Assembling the Tree of Life

Assembling the Tree of Life

EDITED BY Joel Cracraft

Michael J. Donoghue

OXFORD
UNIVERSITY PRESS

2004

OXFORD
UNIVERSITY PRESS

Oxford   New York
Auckland Bangkok Buenos Aires Cape Town Chennai
Dar es Salaam Delhi Hong Kong Istanbul Karachi Kolkata
Kuala Lumpur Madrid Melbourne Mexico City Mumbai Nairobi
São Paulo Shanghai Taipei Tokyo Toronto

Published by Oxford University Press, Inc.,
198 Madison Avenue, New York, New York 10016

www.oup.com

Oxford is a registered trademark of Oxford University Press

Library of Congress Cataloging-in-Publication Data
Assembling the tree of life / edited by Joel Cracraft, Michael J. Donoghue.
p.   cm.
Proceedings of a symposium held at the American Museum of Natural History in New York, 2002.
Includes bibliographical references and index.
ISBN 0-19-517234-5
1. Biology—Classification—Congresses.   I. Cracraft, Joel.   II. Donoghue, Michael J.
QH83.A86 2004
578'.01'2—dc22   2003058012

9 8 7 6 5 4 3 2 1

Printed in the United States of America
on acid-free paper

Contents

Contributors

Marc Allard
Department of Biological Science
The George Washington University
Washington, DC 20052

Robert A. Andersen
Bigelow Laboratory for Ocean
    Sciences
W. Boothbay Harbor, ME 04575

S. L. Baldauf
Department of Biology
University of York
P.O. Box 373
York YO10 5YW
England, UK

F. Keith Barker
James Ford Bell Museum of Natural
    History
University of Minnesota
1987 Upper Buford Circle
St. Paul, MN 55108

Pamela Beresford
Percy FitPatrick Institute
University of Cape Town
Rondebosch 7701
Republic of South Africa

Debashish Bhattacharya
Department of Biological Sciences
University of Iowa
Iowa City, IA 52242–1324

Meredith Blackwell
Department of Biological Sciences
Louisiana State University
Baton Rouge, LA 70803

Elizabeth Borda
Division of Invertebrate Zoology
American Museum of Natural History
Central Park West at 79th Street
New York, NY 10024

Michael Braun
Laboratory of Analytical Biology
Department of Systematic Biology
Smithsonian Institution
4210 Silver Hill Road
Suitland, MD 20746

Rodney A. Bray
Parasitic Worms Division
Department of Zoology
The Natural History Museum
Cromwell Road
London SW7 5BD
England, UK

Thomas D. Bruns
Plant and Microbial Biology
University of California
Berkeley, CA 94720

David Cannatella
Department of Integrative Biology
University of Texas
Austin, TX 78712

M. R. de Carvalho
Departamento de Biologia–FFCLRP
Universidade de São Paulo
Ribeirão Preto
Brazil

Mark W. Chase
Jodrell Laboratory
Royal Botanic Gardens
Kew, Richmond
Surrey TW9 3DS
England, UK

Alice Cibois
Department of Mammalogy and
    Ornithology
Natural History Museum of Geneva
CP 6434
1211 Geneva 6
Switzerland

J. Cockrill
Department of Biology
University of York
P.O. Box 373
York YO10 5YW
England, UK

Jonathan A. Coddington
Department of Systematic Biology
National Museum of Natural History
Smithsonian Institution
Washington, DC 20560

Rita R. Colwell
Director
National Science Foundation
Arlington, VA 22230

Paul Constantino
Department of Anthropology
The George Washington University
2110 G Street NW
Washington, DC 20052

Joel Cracraft
Department of Ornithology
American Museum of Natural History
Central Park West at 79th Street
New York, NY 10024

Peter R. Crane
Royal Botanic Gardens
Kew, Richmond
Surrey TW9 3AB
England, UK

Charles F. Delwiche
Department of Cell Biology and
    Molecular Genetics
University of Maryland College Park
College Park, MD 20742-5815

Michael J. Donoghue
Department of Ecology and Evolution-
    ary Biology
Yale University
New Haven, CT 06520

W. Ford Doolittle
Canadian Institute for Advanced
    Research
Department of Biochemistry and
    Molecular Biology
Dalhousie University
Halifax, Nova Scotia
Canada B3H 4H7

Jerry W. Dragoo
Department of Biology and
Museum of Southwestern Biology
University of New Mexico
Albuquerque, NM 87131

Gareth J. Dyke
Department of Zoology
University College Dublin
Belfield, Dublin 4
Ireland

Gregory D. Edgecombe
Australian Museum
6 College Street
Sydney, New South Wales 2010
Australia

Douglas J. Eernisse
Department of Biological Science
California State University
Fullerton, CA 92834

Peter K. Endress
Institute of Systematic Botany
University of Zurich
Zurich
Switzerland

Julie Feinstein
Department of Ornithology
American Museum of Natural History
Central Park West at 79th Street
New York, NY 10024

Douglas J. Futuyma
Department of Ecology and
    Evolutionary Biology
University of Michigan
Ann Arbor, MI 48109-1079

Jaime García-Moreno
Max Planck Research Centre for
    Ornithology and University of
    Konstanz
Schlossalleé 2
D-78315 Radolfzell
Germany

John Gatesy
Department of Biology
University of California-Riverside
Riverside, CA 92521

David Geiser
Plant Pathology
Pennsylvania State University
University Park, PA 16804

Gonzalo Giribet
Department of Organismic and
    Evolutionary Biology, and Museum
    of Comparative Zoology
Harvard University
16 Divinity Avenue
Cambridge, MA 02138

John Harshman
4869 Pepperwood Way
San Jose, CA 95124

Mark S. Harvey
Department of Terrestrial Inverte-
    brates
Western Australian Museum
Francis Street
Perth, Western Australia 6000
Australia

Gerhard Haszprunar
Zoologischen Staatssammlung
    München
Münchhausenstrasse 27
81247 Munich
Germany

David S. Hibbett
Department of Biology
Clark University
Worcester, MA 01610

David M. Hillis
Section of Integrative Biology and
    Center for Computational Biology
    and Bioinformatics
University of Texas
Austin, TX 78712

P. Hugenholtz
ComBinE Group
Advanced Computational Modelling
    Centre
The University of Queensland
Brisbane 4072
Australia

Timothy James
Department of Biology
Duke University
Durham, NC 27708

G. D. Johnson
Division of Fishes
National Museum of Natural History
Washington, DC 20560

Stefan Koenemann
Institute for Biodiversity and Ecosystem Dynamics
University of Amsterdam
Mauritskade 61
1092 AD Amsterdam
The Netherlands

Robin Lawson
Department of Herpetology
California Academy of Sciences
Golden Gate Park
San Francisco, CA 94118-4599

Michael S. Y. Lee
Department of Environmental Biology,
    University of Adelaide
Department of Palaeontology, South
    Australian Museum
Adelaide, SA 5000
Australia

David R. Lindberg
Museum of Paleontology
1101 Valley Life Science Building
University of California
Berkeley, CA 94720-4780

D. Timothy J. Littlewood
Parasitic Worms Division
Department of Zoology
The Natural History Museum
Cromwell Road
London SW7 5BD
England, UK

François Lutzoni
Department of Biology
Duke University
Durham, NC 27708

Susana Magallón
Departemento de Botánica, Instituto
    de Biología
Universidad Nacional Autónoma de
    México
Circuito Exterior, Anexo al Jardín
    Botánico
AP 70–233
México DF 04510

Richard M. McCourt
Department of Botany
Academy of Natural Sciences
Philadelphia, PA 19103

Jin Meng
Division of Paleontology
American Museum of Natural History
79th Street at Central Park West
New York, NY 10024-5192

David P. Mindell
Department of Ecology and Evolutionary Biology and Museum of
    Zoology
University of Michigan
Ann Arbor, MI 48109-1079

Brent D. Mishler
Department of Integrative Biology
University of California Berkeley
Berkeley, CA 94720

Michael J. Novacek
Division of Paleontology
American Museum of Natural History
79th Street at Central Park West
New York, NY 10024-5192

Kerry O'Donnell
National Center for Agricultural
    Utilization Research
Agriculture Research Service
1815 N. University Street
Peoria, IL 61604

Maureen A. O'Leary
Department of Anatomical Sciences
HSC T-8 (040)
Stony Brook University
Stony Brook, NY 11794-8081

Norman R. Pace
Department of Molecular, Cellular
    and Developmental Biology
Campus Box 0347
University of Colorado
Boulder, CO 80309-0347

J. Pawlowski
Department of Zoology and Animal
    Biology
University of Geneva
1224 Chêne-Bougeries/Geneva
Switzerland

Kevin J. Peterson
Department of Biological Sciences
Dartmouth College
Hanover, NH 03755

Hervé Philippe
Département de Biochimie
Université de Montréal
Pavillon principal—Bureau F-315
C. P. 6128 Succursale Centre-Ville
Montréal, Quebec
Canada H3C 3J7

Winston F. Ponder
Division of Invertebrate Zoology
Australian Museum
Sydney, NSW 2010
Australia

Lorenzo Prendini
Division Invertebrate Zoology
American Museum of Natural History
Central Park West at 79th Street
New York, NY 10024

Kathleen M. Pryer
Department of Biology
Duke University
Durham, NC 27708

Tod W. Reeder
Department of Biology
San Diego State University
San Diego, CA 92182-4614

Joshua S. Rest
Department of Ecology and
    Evolutionary Biology and Museum
    of Zoology
University of Michigan
Ann Arbor, MI 48109-1079

Gregory W. Rouse
South Australian Museum
Adelaide, SA 5000
Australia

Timothy Rowe
Jackson School of Geosciences, C1100
The University of Texas at Austin
Austin, TX 78712

Jorge Salazar-Bravo
Department of Biology and Museum
    of Southwestern Biology
University of New Mexico
Albuquerque, NM 87131

Peter Schikler
Department of Ornithology
American Museum of Natural History
Central Park West at 79th Street
New York, NY 10024

Harald Schneider
Albrecht-von-Haller-Institut für
    Pflanzenwissenschaften
Abteilung Systematische Botanik
Georg-August-Universität Göttingen
Untere Karspüle 2
37073 Göttingen
Germany

Frederick R. Schram
Institute for Biodiversity and
    Ecosystem Dynamics
University of Amsterdam
Mauritskade 61
1092 AD Amsterdam
The Netherlands

Mark E. Siddall
Division of Invertebrate Zoology
American Museum of Natural History
Central Park West at 79th Street
New York, NY 10024

A. G. B. Simpson
Canadian Institute for Advanced
    Research
Department of Biochemistry and
    Molecular Biology
Dalhousie University
Halifax, Nova Scotia
Canada B3H 4H7

Joseph B. Slowinski (deceased)
Department of Herpetology
California Academy of Sciences
Golden Gate Park
San Francisco, CA 94118-4599

Andrew B. Smith
Department of Palaeontology
The Natural History Museum
Cromwell Road
London SW7 5BD
England, UK

Douglas E. Soltis
Department of Botany
University of Florida
Gainesville, FL 32611

Pamela S. Soltis
Florida Museum of Natural History
University of Florida
Gainesville, FL 32611

Michael D. Sorenson
Department of Biology
Boston University
5 Cummington Street
Boston, MA 02215

Joseph Spatafora
Botany and Plant Pathology
Oregon State University
Corvallis, OR 97331

Scott Stanley
411 Cary Pines Drive
Cary, NC 27513

M. L. J. Stiassny
Department of Ichthyology
American Museum of Natural History
Central Park West at 79th Street
New York, NY 10024

John W. Taylor
Department of Plant and Microbial
    Biology
University of California
Berkeley, CA 94720-3102

Maximilian J. Telford
University Museum of Zoology
Department of Zoology
Cambridge University
Downing Street
Cambridge CB2 3EJ
England, UK

Luis P. Villarreal
Department of Molecular Biology and
    Biochemistry, and Center for Virus
    Research
University of California at Irvine
Irvine, CA 92697

David B. Wake
Museum of Vertebrate Zoology and
    Department of Integrative Biology
University of California
Berkeley, CA 94720-3160

David E. Walter
Department of Biological Sciences
University of Alberta
Edmonton, AB
Canada T6G 2E9

Ward C. Wheeler
Division of Invertebrate Zoology
American Museum of Natural History
Central Park West at 79th Street
New York, NY 10024-5192

Michael F. Whiting
Department of Integrative Biology
Brigham Young University
Provo, UT 84042

E. O. Wiley
Ecology and Evolutionary Biology
University of Kansas
Lawrence, KS 66045

Rainer Willmann
Zoologisches Institut der Universität
Georg-August-Universität Göttingen
Berliner Strasse 28
D-37073 Göttingen
Germany

Edward O. Wilson
Department of Organismic and
    Evolutionary Biology and the
    Museum of Comparative Zoology
Harvard University
16 Divinity Avenue
Cambridge, MA 02138

Bernard Wood
Department of Anthropology
The George Washington University
2110 G Street NW
Washington, DC 20052

Gregory Wray
Department of Biology
Duke University
Durham, NC 27708

Terry L. Yates
Department of Biology and Museum
    of Southwestern Biology
University of New Mexico
Albuquerque, NM 87131

Tamaki Yuri
Laboratory of Analytical Biology
Department of Systematic Biology
Smithsonian Institution
4210 Silver Hill Road
Suitland, MD 20746

Assembling the Tree of Life

Michael J. Donoghue

Joel Cracraft

Introduction

Charting the Tree of Life

Many, perhaps even most, people today are comfortable with the image of a tree as a representation of how species are related to one another. The Tree of Life has become, we think, one of the central images associated with life and with science in general, alongside the complementary metaphor of the ecological Web of Life. But this was not always the case. Before Darwin, the reigning view was perhaps that life was organized like a ladder or "chain of being," with slimy "primitive" creatures at the bottom and people (what else!) at the very top. Darwin (1859) solidified in our minds the radically new image of a tree (fig. I.1), within which humans are but one of many (as we now know, millions) of other species situated at the tips of the branches. The tree, it turns out, is the natural image to convey ancestry and the splitting of lineages through time, and therefore is the natural framework for "telling" the genealogical history of life on Earth.

Very soon after Darwin, interest in piecing together the entire Tree of Life began to flourish. Ernest Haeckel's (1866) trees beautifully symbolize this very active period and also, through their artistry, highlight the comparison between real botanical trees and branching diagrams representing phylogenetic relationships (fig. I.2).

However, during this period, and indeed until the 1930s, rather little attention was paid to the logic of inferring how species (or the major branches of the Tree of Life) are related to one another. In part, the lack of a rigorous methodology (especially compared with the newly developing fields of genetics and experimental embryology) was responsible for

a noticeable lull in activity in this area during the first several decades of the 1900s. But, beginning in the 1930s, with such pioneers as the German botanist Walter Zimmermann (1931), we begin to see the emergence of the basic concepts that underlie current phylogenetic research. For example, the central notion of "phylogenetic relationship" was clearly defined in terms of recency of common ancestry—we say that two species are more closely related to one another than either is to a third species if and only if they share a more recent common ancestor (fig. I.3).

This period in the development of phylogenetic theory culminated in the foundational work of the German entomologist Willi Hennig. Many of his central ideas were put forward in German in the 1950s (Hennig 1950), but worldwide attention was drawn to his work after the publication of *Phylogenetic Systematics* in English (Hennig 1966). Hennig emphasized, among many other things, the desirability of recognizing only monophyletic groups (or clades—single branches of the Tree of Life) in classification systems, and the idea that shared derived characteristics (what he called synapomorphies) provided critical evidence for the existence of clades (fig. I.4).

Around this same time, in other circles, algorithms were being developed to try to compute the relatedness of species. Soon, a variety of computational methods were implemented and were applied to real data sets. Invariably, given the tools available in those early days, these were what would now be viewed as extremely small problems.

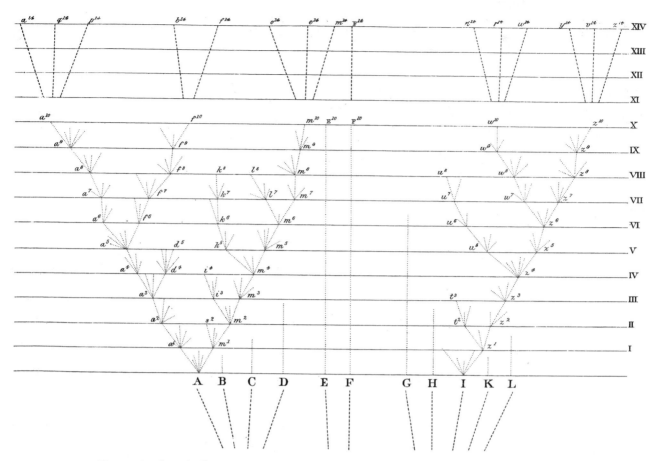

Figure I.1. The only illustration in Darwin's *Origin of Species* (1859), which can be taken to be the beginning of "tree thinking."

Since that time major developments have occurred along several lines. First, although morphological characters were at first the sole source of evidence for phylogenetic analyses, molecular data, especially DNA sequences, have become available at an exponential rate. Today, many phylogenetic analyses are carried out using molecular data alone. However, morphological evidence is crucial in many cases, but especially when the object is to include extinct species preserved as fossils. Ultimately, of course, there are advantages in analyzing all of the evidence deemed relevant to a particular phylogenetic problem—morphological and molecular. And many of our most robust conclusions about phylogeny, highlighted in this volume, are based on a combination of data from a variety of sources.

A second major development has been increasing computational power, and the ease with which we can now manipulate and analyze extremely large phylogenetic data sets. Initially, such analyses were extremely cumbersome and time-consuming. Today, we can deal effectively and simultaneously with vast quantities of data from thousands of species.

Beginning in the 1990s these developments all came together—the image and meaning of a tree, the underlying

conceptual and methodological developments, the ability to assemble massive quantities of data, and the ability to quantitatively evaluate alternative phylogenetic hypotheses using a variety of optimality criteria. Not surprisingly, the number of published phylogenetic analysis skyrocketed (Hillis, ch. 32 in this vol.). Although it is difficult to make an accurate assessment, in recent years phylogenetic studies have been published at a rate of nearly 15 a day.

Where has this monumental increase in activity really gotten us in terms of understanding the Tree of Life? That was the question that motivated the symposium that we organized in 2002 at the American Museum of Natural History in New York, and which yielded the book you have in front of you. Although it may be apparent that there has been a lot of activity, and that a lot can now be written about the phylogeny of all the major lineages of life, it is difficult to convey a sense of just how rapidly these findings have been accumulating. Previously, there was a similar attempt to provide a summary statement across all of life—a Nobel symposium in Sweden in 1988, which culminated in a book titled *The Hierarchy of Life* (Fernholm et al. 1989). That was an exciting time, and the enthusiasm and potential of this en-

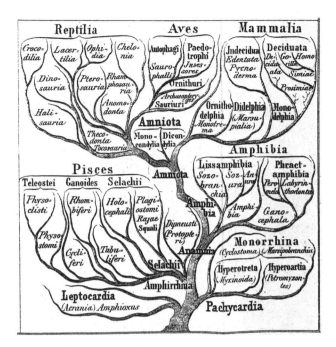

Figure I.2. A phylogenetic tree realized by Haeckel (1866), soon after Darwin's *Origin*.

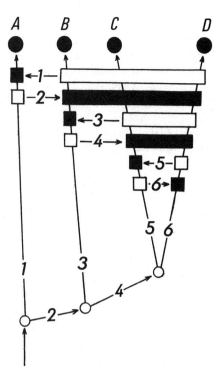

Figure I.4. The conceptual phylogenetic argumentation scheme of Hennig (1966: 91), with solid boxes representing derived (apomorphic) and open boxes representing primitive (plesiomorphic) characters.

deavor were expressed in the chapters of that book. But, in looking back at those pages we are struck by the paucity of data and the minuscule size of the analyses that were being performed at what was surely the cutting edge of research at the time.

It is also clear that so much more of the Tree of Life is being explored today than only a decade ago. Now we can honestly present a picture of the relationships among all of the major branches of the Tree of Life, and within at least some of these major branches we are now able to provide

considerable detail. A decade ago the holes in our knowledge were ridiculously obvious—we were really just getting started on the project. There are giant holes today, which will become increasingly obvious in the years to come (as we learn more about species diversity, and database phylogenetic knowledge), but we believe that it is now realistic to conceive of reconstructing the entire Tree of Life—eventually to include all of the living and extinct species. A decade ago, we could hardly conjure up such a dream. Today we not only can imagine what the results will look like, but we now believe it is attainable.

It also has become increasingly obvious to us just how important it is to understand the structure of the Tree of Life in detail. With the availability of better and better estimates of phylogeny, awareness has rapidly grown outside of systematic biology that phylogenetic knowledge is essential for understanding the history of character change and for interpreting comparative data of all sorts within a historical context. At the same time, phylogeny and the algorithms used to build trees have taken on increasing importance within applied biology, especially in managing our natural resources and in improving our own health and well-being. Phylogenetic trees now commonly appear in journals that had not previously devoted much space to trees or to "tree thinking," and many new tools have been developed to leverage this new information on relationships.

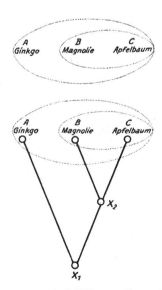

Figure I.3. Zimmermann's (1931) tree, illustrating the concept of "phylogenetic relationship."

In this volume we have tried, with the chapters in the opening and closing sections, to highlight the value of the Tree of Life, and then, in a series of chapters by leading experts, to summarize the current state of affairs in many of its major branches. In presenting this information, we appreciate that many important groups are not covered in sufficient detail, and a few not at all, and we know that in some areas information will already be outdated. This is simply the nature of the progress we are making—new clades are discovered literally every day—and the sign of a healthy discipline. Nevertheless, our sense is that a benchmark of our progress early in the 21st century is a worthy exercise, especially if it can help motivate the vision and mobilize the resources to carry out the mega-science project that the Tree of Life presents. This would surely be one of the most fundamental of all scientific accomplishments, with benefits that are abundantly evident already and surprises whose impacts we can hardly imagine.

Acknowledgments

The rapidly expanding activity in phylogenetics noted above set the stage for a consideration and critical evaluation of our current understanding of the Tree of Life. This juncture in time also coincided with the inception of the International Biodiversity Observation Year (IBOY; available at http://www.nrel.colostate.edu/projects/iboy) by the international biodiversity science program DIVERSITAS (http://www.diversitas-international.org) and its partners. Assembling the Tree of Life (ATOL) was accepted as a key project of IBOY, and a symposium and publication were planned. This volume is the outgrowth of that process.

The ATOL symposium would not have been possible without the participation of many institutions and individuals. Key, of course, was the financial commitment received from the host institutions, the American Museum of Natural History (AMNH) and Yale University, and from the International Union of Biological Sciences (IUBS), a lead partner of DIVERSITAS and convenor of Systematics Agenda 2000 International. Assembling the Tree of Life (ATOL) was accepted as a core project of the DIVERSITAS program, International Biodiversity Observation Year (IBOY). We especially acknowledge the leadership of Ellen Futter (president) and Michael Novacek (senior vice president and provost) of the AMNH and of Alison Richard (provost) of Yale University for making the symposium possible. In addition, a financial contribution from IUBS facilitated international attendance, and we are grateful to Marvalee Wake (president), Talal Younes (executive director), and Diana Wall (director, IBOY) for their support.

The scientific program of the symposium was planned with the critical input of Michael Novacek and many other colleagues, and we are grateful for their suggestions. Ultimately, we tried to cover as much of the Tree of Life as possible in three days and at the same time to include plenary speakers whose charge was to summarize the importance of phylogenetic

knowledge for science and society. We are well aware of the omissions and imbalances that result from an effort such as this one and which are manifest in this volume. Our ultimate goal was to produce a single volume that would broadly cover the Tree of Life and that would be useful to the systematics community as well as accessible to a much wider audience. We challenged the speakers to involve as many of their colleagues as possible and to summarize what we know, and what we don't know, about the phylogeny of each group, and to write their chapters for a scientifically literate general audience, but not at the expense of scientific accuracy. We trust that their efforts will catalyze future research and greatly enhance communication about the Tree of Life.

The symposium itself could not have been undertaken without the tireless effort of numerous people. The staff of the AMNH and its outside symposium coordinator, DBK Events, spent countless hours over many months facilitating arrangements with the speakers and attendees, and not least, making the organizers' lives much easier. It is not possible to identify all of those who contributed, but we would be remiss if we did not mention the following: Senior Vice President Gary Zarr, and especially Ann Walle, Anne Canty, Robin Lloyd, Amy Chiu, and Rose Ann Fiorenzo of the AMNH Department of Communications; Joanna Dales of Events and Conference Services; Mike Benedetto of IT-Network Systems; Frank Rasor and Larry Van Praag of the Audio-Visual Department; and Jennifer Kunin of DBK Events.

Finally, many colleagues helped with production of this volume. Many referees, both inside and outside of our institutions, contributed their time to improve the chapters. Merle Okada and Christine Blake, AMNH Department of Ornithology, helped in many ways with editorial tasks, and Susan Donoghue assisted with the index. Most important, we are grateful to Kirk Jensen of Oxford University Press for believing in the project and facilitating its publication, and to Peter Prescott for seeing it through.

Literature Cited

Darwin, C. R. 1859. On the origin of species. John Murray, London.

Fernholm, B., K. Bremer, and H. Jörnvall (eds.). 1989. The hierarchy of life. Nobel Symposium 70. Elsevier, Amsterdam.

Haeckel, E. 1866. Generelle Morphologie der Organismen: allgemeine Grundzüge der organischen Formen-Wissenschaft, mechanisch begründet durch die von Charles Darwin reformirte Descendenz-Theorie. G. Reimer, Berlin.

Hennig, W. 1950. Grundzüge einer Theorie des phylogenetischen Systematik. Deutscher Zentraverlag, Berlin.

Hennig, W. 1966. Phylogenetic systematics. University of Illinois Press, Urbana.

Zimmermann, W. 1931. Arbeitsweise der botanischen Phylogenetik und anderer Gruppierubgswissenschaften. Pp. 941–1053 in Hanbuch der biologischen Arbeitsmethoden (E. Abderhalden, ed.), Abt. 3, 2, Teil 9. Urban & Schwarzenberg, Berlin.

The Importance of Knowing the Tree of Life

Terry L. Yates
Jorge Salazar-Bravo
Jerry W. Dragoo

The Importance of the Tree of Life to Society

The affinities of all the beings of the same class have
sometimes been represented by a great tree. . . . As buds give
rise by growth to fresh buds, and these, if vigorous, branch
out and overtop on all sides many a feebler branch, so by
generation I believe it has been with the great Tree of Life,
which fills with its dead and broken branches the crust of the
earth, and covers the surface with its ever branching and
beautiful ramifications.
—Charles Darwin, *On the Origin of Species* (1859)

Despite Darwin's vision of the existence of a universal Tree
of Life, assembly of the tree with a high degree of accuracy
has proven challenging to say the least. Generations of systematists have worked on the problem and debated (or
fought) about how to best approach a solution, or questioned
if a solution was even possible. Much of the rest of the biological sciences and medicine either simply accepted decisions of systematists without question or discounted them
entirely as lacking rigor and accuracy. Attempts at solving
the problem met with only limited success and were generally limited to similarity comparisons of various kinds until
the convergence of three important developments: (1) conceptual and methodological underpinnings of phylogenetic
systematics, (2) development of genomics, and (3) rapid
advances in information technology.

Convergence of these three areas makes construction of
a robust tree representing genealogical relationships of all
known species possible for the first time. This, coupled with
the fact that the current lack of a universal tree is severely
hampering progress in many areas of science and limiting the
ability of society to address many important problems and
to capitalize on a host of opportunities, demands that we
undertake this important project now and with conviction.
Although many challenges still stand before us (which themselves represent additional opportunities), constructing a
complete Tree of Life is now conceptually and technologi-
cally possible for the first time. It is relevant to note here that
we still had hundreds of problems to solve when we decided

to land a man on the moon, and their solution produced
hundreds of unexpected by-products. The size of this undertaking and the human resources needed, however, require
an international collaboration instead of a competition. Assembling an accurate universal tree depicting relationships
of all life on Earth, from microbes to mammals, holds enormous potential value for society, and it is imperative that we
start now. This chapter, although not meant to be exhaustive, aims to provide a number of examples where even our
limited knowledge of the tree has provided tangible benefits
to society. The actual value that a fully assembled tree would
hold for society would be limitless.

Enabling Technologies and Challenges

Despite widespread acceptance of phylogenetic systematics
during the 1980s, it was not until the advent of genomics
and modern computer technology, enabled by more efficient
and rapid phylogenetic algorithms in the 1990s, that large-
scale tree assembly became possible. The rapid growth of
genomics, in particular, revolutionized the field of phylogenetic systematics and provided a new level of power to tree
assembly. To reconstruct the evolutionary history of all organisms will require continued advances in computer hardware and development of faster and more efficient algorithms.

The mathematics and computer science communities are
already actively engaged in this challenge, and breakthroughs

are occurring almost daily. For example, researchers working on resolving the relationships of 12 species of bluebells back to a common ancestor have used the 105 genes found in chloroplast DNA from those species (and an outgroup —tobacco) to reconstruct the phylogeny. The resulting analysis examined 14 billion trees. But not only did they reconstruct the phylogeny, they also inferred the gene order of the 105 genes found in the chloroplast genome for each ancestor in the tree, which means 100 billion "genomes" were analyzed. The process took 1 hour and 40 minutes using a 512-processor supercomputer (Moret et al. 2002).

Although this represents a major advancement, additional advancements will be needed for the relationships of the current 1.7 million known species to be reconstructed. Necessary software tools have not been developed to take full advantage of existing data and to permit integration with existing biological databases. The enormous amounts of data being generated by the enabling technologies associated with modern genomics, although posing considerable challenges to the computer world, will allow tree construction at a level of detail far exceeding anything in the past.

Even in groups such as mammals that are well known relative to invertebrates and microbes, the use of genomics in tree construction is increasing our knowledge base at a phenomenal rate and providing important bridges to other fields of knowledge. Recent work by Dragoo and Honeycutt (1997), for example, has revealed that skunks represent a lineage of their own distinct from mustelids (fig. 1.1). Skunks historically have been classified as a subfamily within the Mustelidae (weasels), but genetic data suggest that raccoons are more closely related to weasels than are skunks. Additionally, stink

badgers were classified within a different subfamily of mustelids than skunks. Morphological and genetic data both support inclusion of stink badgers within the skunk clade. The skunk–weasel–raccoon relationship was based on analyses of genes within the mitochondrial genome. However, DNA sequencing of nuclear genes has provided support for this hypothesis as well (Flynn et al. 2000, and K. Koepfli, unpubl. obs.). This discovery is already proving valuable to other fields such as public health and conservation.

These types of advances are producing major discoveries across the entire tree, but nowhere is it more evident than in the microbial world. New discoveries using genomics and phylogenetic analysis have led to the discovery of entire new groups of Archaea (DeLong 1992) that will prove critical to our understanding of the functioning of the world's ecosystems. Others using similar techniques are discovering major groups of important microbes living in extreme environments (Fuhrman et al. 1992) that could lead to discovery of important new classes of compounds. In fact, the number of new species of bacteria being discovered with these methods, as noted by DeLong and Pace (2001), is expanding almost exponentially. It is not only new species that are being discovered but also new kingdoms of organisms within the domains Bacteria and Archaea.

Human Health

Ten people died in April through June 1993 as a result of an unknown disease that emerged in the desert Southwest of the United States. Approximately 70% of the people who ac-

Figure 1.1. Phylogenetic relationship of skunks with relation to weasels as well as other caniform carnivores; modified from Dragoo and Honeycutt (1997). The arrow indicates a sister-group relationship between weasels (Mustelidae) and raccoons (Procyonidae) to the exclusion of skunks. Skunks thus were recognized as a distinct family, Mephitidae.

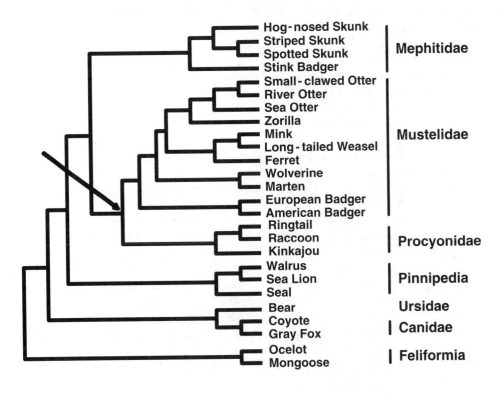

quired this disease died from the symptoms. No known cure or drugs was available to treat this disease, nor was it known if the disease was caused by a virus or bacterium or some other toxin. Later, a previously unknown hantavirus was determined to be the cause and was described as Sin Nombre virus (SNV; Nichol et al. 1993), and it was discovered that the reservoir for this virus was the common deer mouse (Childs et al. 1994).

Phylogenetic analyses of viruses in the genus *Hantavirus* suggested that this new virus was related to Old World hantaviruses. However, the virus was different enough in sequence divergence to suggest that it was not a result of an introduction from the Old World, but rather had evolved in the Western hemisphere. Phylogenetic analyses of both murid rodents and known hantaviruses indicated a high level of agreement between host and virus trees (fig. 1.2), suggesting a long history of coevolution between the two groups (Yates et al. 2002). This information allowed researchers to predict that many of the murid rodent lineages may be associated with other lineages of hantaviruses as well.

Predictions made from analyses of these phylogenetic trees have been supported with the descriptions of at least 25 new hantaviruses in the New World since the discovery of SNV (fig. 1.3). More than half (14) of these newly recognized viruses have been detected in Central and South America. Additionally, many of the viruses are capable of causing human disease. It is likely that many more yet unknown hantaviruses will be discovered in other murid hosts not only in North and South America but also in other countries around the world. The poorly studied regions of such countries as African and Asia quite probably contain many such undescribed viruses.

Further studies enabled by findings of coevolutionary relationships have allowed the development of models that are able to predict areas and times of increased human risk to disease far in advance of any outbreaks (Yates et al. 2002, Glass et al. 2002). Knowledge of phylogenetic relationships of these organisms has thus proven critical for our understanding of diversity of these pathogens and how to predict the risk to humans. An understanding of these relationships also will be critical for us to determine if we are under attack from introduced pathogens.

In 1999 several people were diagnosed with or died from symptoms of a viral infection similar to that caused by the St. Louis encephalitis virus (Flaviviridae). The virus was determined to be transmitted by mosquitoes and not only affected humans but also was killing wild and domestic birds. Phylogenetic analyses using RNA sequencing from this virus as well as other flaviviruses were conducted to determine that the disease causing agent was actually the West Nile virus (Jia et al. 1999, Lanciotti et al. 1999). This virus was determined from those analyses to be closely related to strains found in birds from Israel, East Africa, and Eastern Europe (fig. 1.4; Lanciotti et al. 1999). The information obtained from those studies provided the basic biology needed to allow health officials to effectively treat this new outbreak of West Nile virus as well as make predictions about the spread of the virus using the known potential avian hosts. Advance knowledge of where it might spread next was critical in preventing human and animal infection. West Nile virus has currently spread as far west in the United States as California and has resulted in numerous human and animal deaths.

Conservation

Conservation biology is quite likely the area of science most heavily affected (and will continue to be so) by a better knowledge of the Tree of Life. A more complete Tree of Life will mean that more species are identified. Currently, one of the

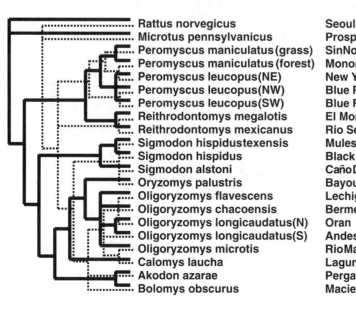

Figure 1.2. Coevolution of New World murid rodents (solid lines) and hantaviruses (dotted lines) based on comparison of each independent phylogeny; modified from Yates et al. (2002).

Rattus norvegicus	Seoul
Microtus pennsylvanicus	Prospect Hill
Peromyscus maniculatus(grass)	SinNombre
Peromyscus maniculatus (forest)	Monongahela
Peromyscus leucopus(NE)	New York
Peromyscus leucopus(NW)	Blue River (IN)
Peromyscus leucopus(SW)	Blue River (OK)
Reithrodontomys megalotis	El Moro Canyon
Reithrodontomys mexicanus	Rio Segundo
Sigmodon hispidustexensis	Muleshoe
Sigmodon hispidus	Black Creek Canal
Sigmodon alstoni	CañoDelgadito
Oryzomys palustris	Bayou
Oligoryzomys flavescens	Lechiguanas
Oligoryzomys chacoensis	Bermejo
Oligoryzomys longicaudatus(N)	Oran
Oligoryzomys longicaudatus(S)	Andes
Oligoryzomys microtis	RioMamore
Calomys laucha	Laguna Negra
Akodon azarae	Pergamino
Bolomys obscurus	Maciel

Figure 1.3. Newly discovered hantaviruses since 1993; modified from Centers for Disease Control and Prevention (2003). Viruses prefixed by an asterisk represent strains known to be pathogenic to humans.

most important issues in conservation biology is the question of how many species are out there (Wheeler 1995). Although no single value can be used with any level of confidence, a figure often cited is 12.5–13 million species (e.g., Singh 2002); Cracraft (2002) estimated (admittedly roughly) that only a very small fraction—in the order of 0.4%—of this figure [or some 50–60 (10^3 taxa)] are included in any sort of phylogenetic analysis. A more developed, inclusive Tree of Life would help identify, catalog, and database elements of biodiversity that may not have been included until now.

A more developed Tree of Life would help incorporate an evolutionary framework with which to base conservation strategies. Two major questions in conservation biology are how variation is distributed in the landscape, and how it came about. Conservation planners, too, need to highlight these spatial components for conservation action. Erwin (1991) convincingly argued for the need to incorporate phylogenies and evolutionary considerations in conservation efforts. Desmet et al. (2002), Barker (2002), and Moritz (2002) have proposed methodological and practical applications for this strategy. For example, Barker (2002) reviewed and expanded on some of the properties of phylogenetic diversity measures to enable capturing both the phylogenetic relatedness of species and their abundances. This measure estimates the relative diversity feature of any nominated set of species by the sum of the lengths of all those branches spanned by the set. These branch lengths reflect patristic or path-length distances of character change. He then used this method to address a number of conservation and management issues (from setting priorities for threatened species management

to monitoring biotic response to management) related to birds at three different levels of analyses: global, New Zealand only, and Waikato specifically.

An improved Tree of Life would allow for rigorous testing of old premises in evolutionary theory. For more than 40 years, the premise that shrinking and expanding of tropical forests in the neotropics and elsewhere has become a paradigmatic force invoked to explain the diversity of species in these biodiverse areas of the world (but see Colinvaux et al. 2001). Research centered on the phylogenies and phylogeographic patterns of various taxa in several tropical areas of the world has now made it clear that the refuge hypothesis (see Haffer 1997, Haffer and Prance 2001) of Amazonian speciation does not explain the patterns of distribution of many taxa. In fact,

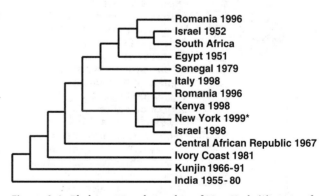

Figure 1.4. Phylogenetic relationship of New York (*) strain of the West Nile virus compared with other strains worldwide; modified from Lanciotti et al. (1999).

Glor et al. (2001), Moritz et al. (2000), and Richardson et al. (2001) have demonstrated that some of the most specious tropical groups have patterns of diversification that resulted during or after the unstable period of the Pleistocene, suggesting a more recent evolutionary history. Phylogenetic patterns indicate that heterogeneous habitats account for more biodiversity than does the accumulation of species through time in an unperturbed environment.

These studies and others (e.g., Moritz 2002) have shown that it is possible to incorporate the knowledge obtained by phylogenetic analyses (i.e., applied phylogenetics of Cracraft 2002) and the distribution of genetic diversity into conservation planning and priority setting for populations within species and for biogeographic areas within regions. Moritz (2002) suggests that the separation of genetic diversity into two dimensions, one concerned with adaptive variation and the other with neutral divergence caused by isolation, highlights different evolutionary processes and suggests alternative strategies for conservation that need to be addressed in conservation planning.

The main tenet in conservation biology is that the "value of biodiversity lies in its option value for the future, the greater the complement of contemporary biodiversity conserved today, the greater the possibilities for future biodiversity because of the diverse genetic resource needed to ensure continued evolution in a changing and uncertain world" (Barker 2002:165). We cannot conserve what we do not know.

Agriculture

The potential value to agriculture of a fully assembled Tree of Life is enormous. The existence of an accurate phylogenetic infrastructure will enable directed searches for useful genes in ancestors of modern-day crop plans, as opposed to the random explorations of the past. Being able to follow individual genes through time armed with knowledge of their ancestral forms will allow a determination of how the function of these genes has changed through time. This knowledge will, in turn, allow selective modification of new generations of plants and animals in a much more precise way than selective breeding alone. For example, a group of researchers working on the Tree of Life for green plants (Oliver et al. 2000) has identified and traced the genes responsible for desiccation tolerance from ancient liverworts to modern angiosperms (fig. 1.5). Given the rate of desertification occurring globally and the rapid increases in human populations, these data may prove invaluable in helping to sustain our global agriculture.

However, our knowledge of the relationships of wild relatives to many important agricultural crops still is limited. Understanding the origins and relationships should help with further improvement of many of the world's crop plants. Recently, however, research on major grain crops such as

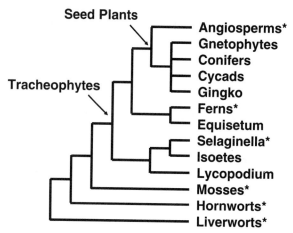

Figure 1.5. Phylogeny of major groups of land plants; modified from Oliver et al. (2000). Asterisks indicate clades that contain desiccation-tolerant species. Oliver et al. (2000) suggest that desiccation tolerance is a primitive state in early land plants that was lost before the evolution of Tracheophytes and then reappeared in at least three major lineages. Additionally, the genes reevolved independently within eight clades found in angiosperms.

wheat, rice, and corn and such other crops as tomatoes and *Manihot* (a major source of starch in South America) has provided insight into the origins of these economically important agricultural products. But, relationships of many other important food and fiber plants, which large parts of our populations worldwide depend on, still remain virtually unknown. These relationships must be understood if we hope to make future genetic improvements, especially because many of the wild progenitors are at risk of extinction and we have yet to study them.

One good example of how phylogenetic relationships may help us to generate an improved crop is seen in corn (*Zea mays mays*). This is a crop of enormous economic importance, and if it is to be used to assist in sustaining human populations, it is imperative that we be able to make continued improvements in disease and/or drought resistance. Corn is a grass with a unique fruiting body commonly referred to as the "corn cob." This is not typically seen in wild grasses, so there have been assorted hypotheses regarding the relationships of corn to other species. Potential relatives to corn are the grasses from Mexico and Guatemala known as teosintes. Recently, Wang et al. (2001) used molecular techniques to conclude that two annual teosinte lineages may actually be the closest relative to corn (fig. 1.6).

These researchers have demonstrated that the origin of this agricultural product probably occurred 9000 years ago in the highlands of Mexico. Additionally, it was determined that the allele responsible for the cob was a result of selection on a regulatory gene rather than a protein-coding gene (Wang et al. 2001). Modern cultivated corn has the poten-

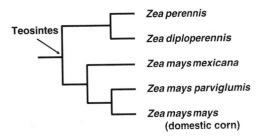

Figure 1.6. Phylogenetic relationship of corn to other teosintes; modified from Wang et al. (2001). This relationship helps explain the morphological variation seen in domestic corncob.

tial to interbreed with several teosinte grasses, so it may be possible to incorporate new traits from these species to improve existing strains of corn crops. These studies illustrate how important it is to protect not only wild species and lineages of teosinte grass but also the habitats in Mexico where they are found.

Invasive Species

Invasive species have become an enormous problem worldwide and cause billions of dollars in damage each year while doing irreparable harm to many native species and ecosystems. Phylogenetic analysis is an important tool in the battle for identifying invasive species and for determining their geographic origin. Recent examples include the West Nile virus example described above and an invasive alga in California. In the latter example, scientists were able to use phylogenetic analysis of DNA sequences to identify the Australian alga species *Caulerpa taxiflora* in California waters. This finding led to an immediate eradication program that, if successful, may save the United States billions of dollars.

In addition, understanding the evolutionary associations of invasive species in the context of closely affiliated groups of species such as host plants or animals is critical for predicting their spread and implementing successful control measures. Wang et al. (1999) performed a phylogenetic analysis to examine relationships of potential pest species of longhorn beetles (Cerambycidae) and found that beetles in certain clades were not likely to become pests, whereas beetles in two other clades could become pests outside of their native Australia. Another clade in this group, the Asian longhorn beetle (*Anoplophora gladripennis*), has been recently introduced into the United States in hardwood packing materials and has already spread from points of introduction to many new areas, killing native hardwood trees as it invades (Meyer 1998). Knowledge of the phylogenetic relationships of trees that this beetle attacks in its native range could prove valuable in predicting the North American trees most likely at risk and could help model its future spread. Likewise, an understanding of the phylogenetic affinities of natural en-

emies of longhorn beetles in Asia will be critical if biological controls for this pest are to be considered in North America.

Invasive ant species have become enormous problems worldwide. The ant *Linepithema humile* has been particularly problematic and has been particularly damaging to native species in Hawaii. Tsutsui et al. (2001) used phylogenetic analyses to trace the origin of this pest to Argentina. Another invasive ant, the fire ant (*Solenopsis invicta*), has caused billions of dollars of damage in the southern United States and has even caused human and animal deaths. Like other eusocial insects, such as Asian termites, fire ants are extremely difficult to control using chemical and other standard methods. Efforts to date in the latter case have been largely ineffective and have led several authors (Morrison and Gilbert 1999, Porter and Briano 2000) to suggest the need for the introduction of biological control agents from the original range of these ants in South America. In particular, these authors have suggested the possible use of host-specific ant-decapitating flies that lay their eggs in the heads of these ants, where the developing larvae eventually kill the ants. Such introductions are always risky but would be extremely so without detailed knowledge of the Tree of Life for the groups in question. According to Rosen (1986), "Reliable taxonomy is the basis for any meaningful research in biology." It is essential also to understand the evolutionary histories of both target pest and natural enemy to predict the possible effects of using one to "control" the other.

Human Land Use

A well-resolved Tree of Life has important implications for disciplines as apparently disparate from biology as the study of human land use patterns, especially when they integrate with other disciplines. For example, phylogenetic analysis was used to discover that two closely related species of rodents in the genus *Calomys* exist in eastern Bolivia (Salazar-Bravo et al. 2002, Dragoo et al. 2003), each harboring a specific arenavirus (fig. 1.7). In the Beni Department of Bolivia, *Calomys* species harbor the Machupo virus (MACV), the etiological agent of Bolivian hemorrhagic fever (BHF), whereas in the Santa Cruz Department, *Calomys callosus* harbors the nonpathogenic Latino virus (LAT). MACV occurs in the Amazon drainage, whereas LAT is found along the drainage of the Parana River. Additionally, it has been found that *Calomys* from each region, despite their genetically based species specificity, will hybridize in the laboratory and create fertile hybrids. It follows that there exists not only the risk of species invasion into a previously isolated ecological zone, but also the risk of hybrids carrying the pathogenic virus into the new region, the possibility of dual arenavirus infection in such rodents, and the chance that virus recombination with unknown consequences might occur.

In the early 1960s MACV produced several outbreaks in northeastern Bolivia, with infection rates of 25% in some towns

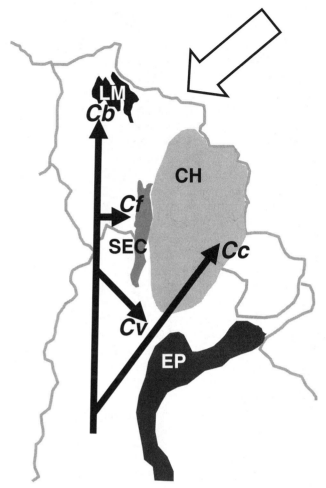

Figure 1.7. Summary cladogram of four closely related taxa of vesper mice (*Calomys*); modified from Salazar-Bravo et al. (2002). Cb, *Calomys* species from the Beni Department of Bolivia; Cf, *C. fecundus*; Cv, *C. venustus*; Cc, *C. callosus*. The white arrow points to the forested area that separates the Llanos de Moxos from the Chaco region. Vegetation is as follows: LM; Llanos de Moxos, SEC; Southeast Coordillera, CH; Chaco, EP Espinal.

and mortality rates approaching 45%. Johnson et al. (1972) noted two distinct phenotypic reactions to infection with MACV and suggested that there may be a genetic component. A *Calomys* species has been reported to express two different immune responses when infected with MACV but not with LAT (Webb et al. 1975). Some individuals become chronically infected, do not produce antibodies, shed large amounts of virus in urine, become infertile, and are the principal vectors of BHF. Others produce an antibody response and all but clear the virus. Although these individuals remain chronically infected, they can reproduce (Justines and Johnson 1969).

There is growing concern in the Bolivian health community about the unintended consequences of an all-weather road connecting Trinidad and Santa Cruz, the capital cities of the Beni and Santa Cruz Departments, respectively, that has been in service for several years. This road breaches a forested natural

barrier between biomes of the respective rodents and viruses. That barrier contains the north–south continental divide of South America (Salazar-Bravo et al. 2002). The new road linking the two home ranges of the virus–rodent pairs is bringing human development to the fringes of both areas along its course. Human populations in both departments are booming. Thirty-five years ago Trinidad and Santa Cruz had about 6000 and 60,000 persons, respectively. Today those numbers have increased 10-fold. Agricultural development has kept pace, especially in the Santa Cruz Department. Therefore, a major concern is whether the rodent and its virus from the north may be now moving, abetted by human commerce, into the southern department. The potential public health risk posed by construction of new roads and new development in the Beni and Santa Cruz Departments makes monitoring this situation essential.

To make predictions about the evolution and spread of arenaviruses, we need to understand the evolutionary history of the rodent reservoirs. The significance of understanding in greater detail evolutionary histories at the population level as well as at the subfamily level goes beyond the importance of prevention and treatment of BHF. The observed patterns of infection and distribution of MACV exhibit a striking number of similarities with not only other arenaviruses but with hantaviruses as well. In addition to the apparent connection to rodent population density and human ecology, these viruses with few exceptions share a common host family of rodents, suggesting a long common evolutionary history.

Economics

Many of the examples presented above will have economic benefits for society. Understanding the Tree of Life also can lead to discovery of new products that can be derived from closely related taxa. These products can be used to affect other areas such as biological control of pest organisms, agricultural productivity, and medicinal necessities. For example, in 1969 a new genus and species of bacterium, *Thermus aquaticus*, was described (Brock and Freeze 1969), which later revolutionized much of the way molecular biology is conducted when the DNA polymerase from this organism was used for the polymerase chain reaction (PCR; Saiki et al. 1988). PCR is a multimillion dollar a year industry that should top $1 billion by the year 2005. This technology has greatly benefited not only systematics and taxonomy but also many other biological sciences, including health and forensics. Discovery of *T. aquaticus* and use of the *Taq* DNA polymerase has spawned many additional technologies. A cursory view of any molecular supply catalog will show numerous chemicals and kits designed for use with PCR technology. Furthermore, such hardware as DNA thermocyclers and automated sequencers also has been developed.

Additionally, DNA polymerases from other closely related thermally stable organisms have been isolated with varying properties such as increased half-life at higher temperatures, decreased activity at lower temperatures, and 3'-5' exonuclease activity. As a result of PCR and the search for new DNA polymerases, many new life forms have been discovered. For example, the thermally stable microbes from which *Taq* was recovered were thought to comprise a tight cluster of a few genera that metabolized sulfur compounds (Woese 1987, Woese et al. 1990). Most of these organisms had to be cultured in the lab in order to be studied (DeLong 1992, Barns et al. 1994). However, PCR technology has allowed for a more in-depth study of these Archaea by using *in situ* amplification of uncultivated organisms that occur naturally in hot springs found in Yellowstone National Park. We now know that the Crenarchaeota display a wide variety of phenotypic and physiological properties in environments ranging from low temperatures in temperate and Antarctic waters to high-temperature hot springs (Barns et al. 1996, and citations therein). In fact, PCR coupled with phylogenetic analysis has allowed the discovery of not only new life forms within the kingdom Crenarchaeota but also new kingdoms within the domain Archaea (fig. 1.8; Barns et al. 1996).

Many new DNA polymerases have been discovered and patented and are now commercially available as a result of some of these discoveries. According to Bader et al. (2001:160), "Simple identification via phylogenetic classification of organisms has, to date, yielded more patent filings than any other use of phylogeny in industry." Patents also have been filed for vaccines associated with various viruses, such as porcine reproductive and respiratory syndrome virus and human immunodeficiency virus, that can target specific closely related virus populations based on phylogenetic analyses (citations within Bader et al. 2001).

Other economically important uses of a well-defined Tree of Life include discovery of biological control organisms as well as chemicals that target specific metabolic pathways of related taxa. Phylogenetic analyses of root-colonizing fungi revealed a group of nonpathogenic fungi that could serve as a biological control against pathogenic fungi (Ulrich et al. 2000). Phylogenetic studies are being conducted on numerous organisms for biological control, including nematodes and associated symbiotic bacteria and target moth, fly, and beetle pests (Burnell and Stock 2000); intracellular bacteria Wolbachia, parasitic wasps, and flies (Werren and Bartos 2001); and insect controls of thistles (Briese et al. 2002). In fact, Briese et al. (2002:149) state, "[G]iven the improved state of knowledge of plant phylogenies and the evolution of host use, it is time to base testing procedure purely on phylogenetic grounds, without the need to include less related test species solely because of economic or conservation reasons."

Other forms of control include using chemicals to attack specific metabolic pathways found in one clade of organisms but not in another. Two such pathways that occur in microbes and/or plants but not mammals are the shikimate pathway and the menevalonant pathway. The chemical glyphosate has been used commercially as an herbicide/pesticide for its ability to disrupt the shikimate pathway in algae, higher plants, bacteria, and fungi but theoretically does not have harmful effects on mammals (Roberts et al. 1998). Another pathway for consideration for an antimicrobial target is the mevalonate pathway. This is one of two pathways that convert isopentenyl diphosphate to isoprenoid found in higher organisms but is the only pathway found in many low-G+C (guanine + cytosine) gram-positive cocci. Phylogenetic analyses indicate that the genes found in these bacteria are more closely related to higher eukaryotic organisms and are likely a result of a very early horizontal gene transfer between eukaryotes and bacteria before the divergence of plants, animals, and fungi (Wilding et al. 2000). This pathway therefore represents a means for control of the gram-positive bacteria.

Another economic value to society may lie in DNA/RNA vaccines. Knowing the phylogenetic relationships of target organisms may allow for the development of broad-scale vaccines or "species"-specific vaccines. DNA vaccines are relatively easy to make and can be produced much quicker than conventional vaccines (Dunham 2002). Although there still are several safety issues to address before wide-scale use of nucleic acid vaccines (Gurunahan et al. 2000), this technology can be used to treat several wildlife diseases (Dunham 2002) and can be used potentially as a defense against a bioterrorist attack.

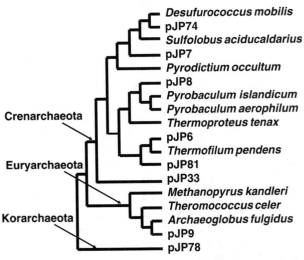

Figure 1.8. Newly discovered organisms of Archaea; modified (reduced tree) from Barns et al. (1996). Taxa labeled "pJP" represent new life forms discovered using ribosomal RNA sequences amplification from uncultured organisms. New taxa were found within two kingdoms representing Crenarchaeota and Euryarchaeota as well as the new kingdom Korarchaeota (pJP78 and other similar rDNA sequences).

Conclusions

Assembling the Tree of Life will be a monumental task and possibly one of the greatest missions we as a society could hope to achieve. It will require numerous collaborations of multiple disciplines within the scientific community. The Tree of Life has already provided many benefits, not only to science but to humanity as well. These benefits are but a small fraction of what a fully assembled tree would have to offer. In many respects, the power of a complete Tree of Life compared with the partial one we have now is analogous to the breakthroughs made possible by a complete periodic table compared with a partial one. Imagine chemists trying to predict the structure and function of new compounds armed with the knowledge of only 10% of the periodic table. The Tree of Life will form the critical infrastructure on which all comparative biology will rest. Once completed, this infrastructure will fuel scientific breakthroughs across all of the life sciences and many other fields of science and engineering and will foster enormous economic development.

Constructing the Tree of Life will create extraordinary opportunities to promote research across interdisciplinary fields as diverse as genomics, computer science and engineering, informatics, mathematics, earth sciences, developmental biology, and environmental biology. The scientific and engineering problem of building the Tree of Life is complex and presents many challenges, but these challenges can be accomplished in our lifetime. Already, the international genomics databases [GenBank (http://www.ncbi.nih.gov/Genbank/index.html), EMBL (http://www.ebi.ac.uk/embl/), and DDBJ (http://www.ddbj.nig.ac.jp/)] grow at an exponential rate, with the number of nucleotide bases doubling approximately every 14 months. Currently, there are more than 17 billion bases from more than 100,000 species listed by the National Center for Biotechnology Information (available at http://www.ncbi.nlm.nih.gov/). Data from nongenomic sources, such as anatomy, behavior, biochemistry, or physiology, also have been collected on thousands of species, and many thousands of phylogenies have been published for groups widely distributed across the tree. To truly benefit industry, agriculture, and health and environmental sciences, the overwhelming amount of data required to construct the Tree of Life must be appropriately organized and made readily available.

Cracraft (2002) considered the question "What is the Tree of Life?" to be one of seven great questions of systematic biology. In many respects, the answer to that question is fundamental to all the others and will enable their resolution. Even fundamental questions such as what a species is and how many there are will be facilitated by assembling the tree. It should be noted that addressing the latter question and assembling the Tree of Life go hand-in-hand and form a positive feedback loop. Discovery of new species will provide new information that will enhance tree assembly, and at the same time tree assembly will provide the information necessary for the discovery of new species.

The other great questions listed by Cracraft (2002) actually require a tree for their resolution. As addressed in this chapter, however, great questions from other disciplines also require a highly resolved tree for their solution. In fact, the answer to few scientific questions offers the potential to fuel as many major discoveries in other disciplines as does resolution of the Tree of Life. Fields such as evolution and development, medicine, and bioengineering will immediately be able to rapidly address questions not before possible without the phylogenetic infrastructure provided by the tree. These discoveries will in turn fuel economic development, inform land management decisions, and protect the environment.

Assembly of the Tree of Life on this scale, however, will require the development of innovative database structures (both hardware and software) that support relational authority files with annotation of both genetic and nongenetic information. Unprecedented levels and methods of computational capabilities will need to be developed as genomic information from the "wet" studies in the laboratory and field is analyzed in the "dry" environments of computers. Already a new field of phyloinformatics and computational phylogenetics is emerging from these efforts that promise to harness phylogenetic knowledge to integrate and transform data held in isolated databases, allowing the invention of new information and knowledge.

What is needed is an international effort to coordinate tree construction, facilitate hardware and software design, promote collaboration among researchers, and facilitate database design and maintenance and the creation of a center to help coordinate and facilitate these activities. Owing to fundamental theoretical advances in manipulating genomic and other kinds of data, to the availability of major new sources of data, and the development of powerful analytical computational tools, we now have the potential (given sufficient resources and coordination) to assemble much of the entire Tree of Life within the next few decades, at least for currently known species. The potential of building a Tree of Life extends far beyond the basic and applied biological sciences and promises to provide much value to society. Building an accurate, complete Tree of Life depicting the relationships of all life on Earth will call for major innovation in many fields of science and engineering similar to those derived from sending a man to the moon or sequencing the entire human genome. The benefits to society from such an undertaking are enormous and may well extend beyond the many provided by these two successful efforts.

Acknowledgments

We thank the Centers for Disease Control and Prevention, the U.S. National Science Foundation, and the National Institutes of Health for previous financial support for many of the discoveries reported here. We especially thank the National Science Foundation for providing the leadership for the initiation of this critical effort. We also thank the Museum of Southwestern Biology of the University of New Mexico (UNM) and the Department of Biology (UNM) for their support.

Literature Cited

Bader, D., B. Moret, and L. Vawter. 2001. Industrial applications of high-performance computing for phylogeny reconstruction. Pp. 159–168 in Proceedings of the SPIE Commercial Applications for High-Performance Computing (H. Siegel, ed.), vol. 4528. Denver, CO.

Barker, G. M. 2002. Phylogenetic diversity: a quantitative framework for measurement of priority and achievement in biodiversity conservation. Biol. J. Linn. Soc. 76:165–194.

Barns, S. M., C. F. Delwiche, J. D. Palmer, and N. R. Pace. 1996. Perspectives on archaeal diversity, thermophily and monophyly from environmental rRNA sequences. Proc. Natl. Acad. Sci. USA 93:9188–9193.

Barns, S. M., R. E. Fundyga, M. W. Jeffries, and N. R. Pace. 1994. Remarkable archaeal diversity detected in a Yellowstone National Park hot spring environment. Proc. Natl. Acad. Sci. USA 91:1609–1613.

Briese, D. T., A. Walker, W. J. Pettit, and J. L. Sagliocco. 2002. Host-specificity of candidate agents for the biological control of Onopordum spp. thistles in Australia: an assessment of testing procedures. Biocontr. Sci. Technol. 12:149–163.

Brock, T. D., and H. Freeze. 1969. Thermus aquaticus gen. n. and sp. n., a nonsporulating extreme thermophile. J. Bacteriol. 98:289–297.

Burnell, A. M., and S. P. Stock. 2000. Heterorhabditis, Steinernema and their bacterial symbionts: lethal pathogens of insects. Nematology 2:31–42.

Centers for Disease Control and Prevention, Special Publications Branch. 2003. Hantavirus Pulmonary Syndrome. On-line slide show. Available: http://www.cdc.gov/ncidod/diseases/hanta/hps/noframes/hpsslideset/hpsslides1–12.htm.

Childs, J. E., T. G. Ksiazek, C. F. Spiropoulou, J. W. Krebs, S. Morzunov, G. O. Maupin, K. L. Gage, P. E. Rollin, J. Sarisky, R. E. Enscore, et al. 1994. Serologic and genetic identification of Peromyscus maniculatus as the primary rodent reservoir for a new hantavirus in the southwestern United States. J. Infect. Dis. 169:1271–1280.

Colinvaux, P. A., G. Irion, M. E. Rasanen, M. B. Bush, and J. de Mello. 2001. A paradigm to be discarded: geological and paleoecological data falsify the Haffer & Prance refuge hypothesis of Amazonian speciation. Amazon. Limnol. Oecol. Reg. Syst. Fluminis Amazon. 16:609–646.

Cracraft, J. 2002. The seven great questions of systematic biology: an essential foundation for conservation and the sustainable use of biodiversity. Ann. Mo. Botan. Gard. 89:127–144

DeLong, E. F. 1992. Archaea in coastal marine environments. Proc. Natl. Acad. Sci. USA 89:5685–5689.

DeLong, E. F., and N. R. Pace. 2001. Environmental diversity of Bacteria and Archaea. Syst. Biol. 50:470–478.

Desmet, P. G., R. M. Cowling, A. G. Ellis, and R. L. Pressey. 2002. Integrating biosystematic data into conservation planning: perspectives from southern Africa's succulent karoo. Syst. Biol. 51:317–330.

Dragoo, J. W., and R. L. Honeycutt. 1997. Systematics of mustelid-like carnivores. J. Mamm. 78:426–443.

Dragoo, J. W., J. Salazar-Bravo, L. J. Layne, and T. L. Yates. 2003. Relationships within the Calomys callosus species group based on amplified fragment length polymorphisms. Biochem. Syst. Ecol. 31:703–713.

Dunham, S. P. 2002. The application of nucleic acid vaccines in veterinary medicine. Res. Vet. Sci. 73:9–16.

Erwin, T. L. 1991. An evolutionary basis for conservation strategies. Science 253:750–752.

Flynn, J. J., M. A. Nedbal, J. W. Dragoo, and R. L. Honeycutt. 2000. Whence the red panda? Mol. Phylogenet. Evol. 17:190–199.

Fuhrman, J. A., K. McCallum, and A. A. Davis. 1992. Novel major archaebacterial group from marine plankton. Nature 356:148–149.

Glass, G. E., T. L. Yates, J. B. Fine, T. M. Shields, J. B. Kendall, A. G. Hope, C. A. Parmenter, C. J. Peters, T. G. Ksiazek, C. S. Li, J. A. Patz, and J. N. Mills. 2002. Satellite imagery characterizes local animal reservoir populations of Sin Nombre virus in the southwestern United States. Proc. Natl. Acad. Sci. USA 99:16817–16822.

Glor, R. E., L. J. Vitt, and A. Larson. 2001. A molecular phylogenetic analysis of diversification in Amazonian Anolis lizards. Mol. Ecol. 10:2661–2668.

Gurunahan, S., D. M. Klinman, and R. A. Seder. 2000. DNA vaccines: immunology, application, and optimization. Annu. Rev. Immunol. 18:927–974.

Haffer, J. 1997. Alternative models of vertebrate speciation in Amazonia: an overview. Biodivers. Conserv. 6:451–476.

Haffer, J., and G. T. Prance. 2001. Climatic forcing of evolution in Amazonia during the Cenozoic: on the refuge theory of biotic differentiation. Amazon.-Limnol. Oecol. Reg. Syst. Fluminis Amazon 16:579–605.

Jia, X. Y., T. Briese, I. Jordan, A. Rambaut, J. C. Chi, J. S. Mackenzie, R. A. Hall, J. Scherret, and W. I. Lipkin. 1999. Genetic analysis of West Nile New York 1999 encephalitis virus. Lancet 354:1971–1972.

Johnson, K. M., P. A. Webb, and G. Justines. 1972. Biology of Tacaribe-complex viruses in lymphocytic choriomeningitis virus and other arenaviruses. Pp. 241–258 in Heinrich-Pette-Institut fur experimentelle Virologie und Immunologie, Universität Hamburg (F. Lehmann-Grube, ed). Springer-Verlag, Berlin.

Justines, G., and K. M. Johnson. 1969 Immune tolerance in Calomys callosus infected with Machupo virus. Nature (Lond.) 222:1090–1091.

Lanciotti, R. S., J. T. Roehrig, V. Deubel, J. Smith, M. Parker, K. Steele, B. Crise, K. E. Volpe, M. B. Crabtree, J. H. Scherret, et al. 1999. Origin of the West Nile virus responsible for and outbreak of encephalitis in the northeastern United States. Science 286:2333–2337.

Meyer, D. A. 1998. Special report: Asian longhorned beetle Anoplophora glabripennis. Hardwood Res. Bull. 500:1–5.

Moret, B. M. E., D. A. Bader, and T. Warnow. 2002. High-performance algorithm engineering for computational phylogenetics. J. Supercomput. 22:99–111.

Moritz, C. 2002. Strategies to protect biological diversity and the evolutionary processes that sustain it. Syst. Biol. 51:238–254.

Moritz, C., J. L. Patton, C. J. Schneider, and T. B. Smith. 2000. Diversification of rainforest faunas: an integrated molecular approach. Annu. Rev. Ecol. Syst. 31:533–563.

Morrison, L. W., and L. E. Gilbert. 1999. Host specificity in two additional *Pseudacteon* spp (Diptera: Phoridae), parasitoids of *Solenopsis* fire ants (Hymentopera: Formicidae). Flor. Entomol. 82:404–409.

Nichol, S. T., C. F. Spiropoulou, S. Morzunov, P. E. Rollin, T. G. Ksiazek, H. Feldmann, A. Sanchez, J. Childs, S. Zaki, and C. J. Peters. 1993. Genetic identification of a hantavirus associated with an outbreak of acute respiratory illness. Science 262:914–917.

Oliver, M. J., Z. Tuba, and B. D. Mishler. 2000. The evolution of vegetative desiccation tolerance in land plants. Plant Ecol. 151:85–100.

Porter, S. D., and J. A. Briano. 2000. Parasitoid-host matching between the little decapitating fly *Pseudacteon curvatus* from Las Flores, Argentina and the black fire ant *Solenopsis richteri*. Fla. Entomol. 83:422–427.

Richardson, J. E., R. T. Pennington, T. D. Pennington, and P. M. Hollingsworth. 2001. Rapid diversification of a species-rich genus of neotropical rain forest trees. Science 293:2242–2245.

Roberts, F., C. W. Roberts, J. J. Johnson, D. E. Kyle, T. Krell, J. R. Coggins, G. H. Coombs, W. K. Milhous, S. Tzipori, D. J. P. Ferguson, D. Chakrabarti, and R. McLeod. 1998. Evidence for the shikimate pathway in apicomplexan parasites. Nature 393:801–805.

Rosen, D. 1986. The role of taxonomy in effective biological control programs. Agric. Ecosyst. Environ. 15:121–129.

Saiki, R. K., D. H. Gelfand, S. Stoffel, S. J. Scharf, R. Higuchi, G. T. Horn, K. B. Mullis, and H. A. Erlich. 1988. Primer-directed enzymatic amplification of DNA with a thermo-stable DNA polymerase. Science 239:487–491.

Salazar-Bravo, J., J. W. Dragoo, M. D. Bowen, C. J. Peters, T. G. Ksiazek, and T. L. Yates. 2002. Natural nadality in Bolivian hemorrhagic fever and the systematics of the reservoir species. Infect. Genet. Evol. 1:191–199.

Singh, J. S. 2002. The biodiversity crisis: a multifaceted review. Curr. Sci. 82:638–647.

Tsutsui, N. D., A. V. Suarez, D. A. Holway, and T. J. Case. 2001. Relationships among native and introduced populations of the Argentine ant (*Linepithema humile*) and the source of introduced populations. Mol. Ecol. 10:2151–2161.

Ulrich, K., C. Augustin, and A. Werner. 2000. Identification and characterization of a new group of root-colonizing fungi within the *Gaeumannomyces-Phialophora* complex. New Phytol. 145:127–135.

Wang, Q., I. W. B. Thornton, and T. R. New. 1999. A cladistic analysis of the Phoracanthine genus *Phoracantha* Newman (Coleoptera: Cerambycidae: Cerambycinae), with discussion of biogeographic distribution and pest status. Ann. Entomol. Soc. Am. 92:631–638.

Wang, R. L., A. Stec. J. Hey, L. Lukens, and J. Doebley. 2001. The limits of selection during maize domestication. Nature 398:236–239.

Webb, P. A., G. Justines, and K. M. Johnson. 1975. Infection of wild and laboratory animals with Machupo and Latino viruses. Bull. WHO 52:493–499.

Werren, J. H., and J. D. Bartos. 2001. Recombination in *Wolbachia*. Curr. Biol. 11:431–435.

Wheeler, Q. D. 1995. Systematics and Biodiversity. Bioscience 45:S21–S28.

Wilding, E. I., J. R. Brown, A. P. Bryant, A. F. Chalker, D. J. Holmes, K. A. Ingraham, S. Iordanescu, C. Y. So, M. Rosenberg, and M. N. Gwynn. 2000. Identification, evolution, and essentiality of the mevalonate pathway for isopentenyl diphosphate biosynthesis in gram-positive cocci. J. Bacteriol. 182:4319–4327.

Woese, C. R. 1987. Bacterial evolution. Microbiol. Rev. 51:221–271.

Woese, C. R., O. Kandler, and M. L. Wheelis. 1990. Towards a natural system of organisms: proposal for the domains Archaea, Bacteria, and Eucarya. Proc. Natl. Acad. Sci. USA 87:4576–4579.

Yates, T. L., J. N. Mills, C. A. Parmenter, T. G. Ksiazek, R. P. Parmenter, J. R. Vande Castle, C. H. Calisher, S. T. Nichol, K. D. Abbott, J. C. Young, et al. 2002. The ecology and evolutionary history of an emergent disease: hantavirus pulmonary syndrome. Bioscience 52:989–998.

Rita R. Colwell

A Tangled Bank

Reflections on the Tree of Life and Human Health

Writing almost 150 years ago, Charles Darwin coined the name "tree of life" to describe the evolutionary patterns that link all life on Earth. His work set a grand challenge for the biological sciences—assembling the Tree of Life—that remains incomplete today. In the intervening years, we have come to understand better the significance of this challenge for our own species. As human activity alters the planet, we depend more and more on our knowledge of Earth's other inhabitants, from microorganisms to mega fauna and flora, to anticipate our own fate. Aldo Leopold, the great naturalist and writer, wrote, "To keep every cog and wheel is the first precaution of intelligent tinkering" (1993:145–146). However, the simple fact is that we do not yet know "what's out there," and we are often unaware of what we have already lost. The total number of species may number between 10 and 100 million, of which approximately 1.7 million are known and only 50,000 described in any detail.

Today, we are in a better position to carry forward Darwin's program. Museums, universities, colleges, and research institutions are invaluable repositories for data painstakingly collected, conserved, and studied over the years. Add a flood of new information from genome sequencing, geographical information systems, sensors, and satellites, and we have the raw material for realizing Darwin's vision.

One of the great challenges we face in assembling the Tree of Life is assembling the talent—bringing together the systematists, molecular biologists, computer scientists, and mathematicians—to design and deploy new computational tools for phylogenetic analysis. Systematists are as scarce as hen's teeth these days. They may be our most endangered species.

The National Science Foundation (NSF) has a long history of supporting the basic scientific research, across all disciplines, that has placed us within reach of achieving this objective. Now, the NSF has begun a new program to help systematists and their colleagues articulate the genealogical Tree of Life. We expect that this tree will do for biology what the periodic table did for chemistry and physics—provide an organizing framework. But advancing scientific understanding is not the sole objective. New knowledge is important for our continued prosperity and well being on the planet. My aim is to explore some of the common ground shared by the Tree of Life project and one important focus of social concern—human health.

My title, "A Tangled Bank," comes from Darwin's *The Origin of Species*, where he invites us to "contemplate a tangled bank" and to reflect on the complexity, diversity, and order found in this commonplace country landscape:

> It is interesting to contemplate a tangled bank, clothed with many plants of many kinds, with birds singing on the bushes, with various insects flitting about, and with worms crawling through the damp earth, and to reflect that these elaborately constructed forms, so different from each other, and dependent upon each other in so complex a manner, have all been produced by laws acting around us. (Darwin 1859)

Darwin understood evolution as the source of complexity and diversity, and his vision radically altered our perspective of life on Earth, past and present. He developed much of his theory in exotic places while sailing on the HMS *Beagle*. Just more than a century later, another voyage, on the Apollo spacecraft, gave us a first view of our blue Earth suspended jewel-like in space. That image is now as familiar as Darwin's country landscape. Awe-inspiring and beautiful, planet Earth appeared to us for the first time as a whole. But above all, we saw it as finite and vulnerable.

Today, another 30 years down the road, we are better able to chart the vast interdependencies that take us from country bank to global systems. We are beginning to understand that abrupt change and what we call "emerging" structures characterize many natural phenomena—from earthquakes to the extinction of some species. We know that the impact of humans on natural systems is increasing, but we don't yet have the full picture of how environmental change—human induced or otherwise—will cascade through natural systems.

There are two themes that intertwine in this chapter. The first is the observation that the health of our species and the health of the planet are inextricably linked. The second is that a new vision of science in the 21st century, biocomplexity, will speed us to a better understanding of those interconnections. I use the term "biocomplexity" to describe the dynamic web of relationships that arise when living things at all levels, from molecules to genes to organisms to ecosystems, interact with their environment.

Early on, we used the term "ecosystems approach" to describe part of what we mean by "biocomplexity." Now, technologies allow us to delve into the structure of the very molecules that compose cells—and simultaneously, to probe the global system that encompasses the biosphere. Advances in DNA sequencing, supercomputing, and computational biology have literally revolutionized our view of the Tree of Life. By comparing genetic sequences from different organisms, we can now chart their genealogy and construct a universal phylogenetic tree.

A cartoon from the British satirical magazine, *Punch*, published shortly after *The Origin of Species*, depicts the evolution of a worm into a human—the human, in this case, being Charles Darwin himself. The caption reads: "Man is but a Worm." The intent, of course, was to ridicule the notion that a human could in any way be related to a lowly worm (Punch 1882). Today, these odd juxtapositions are no longer the subject of satire. In research published in February of 2002, S. Blair Hedges and colleagues from the United States and Japan compared 100 genes shared among three organisms: the human, the fruit fly, and the nematode worm (Blair et al. 2002). The complete genomes of all three organisms have been sequenced; so finding candidate genes was a straightforward exercise in matching. The researchers determined that the human genome is more closely related to the fly than to the worm, clarifying a major branch on the Tree of Life. But it doesn't eliminate the worm from our ancestry.

In the area of genomics, many people are looking at divergent organisms and beginning to realize connections never before imagined. Steven Tanksley and his colleagues at Cornell are exploring the genome of tomatoes to gain insight into how wild strains have evolved into the delicious fruits we find in supermarkets today. A single gene is responsible for "plumping" in tomatoes. He discovered that this gene is similar to a human oncogene—a cancer-causing gene. This match suggests a common mechanism in the cellular processes leading to large, edible fruit in plants and cancers in humans (Frary et al. 2000). This illustrates an important point. Getting the sequence is really only the first step. Functional analysis is needed to confirm the inference of function based on similar (homologous) sequences.

Our current genomic tool kit is a recent development. Research initiated in the late 1920s led scientists to the discovery that an extract from the bacterium that causes pneumonia could change a closely related, but harmless, bacterium into a virulent one in the test tube. A search began for the "transforming factor" responsible for such a change. Both protein and DNA were candidates, but scientific opinion favored protein. The puzzle was solved when Avery et al. (1944) determined that DNA was the transforming factor. Another decade passed before Watson and Crick (1953) described the structure of the DNA molecule and set off a revolution in molecular biology that is still unfolding.

The first genome of a self-replicating, free-living organism—the tiny bacterium *Haemophilus influenzae* strain Rd—was completed in 1995 (Fleischmann et al. 1995). The first genome of a multicellular organism—the nematode worm (*Caenorhabditis elegans*)—was published in 1998 (*Caenorhabditis elegans* Sequencing Consortium 1998), followed by the fruit fly (*Drosophila melanogaster*) genome in 2000 (Adams et al. 2000). The sequencing of the human genome was completed just last year (Venter et al. 2001). Today, we "stand on the shoulders of many giants" who pioneered the revolution in molecular biology and genomics. But all the disciplines have contributed to our progress. From the tiny genome of the first bacterium sequenced with 1.8 million base pairs to the 3.12 billion that comprise the human genome was a leap of enormous magnitude. Researchers from Celera Genomics, who helped sequence the human genome, estimate that assembly of the 3.12 billion base pairs of DNA required 500 million trillion sequence comparisons. Completing the human genome project might have taken years to decades to accomplish without the terascale power of our newest computers and a battery of sophisticated computation tools.

We know that one of the most important tools in modern-day science's arsenal of genetic engineering is PCR—the polymerase chain reaction. This technique was pioneered in the 1980s in the private sector. But first came the discovery of the heat-resistant DNA polymerase needed to untwine the double strands of DNA. Brock and Freeze discovered the source of this heat-resistant enzyme in 1968—a bacterium

(*Thermus aquaticus*), found in a hot spring in Yellowstone National Park (Brock and Freeze 1969).

These new tools have radically changed our perspective of life on Earth and taught us to reorient ourselves on the Tree of Life. DNA sequencing enables researchers to overcome the limitations of culturing microorganisms in the lab and vastly improves our ability to detect and describe microbial species. The surprising feature is the diversity and sheer multitude of microorganisms, which represent the lion's share of Earth's biodiversity. Although microorganisms constitute more than two-thirds of the biosphere, they represent a huge unexplored frontier. Of bacterial species in the ocean, fewer than 1% have been cultured. Just a milliliter of seawater holds about one million cells of these unnamed species and about 10 million viruses. On average, a gram of soil may contain as many as a billion microorganisms.

Research is also revealing phenomenal diversity among microorganisms, especially among prokaryotes. They inhabit a wide range of what we consider extreme environments—hydrothermal vents on the sea floor, the ice floes of polar regions, and the deep, hot, stifling darkness of South African gold mines. Researchers have discovered that these organisms display novel properties and assume novel roles in ecosystems and in Earth's cycles. Many are being investigated for these unique properties and the applications that harnessing them can provide.

In these and other less extreme places, microorganisms have been wildly successful. They adapt very rapidly and evolve very quickly to thrive in novel environments. Among other feats, they have evolved diverse symbiotic relationships with other creatures. The familiar shape of the Tree of Life might appear radically altered if we take into account the intriguing variety of ways that prokaryotes exchange genetic information with other organisms, including lateral gene transfer.

Only a handful of microorganisms are human pathogens. Others infect plants and both domestic and wild animals. But what an impact on human life they have had—both past and present. We know that infectious diseases are a leading cause of death in the world today, including the Americas (WHO 2001). Bacteria play a prominent role, but a wide variety of viruses, protozoa, fungi, and a group of worms, the helminthes, and other parasites also cause infectious diseases. Pathogens—particularly bacteria and viruses—display the same ability to adapt and the same genetic flexibility as their harmless cousins. The increasingly serious problem of drug resistance in pathogens is a direct result of this evolutionary flexibility. Pathogens respond to the excessive and unwarranted use of antibiotics, for example, by developing antibiotic resistance. In many cases, antibiotic genes are linked to heavy metal resistance. Work in my own laboratory in the late 1970s and early 1980s on bacteria in Chesapeake Bay shows a link between genes that encode for metal resistance and genes that encode for antibiotic resistance, notably on plasmids. Other linkages may yet be described.

Knowing how microorganisms have evolved into pathogens and how they differ from less harmful relatives can provide the key in tracking the origin and spread of emerging diseases and their vectors. In 2000 and 2001, several outbreaks of polio were reported from Hispaniola. Phylogenetic analysis showed conclusively that the poliovirus was not the "wild" variety that is the target of eradication efforts worldwide. Where had it come from? The Sabin oral vaccine, a live but weakened poliovirus, is widely used in developing countries. These viruses are shed in the feces of vaccinated individuals. When individuals who have not been vaccinated come into contact with these viruses, possibly in unsanitary food or water, they will become infected. The puzzle in the Hispaniola case is how the attenuated virus reverted to a virulent strain. Genetic sequencing demonstrated that the poliovirus combined with at least four closely related enteroviruses. As the virus spread, one of these variants developed virulence (Kew et al. 2002).

This example demonstrates that human institutions are as much a part of the ecology of infectious disease as recombination on the molecular level. An inadequate vaccination program, combined with poor sanitary conditions, helped to create the environment for the emergence of a new strain of poliovirus.

The rapid increase in cases of dengue fever reported between 1955 and the present day provides another example of a reemerging infectious disease. The World Health Organization (WHO) estimates that as many as 50 million people are infected each year, with an additional 2.5 billion people at risk (WHO 1999). A major epidemic in Brazil caused more than 300,000 cases of dengue in the first three months of 2002 alone. (WHO 2002) Dengue is not a new disease. Major epidemics were recorded in the 18th century in Asia. What caused this infectious disease to reemerge as a major public health problem over the past 50 years? Genetic sequencing has shown that dengue fever and its more deadly form, dengue hemorrhagic fever, are caused by a group of four closely related viruses that infect the mosquito *Aedes aegypti* (Loroño-Pino et al. 1999). Each variant of the dengue virus produces immunity only to itself, so individuals may suffer as many as four infections in a lifetime. Dengue hemorrhagic fever may be caused by these multiple infections. Genetic sequencing is indispensable in tracking the origin and spread of each variant. Knowing which virus type is circulating may be important in determining the potential risk for an outbreak of dengue hemorrhagic fever.

The causes of the current global pandemic are not well understood. But the spread of *Aedes aegypti* is certainly a factor. *Aedes*, a vector for yellow fever, was nearly eradicated in the 1950s and 1960s. After a vaccine for yellow fever became available, mosquito control efforts waned, and *Aedes* has come back with a vengeance to repopulate and even expand its former territory. The Asian tiger mosquito, *Aedes albopictus*, is also a potential vector of epidemic dengue. In the United States, it was first reported in 1995 in Texas, and has since become established in 26 states. It is simply not known whether the tiger mosquito could initiate a major dengue epidemic in the United States. Like *Aedes aegypti*, the tiger mosquito can survive in urban environments. And like *Aedes aegypti*, it is also a possible vector for yellow fever. Once the vector is present, the pathogen may not be far behind.

Genetic sequencing is a critical new tool in the battle to control infectious disease. Sequencing may help to determine the origin of a pathogen, for example, whether it is endemic or imported. And tracking the geographical or ecological origins based on sequencing can also pinpoint natural reservoirs, where health efforts can be focused. We may never be able to eradicate pathogens that are widespread in the environment, but knowledge of how they evolved, their mechanisms of adaptation, and their ecology will help us design effective prevention and control measures.

My own research has focused on the study of how factors combine to cause cholera, a devastating presence in much of the world, although largely controlled in the United States. It is endemic in Bangladesh, for example, where I've done much of my research. My scientific quest to understand cholera began more than 30 years ago, in the 1970s, when my colleagues and I realized that the ocean itself is a reservoir for the bacterium *Vibrio cholerae*, the cause of cholera, by identifying the organism in water samples from the Chesapeake Bay. Copepods, the minute relatives of shrimp that live in salt or brackish waters, are the hosts for the cholera bacterium, which they carry in their gut as they travel with currents and tides. We now know that environmental, seasonal, and climate factors influence copepod populations, and indirectly cholera. In Bangladesh, we discovered that cholera outbreaks occur shortly after sea-surface temperature and height peak. This usually occurs twice a year, in spring and fall, when populations of copepods peak in abundance. Ultimately, we can connect outbreaks of cholera to major climate fluctuations. In the El Niño year of 1991, a major outbreak of cholera began in Peru and spread across South America. Linking cholera with El Niño/Southern Oscillation events provides us with an early warning system to forecast when major cholera outbreaks are likely to occur (Colwell 2002).

Understanding cholera requires us to explore the problem on different scales. We study the relationship between the bacterium *Vibrio cholerae*, which causes the disease, and its copepod host. We look at the ecological factors that affect copepod reproduction and survival. We observe the local and oceanic climatic factors related to currents and sea-surface temperature. On a microscopic level, we look at molecular factors related to the toxin genes in *V. cholerae* to understand the function of genes and how they evolved and adapted in relation to copepods. This in turn may provide new insight into how these pathogens cause disease in humans. Add the economic and social factors of poverty, poor sanitation, and unsafe drinking water, and we begin to see how this microorganism sets off the vast societal traumas of cholera pandemics (Lipp et al. 2002). We cannot eradicate

the cholera bacterium. Understanding *V. cholerae* on the molecular level, tracing the ecology of the disease, forecasting major outbreaks, and controlling them are our only options (Colwell 2002). Other infectious diseases—relayed by vectors, water, food, air, or otherwise—also interact with climate. The El Niño/Southern Oscillation climate pattern has been linked to outbreaks of malaria, dengue fever, encephalitis, and diarrheal disease as well as cholera. Environmental change of all kinds may affect agents of infectious disease. Changes in climate could nudge pathogens and vectors to new regions. Agents of tropical disease could drift toward the polar regions, creating "emerging diseases" at new locales. Because the evolutionary "speed limit" of many pathogens is remarkably high, pathogens might adapt to new ecological circumstances with remarkable ease.

When we look for connections between the Tree of Life and human health, infectious diseases may be the first case that comes to mind. But the nexus among evolution, ecology, genomics, and human health guides us farther afield. When we view our planet through the eyes of complexity, we see motifs that recur with striking constancy. We can often use motifs found in harmless organisms to better understand the mechanisms in their close cousins that cause disease. One case in point is recent research on aphids, the tiny plant pests that cause major agricultural damage. A tiny bacterium, *Buchnera*, lives inside the aphid's cells. It provides essential nutrients to the aphid hosts, and the hosts reciprocate. Over the years, aphids and *Buchnera* have evolved together, so that today, different species of aphids are associated with different species of the bacterium. Baumann and colleagues have traced this cospeciation more than 150–250 million years (Bauman et al. 1997).

The role of these endosymbionts in the adaptation of the aphids to host plants is under investigation as part of the NSF biocomplexity initiative. One of the questions of interest concerns the extent of convergence in the evolution of symbiotic bacteria found within a range of insect groups. *Buchnera* was the first endosymbiont genome to be sequenced. Sequence analysis has shown that *Buchnera* is missing many of the genes required for "independent life"— including the ones that turn off production of the nutrients necessary for the host's survival. Recently, Ochman and Moran (2001) have contrasted the *Buchnera* genome with a hypothetical ancestor of the enteric bacterium *Escherichia coli*, thought to be a relative of *Buchnera*. The comparison shows massive gene reduction in *Buchnera*, a phenomenon also found in many pathogens. Gene loss in both symbionts and pathogens may be key to understanding how human pathogens cause disease. By studying symbionts such as *Buchnera* that live in harmony with their hosts, it may be possible to unravel the adaptive mechanisms that pathogens living inside human cells use to evade the body's defenses. New strategies for combating infections could follow.

Organisms can also shape the physical environment. An example is work by Jillian Labrenz and colleagues (2000) looking at a complex environment: an abandoned and flooded mine. Biofilms here live on the floors of the flooded tunnels. The goal of the work is to understand geomicrobiological processes from the atomic scale up to the aquifer level. Acid drainage from such mines is a severe environmental problem. At one mine being studied, workers accidentally left a shovel in the discharge; the next day half the shovel was eaten away by the acid waste.

We search for ways to remediate the damage in areas like these. Some of the microorganisms in the biofilms play a surprising role (Labrenz et al. 2000). For one, they can clean the zinc-rich waters to a standard better than that of drinking water. At the same time, bacteria in the biofilms are depositing minerals on the tunnel floors. Aggregates of tiny zinc sulfide crystals just 2–5 nm in diameter are formed in very high concentrations by the activity of microorganisms. The work sheds light on an environmental problem, while giving insights into basic science with economic benefit: we are learning how mineral ores of commercial value are formed. Researchers are studying this system on a number of scales—from the early evolution of life on Earth to the nanoscale forces operating inside the microorganisms and in their immediate environment.

Because microorganisms play a central role in the cycling of carbon, nutrients, and other matter, they have large impacts on other life—including humans. Recent research has shed new light on these complex interdependencies in the oceans. The molecule rhodopsin is a photopigment that binds retinal. Activated by sunlight, retinal proteins have been found to serve the energy needs of microorganisms, as well as steer them to light. In people, a different form of the molecule provides the light receptors for vision. Until recently, rhodopsin was thought to occur only in a small number of species, namely, the halobacteria, which thrive in environments 10 times saltier than seawater. Despite the name, they are actually members of the Archaea, one of the three major branches of life and among the oldest forms of life on Earth.

Obed Béjà, Edward DeLong, and colleagues at the Monterey Bay Aquarium Research Institute have now shown that bacteria containing a close variant of this energy-generating, light-absorbing pigment are widespread in the world's oceans (Béjà et al. 2000). This is the first such molecule to be associated with bacteria. The researchers also discovered that genetic variants of these bacteria contain different photopigments in different ocean habitats. The protein pigments appear to be tuned to absorb light of different wavelengths that match the quality of light available (Béjà et al. 2001). These bacteria are present in significant numbers and over a wide geographic range, and may occupy as much as 10% of the ocean's surface. Such abundance may point to a significant new source of energy in the oceans. It is also a startling reminder of what we have yet to discover. We begin to map biocomplexity by tracing the links from the function of a protein to the distribution and variation of bacterial populations to biogeochemical cycles. Human health is ulti-mately linked to the complex dynamics of these vast biogeochemical cycles. Understanding how they function is vital in order to anticipate how disruptions might alter them.

I've taken my examples from the world of microorganisms partly because I'm a microbiologist—but also because this is an emerging frontier. Microorganisms may well be our "canaries in the mineshaft," warning us of subtle environmental changes, from the local to the global. Carl Woese, whose work has done so much to expand our vision of microbial diversity, goes further: "[M]icrobes are the essential, stable underpinnings of the biosphere—without bacteria, other life would not continue to exist" (Woese 1999:263).

This past March, the U.S. Geological Survey published an assessment that sampled 139 waterways across the U.S. for 95 chemicals (Koplin et al. 2002). They found a wide array of substances present in trace amounts in 80% of the waterways sampled. The chemicals ranged from caffeine, to steroids, to antibiotics and other pharmaceuticals. All are bioactive substances—chemicals that interact with organisms at the molecular level. Yet we have very little understanding of how these substances may be affecting microbial communities. Are they altering the structure of microbial ecosystems in soils and water? What are the selective pressures on organisms exposed to these substances? If the composition of microbial communities is seriously altered, or if the abundance or diversity of microorganisms is diminished, what are the implications for the availability of nutrients in ecosystems and for agricultural productivity?

Other organisms may be providing some answers. Research reported recently by Tyrone Hayes and colleagues from the University of California–Berkeley found that atrazine, the nation's top-selling weed killer, turns tadpoles into hermaphrodites with both male and female sexual characteristics. The herbicide also lowers levels of the male hormone testosterone in sexually mature male frogs by a factor of 10, to levels lower than those in normal female frogs. Hayes is now studying how the abnormalities affect the frogs' ability to produce offspring. Although Hayes used the African clawed frog in his research, he and his colleagues found native leopard frogs with the same abnormalities in atrazine-contaminated ponds in the U.S. Midwest (Hayes et al. 2002).

Help in dealing with contaminants in the environment may come from the plant kingdom. Sunflowers have been planted in fields near the Chernobyl nuclear power plant, in what is now Belarus, in an experimental effort to clean the heavily contaminated soils that linger long after the catastrophic accident. One study in 1996 found that the roots of sunflowers floated on a heavily contaminated pond near Chernobyl rapidly adsorbed heavy metals, such as cesium, associated with nuclear contamination (Reuther 1998). The NSF, the U.S. Environmental Protection Agency, and the Office of Naval Research have teamed up to fund new research on plants that can remove organic toxins and heavy metals from contaminated soils. Lena Ma of the University of Florida and colleagues discovered Chinese brake ferns thriving in soils contaminated with arsenic at the site of an abandoned lumber mill (Ma et al. 2001). Arsenic was once widely used as a pesticide in treated wood. Ma found arsenic levels greater than 7,500 parts per million in these samples. Plants fed on a diet of arsenic accumulate more than 2% of total mass in arsenic. Ma is now examining the mechanisms of arsenic uptake, translocation, distribution and detoxification. Other researchers are surveying a wide array of microorganisms for their potential to remove heavy metals and other contaminants from soil and water.

Understanding how organisms respond to change requires that we know what organisms inhabit our world and how they interact. The Tree of Life provides the baseline against which we measure change. In this context, the planned National Ecological Observation Network (NEON; National Science Foundation) will be invaluable. When completed, NEON will be an array of sites across the country furnished with the latest sensor technologies and linked by high-capacity computer lines. The entire system would track environmental change from the microbiological to global scales. Today, we simply do not have the capability to answer ecological questions on a regional to continental scale, whether involving invasive species that threaten agriculture, the spread of disease or bioterrorist agents. Tools such as NEON—which will in time reach international dimensions—will give us a much richer understanding of how organisms react to environmental change.

Eventually, such observatories must be extended to the oceans as well, perhaps with links to the ocean observatories now in the planning stages. The deep sea floor covers nearly 70% of Earth's surface. It may be the most extensive ecosystem on the planet, yet we have only begun to explore its secrets. It may harbor the source of new drugs, or it may be a reservoir for as yet unknown human pathogens. We can only be certain that it will produce surprises. We are all familiar with the submarine vents discovered two decades ago in the deep ocean, marked by the exquisite mineralized chimneys called "black smokers" that form around the hydrothermal vents on the seafloor and tower over dense communities of life. Creatures there live without photosynthesis—relying on microorganisms for sustenance. They exemplify the diversity that we have only recently begun to explore—even in the most extreme environments. These hot springs in the deep sea could have been the wellspring for life on our planet.

The deep sea is a reminder that we stand on the very threshold of a new age of scientific exploration, one that will give us a more profound understanding of our planet and allow us to improve the quality of people's lives worldwide. Yet some of the changes we humans bring about are not for the better. The ozone hole that now appears over Antarctica every year is a reminder that the cumulative effect of billions of individual human actions can have far-reaching, although unintentional, consequences. We understand now that changes in global climate cannot be understood without taking into account the effect that humans have on the environment—the way our individual and institutional actions interact with the atmosphere, the oceans, and the land.

The greatest question of our times may be how we can avoid the pitfalls and still grasp the opportunities that science and technology hold. When we limit our view of human health to problems of disease, diagnosis, and cure, we miss a significant perspective. A larger vision recognizes the evolutionary processes through which we arrived on the scene and the ecological balances that sustain us. We see the vulnerability of the planet and our co-inhabitants on it as *our* vulnerability. The study of biocomplexity science and its essential backbone, the Tree of Life, provide us with a way through and beyond these conundrums. Understanding the relationships among organisms and between organisms and the environment is our surest path to a healthier, more secure future.

Literature Cited

Adams, M. D., S. E. Celniker, R. A. Holt, C. A. Evans, J. D. Gocayne, G. A. Amanatides, S. E. Scherer, P. W. Li, R. A. Hoskins, R. F. Galle, et al. 2000. The genome sequence of *Drosophila melanogaster*. Science 287:2185–2195.

Avery, O. T., C. M. MacLeod, and M. McCarty. 1944. Studies on the chemical nature of the substance inducing transformation of pneumococcal types: induction of transformation by a desoxyribonucleic acid faction isolated from *Pneumococcus* type III. J. Exp. Med. 79(2):137–158.

Baumann, P., N. A. Moran, and L. Baumann. 1997. The evolution and genetics of aphid endosymbionts. Bioscience 47(1):12–20.

Béjà, O., L. Aravind, E. V. Koonin, M. T. Suzuki, A. Hadd, L. P. Nguyen, S. B. Jovanovich, C. M. Gates, R. A. Feldman, J. L. Spudich, E. N. Spudich, and E. F. DeLong. 2000. Bacterial rhodopsin: evidence for a new type of phototrophy in the sea. Science 289:1902–1906.

Béjà, O., E. N. Spudich, J. L. Spudich, M. Leclerc, and E. F. DeLong. 2001. Proteorhodopsin phototrophy in the ocean. Nature 411:786–789.

Blair, J. E., K. Ikeo, T. Gojobori, and S. B. Hedges. 2002. The evolutionary position of nematodes. BMC Evol. Biol. 2:7.

Brock, T. D., and H. Freeze. 1969. *Thermus aquaticus* gen. n. and sp. n., a non-sporulating extreme thermophile. J. Bacteriol. 98:289.

Caenorhabditis elegans Sequencing Consortium. 1998. Genome sequence of the nematode *C. elegans*: a platform for investigating biology. Science 282(5396):2012.

Colwell, R. R. 2002. A voyage of discovery: cholera, climate and complexity. Environ. Microbiol. 4(2):67–69.

Darwin, C. 1859. On the origin of species by means of natural selection, or the preservation of favoured races in the struggle for life. John Murray, London.

Fleischmann, R. D., M. D. Adams, O. White, R. A. Clayton, E. F. Kirkness, A. R. Kerlavage, C. J. Bult, J. F. Tomb, B. A. Dougherty, J. M. Merrick, et al. 1995. Whole-genome random sequencing and assembly of *Haemophilus influenzae* Rd. Science 269(5223):496–512.

Frary, A., T. C. Nesbitt, A. Frary, S. Grandillo, E. van der Knaap, B. Cong, J. Liu, J. Meller, R. Elber, K. B. Alpert, and S. D. Tanksley. 2000. A quantitative trait locus key to the evolution of tomato fruit size. Science 289:85–88.

Hayes, T. B., A. Collins, M. Lee, M. Mendoza, N. Noriega, A. A. Stuart, and A. Vonk. 2002. Hermaphroditic, demasculinized frogs after exposure to the herbicide atrazine at low ecologically relevant doses. Proc. Natl. Acad. Sci. USA 99:5476–5480.

Kew, O., V. Morris-Glasgow, M. Landaverde, C. Burns, J. Shaw, Z. Garib, J. André, E. Blackman, C. J. Freeman, J. Jorba, et al. 2002. Outbreak of poliomyelitis in Hispaniola associated with circulating type 1 vaccine-derived poliovirus. Science 296(5566):356.

Koplin, D. W., E. T. Furlong, M. T. Meyer, E. M. Thurman, S. D. Zaugg, L. B. Barber, and H. T. Buxton. 2002. Pharmaceuticals, hormones, and other organic wastewater contaminants in U. S. streams, 1999–2000: a national reconnaissance. Environ. Sci. Technol. 36(6):1202–1211.

Labrenz, M., G. K. Druschel, T. Thomsen-Ebert, B. Gilbert, S. A. Welch, K. M. Kemner, G. A. Logan, R. E. Summons, G. De Stasio, P. L. Bond, B. Lai, S. D. Kelly, and J. F. Banfield. 2000. Formation of sphalerite (ZnS) deposits in natural biofilms of sulfate-reducing bacteria. Science 290(5497):1744.

Leopold, A. 1993. Round river. Oxford University Press, New York.

Lipp, E. K., A. Huq, and R. R. Colwell. 2002. Effects of global climate on infectious disease: the cholera model. Clin. Microbiol. Rev. 15:757–770.

Loroño-Pino M. A., C. B. Cropp, J. A. Farfán, A. V. Vorndam, E. M. Rodríguez-Angulo, E. P. Rosados-Paedes, L. F. Flores-Flores, B. J. Beaty, and D. J. Gubler. 1999. Common occurrence of concurrent infections by multiple dengue virus serotypes. Am. J. Trop. Med. 61(5):725–730.

Ma, L. Q., K. M. Komar, C. Tu, W. Zhand, Y. Cai, and E. D. Kennelley. 2001. A fern that hyperaccumulates arsenic. Nature 409:579.

NSF. Available: http://www.nsf.gov/bio/bio_bdg03/neon03.htm and http://ibrcs.aibs.org/neon/index.aspl. Last accessed 25 December 2003.

Ochman, H., and N. A. Moran. 2001. Genes lost and genes found: evolution of bacterial pathogenesis and symbiosis. Science 292(5519):1096.

Punch's Almanack, 1882.

Reuther, C. 1998. Growing cleaner: phytoremediation goes commercial, but many questions remain. Academy of Natural Sciences, Philadelphia. Available: http://www.acnatsci.org/research/kye/phyto.html. Last accessed 25 December 2003.

Venter, J. C., M. D. Adams, E. W. Myers, P. W. Li, R. J. Mural, G. G. Sutton, H. O. Smith, M. Yandell, C. A. Evans, R. A. Holt, et al. 2001. The sequence of the human genome. Science 291:1304–1351.

Watson, J. D., and F. H. C. Crick. 1953. A structure for deoxyribose nucleic acid. Nature 171:737–738.

WHO. 1999. Strengthening implementation of the global strategy for dengue fever/dengue haemorrhagic fever. Report of the informal consultation, 18–20 October. World Health Organization, Geneva.

WHO. 2001. The world health report 2001. World Health Organization, Geneva.

WHO. 2002. Communicable disease surveillance and response, disease outbreaks reported. May 8 notice. World Health Organization, Geneva.

Woese, C. 1999. No title. ASM News. 65(5):263.

Douglas J. Futuyma

The Fruit of the Tree of Life

Insights into Evolution and Ecology

A milestone in the history of biology—and indeed of science and of society—was passed in February 2001, when two research groups announced completion of a "draft" of the human genome (International Human Genome Sequencing Consortium 2001, Venter et al. 2001). Even if some biologists felt that this event had rather less scientific significance than the public acclaim might suggest (because, after all, complete genome sequences had already been published for quite a few other species), the social and medical implications are undeniably immense. And for an evolutionary biologist, the most gratifying aspect of this historic event is that the leading publications are pervaded with evolutionary interpretation: "Most human repeat sequence is derived from transposable elements." "The monophyletic LINE1 and Alu lineages are at least 150 and 80 Myr old, respectively." "[M]ost protein domains trace at least as far back as a common animal ancestor." "[C]onservation of gene order [between human and mouse] has been used to identify likely orthologues between the species, particularly when investigating disease phenotypes." [All quotations are taken from International Human Genome Sequencing Consortium (2001).]

An evolutionary perspective has been indispensable for making any sense of the features of the human genome, simply because all the characteristics—genomic and phenotypic alike—of all organisms are the products of evolutionary history. We thus need to understand, as fully as possible, both what that history has been (how old are protein domains?) and what processes have produced it (how did repeat se-

quences arise?). These, indeed, have been the two overarching tasks of the science of evolutionary biology. It should be obvious that studies of history and of processes should each support and illuminate the other. Indeed, they do, and much of the excitement and progress in contemporary evolutionary biology stems exactly from the interpenetration of process-oriented and history-oriented research, a subject of this essay.

The Emergence of a Synthesis

The study of evolutionary history has historically been mostly the task of the "macroevolutionary" fields of paleontology and phylogenetic systematics, whereas evolutionary processes were traditionally viewed through the "microevolutionary" lenses of population and ecological genetics. As recently as 1988, one could bewail the great schism that has divided the two great realms of evolutionary biology for much of its history and urge a meaningful synthesis between them (Futuyma 1988). Historians of science will some day analyze how the synthesis of macroevolutionary and microevolutionary approaches, in which phylogenetic studies play so critical a role and which is still underway, came about. I would like to offer a few historical impressions before sketching some of the ways in which phylogenetics is making indispensable contributions to the broader fields of evolutionary biology and ecology.

Before the Modern Synthesis of evolutionary theory, inferring relationships and erecting classifications in an evolutionary spirit were viewed as major goals for biology and motivated paleontology, morphology, and embryology. Many classifications were developed that were intended to reflect common ancestry (and, in many cases, appear to have achieved that goal remarkably successfully). This work was accompanied by conclusions about the history of character transformations (e.g., the origin of mammalian auditory ossicles). During this period, an "eclipse of Darwinism" in which natural selection suffered ill repute (Bowler 1983), systematic and paleontological research was neither deeply informed by, nor contributed much to, understanding of the causal factors of evolution.

By the 1930s, evolutionary morphology became relegated to the sidelines by the rise of experimental disciplines such as genetics (Bowler 1996), and embryology became an experimental rather than a historically motivated descriptive discipline. Evolutionary biology was transformed by the Modern Synthesis, which arrived at a consensus that genetics supported Darwinism, that natural selection was the most important cause of evolution, and that "macroevolutionary" changes are the consequence of cumulative "microevolutionary" changes. The synthesis could not have occurred without the contributions of systematists such as Mayr, Rensch, and Simpson and of the genetically oriented naturalists Dobzhansky and Stebbins, with their systematic background. Although the systematists drew on earlier phylogenetic studies to support their thesis that macroevolution was explainable by the "neo-Darwinian" synthetic theory [e.g., by pointing out major changes in form associated with changes in function (Mayr 1960)], their contributions to the synthesis arose mostly from their analyses of speciation and intraspecific variation, rather than phylogeny.

The synthesis unquestionably emphasized evolutionary processes rather than evolutionary history as the locus of progress and invigorating challenge, and in this way doubtless joined the growing trend toward experimental biology in marginalizing phylogenetic and historical studies. It is undeniable, however, that few systematists countered by portraying phylogeny as a rigorous discipline (it wasn't; that is why new methods were developed in the 1960s and thereafter) or by demonstrating that it could contribute to conceptual understanding. For example, one of the people who inspired me to study evolution was William L. Brown, Jr., the world's authority on ant systematics. Although he was inspiring in his search to understand evolutionary processes (e.g., Brown and Wilson 1956, Brown 1959), not once, in my memory, did he use ants to illustrate, develop, or test hypotheses about evolutionary processes or history. Many systematists displayed far less interest in evolutionary processes than he, and phylogenetic hypotheses were a less conspicuous part of their work than were description of species and revision of genera. Important though such contributions are, they seldom conveyed intellectual excitement or conceptual progress.

The orthodoxies and preoccupations of a field are often most visible (even if time-lagged) in textbooks, and the few textbooks of evolution published in the 1960s and 1970s illustrate how small a role phylogeny played in evolutionary biology at that time. Both short, elementary paperbacks, whether authored by nonsystematists (Stebbins 1966, Volpe 1970) or systematists (Savage 1963), and longer undergraduate textbooks (Dodson 1960, Eaton 1970) figured at most five phylogenies of real organisms, usually incorporating a fossil record. The Equidae, based on Simpson, and the "reptiles," based on Romer or Colbert, were the usual subjects. Virtually the only conceptual point illustrated was adaptive radiation; certainly no suggestions that phylogeny could inform our understanding of process were made. Perhaps reflecting the senior author's later attitude toward systematics, the major textbook of the 1960s, Ehrlich and Holm's *The Process of Evolution* (1963), contained not a single phylogeny. Although 8 of the 38 short chapters in Grant's *Organismic Evolution* (1977) treat macroevolution, the only two phylogenies depicted accompany a description of the adaptive radiation of Hawaiian honeycreepers and a discussion of the canonical *Hyracotherium*-to-*Equus* "trend."

The virtual invisibility of phylogeny in textbooks was finally ended by Dobzhansky et al. (1977), who included a short discussion of numerical taxonomy and cladistics, several phylogenies illustrating macroevolutionary histories such as the origin of amphibians, several phylogenies based on distance analyses of molecular data (including some of Ayala's own work with electrophoresis), and perhaps most interesting, an illustration of how a phylogeny of Hawaiian *Drosophila*, based on chromosome inversions, supported a postulated history of interisland colonization. This example suggested that phylogenies could be useful for evaluating hypotheses about evolutionary histories. The first edition of my own textbook (Futuyma 1979) described phenetic and cladistic methods, included several phylogenies illustrating the history of diversification, presented several phylogenies as a basis for hypotheses about evolutionary processes (fig. 3.1), and emphasized that "all the examples of rates and directions of evolutionary change discussed [are based] on the assumption that it is possible to infer the phylogenetic history of species correctly."

The resurgence of phylogenetic research and its slow integration into the broader field of evolutionary studies, as reflected by these textbooks, had several causes. First and foremost were attempts to develop rigorous, quantitative methods for erecting classifications (Sokal and Michener 1958) and especially for inferring phylogenies (e.g., Hennig 1950, Edwards and Cavalli-Sforza 1964, Kluge and Farris 1969, Felsenstein 1973). The expectation of greater rigor made the phylogenetic enterprise more optimistic, more conceptually dynamic, and thus more attractive to prospective researchers in the field, and made it potentially more respectable in the view of evolutionary biologists outside the field. [However, I suspect the integration of phylogenetic

Figure 3.1. A rare, early example of a phylogenetic tree used to exemplify an important evolutionary principle. L. H. Throckmorton illustrated parallel evolution of the form of the male ejaculatory bulb in species of the *Drosophila repleta* species group, displaying the morphology on a phylogeny inferred from chromosome inversions. After Throckmorton (1965) and Futuyma (1979).

systematics and other fields of evolutionary study would have happened faster if nonsystematists had not recoiled from the "warfare" among adherents to different systematic doctrines (Hull 1988) and from the astonishingly combative language and behavior of some partisans.]

Second, phylogenetic study became supported by new kinds of data and pursued by individuals trained in a different tradition. Molecular data enabled individuals to do phylogenetic study without apprenticeship in taxon-specific comparative morphology, especially if a molecular clock were valid. Moreover, such data, especially amino acid sequences and electrophoretic allele frequencies, could be interpreted from the perspective not only of systematics but also from that of population genetics. The contributions of individuals whose work embraced both populations genetics and phylogeny (e.g., Felsenstein, Nei, Templeton) may have met resistance from organism-oriented systematists (and to some extent still do), but they did and do form a bridge between phylogenetics and process-oriented evolutionary biology.

Third, the 1970s saw a resurgence of interest in macroevolution, including topics such as developmental constraints (and "evo-devo" generally), punctuated equilibrium and its proposed implication for evolutionary trends, species selection and the differential diversification of clades, and changes in diversity through the Phanerozoic. Such topics could hardly be studied without a phylogenetic framework.

Fourth, some individuals urged a synthesis between phylogenetics and studies of evolutionary processes, and undertook research that required such synthesis. Almost from its inception, the study of molecular evolution depended on a phylogenetic framework, as in the revelation and analysis of gene duplication (e.g., Goodman et al. 1982) and

in tests of rate constancy in sequence evolution (Wilson et al. 1977, Kimura 1983). Some systematists (especially young ones) eagerly sought ways of applying phylogenetic methods to evolutionary questions in areas such as coevolution and character evolution (Brooks and Glen 1982, Mitter and Brooks 1983, Sillén-Tullberg 1988). Felsenstein (1985) offered a method of accounting for phylogeny in comparative studies of adaptation in a paper that elicited more reprint requests than anything else he had published (J. Felsenstein, pers. comm.). In a 1987 address to the Society for the Study of Evolution (SSE), I described ways in which phylogenetic and process-oriented studies could inform each other (Futuyma 1988); later, I organized a symposium on this theme for the 1988 meeting [several of the talks were published in *Evolution* 43(6):1137–1208].

A synthesis slowly developed despite extraordinary *Sturm und Drang* ("storm and stress") in the late 1970s and early 1980s, when it seemed as if "macroevolutionists" and "microevolutionists" were forming increasingly isolated, even hostile, camps (Futuyma 1988). At meetings of the SSE from 1981 through 1988, only about 4% of contributed papers referred to phylogeny (judging from titles in the programs), but this increased to 12% in 1989, when the meeting also included symposia on phylogenies based on ribosomal genes (organized by E. Zimmer and D. Hillis) and on cladistic approaches to evolutionary innovation (organized by C. Mitter and B. Farrell). In 1990, the SSE met with other societies in the fifth International Congress of Systematic and Evolutionary Biology, the theme of which ("the unity of evolutionary biology") was conceived explicitly as a synthesis of historical and process-oriented evolutionary disciplines (C. Mitter, pers. comm.). At this meeting, the Society of Systematic Zoologists decided to become the Society of Systematic Biologists and to meet jointly with the SSE thereafter (Hillis 2001). The joint meetings now include both symposia and a high proportion (about 26% in 2001) of contributed papers with a phylogenetic theme or flavor. Many of the papers explicitly apply phylogenetic methods or information to a wide variety of problems in evolutionary biology. Of course, this growing mutualism between phylogenetic systematics and other subdisciplines of evolutionary biology has also become evident in the contemporary literature.

Phylogenies in Contemporary Evolutionary Biology and Ecology

In the mid-1980s, phylogeny was almost invisible in the pages of *Evolution* and of most other evolutionary journals. Less than two decades later, it pervades the literature on almost every major subject in evolution, to the point at which some have wondered if demands for a phylogenetic framework may even be sometimes excessive (e.g., Westoby et al. 1995; see Silvertown et al. 1997). Moreover, we now seek phylogenies not only of species and higher taxa, but also of genes within genomes and of variant gene sequences within and among species. The same methods can yield trees for organisms and trees for genes, which in turn can shed light on the history and processes that have affected genomes, organisms, and populations.

The many issues in evolution and ecology that are informed by phylogenetic analysis (table 3.1) fall under several major headings, each of which I address briefly below with a few examples. My emphasis is on questions pertaining to the evolution and ecology of organisms and thus, chiefly, on rather traditional questions that phylogenetics can now help answer. I will not treat molecular evolution, in which phylogenetic analysis bears on almost every topic, such as rates of sequence evolution, mutation, and recombination; the evolution of gene families and the homology (paralogy) of functionally different genes; horizontal gene transfer; the time of silencing of pseudogenes; and many others. These topics warrant book-length treatment (e.g., Li 1997) and are far from my areas of competence.

Evolutionary Processes within Species

Phylogenetic methods provide insights into evolutionary processes within species by way of both *phylogenies of genes* and *phylogenies of populations and species*. Traditional population genetic theory deals with the ways in which frequencies of alleles are affected by mutation, genetic drift, gene flow, and natural selection. Coalescent theory expands traditional population genetic theory by analyzing these processes in a history of phylogenetic (or genealogical) relationships among the alleles (Hudson 1990). For example, a population with a constant size of N_e breeding individuals may begin with different gene lineages, each of which diversifies as new mutations occur. If all the sequences are selectively equivalent (neutral), gene lineages become extinct by genetic drift, at a rate inversely proportional to the population size. After about $4N_e$ generations, all except one original lineage will have become extinct, on average, such that all genes are descended from ("coalesce to") one of the original genes. What began as a genetically "polyphyletic" population becomes monophyletic because of genetic drift. The gene tree continues to branch by mutation, but because the tree is continually pruned by genetic drift, only a large population will contain multiple old ("deep") branches that differ by many mutations. Therefore, a gene tree with deep branches indicates a population that has been large or subdivided, and a shallow gene tree signals a small or bottlenecked population (assuming selective neutrality). Given an estimate of the mutation rate (u), in fact, the effective population size can be estimated from the frequency of heterozygotes per site (which is expected to equal the product $4N_e u$ at a diploid locus).

The gene tree can also be affected by selection. For example, balancing selection can maintain different gene lineages, giving rise to much deeper branches in the gene tree

than expected from N_e alone, whereas directional selection that recently fixed an advantageous mutation will have swept away linked neutral variation, resulting in a very shallow gene tree (of sequences that have arisen by mutation since the selective sweep). The effects of selection versus genetic drift can be distinguished by comparing multiple genes that are not closely linked, because genetic drift affects all genes similarly whereas selection affects genes individually.

Among the best-known applications of this approach to date are analyses of human gene trees, which fairly consistently imply that the effective size of the human population has been quite small, on the order of 100,000 or less. (The effective size, which is approximately the harmonic mean of breeding numbers in successive generations, is mostly strongly determined by reductions, or bottlenecks, in size. Therefore, the recent explosive growth of the human population has had little effect on N_e.) Although many basal gene lineages are found in African populations, almost all non-African haplotypes belong to a single nonbasal clade—points that strongly favor the hypothesis that the contemporary human population of the world has been derived from an African population in the very recent past (e.g., Hammer 1995, Ingman et al. 2000). This approach to estimating historical effective population size might also be applied to historical bottlenecks that may have accompanied speciation. In such a study of a pair of sister species of leaf beetles (*Ophraella*), we estimated that N_e was greater than one million, a far cry from a bottleneck (Knowles et al. 1999). However, there may have been enough time since speciation of these beetles for high sequence variation to have been regenerated even if there had been a bottleneck; the method will detect a bottleneck only if divergence has been too recent for coalescence to have occurred in a large population. Similar analyses do indicate small N_e, perhaps due to a speciation bottleneck, in *Drosophila sechellia*, endemic to the Seychelles Islands (Kliman et al. 2000).

In contrast to the very shallow branches of most human gene genealogies, the tree for human genes in the major histocompatibility complex (MHC) shows very deep branches; in fact, different human haplotypes are more closely related to chimpanzee MHC haplotypes than to other human haplotypes. Thus, the MHC polymorphism has been maintained for more than 5 million years, longer than expected for neutral variants if current estimates of human N_e are correct. The gene tree thus provides *prima facie* evidence of balancing selection. It has been suggested that selection by diverse parasites may have maintained variation (Hughes 1999).

Probably the most active area of research in intraspecific phylogeny is phylogeography, the study of the geographic distribution of genealogical lineages (Avise 2000). Often combined with coalescent analysis, such studies are shedding light on histories of population subdivision, gene flow, colonization, and range expansion. For example, the classic studies of Bermingham and Avise (1986) revealed a common history of vicariant differentiation in several species of fresh-

Table 3.1

Some Applications of Phylogenetic Study in Evolutionary Biology and Ecology.

I. Evolutionary processes within species	
1. Isolation, vicariance, and gene flow	Avise (2000), Zink et al. (2000)
2. Colonization and range expansion	Taberlet et al. (1998), Ballard and Sytsma (2000)
3. History of population size	Takahata et al. (1995), Wakeley and Hey (1997)
4. Mutation rates	Kimura (1983), Lynch et al. (1999)
5. Selection on DNA sequences	Hudson (1990)
6. Sexual selection	Basolo (1996), Barraclough et al. (1995)
7. Asexual reproduction vs. recombination	Guttman and Dykhuizen (1994)
II. Character evolution	
1. Meaning and identification of homology and homoplasy	Sanderson and Hufford (1996), Wagner (1989)
2. Rates of evolution	Lynch (1990), Gittleman et al. (1996)
3. Inferring lability and constraint	Gittleman et al. (1996)
4. Comparative method of inferring adaptation	Felsenstein (1985), Martins (1996)
5. Polarity, evolutionary sequences, origin of novelties	Donoghue (1989), Lee and Shine (1998), Wahlberg (2001)
6. Genome evolution (duplications, repeated sequences, etc.)	Fitch (1996), International Human Genome Sequencing Consortium (2001)
7. Locating candidate genes for traits	Crandall and Templeton (1996)
8. Historical framework for experimental analyses	Futuyma et al. (1995), Ryan and Rand (1993)
III. Speciation	
1. Delimiting species	Avise and Ball (1990), Baum and Shaw (1995)
2. Geographic pattern of speciation	Schliewen et al. (1994), Berlocher (1998), Barraclough and Vogler (2000), Coyne and Price (2000)
3. Demography of speciation	Knowles et al. (1999), Hare et al. (2002)
4. Duration of speciation process	McCune and Lovejoy (1998), Avise and Walker (1998)
5. Hybrid speciation, introgression	Rieseberg (1997), Dowling and Secor (1997)
6. Pattern of evolution of reproductive isolation	Coyne and Orr (1989)
7. Dating speciation	Klicka and Zink (1997), Knowles (2000)
IV. Diversity	
1. Hypotheses for diversification (e.g., key adaptations)	Mitter et al. (1988), Sanderson and Donoghue (1996)
2. Estimating speciation and extinction rates	Mooers and Heard (1997), Barraclough and Nee (2001)
3. Estimating number of ghost lineages	Sidor and Hopson (1998)
4. Cospeciation of interacting lineages	Brooks and McLennan (1991), Page and Hafner (1996)
5. Adaptive radiation	Givnish and Sytsma (1997), Schluter (2000)
6. Hypotheses for regional diversity differences	Qian and Ricklefs (1999), Chown and Gaston (2000)
V. Ecology	
1. Community assembly: geographic sources of species	McPeek (1995), Zink et al. (2000)
2. Community assembly: evolution of interactions	Farrell and Mitter (1993), Futuyma and Mitter (1996)
3. Coexistence in relation to phylogenetic affinity	Webb (2000)
4. Convergence in community structure	Losos et al. (1998)
5. Changes in viral infection rates	Holmes et al. (1996)
VI. Conservation	
1. Identifying "management units" and "evolutionarily significant units"	Vane-Wright et al. (1991), Moritz (1994)
2. Conserving "evolutionary history"	Purvis et al. (2000a)
3. Predicting extinction risk	Purvis et al. (2000b)

water fishes in the southeastern United States: The mitochondrial gene tree of each species included two major clades of variant sequences, distributed to the west and east of a probable Pliocene saltwater barrier. Similar studies have revealed the likely sites of refugia for many species during Pleistocene glacial episodes and the routes of postglacial colonization (e.g., Taberlet et al. 1998). Postglacial expansion over broad areas by relatively few colonists appears now to account for lower levels of genetic variation within and among populations at higher latitudes than at lower latitudes. For example, northern populations of MacGillivray's warbler (*Oporornis*

tolmiei) collectively have a shallower mitochondrial gene tree than do southern populations (fig. 3.2; Milá et al. 2000).

Phylogenies of species rather than genes can also help to illuminate evolutionary processes. For example, the relative rate test for constancy of sequence evolution requires phylogenies, and approximate constancy, together with time-calibrated divergence between taxa, is the basis of most estimates of mutation rates at the molecular level (Kimura 1983). A very different example is provided by studies of sexual selection. For instance, Basolo (1996) found that in fishes of the genus *Xiphophorus*, females prefer males with a

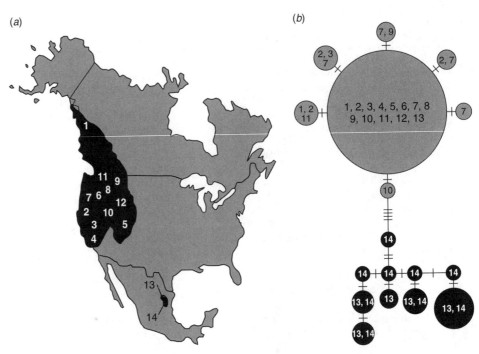

Figure 3.2. An example of inference of historical demography in a phylogeographic analysis. Samples of a mitochondrial cytochrome gene in MacGillivray's warbler (*Oporornis tolmiei*) from localities in western United States and a small region in northern Mexico show high haplotype sequence diversity in Mexico, whereas the northern samples include only a single common haplotype and rare, presumably recently originated variants that differ from the common haplotype by single mutations. The gene tree (or network) is consistent with the hypothesis that northern populations are derived from relatively small numbers of postglacial founders. After Milá et al. (2000).

sword (an elongation of the lower caudal fin rays). Remarkably, females display such a preference not only in those species that have swords (swordtails) but also in species that normally lack them (platies). Although different estimates of phylogenetic relationships within *Xiphophorus* made it ambiguous whether or not the female preference in swordless species reflected a plesiomorphic state (i.e., preference having evolved before the male sword), Basolo showed that the female bias also characterizes *Priapella*, an indisputably primitively swordless sister lineage of *Xiphophorus*. At the time, the idea that preexisting female preferences may play a role in sexual selection was a rather new hypothesis, contending with several other models of sexual selection by female choice.

Speciation

Studies of speciation must have an intimate relation to phylogeny, if for no other reason (obvious now, but perhaps not always so) than that it is often necessary to identify correctly the products of a speciation event, namely, sister species. Even the delimitation of species may depend on phylogenetic data, at least for those who prefer to define species in genealogical terms, such as genetic monophyly (e.g., Baum and Shaw 1995; see also Avise and Ball 1990). A phylogeny is a *sine qua non* for identifying instances in which new species have arisen from interspecific hybrids (e.g., Rieseberg 1997) and for dating speciation events. For example, successive speciation events have apparently occurred within the Pleistocene in montane *Melanoplus* grasshoppers (Knowles 2000). In contrast, many sister species of North American birds that were formerly presumed to have arisen in Pleistocene glacial

refugia appear to have diverged in the Pliocene (Klicka and Zink 1997), although speciation is a continuing process that in some of these cases probably extended into the Pleistocene. This conclusion arises from the suggestion that the minimal duration of the speciation process may be estimated from the difference between the temporal depth of the branch point between sister species and the temporal depth of the deepest nodes within the gene tree of one of those species (McCune and Lovejoy 1998, Avise and Walker 1998). On this basis, Avise and Walker (1998) concluded that speciation in birds and mammals generally takes about 2 Myr (million years), so populations that began diverging in the later Pliocene would have completed speciation in the Pleistocene. McCune and Lovejoy (1998) used this approach to compare the estimated duration of speciation in clades in which allopatric speciation is probable and clades in which they considered sympatric speciation a likely possibility. The results of their analysis were consistent with the hypothesis that sympatric speciation, which cannot occur except by strong selection, should be faster than allopatric speciation.

How to distinguish sympatric from allopatric speciation, and even how to provide convincing evidence that sympatric speciation occurs, have long been vexing questions. Phylogenetic approaches are at last promising answers. Probably the most convincing case of completed sympatric speciation is provided by several apparently monophyletic species groups of cichlids in crater lakes in Cameroon (Schliewen et al. 1994). The lakes are structurally simple and ecologically rather homogeneous, so if speciation occurred within the lakes, as the phylogeny implies, it must have been truly sympatric. In birds, in contrast, monophyletic species groups have evidently not evolved on islands that lack topographic and vegetational barriers, suggesting that bird speciation is usually allopatric, as has long been thought (Coyne and Price 2000). In another approach to the problem, suggested by Berlocher (1998) and Barraclough and Vogler (2000), the degree of range overlap between sister taxa in a clade is plotted against a surrogate for divergence time (e.g., sequence divergence). The overlap between sympatrically originated taxa must remain high or decline (because they start with maximal overlap of the smaller range by the larger), whereas overlap between the ranges of allopatrically originated taxa can only increase with time. Most of the phylogenies analyzed by Barraclough and Vogler (2000) were consistent with allopatric speciation, but two insect phylogenies suggest a role for sympatric speciation.

Character Evolution

Probably all claims about the evolution of characters among species must have a phylogenetic foundation. Historically, this was often not explicitly stated or perhaps even recognized, but clearly phylogenetic assumptions underlie the belief that parasites have "degenerated" in morphology, or that *Hyracotherium* and subsequent equids represent a transformation series. Today, phylogenies are the explicit basis for many, perhaps most, studies of character evolution, whether phenotypic or molecular. They are required to distinguish homology from homoplasy and to estimate rates of character evolution. "Conservative" characters, with low evolutionary rates, provide material for analysis of possible constraints. Homoplasy provides data for the analysis of adaptation by the "comparative method" (Harvey and Pagel 1991), which most practitioners now agree should be based on explicit phylogenies, so that independent evolutionary changes in a trait of interest can be correlated with environmental factors or with other characters.

Phylogeny has long been the (at least implicit) basis for understanding character transformations, such as the origin of novel features (e.g., wings, auditory ossicles, the sting of aculeate Hymenoptera). This enterprise is being rejuvenated as the developmental and genetic bases of such transformations are illuminated in a phylogenetic framework. Both conservation and change in the expression and functional roles of *Hox* genes, for example, provide unprecedented insights into evolutionary changes in body plans (Carroll et al. 2001). We are also better able to evaluate traditional ideas about the polarity of character evolution. For example, the venerable idea that ecological specialists evolve from generalists far more often than the converse has many implications; it might explain, in part, why many clades of herbivorous insects are composed mostly of host-specialized species (Futuyma and Moreno 1988). Only recently, however, has breadth of resource use been mapped onto phylogenies in order to infer the direction of change. In some cases, such as the host range of *Dendroctonus* bark beetles, the traditional hypothesis has been supported (Kelley and Farrell 1998). In quite a few other

phylogenies, however, at least some generalists arise from more specialized ancestors (Nosil 2002), and although it may be premature to conclude that there is "little support for the generalist-to-specialist hypothesis" (Schluter 2000), it is certainly clear that any such trend is far from universal.

To an increasing extent, even experimental studies of character evolution are being designed in a phylogenetic or historical framework. For example, I explicitly conceived my own work on host shifts in *Ophraella* (Coleoptera: Chrysomelidae) as a study of a character that systematics had shown to be interesting, and as an example of mutualism between phylogenetic and population genetic approaches. Insect systematists have long known that host–plant association is a highly conservative character in many groups of phytophagous insects; clades that may date back to the Cretaceous often are restricted to a single plant family (Ehrlich and Raven 1964, Farrell and Mitter 1993). Such features invite the hypothesis that internal constraints may limit evolution (Maynard Smith et al. 1985). Such constraints might manifest by absence or paucity of genetic variation (the prerequisite for any evolution). I posed the hypothesis that the pathways of evolution of host affiliation actually taken by an insect clade may have been more likely, because of constraints on some characters rather than others, than the paths not taken (Futuyma et al. 1993, 1995). Thus, for example, if most host shifts have been between closely related rather than distantly related plants, this hypothesis predicts that features necessary for survival and reproduction on a novel plant would be more genetically variable if the plant is closely related than if it is distantly related to the insect's current host plant.

Most of the 14 currently recognized species of *Ophraella* feed only on one or another genus of plant, in one of four tribes of Asteraceae. Our proposed phylogeny of *Ophraella*, based first on morphological and allozyme characters and later on mitochondrial gene sequences (fig. 3.3; Funk et al. 1995), provides no evidence for cospeciation or codiversification with the host plants but does show that host shifts have been more frequent within than between host tribes (i.e., adaptation to closely related plants has been the norm). Larval and adult beetles feed and survive much better on their own hosts than on those of their congeners, and in some instances the (presumably chemical) barriers to feeding result in almost no feeding at all. Using breeding designs commonly employed in quantitative genetics, we screened large numbers of naive hatchling larvae and newly eclosed adults for their feeding response to and ability to survive on foliage of as many as six species of plants that are hosts of *Ophraella* species, but not of the particular species being screened. We performed such screens for genetic variation in feeding response and survival with four species of *Ophraella*, resulting in a total of 18 combinations of insect and plant species screened for genetic variation in survival and 39 screens of feeding responses (including both larval and adult responses). Overall, we detected genetic variation in survival in only two cases: in both, the plant that supported geneti-

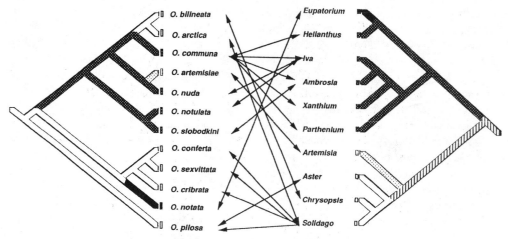

Figure 3.3. The phylogeny of species of *Ophraella* leaf beetles, connected by arrows to their host plants. The hosts belong to four tribes of Asteraceae, indicated by different shading. The poor congruence between the trees is consistent with other evidence that the beetles have shifted among host lineages, for the most part, rather than cospeciating with their host plants. Host shifts have been more frequent within than between plant tribes, illustrating conservatism of diet. The beetle phylogeny is based on mitochondrial DNA sequences, and that of the plants on chloroplast DNA studies (see Funk et al. 1995). A complete plant phylogeny would include many other intercalated genera and tribes. From Futuyma and Mitter (1996).

cally variable survival was very closely related (in the same subtribe) to the beetle species' normal host. Although the correlation was not strong, genetic variation in feeding response was significantly more frequent among tests of species on closely related plants (in the same tribe as the normal host) than on distantly related plants (in a different tribe of the Asteraceae). The results are consistent with the hypothesis that a macroevolutionary pattern of host association revealed by phylogenetic analysis may stem in part from genetic biases revealed by the methods of evolutionary genetics.

Diversity

It seems hardly possible to discuss the origin of organismal diversity without reference to phylogeny. For example, textbook treatments of the subject have usually included phylogenetic diagrams (frequently including reference to stratigraphic distributions, as in classical portrayals of the history of the Equidae). It is only recently, however, that phylogenies have served as explicit tools for testing hypotheses about the history and causes of diversification. For example, parasitologists had proposed phylogenetic hypotheses about parasite–host associations by the 1940s, but only in the early 1980s were phylogenies explicitly used to determine whether the associations were caused by cospeciation and codiversification (one form of coevolution) or by lateral shifts of parasites among preexisting lineages of hosts (e.g., Brooks and Glen 1982, Mitter and Brooks 1983). Subsequent research, including development of methods for distinguish-

ing these hypotheses, has made it clear that different groups of parasites and symbionts (*sensu lato*, including phytophagous insects, microbes, etc.) exemplify both historical patterns (Page and Hafner 1996).

Phylogenies provide by far the most important basis for testing hypotheses about the role of "key innovations" as causes of differences in rates of diversification among clades. The tradition of attributing the high diversity of insects to the evolution of wings, or of Coleoptera to elytra, or of angiosperms to the carpel has been criticized as ad hoc, untestable "storytelling," because each such event is unique (lacking the replication required for any statement about correlation), and each could in principle be attributed to any of the many other apomorphies of these groups (even assuming that their diversity has indeed been caused by any such character). The method of replicated sister-group comparisons introduced by Mitter et al. (1988) provides a more rigorous test by comparing the species diversity of multiple clades in which a putative diversity-enhancing character has independently originated with that of their sister groups that lack the character. The use of sister groups enables diversity differences to be ascribed to differences in rate of speciation and/or extinction rather than differences in age, and the replication provides a basis for statistical test. Mitter et al. (1988) provided evidence that acquisition of the habit of herbivory has enhanced the rate of insect diversification, and Farrell (1998) later used this approach to argue that diversification rate in phytophagous beetles has been greatly increased by shifts from "gymnosperm" to angiosperm hosts. The hypothesis that plant diversification has been enhanced by the evolution of defenses against herbivores—a key element of Ehrlich and Raven's (1964) scenario of coevolution—was supported by the consistently greater species diversity of plant lineages with latex or resin canals in sister-group comparisons—features that have been experimentally shown to deter insect herbivores (fig. 3.4; Farrell et al. 1991). Both key innovations and ecological opportunity offered by "empty ecological space" are associated with enhanced diversification rate, as the many phylogenetic studies of adaptive radiation are demonstrating (Givnish and Sytsma 1997, Schluter 2000).

Differences in species diversity among geographic regions and among environments have attracted attention from both ecologists and evolutionary biologists. Latitudinal gradients in diversity, for example, might represent equilibrial conditions dictated by interactions among species, or might have a more historical explanation based on the history of speciation and extinction (Chown and Gaston 2000). Stebbins (1974), for example, suggested that the tropics might be a "cradle" of new species originating at higher rates than elsewhere, or a "museum" in which extinction rates have been low and species have accumulated over vast spans of time. Phylogenies that provide time depths for many of the clades that contribute to the diversity differences will probably play an important role in resolving this long-persistent controversy. For example, a molecular phylogenetic study of the diverse neotropical tree genus *Inga* (Fabaceae) suggests that most of the approximately 300 species have originated within the last 6 Myr, favoring the "cradle" interpretation (Richardson et al. 2001). On the other hand, diversity at the level of higher taxa (genera, families) may also be highest in tropical latitudes, suggesting that much more comprehensive phylogenies will be needed to compare the distribution of divergence times that would account for differences in diversity between regions.

Geographical variation in species diversity and taxic composition stems in part from the processes that are the subject of historical biogeography (Morrone and Crisci 1995, Humphries and Parenti 1999). This field has always been inseparable from phylogenetic systematics, because the higher taxa that are its subject must have phylogenetic mean-

ing. Part of the "cladistic revolution," in fact, consisted of attempts to establish a more rigorous phylogenetic framework to analyze distributions, in the form of "vicariance biogeography" (Nelson and Platnick 1981). The null hypothesis or guiding principle was that distributions of taxa should be explained by successive disjunctions among regions or areas, resulting in congruent cladograms of taxa and of the areas they occupy. However, the disparagement of "dispersalist" explanations by some early vicariance biogeographers has proven unwarranted, for phylogenetic analyses have been equally powerful in providing evidence of dispersal. For example, a recent analysis of the Chamaeleonidae, based on 644 parsimony-informative molecular, morphological, and behavioral characters, provides strong evidence that chameleons originated in Madagascar after its separation from India, and later dispersed to Africa (at least twice), the Seychelles, and the Comoros archipelago (fig. 3.5; Raxworthy et al. 2002). Chameleons are among many taxa distributed around the Indian Ocean that show more phylogenetic evidence of dispersal than of the Gondwanan vicariance that might have been expected. As more phylogenies are developed, a balanced view of the roles of vicariance and dispersal is emerging (Zink et al. 2000).

Community Ecology

The problems addressed by community ecology include the species diversity and composition of species assemblages, and the structure of their interactions (e.g., food web structure). An evolutionary perspective has been important in community ecology, both by suggesting how evolutionary responses to interspecific interactions may shape community character and by emphasizing the effects of history.

That community composition and structure may be affected by "deep" evolutionary history should be a clear lesson from historical biogeography. The absence of mammals from New Zealand, of sea snakes from the tropical Atlantic, and of bromeliads from Old World tropical forests must count as major differences in community structure, even if a

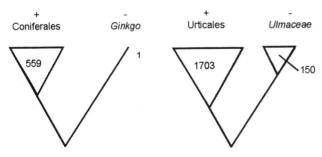

Figure 3.4. Replicated sister-group contrasts can test for effects of apomorphic characters on diversity. These are two of the sister-group pairs of seed plants in which species richness is higher in the clade with apomorphic latex- or resin-bearing canals. Based on Farrell et al. (1991).

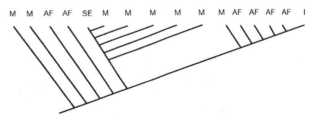

Figure 3.5. A reduced phylogeny of lineages of Chamaeleonidae, showing the pattern of distribution in Madagascar (M), Africa (AF), India (I), and the Seychelles (SE). The phylogeny supports the hypothesis of dispersal from Madagascar and is incompatible with postulated histories of the separation of these land masses. After Raxworthy et al. (2002).

few species play faintly convergent roles (moas, e.g., being possible ungulate vicars). Thirty years ago, the discourse of community ecology made little reference to historical accident, because of a conviction that rapid evolutionary responses to strong ecological interactions should have almost deterministically shaped predictable equilibrial structures. This conviction, or faith, has been shaken, and community ecologists now appear to have a growing appreciation of the importance of history. For example, plant species that we think of as forming coherent assemblages (e.g., maple and hemlock) seem to have undergone quite independent shifts in distribution throughout the Pleistocene (Davis 1976), and differences in the species diversity of trees in Europe, eastern Asia, and eastern North America appear largely attributable to differences in extinctions suffered during glacial episodes, owing to differences in the availability of temperate refuges (Latham and Ricklefs 1993, Qian and Ricklefs 1999).

The impact of much older evolutionary histories has been little analyzed but must be equally significant. For example, many clades of phytophagous insects are so conservative in diet that they have remained associated with the same plant family since the early Cenozoic or earlier (Farrell and Mitter 1993, Farrell 1998). Many genera of leaf beetles (Chrysomelidae) that in New York State feed on only a single plant family include other species that also exist in western North America, Europe, or tropical America. In almost every case, the congeners in those biogeographic regions feed, exclusively or in part, on the same plant families as do their New York relatives (table 3.2; Futuyma and Mitter 1996). Thus, the producer–consumer interface in communities in New York represented by these insect–plant associations must have been shaped in part by sorting among colonizing species from other regions, whose establishment depended on host-related characters that had evolved many millions of years before. (We cannot confidently specify the direction of colonization for most of these genera, because the required phylogenies have not been determined—one example among many in which a more complete Tree of Life would help to describe ecological history.)

The processes that give rise to an assemblage of stably coexisting species include both sorting among colonists from a regional species pool and evolutionary (or coevolutionary) responses to species interactions *in situ*. That is, the characteristics that enable species to coexist may have been "pre-adaptations" that evolved before they came into contact or may have evolved in response to the interaction between them. These processes (which are not mutually exclusive) have been difficult to distinguish, but phylogenetic approaches are providing some resolution. For example, islands in the Lesser Antilles harbor either one species of anole, usually of medium body size, or two species, usually a small and a large one. (Differences in body size are correlated with differences in average prey size, and thus facilitate coexistence.) Although this pattern suggests repeated character displacement between competing species on two-species islands, a phylogeny of the species suggests that the three small spe-

Table 3.2

Fraction (*p*) of New York Genera of Chrysomelidae (Numbering *n*) that Share at Least One Host Plant Family with Congeners in Europe and Tropical America.

	Number of host families in New York					
	1		2		3 or more	
	p	*n*	*p*	*n*	*p*	*n*
Europe	0.94	16	1.00	9	0.97	29
Tropical America	0.93	14	1.00	8	0.80	5

After Futuyma and Mitter (1996).

cies form a monophyletic group and that the two large species that occur on two-species islands likewise are a monophyletic group (Losos 1992). Thus, even if character displacement occurred once, several two-species islands must have been colonized by lineages that already differed in body size, conforming to the hypothesis that preexisting ecological differences are required for species to come into coexistence. Moreover, the independent evolution of large size in one species (*A. ferreus*) on a one-species island suggests that character displacement may not be the sole explanation for evolutionary changes in size.

In contrast to the Lesser Antilles, where coexistence of ecologically different species has been due mostly to species sorting, the Greater Antilles harbor four monophyletic groups of anoles that have undergone strikingly similar adaptive radiations (Losos 1992, Losos et al. 1998). "Ecomorphs" that differ in size, shape, and microhabitat use have evolved in parallel in Cuba, Hispaniola, Jamaica, and Puerto Rico (although the set is slightly incomplete on the latter two islands). It is likely that character displacement among competing species has caused the adaptive divergence. The phylogenetic framework is crucial for showing that the community structure on the several islands has arisen not by sorting ecologically dissimilar from similar species (as in the Lesser Antilles) but by selection stemming from species interactions and the intrinsic functional relationships between anoles and their resources. The belief in predictable evolution of community structure may not be entirely groundless.

Phylogenetic data may also cast light on the processes that affect species assemblages on short time scales (i.e., in ecological time). For example, hypotheses accounting for the high species diversity of trees in many tropical forests include neutral "drift" in the frequencies of ecologically equivalent species (the number of which is ultimately determined by long-term rates of speciation and extinction; Hubbell 2001); greater herbivore- or pathogen-induced mortality of conspecific than allospecific seedlings in the neighborhood of adult trees, resulting in underdispersion of each species (Janzen 1970, Connell 1971); and "niche partitioning" among species, based in part on use of different microhabitats. Webb (2000) reported that tree species within 0.16–hectare plots in lowland dipterocarp forest in West Kalimantan, Indonesia, were phylogenetically closer, on average, than if they had been drawn at random from the entire local pool of 324 species. This pattern is consistent with the hypotheses that phylogenetic affinity is correlated with ecological similarity and that the overall species diversity consists, in part, of assemblages of related species in a mosaic of different microhabitats.

Practical Applications

So many applications of biology depend on taxonomy that we are inclined to forget that phylogenetic assumptions underlie the applications. For instance, a major method of weed management is the use of biological control agents, such as host-specific insects that might be imported from the weed's region of origin. The bulk of research on such insects consists of tests to assure that they will not attack economically important plants such as crops. Most of this effort is devoted to tests of responses to plants in the same higher taxon as the weed, that is, closely related plants. It may seem obvious that a control agent for a weedy species of thistle might be a potential threat to artichoke crops (a member of the same tribe), but of course this rests on an assumption of a phylogenetically sound classification. Conservation biologists have recently raised the concern that biological control agents may attack threatened native species; for example, the weevil *Rhinocyllus conicus* was introduced to North America from Europe to control several adventitious European thistles, but it also attacks several native thistles (Louda et al. 1997). (Advocates of biological control counter criticism by saying that such spread was expected, but that concern for native plant species was not a criterion for introduction when the *Rhinocyllus* program was implemented.) I have suggested that screening of potential biological control include tests for genetic variation in the species' fitness on closely related nontarget species, because genetic adaptation to closely related plants is a common pattern in many clades of herbivorous insects (Futuyma 2000).

Several authors have urged that phylogenetic information be brought to bear in conservation biology (e.g., Vane-Wright et al. 1991, Moritz 1994, Purvis 2000a, 2000b). One might consider giving priority to conserving "phylogenetic history," if, for instance, the choice lay between a species flock of very closely related species and an ecosystem that included endemic or species-poor long phylogenetic branches (e.g., *Sphenodon*, *Welwitschia*). Phylogenetic or phylogeographic information may likewise help to identify "evolutionarily significant units" for management (Moritz 1994).

Conclusions

In an astonishingly short time, phylogenetic methods or frameworks have become integral parts of almost every major area of evolutionary biology, and several parts of ecology as well. A steady stream of papers suggests new uses for phylogenies, with no end of inventiveness in sight. Because it is clear that phylogenetic approaches and data will play an increasingly important role in biological disciplines outside systematics, we might ask how the mutualism between phylogenetic systematics and the other "biodiversity sciences" might best be fostered. I do not presume to offer a deep or even well-informed analysis, but instead a few modest suggestions.

First, systematists and the users of systematics might do well (even for utterly self-serving reasons) to engage in some of their work with an eye toward their mutual or reciprocal benefit. For instance, ecologists engaged in biological inventory projects amass collections that may include huge numbers of species, many rare or even undescribed. The value of these collections for phylogenetic purposes would be enormous if some specimens (or tissues) of each species were cryopreserved for future molecular study. Systematists who are engaged to help identify material from such inventories might consider how future phylogenetic studies of both their "own" taxa and others might be aided if they were to insist that comprehensive samples be donated to frozen tissue collections.

The fruits of phylogenetic studies will be most bountiful if they are presented in ways that will make them most broadly useful, especially in the indefinite future when current methodologies or questions may come to be seen as inadequate or parochial. Most critically, of course, the data themselves must be permanently archived and available (e.g., sequence banks). I would urge, also, that a published phylogenetic study include the results of as many broadly used analytical procedures as possible, including those with which the author strenuously disagrees. One loses nothing by presenting both total-evidence trees and separate trees from, say, morphological and molecular data sets, or trees with both bootstrap and Bremer support values, or indeed, the results of parsimony, maximum likelihood, and other analyses. By all means, an author should assert preference for one or another result, but the interests of scientific understanding—both of the phylogeny of the clade and a broad range of possible evolutionary or ecological questions—will best be served if the "users" of the phylogeny can assess what the range of alternatives might be. (And it is as poor a use of time for the ecologist to rerun alternative analyses from the data bank as for the systematist to revisit remote regions from which the ecologist might have provided a synoptic tissue collection!)

Second, many of the uses to which phylogenies may be put profit from or even require large phylogenies that are as complete as possible. Most published phylogenies are incomplete, for understandable reasons of logistics or convenience. However, inferences about temporal changes in speciation and extinction rates, for example, might be made from phylogenies, but only if all extant taxa are included (Barraclough and Nee 2001). Moreover, tests of many hypotheses, using published phylogenies, are severely limited by the number and reliability of phylogenies suitable to the particular problem at hand; authors frequently use as examples only a few phylogenies, which in some cases are quite controversial. Because many questions in ecology and evolutionary biology are questions of relative frequencies (e.g., the incidence of various modes of speciation), phylogenies of many groups will be needed. Thus, comprehensive phylogenies of large, inclusive clades, such as the ever-growing tree of seed plants, will be useful for many purposes we do not yet envision, especially as these phylogenies become more complete. Although the goal of a complete Tree of Life might not be attainable, the journey toward it will enable us to address ever more hypotheses ever more comprehensively.

Literature Cited

Avise, J. C. 1989. Gene trees and organismal histories: a phylogenetic approach to population biology. Evolution 43:1192–1208.

Avise, J. C. 2000. Phylogeography: the history and formation of species. Harvard University Press, Cambridge, MA.

Avise, J. C., and R. M. Ball, Jr. 1990. Principles of genealogical concordance in species concepts and biological taxonomy. Oxford Surv. Evol. Biol. 7:45–67.

Avise, J. C., and D. Walker. 1998. Pleistocene phylogeographic effects on avian populations and the speciation process. Proc. R. Soc. Lond. B 265:457–463.

Ballard, H. E., Jr., and K. J. Sytsma. 2000. Evolution and biogeography of the woody Hawaiian violets (Viola, Violaceae): arctic origins, herbaceous ancestry and bird dispersal. Evolution 54:1521–1532.

Barraclough, T. G., P. Harvey, and S. Nee. 1995. Sexual selection and taxonomic diversity in passerine birds. Proc. R. Soc. Lond. B 259:211–215.

Barraclough, T. G., and S. Nee. 2001. Phylogenetics and speciation. Trends Ecol. Evol. 16:391–399.

Barraclough, T. G., and A. P. Vogler. 2000. Detecting the geographical pattern of speciation from species-level phylogenies. Am. Nat. 155:419–434.

Basolo, A. L. 1996. The phylogenetic distribution of a female preference. Syst. Biol. 45:290–307.

Baum, D. A., and K. L. Shaw. 1995. Genealogical perspectives on the species problem. Pp. 289–303 in Experimental and molecular approaches to plant biosystematics (P. C. Hoch and A. G. Stephenson, eds.). Monographs in Systematic Botany. Missouri Botanical Garden, St. Louis.

Berlocher, S. H. 1998. Can sympatric speciation via host or habitat shift be proven from phylogenetic and biogeographic evidence? Pp. 99–113 in Endless forms: species and speciation (D. J. Howard and S. H. Berlocher, eds.). Oxford University Press, New York.

Bermingham, E., and J. C. Avise. 1986. Molecular zoogeography of freshwater fishes in the southeastern United States. Genetics 113:939–965.

Bowler, P. J. 1983. The eclipse of Darwinism: anti-Darwinian evolution theories in the decades around 1900. Johns Hopkins University Press, Baltimore.

Bowler, P. J. 1996. Life's splendid drama: evolutionary biology and the reconstruction of life's ancestry 1860–1940. University of Chicago Press, Chicago.

Brooks, D. R., and D. R. Glen. 1982. Primates and pinworms: a case study in coevolution. Proc. Helm. Soc. Wash. 49:76–85.

Brooks, D. R., and D. A. McLennan. 1991. Phylogeny, ecology, and behavior. University of Chicago Press, Chicago.

Brown, W. L., Jr. 1959. General adaptation and evolution. Syst. Zool. 7:157–168.

Brown, W. L., Jr., and E. O. Wilson. 1956. Character displacement. Syst. Zool. 5:49–64.

Carroll, S. B., J. K. Grenier, and S. D. Weatherbee. 2001. From DNA to diversity: molecular genetics and the evolution of animal design. Blackwell Science, Malden, MA.

Chown, S. L., and K. J. Gaston. 2000. Areas, cradles and museums: the latitudinal gradient in species richness. Trends Ecol. Evol. 15:311–315.

Connell, J. H. 1971. On the role of natural enemies in preventing competitive exclusion in some marine animals and in rain forest trees. Pp. 298–310 in Dynamics of populations (P. J. den Boer and G. R. Gradwell, eds.). Proceedings of the Advanced Study Institute in Dynamics of Numbers in Populations, Oosterbeck. Centre for Agricultural Publishing and Documentation, Wageningen, the Netherlands.

Coyne, J. A., and H. A. Orr. 1989. Patterns of speciation in Drosophila. Evolution 43:362–381.

Coyne, J. A., and T. D. Price. 2000. Little evidence for sympatric speciation in island birds. Evolution 54:2166–2171.

Crandall, K. A., and A. R. Templeton. 1996. Applications of intraspecific phylogenies. Pp. 81–99 in New uses for new phylogenies (P. H. Harvey, A. J. Leigh Brown, J. Maynard Smith, and S. Nee, eds.). Oxford University Press, Oxford.

Davis, M. B. 1976. Pleistocene biogeography of temperate deciduous forests. Geosci. Man 13:13–26.

Dobzhansky, T., F. J. Ayala, G. L. Stebbins, and J. W. Valentine. 1977. Evolution. W. H. Freeman, San Francisco.

Dodson, E. O. 1960. Evolution: process and product. Reinhold, New York.

Donoghue, M. J. 1989. Phylogenies and the analysis of evolutionary sequences, with examples from seed plants. Evolution 43:1137–1156.

Dowling, T. E., and C. L. Secor. 1997. The role of hybridization and introgression in the diversification of animals. Annu. Rev. Ecol. Syst. 28:593–619.

Eaton, T. H. 1970. Evolution. W. W. Norton, New York.

Edwards, A. W. F., and L. L. Cavalli-Sforza. 1964. Reconstruction of evolutionary trees. Pp. 67–76 in Phenetic and phylogenetic classification (V. H. Heywood and J. McNeill, eds.). Publ. no. 6. Systematics Association, London.

Ehrlich, P. R., and R. W. Holm. 1963. The process of evolution. McGraw-Hill, New York.

Ehrlich, P. R., and P. H. Raven. 1964. Butterflies and plants: a study in coevolution. Evolution 18:586–608.

Farrell, B. D. 1998. "Inordinate fondness" explained: why are there so many beetles? Science 281:555–559.

Farrell, B. D., D. E. Dussourd, and C. Mitter. 1991. Escalation of plant defense: do latex and resin canals spur plant diversification? Am. Nat. 138:881–900.

Farrell, B. D., and C. Mitter. 1993. Phylogenetic determinants of insect/plant community diversity. Pp. 253–266 in Species diversity in ecological communities: historical and geographical perspectives (R. E. Ricklefs and D. Schluter, eds.). University of Chicago Press, Chicago.

Felsenstein, J. 1973. Maximum likelihood and minimum-steps methods for estimating evolutionary trees from data on discrete characters. Syst. Zool. 22:240–249.

Felsenstein, J. 1985. Phylogenies and the comparative method. Am. Nat. 125:1–15.

Fitch, W. M. 1996. Uses for evolutionary trees. Pp. 116–133 in New uses for new phylogenies (P. H. Harvey, A. J. Leigh Brown, J. Maynard Smith, and S. Nee, eds.). Oxford University Press, Oxford.

Funk, D. J., D. J. Futuyma, G. Ortí, and A. Meyer. 1995. A history of host associations and evolutionary diversification for *Ophraella* (Coleoptera: Chrysomelidae): new evidence from mitochondrial DNA. Evolution 491008–1017.

Futuyma, D. J. 1979. Evolutionary biology. 1st ed. Sinauer, Sunderland, MA.

Futuyma, D. J. 1988. *Sturm und Drang* and the evolutionary synthesis. Evolution 42:217–226.

Futuyma, D. J. 2000. Potential evolution of host range in herbivorous insects. Pp. 42–53 in Host-specificity testing of exotic arthropod biological control agents (R. Van Driesche, T. Heard, A. McClay, and R. Reardon, eds.). U.S. Department of Agriculture Forest Service, Morgantown, WV.

Futuyma, D. J., M. C. Keese, and D. J. Funk. 1995. Genetic constraints on macroevolution: the evolution of host affiliation in the leaf beetle genus *Ophraella*. Evolution 49:797–809.

Futuyma, D. J., M. C. Keese, and S. J. Scheffer. 1993. Genetic constraints and the phylogeny of insect-plant associations: responses of *Ophraella communa* (Coleoptera: Chrysomelidae) to host plants of its congeners. Evolution 47:888–905.

Futuyma, D. J., and C. Mitter. 1996. Insect-plant interactions: the evolution of component communities. Philos. Trans. R. Soc. Lond. B 351:1361–1366.

Futuyma, D. J., and G. Moreno. 1988. The evolution of ecological specialization. Annu. Rev. Ecol. Syst. 19:207–223.

Gittleman, J. L., C. G. Anderson, M. Kot, and H.-K. Luh. 1996. Comparative tests of evolutionary lability and rates using molecular phylogenies. Pp. 289–307 in New uses for new phylogenies (P. H. Harvey, A. J. Leigh Brown, J. Maynard Smith, and S. Nee, eds.). Oxford University Press, Oxford.

Givnish, T. J., and K. J. Sytsma (eds.). 1997. Molecular evolution and adaptive radiation. Cambridge University Press, New York.

Goodman, M., M. L. Weiss, and J. Czelusniak. 1982. Molecular evolution above the species level: branching patterns, rates, and mechanisms. Syst. Zool. 31:376–399.

Grant, V. 1977. Organismic evolution. W. H. Freeman, San Francisco.

Guttman, D. S., and D. E. Dykhuizen. 1994. Clonal divergence in *Escherichia coli* as a result of recombination, not mutation. Science 266:1380–1383.

Hammer, M. F. 1995. A recent common ancestry for human Y chromosomes. Nature 378:376–378.

Hare, M. P., F. Cipriano, and S. R. Palumbi. 2002. Genetic evidence on the demography of speciation in allopatric dolphin species. Evolution 56:804–816.

Harvey, P. H., and M. D. Pagel. 1991. The comparative method in evolutionary biology. Oxford University Press, Oxford.

Hennig, W. 1950. Grundzüge einer Phylogenetischen Theorie der Systematik. Deutscher Zentralverlag, Berlin.

Hillis, D. M. 2001. The emergence of systematic biology. Syst. Biol. 50:301–303.

Holmes, E. C., P. L. Bollyky, S. Nee, A. Rambaut, G. P. Garnett, and P. H. Harvey. 1996. Using phylogenetic trees to reconstruct the history of infectious disease epidemics. Pp. 169–186 in New uses for new phylogenies (P. H. Harvey, A. J. Leigh Brown, J. Maynard Smith, and S. Nee, eds.). Oxford University Press, Oxford.

Hubbell, S. P. 2001. The unified neutral theory of biodiversity and biogeography. Princeton University Press, Princeton, NJ.

Hudson, R. R. 1990. Gene genealogies and the coalescent process. Oxford Surv. Evol. Biol. 7:1–44.

Hughes, A. L. 1999. Adaptive evolution of genes and genomes. Oxford University Press, Oxford.

Hull, D. L. 1988. Science as a process: an evolutionary account of the social and conceptual development of science. University of Chicago Press, Chicago.

Humphries, C. J., and L. R. Parenti. 1999. Cladistic biogeography. Oxford University Press, Oxford.

Ingman, M., H. Kaessmann, S. Pääbo, and U. Gyllensten. 2000. Mitochondrial genome variation and the origin of modern humans. Nature 408:708–713.

International Human Genome Sequencing Consortium. 2001. Initial sequencing and analysis of the human genome. Nature 409:860–921.

Janzen, D. H. 1970. Herbivores and the number of tree species in tropical forests. Am. Nat. 104:501–528.

Kelley, S. T., and B. D. Farrell. 1998. Is specialization a dead end? The phylogeny of host use in *Dendroctonus* bark beetles (Scolytidae). Evolution 52:1731–1743.

Kimura, M. 1983. The neutral theory of molecular evolution. Cambridge University Press, Cambridge.

Klicka, J., and R. M. Zink. 1997. The importance of recent ice ages in speciation: a failed paradigm. Science 277:1666–1669.

Kliman, R. M., P. Andolfatto, J. A. Coyne, F. Depaulis, M. Kreitman, A. J. Berry, J. McCarter, J. Wakeley, and J. Hey. 2000. The population genetics of the origin and divergence of the *Drosophila simulans* complex species. Genetics 156:1913–1931.

Kluge, A. G., and J. S. Farris. 1969. Quantitative phyletics and the evolution of anurans. Syst. Zool. 18:1–32.

Knowles, L. L. 2000. Tests of Pleistocene speciation in montane grasshoppers (genus *Melanoplus*) from the sky islands of western North America. Evolution 54:1337–1348.

Knowles, L. L., D. J. Futuyma, W. F. Eanes, and B. Rannala. 1999. Insight into speciation from historical demography in the phytophagous beetle genus *Ophraella*. Evolution 53:1846–1856.

Latham, R. E., and R. E. Ricklefs. 1993. Continental comparisons of temperate-zone tree species diversity. Pp. 294–314 in Species diversity in ecological communities (R. E. Ricklefs and D. Schluter, eds.). University of Chicago Press, Chicago.

Lee, M. Y. S., and R. Shine. 1998. Reptilian viviparity and Dollo's law. Evolution 52:1441–1450.

Li, W.-H. 1997. Molecular evolution. Sinauer, Sunderland, MA.

Losos, J. B. 1992. The evolution of convergent structure in Caribbean *Anolis* communities. Syst. Biol. 41:403–420.

Losos, J. B., T. R. Jackman, A. Larson, K. de Queiroz, and L. Rodríguez-Schettino. 1998. Contingency and determinism in replicated adaptive radiations of island lizards. Science 279:2115–2118.

Louda, S. M., D. Kendall, J. Connor, and D. Simberloff. 1997. Ecological effects of an insect introduced for biological control of weeds. Science 277:1088–1090.

Lynch, M. 1990. The rate of morphological evolution from the standpoint of the neutral expectation. Am. Nat. 136:727–741.

Lynch, M., J. Blanchard, D. Houle, T. Kibota, S. Schultz, L. Vassilieva, and J. Willis. 1999. Perspective: spontaneous deleterious mutation. Evolution 53:645–663.

Martins, E. P. (ed.) 1996. Phylogenies and the comparative method in animal behavior. Oxford University Press, Oxford.

Maynard Smith, J., R. Burian, S. Kauffman, P. Alberch, J. Campbell, B. Goodwin, R. Lande, D. Raup, and L. Wolpert. 1985. Developmental constraints and evolution. Quart. Rev. Biol. 60:265–287.

Mayr, E. 1960. The emergence of evolutionary novelties. Pp. 349–380 *in* The evolution of life (S. Tax, ed.). University of Chicago Press, Chicago.

McCune, A. R., and N. J. Lovejoy. 1998. The relative rate of sympatric and allopatric speciation in fishes: tests using DNA sequence divergence between sister species and among clades. Pp. 172–185 *in* Endless forms: species and speciation (D. J. Howard and S. H. Berlocher, eds.). Oxford University Press, New York.

McPeek, M. A. 1995. Morphological evolution mediated by behavior in the damselflies of two communities. Evolution 49:749–769.

Milá, B., D. J. Girman, M. Kimura, and T. B. Smith. 2000. Genetic evidence for the effect of a postglacial population expansion on the phylogeography of a North American songbird. Proc. R. Soc. Lond. B 267:1033–1040.

Mitter, C., and D. R. Brooks. 1983. Phylogenetic aspects of coevolution. Pp. 65–98 *in* Coevolution (D. J. Futuyma and M. Slatkin, eds.). Sinauer, Sunderland, Mass.

Mitter, C., B. Farrell, and B. Wiegmann. 1988. The phylogenetic study of adaptive zones: has phytophagy promoted insect diversification? Am. Nat. 132:107–128.

Mooers, A. Ø., and S. B. Heard. 1997. Inferring evolutionary process from phylogenetic tree shape. Q. Rev. Biol. 72:31–54.

Moritz, C. 1994. Defining evolutionary significant units for conservation. Trends Ecol. Evol. 9:373–375.

Morrone, J. J., and J. V. Crisci. 1995. Historical biogeography: introduction to methods. Annu. Rev. Ecol. Syst. 26:373–401.

Nelson, G., and N. I. Platnick. 1981. Systematics and biogeography: cladistics and vicariance. Columbia University Press, New York.

Nosil, P. 2002. Transition rates between specialization and generalization in phytophagus insects. Evolution 56:1701–1706.

Page, R. D. M., and M. S. Hafner. 1996. Molecular phylogenies and host-parasite cospeciation: gophers and lice as a model system. Pp. 255–270 *in* New uses for new phylogenies (P. H. Harvey, A. J. Leigh Brown, J. Maynard Smith, and S. Nee, eds.). Oxford University Press, Oxford.

Purvis, A., P.-M. Agapow, J. L. Gittleman, and G. M. Mace. 2000a. Non-random extinction and the loss of evolutionary history. Science 288:328–330.

Purvis, A., J. L. Gittleman, G. Cowlishaw, and G. M. Mace. 2000b. Predicting extinction risk in declining species. Proc. R. Soc. Lond. B 267:1947–1952.

Qian, H., and R. E. Ricklefs. 1999. A comparison of the taxonomic richness of vascular plants in China and the United States. Am. Nat. 154:160–181.

Raxworthy, C. J., M. R. J. Forstner, and R. A. Nussbaum. 2002. Chameleon radiation by oceanic dispersal. Nature 415:784–787.

Richardson, J. E., R. T. Pennington, T. D. Pennington, and P. M. Hollingsworth. 2001. Rapid diversification of a species-rich genus of neotropical rain forest trees. Science 293:2242–2245.

Rieseberg, L. H. 1997. Hybrid origins of plant species. Annu. Rev. Ecol. Syst. 28:359–389.

Ryan, M. J., and A. S. Rand. 1993. Sexual selection and signal evolution—the ghost of biases past. Philos. Trans. R. Soc. Lond. B 340:187–195.

Sanderson, M. J., and M. J. Donoghue. 1996. Reconstructing shifts in diversification rate on phylogenies. Trends Ecol. Evol. 11:15–20.

Sanderson, M. J., and L. Hufford (eds.). 1996. Homology: the recurrence of similarity in evolution. Academic Press, San Diego.

Savage, J. M. 1963. Evolution. Holt, Rinehart and Winston, New York.

Schliewen, U. K., D. Tautz, and S. Pääbo. 1994. Sympatric speciation suggested by monophyly of crater lake cichlids. Nature 368:629–632.

Schluter, D. 2000. The ecology of adaptive radiation. Oxford University Press, Oxford.

Sidor, C. A., and J. A. Hopson. 1998. Ghost lineages and "mammalness": assessing the temporal pattern of character acquisition in the Synapsida. Paleobiology 24:254–273.

Sillén-Tullberg, B. 1988. Evolution of gregariousness in aposematic butterfly larvae: a phylogenetic analysis. Evolution 42:293–305.

Silvertown, J., M. Franco, and J. L. Harper (eds.). 1997. Plant life histories: ecology, phylogeny and evolution. Cambridge University Press, Cambridge.

Sokal, R. R., and C. D. Michener. 1958. A statistical method for evaluating systematic relationships. Univ. Kansas Sci. Bull. 38:1409–1438.

Stebbins, G. L. 1966. Processes of organic evolution. Prentice-Hall, Englewood Cliffs, NJ.

Stebbins, G. L. 1974. Flowering plants: evolution above the species level. Harvard University Press, Cambridge, MA.

Taberlet, P., L. Fumagalli, A.-G. Wust-Saucy, and J.-F. Cosson, 1998. Comparative phylogeography and postglacial colonization routes in Europe. Mol. Ecol. 7:453–464.

Takahata, N., Y. Satta, and J. Klein. 1995. Divergence time and population size in the lineage leading to modern humans. Theor. Popul. Biol. 48:198–221.

Throckmorton, L. H. 1965. Similarity versus relationship in *Drosophila*. Syst. Zool. 14:221–236.

Vane-Wright, R. I., C. J. Humphries, and P. H. Williams. 1991. What to protect—systematics and the agony of choice. Biol. Conserv. 55:235–254.

Venter, J. C., M. D. Adams, E. W. Myers, P. W. Li, R. J. Mural, G. G. Sutton, H. V. Smith, M. Yandell, C. A. Evans, et al. 2001. The sequence of the human genome. Science 291:1304–1351.

Volpe, E. P. 1970. Understanding evolution, 2nd ed. Wm. C. Brown, Dubuque, IA.

Wagner, G. P. 1989. The biological homology concept. Annu. Rev. Ecol. Syst. 20:51–69.

Wahlberg, N. 2001. The phylogenetics and biochemistry of host-plant specialization in melitaeine butterflies (Lepidoptera: Nymphalidae). Evolution 55:522–537.

Wakeley, J., and J. Hey. 1997. Estimating ancestral population parameters. Genetics 145:847–855.

Webb, C. O. 2000. Exploring the phylogenetic structure of ecological communities: an example for rain forest trees. Am. Nat. 156:145–155.

Westoby, M., M. R. Leishman, and J. M. Lord. 1995. On misinterpreting the "phylogenetic correction." J. Ecol. 83:531–534.

Wilson, A. C., S. S. Carlson, and T. J. White. 1977. Biochemical evolution. Annu. Rev. Biochem. 46:573–639.

Zink, R. M., R. C. Blackwell-Rago, and F. Ronquist. 2000. The shifting roles of dispersal and vicariance in biogeography. Proc. R. Soc. Lond. B 267:497–503.

The Origin and Radiation of Life on Earth

S. L. Baldauf

D. Bhattacharya

J. Cockrill

P. Hugenholtz

J. Pawlowski

A. G. B. Simpson

The Tree of Life

An Overview

Most of life, for most of life's history, is about single-celled organisms, which come in one of two types, eukaryotic and prokaryotic. Most of life is probably prokaryotic, in terms of numbers of cells, numbers of species, and time on Earth. Two of the three domains of life are prokaryotic, the Archaea and the Bacteria, and theirs are the oldest fossils, found in the oldest unmetamorphosed rock [3.5 Byr (billion years) old; Schopf et al. 2002; but see Van Zuillen et al. 2002]. Therefore, the last universal common ancestor of all life (LUCA) was probably prokaryotic, that is, a small cell (1–5 μm diameter), with a small genome [~1–10 megabases (million bases)], few or no internal membrane-bound structures, and able to meet all its living requirements using simple compounds (autotrophic).

Eukaryotes were almost certainly derived from prokaryotes (but see Philippe, ch. 7 in this vol.). The oldest even arguably eukaryotic fossils are only ~1.8 Byr old (Brocks et al. 1999). All well-studied eukaryotes have cells that are at least an order of magnitude larger than those of prokaryotes with genomes (100–10,000 megabases). However, we now know that bacterial-sized eukaryotes, probably with nearly bacterial-sized genomes (picoeukaryotes; described below), are common (Moon-van der Staay et al. 2001, Lopéz-Garcia et al. 2001), but even these are clearly distinct from prokaryotic cells. Thus, eukaryotic cells are more structurally complex than those of prokaryotes, having various internal membrane-bound organelles, such as a nucleus, and are, for the most part, energetically dependent on endosymbiotic bacteria, that is, mitochondria and chloroplasts.

Until the 1980s, universal trees of life were based on a combination of structural and biochemical data characters, but these generally have either too much or too little variation to reflect reliably ancient evolutionary relationships. Therefore, before the advent of molecular biology, constructing an evolutionarily meaningful tree of life was a dubious undertaking, at best. It was the discovery of the conservative, ubiquitous nature of ribosomal RNAs that changed this.

All living cells make protein and in pretty much the same way using ribosomes that consist of a large and small subunit (LSU and SSU). The catalytic core of each ribosomal subunit is an RNA molecule, the ribosomal RNAs (rRNAs). LSU and SSU rRNAs are large molecules, highly conserved across all life, and extremely abundant. It is these characteristics that make them such excellent "molecular phylogenetic markers," particularly SSU rRNA (also known by its sedimentation coefficient of 12S for mitochondrial or 16S–18S for nuclear SSU or rRNA; Green and Noller 1997).

The highly conserved nature of SSU rRNAs allows these sequences to be obtained relatively easily from most living organisms and meaningfully compared with each other. Thus, SSU rRNA data provided, for the first time, large numbers of clearly homologous characters across all life and led to the first universal evolutionary trees derived by objective, quantitative criteria. The most startling early discovery was that prokaryotic cells are actually two fundamentally different groups of organisms, archaebacteria (Archaea) and true bacteria (Eubacteria or simply "Bacteria"), as different from each other as either is from eukaryotes (Eucarya; Woese and Fox 1977).

There are now more than 40,000 SSU rRNA sequences in the public domain (Benson et al. 2004). These clearly identify many (but likely not all) major taxonomic groups, some previously only guessed at or entirely unknown. Parts of the molecule are so highly conserved that they can be used as primers to determine SSU rRNA sequences from even trace amounts of DNA using polymerase chain reaction (PCR). This technology has recently been adapted to allow sequencing of SSU rRNA from uncultured organisms or even from mixed pools of total environmental DNA, an approach called environmental or culture-independent PCR (ciPCR; Amann et al. 1995, Moreira and Lopéz-Garcia 2001, Hugenholtz et al. 1998; see also Pace, ch. 5 in this vol.). This has revealed a tremendous diversity of previously unknown organisms at all taxonomic levels.

SSU rRNA data first defined the universal Tree of Life and remain the cornerstone of molecular systematics. Although protein genes trees have revealed important discrepancies in the SSU rRNA tree, each protein gene tree seems to have its own, unique inaccuracies as well. Nonetheless, on the whole, there is a general consensus on most branches among most molecules, although no single gene seems able to accurately reconstruct them all (Baldauf et al. 2000). Individual genes also seem to lack sufficient information to resolve the deepest branches in the tree. For this reason, most studies of deep phylogeny now employ multigene "concatenated" data sets (CDSs). However, even this may not work for bacteria and archaeans because of frequent trading of genes among even very distantly related taxa [lateral gene transfer (LGT); see Doolittle, ch. 6 in this vol.).

The following is a summary of the major groups of life as we currently see them, and our best guesses as to how they are related to each other. We have tried to provide a brief description of each of the major groups, a summary of their likely higher order relationships, and the nature of the supporting data, both molecular and nonmolecular. The reader should keep in mind that the deepest divergences in these trees require large CDSs to test them, and only a few of these are yet available. Furthermore, most habitats remain unsampled by ciPCR studies, and the identities of these new "ciPCR taxa" need to be confirmed with other data. Therefore, the following is very much a summation of a work in progress, but, with a little luck, one we can continue to build on for a while.

Overview of the Tree

Figure 4.1 summarizes our current best guess as to the composition of and relationships among the major groups of living organisms based on a large number of independent, partially overlapping studies. Emphasis is placed on SSU rRNA trees, because these are the most comprehensive, and

on CDS trees, because these are the most accurate. The integrity of the three domains of life, Archaea, Bacteria, and Eucarya, is now confirmed by a tremendous body of data, including nearly 100 completely sequenced genomes. The identities of most of the major groups within these domains are also confirmed by many different data, both molecular and nonmolecular. Some of the relationships among the major groups ("deep branches") are also well resolved by substantial bodies of data, but the majority of these deepest branches are still only tenuously supported (shaded bars on figure 4.1).

Arguably the single most outstanding question in the Tree of Life is the position of the root. This can theoretically be tested using ancient gene duplications that occurred before the origin of the last common ancestor of all life. A number of these duplications are known, and all seem to tell the same story, that the root of the universal tree lies within Bacteria, making Archaea and Eucarya sister taxa (Gogarten et al. 1989, Iwabe et al. 1989). This agrees with the striking similarities between Archaea and Eucarya in nearly all aspects of cellular information processing. Nonetheless, these are still only a handful of genes each with only a small number of universally alignable positions. These limitations, together with the immense evolutionary distance involved (2–4 Byr), make this an extremely difficult phylogenetic problem (see Philippe, ch. 7 in this vol.), and this location for the universal root still needs to be regarded with caution.

Domain Bacteria

Bacteria are highly variable, and there are few general rules about them that are not violated somewhere. Sizes average 1–5 γm but range from 0.1 to 660 γm. Most have a peptidoglycan cell wall sandwiched between an inner and outer cell membrane composed of ester-linked lipids, but the cell wall, the outer membrane, or both may be absent. A variety of internal and external structures are found in bacteria, but these are rarely membrane bound. Multicellular assemblages are common, sometimes with terminally differentiated cell types, and complex life cycles are found, sometimes including several developmental stages. Motility is by means of flagella, gliding, or adjustable buoyancy using gas-filled vacuoles, and warfare is waged using a wide assortment of "antibiotics." Habitats seem to be any where there is water, even small or sporadic amounts. These include everything from deep crustal groundwater to natural gas deposits, volcanoes, oil spills, clouds, and many, many more (Madigan et al. 1997, Paustian 2003).

Bacterial genomes are most commonly organized into a single circular chromosome with a single origin of replication, very little repetitive DNA, many genes organized into operons, and introns extremely rare. The chromosome is located in a nuclear region (nucleoid) that is rarely membrane bound, and proteins are synthesized nearby on 70S ribo-

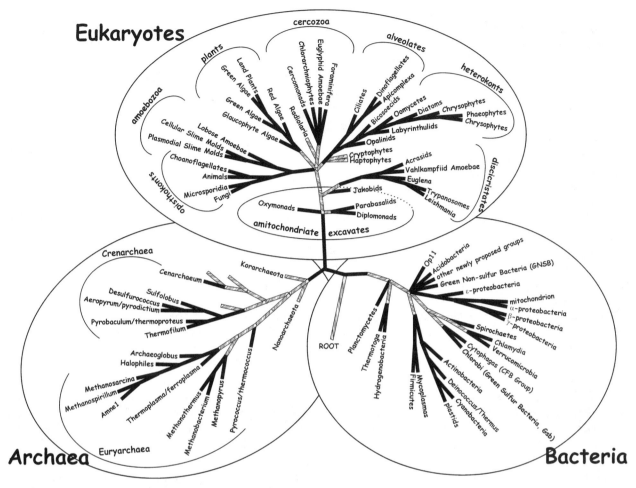

Figure 4.1. The Tree of Life. The tree shown is our current best guess on the major groups of life and their relationships to each other. Solid bars indicate groupings for which there is considerable molecular phylogenetic support. Shaded bars indicate tentative groupings with moderate, weak, or purely ultrastructural support.

somes such that transcription and translation are simultaneous (coupled). Extrachromosomal DNA minicircles (plasmids) are common, carry a variety of genes often including ones for antibiotic synthesis and resistance, and vary widely in size. Gene expression is regulated by diffusible RNA polymerase subunits, called sigma factors, that bind directly to specialized promoter elements immediately preceding their genes. Cells are generally haploid in lab culture, but most are probably haplodiploid in nature, with large stretches of the chromosome existing in multiple copies (Madigan et al. 1997).

Photosynthesis is common and usually anoxygenic, using photosystem I or II (PSI, PSII). Only cyanobacteria use both PSI and PSII, which, when coupled, can split water and release oxygen (i.e., perform oxygenic photosynthesis). A wide diversity of bacteria are thermophilic (prefer or require high temperatures). It therefore appears that thermophily must have evolved multiple times, probably aided by lateral transfer of critical genes such as DNA gyrase (Forterre 2002). Adaptations to thermophily include positive supercoiling of

DNA and its packaging with histone-like proteins, increased guanine + cytosine (G+C) content in catalytic RNAs (but not in protein-encoding DNA), and on-demand production of heat-labile small molecules. Parasitism and symbioses are widespread, mostly with eukaryotes. However, bacteria can parasitize other bacteria or members of Archaea and also form sometimes extremely complex symbioses or highly coordinated commensal relationships with them (Madigan et al. 1997).

Because Bacteria are too biochemically and morphologically plastic to be classified by such characters, their higher order systematics and the entire field of bacterial evolution did not really exist before molecular phylogeny. The first true phylogenetic treatment of bacteria was Carl Woese's now classic 1987 paper in which he placed all the major groups of cultured taxa into 12 "classical" groups, some predicted and others still without phenotypic justification. More are being added from existing culture collections—SSU rRNA sequences exist for fewer than half of these taxa—and the exploration of new habitats. However, the biggest revolution in our appreciation of bacterial diversity has come from ciPCR studies (Hugenholtz et al. 1998). These suggest the possible existence of 10–20 or more new groups, some of them widespread and diverse and probably important components of a variety of ecosystems (see Pace, ch. 5 in this vol.). The study of bacterial evolution, and bacteriology in general, has been further revolutionized by the advent of rapid whole-genome sequencing, revealing the entire genetic inventory of diverse bacterial species. There are ~70 completed bacterial genomes listed at the National Center for Biotechnology Information genomics server (Benson et al. 2004) and severalfold more in progress (Bernal et al. 2001). There are also probably as many again in the private domain, particularly from medically and commercially important taxa.

Molecular phylogenies of SSU rRNA and other universal genes seem, for the most part, to define the major groups of bacteria but not the relationships among them. This is because of an assortment of problems, including the antiquity of these relationships, lack of sequence data from important taxa, and LGT (see Doolittle, ch. 6 in this vol.). The extent of LGT in bacterial evolution has only recently been recognized, and although informational (transcription and translation component encoding) genes seem less susceptible (Jain et al. 1999), no gene appears to be immune (see Doolittle, ch. 6 in this vol.; see also Asai et al. 1999). Analyses of multigene CDSs now show some consistent strong resolution of major deep branches. However, these studies are still few in number, only include taxa with completely sequenced genomes, are somewhat overlapping in gene content, and may not always be free of LGT-induced artifacts. Therefore, figure 4.2 shows a somewhat optimistic view of higher order bacterial systematics, and many newly described lineages are missing because of a general lack of information on them. The following adheres to the standardized bacterial nomenclature as proposed in the most recent edition of *Bergey's Manual of Systematic Bacteriology* (2001).

Hyperthermophiles: Thermotogae and Aquificae

Thermotogae

Thermotogae (fig. 4.2, node 2) are nonphotosynthetic rod-shaped hyperthermophilic (65–95°C) anaerobes that consume organic compounds and generate hydrogen gas and hydrogen sulfide. Besides being phenotypically narrow, the group as a whole is not particularly large or widespread, as ciPCR studies indicate, and so far they are almost exclusively restricted to geothermal habitats (Hugenholtz 1998). The only well-characterized taxon is *Thermotoga maritima*, originally isolated from geothermal marine sediments and named for its loose "toga-like" outer membrane. This taxon is usually among the deepest, if not the deepest, branch in phylogenetic trees within Bacteria. This seemed to be supported by initial analyses of its completed genome sequence, which

showed that 24% of its genes are more similar to homologs in Archaea than to those in Bacteria (Nelson et al. 1999). However, this number appears to be considerably overestimated (Ochman 2001) because it is based on a simple database search strategy ("blastology"; see Doolittle, ch. 6 in this vol.), and in-depth analyses of the remaining archaea-like genes show that at least some are probably the result of relatively recent LGT (Nesbo et al. 2001).

Aquificae (Aquifex/Hydrogenobacter Group)

Isolated from hot springs, volcano calderas, and marine hydrothermal vents, Aquificae (fig. 4.2, node 2) thrive at 86–95°C, making them some of the most thermophilic bacteria known. Like the Thermotogae, members of this group appear to be restricted to geothermal habitats (Hugenholtz et al. 1998), where they live by splitting hydrogen gas or hydrogen sulfide and fixing carbon dioxide for carbon, all abundant in geothermal volcanic gases (Hjorleifsdottir et al. 2002). The Aquificae are more diverse than are Thermotogae and include halophiles, isolated from saline hot springs, and an acidophile, isolated from an acidic solfatar (sulfur deposits, e.g., volcanoes; Takacs et al. 2001). The best characterized are species of *Aquifex*, a blue filament and currently the most thermophilic bacteria known. The completely sequenced, relatively small (1.55 megabases) genome of *A. aeolicus* lacks many metabolic pathways, consistent with the organism's obligate chemolithotrophic lifestyle (Deckert et al. 1998). New genera belonging to this group have recently been described (Huber et al. 2002a). These new taxa significantly extend the phylogenetic diversity of the group (according to SSU rRNA divergence) but not particularly its physiological diversity, because all are hyperthermophilic chemolithoautotrophs.

Phylogeny

Thermotogae and Aquificae are the most consistently basal branches in bacterial trees, both in CDS and single gene analyses (fig. 4.2, node 2). However, they are found in a variety of arrangements either together as a group (Olsen et al. 1994, Bocchetta et al. 2000, Wolf et al. 2001, Daubin et al. 2002) or as adjacent branches and in alternating order (Olsen et al. 1994, Brown et al. 2001). On the other hand, Brochier and Philippe (2002) suggest that the basal branching of these two taxa in SSU rRNA trees at least is due to a long-branch attraction artifact. This is the tendency in phylogenetic trees for highly divergent sequences, that is, those with long terminal branches, to group together and/or be drawn toward the base of the tree when a distant outgroup is used to root it.

Green Nonsulfur Bacteria (Chloroflexi)

The green nonsulfur (GNS) bacterial group is currently defined solely on the basis of SSU rRNA phylogeny. Members of the group are found diverse habitats, sometimes in abundance (Hugenholtz et al. 1998, Bjornsson et al. 2002), and

A

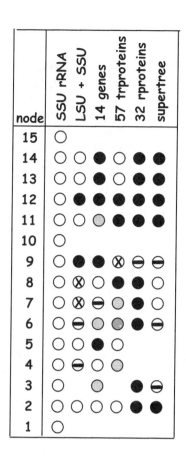

B

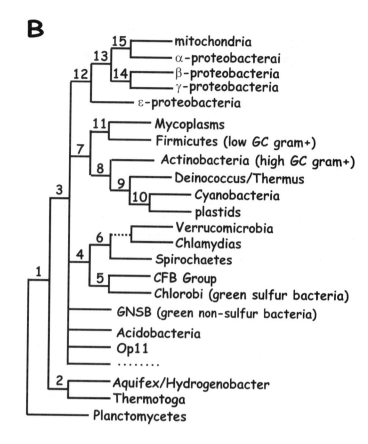

Figure 4.2. Support for deep branches in the bacterial tree. (A) shows data supporting the consensus phylogeny of major bacterial groups shown in (B). Bootstrap values (% BP) from individual data sets supporting the numbered nodes are indicated by circles in black (75–100% BP), gray (60–75% BP), or white (<60% BP). Rejection of nodes by individual data sets is measured by the strongest bootstrap support for any conflicting grouping and is indicated by a circled cross (65–100% BP) or bar (<65% BP). For SSU rRNA phylogeny, white or black circles indicate only presence or absence in trees, respectively. Data sets used are SSU rRNA, combined SSU and LSU (Brochier et al. 2002), 14 assorted conserved genes (Brown et al. 2001), 57 translational proteins (Brochier et al. 2002), 32 ribosomal proteins (Wolf et al. 2001), and supertree analysis of 121 genes (Daubin et al. 2002). Most of the new ciPCR-only groups are not included because of space limitations and their current omission from combined data sets, as indicated by the dotted line below Op11.

the group appears as an early branch of Bacteria in some phylogenetic trees (Oyaizu et al. 1987). Four major subdivisions are defined, with most described taxa falling within a single subdivision, Chloroflexi. Two entire subdivisions, including Subdivision 1, the most divergent (by SSU rRNA analysis), are known only from ciPCR. Recently an isolate belonging to Subdivision 1 has been obtained from activated sludge (Sekiguchi et al. 2001), although it has not been characterized in detail.

Contrary to the group name, not all members are green and sulfide intolerant. The group is metabolically diverse and includes thermophilic sulfur-intolerant green phototrophs (Chloroflexi), thermophilic red phototrophs (Heliothrix), mesophilic sulfur-tolerant green phototrophs (Oscillochloris), thermophilic heterotrophs (Herpetosiphon, Dehalococcoides, Thermoleophilum), *Thermomicrobium*, *Sphaerobacter*, and probably many more. Morphologically, the group appears to

be rich in filamentous representatives (Bjornsson et al. 2002), including all of the described species with the exceptions of the rod-shaped *Thermomicrobium* and *Sphaerobacter*, and coccus-shaped *Dehalococcoides*.

Chloroflexi

Chloroflexi contain the best characterized of the GNS bacteria. They are superficially similar but apparently unrelated to the green sulfur bacteria (GSB; described below). Members of Chloroflexi are moderately thermophilic (35–72°C) gliding green filaments that leave a characteristic slime trail. They are metabolically versatile but generally act as facultative anoxygenic phototrophs and are common in microbial mats, sometimes forming massive accumulations in hot springs. In fact, it may have been GNS bacteria rather than cyanobacteria that formed the large >3 Byr old continuous microbial mats known as stromatolites (Oyaizu et al. 1987).

Their photosynthesis is a hybrid of features of both GSB (light is harvested in cylindrical organelles called chlorosomes using bacteriochlorophyll) and proteobacteria (electron transport occurs across the cell membrane using PSI).

Other GNS Bacteria

The other GNS bacteria exhibit a wide variety of phenotypes. *Herpetosiphon* is a mesophilic gliding bacterium that is poorly characterized but maybe very common in soil. Other non-photosynthetic members of the group have very unusual growth substrates. *Thermoleophilum* lives at 60°C and can grow on wax using a novel respiratory naphthoquinone. Species of *Dehalococcoides* can thrive on chlorohydrocarbons such as tetrachloroethane, which are toxic, highly persistent, and ubiquitous environmental contaminants (Adrian et al. 2002). This makes them potential bioremediation agents. However, little else is known about these taxa.

Phylogeny

The chimeric nature of the chloroflexan photosynthetic apparatus led to early speculation that the GNS bacteria might be very ancient and represent the ancestral bacterial photosynthetic apparatus. This is consistent with SSU rRNA phylogeny, which places them among the deepest branches in the bacterial tree (fig. 4.2; Woese 1987, Pace 1997). However, with the addition of environmental clades, the relative branching order among major bacterial groups now appears to be unresolved, and no deep branches can be identified with certainty (Hugenholtz et al. 1998). There are very few other molecular data on the group, and no completed genome sequences, so they have not yet been included in CDS or supertree analyses. However, sequencing of at least two GNS bacterial genomes is in progress (Bernal et al. 2001).

Planctomycetes

Planctomycetes (fig. 4.2, node 1) are perhaps the most phenotypically unique group of bacteria known, with a whole series of unusual although possibly interrelated features. They are aquatic, appendage forming (prosthecate) aerobes. Cells anchor themselves to various substrates using stalks, and when anchored to each other, they form rosettes. Probably because of this morphology, they divide by asymmetrical budding rather than binary fission, as is also the case in other stalk-forming bacteria such as β-proteobacteria (described below; Hallbeck et al. 1993). This means that cells are dimorphic with distinct mother and daughter cells, resulting in a colonial genealogy. Daughter cells or "swarmers" are motile by means of flagella that are often lost when the cells develop stalks and settle down to become mother cells.

Unique among bacteria, Planctomycetes have peptidoglycan-free proteinaceous cell walls that are covered with distinctive pits of unknown function. Most striking of all is the presence in many of the Planctomycetes of a single- or double-membrane-bound nucleoid, reminiscent of the membrane-bound nuclei of eukaryotes. The recently described taxon *Candidatus* "*Brocadia anammoxidans*" also has a membrane-bound anammoxosome region that performs anaerobic ammonia oxidation (anammox; Lindsay et al. 2001). The membrane of this "organelle" consists of biologically unprecedented ladderane lipids that make the anammoxosome highly impermeable and thus protect the cell from the anammox intermediates (Sinninghe Damste et al. 2002).

Because of their distinctive morphology, many Planctomycetes-like species have been described, but few have been successfully cultured. They are well represented in the majority of ciPCR studies, including those of geothermal vents, fresh and marine waters, and deep subsurface habitats (Hugenholtz et al. 1998). Their SSU rRNA sequences are highly usual; although clearly a monophyletic group, they exhibit very low sequence similarity to all other bacteria. This is consistent with their being either very odd (i.e., rapidly evolving) or very old (Woese 1987). Speculation has arisen, at various times, that they might represent an extremely early divergence from the common bacterial root. This idea is not supported by most SSU rRNA phylogenies, where they do not branch particularly deeply and tend to have an affinity for *Chlamydia* and *Verrucomicrobia*, although without strong statistical support. However, a recent reanalysis of SSU rRNA data using only the most slowly evolving positions places them back at the base of the Bacteria with moderate support (Brochier and Philippe 2002), although the merits of this approach are not proven. Additional molecular data are needed to test this possibility.

CFB Group (Bacteroidetes) and GSB (Chlorobi)

CFB Group (Cytophagas, Flexibacteria, Flavobacteria, and Bacteroides)

The CFB group (fig. 4.2, node 5) is a group without phenotypic justification, that is, lacking common defining features, perhaps in part because the different taxa have been studied in very different ways (Woese 1987). They are generally rod shaped but pleomorphic (variable), may form sheets, and may move, either by gliding or with flagella. They tend to have peptidoglycan-free cell walls and unusual membrane lipids. All are organochemotrophs, that is, are nonphotosynthetic and non-carbon-fixing. Most can degrade large complex macromolecules such as cellulose and chitin, and they are common animal commensals. *Bacteroides thetamicrobium* is the most abundant organism in human gut (~10^{10} cells/g body weight), and in its membranes it has sphingolipids, lipids that are otherwise largely restricted to mammalian nerve cells.

Flavobacteria and Cytophagas

Flavobacteria, named for their characteristic carotene-induced yellow color, are obligately aerobic nonmotile rods with a mitochondria-like electron transport chain. They are found in soils and aquatic environments and receive considerable attention as opportunistic pathogens of fish in crowded conditions such as farms or aquaria. Obligate intracellular parasites/symbionts of the amoebozoan *Acanthamoeba* are also known. The cytophagas, including flexibacteria, are essentially gliding flavobacteria. They occur in similar habitats and are especially noted for their ability to degrade complex macromolecules such as DNA, cellulose, chitin, and agar, suggesting that they may have important roles in natural nutrient recycling.

Bacteroidetes

The *Bacteroides* constitute the third major group within the CFB phylum. All are obligate anaerobes and capable of living freely, but they are most commonly encountered in the mammalian gut. Here they are extremely abundant and highly diverse (Ramsaka et al. 2000). They possess thick polysaccharide coats that permit them to survive in these environments, where they break down host-indigestible materials such as cellulose and pectin. Some of these breakdown products may be absorbed by the host, but their primary benefit to humans may be in rendering the gut inhospitable to potential pathogens and also, by sheer force of numbers, physically blocking the latter from attaching. *Porphyromonas* species are associated with dental disease in humans, although whether as cause or effect is not known. ciPCR has identified a complex assemblage of *Prevotolla* species in the guts of ruminants (Ramsaka et al. 2000).

CFB Group Phylogeny

The CFB group of bacteria as a whole appear to be ubiquitous. ciPCR studies find them in every habitat examined so far and often in abundance (Hugenholtz et al. 1998). Phylogenetically the group is diverse and poorly understood. No completed genome sequences exist at this time, so they are not included in current CDS studies, but the genomes of *Bacteroides fragilis*, *Porphyromonas gingivalis*, and *Cytophaga hutchinsonii* will be completed soon. Although SSU rRNA phylogeny strongly supports the group as a whole, it does not support the integrity of any of the three subgroups, and major reclassification is underway (Olsen et al. 1994, Maidak et al. 2001). In addition, new, apparently basal lineages within the group have been described recently such as *Rhodothermus* (Andresson and Fridjonsson 1994).

Green Sulfur Bacteria (GSB) Group (Chlorobi)

The cultivated members of the GSB (fig. 4.2, node 5) are obligately anaerobic, sometimes gliding but mostly nonmotile, green or brown phototrophs. They thrive in high-sulfur, low-light habitats such as sulfur-rich muds, where they oxidize sulfur and excrete sulfate. This gives rise to their characteristic large, iridescent extracellular sulfate globules. The cultivated GSB tend to be rod-shaped, often twisted into a variety of shapes, including crescents, rings, ovals, or spheres, or aggregated into long, sometimes spiral chains.

Chlorobi are extremely efficient photosynthesizers, requiring approximately one-quarter the light intensity required by other phototrophs. They include *Chlorochromatium aggregatum*, which is a consortium consisting of a *Chlorobium* and an unidentified host that is heterotrophic (i.e., obtains nutrients by nonphotosynthesis means; Overmann and van Gemerden 2000). The large, motile, light-seeking host provides transport for the many small chlorobi attached to its surface. These may in turn provide their host with photosynthate. The whole consortium divides synchronously, although the relationship appears to be obligate only for the host (Overmann and van Gemerden 2000).

GSB photosynthesis superficially resembles that of GNS bacteria, with whom they were once thought to be closely related. Both use similar pigments (bacteriochlorophyll b, c, or d) and have at least superficially similar cylindrical light harvesting organelles (chlorosomes). However, the structures of these organelles may be substantially different between the two groups. Also, GSB use PSII and fix carbon dioxide with a reverse Krebs (tricarboxylic acid) cycle, whereas GNS bacteria use PSI and the Calvin cycle, at least primarily (Hanson and Tabita 2001). The two groups are also clearly separated in SSU rRNA phylogenies (e.g., Woese 1987, Hugenholtz et al. 1998; see also Pace, ch. 5 in this vol.).

The cultivated representatives of the GSB appear as a closely related group in SSU rRNA trees. The majority of sequence diversity in the group appears to be represented by as yet uncultivated lineages detected in environmental SSU rRNA surveys. These ciPCR studies also indicate GSB-type bacteria in diverse habitats, including subsurface layers completely devoid of light (Hugenholtz et al. 1998). The latter suggests that the phenotypic diversity of the group may also be much broader than the currently recognized anaerobic phototrophy and may include nonphotosynthetic members.

Phylogeny

The robust grouping of the GSB and CFB groups (fig. 4.2, node 5), always excluding the GNS group, was first noted based on SSU rRNA sequence signatures, structural features, and phylogeny (Woese 1987). This relationship has continued to hold strongly throughout the massive expansion of the SSU rRNA database (Pace 1997). Because the genome sequence of *Chlorobium tepidum* was only released this year, representatives of the CFB and GSB groups have been included in only two CDS studies to date (fig. 4.2A, line 5). These both strongly support a GSB + CFB clade and suggest that ultimately the two should be combined as a single major bacterial group.

Chlamydiae and Spirochaetes

Chlamydiae

The few described species of Chlamydiae (fig. 4.2, node 6) are a closely related, highly specialized, medically important group of obligate intracellular parasites. They cause a number of significant human diseases such as pelvic inflammatory disease and trachoma, the leading cause of preventable blindness. Their cell walls lack peptidoglycan, but they retain the necessary enzymes to make it and are therefore sensitive to β-lactam antibiotics (which target cell-wall biosynthetic enzymes; Stephens et al. 1998). The life cycle consists of a desiccation-resistant infective form (elementary bodies) that "germinates" upon entering the host to form reticulate bodies sequestered in intracellular vacuoles or "inclusions." Upon maturing, they revert to elementary bodies that escape by lysing the host cell.

Known clinical isolates probably account for only a small subset of the Chlamydiae, and even many clinical isolates remain uncharacterized. They have also recently been identified as intracellular parasites of amoebae. ciPCR has indicated as many as four additional subgroups of Chlamydiae (Horn and Wagner 2001), although they have not been found in many habitats other than soil (Hugenholtz et al. 1998). Four *Chlamydia* genomes have been completely sequenced. They are extremely reduced with around 1000 protein-coding genes and lack many biosynthetic pathways, including those for basic small molecules. Rough sequence matching (blastology) suggests that they have acquired an unprecedented ~35 genes from their hosts (Stephens et al. 1998).

Spirochaetes

Most Spirochaetes (fig. 4.2, node 6) are free-living or harmless commensals, part of the normal host bioflora, but a number are important obligate intracellular parasites. Known free-living species are chemorganotrophs, obtaining both carbon and energy from organic compounds. Parasitic species include the causative agents of syphilis, Lyme disease, leptospirosis, and relapsing fever. Spirochaetes have a very distinctive spiral morphology and corkscrew-like movement. This is the result of paired polar flagella that extend toward each other and intertwine along the midline of the cell. Unusually, the flagella lie within the periplasmic space rather than outside the cell. Therefore, when they beat they turn like a rotor, spinning the cell within its outer membrane sheath and propelling it forward.

ciPCR studies show that members of the Spirochaetes occur in a wide variety of habitats, including thermophilic, but apparently not marine environments (Hugenholtz et al. 1998). A large diversity of Spirochaetes, mostly of the genus *Treponema*, have been found by ciPCR to the hindgut of termites (Lilburn et al. 1999). Their role here appears to be in fixing nitrogen for their hosts and the parabasalid protists, also found exclusively in this habitat (Lilburn et al. 2001; described below).

Phylogeny

There is no phenotypic resemblance between Chlamydiae and Spirochaetes (fig. 4.2, node 6). Although both include a number of obligate intracellular parasites with highly reduced genomes, these are undoubtedly correlated characters that have evolved independently many times. Nonetheless, the two taxa group together in nearly all CDS analyses with moderate to strong statistical support (fig. 4.2A, line 6). Their grouping is also suggested weakly (Pace 1997), although not consistently (Brochier et al. 2002), in SSU rRNA trees. Although Chlamydiae may have heightened levels of LGT, acquiring genes particularly from their hosts (Subramanian et al. 2000, Stephens et al. 1998), Spirochaetes may have much lower levels of LGT (Dykhuizen and Baranton 2001). The postulated origin of eukaryotic flagella from endosymbiotic Spirochaetes (Sagan and Margulis 1987) has found no molecular support.

Deinococcus-Thermus Group and Cyanobacteria

Deinococcus-Thermus Group

Deinococci (fig. 4.2, node 9) are aerobic, nonmotile, red-pigmented, tetrad-forming chemorganotrophic rods or cocci. They are extremely "tough" and occur in some of the most inhospitable environments known: Antarctic dry valleys, dust, cloud droplets, irradiated food, and medical instruments. They can tolerate, among other things, high levels of ultraviolet and gamma-irradiation (up to 1500 kilorads), extreme desiccation and starvation, and mutagens such as hydrogen peroxide. All of these conditions can cause double-stranded breaks in DNA. In *Escherichia coli*, two or three such lesions are lethal, but *Deinococcus radiodurans* can rapidly and accurately repair 1000 or more. It does this by encoding every pathway for DNA protection and accurate repair known and maintaining its genome in multiple copies (White et al. 1999, Makarova et al. 2001). Although unrelated to "true" gram-positive bacteria, Deinococci have thickened gram-positive-staining cell walls. All of these characters make them attractive targets to engineer for bioremediation, and variants of these bacteria can clean up mercury and toluene. The National Aeronautics and Space Administration also plans to use *D. radiodurans* as a model in simulations used to guide the search for life on Mars.

The Thermus group includes three described genera, all hyperthermophiles isolated from hot springs. *Thermus aquaticus* is particularly noteworthy as the source of *Taq* polymerase (which is named after it), the most widely used enzyme for DNA amplification by thermocycling (PCR). The Deinococcus–Thermus grouping was unsuspected before SSU rRNA phylogeny (Woese 1987), and there is still no phenotypic justification for the group. However, it is unambiguously supported by a large body of molecular sequence data (White et al. 1999). Although still not publicly available, the genome sequence of *T. aquaticus* was completed probably years ago by private industry hoping to mine it for more heat-stable enzymes.

Cyanobacteria

Formerly known as the blue-green algae, cyanobacteria (fig. 4.2, node 9) comprise a large, distinct, well-characterized group. They are ubiquitous, occurring anywhere there is light and even tiny amounts of transient moisture and can survive long periods of desiccation and dormancy. Habitats include between ice crystals in frozen water, in hot springs up to 70°C (the photosynthetic limit), and on or within desert rocks and soil. They are the only oxygenic photosynthetic bacteria and use a variety of pigments to trap (harvest) light, resulting in a range of colors from blue-green to red-brown. Many also fix nitrogen, often in separate terminally differentiated thick-walled cells (heterocysts). Morphologies range from single cells or small colonies to macroscopic filaments and mats.

Cyanobacteria can be extremely abundant and form large macroscopic filaments and mats. On the other hand, tiny *Prochloron* (0.2–0.7 γm) may be the most abundant creature on the planet and our single greatest source of oxygen (Chisholm et al. 2002). Cyanobacteria are also the most frequent photosynthetic component of lichens and a frequent source of color in reef animals. The oldest recognizable fossils appear to be cyanobacteria, and they are the original source of atmospheric oxygen. They also probably at least helped build the oldest known living structures, the stromatolites, although recent evidence suggests these may consist largely of GNS bacteria (Oyaizu et al. 1987; described above).

Cyanobacteria are the only bacteria with chlorophyll a and both photosystems (PSI and PSII), which allows them to generate enough energy to split water and thereby release oxygen in the form of O_2. The accessory pigments and proteins for capturing the light to do this vary among species. This led to theories that eukaryotic photosynthesis originated multiple times, with the differently pigmented eukaryotic algae acquiring their plastids from different cyanobacteria (Urbach et al. 1992). However, there is now considerable molecular data on eukaryotic plastids, and these strongly support a single common endosymbiotic origin for all of them (Douglas 1998).

Phylogeny

A large supergroup consisting of Deinococcus-Thermus, Cyanobacteria, and Actinobacteria is found in all CDS analyses except those using SSU + LSU (large ribosomal subunit), often with strong statistical support (65–100% bootstrap; fig. 4.2, node 8). Within this supergroup, Deinococcus-Thermus and Cyanobacteria are most often found together, a grouping that is also found by SSU + LSU (fig. 4.2A, line 8). As further support of this relationship, these two taxa also appear to exclusively share a large insertion in protein synthesis elongation factor Tu genes (Gupta 2001). CDS analyses are currently limited by the fact that there is only a single published genomic sequence each for Deinococcus-Thermus and Cyanobacteria. Better resolution should be possible with the five or more cyanobacterial genomes currently in progress (CyanoBase 2003) and release of any completed *Thermus* sequences.

Actinobacteria and Firmicutes (High- and Low-G+C Gram-Positive Bacteria)

Actinobacteria (High-G+C Gram-Positive Bacteria)

The Actinobacteria consist of five major subdivisions: Actinobacterae, Acidimicrobidae, Rubrobacteridae, and Coriobacteridae. The Actinobacterae include most of the well-characterized taxa. These are chemorganotrophic, often filamentous, mostly aerobic bacteria with ~70% G+C in their genomes. They are speciose and often highly abundant. ciPCR studies find them in every habitat sampled and particularly plentiful in soil and freshwater (Hugenholtz et al. 1998). Shapes vary from rods to straight or branching filaments and mycelia, and many form highly resistant, potentially long-lived spores. *Streptomycetes* and *Actinomycetes* were once mistaken for fungi because of their branching aerial hyphae.

Most Actinobacteria are free-living or harmless animal commensals or, at the most, opportunistic pathogens. However, the group also includes *Corynebacterium diptheriae* (diphtheria), *Mycobacterium leprae* (leprosy), and *M. tuberculosis*, the single most lethal infectious agent of humans (Cole 2002). Mycobacteria are particularly problematic because they have complex, lipid-rich cell walls resistant to various environmental insults, including many antibiotics. Important beneficial species include *Proprionibacteria*, used in cheese production, and *Streptomycetes* and *Actinomycetes*, producers of more than two-thirds of all naturally occurring antibiotics. *Arthrobacter* is possibly the single most common cultivated soil organism and an important natural herbicide degrader, as are some *Actinomycetes*.

Firmicutes (low-G+C Gram-Positive Bacteria)

Sometimes referred to as the *Bacillus-Clostridium* group, members of the Firmicutes (fig. 4.2, node 11) are chemorganotrophic, often anaerobic, non-filament-forming taxa with ~30% G+C in their genomic DNA. Like the Actinobacteria, they are widespread, found so far in all but geothermal habitats, and predominate in both soil and wastewater (Hugenholtz et al. 1998). There are three subgroups; endospore formers, lactic acid bacteria (anaerobic fermenters), and cocci.

Endospore formers are primarily soil inhabitants. Notable members include *Bacillus anthrasis* (anthrax), *Clostridia* (tetanus, botulism, gas gangrene), and *Bacillus thuringiensis* (commercial source of the powerful insecticide Bt toxin). Other *Bacilli* are important sources of industrial enzymes such as amylases and proteases. Endospores are formed from the entire cell contents and have a dense outer coating. This makes them highly resistant and potentially very long-lived, possibly surviving many millions of years (Cano and Borucki 1995, Vreeland et al. 2000) and perhaps even space travel.

Lactic acid fermenters are anaerobic but oxygen tolerant. They produce vast quantities of lactic acid, probably the earliest preservative, and various *Lactobacilli* are still used in the production of buttermilk, yogurt, and pickles. The cocci include *Heliobacterium* and *Mycoplasma*. *Heliobacteria*, the only photosynthetic members of Firmicutes, occur in rice paddy soils, where they are important fixers of nitrogen. The cell-wall-free mycoplasmas are among the smallest independent-living organisms known in terms of both physical size and the size of their genomes (Fraser et al. 1995). They are metabolically simple, often parasitic, and the only non-eukaryote with cholesterol, which they use to strengthen their membranes. *Mycoplasma pneumoniae* is the causative agent of "walking pneumonia," which has a slower onset than other bacterial forms (*Chlamydia pneumoniae*, *Streptococcus pneumoniae*, *Klebsiella pneumoniae*).

Sporomusa

This is an intriguing, relatively new taxon containing what were previously considered to be a diversity of species (Willems and Collins 1995, Janssen and O'Farrell 2002). All possess classical gram-negative cell walls and an outer cell membrane (Kuhner et al. 1997). They are sometimes listed as a third division of "gram-positive" bacteria, but SSU rRNA trees generally place them at the base of the Firmicutes group (Willems and Collins 1995, Janssen and O'Farrell 2002). This suggests the possibility that the gram-positive cell wall could have evolved independently in Firmicutes and Actinobacteria. This possibility would, in turn, lend further support to the growing idea that the "gram-positive bacteria" may not be a true phylogenetic group, as indicated by all current molecular trees (fig. 4.2, node 7).

Phylogeny

The two groups of traditional gram-positive bacteria are united by the shared absence of an outer cell membrane and, except for mycoplasmas, the presence of a thick gram-stain-retaining cell wall. However, a thickened cell wall is not a complex structure, leaving open the possibility that it could have evolved twice independently. Pressure for this could be common because it probably helps cells resist high salt and desiccation. This is consistent with its independent presence in the highly desiccation-resistant Deinococci and *Methylbacterium*, a proteobacterium. Therefore, the only unique character uniting the Firmicutes and Actinobacteria groups is a single shared loss, that of the outer cell membrane. However, although phylogenetic trees show no tendency to unite these two groups, these analyses are hampered by the extreme differences in the G+C content of these taxa, which affects even amino acid substitution patterns in proteins (Cole 2002).

Proteobacteria (Purple Bacteria)

Proteobacteria (fig. 4.2, node 12) are, more or less, the traditional "gram-negative" bacteria and the single largest group of described bacteria. The α, β, and γ subgroups each have more described taxa than all other bacteria groups combined except cyanobacteria and are found in every habitat type sampled, predominating in many (Hugenholtz et al. 1998). The group is highly diverse and difficult to define but unambiguously monophyletic in molecular trees (fig. 4.2, node 12). Nearly every described bacterial morphology is found and in nearly every subgroup, apparently switching rapidly over evolutionary time and with changing growth conditions. Purple photosynthesis is dispersed throughout and was probably the ancestral state, but with multiple losses (Woese 1987). Many also fix carbon dioxide. The five subgroups were originally identified on the basis of SSU rRNA and given provisional Greek letter names, which seem to have stuck (Woese 1987, Olsen et al. 1994).

γ-Proteobacteria

This is a large, diverse, metabolically rich group and, together with the γ subdivision, the most widespread (Hugenholtz et al. 1998). The group abounds with symbionts, commensals, and parasites, including many complex symbioses, most notably the eukaryotic mitochondrion. Well-known species include *Agrobacterium* (used in plant genetic engineering), *Rhizobium* and *Bradyrhizobium* (nitrogen-fixing symbionts of legumes), and *Rickettsia* (typhus, Rocky Mountain spotted fever). *Rickettsia*-like proteobacteria are probably the closest living relatives of the mitochondrion. Morphologies vary from rods to spirals to budding stalks, the latter being complex extensions of the cytoplasm. Similar to the Planctomycetes, α-proteobacteria may form stalks anchored to the substrate or each other (rosettes) and reproduce by asymmetrical budding. Most are chemorganotrophic, but there are also purple nonsulfur (high-sulfur-intolerant) phototrophs and extracellular and obligate intracellular parasites (*Rickettsia*).

β-Proteobacteria

Also morphologically and biochemically diverse and widely distributed (Hugenholtz et al. 1998), the group includes *Neisseria gonorrhoea*, *Neisseria meningitidis*, *Bordetella pertussins* (whooping cough), and *Thiobacillus*. *Nitrosomonas* plays an important ecological role by completing the final step in nitrogen recycling. Recent molecular phylogenetic data show these taxa to be a subgroup within the γ-proteobacteria (e.g., Brochier et al. 2002).

γ-Proteobacteria

This group consists of another bewildering array of phenotypes, and representatives have been identified in most ciPCR-sampled habitats. The group includes purple sulfur phototrophs (e.g., *Chromatium*); enterics such as *Escherichia coli* and *Salmonella* (mild to severe food poisoning, typhoid); human pathogens such as *Legionnella* (Legionnaire's disease), *Vibrio cholerae* (cholera), *Haemophilus influenza*, *Yersinia pestis* (plague), and *Proteus vulgaris* (cystitis); and a whole host of bizarre phenotypes, including bioluminescent (*Vibrio*), fluorescent (*Pseudomonas*), metal reducing (*Shewanella*), methane consuming (*Methylomonas*), nitrogen fixing (e.g., *Azotobacter*), plant pathogenic (*Xanthomonas*), magnetotrophic, and many more.

δ-*Proteobacteria*

These include two major subdivisions, Myxococcus and Desulfovibrio. Members of Desulfovibrio are morphologically diverse, aquatic or moist soil inhabiting chemolithotrophs that are capable of oxidizing metals such as underground pipes. Bdellovibrios include bacterial parasites that invade and reproduce within the periplasm of their bacterial host. Members of Myxococcus also prey on other bacteria. Under conditions of nutrient starvation they aggregate to form motile multicellular "slugs." These mature into stalked fruiting bodies carrying a head of spores, a sort of miniature cellular slime mold (described below).

ε-*Proteobacteria*

This is the most restricted subgroup, mostly inhabiting extreme environments such as hydrothermal vents and acid lakes. They include *Helicobacter pylori* (causative agent of peptic ulcers), *Campylobacter jejuni* (gastroenteritis), and *Thiovolum* (symbiont of hydrothermal vent invertebrates).

Phylogeny

This huge, highly diverse, and well-studied taxon is probably more appropriately treated as a supergroup. Nonetheless, its monophyly is strongly supported in all CDS trees (fig. 4.2A, line 12). Only the δ division, which SSU rRNA trees place as the deepest branch in the group, are omitted so far from the CDS trees because of the lack of a complete genome sequence.

Bacteria: Recent Additions

The preceding are the "classic" bacterial groups, originally defined by SSU rRNA analyses in 1987 (Woese 1987) and including all well-studied taxa. What follows are descriptions of the largest and most robust of the newly identified major groups. Pace (ch. 5 in this vol.) now estimates a total of about 40 "phylum-level" groups of bacteria. This is based on continued SSU rRNA characterization of already-collected strains, isolation of new strains using traditional and new advanced culturing techniques, and, especially, ciPCR. The latter indicate that some of these new groups may be diverse, widespread, and abundant (Hugenholtz et al. 1998). As many as 20 of them are identified solely by their ciPCR sequences, although some have been subsequently confirmed by fluorescence *in situ* hybridization or even isolated and cultured. A word of warning: any classification based on a single gene must be treated with caution. Even the best phylogenetic methods cannot always distinguish genuinely distinctive sequences from rapidly evolving ones that may nonetheless belong to well-established groups.

Acidobacteria are currently the largest and most diverse newly recognized group with at least eight major subdivisions (Hugenholtz et al. 1998). They were first identified in acid environments such as acid bogs and mine drainage. However, ciPCR studies show all subdivisions represented in 43 separate soil samples (Barns et al. 1999), and fluorescent ciPCR-based probes indicate a diversity of morphological types (Ludwig et al. 1997). Because the group is genetically and metabolically diverse and environmentally widespread, it is probably of significant ecological importance (Barns et al. 1999). Only three cultured members are described, *Acidobacterium capsulatum*, *Holophage foetida*, and *Geothrix fermentans*, representing two subdivisions of the Acidobacteria. Recently, representatives of a third Acidobacteria subdivision have been isolated from soil (Sait et al. 2002).

Verrucomicrobia are a large, diverse, and widespread group with five or six subdivisions currently indicated (Hugenholtz et al. 1998). Division 1 includes most of the cultured taxa, such as *Verrucomicrobia* and the appendaged *Prosthecobacter* species (Hedlund et al. 1996). Division 2 includes a ciliate ectosymbiont, which defends its host by ejecting proteinaceous "spines." These spines appear to contain the closest bacterial homologs yet found of the eukaryotic protein tubulin (Petroni et al. 2002). Division 3 includes the ciPCR taxon EA25, thought to constitute up to 10% of the bacteria in some soils. Division 4 includes *Ultramicrobium*, the smallest bacterium known (0.1 μm^3 in volume) and able to pass through the normal bacteria-excluding 0.2 μm filters. This taxon may account, at least in part, for earlier reports of a surprising abundance of viruses in the open ocean.

OP11 is a purely "environmental lineage," a major group of the bacteria with no known cultivated members. There are five subdivisions indicated, all known entirely from ciPCR sampling. OP11 "phylotypes" seem to be present and abundant in most habitats and a major constituent of subsurface environments. All the OP11 phylotype sequences have long branches in SSU rRNA trees, which suggests that they have undergone accelerated evolution. If so, this would make it difficult to identify cultured relatives if they exist (Hugenholtz et al. 1998).

Other Newly Proposed Groups

These tend to have fewer representatives. However, most contain of at least two subdivisions and have been identified in a number of ciPCR studies. Others consist of recently characterized species that have been successfully cultured but seem to lack close relatives in SSU rRNA trees. All of these small possible new groups should be considered provisional until further data are available because they could simply be taxa with highly divergent SSU rRNA sequences obscuring their true affinities. This list does not include the roughly one-third of all ciPCR groups that still lack any cultivated representatives and have not yet been formally named.

Coprothermobacter is a moderate thermophile (55°C). Its SSU rRNA sequence shows some affinity for the Hyperthermophiles (described above), which would make it the first mesophilic member of the group (Etchebehere and Muxi 2000). The latter may also include *Dictyoglomus*. This hyperthermophile is popular as a source of proteins for crystallography and new thermostable enzymes (Ding et al. 1999).

Fibrobacter is one of the two most abundant genera of cellulose degraders within the complex biota in the rumen of grazing animals. It may be related to the GSB group of bacteria (described above; Gordon and Giovannoni 1996; but see also Pace 1997). The flexistipes include mesophiles and mild thermophiles that nonetheless seem to be largely restricted to geothermal habitats.

The Fusobacteria include major constituents of tooth plaque, with highly distinct phylotypes seeming to specialize on different mammalian taxa (Foster et al. 2002). The Nitrospira, including *Leptospirillum*, *Nitrospira*, *Magnetobacterium*, and *Thermodesulfovibrio*, were previously assigned to the δ-proteobacteria but have highly distinct SSU rRNA sequences (Pace 1997). As the names imply, this a metabolically very diverse collection of taxa. *Nitrospira* species are nitrite oxidizers and include the principle natural detoxifiers of freshwater aquaria (Hovanec and DeLong 1996). The synergistes include six genera, all strictly anaerobic. *Synergistes* species are relatively scarce rumen bacteria that can break down dihydroxypyridine, the compound that renders legumes poisonous to grazing mammals.

The Thermodesulfobacteria are low-G+C thermophiles isolated from hot springs, hydrothermal vents, and oil platforms. They are the only bacteria that appear to have archaea-like membrane lipids. *Gemmatimonadetes* was formerly classified as candidate division "BD" or "KS-B" and consisted solely of ciPCR sequences. A single strain has recently been isolated from sludge processed under enhanced biological phosphorus removal conditions (Zhang et al. 2003) and appears to accumulate polyphosphate. Other members of the group have been identified in soil and marine samples (Zhang et al. 2003). TM7 is known only from ciPCR but has been partially characterized nonetheless as streptomycin-resistant, sheathed filaments. It also appears to have a typical gram-positive cell envelope and may have some bearing on discussions of the mono- versus polyphyletic origin of this trait (Hugenholtz et al. 2001). Full taxon lists for each of these groups can be found at the National Center for Biotechnology Information taxonomy server (available at http://www.ncbi.nlm.nih.gov/Taxonomy/).

Domain Archaea

The Archaea are easily the least understood of the three domains, with orders of magnitude fewer described species compared with members of Bacteria or Eucarya. Most of the characterized taxa are extremophiles inhabiting some of the harshest environments imaginable, but ciPCR suggests there are as many or more mesophiles. Before 1970 bacteriologists dispersed them across various classical bacterial taxa, and it was not suspected that such diverse organisms could be related to each other. However, even early SSU rRNA analyses showed clearly and strikingly both that Archaea form a coherent group and that they constitute a third domain of life (Woese and Fox 1977).

Circumscription of Archaea is qualified by the fact that very few taxa have been studied in culture, and there is almost no information on the many, sometimes major ciPCR-indicated subdivisions. Even most cultivated taxa are poorly characterized, partly because of technical problems such as working with organisms that cannot survive below the melting point of agar or prefer corrosive media such as 0.1 M sulfuric acid. Therefore, there are no archaeal genetic systems, and much of what we know comes from recent genomic sequencing (Bernal et al. 2001). There is also a lack of incentive: Archaea includes no known pathogens, and their ecological roles and economic potential are largely unknown.

Members of Archaea are perhaps best described as a mix of bacterial and eukaryotic features: essentially eukaryotic "brains" in bacterial cells, living bacterial lives (Coulson et al. 1991). Morphologically and metabolically they are bacterial, with small (0.5–5 μm) cells, lacking internal membrane-bound organelles, and usually surrounded by rigid cell walls, albeit ones made of protein rather than of peptidoglycan. Their genomes are small (~ 1.5–3×10^6 base pairs), mostly closed circles, probably with a single origin of replication. Genes tend to be in operons often structurally identical to those of bacteria. Known taxa are autotrophs or chemorganotrophs, sometimes photosynthetic, often sulfur-dependent, frequently fixing carbon dioxide. A variety of symbioses and commensalisms are also known (Madigan et al. 1997).

On the other hand, information processing (i.e., RNA transcription and translation) is eukaryotic both in overall organization and in individual component sequences (Kyrpides and Woese 1998, Edgell and Doolittle 1997). Even before the genes were sequenced, it was noted that archaeal RNA polymerases had a subunit composition similar to those of eukaryotes (Huet et al. 1983). This was later confirmed by striking similarities in their protein sequences and structures. Archaeal RNA polymerases also require the eukaryotic-type transcription factors TBP, TFIIB, and TFIID to bind their promoters (Bell and Jackson 2001). This is unlike the much smaller bacterial RNA polymerase that can bind DNA on its own and uses small exchangeable subunits called sigma factors to identify its promoters.

Likewise, the components of archaeal protein synthesis are mostly either uniquely shared with eukaryotes or are eukaryotic versions of universal ones (Kyrpides and Woese 1998). Many are encoded in canonical bacterial operons but have strikingly eukaryotic sequences and structures (Garrett et al. 1991) and are functionally interchangeable with eukaryotic factors *in vitro*. Members of Archaea also have fundamentally eukaryotic DNA replication (Myllykallio et al. 2000, She et al. 2001, Bohlke et al. 2002), and euryarchaeotes (but no known crenarchaeotes) use histones to package their DNA in nucleosome-like structures (Sandman and Reeve 2001).

The possibility of members of Archaea having given rise to either Bacteria or Eucarya or both has long been a contentious issue (Rivera and Lake 1992, Baldauf et al. 1996). However, with more data now available, most analyses, including CDS trees, unambiguously support their being a monophyletic group (Brown et al. 2001, and S. L. Baldauf and J. Cockrill, unpubl. obs.). Members of Archaea are also distinct in possessing highly unique membrane lipids that appear to be restricted to them, with the possible exception of the thermodesulfobacteria (described above). These membranes consist of isoprenyl lipids ether-linked to D-glycerol, distinctly different from the ester-linked fatty acid lipids and L-glycerol of members of Eucarya and Bacteria. Archaeans are also the only known organisms with lipid monolayer membranes. These are common in hyperthermophiles, where they probably provide membrane stability at high temperatures and account for these cells' inability to live at lower temperatures (Coulson et al. 2001).

All characterized members of Archaea and most ciPCR phylotypes fall cleanly into two distinct groups, the Crenarchaeota and the Euryarchaeota (fig. 4.3). The cultured Crenarchaeota or "thermoacidophiles" are phenotypically narrow, whereas the Euryarchaeota are extremely diverse. However, ciPCR suggests that both groups are much broader, particularly the Crenarchaeota. ciPCR also indicates the possible existence of two additional major divisions of Archaea (fig. 4.3), the Korarchaeota, known only from ciPCR studies (Barns et al. 1996), and the extremely small (0.4 µm diameter) Nanoarchaeota (Huber et al. 2002b). However, the classification of these taxa as major new archaeal lineages is based entirely on SSU rRNA trees and is not unambiguously supported even by them.

Crenarchaeota

All cultured Crenarchaeota (fig. 4.3, nodes 13 and 14) are hyperthermophiles, including some of the most thermophilic

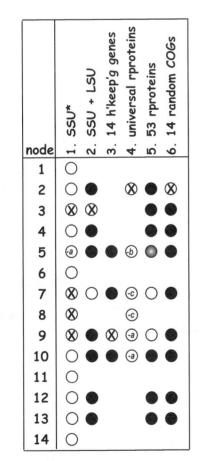
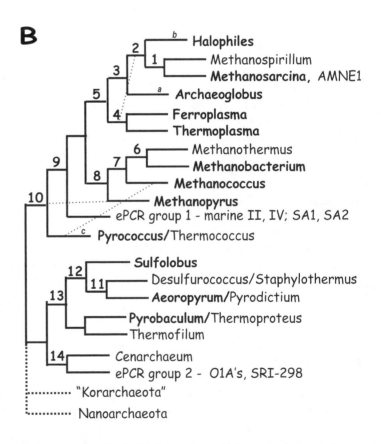

Figure 4.3. Support for deep branches in the archaeal tree. (A) shows support for the labeled nodes in (B) (same symbols as in fig. 4.2A). Data sets used are SSU, SSU + LSU and 53 ribosomal proteins (Matte-Tailliez et al. 2002), 14 housekeeping genes (Brown et al. 2001), universal ribosomal proteins (Wolf et al. 2001), 14 random clusters of orthologous groups of proteins (COGs) (J. Cockrill and S. L. Baldauf, unpubl. obs.). Lowercase letters within open circles indicate where a single taxon interrupts an otherwise strongly supported grouping, as indicated in (B). In (B), boldface is used to indicate the taxa used in combined analyses (i.e., taxa for which complete genome sequence data are available). Major differences between SSU and all other analyses are indicated by lightly dotted lines. Two potential major new archaeal lineages are attached to the base of the tree by dashed lines.

and acidophilic organisms known. Optimum growth temperatures range from 75°C to 100°C, with some cells surviving temperatures as high as 115°C and most unable to survive below 70°C. They have a wide range of pH tolerances but may flourish at pH 1–2 and still live and grow at pH 0. Most cells are flagellate, but shapes vary widely from simple disks (*Thermodiscus*), nearly rectangular rods (*Thermoproteus*, *Pyrobaculum*), irregularly lobed cocci (*Sulfolobus*), or extremely long thin filaments (*Thermofilum*), to grapelike aggregates (*Staphylothermus*) and large fibrous networks (*Pyrodictium*; Barns and Burggraf 1997). Many species are acidophilic, using sulfur or sulfur compounds as electron donors, acceptors, or both. Carbon may be acquired from organic compounds (chemorganotrophy) or by fixing carbon dioxide using ribulose biphosphate carboxylase or a reverse tricarboxylic acid cycle.

Although all the cultured members of Crenarchaeota are hyperthermophilic and often anaerobic, ciPCR indicates the existence of mesophilic aerobic taxa throughout the group. Unlike the crenarchaeal thermophiles, many of the mesophiles appear to be widespread. Their habitats range from shallow sediments (Hershberger et al. 1996) to the open ocean (DeLong 1992, Fuhrman et al. 1992). Some may also be extremely abundant, including dominating the ocean's interior, "the world's largest biome" (Karner et al. 2001). They are found throughout the water column, increasing in abundance with depth until they constitute ~39% of all the microbial cells (DeLong 1992, Fuhrman et al. 1992).

The Crenarchaeota are the least-characterized major division of living organisms, and no taxonomic group has been more fundamentally revised by ciPCR data. These now indicate that there are two major divisions of Crenarchaeota, referred to in the following as Divisions I and II. Division II was discovered only in the last 5 years and consists entirely of ciPCR phylotypes plus *Cenarchaeum*, an obligate symbiont of a marine sponge. General references for the following are Madigan et al. (1997), Barns and Burggraf (1997), and Brown (2002).

Crenarchaeota Division I

Thermoproteales. Thermoproteales (fig. 4.3, node 13) currently include six described genera: *Caldivirga*, *Pyrobaculum*, *Thermocladium*, *Thermofilum*, *Thermoproteus*, and *Vulcanisaeta*. These are generally rod shaped thermophiles living at near neutral pH. *Thermoproteus* (60–96°C, optimum 85°C) cells are long rods that reproduce by budding and are very common in solfatars (sulfur deposits). *Thermofilum* species (70–95°C) form extremely long, thin (1–100 (0.15–0.3 μm) filaments and are common in deep-sea hydrothermal vents, as are *Pyrobaculum* species (74–115°C, optimum 100°C), which are rods or flattened cocci. The latter can grow well at temperatures up to 115°C, making them, together with *Methanopyrus* (described below), the most thermophilic organisms known (note: at these depths 115°C is well below the boiling point of water). *Pyrobaculum* species

are high-sulfur-intolerant denitrifiers, using oxygen instead of sulfur as a terminal electron acceptor (aerobic respiration). This is unusual for crenarchaeotes, which are generally obligate or at least facultative anaerobes.

Sulfolobales. This group currently includes five genera: *Acidianus*, *Methanosphaera*, *Sulfolobus*, *Sulfurisphaera*, and *Stygiolobus*. This is an entire group of organisms whose natural habitat is essentially boiling sulfuric acid. All are coccoid-shaped thermophilic acidophiles with growth optima of 75–95°C and pH 1.0–3.5. The most extreme is *Acidianus*, which can grow at pH 0. *Sulfolobus* species (55–87°C, optima 75–85°C, pH 2–3) thrive in thermoacidic environments such as solfatars, boiling mud pots, and hot acid mine drainage, where they often grow in great abundance and in near monoculture. They reproduce by budding, which produces characteristic irregularly shaped lobes that may also function in adhesion and for which they get their name. Both *Sulfolobus* and *Acidianus* (65–95°C, optima 85–90°C pH 2) have extremely low genomic G+C content (37% and 31%, respectively), demonstrating once again that thermophilic adaptation does not require elevated genomic percent guanine + cytosine (G+C) (She et al. 2001).

Desulfurococcales. Desulfurococcales (fig 4.3, node 11) is currently the largest group of crenarchaeotes, with 11 described genera divided into two groups. All are coccoid or rod shaped, neutrophilic (living at neutral pH) hyperthermophiles. Group 1 includes *Aeropyrum*, *Desulfurococcus*, *Ignicoccus*, *Staphylothermus*, *Stetteria*, *Sulfophobococcus*, *Thermodiscus*, and *Thermosphaera*. *Staphylothermus* species (65–98°C, optimum 92°C) are cocci that grow in grapelike clusters. *Thermodiscus* (75–98°C, optimum 90°C) cells are, not surprisingly, disk-shaped. *Ignicoccus* includes the host of a newly discovered symbiont thought to represent a new subdivision of Archaea, the Nanoarchaeota (described below). *Aeropyrum* are unusual in being strictly aerobic hyperthermophiles (optimum 90–95°C).

Members of Group 2 (*Hyperthermus*, *Pyrodictium*, and *Pyrolobus*) are all found in shallow submarine volcanic habitats. *Pyrodictium* species (62–110°C, optimum 100°C) are disk-shaped cells held in networks by hollow, proteinaceous fibers growing in "moldlike" layers on suspended sulfur crystals. The fibers maybe arranged in regular patterns similar to the protein in bacterial flagella (Madigan et al. 1997). Their mature rRNA molecules have an unusually high percentage of modified bases, presumably to stabilize their structure and hence their activity at high temperatures. The other cultured members of the group are *Pyrolobus* (90–113°C, optimum 105°C), with lobed cocci, and *Hyperthermus* (95–106°C, optimum 108°C).

Crenarchaeota Division II (Cenarchaeum Group)

This second major division of the Crenarchaeota (fig. 4.3, node 14) includes a wide variety of ciPCR phylotypes indicating mesophiles, thermophiles, and hyperthermophiles.

Some of the mesophiles are also widespread and/or extremely abundant. The only characterized member of the division is *Cenarchaeum symbiosum*, an obligate symbiont of the marine sponge *Axinella* species (Preston et al. 1996). The association between the sponge and symbiont appears to be stable, widespread, and highly specific, and the symbiont is highly abundant within its host (Preston et al. 1996). This is the only known eukaryote–archaean symbiosis involving a crenarchaeote. Although the symbiont cannot be separated from its host, large fragments if its genome have been characterized using whole-organism genomic libraries (Schleper et al. 1998).

Crenarchaeote Phylogeny

ciPCR data suggest the need for major revision of the Crenarchaeota, but further information is needed on at least some of these "taxa." Nonetheless, it now appears clear that there are at least two deeply separated subdivisions within the group, referred to here as Divisions I and II. Division I includes all the cultured hyperthermophiles, which form a tight cluster nested within several layers of deeply diverging ciPCR "groups" (fig. 4.3, node 13; see Pace, ch. 5 in this vol.). Division II is almost entirely composed of ciPCR phylotypes, including many mesophiles. There are also several groups of ciPCR phylotypes that do not clearly belong to either division, and these may indicate additional distinct lineages. Because the latter are mostly hyperthermophiles, this is still thought to be the ancestral condition of the group, with mesophily derived multiple times within it.

Within Division I, the major branching patterns are fairly well resolved for the taxa with completed genome sequences: *Sulfolobus solfataricus*, *Pyrobaculum aerophilum* (Thermoproteales), and *Aeropyrum pernix* (Desulfurococcales). CDS trees for these taxa agree with SSU rRNA trees, grouping *Sulfolobus* and *Aeropyrum* together (fig. 4.3, node 12) to the exclusion of *Pyrobaculum*. Because *Sulfolobus* is the only acidophile of the three, this suggests that acidophily was derived from neutrophily, rather than ancestral as originally postulated. However, these studies are starkly taxonomically narrow, which makes any broad conclusions premature. Because Division II, with the exception of *Cenarchaeum*, is known only from ciPCR phylotypes (fig. 4.3, node 14), all information on the group comes from SSU rRNA trees, which support it strongly.

Euryarchaeota

The Euryarchaeota (fig. 4.3, node 10) are an extremely diverse group including mesophilic, thermophilic, and hyperthermophilic methanogens; thermoacidophiles; sulfur-reducing thermophiles; and extreme halophiles. These tend to form roughly seven robust subgroups, three of which are methanogenic (fig. 4.3). All of these have a broad sampling of cultured representatives. Euryarchaeotes include a number of environmentally important organisms, both beneficial and harmful.

Archaeoglobi

Archaeoglobus species are sulfate-reducing, obligately anaerobic hyperthermophiles (60–90°C, optimum 83°C) with irregularly spherical, flagellated cells. They are found in hydrothermal environments and subsurface oil fields, where the iron sulfide they produce may corrode oil and gas mining equipment (Klenk et al. 1997). None of the current molecular phylogenetic data place them even close to the Thermococci, the other thermophilic sulfate-reducing Euryarchaeotes (fig. 4.3). Thus, the thermoacidophilic habit appears to have arisen in Euryarchaeota at least twice independently.

Halobacteria

These extreme halophiles require a minimum of 1.5–2.0 M NaCl (or equivalent) and many can survive in saturated or near-saturated brine (up to 5 M NaCl). They are probably present in all high-salt environments, including salted fish, and are common in hypersaline seas, salterns (shallow salt evaporation pools), and subterranean salt deposits. They photosynthesize using the purple retinal pigment conjugated to bacteriorhodopsin. This results in the pink hue of salt evaporation pools, the purple of the Dead Sea, and the red in salted herring (the famous "red herring").

Unlike other halophilic organisms, members of Halobacteria are isosmotic with their environment, so there is no osmotic pressure on their cell walls. This allows for unusual morphologies such as *Haloarcula*, which has ultrathin, 0.1–μm-thick cells that form perfect squares, rectangles, and even triangles, mostly with tufts of flagella at their apexes. These are the only phototropic archaeans, and they use a mechanism fundamentally different from any bacterial photosynthesis. The light-harvesting machinery is embedded in the cell membrane itself, and absorbed light is used directly to pump protons across the membrane. This creates a proton-motive force that is then coupled to ATP synthesis (Lanyi and Luecke 2001).

Methanogens

The methanogenic members of Archaea include at least six major groups (fig. 4.3, nodes 1, 6, 7, and 8), and they are almost certainly not monophyletic. Nonetheless, all are obligately anaerobic and methanogenic (methanogenic enzymes are oxygen intolerant) and possess a unique fluorescent cytochrome, F420, that is found nowhere else. They are widely dispersed in nature and are found in sediment, soil, wastewater treatment ponds, landfills, subterranean oil deposits, and animal intestinal flora. They are the source of swamp gas and intestinal methane and are important components of the rhizosphere, the plant root environment.

Three of the methanogen groups, represented by *Methanosarcina*, *Methanospirilla*, and *Methanococcus*, are currently placed together in the Methanococcales. However, CDS trees tend to split them up, placing the first two taxa with the Halobacteria (fig. 4.3, node 2) and the latter some distance away with Methanobacteria and *Methanothermus* (fig. 4.3, node 7). Because *Methanopyrus* also appears to be closely associated with the latter group (Slesarev et al. 2002), this means that all currently known methanogenic archaeans can probably be assigned to two well-defined groups.

Methanogen Group 1 (MG1): Methanosarcina + Methanospirillum.   Methanosarcina + Methanospirillum (fig. 4.3, node 1) produce methane from a variety of substrates, including acetate, methylamines, and methanol, an unusual metabolic diversity for methanogens. They occur in diverse habitats from anaerobic lake bottoms and muds to cattle rumen, where they are responsible for methane production, a significant source of global greenhouse gas (Deppenmeier et al. 2002). *Methanosarcina* species are mildly thermophilic (40–55°C) and common in soils, sediment, swamps and wastewater treatment sludge, where they play an essential role in the early stages of sewage processing. The group also includes cold-adapted species that survive temperatures as low as −10°C and are found in Antarctic lakes and cold deep-marine sediments (Thomas and Cavicchioli 1998).

Methanogen Group 2 (MG2).   MG2 includes Methanococci + Methanobacteria + Methanopyri (fig. 4.3, nodes 7 and 8). This tentative group includes mesophiles, thermophiles, and hyperthermophiles. Most are highly self-sufficient, metabolically speaking, and some are pure prototrophs, that is, capable of living on only hydrogen gas, carbon dioxide, and either nitrogen gas or ammonium ions. Methanococci are irregular flagellated cocci. They include borderline mesophiles to hyperthermophiles (48–98°C) isolated from marine and freshwater sediments and the deep-sea hydrothermal vents known as "black smokers" (described below). Methanobacteria are nonmotile rods or filaments. They include mesophilic to moderately thermophilic (40–70°C, optimum 65°C) pure prototrophs common in animal colon and rumen and also isolated from sewage sludge and sea sediments and as symbionts of animals, plants, and protists. In the termite hindgut, they form symbioses or endosymbioses with cellulose-digesting protists, from whom they get hydrogen gas (Tokura et al. 2000). *Methanopyrus kandlerii*, the sole cultivated member of the Methanopyri, is a "gram-positive" hyperthermophile (80–110°C) first isolated from a 2000-m-deep black smoker. It has an internal salt concentration of 1.1 M, probably part of the means by which they maintain enzymatic activity in extreme heat.

Methanogen Phylogeny.   Most of the phylogeny of archaeal methanogens is based on SSU rRNA trees, which split them into numerous separate groups and tend to place *Methanopyrus* as the deepest euryarchaeote branch. However current CDS analyses tend to restrict the methanogen to only two distinct groups (MG1 and MG2; fig. 4.3, nodes 1 and 8), including *Methanopyrus* as sister to a *Methanococcus + Methanobacterium* clade (MG2; fig. 4.3, node 8; Slesarev et al. 2002).

Furthermore, neither methanogen group appears to be among the deepest branches in the euryarchaeote tree. On the other hand, CDS data strongly reject recent claims that the archaeal methanogens are monophyletic; MG1 is nested within a substantial group of nonmethanogens, at some distance from MG2 (fig. 4.3, nodes 2–5).

Thermococci

These include *Thermococcus* and *Pyrococcus*, which are thermophilic/hyperthermophilic (75–100°C, pH ~7) flagellated cells commonly found, along with other thermophilic members of Archaea, in and around the "black smokers" formed by deep-sea hydrothermal vents. Black smokers are mineral chimneys formed by the buildup of sulfides deposited by the mineral-rich waters spewing from the vents and giving the appearance of belching black smoke. The warm hydrogen sulfide–rich habitat around these vents supports a rich fauna, essentially oases of life along the otherwise largely barren seafloor. These communities are dependent on energy from the oxidation of sulfide rather than light; that is, they are lithotrophic rather than phototrophic.

Thermococci and Pyrococci together form a distinct, tight phylogenetic group in all phylogenetic trees. The group appears to be quite shallow, but this may reflect a slow rate of molecular sequence evolution, which is seen in many hyperthermophiles. This may be due to the restrictive amino acid requirements of thermostable proteins; thermococcalean proteins tend to have highly biased amino acid use and favor nonpolar amino acids over polar ones by a ratio of around 3:1 (Howland 2000). Unlike most other major subdivisions of Archaea, uncultured Thermococci do not display a large diversity in any of the habitats sampled by ciPCR analyses (Maidak et al. 2001).

Thermoplasmata (Thermoplasma, Ferroplasma, and Picrophilus)

These are all thermophilic extreme acidophiles (growth optima 40–60°C, pH 0.5–2.0), the only organisms able to survive, never mind thrive, at pH < 0 (Ruepp et al. 2000, Edwards et al. 2000, Schleper et al. 1995). Cells are small (0.2–5 μm), spherical, and sometimes flagellated and, unlike all other archaeans, lack a cell wall. Despite this, they survive external pHs of 0–4 while maintaining an internal pH ~7 (Ruepp et al. 2000). Natural habitats include hot solfatars and coal refuse, which are rich in highly toxic metals such as copper, arsenic, cadmium, and zinc. It appears that they make their "living" by scavenging complex organics released by cells that are killed by these extreme conditions. *Ferroplasma* is responsible for the acidification of coal mine drainage, the primary environmental problem associated with mining (Edwards et al. 2000).

Thermoplasma acidophilum has an extremely small genome (~1.6 megabases), and, in an apparent case of massive LGT, it shares 17% of it exclusively with the crenarchaeote *Sulfolobus* (Ruepp et al. 2000). The shared sequence resides in approximately five large blocks and codes for many of the transport and metabolic pathways needed for this unique lifestyle, which requires importing a variety of complex organic compounds. Consistent with this, Thermoplasmata and Sulfolobales often co-occur in habitats that they share almost exclusively except for a few species of *Bacillus*. ciPCR indicates that Thermoplasmata is much larger and broader than the currently known taxa, although it still appears to be restricted to hot acid environments (Maidak et al. 2001). The lack of a cell wall in thermoplasmas has led to speculation that the group might include the direct ancestor of Eucarya. However, a large body of molecular phylogenetic data now soundly reject this (fig. 4.3, node 5).

Euryarchaeote Phylogeny

Euryarchaeota is an ancient, large, and extremely diverse group, and resolving relationships within it will be difficult. This is also complicated not only by LGT but also, perhaps more important, by the strong biases in the amino acid composition of their proteins. These are required for adaptation to extremes of salt, pH, and temperature. Therefore, inclusion of sequence data from the mesophilic taxa indicated by ciPCR phylotypes could potentially improve resolution considerably.

Nonetheless, certain trends can be identified at this point with some confidence. The Euryarchaeota are almost certainly a monophyletic group (fig. 4.3, node 10); theories that bacteria might have originated from euryarchaeote ancestors have been largely abandoned (Lake 1988). Within the euryarchaeotes it appears unlikely that the methanogens form a monophyletic group. Recent claims to the contrary are not supported by analyses with fuller taxonomic representation (Slesarev et al. 2002).

The earliest branch of the euryarchaeotes appears to be the Thermococcoides (fig. 4.3, node 9). This is followed by a number of ciPCR lineages that may or may not form a single group. Data beyond SSU rRNA sequences are needed before any more conclusions on these taxa can be drawn. The remaining euryarchaeotes appear to split into two groups: MG1 and their allies (MG1+; fig. 4.3, node 7) and MG2 (fig. 4.3, node 8). The inclusion of *Methanopyrus* in MG2 is still tentative; the grouping is only weakly supported by the only CDS study with *Methanopyrus* in it, and it is strongly rejected by SSU rRNA trees, in which *Methanopyrus* SSU shows no clear affinity for any other euryarchaeote sequence and tends to fall toward the base of the tree.

MG1+ (fig. 4.3, node 5) is a surprisingly robust clade. The group is further supported by the shared presence of cytochrome b and/or c, which are found among members of Archaea only in Methanomicrobiales, Halobacteria, and Thermoplasmas. The most problematic taxon within MG1+ is the Halobacteria, probably because of their extreme, uniquely biased amino acid use (Ng et al. 2000). Nonetheless, they group together strongly with MG1 in a number of trees (fig. 4.3, node 5), and when they do not it is often because they are found in highly unlikely positions, such as at the base

of the entire Archaea domain (Slesarev et al. 2002). A grouping of Halobacteria + MG1 is also consistent with the fact that the latter includes *Methanohalophilus* species, the only known halophilic methanogens. The exact position of *Archaeoglobus* within MG1+ is also very unstable, suggesting that there may have been considerable LGT during its evolution (fig. 4.3A).

Korarchaeota and Nanoarchaeota

Two additional major subdivisions of Archaea have been suggested recently, the Korarchaeota and Nanoarchaeota. The Korarchaeota were originally identified in a ciPCR study of a Yellowstone Park hot spring (74–93°C). Two phylotypes were found that formed a distinct group that was clearly archaeal but not specifically related to either Crenarchaeota or Euryarchaeota. The group, provisionally named "Korarchaeota" (Barns et al. 1996), has since been detected in Icelandic sulfide hot springs (Hjorleifsdottir et al. 2002) and geothermal effluent (Marteinsson et al. 2001), sometimes in abundance (Hjorleifsdottir et al. 2002). This indicates that the group at least is real, but additional data are still needed to test their classification as a unique archaeal subdivision. Caution is warranted also by the fact that their position as a unique branch among archaeal SSU rRNA sequences varies depending on the taxon composition of the data set and the analytical method used. Nonetheless, korarchaeote SSU rRNA sequences lack features generally associated with phylogenetic artifact; that is, they do not form long branches and lack strong percentage G+C bias.

The "Nanoarchaeota" were described even more recently and have been encountered so far only once (Huber et al. 2002b). They are hyperthermophiles (70–98°C) from an Icelandic coastal hot submarine vent and were found attached to cells of *Igniococcus*, a desulfurococcalean crenarchaeote (described above). Everything about them is small, including their cells (0.4 μm diameter) and their genomes (500 kilobases), which is near the theoretical limit for a "free-living organism" (Huber et al. 2002b). So far, they can be cultured only when attached to a live host. However, they are probably not parasites because the host grows equally well with or without them. *Nanoarchaeum* SSU rRNA is clearly archaeal but otherwise highly divergent and shows no specific affinity for any currently known archaeal group (Huber et al. 2002b). However, unlike the korarchaeote SSU rRNAs, these sequences do possess features associated with phylogenetic artifact; that is, they are extremely divergent and lack a number of otherwise universally conserved nucleotides. This suggests they may belong to a rapidly evolving lineage rather than an ancient one.

Summary of Archaebacteria

The Archaea include the most extremophilic organisms known, and more than for any other group of taxa, our understanding of them is being fundamentally rewritten by ciPCR and whole-genome data. Genomic sequencing is the only way to study many of them in any detail, and ciPCR studies indicate major mesophilic components and additional new groups at all taxonomic levels. Some consistent resolution seems to be emerging from the still very limited CDS trees, but little can be said with confidence. Resolving the archaeal tree will require more genes and more taxa representing the true diversity of the group, as well as careful attention to confounding factors such as LGT. Protein gene sequences from mesophilic taxa may be the key to circumventing the systematic phylogenetic artifact caused by the highly skewed amino acid composition of extremophile sequences. Although these taxa have largely escaped cultivation, recent progress in genomic analyses of uncultured taxa could circumvent this limitation (Schleper et al. 1998).

Domain Eucarya

Eukaryotes are defined first and foremost by the presence of a nucleus surrounded by a double membrane punctuated with large highly complex pores. The nuclear membrane is part of a larger endo-membrane system that also includes the endoplasmic reticulum and Golgi apparatus, which synthesizes membranes and processes, sorts, and packages proteins for distribution or export. Other organelles are also usually present, most notably mitochondria and chloroplasts (more correctly "plastids"), both of which are descended from bacterial endosymbionts (Alberts et al. 2002). Mitochondria in particular, but possibly also plastids (Andersson and Roger 2002), originated early in eukaryotic evolution, and no premitochondrial eukaryotes appear still to exist. However, mitochondria have been lost, reduced, or converted to fermentative, hydrogen-gas-producing organelles (hydrogenosomes; Dyall and Johnson 2000) several times independently over the course of eukaryote evolution.

Eukaryotic cells vary widely in size (from <1.0 to 100 μm in diameter) and often form colonies or multicellular structures. They have numerous other unique features, many probably correlated with the advent of membrane-bound nuclei and the invention of endocytosis, such as the actin cytoskeleton probably derived from bacterial cell-division protein ftsA (van den Ent and Lowe 2000). Eukaryotic flagella are large, complex multiprotein structures unrelated to bacterial flagella, which are composed almost exclusively of flagellin. The eukaryotic flagellum was probably derived early in eukaryotic evolution, possibly from the cytoskeleton, and is clearly not of endosymbiotic origin (Cavalier-Smith 2002). Sequestering DNA into a membrane-bound nucleus spatially separates transcription (copying of DNA into RNA) from translation (decoding of RNA into protein), unlike in bacteria, which have the two processes coupled and possibly coregulated. However, it now appears that a significant amount of eukaryotic translation may occur in the nucleus, coupled with transcription as in bacteria (Iborra et al. 2002).

Eukaryotic information processing is essentially an expanded version of the archaeal system. Transcription uses archaea-like RNA polymerases, and gene expression is controlled with the same basic machinery, although with many eukaryote-unique factors layered on top (Bell and Jackson 2001). Eukaryotes are still unique in having large operon-free genomes on multiple linear chromosomes, packaged around histones, usually containing large amounts of repetitive DNA. Introns tend to be much more common, to the point of being highly abundant in some plants, animals, fungi, and amoebozoans. These introns are mostly of the spliceosomal type; that is, they require a large multiprotein complex (the "spliceosome") to remove them from the pre-messenger RNA transcript, unlike the self-splicing introns of the members of Bacteria and Archaea (Logsdon 1998). Eukaryotes are also the only organisms known to have true diploidy (and polyploidy) and sex (meiosis). However, these rules are nearly all broken somewhere among eukaryotes, and more exceptions will undoubtedly be found.

Eukaryotes are a highly derived, unquestionably monophyletic group. More so than for members of either Archaea or Bacteria, most of what we know about them is based on SSU rRNA trees. These define most of the major groups, some of which were not previously suspected. The overall structure of this SSU rRNA tree also led to the influential "Archezoa" (Cavalier-Smith 1987) and "Crown Radiation" (Sogin and Gunderson 1987, Sogin 1991) hypotheses, which have since been disproved. The former was based on the observation that the deepest eukaryote SSU branches, admittedly largely parasitic, lacked mitochondria and most other internal structures, and therefore might represent primitive pre-mitochondrial lineages. The latter hypothesis suggested that the clustering of most of the other "more advanced" lineages meant that they arose comparatively late in eukaryote evolution, perhaps in a single explosive radiation (see Philippe, ch. 7 in this vol.). A fairly large body of data now agrees that both phenomena are different aspects of the same artifact: fast-evolving (long-branched) taxa (members of the Archezoa) being drawn toward the base of the tree and causing the remaining taxa (the crown) to appear as a dense cluster (Morin 2000, Philippe and Germot 2000).

Fourteen major eukaryotic groups are currently defined based on molecular phylogenetic data (fig. 4.4; Cavalier-Smith 1998, Baldauf et al. 2000). These include most of Patterson's (1999) 60 "ultrastructural types." The most thorough reference on eukaryote morphology and fine-level taxonomic diversity is *The Illustrated Guide to the Protozoa* (Lee et al. 2000). All unreferenced material in the following sections is derived from that book, which relies heavily on the work of Patterson, Brugerolle, and colleagues or of Hausmann and Hülsmann (1996).

However, it is now apparent that this description is far from representing the true diversity of eukaryotes. In culture collections alone, more than 200 taxa are without known relatives, and several major groups of amoebae lack even SSU

A

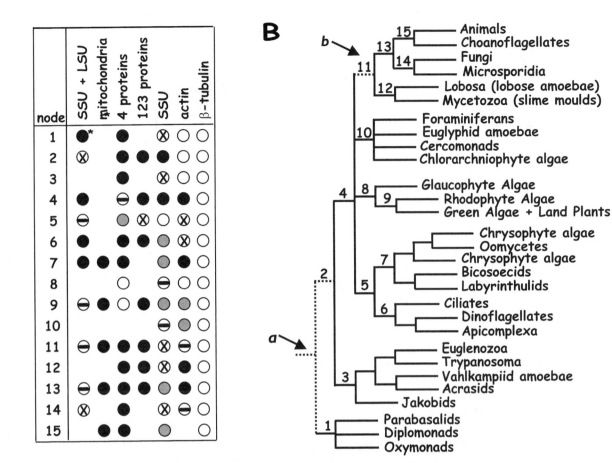

B

Figure 4.4. Support for deep branches in the eukaryote tree. (A) shows support for the labeled nodes in (B) (same symbols as in fig. 4.2A). Data sets used are combined SSU and LSU rRNA (Van der Auwera et al. 1998), combined mitochondrial proteins (Burger et al. 1999), four combined proteins (Baldauf et al. 2000), 123 combined proteins (Bapteste et al. 2002), and individual gene phylogenies for SSU rRNA (Van de Peer and De Wachter 1997, Sogin 1991), actin (Bhattacharya and Weber 1997, Keeling 2001), and β-tubulin (Keeling and Doolittle 1996, Keeling et al. 2000). Alternative rootings of the tree are indicated by dashed lines and arrows: "a," for the molecular phylogenetic root using archaeal outgroup sequences, and "b," as indicated by the fusion of the genes for DHFR and TS (described in text; Stechmann and Cavalier-Smith 2002). The asterisk (*) indicates that nodes 1 and 4 in the SSU + LSU CDS are interrupted by aberrant deep branching of microsporidian and lobosan sequences.

rRNA sequences (Patterson 1999). More important, recent ciPCR studies suggest the existence of major undiscovered eukaryotic lineages (Amaral-Zettler et al. 2002, Dawson and Pace 2002, Moriera and Lopéz-Garcia 2003). These "nanoeukaryotes," cells less than 2–3 μm in diameter, have previously escaped detection because they are all but indistinguishable from bacteria under the light microscope. Some of the new taxa appear to represent major new subdivisions of established groups (e.g., Alveolates; fig. 4.4) or perhaps even the first known representatives of entire new lineages. Major revisions in the eukaryotic tree are to be expected in the very near future (Moriera and Lopéz-Garcia 2002).

Excavates

Excavates 1: Amitochondriate Excavates (fig. 4.4, node 1)

Among the best candidates for the earliest diverging Eukaryotes are the group of taxa recently united as the Excavata.

This is a diverse assemblage of single-celled organisms most of which possess a conspicuous "excavated" ventral feeding groove (Cavalier-Smith 2002, Simpson and Patterson 1999, 2001). However, the group as a whole lacks material molecular phylogenetic support. For convenience they are treated as two somewhat arbitrary subgroups, (1) "amitochondriate" excavates, which lack classical mitochondria, and (2) "mitochondriate" excavates.

The best known amitochondriate excavates are diplomonads. These typically exhibit a "doubled" morphology, with duplicate nuclei, sets of flagella, and cytoskeletons arranged back to back in each cell. The intestinal parasite *Giardia intestinalis* is a major human diarrheal agent, whereas *Spironucleus* includes some serious fish parasites. Some other diplomonads are free-living and are common in low-oxygen habitats (Bernard et al. 2000). Retortamonads are broadly similar to diplomonads but have a single nucleus, flagellar cluster, and feeding groove per cell. Most are intestinal commensals.

Oxymonads are flagellated symbionts from the intestinal tracts of animals, mostly termites. Some attach to the gut wall by stalk or "holdfast," others squirm using an internal motile cytoskeleton, and still others are free-swimming cells. Diplomonads, retortamonads, and oxymonads all seem to lack any cellular structure that may be homologous to mitochondria.

Parabasalids are a diverse group almost entirely comprised of parasites and symbionts united by the presence of a parabasal apparatus, which is a complex of Golgi stacks and striated cytoskeletal elements. In place of mitochondria, parabasalids have organelles called hydrogenosomes that anaerobically generate ATP from pyruvate, liberating hydrogen gas in the process (Rotte et al. 2000). Some parabasalids from termites, for example, hypermastigids, are huge multiflagellated cells, hundreds of micrometers long and covered in ectosymbiotic bacteria, whereas most "trichomonad" parabasalids are small teardrop-shaped cells with four to six flagella. Trichomoniasis, caused by the trichomonad parabasalid *Trichomonas vaginalis*, is the most common human sexually transmitted infection affecting ~170 million people worldwide (Müller 1988). *Trimastix* and *Carpediemonas* are free-living, groove-bearing, bacterivorous flagellates that inhabit low-oxygen environments (Bernard et al. 2000). Although neither has classical mitochondria, both have small organelles that superficially resemble the hydrogenosomes of parabasalids.

Phylogeny. The amitochondriate excavates have been central to exploring the origin and early diversification of eukaryotic cells. On the strength of early SSU rRNA phylogenies, diplomonads, retortamonads, oxymonads, and parabasalids were widely thought to be among the earliest branching eukaryotes, diverging before the acquisition of the bacterial symbiont that became the mitochondrion (Cavalier-Smith 1987, Sogin 1991). This deep-branching placement is also seen with protein-coding genes (e.g., Baldauf et al. 1996, Roger 1999, Bapteste et al. 2002). However, several genes of mitochondrial origin have since been found in parabasalid and diplomonad nuclear genomes (Roger 1999, Tachezy et al. 2001), suggesting that both groups originally had a mitochondrial symbiont (in parabasalids, this symbiont is preserved as the hydrogenosome).

Recent phylogenetic evidence also demonstrates that diplomonads and retortamonads are very closely related to *Carpediemonas* and parabasalids, whereas oxymonads are close to *Trimastix* (Dacks et al. 2001, Simpson and Patterson 2001, Silberman et al. 2002, Embley and Hirt 1998, Simpson et al. 2002). Although mitochondrial origins of the hydrogenosome-like organelles of *Trimastix* and *Carpediemonas* have not been proven, it seems very likely that all amitochondriate excavates have ancestors that bore mitochondrial symbionts. It is also argued that the basal placements of diplomonads and parabasalids in many molecular phylogenetic trees could be analysis artifacts caused by aberrant (especially accelerated) gene sequence evolution in these groups (Embley and Hirt 1998, Philippe and Adoutte 1998). Therefore, the relevance of amitochondriate excavates

to understanding early eukaryotic history is now uncertain, although they remain fascinating organisms for exploring the biochemical diversity and potential of eukaryotic cells.

Excavates 2: Mitochondriate Excavates (Discicristates, Jakobids, and Malawimonas)

The most important and best known mitochondrion-bearing excavates are the Discicristates (fig. 4.4, node 3). They are among the most recent of major eukaryotic groups to be confirmed by strong molecular phylogenetic support (Baldauf et al. 2000). Discicristates include the Euglenozoa and the Heterolobosea, which share the unusual characteristic of having mitochondria whose cristae are discoid in shape. These infoldings of the inner mitochondrial membrane are the site of electron transport and ATP production. Other mitochondriate excavates are the more obscure jakobids and *Malawimonas*.

Euglenozoa contain two major supergroups: kinetoplastids and euglenids. Kinetoplastids are small uni- or biflagellated cells with a distinctive and baroque mitochondrial genome organization. The mitochondrial DNA is condensed into a large mass or masses called the kinetoplast, and many of the messenger RNAs for mitochondrial genes require extensive RNA editing (mediated by other smaller RNA molecules called guide RNAs) before they encode functional proteins (Sollner-Webb 1996). The kinetoplastids include the trypanosomatid parasites, among which are the agents of several deadly human diseases: sleeping sickness, Chagas disease, and leishmaniases. Many other kinetoplastids are also commensals or parasites, but free-living forms are abundant consumers of bacteria and small eukaryotes.

Euglenids are usually free-living uni- or biflagellate cells enclosed by a thickened pellicle made longitudinal proteinaceous strips. Most of the diversity of euglenids are free-living osmotrophs, or phagotrophs that are often able to consume large eukaryotic cells, although the most famous euglenids are the photosynthetic forms, such as *Euglena*. The photosynthetic euglenids have chloroplasts that are of secondary origin—they are derived from an eukaryotic green algal cell that was ingested by a nonphotosynthetic euglenid ancestor.

Heterolobosea (fig. 4.4, node 3). These are mostly amoebae, although many have flagellate phases in their life cycles (Patterson and Sogin 2000). Heteroloboseids differ in appearance from lobose amoebae in their "eruptive" formation of pseudopodia. Most are soil or freshwater bacterivores, although one, *Naegleria fowleri*, is a rare but often fatal facultative human pathogen. A subgroup, the acrasids, are slime molds that form fruiting bodies, but they are unrelated to the "true" mycetozoan slime molds (Roger et al. 1996).

Jakobids (i.e., core jakobids) are small free-living bacterivores. They have the most bacteria-like mitochondrial genomes known, having retained genes apparently lost, relocated, or replaced in other studied eukaryotes (Lang et al. 1997). A final small group, *Malawimonas*, is superficially similar to jakobids but might be more closely related to some or all amitochondriate excavates (O'Kelly and Nerad 1999, Simpson et al. 2002).

Jakobids are also interesting because their bacteria-like mitochondrial genomes may represent a primitive state for living eukaryotes (Lang et al. 1997). However, the recent resolution of the broad-scale eukaryotic tree using the dihydrofolate reductase–thymidylate synthase gene fusion suggests that Excavata might be closer to Plantae than to Opisthokonta (Stechmann and Cavalier-Smith 2002) and thus not especially deeply branching after all (assuming that Excavata is, in fact, a natural group).

Phylogeny. The monophyly of excavates is currently contentious. The cytoskeleton supporting the cell is distinctively similar in all of them. The exceptions, most notably parabasalids, are convincingly related to at least one "good" excavate in molecular trees (e.g., Baldauf et al. 2000). Thus, morphology suggests that excavates descend from a similar common ancestor (Simpson and Patterson 1999, Simpson et al. 2002). By contrast, almost all molecular analyses place excavates as multiple separated clusters distributed across the diversity of eukaryotes (Simpson et al. 2002). However, different groups of excavates exhibit drastic differences in evolutionary rate for commonly used molecular markers, a property known to complicate and confound phylogenetic analysis (Philippe and Adoutte 1998). Resolving whether excavates are a natural group using molecular markers promises to be a difficult problem in eukaryotic phylogeny. One theoretical possibility is that excavates currently represent an ancestral grade for most or all living eukaryotes, rather than a natural group (O'Kelly 1993).

Chromalveolates

Chromalveolates (fig. 4.4, node 5) are a broadly diverse group of protists that includes the Chromista (fig. 4.4, node 7), comprising the cryptophytes, haptophytes, and stramenopiles (heterokonts) and the Alveolata (fig. 4.4, node 6), which include the parasitic apicomplexans, ciliates, and dinoflagellates. The chromalveolates were postulated primarily on the basis of molecular phylogenetic analyses that unite particular members of these disparate lineages (described below), and the hypothesis that all taxa containing a chromophytic plastid (i.e., containing chlorophyll c) share a common origin (Cavalier-Smith 2000).

Chromalveolates 1: Chromists

The Chromista (Cavalier-Smith 1986) are a provisional group including the cryptophyte, haptophyte, and stramenopiles. These are largely marine, unicellular algae that are some of the most important photosynthetic forms on the planet. The stramenopiles have unambiguous molecular and ultrastructural justification. Almost all groups within Heterokonta include organisms with a "tinsillated" flagellum, and most have a second, shorter, smooth flagellum. The shorter flagellum is posteriorly directed and often associated with an eyespot.

The tinsillated flagellum is anteriorly directed and bears two rows of stiff, tripartite hairs along its length. These hairs reverse the flow around the flagellum so that the cell is dragged forward although the medium, rather than pushed along. The group is named for the structure of the flagellar hairs (stramenopiles), but they are also often referred to as heterokonts, which means "different flagella." These characters are not found in the cryptophytes or haptophytes, and their phylogenetic affinity has so far been hard to resolve.

Stramenopiles (Diatoms, Kelps, Oomyetes, Labyrinthulids). Stramenopiles are possibly the largest and most diverse group of eukaryotes. They include opalinids (endocommensals, mostly in cold-blooded vertebrates), oomycetes (including water molds and downy mildews, previously classified as fungi), bicosoecids (small heterotrophic biflagellates), labyrinthulids (slime nets), and all the diverse types of chlorophyl a and c algae. The latter include the diatoms, dominant marine photoautotrophs that reside in lidded boxes made of silica (glass) called frustules. There are ~11,000 recognized species, and millions of undescribed ones by some estimates (Norton et al. 1996). Other stramenopile algae include the multicellular kelps, which are particularly widespread in temperate intertidal and subtidal zones. These have true parenchyma and build "forests" in nearshore environments that support complex ecosystems including fish and marine mammals.

Oomycetes are important group of parasites or saprobes (e.g., *Phytophthora infestans*, the cause of potato blight and the great famine of Ireland). Although lacking a plastid, recent sequence data suggest that these taxa may have once been photosynthetic (Andersson and Roger 2002). There is also a huge diversity of very small free-swimming phototrophic, mixotrophic, and heterotrophic stramenopiles in most planktonic systems (Moriera and Lopéz-Garcia 2002), for example, the bicosoecid *Cafeteria*, perhaps the world's most abundant predator. Others, such as *Blastocystis* and opalinids, are commensals in the guts of animals. Although lacking a plastid, recent sequence data suggest that they may have once been photosynthetic because they retain nuclear-encoded genes of apparent cyanobacterial origin (Andersson and Roger 2002).

Haptophytes. Haptophytes get their name from the presence of a unique anterior appendage, the haptonema, used for adhesion and capturing prey. The group includes the coccolithophorids, which build external coverings of calcium carbonate scales (coccoliths) and tend to dominate open oceanic waters worldwide. *Emiliana huxleyi*, in particular, has received considerable attention because of its important role in cloud production through dimethyl sulfoxide release, the effects of its "blooms" on temperature and optical quality of oceanic waters, and its role as a major carbon sink (Buitenhuis et al. 1996). Coccoliths from dead cells accumulate as limestone deposits on the ocean bottom, forming the largest inorganic reservoir of carbon on Earth. The haptophyte *Chrysochromulina* is an important source of toxic blooms.

Cryptophytes. The cryptophytes are perhaps the least known of the chromists, being relatively small (mostly 2–10 μm diameter) unicells and primarily found in cold or deep aquatic environments. The group has been critical to our understanding of plastid secondary endosymbiosis because they have retained an intermediate stage in the process. Current theory holds that all chromalveolates acquired their plastid by a single event (Cavalier-Smith 2000), in which a common ancestor of the group ingested a single-celled photosynthetic eukaryote, in this case a red alga (vs. a green alga in the case of euglenids). The host would have then transferred the red algal nuclear genes required for plastid maintenance into their own nuclear genome, and the original red algal nucleus would have been lost. However, in the case of cryptophytes a remnant of the red algal nucleus persists as a "nucleomorph" that resides together with the plastid surrounded by a double membrane—a kind of cell within a cell. Analysis of the nucleomorph genome (e.g., Douglas et al. 1999, 2001) provided the first phylogenetic evidence for the chimeric nature of algal cells by confirming the red algal origin of the cryptophyte plastid.

Chromalveolates 2: Alveolates

The alveolates (fig. 4.4, node 6) represent another large assemblage of protists with strong molecular and ultrastructural justification. The group includes the dinoflagellates, many of which are algae, the parasitic apicomplexans, and the ciliates (Gajadhar et al. 1991). All members of the group possess sacs or alveoli under the plasma membrane. The alveoli form the pellicle in ciliates and surround the peripheral armor plates in dinoflagellates.

Ciliates.   Ciliates are mostly free-living aquatic unicells. These well-known protists (e.g., *Paramecium tetraurelia*) are characterized by an abundance of cilia on their body surface, nuclear dualism, and the presence of a conjugation stage during the sexual phase of the life cycle (Hausmann and Hülsmann 1996). Nuclear dualism refers to the maintenance of two different types of nuclei in each cell. The smaller micronucleus contains the diploid germ nucleus, whereas the second much larger macronucleus contains thousands of copies of only the physiologically active genes. Ciliate nuclear genome organization is truly remarkable; genes are not only fragmented by introns and short intervening sequences, but the order of the gene fragments themselves may be scrambled (Prescott 2000). Therefore, extensive editing can be required during generation of the macronucleus in order to produce the active working copy of the gene.

Dinoflagellates.   This is a diverse, predominantly unicellular group, characterized by having one transverse and one longitudinal flagellum, resulting in a unique rotatory swimming motion. Most are covered by often elaborate plates or armor. Although the group was probably primitively photosynthetic (described below), only about half of the extant dinoflagellates still are, and many of these species are mixotrophs. These ingest bacteria and other eukaryotes and are notorious for acquiring temporary endosymbionts from them, particularly plastids from a variety of algae. In fact, the group appears to include the first known example of tertiary endosymbiosis involving the secondary endosymbiosis of a haptophyte, itself already secondarily endosymbiotic (described above; Yoon et al. 2002a). Others, such as *Symbiodinium* species, are themselves endosymbionts of corals, and these and other dinoflagellates are a common source of phosphorescence in marine waters. Under all trophic condition, the dinoflagellates are an important component of marine ecosystems as symbionts and primary producers. They also produce some of the most potent neurotoxins known and are the main source of toxic red tides and other forms of fish and shellfish poisoning.

Apicomplexa.   Closely related to the dinoflagellates are the apicomplexans, which formerly constituted the bulk of the "sporozoa." They include some of the most important protozoan disease agents of both invertebrates and vertebrates and are the causative agents of malaria and toxoplasmosis. All are obligate, mostly intracellular parasites characterized by the presence of an intricate apical complex. This is a system of organelles and microtubules situated at the posterior of the cell that functions in the attachment and initial penetration of the host. Their complex life cycles are completed entirely within the host, and they exist outside it only as spores or oocysts. The group appears to have been derived from photosynthetic ancestors, and recent data show that they retain a vestigial plastid (apicoplast) most likely of red algal origin (Fast et al. 2001). Much research in malaria is now being directed at finding drugs that target potential functions of this organelle.

Phylogeny

Chromalveolates (fig. 4.4, node 5) are a broadly diverse group of protists postulated primarily on the hypothesis that all taxa containing a chromophytic (chlorophyll c) plastid share a common origin (Cavalier-Smith 2000). Molecular phylogenetic support for the heterokont and alveolate groupings is generally strong, although these data sets are mostly very taxon limited. However, support for the entire chromalveolates grouping is still slight, although some analyses of nuclear-encoded genes have shown moderate to moderately strong support for a stramenopile + alveolate clade (e.g., Van de Peer and De Wachter 1997, Baldauf et al. 2000; see also Philippe, ch. 7 in this vol.).

More recently, a much more inclusive five-gene plastid data set shows the first robust support for a single common origin of chromist plastids, implying a monophyletic origin for the chromalveolates (Yoon et al. 2002b). These trees show the cryptophytes as the deepest branch in the group, implying that the nucleomorph and phycobilin pigments are ancestral characters lost before the divergence of haptophytes and stramenopiles. Plastid loss after secondary endosymbiosis must also have been common (e.g., the oomycetes). Molecular clock analyses place the earliest date for the origin of chromists at ~1.26 Byr ago. Thus, a single, ancient event, the secondary endosymbiosis of a red algal plastid, appears to have been a fundamental one in eukaryote evolution, giving rise to an entire protist superassemblage, the chromalveolates.

Plantae

The Plantae (fig. 4.4, node 8) consists of the rhodophytes (red algae), glaucophytes (glaucophyte algae), and Viridiplantae or green plants (green algae + land plants). Rhodophytes vary from large seaweeds to crustose mats that look more like rocks than living plants. Their plastids have two membranes and unstacked thylakoids. Light is harvested primarily with chlorophyll a and phycoerythrins (red chromophores) conjugated to phycobiliproteins. There are two major subgroups, bangiophytes and florideophytes; the former appears to be older and may have given rise to the latter. Glaucophytes are a small but distinct group of unicellular flagellates. They harvest light energy in plastids called "cyanelles" using rhodophyte-like proteins and pigments. Cyanelles have two membranes, unstacked thylakoids and, most remarkably, bacteria-like peptidoglycan walls. Viridiplants vary from single-celled flagellates to large marine filaments to redwoods. Their plastids have two membranes and stacked thylakoids, and they harvest light with chlorophylls a and b attached to chlorophyll–a-b–binding proteins. Virdiplantae includes the chlorophyte, ulvophyte, trebouxiophytes, charophyte, and "prasinophyte" algae (see Delwiche et al., ch. 9 in this vol.). Land plants were clearly derived from charophyte algae, and the single-celled "prasinophytes" are almost certainly para- or even polyphyletic (Turmel et al. 2002).

Phylogeny

Plantae are probably the only eukaryotic group to acquire photosynthesis directly from cyanobacteria (primary endosymbiosis). That this only happened once is most strongly supported by the fact that their plastid genomes have a similar, derived gene order and composition (Douglas 1998). It is also consistent with the fact that these are all the eukaryotes whose plastids have only two membranes, thought to correspond to the inner and outer membrane of the original cyanobacterial endosymbiont (Archibald and Keeling 2002). All other algae have three or four outer plastid membranes, believed to be the result of additional endosymbioses (see Delwiche et al., ch. 9 in this vol.).

If their primary endosymbiosis only happened once in eukaryotic evolution, then red green and glaucophyte plants would be expected to form a clade. However, there are still very few molecular data to test this; there are few nuclear gene sequences for very few rhodophytes, and even fewer nuclear and no mitochondrial data for glaucophytes. Actin and mitochondrial sequence trees strongly support a monophyletic red–green clade (Burger et al. 1999), as do some nuclear markers (Hilario and Gogarten 1998). However, others still appear to reject it strongly (Stiller et al. 2001). Actin trees also tentatively place all three plant lineages together

(Bhattacharya and Weber 1997), as do combined sequence data (Moreira et al. 2000, Baldauf et al. 2000). Clearly, more data are needed from a more representative sampling of algal lineages.

Cercozoa, Foraminifera, and Radiolaria

This is a heterogeneous assemblage of morphologically and ecologically diverse forms (fig. 4.4, node 10), including cercomonads, thaumatomonads, cryothecomonads, *Spongomonas*, chlorarachniophytes, euglyphids, *Gromia*, plasmodiophorids, and haplosporids, probably also the foraminiferans, and possibly also the radiolarians. With the exception of the latter two taxa, each group is currently represented by just a few genera. The foraminiferans and radiolarians, on the other hand, are large and well-characterized groups. Most members of the Cercozoa produce filose pseudopodia (or axopodia) that are used to capture food particles.

Chlorarachniophytes (genera *Chlorarachnion*, *Lotharella*, *Gymnochlora*) are photosynthetic marine amoebae with reticulate (anastomosing, networklike) pseudopodia and a uniflagellate dispersal stage. Theirs is another example of secondary endosymbiosis, and, similar to Cryptophytes (described above), they retain a remnant of the primary endosymbiont nucleus (nucleomorph). Euglyphids are testate amoebae with filose pseudopodia, found commonly in freshwater and in mosses. Their silica hard outer shells (test) are composed of regularly arranged, secreted plates, which are also used as characters for species identification. The gromids are widespread marine protists characterized by filose pseudopodia and a large (up to 5 mm) spherical to ovoid organic test with a characteristic layer of honeycomb membranes. It has a complex life cycle with a well-documented gamontic phase.

Cercomonads (genera *Cercomonas*, *Heteromita*, *Massisteria*) are common, heterotrophic flagellates, with two naked flagella, usually able to produce pseudopodia in their trophic stage. Thaumatomonads (e.g., *Protaspis*, *Thaumatomonas*) are biflagellate heterotrophic, mostly benthic flagellates. They maintain a rigid cell profile but feed with ventral pseudopodia. Cryothecomonads are flagellated planktonic predators known mostly from polar oceans. *Spongomonas* are sessile flagellates that embed into a spongy-walled matrix. Plasmodiophorids and haplosporids are typically plasmodial endoparasites of other eukaryotes. The Plasmodiophora members (10 genera) are plant parasites, sometimes treated as fungi. They are characterized by multinucleated plasmodia, unusual cruciform nuclear division and zoospores with two anterior flagella. The Haplosporidia members (three genera) cause diseases in freshwater and marine invertebrates. They form large multinucleate plasmodia with unusual organelles, called haplosporosomes, of unknown function.

Foraminifera and Radiolaria

Compared with Cercozoa, Foraminifera and Radiolaria are morphologically well-defined, large groups, composed of about 940 and 140 modern genera, respectively. Foraminiferans are widely distributed in all types of marine environment, but some also occur in freshwater and terrestrial habitats. They are characterized by finely granular reticulated pseudopodia (granuloreticulopodia) with bidirectional cytoplasmic flow. Most members of Foraminifera possess a test, which may be organic, agglutinated or calcareous, and composed of single or multiple chambers. Many foraminiferans have complex life cycles consisting of alternation of sexual and asexual generations. Nuclear dimorphism has been observed in a few species. Some calcareous foraminiferans live in endosymbiosis with dinoflagellates, diatoms, green algae, or red algae.

Radiolaria are characterized by the combination of internal mineralized "skeletons" and axopodia—long, radiating, unbranched processes stiffened by arrays of microtubules. All are marine and pelagic, solitary or colonial. Some live in symbiosis with different types of algae. Radiolarians consist of three distinct classes: Acantharea, Phaeodarea, and Polycystinea. Acantharia are characterized by delicate skeletons that consists of radial spicules, composed of strontium sulfate, joined at the center of the cell and emerging from the cell surface in a regular pattern. Phaeodaria are characterized by siliceous skeletons formed of hollow radial spines (not always present) and a very thick capsular membrane. Polycystinea is divided into Spumellaria, whose members possess a spherical cell body plan, and Nassellaria, with members having a nonspherical body plan and skeletons varying from simple spicules to complex helmet-shaped structures. Both foraminiferan and radiolarian (polycystine and phaeodarian) skeletons contribute substantially to the microfossil record in marine sediments extending back to the Cambrian. Their fossilized tests are used in micropaleontology as biostratigraphic markers and as paleoceanographic indicators to determine ancient water temperature, ocean depths, circulation patterns, and the age of water masses.

Phylogeny

The grouping of Cercozoa, Foraminifera, and Radiolaria is based almost exclusively on molecular phylogenetic data. Although the majority of the protists belonging to these groups possess pseudopodia, this character is also present in Amoebozoa (described below) and other, now clearly unrelated groups. The cercozoan clade was originally demonstrated by a series of SSU rRNA analyses progressively adding more of the unusual members of the group (Bhattacharya et al. 1995, Bulman et al. 2001, Wylezich et al. 2002), the most recent addition being the enigmatic soil flagellate *Proleptomonas faecicola* (Vickerman et al. 2003) and a marine filosean *Gromia oviformis* (Burki et al. 2002). Cercozoan affinity was also suggested for Haplosporidia and *Marteilia* (Paramyxea; Cavalier-Smith 2000), despite earlier molecular study, which considered them as independent eukaryotic phyla (Berthe et al. 2000). SSU rRNA trees always place *Plasmodiophora* as the deepest "reliably placed" branch

in the clade and cercomonads *sensu stricto* as para- or polyphyletic within the group.

The grouping of *Chlorarachnion* and *Cercomonas* has been confirmed by α-tubulin (Keeling et al. 1998) and actin (Keeling 2001) gene phylogeny. The latter also first revealed the close relation between Cercozoa and Foraminifera, contradicting the previous rRNA-based analyses (Pawlowski et al. 1996), and is now confirmed by polyubiquitine structure (Archibald et al. 2003) and analysis of RNA polymerase II subunit 1 sequences (Longet et al. 2003). The latter also shows that, among the Cercozoa, *Gromia* appears to be the closest relative to the foraminiferans.

Neither the composition nor the overall phylogenetic position of Radiolaria is well established. Early SSU rRNA analyses including Acantharia and Polycystina suggested the group was polyphyletic (Amaral-Zettler et al. 1997). However, later analyses of the same data (Pawlowski et al. 1999) or with the addition of new ciPCR sequences (Lopéz-Garcia et al. 2003) showed that these two groups are actually closely related, and the group as a whole to be related to the Cercozoa (Cavalier-Smith 2002). There are currently no molecular data for Phaedaria and their morphological distinctiveness has led to suggestions that they might have an origin independent from the other two groups.

There are also currently no published SSU rRNA data on Heliozoa, and their inclusion in a Cercozoa + Foraminifera + Radiolaria clade is contradicted by morphological data indicating that at least some of them (actinophryids) are related to stramenopiles (Mikryukov and Patterson 2001). However, Radiolaria and Cercozoa appear adjacent to each other in some SSU rRNA trees (Lopéz-Garcia et al. 2003, Cavalier-Smith 2002) and may form a very weak clade when only the shortest branches are analyzed (A. G. B. Simpson, unpubl. obs.).

Amoebozoa

Lobosa (Lobose Amoebae)

Approximately 14 amoeboid types (fig. 4.4, node 12) are recognized. They appear to be scattered across the eukaryote tree and may have arisen independently from flagellate ancestors a number of times (Patterson et al. 2000, Cavalier-Smith 1998, 2002). Traditional taxonomy of amoebae relies mainly on pseudopodial morphology and, where present, the morphology and composition of extracellular scales or shells (tests). There are very few, if any, molecular data on most of them and generally little indication of their place within the larger eukaryote tree. Nonetheless, the phylogenetic positions of the heterolobosean, foraminiferan, and euglyphid (all described above) amoebae seem to be resolved, as is that of the lobose amoebae.

The nontestate (naked) lobose amoebae (also known as ramicristate or gymnamoebae) are now clearly placed with the Mycetozoa based on mitochondrial genome synapomorphies and phylogenetic trees (Bhattacharya and Weber

1997, Baldauf et al. 2000). They generally have one to many lobose or tubelike pseudopods (lobopodia), usually a single nucleus, and mitochondrial cristae that are tubular and branched. Sizes range from a few micrometers to several millimeters, and many smaller forms probably remain to be discovered. They are cosmopolitan in distribution and important as major bacterial predators. Some form cysts to survive desiccation or other harsh conditions or to invade hosts. The group consists largely of widespread free-living species, but they also include animal commensals and opportunistic pathogens, such as *Acanthamoeba*, which causes eye infections in contact lens wearers. The naked lobose amoebae may be related to the testate lobose amoebae (Arcinellinids), but there are no molecular data on them.

Pelobionts and Entamoebae

Pelobionts are amoeboflagellates (possessing both amoeboid and flagellate morphologies) that mostly live in low-oxygen environments. They have one or many apical flagella and vary widely in size; the most famous, *Pelomyxa*, is a massive amoeba as long as 3 mm with numerous nonmotile flagella on its surface. Entamoebae are small aflagellate amoebae, and almost all are small commensals or parasites of animals. Several species live in the mouth and intestinal tract of humans, causing amoebic dysentery, and they sometimes invade the liver (*Entamoeba histolytica*), resulting in serious illness. Both pelobionts and entamoebas lack mitochondria and, partly on this basis, were widely thought to represent very early diverging eukaryotes (Cavalier-Smith 1987). However, they have recently been shown to have mitosomes, small organelles of mitochondrial origin (Tovar et al. 1999).

Mycetozoa (True Slime Molds)

The Mycetozoa contain the myxogastrid, dictyostelid, and protostelid slime molds, although the latter may be paraphyletic with respect to either or both of the former (Olive and Stoianovitch 1975). Members of the three groups have very different trophic (feeding stage) morphologies, as described below. This has led to a long-running debate as to whether they are related or not, and the myxogastrids and dictyostelids have been variously classified as plants, animals, and fungi in the ~150 years since they were first described. However, they all have distinctly similar fruiting bodies consisting of a cellulosic stalk supporting spore-bearing sori, albeit of widely varying size, form, and complexity (Olive and Stoianovitch 1975).

The myxogastrids (Myxogastridae) are also known as the plasmodial, true, or acellular slime molds. The best known is *Physarum polycephalum*, easily grown on agar plates in the lab. These are amoeboflagellates, switching between amoeboid and flagellate morphologies early in their life cycle before maturing into large plasmodia with 10,000 or more nuclei. Plasmodia are capable of a slow, creeping movement propelled by cytoplasmic pulsations, even though they can be 100 cm or more in diameter. The trophic stage of dictyo-

stelids (Dictyostelidae), on the other hand, is strictly amoeboid. Under appropriate conditions the amoebae can aggregate, although cells never fuse to form true plasmodia. As many as 10,000 or more cells stream together to form a "slug," which surrounds itself with a single outer covering and acts much like a very simple multicellular organism in that it has a defined head and tail region and is mobile. In *Dictyostelium discoideum*, cell fate is determined in the slug, and only the cells in the tail region can form spores. Protostelids (Protostelidae) were first described in 1960 and are almost entirely microscopic. They can be either amoeboflagellate or strictly amoeboid, sometimes among apparently closely related taxa (Olive and Stoianovitch 1975). Thus, they seem to bridge the "gap" between the dictyostelid and myxogastrid morphologies (Olive and Stoianovitch 1975), and may be paraphyletic with respect to them.

The possible monophyly of Mycetozoa has been debated since their discovery in the late 1800s, based on the striking differences in their trophic stages and striking similarities in their fruiting bodies. This was not helped by early rRNA trees, which separated *D. discoideum* (Dictyostelidae) and *Physarum polycephalum* (Myxogastridae) widely. However, all other molecular data tend to place them together, mostly with very strong support (Baldauf et al. 2000). Only a single molecular study includes a protostelid sequence (*Planoprotostelium aurantium*), placing it as the sister group to a strong *Physarum polycephalum* + *Dictyostelium discoideum* clade. This suggests that not only are the myxgastrids and dictyostelids related, but they are in fact only a subgroup of Mycetozoa (Baldauf and Doolittle 1997).

Phylogeny

Although there are few molecular data from lobose amoebae, and all from *Acanthamoeba castellanii*, based on these sequences the monophyly of the Lobosa + Mycetozoa (fig. 4.4, node 12) is strongly supported by actin (e.g., Bhattacharya et al. 1995) and combined (Baldauf et al. 2000) data. Lack of support for this grouping from SSU rRNA is not surprising because these data rarely even bring the Mycetozoa together, much less support them strongly as a group. Combined sequence data also support a strong grouping of pelobionts and entamaebids and place together with the Mycetozoa (Bapteste et al. 2002). Although *Entamoebae histolytica* is represented in many single gene trees, it is almost never united with the Mycetozoa. However, sequences from this taxon also tend to be very divergent and to form highly unstable long branches in phylogenetic trees (e.g., Keeling and Doolittle 1996).

Perhaps the most convincing data for Lobosa + Mycetozoa are shared unique similarities in their mitochondrial genomes (Ogawa et al. 2000), but because pelobionts and entamaebids lack mitochondria, these data cannot be extended to them. Morphologically, the Amoebozoa are united by the presence of lobose pseudopodia moving in a smooth, noneruptive manner and tubular mitochondrial cristae.

Acrasids were until recently grouped with the dictyostelids because they form similar-looking fruiting bodies. However, they have discoidal mitochondrial cristae, form eruptive filose pseudopodia, do not aggregate, and have now been unambiguously reclassified as Heterolobosea (described above), and molecular phylogeny seems to firmly link both groups to each other and to the Mycetozoa (Bapteste et al. 2002).

Opisthokonta (Animalia and Fungi)

Animalia

Animals (fig. 4.4, node 15) are defined as multicellular heterotrophs capable of complex and relatively rapid movement, acquiring food by ingestion and digesting it in an internal cavity. Their cells lack rigid cell walls, and all except sponges are made up of cells organized into specialized tissues, which are mostly further organized into specialized organs. Most are diploid and reproduce sexually by means of differentiated eggs and sperm. Animal development is characterized by distinctive stages including a zygote, blastula, and gastrula (see Eernisse and Peterson, ch. 13 in this vol.).

Fungi

Fungi (fig. 4.4, node 14) are single or multicellular heterotrophs, acquiring their food by absorption after first digesting it extracellularly with secreted hydrolytic enzymes. Cell walls are generally present and composed of chitin; multicellular forms consist of multinucleate filamentous tubes, termed hyphae. There are five major subtypes: chytrids, zygomycetes, ascomyetes, basidiomycetes, and microsporidians. Thraustochytrids, oomycetes, mycetozoa, plasmodiophorids, and labyrinthulids have all been removed from the group, mostly to the heterokonts. The earliest branches of true fungi are clearly chytrids, although neither they nor the members of Zygomycetes are monophyletic (see Taylor et al., ch. 12 in this vol.). The microsporidia are often depicted as extremely early-diverging lineages in molecular trees, but this is now known to an artifact of their fast evolutionary rates for most genes (fig. 4.4, Baldauf et al. 2000).

Animal–Fungus Allies ("Choanozoa")

A diverse group of taxa have been recently assigned to the opisthokont clade, although their various branching positions within it are not generally well resolved. These include choanoflagellates (aquatic uniflagellates), ichthyosporeans (obligate intracellular parasites of aquatic animals), corallochytreans (free-living saprophytes), and nucleariids (cristidiscoidean amoebae). This is a diverse collection of single-celled taxa with seemingly little in common. Ichthyosporeans and corallochytreans are highly reduced morphologies. Nucleariids lack both "diagnostic" features of opisthokonts, that is, have no flagella, much less a single basal one, and their mitochondrial cristae appear to be discoidal rather than flattened (Zettler et al. 2001). Only choanoflagellates are long-standing candidates for the sis-

ter group to animals, because of their strong resemblance to the collar cells of sponges. However, the reassignment of these taxa to Opisthokonta, originally based on SSU rRNA trees, has been confirmed for choanoflagellates (*Monosiga*) and ichthyosporeans (*Amoebidium*) based on combined mitochondrial gene trees (Burger et al. 2003). These trees strongly place the choanoflagellate as the closest sister group to animals and *Amoebidium* as a sister group to the choanoflagellate–animal clade.

Phylogeny

The sisterhood of animals and fungi is now well accepted among evolutionary protistologists (Cavalier-Smith 1998, Patterson 1999) and is supported by all large, broadly taxonomically sampled molecular data sets, including SSU rRNA, LSU rRNA, HSP70 (70 KD heat shock protein), EF-1α (protein synthesis elongation factor 1-α), α-tubulin, β-tubulin, and actin, by combined analysis of 23 proteins using the sum of likelihood scores method, and by all CDS trees (Baldauf et al. 2000, Moreira et al. 2000, Bapteste et al. 2002). A small number of morphologically synapormophies have been defined—the unique combination of flattened mitochondrial cristae and, when flagellate, the presence of a single basal flagellum on reproductive cells (Cavalier-Smith 1998, 2002) and similarities in the flagellar anchorage system (Patterson 1999). However, these characters are only sporadically found among the various opisthokont allies (described above). Nonetheless, the grouping is often not found in small, poorly taxonomically sampled single-gene trees, probably because of long-branch problems and hidden paralogy (Baldauf and Palmer 1993).

Possible New Additions

There are more than 200 poorly known but distinct groups of eukaryotes whose affinities are unclear (Patterson 1999). Most of these are small free-living heterotrophic flagellates or amoebae or are parasites of various kinds. Many will doubtless turn out to fall within one or more of the groups described above, but there are reasons for guessing that some form distinct major groups. Apusomonads and *Ancyromonas* are probably closely related, small gliding flagellates supported by submembranous thecae. Some SSU rRNA trees weakly suggest that they are closely related to opisthokonts. Collodictyonids are free-swimming predators of other eukaryotic cells that form no close relationships in SSU rRNA trees (Brugerolle et al. 2002).

Heliozoa ("sun animals") are a large, diverse collection of cells that capture food particles using radiating stiffened pseudopodia. They form at least four distinct groups and are widely assumed to be polyphyletic, although actinophryid heliozoa alone are thought to be descended from heterokonts. Tenuous morphological considerations suggest pivotal roles for *Phalansterium* and *Multicilia* (small flagellates) for understanding the evolution of Amoebozoa, and possibly all eukaryotes, but this is not confirmed with detailed examinations or molecular data. Little is known of the positions of kathablepharids, spironemids, or *Telonema*, to name just a few.

The Eukaryote Root

Probably the single most outstanding question in eukaryote evolution is the location of the root of tree. The predominant theory until recently has been the Archezoa hypothesis based on the observation that the deepest branches in the eukaryotic SSU rRNA tree were mitochondrion-lacking organisms, that is, microsporidia, diplomonads, and parabasalids (described above; Cavalier-Smith 1987, Sogin 1991). This led to the suggestion that these taxa diverged before unique acquisition of the mitochondrial symbiont in eukaryotes. However, nuclear-encoded mitochondria-like genes have been found in representatives of each of these groups, suggesting they once had at least the precursor of this organelle (Roger 1999, Tachezy et al. 2001). Analyses of protein-encoding genes also showed microsporidia to be members of Fungi (Keeling et al. 2000, Hirt et al. 1999, Baldauf et al. 2000). The deep placement of their sequences in SSU rRNA trees is an extreme case of long-branch attraction (Embley and Hurt 1998; see also Philippe, ch. 7 in this vol.).

Most other molecular phylogenies, including CDS trees, still place diplomonads and/or parabasalids as the most basal eukaryote branches (Hashimoto et al. 1994, Philippe and Adoutte 1998, Bapteste et al. 2002). However, these sequences still tend to form very long branches in these trees, and it can still be argued that their deep placement is simply a long-branch artifact (see Philippe, ch. 7 in this vol.). Methods designed to compensate for long-branch attraction, such as transversion parsimony or covarion analyses, tend to show the eukaryote tree without any deep resolution, which may indicate that the major eukaryote groups arose by explosive radiation (see Philippe, ch. 7 in this vol.), or simply that these methods remove most of the information from a data set so that nothing can be resolved.

A radically different placement of the eukaryote root is suggested by a recently investigated a gene fusion involving dihydrofolate reductase (DHFR) and thymidylate synthase (TS). The genes for these proteins are separate in bacteria and opisthokonts but fused in plants, cercozoans, chromalveolates, apusomonads, centrohelids, and discicristates (Stechmann and Cavalier-Smith 2002). If this root is correct, it places the opisthokonts as one of, if not the, first extant branch off the main line of eukaryote descent. If the excavates are then taken as monophyletic, for which there is currently no strong molecular phylogenetic support, this shift in the root makes the amitochondriate excavates a relatively recently derived group (sister group to discicristates). The strength of this character rests on the assumption that gene fusions are highly irreversible, which is not true and difficult to evaluate here. The scenario is further complicated by

the fact that *Dictyostelium*, a pivotal taxon in this scheme, lacks these genes entirely (Myllykallio et al. 2002), as do members of the alternative deepest eukaryote branch, amitochondriate excavates. Further data are clearly needed.

Literature Cited

Bergey's Manual Trust. 2001. Bergey's manual of systematic bacteriology. Available: http://www.cme.msu.edu/bergeys/. Last accessed 15 December 2003.

Adrian, L., U. Szewzyk, J. Wecke, and H. Gorisch. 2002. Bacterial dehalorespiration with chlorinated benzenes. Nature 408:580–83

Alberts, B., A. Johnson, J. Lewis, M. Raff, K. Roberts, and P. Walter. 2002. Molecular biology of the cell. 4th ed. Taylor and Francis, New York.

Amann, R. I., W. Ludwig, and K. H. Schleifer. 1995. Phylogenetic identification and in situ detection of individual microbial cells without cultivation. Microbiol. Rev. 59:143–169.

Amaral-Zettler, L., M. L. Sogin, and D. A. Caron. 1997. Phylogenetic relationship between the Acantharea and the Polycystinea: a molecular perspective of Haeckel's Radiolaria. Proc. Natl. Acad. Sci. USA 94:11411–11416.

Amaral-Zettler, L. A., F. Gomez, E. Zettler, B. G. Keenan, R. Amils, and M. L. Sogin. 2002. Eukaryotic diversity in Spain's River of Fire. Nature 417:137.

Andersson, J. O., and A. J. Roger. 2002. A cyanobacterial gene in nonphotosynthetic protists—an early chloroplast acquisition in eukaryotes? Curr. Biol. 12:115–119.

Andresson, O. S., and O. H. Fridjonsson. 1994. The sequence of the single 16S rRNA gene of the thermophilic eubacterium *Rhodothermus marinus* reveals a distant relationship to the group containing *Flexibacter*, *Bacteroides*, and *Cytophaga* species. J. Bacteriol. 176:6165–6169.

Archibald, J. M., and P. J. Keeling. 2002. Recycled plastids: a "green movement" in eukaryotic evolution. Trends Genet. 18:577–584.

Archibald, J. M., D. Longet, J. Pawlowski, and P. J. Keeling. 2003. A novel polyubiquitin structure in Cercozoa and Foraminifera: evidence for a new eukaryotic supergroup. Mol. Biol. Evol. 20:62–66.

Asai, T., D. Zaporojets, C. Squires, and C. L. Squires. 1999. An *Escherichia coli* strain with all chromosomal rRNA operons inactivated: complete exchange of rRNA genes between bacteria. Proc. Natl. Acad. Sci. USA 96:1971–1976.

Baldauf, S. L., and W. F. Doolittle. 1997. Origin and evolution of the slime molds (Mycetozoa). Proc. Natl. Acad. Sci. USA 94:12007–12012.

Baldauf, S. L., and J. D. Palmer. 1993. Animals and fungi are each other's closest relatives: congruent evidence from multiple proteins. Proc. Natl. Acad. Sci. USA 90:11558–11562.

Baldauf, S. L., J. D. Palmer, and W. F. Doolittle. 1996. The root of the universal tree and the origin of eukaryotes based on elongation factor phylogeny. Proc. Natl. Acad. Sci. USA 93:7749–54.

Baldauf, S. L., A. J. Roger, I. Wenk-Siefert, and W. F. Doolittle.

2000. A kingdom-level phylogeny of eukaryotes based on combined protein data. Science 290:972–977.

Bapteste, E., H. Brinkmann, J. A. Lee, D. V. Moore, C. W. Sensen, P. Gordon, L. Durufle, T. Gaasterland, P. Lopez, M. Muller, and H. Philippe. 2002. The analysis of 100 genes supports the grouping of three highly divergent amoebae: *Dictyostelium*, *Entamoeba*, and *Mastigamoeba*. Proc. Natl. Acad. Sci. USA 99:1414–19.

Barns, S. M., and S. Burggraf. 1997. Crenarchaeota. *In* The Tree of Life Web Project (D. R. Maddison, ed.). Available at http://tolweb.org/tree?group=Crenarchaeota&contgroup=Archaea. Last accessed 15 December 2003.

Barns, S. M., C. F. Delwiche, J. D. Palmer, and N. R. Pace. 1996. Perspectives on archaeal diversity, thermophily and monophyly from environmental rRNA sequences. Proc. Natl. Acad. Sci. USA 93:9188–9193.

Barns, S. M., S. L. Takala, and C. R. Kuske. 1999. Wide distribution and diversity of members of the bacterial kingdom Acidobacterium in the environment. Appl Environ. Microbiol. 65:1731–1737.

Bell, S. D., and S. P. Jackson. 2001. Mechanism and regulation of transcription in Archaea. Curr. Opin. Microbiol. 4:208–213.

Benson, D. A., L. Karsch-Mizrachi, D. J. Lipman, J. Ostell, and D. L. Wheeler. 2004. GenBank. Nucleic Acids Res. 32:D23–D26.

Bernal, A., U. Ear, and N. Kyrpides. 2001. Genomes OnLine Database (GOLD): a monitor of genome projects worldwide. Nucleic Acids Res. 29:126–127.

Bernard, C., A. G. B. Simpson, and D. J. Patterson. 2000. Some free-living flagellates (Protista) from anoxic habitats. Ophelia 52:113–142.

Berthe, F. C. J., F. Le Roux, E. Peyretaillade, P. Peyret, D. Rodriguez, M. Gouy, and C. P. Vivares. 2000. Phylogenetic analysis of the small subunit ribosomal RNA of *Marteilia refringens* validates the existence of phylum Paramyxea. J. Eukaryot. Microbiol. 47:288–293.

Bhattacharya, D., T. Helmchen, and M. Melkonian. 1995. Molecular evolutionary analyses of nuclear-encoded small subunit ribosomal RNA identify an independent Rhizopod lineage containing the Euglyphida and the Chlorarachniophyta. J. Eukaryot. Microbiol. 42:65–69.

Bhattacharya, D., and K. Weber. 1997. The actin gene of the glaucocystophyte *Cyanophora paradoxa*: analysis of the coding region and introns, and an actin phylogeny of eukaryotes. Curr. Genet. 31:439–46.

Bjornsson, L., P. Hugenholtz, G. W. Tyson, and L. L. Blackall. 2002. Filamentous Chloroflexi (green non-sulfur bacteria) are abundant in wastewater treatment processes with biological nutrient removal. Microbiology 148:2309–2318.

Bocchetta, M., S. Gribaldo, A. Sanangelantoni, and P. Cammarano. 2000. Phylogenetic depth of the bacterial genera *Aquifex* and *Thermotoga* inferred from analysis of ribosomal protein, elongation factor, and RNA polymerase subunit sequences. J. Mol. Evol. 50:366–380.

Bohlke, K., F. M. Pisani, M. Rossi, and G. Antranikian. 2002. Archaeal DNA replication: spotlight on a rapidly moving field. Extremophiles. 6:1–14.

Brochier, C., E. Bapteste, D. Moreira, and H. Philippe. 2002. Eubacterial phylogeny based on translational apparatus proteins. Trends Genet. 18:1–5.

Brochier, C., and H. Philippe. 2002. Phylogeny: a non-hyperthermophilic ancestor for bacteria. Nature 417:244.

Brocks, J. J., G. A. Logan, R. Buick, and R. E. Summons. 1999. Archean molecular fossils and the early rise of eukaryotes. Science 13:1033–1036.

Brown, J. R., C. J. Douady, M. J. Italia, W. E. Marshall, and M. J. Stanhope. 2001. Universal trees based on large combined protein sequence data sets. Nat. Genet. 28:281–285.

Brugerolle, G., G. Bricheux, H. Philippe, and G. Coffea. 2002. *Collodictyon triciliatum* and *Diphylleia rotans* (= *Aulacomonas submarina*) form a new family of flagellates (Collodictyonidae) with tubular mitochondrial cristae that is phylogenetically distant from other flagellate groups. Protist 153:59–70.

Buitenhuis, E., J. Bleijswijk, D. van Bakker, and M. Veldhuis. 1996. Trends in inorganic and organic carbon in a bloom of *Emiliania huxleyi* in the North. Sea. Mar. Ecol. Prog. Series. 143:271–282.

Bulman, S. R., S. F. Kühn, J. W. Marshall, and E. Schnepf. 2001. A phylogenetic analysis of the SSU rDNA from members of the Plasmodiophorida and Phagomyxida. Protist 152:43–51.

Burger, G., L. Forget, Y. Zhu, M. W. Gray, and B. F. Lang. 2003. Unique mitochondrial genome architecture in unicellular relatives of animals. Proc. Natl. Acad. Sci USA 100:892–897.

Burger, G., D. Saint-Louis, M. W. Gray, and B. F. Lang. 1999. Complete sequence of the mitochondrial DNA of the red alga *Porphyra purpurea*: cyanobacterial introns and shared ancestry of red and green algae. Plant Cell. 11:1675–1694.

Burki, F., C. Berney, and J. Pawlowski. 2002. Phylogenetic position of *Gromia oviformis* Dujardin inferred from nuclear-encoded small subunit ribosomal DNA. Protist 153:251–260.

Cano, R. J., and M. K. Borucki. 1995. Revival and identification of bacterial spores in 25– to 40–million-year-old Dominican amber. Science 268:1060–1064.

Cavalier-Smith, T. 1986. The kingdoms of organisms. Nature 324:416–417.

Cavalier-Smith, T. 1987. Eukaryotes with no mitochondria. Nature 326:332–333.

Cavalier-Smith, T. 1998. A revised six-kingdom system of life. Biol. Rev. 73:203–266.

Cavalier-Smith, T. 2000. Membrane heredity and early chloroplast evolution. Trends Plant Sci. 5:174–182.

Cavalier-Smith, T. 2002. The phagotrophic origin of eukaryotes and phylogenetic classification of Protozoa. Int. J. Syst. Evol. Microbiol. 52:297–354.

Chisholm, S. W., R. J. Olson, E. R. Zettler, R. Goericke, J. B. Waterbury, and N. A. Welschmeyer. 2002. A novel free-living prochlorophyte abundant in the oceanic euphotic zone. Nature 334:340–343.

Cole, S. T. 2002. Comparative and functional genomics of the *Mycobacterium* tuberculosis complex. Microbiology 148:2919–2928.

Coulson, R. M., A. J. Enright, and C. A. Ouzounis. 1991. Functional classes in the three domains of life. J. Mol. Evol. 49:551–557.

Coulson, R. M., A. J. Enright, and C. A. Ouzounis. 2001.

Transcription-associated protein families are primarily taxon-specific. Bioinformatics 17:95–97.

CyanoBase. 2003. The genomic database for Cyanobacteria. Kazusa DNA Research Institute, Kisarazu, Japan. Available: http://www.kazusa.or.jp/cyano/index.html. Last accessed 15 December 2003.

Dacks, J. B., J. D. Silberman, A. G. Simpson, S. Moriya, T. Kudo, M. Ohkuma, and R. J. Redfield. 2001. Oxymonads are closely related to the excavate taxon *Trimastix*. Mol. Biol. Evol. 18:1034–1044.

Daubin, V., M. Gouy, and G. Perriere. 2002. A phylogenomic approach to bacterial phylogeny: evidence of a core of genes sharing a common history. Genome Res. 12:1080–1090.

Dawson, S. C., and N. R. Pace. 2002. Novel kingdom-level eukaryotic diversity in anoxic environments. Proc. Natl. Acad. Sci. USA 99:8324–8329.

Deckert, G., P. V. Warren, T. Gaasterland, W. G. Young, A. L. Lenox, D. E. Graham, R. Overbeck, M. A. Snead, M. Keller, M. Aujay, et al. 1998. The complete genome of the hyperthermophilic bacterium *Aquifex aeolicus*. Nature 392:353–358.

DeLong, E. F. 1992. Archaea in coastal marine environments. Proc. Natl. Acad. Sci. USA 89:5685–5689.

Deppenmeier, U., A. Johann, T. Hartsch, R. Merkl, R. A. Schmitz, R. Martinez-Arias, A. Henne, A. Wiezer, S. Baumer, C. Jacobi, et al. 2002. The genome of *Methanosarcina mazei*: evidence for lateral gene transfer between bacteria and archaea. J. Mol. Microbiol. Biotechnol. 4:453–461.

Ding, Y. H., R. S. Ronimus, and H. W. Morgan. 1999. Purification and properties of the pyrophosphate-dependent phosphofructokinase from *Dictyoglomus thermophilum* Rt46 B. 1. Extremophiles. 3:131–137.

Douglas, S. E. 1998. Plastid evolution: origins, diversity, trends. Curr. Opin. Genet. Dev. 8:655–661.

Douglas, S. E., C. A. Murphy, D. F. Spencer, and M. W. Gray. 1999. Cryptomonad algae are evolutionary chimaeras of two phylogenetically distinct unicellular eukaryotes. Nature 350:148–151.

Douglas, S. E., S. Zauner, M. Fraunholtz, M. Beaton, S. Penny, L. T. Deng, W. Wu, M. Reith, T. Cavalier-Smith, and U. G. Maier. 2001. The highly reduced genome of an enslaved algal nucleus. Nature 410:1091–1096.

Dyall, S. D., and P. J. Johnson. 2000. Origins of hydrogenosomes and mitochondria: evolution and organelle biogenesis. Curr. Opin. Microbiol. 3:404–411.

Dykhuizen, D. E., and G. Baranton. 2001. The implications of a low rate of horizontal transfer in *Borrelia*. Trends Microbiol. 9:344–350.

Edgell, D. R., and W. F. Doolittle. 1997. Archaea and the origin(s) of DNA replication proteins. Cell 89:995–998.

Edwards, K. J., P. L. Bond, T. M. Gihring, and J. F. Banfield. 2000. An archaeal iron-oxidizing extreme acidophile important in acid mine drainage. Science 287:1796–1799.

Embley, T. M., and R. P. Hirt. 1998. Early branching eukaryotes? Curr. Opin. Genet. Dev. 8:624–629.

Etchebehere, C., and L. Muxi. 2000. Thiosulfate reduction and alanine production in glucose fermentation by members of the genus *Coprothermobacter*. Antonie Van Leeuwenhoek. 77:321–327.

Fast, N. M., J. C. Kissinger, D. S. Roos, and PJ. Keeling. 2001. Nuclear-encoded, plastid-targeted genes suggest a single common origin for apicomplexan and dinoflagellate plastids. Mol. Biol. Evol. 18:418–426.

Forterre, P. 2002. A hot story from comparative genomics: reverse gyrase is the only hyperthermophile-specific protein. Trends Genet. 18:36–37.

Foster, G., H. M. Ross, R. D. Naylor, M. D. Collins, C. P. Ramos, F. Fernandez Garayzabal, R. J. Reid, et al. 2002. *Cetobacterium ceti* gen. nov., sp. nov., a new gram-negative obligate anaerobe from sea mammals. Lett. Appl. Microbiol. 21202–21206.

Fraser, C. M., J. D. Gocayne, O. White, M. D. Adams, R. A. Clayton, R. D. Fleischmann, C. J. Bult, A. R. Kerlavage, G. Sutton, J. M. Kelley, et al. 1995. The minimal gene complement of *Mycoplasma genitalium*. Science 270:397–403.

Fuhrman, J. A., K. McCallum, and A. A. Davis. 1992. Novel major archaebacterial group from marine plankton. Nature 356:148–149.

Gajadhar, A. A., W. C. Marquardt, R. Hall, J. Gunderson, E. V. Ariztia-Carmona, and M. L. Sogin. 1991. Ribosomal RNA sequences of *Sarcocystis muris*, *Theileria annulata* and *Crypthecodinium cohnii* reveal evolutionary relationships among apicomplexans, dinoflagellates, and ciliates. Mol. Biochem. Parasitol. 45:147–154.

Garrett, R. A., J. Dalgaard, N. Larsen, J. Kjems, and A. S. Mankin. 1991. Archaeal rRNA operons. Trends Biochem. Sci. 16:22–26.

Gogarten, J. P., H. Kibak, P. Dittrich, L. Taiz, E. J. Bowman, B. Bowman, M. Manolson, R. Poole, T. Date, T. Oshima, et al. 1989. Evolution of the vacuolar H+-ATPase: implications for the origin of eukaryotes. Proc. Natl. Acad. Sci. USA 86:6661–6665.

Gordon, D. A., and S. J. Giovannoni. 1996. Detection of stratified microbial populations related to *Chlorobium* and *Fibrobacter* species in the Atlantic and Pacific oceans. Appl. Environ. Microbiol. 62:1171–1177.

Green, R., and H. F. Noller. 1997. Ribosomes and translation. Annu. Rev. Biochem. 66:679–716.

Gupta, R. S. 2001. The branching order and phylogenetic placement of species from completed bacterial genomes, based on conserved indels found in various proteins. Int. Microbiol. 4:187–202.

Hallbeck, L., F. Stahl, and K. Pedersen. 1993. Phylogeny and phenotypic characterization of the stalk-forming and iron-oxidizing bacterium *Gallionella ferruginea*. J. Gen. Microbiol. 139:1531–1535.

Hanson, T. E., and F. R. Tabita. 2001. A ribulose-1,5-bisphosphate carboxylase/oxygenase (RubisCO)-like protein from *Chlorobium tepidum* that is involved with sulfur metabolism and the response to oxidative stress. Proc. Natl. Acad. Sci. USA 98:4397–4402.

Hashimoto, T., Y. Nakamura, F. Nakamura, T. Shirakura, J. Adachi, N. Goto, K. Okamoto, and M. Hasegawa. 1994. Protein phylogeny gives a robust estimation for early divergences of eukaryotes: phylogenetic place of a mitochondria-lacking protozoan, *Giardia lamblia*. Mol. Biol. Evol. 11:65–71.

Hausmann, K., and N. Hülsmann. 1996. Protozoology. Thieme Medical Publishers, New York.

Hedlund, B. P., J. J. Gosink, and J. T. Staley. 1996. Phylogeny of *Prosthecobacter*, the fusiform caulobacters: members of a recently discovered division of the bacteria. Int. J. Syst. Bacteriol. 46:960–966.

Hershberger, K. L., S. M. Barns, A. L. Reysenbach, S. C. Dawson, and N. R. Pace NR. 1996. Wide diversity of Crenarchaeota. Nature 384:420.

Hilario, E., and J. P. Gogarten. 1998. The prokaryote-to-eukaryote transition reflected in the evolution of the V/F/A-ATPase catalytic and proteolipid subunits. J. Mol. Evol. 46:703–715.

Hirt, R. P., J. M. Logsdon, Jr., B. Healy, M. W. Dorey, W. F. Doolittle, and T. M. Embley. 1999. Microsporidia are related to Fungi: evidence from the largest subunit of RNA polymerase II and other proteins. Proc. Natl. Acad. Sci. USA 96:580–585.

Hjorleifsdottir, S., S. Skirnisdottir, G. O. Hreggvidsson, O. Holst, and J. K. Kristjansson. 2002. Species composition of cultivated and noncultivated bacteria from short filaments in an Icelandic hot spring at 88 degrees C. Microb. Ecol. 42:117–125.

Horn, M., and M. Wagner. 2001. Evidence for additional genus-level diversity of Chlamydiales in the environment. FEMS Microbiol. Lett. 204:71–74.

Hovanec, T. A., and W. F. DeLong. 1996. Comparative analysis of nitrifying bacteria associated with freshwater and marine aquaria. Appl. Environ. Microbiol. 62:2888–2896.

Howland, J. L. 2000. The surprising Archaea. Oxford University Press, Oxford

Huber, H., S. Diller, C. Horn, and R. Rachel. 2002a. *Thermovibrio ruber* gen. nov., sp. nov., an extremely thermophilic, chemolithoautotrophic, nitrate-reducing bacterium that forms a deep branch within the phylum Aquificae. Int. J. Syst. Evol. Microbiol. 52:1859–1865.

Huber, H., M. J. Hohn, R. Rachel, T. Fuchs, V. C. Wimmer, and K. O. Stetter. 2002b. A new phylum of Archaea represented by a nanosized hyperthermophilic symbiont. Nature 417:63–67.

Huet, J., R. Schnabel, A. Sentenac, and W. Zillig. 1983. Archaebacteria and eukaryotes possess DNA-dependent RNA polymerases of a common type. EMBO J. 2:1291–1294.

Hugenholtz, P., B. M. Goebel, and N. R. Pace. 1998. Impact of culture-independent studies on the emerging phylogenetic view of bacterial diversity. J. Bacteriol. 180:4765–4774.

Hugenholtz, P., G. W. Tyson, R. I. Webb, A. M. Wagner, and L. L. Blackall. 2001. Investigation of candidate division TM7, a recently recognized major lineage of the domain Bacteria with no known pure-culture representatives. Appl. Environ. Microbiol. 67:411–419.

Iborra, F. J., D. A. Jackson, and P. R. Cook. 2002. Coupled transcription and translation within nuclei of mammalian cells. Science 293:1139–1142.

Iwabe, N., K. Kuma, M. Hasegawa, S. Osawa, and T. Miyata. 1989. Evolutionary relationship of archaebacteria, eubacteria, and eukaryotes inferred from phylogenetic trees of duplicated genes. Proc. Natl. Acad. Sci. USA 86:9355–9359.

Jain, R., M. C. Rivera, and J. A. Lake. 1999. Horizontal gene transfer among genomes: the complexity hypothesis. Proc. Natl. Acad. Sci. USA 96:3801–3806.

Janssen, P. H., and K. A. O'Farrell. 2002. *Succinispira mobilis* gen. nov., sp. nov., a succinate-decarboxylating anaerobic bacterium. Int. J. Syst. Bacteriol. 49:1009–1013.

Karner, M. B., E. F. DeLong, and D. M. Karl. 2001. Archaeal dominance in the mesopelagic zone of the Pacific Ocean. Nature 409:507–510.

Keeling, P. J. 2001. Foraminifera and Cercozoa are related in actin phylogeny: two orphans find a home? Mol. Biol. Evol. 18:1551–1557.

Keeling, P. J., J. A. Deane, and G. J. McFadden. 1998. The phylogenetic position of alpha- and beta-tubulins from the *Chlorarachnion* host and *Cercomonas* (Cercozoa). J. Eukaryot. Microbiol. 45:561–570.

Keeling, P. J., and W. F. Doolittle. 1996. Alpha-tubulin from early-diverging eukaryotic lineages and the evolution of the tubulin family. Mol. Biol. Evol. 13:1297–1305.

Keeling, P. J., M. A. Luker, and J. D. Palmer. 2000. Evidence from beta-tubulin phylogeny that microsporidia evolved from within the fungi. Mol. Biol. Evol. 17:23–31.

Klenk, H. P., R. A. Clayton, J. F. Tomb, O. White, K. E. Nelson, K. A. Ketchum, R. J. Dodson, M. Gwinn, E. K. Hickey, J. D. Peterson, et al. 1997. The complete genome sequence of the hyperthermophilic, sulphate-reducing archaeon *Archaeoglobus fulgidus*. Nature 390:364–370.

Kuhner, C. H., C. Frank, A. Griesshammer, M. Schmittroth, G. Acker, A. Gossner, and H. L. Drake. 1997. *Sporomusa silvacetica* sp, nov., an acetogenic bacterium isolated from aggregated forest soil. Int. J. Syst. Bacteriol. 47:352–358.

Kyrpides, N. C., and C. R. Woese. 1998. Archaeal translation initiation revisited: the initiation factor 2 and eukaryotic initiation factor 2B alpha-beta-delta subunit families. Proc. Natl. Acad. Sci. USA 95:3727–3730.

Lake, J. A. 1988. Origin of the eukaryotic nucleus determined by rate-invariant analysis of rRNA sequences. Nature 331:184–186.

Lang, B. F., G. Burger, C. J. O'Kelly, R. Cedergren, G. B. Golding, C. Lemieux, D. Sankoff, M. Turmel, and M. W. Gray. 1997. An ancestral mitochondrial DNA resembling a eubacterial genome in miniature. Nature 387:493–497.

Lanyi, J. K., and H. Luecke. 2001. Bacteriorhodopsin. Curr. Opin. Struct. Biol. 11:415–419.

Lee, J., G. F. Leedale, and P. Bradbury (eds.). 2000. Illustrated guide to the Protozoa, 2nd ed. Society of Protozoologists, Lawrence, KS.

Lilburn, T. G., K. S. Kim, N. E. Ostrom, K. R. Byzek, J. R. Leadbetter, and J. A. Breznak. 2001. Nitrogen fixation by symbiotic and free-living spirochetes. Science 292:2495–2498.

Lilburn, T. G., T. M. Schmidt, and J. A. Breznak. 1999. Phylogenetic diversity of termite gut spirochaetes. Environ. Microbiol. 1:331–345.

Lindsay, M. R., R. I. Webb, M. Strous, M. S. Jetten, M. K. Butler, R. J. Forde, and J. A. Fuerst. 2001. Cell compartmentalisation in planctomycetes: novel types of structural organisation for the bacterial cell. Arch. Microbiol. 175:413–429.

Logsdon, J. M., Jr. 1998. The recent origins of spliceosomal introns revisited. Curr. Opin. Genet. Dev. 8:637–648.

Longet, D., J. M. Archibald, P. J. Keeling, and J. Pawlowski. 2003. Foraminifera and Cercozoa share a common origin according to RNA polymerase II phylogenies. Int. J. Syst. Evol. Microbiol. 53:1735–1739.

Lopéz-Garcia, P., F. Rodriguez-Valera, C. Pedros-Alio, and D. Moreira. 2001. Unexpected diversity of small eukaryotes in deep-sea Antarctic plankton. Nature 409:603–607.

Lopéz-Garcia, P., H. Philippe, F. Gail, and D. Moreira. 2003. Autochthonous eukaryotic diversity in hydrothermal sediment and experimental microcolonizers at the Mid-Atlantic Ridge. Proc. Natl. Acad. Sci. USA 100:697–702.

Ludwig, W., S. H. Bauer, M. Bauer, L. Held, G. Kirchhof, R. Schulze, L. Huber, S. Spring, A. Hartmann, and K. H. Schleifer. 1997. Detection and in situ identification of representatives of a widely distributed new bacterial phylum. FEMS Microbiol. Lett. 153:181–190.

Madigan, M. T., J. M. Martinko, and J. Parker. 1997. Brock biology of microorganisms. 8th ed. Prentice Hall, Upper Saddle River, NJ.

Maidak, B. L., J. R. Cole, T. G. Lilburn, C. T. Parker, Jr., P. R. Saxman, R. J. Farris, G. M. Garrity, G. J. Olsen, T. M. Schmidt, and J. M. Tiedje. 2001. The RDP-II (Ribosomal Database Project). Nucleic Acids Res. 29:173–174.

Makarova, K. S., I. Aravind, Y. L. Wolf, R. L. Tatusov, K. W. Minton, E. V. Koonin, and M. J. Daly. 2001. Genome of the extremely radiation-resistant bacterium Deinococcus radiodurans viewed from the perspective of comparative genomics. Microbiol. Mol. Biol. Rev. 65:44–79.

Marteinsson, V. T., J. K. Kristjansson, H. Kristmannsdottir, M. Dahlkvist, K. Saemundsson, M. Hannington, S. K. Petursdottir, A. Geptner, and P. Stoffers. 2001. Discovery and description of giant submarine smectite cones on the seafloor in Eyjafjordur, northern Iceland, and a novel thermal microbial habitat. Appl. Environ. Microbiol. 67:827–833.

Matte-Tailliez, O., C. Brochier, P. Forterre, and H. Philippe. 2002. Archaeal phylogeny based on ribosomal proteins. Mol. Biol. Evol. 19:631–639.

Mikryukov, K. A., and D. J. Patterson. 2001. Taxonomy and Phylogeny of Heliozoa. III. Actinophryids. Acta Protozool. 40:3–25.

Möller, B., R. Oßmer, B. H. Howard, G. Gottschalk, and H. Hippe. 1984. Sporomusa, a new genus of gram-negative anaerobic bacteria including Sporomusa sphaeroides spec. nov. and Sporomusa ovata spec. nov. Arch. Microbiol. 139:388–396.

Moon-van der Staay, S. Y., R. De Wachter, and D. Vaulot. 2001. Oceanic 18S rDNA sequences from picoplankton reveal unsuspected eukaryotic diversity. Nature 409:607–610.

Moreira, D., H. Le Guyader, and H. Philippe. 2000. The origin of red algae and the evolution of chloroplasts. Nature 405:69–72.

Morin, L. 2000. Long branch attraction effects and the status of basal eukaryotes. J. Eukaryot. Microbiol. 47:167–177.

Müller, M. 1988. Energy metabolism of protozoa without mitochondria. Annu. Rev. Microbiol. 42:465–488.

Myllykallio, H., G. Lipowski, D. Leduc, J. Filee, P. Forterre, and U. Liebl. 2002. An alternative flavin-dependent mechanism for thymidylate synthesis. Science 297:105–107.

Myllykallio, H. P., P. Lopez, P. Lopez-Garcia, R. Heilig, W. Saurin, Y. Zivanovic, H. Philippe, and P. Forterre. 2000. Bacterial mode of replication with eukaryotic-like machinery in a hyperthermophilic archaeon. Science 288:2212–2215.

Nelson, K. E., G. Lipowski, D. Leduc, J. Filee, P. Forterre, and U. Liebl. 1999. Evidence for lateral gene transfer between Archaea and bacteria from genome sequence of Thermotoga maritima. Nature 399:323–329.

Nesbo, C. L., S. L'Haridon, K. O. Stetter, and W. F. Doolittle. 2001. Phylogenetic analyses of two archaeal genes in thermotoga maritima reveal multiple transfers between archaea and bacteria. Mol. Biol. Evol. 18:362–375.

Ng, W. W., S. P. Kennedy, G. G. Mahairas, B. Berquist, M. Pan, H. D. Shukla, S. R. Lasky, N. S. Baliga, V. Thorsson, J. Sbrogna, et al. 2000. Genome sequence of Halobacterium species NRC-1.v. Proc. Natl. Acad. Sci. USA 97:12176–12181.

Norton, T. A., M. Melkonian, and R. A. Andersen. 1996. Algal biodiversity. Phycologia 35:308–326.

O'Kelly, C. J. 1993. The jakobid flagellates: Structural features of Jakoba, Reclinomonas and Histiona and implications for the early diversification of eukaryotes. J. Eukaryot. Microbiol. 40:627–636.

O'Kelly, C. J., and T. A. Nerad. 1999. Malawimonas jakobiformis n. gen., n. sp. (Malawimonadidae fam. nov.): a Jakoba-like heterotrophic nanoflagellate with discoidal mitochondrial cristae. J. Eukaryot. Microbiol. 46:522–531.

Ochman, H. 2001. Lateral and oblique gene transfer. Curr. Opin. Genet. Dev. 11:616–619.

Ogawa, S. R., R. Yoshino, K. Angata, M. Iwamoto, M. Pi, K. Kuroe, K. Matsuo, T. Morio, H. Urushihara, K. Yanagisawa, and Y. Tanaka. 2000. The mitochondrial DNA of Dictyostelium discoideum: complete sequence, gene content and genome organization. Mol. Gen. Genet. 263:514–519.

Olive, L., and D. Stoianovitch. 1975. The mycetozoans. Academic Press, New York.

Olsen, G. J., C. R. Woese, and R. Overbeek. 1994. The winds of (evolutionary) change: breathing new life into microbiology. J. Bacteriol. 176:1–6.

Overmann, J., and H. van Gemerden. 2000. Microbial interactions involving sulfur bacteria: implications for the ecology and evolution of bacterial communities. FEMS Microbiol. Rev. 24:591–599.

Oyaizu, H., B. Debrunner-Vossbrinck, L. Mandelco, J. A. Studier, and C. R. Woese. 1987. The green non-sulfur bacteria: a deep branching in the eubacterial line of descent. Syst. Appl. Microbiol. 9:47–53.

Pace, N. R. 1997. A molecular view of microbial diversity and the biosphere. Science 276:734–740.

Patterson, D. J. 1999. The diversity of eukaryotes. Am. Nat. 154:S96–S124.

Patterson, D. J., A. Rogerson, and A. G. Simpson. 2000. Amoebae of Uncertain Affinity. Pp. 804–827 in The illustrated guide to the protozoa. 2nd ed. (J. J. Lee, G. F. Leedale, and P. Bradbury, eds.). Society of Protozoologists, Lawrence, KS.

Patterson, D. J., and M. L. Sogin. 2000. Eukaryotes in The Tree of Life Web Project (D. R. Maddison, ed.). Available: http://tolweb.org/tree?group=Eukaryotes&contgroup=Life_on_Earth. Last accessed 15 December 2003.

Paustian, T. 2003. Microbiology Webbed Out. Available: http://www.bact.wisc.edu/MicrotextBook/. Last accessed 25 December 2003.

Pawlowski, J., I. Bolivar, J. F. Fahrni, T. Cavalier-Smith, and M. Gouy. 1996. Early origin of Foraminifera suggested by SSU rRNA gene sequences. Mol. Biol. Evol. 13:445–450.

Pawlowski, J., I. Bolivar, J. F. Fahrni, C. DeVargas, and S. Bowser. 1999. Molecular evidence that *Reticulomyxa filosa* is a freshwater naked foraminifer. J. Eukaryot. Microbiol. 46:612–617.

Petroni, G., S. Spring, K. H. Schleifer, F. Verni, and G. Rosati. 2002. Defensive extrusive ectosymbionts of *Euplotidium* (Ciliophora) that contain microtubule-like structures are bacteria related to *Verrucomicrobia*. Proc. Natl. Acad. Sci. USA 97:1813–1817.

Philippe, H., and A. Adoutte. 1998. The molecular phylogeny of protozoa: solid facts and uncertainties. Pp. 25–36 in Evolutionary relationships among protozoa (G. H. Coombs, K. Vickerman, M. A. Sleigh, and A. Warren, eds.). Kluwer Academic Publishers, Dordrecht, the Netherlands.

Philippe, H., and A. Germot. 2000. Phylogeny of eukaryotes based on ribosomal RNA: long-branch attraction and models of sequence evolution. Mol. Biol. Evol. 17:830–834.

Prescott, D. M. 2000. Genome gymnastics: unique modes of DNA evolution and processing in ciliates. Nat. Rev. Genet. 1:191–198.

Preston, C. M., K. Y. Wu, T. F. Molinski, and E. F. DeLong. 1996. A psychrophilic crenarchaeon inhabits a marine sponge: *Cenarchaeum symbiosum* gen. nov., sp. nov. Proc. Natl. Acad. Sci. USA 93:6241–6246.

Ramsaka, A., M. Peterkaa, K. Tajimab, J. C. Martin, J. Wood, M. E. Johnston, R. I. Aminovd, H. J. Flint, and G. Avgustina. 2000. Unravelling the genetic diversity of ruminal bacteria belonging to the CFB phylum. FEMS Microbiol. Ecol. 33:69–79.

Rivera, M. C., and J. A. Lake. 1992. Evidence that eukaryotes and eocyte prokaryotes are immediate relatives. Science 257:74–76.

Roger, A. J. 1999. Reconstructing early events in eukaryotic evolution. Am. Nat. 154:S146–S163.

Roger, A. J., M. W. Smith, R. F. Doolittle, and W. F. Doolittle. 1996. Evidence for the Heterolobosea from phylogenetic analysis of genes encoding glyceraldehyde-3-phosphate dehydrogenase. J. Eukaryot. Microbiol. 43:475–485.

Rotte, C., K. Henze, M. Muller, and W. Martin. 2000. Origins of hydrogenosomes and mitochondria. Curr. Opin. Microbiol. 3:481–486.

Ruepp, A., W. Graml, M. L. Santos-Martinez, K. K. Koretke, C. Volker, H. W. Mewes, D. Frishman, S. Stocker, A. N. Lupas, and W. Baumeister. 2000. The genome sequence of the thermoacidophilic scavenger *Thermoplasma acidophilum*. Nature 407:508–513.

Sagan, D., and L. Margulis. 1987. Bacterial bedfellows: a microscopic menage a trois may be responsible for a major step in evolution. Nat. Hist. 96:26–33.

Sait, M., P. Hugenholtz, and H. H. Janssen. 2002. Cultivation of globally distributed soil bacteria from phylogenetic lineages previously only detected in cultivation-independent surveys. Environ Microbiol. 4:654–666.

Sandman, K., and J. N. Reeve. 2001. Chromosome packaging by archaeal histones. Adv. Appl. Microbiol. 50:75–99.

Schleper, C., E. F. DeLong, C. M. Preston, R. A. Feldman, K. Y. Wu, and R. V. Swanson. 1998. Genomic analysis reveals chromosomal variation in natural populations of the uncultured psychrophilic archaeon *Cenarchaeum symbiosum*. J. Bacteriol. 180:5003–5009.

Schleper, C., G. Puehler, I. Holz, A. Gambacorta, D. Janekovic, U. Santarius, H. P. Klenk, and W. Zillig. 1995. *Picrophilus* gen. nov., fam. nov.: a novel aerobic, heterotrophic, thermoacidophilic genus and family comprising Archaea capable of growth around pH 0. J. Bacteriol. 177:7050–59.

Schopf, J. W., A. B. Kudryavtsev, D. G. Agresti, T. J. Wdowiak, and A. D. Czaja. 2002. Laser-Raman imagery of Earth's earliest fossils. Nature 416:73–76.

Sekiguchi, Y., H. Takahashi, Y. Kamagata, A. Ohashi, and H. Harada. 2001. *In situ* detection, isolation, and physiological properties of a thin filamentous microorganism abundant in methanogenic granular sludges: a novel isolate affiliated with a clone cluster, the green non-sulfur bacteria, subdivision I. Appl. Environ. Microbiol. 67:5740–5749.

She, Q., R. K. Singh, F. Confalonieri, Y. Zivanovic, G. Allard, M. J. Awayez, C. C. Chan-Weiher, I. G. Clausen, B. A. Curtis, A. De Moors, et al. 2001. The complete genome of the crenarchaeon *Sulfolobus solfataricus* P2. Proc. Natl. Acad. Sci. USA 98:7835–7840.

Silberman, J. D., A. G. Simpson, J. Kulda, I. Cepicka, V. Hampl, P. J. Johnson, and A. J. Roger. 2002. Retortamonad flagellates are closely related to diplomonads—implications for the history of mitochondrial function in eukaryote evolution. Mol. Biol. Evol. 19:777–786.

Simpson, A. G. B., and D. J. Patterson. 1999. The ultrastructure of *Carpediemonas membranifera* (Eukaryota) with reference to the excavate hypothesis. Eur. J. Protistol. 35:353–370.

Simpson, A. G. B., and D. J. Patterson. 2001. On core jakobids and excavate taxa: the ultrastructure of *Jakoba incarcerata*. J. Eukaryot. Microbiol. 48:480–492.

Simpson, A. G. B., A. J. Roger, J. D. Silberman, D. D. Leipe, V. P. Edgcomb, L. S. Jermiin, D. J. Patterson, and M. L. Sogin. 2002. Evolutionary history of 'early diverging' eukaryotes: the excavate taxon *Carpediemonas* is closely related to *Giardia*. Mol. Biol. Evol. 19:1782–1791.

Sinninghe Damste, J. S., M. Strous, W. I. Rijpstra, E. C. Hopmans, J. A. Geenevasen, A. C. van Duin, L. A. van Niftrik, and M. S. Jetten. 2002. Linearly concatenated cyclobutane lipids form a dense bacterial membrane. Nature 419:708–712.

Slesarev, A. I., K. V. Mezhevaya, K. S. Makarova, N. N. Polushin, O. V. Shcherbinina, V. V. Shakhova, G. I. Belova, L. Aravind, D. A. Natale, I. B. Rogozin, et al. 2002. The complete genome of hyperthermophile *Methanopyrus kandleri* AV19 and monophyly of archaeal methanogens. Proc. Natl. Acad. Sci. USA 99:4644–4649.

Sogin, M. L. 1991. Early evolution and the origin of eukaryotes. Curr. Opin. Genet. Dev. 1:457–463.

Sogin, M. L., and J. H. Gunderson. 1987. Structural diversity of eukaryotic small subunit ribosomal RNAs. Evolutionary implications. Ann. N.Y. Acad. Sci. 503:125–139.

Sollner-Webb, B. 1996. Trypanosome RNA editing: resolved. Science 273:1182–1183.

Stechmann, A., and T. Cavalier-Smith. 2002. Rooting the eukaryote tree by using a derived gene fusion. Science 297:89–91.

Stephens, R. S., S. Kalman, C. Lammel, J. Fan, R. Marathe, L. Aravind, W. Mitchell, L. Olinger, R. I. Tatusov, Q. Zhao, E. V. Koonin, and R. W. Davis. 1998. Genome sequence of an obligate intracellular pathogen of humans: *Chlamydia trachomatis*. Science 282:754–759.

Stiller, J. W., J. Riley, and B. D. Hall. 2001. Are red algae plants? A critical evaluation of three key molecular data sets. J. Mol. Evol. 52:527–539.

Subramanian, G., E. V. Koonin, and L. Aravind. 2000. Comparative genome analysis of the pathogenic spirochetes *Borrelia burgdorferi* and *Treponema pallidum*. Infect Immun. 68:1633–1648.

Tachezy, J., L. B. Sanchez, and M. Müller. 2001. Mitochondrial type iron-sulfur cluster assembly in the amitochondriate eukaryotes *Trichomonas vaginalis* and *Giardia intestinalis*, as indicated by the phylogeny of IscS. Mol. Biol. Evol. 18:1919–1928.

Takacs, C. D., M. Ehringer, R. Favre, M. Cermola, G. Eggertsson, A. Palsdottir, and A. Reysenbach. 2001. Phylogenetic characterization of the blue filamentous bacterial community from an Icelandic geothermal spring. FEMS Microbiol. Ecol. 35:123–128.

Thomas, T., and R. Cavicchioli. 1998. Archaeal cold-adapted proteins: structural and evolutionary analysis of the elongation factor 2 proteins from psychrophilic, mesophilic and thermophilic methanogens. FEBS Lett. 439:281–286.

Tokura, M., M. Ohkuma, and T. Kudo. 2000. Molecular phylogeny of methanogens associated with flagellated protists in the gut and with the gut epithelium of termites. FEMS Microbiol. Ecol. 33:233–240.

Tovar, J., A. Fischer, and C. G. Clark. 1999. The mitosome, a novel organelle related to mitochondria in the amitochondrial parasite *Entamoeba histolytica*. Mol. Microbiol. 32:1013–1021.

Turmel, M., C. Otis, and C. Lemieux. 2002. The complete mitochondrial DNA sequence of *Mesostigma viride* identifies this green alga as the earliest green plant divergence and predicts a highly compact mitochondrial genome in the ancestor of all green plants. Mol. Biol. Evol. 19:24–38.

Urbach, E., D. L. Robertson, and S. W. Chisholm. 1992. Multiple evolutionary origins of prochlorophytes within the cyanobacterial radiation. Nature 355:267–270.

Van de Peer, Y., and R. De Wachter. 1997. Evolutionary relationships among the eukaryotic crown taxa taking into account site-to-site rate variation in 18S rRNA. J. Mol. Evol. 45:619–630.

van den Ent, F., and J. Lowe. 2000. Crystal structure of the cell division protein FtsA from *Thermotoga maritima*. EMBO J. 19:5300–5307.

Van der Auwera, G., C. J. Hofmann, P. De Rijk, and R. De Wachter. 1998. The origin of red algae and cryptomonad nucleomorphs: a comparative phylogeny based on small and large subunit rRNA sequences of *Palmaria palmata*, *Gracilaria verrucosa*, and the *Guillardia theta* nucleomorph. Mol. Phylogenet. Evol. 10:333–342.

Van Zuillen, M. A., A. Lepland, and G. Arrhenius. 2002. Reassessing the evidence for the earliest traces of life. Nature 418:627–630.

Vickerman, K., D. Le Ray, and J. Hoef-Emden De Jonckheere. 2003. The soil flagellate *Proleptomonas faecicola*: cell organisation and phylogeny suggest that the only described free-living trypanosomatid is not a kinetoplastid but has cercomonad affinities. Protist 153:9–24.

Vreeland, R. H., W. D. Rosenzweig, and D. W. Powers. 2000. Isolation of a 250 million-year-old halotolerant bacterium from a primary salt crystal. Nature 407:897–900.

White, O., J. A. Eisen, J. F. Heidelberg, E. K. Hickey, J. D. Peterson, R. J. Dodson, D. H. Haft, M. L. Gwinn, W. C. Nelson, D. L. Richardson, et al. 1999. Genome sequence of the radioresistant bacterium *Deinococcus radiodurans* R1. Science 286:1571–1577.

Willems, A., and M. D. Collins. 1995. Phylogenetic placement of *Dialister pneumosintes* (formerly *Bacteroides pneumosintes*) within the Sporomusa subbranch of the Clostridium subphylum of the gram-positive bacteria. Int. J. Syst. Bacteriol. 45:405.

Woese, C. R. 1987. Bacterial evolution. Microbiol. Rev. 51:221–271.

Woese, C. R., and G. E. Fox. 1977. Phylogenetic structure of the prokaryotic domain: the primary kingdoms. Proc. Natl. Acad. Sci. USA 74:5088–5090.

Wolf, Y. I., I. B. Rogozin, N. V. Grishin, R. L. Tatusov, and E. V. Koonin. 2001. Genome trees constructed using five different approaches suggest new major bacterial clades. BMC Evol. Biol. 1:8–29.

Wylezich, C., R. Meisterfeld, S. Meisterfeld, and M. Schlegel. 2002. Phylogenetic analyses of small subunit ribosomal RNA coding regions reveal a monophyletic lineage of euglyphid testate amoebae (order Euglyphida). J. Eukaryot. Microbiol. 49:108–118.

Yoon, H. S., J. Hackett, and D. Bhattacharya. 2002a. A single origin of the peridinin-, and fucoxanthin-containing plastids in dinoflagellates through tertiary endosymbiosis. Proc. Natl. Acad. Sci. USA 99:11724–11729.

Yoon, H. S., J. Hackett, G. Pinto, and D. Bhattacharya. 2002b. A single, ancient origin of chromist plastids. Proc. Natl. Acad. Sci. USA 99:15507–15512.

Zettler, L. A. A., T. A. Nerad, C. J. O'Kelly, and M. L. Sogin. 2001. The nucleariid amoebae: more protists at the animal-fungal boundary. J Eukaryot. Microbiol. 48:293–297.

Zhang, H., Y. Sekiguchi, S. Hanada, P. Hugenholtz, H. Kim, Y. Kamagata, and K. Nakamura. 2003. *Gemmatimonas aurantiaca* gen. nov., sp. nov., a gram-negative aerobic polyphosphate-accumulating microorganism, the first cultured representatives of the new bacterial phylum *Gemmatimonadetes* phy. nov. Int. J. Syst. Evol. Microbiol. 53:1155–1163.

Norman R. Pace

The Early Branches in the Tree of Life

The development of DNA sequencing technology in the last decades of the 20th century revolutionized biology, including the ways in which we can study the history of life. Before the availability of gene sequences, relationships of fossils were the main hope to chart the evolution of life. The character traits used to relate organisms in evolution were primarily morphological and could not be applied to microbial organisms. With gene sequences, contemporary organisms are related quantitatively in terms of nucleotide differences. Variation in sequences among modern organisms is a measure of the extent of biodiversity. Gene and now whole-genome sequences also allow the inference of maps of the history of evolution, in the form of phylogenetic trees. The results are illuminating and provide grist for conjecture and controversy on the evolutionary process. The purpose of this article is to tour the large-scale structure of the phylogenetic Tree of Life and to provide some interpretation of this emerging view of life's history. I emphasize how our understanding of the extent of the tree has expanded because of recent molecular studies of microbial diversity in the environment.

Molecular Phylogeny: Inference of Phylogenetic Trees

Ancestral relationships of modern organisms are derived using the techniques of "molecular phylogeny." The basic notion of molecular phylogeny is simple. Sequences of ho-

mologous (more properly, orthologous) genes, genes with common ancestry and function, from different organisms are aligned so that corresponding DNA bases can be compared. The number of differences between pairs of sequences is counted, which is considered to be some measure of the evolutionary distance that has separated the pairs of organisms. Just as geographical maps can be constructed from distances between land features, evolutionary maps—"phylogenetic trees"—can be inferred from evolutionary distances (sequence changes) between homologous genes. Calculations of the path of evolution are fraught with statistical uncertainties, however.

The process of inferring the best relatedness trees from pairwise sequence counts is complex and dependent on models of evolution used to calculate such trees (Swofford et al. 1996). One complexity that vexes attempts to infer the deeper relationships in the universal phylogenetic tree with certainty is that the actual number of sequence changes was greater than the observed number. This is because of the probabilities of back mutations, where no change is counted, and multiple past mutations, which are counted as only single changes. Numbers of mutational events per observed mutation can be estimated statistically, but a significant amount of the information used to build trees then becomes inferential, not directly observed. The mathematics of estimating actual changes from observed change are such that deeper branch points in phylogenetic trees are accompanied by greater statistical uncertainty as to their position. Still another complexity is that different lines of descent have evolved at different rates, which confuses tree-building algorithms.

Current advanced methods for inference of phylogenetic relationships are well developed to cope with the problems mentioned and others, but statistical vagaries are inescapable. The methods in common use are dependent on different models for reconstructing relationships, and this can influence the topological outcome of phylogenetic calculations. Popular methods for inferring phylogenetic trees from sequence relationships include evolutionary distance (ED), maximum parsimony (MP), and maximum likelihood (ML). ED uses corrected sequence differences directly as distances to calculate the pattern of ancestral connections. MP presumes that the fewest changes make the best trees, so optimal relatedness patterns are estimated by the minimum number of changes required to generate the topology. ML is a statistical method that calculates the likelihood of a particular topology given the sequence differences. In each case, statistical uncertainties in the calculations render any particular result questionable. Consequently, nodes in trees are tested many times using the same method and with subsets of the sequence collection, so-called "bootstrap analysis." The reliability of a particular result, for instance, a branch point in a tree or the composition of a relatedness group, is tested by the frequency with which the result occurs in the set of bootstrap trees. At the current state of their development, the different methods for calculating phylogenetic trees usually give generally comparable results. Nonetheless, intrinsic uncertainties in any tree must be acknowledged, particularly in the placement of deeper branches.

What Gene for Deep Phylogeny?

Any collection of homologous gene sequences can be used to infer phylogenetic relationships among those genes, but genes used to infer the overall structure of evolution, a universal phylogenetic tree, have special constraints on their properties (Woese 1987). One is that the gene must occur in all forms of life, so all can be related to one another. The hemoglobin gene, for instance, would not be useful for large-scale phylogeny because many groups do not contain the gene. A second constraint is that the gene must have resisted, over the ages, lateral transfer between genetic lines of descent. Genomic studies have shown clearly that many kinds of genes, for example, metabolic genes, have experienced extensive lateral transfer during the course of their evolution (Koonin et al. 2001, Woese 2000). Use of such genes for phylogenetic reconstructions produces conflicting results. A third constraint on genes that can be used to infer global phylogenetic trees is that they contain sufficient information, numbers of homologous nucleotides, so that relationships can be established with the best statistical reliability. There are not many genes that meet all these requirements. Most genes occur in only a limited diversity of organisms, and

many have undergone lateral transfer. The most generally accepted large-scale phylogenetic results are based on the use of ribosomal RNA (rRNA) gene sequences, those of the large subunits and small subunits (SSUs) of rRNAs. Ribosomes are present in all cells and major organelles, and phylogenetic trees inferred with these gene sequences are congruent with trees constructed using other elements of the cellular nucleic acid–based, information-processing machinery. Therefore, changes in the rRNA sequences seem to reflect the evolutionary path of the genetic machinery.

SSU rRNA sequences were first used for phylogenetic studies by Carl Woese, even before it was possible to determine gene sequences rapidly. Woese painstakingly prepared radioactive rRNAs from many diverse organisms, mostly microbes, and compared their content of short patches of sequences, fragments called oligonucleotides. The prevailing notion of life's evolutionary diversity at the time was framed in the context of two kinds of organisms, procaryote or eucaryote. Consequently, it was unexpected when the rRNA sequences from diverse organisms fell into three, not two, fundamentally distinct groups (Woese and Fox 1977). There had to be three primary lines of evolutionary descent, phylogenetic "domains," now termed Archaea (formerly archaebacteria), (eu)Bacteria, and Eucarya (eucaryotes; Woese et al. 1990). Woese's 1977 paper reporting the discovery of Archaea sparked publicity and controversy (Woese and Fox 1977). The concept of three primary relatedness groups of life touched off a flurry of refutations defending the procaryote–eucaryote or the five-kingdoms notions to account for biological organization. These familiar notions had never previously been tested, however, and the analysis of rRNA sequences proved them fundamentally incorrect. The shift in public and textbook treatment of the large organization of life is ongoing.

The Three Phylogenetic Domains of Life

Figure 5.1 is derived from a tree calculated using the particular set of rRNA sequences (Barns et al. 1996). The figure is a rough map of the course of evolution of the genetic core of cells, the collection of genes that propagates replication and gene expression. The dimension along the lines is sequence change, not time. Estimated evolutionary change that separates contemporary sequences (organisms) is read along line segments. The "root" of the universal tree, the point of origin for modern lineages, cannot be established using sequences of only one type of molecule. However, phylogenetic studies of gene families that originated before the last common ancestor of the three domains have placed the root on the bacterial line (Gogarten et al. 1989, Iwabe et al. 1989). This means that Eucarya and Archaea had a common history that excluded the descendants of the bacterial line. This period of evolutionary history shared by Eucarya and Archaea was an important time in the evolution of cells, during which the refinement of the primordial information-processing mechanisms occurred. Thus,

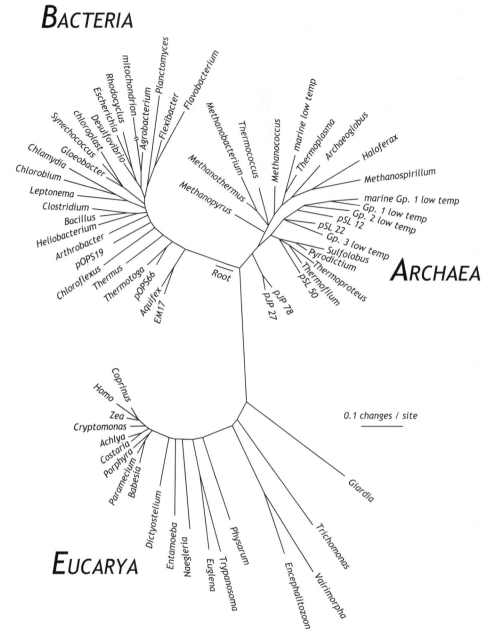

Figure 5.1. Universal tree based on SSU rRNA sequences. Sixty-four rRNA sequences representative of all known phylogenetic domains were aligned, and a tree was produced with an ML method (Barns et al. 1996). That tree was modified, resulting in the composite one shown, by trimming and adjusting branch points to incorporate the results of other analyses. The scale bar corresponds approximately to 0.1 changes per nucleotide (Pace 1997).

modern representatives of Eucarya and Archaea share many properties that differ from bacterial cells in fundamental ways. One example of similarities and differences is in the nature of the transcription machinery. The RNA polymerases of Eucarya and Archaea resemble each other far more than either resembles the bacterial type of polymerase. Moreover, whereas all bacterial cells use sigma factors to regulate the initiation of transcription, eucaryal and archaeal cells use TATA-binding proteins (Marsh et al. 1994, Rowlands et al. 1994). The shared evolutionary history of Eucarya and Archaea suggests that we may be able to recognize fundamental elements of our own cells through study of the far simpler archaeal version.

The rRNA sequence information, along with other molecular data, solidly confirms the century-old notion that mito-

chondria and chloroplasts are derived from bacterial symbionts. The sequence comparisons establish that mitochondria are representatives of the Proteobacteria, the group indicated by *Escherichia* and *Agrobacterium* in figure 5.1. Chloroplasts derived from cyanobacteria, represented by *Synechococcus* and *Gloeobacter* in figure 5.1. Thus, all of the respiratory and photosynthetic capacity of eucaryotic cells was obtained from bacterial symbionts. The nuclear component of the modern eucaryotic cell did not derive from an ancient bacterial or archaeal symbiosis, however. Molecular trees based on rRNA and other reliable genes show unequivocally that the Eucarya are as old as the Archaea. The mitochondrion and chloroplast came in relatively late in the sense of sequence change in rRNA, but early in the chronological history of life (described below). This later evolution of the major organelles is evidenced by the fact that mitochondria and chloroplasts diverged from peripheral branches in the molecular trees (fig. 5.1). Moreover, the most deeply divergent eucaryotes in phylogenetic trees even lack mitochondria. These latter kinds of organisms, little-studied but sometimes troublesome anaerobic creatures such as *Giardia*, *Trichomonas*, and *Vairimorpha*, nonetheless contain at least a few bacteria-type genes (Sogin and Silberman 1998). These genes may be evidence of an earlier symbiosis that was lost, or perhaps a gene transfer event between the evolutionary domains.

A Microbial World

A sobering aspect of large-scale phylogenetic trees such as shown in figure 5.1 is the graphical realization that most of our knowledge in biological sciences has focused on but a small slice of biological diversity. The organisms most represented in our textbooks of biology, animals (*Homo* in fig. 5.1), plants (*Zea*), and fungi (*Coprinus*), constitute only peripheral branches even of eucaryotic cellular diversity. Life's genetic diversity is mainly microbial in nature. Although the biosphere is absolutely dependent on the activities of microorganisms, our understanding of the makeup and natural history of microbial ecosystems is, at best, rudimentary. One reason for the paucity of information is that microbiologists traditionally have relied on laboratory cultures for the detection and identification of microbes. Yet, more than 99% of natural microbes are not cultured using standard techniques. Consequently, most environmental microbes have remained largely unknown.

The development of cloning and sequencing technology, coupled with the relational perspective afforded by phylogenetic trees, made it possible to identify environmental microbes without the requirement for culture (Pace 1997). The occurrence of phylogenetic types of organisms, "phylotypes," and their distribution in natural communities can be surveyed by sequencing rRNA genes obtained directly from environmental DNA by cloning. This sidesteps the need to culture organisms in order to learn something about them. A sequence-based phylogenetic assessment of an uncultivated organism can provide insight into many of the properties of the organism through comparison with its studied relatives. On the other hand, many of the phylotypes detected in the environment have no close relatives in the culture collections, so little can be inferred about the properties of the organisms that correspond to the sequences. The sequences, however, can be used to devise experimental tools, for instance, molecular hybridization probes, that can be used identify and study the inhabitants of microbial ecosystems. Regardless of the properties of the organisms they represent, the novel rRNA sequences have provided additional perspective on the topology of the universal tree. The following sections discuss the evolutionary structures of the three domains.

Bacteria

Most knowledge of microorganisms has derived from the study of only a few kinds of bacteria, mainly cultured organisms and in the context of disease or industrial products. Any general census of bacteria that make up naturally occurring microbial communities was not possible until the development of the molecular methods that identify rRNA sequence-based phylotypes without culture. As rRNA sequences have accumulated in the databases, now numbering more than 80,000, it is apparent that the heavily studied species represent only a fraction of bacterial diversity.

The phylogenetic tree shown in figure 5.1 is based on a calculated result with the sequences included. Trees inferred with such a diversity of sequences can accurately portray relationships between the domains, but the order of branches within the domains is likely to be inaccurate because of the small number of taxa selected for the analysis. A summary of the results of tree calculations with different methods and different suites of bacterial rRNA sequences is diagrammed in figure 5.2 (Hugenholtz et al. 1998a). The wedges indicate the radiations of the major clades, relatedness groups. These are termed "phylogenetic divisions," or "phyla." The number of known bacterial divisions has expanded substantially in recent years. The first compilation, by Woese in 1987 (fig. 5.2 inset), could include only about 12 divisions. About 40 such deeply related groups of bacteria have now been identified by rRNA sequences. Only about two-thirds of the bacterial divisions have cultured representatives (filled wedges in fig. 5.2). The remaining (open wedges) have been detected only in molecular surveys of environmental rRNA genes. Organisms that belong to these bacterial divisions without cultured members sometimes are abundant in their respective environments, and therefore, their activities are likely significant in the local biogeochemistry. Sequences that identify members of the WS6 division, for instance, are conspicuous in hydrocarbon bioremediation sites and so likely are important for that process (Dojka et al. 1998). OP11 sequences, first detected in a hot spring in Yellowstone National Park (Hugenholtz et al. 1998b), commonly are abundant in anaerobic environments (J. K. Harris, S. T. Kelley, and N. R. Pace, unpubl. obs.). The rRNA sequences thus point to areas for investigation by microbiologists.

Phylogenetic analyses of available molecular sequences, rRNA and protein, have failed to resolve convincingly any specific branching orders of the bacterial divisions. Trees produced using rRNA sequences (e.g., figs. 5.1 and 5.2) often indicate that a few of the division lineages (e.g., Aquificales, Thermotogales) branch more deeply than the main radiation, but this is possibly an artifact of the high-temperature nature of those organisms and their rRNAs. The base of the bacterial tree is best seen as a polytomy, an expansive radiation that is not resolved with the current data. It is possible that future studies will draw together some of the groups that

Figure 5.2. Diagrammatic representation of the phylogenetic divisions of Bacteria. Phylogenetic trees containing sequences from the indicated organisms or groups of organisms, chosen to represent the broad diversity of Bacteria, were used as the basis of the figure. Wedges indicate that several representative sequences fall within the indicated depth of branching. Solid wedges are represented by cultured organisms. Open wedges are represented only by environmental sequences and are named after rRNA gene clone libraries (OP, WS, TM, OS). The smaller or larger areas of the sectors correspond to smaller or larger numbers of sequences available. The scale corresponds approximately to 0.1 changes per nucleotide (Hugenholtz et al. 1998a). The inset shows the bacterial tree of the 12 phylogenetic divisions known in 1987 (Woese 1987).

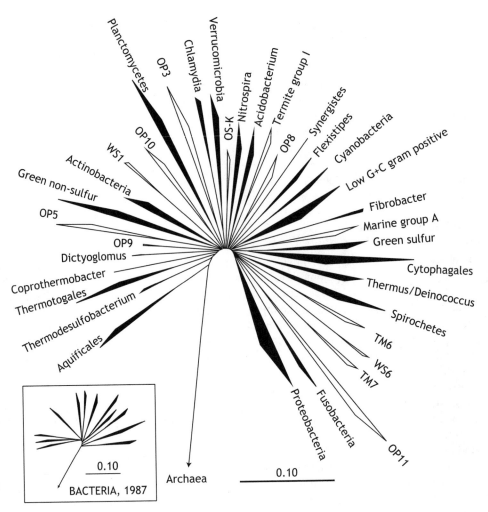

now seem to constitute division-level diversity. An important direction in this regard is the accumulation of additional sequences, particularly those that represent the entire diversity of the bacterial divisions. Broad taxon representation of sequences is required to produce the most accurate phylogenetic trees (Hillis 1998). Currently, however, as illustrated in figure 5.3, most rRNA sequences are from only a few of the bacterial divisions. Further environmental surveys with molecular methods will be the most efficient way, possibly the only way, to gather a broader information base on bacterial diversity. It is also likely that genomic studies will contribute to the resolution of the bacterial tree. For instance, the common occurrence of gene families could be evidence for a specific relationship between divisions that are not convincingly relatives within the accuracy of the rRNA trees. Although the understanding of the fine structure of the bacterial tree will improve, the current picture of the base of the tree as an expansive radiation of independent lines of genetic descent is unlikely to change.

This overall structure of the bacterial phylogenetic tree (fig. 5.2), a line of descent with no (surviving) branches and then a burst of diversifying genetic lineages, is intriguing. This evolutionary radiation surely was one of the great landmarks in biology, and the consequences of that diversification included profound modification of this planet, through the metabolic activities of the resulting organisms. What could have sparked such a spectacular radiation in the bacterial tree? One possibility is that the expansive genetic differentiation resulted when early life developed sufficient sophistication that stable, independent lines of descent could be established. Before that, the rudimentary nature of biochemical processes may have precluded the establishment of independent genetic lines of descent. Genes would have been shared by communities of replicating entities. Woese has discussed the transition between early biochemistry and the establishment of the cellular lines of descent as analogous to an annealing process (Woese 1998, 2000). Initially, mutation rates and lateral transfer would have been high. As increasingly complex and specific structures accumulated, both mutation rates and lateral transfer would have tapered off, and discrete genetic lines of descent could be established.

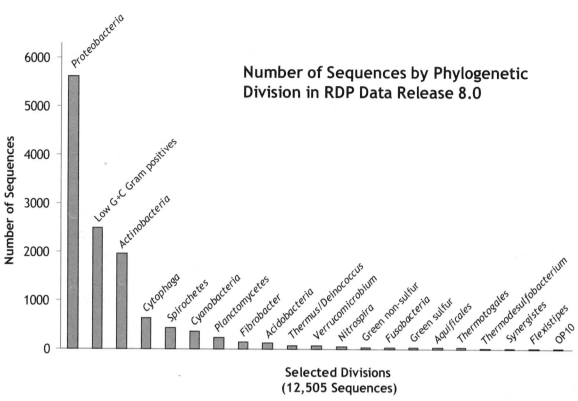

Figure 5.3. Phylogenetic distribution of SSU rRNA sequences > 500 nucleotides in length in the RDP-ARB database (http://rdp.cme.msu.edu/html/). Figure compiled by Kirk Harris.

Archaea

In 1977, at the time of the recognition that archaeans are fundamentally distinct from both bacteria and eucaryotes, only a few species of those organisms had been cultured and studied. The properties of these organisms seemed unusual. Some of the cultured species were highly anaerobic methanogens, using molecular hydrogen as an energy source and respiring with carbon dioxide, to make methane. Others thrived in saturated brine, for instance, Israel's Dead Sea, and produced a rhodopsin-like pigment akin to that in our own eyes. A third type of what became known as members of Archaea were acidophilic thermophiles, found in acidic geothermal springs. Most examples of Archaea that have been cultured since their recognition also have been obtained from those environments. Consequently, archaeans popularly have been considered restricted to environments that are "extreme" by human standards. Molecular studies have shown, however, that this perception is seriously distorted. Archaeal rRNA genes belonging to uncultured organisms are widely distributed in environments that are not necessarily extreme. Our understanding of the structure of the archaeal phylogenetic tree rests on only about 1000 rRNA sequences, about half from cultured organisms and the others from environmental surveys of rRNA genes. Relatively few environments have been analyzed for Archaea, however, so the extent of

diversity that makes up that phylogenetic domain surely is far broader than we know.

Figure 5.4 is a diagram of the known phylogenetic makeup of the domain Archaea. There are two main relatedness groups, Euryarchaeota and Crenarchaeota. A potential third deeply divergent lineage of Archaea, Korarchaeota, is represented only by environmental rRNA gene sequences, so the status of this group needs to be tested and consolidated by further studies of gene sequences and descriptions of organismal properties (Barns et al. 1996). The branch between these main evolutionary clades of Archaea are the deepest within any of the three domains. The depth of separation of Euryarchaeota and Crenarchaeota also is indicated by many biochemical properties and genomic features. For instance, even DNA is packaged differently in these two kinds of organisms: euryarchaeotes use histones to package chromatin, much as do eucaryotes, whereas crenarchaeal genomes evidently lack histone genes (Pereira et al. 1997). The mode of packaging DNA by the latter organisms is not known.

There are cultured representatives of most of the main lineages of Euryarchaeota. Molecular analyses of environmental sequences have revealed no new groups that diverge deeply in the euryarchaeal tree. In contrast, most of the known rRNA diversity of Crenarchaeota is known only from environmental sequences. All cultured crenarchaea are thermophilic and often are obtained from geothermal environ-

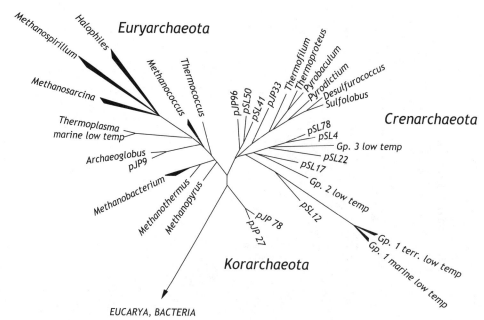

Figure 5.4. Diagrammatic representation of the phylogeny of Archaea. Wedges indicate that several representative sequences fall within the indicated depth of branching. Names correspond to organisms or groups of organisms, or environmental clones (Dawson 2000.

ments. The properties of these organisms did much to popularize the notion of archaeans as exclusively "extremophiles." It came as a surprise, then, when abundant, phylogenetically diverse crenarchaeal rRNA gene sequences were discovered in more moderate habitats ranging from shallow and deep marine waters, soils, sediments, and rice paddies, to symbionts in some invertebrates (DeLong and Pace 2001). As shown in figure 5.4, only one of the main relatedness groups in Crenarchaeota is composed of named organisms. The other groups consist of environmental organisms represented only by sequences. These otherwise largely unknown organisms are some of the most abundant creatures on Earth. In the oceans, for instance, low-temperature crenarchaea occur at concentrations of 10^7 to 10^8 cells per liter throughout the water column at all latitudes, and typically constitute 20–50% of the cells present. The niche in the global ecosystem that these organisms fill is not known. Cultured crenarchaea commonly use hydrogen as an energy source, and molecular hydrogen is pervasive in the environment at very low levels (Morita 2000). Perhaps the low-temperature crenarchaea tap this ubiquitous fuel. Although low-temperature crenarchaea have so far eluded pure culture for laboratory studies, recent developments in genome science are being exploited to learn more about them. Environmental DNA is cloned as large pieces that can be linked together and sequenced to gain further information on the organisms identified by the rRNA sequences (DeLong et al. 1999).

Eucarya

Molecular evolutionary studies of eucaryotes have relied generally on a sparse collection of gene sequences that do not represent the full range of eucaryotic diversity in nature.

As shown in figure 5.1, the most diverse eucaryotic rRNA sequences are derived from microbes. Yet, such organisms are the least known of eucaryotes and have received the least attention from molecular phylogenetic studies. More than 100,000 microbial eucaryotes, "protists," have been described (Patterson and Sogin 1993), but only a few thousand have been investigated for rRNA sequence (Sogin and Silberman 1998). Moreover, as with the collection of bacterial rRNA sequences, the collection of eucaryal sequences is heavily biased toward only a few relatedness groups. The recent addition of environmental rRNA gene sequences to phylogenetic calculations has improved the resolution of the eucaryotic tree by providing additional diversity Dawson and Pace 2002). A diagram that summarizes the phylogeny of the eucaryotic taxonomic kingdoms from the rRNA perspective is shown in figure 5.5. There is no convention for the taxonomic organization of sequence-based relatedness groups of eucaryotes. Based on various traditional or molecular classification schemes, eucaryotes have been categorized into anywhere from three to more than 70 major kingdoms. Eucaryal sequences available in the databases fall into about 30 independent relatedness clusters, the known kingdom-level relatedness groups (Dawson 2000; not all shown in fig. 5.5).

From the perspective of rRNA sequences, the overall topology of the eucaryal tree is seen as a basal radiation of independent lines of descent, one of which gave rise to other main lines, one of which culminated in the "crown radiation" of the familiar taxonomic kingdoms such as animals, plants, stramenopiles, and so forth (fig. 5.5). The specific positions of intermediate branches in the rRNA tree are only approximate, but the successive branching order is indicated by several kinds of analyses (Dawson and Pace 2002, Sogin et al. 1989). The accuracy with which the kingdom-level lines can be resolved will improve as the sequence collection available

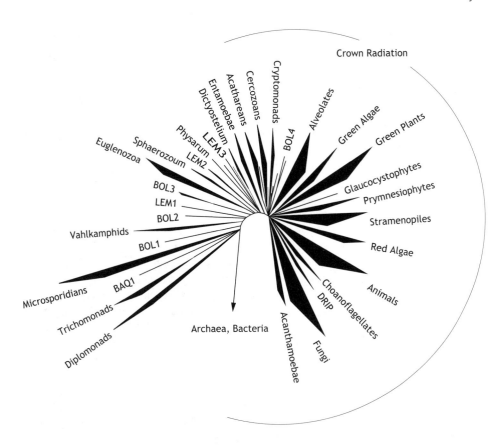

Figure 5.5. Schematic diagram of the evolution of Eucarya. The branch points of these kingdom-level groups are based on trees inferred with ED, MP, and ML and representative sequences. The areas of the wedges reflect nonlinearly the relative numbers of SSU rRNA sequences of these groups in GenBank. Groups named LEM, BOL, and BAQ are represented only by environmental rRNA gene clones (Dawson and Pace 2002).

for analysis grows. This view of successive branching in the eucaryotic tree contrasts with the results of some comparisons of protein-encoding genes, with limited phylogenetic representation (Philippe et al. 2000). Those results have been interpreted to indicate that there is no particular branching order, that the contemporary kingdom-level lines derive from a single expansive radiation analogous to the bacterial radiation (fig. 5.2). Proponents of this view have argued that extensive sequence differences between basal-derived and crown-group rRNA genes do not reflect great evolutionary distances, but rather are a consequence of relatively rapid evolution in the basal lines. Some of the environmental rRNA gene sequences branch more deeply in the tree than the crown radiation, however, and are not rapidly evolving lines. These environmental sequences punctuate the long lines between the crown and the previously identified basal divergences. The occurrence of deeply divergent eucaryotic lines with slow substitution rates (short lines) indicates that the high rates (long lines) previously ascribed to the basal divergences in rRNA trees are not the norm. Phylogenetic trees based on a single gene, SSU rRNA in this case, of course cannot reflect the genealogies of all the genes that specify organisms because of the potential influence of lateral transfer. Genes with phylogenies that are not congruent with the rRNA tree possibly have undergone lateral transfer in their evolution.

The successive radiations of the main lines of descent are significant landmarks in eucaryotic history. Correlation of cellular properties or genomic sequences with rRNA trees

may provide clues regarding the biological innovations that sparked these deep radiations. One noteworthy correlation may be the phylogenetic distribution of the major organelles, chloroplasts and mitochondria. All characterized representatives of the basal lineages of eukaryotes lack mitochondria and chloroplasts, whereas organisms of more peripherally branching groups have those organelles. As diagrammed in figure 5.6, the distribution of these organelles indicates that much of the modern diversity of eukaryotes may have been made possible by the metabolic power and light-harvesting capacity of bacteria.

Time and the Tree of Life

Because sequences of genes change with time, it seems natural to try to infer the times of branch points in evolutionary history by the extents of sequence divergence between modern genes. Indeed, molecular phylogenetic trees often are interpreted in the context of time since the divergence of particular branches. This simple correlation between time and sequence change is not well founded, however, because different lines of descent can change at different rates. This is seen in the lengths of line segments (extents of sequence change) in the three-domain tree in figure 5.1. Thus, lines leading to modern-day members of Archaea are systematically short compared with the lines leading to their sister group, modern eucaryotes. Moreover, the rate of change in sequences is

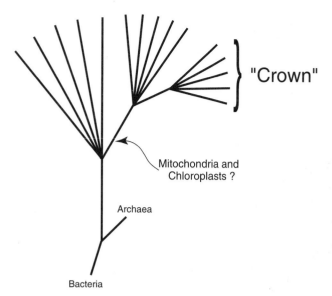

Figure 5.6. Possible pattern of eukaryotic rRNA diversification. The diagram shows the pattern of eukaryotic evolution and the incorporation of the major organelles, chloroplasts, and mitochondria. As described in the text, the organelles would have been in place more than 3.5 Byr ago.

not constant with time. This is seen in the mitochondria, which have undergone many more sequence (and other) changes than has their sister line in this tree, the line leading to the proteobacterium *Agrobacterium tumefaciens* (fig. 5.1). Thus, a sequence-based phylogenetic tree cannot be used to date events unless the tree can be calibrated by correlating a historical occurrence with some feature in the tree.

The deep evolutionary branches that gave rise to the phylogenetic domains blur into the origin of life, and their subbranches probably happened early, as well. A geological and biological correlation that may estimate one time point in the Tree of Life is the occurrence of molecular oxygen and the phylogenetic radiation of the only organisms that produce oxygen, the cyanobacteria. Although oxygen did not become abundant until 2–2.5 billion years (Byr) ago, there is evidence for oxidized iron in 3.5-Byr-old rocks (Sleep 2002). The occurrence of stromatolites in those rocks indicates that complex microbial communities had developed by that time. Moreover, the shapes of ostensible microfossils in cherts of the same age are proposed to resemble morphologically conspicuous, modern-day cyanobacteria (Schopf 1994). This presence of oxygen, bolstered by the fossil record, suggests that the cyanobacterial radiation (indicated by *Gloeobacter*, *Synechococcus*, and chloroplast in fig. 5.1) had already occurred by 3.5 Byr ago. The main bacterial divergences must have occurred even before the time of the cyanobacterial radiation. Because the phylogenetic line that led to chloroplasts originated at the base of the cyanobacterial radiation, it seems likely that chloroplasts, as well, were derived early. The branch point of a mitochondrial lineage from proteobacteria is consistent with the early appearance of that

organelle, too. Therefore, the modern kind of eucaryotic cell, with organelles, probably also arose early, more than 3.5 Byr ago. The eucaryotic nuclear line of descent is even more ancient, as old as the archaeal line.

Conclusion and Prospects

The general outlines of a universal phylogenetic tree are now in place. It is clear, however, that it incompletely portrays the breadth of biological diversity. A main reason that it is incomplete is because our understanding of microbial diversity is rudimentary. Molecular studies of environmental organisms continue to reveal major relatedness groups that were not suspected. Are there still other primary domains to be discovered? Perhaps. The methods used to hunt organisms in the environment are heavily dependent on the microbial diversity that we already know about. Are there other new bacterial divisions and eukaryotic kingdoms to be discovered? Almost certainly. Even the limited studies of microbial ecosystems so far have turned up remarkable novelty, and the complexity of those ecosystems indicates that much broader diversity will be encountered.

The complexity of the microbial world does not fit well into the call of many biologists to enumerate all of Earth's species. Microbial diversity is too broad, far too complex to be accommodated by species counts. On the other hand, a sampling and an articulation of the extent of cellular diversity can be accomplished by sequence surveys of environmental rRNA genes. The sequences reflect the kinds of organisms that they represent, and the frequencies of the phylotypes are a rough census of the microbial world. An expanded sequence representation of life's diversity also will afford more accurate molecular phylogenetic reconstructions and bring us to a closer understanding of our earliest beginnings.

Dedication

This article is dedicated to Roy Chapman Andrews, who knew that there were things to be discovered; and to the American Museum of Natural History, which gave him the opportunity to go find them.

Acknowledgments

I thank colleagues in my lab for comments that improved this article. My research activities are supported by the National Institutes of Health, the National Science Foundation, and the NASA Astrobiology Institute.

Literature Cited

Barns, S. M., C. F. Delwiche, J. D. Palmer, and N. R. Pace. 1996. Perspectives on archaeal diversity, thermophily and monophyly from environmental rRNA sequences. Proc. Natl. Acad. Sci. USA 93:9188–9193.

Dawson, S. C. 2000. Evolution of the Eucarya and Archaea: perspectives from natural microbial assemblages. Thesis, University of California, Berkeley.

Dawson, S. C., and N. R. Pace. 2002. Novel kingdom-level eukaryotic diversity in anoxic environments. Proc. Natl. Acad. Sci. USA. 99:8324–8329.

DeLong, E. F., and. N. R. Pace. 2001. Environmental diversity of Bacteria and Archaea. Syst. Biol. 50:470–478.

DeLong, E. F., C. Schleper, R. Feldman, and R. V. Swanson. 1999. Application of genomics for understanding the evolution of hyperthermophilic and nonthermophilic Crenarchaeota. Biol. Bull. 196:363–366.

Dojka, M. A., P. Hugenholtz, S. K. Haack, and N. R. Pace. 1998. Microbial diversity in a hydrocarbon- and chlorinated-solvent-contaminated aquifer undergoing intrinsic bio-remediation. Appl. Environ. Microbiol. 64:3869–3877.

Gogarten, J. P., H. Kibak, P. Dittrich, L. Taiz, E. J. Bowman, B. J. Bowman, M. F. Manolson, R. J. Poole, T. Date, T. Oshima, J. Konishi, K. Denda, and M. Yoshida. 1989. Evolution of the vacuolar H^+-ATPase: implications for the origin of eukaryotes. Proc. Natl. Acad. Sci. USA 86:6661–6665.

Hillis, D. M. 1998. Taxonomic sampling, phylogenetic accuracy, and investigator bias. Syst. Biol. 47:3–8.

Hugenholtz, P., B. M. Goebel, and N. R. Pace. 1998a. Impact of culture-independent studies on the emerging phylogenetic view of bacterial diversity. J. Bacteriol. 180:4765–4774.

Hugenholtz, P., C. Pitulle, K. L. Hershberger, and N. R. Pace. 1998b. Novel division level bacterial diversity in a Yellowstone hot spring. J. Bacteriol. 180:366–376.

Iwabe, N., K. Kuma, M. Hasegawa, S. Osawa, and T. Miyata. 1989. Evolutionary relationship of archaebacteria, eubacteria, and eukaryotes inferred from phylogenetic trees of duplicated genes. Proc. Natl. Acad. Sci. USA 86:9355–9359.

Koonin, E. V., K. S. Makarova, and L. Aravind. 2001. Horizontal gene transfer in prokaryotes: quantification and classification. Annu. Rev. Microbiol. 55:709–742.

Marsh, T. L., C. I. Reich, R. B. Whitelock, and G. J. Olsen. 1994. Transcription factor IID in the Archaea: sequences in the *Thermococcus celer* genome would encode a product closely related to the TATA-binding protein of eukaryotes. Proc. Natl. Acad. Sci. USA 91:4180–4185.

Morita, R. Y. 2000. Is H_2 the universal energy source for long-term survival? Microb. Ecol. 38:307–320.

Pace, N. R. 1997. A molecular view of microbial diversity and the biosphere. Science 276:734–740.

Patterson, D. J., and Sogin, M. L. 1993. Eukaryote origins and protistan diversity. Pp. 13–46 *in* The origin and evolution of prokaryotic and eukaryotic cells (H. Hartman and K. Matsuno, eds.). World Scientific, River Edge, NJ.

Pereira, S. L., R. A. Grayling, R. Lurz, and J. N. Reeve. 1997. Archaeal nucleosomes. Proc. Natl. Acad. Sci. USA 94:12633–12637.

Philippe, H., P. Lopez, H. Brinkmann, K. Budin, A. Germot, J. Laurent, D. Moreira, M. Muller, and H. Le Guyader. 2000. Early-branching or fast-evolving eukaryotes? An answer based on slowly evolving positions. Proc. R. Soc. Lond B 267:1213–1221.

Rowlands, T., P. Baumann, and S. P. Jackson. 1994. The TATA-binding protein: a general transcription factor in eukaryotes and archaebacteria. Science 264:1326–1329.

Schopf, J. W. 1994. The oldest known records of life: early archaean stromatolites, microfossils, and organic matter. Pp. 193–207 *in* Early life on Earth (S. Bengston, ed.). Columbia University Press, New York.

Sleep, N. 2002. Oxygenating the atmosphere. Nature 410:317–319.

Sogin, M. L., J. H. Gunderson, H. J. Elwood, R. A. Alonso, and D. A. Peattie. 1989. Phylogenetic meaning of the kingdom concept: an unusual ribosomal RNA from *Giardia lamblia*. Science 243:75–77.

Sogin, M. L., and J. D. Silberman. 1998. Evolution of the protists and protistan parasites from the perspective of molecular systematics. Int. J. Parasitol. 28:11–20.

Swofford, D. L., G. J. Olsen, P. J. Waddell, and D. M. Hillis. 1996. Phylogenetic inference. Pp. 407–514 *in* Molecular systematics (D. M. Hillis, C. Moritz, and B. K. Mable, eds.). Sinauer Associates, Sunderland, MA.

Woese, C. R. 1987. Bacterial evolution. Microbiol. Rev. 51:221–271.

Woese, C. R. 1998. The universal ancestor. Proc. Natl. Acad. Sci. USA 95:6854–6859.

Woese, C. R. 2000. Interpreting the universal phylogenetic tree. Proc. Natl. Acad. Sci. USA 97:8392–8396.

Woese, C. R., and G. E. Fox. 1977. Phylogenetic structure of the prokaryotic domain: the primary kingdoms. Proc. Natl. Acad. Sci. USA 74:5088–5090.

Woese, C. R., O. Kandler, and M. L. Wheelis. 1990. Towards a natural system of organisms: proposal for the domains Archaea, Bacteria, and Eucarya. Proc. Natl. Acad. Sci. USA 87:4576–4579.

W. Ford Doolittle

Bacteria and Archaea

The Triumph of Molecular Phylogeny

The collection of chapters in this volume and the symposium for which they were assembled celebrate one of the signal achievements of 20th century biology: the integration of molecular sequence analyses with more traditional comparative and paleontological approaches in the construction of a universal Tree of Life. *Integration* is one of the key words here. Without molecular data, we would still find it easy to tell birds from bees or to distinguish any bird or bee from broccoli, brewer's yeast, or bacteria. But we would have no strong basis for deciding, as we have (see Baldauf et al., ch. 4 in this vol.), that all birds and bees are closer kin to yeast than to broccoli. Nor would we have much reason to be as confident as we are that, despite the manifest differences in size, shape, and lifestyle, organisms in the first four groups—all eukaryotes, with nucleated cells—share a common ancestor with the nonnucleated prokaryotes (Bacteria and Archaea). For all the very deep branchings, only molecular data—in the form of DNA or protein sequence, or sometimes three-dimensional protein structure—can provide unarguable evidence for common ancestry and define lines of descent.

Unarguable is another key word. Of course, biologists have never been at a loss for theories about how one type of living thing *might* be evolutionarily related to another, and what features *might* be important for deciding this. I remember being taught in high school that brewer's yeast and other fungi were really a complex kind of bacterium, because of their shared absorptive mode of taking nutrients, cell walls, and general cellular simplicity, for instance. By the time I started college, this view had been replaced by the synthesis known as Whittaker's Five Kingdoms (Animals, Plants, Fungi, Protozoa, and Bacteria each as separate assemblages). Such theories were always fluid and arguable, because there were few commonly agreed upon grounds for formulating or proving them. One difficulty was in knowing which shared features are truly homologous (similar because they derive from such a feature in a common ancestor) and which are analogous (independently evolved for similar purposes, e.g., the wings of birds and bats, or the aquatic habits of fishes and whales). Claims for evolutionary relatedness can only be made on the basis of homologous traits. Another difficulty was in converting data about shared features (if homologous) into quantifiable measures of overall organismal similarity. How do we combine data about biochemical pathways, cellular ultrastructure, and behavior, which are so profoundly different in quality, into a single quantity measuring relatedness?

Molecular sequence data, at least at first blush, obviate both problems. There are 20^{100} possible proteins 100 amino acids long. Anything more than about 15% sequence identity between two proteins cannot be mere coincidence and is unlikely to be the result of evolution independently rediscovering the same solution twice (convergence), because one of evolutionary biology's best-learned lessons is that there are many different ways to solve the same challenge. So significant sequence similarity *can* be taken as significant evidence of homology. It is also eminently quantifiable: we have only to line two sequences up so as to optimize the match, and count the identical amino acids (or nucleotides, for an RNA or DNA sequence). These advantages of molecular sequence data were first recognized by Emile Zuckerkandl and Linus Pauling, whose 1965 papers founded the now flourishing discipline of molecular phylogeny (Zuckerkandl and Pauling 1965). Further, Zuckerkandl and Pauling argued that gene sequence data (or its direct read-out in RNA or protein sequence) deserve our attention more than features of organismal form and function, because they are *more fundamental*. DNA sequence *determines* organismal form and function, and not the other way round. Indeed, the latter contain no evolutionary information that is not encoded in the former.

Implicit in Zuckerkandl and Pauling's arguments, and embodied in the molecular phylogenetic work they inspired, was the assumption that, in picking a gene to do phylogeny with, all we needed to worry about was the ease with which it (or its RNA or protein product) can be isolated and sequenced, and the breadth of its distribution. (Hemoglobins are marvelous for doing vertebrate phylogeny, but plants and bacteria don't have them.) What we didn't have to concern ourselves with was the possibility that *different genes* in a genome might have *different phylogenetic histories*. This assumption is depicted in figure 6.1A and could be summarized as individual gene trees = genome tree = organism tree. Carl Woese made something like this assumption near the end of the 1960s, when he chose small subunit (SSU) ribosomal RNA (rRNA) as a "universal molecular chronometer,"

a stand-in for all genes. SSU rRNA was one of the few ubiquitously distributed gene products that could be easily isolated and (at least partially) sequenced at that time (Woese 1987). It still would be one of the best all around choices (see Pace, ch. 5 in this vol.).

Ironically, a strong violation of the principle illustrated by figure 6.1A was proposed by Lynn Margulis, at very nearly the same time, and provided one of the first hypotheses about deep phylogeny that the infant discipline of molecular phylogeny could cut its teeth on (Margulis 1970). She dusted off and made modern the endosymbiont hypothesis for the origin of chloroplasts and mitochondria, first proposed by Mereschowsky in the late 19th century. According to this notion, these energy-generating organelles (the first responsible for photosynthesis in all plants and algae, and the second for respiration in almost all eukaryotes) were once free-living bacteria that had become trapped in the cytoplasm of ancient eukaryotic cells, as permanent endosymbionts (fig. 6.2). In this sheltered and nutrient-rich environment, many genes useful only for independent life were lost, whereas many producing proteins still needed for photosynthesis or respiration were transferred to the nucleus (so that their products would thenceforth have to be transported back into the organelle). A few genes were retained on the tiny residual genomes found in mitochondria and plastids, however, and these could unequivocally be used to trace the evolutionary origins of these organelles.

Among such retained genes were those for organellar versions of SSU rRNA. By the mid 1970s, several groups had shown that chloroplast and mitochondrial SSU rRNA genes

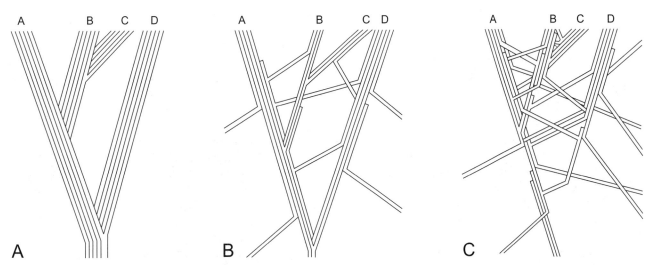

Figure 6.1. Three models for the relationships between organismal, genome, and gene phylogenies, for four imaginary species (labeled A, B, C, and D). (A) shows the "standard model": no genes are exchanged between genomes, so the gene complements of any genome can change only through loss of genes or duplication of genes, followed by divergence in sequence and function. (B) shows the "stable core": some, possibly even most, genes can be exchanged between genomes over evolutionary time, but a core of genes is immune to this process, and the (congruent) phylogenies of these genes can be used to trace organismal phylogeny, and construct the true Tree of Life. In (C), the "shifting core" model, no two genes need have the same phylogeny throughout all of life's history. Nevertheless, within restricted regions of the tree, most genes might evolve in a coherent fashion, showing congruent phylogenies.

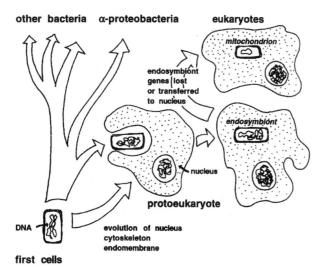

other bacteria   **α-proteobacteria**   **eukaryotes**

mitochondrion

endosymbiont
genes lost
or transferred
to nucleus

endosymbiont

nucleus

protoeukaryote

DNA

evolution of nucleus
cytoskeleton
endomembrane

first cells

Figure 6.2. The endosymbiont hypothesis for the origin of mitochondria. A respiring α-proteobacterium was acquired by a nonrespiring host (the protoeukaryote) as an endosymbiont, conferring the benefits of respiration (efficient metabolism). The endosymbiont lost genes needed for independent growth and transferred many other genes to the nucleus. A small mitochondrial genome (sometimes only a dozen genes) remains in the organelle. A similar hypothesis would have chloroplasts derive from cyanobacteria (blue-green algae). Both hypotheses are considered proven (Gray and Doolittle 1982).

were indeed of independent bacterial origin (cyanobacteria and α-proteobacteria, respectively), exhibiting phylogenies clearly different from each other (Gray and Doolittle 1982). More to the point, their phylogenies also differed from that of the nuclear-gene-encoded SSU rRNA of cytoplasmic ribosomes—a marker for the evolutionary history of the protoeukaryotic host that first harbored the symbionts (fig. 6.2). So this very important idea about cellular evolution was also the first serious counterexample to the assumption that all of an organism's genes should have the same phylogeny. Indeed, it was the fact that they don't that *proved* the endosymbiont hypothesis.

In the rest of this chapter I show that there are very many other genes like this, genes that show different phylogenies from SSU rRNA and from each other (and have nothing to do with the endosymbiont hypothesis). Within the prokaryotic domains (Bacteria and Archaea), in particular, much coding DNA can be and demonstrably has been exchanged across species, genus, phylum, or even domain boundaries—so many genes, indeed, that the pattern of relationships defined by SSU rRNA genes may not be exhibited by the majority of the genes in any genome. For prokaryotes, the appropriate model for typical relationships between gene phylogenies might look more like B or C than A in figure 6.1. This is probably not so much a problem for eukaryotes, especially complex multicellular ones, and I *will* confine myself to the topic assigned me, Bacteria and Archaea. But because there seems to be so much gene sharing between the two, my title might more appropriately have been Bacteriandarchaea.

None of this necessarily means that Darwin was fundamentally wrong, or that the concept of a unique and universal *organismal* Tree of Life is passé, or that—if certain assumptions hold—rRNA does not track this tree best. But there is not a unique universal *genomic* tree, and we need to develop more sophisticated (but also much more interesting and exciting) ways of thinking about what we mean by *the* Tree of Life.

Superbugs, Drugs, and Lateral Gene Transfer

The mid 1960s also saw the discovery of lateral gene transfer (LGT), the process (or rather, collection of processes) underlying microbial gene sharing. Infectious disease microbiologists, mostly in the United States and Japan, found that the rapid rise of resistance to commonly (and often excessively) used antibiotics among human pathogens (especially in hospitals) was not due to the expected Darwinian mechanism of random mutation followed by natural selection (Falkow 1975). Instead, genes determining resistance to antibiotics (by a variety of mechanisms) had been recruited from preexisting natural reservoirs and were being passed around among pathogens on small circular DNA molecules (plasmids), themselves well adapted to spreading infectiously between bacterial species (fig. 6.3). *Selection* is still involved— pathogens receiving the resistance-conferring plasmids produce more progeny because they have them. So the process is Darwinian. But it was not mutations occurring within genes within species, but whole genes (or suites of genes) transferred across species boundaries, on which selection was acting. Indeed, we now know that plasmids can carry several different genes for resistance to several different kinds of antibiotics simultaneously, and that special mechanisms and genetic devices (insertion sequences, transposons, and integrons, to give some names) have evolved to facilitate the assembly and transmission of such genes (Bushman 2002).

We were also soon to learn that antibiotic resistance determinants were not the only kinds of coding sequences that plasmids could carry. Clusters of several genes involved in the synthesis of unusual and inessential metabolites or the degradation of unusual and rarely available substrates were also exchanged in this way. Two Canadians (Sorin Sonea and Maurice Panniset) and an Australian (Darryl Reanney) soon constructed a bold if inchoate theory on this foundation (Sonea and Panniset 1976, Reanney 1976). They asserted that because of between-species gene transfer—mediated not only by plasmids but also by bacterial viruses (phages) and through cell-to-cell contact (conjugation) or DNA uptake (transformation)—all bacteria might be viewed as one species, responding to environmental challenges (over evolutionary time) as a single "global superorganism." As I recall it, these claims were widely dismissed during the 1970s and

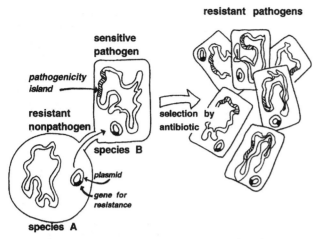

Figure 6.3. Bacterial antibiotic resistance genes found on plasmids have been the major cause of the rise in drug-resistant "superbugs." Their spread is one form of LGT. Also, genes for many functions related to pathogenicity are clustered in transferrable regions of bacterial chromosomes.

1980s—they were so hopelessly radical! Most of the genes then known to be transferred by plasmids could be viewed as somehow "specialized" and, under most circumstances, dispensable. Genes for core informational functions (replication, transcription, and translation) were not known to be subject to LGT, nor were genes of basic and widely conserved metabolic pathways. So LGT was seen as a genetic add-on, not a fundamental evolutionary force. It might even have appeared on the scene recently, as the microbes' way of coping with human activity, namely, antibiotic use and the flooding of microbial environments with many unusual pollutants, some highly toxic but some of novel nutritional value (for bacteria).

Pathogenicity (and Other) Islands

As we acquired the ability to characterize and especially to sequence longer and longer stretches of DNA, however, we could begin to see that still much more complex genetic packages could be delivered across species boundaries by LGT. And chromosome as well as plasmids could harbor the transferred genes. In particular, pathogenic bacteria often differ from harmless relatives by the possession of large functionally specialized clusters, called pathogenicity islands, some containing more than 100 genes (Hacker and Kaper 2000). These include virulence factors of many sorts, facilitating survival within, protection from, or attack on the host, as well as genes promoting the islands' transfer as units. Often, pathogenicity islands are inserted within a particular type of chromosomal sequence (a gene for transfer RNA) and have different compositional characteristics (relative composition of G, C, A, and T) than the surrounding genes (fig. 6.3). Most cogently, the genes of which they are composed may be found

in very similar form in very distantly related bacterial (or even archaeal) genomes, but not in the pathogen's closest relatives. Clearly, they have been transferred into the genomes in which we find them, although we don't generally know the transfer mechanism. So, very complex and important (for bacteria and for us) suites of biochemical/physiological/behavioral characteristics can be acquired in "one fell swoop" by LGT. And recently, we've come to realize that there are also "symbiosis islands" (promoting cooperation with hosts), "saprophytic islands" (facilitating decay), and "ecological islands" (metabolism in unusual circumstances).

Genomic Diversity: The Iceberg of Which Phylotypic Diversity Is but the Tip

Still, resistance factors and complex multigene determinants of interactions (benign or malign) with hosts and environments might be seen as "specialized." Surely, they constitute no serious threat to our understandings of the evolutionary histories of the everyday genes comprising the bulk of most genomes, or to our ability to reconstruct the universal tree using a nontransferrable marker, like SSU rRNA.

Genomics and, in particular, the appearance of complete bacterial and archaeal genomic sequences now call even this view into question. More than 100 such sequences will soon be publicly available, and these will demolish the notion that genomes in general contain just a few genes (or gene clusters) of foreign origin, and these only for specialized functions. Particularly striking are the comparisons that can be drawn between different isolates of the very same bacterial species. Consider for instance *Escherichia coli*, the laboratory workhorse of molecular biologists and biotechnologists for the last five decades. The complete genome sequence of K12, their favorite strain, was reported in 1997 (Blattner et al. 1997). Many of its 4405 genes were already familiar from genetic experiments or piecemeal gene sequencing studies. The community therefore thought that it had this species under wraps, genomically—until four years later, when the genome of another *E. coli* isolate, O157:H7, was completed (Perna et al. 2001) This is the strain that first attracted popular attention in 1993 through the death of three young customers of a fast-food restaurant in California, and two years ago killed seven drinking from contaminated wells in Ontario. The sequencing showed that it has 1387 genes that K12 doesn't have, whereas K12 itself has 528 genes not found in O157:H7—numbers corresponding to 26% of the genome of O157 and 12% of K12's. Many of these differences can only be explained by LGT, verifiable through similarity to homologous genes in evolutionarily distant bacteria (or even archaea) and, most persuasively, through the construction of phylogenetic trees for each gene. These many differences are also clearly the consequence of many different LGT events, not just the acquisition of a few large pathogenicity islands. In fact there are 177 physically separated "O islands" (genes or gene clusters present in O157 but not K12) and about 234 "K islands." Although many of the strain-specific genes of O157:H7 are likely to be specialized determinants of virulence, many are not. They encode seemingly pedestrian microbial functions (e.g., carbohydrate transfer, glutamate fermentation, or aromatic compound degradation).

Preliminary data for other *E. coli* strains show the O157:H7 versus K12 difference to be typical, not aberrant. Similar studies based on similar information on other pathogens produce similar results. Strains of the same "species" often differ from each other by up to 25% in gene content. Simple logic (with the assumption that, on average, bacterial genomes are getting neither larger nor smaller) dictates that about half of this difference can be attributed to acquisition of new genes by one or the other strain, after their joint separation from a common ancestor. (The other half could be explained by loss, from one or the other strain, of genes present in that ancestor.)

We know about genomic variability in pathogens because it is easy to obtain funding to study the biology of pathogens. Data on nonpathogens are scant. Recently Camilla Nesbø in my lab, with Karen Nelson at The Institute for Genomic Research, has been looking at genomic diversity within *Thermotoga maritima*, a nonpathogen *par excellence*. This hyperthermophilic bacterium grows best at 80°C and was isolated from the seafloor in a geothermal area near Vulcano, Italy. Preliminary data suggest that here, too, there will be something like 20% variability in gene content, between otherwise very similar isolates. If this turns out to be generally true for "environmental microbes" (including Archaea), then we cannot explain away within-species genomic variation as a by-product of intense host–parasite warfare: we must accept it as a fact of prokaryotic life. We must also accept, then, that the microbial world is *even more* wildly diverse than those who use "phylotyping" (amplification and sequencing of SSU rRNA genes from environmental DNA samples; see Pace, ch. 5 in this vol.) have already told us. Such studies have revealed, through a plethora of new twigs on the branches of the SSU rRNA tree, a hitherto unimaginable diversity of relatives of known groups. They have also led to the discovery of completely new groups, without previously known relatives. For each isolate identified by a single SSU rRNA sequence ("phylotype"), however, there may now be many more genomic variants, differing in their content of truly different (nonhomologous) genes by more than, say, *the genomes of all the animals*. (Animals *do*, of, course vary in gene content, but through duplication and functional divergence of genes they already have, or through gene loss—scarcely ever through the introduction of genuinely novel genes by LGT.)

How Much Exchange over Life's Whole History?

There is no easy way to know how old any bacterial species is, or (which is almost the same question) how long strains within a species have been diverging—and surely there is no uniform age. Howard Ochman and Isaac Jones estimate that various *E. coli* strains began to diverge about 25–40 Myr (million years) ago, based on an often quoted but largely unverified estimate of the divergence of *Escherichia* from *Salmonella* at 100–150 Myr ago (Ochman and Jones 2000). In contrast, *Yersinia pestis*, the cause of plague, may be only a few thousand years old (Achtman et al. 1999)! But however ancient bacterial species in general may be, their ages *will* be dwarfed by that of life itself. So, if 10–20% of a genome can "turn over" because of LGT and gene loss within (generously) 100 Myr, what fraction would we expect to have been affected by LGT over 3.8 billion years? No one thinks that all genes are equally exchangeable, but still it *is* reasonable to ask what fraction of any contemporary genomes' genes has been affected by LGT. There are several ways one might try to do this.

Ochman and Jeff Lawrence look at basic compositional features of genes, in particular, the relative frequencies of A, T, G, and C and the choice among alternative codings for the same amino acids (Ochman et al. 2000). Prokaryotic species differ significantly in these parameters, which tend to be similar within a genome. Thus, a *recently transferred* gene might "stick out like a sore thumb" from the surrounding long-term residents. (With time—perhaps a few hundred million years—genome-specific mutational and selectional pressures will attenuate and ultimately erase the differences.) With analyses based on these premises, Ochman and collaborators find foreign gene contents from 0.0% (for *Mycoplasma genitalium* or *Rickettsia prowazecki*, intracellular human parasites) to 16.6% for the cyanobacterium *Synechocystis*, with *E. coli* boasting 12.8% transfers.

Eugene Koonin and his colleagues employ a completely different method (called BLAST) that makes all possible pairwise comparisons between each of a genome's genes and all homologous genes in other genomes (or the larger databases), and calculates sequence similarity (Koonin et al. 2001). Genes that have greatest sequence similarity to genes in species that are distant on the rRNA tree (rather than to genes in species that are close) are likely transfers. The most easily detected transfers would be those involving the greatest distances: genes in an archaeal genome that are most similar to homologs in the bacterial domain, and vice versa. Koonin finds up 15.6% interdomain transfer (for an archaean, *Halobacterium salinarum*). Rumor in the field now has it that similar analyses will show that one-third of the genes in the yet-to-be-published genome sequence of the methane-producing archaean *Methanosarcina mazei* are of bacterial provenance—an astonishing result!

The third and best way to assess a genome's origins is to construct phylogenetic trees for each of its genes, by state-of-the-art methods. For many individual genes, compelling cases can be developed. My favorite example is the gene for HMGCoA reductase (3-hydroxy-3-methylglutaryl coenzyme A reductase), a key enzyme in the synthesis of isoprenoid compounds (sterols, e.g.) in all three domains (and the target of the statins that many people take to reduce endogenous cholesterol synthesis). Our attention was first drawn to HMGCoA reductase because BLAST analyses showed that the version of this gene in *Archaeoglobus fulgidis* (a hyperthermophilc archaean sometimes found in undersea oil wells) was very like homologous genes in bacteria and unlike the versions found in other Archaea. In fact, most Archaea have an HMGCoA reductase very similar to that of eukaryotes, so for them statins are antibiotics! A tree prepared by Yan Boucher for HMGCoA reductases (fig. 6.4) not only confirmed this result but identified other transfers—Bacteria to *Giardia intestinalis* (a single-celled pathogenic eukaryote), Archaea to *Vibrio cholerae* (a bacterial pathogen), and Archaea to *Streptomyces* species (bacteria that produce antibacterial antibiotics). Gene-by-gene analyses are time consuming, because human judgment is still often required. Less reliable but very rapid programs for preparing, by simple automatic methods, all the trees for all the genes in a genome are being developed. That of Thomas Sicheritz-Ponten and Siv Andersson shows, not unlike Koonin's BLAST studies, interdomain (Bacteria to Archaea or Archaea to Bacteria) transfers amounting to up to about 20% of a genome (Sicheritz-Ponten and Andersson 2001).

Is this about the limit? Are 70–80% of most genomes well behaved in the long-term evolutionary sense, as well as the short? Probably not. Foreign gene estimates are all likely to be underestimates. Ochman's analyses, for instance, can only look back a few hundred million years. Koonin's and Sicheritz-Ponten's results described interdomain transfers (Bacteria to Archaea or vice versa). Because Bacteria and Archaea have dissimilar gene expression machinery and control signals, genes transferred between them should often be poorly read. Harder to detect, intradomain transfers should be much more frequent.

Hunting Down the Core

There is another way to skin this cat. Instead of asking what fraction of genes in a given genome have clearly *different histories* than the majority (or than SSU rRNA), we can ask if we can find, by comparing all genomes, a stable core of shared genes (fig. 6.1B) that have the *same history*. There is a general belief that such a core should exist, based on a hypothesis and an observation.

The *hypothesis*, first articulated by Woese when he decided to settle on SSU rRNA as a "universal molecular chronometer," has come to be called the "complexity hypothesis" (Jain et al. 1999). The idea is simple: genes whose protein (or RNA) products must interact in the cell will *coevolve*. Mutations that affect the structure of one gene product (call it A) will be compensated by mutations that affect another, interacting, gene product (B) in a compensatory way, so that the essential interactions between A and B are preserved throughout the evolutionary history of a species or lineage.

Meanwhile, in another, related lineage, the homologous gene products A' and B' will also be coevolving, but likely along a somewhat different path. If the B gene of the first lineage were replaced by the B'> of the second lineage, there might be problems: the A gene product might not interact as effectively with the B' product (and similarly, A' might not be effective with B). This seems a very reasonable conjecture, and the corollary—that genes involved in even more complex interactions (A + B + C + D + E . . .) should be very hard to exchange for homologous genes in different lineages, without detriment to growth—seems inescapable.

SSU rRNA is the central part of an enormously complex structure, the ribosome. This factory for translation (the RNA → protein part of DNA → RNA → protein) also requires two other RNAs and more than 50 proteins, in order to do its vital and always essential job. The complexity hypothesis would predict that the genes encoding these RNAs and proteins could not be transferred across even very short evolutionary distances. Similarly, the various genes encoding the machineries of transcription (DNA → RNA) and replication (copying of DNA) should be hard to transfer. Certainly, it is the case that the genes identified as foreign in individual sequenced bacterial or archaeal genomes are seldom genes of these informational classes. But there *are* now several reliable reports of transfer of "informational genes," especially those involved in translation and (in a few cases) SSU rRNA itself (Yap et al. 1999)!

The *observation* on which confidence in a stable core rests is what some of us call "coherence." *Many* individual genes, when known from a sufficient number of species, do re-create the same major groups—Archaea (and within them euryarchaeotes and crenarchaeotes) or Bacteria (and within them the known bacterial phyla, such as cyanobacteria, α-, β- and γ-proteobacteria and so forth). There is no published systematic survey that says how many "many" is, however, or that compares a large number of well-resolved trees for congruence of topology. And few genes agree on branching order of bacterial phyla (even though they do distinguish Bacteria and Archaea). Pace (ch. 5 in this vol.) suggests that the poor resolution at the base of the bacteria bespeaks a rapid radiation some 3.5 billion years ago, perhaps caused by a key innovation. This is one explanation but not the only.

Surely the most rigorous test of the stable core idea would be to compare all bacterial *and* archaeal genomes, distill out the set of genes of which all genomes have a copy, make trees, and tally up how many subscribe to which topology. Efforts to do this have failed: there are very few genes shared by all genomes (even all bacterial or all archaeal genomes)—perhaps 50 or fewer (Teichman and Mitchison 1999). Few of these genes give statistically robust trees, so we simply cannot say whether their topologies are congruent or not. The assumption that there might be a stable core of genes for all prokaryotes is not disproved by this, but neither is it proven: it remains a hypothesis. In an effort to test the stable core idea on a more limited basis, we looked at the core of genes

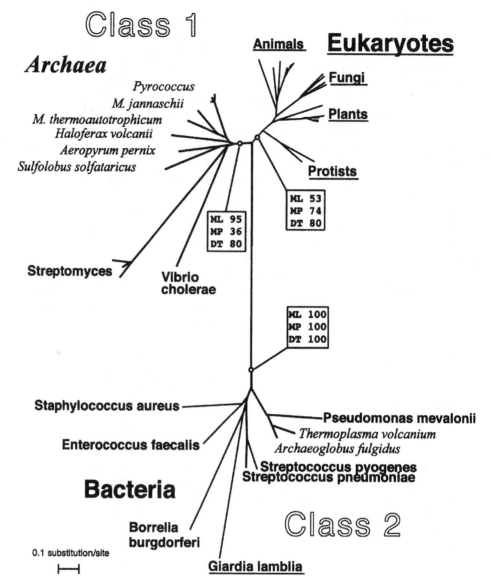

Figure 6.4. Phylogeny of genes encoding HMGCoA reductase, a key enzyme in the synthesis of sterols and related lipids. The predominant bacterial form (class 2) and predominant eukaryotic/archaeal form (class 2) are unquestionably homologous but with different functional characteristics. Four LGT events are very strongly supported by the phylogenetic analysis. The boxed numbers are bootstrap values, measures of statistical robustness, for a tree obtained by maximum likelihood, maximum parsimony, and distance methods. Archaeal names are italicized, eukaryotic names are underlined, and bacterial names are in regular letters.

shared by four sequenced eukarychaeotes, asking if these all produced the same tree (Nesbø et al. 2001). Several hundred genes could be looked at and, because there are only three unrooted phylogenetic trees for four taxa, easily scored for agreement or disagreement. It turns out that each of the three possible trees is significantly represented among the 263 shared genes we looked at. In other words, although there is a core of genes shared by the four genomes, it does not seem to be a *stable* core. The shared genes often appear to have different phylogenetic histories. This could mean that genes are not infrequently replaced by homologous but possibly quite different versions of themselves, transferred in across species lines.

So it is not possible to prove that there is any sizable stable core, even within a relatively restricted group such as the euryarchaeotes. Hervé Philippe and collaborators have tried another approach (Brochier et al. 2002). Individual trees constructed for 57 translational proteins shared by 45 bac-

terial species mostly disagree, as expected: there is too much noise and too little phylogenetic signal. But if they strung all gene sequences together to obtain one concatenated sequence, then a statistically robust tree could be obtained, and 44 of the 57 genes did not significantly contradict this result. (The 13 others showed significant evidence for transfer.) So perhaps these comprise a true core for all of Bacteria. But 44 is but a few percent of the number of genes in a typical bacterial genome. And when Brown and collaborators (2001) included members of Archaea in a similar study, they were obliged to reduce the apparent stable core even further, to only 14 genes. Woese may be correct in asserting, "An organismal genealogical trace of some kind does seem to exist . . . but that trace is carried clearly almost exclusively in the componentry of the cellular information processing systems" (Woese 2000:8393). However, when it comes to prokaryotes, and the deepest branches of the universal tree, proving even this modest claim is surprisingly difficult!

Other Models

Absence of evidence is not evidence of absence. A conservative summary of what I've said so far is that the existence of a stable core is *hard to prove*. The signal-to-noise ratio in the data we need to decide about events occurring three and more billion years ago is too low, and our methods are still too crude. "Hard to prove" is not "disproven." But all parties to the debate now accept that the core of genes that has been stably associated in all prokaryotic genomes since the first genome is far smaller than we used to think. And, just maybe, there might be no such core.

What if there weren't? Could there be some other model than those depicted in figure 6.1, A and B, to explain the undeniable fact that we can classify bacteria and archaea into groups that have many shared defining features—that the entire edifice of Linnaean hierarchical classification has been more or less successfully imposed on microbial systematics? Jeff Lawrence, Peter Gogarten, and I have been working on such a model, which is still in the verbal stages (no formal mathematics) and has as yet no fixed name (Gogarten et al. 2002). Here I call it the model of the "shifting core" or, alternatively, the model of "nested gene pools." In fact, it's not much different from what Woese himself now believes (Woese 2000), although we would probably disagree on the values of its parameters.

Imagine that all genes are potentially exchangeable but that the frequency or likelihood of exchange varies tremendously. Many factors would affect this. Complexity of interactions of the gene's product, and whether or not it was genetically linked (and so could be co-transferred) with other interacting genes would be important factors, related to the genes themselves. So would essentiality: genes that must always be present can only be replaced through an intermediate stage in which both the originally resident and the incoming foreign gene are found in the same genome. (Such intermediates are well known.) Biochemistry of the donor and recipient organism would be a key determinant. Transferred genes for various components of the photosynthetic apparatus are only likely to be of any use to species that already do photosynthesis. If of no use, transferred genes will soon be lost and we will never know that a transfer occurred. Similarly, the differences in gene expression systems between Bacteria and Archaea must reduce the frequency of successfully fixed transfers between them. Environmental niche matters, too: genes from thermophiles make proteins that work best in other thermophiles. Finally, donors and recipients must be found in close proximity in nature, and physical and genetic mechanisms to pass DNA between them (including "accidental" mechanisms) must exist.

Imagine that we ourselves create hundreds of different bacterial species, with genes and genomes made from scratch by machine, and then set them up in various niches and allow them to transfer genes according to such rules. Although there would initially have been no deep "phylogenetic" relationships between these human-made species or their genes, patterns of shared genes and similarities in sequences would eventually emerge, because of recurring transfers at different frequencies. In other words, LGT itself can create and maintain the patterns we seek to explain by the model depicted in figure 6.1A, but the underlying process would be as shown in figure 6.1C. According to this model, organisms that exchange genes most frequently would comprise "species." Different species whose organisms share genes somewhat less frequently would comprise genera, and so on up the Linnaean ladder. Bacteria are coherent as a domain because they more frequently exchange genes with other bacteria than with members of Archaea (and vice versa), but still, interdomain transfer does occasionally happen.

This model may not be correct in its extreme form (no stable core at all), but something like it must apply in the long run to most of the genes that make up prokaryotic genomes. In the short run (corresponding to the divergence of strains in a species or species in a genus, perhaps), it may most accurately describe only the 20% of a genome's worth of genes that are found in some genomes but not others. [However, recombination within genes—which I have not discussed—may have a similarly confounding effect, at this level (see Maynard Smith et al. 2000).]

The One True Tree

Darwin did describe the relationships of all organisms as a tree and thought that the patterns of similarities and differences between all contemporary species could be explained as the result of successive bifurcative speciation events, going back to one, or just a few first living things. If we had a videotape of all that (and 3.8 billion years to sit down and watch it!), we could trace all the bifurcations, and that tracing would be the universal Tree of Life. But there is no video, so we have been trying to reconstruct these bifurcations by comparing the sequences of genes, initially on the assumption that any gene would in principle do, but more recently with the belief that only some genes will tell the true story. But even if none do, and figure 6.1C shows how genomes truly evolve, the situation need not be seen as hopeless. Some kind of consensus of the phylogenies of all genes of all genomes, weighted perhaps in favor of those least frequently transferred, might still have a good chance of recreating the pattern of speciation events recorded on our imaginary videotape. We don't yet know how best to make such consensus phylogenies. Some investigators want to call them "genome phylogenies," a misleading term, I believe. Frequent LGT does not mean there is no single true universal Tree of Life for organisms, only that reconstructing this tree has become more problematic. But frequent LGT does mean that there is no single true universal *tree of genomes*, because these are made up of parts that have different phylogenies!

Cold Comfort to Creationism

Advocates of Biblical interpretations of life's history and proponents of "intelligent design" like to cite disagreement within the evolutionary community and, in particular, claims to have "overthrown Darwin" as support for their views. Therefore, early publications asserting that evidence for extensive LGT was "uprooting the Tree of Life" have found popularity with them. Perhaps some of us (especially me) were not careful enough in stating that what was being uprooted was the tree of *genomes*. Our acceptance of the video version of the *organismal* tree remains steadfast, regardless of problems in constructing it.

Even so, there is a challenge to Darwinism, as it has itself evolved over the last century. Darwinists (more properly, neo-Darwinists) see adaptation happening as the result of selection among mutations that have arisen in genes within populations of species, and speciation as most commonly the result of divergent (and ultimately incompatible) adaptations being fixed in different populations. Explicitly or implicitly, figure 6.1A is the model of genome evolution *most* compatible with this neo-Darwinian view. This, I assert, is what Darwin himself would have expected, had he lived to see the centenary of the publication of *The Origin of Species*. If adaptations are instead often due to acquisition of genes from different species, then figure 6.1C might the more relevant model. I'd hope that Darwin, had he hung on for still another half century, would have found this at least amusing and recognized the profound difference.

In any case, what does it matter what Darwin would think? Evolutionary biologists are committed to materalistic, nonsupernatural explanations of the patterns of similarity and difference we see in the living world, not to the correctness of Darwin's own particular explanations. If we substitute one materalistic, nonsupernatural explanation for another, this is a sign of paradigmatic health, not weakness. Sometimes I think we ourselves forget this, and defend Darwin and neo-Darwinism (and, indeed, the gene-based Tree of Life) as if they were received truth, not provisional interpretations of a fascinatingly complex world. We should stop doing that!

Literature Cited

Achtman, M., K. Zurth, G. Morelli, G. Torrea, A. Guiyole, and E. Carniel. 1999. *Yersinia pestis*, the cause of plague, is a recently emerged clone of *Yersinia tuberculosis*. Proc. Natl. Acad. Sci. USA 96:14043–14048.

Blattner, F. R., G. Plunkett, III, C. A. Bloch, N. T. Perna, V. Burland, M. Riley, J. Collado-Vides, J. D. Glasner, C. K. Rode, G. F. Mayhew, et al. 1997. The complete genome sequence of *Escherichia coli* K-12. Science 277:1453–1474.

Brochier, C., E. Bapteste, D. Moreira, and H. Philippe. 2002. Eubacterial phylogeny based on translational apparatus proteins. Trends Genet. 18:1–5.

Brown, J. R., C. J. Douady, M. J. Italia, W. E. Marshall, and M. J. Stanhope. 2001. Universal trees based on large combined protein sequence datasets. Nat. Genet. 28:281–285.

Bushman, F. 2002. Lateral DNA transfer. Cold Spring Harbor Laboratory Press, Cold Spring Harbor, NY.

Falkow, S. 1975. Infectious multiple drug resistance. Pion Ltd., London.

Gogarten, J. P., W. F. Doolittle, and J. G. Lawrence. 2002. Prokaryotic evolution in the light of gene transfer. Mol. Biol. Evol. 19:2226–2238.

Gray, M. W., and W. F. Doolittle. 1982. Has the endosymbiont hypothesis been proven? Microbiol. Rev. 46:1–42.

Hacker, J., and J. B. Kaper. 2000. Pathogenicity islands and the evolution of microbes. Annu. Rev. Microbiol. 54:641–679.

Jain, R. C., M. C. Rivera, and J. A. Lake. 1999. Horizontal gene transfer among genomes: the complexity hypothesis. Proc. Natl. Acad. Sci. USA 96:3801–3806.

Koonin, E. V., K. S. Marakova, and L. Aravind. 2001. Horizontal gene transfer in prokaryotes: quantification and classification. Annu. Rev. Microbiol. 55:709–742.

Margulis, L. 1970. Origin of eukaryotic cells. Yale University Press, New Haven, CT.

Maynard Smith, J., E. J. Feil, and N. H. Smith. 2000. Population structure and evolutionary dynamics of pathogenic bacteria. Bioessays 22:1115–1122.

Nesbø, C. L., Y. Boucher, and W. F. Doolittle. 2001. Defining the core of nontransferable prokaryotic genes: the euryarchaeal core. J. Mol. Evol. 53:340–350.

Ochman, H., and I. B. Jones. 2000. Evolutionary dynamics of full genome content in *Escherichia coli*. EMBO J. 19:6637–6643.

Ochman, H., J. G. Lawrence, and E. A. Groisman 2000. Lateral gene transfer and the nature of bacterial innovation. Nature 405:299–304.

Perna, N. T., G. Plunkett, III, V. Burland, B. Mau, J. D. Glasner, D. J. Rose, G. F. Mayhew, P. S. Evans, J. Gregor, et al. 2001. Genome sequence of enterohaemorrhagic *Escherichia coli* O157:H7. Nature 409:529–532.

Reanney, D. C. 1976. Extrachromosomal elements as possible elements of adaptation and development. Bacteriol. Rev. 40:552–590.

Sicheritz-Ponten, T., and S. G. Andersson. 2001. A phylogenomic approach to microbial evolution. Nucleic Acids Res. 29:545–552.

Sonea, S., and M. Panniset. 1976. Manifesto for a new bacteriology. Rev. Can. Biol. 35:103–167.

Teichman, S. A., and G. Mitchison. 1999. Is there phylogenetic signal in prokaryotic proteins? J. Mol. Evol. 49:98–107.

Woese, C. R. 1987. Bacterial evolution. Microbiol. Rev. 51:221–271.

Woese, C. R. 2000. Interpreting the universal phylogenetic tree. Proc. Natl. Acad. Sci. USA 97:8392–8396.

Yap, W. H., Z. Zhang, and Y. Wang. 1999. Distinct types of rRNA operons exist in the genomes of the actinomycete *Thermomonospora chromogena* and evidence for horizontal transfer of an entire rRNA operon. J. Bacteriol. 181:5201–5209.

Zuckerkandl, E., and L. Pauling. 1965. Evolutionary divergence and convergence in proteins. Pp. 97–166 *in* Evolving genes and proteins (V. Bryson and H. J. Vogel, eds.). Academic Press, New York.

Hervé Philippe

The Origin and Radiation of Eucaryotes

The inference of the universal Tree of Life has been a major quest in biology since the publication of the theory of evolution by Charles Darwin in 1859 (Darwin 1859). The first attempt was done by Haeckel seven years later (Haeckel 1866). Yet, although this early phylogeny still appears reasonable, progress toward the resolution of the universal tree remained elusive for decades. This was in part because of the lack of rigorous method (the famous "art" of taxonomy) but was greatly resolved by the German entomologist Willy Hennig through the development of the so-called cladistic method (Hennig 1966). Indeed, the main difficulty was the scarcity of morphological characters (*sensu lato*, e.g., including ultrastructural or biochemical). The best example of this difficulty is provided by the study of prokaryotes. After many years of trials, Stanier and Van Niel were forced to conclude that "any systematic attempt to construct a detailed scheme of natural relationships becomes the purest speculation . . . the ultimate scientific goal of biological classification cannot be achieved in the case of bacteria" (Van Niel 1955:5). Similar difficulties, albeit to a lesser extent, were encountered for the phylogeny of unicellular eucaryotes (protists; Taylor 1978).

The discovery that molecular data (protein, and later, DNA sequences) contained information about the history of the organisms harboring them has revolutionized the field of phylogeny (Zuckerkandl and Pauling 1965). Until the 1980s, sequencing remained a limiting factor and reduced the impact of molecular phylogeny. Only the study of ribosomal RNA (rRNA), first through oligonucleotide catalogs

and then through sequencing, allowed the construction of the universal Tree of Life (Woese 1987, Woese and Fox 1977). The main achievement was the proposal that prokaryotes should be divided into two groups, called domains, the Bacteria (Eubacteria) and the Archaea (Archaebacteria). A short time later, following the suggestion of Schwartz and Dayhoff (1978), two groups located the root of the universal Tree of Life through the use of anciently duplicated genes [i.e., elongation factors (Iwabe et al. 1989) and ATPases (Gogarten et al. 1989)]. The root fell within the bacterial branch, making Archaea and Eucarya sister groups, rendering the prokaryotes paraphyletic. Quite surprising, the quest for the universal Tree of Life, which has been very elusive for more than a century, was considered as generally solved thanks to the molecular phylogenetic studies of the 1980s. In 1990, a rooted universal tree was published (Woese et al. 1990), and since then it has generally been used as the reference tree in textbooks and review papers.

The fact that scientists consider this question as fairly solved is very peculiar. Indeed, microbiologists have shown that the majority of biochemical, physiological, or morphological characters each tell a different story about the relationships among prokaryotes (Van Niel 1955). This is to be expected for organisms that evolved over billions of years, given it is also true for organisms that diversified much more recently (e.g., mammals, birds, or angiosperms). The use of molecular data clearly allowed systematists to increase the number of informative characters, but not to avoid the in-

95

herent difficulty of inferring ancient events. The first molecular phylogenies, which are often quoted for showing the efficiency of the method, contain serious and indisputable errors. I will discuss only the most famous example: the phylogeny of eucaryotes based on cytochrome c (Fitch and Margoliash 1967). In this tree, primates emerge at the base of the mammals, well before the marsupials, and snakes at the base of amniotes, far from their generally accepted position (diapsids, represented by turtle and birds). Thus, despite the known theoretical and practical difficulties of inferring the universal Tree of Life, a phylogeny based on very few data (mainly 1000 positions for rRNA) was perceived as an accurate estimate.

At least three major problems have recently challenged this universal tree. First, the discovery of many uncultured organisms through molecular ecology techniques has generated many new phyla, especially in prokaryotes (see Pace, ch. 5 in this vol.). Second, lateral gene transfer (LGT) between distantly related organisms has been revealed as a much more common phenomenon than previously thought (Koonin et al. 2001). Even if one can demonstrate that tens of genes share the same historical pattern within Bacteria (Brochier et al. 2002) and Archaea (Matte-Tailliez et al. 2002), LGT raises serious questions about our view of prokaryotic evolution (see Doolittle, ch. 6 in this vol.). Third, the impact of tree reconstruction artifacts is not negligible, and in this chapter I focus on this problem. After a brief overview of the Tree of Life based on rRNA (Woese et al. 1990), I discuss the most frequent artifacts and provide a brief explanation of their causes. Then, I will detail the case of the bacterial phylogeny based on rRNA. This will allow pinpointing the sections of the current universal Tree of Life that are likely incorrect. After summarizing recent progress toward their resolution, I present my personal view of the universal Tree of Life and its implication for the origin of eucaryotes.

The rRNA Tree

The rRNA tree (fig. 7.1) is so well known that I will only discuss a few points. The advantages of rRNA as a universal marker are enormous (Woese 1987): (1) universality, (2) large size (a few thousand nucleotides), (3) high degree of conservation, and (4) extremely low probability of being affected by LGT. These advantages were empirically confirmed because clades well established through morphological analysis (e.g., spirochaetes, cyanobacteria, animals, red algae, ciliates) were recovered with rRNA. Moreover, rRNA phylogenies also disclosed a number of assemblages that are not expected, based on previous morphological analysis. For example, an ensemble containing the morphologically very diverse ciliates, dinoflagellates, and apicomplexans emerged (Gajadhar et al. 1991). Indeed, when looking for a derived morphological character that may be shared by these three phyla, the only one that emerged was the presence of submembranar vesicles,

closely apposed to the plasma membrane and known as alveoli in ciliates. Some very curious eucaryotic organisms were unambiguously located within well-known clades [e.g., *Pneumocystis* within Fungi (Edman et al. 1988), *Dientamoeba* within trichomonads (Silberman et al. 1996a), *Blastocystis* within stramenopiles (Silberman et al. 1996)]. Let me discuss now the phylogenetic pattern related to the early evolution of eucaryotes.

The location of the root between Bacteria and a clade containing Archaea and Eucarya, which is based on the analysis of a few anciently duplicated genes (Brown and Doolittle 1997), has profound implications about the nature of the "last universal common ancestor" (LUCA). The most parsimonious interpretation is that LUCA was a prokaryote-like organism, because a eucaryote-like LUCA implies two major transitions from eucaryotes to prokaryotes, one to Bacteria, the other to Archaea. It should nevertheless be noted that, because of the RNA-world hypothesis, this possibility has been envisioned (Poole et al. 1999). The RNA-world hypothesis predicts a biota antecedent to our own that used an RNA-like molecule for a variety of tasks today performed by RNA, DNA, and proteins together (Yarus 2002). This hypothesis is widely accepted as a probable stage in the early evolution of life. Accordingly, proteins have gradually replaced RNA as the main biological catalysts. Therefore, the numerous RNA-based mechanisms of eucaryotes would be remnants of the RNA world, suggesting that prokaryotes derived from a eucaryotic-like organism (Poole et al. 1999). According to the tree in figure 7.1, LUCA was a prokaryote-like organism and had a circular chromosome with a single origin of replication, and many genes organized with operons. Yet, contrary to a frequent belief (e.g., Gupta and Singh 1994, Martin and Müller 1998, Slesarev et al. 1998), nothing can be said about the machinery of replication, transcription, and translation. It is clear that this machinery is more similar between Archaea and Eucarya. However, even with a root in the bacterial branch, the ancestral state can be equally parsimoniously similar to the bacterial one or to the eucaryotic one. In both cases, a transition from one type to another is required. Thus, the similarity between Archaea and Eucarya for the informational genes cannot be considered as a synapomorphy supporting the monophyly of this clade.

A second point is that, in the bacterial portion of the tree (fig. 7.1), the first two lineages to emerge are the Aquificales and the Thermotogales (Burggraf et al. 1992, Woese 1987). Because these two phyla mainly contain hyperthermophilic organisms (e.g., *Aquifex* and *Thermotoga*), and because most of the basal lineages within Archaea are also hyperthemophilic, the most parsimonious explanation is that LUCA was a hyperthermophilic organism (Stetter 1996). This implies that adaptation to life at low temperatures (below 60°C) occurred many times independently. In particular, in classical scenarios of eucaryotic origin, the archaeal lineage at the origin of eucaryotic cells must have become mesophilic. Moreover, the hyperthermophilic nature of LUCA led to the

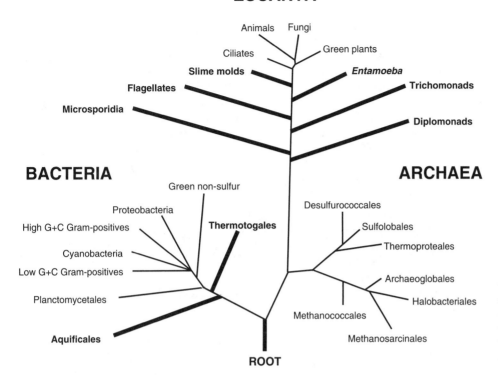

EUCARYA

Animals Fungi

Ciliates Green plants

Slime molds *Entamoeba*

Flagellates **Trichomonads**

Microsporidia **Diplomonads**

BACTERIA **ARCHAEA**

Green non-sulfur

Desulfurococcales

Proteobacteria

Sulfolobales

High G+C Gram-positives **Thermotogales**

Thermoproteales

Cyanobacteria

Low G+C Gram-positives Archaeoglobales

Planctomycetales Halobacteriales

Methanococcales

Aquificales Methanosarcinales

ROOT

Figure 7.1. Universal Tree of Life based on rRNA and rooted with anciently duplicated genes, modified from Stetter (1996). The thick branches with boldface names are likely misplaced by LBA artifact.

hypothesis of a hyperthermophilic origin of life, most likely in hydrothermal ecosystems (Nisbet and Sleep 2001, Pace 1991, Reysenbach and Shock 2002, Russell and Hall 1997, Stetter 1996, Woese 1987). Although elongation of oligopeptides (Imai et al. 1999) and synthesis of amino acids (Amend and Shock 1998) are favored at high temperature, the degradation of RNA at such temperature argues against a hot origin of life if one accepts the RNA-world hypothesis (Levy and Miller 1998, Moulton et al. 2000).

Finally, within eucaryotes, the first three lineages to emerge (diplomonads, microsporidia, and trichomonads) are all devoid of mitochondria (Sogin 1991). This seems to strongly confirm the Archezoa hypothesis (Cavalier-Smith 1987) that these three lineages are primitively devoid of mitochondria and that the mitochondrial endosymbiosis from an α-proteobacteria occurred relatively late during eucaryotic evolution, after the emergence of these three groups. However, the discovery of genes of mitochondrial origin (e.g., those encoding cpn60, HSP70, and Val-tRNA synthetase) in all the amitochondriate organisms in which they have been looked for (e.g., *Entamoeba*, *Trichomonas*, *Nosema*, *Encephalitozoon*, *Giardia*, *Neocallimastix*) suggests a secondary loss of mitochondria (for a review, see Embley and Hirt 1998). In *Entamoeba*, trichomonads, and microsporidia, several such genes have been found, and their products have been shown to be located in a double-bound organelle (hydrogenosome and mitosome/crypton; Bui et al. 1996, Mai et al. 1999, Tovar et al. 1999, Williams et al. 2002). Similarly, the diplomonad *Giardia intestinalis* has specialized mem-

branes with electron transport and membrane-potential-generating functions (Lloyd et al. 2002). This further indicates that these organisms have lost their mitochondria. Yet, at least one gene, *Val-tRNA synthetase*, which was first believed to be of mitochondrial origin (Hashimoto et al. 1998), has probably been acquired by LGT from γ-proteobacteria (Gribaldo and Philippe 2002). This is not unexpected because LGTs are frequent, especially for amitochondriate eucaryotes (Andersson et al. 2003). Because only a few genes of mitochondrial origin were found in the genome of a microsporidia (*Encephalitozoon cuniculi*; Katinka et al. 2001) and of a diplomonad (*Giardia lamblia*; McArthur et al. 2000), it is not impossible that these genes have also been acquired by LGT from other eucaryotes (Sogin 1997), and therefore it is not possible on these grounds to completely reject the hypothesis that at least some of the amitochondriate eucaryotes never did harbor a mitochondrion.

Tree Reconstruction Artifacts

The information that is used to infer molecular phylogeny consists of the mutations that have been fixed in an ancestral species, which are called substitutions. If, for a given position, a substitution occurred only once over the phylogenetic tree under study, then an unambiguous signal would be provided: a partition of the species into the ones possessing a given new character state (e.g., a change to A) and the ones possessing the alternative primitive state (e.g., G) would provide support for one node on the phylogeny. If many characters of this type are available, they will define many different compatible partitions that will allow inferring the correct phylogenetic tree. Unfortunately, in real sequences, such perfect characters with a single substitution are extremely rare, and almost all base positions have undergone many more than one substitution. If, for example, a base position has undergone 25 substitutions across a tree connecting 50 species, the taxon partitions suggested by the sharing of the various nucleotides will almost certainly be at odds with the correct phylogeny. This base position, therefore, has evolved too fast for the phylogeny under study and will contribute more noise than signal (such a position is said to be saturated).

In practice, an alignment of homologous sequences contains a mixture of slow- and fast-evolving positions (the situation is indeed more complicated because of heterotachy; see below). If there were no bias, fast-evolving positions will contribute random noise that will not favor any specific phylogeny, and the correct phylogeny will be inferred primarily on the basis of the slow-evolving positions. Unfortunately, several biases exist that can confound phylogenetic inference. The easiest biases to understand are those of nucleotide or amino acid composition. Assume that two lineages increased the G+C (guanosine + cytosine) content of their sequences independently. In that case, the noise contributed by fast-evolving positions will not be random but will favor the grouping of two G+C-rich lineages (Hasegawa and Hashimoto 1993, Lockhart et al. 1992). Another very important bias is the existence of unequal evolutionary rate among lineages. In the case of four species in which two are slowly evolving and two are fast evolving, the noise will favor the grouping of the two slowly evolving lineages because they share many ancestral characters. As a result, the two fast-evolving species will be grouped together, a phenomenon called the long-branch attraction (LBA) artifact (Felsenstein 1978).

These problems are known since the beginning of molecular phylogeny, and many attempts have been made to develop methods of inference less sensitive to nonrandom bias (for a review, see Swofford et al. 1996). To deal with the noise created by fast-evolving positions, it is necessary to have a model of sequence evolution as realistic as possible in order to infer the existence of multiple substitutions. Starting from the very simple model of Jukes and Cantor (1969), researchers have developed very complex models such as the general time-reversible model (Waddell and Steel 1997) or the Γ model that deals with among-site rate variation (Yang 1996). Other models that are not reversible have been implemented, particularly to avoid the bias due to nucleotide composition (Galtier and Gouy 1998). Nevertheless, even the most complex model is far from biological reality. One of the most important phenomena that is just beginning to be considered (Galtier 2001, Huelsenbeck 2002, Penny et al. 2001) is heterotachy, the variation of evolutionary rate of a given

position over time (i.e., fast in one part of the tree and slow in another one). Many studies have shown that this phenomenon is quite common (Galtier 2001, Huelsenbeck 2002, Lockhart et al. 2000, Lopez et al. 1999, Miyamoto and Fitch 1995, Penny et al. 2001); for example, up to 95% of the variable positions cytochrome b are heterotachous for a sample of ~2000 vertebrate sequences (Lopez et al. 2002). Heterotachy can increase the impact of LBA artifacts when two fast-evolving lineages display a higher number of variable positions (Germot and Philippe 1999). In fact, when a distant outgroup is used, the fast-evolving species and the outgroup have long branches that often attract each other. This leads to a very simple principle: early-emerging lineages are often fast-evolving ones misplaced by the LBA artifact. On the universal tree based on rRNA, all the basal branches (indicated in bold in fig. 7.1) are thus potentially erroneous.

The Case of the Bacterial Phylogeny Based on rRNA

The first two lineages to emerge in eubacterial phylogeny (Aquificales and Thermotogales) display rather short branches and for this reason are generally assumed to not be misplaced because of LBA (Burggraf et al. 1992, Stetter 1996). We recently reanalyzed the rRNA based phylogeny of Bacteria using a large data set, 95 species and 1147 positions (Brochier and Philippe 2002). If one examines the distribution of the number of substitutions per site (solid bars in fig. 7.2), it appears that most of the changes are contributed by fast-evolving positions. More precisely, there are many slowly evolving positions (e.g., 373 without changes, 154 with a single substitution) and relatively few fast-evolving positions (e.g., only 154 positions with more than 16 substitutions). This distribution of the observed substitutions is expected when the substitution rate is distributed according to a Γ law with a low α parameter (0.4 here). However, the point that is rarely discussed is the relative contributions of the slowly and fast-evolving positions to tree selection. Within a parsimony framework, the criterion to select the best phylogeny is the minimum total number of steps. Yet, as shown by the shaded bars in figure 7.2, the importance of slow- and fast-evolving sites is completely the reverse of the distribution of these sites. In fact, the slowly evolving sites (fewer than five changes) contribute very few of the total number of changes (~900 steps), whereas the fast-evolving ones are the major contributors (~3800 steps). As a result, the fast-evolving sites are the most influent in the selection of the tree topology, whereas the slowly evolving ones contain the most reliable signal.

To investigate this fundamental issue of molecular phylogeny, we used the Slow-Fast (SF) method (Brinkmann and Philippe 1999), which evaluates the evolutionary rate of positions in terms of the sum of the number of steps in pre-

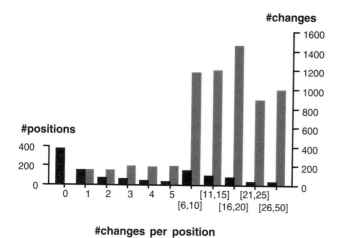

#changes

#positions

#changes per position

Figure 7.2. Distribution of the number of substitutions per position for the rRNA of 95 prokaryotic species (solid bars). The number of substitutions brought by each class of positions is indicated by the shaded bars.

defined monophyletic groups (here, the bacterial phyla) and thus allows study of the phylogenetic relationships among these groups. Interestingly, the first bacteria to emerge in the tree based on the most reliable positions (fewer than five substitutions) are, with a reasonable statistical support, Plancto-mycetes (Brochier and Philippe 2002). This phylum is a major division of Bacteria, whose members share several original features such as the lack of peptidoglycan in their cell walls or a budding mode of reproduction (Fuerst 1995). The most intriguing feature is the existence of a single or double membrane around the bacterial chromosome in *Gemmata* and *Pirellula* species, which has been compared with the eucaryotic nucleus (Fuerst 1995). Yet, evolutionary homology with the eucaryotic nucleus has not been proved. Despite these unique characteristics, this group remains little studied, although it was recently implied in anaerobic ammonia oxidation (Strous et al. 1999). If the early emergence of Planctomycetales were confirmed by genomic data (Jenkins et al. 2002), the early emergence of the most "eucaryote-like" bacteria at the base of the tree would challenge the current view on the nature of LUCA. In contrast, the hyperthermophilic bacteria robustly emerged late in the tree based on slowly evolving positions (Brochier and Philippe 2002). This is in agreement with the growing evidence that they secondarily adapted to high temperature (Aravind et al. 1998, Forterre et al. 2000, Galtier et al. 1999, Nelson et al. 1999), which seriously weakened the hypothesis that LUCA was hyperthermophile. Finally, in this tree, hyperthermophylic bacteria show a very high evolutionary rate, which was masked in standard analysis by the fast-evolving positions (Bromham et al. 2000, Philippe and Laurent 1998, Philippe et al. 1994). Therefore, contrary to recent claims (Dawson and Pace 2002), apparently slowly evolving lineages (e.g., Aquificales and Thermotogales; fig. 7.1) can be misplaced by the LBA artifact.

Recent Advances into the Eucaryotic Phylogeny

The impact of LBA artifact is not limited to the bacterial phylogeny but applies to all the branches indicated in bold in figure 7.1 (Brinkmann and Philippe 1999, Philippe et al. 2000b). This is especially dramatic in the case of eucaryotes, for which more than 10 early-branching lineages could be artificially located (Philippe and Adoutte 1998). Indeed, the eucaryotic tree was previously divided into two parts: (1) the so-called crown, in which the branching order between phyla was very poorly resolved, which is interpreted as the result of an adaptive radiation (Knoll 1992); and (2) the base, which contains "primitive" eucaryotes, especially the amitochondriate ones. We have proposed that all the lineages of the classical base are very likely misplaced and in fact belong to the crown, what we called the "big bang" hypothesis (Philippe and Adoutte 1998).

As recently reviewed (Philippe et al. 2000a), many lines of evidence are in agreement with the hypothesis that the eucaryotes branching early in the rRNA are misplaced because of LBA. First, the evolutionary rates of different eucaryotic phyla have been estimated for several genes, and it has been shown that the faster a phylum evolves, the earlier it emerges (e.g., euglenozoans for rRNA and ciliates for actin; Moreira et al. 2002, Philippe and Adoutte 1998). Second, the addition of new sequences in phylogenetic analyses, which is known to reduce the impact of the LBA artifact (Hendy and Penny 1989), results in an upward movement of the early-branching species in the tree (Moreira et al. 1999). Third, the use of more realistic models of sequence evolution, also known to attenuate the impact of LBA (Huelsenbeck 1998), leads, in rRNA trees, to a later emergence of euglenozoans (Peyretaillade et al. 1998, Tourasse and Gouy 1998), microsporidia (Peyretaillade et al. 1998, Van de Peer et al. 2000), *Physarum* (Peyretaillade et al. 1998), and trichomonads and heteroloboseans (Silberman et al. 1999). In fact, the most recent analyses that used a Γ law to model the rate heterogeneity among sequence sites showed that the "classical" tree cannot be statistically differentiated from the ones that locate all the lineages within the crown (Philippe and Germot 2000, Simpson et al. 2002). Fourth, several characteristics [highly heterogeneous rRNA length, large number of unique substitutions, attraction by artificial random sequences, and high Relative Apparent Synapomorphy Analysis (RASA) taxon variance] suggested that the basal lineages of the rRNA tree are fast evolving (Stiller and Hall 1999). Fifth, if a basal emergence in the rRNA tree is correct, one expects that the slowly evolving positions, which contain most of the ancient phylogenetic information, will provide strong support for the basal branching. Yet, as for Bacteria, when using the S-F method, the basal taxa in the standard rRNA tree do not emerge early when only slow-evolving positions are used, but display very long branches (Philippe et al. 2000b). Sixth, phylogenies based on protein sequences generally suggest a late emergence for the taxa emerging early in the rRNA tree. A clear example is provided by microsporidia, which are located very close to the base of eucaryotes in rRNA tree (fig. 7.1) but are indeed highly derived fungi (for review, see Keeling and Fast 2002).

The phylogenetic relationships within the crown of the eucaryotic rRNA tree are known to be difficult quite to resolve, possibly because of a rapid diversification (Knoll 1992, Sogin 1991). Indeed, eucaryotic rRNA phylogenies inferred with a comprehensive taxonomic sampling and a Γ law model are very poorly resolved, the bootstrap values for the nodes connecting the major phyla being almost all below 50% (Brugerolle et al. 2002, Cavalier-Smith 2002, Simpson et al. 2002). Because many more lineages than first acknowledged (the artifactually early-branching phyla and the newly discovered, uncultured groups; Dawson and Pace 2002, Lopez-Garcia et al. 2001) belong to the already poorly resolved crown, The complete resolution of the eucaryotic phylogeny constitutes a great challenge.

Two quite different approaches can be used, which we have called statistician and Hennigian (Philippe and Laurent 1998). The statistician approach consists in the analysis of very large data sets, with tree reconstruction methods as refined as possible. The underlying idea is that the resolving power will increase and that the biases brought by different genes will be different and thus will be minimized. The Hennigian approach consists in the use of very slowly evolving characters, such as insertion/deletion or gene fusion events [also called rare genomic events (Rokas and Holland 2000)]. The assumption is that these characters are less homoplastic, and therefore the most simple tree reconstruction method (i.e., maximum parsimony) will provide a good estimate of the good phylogeny. These two approaches have been applied to the case of eucaryotes, with both more and less success.

In the statistician approach, because of the limited amount of available sequences, one has to choose between many genes/few species (13/12; Moreira et al. 2000) and few genes/many species (4/60; Baldauf et al. 2000). As expected (Graybeal 1998, Lecointre et al. 1993, 1994), the first approach provided a fully resolved tree (Moreira et al. 2000) but is very sensitive to LBA, whereas the second is not severely affected by LBA but is very poorly resolved. For example, the Euglenozoa and the Apicomplexa emerge strongly but artificially at the base when few species are used (Moreira et al. 2000). On the contrary, they belong to a large group of protists (including also stramenopiles and heteroloboseans) when many species are used (Baldauf et al. 2000), but with a weak support (bootstrap value around 50%). In contrast, red algae and green plants strongly group together in the clade Plantae in the first analysis but very weakly (bootstrap value below 50%) in the second one. The monophyly of Plantae found with nuclear genes strongly suggests the hypothesis of a unique primary endosymbiosis of a cyanobacteria at the origin of chloroplast, as already proposed by plastid and mitochondrial data (Palmer 2000).

We recently tried to make a compromise between these two extremes in order to increase simultaneously both accuracy and resolving power (Bapteste et al. 2002). We used 123 genes for 30 species, representing about 25,000 unambiguously aligned positions. The corresponding phylogeny is shown in figure 7.3. Not surprisingly, the results are in between the previous ones (with 12 and 60 species, respectively; Baldauf et al. 2000, Moreira et al. 2000), which is illustrated by three examples. (1) One fast-evolving species, a parasitic amitochondriate amoeba *Entamoeba histolytica*, is strongly grouped with a free-living amitochondriate amoeba (*Mastigamoeba*), this clade being a sister group of Mycetozoa, represented here by *Dictyostelium*. The monophyly of this large clade of amoeboid organisms contrasts with their pronounced polyphyly on classical rRNA trees (Sogin 1991). The statistician approach has provided convincing evidence for a difficult phylogenetic question. (2) The early emergence of diplomonads and Euglenozoa (fig. 7.3) is very likely due to LBA. In fact, when we added microsporidia to our data set, we found very strong support for their early emergence (H. Brinkman, M. van der Giezen, T. M. Embley, and H. Philippe, unpubl. obs.). However, the evidence for considering microsporidia as derived fungi is very strong (Keeling and Fast 2002), but many of the genes used evolved very fast in this group, thus generating LBA. The number of species used in our study is thus insufficient to eliminate LBA, all the more so because a very distant outgroup (Archaea) is used. It is likely that the use of genes of mitochondrial origin, with a very close α-proteobacterial outgroup, will be a good way to avoid this problem (Philippe 2000). (3) Several nodes (e.g., the grouping of stramenopiles and alveolates) are weakly supported. This indicates that the number of genes used is still insufficient, and/or, as proposed by the big bang hypothesis, the time between speciation events is too short to discriminate branching orders. In summary, the statistician approach has allowed, and will allow, progress in the resolution of the phylogeny of eucaryotes. However, because it is very sensitive to the inconsistency of the methods, it is of prime importance to improve the tree reconstruction methods, especially by taking into account heterotachy (Galtier 2001, Huelsenbeck 2002, Penny et al. 2001).

In the Hennigian approach, very few characters useful for resolving the phylogeny of eucaryotes have been discovered. First, a few insertion/deletions have been proposed. In particular, an insertion of about 12 amino acids in the elongation factor EF-1α is shared only by animals and fungi (Baldauf and Palmer 1993), and also by microsporidia (Van de Peer et al. 2000), suggesting the monophyly of this clade, called Opisthokonta. However, the same insertion is also present in some green algae but not in land plants (H. Philippe, unpubl. obs.). Similarly, two small indels of one amino acid in enolase are shared by trichomonads and prokaryotes, suggesting that trichomonads constitute the first lineage to emerge within eucaryotes (Keeling and Palmer 2000). However, the same indels are also present in several independent

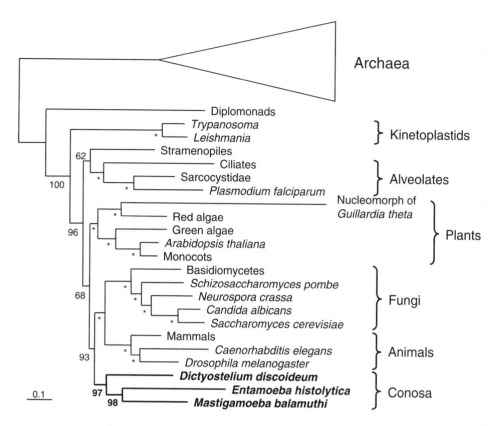

Figure 7.3. Phylogenetic tree based on 123 genes, redrawn from Bapteste et al. (2000). The tree was inferred by a separate maximum likelihood analysis, taking into account among-sites rate variation (JTT + Γ model). For reducing computational time, several nodes, which were recovered through preliminary analyses, were constrained (indicated by asterisks). The bootstrap values were obtained by bootstrapping the 123 genes, a modification of the RELL method (Kishino et al. 1990).

lineages (e.g., in several members of Archaea and in a few of Bacteria; Bapteste and Philippe 2002, Hannaert et al. 2000), casting doubts on the use of this character as a phylogenetic marker. In fact we have found in enolase, IMPDH, and Val tRNA synthetase several large indels that contradict each other and also the phylogeny inferred from the very same gene containing the indel (Bapteste and Philippe 2002, Gribaldo and Philippe 2002). This indicates that indels are not always very good characters, because they are prone to convergence and that they are very sensitive to LGT (with or without recombination; see Bapteste and Philippe 2002). It is thus very hazardous to base phylogenetic inference on a single indel. Finally, an insertion in a very highly conserved gene (ubiquitin) for which a comprehensive taxonomic sampling is available provide convincing evidence for the sister-group relationship of Cercozoa and Foraminifera (Archibald et al. 2003).

Other rare genomic events are more promising. The first case is the nonhomologous replacement of the mitochondrial RNA polymerase by the T3/T7-like one. In all the mitochondriate eucaryotes, except the jakobids (e.g., *Reclinomonas americana*), the original bacterial polymerase encoded in the mitochondria has been replaced by T3/T7 polymerase (Cermakian et al. 1996, Lang et al. 1997). This replacement suggests that jakobids are the first eucaryotic lineage to emerge. However, in the plastid of land plants, the bacterial and the T3/T7-like RNA polymerases are known to have coexisted for several hundred of millions years (Gray and

Lang 1998), and the bacterial form has been lost in one parasitic nonphotosynthetic plant (Wolfe et al. 1992). It is therefore quite possible that different lineage sorting has affected the RNA polymerase of mitochondria. Nevertheless, jakobids are good candidates for being the first emerging eucaryotes. A second case of a rare genomic event is the fusion of the dihydrofolate reductase and thymidylate synthase genes. These two genes are separated in all the bacteria and all the opistokonts, but are fused, when present, in the other eucaryotes (Philippe et al. 2000b, Stechmann and Cavalier-Smith 2002). This is a strong argument to locate the root of the eucaryotic tree between opistikonts and all the other eucaryotes. Yet, it should be noted that these genes have been lost in several lineages (e.g., *Entamoeba* and *Giardia*) and replaced by nonhomologous genes in some others (e.g., *Dictyostelium*; Dynes and Firtel 1989). This gene fusion suggests that opistokonts are also very good candidates for being the first emerging eucaryotes. In summary, the use of rare genomic events has provided some interesting hypotheses for rooting the eucaryotic tree. If such a root is reliably inferred, it will be possible to construct eucaryotic phylogenies without the need of a non-eucaryotic outgroups, thus seriously reducing the importance of LBA.

As expected from the results based on rRNA, the eucaryotic phylogeny turned to be a very difficult question. The very large amount of new molecular data has recently allowed resolving several nodes (fig. 7.4). The resolution will continue to be improved thanks to the sequencing of complete genomes and of a large sample of cDNAs (http://megasun.bch.umontreal.ca/pepdb/pep_main.html) for many protists.

A Personal Point of View on the Universal Tree of Life

In conclusion, several basal branches of the universal Tree of Life based on rRNA (indicated in bold in fig. 7.1), which may be misplaced because of LBA artifact, have been relocated upper in the tree (e.g., hyperthermophilic bacteria and microsporidia). For some others (e.g., diplomonads and the root of the Tree of Life), it appeared that their high evolutionary rates for numerous genes prevented their reliable placement, because current tree reconstruction methods are still sensitive to LBA. The support in favor of their early emergence has thus been weakened. Nevertheless, the global picture provided by rRNA remains correct, and one can still consider rRNA as one of the best phylogenetic markers, despite some weaknesses. The progresses to fix the potential errors highlighted in figure 7.1 are summarized in figure 7.4. It should be noted that several nodes are supported with little support (e.g., a single gene) and reflect my working hypothesis rather than a robust and widely accepted consensus.

I would like to emphasize two general issues that are especially relevant to the origin and evolution of eucaryotes. The first is that we are strongly influenced by the Aristotelian view that simple organisms are primitive organisms (the famous *scala natura*). It is for this reason that we easily believe that prokaryotes precede eucaryotes and that amitochondriate eucaryotes predate the mitochondrial endosymbiosis. Yet, the study of eucaryotic phylogeny (Embley and Hirt 1998, Philippe et al. 2000a) has shown that simplification is a major evolutionary trend. As brilliantly argued more than 50 years ago (Lwoff 1943), we have a major psychological reluctance to accept the importance of simplification, because we associate evolution, progress, and complexity (Gould 1996). The second is that molecular phylogeneticists, because of the constraint of having to study extant organisms, often forget extinct organisms. In fact, extinction is a very common phenomenon, and one should take extinct organisms into account for every evolutionary scenario. Even if a lot of speculations are required to infer the characteristics of past microorganisms, the numerous extinct organisms quite different from extant eucaryotes and prokaryotes should not be ignored (e.g., the organisms thriving during the hypothetical RNA world). As a result, the absence of early-branching eucaryotes proposed by the "big bang" hypothesis does not imply that complex eucaryotes suddenly evolved from scratch. As shown in figure 7.4, this can just be due to the extinction of all the intermediary forms, as is well known for mammals and birds.

Finally, as explained in detail elsewhere (Forterre and Philippe 1999), we favor the hypothesis that LUCA was an eucaryote-like organism that would have evolved through simplification into a prokaryote-like form. The main argument is that many RNA-based mechanisms inherited from the RNA world have been replaced by protein-based mechanisms in prokaryotes (Poole et al. 1999). Nevertheless, this argument is not decisive, because RNA-based mechanisms can appear in prokaryotes (e.g., transfer-messenger RNA in Bacteria).

Acknowledgments

This chapter is dedicated to the memory of André Adoutte (1947–2002), who was my Ph.D. supervisor and, as early as 1987, was concerned by the limitation of tree reconstruction methods. Most of the work discussed in this chapter was due to his brilliant intuitions. I also dedicate this chapter to the memory of Stephen J. Gould (1941–2002), whose books motivated me to move from mathematics to evolutionary biology. I thank Simonetta Gribaldo for careful reading of the manuscript and Joel Cracraft for many helpful suggestions.

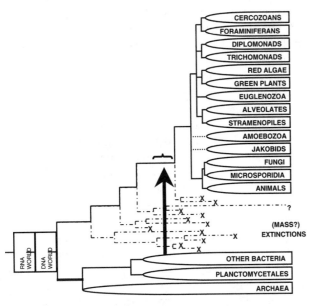

Figure 7.4. Simplified universal Tree of Life. The root was located in the eukaryotic branch based on the analysis of slowly evolving positions (Brinkmann and Philippe 1999). The mitochondrial endosymbiosis (arrow) is supposed to have occurred before the diversification of extant eukaryotes and before the major increase of atmospheric oxygen (Canfield and Teske 1996). A few extinct lineages are indicated in broken lines, to show that the diversity of extant organisms provides a very sparse sampling of ancient diversity.

Literature Cited

Amend, J. P., and E. L. Shock. 1998. Energetics of amino acid synthesis in hydrothermal ecosystems. Science 281:1659–1662.

Andersson, J. O., A. M. Sjogren, L. A. Davis, T. M. Embley, and A. J. Roger. 2003. Phylogenetic analyses of diplomonad genes reveal frequent lateral gene transfers affecting eukaryotes. Curr. Biol. 13:94–104.

Aravind, L., R. L. Tatusov, Y. I. Wolf, D. R. Walker, and E. V. Koonin. 1998. Evidence for massive gene exchange between archaeal and bacterial hyperthermophiles. Trends Genet. 14:442–444.

Archibald, J. M., D. Longet, J. Pawlowski, and P. J. Keeling. 2003. A novel polyubiquitin structure in cercozoa and foraminifera: evidence for a new eukaryotic supergroup. Mol. Biol. Evol. 20:62–66.

Baldauf, S. L., and J. D. Palmer. 1993. Animals and fungi are each other's closest relatives: congruent evidence from multiple proteins. Proc. Natl. Acad. Sci. USA 90:11558–11562.

Baldauf, S. L., A. J. Roger, I. Wenk-Siefert, and W. F. Doolittle. 2000. A kingdom-level phylogeny of eukaryotes based on combined protein data. Science 290:972–977.

Bapteste, E., H. Brinkmann, J. A. Lee, D. V. Moore, C. W. Sensen, P. Gordon, L. Durufle, T. Gaasterland, P. Lopez, M. Muller, and H. Philippe. 2002. The analysis of 100 genes supports the grouping of three highly divergent amoebae: *Dictyostelium*, *Entamoeba*, and *Mastigamoeba*. Proc. Natl. Acad. Sci. USA 99:1414–1419.

Bapteste, E., and H. Philippe. 2002. The potential value of indels as phylogenetic markers: position of trichomonads as a case study. Mol. Biol. Evol. 19:972–977.

Brinkmann, H., and H. Philippe. 1999. Archaea sister group of Bacteria? Indications from tree reconstruction artifacts in ancient phylogenies. Mol. Biol. Evol. 16:817–825.

Brochier, C., E. Bapteste, D. Moreira, and H. Philippe 2002. Eubacterial phylogeny based on translational apparatus proteins. Trends Genet. 18:1–5.

Brochier, C., and H. Philippe. 2002. Phylogeny: a non-hyperthermophilic ancestor for bacteria. Nature 417:244.

Bromham, L., D. Penny, A. Rambaut, and M. D. Hendy. 2000. The power of relative rates tests depends on the data. J. Mol. Evol. 50:296–301.

Brown, J. R., and W. F. Doolittle. 1997. Archaea and the prokaryote-to-eukaryote transition. Microbiol. Mol. Biol. Rev. 61:456–502.

Brugerolle, G., G. Bricheux, H. Philippe, and G. Coffe. 2002. *Collodictyon triciliatum* and *Diphylleia rotans* (= *Aulacomonas submarina*) form a new family of flagellates (Collodictyonidae) with tubular mitochondrial cristae that is phylogenetically distant from other flagellate groups. Protist 153:59–70.

Bui, E. T., P. J. Bradley, and P. J. Johnson. 1996. A common evolutionary origin for mitochondria and hydrogenosomes. Proc. Natl. Acad. Sci. USA 93:9651–9656.

Burggraf, S., G. J. Olsen, K. O. Stetter, and C. R. Woese. 1992. A phylogenetic analysis of Aquifex pyrophilus. Syst. Appl. Microbiol. 15:352–356.

Canfield, D. E., and A. Teske. 1996. Late proterozoic rise in atmospheric oxygen concentration inferred from phylogenetic and sulphur-isotope studies. Nature 382:127–132.

Cavalier-Smith, T. 1987. Eukaryotes with no mitochondria. Nature 326:332–333.

Cavalier-Smith, T. 2002. The phagotrophic origin of eukaryotes and phylogenetic classification of Protozoa. Int. J. Syst. Evol. Microbiol. 52:297–354.

Cermakian, N., T. M. Ikeda, R. Cedergren, and M. W. Gray. 1996. Sequences homologous to yeast mitochondrial and bacteriophage T3 and T7 RNA polymerases are widespread throughout the eukaryotic lineage. Nucleic Acids Res. 24:648–654.

Darwin, C. 1859. The origin of species by means of natural selection. Murray, London.

Dawson, S. C., and N. R. Pace. 2002. Novel kingdom-level eukaryotic diversity in anoxic environments. Proc. Natl. Acad. Sci. USA 99:8324–8329.

Dynes, J. L., and R. A. Firtel. 1989. Molecular complementation of a genetic marker in Dictyostelium using a genomic DNA library. Proc. Natl. Acad. Sci. USA 86:7966–7970.

Edman, J. C., J. A. Kovacs, H. Masur, D. V. Santi, H. J. Elwood, and M. L. Sogin. 1988. Ribosomal RNA sequence shows Pneumocystis carinii to be a member of the fungi. Nature 334:519–522.

Embley, T. M., and R. P. Hirt. 1998. Early branching eukaryotes? Curr. Opin. Genet. Dev. 8:624–629.

Felsenstein, J. 1978. Cases in which parsimony or compatibility methods will be positively misleading. Syst. Zool. 27:401–410.

Fitch, W. M., and E. Margoliash. 1967. Construction of phylogenetic trees. Science 155:279–284.

Forterre, P., C. Bouthier De La Tour, H. Philippe, and M. Duguet. 2000. Reverse gyrase from hyperthermophiles: probable transfer of a thermoadaptation trait from archaea to bacteria. Trends Genet. 16:152–154.

Forterre, P., and H. Philippe. 1999. Where is the root of the universal tree of life? Bioessays 21:871–879.

Fuerst, J. A. 1995. The planctomycetes: emerging models for microbial ecology, evolution and cell biology. Microbiology 141:1493–1506.

Gajadhar, A. A., W. C. Marquardt, R. Hall, J. Gunderson, E. V. Ariztia-Carmona, and M. L. Sogin. 1991. Ribosomal RNA sequences of Sarcocystis muris, Theileria annulata and Crypthecodinium cohnii reveal evolutionary relationships among apicomplexans, dinoflagellates, and ciliates. Mol. Biochem. Parasitol. 45:147–154.

Galtier, N. 2001. Maximum-likelihood phylogenetic analysis under a covarion-like model. Mol. Biol. Evol. 18:866–873.

Galtier, N., and M. Gouy. 1998. Inferring pattern and process: maximum-likelihood implementation of a nonhomogeneous model of DNA sequence evolution for phylogenetic analysis. Mol. Biol. Evol. 15:871–879.

Galtier, N., N. Tourasse, and M. Gouy. 1999. A nonhyperthermophilic common ancestor to extant life forms. Sciences 283:220–221.

Germot, A., and H. Philippe. 1999. Critical analysis of eukaryotic phylogeny: a case study based on the HSP70 family. J. Eukaryot. Microbiol. 46:116–124.

Gogarten, J. P., H. Kibak, P. Dittrich, L. Taiz, E. J. Bowman, B. J. Bowman, M. F. Manolson, R. J. Poole, T. Date, T. Oshima, et al. 1989. Evolution of the vacuolar H+-ATPase: implications for the origin of eukaryotes. Proc. Natl. Acad. Sci. USA 86:6661–6665.

Gould, S. J. 1996. Full house: the spread of excellence from Plato to Darwin. Harmony Books, New York.

Gray, M. W., and B. F. Lang. 1998. Transcription in chloroplasts and mitochondria: a tale of two polymerases. Trends Microbiol. 6:1–3.

Graybeal, A. 1998. Is it better to add taxa or characters to a difficult phylogenetic problem? Syst. Biol. 47:9–17.

Gribaldo, S., and H. Philippe. 2002. Ancient phylogenetic relationships. Theor. Pop. Biol. 61:391–408.

Gupta, R. S., and B. Singh. 1994. Phylogenetic analysis of 70 kD heat shock protein sequences suggests a chimeric origin for the eukaryotic cell nucleus. Curr. Biol. 4:1104–1114.

Haeckel, E. 1866. Generelle Morphologie der Organismen: Allgemeine Grundzüge der organischen Formen-Wissenschaft, mechanisch begründet durch die von Charles Darwin reformirte Descendenz-Theorie. 2 vols. Georg Reimer, Berlin.

Hannaert, V., H. Brinkmann, U. Nowitzki, J. A. Lee, M.-A. Albert, C. W. Sensen, T. Gaasterland, M. Müller, P. Michels, and W. Martin. 2000. Enolase from *Trypanosoma brucei*, from the amitochondriate protist *Mastigamoeba balamuthi*, and from the chloroplast and cytosol of *Euglena gracilis*: pieces in the evolutionary puzzle of the eukaryotic glycolytic pathway. Mol. Biol. Evol. 17:989–1000.

Hasegawa, M., and T. Hashimoto. 1993. Ribosomal RNA trees misleading? Nature 361:23.

Hashimoto, T., L. B. Sanchez, T. Shirakura, M. Muller, and M. Hasegawa. 1998. Secondary absence of mitochondria in *Giardia lamblia* and *Trichomonas vaginalis* revealed by valyl-tRNA synthetase phylogeny. Proc. Natl. Acad. Sci. USA 95:6860–6865.

Hendy, M., and D. Penny. 1989. A framework for the quantitative study of evolutionary trees. Syst. Zool. 38:297–309.

Hennig, W. 1966. Phylogenetic systematics. University of Illinois Press, Urbana.

Huelsenbeck, J. P. 1998. Systematic bias in phylogenetic analysis: is the Strepsiptera problem solved? Syst. Biol. 47:519–537.

Huelsenbeck, J. P. 2002. Testing a covariotide model of DNA substitution. Mol. Biol. Evol. 19:698–707.

Imai, E., H. Honda, K. Hatori, A. Brack, and K. Matsuno. 1999. Elongation of oligopeptides in a simulated submarine hydrothermal system. Science 283:831–833.

Iwabe, N., K. Kuma, M. Hasegawa, S. Osawa, and T. Miyata. 1989. Evolutionary relationship of archaebacteria, eubacteria, and eukaryotes inferred from phylogenetic trees of duplicated genes. Proc. Natl. Acad. Sci. USA 86:9355–9359.

Jenkins, C., V. Kedar, and J. A. Fuerst. 2002. Gene discovery within the planctomycete division of the domain Bacteria using sequence tags from genomic DNA libraries. Genome Biol. 3: research0031.1–0031.11.

Jukes, T. H., and C. R. Cantor. 1969. Evolution of protein molecules. Pp. 21–132 *in* Mammalian protein metabolism (H. N. Munro, ed.). Academic Press, New York.

Katinka, M. D., S. Duprat, E. Cornillot, G. Metenier, F. Thomarat, G. Prensier, V. Barbe, E. Peyretaillade, P. Brottier, P. Wincker, et al. 2001. Genome sequence and gene compaction of the eukaryote parasite *Encephalitozoon cuniculi*. Nature 414:450–453.

Keeling, P. J., and N. M. Fast. 2002. Microsporidia: biology and evolution of highly reduced intracellular parasites. Annu. Rev. Microbiol. 56:93–116.

Keeling, P. J., and J. D. Palmer. 2000. Parabasalian flagellates are ancient eukaryotes. Nature 405:635–637.

Kishino, H., T. Miyata, and M. Hasegawa. 1990. Maximum likelihood inference of protein phylogeny, and the origin of chloroplasts. J. Mol. Evol. 31:151–160.

Knoll, A. H. 1992. The early evolution of eukaryotes: a geological perspective. Science 256:622–627.

Koonin, E. V., K. S. Makarova, and L. Aravind. 2001. Horizontal gene transfer in prokaryotes: quantification and classification. Annu. Rev. Microbiol. 55:709–742.

Lang, B. F., G. Burger, C. J. O'Kelly, R. Cedergren, G. B. Golding, C. Lemieux, D. Sankoff, M. Turmel, and M. W. Gray. 1997. An ancestral mitochondrial DNA resembling a eubacterial genome in miniature. Nature 387:493–497.

Lecointre, G., H. Philippe, H. L. V. Le, and H. Le Guyader. 1993. Species sampling has a major impact on phylogenetic inference. Mol. Phylogenet. Evol. 2:205–224.

Lecointre, G., H. Philippe, H. L. V. Le, and H. Le Guyader. 1994. How many nucleotides are required to resolve a phylogenetic problem? The use of a new statistical method applicable to available sequences. Mol. Phylogenet. Evol. 3:292–309.

Levy, M., and S. L. Miller. 1998. The stability of the RNA bases: implications for the origin of life. Proc. Natl. Acad. Sci. USA 95:7933–7938.

Lloyd, D., J. C. Harris, S. Maroulis, R. Wadley, J. R. Ralphs, A. C. Hann, M. P. Turner, and M. R. Edwards. 2002. The "primitive" microaerophile *Giardia intestinalis* (syn. lamblia, duodenalis) has specialized membranes with electron transport and membrane-potential-generating functions. Microbiology 148:1349–1354.

Lockhart, P. J., C. J. Howe, D. A. Bryant, T. J. Beanland, and A. W. Larkum. 1992. Substitutional bias confounds inference of cyanelle origins from sequence data. J. Mol. Evol. 34:153–162.

Lockhart, P. J., D. Huson, U. Maier, M. J. Fraunholz, Y. Van De Peer, A. C. Barbrook, C. J. Howe, and M. A. Steel. 2000. How molecules evolve in Eubacteria. Mol. Biol. Evol. 17:835–838.

Lopez, P., D. Casane, and H. Philippe. 2002. Heterotachy, an important process of protein evolution. Mol. Biol. Evol. 19:1–7.

Lopez, P., P. Forterre, and H. Philippe. 1999. The root of the tree of life in the light of the covarion model. J. Mol. Evol. 49:496–508.

Lopez-Garcia, P., F. Rodriguez-Valera, C. Pedros-Alio, and D. Moreira. 2001. Unexpected diversity of small eukaryotes in deep-sea Antarctic plankton. Nature 409:603–607.

Lwoff, A. 1943. L'évolution physiologique. Etude des pertes de fonctions chez les microorganismes. Hermann et Cie, Paris.

Mai, Z., S. Ghosh, M. Frisardi, B. Rosenthal, R. Rogers, and J. Samuelson. 1999. Hsp60 is targeted to a cryptic mitochondrion-derived organelle ("crypton") in the microaerophilic protozoan parasite *Entamoeba histolytica*. Mol. Cell. Biol. 19:2198–2205.

Martin, W., and M. Müller. 1998. The hydrogen hypothesis for the first eukaryote. Nature 392:37–41.

Matte-Tailliez, O., C. Brochier, P. Forterre, and H. Philippe. 2002. Archaeal phylogeny based on ribosomal proteins. Mol. Biol. Evol. 19:631–639.

McArthur, A. G., H. G. Morrison, J. E. Nixon, N. Q. Passamaneck, U. Kim, G. Hinkle, M. K. Crocker, M. E. Holder, R. Farr, C. I. Reich, et al. 2000. The *Giardia* genome project database.

Federation European Microbiological Societies Microbiology Letters 189:271–273.

Miyamoto, M. M., and W. M. Fitch. 1995. Testing the covarion hypothesis of molecular evolution. Mol. Biol. Evol. 12:503–513.

Moreira, D., S. Kervestin, O. Jean-Jean, and H. Philippe. 2002. Evolution of eukaryotic translation elongation and termination factors: variations of evolutionary rate and genetic code deviations. Mol. Biol. Evol. 19:189–200.

Moreira, D., H. Le Guyader, and H. Philippe. 1999. Unusually high evolutionary rate of the elongation factor 1 alpha genes from the Ciliophora and its impact on the phylogeny of eukaryotes. Mol. Biol. Evol. 16:234–245.

Moreira, D., H. Le Guyader, and H. Philippe. 2000. The origin of red algae: implications for the evolution of chloroplasts. Nature 405:69–72.

Moulton, V., P. P. Gardner, R. F. Pointon, L. K. Creamer, G. B. Jameson, and D. Penny. 2000. RNA folding argues against a hot-start origin of life. J. Mol. Evol. 51:416–421.

Nelson, K. E., R. A. Clayton, S. R. Gill, M. L. Gwinn, R. J. Dodson, D. H. Haft, E. K. Hickey, J. D. Peterson, W. C. Nelson, K. A. Ketchum, et al. 1999. Evidence for lateral gene transfer between archaea and bacteria from genome sequence of *Thermotoga maritima*. Nature 399:323–329.

Nisbet, E. G., and N. H. Sleep. 2001. The habitat and nature of early life. Nature 409:1083–1091.

Pace, N. R. 1991. Origin of life—facing up to the physical setting. Cell 65:531–533.

Palmer, J. D. 2000. A single birth of all plastids? Nature 405:32–33.

Penny, D., B. J. McComish, M. A. Charleston, and M. D. Hendy. 2001. Mathematical elegance with biochemical realism: the covarion model of molecular evolution. J. Mol. Evol. 53:711–723.

Peyretaillade, E., C. Biderre, P. Peyret, F. Duffieux, G. Metenier, M. Gouy, B. Michot, and C. P. Vivares. 1998. Microsporidian *Encephalitozoon cuniculi*, a unicellular eukaryote with an unusual chromosomal dispersion of ribosomal genes and a LSU rRNA reduced to the universal core. Nucleic Acids Res. 26:3513–3520.

Philippe, H. 2000. Long branch attraction and protist phylogeny. Protist 51:307–316.

Philippe, H., and A. Adoutte. 1998. The molecular phylogeny of Eukaryota: solid facts and uncertainties. Pp. 25–56 *in* Evolutionary relationships among protozoa (G. Coombs, K. Vickerman, M. Sleigh and A. Warren, eds.). Kluwer, Dordrecht.

Philippe, H., and A. Germot. 2000. Phylogeny of eukaryotes based on ribosomal RNA: long-branch attraction and models of sequence evolution. Mol. Biol. Evol. 17:830–834.

Philippe, H., A. Germot, and D. Moreira. 2000a. The new phylogeny of eukaryotes. Curr. Opin. Genet. Dev. 10:596–601.

Philippe, H., and J. Laurent. 1998. How good are deep phylogenetic trees? Curr. Opin. Genet. Dev 8:616–623.

Philippe, H., P. Lopez, H. Brinkmann, K. Budin, A. Germot, J. Laurent, D. Moreira, M. Müller, and H. Le Guyader. 2000b. Early branching or fast evolving eukaryotes? An answer based on slowly evolving positions. Philos. Trans. R. Soc. Lond. B 267:1213–1221.

Philippe, H., U. Sörhannus, A. Baroin, R. Perasso, F. Gasse, and A. Adoutte. 1994. Comparison of molecular and paleontological data in diatoms suggests a major gap in the fossil record. J. Evol. Biol. 7:247–265.

Poole, A., D. Jeffares, and D. Penny. 1999. Early evolution: prokaryotes, the new kids on the block. Bioessays 21:880–889.

Reysenbach, A. L., and E. Shock. 2002. Merging genomes with geochemistry in hydrothermal ecosystems. Science 296:1077–1082.

Rokas, A., and P. W. H. Holland. 2000. Rare genomic changes as a tool for phylogenetics. Trends Ecol. Evol. 15:454–459.

Russell, M. J., and A. J. Hall. 1997. The emergence of life from iron monosulphide bubbles at a submarine hydrothermal redox and pH front. J. Geol. Soc. Lond. 154:377–402.

Schwartz, R. M., and M. O. Dayhoff. 1978. Origins of prokaryotes, eukaryotes, mitochondria, and chloroplasts. Science 199:395–403.

Silberman, J. D., C. G. Clark, L. S. Diamond, and M. L. Sogin. 1999. Phylogeny of the genera *Entamoeba* and *Endolimax* as deduced from small-subunit ribosomal RNA sequences. Mol. Biol. Evol. 16:1740–1751.

Silberman, J. D., C. G. Clark, and M. L. Sogin. 1996a. *Dientamoeba fragilis* shares a recent common evolutionary history with the trichomonads. Mol. Biochem. Parasitol. 76:311–314.

Silberman, J. D., M. L. Sogin, D. D. Leipe, and C. G. Clark. 1996b. Human parasite finds taxonomic home. Nature 380:398.

Simpson, A. G., A. J. Roger, J. D. Silberman, D. D. Leipe, V. P. Edgcomb, L. S. Jermiin, D. J. Patterson, and M. L. Sogin. 2002. Evolutionary history of "early-diverging" eukaryotes: the excavate taxon Carpediemonas is a close relative of Giardia. Mol. Biol. Evol. 19:1782–1791.

Slesarev, A. I., G. I. Belova, S. A. Kozyavkin, and J. A. Lake. 1998. Evidence for an early prokaryotic origin of histones H2A and H4 prior to the emergence of eukaryotes. Nucleic Acids Res. 26:427–430.

Sogin, M. 1997. History assignment: when was the mitochondrion founded? Curr. Opin. Genet. Dev. 7:792–799.

Sogin, M. L. 1991. Early evolution and the origin of eukaryotes. Curr. Opin. Genet. Dev 1:457–463.

Stechmann, A., and T. Cavalier-Smith. 2002. Rooting the eukaryote tree by using a derived gene fusion. Science 297:89–91.

Stetter, K. O. 1996. Hyperthermophiles in the history of life. Ciba Found. Symp. 202:1–10.

Stiller, J., and B. Hall. 1999. Long-branch attraction and the rDNA model of early eukaryotic evolution. Mol. Biol. Evol. 16:1270–1279.

Strous, M., J. A. Fuerst, E. H. Kramer, S. Logemann, G. Muyzer, K. T. van de Pas-Schoonen, R. Webb, J. G. Kuenen, and M. S. Jetten. 1999. Missing lithotroph identified as new planctomycete. Nature 400:446–449.

Swofford, D. L., G. J. Olsen, P. J. Waddell, and D. M. Hillis. 1996. Phylogenetic inference. Pp. 407–514 *in* Molecular systematics (D. M. Hillis, C. Moritz and B. K. Mable, eds.). Sinauer Associates, Sunderland, MA.

Taylor, F. J. R. 1978. Problem in the development of an explicit hypothetical phylogeny of the Lower Eukaryotes. Biosystems 10:67–89.

Tourasse, N. J., and M. Gouy. 1998. Evolutionary relationships between protist phyla constructed from LSU rRNAs accounting for unequal rates of substitution among sites. Pp. 57–75 in Evolutionary relationships among protozoa (G. Coombs, K. Vickerman, M. Sleigh and A. Warren, eds.). Chapman and Hall, London.

Tovar, J., A. Fischer, and C. G. Clark. 1999. The mitosome, a novel organelle related to mitochondria in the amitochondrial parasite *Entamoeba histolytica*. Mol. Microbiol. 32:1013–1021.

Van de Peer, Y., A. Ben Ali, and A. Meyer. 2000. Microsporidia: accumulating molecular evidence that a group of amito-chondriate and suspectedly primitive eukaryotes are just curious fungi. Gene 246:1–8.

Van Niel, C. B. 1955. The microbe as a whole. Pp. 3–12 in Perspectives and horizons in microbiology (S. A. Waskman, ed.). Rutgers University Press, New Brunswick, NJ.

Waddell, P. J., and M. A. Steel. 1997. General time-reversible distances with unequal rates across sites: mixing gamma and inverse Gaussian distributions with invariant sites. Mol. Phylogenet. Evol. 8:398–414.

Williams, B. A., R. P. Hirt, J. M. Lucocq, and T. M. Embley.

2002. A mitochondrial remnant in the microsporidian Trachipleistophora hominis. Nature 418:865–869.

Woese, C. R. 1987. Bacterial evolution. Microbiol. Rev. 51:221–271.

Woese, C. R., and G. E. Fox. 1977. Phylogenetic structure of the prokaryotic domain: the primary kingdoms. Proc. Natl. Acad. Sci. USA 74:5088–5090.

Woese, C. R., O. Kandler, and M. L. Wheelis. 1990. Towards a natural system of organisms: proposal for the domains Archaea, Bacteria, and Eucarya. Proc. Natl. Acad. Sci. USA 87:4576–4579.

Wolfe, K. H., C. W. Morden, and J. D. Palmer. 1992. Function and evolution of a minimal plastid genome from a non-photosynthetic parasitic plant. Proc. Natl. Acad. Sci. USA 89:10648–10652.

Yang, Z. 1996. Among-site rate variation and its impact on phylogenetic analyses. Trends Ecol. Evol. 11:367–370.

Yarus, M. 2002. Primordial genetics: phenotype of the ribocyte. Annu. Rev. Genet. 36:125–151.

Zuckerkandl, E., and L. Pauling. 1965. Evolutionary divergence and convergence in proteins. Pp. 97–166 in Evolving genes and proteins 9 V. Bryson and H. J. Vogel, eds.). Academic Press, New York.

David P. Mindell
Joshua S. Rest
Luis P. Villarreal

Viruses and the Tree of Life

Viruses, Taxa, and Life

Viruses are rarely included in syntheses regarding the common origin and history for all life forms. There are many reasons for this, including our ignorance of their deep history, an earlier reluctance to consider them as living organisms, and their extreme changeability. However, increasing amounts of molecular sequence data enable more comparisons among viruses and between viruses and other organisms, and we attempt here a brief perspective on the integration of viruses and the Tree of Life. At the outset, we wish to emphasize that viruses have arisen on multiple, independent occasions, being a grade rather than a single clade, and to alert readers to the limitations of the "tree of life" metaphor when applied to virus histories.

Viruses are obligate intracellular parasites averaging 30 nm long, or 1/100th the size of many bacteria. They are the last major kind of organisms to be described, and may represent the last and broadest organismal frontier. Many viruses, when in reproductive mode, can produce thousands of offspring per hour in each of the hundreds or thousands of cells infected in a single host individual. This provides copious grist for the evolutionary mill, in producing a multitude of "winning" virus forms and lifestyles that have ultimately succeeded in colonizing all other organisms, from bacteria to algae, fungi, plants, and animals, and moving with them to all regions and habitats on Earth. The associations between viruses and their hosts range from ephemeral one-time visits without consequence to chronic, fatal associations. In a longer time frame, the associations range from a possibly crucial, transformational role for life's earliest forms, to extinctions of host populations, to an ongoing and deeply integrated role in the evolution of host organisms and their genomes. Success in being small requires great economy in structure and content. Whereas the human nuclear genome includes roughly three billion bases of DNA and about 35,000 genes, many common viruses, such as HIV, carry a mere 10,000 bases of RNA or DNA and nine or so genes. Therefore, an HIV genome is only 0.0003% the size of a human's genome.

The International Committee on Taxonomy of Viruses has published a series of reports seeking to bring order to the expanding catalog of known virus diversity using the familiar nested taxonomic categories of species, genus, and family. The most recent report (Van Regenmortel et al. 2000) names roughly 3600 species and estimates that at least 30,000 viruses, strains, and subtypes are being actively studied in research labs around the world. There is a sense that a "significant fraction" of the primary kinds of viruses are now known, based on the low frequency for discovery of viruses that do not fit into existing families. However, the lower level viral taxa described represent just the tip of the iceberg, because little survey work has been done for viruses outside of those infecting humans and our domestic animals and plants. We have no idea how many different viruses with unique capabilities infect archaebacteria, whales, slime molds, or other of the myriad forms of life.

Early classification for viruses centered on the similarity in diseases or symptoms caused, the means of transmission, or the kinds of organisms or even body organs infected. For example, viruses able to induce swelling of the liver with accompanying fever and yellowing of the skin (jaundice) caused by buildup of a bile pigment were classified together as the "hepatitis viruses." This included what are now seen as distantly related groups such as hepatitis A virus, hepatitis B virus, yellow fever virus, and Rift Valley fever virus. Biochemical and molecular studies in the 1960s and early 1970s facilitated classification of viruses based on the nature of their genetic material, whether RNA or DNA, and whether the genome was double or single stranded and, if single stranded, whether that strand was identical to the messenger RNA (mRNA) transcript (positive-stranded) or complementary to it (negative-stranded; Baltimore 1971). About this time, an approach to classification of viruses was widely adopted in which as many characteristics as possible were considered, and weighted as criteria for classifying viruses into families, genera, and species. The relative weight accorded to different characteristics was arbitrary and potentially biased toward maintenance of groupings that fit preconceived notions of relationships. Beginning in the 1980s and 1990s, biologists sought to develop a taxonomy for viruses based on phylogenetic analyses of shared traits, primarily DNA sequences, although this is a work very much in progress with no guarantee of advance after the most obvious relationships are determined. Based on similarity in the nature of the viral genome, strandedness [(+)sense or (–)antisense] of the viral genome, capacity (or not) for reverse transcription, and polarity of the viral genome, six primary groups are generally recognized, composed of at present 62 families and 233 genera (Van Regenmortel et al. 2000; table 8.1).

Because viruses reproduce asexually, the "biological" species definition, with species recognized on the basis of reproductive isolation among sexually reproducing individuals, is not relevant. This is also the case with the vast majority of other life forms, including bacteria and many eukaryotes, where species and higher level taxa are recognized on the basis of common descent and either relative age of divergence or degree of differentiation. The concept of "quasi-species" was initially developed to describe a wild-type genome of RNA molecules accompanied by a distribution of its mutants in studies of the origin of life (Eigen 1971), and has been extended to RNA viruses. However, the term "quasi-species" is derived from chemistry, in which "species" refers to an assembly of identical molecules, rather than being derived from evolutionary biology in which "species" generally refers to gene flow among individuals or diagnosable evolutionary units. Although the quasi-species concept has been useful as a population genetic model, it has no direct application to systematics and taxonomy. As an indication of this, any particular RNA sequence may belong to more than one quasi-species, depending on which traits the wild-type selected for study is intended to model. A recent definition

explicitly for virus species is as "a polythetic class of viruses that constitute a replicating lineage and occupy a particular ecological niche" (Van Regenmortel 2000). A polythetic class is one in which no single feature is essential for membership.

Viruses have traditionally been excluded from considerations of the Tree of Life. Initially, some biologists balked at recognizing them as life forms and did not consider them to be taxa (a term used loosely to designate any evolutionary lineage), because they depend on their hosts for replication of their own DNA or RNA. In retrospect, this view appears arbitrary and unnecessarily restrictive. Viruses exhibit many features common to other life forms, including structural organization based on heritable nucleic acid sequences, reproduction, use of material resources from their environment, internal homeostatic controls within individuals (virions) to promote survival in changing environments, diversity in form and function of parts, and the capacity to adjust to changing conditions over time and to evolve. There are many obligate parasites that we do not hesitate to call "alive" or recognize as taxa, including the specialized and entirely dependent *Escherichia coli* in our digestive tracts and the many forms of mycorrhizal fungi dependent on and restricted to life on plant roots, as just two examples. Although viruses closely resemble mobile genetic elements, including plasmids, episomes, transposons, and retrotransposons, viruses differ in having individuals mature within proteinaceous capsid and envelope structures that permit efficient target cell receptor specificity and transmission among cells and among host individuals.

Many biological terms, units and concepts defy exact definition and application, due in part to the dynamic processes involved in evolution and the existence of variable intermediates between so many of the recognized units. Consider the difficulty in defining some of the most frequently invoked biological units such as "species" and "gene." "Life" may be seen as similarly difficult to define, and ultimately, its definition is a matter of human convention. T. Dobzhansky famously remarked that nothing in biology makes sense except in the light of evolution, and by extension, it is now widely recognized that nothing in evolution makes sense except in the light of phylogeny. Thus, understanding virus evolution, which is often distinct from that of their hosts, requires a phylogenetic perspective and, ultimately, inclusion in the phylogenetic Tree of Life. Evolution of viruses is increasingly seen as a key component in the history of life.

The more substantive, empirical reason that viruses have been excluded from Tree of Life discussions in the past involves the difficulty and frequent impossibility of finding homologous traits suitable for phylogenetic analyses relating diverse viruses and relating viruses to other organisms (Holland and Domingo 1998), as well as widespread recombination among lineages (see Worobey and Holmes 1999). The shortage of homologous traits will be a lasting impediment to direct comparisons and phylogenetic analyses for

Table 8.1
Six Classes and 62 Recognized Families of Viruses.

Virus family[a]	Representative common name(s)	Known Hosts[b]
Double-strand DNA viruses		
Myoviridae (6)	Phage T4	Arc, Eub
Siphoviridae (6)	Phage 1	Arc, Eub
Podoviridae (3)	Phage T7	Eub
Tectiviridae (1)	Phage PRD1	Eub
Corticoviridae (1)	Phage PM2	Eub
Plasmaviridae (1)	Phage L2	Eub
Lipothrixviridae (1)	Thermoproteus virus 1	Arc
Rudiviridae (1)	Sulfolobus virus SIRV-1	Arc
Fuselloviridae (1)	Sulfolobus virus SSV-1	Arc
Poxviridae (13)	Vaccinia virus, cowpox	Inv, Ver
Asfarviridae (1)	African swine fever virus	Ver
Iridoviridae (4)	Lymphocystis disease virus 1	Inv, Ver
Phycodnaviridae (3)	*Paramecium bursaria* Chlorella virus 1	Alg
Baculoviridae (2)	*Cydia pomonella* granulovirus (CpGV)	Inv
Herpesviridae (9)	Human herpesvirus 1, bald eagle herpesvirus	Ver
Adenoviridae (2)	Human adenovirus A, snake adenovirus	Ver
Polyomaviridae (1)	Simian virus 40 (SV-40), bovine polyomavirus	Ver
Papillomaviridae (1)	Human papillomavirus, canine oral papillomavirus	Ver
Polydnaviridae (2)	*Campoletis aprilis* ichnovirus	Inv
Ascoviridae (1)	*Diadromus pulchellus* ascovirus	Inv
Single-strand DNA viruses		
Inoviridae (2)	Phage M13, *Vibrio* phage v6	Eub
Microviridae (4)	Phage fX174, *Chlamidia* phage 1 (Ch-1)	Eub
Geminiviridae (3)	Maize streak virus (MSV), beet curly top virus	Pla
Circoviridae (1)	Chicken anemia virus, porcine circovirus	Ver
Parvoviridae (6)	Canine parvovirus, *Aedes aegypti* densovirus	Inv, Ver
DNA–RNA reverse-transcribing viruses		
Hepadnaviridae (2)	Hepatitis B virus	Ver
Caulimoviridae (6)	Petunia vein clearing-like virus	Pla
Pseudoviridae (2)	*Saccharomyces cerevisiae* Ty-1 virus	Fun, Inv, Pla
Metaviridae (2)	*Drosophila melanogaster* gypsy virus	Fun, Inv, Pla
Retroviridae (7)	HIV-1, avian leukosis virus	Ver
Double RNA viruses		
Cystoviridae (1)	Phage f6	Eub
Reoviridae (9)	Mammalian orthoreovirus, rice dwarf virus	Inv, Pla, Ver
Birnaviridae (3)	Infectious pancreatic necrosis virus	Inv, Ver
Totiviridae (3)	*Giardia lamblia* virus	Fun, Pro
Partitiviridae (4)	*Penicillium chrysogenum* virus	Fun, Pla
Hypoviridae (1)	*Cryphonectria hypovirus* 1–EP713	Fun
(−) Sense single-strand RNA viruses		
Bornaviridae (1)	Borna disease virus	Ver
Filoviridae (2)	Marburg virus, Zaire ebola virus	Ver
Paramyxoviridae (11)	Mumps virus, measles virus	Pla, Ver
Rhabdoviridae (6)	Rabies virus, potato yellow dwarf virus	Ver, Pla
Orthomyxoviridae (4)	Influenza A virus	Ver
Bunyaviridae (5)	Hantaan virus, tomato spotted wilt virus	Pla, Ver
Arenaviridae (1)	Hepatitis delta virus	Ver
(+) Sense single-strand RNA viruses		
Leviviridae (2)	Phage MS2	Eub
Narnaviridae (2)	*Saccharomyces cerevisiae* narnavirus 20S	Fun
Picornaviridae (6)	Poliovirus, hepatitis A virus	Ver
Sequiviridae (2)	Parsnip yellow fleck virus	Pla
Comoviridae (3)	Tobacco ringspot virus	Pla
Potyviridae (6)	Ryegrass mosaic virus	Pla
Caliciviridae (4)	Rabbit hemorrhagic disease virus	Ver
Astroviridae (1)	Human astrovirus 1	Ver
Nodaviridae (2)	Striped jack nervous necrosis virus	Inv, Ver

(continued)

Table 8.1

(*continued*)

Virus family[a]	Representative common name(s)	Known Hosts[b]
Tetraviridae (2)	Southern bean mosaic virus	Inv
Luteoviridae (3)	Barley yellow dwarf virus-PAV	Pla
Tombusviridae (8)	Oat chlorotic stunt virus	Pla
Coronaviridae (2)	Equine torovirus	Ver
Arteviridae (1)	Equine arteritis virus	Ver
Flaviviridae (3)	Hepatitis C virus, dengue virus	Ver
Togaviridae (2)	Rubella virus, tobacco mosaic virus	Ver, Pla
Bromoviridae (5)	Cucumber mosaic virus	Pla
Closteroviridae (2)	Grapevine virus A	Pla
Barnaviridae (1)	Mushroom bacilliform virus	Fun

[a]Numbers in parentheses denote number of recognized genera. Some genera are currently unassigned to a family and are not included here.
[b]Arc, Archaea; Eub, Eubacteria; Fun, fungi; Inv, invertebrates; Pro, protists; Ver, vertebrates; Pla, plants; Alg, lower animals.
Follows Van Regenmortel et al. (2000).

many virus groups, particularly higher level taxa. However, as new molecular data for both viruses and their hosts are collected, and as comparative evolutionary analyses proceed, an increasing number of explicit hypotheses regarding virus relationships, especially among close relatives, are being developed. Minimally, these provide hypotheses for further testing. In the following sections we provide a brief overview of existing hypotheses regarding virus evolutionary history, recognizing them to be speculative and in many cases only weakly supported.

Virus Origins

Our understanding of ancient virus origins is extremely limited, because of their fast pace of evolutionary change, recombination among lineages, and the very small number of homologous characters available, if any, for comparison between viruses and other organisms. Despite these severe limitations, three general hypotheses for the mechanism of viral origins have been identified and can be referred to as (1) the primordial, (2) the escaped transcript, and (3) the regressive hypotheses (reviewed in Strauss et al. 1996, DeFilippis and Villareal 2001; fig. 8.1). These rely on the same evolutionary mechanisms, including mutation, recombination, and natural selection, known to operate in more recent times and throughout the history of life. These three hypotheses are not mutually exclusive, and more than one may apply in any particular case. These hypotheses of virus origins are distinct from hypotheses of phylogenetic relationship showing patterns of common ancestry among virus lineages subsequent to their origins from nonviruses.

The primordial hypothesis holds that some RNA viruses have been present since the beginnings of life on Earth about 3.8 billion years ago. In this primordial hypothesis, simple RNA molecules, with strings of concatenated nucleotides, arose from pools of free nucleotides as a result of the chemi-

cal and physical attractions among singleton nucleotides. Simple RNA molecules have now been shown to be capable of copying themselves by serving as a polymerase enzyme. They are also able to cut other nucleotide strings and successfully integrate themselves into the cut site. Discovery of these abilities, together with the observation that RNA sequences self-assemble more readily but are less stable over time compared with DNA sequences, have fueled the view of early life being encoded by RNA. Eventually, information storage by reactive RNA molecules was replaced, via reverse transcriptase (RT) activity, by information storage in more stable DNA molecules (reviewed in Joyce 2002). Although some of these early self-replicating molecules eventually collected and organized into duplicating units that we can call "host cells," other molecules were packaged into virus particles that coevolved with host cells and parasitized them. The fact that viruses and their related genetic elements are ubiquitous within the cells or genomes of all life forms also suggests an early origin (fig. 8.1, upper panel). Evidence and scenarios for evolution at the RNA level that may have taken place in simple viral or previral systems are reviewed in Robertson (1992) and Robertson and Neel (1999).

The escaped transcript hypothesis posits that viruses arose from mRNAs or other host-cell RNA or DNA molecules that acquired the ability to be replicated and packaged in a proteinaceous coat, enabling an escape from their cellular confines. mRNAs routinely pass through the membrane of the nucleus, on their way to the ribosomes in the cellular cytoplasm, where they are translated into amino acids. Successful passage through the nuclear membrane makes navigation of the cell wall seem feasible as well, although the mechanisms differ significantly. In this scenario, viruses evolved through a series of intermediate forms, from an obligate intracellular progenitor. Figure 8.1's lower panel illustrates the escaped transcript hypothesis, with the dashed line indicating viral origin from a set of characters that eventually obtained features (additional genes) enabling survival and

Primordial hypothesis

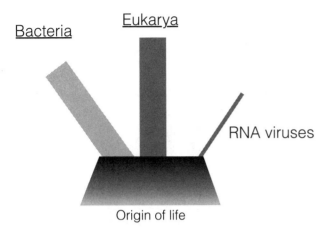

Origin of life

Escaped transcript hypothesis

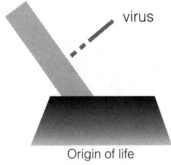

Origin of life

Figure 8.1. Hypotheses for virus origins. (Upper panel) Primordial hypothesis: RNA viruses arise early in the history of life, concomitant with evolution of first cells; dark shading for Eucarya lineage denotes viral genetic contribution to early evolution of Eucarya. (Lower panel) Escaped transcript hypothesis: RNA viruses arise from mRNAs or other host-cell RNA or DNA molecules that acquired the ability to be replicated and packaged in a proteinaceous coat. The polygon base of the diagram denotes early history of life before and including evolution of first cells and horizontal transfer of genetic material.

evolution as a distinct biological entity. Initially developed by Lwoff (1957) and Temin (1980), this hypothesis is widely held for DNA viruses and retroviruses. Despite its appeal, no virus family can be firmly linked to an origin of this kind at present.

The regressive hypothesis supposes that viruses are descended from formerly free-living bacteria that have lost functions and the DNA and structures associated with them. This seemed plausible in the past, given the existence of parasitic,

intracellular bacteria that are entirely dependent on their hosts for energy and synthesis of proteins. However, with the advent of molecular data, this model now appears untenable, given the many structural, functional, and molecular sequence traits known to be shared between viruses and various nonbacterial genetic elements, and as the many disparities between viruses and bacteria become better known.

Virus Phylogenies

Sixty-two different virus families have been recognized (table 8.1), and support for them as monophyletic groups varies from strong to limited. However, only a small number of these families have been related to each other in higher level taxonomic groupings based on phylogenetic considerations, and these include the only three currently recognized orders: Caudovirales, Mononegavirales, and Nidovirales. Other hypothesized relationships among families exist, although the hypothesized clades have not been named. In the following section, we briefly review some of the phylogenetic hypotheses among as well as within virus families.

The earliest classification encompassing all viruses is phenetic, being based on the nature of their genetic material, as mentioned above. These fundamental differences in the genomes, and the associated differences in their molecular biology, suggest the hypothesis that these groups stem from independent and mechanistically different origins. In addition, sets of viruses within the three primary groups (DNA viruses, RNA viruses, reverse-transcribing viruses) mentioned have basic differences from each other [e.g., (+)strand = sense strand vs. (−)strand = sense strand, segmented vs. nonsegmented genomes) that might also be the result of independent origins. Based on these differences in form and function and the apparent feasibility of repeated, independent origins, most researchers would agree that the viral lifestyle has arisen on multiple occasions. If this is the case, viruses as a group comprise a grade, rather than a clade. Grades share a particular lifestyle or form of organization, rather than common ancestry, and that makes them a group sharing convergent similarity, as opposed to a clade, which denotes a monophyletic group representing all and only the descendents of a particular common ancestor. Recognizing viruses as a grade underscores their potential for future independent origins.

RNA Viruses

RNA viruses have RNA genomes and do not replicate via a DNA intermediate as in the reverse-transcribing viruses. The taxonomic majority have single-strand positive [(+)strand = sense strand] genomes, others have single-strand negative (or antisense) genomes, and the rest have double-stranded genomes. Phylogenetic analyses using conserved RNA-dependent RNA polymerase (RdRp) amino acid sequences for representatives of all three RNA virus groups mentioned have been controversial. Zanotto et al. (1996) found that RdRp sequences cannot be used for simultaneous phylogenetic analysis of all RNA viruses based on a lack of sequence similarity and reliable phylogenetic signal, with alternative alignments and phylogenetic methods yielding incongruent topologies and none of the hypothesized multifamily supergroups (described below) receiving significant support. More recently, Gibbs et al. (2000) present analyses supporting monophyly of RdRp sequences from the postulated alpha-like virus supergroup of single-strand positive RNA (ss+RNA) viruses (including alfamoviruses and closteroviruses, among others), although their analyses also do not support simultaneous analysis of all RdRp sequences.

Previously, a single, common origin for this RdRp in all RNA viruses had been postulated (Gorbalenya 1995), consistent with the notion of a single origin for RNA viruses (Strauss et al. 1996; fig. 8.2, upper panel). Analyses of RdRp together with helicase and chymotrypsin-like proteases had suggested that each of the three primary RNA virus genomic classes [ss+, single-strand negative (ss–), double-strand (ds)] represents a monophyletic group (Gorbalenya 1995). Some researchers had suggested that dsRNA viruses originated multiple times independently from ss+RNA viruses (Koonin and Dolja 1993, Ward 1993), which comprise about 80% of known RNA viruses. Others interpreted phylogenetic evidence to suggest that dsRNA viruses gave rise to ss+RNA viruses, which gave rise, in turn, to ss–RNA viruses (Bruenn 1991, Goldbach and De Haan 1994). There is no consensus on this, and utility of RdRp at this level is problematic. Further, RNA viruses had been classified into six "supergroups" (Carmo-like, Sobemo-like, Picorna-like, Flavi-like, Alpha-like, and Corona-like viruses), each including multiple families, based on morphologic and genomic characteristics as well as phylogenetic analysis of conserved protein sequences (Gorbalenya and Koonin 1989, Gorbalenya 1995). Among the ss+RNA viruses, the families Coronaviridae and Arteriviridae were placed together as the only two members of the order Nidovirales. An explicit hypothesis for phylogeny among ss+RNA Picorna-like viruses is presented in figure 8.3, upper panel, and among Tombusviridae taxa, in figure 8.3, lower left panel. Among the ss–RNA viruses, four families of enveloped, linear, nonsegmented viruses (Bornaviridae, Filoviridae, Paramyxoviridae, and Rhabdoviridae) were placed together in the order Mononegavirales (fig. 8.3, lower right panel). Bornaviridae differs from the others in having a unique pattern of mRNA processing. These high-level groupings remain speculative.

Although both the RNA viruses and the reverse-transcribing viruses have RNA genomes, their use of different virally encoded polymerases (RdRp and RT, respectively) suggests separate origins for them. However, an alternative view, which assumes a common ancestor for RNA viruses and the reverse-transcribing viruses, or at least their polymerases, has been used in rooting phylogeny for RT sequences with RdRp (e.g., Eickbush 1997). The structures of two RTs and three

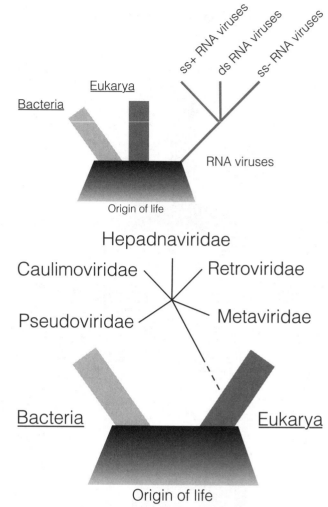

Figure 8.2. Hypotheses for phylogeny and origins among RNA viruses showing potential monophyly after a primordial origin (upper panel) and reverse-transcribing viruses showing potential monophyly and an escaped transcript origin (lower panel). ds, double strand; ss, single strand.

RdRps have been determined, and the similarity between these structures, in configuration and order of domains, is consistent with the view that RNA-dependent polymerases of picornaviruses, flaviviruses, and retroviruses share a common ancestor (e.g., Bressanelli et al. 1999, Ago et al. 1999). However, alignments for RdRp and RT must still be viewed cautiously because of relatively low similarity between RT and RdRp sequences, and the possibility that their similarity might be due to similar functions and convergent evolution.

Reverse-Transcribing Viruses

The five families in this group (table 8.1, fig. 8.2, lower panel) all replicate by reverse transcription and encode the enzyme RT. All five families are thought to share common ancestry, possibly via descent from host genomic elements with RT known as long-terminal-repeat (LTR) retrotransposons, and

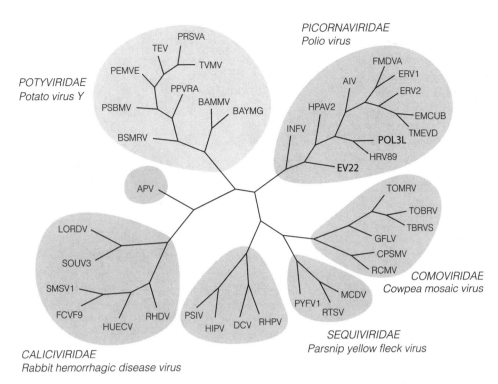

Figure 8.3. (Upper panel) Phylogenetic hypothesis for ss+RNA viruses of the Picorna-like supergroup based on RNA polymerase 3D^{pol} (Gromeier et al. 1999). Two provisional groups are unassigned to a family. (Lower left panel) Phylogenetic hypothesis for select Tombusviridae genera based on DNA polymerase. (Lower right panel) Phylogenetic hypothesis for order Mononegavirales based on DNA polymerase (Pringle and Easton 1997). Note the non-monophyly for Paramyxoviridae. Common names are given for family representatives.

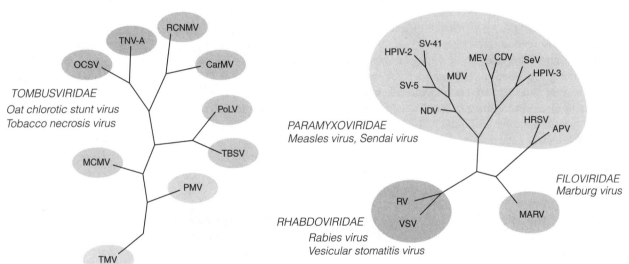

to comprise a monophyletic group. Position of the root is not known, and correspondingly, relationships among families remain uncertain. It is also possible, however, that two or more of the five families denote independent origins (see Temin 1980, Xiong and Eickbush 1990, Eickbush 1997, McClure 1999, Boeke et al. 2000). Retroviridae, Metaviridae, and Pseudoviridae have RNA genomes, whereas Caulimoviridae and Hepadnaviridae (including hepatitis B virus) have DNA genomes, transcribed by host DNA polymerase, and then reverse transcribed by the virus's own RT. A phylogenetic hypothesis for seven genera within the best-known family, Retroviridae, is presented in figure 8.4, left panel.

Phylogenetic analyses of conserved RT domains unite an impressive array of elements, including RT from reverse-transcribing virus families, numerous cellular and organellar retroelements, and the cellular gene telomerase, which performs elongation of telomeres (repeated DNA sequences capping chromosome ends) in eukaryotes. RT analyses rooted with RdRp indicate monophyly for a set of RT sequences from prokaryotic and mitochondrial genomes, including group II introns and retrons as sister groups, with successively basal divergences for non-LTR retrotransposons, telomerases, and LTR retrotransposons, which include retroviruses (Eickbush 1997). Analyses excluding RdRps and using the prokaryotic retroelements as the outgroup yield a

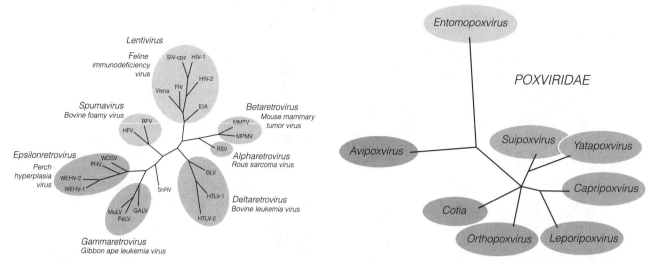

Figure 8.4. (Left panel) Phylogenetic hypothesis for the seven Retroviridae genera based on RT sequences (Hunter et al. 2000, Dimmic et al. 2002). Common names are given for genus representatives. (Right panel) Phylogenetic hypothesis for select Poxviridae genera based on thymidine kinase DNA sequences (Moyer et al. 2000).

different topology, with LTR retrotransposons and telomerases as sister taxa and non-LTR retrotransposons as sister to them. This difference in topology implies different scenarios for the relative timing of origin for telomerase, retrotransposons, and reverse-transcribing viruses. Telomerases and non-LTR retrotransposons have similar catalytic mechanisms, in which the 3' hydroxyl group of a DNA end is used to prime reverse transcription. Their functional similarity is demonstrated even more dramatically by the finding that non-LTR retrotransposons (TART and HeT-A) appear to have replaced telomerase for telomere replication in *Drosophila melanogaster* (Levis et al. 1993). Regardless of which topology for the vast array of RT sequences is correct, gene trees like those described above indicate the dynamic nature of RT and reverse-transcribing virus evolution, and the important role of RT in evolutionary history.

DNA Viruses

The DNA viruses are a heterogeneous group. Some have double-stranded genomes, and others have single-stranded genomes. Some are enveloped, and others are not; some encode polymerase, and some others do not. They vary in size from <2 to >670 kilobases. There is no evidence indicating monophyly for DNA viruses overall, and it appears likely that DNA viruses have had multiple origins, possibly via the hypothesized escaped element mechanism outlined above. Like RT, all DNA-dependent DNA polymerases (DdDps), whether from DNA viruses or from the genomes of eukaryotes and prokaryotes, appear to have evolved from a single common ancestor (Knopf 1998, Wang 1991). The ordering of functional domains for these proteins appears well conserved. However, DNA viruses with DdDp (including phycodnaviruses, poxviruses, baculoviruses, and mycobac-

teriophages, among others) are highly divergent and cannot be linked by evidence to form a monophyletic group. Filée et al. (2002) present phylogenetic analyses for five different DNA polymerase families, also indicating a complex history of lateral gene transfer among viruses, plasmids, and their diverse hosts. Among the dsDNA viruses, three diverse families of tailed viruses infecting bacteria (Myoviridae, Siphoviridae, and Podaviridae) are placed together in the order Caudovirales. The ssDNA viruses all use a protein-primed DNA replication mechanism that is distinct from that of other viruses. Poxviridae is an example of a large and well-known DNA virus family with well-supported phylogenetic structure (fig. 8.4, right panel).

Why Try to Integrate Viruses in the Tree of Life?

Efforts to determine the phylogenetic origins and subsequent pattern of evolution for viruses, obscured as they are, can be justified on the same basis as all Tree of Life research: we desire a comprehensive understanding of life's history. This comprehensive understanding entails inclusion of all taxa, to whatever extent possible, for two reasons: first, so all major groups are accounted for (i.e., so the vastness of our ignorance is appropriately exposed, and not hidden for convenience), and second, so the record of character and organismal change can be recovered as accurately as possible. One of the lessons of phylogenetics is that our understanding of the record of evolutionary change generally improves as we integrate more taxa and more characters into our analyses. Although most events in the long and varied evolutionary histories for the grade we call "viruses" are unrecoverable, viruses are not unique in this regard. As one example, paleontologists also work with small amounts of fragmentary data to reconstruct history based on one or a few representatives of diverse (and in their case often extinct) clades. The unique and significant role of viruses (see below) in the evolution of life makes the effort of placing them in the context of the Tree of Life particularly compelling.

Reverse Transcriptase and Transition from an RNA to a DNA World

An early difficulty in studies of the origin and evolution of life had been in explaining DNA synthesis. DNAs are synthesized with the help of enzymes, which are themselves encoded by DNA. This leaves one wondering how those early DNA-synthesizing proteins came into being. Beginning in the late 1960s a series of hypotheses and, later, discoveries were made that led to our current view of an early RNA world as a precursor to our current DNA world, where all organisms other than viruses have DNA genomes. The ribonucleotides in RNA were found to be more readily synthesized than the deoxyribonucleotides in DNA, and most important, some RNAs (ribozymes) were indeed capable of self-replication. The finding that RNAs are less stable over time than DNAs provided the underlying pressure for natural selection to effect a change from RNA to DNA as the heritable material for storing information that encodes organisms. RT is the only known enzyme capable of synthesizing DNA from RNA templates and has apparently played a pivotal role in the transition between RNA and DNA worlds. This enzyme is the defining feature of the reverse-transcribing viruses (table 8.1) and for a larger, encompassing group of genetic elements (retroids, e.g., retrons, retrotransposons, retroplasmids). As a consequence, understanding the history of RT evolution, in the reverse-transcribing families of viruses (table 8.1) and other retroids, gives us a fuller picture of the capabilities and past activities of this apparently seminal agent. The extent to which retroids have been involved in ancient and recent events of genome evolution is just beginning to be assessed (e.g., McClure 1999, Moran et al. 1999, Kidwell and Lisch 2000).

Viruses and Eukaryotic Genomes

Phylogeneticists are silent regarding diversification among RNA world entities, because none survive as such, with the possible exception of some RNA viruses, as mentioned above. The three extant, primary lineages of DNA-based organisms are recognized as Bacteria, Archaea, and Eucarya (Woese 1987). Hypotheses regarding the origin of eukaryotic cells generally invoke symbioses between eubacterial and methanogenic archaeal taxa (e.g., Lake and Rivera 1994, Martin and Muller 1998, Moreira and Lopez-Garcia 1998), although this view has been questioned recently, with emphasis given to "communal" genomic evolution and horizontal gene transfer as a primary force (Woese 2002). There is

limited evidence suggesting a possible role for horizontal gene transfer from some dsDNA viruses, in the early evolution of Eucarya. Phylogenetic evidence suggesting a viral contribution to eukaryotic cellular evolution entails finding of sister relationships for orthologous viral and eukaryotic (nuclear) genes, which are preceded by divergences among virus orthologs. Such interpretations are, of course, critically dependent on assumptions regarding position of the phylogenetic root. For example, combined analyses of guanyltransferases and related ATP-dependent ligases from diverse Poxviridae and Asfarviridae taxa (e.g., African swine fever virus) and diverse eukaryotes (including *Homo*, *Saccharomyces*, and *Methanococcus*) support earlier divergence among virus orthologs relative to divergence among eukaryotic orthologs (Bell 2001). Similar phylogenetic patterns have been found for various DNA polymerases (Knopf 1998, Villarreal 1999), DNA topoisomerase (Garcia-Beato et al. 1992), and possibly RNA polymerase large subunit (Sonntag and Darai 1996). Similar phylogenetic patterns relating these viral and eukaryotic sets of orthologs is consistent with a common evolutionary history for each set, and their presence in an ancestral virus, possibly residing within an archaeal host, before the emergence of eukaryotes. Horizontal transfer can be multidirectional, and phylogenetic analyses are revealing instances of eukaryotic gene capture by viruses as well (e.g., Hughes 2002). As more eukaryotic genes and genomes are sequenced, more evidence for past colonization events by viruses is coming to light (especially for retroviruses; e.g., Dimcheff et al. 2000).

Applications to Individual and Public Health

Traditionally, viral pathogens are identified on the basis of disease symptoms and in the context of epidemiological (population) analyses. However, as molecular sequencing becomes routine and databases grow, rapid identification of viral isolates can often be done based on explicit sequence comparisons of unknown isolates with known sequences. Quick characterizations based on presence or absence of particular sequences often suffice for basic diagnosis, but phylogenetic analyses allow much greater detail. For some viruses, phylogenetic identification is particularly important for identifying particular strains or subtypes (as for HIV-1) having a small number of unique changes that can underlie significant differences in virulence, transmissibility, drug resistance, or other traits of interest. Further, phylogenetic analyses ensure that identification is based on evolutionary relatedness rather than just similarity, which can reflect convergence. Thus, having virus phylogenies available, in as much detail as possible, helps in rapid, accurate identification of unknown viral isolates and in understanding of the health risks and preventative measures that might be taken.

We can better understand a virus epidemic's origin and work more effectively to reduce future epidemics, if we understand the pathogen's phylogenetic history, host species range, and the geographic ranges of both host and pathogen. For example, understanding phylogeny of Lentiviridae taxa, including HIV and other primate immunodeficiency viruses (e.g., Sharp et al. 2001), informs us about the importance of avoiding direct contact with blood or other infected tissues from other primates, particularly chimpanzees harboring a closely related SIV (simian immunodeficiency virus). Detailed phylogeny for HIV-1 taxa helps in tracking the spread of the most virulent lineages and understanding which sequence-level changes are associated with enhanced transmissibility and virulence, and which particular sequence sites are subject to accelerated rates of change due to selection pressure imposed by hosts' immune systems. Similarly, understanding the phylogenetic position for West Nile viruses (Flaviviridae) can potentially help in determining the source and the cause for its recent spread to the Western Hemisphere as well as its history of change (e.g., Anderson et al. 2001). Accurate phylogeny for pathogens is important in understanding any zoonosis (disease transmitted from nonhuman to human hosts). If we can determine phylogeny for the viral lineages we can potentially infer the molecular changes that are associated with cross-species transmission and increased virulence and can potentially enhance remediation efforts, including, in some cases, development of antiviral medications.

Phylogeny can contribute to improved vaccine development, because identification of viruses best suited for development of host immunity generally entails choice of the same lineage as, or one closely related to, that in circulation. Information on relatedness is also relevant in constructing chimeric (recombinant) virus vaccines. Attenuated (weakened) chimeric viruses used as vaccines may include the genes whose products elicit development of the desired antibodies, as well as including other sequence regions bearing mutations that keep the virus benign. Further, consensus sequences or even phylogenetically inferred ancestral sequences could be used in vaccine design to minimize the differences between engineered vaccine strains and diverse strains in circulation (e.g., Gaschen et al. 2002).

Recent work on wildlife infectious diseases indicates that the majority are viral in origin and that their spread into new wildlife species is often mediated by human disturbances (Dobson and Foufopoulos 2001). Understanding the virus phylogeny can help inform enlightened management practices. This may include reducing human disturbances that foster cross-species transmission for viruses related to the known pathogen, restricting introductions of species associated with viruses closely related to those known to cross host-species boundaries, and restricting the handling of live individuals or of tissues harboring similarly related viruses.

Gene therapy is a novel form of molecular medicine attempting to correct genetic disorders and inhibit disease progression. Functional copies of human genes are inserted into viral expression vectors and carried by them into cells, where they are integrated into the host's genome or maintained as autonomous units (Pfeifer and Verma 2001). The potential exists to influence the outcome of many diseases, ranging from birth defects, to cancer, to neurological disorders. Most work to date has focused on a small set of animal viruses, including SV40 (Polyomaviridae), murine lukemia virus, HIV (Retroviridae), adenovirus (Adenoviridae), and adeno-associated virus (Parvoviridae). As suitable viruses and viral components are identified, knowledge of their phylogenetic relationships may crucially inform the search for additional candidates, given that the desired traits are more likely to be shared with closely related groups than with distantly or unrelated groups.

Outlook

The problems faced by biologists working on the origins and phylogeny of viruses are severe and quantitatively, although not qualitatively, different from those faced by systematists working on other taxa. The two primary challenges may be summarized as (1) identifying as many homologous traits (Mindell and Meyer 2001) as possible for comparisons among viruses and between viruses and other organisms, and (2) identifying recombination among lineages and its role in diversification of taxa. Shortages of homologous characters are inherent in the study of viruses, because of small genome sizes, apparent independent origins for multiple groups, rapid rates of sequence evolution (for RNA viruses in particular) confounding alignments, and high levels of viral lineage extinction. Frequent recombination is also inherent among and within viral lineages, stemming from the ability of multiple viruses to coinfect individual host cells and their general capacity for dramatic change. Although problematic for systematists, the capability for recombination is a key feature in the evolutionary success of viruses. One form of recombination (reassortment) is particularly well known as a successful strategy for influenza A viruses (Orthomyxoviridae), mixing genome segments from different parental lineages in progeny, yielding novel genotypes not recognized by hosts' immune systems. Recombination among viral lineages, due to template switching, is also common in the proliferation and spread of HIV-1 among human populations (Robertson et al. 1995) and dengue fever viruses (Flaviviridae) as well (Worobey et al. 1999).

As a consequence of these inherent difficulties, much of the complex evolutionary history for viruses is unrecoverable. However, in assembling the Tree of Life, we seek a maximally comprehensive understanding of life's history, which means that all life forms, including viruses, must still be considered. Continued study of virus evolution has important applied uses as well, for individual health, public health, and environmental health. Despite limitations, increasingly sophisticated methods for sequence alignments and phylogenetic analyses, combined with an expanding molecular sequence database for diverse viral taxa, will allow systematists to improve resolution of some, although by no means all, ancient relationships. Secondary and tertiary structure of proteins are a promising source of conserved characters, and additional phylogenetic insights for ancient events are likely to be found as structural databases grow and are used in comparative analyses. Increased understanding of viral history, for both virus lineages and virus genes, has begun and will continue to transform our view of the shape, the shaping, and the interconnectedness of the Tree of Life.

Finally, we can ask how well the "tree of life" metaphor, coined by Darwin, describes complex virus histories that include recombination among lineages, occasional horizontal transfer of genes with hosts, and possible origination from sets of escaped genetic characters (rather than the usual mode of whole organismal population divergence and lineage splitting). Trees as phylogenetic diagrams give the impression of organismal diversification resulting from a series of nearly instantaneous lineage bifurcations, with single lines dividing neatly into two, and continuing in splendid genetic isolation from each other. Although there are many well-defined monophyletic viral groups, one can only conclude that the overall fit of the metaphor is poor. Nonetheless, the metaphor of the Tree of Life is useful and deeply entrenched in biological discourse, even if simplistic or misleading in some ways. Interestingly, before settling on the phrase "tree of life," Darwin wrote of a "coral of life" (Barrett et al. 1987; see Gould 2002). With occasional connections among branches for some forms, corals may provide a better depiction of viral origins and diversification.

Acknowledgments

We thank Eddie Holmes for valuable comments on an earlier draft of the manuscript, and we thank the editors of this book for their willingness to try something new. D.P.M. was supported by National Science Foundation grant DBI 9974525.

Literature Cited

Ago, H., T. Adachi, A. Yoshida, M. Yamamoto, N. Habuka, K. Yatsunami, and M. Miyano. 1999. Crystal structure of the RNA-dependent RNA polymerase of hepatitis C virus. Structure 7:1417–1426.

Anderson, J. F., C. R. Vossbrinck, T. G. Andreadis, A. Iton, W. H. Beckwith, and D. R. Mayo. 2001. A phylogenetic approach to following West Nile virus in Connecticut. Proc. Natl. Acad. Sci. USA 98:12885–12889.

Baltimore, D. 1971. Expression of animal virus genomes. Bacteriol. Rev. 35:235–241.

Barrett, P. H., P. J. Gautey, S. Herbert, D. Kohn, and S. Smith. 1987. Charles Darwin's notebooks, 1836–1844. Cambridge University Press, Cambridge.

Bell, P. J. L. 2001. Viral eukaryogenesis: was the ancestor of the nucleus a complex DNA virus? J. Mol. Evol. 53:251–256.

Boeke, J. D., T. H. Eickbush, S. B. Sandmeyer, and D. F. Voytas.

2000. Pp. 349–357 in Virus taxonomy (M. H. V. Van Regenmortel, C. M. Fauquet, D. H. L. Bishop, E. B. Carstens, M. D. Estes, S. M. Lemon, J. Maniloff, M. A. Mayo, D. J. McGeoch, C. R. Pringle, and R. B. Wickner, eds.). Academic Press, San Diego.

Bressanelli, S., L. Tomei, A. Roussel, I. Incitti, R. L. Vitale, M. Mathieu, R. De Francesco, and F. A. Rey. 1999. Crystal structure of the RNA-dependent RNA polymerase of hepatitis C virus. Proc Natl. Acad. Sci. USA. 96:13034–13039.

Bruenn, J. A. 1991. Relationships among the positive strand and double-strand RNA viruses as viewed through their RNA-dependent RNA polymerases. Nucleic Acids Res. 19:217–226.

DeFilippis, V. R., and L. P. Villarreal. 2001. Virus evolution. Pp. 353–370 in Field's virology, 4th ed. (D. M. Knipe and P. M. Howley, eds.), vol. 1. Lippincott, Williams and Wilkins, New York.

Dimcheff, D. E., S. V. Drovetski, M. Krishnan, and D. P. Mindell. 2000. Cospeciation and horizontal transmission of avian sarcoma and leukosis virus *gag* genes in galliform birds. J. Virol. 74:3984–3995.

Dimmic, M. W., J. S. Rest, D. P. Mindell, and R. A. Goldstein. 2002. rtREV: a substitution matrix for inference of retrovirus and reverse transcriptase phylogeny. J. Mol. Evol. 55:65–73.

Dobson, A., and J. Foufopoulos. 2001. Emerging infectious pathogens of wildlife. Philos. Trans. R. Soc. Lond. B 356:1001–1012.

Eickbush, T. H. 1997. Telomerase and retrotransposons: which came first? Science 277:911–912.

Eigen, M. 1971. Self-organization of matter and the evolution of biological macromolecules. Naturwissenschaften 58:465–523.

Filée, J., P. Forterre, T. Sen-Lin, and J. Laurent. 2002. Evolution of DNA polymerase families: evidences for multiple gene exchange between cellular and viral proteins. J. Mol. Evol. 54:763–773.

Garcia-Beato, R., J. M. P. Freue, C. Lopez-Otin, R. Blasco, E. Vinuela, and M. L. Salas. 1992. A gene homologous to topoisomerase II in African swine fever virus. Virology 188:938–947.

Gaschen, B., J. Taylor, D. Yusim, B. Foley, F. Gao, D. Lang, V. Novitsky, B. Haynes, B. Hahn, T. Bhattacharya, and B. Korber. 2002. Diversity considerations in HIV-1 vaccine selection. Science 296:2354–2360.

Gibbs, M. J., R. Koga, H. Moriyama, P. Pfeiffer, and T. Fukuhara. 2000. Phylogenetic analysis of some large double-stranded RNA replicons from plants suggests they evolved from a defective single-stranded RNA virus. J. Gen. Virol. 81:227–233.

Goldbach, R., and P. De Haan. 1994. RNA viral supergroups and the evolution of RNA viruses. Pp. 105–119 in The evolutionary biology of viruses (S. S. Morse, ed.). Raven Press, New York.

Gorbalenya, A. E. 1995. Origin of RNA viral genomes: approaching the problem by comparative sequence analysis. Pp. 49–66 in Molecular basis of virus evolution (A. J. Gibbs, C. H. Calisher, and F. Garacia-Arenal, eds.). Cambridge University Press, Cambridge.

Gorbalenya, A. E., and E. V. Koonin. 1989. Viral-proteins containing the purine ntp-binding sequence pattern. Nucleic Acids Res. 17:8413–8440.

Gould, S. J. 2002. The structure of evolutionary theory. Harvard University Press, Cambridge, MA.

Gromeier, M., E. Wimmer, and A. E. Gorbalenya. 1999. Genetics, pathogenesis and evolution of picornaviruses. Pp. 287–343 in Origin and evolution of viruses (Domingo, E., R. G. Webster, and J. J. Holland, eds.). Academic Press, San Diego.

Holland, J. J., and E. Domingo. 1998. Origin and evolution of viruses. Virus Genes 16:13–21.

Hughes, A. L. 2002. Origin and evolution of viral interleukin-10 and other DNA virus genes with vertebrate homologues. J. Mol. Evol. 54:90–101.

Hunter, E., J. Casey, B. Hahn, M. Hayami, B. Korber, R. Kurth, J. Neil, A. Rethwilm, P. Sonigo, and J. Stoye. 2000. Pp. 369–387 in Virus taxonomy (M. H. V. Van Regenmortel, C. M. Fauquet, D. H. L. Bishop, E. B. Carstens, M. D. Estes, S. M. Lemon, J. Maniloff, M. A. Mayo, D. J. McGeoch, C. R. Pringle, and R. B. Wickner, eds.). Academic Press, San Diego.

Joyce, G. F. 2002. The antiquity of RNA-based evolution. Nature 418:214–221.

Kidwell, M. G., and D. R. Lisch. 2000. Transposable elements and host genome evolution. Trends Ecol. Evol. 15:95–99.

Knopf, C. W. 1998. Evolution of viral DNA-dependent polymerases. Virus Genes 16:47–58.

Koonin, E. V., and V. V. Dolja. 1993. Evolution and taxonomy of positive-strand RNA viruses—implications of comparative-analysis of amino-acid-sequences. Crit. Rev. Biochem. Mol. 28:375–430.

Lake, J. A., and M. C. Rivera. 1994. Was the nucleus the 1st endosymbiont? Proc. Natl. Acad. Sci. USA 91:2880–2881.

Levis, R. W., R. Ganesan, K. Houtchens, L. A. Tolar, and F. Sheen. 1993. Transposons in-place of telomeric repeats at a Drosophila telomere. Cell 75:1083–1093.

Lwoff, A. 1957. The concept of virus. J. Gen. Microbiol. 17:239–253.

Martin, W., and M. Muller. 1998. The hydrogen hypothesis for the first eukaryote. Nature 392:37–41.

McClure, M. A. 1999. The retroid agents: disease, function and evolution. Pp. 163–195 in Origin and evolution of viruses (E. Domingo, R. G. Webster, and J. J. Holland, eds.). Academic Press, San Diego.

Mindell, D. P., and A. Meyer. 2001. Homology evolving. Trends Ecol. Evol. 16:434–440.

Moran, J. V., R. J. De Barardinis, and H. H. Kazazian. 1999. Exon shuffling by L1 retrotransposition. Science 283:1530–1534.

Moreira, D., and P. Lopez-Garcia. 1998. Symbiosis between methanogenic archaea and delta-proteobacteria as the origin of eukaryotes: the syntrophic hypothesis. J. Mol. Evol. 47:517–530.

Moyer, R. W., B. M. Arif, D. N. Black, D. B. Boyle, R. M. Buller, K. R. Dumbell, J. J. Esposito, G. McFadden, B. Moss, A. A. Mercer, et al. 2000. Poxviridae. Pp. 137–157 in Virus taxonomy (M. H. V. Van Regenmortel, C. M. Fauquet, D. H. L. Bishop, E. B. Carstens, M. D. Estes, S. M. Lemon, J. Maniloff, M. A. Mayo, D. J. McGeoch, C. R. Pringle, and R. B. Wickner, eds.). Academic Press, San Diego.

Pfeifer, A., and I. M. Verma. 2001. Virus vectors and their applications. Pp. 469–491 in Field's virology, 4th ed. (D. M. Knipe and P. M. Howley, eds.), vol. 1. Lippincott, Williams and Wilkins, New York.

Pringle, C. R., and A. J. Easton. 1997. Monopartite negative strand RNA genomes. Semin. Virol. 8:49–57.

Robertson, D. L., P. M. Sharp, F. E. McCutchan, and B. H. Hahn. 1995. Recombination in HIV-1. Nature 374:124–126.

Robertson, H. D. 1992. Replication and evolution of viroid-like pathogens. Curr. Top. Microbiol. Immunol. 176:214–219.

Robertson, H. D., and O. D. Neel. 1999. Virus origins: conjoined RNA genomes as precursors to DNA genomes. Pp. 25–35 in Origin and evolution of viruses (E. Domingo, R. Webster, and J. Holland, eds.). Academic Press, San Diego.

Sharp, P. M., E. Bailes, R. R. Chaudhuri, C. M. Rodenburg, M. O. Santiago, and B. H. Hahn. 2001. The origins of acquired immune deficiency syndrome viruses: where and when? Philos. Trans. R. Soc. Lond. B 356:867–876.

Sonntag, K-C., and G. Darai. 1996. Evolution of viral DNA-dependent RNA polymerases. Virus Genes 11:271–284.

Strauss, E. G., J. H. Strauss, and A. J. Levine. 1996. Virus evolution. Pp. 153–171 in Field's virology, 3rd ed. (B. N. Field, D. M. Knipe, P. M. Howley, et al., eds.). Lippincott-Raven, Philadelphia.

Temin, H. M. 1980. Origin of retroviruses from cellular moveable genetic elements. Cell 21:599–600.

Van Regenmortel, M. H. V. 2000. Introduction to the species concept in virus taxonomy. Pp. 3–16 in Virus taxonomy (M. H. V. Van Regenmortel, C. M. Fauquet, D. H. L. Bishop, E. B. Carstens, M. D. Estes, S. M. Lemon, J. Maniloff, M. A. Mayo, D. J. McGeoch, C. R. Pringle, and R. B. Wickner, eds.). Academic Press, San Diego.

Van Regenmortel, M. H. V., C. M. Fauquet, D. H. L. Bishop, E. B. Carstens, M. D. Estes, S. M. Lemon, J. Maniloff, M. A. Mayo, D. J. McGeoch, C. R. Pringle, and R. B. Wickner, (eds.). 2000. Virus taxonomy. Academic Press, San Diego.

Villarreal, L. P. 1999. DNA virus contribution to host evolution. Pp. 391–420 in Origin and evolution of viruses (E. Domingo, R. Webster, and J. Holland, eds.). Academic Press, New York.

Wang, T. S. 1991. Eukaryotic DNA polymerases. Annu. Rev. Biochem. 60:513–552.

Ward, C. W. 1993. Progress towards a higher taxonomy of viruses. Res. Virol. 144:419–453.

Woese, C. R. 1987. Bacterial evolution. Microbiol. Rev. 51:221–271.

Woese, C. R. 2002. On the evolution of cells. Proc. Natl. Acad. Sci. USA 99:8742–8747.

Worobey, M., and E. C. Holmes. 1999. Evolutionary aspects of recombination in RNA viruses. J. Gen. Virol. 80:2535–2543.

Worobey, M., A. Rambaut, and E. C. Holmes. 1999. Widespread intra-serotype recombination in natural populations of dengue virus. Proc. Natl. Acad. Sci. USA 96:7352–7357.

Xiong, Y., and T. H. Eickbush. 1990. Origin and evolution of retroelements based upon their reverse transcriptase sequences. EMBO J. 9:3353–3362.

Zanotto, P. M. de A., M. J. Gibbs, E. A. Gould, and E. C. Holmes. 1996. A reevaluation of the higher taxonomy of viruses based on RNA polymerases. J. Virol. 70:6083–6096.

The Relationships of Green Plants

Algal Evolution and the Early Radiation of Green Plants

Charles F. Delwiche
Robert A. Andersen
Debashish Bhattacharya
Brent D. Mishler
Richard M. McCourt

Eukaryotes perform photosynthesis thanks to a specialized organelle that is derived from once free-living cyanobacteria (i.e., blue-green algae). In land plants and green algae this organelle is called the "chloroplast" in reference to its green pigmentation, and by analogy the photosynthetic organelles of groups with other pigmentation patterns have been called "rhodoplasts" (red algae), "chromoplasts" (brown algae), and so forth. This terminology is confused by the fact that the chloroplasts of land plants exist in a number of developmental forms that have sometimes been given names redundant with those of the different algal lineages (e.g., "chromoplast" for the carotenoid-rich form that gives color to some ripe fruit). For simplicity, and because these organelles seem to share a common ancestry (Delwiche et al. 1995), we use the term "plastid" to refer to all such organelles. The hallmark of the plastid is its reduced genome and concomitant complete dependence upon the nuclear genome of the host cell. Like mitochondria, the other clearly endosymbiotic organelle, plastids have genomes that are greatly reduced in size and complexity from those of their free-living cyanobacterial relatives (Glöckner et al. 2000). For example, the fully sequenced genome of the free-living cyanobacterium *Nostoc* sp. PCC 7120 is 6.4 Mb (million bases) in size, and encodes about 6626 genes (i.e., protein- or RNA-coding regions), and that of *Synechocystis* sp. PCC 6803 is 3.5 Mb and encodes about 4003 genes. By contrast, well-characterized plastid genomes range from 136 Kb and 191 genes in the glaucocystophyte *Cyanophora paradoxa* to 120 Kb and 108 genes in the land plant *Pinus thunbergii*, with smaller and less complex genomes in some species with nonphotosynthetic plastids. Some larger plastid genomes are known, but these seem to be special cases, and a typical plastid genome encodes ≤5% of the number of genes found in a free-living cyanobacterium (Palmer and Delwiche 1998). The number of proteins expressed in a typical plastid is, however, much larger than the number encoded in the plastid genome. This is accounted for by the massive transfer of former plastid genes to the nucleus (Martin et al. 2002). Because plastids still need the products of these transferred genes, they are utterly dependent upon the nuclear genome of their host cell. Thus, plastids are tightly integrated into the host cell and show close genetic, physiological, and developmental coordination with the host.

With the exception of land plants, all eukaryotes with plastids are called "algae." The land plants or "embryophytes," are a monophyletic group of green algae that are characterized by a life cycle that involves multicellular haploid and diploid phases (the "alternation of generations") and a suite of distinctive ultrastructural and biochemical features, notably, phragmoplastic cell division and plasmodesmata (described below). The term "alga" has fallen out of favor in recent years, in large part because of the belief that it encompasses a polyphyletic set of lineages. This is accurate with reference to the nuclear phylogeny, but neglects the plastid component of the cell. Because there is substantial (although not conclusive) evidence that plastids are monophyletic, in this chapter we use "algae" to refer to a diverse assemblage of eukaryotic autotrophic organisms, usually aquatic, that are polyphyletic with respect to their nuclear genomes but monophyletic with respect to their plastids (described below). In a strict sense, the land plants should be viewed as a specific, largely terrestrial, lineage of green algae. With renewed interest in algae that poison, kill, pollute, or aggressively invade new habitats, the general public is regaining an interest and appreciation of this diverse assemblage of organisms. And that interest extends to the more benign and beautiful algae, as well as those species used for food by a large number of human societies, for example, nori (*Porphyra*), kombu (*Laminaria*), and many others in Japan and throughout Asia; chu rhu (*Monostroma*) in Bhutan; dulse (*Palmaria*) in Canada and the United States, to name a few. The oceans cover approximately 71% of Earth's surface, and cyanobacteria and algae account for nearly all of the primary production in oceans, directly supporting nearly all marine animal life.

The plastids of three distinct algal lineages are directly derived from free-living cyanobacteria. Characterized by the presence of only two unit membranes, such primary plastids are most familiar in the green algae (chlorophytes) and their derived, terrestrial subgroup, the land plants (embryophytes), but are also found in the red algae (rhodophytes) and in the glaucocystophytes, a small and relatively obscure group of organisms whose plastids retain the peptidoglycan cell wall of the cyanobacterial endosymbiont. All other photosynthetic eukaryotes rely on plastids that were acquired when the host cell ingested another eukaryote that already had plastids. Termed "secondary" plastids because their evolution required (at least) two sequential endosymbiotic relationships to be established, these organelles are always surrounded by more than two unit membranes, and in some cases are part of complex endomembrane systems that include tiny residual eukaryotic nuclei (nucleomorphs). Table 9.1 lists the principal groups of photosynthetic eukaryotes discussed here, along with the characteristics of their plastids. A summary view of our current knowledge of plastid and host ancestries is shown in figure 9.1, which can be compared with Delwiche (1999), which does not incorporate the chromalveolate hypothesis. For more information on the evolution of plastids and algae, the reader is referred to Delwiche (1999) and Palmer (2003).

Glaucocystophytes, Red Algae, and the Relationships among Taxa with Primary Plastids

Two groups of algae have plastids with pigmentation that resembles that of typical cyanobacteria: the glaucocystophytes and the red algae. The glaucocystophytes are an unfamiliar and relatively rare group of mostly freshwater algae with plastids that have often been confused with cyanobacteria. Unlike red algae, they are flagellate, although in some cases the flagella are reduced to vestigial appendages inside of a cell wall (Kies

and Kramer 1989, Bhattacharya et al. 1995). Glaucocystophytes are pigmented with chlorophyll a and light-harvesting protein structures termed "phycobilisomes" that are also found in most cyanobacteria and in red algae. These light-harvesting protein complexes form distinctive knobs studding the surface of the thylakoid membranes, where photosynthesis occurs. The phycobilisomes also prevent the thylakoid membranes from forming stacks of the type seen in green plants and other organisms without phycobilisomes. Because phycobilisomes have a characteristic absorption spectrum and result in a distinctive ultrastructure, the glaucocystophyte plastid bears a striking superficial resemblance to cyanobacteria. This similarity was reinforced by the retention of a thin peptidoglycan cell wall on the plastid, and long after the endosymbiotic origin of plastids was recognized, the plastids of glaucocytophytes were a source of confusion. Even today many authors fail to distinguish these organelles from their free-living relatives. In fact, these structures are authentic plastids, with genome size and complexity that are markedly smaller than those of free-living cyanobacteria and comparable with those of other plastids (Stirewalt et al. 1995), and molecular phylogenetic analyses firmly place them with other plastids (Bhattacharya and Schmidt 1997).

Glaucocystophytes are rare inhabitants of clean freshwater lakes, streams, and ditches and are not usually found in high population densities. Only a handful of genera are known, with only rudimentary knowledge for several, and only about 13 species in three genera have been described with reasonable confidence. Although they are unlikely to be of any great environmental or ecological significance, the glaucocystophytes seem to occupy a key position in the evolution of eukaryotes as one of the earliest-diverging lineages of algae (Martin et al. 1998). The relationships among the major algal groups remain unresolved, but the most likely placement for the glaucocystophytes is either as the earliest-diverging lineage of algae with primary plastids and the sister group to red + green algae, or as the sister group to red algae alone. In either case, these organisms display key ancestral characters that have been lost in related lineages, most notably, the peptidoglycan plastid wall, which is absent in all other lineages, and flagella, which are absent in the red algae.

By contrast with the glaucocystophytes, the red algae are a diverse and widespread group that dominates many temperate and tropical marine intertidal environments. Red algae can also be found growing at depths where the incident light is a tiny fraction of that at the surface, in freshwater environments, and even occasionally in soil crusts. But red algae are rare in the open ocean, presumably because they lack flagella at any stage of the life history. They are environmentally important primary producers and provide food and industrial chemicals, including the polysaccharide agarose that is a staple of bacteriology and molecular biology. There are three fundamental lineages of red algae, the subclasses Florideophycidae and Bangiophycidae and the order Cyani-

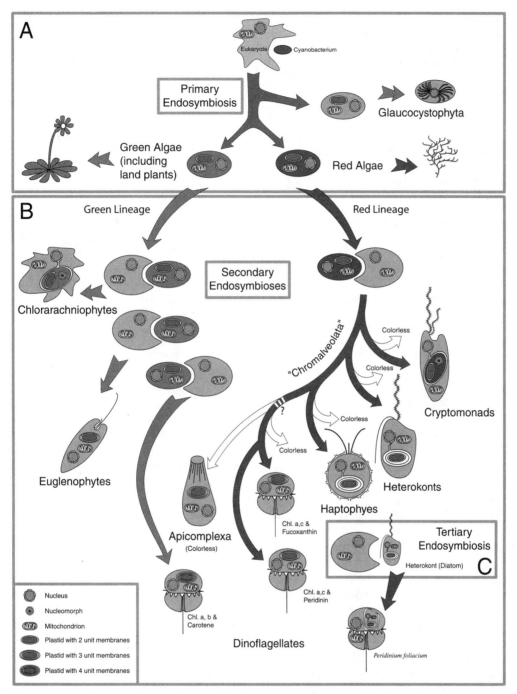

Figure 9.1. Another hypothesis for endosymbiotic events in the evolution of plastids. This hypothesis should be compared to that presented by Delwiche (1999). This scenario includes the "chromalveolate hypothesis" presented by Cavalier-Smith (1999), which proposes that the chlorophyll a/b taxa (cryptomonads, heterokonts, haptophytes, and dinoflagellates), as well as a number of colorless taxa, are descendants of a single endosymbiotic event. Under this scenario, there are relatively few primary endosymbiotic events but more losses of pigmentation and plastids. Some nonphotosynthetic lineages, for example, members of Apicomplexa, retain unpigmented plastids and plastid genomes, but such direct evidence of a photosynthetic past has not been documented from all of the colorless lineages that would be implied by this hypothesis. The "cabozoan hypothesis" is not shown here (Cavalier-Smith 1999).

Table 9.1

Major Lineages of Algae, along with the Characteristics of Their Plastids and Estimate of Their Biological Diversity.

Lineage	Membranes	Pigmentation	Diversity[a]
Primary plastids			
Glaucocystophytes	2	a, PB	50
Rhodophytes	2	a, PB	5500–20,000
Chlorophytes	2	a, b	13,500–100,000
Charophytes (excluding land plants)	2	a, b	20,000
Charophytes (including land plants)	2	a, b	500,000–1,000,000
Secondary or tertiary plastids			
Cryptomonads	4 w/ nucleomorph	a, c, PB	1200
Heterokonts (= Stramenopiles)	4	a, c	107,500–10,000,000
Haptophytes (= Coccolithophorids)	4	a, c	2000
Dinoflagellates	3 (4 in some)	Various	3500–11,000
Apicomplexa	4	None	4800–4,800[b]
Chlorarachniophytes	4 w/ nucleomorph	a, b	~20
Euglenoids	3	a, b	800

[a]Estimated number of species, based on Norton et al. (1996) and Van den Hoek et al. (1995).
[b]Perkins et al., (2000).

diales, each of which is thought to be monophyletic (Oliveira and Bhattacharya 2000). The florideophytes are primarily found in marine environments and are known for complex life histories that fascinate some life science students and torment the remainder. The bangiophytes are found in both marine and freshwater environments. Although they are typically less structurally and developmentally complex than the better studied florideophytes, bangiophytes do show considerable phylogenetic diversity and seem to be key to understanding the evolution of the group (Oliveira and Bhattacharya 2000, Müller et al. 2001). The Cyanidiales are a small group whose members occur primarily in acidic hot springs and differ markedly from other red algae in a number of key properties (Albertano et al. 2000), and may be an outgroup to the remainder of the red algae. The plastids of red algae are pigmented with chlorophyll a and phycobilisomes similar to those of glaucocystophytes and cyanobacteria.

It is interesting to note that although the red algae are generally viewed as a marine group, many of the earliest-branching bangiophytes and members of Cyanidiales are not marine inhabitants. Because the glaucocystophytes and many green algae are freshwater forms, this raises the possibility that the earliest photosynthetic eukaryotes were freshwater organisms.

A key problem in algal evolution, and one that has been the topic of much debate, is whether all primary plastids constitute a monophyletic group, and if so, whether they are the result of a single endosymbiotic event. Most molecular phylogenetic analyses of plastid genes show red, green, and glaucocystophyte plastids to be a monophyletic group (Delwiche et al. 1995, Delwiche 1999, McFadden 2001), but analyses of nuclear genes are more equivocal. Early analyses of nuclear ribosomal RNA (rRNA) genes typically did not place red and green algae together in a monophyletic group,

albeit with relatively little support for their relative positions, and more recent analyses of RNA polymerase genes also show these as two distinct groups (Stiller and Hall 1997). However, analyses of nuclear-encoded protein-coding genes have more consistently and strongly supported monophyly of the three lineages with primary plastids (Baldauf et al. 2000, Moreira and Philippe 2001). Thus, glaucocystophytes, red algae, and green algae constitute a single monophyletic group in many molecular phylogenetic analyses of both plastid and nuclear genes. On the surface this would imply that all primary plastids are derived from a single endosymbiotic event.

It is important to note, however, that even if both host and endosymbiont lineages form a monophyletic group, this does not guarantee that the plastids are the result of a single endosymbiotic event. It is always possible that closely related host lineages independently acquired closely related endosymbionts (much as dinoflagellates inhabit lineages of related animals in the modern world). Indeed, although there may have been a single, and singularly momentous case of indigestion leading to the development of endosymbiotic plastids, it seems more likely that the acquisition of plastids was the result of a gradual adaptation of the ingesting host lineage to the retention of ingested cyanobacteria for increasing lengths of time. This would eventually lead to retention of the endosymbiont through the complete cell cycle. Once the endosymbiont became heritable, then the door would have been open to permanent and obligate symbiosis, but there is a good chance that the host cell had significant adaptations to the presence of an endosymbiont long before the endosymbiont became permanent. Consequently, evidence that can distinguish between single and multiple origins of primary plastids would have to come from properties that are distinctive to the phenomenon of endosymbiosis itself. Among such properties would be the characteristics of the transit peptides that target nuclear-encoded gene products into the plastid, and the content of the plastid genome after its reduction to that of an organelle (Löffelhardt et al. 1997). However, although gene content and arrangement are superficially similar in all three lineages of primary plastids, the importance of such similarity can be interpreted only with knowledge of the degree of similarity that would be expected from independent endosymbiotic events, and it is entirely possible that the similar content of plastid genomes is the product of convergent evolution (Stiller et al. 2003)

Recent study of plastid retention in sea slugs provides support for the view that the endosymbiotic origin of plastids is likely to have involved a long period of predation and facultative retention of plastids before the permanent and obligate symbiosis. These remarkable animals eat algae and selectively retain the chloroplasts, which are ingested by cells and retained in a highly branched digestive tract that extends nearly to the surface of the mantle (Rumpho et al. 2000). In some cases the plastids are thought to serve primarily as a form of camouflage, but in others (e.g., *Elosia chlorotica*) the sea slug is able to survive indefinitely with light as its sole energy source. These "solar-powered" sea slugs very nearly qualify as algae in the definition given above, except that the plastids are not retained through the complete life cycle and have to be obtained each generation by eating an alga. However, the length of time that the plastids are retained and their continued functionality despite the photodegradation that takes place in a normal functioning plastid suggest that the sea slugs have sophisticated adaptations for plastid maintenance. Recent work indicates that there are genes resident within the sea slug's nuclear genome that serve the specific purpose of maintaining the plastid (Green et al. 2000, Hanten and Pierce 2001), and it may well be that these genes were derived from the prey genome via a process of horizontal gene transfer. Apparently you really are what you eat (Doolittle 1998).

Cryptomonads, Heterokonts, and Haptophytes: Secondary Plastids Derived from Red Algae

The phenomenon of secondary endosymbiosis—in which eukaryotes have acquired plastids by establishing a symbiotic relationship with other eukaryotes that already had plastids—has created a great deal of confusion. Because organisms with secondary plastids are chimeric (i.e., composed of tissues of two distinct evolutionary ancestries), they present a bewildering mixture of characters from seemingly unrelated organisms. Ultrastructural observations by Gibbs (1962, 1981) led her to propose that the plastids of several groups were acquired by secondary endosymbiosis. The most spectacular form of secondary plastids is exemplified by the cryptomonads, which are thought to have acquired a red algal endosymbiont (Gillott and Gibbs 1980).

In these organisms, a set of four membranes surrounds the plastid stroma and thylakoids. The two innermost envelopes correspond to the plastid envelope of the primary plastid, and the two outer membranes presumably correspond to the red algal plasma membrane and a food vacuole of the host cell. Remarkably, in the space that would have been the cytoplasm of the red algal endosymbiont, there are ribosomes and a degenerate eukaryotic nucleus, or "nucleomorph." The nucleomorph is a greatly reduced red algal nucleus, with three chromosomes (Douglas et al. 2001). In a striking example of convergent evolution, a very similar overall genome structure is seen in the green-pigmented secondary plastids of chlorarachniophytes (described below).

The plastids of cryptomonads, like those of red algae, contain phycobiliproteins, but these proteins are located in the thylakoid lumen and are not organized into phycobilisomes. However, unlike typical red algal plastids, two chlorophylls, a and c, are present. Chlorophyll c has been taken as a character linking a putative group of algae referred to as "chromophytes" or, more recently, Chromalveolates (Cavalier-Smith 2000), including cryptomonads, heterokonts, haptophytes, and dinoflagellates, all of which share chlorophyll c and many of which have light-harvesting carotenoids. The chromophyte clade was originally proposed by Chadefaud (1950) as an algal lineage to stand equal with the blue-green algae (= cyanobacteria), red algae, and green algae. The chromophytes included the cryptophytes, heterokont algae (including haptophytes), dinoflagellates, and euglenoid algae in Chadefaud's definition, but the euglenoids were removed when the division Chromophyta was formally described (Christensen 1989). Molecular studies during the 1990s suggested that the host-cell lineages of these organisms do not constitute a monophyletic group, and the concept of chromophytes fell into disfavor, although it has retained fairly widespread use because these organisms do share a number of ecological and structural similarities. However, recent analyses of plastid-encoded genes (Fast et al. 2001, Yoon et al. 2002) provide support for the hypothesis that the chromophytes, and in particular, the clade defined by the cryptomonads, heterokonts, and haptophytes (Chromista; Cavalier-Smith 1986) may in fact be a monophyletic group, or at least the product of symbiotic events involving organisms with closely related plastids (fig. 9.1). This is still very much an area of active investigation, and a recent extensive analysis of plastid genes and genomes did not find support for the chromalveolate hypothesis (Martin et al. 2002).

The heterokonts, which are also known as "stramenopiles" for the bristles on their anterior flagellum, constitute one of the great lineages of eukaryotic diversity and include many protists once classified as algae, protozoa, and aquatic fungi. These organisms have not received the measure of study given to other eukaryotic lineages with comparable diversity and age (i.e., green plants, animals, and fungi). Space limitations will not allow this group to be given full justice here, but they are among the dominant primary producers in most marine environments and, as such, lie near the base of the food chain for two-thirds of the planet. As might be expected for a group of such global significance, the heterokonts show tremendous biological diversity in terms of number of species, molecular sequence divergence, and structural variation (Andersen 1992, 1998, Potter et al. 1997). The term "heterokont" was first proposed by Luther (1899) for the xanthophytes and freshwater raphidophytes, but today the term refers to a larger phylogenetic lineage characterized (in most cases) by organization of two flagella, the anteriorly directed one being decorated with minute but elaborate flagellar hairs (Van den Hoek et al. 1995). Such characteristic flagella occur at some stage of the life cycle of many, but by no means all, organisms in the lineage.

Interestingly, many early-branching heterokonts such as the oomycetes, thraustochytrids, and bicosoecids are colorless and apparently lack plastids. This may, on the surface, suggest a nonphotosynthetic ancestry for this group, but a recent analysis of the 6-phosphogluconate dehydrogenase (*gnd*) gene from cyanobacteria and different protists suggests otherwise. Phylogenetic analysis of *gnd* indicates that the parasitic heterokont *Phytophtora infestans* was likely once photosynthetic because it retains a *gnd* gene (of cyanobacterial affinity) that is closely related to the homologue in photosynthetic members of this lineage (Andersson and Roger 2002). Weaker evidence for a photosynthetic ancestry comes from a phylogenetic analysis of plastid-targeted *GAPDH* (the gene encoding glyceraldehyde phosphate dehydrogenase) that groups heterokonts, dinoflagellates, and apicomplexans in one clade, consistent with these groups having once shared a common plastid, with losses occurring in ciliates and in nonphotosynthetic heterokonts (Fast et al. 2001).

The photosynthetic heterokonts have a chromophyte pigmentation and include such ecologically important groups as the diatoms and brown algae. The chloroplast typically has a girdle lamella, a saclike structure that encloses all remaining lamellae. The chloroplast is almost without exception connected to the nucleus; that is, the nuclear envelope is continuous with the plastid endoplasmic reticulum. The storage product is a β-1,3-linked glucan consisting of only about 25–35 residues, and because of the small molecular size, the storage product is maintained in a cytoplasmic vacuole. A large and diverse number of microscopic algal groups make up the heterokont algae, along with the brown algae, which are as large and structurally complex as are animals or plants. The microalgae include the diatoms, which produce silica cell walls of opaline glass, like that found in glass windows. The diatoms (with as many as 10 million species!) are the "insects" of the microbial world (Norton et al. 1996), and diatoms are probably the major original carbon source of many petroleum deposits (crude oil, natural gas). The heterokont algae, with a few noteworthy exceptions (e.g., the diatom *Pseudo-nitzschia* and the raphidophyte *Chattonella*), are rarely toxic or harmful.

Haptophytes are predominately marine phytoplankters, including the calcarious-scaled coccolithophores that are famous for having formed the White Cliffs of Dover. Although there are probably only a few thousand species of haptophytes worldwide, these organisms are of great environmental significance (see Tyrrell 2003). Some species (e.g., *Emiliania huxleyi*, *Gephyrocapsa oceanica*) can occur in vast populations, particularly in temperate and polar seas, and are responsible for substantial primary productivity. As a consequence, they make a noticeable contribution to the global carbon cycle and are thought, for example, to have given rise to the North Sea oil fields. The pigmentation of haptophytes is similar or identical to that of some heterokont algae, they also store small β-1,3-linked glucans in cytoplasmic vacuoles, and there is a membrane continuity between the nuclear envelope and the plastid endoplasmic reticulum. However, haptophytes lack flagellar hairs like those found in the heterokonts. A few (e.g., *Chrysochromulina*) are harmful or toxic, killing fish when they occur in bloom conditions.

Dinoflagellates and Apicomplexans: A Confusion of Plastids

Dinoflagellates show the greatest diversity of plastids of any eukaryotic group (Delwiche 1999). They are members of the alveolates, another major eukaryotic lineage comparable with the heterokonts, plants, animals, and fungi. Many dinoflagellates (e.g., *Alexandrium*, *Prorocentrum*) produce deadly toxins such as saxitoxin or okadaic acid, and they cause shellfish poisoning and other types of death or illness to humans and marine life.

Only about one-half of all dinoflagellates are photosynthetic, and like the heterokonts, many of the basal branching lineages are colorless and presently show no structural sign of having had a plastid in the past. Curiously, members of the closely related Apicomplexa, a nonphotosynthetic group of obligate parasites, have a spherical structure composed of four nested membranes that has been shown to be a remnant, colorless plastid (Köhler et al. 1997), which raises the obvious possibility that the common ancestor of both apicomplexans and dinoflagellates was equipped with a plastid. There is at present only scanty evidence that the nonphotosynthetic basal lineages of dinoflagellates were ever equipped with a plastid, although tantalizing evidence from GAPDH suggests dinoflagellates may share a common plastid origin with heterokonts and haptophytes (Fast et al. 2001). These data should be viewed with caution, however, both because GAPDH has been notoriously difficult to interpret and because (as discussed above in the context of primary plastids), it can be difficult to infer the number of endosymbiotic events, particularly from limited information.

Most dinoflagellates that are photosynthetic rely on a characteristic peridinin-pigmented plastid that is surrounded by three unit membranes. There are, however, a number of photosynthetic dinoflagellates that show other pigmentation types (Delwiche and Palmer 1997) and seem to have acquired their plastids via independent endosymbiotic events involving green algae, diatoms, cryptophytes, haptophytes, and other organisms (Delwiche 1999, Tengs et al. 2000, Yoon et al. 2002). Curiously, some dinoflagellates with plastids other than the typical peridinin-type plastid are phylogenetically dispersed as subclades among the peridinin-containing taxa. This may imply that dinoflagellates with peridinin-type plastids are less tightly bound to their endosymbiont than are most algae. There are at least two possible explanations for this phenomenon. The first involves the peculiar type of rubisco that performs photosynthesis in peridinin-containing dinoflagellates. This form II rubisco is probably far more sensitive to oxygen than is form I rubisco, which is used by all other oxygenic phototrophs, and may mean that the quantum yield of photosynthesis in dinoflagellates is lower than in other organisms. In practice, dinoflagellates can be very difficult to transport, because they can only survive a short time in the dark, and this may reflect a relatively low ratio between photosynthesis and respiration attributable in part to the type of rubisco they use for photosynthesis. A second possible explanation involves the unusual genomic structure of dinoflagellate plastid genomes, which seem to be coded entirely as single-gene minicircles (Zhang et al. 1999, 2001). This genomic organization might be less stable than a more typical chromosomal organization, and this could in turn lead to more frequent loss of plastids than in other groups. These and other hypotheses remain to be tested.

A recent study provides a surprising view of dinoflagellate plastid evolution, suggesting that this lineage may have undergone a tertiary plastid replacement (i.e., the uptake of an alga containing a secondary endosymbiont) involving a haptophyte (Yoon et al. 2002). These data suggest that the dinoflagellates once likely contained a secondary plastid of red algal origin (see GAPDH data in Fast et al. 2001) that may have been shared by all chromalveolates and that was subsequently replaced by a haptophyte plastid before the radiation of the photosynthetic dinoflagellates. If this is correct, and if the chromalveolate hypothesis is correct (Cavalier-Smith 1999), then the implication would be that there was a single ancient endosymbiotic event that gave rise to the plastids of all chromophytes, but that the dinoflagellates lost this red algal endosymbiont and later reacquired a haptophyte endosymbiont. Under this scenario, the fucoxanthin-pigmented dinoflagellates, which have pigmentation and chloroplast morphology very similar to those of haptophytes (Tengs et al. 2000), would represent the primitive condition among plastid-containing dinoflagellates, whereas those with peridinin-type plastids would represent a derived condition (Yoon et al. 2002). Together, these data are potentially valuable for resolving long-standing questions about plastid evolution and the number of secondary and tertiary endo-

symbioses, but they clearly need to be corroborated with resolved host-cell trees using either or both nuclear and mitochondrial genes. Several potentially serious analytical problems exist with the data that have been examined to date, and at present several competing hypotheses are plausible.

Chlorarachniophyes and Euglenoids: Secondary Plastids Derived from Green Algae

Although the greatest diversity of organisms with secondary plastids is found among the organisms that acquired secondary plastids from red algae, there are also organisms with secondary plastids derived from green algae (none have yet been shown to be derived from glaucocystophytes). In a remarkable display of parallel evolution, the amoeboid chlorarachniophytes have a green algal plastid in a four-membrane compartment similar to that of cryptomonads, complete even to the presence of a nucleomorph (McFadden et al. 1994).

The euglenoids are common in eutrophic freshwater or estuarine habitats. Like dinoflagellates, only about one-half of all species are photosynthetic, and the dependence of the host cell upon the plastid seems to be relatively weak. When grown at high temperatures and with an external carbon source, some euglenoids will undergo cell division more rapidly than the plastid can divide, and the host cell will be "cured" of its plastid, which will, of course, never regenerate (Gibbs 1978). This observation was important in early discussions of the endosymbiotic origin of plastids.

Green Algal Diversity

Green algae are not the most diverse of algal groups (table 9.1), but their presumed close relationship to higher plants and frequent occurrence in freshwater habitats used by humans make them one of the more familiar and well-studied groups of algae. Their plastids are primary and contain chlorophylls a and b and a variety of accessory pigments such as carotenes and xanthophylls. Most green algae are microscopic, but within this diminutive realm they manifest a relatively broad variety of growth habits, from unicells (e.g., *Chlamydomonas*), to colonies (e.g., *Volvox*), to unbranched (e.g., *Ulothrix*) and branched filaments (e.g., *Cladophora*), to true parenchymatous forms (e.g., *Coleohaete* and *Chara*). This variety of morphology was once considered to represent a progression of forms from simple to complex (e.g., unicells to large colonies of unicells; unbranched to branched filaments; small branched clumps to large, plantlike forms; Smith 1955, Fritsch 1965, Bold and Wynne 1985). It appears, however, that complex forms have arisen numerous times from simple ancestors, and numerous reversions to simple forms further complicate matters (e.g., McCourt et al. 2000). Some green algae are macroscopic, particularly those that occur in intertidal or subtidal marine benthic habitats. In many cases, these larger forms are coenocytic; that is, their thalli or plant bodies are composed of ramified networks of tubes containing a cytoplasm with many nuclei and plastids but not divided into discrete cells. These tubular modules may be compressed into spongy large thalli (e.g., *Codium*) or be elaborate structures more than a meter in length with rootlike processes, leaflike assimilators, and spreading connectors analogous to stolons (e.g., *Caulerpa*). That such structural complexity is possible in a thallus that is, in effect, a single cell challenges conventional thought on the role of multicellularity in plant development.

Other green algae are microscopic and have cells of more familiar form, but even these can show a fair measure of complexity and tissue differentiation. To give one example, *Volvox*, which is widely familiar to biology students, is organized into a sphere of biflagellate cells linked by thin cytoplasmic strands and often with new thalli developing on the inside of the sphere. Although often described as a colony, most species of *Volvox* show clear functional differentiation among cells, and development of the thallus includes an elaborate process of inversion reminiscent of gastrulation, although these processes are certainly not homologous (Kelland 1977, Kirk 1999, 2001). Highly complex algae are also found in the lineage that includes land plants, the Charophyta (or Streptophyta; described below).

Green algae are nearly ubiquitous, albeit not terribly abundant in many habitats. Found most often in aquatic environments from freshwater to marine and hypersaline, green algae are common and sometimes abundant phytoplankton in lakes and streams. Many grow attached to rocks or other hard substrata, although large free-floating mats of "pond scum" are widely distributed in quiet fresh waters. Certain groups have colonized subaerial or truly terrestrial habitats, such as soil interstices, within limestone rocks, on the surface of desert soils, on bark and leaf surfaces of some seed plants, or as photosynthetic symbionts (not endosymbionts) in lichens. Green algae possess a mirror image distribution compared with two other large groups of algae, the red and brown algae: the latter two are the dominant macrophytes in the oceans, with a relative few species occupying freshwater habitats, whereas green algae are far more abundant in freshwater. In all these groups there are marked size differences between marine and freshwater species: marine greens, like reds and browns, are generally large; freshwater taxa of these three groups tend to be smaller and most are microscopic. As is the case with many marine and freshwater organisms capable of aestivation or dormancy, freshwater green algae frequently form resistant spores or bodies through sexual or vegetative means; these structures are much less common in marine algae.

Large marine green algae include several abundant and widespread taxa such as the sea lettuce (*Ulva*) and dead-man's fingers (*Codium*). Members of the Caulerpales can at times be conspicuous members of reef communities and are raised by aquarium enthusiasts. It should be noted that a semi-domesticated variety of *Caulerpa*, of recent infamy, has wreaked havoc in the Mediterranean as an aggressive exotic (Meinesz et al. 1993), and it now threatens North America (Jousson et al. 2000). Although some planktonic green algal unicells (e.g., *Dunaliella*) are known in the world's seas, the oceanic phytoplankton is primarily dominated by other groups of algae, notably diatoms and haptophytes (Graham and Wilcox 2000), along with cyanobacteria.

Instead of basing phylogeny on growth habit (Fritsch 1965), modern approaches have discovered ultrastructural characters of flagellar structure, particularly the anchorage of flagella in the cell (O'Kelly and Floyd 1983) and cell division (Pickett-Heaps 1975) that are morphological synapomorphies for groups composed of a variety of body forms. This ultrastructural anatomical consistency contrasts with a diversity of thallus types, from unicells to colonies to branched and unbranched filaments. The ultrastructural approach was pioneered by Pickett-Heaps and Marchant (1972) and others but was codified by Mattox and Stewart (1984) in a dramatic restructuring of the systematics of the green algae. Mattox and Stewart's hypothesis was that there are four or five major lineages of green algae, each characterized by a particular type of motile unicell, and all showing some degree of morphological convergence of body types that had previously been considered of overriding taxonomic importance. This radical new classification that placed emphasis on ultrastructural features led the way to other studies of biochemistry (e.g., glycolate metabolism enzymes) and cell division (e.g., mode of cell wall formation at cytokinesis) that corroborated Mattox and Stewart's hypothesis. Although new data and analyses have led to substantial revision of the system established by Mattox and Stewart, their treatment marked a turning point in green algal systematics, and most modern classifications rely heavily on it.

A new comprehensive treatment of green algal systematics is badly needed. Although Mattox and Stewart's (1984) system established a baseline for modern classification, it is now nearly 20 years old and was developed in the absence of any molecular systematic data and without formal phylogenetic analysis. Molecular data both have confirmed many elements of their system and have helped reveal a series of additional surprising arrangements (e.g., a monophyletic class Trebouxiophyceae, containing many photobionts of lichens along with other forms). Recent comprehensive treatments have been presented in response to the need to organize general textbooks (e.g., Van den Hoek et al. 1995, Graham and Wilcox 2000). Although these systems include more recent information, including some molecular phylogenetic data, and are in some respects excellent, they are fundamentally an afterthought in the context of broader texts and fail to make full use of modern data and analytical methods.

In Mattox and Stewart's (1984) system, there were five classes of green algae: Charophyceae, Micromonadophyceae (now known as Prasinophyceae), Ulvophyceae, Pleurastrophyceae (now Trebouxiophyceae; Friedl 1995), and Chlorophyceae. Although considerable uncertainty remains as to the relationships among these groups, most of them seem to more or less correspond to monophyletic groups, with the exception of the Prasinophyceae, which are probably paraphyletic (Fawley et al. 2000), and the Charophyceae, from which Mattox and Stewart omitted the land plants. The status of the Ulvophyceae is uncertain, with some data indicating that there are two or more unrelated elements submerged within this group (Van den Hoek et al. 1995).

Molecular phylogenetic data indicate that the chlorophytes as a whole are divided into two primary lineages (fig. 9.2A; Graham and Wilcox 2000). The first of these is a clade composed of Mattox and Stewart's Charophyceae plus the land plants, a group termed "Streptophyta" by Bremer (1985) and "Charophyta" by Karol et al. (2001). The latter term will be used here. This group is characterized by asymmetric placement of flagella (if present), along with several other ultrastructural, biochemical, and molecular features. A few (but not all) charophyte orders perform cell division in a manner that is strikingly similar to the way it occurs in land plants. The charophyte lineage will be considered in more detail below. The second lineage seems to include all of the other classes recognized by Mattox and Stewart and thus includes the bulk of the green algae. We refer to that lineage here as the "Chlorophyta *sensu stricto.*" The branching order among these groups is still under investigation.

The Prasinophycae may be the least natural of the classes recognized by Mattox and Stewart (1984). They are scaly, unicellular flagellates, a cell morphology that probably represents the ancestral condition in green algae (Van den Hoek et al. 1995). In the absence of clear synapomorphic characters that define the group, it is not surprising that at least some organisms classified in the Prasinophyceae on the basis of morphology would be best placed elsewhere. Nonetheless, the majority of prasinophytes fall within the Chlorophyta *sensu stricto*, where they form a paraphyletic grade at the base of the group (fig. 9.2C; Fawley et al. 2000). It may well be that there is a great deal of unrecognized biological diversity among these organisms.

An organism of particular importance is *Mesostigma viride*, a unicellular and scaly flagellate that was placed by Mattox and Stewart (1984) and others among the prasinophytes but that has some (possibly plesiomorphic) ultrastructural similarities to charophytes (Rogers et al. 1981). Molecular phylogenetic analyses support the distinctive nature of *Mesostigma* but differ on whether it is sister to the charophytes (Karol et al. 2001) or to all of known green algal diversity (i.e., to the clade comprising charophytes and chlorophytes *sensu stricto*; fig. 9.2A; Lemieux et al. 2000). Resolving this issue will require phylogenetic analysis of a rather rich data set, including genome-scale data from a substantial number of organisms, but the potential rewards of such a study are great. Whichever phylogenetic position is correct, *Mesostigma* is clearly a pivotal organism in green plant evolution.

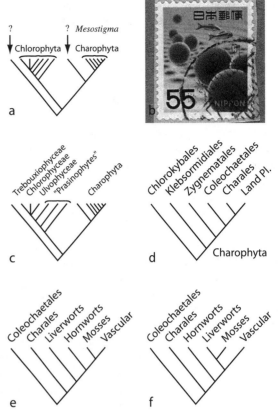

Figure 9.2. Phylogenetic relationships among the green plants. (A). All green plants, including Charophyta, Chlorophyta, and two possible placements of *Mesostigma viride*. (B) A Japanese stamp commemorating *Aegagropila linnaei* ("marimo balls"). (C) Primary lineages in Chlorophyta. (D) Primary lineages in Charophyta. (E) The "liverworts basal" hypothesis for the branching order among the four early-diverging lineages of land plants. (F) The competing "hornworts basal" hypothesis (by C. F. Delwiche).

The Ulvophyceae have a cruciate flagellar root system with basal bodies that are offset counterclockwise, and neither a phycoplast nor a phragmoplast is formed during cell division. These are likely to be plesiomorphic conditions, and recent classifications have suggested that these organisms should be divided into two or more separate groups (Van den Hoek et al. 1995, Watanabe et al. 2001). The ulvophytes include both marine and freshwater forms, are highly varied in form, and include some ecologically and economically important organisms. Familiar ulvophytes are the sheetlike *Ulva* that covers riprap worldwide, and the filamentous *Ulothrix* (another polyphyletic genus) that is common in cold-water environments. *Cladophora*, a small coenocytic branched filament, occurs in both freshwater and marine environments. *Cladophora* and its close relatives are extremely widespread, are both economically and culturally important, and would benefit from additional systematic analysis (Hanyuda et al. 2002). A close relative of *Cladophora*,

Aegagropila, grows into the famous marimo balls of Lake Akan in Japan and is probably the only alga to have been designated a national treasure and commemorated on a postage stamp (fig. 9.2B). Also classified among the ulvophytes are huge and elaborate coenocytes of *Bryopsis*, *Caulerpa*, and their relatives, and the microscopic but highly complex and fully terrestrial Trentepohliales (Chapman et al. 2001, Thompson and Wujek 1997).

Members of Chlorophyceae are mostly freshwater, have a cruciate microtubular root system with basal bodies that are either directly opposed or offset clockwise, and at least in most cases form a distinctive structure called a "phyco-plast" during cell division. The chlorophytes include such familiar taxa as *Chlamydomonas*, an important model organism (Harris 1989), and *Volvox*, also a model system and famous for its beautiful spherical form. Also in this group are *Hydrodictyon*, an elaborate net-shaped organism that was mentioned in Chinese literature nearly 2000 years ago (Tilden 1937); *Characiosiphon*, a multinucleate coenocyte; and highly differentiated filamentous algae such as *Stigeoclonium*, *Draparnaldia*, and *Fritschiella*.

The Trebouxiophyceae are mostly small unicells that live in freshwater or terrestrial environments, but some are filamentous or even organized into bladelike sheets (e.g., *Prasiola*). A key model system in the study of the physiology and biochemistry of photosynthesis was a trebouxiophycean species of *Chlorella*, a genus that molecular phylogenetic studies indicate is grossly polyphyletic and is currently undergoing revision (Huss et al. 1999). Lichen phycobionts are often trebouxiophytes, and they are rather common in terrestrial habitats both in lichenized and unlichenized states (Friedl 1995, Lewis and Flechtner 2002).

Green Algae and the Colonization of the Land

Although a great diversity of algae, green and otherwise, inhabit the terrestrial environment (Lewis and Flechtner 2002), a single monophyletic group characterized by a life cycle that involves an alternation of multicellular diploid and haploid generations and a syndrome of ultrastructural and biochemical features. This group dominates the land in terms of biomass, primary production, ground coverage, and known biological diversity among phototrophs. This lineage, referred to here as "land plants" and known more formally as "embryophytes" in reference to their life cycle, are from a phylogenetic perspective green algae that have become adapted to life in the terrestrial environment (Karol et al. 2001).

Green algae as a whole contain chlorophylls a and b, store starch inside the chloroplast, have cellulosic cell walls, and possess a unique star-shaped structure at the flagellar transition zone (Graham 1993). These features are shared with land plants, and consequently even before biochemical and ultrastructural similarities were known in detail, the green algae were considered ancestral to land plants (Bower 1908,

Fritsch 1965). However, thought on the identity of the closest relatives of land plants has changed considerably as knowledge of cell biology, ecology, and phylogeny of the green algae has developed (Van den Hoek et al. 1995). In the absence of phylogenetic information that is independent of morphology, heterotrichous forms such as *Fritschiella*, with prostrate rhizoids and upright, branched structures, were thought to represent a stage in the evolutionary series from unicells to land plants (Singh 1941). At the same time it was recognized that there was ample opportunity for convergent evolution, and all classifications were viewed as tentative, and probably artificial (Tilden 1937, Fritsch 1965). The classical sequence of increasing grades of developmental complexity (Smith 1955, Bold and Wynne 1985), which runs from permanently flagellate unicells, through coccoid forms with motile stages, unbranched and branched filaments, to complex thalli composed of layers of cells organized in three dimensions has proven to have some truth to it, but various forms of complexity have evolved independently in several lineages. Consequently, study with the light microscope could identify a number of candidate taxa that seemed to be relevant to the origin of land plants, but was not effective at choosing among these.

With the rapid accumulation of ultrastructural data in the 1970s and molecular data in the 1980s and 1990s, the picture that has emerged places land plants firmly within the Charophyta (Mishler and Churchill 1985, Bremer et al. 1987, Graham et al. 1991, McCourt 1995, Graham and Wilcox 2000, Karol et al. 2001). The charophytes are remarkably diverse structurally and display the full range of classical grades of complexity. Interestingly enough, the phylogeny within this group seems to follow the classical developmental sequence rather well, albeit with some reversals and parallelism (McCourt et al. 2000, Karol et al. 2001, Delwiche et al. 2002). Because of their structural diversity, although several charophytes had been discussed with respect to the origin of land plants (e.g., Zygnematales, Coleochaetales, Charales), these taxa were rarely classified together before the advent of ultrastructural and molecular phylogentic data, and even today some authorities shy away from treating them as an integrated whole (e.g., Wehr and Sheath 2003).

Graham (1993) termed these green algae "charophyceans," derived from Mattox and Stewart's (1984) class Charophyceae. Rather than substitute a class-based name for a paraphyletic group of algae, however, we use the term "charophytes" informally, or the division name "Charophyta" to refer to the whole clade including land plants (Karol et al. 2001). These terms have their own limitations—historically "charophytes" has referred to the Charales and related fossil forms (Tappan 1980). Bremer and Wanntorp (1981) recognized the importance of naming monophyletic groups and used Jeffrey's (1967) term "Streptophyta." This term refers to the twisted shape of the sperm and was originally used to refer to the clade composed of embryophytes and the Charales (Jeffrey 1967). More recently, the streptophytes have been taken to include other algae on the plant lineage as well, but we prefer the term in its original sense and use "Charophyta" to refer to the more inclusive group. The groups that make up the charophytes include several previously recognized orders, each of which is monophyletic (fig. 9.2D): Chlorokybales, Klebsormidiales, Zygnematales (including Desmidiaceae), Coleochaetales, and Charales, plus, of course, land plants.

As noted above, the unicellular flagellate *Mesostigma* (fig. 9.3F) is apparently a basal branch within the charophytes (Karol et al. 2001), although other analyses question this genus's placement (Lemieux et al. 2000, Turmel et al. 2001). To further complicate matters, analyses of rRNA weakly support a topology that places *Mesostigma* in a clade with the genus *Chaetosphaeridium* (Marin and Melkonian 1999). This topology differs from that found in analyses of protein-coding genes (Karol et al. 2001, Bhattacharya et al. 1998) but does not seem to be a spurious result. The Chlorokybales (fig. 9.3E) are a monotypic order consisting of the species *Chlorokybus atmosphyticus*, a rare soil alga that forms small packets of cells embedded in mucilage. The Klebsormidiales (fig. 9.3D) include the genera *Klebsormidium*, *Interfilum*, and *Entransia*; they are unbranched filaments and are common in freshwater and terrestrial environments. *Klebsormidium* is frequently a component of the green film that accumulates on sheltered walls and structures in warm, moist climates and of desert crusts (Lockhorst 1996, Lewis and Flechtner 2002).

Members of Zygnematales (fig. 9.3C) are common freshwater algae including both filamentous (e.g., *Spirogyra*) and unicellular (e.g., *Micrasterias*) forms. In this group the unicellular form seems to be a derived condition, with some species having independently reacquired a filamentous growth form (e.g., *Desmidium*; McCourt et al. 2000). No member of Zygnematales has any flagella stage, and consequently they indulge in a distinctive form of sexual reproduction termed "conjugation" that does not require motile cells. Although the filamentous Zygnematales are often considered to be unbranched, some branching is associated with holdfast formation, and the conjugation tube is developmentally similar to a branch (Fritsch 1965).

Coleochaetales (fig. 9.3B) include the genera *Coleochaete*, *Chaetosphaeridium*, and the exquisitely rare *Awadhiella* (Delwiche et al. 2002, Nandan Prasad and Kumar Asthana 1979). They form complex branched thalli that are found living on submerged rocks and vegetation worldwide. Reproduction is oogamous, and in *Coleochaete* there are elaborate developmental changes that occur in response to the fertilization of the egg. Members of the genus *Coleochaete* show many structural and biochemical features that resemble land plants, including phragmoplastic cell division, plasmodesmata, plantlike peroxysomes, and the production of more than four meiotic products from the zygote (Graham 1993, Delwiche et al. 2002). The thalli have a distinct three-dimensional organization, and in some the laterally adjacent cell files are so tightly adjoined that the tissue organization has

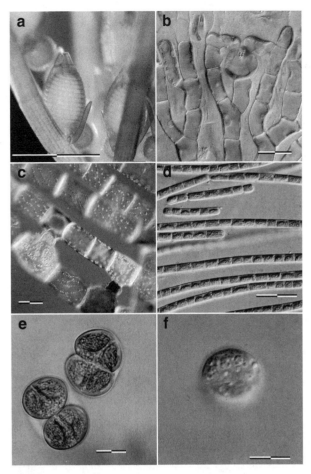

Figure 9.3. Representative species in Charophyta. (A) *Chara globularis* (Charales) KGK0044, showing cortication of developing oogonium and antheridium in the background. Freshwater. Scale bar, 1 mm. (B) *Coleochaete pulvinata* (Coleochaetales) CFD 56a6, showing early developmental stage of zygote cortication. Freshwater. Scale bar, 30 μm. (C) *Spirogyra maxima* (Zygnematales) UTEX 2495, showing conjugation tubes and partially developed zygotes. Freshwater. Scale bar, 100 μm. (D) *Klebsormidium nitens* (Klebsormidiales) SAG 335–2b, showing a single parietal chloroplast per cell. Moist soils or freshwater. Scale bar, 30 μm. (E) *Chlorokybus atmosphyticus* (Chlorokybales) UTEX 2591 growing in characteristic sarcinoid packets of cells. Moist soils. Scale bar, 10 μm. (F) *Mesostigma viride* (Mesostigmatales) SAG 50-1. Note surface scales visible on upper portion of cell; flagella are not visible, but would emerge from the medial groove in direction of viewer. Freshwater. Scale bar, 10 μm.

been viewed as a simple parenchyma (Graham 1993). Vegetative growth occurs by cell division in very specific locations in the thallus, depending upon the species (Delwiche et al. 2002). Typically cell divisions that increase the length of a filament occur in the apical cell of the filament, and branching occurs either in the apical cell or in the second or third cell from the apex. Because of these similarities to land plants, *Coleochaete* has long been discussed in the context of the origin of land plants (e.g., Bower 1908), but many of these characters are also found in the Charales and not all are uniformly present in Coleochaetales.

Charales (fig. 9.3A) are large and complex organisms with a conspicuous node/internode organization reminiscent of the more developmentally complex land plants. However, the thallus structure is fundamentally different from that of land plants, with the internodes formed from a single giant cell, in some cases secondarily covered by corticating filaments (Graham and Wilcox 2000). Reproduction is oogamous, and the zygote, which may be a millimeter or more in diameter, is surrounded by a thick sporopollenin wall, which, in addition to heavy calcification in many forms, ensures that oopores and associated structures often form well-preserved fossils (Delwiche et al. 1989, Feist and Feist 1997). As a consequence, the Charales have a rich fossil record extending back well more than 400 million years (Grambast 1974, Tappan 1980). The Charales have a distinctive and complex form, with elaborate multicellular antheridia and oogonia covered with sterile jacket cells, early development consisting of protonemal filaments with subsequent formation of nodes and internodes, well-differentiated rhizoids, stolonlike growth, and a variety of other developmental responses to the environment. Because of these striking apomorphies, many authors have separated them into a division or class of their own. Although many of these features are highly specialized and clearly evolved independently of their analogs in land plants (Fritsch 1965, Graham 1993), there are enough features shared with land plants that, like Coleochaetales, Charales has long featured in the search for the sister taxon to land plants.

As noted above, members of Charales are now thought to be the closest living relatives of land plants, with a recent multigene analysis supporting a strong sister-group relationship between these two groups (Karol et al. 2001). This analysis is the first to provide robust support for a sister taxon to land plants. If accurate, this arrangement suggests that the common ancestor of Charales and land plants exhibited several traits: branching thallus, oogamy, branched rhizoidal structures, a complex sperm, and a freshwater habitat. The Charales have clear developmental patterning in three dimensions, and the organization of cells in the nodes is often considered to be parenchyma, but it does not have the degree of complexity and tissue differentiation found in land plants. Growth occurs by division of an apical cell with a single cutting face.

A freshwater origin for land plants is perhaps not surprising, given that terrestrial animals also likely originated there. The availability of a robust phylogeny for the charophytes also helps resolve a long-standing issue in botany, the origin of the life history of land plants (e.g., Bower 1908, Tilden 1937, Fritsch 1965, Mishler and Churchill 1985, Graham 1993). All charophytes except for land plants have a life cycle in which the vegetative cells are haploid and the only diploid stage is the zygote (i.e., haplontic), so the life cycle of land plants almost certainly arose by intercalation of a multicellular diploid phase in a haplontic life cycle.

What other conclusions can be made about the origin of land plants from an understanding of their placement within Charophyta? Graham (1993) has discussed in detail the features that unite Charales, Coleochaetales, and other charophyte algae with land plants, as well as those features that are distinctive to land plants. Among the characters that are unique to land plants (and fairly universal among them) are an alternation of generations in the life cycle; tissue composed of cells organized in three dimensions ("parenchyma") and with fairly small cells; large size overall, which may be dependent upon the preceding; extensive tissue differentiation and specialization; a well-developed cuticle; distinctive and complex multicellular antheridia and archegonia; numerous spores with a thick sporopollenin coat; and a number of technical biochemical and ultrastructural characters. Although the gap between the "algal" charophytes and land plants is smaller than some have suggested, there are clearly a number of features that are unique to the land plants. These features, and probably some combination of them, permitted this clade to undergo dramatic diversification and to inhabit a wide range of habitats. This is particularly interesting in view of the fact that several other charophytes are partially or wholly terrestrial (e.g., *Chlorokybus*, *Klebsormidium*), and yet the embryophytes overwhelmingly dominate the land. The embryophytes have a known fossil record that is only slightly deeper than that of Charales and show a degree of molecular divergence that is comparable with the other charophycean orders. Perhaps it is primarily their ability to grow large that has given them command of the terrestrial environment.

The Land Plants: Terrestrial Green Algae

Land plants are the unchallenged masters of the terrestrial environment. They are found in nearly every terrestrial environment with the exception of the high alpine and polar regions and severe deserts. Because of a high degree of tissue differentiation and sophisticated adaptations to water management, some land plants are able to survive even in very dry environments by relying on water storage, tap roots that can mine deep water, or life cycles that involve long periods of dormancy. Algae that are not part of the land plant clade can be found living in similarly dry environments (e.g., *Klebsormidium*; Lewis and Flechtner 2002), but these survive by dormancy and remarkable desiccation tolerance (Oliver et al. 2000) and do not achieve the biomass found in land plants.

Although the most conspicuous members of the terrestrial flora are large vascular plants, these are relatively derived members of the clade, and their extreme adaptations to the terrestrial environment obscure some of the fundamental similarities of land plants to other charophytes. Early branches in the land plant lineage include a number of small and inconspicuous organisms without fully developed vascular tissue and with reproductive mechanisms that require the availability of liquid water. Although the early-diverging land plants have a life cycle involving an alternation of multicellular haploid and diploid generations that is characteristic of land plants, there are three lineages, sometimes artificially lumped together as "bryophytes," in which the conspicuous, long-lived, and vegetatively spreading generation is haploid (the "gametophyte" after its ability to produce gametes), and the diploid stage (the "sporophyte" after its ability to produce spores) is simple, unbranched, generally short-lived, and incapable of surviving independently of the gametophyte.

The liverworts have relatively simple gametophytes, which may be flattened thalli with dichotomous branching and no leaves or dorsiventral stems with filmy leaves that are only a single layer of cells thick. The cells have multiple, discoidal chloroplasts without pyrenoids, similar to those of Charales and all other land plants except the hornworts. In the complex thalloid liverworts, internal air chambers communicate with the atmosphere by means of a pore that is surrounded by a ring of specialized cells but that does not actively regulate the flow of gasses into the chamber. A cuticle is present but is typically very thin and does not provide robust protection against drying. The tissue is well organized in three dimensions, and growth occurs by division of an apical cell with three or more cutting faces. The sporophyte is very simple, without pores or stomata, and it follows a fixed developmental trajectory (i.e., determinate). Liverworts are fairly common in wet environments around the world but are rarely conspicuous. There are roughly 8000 species and 330 genera, and they show considerable diversity of structure and ecology, but all are small plants that require at least periodically wet conditions to thrive (Schofield 1985).

The hornworts are superficially similar to the thalloid liverworts and have often been confused with them. However, the hornworts differ from liverworts in a number of key characters and are almost certainly a monophyletic group. Like thalloid liverworts, the gametophytes are flattened structures without a distinct stem and leaves and radiate with irregular branching from the point of germination to form a more or less disk-shaped structure. The cells of most species have a single large chloroplast with a pyrenoid, similar to Coleochaetales and some other charophytes, but quite unlike Charales or other land plants (Graham and Kaneko 1991). Many species have a permanent symbiotic relationship with a nitrogen-fixing cyanobacterium, *Anabaena*. The sporophytes are dramatically different from those of the liverworts, with tracheophyte-like stomata capable of opening and closing, tissue specialized for water conduction, and indeterminate growth. In some cases the sporophyte seems to become largely independent of the gametophyte, and it may survive for a considerable period of time (a year or more). Hornworts are moderately rare but can be found on bare soil or rocks on stream margins as well as wet cliffs and road cuts.

There are probably fewer than 500 species in six or seven genera (Schofield 1985).

The mosses are a large, widespread, and familiar group. The gametophyte is organized into distinct stems and leaves with an elaborate and precisely coordinated architecture. Many species have a fairly well-developed vascular system. Taken together, these features give the gametophyte of mosses a structure that is strongly reminiscent of the architecture of vascular plants, but it is important to remember that the gametophytes of vascular plants do *not* have a similar form. Consequently, this similarity is very likely the result of convergent evolution or, possibly, a heterochronic shift. The sporophyte of mosses is unbranched and determinate but is structurally quite complex, with stomata, a specialized aperture to release the spores, conducting tissue, and, depending upon the species, a variety of teeth and other specialized structures. Mosses are often inconspicuous but are nearly ubiquitous, extending their range far into alpine, arctic, and desert regions where few other land plants occur. The species that occupy these extreme environments rely on great desiccation tolerance and rapid recovery to take advantage of brief periods of moisture availability and moderate temperatures to complete their life cycles. There are about 10,000 species of moss in roughly 700 genera (Schofield 1985).

Although the liverworts, hornworts, and mosses each almost certainly constitute a monophyletic group, the branching order among these groups in relation to the fourth land plant lineage, the tracheophytes (for discussion of this major group of land plants, see Pryer et al., ch. 10 in this vol.), remains a topic of active debate. Classical botanical thought considered a wide range of possibilities and often declined to speculate on the relationships among them (e.g., Bold 1967), or arbitrarily grouped them into a single heterogeneous taxon. Cladistic analyses based on morphological data suggested that the three lineages formed a ladderlike grade, with liverworts most basal, hornworts next, and mosses most closely related to a monophyletic tracheophyte clade (fig. 9.2E), although a few analyses reversed the branching order of liverworts and hornworts, placing hornworts most basal (reviewed in Bremer et al. 1987). Early molecular phylogenetic analyses were mostly based either on a single rRNA gene or on the plastid gene *rbcL*, and although there was strong support from many sources for monophyly of land plants as a whole and of vascular plants, the branching order among the three "bryophyte" lineages varied greatly from analysis to analysis, with almost every possible branching order represented (Chapman and Buchheim 1992, Mishler et al. 1994, Kranz et al. 1995). For a time the matter appeared settled, with most authorities accepting the liverwort/hornwort/moss sequence (fig. 9.2E), but Nickrent et al. (2000) presented an analysis based on four genes (*rbcL*, and SSU rDNA from the chloroplast, mitochondrial, and nuclear genomes) that reopened this question. In this analysis they investigated the contributions of each of the individual genes to the analysis, as well as differences in rate at different codon positions, evidence for saturation, and the results of several analytical methods. From this they concluded that rRNA genes supported a "hornworts basal" topology (fig. 9.2F) and suggested that prior analyses based on *rbcL* had favored a "liverworts basal" topology (fig. 9.2E) because of analytical artifacts traceable to saturation at third-codon positions. This is, however, unlikely to be the last word on the matter. In addition to several potential morphological synapomorphies such as conducting tissue, mosses and tracheophytes alone share isoprene emission (Hanson et al. 1999). Mosses, hornworts, and tracheophytes share an apparently derived ability to conjugate auxin (Sztein et al. 1995), as well as three synapomorphic mitochondrial introns (Qiu et al. 1998). Furthermore, a recent analysis based on 11,518 amino acid sites from the 52 proteins encoded by all green plant chloroplast genomes placed liverworts as the first diverging group of land plants (Kugita et al. 2003). Although this latter analysis is based solely on chloroplast genes and would benefit greatly from availability of chloroplast genome sequences from a larger number of organisms, it suggests that the topology shown in figure 9.2E is not simply an artifact peculiar to the gene *rbcL*. In the future, "total evidence" analyses are needed to combine all relevant genomic, biochemical, and morphological data, from a sufficient sample of taxa representing all major lineages (Mishler 2000).

Fortunately, with the rapid collection of genome-scale DNA sequence data from diverse organisms, it is likely that it will be possible to directly address this question and many others with large and well-curated molecular data sets in the near future. High-throughput sequencing projects have begun to take advantage of phylogenetic information to select organisms for study, and it appears that substantial quantities of sequence information will soon become available from some of the groups that have been largely neglected to date. It is important to recognize that some unfamiliar organisms are of great importance and that there is potentially a great deal to be gained by studying such organisms.

Acknowledgments

This work was supported in part by National Science Foundation grants DEB-9978117 and MCB-9984284 and Green AToL grant DEB-0228729. Tsetso Bachvaroff, John Hall, Ken Karol, Jeff Lewandowski, and other members of the Delwiche lab provided useful commentary. Dan Nickrent participated in useful discussions and provided access to unpublished data. Tashi Wangchuk provided "chu rhu" for analysis and snacking.

Literature Cited

Albertano P., C. Ciniglia, G. Pinto, and A. Pollio. 2000. The taxonomic position of *Cyanidium*, *Cyanidioschyzon* and *Galdieria*: an update. Hydrobiology 433:137–143.

Andersen, R. A. 1992. Diversity of eukaryotic algae. Biodivers. Conserv. 1:267–292.

Andersen, R. A. 1998. What to do with the protists? Aust. Syst. Bot. 11:185–201.

Andersson, J. O., and A. J. Roger. 2002. A cyanobacterial gene in nonphotosynthetic protists—an early chloroplast acquisition in eukaryotes? Curr. Biol. 12:115–119.

Baldauf, S. L., A. J. Roger, I. Wenk-Siefert, and W. F. Doolittle. 2000. A kingdom-level phylogeny of eukaryotes based on combined protein data. Science 290:972–977.

Bhattacharya, D., T. Helmchen, C. Bibeau, and M. Melkonian. 1995. Comparisons of nuclear-encoded small-subunit ribosomal RNAs reveal the evolutionary position of the Glaucocystophyta. Mol. Biol. Evol. 12:415–420.

Bhattacharya, D., and H. A. Schmidt. 1997. Division Glaucocystophyta. Plant Syst. Evol. 11:139–148.

Bhattacharya, D., K. Weber, S. An Seon, and W. Berning-Koch. 1998. Actin phylogeny identifies *Mesostigma viride* as a flagellate ancestor of the land plants. J. Mol. Evol. 47:544–550.

Bold, H. C. 1967. Morphology of plants. 2nd ed. Harper & Row, New York.

Bold, H. C., and M. J. Wynne. 1985. Introduction to the algae. Prentice-Hall, Englewood Cliffs, NJ.

Bower, F. O. 1908. The origin of a land flora; a theory based on the facts of alternation. Macmillan, London.

Bremer, K. 1985. Summary of green plant phylogeny and classification. Cladistics 1:369–385.

Bremer, K., C. J. Humphries, B. D. Mishler, and S. P. Churchill. 1987. On cladistic relationships in green plants. Taxon 36:339–349.

Bremer, K., and H.-E. Wanntorp. 1981. A cladistic classification of green plants. Nord. J. Bot. 1:1–3.

Cavalier-Smith, T. 1986. The kingdom Chromista: origins and systematics. Pp. 309–347 in Progress in phycological research (F. E. Round and D. J. Chapman, eds.), vol. 4. Biopress, Bristol.

Cavalier-Smith, T. 1999. Principles of protein and lipid targeting in secondary symbiogenesis: euglenoid, dinoflagellate, and sporozoan plastid origins and the eukaryote family tree. J. Eukaryot. Microbiol. 46:347–366.

Cavalier-Smith, T. 2000. Membrane heredity and early chloroplast evolution. Trends Plant Sci. 5:174–82.

Chadefaud, M. 1950. Les cellules nageuses des algues dans lembranchement des chlorophycees. Cr. Herbd. Acad. Sci. 231:988–990.

Chapman, R. L., O. Borkhsenious, R. C. Brown, M. C. Henk, and D. A. Waters. 2001. Phragmoplast-mediated cytokinesis in Trentepohlia: results of TEM and immunofluorescence cytochemistry. Int. J. Syst. Evol. Microbiol. 51:759–765.

Chapman, R. L., and M. A. Buchheim. 1992. Green algae and the evolution of land plants: inferences from nuclear-encoded rRNA gene sequences. Biosystems 28:127–137.

Chesnick, J. M., C. W. Morden, and A. M. Schmieg. 1996. Identity of the endosymbiont of *Peridinium foliaceum* (Pyrrophyta): analysis of the rbcLS operon. J. Phycol. 32:850–857.

Chretiennotdinet, M. J., C. Courties, A. Vaquer, J. Neveux, H. Claustre, J. Lautier, and M. C. Machado. 1995. A new marine picoeucaryote—*Ostreococcus tauri* Gen Et Sp-Nov (Chlorophyta, Prasinophyceae). Phycologia 34:285–292.

Christensen, T. 1989. The chromophyta, past and present. Pp. 1–12 *in* The chromophyte algae: problems and perspectives (J. C. Green, B. S. C. Leadbeater, and W. L. Diver, eds.). Clarendon Press, Oxford.

Delwiche, C. F. 1999. Tracing the thread of plastid diversity through the tapestry of life. Am. Nat. 154:S164–S177.

Delwiche, C. F., L. E. Graham, and N. Thomson. 1989. Lignin-like compounds and sporopollenin in *Coleochaete*, an algal model for land plant ancestry. Science 245:399–401.

Delwiche, C. F., K. G. Karol, M. T. Cimino, and K. J. Sytsma. 2002. Phylogeny of the genus *Coleochaete* (Charophyceae, Chlorophyta) and related taxa inferred by analysis of the chloroplast gene *rbcL*. J. Phycol. 38:394–403.

Delwiche, C. F., M. Kuhsel, and J. D. Palmer. 1995. Phylogenetic analysis of *tufA* sequences indicates a cyanobacterial origin of all plastids. Mol. Phylogenet. Evol. 4:110–128.

Delwiche, C. F., and J. D. Palmer. 1997. The origin of plastids and their spread via secondary symbiosis. Plant Syst. Evol. 11:53–86.

Doolittle, W. F. 1998. You are what you eat: a gene transfer ratchet could account for bacterial genes in eukaryotic nuclear genomes. Trends Genet. 14:307–311.

Douglas, S., S. Zauner, M. Fraunholz, M. Beaton, S. Penny, L.-T. Deng, X. Wu, M. Reith, T. Cavalier-Smith, and U.-G. Maier. 2001. The highly reduced genome of an enslaved algal nucleus. Nature 410:1091–1098.

Fast, N. M., J. C. Kissinger, D. S. Roos, and P. J. Keeling. 2001. Nuclear-encoded, plastid targeted genes suggest a single common origin for Apicomplexan and dinoflagellate plastids. Mol. Biol. Evol. 18:418–426.

Fawley Marvin, W., Y. Yun, and M. Qin. 2000. Phylogenetic analyses of 18S rDNA sequences reveal a new coccoid lineage of the Prasinophyceae (Chlorophyta). J. Phycol. 36:387–393.

Feist, M., and R. Feist. 1997. Oldest record of a bisexual plant. Nature 385:401.

Friedl, T. 1995. Inferring taxonomic positions and testing genus level assignments in coccoid green lichen algae—a phylogenetic analysis of 18S ribosomal-RNA sequences from *Dictyochloropsis reticulata* and from members of the genus *Myrmecia* (Chlorophyta, Trebouxiophyceae Cl-Nov). J. Phycol. 31:632–639.

Fritsch, F. E. 1965. The structure and reproduction of algae. Cambridge University Press, Cambridge.

Gibbs, S. P. 1962. Nuclear envelope-chloroplast relationships in algae. J. Cell Biol. 14:433–444.

Gibbs, S. P. 1978. The chloroplasts of Euglena may have evolved from symbiotic green algae. Can. J. Bot. 56:2883–2889.

Gibbs, S. P. 1981. The chloroplast endoplasmic reticulum: structure, function, and evolutionary significance. Int. J. Cytol. 72:49–99.

Gillott, M. A., and S. P. Gibbs. 1980. The cryptomonad nucleomorph: its ultrastructure and evolutionary significance. J. Phycol. 16:558–568.

Glöckner, G., A. Rosenthal, and K. Valentin. 2000. The structure and gene repetoire of an ancient red algal plastid genome. J. Mol. Evol. 51:382–390.

Graham, L. E. 1993. Origin of land plants. John Wiley and Sons, New York.

Graham, L. E., C. F. Delwiche, and B. D. Mishler. 1991. Phylogenetic connections between the "green algae" and the "bryophytes." Pp. 3–443 *in* Bryophyte systematics (N. G. Miller, ed.). J. Cramer, Stuttgart.

Graham, L. E., and Y. Kaneko. 1991. Subcellular structures of relevance to the origin of land plants embryophytes from green algae. Crit. Rev. Plant Sci. 10:323–342.

Graham, L. E., and L. W. Wilcox. 2000. Algae. Prentice-Hall, Upper Saddle River, NJ.

Grambast, L. 1974. Phylogeny of the Charophyta. Taxon 23:463–481.

Green, B. J., W. Y. Li, J. R. Manhart, T. C. Fox, E. J. Summer, R. A. Kennedy, S. K. Pierce, and M. E. Rumpho. 2000. Mollusc-algal chloroplast endosymbiosis. Photosynthesis, thylakoid protein maintenance, and chloroplast gene expression continue for many months in the absence of the algal nucleus. Plant Physiol. 124:331–342.

Hanson, D. T., S. Swanson, L. E. Graham, and T. D. Sharkey. 1999. Evolutionary significance of isoprene emission from mosses. Am. J. Bot. 86:634–639.

Hanten, J. J., and S. K. Pierce. 2001. Synthesis of several light-harvesting complex I polypeptides is blocked by cyclohex-imide in symbiotic chloroplasts in the sea slug, Elysia chlorotica (Gould): a case for horizontal gene transfer between alga and animal? Biol. Bull. 201:34–44.

Hanyuda, T., I. Wakana, S. Arai, K. Miyaji, Y. Watano, and K. Ueda. 2002. Phylogenetic relationships within Cladophorales (Ulvophyceae, Chlorophyta) inferred from 18S rRNA gene sequences, with special reference to *Aegagropila linnaei*. J. Phycol. 38:564–571.

Harris, E. H. 1989. The Chlamydomonas sourcebook. Academic Press, San Diego.

Huss, V. A. R., C. Frank, E. C. Hartmann, M. Hirmer, A. Kloboucek, B. M. Seidel, P. Wenzeler, and E. Kessler. 1999. Biochemical taxonomy and molecular phylogeny of the genus *Chlorella* sensu lato (Chlorophyta). J. Phycol. 35:587–598.

Jeffrey, C. 1967. The origin and differentiation of the archegoniate land plants: a second contribution. Kew Bull. 21:335–349.

Jousson, O., J. Pawlowski, L. Zaninetti, F. W. Zechman, F. Din, G. Di Guiseppe, R. Woodfield, R. A. Millar, and A. Meinesz. 2000. Invasive alga reaches California. Nature 408:157–158.

Karol, K. G., R. M. McCourt, M. T. Cimino, and C. F. Delwiche. 2001. The closest living relatives of plants. Science 294:2351–2353.

Kelland, J. L. 1977. Inversion in *Volvox* (Chlorophyceae). J. Phycol. 13:373–378.

Kies, L., and B. P. Kramer. 1989. Phylum Glaucocystophyta. Pp. 152–166 *in* Handbook of Protoctista (L. Margulis, J. O. Corliss, M. Melkonian, and D. J. Chapman, eds.). Jones and Bartlett Publishers, Boston.

Kirk, D. L. 1999. Evolution of multicellularity in the volvocine algae. Curr. Opin. Plant Biol. 2:496–501.

Kirk, D. L. 2001. Germ-soma differentiation in *Volvox*. Dev. Biol. 238:213–223.

Köhler, S., C. F. Delwiche, P. W. Denny, L. G. Tilney, P. Webster, R. J. M. Wilson, J. D. Palmer, and D. S. Roos. 1997. A plastid of probable green algal origin in apicomplexan parasites. Science 275:1485–1489.

Kranz, H. D., D. Miks, M.-L. Siegler, I. Capesius, C. Sensen, W., and V. A. R. Huss. 1995. The origin of land plants: phylogenetic relationships among charophytes, bryophytes, and vascular plants inferred from complete small-subunit ribosomal RNA gene sequences. J. Mol. Evol. 41:74–84.

Kugita, M., A. Kaneko, Y. Yamamoto, Y. Takeya, T. Matsumoto, and K. Yoshinaga. 2003. The complete nucleotide sequence of the hornwort (Anthoceros formosae) chloroplast genome: insight into the earliest land plants. Nucleic Acids Res. 31:716–721.

Lemieux, C., C. Otis, and M. Turmel. 2000. Ancestral chloroplast genome in Mesostigma viride reveals an early branch of green plant evolution. Nature 403:649–652.

Lewis, L. A., and V. R. Flechtner. 2002. Green algae (Chlorophyta) of desert microbiotic crusts: diversity of North American taxa. Taxon 51:443–451.

Lockhorst, G. M., ed. 1996. Comparative taxonomic studies on the genus Klebsormidium (Charophyceae) in Europe. Gustav Fischer Verlag, Stuttgart.

Löffelhardt, W., H. Bohnert, and D. Bryant. 1997. The complete sequence of the Cyanophora paradoxa cyanelle genome (Glaucocystophyceae). Plant Syst. Evol. 11:S149–S162.

Luther, A. 1899. Über Chlorosaccus eine neue Gattung der Süsswasseralgen, nebst Bemerkungen zur Systmatick verwandter algen. Bihang till Kongliga Svenska Ventenskaps-Academiens Handlingar 24:1–22.

Marin, B., and M. Melkonian. 1999. Mesostigmatophyceae, a new class of streptophyte green algae revealed by SSU rRNA sequence comparisons. Protist 150:399–417.

Martin, W., T. Rujan, E. Richly, A. Hansen, S. Cornelsen, T. Lins, D. Leister, B. Stoebe, M. Hasegawa, and D. Penny. 2002. Evolutionary analysis of Arabidopsis, cyanobacterial, and chloroplast genomes reveals plastid phylogeny and thousands of cyanobacterial genes in the nucleus. Proc. Natl. Acad. Sci. USA 99:12246–12251.

Martin, W., B. Stoebe, V. Goremykin, S. Hansmann, M. Hasegawa, and K. Kowallik. 1998. Gene transfer to the nucleus and the evolution of chloroplasts. Nature 393:162–165.

Mattox, K. R., and Stewart, K. D. 1984. A classification of the green algae: a concept based on comparative cytology. Pp. 29–72 in Systematics of the green algae (D. E. G. Irvine and D. M. John, eds.). Academic Press, London.

McCourt, R. M. 1995. Green Algal Phylogeny. Trends Ecol. Evol. 10:159–163.

McCourt, R. M., K. G. Karol, J. Bell, K. M. Helm-Bychowski, A. Grajewska, M. F. Wojchiechowski, and R. W. Hoshaw. 2000. Phylogeny of the conjugating green algae (Zygnemophyceae) based on rbcL sequences. J. Phycol. 36:747–758.

McFadden, G. I. 2001. Primary and secondary endosymbiosis and the origin of plastids. J. Phycol. 37:951–959.

McFadden, G. I., P. R. Gilson, C. J. B. Hofmann, G. J. Adcock, and U.-G. Maier. 1994. Evidence that an amoeba acquired a chloroplast by retaining part of an engulfed eukaryotic alga. Proc. Natl. Acad. Sci. USA 91:3690–3694.

Meinesz, A., J. Devaugelas, B. Hesse, and X. Mari. 1993. Spread of the introduced tropical green alga Caulerpa-taxifolia in northern Mediterranean waters. J. Appl. Phycol. 5:141–147.

Mishler, B. D. 2000. Deep phylogenetic relationships among "plants" and their implications for classification. Taxon 49:661–683.

Mishler, B. D., and S. P. Churchill. 1985. Transition to a land flora: phylogenetic relationships of the green algae and bryophytes. Cladistics l:305–328.

Mishler, B. D., L. A. Lewis, M. A. Buchheim, K. S. Renzaglia, D. J. Garbary, C. F. Delwiche, F. W. Zechman, T. S. Kantz, and R. L. Chapman. 1994. Phylogenetic-Relationships of the Green Algae and Bryophytes. Ann. Mo. Bot. Gard. 81:451–483.

Moreira, D., and H. Philippe. 2001. Sure facts and open questions about the origin and evolution of photosynthetic plastids. Res. Microbiol. 152:771–780.

Müller, K. M., M. C. Oliveira, R. Sheath, and D. Bhattacharya. 2001. Ribosomal DNA phylogeny of the Bangiophycidae (Rhodophyta) and the origin of secondary plastids. Am. J. Bot. 88:1390–1400.

Nandan Prasad, B., and D. Kumar Asthana. 1979. Awadhiella—a new genus of Coleochaetaceae from India. Hydrobiologia 62:131–135.

Nickrent, D. L., C. L. Parkinson, J. D. Palmer, and R. J. Duff. 2000. Multigene phylogeny of land plants with special reference to bryophytes and the earliest land plants. Mol. Biol. Evol. 17:1885–1895.

Norton, T. A., M. Melkonian, and R. A. Andersen. 1996. Algal biodiversity. Phycologia 35:308–326.

O'Kelly, C. J., and G. L. Floyd. 1983. Flagellar apparatus absolute orientations and the phylogeny of the green algae. Biosystems 16:227–251.

Oliveira, M. C., and D. Bhattacharya. 2000. Phylogeny of the Bangiophycidae (Rhodophyta) and the secondary endosymbiotic origin of algal plastids. Am. J. Bot. 87:482–492.

Oliver, M. J., Z. Tuba, and B. D. Mishler. 2000. The evolution of vegetative desiccation tolerance in land plants. Plant Ecol. 151:85–100.

Palmer, J. D. 2003. The symbiotic birth and spread of plastids: how many times and whodunit? J. Phycol. 39:4–11.

Palmer, J. D., and C. F. Delwiche. 1998. The origin and evolution of plastids and their genomes. Pp. 375–409 in Molecular systematics of plants II: DNA sequencing (P. S. Soltis, D. E. Soltis, and J. J. Doyle, eds.). Kluwer Academic Publishers, Boston.

Pickett-Heaps, J. D. 1975. Green algae: structure, reproduction, and evolution of selected genera. Sinauer Associates, Sunderland, MA.

Pickett-Heaps, J. D., and H. J. Marchant. 1972. The phylogeny of the green algae: a new proposal. Cytobios 6:255–264.

Pimm, S. L., G. J. Russell, J. L. Gittleman, and T. M. Brooks. 1995. The future of biodiversity. Science 269:347–350.

Potter, D., T. C. Lajeunesse, G. W. Saunders, and R. A. Anderson. 1997. Convergent evolution masks extensive biodiversity among marine coccoid picoplankton. Biodivers. Conserv. 6:99–107.

Qiu, Y.-L., Y. Cho, J. C. Cox, and J. D. Palmer. 1998. The gain of three mitochondrial introns identifies liverworts as the earliest land plants. Nature 394:671–674.

Rogers, C. E., D. S. Domozych, K. D. Stewart, and K. R. Mattox. 1981. The flagellar apparatus of Mesostigma viride (Prasinophyceae): multilayered structures in a scaly green flagellate. Plant Syst. Evol. 138:247–258.

Rumpho, M. E., E. J. Summer, and J. R. Manhart. 2000. Solar-powered sea slugs. Mollusc/algal chloroplast symbiosis. Plant Physiol. 123:29–38.

Schofield, W. B. 1985. Introduction to bryology. Macmillan, New York.

Singh, R. N. 1941. On some phases in the life history of the terrestrial alga, *Fritschiella tuberosa* Iyeng., and its autecology. New Phytol. 40:170–182.

Smith, G. M. 1955. Cryptogamic botany. McGraw-Hill, New York.

Stiller, J. W., and B. D. Hall. 1997. The origins of red algae: implications for plastid evolution. Proc. Natl. Acad. Sci. USA 94:4520–4525.

Stiller, J. W., D. C. Reel, and J. C. Johnson. 2003. A single origin of plastids revisited: convergent evolution in organellar genome content. J. Phycol. 39:95–105.

Stirewalt, V., C. Michalowski, W. Löffelhardt, H. Bohnert, and D. Bryant. 1995. Nucleotide sequence of the cyanelle genome from *Cyanophora paradoxa*. Plant Mol. Biol. Rep. 13:327–332.

Sztein, A. E., J. D. Cohen, J. P. Slovin, and T. J. Cooke. 1995. Auxin metabolism in representative land plants. Am. J. Bot. 82:1514–1521.

Tappan, H. 1980. The paleobiology of plant protists. W. H. Freeman and Co., San Francisco.

Tengs, T., O. J. Dahlberg, K. Shalchian-Tabrizi, D. Klaveness, K. Rudi, C. F. Delwiche, and K. S. Jakobsen. 2000. Phylogenetic analyses indicate that the 19'-hexanoyloxy-fucoxanthin-containing dinoflagellates have tertiary plastids of haptophyte origin. Mol. Biol. Evol. 17:718–729.

Thompson, R. H., and D. E. Wujek. 1997. *Trentepohliales*: *Cephaleuros*, *Phycopeltis*, and *Stomatochroon*. Science Publishers, Inc., Einfield, NH.

Tilden, J. E. 1937. The algae and their life relations. 1968 repr. ed. Hafner Publishing, New York.

Turmel, M., C. Otis, and C. Lemieux. 2001. The complete mitochondrial DNA sequence of Mesostigma viride identifies this green alga as the earliest green plant divergence and predicts a highly compact mitochondrial genome in the ancestor of all green plants. Mol. Biol. Evol. 19:24–38.

Tyrrell, T. 2003. *Emiliana huxleyi* Home Page. Available: http://www.soes.soton.ac.uk/staff/tt/en. Last accessed 25 December 2003.

Van den Hoek, C., D. G. Mann, and H. M. Jahns. 1995. Algae: an introduction to phycology. Cambridge University Press, Cambridge.

Watanabe, S., N. Kuroda, and F. Maiwa. 2001. Phylogenetic status of *Helicodictyon planctonicum* and *Desmochloris halophila* gen. et comb. nov and the definition of the class Ulvophyceae (Chlorophyta). Phycologia 40:421–434.

Wehr, J. D., and R. G. Sheath (eds.). 2003. Freshwater algae of North America. Academic Press, San Diego.

Yoon, H. S., J. Hackett, and D. Bhattacharya. 2002. A single origin of the peridinin- and fucoxanthin- containing plastids in dinoflagellates through tertiary endosymbiosis. Proc. Natl. Acad. Sci. USA 99:11724–11729.

Zhang, Z., T. Cavalier-Smith, and B. R. Green. 2001. A family of selfish minicircular chromosomes with jumbled chloroplast gene fragments from a dinoflagellate. Mol. Biol. Evol. 18:1558–1565.

Zhang, Z., B. R. Green, and T. Cavalier-Smith. 1999. Single gene circles in dinoflagellate chloroplast genomes. Nature 400:155–159.

Kathleen M. Pryer
Harald Schneider
Susana Magallón

The Radiation of Vascular Plants

Vascular plants include our major food resources in the form of leaves, stems, roots, fruits, and seeds. They further sustain human life by providing other essentials such as wood, fibers, and medicines. Plants are the dominant primary producers in terrestrial habitats, and by the process of photosynthesis, they actively convert solar energy, water, and carbon dioxide into carbohydrates (sugars) and oxygen, which is vital to all living things. The rise and spread of vascular plants resulted in a dramatic drop in atmospheric carbon dioxide (CO_2) about 400 million years ago during the mid-Paleozoic (Algeo et al. 2001, Berner 2001, Driese and Mora 2001, Raven and Edwards 2001). This decline in atmospheric CO_2 triggered the evolution of vascular plants with more complex body plans, including such organs as leaves, specialized for optimizing photosynthesis (Beerling et al. 2001, Pataki 2002, Shougang et al. 2003). Vascular plants therefore both caused and reacted to global changes in their physical environment early in their evolution. The earliest radiation of vascular plants has been interpreted as one in which rapidly diversifying lineages colonized and shaped different terrestrial habitats (DiMichele et al. 2001). Repeated reciprocation between climatic change and vascular plant radiation is noted throughout the fossil record, with particularly marked changes in floristic patterns occurring at the end of the Permian (Looy et al. 2001), at the Triassic/Jurassic (McElwain et al. 1999) and Paleocene/Eocene boundaries (Tiffney and Manchester 2001), and in the Cretaceous (Friis et al. 2001b).

The advent of terrestrial primary producers capable of forming a huge biomass correlates with the simultaneous rise of terrestrial animals in the Paleozoic, including various groups of arthropods and tetrapods (Coates 2001, Shear and Selden 2001, Carroll 2002). For example, the first known mites are found together with the first vascular plants in Devonian Rhynie Chert beds in Aberdeenshire in the north of Scotland. According to a recent molecular clock estimate, basal groups of insects originated during the Late Devonian (Gaunt and Miles 2002), coinciding with the diversification of vascular plants. The wide spectrum of fossilized insects observed in the Late Carboniferous (Labandeira 2001), including herbivorous groups, also suggests a simultaneous adaptive radiation of vascular plants and insects. The establishment of vascular plants with large and complex body plans in the Carboniferous and Permian resulted in an increased amount and diversity of vegetative biomass that favored the diversification of herbivorous tetrapods in the Permian (Sues and Reisz 1998, Coates 2001). Vascular plants are not only the major nutrient source for the consumers in their ecosystems, but they also play an important role in symbiotic associations with fungi (Brundrett 2002). Rhynie Chert fossils of glomalean mycorrhizal fungi discovered in association with preserved plant shoots exquisitely document complex plant–fungi interactions by the Early Devonian, suggesting that mycorrhizal associations were a critical factor in the early and successful colonization of land by terrestrial plants (Taylor et al. 1995, Blackwell 2000, Cairney 2000,

Hibbett et al. 2000, Redecker et al. 2000, Brundrett 2002; see also ch. 12 in this vol.).

In this review, we summarize the results of various recent studies that have used morphological and/or molecular evidence to infer the phylogeny of living vascular plants, and those that have used morphological/anatomical evidence to understand relationships of fossil plants. These studies differ widely in their taxon sampling, in the parts of the green branch of the Tree of Life they focus on, and also in their methodology. It is a challenging exercise, therefore, to distill from them not only a summary but also a fresh look at our current understanding of the evolution and relationships among both living and extinct vascular plants. It should be noted at the outset that we view the continued traditional application of several taxonomic names and ranks, especially to fossil groups that are clearly not monophyletic (e.g., Rhyniophyta), as hampering progress in our understanding and discussions of vascular plant evolution. Rather than abandon these names entirely, we retain most of them as common names in quotation mark (e.g., "rhyniophytes") to clarify historical usage. In our phylogenetic figures, we attempt to illustrate progress that has been made in discerning the relationships of members of these groups and our best sense of where they "fit in." Also, where we integrate fossils together with living taxa, we try to distinguish between stem and crown groups (see Smith 1994:94–98). Stem groups include taxa that are in fact part of a particular lineage but that lack some character(s) (synapomorphy) that distinguishes the crown group.

We were especially fortunate to be able to build on several thorough reviews that have been published in recent years. For additional information and different perspectives, the reader is referred to Kenrick and Crane (1997b), Bateman et al. (1998), Doyle (1998b), Rothwell (1999), Renzaglia et al. (2000), Donoghue (2002), Judd et al. (2002: ch. 7), and Schneider et al. (2002).

What Are Vascular Plants?

Vascular plants make up the bulk of all the land plant lineages. They are a monophyletic group characterized by the presence of specialized cells, tracheids and sieve elements, which conduct water and nutrients throughout the plant body and provide structural support (Kenrick and Crane 1997a, 1997b, Bateman et al. 1998, Schneider et al. 2002). Land plants are typified by an alternation of generation phases, whereby heteromorphic haploids (gametophytes) and diploids (sporophytes) alternate throughout the plant's life cycle (Kenrick 1994, 2000, 2002b, Mable and Otto 1998, Renzaglia et al. 2000). The gametophytes of nonvascular land plant lineages (mosses, liverworts, and hornworts) are the dominant or more visible phase that bears a comparatively tiny sporophyte with a single sporangium. The recent confirmation of Charales as the green algal lineage most closely related to land plants (Karol et al. 2001) supports the view

that a dominant gametophyte phase is the plesiomorphic (ancestral) condition, whereas a predominant sporophyte phase, which is found in vascular plants, is derived (Mable and Otto 1998, Kenrick 2000). The gametophyte phase is diminutive in vascular plants compared with the highly branched sporophyte that bears more than a single sporangium (polysporangiate). Figure 10.1 contrasts these major differences in morphology and life cycle between nonvascular and vascular plants. Observations from the fossil record (Kenrick 2002b) the reconstruction of life cycle evolution based on living taxa (Schneider et al. 2002: fig. 17.2a) converge on a scenario suggesting that over time there was a trend from a short-lived sporophyte phase ("bryophytes") to one whereby both the gametophyte and sporophyte phases were essentially codominant (putatively isomorphic in "rhyniophytes") and that eventually the sporophyte phase came to dominate the life cycle in vascular plants. Heterosporous lineages (those that produce two spore types), and especially seed plants, demonstrate this trend most clearly, with the gametophyte phase becoming extremely reduced both in size and in duration.

Unequivocal evidence for the earliest polysporangiate plants dates back to the Rhynie Chert beds of the Late Silurian. Representatives such as *Aglaophyton* (fig. 10.2 inset), *Horneophyton*, and *Rhynia* all had simple and diminutive body plans with dichotomously branched axes and no roots or leaves (Kenrick and Crane 1997a, 1997b, Crane 1999, Edwards and Wellman 2001). The erect axes were terminated by round or ovoid sporangia and possessed water-conducting cells that were either unthickened and unornamented (e.g., *Aglaophyton*) or well-developed tracheids (e.g., *Rhynia*) organized in centrarch protosteles (protoxylem is centrally located in the vascular cylinder and xylem maturation is centrifugal—toward the periphery of the axis. These plants are often referred to as "rhyniophytes," and they were never very diverse either in species number or in morphology and quickly became replaced by plants with a more complex organization. The descendants of these "rhyniophytes" diversified rapidly in the Early Devonian, resulting in a split into two major groups (fig. 10.2), the lycophytes (Lycophytina) and the euphyllophytes (Euphyllophytina). The primary feature that unites these two groups to distinguish them from the "rhyniophytes" is the differentiation of the plant body into aerial (shoot) and subterranean (root) components (Gensel and Berry 2001, Gensel et al. 2001, Schneider et al. 2002), which argues for a single origin of roots rather than several independent origins (Raven and Edwards 2001, Kenrick 2002a).

Lycophytes and Zosterophytes (Lycophytina)

The earliest lycophyte lineages diversified in the Early Devonian and are referred to here as "protolycophytes" (fig. 10.3), plants characterized by mostly dichotomously branching axes that are either naked or covered by spiny appendages. The aerial axes mostly possess an exarch pro-

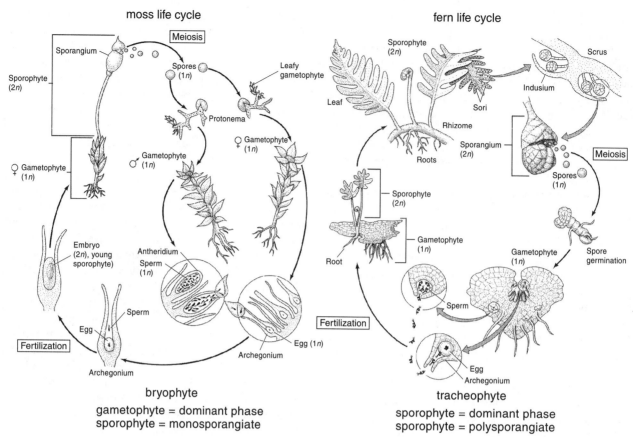

Figure 10.1. Comparison of alternation of generation phases between representative "bryophyte" and tracheophyte life cycles to illustrate the evolutionary transition from a dominant autotrophic gametophyte and a nutritionally dependent monosporangiate sporophyte in "bryophytes" to a dominant autotrophic polysporangiate sporophyte in tracheophytes. In ferns, the gametophytes are independent of the sporophytes; in seed plants, the microgametophytes are independent but the megagametophytes are retained on the sporophyte. Figure modified from Singer (1997).

tostele (protoxylem is located at the edge of the vascular cylinder and xylem maturation is centripetal—toward the center of the axis) and bear dorsiventral sporangia (often kidney-shaped) that open into two equal-sized valves via transverse dehiscence. These sporangia are laterally inserted either on terminate or nonterminate axes (Kenrick and Crane 1997a). Zosterophytes [eg., *Zosterophyllum* (fig. 10.3 inset), *Sawdonia*] and prelycophytes (e.g., *Asteroxylon*, *Drepanophycus*) were a dominant component of the landscape until they became extinct in the Early Carboniferous. Their descendants, which include three extant lineages of lycophytes—Lycopodiales, Selaginellales, and Isoëtales (fig. 10.3)—bear lycophylls, leaves that develop exclusively by intercalary growth (Crane and Kenrick 1997, Kenrick 2002b; Schneider et al. 2002). Intercalary growth is characterized by meristematic activity that is not apical but rather is more diffusely organized toward the base of the lycophylls.

These three lineages diversified in the Late Devonian and Carboniferous and can be easily distinguished by both reproductive and vegetative features (Kenrick and Crane 1997a, Judd et al. 2002). The Lycopodiales are homosporous (pro-

duce spores of a single type) and are sister to a clade of heterosporous (producing two spore types) lycophytes, Selaginellales and Isoëtales, which bear a sterile leaflike appendage (ligule) on the adaxial (facing toward main axis) leaf surface. The heterosporous lycophytes were morphologically and ecologically diverse throughout the Carboniferous, including arborescent forms with a unique type of secondary xylem, such as the isoetalean lycophyte *Lepidodendron* (fig. 10.3, inset), but declined drastically starting in the Upper Carboniferous (Pigg 2001, DiMichele and Phillips 2002) and continuing throughout the Mesozoic. Today the lycophyte representatives that remain are diminutive in stature and diminished in diversity (<1% of extant vascular plants).

"Trimerophytes" and Euphyllophytes (Euphyllophytina)

Euphyllophytes are the sister group to the lycophytes. Although the euphyllophytes encompass an astonishing morphological diversity, they all share several features in common,

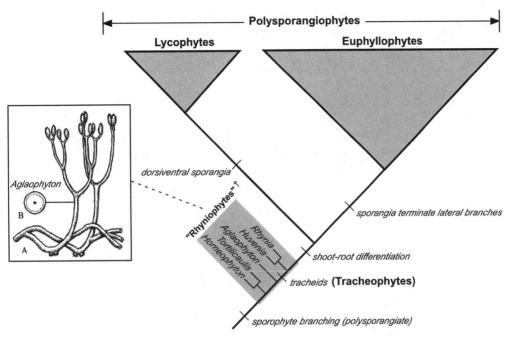

Figure 10.2. Vascular plant phylogeny: relationships of early polysporangiate taxa. Gray triangles indicate extant lycophyte and euphyllophyte crown groups; shaded box highlights extinct (†) "rhyniophyte" stem group with some representative taxa. Critical synapomorphies are indicated on the branches. Phylogeny based largely on Kenrick and Crane (1997:129, fig. 4.31) and Meyer-Berthaud and Gerrenne (2001). Inset, Sketch of a representative early Devonian "rhyniophyte," *Aglaophyton*: A, dichotomously branched creeping and erect axes, the latter terminated by ovoid sporangia; B, tiny central strand of unthickened water-conducting cells. Plant drawing from Fischer et al. (1998).

such as sporangia that terminate some lateral branches (figs. 10.2 and 10.4), a distinctively lobed primary xylem strand (Stein 1993, Kenrick and Crane 1997a) and a 30-kilobase chloroplast inversion (Raubeson and Jansen 1992a). Early members of the euphyllophyte lineage, such as *Psilophyton* (fig. 10.4 inset), are referred to as "trimerophytes" and were homosporous and leafless plants restricted to the Devonian. They exhibited pseudomonopodial branching (overtopping) resulting in a differentiation of the shoot system whereby one axis is dominant with indeterminate growth (main axis continues to grow) and the lateral axes were determinate (terminated by sporangia). Later during the evolution of this lineage the determinate axes were transformed into euphylls—leaves that develop with an apical and/or marginal meristem resulting in a gap being formed in the stele (stem vascular cylinder) above the point of leaf insertion (Schneider et al. 2002), which argues for a single origin of euphylls, rather than several independent origins (Boyce and Knoll 2002). Therefore, as early as the Devonian there was a major transition from vascular plants without leaves to those that possessed euphylls (Beerling et al. 2001, Shougang et al. 2003).

During the evolution of the Euphyllophytina there was a split in the early-mid Devonian into two major clades, monilophytes and lignophytes (Pryer et al. 2001). The monilophytes (= Infradivision Moniliformopses, *sensu* Kenrick and Crane

1997a; Judd et al. 2002: ch. 7) include horsetails, eusporangiate and leptosporangiate ferns, and whisk ferns (*Psilotum* and relatives). The lignophytes include all seed plants and their closest relatives (Doyle 1998b). The ancient radiation of these two divergent lineages gave rise to what now is 99% of extant vascular plant diversity.

Monilophytes

The monilophytes comprise five major extant lineages (fig. 10.5A): Equisetopsida (horsetails), Polypodiidae (leptosporangiate ferns), Psilotidae (whisk ferns), Marattiidae (marattiaceous ferns), and Ophioglossidae (moonwort ferns). Previous assessments of relationships among these lineages were contradictory and often placed one or more of them as sister to the seed plants, implying that vascular plant evolution had proceeded in a progressive and steplike fashion. Recent recognition that these lineages are clustered together in a single clade that is sister to seed plants has helped to stabilize this pivotal region of the vascular plant phylogeny (Kenrick and Crane 1997a, 1997b, Nickrent et al. 2000, Renzaglia et al. 2000, Pryer et al. 2001, Rydin et al. 2002).

Among the extant monilophytes, the earliest-diverging lineages are those with the poorest fossil record—the Psiloti-

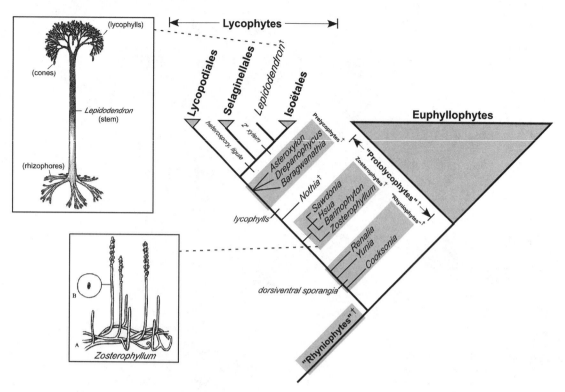

Figure 10.3. Vascular plant phylogeny: relationships of lycophytes. Phylogeny based largely on Kenrick and Crane (1997a:129, fig. 4.31) and Meyer-Berthaud and Gerrienne (2001); extinct taxa are indicated with a dagger (†). Early in the evolution of the lycophyte lineage there was a transition to sporangia that had a dorsoventral organization and that opened with transverse dehiscence. The crown group of lycophytes shares a single origin of leaves (lycophylls) that develop by an intercalary meristem. Bottom inset, Sketches of a representative early Devonian lycophyte, *Zosterophyllum*: A, dichotomously branched axes, erect axes bearing lateral dorsoventral sporangia; B, Tiny central vascular strand (protostele) composed of tracheids. Plant drawing from Arens et al. (1998a). Top inset, *Lepidodendron*, typical arborescent lycopsid. Plant drawing from Arens et al. (1998b). Taxonomic issues: *Hsua* = "rhyniophyte" *sensu* Banks (1975, 1992) = putative zosterophyte *sensu* Kenrick and Crane (1997a); *Nothia* = "rhyniophyte" *sensu* Banks (1975, 1992) = putative zosterophyte *sensu* Kenrick and Crane 1997a); *Barinophyton* = *incertae sedis sensu* Banks (1975, 1992) = zosterophyte *sensu* Kenrick and Crane (1997a); *Cooksonia* = "rhyniophyte" *sensu* Banks (1975, 1992) = polyphyletic, with some species part of Lycophytina stem group *sensu* Kenrick and Crane (1997a).

dae and Ophioglossidae (fig. 10.5B)—eusporangiate ferns (produce thick-walled sporangia containing numerous spores) with such radically different phenotypes that their recognition as sister taxa became apparent only after the accumulation of data from a number of molecular markers (Nickrent et al. 2000, Pryer et al. 2001). Morphological characters that support this relationship are exceedingly difficult to discern given the extreme simplification that one observes in both their vegetative and reproductive structures. However, these two monilophyte lineages share a reduction in root systems, whereby Ophioglossidae have no root hairs and Psilotidae have lost roots altogether, with both lineages relying on endomycorrhizal associations for nutrient absorption (Schneider et al. 2002).

How the remaining lineages of extant monilophytes (Equisetopsida, Polypodiidae, and Marattiidae) are related to one another is still unclear. Extant Equisetopsida (15 species; Des Marais et al. 2003) and Marattiidae (300 species; Hill and Camus 1986) are relatively species poor, but both these groups have very rich fossil records in the Late Paleozoic and Early Mesozoic (fig. 10.5B; Bateman et al. 1998, Rothwell 1999, Liu et al. 2000, Berry and Fairon-Demaret 2001). In contrast, the Polypodiidae have had a rich fossil record from the Late Paleozoic until the Recent period, with extant taxa numbering greater than 10,000 species (Collinson 1996, Skog 2001). Polypodiidae share the notable characteristic of being leptosporangiate—having sporangia with a wall that is a single cell layer thick and containing relatively few meiospores (<1000, usually 64).

Continued emphasis on increasing the availability of molecular markers will likely improve resolution among these deep branches in the monilophyte clade. However, it

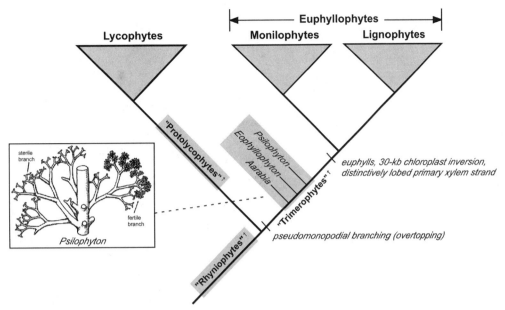

Figure 10.4. Vascular plant phylogeny. Relationships of "trimerophytes" and euphyllophytes (= monilophytes + lignophytes). Phylogeny based largely on Kenrick and Crane (1997a:240, fig. 7.10) and Meyer-Berthaud and Gerrienne (2001). Extinct (†) taxa are in shaded boxes. The transition to "trimerophytes" is marked by a change to pseudomonopodial branching, whereby a main indeterminate axis develops and overtops lateral determinate axes. Euphyllophytes share a common origin of leaves (euphylls) that develop by an apical and/or marginal meristem, a 30-kilobase inversion in the chloroplast genome organization, and a distinctively lobed primary xylem strand. Inset, sketch of *Psilophyton* shows sterile axes with forked tips and fertile lateral axes terminated by sporangia. Plant drawing from Arens et al. 1998c).

will also be critical for future studies to incorporate morphological data from fossil members pertinent to this clade if we really are going to improve our understanding of the evolution of monilophyte lineages through time and clarify ideas concerning homology. These include arborescent relatives of Equisetopsida, such as *Archaeocalamites* and *Calamites* (fig. 10.5A); fossil relatives of Marattiidae, such as *Psaronius* (fig. 10.5A), which was an important tree-fern-like component of Carboniferous landscapes; and extinct Polypodiidae, such as *Botryopteris*, which although less abundant and more diminutive (Bateman et al. 1998, DiMichele et al. 2001), were opportunistic and scandent members of the terrestrial ecosystem, much as their living relatives are today.

As to the relationships of other fossil monilophytes of the Devonian with highly divergent morphologies, such as "iridopterid" (*Ibykya*), "cladoxylopsid" (*Calamophyton*, *Pseudosporochnus*), and "zygopterid" ferns (*Rhacophyton*), much work remains to be done (Bateman et al. 1998, Rothwell 1999, Berry and Farion-Dermaret 2001). These fossil plants have been discussed as relatives to horsetails and ferns in the broad sense, but with no clear picture emerging as to which are stem group or crown group members (fig. 10.5A). Integrating these taxa into a phylogeny of living members will certainly improve our understanding of evolutionary transitions in morphology in this group.

Within the leptosporangiate ferns (Polypodiidae), our knowledge of extant fern relationships has dramatically improved over the last 10 years (Hasebe et al. 1995, Pryer et al.

1995, 2001, Schneider 1996, Wolf et al. 1998). A few highlights include the determination that Osmundaceae is the earliest-diverging leptosporangiate family; gleichenioid and dipteroid ferns together with *Matonia* are a monophyletic early-diverging group and not, as was once thought, a paraphyletic grade of basal ferns; the heterosporous ferns (Marsileaceae and Salviniaceae) are sister group to a large clade of derived homosporous ferns that includes tree ferns and the species-rich "polypodiaceous" ferns; the most derived lineage of ferns including dennstaedtioid, ptendoid, dryopteridoid, and polypodioid ferns, once thought to be polyphyletic (Smith 1995), are now known to be monophyletic. Extant lineages differ enormously in their diversity and history, including their time of origin, time of greatest diversity, and time of decline (fig. 10.5B). Osmundaceous ferns, the most basal lineage of leptosporangiate ferns, are a small group today, but they were highly diverse from the Permian and throughout the Mesozoic until they began to decline in the Upper Cretaceous (Skog 2001). Other basal Polypodiidae lineages, such as the gleichenioid and schizaeoid ferns, followed a similar pattern (fig. 10.5B). In stark contrast, the clade of ferns (Polypodiales; fig. 10.5B) with the greatest diversity today (>80% of all extant leptosporangiate ferns) might have originated as early as the Cretaceous (Skog 2001; but see Collinson 1996) and has diversified throughout the Cenozoic (Collinson 2001). The origin and diversification pattern observed in this clade of ferns parallels that observed in the angiosperms, albeit at a relatively smaller scale

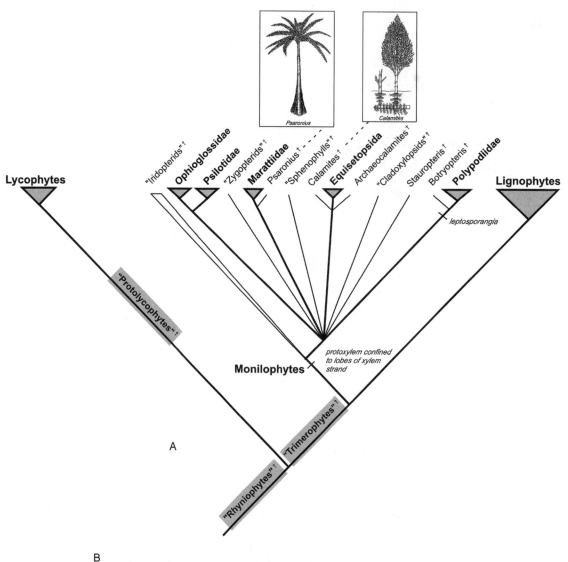

Lycophytes

"Iridopterids"†
Ophioglossidae
Psilotidae
"Zygopterids"†
Marattiidae
Psaronius†
"Sphenophylls"†
Calamites†
Equisetopsida
Archaeocalamites
"Cladoxylopsids"†
Stauropteris†
Botryopteris†
Polypodiidae
Lignophytes

Psaronius

Calamites†

leptosporangia

"protolycophytes"†

Monilophytes

protoxylem confined
to lobes of xylem
strand

A

"Trimerophytes"†

"Rhyniophytes"†

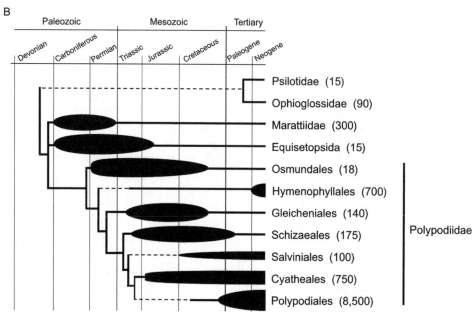

B

	Paleozoic			Mesozoic			Tertiary	
	Devonian	Carboniferous	Permian	Triassic	Jurassic	Cretaceous	Paleogene	Neogene

Psilotidae (15)

Ophioglossidae (90)

Marattiidae (300)

Equisetopsida (15)

Osmundales (18)

Hymenophyllales (700)

Gleicheniales (140)

Schizaeales (175)

Salviniales (100)

Cyatheales (750)

Polypodiales (8,500)

Polypodiidae

(H. Schneider, E. Schuettpelz, K. M. Pryer, R. Cranfill, S. Magallón, and R. Lupia, unpubl. obs.).

Lignophytes

The lignophytes (Doyle and Donoghue 1986, Rothwell and Serbet 1994, Bateman et al. 1998, Doyle 1998b) include all plants that reproduce via seeds (spermatophytes), together with their immediate "seed-free" precursors (fig. 10.6). Spermatophytes are the only living lignophytes, and with more than 260,000 species, they constitute the most diverse group of extant plants. The overwhelming majority of this diversity belongs to angiosperms (flowering plants), which produce their seeds enclosed within carpels (modified leaves). The remaining spermatophytes are "gymnosperms," represented by four extant lineages (fig. 10.6): Cycadophyta (cycads, ~130 species), Gnetophyta (gnetophytes or gnetales, ~70 species), Ginkgophytes (*Ginkgo biloba*, a single living species), and Coniferophyta (conifers, ~550 species). All "gymnosperms" produce naked seeds, that is, not enclosed within a carpel.

Lignophytes share the capability of forming wood by means of a bifacial cambium—a region of persistent cell division in their stems that produces secondary phloem toward the outside and secondary xylem toward the inside. At maturity, these secondary xylem cells form wood. Lignophyte precursors share a tetrastichous branch arrangement and a distinctive form of protoxylem ontogeny (= Infradivision Radiatopses, *sensu* Kenrick and Crane 1997a; Schneider et al. 2002; fig. 10.6). *Pertica*, formerly regarded as a "trimerophyte" (*sensu* Banks 1975, 1992), and *Tetraxylopteris*, a "progymnosperm," have been tentatively identified as lignophyte precursors (Kenrick and Crane 1997a; fig. 10.6). The earliest lignophytes had a gymnospermous wood-producing stem anatomy, and some were large trees. Unlike gymnosperms, however, these woody plants did not produce seeds, but rather were free-sporing. Collectively, these plants are known as "progymnosperms" and were important components of the mid-Paleozoic vegetation (Meyer-Berthaud et al. 1999). Early representatives of this group, such as the Middle Devonian (Eifelian) *Aneurophyton*, produced a single type of spore (homospory). Younger "progymnosperms" produced two different types of spores (heterospory), microspores and megaspores, which gave rise to microgametophytes/sperm cells and megagametophytes/egg cells, respectively. The megagametophyte and microgametophyte phases of the life cycle of these fossils are believed to have been retained within the walls of the megaspore and the microspore, respectively (endospory), which is the condition observed in all known living heterosporous plants. The heterosporous "progymnosperms," including the Late Devonian *Archaeopteris* (fig. 10.6, inset), are considered to be the closest relatives to the seed plant lineage (spermatophytes).

Although heterospory evolved several times in different tracheophyte lineages, only in the lineage leading to spermatophytes was it accompanied by a sophisticated suite of innovations and modifications involving the structure and function of megagametophytes, microgametophytes, and associated sporophytic tissues, giving rise to the complex structures that are seeds (Bateman and DiMichele 1994). The fossil record indicates that the series of steps leading from heterospory to seeds occurred a single time in the evolution of plants (Crane 1985, Doyle and Donoghue 1986, Nixon et al. 1994, Rothwell and Serbet 1994). Several Late Devonian (Fammenian) reproductive structures, such as in *Elkinsia* and *Archaeosperma* (fig. 10.6), exhibit some of the early steps in the evolution of the seed, but lack several critical attributes found in later forms. These structures consisted of an unopened (indehiscent) megasporangium that retained within its walls a single functional megaspore with an endosporic megagametophyte. Partially fused protective lobes of sporophytic origin (integumentary lobes) enveloped the indehiscent megasporangium, thereby retaining the megasporangium on the sporophyte parent plant.

Subsequent spermatophyte lineages had seeds with a completely fused envelope (integument) that enclosed the

Figure 10.5. Vascular plant phylogeny: relationships of ferns and horsetails (monilophytes). (A) Phylogeny based largely on Pryer et al. (2001) and Kenrick and Crane (1997a). The monilophytes share the positional and ontogenetic characteristic of having their protoxylem confined to the outer lobed ends of the xylem strand ("necklacelike," L. *moniliformis*). The greatest species diversity within the monilophytes (~12,000 species) is found in the Polypodiidae clade, which shares the derived leptosporangiate condition: thin-walled sporangia that produce a low number of spores (generally 64). Sketch of representative extinct crown group monilophytes: left inset, *Psaronius*, Pennsylvanian marratialean "tree fern." Plant drawing from Arens et al. (1998d): right inset, *Calamites*, Carboniferous arborescent relative of the modern horsetail, *Equisetum*. Plant drawing from Arens et al. (1998e) By integrating extant taxa (Pryer et al. 2001) together with their fossil (†) relatives (Kenrick and Crane 1997a, Berry and Fairon-Demaret 2001, Meyer-Berthaud and Gerrienne 2001) in this phylogeny, we hope to demonstrate that much of the morphological diversity that once existed in this clade is not represented in studies that consider only the living taxa. The representative fossils (†) encompass several groups: "cladoxylopsids" (*Calamophyton*, *Pseudosporochnus*), "iridopterids" (*Hyenia*, *Ibyka*), "sphenophylls" (*Sphenophyllum*, *Bowmanites*), "zygopterids" (*Rhacophyton*, *Zygopteris*), and *Stauropteris*. (B) Phylogeny of extant monilophytes (Pryer et al. 2001) plotted onto a geological time scale (Geological Society of America 1999) to illustrate the diversification of leptosporangiate ferns through time (Skog 2001). Dashed lines indicate ghost lineages—lineages without a corresponding fossil record (a striking example is the branch that unites Psilotidae and Ophioglossidae); continuous lines indicate congruence between fossil record and phylogeny. Thickened areas only generally approximate the relative diversity of groups through time.

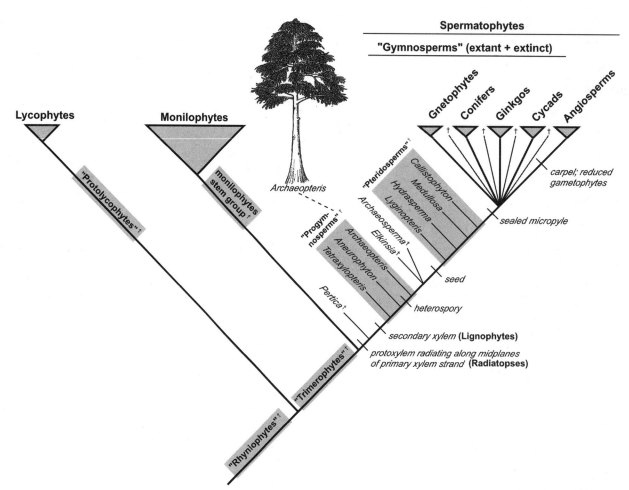

Figure 10.6. Vascular plant phylogeny: relationships of seed plants and the extinct "progymno-sperm," "pteridosperm," and derived "trimerophyte" stem group lineages. Phylogeny based primarily on Kenrick and Crane (1997a). Members of this clade share the positional and ontoge-netic characteristic of having protoxylem with multiple strands occurring along the midplanes of the lobed primary xylem ribs, corresponding to the "radiate protoxylem" group of Stein (1993) (= Infradivision Radiatopses in Kenrick and Crane 1997a). The greatest species diversity within this clade is found in the angiosperms (~260,000 species), which share several derived characters, including a carpel that encloses the seed and highly reduced male and female gametophytes. Critical synapormophies (e.g., secondary xylem, heterospory, seed, sealed micropyle) are plotted onto the topology at positions we believe best reflect our current understanding of the evolution of these features. A "seed" is a complex structure and is defined here as a megasporangium containing a single functional megaspore enclosed in one or more integuments of sporophytic origin. Inset, Sketch of representative extinct "progymnosperm": *Archaeopteris*. Plant drawing from Arens et al. (1998f). Relationships among all five major extant seed plant lineages remains elusive with no consensus as to the closest relative to the angiosperms. The figure attempts to illustrate that taking into account all the lineages (extinct and extant) that produce naked seeds ("gymnosperms") as a whole, results in "gymnosperms" being a paraphyletic assemblage, regardless of how the modern groups turn out to be related (i.e., even if all four living lineages are a monophyletic sister group to the angiosperms). Extinct (†) lineages interspersed among the five extant lineages of seed plants represent such groups as Bennettitales, Pentoxylales, Caytoniales, Corystospermales, and Cordaitales.

megasporangium except at its apex, where a small opening remained. Through this aperture (the micropyle), pollen grains entered the pollen chamber and released either sperm cells or formed pollen tubes to establish contact with the megagametophyte and, eventually, with the egg cells. The earliest seeds with a completely fused integument and a well-defined micropyle are known from the lowermost Carboniferous (e.g., *Stamnostoma*; Long 1960). Spermatophyte diversity increased dramatically during the Carboniferous, giving rise to several lineages with fernlike foliage, collectively known as "pteridosperms" or "seed ferns" (fig. 10.6). These plants encompass an extremely broad array of seed morphology and reproductive biology, but they all have seeds with a micropyle that did not seal following pollen grain capture. How "pteridosperm" lineages are related is incompletely understood (fig. 10.6); however, it appears that the earliest-diverging lineages were composed of forms such as *Lyginopteris*, in which pollen reception involved sophisticated elaborations of the megasporangium wall apex, but in later forms, such as in *Medullosa* and *Callistophyton*, the function of pollen reception was taken up by the micropyle formed by the integuments. The Mesozoic "glossopterids" are putative "pteridosperms," although some authors have indicated that they may be more closely allied with members of the seed plant crown group (Doyle 1998b, Willis and McElwain 2002).

Plants in which the micropyle became sealed after pollen grain capture gave rise to the clade that includes the five major groups of living spermatophytes, as well as many other lineages that are now extinct (fig. 10.6). Included among these are some Mesozoic plants that various authors have previously called "seed ferns." In this chapter, we restrict the use of "seed ferns" to seed plants without a sealed micropyle; therefore, taxa previously regarded as "seed ferns" that have a sealed micropyle, such as the Caytoniales, no longer fit this definition and are regarded here as part of the seed plant crown group. Discerning relationships among major living spermatophyte clades and their extinct relatives has proven to be extremely problematic, due, at least in part, to the old age of most of the lineages involved (except probably for the angiosperm crown group), and to the scant proportion of overall spermatophyte diversity represented by the living members. The use of morphological and molecular data in phylogenetic studies has resulted in dramatically different views. Morphological studies benefit from incorporating information about extinct clades but are affected by the problematic interpretation of homologies for insufficiently known characters. Molecular studies are severely impacted by the relatively meager taxonomic representation of overall spermatophyte diversity that is provided by living representatives.

Analyses of morphological data have recognized a clade (the anthophytes) that includes angiosperms and gnetophytes, together with the extinct Bennettitales and Pentoxylales (Crane 1985, Doyle and Donoghue 1986, Rothwell and Serbet 1994). As a result, the idea that, among living spermatophytes, angiosperms and gnetophytes are most closely related (anthophyte

hypothesis, fig. 10.7A) prevailed for more than a decade (Donoghue and Doyle 2000). However, increasing evidence from studies based on molecular data has now rejected the phylogenetic closeness between angiosperms and gnetophytes (Donoghue and Doyle 2000), although, at this writing, none of these studies have yet converged on an alternative, well-supported scheme of relationships among the living major clades of spermatophytes. The conflict spans not only analyses based on morphological versus molecular data, but also analyses based on different types of molecular data and on different approaches to analytical methods and taxon sampling (e.g., Sanderson et al. 2000, Magallón and Sanderson 2002, Rydin and Källersjö 2002, Rydin et al. 2002).

Several studies based on different genes and gene combinations lace angiosperms as the sister to all other living spermatophytes (gymnosperm hypothesis; fig 10.7B), suggesting that angiosperms are not closely related to any one of the extant groups of gymnosperms. Most molecular-based studies indicate a close association between gnetophytes and conifers, some even placing gnetophytes *within* conifers, thus rendering the conifers a paraphyletic assemblage (e.g., Chaw et al. 2000, Gugerli et al. 2001, Magallón and Sanderson 2002, Soltis et al. 2002). The suggestions of extant gymnosperm monophyly and conifer paraphyly are unexpected and should be viewed as provisional. Still other studies (e.g.,

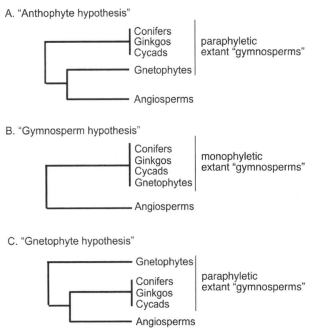

Figure 10.7. Alternative hypotheses of relationships (A, B, C) among five major extant lineages of seed plants. The anthophyte hypothesis places the gnetophytes as sister to the angiosperms. This hypothesis is based mostly on morphological evidence (Crane 1985, Doyle and Donoghue 1986), but most recent molecular studies (e.g., Barkman et al. 2000) have not supported any evidence for an anthophyte clade (but see Rydin et al. (2002).

Graham and Olmstead 2000, Sanderson et al. 2000) have shown that sometimes gnetophytes can be placed as sister to all other extant seed plant groups (gnetophyte hypothesis; fig. 10.7C).

A close proximity between gnetophytes and conifers has been proposed previously on the basis of various anatomical and morphological similarities (Coulter and Chamberlain 1917, Bailey 1953, Bierhorst 1971, Carlquist 1996), but the placement of gnetophytes *within* the conifers has disturbing implications from a traditional perspective on conifer evolution. Conifers show remarkable homogeneity in their vegetative and reproductive morphological attributes, including a growth form that is nearly always a monopodial tree, mostly needle-shaped leaves, gymnospermous wood, simple pollen cones, and usually compound seed cones with a distinctive organization, whereas gnetophytes display extraordinary variability in each of these characters. A molecular character often cited in support of conifer monophyly is the loss of one of the inverted repeat (IR) copies of the chloroplast genome, which is shared exclusively by all conifers (Raubeson and Jansen 1992b).

Although it is certainly possible that angiosperms are not closely related to any one lineage of living gymnosperms, it is important to keep in mind that molecular evidence alone simply cannot provide information regarding the relationship of angiosperms to any of the extinct groups of gymnosperms. Regardless of how the issue of relationships among the five extant seed plant lineages is finally resolved, "gymnosperms" in the broad sense, which include the early-diverging fossil lineages, are not monophyletic (fig. 10.6). It is highly likely that at the base of the lineage leading to modern angiosperms, there were some gymnosperms that are now extinct.

Taken as a whole, there have been remarkable improvements in our understanding of relationships within the major living spermatophyte lineages. Well-supported examples include the determination that *Cycas* is the sister to all other cycads; among living gnetophytes, *Gnetum* and *Welwitschia* are more closely related to one another than either is to *Ephedra*; Pinaceae is the earliest-diverging clade among living conifers; and Araucariaceae plus Podocarpaceae is sister to a clade that includes Taxaceae (yew), Taxodiaceae (redwood), and Cupressaceae (cypress) (Barkman et al. 2000, Chaw et al. 2000, Gugerli et al. 2001, Magallón and Sanderson 2002, Rydin et al. 2002).

A number of significant innovations originated on the lineage leading to angiosperms, including the carpel, which encloses the seeds, a second integumentary layer around the seed, and an extreme reduction of the megagametophyte (Bateman et al. 1998, 1998b, Theissen et al. 2002). Although our understanding of relationships within the angiosperms has improved dramatically over recent years (Qiu et al. 1999, 2000, Soltis et al. 1999, 2000; see ch. 11 in this vol.), the nature and homology of several characters unique to angiosperms are still unclear.

Vascular Plants, the Phylogenetic and Genomic Revolutions, and Fossils

Our understanding of the phylogeny of vascular plants has changed tremendously in the last 20 years due to the introduction of molecular techniques (Soltis and Soltis 2000) into plant sciences and the concomitant application of explicit phylogenetic methods to both molecular and morphological data. Before that time (and to some extent even in the present), an Aristotelian interpretation of relationships prevailed, one that promoted a linear and unidirectional transition in vascular plant evolution from simple to complex organization. For example, it was commonly thought that the whisk fern *Psilotum* was a "living fossil" or remnant of the earliest lineage of vascular plants, given its remarkable superficial resemblance to the dichotomously branched "rhyniophyte" fossils (Parenti 1980, Gifford and Foster 1989, Rothwell 1996, 1999, DiMichele et al. 2001). We now know that *Psilotum* is well embedded within the euphyllophytes and that its scalelike leaves and lack of roots do not indicate an ancient origin, but are rather the result of morphological simplification during the evolution of these plants (Schneider et al. 2002).

Although there has been remarkable progress in our understanding of plant evolution, some relationships are still enigmatic. For example, relationships among the major seed plant lineages, recently thought to be close to resolution (Donoghue and Doyle 2000), are now under renewed scrutiny, and we are almost no farther along than we were 20 years ago in identifying the closest relatives to the angiosperms. Molecular data appear to have rejected the anthophyte hypothesis (fig. 10.7A)—gnetophytes sister to angiosperms—but they continue to be ambiguous about the position of gnetophytes: either within the putatively monophyletic extant gymnosperms (gymnosperm hypothesis; fig. 10.7B), or sister to all other living seed plants (gnetophyte hypothesis; fig. 10.7C; (Goremykin et al. 1997, Doyle 1998, Barkman et al. 2000, Frohlich and Parker 2000, Sanderson et al. 2000, Magallón and Sanderson 2000, Rydin and Källersjö 2002, Rydin et al. 2002).

Papers on vascular plant phylogeny are now being published that include in excess of eight or more genes (Graham and Olmstead 2000, Soltis et al. 2002), but a clear picture of branching relationships resulting from the deep seed plant radiation is still not emerging. Two approaches are currently favored to resolve these persistently stubborn questions. The first promises to take advantage of the exceptional progress in our ability to sequence large pieces of whole genomes under the assumption that the accumulation of large amounts of genetic information and, in particular, data about structural mutations within the genome may provide a breakthrough. The second favors the integration of more than 100 years of accumulated knowledge of fossils into a modern phylogenetic framework. Exactly how to go about doing this, integrating data from molecules together with morphological characters from both living and extinct taxa, is one of the exciting challenges now facing us (Doyle and Donoghue 1987, Wilkinson 1995, Nixon 1996, Wiens 1998, O'Leary 2000, Kearney 2002). The latter approach reflects a rather surprising renaissance in how to view morphological data in modern phylogenetic studies. Pushed aside in the early days of DNA sequencing, when molecules were thought to be the holy grail for sorting out all questions on early land plant evolution, they are now back in favor once again as a valuable resource in more synthetic approaches. In addition, there has been a recent notable increase in the description of exciting new plant fossils (e.g., Sun et al. 1998, 2002, Friis et al. 2001a).

Because recent advances in our understanding of vascular plant relationships are due mostly to the introduction of molecular data, our interpretations are nearly exclusively restricted to living taxa. When one stops to ponder the vascular plant tree through hundreds of millions of years, one is struck not only by the large number of taxa that have come before and that are no longer extant (and that are not available for DNA sequencing studies), but also the extent of morphological diversity that is no longer represented in living plants. Extinctions have wiped out major parts of whole lineages that contributed heavily to plant diversity in the Paleozoic and Mesozoic. For example, extant moniliophytes consist of five distinct and ancient lineages. With the exception of leptosporangiate ferns (Polypodiidae), these lineages are not rich in either species number or morphological diversity. However, some of these lineages, such as the horsetails, were among the more diverse and dominant groups in the Upper Paleozoic and Early Mesozoic. Although horsetails have managed to survive until today with one species-poor lineage, *Equisetum*, other groups of monilophytes, such as Cladoxydopsidales and Zygopteridales, have gone completely extinct. On the surface, this would seem to support the idea that terrestrial ecosystems have witnessed a sequential replacement of lineages through time whereby, for example, such groups as the lycophytes, which were dominant in the Paleozoic, came to be superseded in diversity by the euphyllophytes, especially seed plants (Niklas et al. 1985). This has led to the seemingly popular notion that once such lineages "crash": they either go extinct or experience a prolonged period of stasis. However, some of these "superseded" lineages have undergone subsequent radiations, as observed in lycophytes (Late Tertiary; Wikström and Kenrick 2001) and derived leptosporangiate ferns (Late Cretaceous-Early Tertiary; H. Schneider, E. Schuettpelz, K. M. Pryer, R. Cranfill, S. Magallón, and R. Lupia, unpubl. obs.).

The recent implementation of highly sophisticated genetic tools to study the plant genome and the expression of its genes has generated a new breed of studies that integrate the study of plant development with evolution (Cronk 2001, Cronk et al. 2002, Schneider et al. 2002). This approach can be used to explore the evolution of critical morphological characters, such as the origin of leaves, for example, that have been the subject of long-standing controversies (Langdale et al. 2002, Schneider et al. 2002). Incorporating data from fossils and plant development in integrative and comparative studies promises to help us to overcome our currently incomplete knowledge of vascular plant relationships through time. The results of such studies will inform our understanding of the evolution of these extinct taxa, will afford us clearer insights into the morphological evolution of extant plants, and will even permit us to interpret fundamental changes in global ecology—including climate—throughout the last 450 million years (McElwain et al. 1999, Beerling et al. 2001, Berner 2001, Driese and Mora 2001, Willis and McElwain 2002).

Acknowledgments

We thank Joel Cracraft and Michael J. Donoghue for inviting us to participate in this symposium and for their patience and encouragement during manuscript preparation. Support from the National Science Foundation to the "Deep Time" Research Coordination Network initiative has been helpful in bringing the coauthors and others together to discuss several points made in this chapter. K.M.P. and H.S. gratefully acknowledge grant support from the NSF (DEB-0089909). An exceptional website on early land plants, developed by Nan C. Arens and Caroline Strömberg, was useful to us (available at http://www.ucmp.berkeley.edu/IB181/HpageIB181.html). It was the source of several plant drawings by C. Strömberg, reproduced here courtesy of the University of California Museum of Paleontology.

Literature Cited

Algeo, T. J., S. E. Scheckler, and J. B. Maynard. 2001. Effects of the Middle to Late Devonian spread of vascular plants on weathering regimes, marine biotas, and global climate. Pp. 213–236 in Plants invade the land: evolutionary and environmental perspectives (P. G. Gensel and D. Edwards, eds.). Columbia University Press, Cambridge, MA.

Arens, N. C., C. Strömberg, and A. Thompson. 1998a. Lab V—lycophytes. In Virtual paleobotany laboratory. Available: http://www.ucmp.berkeley.edu.IB181.VPL/Lyco/Lyco1.html. Last accessed 10 March 2003.

Arens, N. C., C. Strömberg, and A. Thompson. 1998b. Lab V—lycophytes. In Virtual paleobotany laboratory. Available: http://www.ucmp.berkeley.edu/IB181/VPL/Lyco/Lyco2.html. Last accessed 10 March 2003.

Arens, N. C., C. Strömberg, and A. Thompson. 1998c. Lab VII—the origin of seed plants. In Virtual paleobotany laboratory. Available: http://www.ucmp.berkeley.edu/IB18l/VPL/Osp/Osp1.html. Last accessed 10 March 2003.

Arens, N. C., C. Strömberg, and A. Thompson. 1998d. Lab VI—sphenopsids and ferns. In Virtual paleobotany laboratory. Available: http://www.ucmp.berkeley.edu/IB181/VPL/SpheFe/SpheFe4.htmlp. Last accessed 10 March 2003.

Arens, N. C., C. Strömberg, and A. Thompson. 1998e. Lab VI—sphenopsids and ferns. *In* Virtual paleobotany laboratory. Available: http://www.ucmp.berkeley.edu/IB181/VPL/SpheFe/SpheFe2.html. Last accessed 10 March 2003.

Arens, N. C., C. Strömberg, and A. Thompson. 1998f. Lab VII—the origin of seed plants. *In* Virtual paleobotany laboratory. Available: http://www.ucmp.berkeley.edu/IB181/VPL/Osp/Osp2.html. Last accessed 10 March 2003.

Bailey, I. W. 1953. Evolution of the tracheary tissue of land plants. Am. J. Bot. 40:4–8.

Banks, H. P. 1975. Reclassification of Psilophyta. Taxon 24:401–413.

Banks, H. P. 1992. The classification of early land plants—revisited. Geophytology 22:49–64.

Barkman, T. J., G. Chenery, J. R. McNeal, L. Lyons-Weiler, W. J. Ellisens, G. Moore, A. D. Wolfe, and C. W. dePamphilis. 2000. Independent and combined analyses of sequences from all three genomic compartments converge on the root of flowering plant phylogeny. Proc. Natl Acad. Sd. USA 97:13166–13171.

Bateman, R. M., P. R. Crane, W. A. DiMichele, P. Kenrick, N. P. Rowe, T. Speck, and W. E. Stein. 1998. Early evolution of land plants: phylogeny, physiology, and ecology of the primary terrestrial radiation. Annu. Rev. Ecol. Syst. 29:263–292.

Bateman, R. M., and W. A. DiMichele. 1994. Heterospory: the most iterative key innovation in the evolutionary history of the plant kingdom. Biol. Rev. 69:345–417.

Beerling, D. J., C. P. Osborne, and W. G. Chaloner. 2001. Evolution of leaf-form in land plants linked to atmospheric CO_2 decline in the Late Palaeozoic era. Nature 410:352–354.

Berner, R. A. 2001. The effect of the rise of land plants on atmospheric CO_2 during the Paleozoic. Pp. 173–178 *in* Plants invade the land: evolutionary and environmental perspectives (P. G. Gensel and D. Edwards, eds.). Columbia University Press, Cambridge, MA.

Berry, C. M., and M. Fairon-Demaret. 2001. The Middle Devonian flora revisited. Pp. 120–139 *in* Plants invade the land: evolutionary and environmental perspective (P. G. Gensel and D. Edwards, eds.). Columbia University Press, Cambridge, MA.

Bierhorst, D. W. 1971. Morphology of vascular plants. Macmillan Press, New York.

Blackwell, M. 2000. Terrestrial life—fungal from the start? Science 289:1884–1885.

Boyce, C. K., and A. H. Knoll. 2002. Evolution of developmental potential and the multiple independent origins of leaves in Paleozoic vascular plants. Paleobiology 28:70–100.

Brundrett, M. C. 2002. Coevolution of roots and mycorrhizas of land plants. New Phytol. 154:275–304.

Cairney, J. W. G. 2000. Evolution of mycorrhiza systems. Naturwissenschaften 87:467–475.

Carlquist, S. 1996. Wood, bark, and stem anatomy of Gnetales: a summary. Int. J. Plant Sci. 157(suppl.):S58–S76.

Carroll, R. 2002. Early land vertebrates. Nature 418:35–36.

Chaw, S.-M., C. L. Parkinson, Y. Cheng, T. M. Vincent, and J. D. Palmer. 2000. Seed plant phylogeny inferred from all three plant genomes: monophyly of extant gymnosperms and origin of Gnetales from conifers. Proc. Natl. Acad. Sci. USA 97:4086–4091.

Coates, M. I. 2001. Origin of tetrapods. Pp. 74–79 *in* Palaeobiology II (D. E. G. Briggs and P. R. Crowther, eds.). Blackwell Science, London.

Collinson, M. E. 1996. "What use are fossil ferns?" twenty years on: with a review of the fossil history of extant pteridophyte families and genera. Pp. 349–394 *in* Pteridology in perspective (J. M. Camus, M. Gibby, and R. J. Johns, eds.). Royal Botanic Gardens, Kew.

Collinson, M. E. 2001. Cainozoic ferns and their distribution. Brittonia 53:173–235.

Coulter, J. M., and C. J. Chamberlain. 1917. Morphology of gymnosperms. University of Chicago Press, Chicago.

Crane, P. R. 1985. Phylogenetic analysis of seed plants and the origin of angiosperms. Ann. Mo. Bot. Gard. 72:716–793.

Crane, P. R. 1999. Major patterns in botanical diversity. Pp. 171–187 *in* Evolution: investigating the evidence. (J. Scotchmoor and D. A. Springer, eds.). Palaeontological Society Special Publications 11.

Crane, P. R., and P. Kenrick. 1997. Diverted development of reproductive organs: a source of morphological innovation in land plants. Plant Syst. Evol. 206:161–174.

Cronk, Q. C. B. 2001. Plant evolution and development in a post-genomic context. Nat. Rev. Genet. 2:607–620.

Cronk, Q. C. B., R. M. Bateman, and J. A. Hawkins (eds.). 2002. Developmental genetics and plant evolution. Taylor and Francis, London.

Des Marais, D. L., K. M. Pryer, D. M. Britton, and A. R. Smith. 2003. Phylogenetic relationships and evolution of extant horsetails, *Equisetum*, based on chloroplast DNA sequence data (*rbcL* and *trnL-F*). mt. J. Plant Sci. 164:737–751.

DiMichele, W. A., and T. L. Phillips. 2002. The ecology of Paleozoic ferns. Rev. Palaeobot. Palynol. 119:143–159.

DiMichele, W. A., W. E. Stein, and R. M. Bateman. 2001. Ecological sorting of vascular plant classes during the Paleozoic evolutionary radiation. Pp. 285–335 *in* Evolutionary paleoecology: the ecological context of macroevolutionary change (W. D. Allmon and D. J. Bottjer, eds.). Columbia University Press, New York.

Donoghue, M. J. 2002. Plants. Pp. 911–918 *in* Encyclopedia of evolution (M. Pagel, ed.), vol. 2. Oxford University Press, Oxford.

Donoghue, M. J., and J. A. Doyle. 2000. Seed plant phylogeny: demise of the anthophyte hypothesis? Curr. Biol. 10:R106–R109.

Doyle, J. A. 1998a. Molecules, morphology, fossils, and the relationships of angiosperms and Gnetales. Mol. Phylogenet. Evol. 9:448–462.

Doyle, J. A. 1998b. Phylogeny of vascular plants. Annu. Rev. Ecol. Syst. 29:567–599.

Doyle, J. A., and M. J. Donoghue. 1986. Seed plant phylogeny and the origin of angiosperms: an experimental cladistic approach. Bot. Rev. 52:321–431.

Doyle, J. A., and M. J. Donoghue. 1987. The importance of fossils in elucidating seed plant phylogeny and macroevolution. Rev. Palaeobot. Palynol. 50:63–95.

Driese, S. G., and C. I. Mora. 2001. Diversification of Siluro-Devonian plant traces in paleosols and influence on estimates of paleoatmospheric CO_2 levels. Pp. 237–253 *in* Plants invade the land: evolutionary and environmental perspectives (P. G. Gensel and D. Edwards, eds.). Columbia University Press, Cambridge, MA.

Edwards, D., and C. Wellman. 2001. Embryophytes on land: the Ordovician to Lochkovian (Lower Devonian) record. Pp. 3–28 *in* Plants invade the land: evolutionary and environmental perspectives (P. G. Gensel and D. Edwards, eds.). Columbia University Press, Cambridge, MA.

Fischer, D., T. Liu, E. Yip, and K. Yu. 1998. Localities of the Devonian: Rhynie Chert, Scotland. *In* Virtual paleobotany, laboratory. Available: http://www.ucmp.berkeley.edu/devonian/rhynie.html. Last accessed 10 March 2003.

Friis, E. M., K. R. Pedersen, and P. R. Crane. 2001a. Fossil evidence of water lilies (Nymphaeales) in the Early Cretaceous. Nature 410:357–360.

Friis, E. M., K. R. Pedersen, and P. R. Crane. 200lb. Origin and radiation of angiosperms. Pp. 97–102 *in* Palaeobiology II (D. E. G. Briggs and P. R. Crowther, eds.). Blackwell Science, London.

Frohlich, M. W., and D. S. Parker. 2000. The mostly male theory of flower evolutionary origins: from genes to fossils. Syst. Bot. 25:155–170.

Gaunt, M. W., and M. A. Miles. 2002. An insect molecular clock dates the origin of the insects and accords with palaeontological and biogeographic landmarks. Mol. Biol. Evol. 19:748–761.

Gensel, P. G., and C. M. Berry. 2001. Early lycophyte evolution. Am. Fern J. 91:74–98.

Gensel, P. G., M. E. Kotyk, and J. F. Basinger. 2001. Morphology of above- and below-ground structures in Early Devonian (Pragian-Emsian) plants. Pp. 83–102 *in* Plants invade the land: evolutionary and environmental perspectives (P. G. Gensel and D. Edwards, eds.). Columbia University Press, New York.

Geological Society of America. 1999. Geologic time scale. Product code CTS004. (A. R. Palmer and J. Geissman, comps.).

Gifford, E. M., and A. S. Foster. 1989. Morphology and evolution of vascular plants. 3rd ed. Freeman, New York.

Goremykin, V., V. Bobrova, J. Pahnke, A. Troitsky, A. Antonov, and W. Martin. 1997. Noncoding sequences from the slowly evolving chloroplast inverted repeat in addition to *rbcL* data do not support Gnetalean affinities of angiosperms. Mol. Biol. Evol. 13:383–396.

Graham, S. W., and R. G. Olmstead. 2000. Utility of 17 chloroplast genes for inferring the phylogeny of the basal angiosperms. Am. J. Bot. 87:1712–1730.

Gugerli, F., C. Sperisen, U. Büchler, I. Brunner, S. Brodbeck, J. D. Palmer, and Y.-L. Qiu. 2001. The evolutionary split of Pinaceae from other conifers: evidence from an intron loss and a multigene phylogeny. Mol. Phylogenet. Evol. 21:167–175.

Hasebe, M., P. G. Wolf, K. M. Pryer, K. Ueda, M. Ito, R. Sano, G. J. Gastony, J. Yokoyama, J. R. Manhart, N. Murakami, E. H. Crane, C. H. Haufler, and W. D. Hauk. 1995. Fern phylogeny based on *rbcL* nucleotide sequences. Am. Fern J. 85:134–181.

Hibbett, D. S., L.-B. Gilbert, and M. J. Donoghue. 2000. Evolutionary instability of ectomycorrhizal symbioses in basidomycetes. Nature 407:506–508.

Hill, C. R., and J. M. Camus. 1986. Evolutionary cladistics of marattialean ferns. Bull. Br. Mus. Nat. Hist. (Bot.) 14:219–300.

Judd, W. S., C. S. Campbell, E. A. Kellogg, P. F. Stevens, and M. J. Donoghue. 2002. Plant systematics: a phylogenetic approach. 2nd ed. Sinauer, Sunderland, MA.

Karol, K. G., R. M. McCourt, M. T. Cimino, and C. F. Delwiche. 2001. The closest living relatives of land plants. Science 294:2351–2353.

Kearney, M. 2002. Fragmentary taxa, missing data, and ambiguity: mistaken assumptions and conclusions. Syst. Biol. 51:369–381.

Kenrick, P. 1994. Alternation of generations in land plants: new phylogenetic and morphological evidence. Biol. Rev. 69:293–330.

Kenrick, P. 2000. The relationships of vascular plants. Philos. Trans. R. Soc. Lond. B 355:847–855.

Kenrick, P. 2002a. The origin of roots. Pp. 1–13 *in* Plant roots: the hidden half. 3rd ed. (Y. Waisel, A. Eshel, and U. Kafkafi, eds.). Dekker, New York.

Kenrick, P. 2002b. The telome theory. Pp. 365–387 *in* Developmental genetics and plant evolution (Q. C. B. Croak, R. M. Bateman, and J. A. Hawkins, eds.). Taylor and Francis, London.

Kenrick, P., and P. R. Crane. 1997a. The origin and early diversification of land plants: a cladistic study. Smithsonian Institution Press, Washington, DC.

Kenrick, P., and P. R. Crane. 1997b. The origin and early evolution of plants on land. Nature 389:33–39.

Labandeira, C. C. 2001. The rise and diversification of insects. Pp. 82–88 *in* Palaeobiology II (D. E. G. Briggs and P. R. Crowther, eds.). Blackwell Science, London.

Langdale, J. A., R. W. Scotland, and S. B. Corley. 2002. A developmental perspective on the evolution of leaves. Pp. 388–394 *in* Developmental genetics and plant evolution (Q. C. B. Cronk, R. M. Bateman, and J. A. Hawkins, eds.). Taylor and Francis, London.

Liu, Z.-H., J. Hilton, and C.-S. Li. 2000. Review on the origin, evolution and phylogeny of Marattiales. Chin. Bull. Bot. 17:39–52.

Long, A. G. 1960. "*Stamnostoma huttonense*" gen. et sp. nov.— pteridosperm seed and cupule from the calciferous sandstone series of Berwickshire. Trans. R. Soc. Edinb. 64:201–215.

Looy, C. V., R. J. Twitchett, D. L. Dilcher, J. H. A. Van Konijnenburg-Van Cittert, and H. Visscher. 2001. Life in the end-Permian dead zone. Proc. Natl Acad. Sci. USA 98:7879–7883.

Mable, B. K., and S. P. Otto. 1998. The evolution of life cycles with haploid and diploid phases. Bioessays 20:453–462.

Magallón, S., and M. J. Sanderson. 2002. Relationships among seed plants according to highly conserved genes: sorting conflicting phylogenetic signals among ancient lineages. Am. J. Bot. 89:1991–2006.

McElwain, J. C., D. J. Beerling, and F. I. Woodward. 1999. Fossil plants and global warming at the Triassic-Jurassic boundary. Science 285:1386–1390.

Meyer-Berthaud, B., and P. Gerrienne. 2001. *Aarabia*, a new Early Devonian vascular plant from Africa (Morocco). Rev. Palaeobot. Palynol. 116:39–53.

Meyer-Berthaud, B., S. E. Scheckler, and J. Wendt. 1999. *Archaeopteris* is the earliest known modern tree. Nature 398:700–701.

Nickrent, D. L., C. L. Parkinson, J. D. Palmer, and R. J. Duff. 2000. Multigene phylogeny of land plants with special reference to bryophytes and earliest land plants. Mol. Biol. Evol. 17:1885–1895.

Niklas, K. J., B. H. Tiffney, and A. H. Knoll. 1985. Patterns in vascular land plant diversification: an analysis at the species level. Pp. 97–128 in Phanerozoic diversity patterns: profiles in macroevolution (J. W. Valentine, ed.). Princeton University Press, Princeton University Press, Princeton, NJ.

Nixon, K. C. 1996. Paleobotany in cladistics and cladistics in paleobotany: enlightenment and uncertainty. Rev. Palaeobot. Palynol. 90:361–373.

Nixon, K. C., W. L. Crepet, D. Stevenson, and E. M. Friis. 1994. A reevaluation of seed plant phylogeny. Ann. Mo. Bot. Gard. 81:484–533.

O'Leary, M. A. 2000. Operational obstacles to total evidence analyses considering that 99% of life is extinct. J. Vert. Paleontol. 20(suppl.):61A.

Parenti, L. R. 1980. A phylogenetic analysis of the land plants. Biol. J. Linn. Soc. 13:225–242.

Pataki, D. E. 2002. Atmospheric CO_2, climate and evolution— lessons from the past. New Phytol. 154:1–14.

Pigg, K. B. 2001. Isoetalean lycopsid evolution: from the Devonian to the present. Am. Fern J. 91:99–114.

Pryer, K. M., H. Schneider, A. R. Smith, R. Cranfill, P. G. Wolf, J. S. Hunt, and S. D. Sipes. 2001. Horsetails and ferns are a monophyletic group and the closest living relatives to seed plants. Nature 409:618–622.

Pryer, K. M., A. R. Smith, and J. E. Skog. 1995. Phylogenetic relationships of extant ferns based on evidence from morphology and *rbcL* sequences. Am. Fern J. 85:205–282.

Qiu, Y.-L., J. Lee, F. Berasconi-Quadroni, D. E. Soltis, P. S. Soltis, M. Zanis, E. A. Zimmer, Z. Chen, V. Savolainen, and M. W. Chase. 1999. The earliest angiosperms. Nature 402:404–407.

Qiu, Y.-L., J. Lee, F. Berasconi-Quadroni, D. E. Soltis, P. S. Soltis, M. Zanis, E. A. Zimmer, Z. Chen, V. Savolainen, and M. W. Chase. 2000. Phylogeny of basal angiosperms: analyses of five genes from three genomes. Int. J. Plant Sci. 161(suppl.):S3–S27.

Raubeson, L. A., and R. K. Jansen. 1992a. Chloroplast DNA evidence on the ancient evolutionary split in vascular plants. Science 255:1697–1699.

Raubeson, L. A., and R. K. Jansen. 1992b. A rare chloroplast-DNA structural mutation is shared by all conifers. Biochem. Syst. Ecol. 20:17–24.

Raven, J. A., and D. Edwards. 2001. Roots: evolutionary origin and biogeochemical significance. J. Exp. Bot. 22:381–401.

Redecker, D., R. Kodner, and L. E. Graham. 2000. Glomalean fungi from the Ordovician. Science 289:1920–1921.

Renzaglia, K. S., R. J. Duff, D. L. Nickrent, and D. J. Garbary. 2000. Vegetative and reproductive innovations of early land plants: implications for a unified phylogeny. Philos. Trans. R. Soc. Lond. B 355:768–793.

Rothwell, G. W. 1996. Phylogenetic relationships of ferns: a paleobotanical perspective. Pp. 395–404 in Pteridology in perspective (J. M. Camus, M. Gibby, and R. J. Johns, eds.). Royal Botanic Gardens, Kew.

Rothwell, G. W. 1999. Fossils and ferns in the resolution of land plant phylogeny. Bot. Rev. 65:188–218.

Rothwell, G. W., and R. Serbet. 1994. Lignophyte phylogeny and the evolution of spermatophytes. Syst. Bot. 19:443–482.

Rydin, C., and M. Källersjö. 2002. Taxon sampling and seed plant phylogeny. Cladistics 18:485–513.

Rydin, C., M. Källersjö, and E. M. Friis. 2002. Seed plant relationships and the systematic position of Gnetales based on nuclear and chloroplast DNA: conflicting data, rooting problems, and the monophyly of conifers. Int. J. Plant Sci. 163:197–214.

Sanderson, M. J., M. F. Wojciechowski, J.-M. Hu, T. Ser Khan, and S. G. O'Brady. 2000. Error, bias, and long-branch attraction in data for two chloroplast photosystem genes in seed plants. Mol. Biol. Evol. 17:782–797.

Schneider, H. 1996. Vergleichende Wurzelanatomie der Farne. Shaker, Aachen.

Schneider, H., K. M. Pryer, R. Cranfill, A. R. Smith, and P. G. Wolf. 2002. Evolution of vascular plant body plans: a phylogenetic perspective. Pp. 330–363 in Developmental genetics and plant evolution (Q. C. B. Croak, R. M. Bateman, and J. A. Hawkins, eds.). Taylor and Francis, London.

Shear, W. A., and P. A. Selden. 2001. Rustling in the under-growth: animals in early terrestrial ecosystems. Pp. 29–51 in Plants invade the land: evolutionary and environmental perspectives (P. G. Gensel and D. Edwards, eds.). Columbia University Press, New York.

Shougang, H., C. B. Beck, and W. Deming. 2003. Structure of the earliest leaves: adaptations to high concentrations of atmospheric CO_2. Int. J. Plant Sci. 164:71–75.

Singer, S. R. 1997. Plant life cycles and angiosperm development. Pp. 493–513 in Embryology: constructing the organism (S. F. Gilbert and A. M. Raunio, eds.). Sinauer Associates, Sunderland, MA.

Skog, J. E. 2001. Biogeography of Mesozoic leptosporangiate ferns related to extant ferns. Brittonia 53:236–269.

Smith, A. B. 1994. Systematics and the fossil record: document-ing evolutionary patterns. Blackwell, London.

Smith, A. R. 1995. Non-molecular phylogenetic hypotheses for ferns. Am. Fern J. 85:104–122.

Soltis, D. E., P. S. Soltis, M. W. Chase, M. E. Mort, D. C. Albach, M. Zanis, V. Savolainen, W. H. Hahn, S. B. Hoot, M. F. Fay, M. Axtell, S. M. Swensen, L. M. Prince, W. J. Kress, K. C. Nixon, and J. S. Farris. 2000. Angiosperm phylogeny inferred from 18S rDNA, *rbcL*, and *atpB* sequences. Bot. J. Linn. Soc. 133:381–461.

Soltis, D. E., P. S. Soltis, and M. J. Zanis. 2002. Phylogeny of seed plants based on evidence from eight genes. Am. J. Bot. 89:1670–1681.

Soltis, P. S., and D. E. Soltis. 2000. Contributions of plant molecular systematics to studies of molecular evolution. Plant Mol. Biol. 24:45–75.

Soltis, P. S., D. E. Soltis, and M. W. Chase. 1999. Angiosperm phylogeny inferred from multiple genes as a tool for comparative biology. Nature 402:402–404.

Stein, W. E. 1993. Modelling the evolution of the stelar architecture in vascular plants. Int. J. Plant Sci. 154:229–263.

Sues, H.-D., and R. R. Reisz. 1998. Origins and early evolution of herbivory in tetrapods. Trends Ecol. Evol. 13:141–145.

Sun, G., D. L. Dilcher, S. Zheng, K. C. Nixon, and X. Wang. 2002. Archaefructaceae, a nev basal angiosperm family. Science 296:899–904.

Sun, G., D. L. Dilcher, S. Zheng, and Z. Zhou. 1998. In search of the first flower: a Jurassic angiosperm, *Archaefructus*, from northeast China. Science 282:1692–1695.

Taylor, T. N., W. Remy, H. Haas, and H. Kerp. 1995. Fossil arbuscular mycorrhizae from the early Devonian. Mycologia 87:560–573.

Theissen, G. A. Becker, K.-U. Winter, T. Münster, C. Kirchner, and H. Saedler. 2002. How the land plants learned their floral ABCs: the role of MADS-box genes in the evolutionary origin of flowers. Pp. 173–205 *in* Developmental genetics and plant evolution (Q. C. B. Cronk, R. M. Bateman, and J. A. Hawkins, eds.). Taylor and Francis, London.

Tiffney, B. H., and S. R. Manchester. 2001. The use of geological and paleontological evidence in evaluating plant phylogeographic hypotheses in the northern hemisphere Tertiary. Int. J. Plant Sci. 162(suppl.):S3–S17.

Wiens, J. J. 1998. Does adding characters with missing data increase or decrease phylogenetic accuracy? Syst. Biol. 47:625–640.

Wikström, N., and P. Kenrick. 2001. Evolution of Lycopodiaceae (Lycopsida): estimating divergence times from *rbcL* gene sequences by use of nonparametric rate smoothing. Mol. Phylogenet. Evol. 19:177–186.

Wilkinson, M. 1995. Coping with abundant missing entries in phylogenetic inference using parsimony. Syst. Biol. 44:501–514.

Willis, K. J., and J. C. McElwain. 2002. The evolution of plants. Oxford University Press, New York.

Wolf, P. G., K. M. Pryer, A. R. Smith, and M. Hasebe. 1998. Phylogenetic studies of extant pteridophytes. Pp. 541–556 *in* Molecular systematics of plants II. DNA sequencing (D. E. Soltis, P. S. Soltis, and J. J. Doyle, eds.). Kluwer, Boston.

Pamela S. Soltis

Douglas E. Soltis

Mark W. Chase

Peter K. Endress

Peter R. Crane

The Diversification of Flowering Plants

In this chapter, we provide an overview of the phylogeny of flowering plants, with special emphasis on the root and major clades of the angiosperms, and patterns of radiation in the evolutionary history of angiosperms. Given the size of the angiosperm clade, we will not examine relationships within major clades in any detail; instead, we refer the reader to publications that focus on those clades or grades [e.g., basal angiosperms (Zanis et al. 2002), monocots (Chase et al. 2000), early-diverging eudicots (Hoot et al. 1999), asterids (Albach et al. 2001, Bremer et al. 2002)]. After this overview, we use the phylogeny to examine patterns of evolution in three important features of flowering plants: double fertilization and endosperm formation, closed carpels, and perianth structure and organization.

The flowering plants are one of five clades of extant seed plants, and they are by far the largest, most diverse, and most important ecologically of all living embryophytes (land plants). There are at least 260,000 (Takhtajan 1997) species of flowering plants (i.e., five to six times the number of living species of vertebrates), classified in approximately 450 families (e.g., 453, as listed in Angiosperm Phylogeny Group II 2003). The clade has a fossil history that extends back at least to the early Cretaceous, conservatively approximately 130 million years ago (Mya).

Several features have been identified as synapomorphies of angiosperms (see Doyle and Donoghue 1986). Perhaps foremost among these is double fertilization, with its joint processes of zygote production, endosperm formation, and angiospermy (i.e., presence of closed carpels). In angiosperms, one sperm unites with the egg to form a diploid zygote. Then, typically a second sperm unites with two additional nuclei, the polar nuclei, to produce a triploid endosperm. Endosperm tissue provides a source of nutrients for the developing embryo. Although double fertilization has also been reported in the gnetophytes (Friedman 1990, Carmichael and Friedman 1996), the process of double fertilization and its consequent formation of endosperm are unique to angiosperms. We examine variation on the basic theme of fertilization and endosperm formation in angiosperms later in this chapter.

A second synapomorphy of the angiosperms is the carpel, the floral structure that contains the ovule(s), which after fertilization will become the seed(s). The closure of the carpel provides additional protection for the seeds and may be accomplished by secretion or fusion of the carpel margins or flanks. We consider the evolution of the closed carpel later in this chapter.

Additional synapomorphies of flowering plants include phloem tissue composed of sieve tubes and companion cells, stamens with two pairs of pollen sacs, and aspects of gametophyte development and structure. Although important characteristics, we do not examine these features further in this chapter.

Relationships among clades of extant seed plants remain unclear, despite considerable recent attention (see Pryer et al., ch. 10 in this vol.). Nearly all morphological analyses of seed plants place the gnetophytes (or at least some of them) as the sister group to the angiosperms (e.g., Crane 1985, Doyle and Donoghue 1986, 1992, Loconte and Stevenson 1990, Nixon et al. 1994, Rothwell and Serbet 1994; for reviews, see Doyle 1996, 1998, Donoghue and Doyle 2000); however, molecular analyses have found a variety of topologies, depending on the gene(s) used and the taxa sampled (e.g., Hamby and Zimmer 1992, Hasebe et al. 1992, Albert et al. 1994, Chaw et al. 1997, 2000, Goremykin et al. 1996, Malek et al. 1996, Hansen et al. 1999, Winter et al. 1999, Soltis et al. 1999b, Bowe et al. 2000, Sanderson et al. 2000, Rydin et al. 2002, D. Soltis et al. 2002). Most multigene analyses have found a clade of extant gymnosperms, consisting of cycads, *Ginkgo*, conifers, and gnetophytes, as the sister to the angiosperms. However, given that the number of extinct seed plant lineages nearly equals the number of extant groups, analyses of living seed plants only are likely to provide inadequate inferences of phylogeny. Resolution of seed plant relationships, including the sister group of the angiosperms, will require careful phylogenetic analyses that integrate data for fossil and extant groups (see Donoghue and Doyle 2000).

Overview of Angiosperm Phylogeny

The phylogenetic overview of flowering plants that follows (summarized in fig. 11.1) is drawn from several sources. The backbone of the tree comes from the collaborative study by Soltis et al. (2000), which included 560 species of angiosperms, seven gymnosperms as outgroups, and data from two plastid genes (*rbcL* and *atpB*) and one nuclear gene (18S ribosomal DNA), for a total of more than 4700 aligned nucleotides. The *rbcL* analysis of nearly all families of eudicots (Savolainen et al. 2000) helped to place several groups not sampled in the three-gene analysis. In addition, analyses of specific groups provided information on relationships of basal angiosperms (Zanis et al. 2002), monocots (Chase et al. 2000), core eudicots (D. Soltis et al. 2002), and asterids (Albach et al. 2001, Bremer et al. 2002).

The Root of the Angiosperms

Phylogeny of Extant Angiosperms

Most analyses focusing on the root of the angiosperms concur in finding *Amborella* as the sister to all other extant flowering plants (D. Soltis et al. 1997, 2000, P. Soltis et al. 1999a, Qiu et al. 1999, Mathews and Donoghue 1999, Graham and Olmstead 2000, Zanis et al. 2002; fig. 11.1). Nymphaeaceae (water lilies) and Austrobaileyales (composed of *Austrobaileya*, *Trimenia*, *Schisandra*, *Kadsura*, and *Illicium*) occupy the next branches in this basal grade; all other angiosperm species form a clade that is sister to Austrobaileyales. This large clade of all other angiosperms has been referred to as the euangiosperms (Qiu et al. 1999), but no consensus has

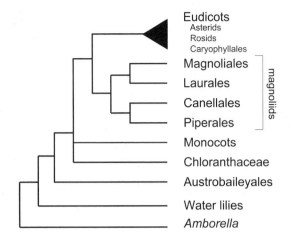

Figure 11.1. Overview of angiosperm phylogeny, showing eudicots and major clades of basal angiosperms. The branch uniting the eudicots with the magnoliids is only weakly supported.

yet been reached on a name for this clade. In contrast to those studies that show *Amborella* as the sister group to the rest of the angiosperms, a few analyses have found alternative rootings, the more strongly supported of which shows *Amborella* together with the water lilies as the sister to all other living flowering plants (e.g., Barkman et al. 2000). Statistical analyses of alternative rootings, however, generally favor the tree with *Amborella* as sister to the rest of the angiosperms, although the *Amborella* + water lilies tree cannot be conclusively rejected (Zanis et al. 2002). However, more recent analyses of rapidly evolving genes also place *Amborella* alone as sister to all other extant flowering plants (Borsch et al. 2003, Hilu et al. 2003). Furthermore, *Amborella* is unequivocally reconstructed as the sister group to all other angiosperms based on both sequence and structural features of the floral genes *AP3* and *PI* (S. Kim et al., unpubl. obs.). Regardless of which rooting is correct, tremendous progress has been made in a short time, due in large part to large-scale collaborations among angiosperm systematists.

Although *Amborella* has received considerable attention since its noteworthy position was reported, we summarize some of its basic attributes here. *Amborella* is monotypic, and *A. trichopoda* is restricted to cloud forests on New Caledonia. The plants are dioecious shrubs with vesselless xylem (but see Feild et al. 2000). The flowers, although functionally unisexual, have a bisexual organization: at least in the female flowers, sterile stamens are present between tepals and carpels (Endress and Igersheim 2000b; for additional studies of floral development in *A. trichopoda*, see Posluszny and Tomlinson 2003). The small (generally <0.5 cm in diameter) flowers are composed of a moderate number of spirally arranged floral organs. The perianth is undifferentiated; that is, there is no clear distinction into sepals and petals (described below). The carpels are completely ascidiform (i.e., without a conduplicate part) and are closed by secretion and not by postgenital fusion (Endress and Igersheim 2000b; (see section below titled Closed Carpels).

The Fossil Record of Early Angiosperms

The fossil record of angiosperms from the early Cretaceous portrays tremendous diversity in size, structure, and organization of flowers. For example, the fossil *Archaeanthus* from the mid-Cretaceous (uppermost Albian to mid-Cenomanian) supported the long-standing view (e.g., Cronquist 1968) that the first flowers were large and *Magnolia*-like in size and structure (Dilcher and Crane 1984; fig. 11.2). This view of the early angiosperm flower prevailed among most paleobotanists and systematists for at least the latter half of the twentieth century.

During the past several years, views of early flowers have changed dramatically, because of new paleobotanical techniques of studying charcoalified mesofossils from new fossil sites, most notably in Portugal and eastern North America (for review, see Friis et al. 2000). These fossil deposits harbor abundant diversity in floral morphology. Furthermore, many extant lineages of flowering plants were established by 100–90 Mya (see Magallón and Sanderson 2001), and many fossils that do not appear to fit into extant groups were also present. Despite this morphological and phylogenetic diversity, these fossils are uniformly small, all less than 1 cm in diameter. Among these fossils is a water lily from approximately 125 Mya (Friis et al. 2001; fig. 11.3), consistent with the near-basal position of the extant water lily clade in molecular phylogenetic trees.

Recent discoveries of two species described in the fossil genus *Archaefructus*, *A. sinensis* and *A. liaoningensis*, have provided new information on early angiosperms (Sun et al. 1998, 2002). The recently discovered fossils of *A. sinensis* are beautifully preserved specimens at reproductive maturity (fig. 11.4) and come from the lower part of the Yixian Formation in Beipiao and Lingyuan of western Liaoning, China, dated to the early Cretaceous. Although the date for this site is not clear, the minimum age certainly places *Archaefructus* as one of the oldest unambiguous angiosperm fossils. *Archaefructus sinensis* has spirally arranged to whorled carpels and

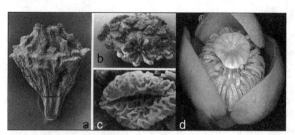

Figure 11.3. Fossil and modern water lilies. (A–C) Early Cretaceous water lily fossils (from Friis et al. 2001); the flower is inferred to have been no more than 1 cm in diameter. (A) Lateral view, showing numerous scars from attachment of stamens (24) and perianth parts (6). (B) Apical view, showing 12 carpels in the center, surrounded by rhomboidal stamen scars and narrow elliptic perianth scars. (C) Monocolpate pollen grain with reticulate pollen wall. (D) Modern *Nuphar*; note the numerous stamens and four of six perianth parts (two additional small, sepaloid perianth parts are not visible from this view).

paired stamens. From the dissected leaf morphology and the abundance of fish fossils found in the same deposit, *Archaefructus* is inferred to have been aquatic (Sun et al. 2002). A phylogenetic analysis that included *Archaefructus* and 173 extant taxa, molecular characters from three genes (taken from Soltis et al. 2000), and 17–108 morphological characters (taken from Doyle and Endress 2000) concluded that *Archaefructus* is the sister to all extant angiosperms (Sun et al. 2002). However, a reanalysis of the data, including additional material, suggests other potential placements (Friis et al. 2003).

Based on early Cretaceous fossils such as *Archaefructus* and the abundance of early to mid-Cretaceous fossils from Portugal and eastern North America, the fossil record is most consistent with a hypothesis of early flowers having been fairly small. However, the diversity of form suggests an early

a   b

Figure 11.2. (A) Reconstruction of fossil genus *Archaeanthus* from the mid-Cretaceous (Dilcher and Crane 1984). (B) Photograph of modern *Magnolia grandiflora*. Note the similarity in floral structure between *Archaeanthus* and *Magnolia*; *Archaeanthus* fits contemporary models of the ancestral flower.

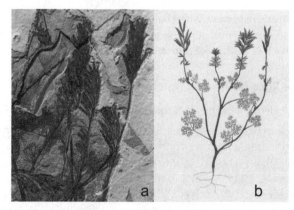

Figure 11.4. *Archaefructus sinensis* from the Early Cretaceous (courtesy of D. L. Dilcher). (A) Reproductive shoots showing numerous carpels and dissected leaves; note the fossil fish skeleton at top left, indicating aquatic habitat of *Archaefructus*. (B) Reconstruction, clearly showing dissected leaves and paired stamens subtending carpels.

radiation of angiosperms, with associated diversification in floral structure (e.g., Friis et al. 2000). A more complete understanding of early floral evolution will require integrated phylogenetic analyses of extant and fossil species.

Molecular versus Fossil Ages of Angiosperms

Estimates of the age of the angiosperms and the timing of important divergences based on molecular data do not generally agree with each other or with dates determined from the fossil record. For example, the age of the angiosperms has been estimated as 350–420 Mya (Ramshaw et al. 1972), > 319 Mya (Martin et al. 1989, 1993), 200 Mya (Wolfe et al. 1989), 160 Mya (Goremykin et al. 1996), 158–179 Mya (Wikström et al. 2001), 140–190 Mya (Sanderson and Doyle 2001), and 126.9–134.5 Mya (P. Soltis et al. 2002) using different genes and different methods. Although some of these estimates fall close to or slightly older than the age implied by the fossil record (i.e., 125–135 Mya), many molecular-based estimates of the age of the angiosperms published to date greatly exceed this age. Many sources of error can lead to poor DNA-based estimates of divergence times (e.g., Sanderson and Doyle 1991, P. Soltis et al. 2002). Unequal rates of evolution among lineages, especially when combined with inadequate sampling of taxa, can distort estimates of divergence times based on molecular data. A similar pattern of older molecular-based estimates than fossil dates has been observed for other groups of organisms, including fungi, animals, and the divergence of the crown-group eukaryotes (e.g., Heckman et al. 2001). This pattern has been attributed to a methodological bias such that clock-based methods will overestimate ages (Rodríguez-Trelles et al. 2002). However, in at least the cases of angiosperms and land plants (Sanderson and Doyle 2001, P. Soltis et al. 2002, Sanderson 2003), sufficient data and appropriate sampling have produced estimates that are generally in line with the fossil record.

Within the angiosperms, estimated divergence times are also generally older than indications from the fossil record (e.g., molecular dates from Wikström et al. 2001, compared with fossil dates from Magallón et al. 1999). However, these discrepancies are on a much smaller scale than those reported for the age of the angiosperms, typically differing by tens rather than hundreds of millions of years. Further refinement of analytical methods is needed to allow accurate estimation of divergence times for those many groups of flowering plants that lack a fossil record.

Major Clades of Angiosperms

Apart from the basal grade of *Amborella*, water lilies, and Austrobaileyales, relationships among other basal clades of angiosperms are less clear. The monocots and magnoliids represent two large clades that diversified early in angiosperm history, but their exact placements, as well as those of smaller clades such as Ceratophyllaceae and Chloranthaceae, are not well supported, even in analyses based on 11 genes and more

than 15,000 aligned nucleotides (Zanis et al. 2002). Ceratophyllaceae and Chloranthaceae have fossil records that extend back at least 125 Mya (e.g., Couper 1958, Walker and Walker 1984, Friis et al. 2000, Dilcher 1989; for review, see Endress 2001), placing them among the oldest angiosperm fossils.

Monocots

The monocots are one of the largest clades of angiosperms, with an estimated 65,000 or more species (Takhtajan 1997) and approximately 20% of all angiosperms. *Acorus* is the sister to all other monocots, and the alismatid families with a large number of aquatic species follow *Acorus* as the next successive sister group to the rest of the monocots. Key lineages within monocots are the graminoid families (restios, sedges, and grasses), palms, yams, gingers, lilies, and Asparagales, a large clade that includes the orchids, irises, and hyacinths, among other groups (fig. 11.5). Relationships among major clades of monocots remain largely unresolved (e.g., Chase et al. 2000), but only three genes have been sampled to infer these relationships; ongoing collaborative research using several additional genes should help resolve monocot phylogeny. Based on both the fossil record and molecular clock estimates, many lineages of monocots extend back in the fossil record at least 80–100 Mya (e.g., Bremer 2000, Wikström et al. 2001, Gandolfo et al. 2002).

Magnoliids

The magnoliid clade, which comprises fewer than 5% of all living species of flowering plants, contains most of the groups considered to be "primitive angiosperms" by many previous authors (e.g., Cronquist 1981, 1988, Takhtajan 1997) and consists of four subclades: Magnoliales and Laurales are sisters, and Piperales and Canellales are sisters (fig. 11.1).

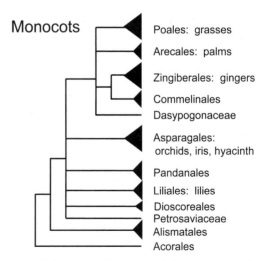

Figure 11.5. Summary of monocot phylogeny. *Acorus* (Acorales) and Alismatales are successive sisters to the rest of the clade. Despite intensive study, relationships among major clades of monocots are mostly unresolved.

Magnoliales, long considered the most ancient group of living angiosperms (although phylogenetic analyses now indicate otherwise; see above), are composed of six families of woody plants from tropical to warm-temperate habitats. The *Magnolia* family, with the *Magnolia* and tulip poplar (*Liriodendron*), may be the most familiar family in the order, but Magnoliales also contain Annonaceae (paw paw family), Myristicaceae (nutmeg family), and three small families, Degeneriaceae, Himantandraceae, and Eupomatiaceae. The largest family of Laurales is the laurel family (Lauraceae), and the order also contains a number of small families (Calycanthaceae, Monimiaceae, Gomortegaceae, Atherospermataceae, and Hernandiaceae; for review of the phylogeny of Laurales, see Renner 1999). Canellales consist of only two families, Canellaceae and Winteraceae; and Piperales contain only five, Aristolochiaceae, Lactoridaceae, Piperaceae, Hydnoraceae (Nickrent et al. 2002), and Saururaceae. The magnoliids are weakly supported as sister to the eudicots (Zanis et al. 2002). Although phylogenetic evidence argues against members of the magnoliid clade as being among the most ancient extant flowering plants, this clade does extend back into the mid-Cretaceous (*Archaeanthus*; Dilcher and Crane 1984). Furthermore, this clade may represent an early radiation in the history of flowering plants. Despite the small number of extant species, floral diversity in magnoliids is extensive, with flowers ranging from the large, showy flowers of *Magnolia* to the simple, perianthless flowers of Piperaceae.

Most clades of basal angiosperms are characterized by generally uniaperturate or uniaperturate-derived pollen grains (fig. 11.6; see Sampson 2000). This type of pollen is also produced by all extant and fossil gymnosperms. In contrast, triaperturate (or triaperturate-derived) pollen is produced by a single large clade of angiosperms, the eudicots (see below; Donoghue and Doyle 1989, Doyle and Hotton 1991). In fact, this single pollen character is the only nonmolecular synapomorphy identified for this clade that contains approximately 75% of all angiosperm species (Drinnan et al. 1994). The distinction between uniaperturate and triaperturate pollen is clear in the fossil record, making assignment of fossil specimens to the eudicot clade unambiguous. Furthermore, the pollen record clearly shows that the earliest triaperturate pollen appeared 125 Mya. Moreover, the richness of the pollen record makes the age of 125 Mya one of the most secure dates in the paleobotanical record. The origin of the eudicots at least 125 Mya indicates that this clade arose early in angiosperm evolution and is nearly as old as the angiosperms themselves.

Eudicots

The eudicot clade consists of a basal grade of five main lineages and a large clade that contains most species of eudicots (fig. 11.7). Ranunculales, which include Ranunculaceae (buttercups, columbines, and larkspurs) and Papaveraceae (poppies), among others, are the sister group to all other eudicots

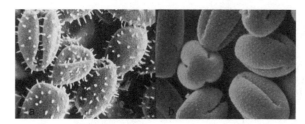

Figure 11.6. (A) Typical uniaperturate pollen grains of gymnosperms and noneudicots; note the single groove. (B) Triaperturate pollen grains of eudicots; note three grooves.

(e.g., Hoot et al. 1999, Soltis et al. 2000, Kim et al. 2003). Other basal lineages are Proteales (proteas, sycamores, and the water lotus, *Nelumbo*), Sabiaceae, Trochodendraceae, and Buxaceae (boxwoods), but relationships among these groups and their relationships to the core eudicots are not completely clear.

Most eudicots fall in the core eudicot clade (fig. 11.7). Within the core eudicots, Gunnerales are sister to the rest of the clade (Soltis et al. 2003), but the interrelationships among

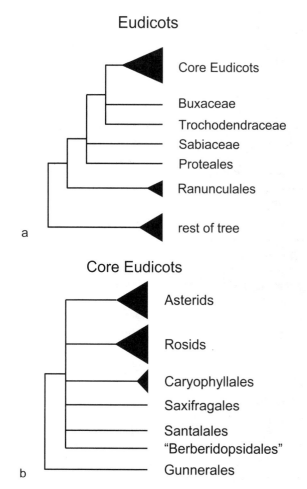

Figure 11.7. Summary of phylogenetic relationships among major clades of core eudicots. (A) Basal eudicots. (B) Core eudicots.

the remaining six main clades of core eudicots are not resolved. The most prominent clades of core eudicots are the rosids and asterids, each with thousands of species, along with the smaller Caryophyllales, Santalales, and Saxifragales clades and the very small clade of *Berberidopsis* and *Aextoxicon*. Because the core eudicots contain more than 70% of all angiosperm species and most of the morphological and physiological diversity of flowering plants, resolution of relationships among major clades of core eudicots is needed to clarify major patterns of evolution.

Eudicots: Gunnerales.   Gunnerales consist of only two families, Gunneraceae with the single genus *Gunnera* (40 species) and Myrothamnaceae with only two species of *Myrothamnus*. Despite strong molecular support for the relationship between the two families, they were not previously considered to be close relatives, because they differ dramatically in morphology. Plants of *Myrothamnus* are small, xerophytic shrubs, whereas plants of *Gunnera* are small or immense perennial herbs often from moist or humid habitats in the Southern Hemisphere.

Eudicots: "Berberidopsidales."   This clade also contains only two small families, Berberidopsidaceae with the genera *Berberidopsis* from South America and *Streptothamnus* from southeastern Australia (although *Streptothamnus* is sometimes considered part of *Berberidopsis*) and Aextoxicaceae with a single species of *Aextoxicon*. Again, despite strong molecular support, no obvious morphological characters unite these groups, and this clade is recognized solely on the basis of molecular data. Both families have encyclocytic stomata, a rare feature in angiosperms and an apparent synapomorphy for these families. The clade has not been formally recognized as an order (Angiosperm Phylogeny Group II 2003), despite its strong molecular support.

Eudicots: Santalales.   This clade consists of seven families, many species of which are parasites, although some plants photosynthesize during part of their life cycle and obtain mostly water and dissolved nutrients from their hosts. Untangling the relationships of parasitic plants has long been difficult because they often exhibit morphologies that appear to have been highly modified by their adaptation to the parasitic habit. For example, many parasites appear to have lost leaves, perianth parts, integuments, and chlorophyll relative to their nonparasitic relatives (see Nickrent et al. 1998). Aerial parasites, such as mistletoes, appear to have arisen multiple times independently in Santalales.

Eudicots: Caryophyllales.   The limits of Caryophyllales extend beyond those of previous circumscriptions (e.g., Cronquist 1988, Takhtajan 1987, 1997). "Traditional Caryophyllales" include several well-known families such as Cactaceae (cactus family), Caryophyllaceae (pink family), and Amaranthaceae (spinach family). The sister clades to this traditional group include carnivorous plants in Droseraceae (sundews) and Nepenthaceae (Old World pitcher plants) and are now also included in Caryophyllales (Angiosperm Phylogeny Group II 2003). Many members of Caryophyllales are

adapted to extremely harsh environmental conditions, such as high-alkaline soils, high-salt conditions, extreme aridity, and nutrient-poor soils. They have conquered these habitats through a variety of adaptations, such as unusual photosynthetic pathways [Crassulacean Acid Metabolism (CAM) and C_4], unusual morphologies (e.g., succulence), unusual methods of nutrient uptake (e.g., carnivory), and secretion of excessive salt by special glands.

Eudicots: Rosids.   The rosids and asterids are by far the two largest clades of core eudicots. The rosids are a clade of extremely well-supported groups, the interrelationships of which are not clear. This pattern is true at both the deep nodes within the rosids and within the two large eurosid I and II clades (fig. 11.8). Eurosid I consists of the large order Malpighiales (examples of which are poplars, willows, passion flowers, violets, St. John's wort, and flax), Oxalidales (e.g., *Oxalis*), and orders corresponding to the melon, oak/hickory/walnut, legume, and rose clades. Eurosid II contains Malvales (mallows, cotton, basswood, chocolate), Sapindales (citrus, maples, horse chestnuts), and Brassicales (mustards, capers, papaya, nasturtium). Three additional orders form part of a basal split in the rosids: Myrtales (myrtles), Geraniales (geranium), and Crossosomatales.

Despite the lack of resolution among major groups of rosids, the rosids offer some interesting cases of chemical and physiological evolution. One of these concerns the origin of symbioses with nitrogen-fixing bacteria. The legumes (Fabaceae), a prominent clade in eurosid I, are well known for their symbiotic associations with rhizobial bacteria. However, nodular symbioses with nitrogen-fixing bacteria are also found in nine other families of angiosperms, and nearly all of these symbioses are with actinomycetes rather than rhizobia. Traditional classifications of angiosperms indicated that these 10 families were distantly related. This inference in turn led crop geneticists and breeders to view the genetic machin-

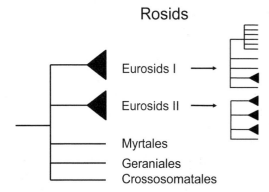

Figure 11.8. Pattern of radiation in the rosids. Myrtales, Geraniales, and Crossosomatales are basal branches of rosids. Eurosids I and II are large clades, each of which consists of multiple lineages, shown diagrammatically to the right. Note repeated pattern of radiation, from basal relationships in rosids to basal relationships within eurosids I and II to additional radiations nearer the tips.

ery for nitrogen-fixing symbioses to be quite simple, and perhaps transferable among distantly related species, for example, from a bean to cereal grasses. Early molecular phylogenetic studies (e.g., Chase et al. 1993) indicated, however, that these families might be fairly closely related, and more focused analyses confirmed these ideas (Soltis et al. 1995, 1997, 2000). In fact, all of the 10 families with nodular symbiotic associations fall in a single clade within the eurosid I clade, along with several families that lack these symbioses. The focused placement of all 10 families within a single clade supports the hypothesis that there was likely a single origin of the predisposition for symbiotic associations with nitrogen-fixing bacteria followed by multiple refinements of symbiosis within this clade. This finding suggests, contrary to previous hypotheses, that the genetic transfer of the needed machinery from a legume to a cereal may be difficult.

A second example of complex chemical evolution in the rosids involves the origin and diversification of glucosinolate compounds. These compounds are generally considered to be important plant defense compounds and are perhaps best known in the mustards and their close relatives, all classified in Brassicaceae. However, as with nitrogen-fixing symbioses, glucosinolates have been reported in families outside Brassicaceae, and these groups were considered distantly related based on traditional classifications. On the contrary, phylogenetic analyses, based initially on morphology and ultimately molecular data (e.g., Rodman 1991, Rodman et al. 1993, 1998), found that all but one of these families form a single clade, now referred to as Brassicales (Angiosperm Phylogeny Group 1998, Angiosperm Phylogeny Group II 2003), nested well within the eurosid II clade. The only exception is Putranjivaceae; *Drypetes* and *Putranjiva* are both reported to produce glucosinolates. Putranjivaceae are also in the rosids, but distantly related to Brassicales; this phylogenetic placement suggests that glucosinolate production in *Drypetes* and *Putranjiva* arose independently from that in Brassicales and that glucosinolates in Putranjivaceae may be produced through a different biosynthetic pathway (Rodman et al. 1998).

Eudicots: Saxifragales. The sister group to the rosids may be Saxifragales, a clade of 14 families—including Saxifragaceae (containing coral bells), Crassulaceae (stonecrops), Grossulariaceae (gooseberries, currants), and several groups previously not considered at all closely related to these families, such as Altingiaceae (sweet gum), Cercidiphyllaceae, Daphniphyllaceae, Hamamelidaceae (witch hazel), and Paeoniaceae (peonies). These families were previously classified in three different subclasses of angiosperms (e.g., Cronquist 1981, Takhtajan 1997), reflecting their morphological diversity in habit, size, life history, and flowers. For example, the clade includes trees, shrubs, lianas, annual and perennial herbs, succulents, and aquatics, with further differences in number of floral parts and the degree of fusion of floral organs. Because of this morphological diversity, nonmolecular synapomorphies have not yet been identified, although features of wood anatomy and leaf venation are similar in the woody members of the clade.

Despite strong support for Saxifragales as a clade of core eudicots, their position is uncertain. They have variously appeared as sister to the rosids or sister to the rest of (or most of) the core eudicots. The diversification of Saxifragales appears to have been contemporaneous with the initial radiations of the eudicots, magnoliids, and monocots (Fishbein et al. 2001). Furthermore, the oldest confirmed fossils of Saxifragales are dated to 89.5 Mya, which is comparable with the oldest fossils of core eudicots (Magallón et al. 1999).

Eudicots: Asterids. The final clade of core eudicots is the asterids, a huge clade consisting of nearly 80,000 species classified into approximately 4700 genera and 100 families (Thorne 1992). This clade is composed of four subclades (e.g., Albach et al. 2001, Bremer et al. 2002): Cornales (dogwoods and hydrangeas), Ericales (blueberries, cranberries, azaleas, camellias, and phlox), euasterids I (= lamiids; see Bremer et al. 2002; e.g., mints, snapdragons, tomato, and potato), and euasterids II (= campanulids; Bremer et al. 2002; e.g., sunflowers, carrot family, and honeysuckles and relatives). Most analyses indicate that Cornales is a sister group to a clade of Ericales and euasterids (e.g., Soltis et al. 2000, Bremer et al. 2002). Relationships within clades of asterids have been addressed by Xiang et al. (1998) for Cornales, Judd and Kron (1993) and Anderberg et al. (2002) for Ericales, B. and K. Bremer and their students (B. Bremer et al. 2002, K. Bremer et al. 2001) for euasterids, Plunkett et al. (1997) and Plunkett and Lowrey (2001) for Apiales, and Donoghue et al. (2001, 2003) for Dipsacales.

Radiations in Angiosperm Phylogeny

A recurrent pattern in the angiosperm trees is that of radiations. This pattern is clearly evident in the rosids, but it is also present within the asterids, within the core eudicots, near the base of the eudicots, within the magnoliids, within the monocots, and even earlier in angiosperm phylogeny. Undoubtedly, some of these radiations may be resolved by additional data, as for basal angiosperms (e.g., Zanis et al. 2002), but some of these starburst patterns remain even after the analysis of several genes totaling several thousand nucleotides (e.g., Saxifragales; Fishbein et al. 2001). Given that the eudicots themselves originated shortly after the fossil record discloses the origin of angiosperms as a whole, many of these radiations actually trace back to an early point in angiosperm history. The evolutionary history of flowering plants seems to be one of repeated radiations, perhaps associated with innovations and the opening up of new habitats.

Gaps in Our Knowledge of Angiosperm Phylogeny

Although many aspects of angiosperm phylogeny have been clarified, areas of uncertainty remain. For example, are the radiations described above true radiations, or do they simply appear as radiations because we lack the information to discriminate the true branching patterns of history? Additional study should be devoted to those putative points of major radiation, such as the core eudicots, the rosids, the asterids, and the lilioid monocots. Relationships among the basal nodes of the eudicot clade also need clarification. Thus, although many of the major lineages of angiosperms and their interrelationships have been identified in recent studies, further study is required to resolve the topology of the angiosperm branch of the Tree of Life and to interpret patterns of diversification across the angiosperm clade.

Evolution of Key Angiosperm Features

The major traits that distinguish angiosperms from their gymnospermous ancestors are all characters that have presumably made the reproductive process more efficient in angiosperms. Modifications occurred both in the structure of the reproductive organs and in the set of processes that collectively result in sexual reproduction. In this section, we examine the evolution of three of these important features: double fertilization and endosperm formation, closed carpels, and the structure and organization of the perianth.

Double Fertilization and Endosperm Formation

The typical process of double fertilization in angiosperms involves (1) the union of an egg and a single sperm nucleus to form the zygote and (2) the union of two polar nuclei and a second sperm nucleus to form triploid endosperm, the nutritive material for the developing embryo (fig. 11.9).

This process involves the formation of an eight-nucleate embryo sac.

Although double fertilization and embryo formation occur in this manner in the vast majority of angiosperms, some species show variation on this general theme. One important variant is the formation of endosperm through the union of the second sperm nucleus with a single haploid central cell of the embryo sac, producing a diploid endosperm. This process involves a four-nucleate embryo sac rather than the typical eight-nucleate embryo sac produced by most flowering plants. This variation on the general process of sexual reproduction has been documented in detail in *Nuphar*, a water lily, one of the basal lineages of flowering plants (Williams and Friedman 2002). The same process appears to occur in some other basal angiosperms, such as *Illicium* (Friedman and Williams 2003). This phylogenetic distribution of a four-nucleate embryo sac and diploid endosperm formation suggests the possibility that these features are ancestral in the angiosperms. However, there is at least one report of an eight-nucleate embryo sac in *Amborella*, which sits at the pivotal position of sister to all other angiosperms. If *Amborella* indeed exhibits the typical angiosperm processes of eight-nucleate embryo sac formation and triploid endosperm production, then multiple changes in embryo sac structure and endosperm formation occurred early in the history of angiosperms. Possibilities include (1) parallel development of the *Nuphar-Illicium* type of reproduction in the water lilies and Austrobaileyales, (2) a reversion to the *Amborella* type in the majority of angiosperms after the

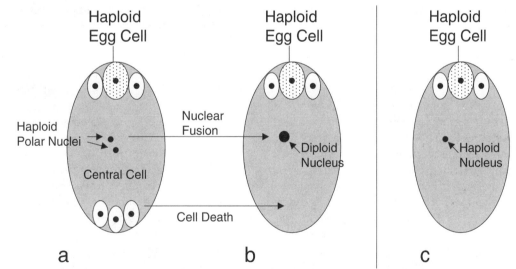

Figure 11.9. Pattern of double fertilization and endosperm formation (redrawn from Williams and Friedman 2002). (A and B) A seven-celled, eight-nucleate female gametophyte (embryo sac), typical of most angiosperms. In double fertilization, one sperm unites with the egg to form a diploid zygote, and a second sperm nucleus unites with the fused polar nuclei of the central cell to form triploid endosperm. The three cells opposite the egg and its adjacent cells disintegrate. (C) Four-celled female gametophyte of the water lily *Nuphar*. In *Nuphar*, and some other basal angiosperms, one sperm fuses with the egg to form the zygote, and the second sperm unites with the haploid nucleus of the central cell to form diploid endosperm. Redrawn and modified from Williams and Friedman (2002).

development of the *Nuphar-Illicium* type, or (3) parallel development of the eight-nucleate type in *Amborella* and the majority of angiosperms from a four-nucleate ancestor. This fundamental aspect of angiosperm embryology requires further study.

Closed Carpels

The closed carpel provides protection for both the ovule before fertilization and the developing embryo and seed after fertilization and thus represents a tremendously important innovation in the history of plants. In addition, the closed carpel allows for competition among pollen grains and thus selection at the gametic level (Mulcahy 1979).

Numerous hypotheses have been presented for the origin of the carpel, but most relate to the folding or tubular development of a fertile leaf bearing ovules that become tucked inside the new structure. The closure of the newly formed carpel may have been initially by secretions. In fact, a number of angiosperm groups have carpels that are closed not by fusion of adjacent surfaces but by mucilaginous secretions (Endress and Igersheim 2000a; fig. 11.10). The carpels of *Amborella*, the small-flowered water lilies (*Cabomba*), *Austrobaileya*, *Trimenia*, *Schisandra*, and *Kadsura* are fused entirely by secretions, whereas those of *Illicium* and the large-flowered water lilies (e.g., *Nymphaea*) have carpels that are closed at least partly by fusion and partly by secretion. Within the magnoliids, the degree of fusion increases, with less of a role played by secretion. The carpels of the eudicots are nearly all closed by fusion of adjacent tissues. Therefore, the ancestral condition appears to have been closure by secretion, with fusion evolving later, probably independently in a number of lineages.

Evolution of a Differentiated Perianth: Morphology and Floral Genes

Perianth is the collective term for the sepals and petals of a flower (or the tepals, if sepals and petals are undifferentiated from each other), and this structure plays a tremendously important role in plant reproduction. Perianth parts provide protection for the developing reproductive structures of the flower when the flower is in bud. Further, a showy perianth, typically the corolla (the collective term for the petals), is an important attractant for pollinators. In most angiosperms, the perianth is clearly differentiated into an outer whorl (series) of typically green sepals and an inner whorl of typically colored petals. However, many basal angiosperms and some early-diverging eudicots lack a differentiated perianth, with all perianth parts appearing identical. Multiple hypotheses have been proposed to explain the origin of the differentiated, or bipartite, perianth, with alternative hypotheses seeming more likely for different groups of species (e.g., Takhtajan 1991). The form of the perianth of the original flower has also been debated, with alternative hypotheses ranging from

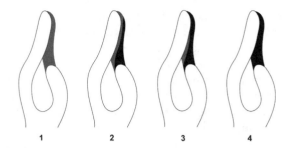

Figure 11.10. Carpel closure in basal angiosperms (redrawn from Endress and Igersheim 2001a). (1) Carpel is closed by secretions only, as indicated by gray shading. This method of carpel closure occurs in *Amborella*, the water lily *Cabomba*, and *Austrobaileya*, *Trimenia*, and *Schisandra* of Austrobaileyales. (2 and 3) Increasing role of congenital fusion (black shading) and corresponding reduced role of secretions. (4) Complete congenital fusion, as in Magnoliales, Canellales, monocots, and eudicots.

a large, showy, undifferentiated perianth, as in *Magnolia* (e.g., Cronquist 1968), to a small, inconspicuous, or absent perianth (e.g., Friis et al. 1986).

The distribution of differentiated and undifferentiated perianths across the phylogenetic tree for angiosperms clearly shows that a differentiated perianth arose multiple times in the basal angiosperms and early eudicots (fig. 11.11; see also Albert et al. 1998, Ronse DeCraene et al. 2003). Petals are clearly not phylogenetically homologous across the angiosperms. Furthermore, petals appear to have arisen via different mechanisms in different groups (Takhtajan 1991) and thus are not ontogenetically or structurally homologous either.

In contrast to stamens and carpels, sepals and petals cannot be distinguished unambiguously by their structures and functions; therefore, genetic data may be useful for clarifying structural identities and homologies of perianth organs among groups (see Kramer et al. 1998, Kramer and Irish 2000, Endress 2001, Theissen 2001). Floral organ identity in the model angiosperm *Arabidopsis thaliana* (Brassicaceae) is controlled by overlapping expression of three classes of genes (A, B, and C class) in adjacent "whorls" of the flower (Coen and Meyerowitz 1991; fig. 11.12). Most of the ABC genes are members of the large MADS-box gene family that occurs throughout plants, animals, and fungi. Expression of A-class genes alone produces sepals, coexpression of A and B genes yields petals, coexpression of B and C genes specifies stamens, and C-class expression alone leads to carpel formation (for modifications to the model, see Theissen and Saedler 2001, Honma and Goto 2001; for review, see Theissen 2001). This pattern of gene action in a core eudicot with a clearly bipartite perianth can be used to evaluate the genetic nature of undifferentiated perianth organs—tepals—in other plant groups. Do tepals share expression patterns with sepals or with petals, or do they exhibit their own combination of MADS-box gene expression (e.g., Albert et al. 1998)? Conversely, can patterns of gene expression be used

Figure 11.11. Parsimony reconstruction of the evolution of differentiated versus undifferentiated perianth in basal angiosperms using MacClade (Maddison and Maddison 1992) and a tree based on analyses by Zanis et al. (2002), Renner (1999), Karol et al. (2000), Hoot and Crane (1995), and Hoot (1995). A differentiated perianth has clearly distinguishable sepals and petals. Here, "perianth differentiation" includes morphological and/or positional differentiation.

to determine whether a structure is fundamentally a sepal or a petal? To date, most studies of gene expression in flowers with undifferentiated perianths (e.g., Kramer et al. 1998, 2003, Kramer and Irish 2000, Tzeng and Yang 2001, Kanno et al. 2003) have focused on B-class genes because of their role in petal formation in core eudicots. These studies have generally demonstrated B-class gene expression throughout the perianth, suggesting, perhaps, that these structures are "petals." However, this pattern of B-class expression even extends to monocots such as lilies (Tzeng and Yang 2001) and tulips (van Tunen et al. 1993, Kanno et al. 2003) with outer perianth segments that correspond positionally to sepals. Thus, most authors agree on a "modified" ABC model of floral organ identity in basal angiosperms, monocots, and basal eudicots (Kramer et al. 1998, 2003, Kramer and Irish 2000, Tzeng and Yang 2001, Kanno et al. 2003; fig. 11.12) and suggest that gene expression patterns alone cannot be used to infer homology of floral organs. Furthermore, because B-class genes are expressed throughout the perianths of both *Amborella* and *Nuphar* (a water lily; S. Kim et al., unpubl. obs.), it appears that early angiosperms may have exhibited diffuse expression of these organ-determining genes throughout the flower. Later in angiosperm evolution, expression of these genes became localized, resulting

in the uniform, predictable, synorganized flower of the core eudicots.

Conclusions

The diversification of flowering plants has been phenomenal, generating upward of 300,000 extant species in less than 150 million years. Flowering plants have thrived on all land masses, and they continue to dominate, and form the basis of, all terrestrial ecosystems. They also play crucial roles in many aquatic, including some marine, habitats. Their evolution has been closely tied to diversification in many other groups of organisms, such as fungi, beetles, butterflies, flies, and mammals. Clear understanding of all of these branches of the Tree of Life will allow formulation and tests of hypotheses of codiversification and coevolution.

Within angiosperms, information on phylogeny has already guided research as diverse as ecology and genomics. Phylogenetic information may also be crucial for conserving rare species, eliminating invasive species, and improving crops. Continued efforts to include all 300,000 "leaves" in the "Tree of Flowering Plants" will ultimately generate unprecedented and unforeseeable benefits to organismal biology and society.

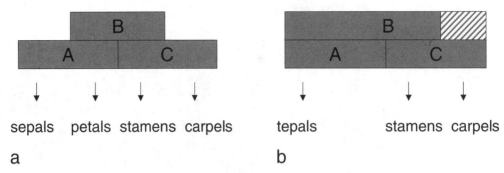

Figure 11.12. (A) The ABC model of floral organ identity in *Arabidopsis thaliana*, a rosid (Coen and Meyerowitz 1991). Expression of A-class gene(s) alone specifies sepal formation, coexpression of A and B specifies petals, coexpression of B and C specifies stamens, and expression of C alone specifies carpels. (B). "Modified" ABC model, based on work on monocots and basal angiosperms. Note that B-class gene expression occurs throughout all major organs of the flower. Coexpression of A and B, even in the outer whorl of perianth segments in tulip, produces morphologically identical tepals in two positional series. The hatched area indicates that weak B-class gene expression is also observed in carpels. Modified from Kanno et al. (2003).

Acknowledgments

We thank Joel Cracraft and Michael Donoghue for inviting us to participate in the Tree of Life Symposium and to contribute this chapter. This work was supported in part by NSF grants DEB-0090283 and PGR-0115684.

Literature Cited

Albach, D. C., P. S. Soltis, D. E. Soltis, and R. G. Olmstead. 2001. Phylogenetic analysis of the Asteridae s.l. using sequences of four genes. Ann. Mo. Bot. Gard. 88:163–212.

Albert, V. A., A. Backlund, K. Bremer, M. W. Chase, J. R. Manhart, B. D. Mishler, and K. C. Nixon. 1994. Functional constraints and *rbcL* evidence for land plant phylogeny. Ann. Mo. Bot. Gard. 81:534–567.

Albert, V. A., M. H. G. Gustafsson, and L. DiLaurenzio. 1998. Ontogenetic systematics, molecular developmental genetics, and the angiosperm petal. Pp. 349–374 *in* Molecular systematics of plants II (D. E. Soltis, P. S. Soltis, and J. J. Doyle, eds.). Kluwer, Boston.

Anderberg, A. A., C. Rydin, and M. Källersjö. 2002. Phylogenetic relationships in the order Ericales s.l.: analyses of molecular data from five genes from the plastid and mitochondrial genomes. Am. J. Bot. 89:677–687.

Angiosperm Phylogeny Group. 1998. An ordinal classification for the families of flowering plants. Ann. Mo. Bot. Gard. 85:531–553.

Angiosperm Phylogeny Group II. 2003. An update of the Angiosperm Phylogeny Group classification for the orders and families of flowering plants: APG II. Bot. J. Linn. Soc. 141:399–436.

Barkman, T. J., G. Chenery, J. R. McNeal, J. Lyons-Weiler, and C. W. dePamphilis. 2000. Independent and combined analyses of sequences from all three genomic compartments converge on the root of flowering plant phylogeny. Proc. Natl. Acad. Sci. USA 97:13166–13171.

Borsch, T., K. W. Hilu, D. Quandt, V. Wilde, C. Neinhuis, and W. Barthlott. 2003. Non-coding plastid *trnT-trnF* sequences reveal a highly supported phylogeny of basal angiosperms. J. Evol. Biol. 16:558–576.

Bowe, L. M., G. Coat, and C. W. dePamphilis. 2000. Phylogeny of seed plants based on all three genomic compartments: extant gymnosperms are monophyletic and Gnetales' closest relatives are conifers. Proc. Natl. Acad. Sci. USA 97:4092–4097.

Bremer, B., K. Bremer, N. Heidari, P. Erixon, R. G. Olmstead, A. A. Anderberg, M. Källersjö, and E. Barkhordarian. 2002. Phylogenetics of asterids based on 3 coding and 3 non-coding chloroplast DNA markers and the utility of non-coding DNA at higher taxonomic levels. Mol. Phylogenet. Evol. 24:274–301.

Bremer, K. 2000. Early Cretaceous lineages of monocot flowering plants. Proc. Natl. Acad. Sci. USA 97:4707–4711.

Bremer, K., A. Backlund, B. Sennblad, U. Swenson, K. Andreasen, M. Hjertson, J. Lundberg, M. Backlund, and B. Bremer. 2001. A phylogenetic analysis of 100+ genera and 50+ families of euasterids based on morphological and molecular data with notes on possible higher level morphological synapomorphies. Plant Syst. Evol. 229:137–169.

Carmichael, J. S., and W. E. Friedman. 1996. Double fertilization in *Gnetum gnemon* (Gnetaceae): its bearing on the evolution of sexual reproduction within the Gnetales and the Anthophyte clade. Am. J. Bot. 83:767–780.

Chase, M. W., D. E. Soltis, R. G. Olmstead, D. Morgan, D. H. Les, B. D. Mishler, M. R. Duvall, R. A. Price, H. G. Hills, Y.-L. Qiu, et al. 1993. Phylogenetics of seed plants: an analysis of nucleotide sequences from the plastid gene *rbcL*. Ann. Mo. Bot. Gard. 80:528–580.

Chase, M. W., D. E. Soltis, P. S. Soltis, P. J. Rudall, M. F. Fay, W. J. Hahn, S. Sullivan, J. Joseph, M. Molvray, P. J. Kores, T. J. Givnish, K. J. Sytsma, and J. C. Pires. 2000. Higher-level systematics of the monocotyledons: an assessment of current knowledge and a new classification. Pp. 3–16 *in* Monocots: systematics and evolution (K. L. Wilson and D. A. Morrison, eds.). CSIRO Publishing, Collingwood, Victoria, Australia.

Chaw, S.-M., C. L. Parkinson, Y. Cheng, T. M. Vincent, and J. D. Palmer. 2000. Seed plant phylogeny inferred from all three plant genomes: monophyly of extant gymnosperms and origin of Gnetales from conifers. Proc. Natl. Acad. Sci. USA 97:4086–4091.

Chaw, S.-M., A. Zharkikh, H.-M. Sung, T.-C. Lau, and W.-H. Li. 1997. Molecular phylogeny of extant gymnosperms and seed plant evolution: analysis of nuclear 18S rDNA sequences. Mol. Biol. Evol. 14:56–68.

Coen, E. S., and E. M. Meyerowitz. 1991. The war of the whorls: genetic interactions controlling flower development. Nature 353:31–37.

Couper, R. A. 1958. British Mesozoic microspores and pollen grains: a systematic and stratigraphic study. Palaeontogr. Abt. B 103:75–179.

Crane, P. R. 1985. Phylogenetic analysis of seed plants and the origin of angiosperms. Ann. Mo. Bot. Gard. 72:716–793.

Cronquist, A. 1968. The evolution and classification of flowering plants. Houghton Mifflin, Boston.

Cronquist, A. 1981. An integrated system of classification of flowering plants. Columbia University Press, New York.

Cronquist, A. 1988. The evolution and classification of flowering plants. 2nd ed. New York Botanical Garden, Bronx, NY.

Dilcher, D. L. 1989. The occurrence of fruits with affinities to Ceratophyllaceae in lower and mid-Cretaceous sediments. Am. J. Bot. 76:162.

Dilcher, D. L., and P. R. Crane. 1984. *Archaeanthus*: an early angiosperm from the Cenomanian of the western interior of North America. Ann. Mo. Bot. Gard. 71:351–383.

Donoghue, M. J., C. D. Bell, and R. C. Winkworth. 2003. The evolution of reproductive characters in Dipsacales. Int. J. Plant Sci. 164(5 suppl.):S453–S464.

Donoghue, M. J., and J. A. Doyle. 1989. Phylogenetic analysis of angiosperms and the relationships of Hamamelidae. Pp. 17–45 *in* Evolution, systematics, and fossil history of the Hamamelidae, Vol. 1: Introduction and "lower" Hamamelidae (P. R. Crane and S. Blackmore, eds.). Clarendon Press, Oxford.

Donoghue, M. J., and J. A. Doyle. 2000. Seed plant phylogeny: demise of the anthophyte hypothesis? Curr. Biol. 10:R106–R109.

Donoghue, M. J., T. Eriksson, P. A. Reeves, and R. G. Olmstead. 2001. Phylogeny and phylogenetic taxonomy of dipsacales, with special reference to *Sinadoxa* and *Tetradoxa* (Adoxaceae). Harv. Pap. Bot. 6:459–479.

Doyle, J. A. 1996. Seed plant phylogeny and the relationships of Gnetales. Int. J. Plant Sci. 157:S3–S39.

Doyle, J. A. 1998. Molecules, morphology, fossils, and the relationships of angiosperms and Gnetales. Mol. Phylogenet. Evol. 9:448–462.

Doyle, J. A., and M. J. Donoghue. 1986. Seed plant phylogeny and the origin of the angiosperms: an experimental cladistic approach. Bot. Rev. 52:321–431.

Doyle, J. A., and M. J. Donoghue. 1992. Fossils and seed plant phylogeny reanalyzed. Brittonia 44:89–104.

Doyle, J. A., and P. K. Endress. 2000. Morphological phylogenetic analyses of basal angiosperms: comparison and combination with molecular data. Int. J. Plant Sci. 161:S121–S153.

Doyle, J. A., and C. L. Hotton. 1991. Diversification of early angiosperm pollen in a cladistic context. Pp. 169–195 *in* Pollen and spores: patterns of diversification (S. Blackmore and S. H. Barnes, eds.). Clarendon Press, Oxford.

Drinnan, A. N., P. R. Crane, and S. B. Hoot. 1994. Patterns of floral evolution in the early diversification of non-magnoliid dicotyledons (eudicots). Plant Syst. Evol. 8(suppl.):93–122.

Endress, P. K. 2001. The flowers in extant basal angiosperms and inferences on ancestral flowers. Int. J. Plant Sci. 162:1111–1140.

Endress, P. K., and A. Igersheim. 2000a. Gynoecium structure and evolution in basal angiosperms. Int. J. Plant. Sci. 161:S211–S223.

Endress, P. K., and A. Igersheim. 2000b. The reproductive structures of the basal angiosperm *Amborella trichopoda* (Amborellaceae). Int. J. Plant Sci. 161:S237–S248.

Feild, T. S., M. A. Zweiniecki, T. Brodribb, T. Jaffré, M. J. Donoghue, and N. M. Holbrook. 2000. Structure and function of tracheary elements in *Amborella trichopoda*. Int. J. Plant. Sci. 161:705–712.

Fishbein, M., C. Hibsch-Jetter, D. E. Soltis, and L. Hufford. 2001. Phylogeny of Saxifragales (angiosperms, eudicots): analysis of a rapid, ancient radiation. Syst. Biol. 50:817–847.

Friedman, W. E. 1990. Double fertilization in *Ephedra*, a nonflowering seed plant: its bearing on the origin of angiosperms. Science 247:951–954.

Friedman, W. E., and J. H. Williams. 2003. Modularity of the angiosperm female gametophyte and its bearing on the early evolution of endosperm in flowering plants. Evolution 57:216–230.

Friis, E. M., P. R. Crane, and K. R. Pedersen. 1986. Floral evidence for Cretaceous chloranthoid angiosperms. Nature 320:163–164.

Friis, E. M., J. A. Doyle, P. K. Endress, and Q. Leng. 2003. *Archaefructus*—angiosperm precursor or specialized early angiosperm? Trends Plant Sci. 8:369–373.

Friis, E. M., K. R. Pedersen, and P. R. Crane. 2000. Reproductive structure and organization of basal angiosperms from the early Cretaceous (Barremian or Aptian) of western Portugal. Int. J. Plant Sci. 161:S169–S182.

Friis, E. M., K. R. Pedersen, and P. R. Crane. 2001. Fossil evidence of water lilies in the early Cretaceous. Nature 410:357–360.

Gandolfo, M. A., K. C. Nixon, and W. L. Crepet. 2002. Triuridaceae fossil flowers from the Upper Cretaceous of New Jersey. Am. J. Bot. 89:1940–1957.

Goremykin, V., V. Bobrova, J. Pahnke, A. Troitsky, A. Antonov, and W. Martin. 1996. Noncoding sequences from the slowly evolving chloroplast inverted repeat in addition to *rbcL* data do not support Gnetalean affinities of angiosperms. Mol. Biol. Evol. 13:383–396.

Graham, S. W., and R. G. Olmstead. 2000. Utility of 17 chloroplast genes for inferring the phylogeny of the basal angiosperms. Am. J. Bot. 87:1712–1730.

Hamby, R. K., and E. A. Zimmer. 1992. Ribosomal RNA as a phylogenetic tool in plant systematics. Pp. 50–91 *in* Molecular systematics of plants (P. S. Soltis, D. E. Soltis, and J. J. Doyle, eds.). Chapman and Hall, New York.

Hansen, A., S. Hansmann, T. Samigullin, A. Antonov, and W. Martin. 1999. *Gnetum* and the angiosperms: molecular evidence that their shared morphological characters are convergent, rather than homologous. Mol. Biol. Evol. 16:1006–1009.

Hasebe, M., R. Kofuji, M. Ito, M. Kato, K. Iwatsuki, and K. Ueda. 1992. Phylogeny of gymnosperms inferred from *rbcL* gene sequences. Bot. Mag. Tokyo 105:673–679.

Heckman, D. S., D. M. Geiser, B. R. Eidell, R. L. Stauffer, N. L. Kardos, and S. B. Hedges. 2001. Molecular evidence for the early colonization of land by fungi and plants. Science 293:1129–1133.

Hilu, K. W., T. Borsch, K. Müller, D. E. Soltis, P. S. Soltis, V. Savolainen, M. W. Chase, M. Powell, L. A. Lawrence, R. Evans, et al. 2003. Angiosperm phylogeny based on *matK* sequence information. Am. J. Bot. 90:1758–1776.

Honma, T., and K. Goto. 2001. Complexes of MADS-box proteins are sufficient to convert leaves into floral organs. Nature 409:525–529.

Hoot, S. B. 1995. Interfamilial relationships in the Ranunculidae based on molecular systematics. Plant Syst. Evol. 9(suppl.): 119–131.

Hoot, S. B., and P. R. Crane. 1995. Phylogeny of the Ranunculaceae based on preliminary *atpB*, *rbcL*, and 18S nuclear ribosomal DNA sequence data. Plant Syst. Evol. 9(suppl.): 241–251.

Hoot, S. B., S. Magallón, and P. R. Crane. 1999. Phylogeny of basal eudicots based on three molecular datasets: *atpB*, *rbcL*, and 18S nuclear ribosomal DNA sequences. Ann. Mo. Bot. Gard. 86:1–32.

Judd, W. S., and K. A. Kron. 1993. Circumscription of Ericaceae (Ericales) as determined by preliminary cladistic analyses based on morphological, anatomical, and embryological features. Brittonia 45:99–114.

Kanno, A., H. Saeki, T. Kameya, H. Saedler, and G. Theissen. 2003. Heterotopic expression of class B floral homeotic genes supports a modifed ABC model for tulip (*Tulipa gesneriana*). Plant Mol. Biol. 52:831–841.

Karol, K. G., Y. Suh, G. E. Schatz, and E. A. Zimmer. 2000. Molecular evidence for the phylogenetic position of *Takhtajania* in the Winteraceae: inference from nuclear ribosomal and chloroplast gene spacer sequences. Ann. Mo. Bot. Gard. 87:414–432.

Kim, S., D. E. Soltis, P. S. Soltis, M. J. Zanis, and Y. Suh. 2003. Phylogenetic relationships among early-diverging eudicots based on four genes: were the eudicots ancestrally woody? Mol. Phylogenet. Evol.

Kramer, E. M., V. S. DiStilio, and P. M. Schlüter. 2003. Complex patterns of gene duplication in the *APETALA3* and *PISTILLATA* lineages of the Ranunculaceae. Int. J. Plant Sci. 164:1–11.

Kramer, E. M., R. L. Dorit, and V. F. Irish. 1998. Molecular evolution of genes controlling petal and stamen development: duplication and divergence within the APETALA3 and PISTILLATA MADS-box gene lineages. Genetics 149:765–783.

Kramer, E. M., and V. F. Irish. 2000. Evolution of the petal and stamen developmental programs: evidence from comparative studies of the lower eudicots and basal angiosperms. Int. J. Plant Sci. 161:S29–240.

Loconte, H., and D. W. Stevenson. 1990. Cladistics of the Spermatophyta. Brittonia 42:197–211.

Maddison, W. P., and D. R. Maddison. 1992. MacClade: analysis of phylogeny and character evolution, ver. 3. Sinauer, Sunderland, MA.

Magallón, S., P. R. Crane, and P. S. Herendeen. 1999. Phylogenetic pattern, diversity, and diversification of eudicots. Ann. Mo. Bot. Gard. 86:297–372.

Magallón, S., and M. J. Sanderson. 2001. Absolute diversification rates in angiosperm clades. Evolution 55:1762–1780.

Malek, O., K. Lattig, R. Hiesel, A. Brennicke, and V. Knoop. 1996. RNA editing in bryophytes and a molecular phylogeny of land plants. EMBO J. 14:1403–1411.

Martin, W., A. Gierl, and H. Saedler. 1989. Molecular evidence for pre-Cretaceous angiosperm origins. Nature 339:46–48.

Martin, W., D. Lydiate, H. Brinkmann, G. Forkmann, H. Saedler, and R. Cerff. 1993. Molecular phylogenies in angiosperm evolution. Mol. Biol. Evol. 10:140–162.

Mathews, S., and M. J. Donoghue. 1999. The root of angiosperm phylogeny inferred from duplicate phytochrome genes. Science 286:947–949.

Mulcahy, D. 1979. The rise of the angiosperms: a genecological factor. Science 206:20–23.

Nickrent, D. L., A. Blarer, Y.-L. Qiu, D. E. Soltis, P. S. Soltis, and M. J. Zanis. 2002. Molecular data place Hydnoraceae with Aristolochiaceae. Am. J. Bot. 89:1809–1817.

Nickrent, D. L., R. J. Duff, A. Colwell, A. D. Wolfe, N. D. Young, K. E. Steiner, and C. W. dePamphilis. 1998. Molecular phylogenetic and evolutionary studies of parasitic plants. Pp. 211–241 *in* Molecular systematics of plants II (D. E. Soltis, P. S. Soltis, and J. J. Doyle, eds.). Kluwer, Boston.

Nixon, K. C., W. L. Crepet, D. Stevenson, and E. M. Friis. 1994. A reevaluation of seed plant phylogeny. Ann. Mo. Bot. Gard. 81:484–533.

Plunkett, G. M., and P. P. Lowrey II. 2001. Relationships among "ancient araliads" and their significance for the systematics of Apiales. Mol. Phylogenet. Evol. 19:259–276.

Plunkett, G. M., D. E. Soltis, and P. S. Soltis. 1997. Clarification of the relationship between Apiaceae and Araliaceae based on *matK* and *rbcL* sequence data. Am. J. Bot. 84:567–580.

Posluszny, U., and P. B. Tomlinson. 2003. Aspects of inflorescence and floral development in the putative basal angiosperm *Amborella trichopoda* (Amborellaceae). Can. J. Bot. 81:28–39.

Qiu, Y.-L., J. Lee, F. Bernasconi-Quadroni, D. E. Soltis, P. S. Soltis, M. Zanis, E. A. Zimmer, Z. Chen, V. Savolainen, and M. W. Chase. 1999. The earliest angiosperms: evidence from mitochondrial, plastid and nuclear genomes. Nature 402:404–407.

Ramshaw, J. A. M., D. L. Richardson, B. T. Meatyard, R. H. Brown, M. Richardson, E. W. Thompson, and D. Boulter. 1972. The time of origin of the flowering plants determined using amino acid sequence data of cytochrome *c*. New Phytol. 71:773–779.

Renner, S. S. 1999. Circumscription and phylogeny of the Laurales: evidence from molecular and morphological data. Am. J. Bot. 86:1301–1315.

Rodman, J. E. 1991. A taxonomic analysis of glucosinolate-producing plants. II. Cladistics. Syst. Bot. 16:619–629.

Rodman, J. E., R. A. Price, K. Karol, E. Conti, K. J. Sytsma, and J. D. Palmer. 1993. Nucleotide sequences of the *rbcL* gene indicate monophyly of mustard oil plants. Ann. Mo. Bot. Gard. 80:686–699.

Rodman, J. E., P. S. Soltis, D. E. Soltis, K. J. Sytsma, and K. G. Karol. 1998. Parallel evolution of glucosinolate biosynthesis inferred from congruent nuclear and plastid gene phylogenies. Am. J. Bot. 85:997–1006.

Rodríguez-Trelles, F., R. Tarrío, and F. J. Ayala. 2002. A methodological bias toward overestimation of molecular evolutionary time scales. Proc. Natl. Acad. Sci. USA 99:8112–8115.

Ronse DeCraene, L. P., P. S. Soltis, and D. E. Soltis. 2003. Evolution of floral structures in basal angiosperms. Int. J. Plant. Sci. 164(suppl.):S329–S363.

Rothwell, G. W., and R. Serbet. 1994. Lignophyte phylogeny and the evolution of spermatophytes: a numerical cladistic analysis. Syst. Bot. 19:443–482.

Rydin, C., M. Källersjö, and E. M. Friis. 2002. Seed plant relationships and the systematic position of Gnetales based on nuclear and chloroplast DNA: conflicting data, rooting problems, and the monophyly of conifers. Int. J. Plant Sci. 163:197–214.

Sampson, F. B. 2000. Pollen diversity in some modern magnoliids. Int. J. Plant Sci. 161:S193–S210.

Sanderson, M. J. 2003. Molecular data from 27 proteins do not support a Precambrian origin of land plants. Am. J. Bot. 90:954–956.

Sanderson, M. J., and J. A. Doyle. 2001. Sources of error and confidence intervals in estimating the age of angiosperms from *rbcL* and 18S rDNA data. Am. J. Bot. 88:1499–1516.

Sanderson, M. J., M. F. Wojciechowski, J.-M. Hu, T. Sher Khan, and S. G. Brady. 2000. Error, bias, and long-branch attraction in data for two chloroplast photosystem genes in seed plants. Mol. Biol. Evol. 17:782–797.

Savolainen, V., M. F. Fay, D. C. Albach, A. Backlund, M. van der Bank, K. M. Cameron, S. A. Johnson, M. D. Lledó, J.-C. Pintaud, M. Powell, et al. 2000. Phylogeny of the eudicots: a nearly complete familial analysis based on *rbcL* gene sequences. Kew Bull. 55:257–309.

Soltis, D. E., A. E. Senters, M. J. Zanis, S. Kim, J. D. Thompson, P. S. Soltis, L. P. Ronse DeCraene, P. K. Endress, and J. S. Farris. 2003. Gunnerales are sister to other core eudicots: implications for the evolution of pentamery. Am. J. Bot. 90:461–470.

Soltis, D. E., P. S. Soltis, M. W. Chase, M. E. Mort, D. C. Albach, M. Zanis, V. Savolainen, W. J. Hahn, S. B. Hoot, M. F. Fay, et al. 2000. Angiosperm phylogeny inferred from 18S rDNA, *rbcL*, and *atpB* sequences. Bot. J. Linn. Soc. 133:381–461.

Soltis, D. E., P. S. Soltis, D. R. Morgan, S. M. Swensen, B. C. Mullin, J. M. Dowd, and P. G. Martin. 1995. Chloroplast gene sequence data suggest a single origin of the predisposition for symbiotic nitrogen fixation in angiosperms. Proc. Natl. Acad. Sci. USA 92:2647–2651.

Soltis, D. E., P. S. Soltis, D. L. Nickrent, L. A. Johnson, W. J. Hahn, S. B. Hoot, J. A. Sweere, R. K. Kuzoff, K. A. Kron, M. W. Chase, et al. 1997. Angiosperm phylogeny inferred from 18S ribosomal DNA sequences. Ann. Mo. Bot. Gard. 84:1–49.

Soltis, D. E., P. S. Soltis, and M. J. Zanis. 2002. Phylogeny of seed plants based on evidence from eight genes. Am. J. Bot. 89:1670–1681.

Soltis, P. S., D. E. Soltis, and M. W. Chase. 1999a. Angiosperm phylogeny inferred from multiple genes as a tool for comparative biology. Nature 402:402–404.

Soltis, P. S., D. E. Soltis, V. Savolainen, P. R. Crane, and T. G. Barraclough. 2002. Rate heterogeneity among lineages of tracheophytes: integration of molecular and fossil data and evidence for molecular living fossils. Proc. Natl. Acad. Sci. USA 99:4430–4435.

Soltis, P. S., D. E. Soltis, P. G. Wolf, D. L. Nickrent, S.-M. Chaw, and R. L. Chapman. 1999b. The phylogeny of land plants inferred from 18S rDNA sequences: pushing the limits of rDNA signal? Mol. Biol. Evol. 16:1774–1784.

Sun, G., D. L. Dilcher, S. Zheng, and Z. Zhou. 1998. In search of the first flower: a Jurassic angiosperm, *Archaefructus*, from northeast China. Science 282:1692–1695.

Sun, G., Q. Ji, D. L. Dilcher, S. Zheng, K. C. Nixon, and X. Wang. 2002. Archaefructaceae, a new basal angiosperm family. Science 296:899–904.

Takhtajan, A. 1987. System of Magnoliophyta. Academy of Sciences, Leningrad.

Takhtajan, A. 1991. Evolutionary trends in flowering plants. Columbia University Press, New York.

Takhtajan, A. 1997. Diversity and classification of flowering plants. Columbia University Press, New York.

Theissen, G. 2001. Development of floral organ identity: stories from the MADS house. Curr. Opin. Plant Biol. 4:75–85.

Theissen, G., and H. Saedler. 2001. Floral quartets. Nature 409:469–471.

Thorne, R. F. 1992. Classification and geography of the flowering plants. Bot. Rev. 58:225–348.

Tzeng, T. Y., and C. H. Yang. 2001. A MADS box gene from lily (*Lilium longiflorum*) is sufficient to generate dominant negative mutation by interacting with PISTILLATA (PI) in *Arabidopsis thaliana*. Plant Cell Physiol. 42:1156–1168.

van Tunen, A. J., W. Eikelboom, and G. C. Angenent. 1993. Floral organogenesis in *Tulipa*. Flower. News Lett. 16:33–38.

Walker, J. W., and A. G. Walker. 1984. Ultrastructure of Lower Cretaceous angiosperm pollen and the origin and early evolution of flowering plants. Ann. Mo. Bot. Gard. 71:464–521.

Wikström, N., V. Savolainen, and M. W. Chase. 2001. Evolution of the angiosperms: calibrating the family tree. Proc. R. Soc. Lond. B 268:2211–2220.

Williams, J. H., and W. E. Friedman. 2002. Identification of diploid endosperm in an early angiosperm lineage. Nature 415:522–526.

Winter, K.-U., A. Becker, T. Munster, J. T. Kim, H. Saedler, and G. Theissen. 1999. The MADS-box genes reveal that gnetophytes are more closely related to conifers than to flowering plants. Proc. Natl. Acad. Sci. USA 96:7342–7347.

Wolfe, K. H., M. Gouy, Y.-W. Yang, P. M. Sharp, and W.-H. Li. 1989. Date of the monocot-dicot divergence estimated from chloroplast DNA sequence data. Proc. Natl. Acad. Sci. USA 86:6201–6205.

Xiang, Q.-Y., D. E. Soltis, and P. S. Soltis. 1998. Phylogenetic relationships of Cornaceae and close relatives inferred from *matK* and *rbcL* sequences. Am. J. Bot. 85:285–297.

Zanis, M. J., D. E. Soltis, P. S. Soltis, S. Mathews, and M. J. Donoghue. 2002. The root of the angiosperms revisited. Proc. Natl. Acad. Sci. USA 99:6848–6853.

IV

The Relationships of Fungi

John W. Taylor

Joseph Spatafora

Kerry O'Donnell

François Lutzoni

Timothy James

David S. Hibbett

David Geiser

Thomas D. Bruns

Meredith Blackwell

The Fungi

The fungi contain possibly as many as 1.5 million species (Hawksworth 1991, 2001), ranging from organisms that are microscopic and unicellular to multicellular colonies that can be as large as the largest animals and plants (Alexopoulos et al. 1996). Phylogenetic analyses of nuclear small subunit (nSSU) ribosomal DNA (rDNA) put fungi and animals as sister clades that diverged 0.9 to 1.6 billion years ago (Wainright et al. 1993, Berbee and Taylor 2001, Heckman et al. 2001). The grouping of fungi and animals as sister taxa is controversial, with some protein-coding genes supporting the association and others not (Wang et al. 1999, Loytynoja and Milinkovitch 2001, Lang et al. 2002). Assuming that fungi and animals are sister taxa, a comparison of basal fungi (Chytridiomycota) with basal animals and associated groups (e.g., choanoflagellates and mesomycetozoa) should shed light on the nature of the last common ancestor of animals and fungi (fig. 12.1). It must have been unicellular and motile, indicating that multicellularity evolved independently in the two clades, and again in the several differently pigmented plant clades (M. Medina, A. C. Collins, J. W. Taylor, J. W. Valentine, J. H. Lips, L. Amaral-Zettler, and M. L. Sogin, unpubl. obs.). Fungi, like animals, are heterotrophs but, unlike animals, fungi live in their food. They do so as unicellular yeasts or as thin, filamentous tubes, termed hyphae (hypha, singular), which absorb simple molecules and export hydrolytic enzymes to make more simple molecules out of complex polymers, such as carbohydrates, lipids, proteins, and nucleic acids. Fungi have been spectacularly successful in the full range of heterotrophic interactions—decomposition, symbiosis, and parasitism. Fungi are well known to decay food stored too long in the refrigerator, wood in homes that have leaky roofs, and even jet fuel in tanks where condensation has accumulated. In nature, apart from fire, almost all biological carbon is recycled by microbes. The hyphae of filamentous fungi do the hard work in cooler climes and wherever invasive action is needed, as in the decay of wood.

Fungi enter into many symbioses, three of the most widespread and enduring are with microbial algae and cyanobacteria as lichens, with plants as mycorrhizae, and again with plants as endophytes. These symbioses are anything but rare. Nearly one-fourth of all described fungi form lichens, and lichens are the last complex life forms seen as one travels to either geographic pole (Brodo et al. 2001). Almost all plant species form mycorrhizae, and there is good fossil and molecular phylogenetic evidence that the first land plants got there with fungi in their rhizomes (Smith and Read 1997). There probably is not a plant that lacks a fungal endophyte, and there is good evidence that the endophytes improve plant fitness by deterring insect and mammalian herbivores and affect plant community structure (Clay 2001). Fungi are not limited to symbioses with autotrophs. Symbioses with animals are also prevalent, with partners ranging from ants and other insects to the gut of many ruminate animals and other herbivores (Blackwell 2000). Many insects may have been able to occupy new habitats due to associations with gut yeasts that provide digestive enzymes (Suh et al. 2003).

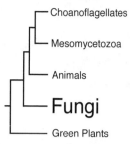

Figure 12.1. Phylogenetic tree showing relationships of the fungi, animals, and green plants based on nSSU rDNA.

Fungi also are well-known parasites. The stories of the spread of plant pathogens such as wheat rust, chestnut blight, and Dutch elm disease are biological and social tragedies, often initiated by intercontinental transport of pathogenic fungi (Agrios 1997). Fungi also plague humans, with athlete's foot and ringworm being the relatively benign end of a spectrum that ends in coccidioidomycosis, histoplasmosis, and other systemic and sometimes fatal diseases (Kwon-Chung and Bennett 1992). In the era of immune suppression, many yeasts and filamentous fungi, heretofore considered not to be serious human pathogens, have been found to cause grave systemic disease, among them *Aspergillus fumigatus* and *Candida albicans*. The close relationship of fungi and animals brings with it a similarity in metabolism that has made it difficult to find pharmaceuticals that attack the fungus and not the host.

Fungi have life histories that are far more interesting than those of most animals. Typically, fungi can mate and use meiosis to make progeny that have recombined genotypes, and they also can reproduce clonally via mitosis to make progeny with identical genotypes (Alexopoulos et al. 1996). Reproduction involves spore formation, with both mitotic and meiotic spores often facilitating long-distance transport and resistance to adverse environmental conditions. Huge numbers of spores can be produced, with the record annual spore release of several trillion being held by giant puffballs and the large fruiting bodies of wood-rotting Basidiomycota. Reproduction often is triggered by exhaustion of the food supply. Before mating, individuals find partners by chemical communication via pheromones, which range from complex organic compounds in Chytridiomycota and Zygomycota to oligopeptides in Ascomycota and Basidiomycota. Spores germinate to produce hyphae or germinate by budding to produce yeasts; in both cases the cell wall is composed of glucose polymers, the best known being chitin, a polymer of N-acetylglucosamine. Most fungi are not self-motile, the exception being the Chytridiomycota, which produce unicellular zoospores that have one typical eukaryotic flagellum inserted posteriorly.

Humans have domesticated yeasts to make bread, beer, wine, and fermentations destined for distillation. They have done the same with a number of filamentous fungal species, with species of *Penicillium* being the best known because of their role in making the camembert and roquefort families of cheese, dry-cured sausage, and the life-saving antibiotic penicillin. Biologists also have exploited several fungi as model organisms for genetics, biochemistry, and molecular biology, among them, *Neurospora crassa*, *Saccharomyces cerevisiae*, and *Schizosaccharomyces pombe*—Nobel Prize winners all.

Within the monophyletic Fungi, four major groups generally are recognized: Chytridiomycota, Zygomycota, Basidiomycota, and Ascomycota (fig. 12.2). Analysis of nSSU rDNA shows the Ascomycota and Basidiomycota to be monophyletic, but the Zygomycota and Chytridiomycota are not easily made into monophyletic groups, and their monophyly, or lack thereof, is controversial (Nagahama et al. 1995). The earliest divergences within Fungi involve certain Chytridiomycota and Zygomycota. The hyphae of these fungi typically lack the regularly spaced, cross walls (septa) typical of Ascomycota and Basidiomycota. In Chytridiomycota and Zygomycota, haploid nuclei are brought together by mating and fuse without delay. One of the clades radiating among the Chytridiomycota and Zygomycota leads to the Glomales + Ascomycota + Basidiomycota clade. Again, the placement of the Glomales on this branch may be controversial (James et al. 2000). Together, the Ascomycota and Basidiomycota form an informal group, the dikaryomycetes, which have regularly spaced cross walls in their hyphae, oligopeptide mating pheromones, and, because of an extended period between mating and nuclear fusion, pairs of genetically dissimilar nuclei in mated hyphae (i.e., a dikaryon). In the following sections, each of these groups is discussed, beginning with the largest and most familiar ones: Ascomycota, Basidiomycota, Zygomycota, and Chytridiomycota. Mycologists study more organisms than are found in the monophyletic Fungi, but inclusion of these organisms is beyond the scope of this chapter; some are covered elsewhere in this volume and are treated in mycology textbooks (Alexopoulos et al. 1996). These "fungal" groups include the water molds (Oomycota, Straminipila), home of the infamous plant pathogen *Phytophthora infestans*, cause of late blight of potato; the cellular slime molds (Dictyosteliomycota), home of the model social microbe *Dictyostelium discoideum*; the plasmodial slime molds (Myxomycota), home of the cell biology model organism *Physarum polycephalum*; and a myriad of other myxomycetes having beautiful sporangia. Conversely, some organisms not presently classified as Fungi may belong there, especially the microsporidia, a group of obligate animal parasites that branch deeply on the eukaryote branch in rDNA trees, but close to, or within, the fungi in some protein gene trees (Keeling et al. 2000, Tanabe et al. 2002).

Ascomycota

The Acomycota, or sac fungi (Gr. *ascus*, sac; *mycetos*, fungi), are the largest of the four major groups of Fungi in terms of number of taxa. With approximately 45,000 sexual and

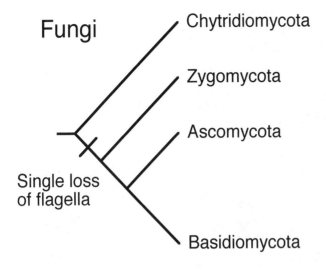

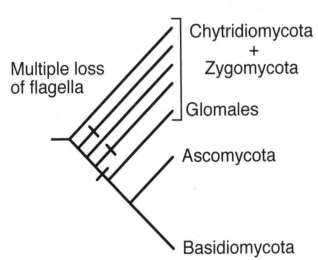

Figure 12.2. Alternative phylogenetic trees showing the relationships among the major groups of fungi. Each branch is monophyletic if flagella have been lost just once in the evolution of fungi, but both Zygomycota and Chytridiomycota are non-monophyletic if flagella have been lost independently.

asexual species, it accounts for about 65% of all described fungi (Hawksworth et al. 1995, Kirk et al. 2001). This group is characterized by the production of meiospores (ascospores) within sac-shaped cells (asci). It includes more than 98% of the fungi that combine with green algae or cyanobacteria or both to form lichens, as well as the majority of fungi that lack morphological evidence of sexual reproduction (mitosporic fungi). Ascomycota include many well-known fungi that have transformed civilization through food and medicine and that serve as model organisms through which major advancements in science have been made (Taylor et al. 1993). Some examples of these fungi include *Saccharomyces cerevisiae* (the yeast of commerce and foundation of the baking and brewing industries, not to mention molecular genetics), *Penicillium chrysogenum* (producer of the antibiotic penicillin), *Tolypocladium inflatum* (producer of the immunosuppressant

drug cyclosporin A, which revolutionized the field of organ transplantation), *Morchella esculenta* (the edible morel), and *Neurospora crassa* (the "one-gene-one-enzyme" organism). There are also many notorious members of Ascomycota that cause disease in humans and in many ecologically and economically important organisms. Some of these examples include *Aspergillus flavus* (producer of aflatoxin, the fungal contaminant of nuts and stored grain that is both a toxin and the most potent known natural carcinogen), *Candida albicans* (cause of thrush, diaper rash, and vaginitis), *Pneumocystis carinii* (cause of a pneumonia in people with compromised immune systems), *Magnaporthe grisea* (cause of rice blast disease), and *Cryphonectria parasitica* (responsible for the demise of 4 billion chestnut trees in the eastern United States; Alexopoulos et al. 1996).

Characteristics

The shared derived character state that defines members of the Ascomycota is the ascus (fig. 12.3). It is within the ascus that nuclear fusion (karyogamy) and meiosis ultimately take place. In the ascus, one round of mitosis typically follows meiosis to produce eight nuclei, and eventually eight ascospores; however, numerous exceptions exist that result in asci containing from one to more than 100 ascospores, depending on the species. Ascospores are formed within the ascus by the enveloping membrane system, a second shared derived character unique to Ascomycota. This double membrane system packages each nucleus with its adjacent cytoplasm and organelles and provides the site for ascospore wall formation. These membranes apparently are derived from the ascus plasma membrane in the majority of filamentous species, and the nuclear membrane in the majority of "true yeasts," and are assumed to be homologous (Wu and Kimbrough 1992, Raju 1992).

Within Ascomycota, two major growth forms exist. Species that form a mycelium consist of filamentous, often branching, hyphae. Hyphae exhibit apical growth and in Ascomycota are compartmentalized by evenly spaced septations that originate by centripetal growth from the cell wall. These septations are relatively simple in morphology and possess a single pore through which cytoplasmic connectivity may exist between hyphal compartments. Numerous examples exist, however, in which the pores become plugged, preventing or at least regulating movement between adjacent hyphal compartments. Hyphae also are the basic "cellular" building blocks for the different types of fungal tissues (e.g., the meiosporangia or fruiting bodies termed ascomata). The second major type of growth form found within Ascomycota is the yeast, a single-celled growth form that multiplies most commonly by budding. Both yeasts and hyphae have cell walls made of varying proportions of chitin and β-glucans (Wessels 1994). It is important to note that neither the hyphal (filamentous) morphology nor the yeast morphology is indicative of phylogenetic relationships. In fact, many spe-

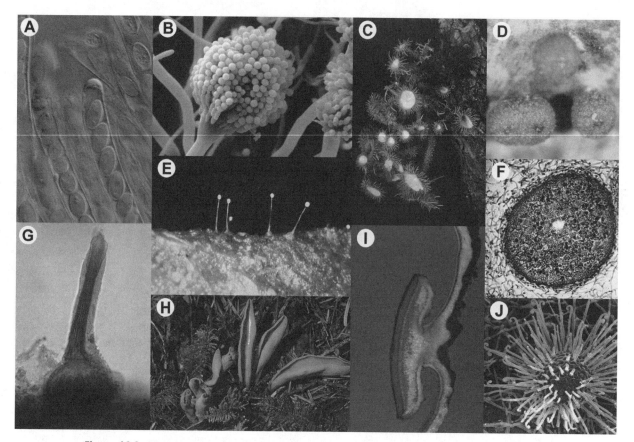

Figure 12.3. Macroscopic and microscopic images of meiotic and mitotic stages of Ascomycota. (A) Young asci and ascospores of *Otidea* (courtesy of J. W. Spatafora). (B) Scanning electron micrograph of conidia and conidiophores of *Aspergillus* (courtesy of C. W. Mims). (C) Lichen thallus of *Usnea* showing apothecia (courtesy of S. Sharnoff). (D) Perithecia of *Nectria* (courtesy of J. W. Spatafora). (E) Dungscape showing perithecial necks of *Sphaeronaemella fimicola* emerging from dung substrate (courtesy of D. Malloch and M. Blackwell). (F) Cross section of cleistothecium of *Talaromyces* with asci dispersed throughout central cavity of cleistothecium (courtesy of T. Volk). (G) *Kathistes calyculata* perithecium with basal asci and terminal, incurved setae (courtesy of D. Malloch and M. Blackwell). (H) Ear-shaped apothecia of *Otidea* (courtesy of W. Colgan III). (I) Cross section of *Lobaria* thallus showing arrangement of green algal layer (courtesy of S. Sharnoff). (J) Scanning electron micrograph of cleistothecium of *Uncinula* with hooked appendages (courtesy of C. W. Mims).

cies of Ascomycota are dimorphic, producing both hyphal and yeast stages at certain points in their life cycle. Regardless of the growth form, all members of Ascomycota are eukaryotes, typically possessing a single haploid nucleus, or several identical haploid nuclei, per hyphal compartment or yeast cell, although examples exist of diploid species of Ascomycota (e.g., *Candida albicans*) or species possessing long-lived diploid stages (e.g., *Saccharomyces cerevisiae*).

Reproduction and Life Cycle

Like much of life apart from the vertebrates, fungi have more than one reproductive option, a phenomenon termed pleomorphy (Sugiyama 1987). This phenomenon is arguably most pronounced among members of Ascomycota. The textbook Ascomycota example can make spores sexually (ascospores or meiospores) and asexually (conidia or mitospores;

fig. 12.4), although many species are known to reproduce only by ascospores, and many more are known to reproduce only by conidia. After meiosis, the ascospores take shape inside the ascus with new cell walls synthesized de novo in association with the aforementioned enveloping membrane system. Conidia contain mitotic nuclei, and their cell wall is a modification or extension of a preexisting hyphal or yeast wall. In hyphal Ascomycota, conidia may be produced by specialized hyphae that range from structures scarcely differentiated from vegetative mycelium (*Geotrichum candidum*) to hyphae consisting of elaborate heads of ornamented condida (*Aspergillus niger*; Cole and Kendrick 1981). Classification of Ascomycota is based on characteristics of sexual reproduction (i.e., ascomata and asci), and for this reason species that reproduce only asexually have been problematic in their integration into the classification of Ascomycota. In older systems of classification, all asexual members of

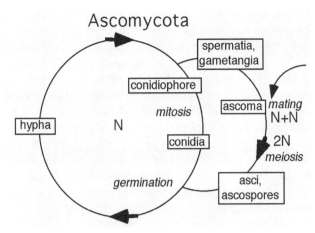

Figure 12.4. Generalized Ascomycota life cycle. The thallus (body) typically is hyphal and haploid. Vegetative hyphae can differentiate into reproductive structures for clonal (conidiophores, conidia) or sexual reproduction (spermatia, gametangia) or both. Sexual reproduction involves mating to produce, in a limited set of hyphae, a short-lived dikaryotic phase (N+N). Typically, the dikaryon is surrounded by a developing haploid ascoma. Karyogamy produces a zygote and is followed immediately by meiosis to produce ascospores. Both ascospores and conidia germinate to produce haploid hyphae.

Ascomycota were placed in the admittedly artificial Deuteromycota. This classification scheme has since been abandoned, and with the advent of molecular phylogenetics, sexual and asexual taxa can be integrated into a common system of classification based on comparison of gene sequences that are ubiquitously distributed across their genomes (Taylor 1995).

Ascospores and conidia are propagules whose main functions are dispersal to and colonization of appropriate substrates or hosts. Ascospores may or may not be forcibly ejected from an ascus. With forcible ejection, turgor pressure builds within the ascus, resulting in the eventual violent eruption of the ascospores from the ascus. In these systems, wind is the primary dispersal agent. Other members of Ascomycota do not forcibly eject their ascospores. In these systems the ascus wall breaks down, passively releasing the ascospores into the environment. This latter mechanism is especially common among Ascomycota that rely on arthropods and water to disperse their ascospores (Ingold 1965). In an analogous manner, conidia also may be produced in a relatively dry mass and be dispersed by wind, or may be produced in wet or sticky heads and be dispersed by water or arthropods (fig. 12.3). In most species, both ascospores and conidia are capable of germination, restoring the dominant haploid mycelial stage (fig. 12.4).

Species of Ascomycota may be either self-fertile (homothallic) or self-sterile (heterothallic), with the latter form requiring a separate and mating-compatible partner for sexual reproduction. Genetic regulation of sex expression and mating is well understood in several model members of Ascomycota, such as budding yeast (*Saccharomyces cerevisiae*), fission yeast (*Schizosaccharomyces pombe*), and *Neurospora crassa*; there are two sexes, and mating is coordinated by the aforementioned oligopeptide pheromones (Marsh 1991, Glass and Lorimer 1991). In yeast species, individual yeast cells function as gametangia and fuse to form the zygote, which eventually becomes the ascus after karyogamy and meiosis. In hyphal species, female gametangia (ascogonia) are produced and are fertilized either by male gametangia (antheridia) or by minute conidia that function as spermatia. In this latter example, cytoplasmic fusion (plasmogamy) may not be immediately followed by karyogamy, leading to a short phase where two genetically different nuclei occupy the same hyphal segment, as mentioned in the introductory remarks. These dikaryotic hyphae may be protected and nourished by differentiated haploid hyphae, which form a fruiting body (the ascoma; plural, ascomata; fig. 12.3). It is within the ascomata that asci eventually are produced from the dikaryotic hyphae originating from sexual reproduction. Asci exhibit a range of morphologies across Ascomycota with unitunicate asci possessing a single functional wall layer and bitunicate asci possessing two functional wall layers that operate much like a "jack-in-the-box" (Luttrell 1951, 1955). Unitunicate asci may be operculate and possess an apical lid (operculum) through which ascospores are released, or they may be inoperculate and release their ascospores through an apical pore or slit. As discussed below, ascus morphology does correlate with phylogeny. Ascospores are released from the asci as described above and germinate to form a new haploid mycelium, which will go on to produce hyphae, conidia, and ascospores that are characteristic of the species.

Nutrition, Symbioses, and Distribution

Like other fungi, members of Ascomycota are heterotrophs and obtain nutrients from dead (saprotrophism) or living (ranging from mutualism through parasitism) organisms (Griffin 1994, Carroll and Wicklow 1992). If water is present, as saprotrophs they can consume almost any carbonaceous substrate, including jet fuel (*Amorphotheca resinae*) and wall paint (*Aureobasidium pullulans*), and play their biggest role in recycling dead plant material. As symbionts, they may form obligate mutualistic associations with photoautotrophs such as algae and cyanobacteria (lichens; Brodo et al. 2001, Lutzoni et al. 2001, Nash 1996; fig. 12.3), plant roots (mycorrhizae; Varma and Hock 1999), and the leaves and stems of plants (endophytes; Arnold et al. 2001, Carroll 1988, 1995). Other Ascomycota form symbiotic associations with an array of arthropods, where they can line beetle galleries and provide nutrition for the developing larvae (*Ceratocystis* and *Ophiostoma*) or inhabit the gut of insects to participate in sterol and nitrogen metabolism (*Symbiotaphrina* and other yeasts and yeastlike symbionts). In return, the insects maintain pure cultures of the fungi and provide for their transport (Benjamin et al. in press, Currie et al. 2003). As para-sites and pathogens, ascomycetes account for most of the animal and plant pathogenic fungi, including those mentioned in the introduction to the Ascomycota section and many others, such as *Ophiostoma ulmi*, the Dutch elm disease fungus that is responsible for the demise of elm trees in North America and Europe (Agrios 1997). Numerous species are known from marine and aquatic ecosystems, where they are most frequently encountered on plant debris but may also be parasites of algae and other marine organisms (Kohlmeyer and Kohlmeyer 1979, Spatafora et al. 1998).

Ascomycota can be found on all continents and many genera and species display a cosmopolitan distribution (*Candida albicans* or *Aspergillus flavus*). Others are found on more than one continent (*Ophiostoma ulmi* or *Cryphonectria parasitica*), but many are known from only one narrowly restricted location. For example, the white piedmont truffle (*Tuber magnatum*) is known from only one province of northern Italy.

Relationships of Ascomycota to Other Fungi

The Ascomycota are the sister group to Basidiomycota. This relationship is supported by the aforementioned presence in members of both groups of regularly septate hyphae, and pairs of unfused haploid nuclei present in some stage of the thallus after mating and before nuclear fusion (dikaryons). Further support comes from the apparent homology between structures that coordinate simultaneous mitosis of dikaryotic nuclei (Ascomycota croziers and Basidiomycota clamp connections). Finally, numerous molecular phylogenetic studies all support the hypothesis that Ascomycota and Basidiomycota share a more recent common ancestor with one another than with any other major group (e.g., Zygomycota, Chytridiomycota) in Fungi (e.g., Bruns et al. 1992, Berbee and Taylor 1993, Tehler et al. 2000).

Phylogenetic Relationships within Ascomycota

Comparison of the genes that encode for the nuclear ribosomal RNAs (rRNAs) and the gene family of RNA polymerase, especially RNA polymerase II subunit B, supports a monophyletic Ascomycota that possesses three major subgroups (fig. 12.5; Berbee and Taylor 1993, Bruns et al. 1992, Spatafora 1995, Liu et al. 1999, Lutzoni et al. 2001). In the most recent classification (Eriksson et al. 2003), the three groups are designated subphylum Taphrinomycotina (= class Archiascomycetes), subphylum Saccharomycotina (= class Hemiascomycetes), and subphylum Pezizomycotina (= class Euascomycetes).

Taphrinomycotina are a group recently discovered from comparison of nucleic acid sequences and contains several species previously thought to be Saccharomycotina (Nishida and Sugiyama 1994). Some species, such as the fission yeast, *Schizosaccharomyces pombe*, are unicellular, but others grow as hyphae as well as single cells (e.g., *Taphrina* species).

Members of Taphrinomycotina do not produce ascomata with the exception of the genus *Neolecta*. *Neolecta* produces stipitate, club-shaped ascomata and once was classified among the Pezizomycotina. Recent molecular phylogenetic studies of independent gene data sets do not support the placement of *Neolecta* within the Pezizomycotina (Landvik 1996, Landvik et al. 2001). Rather all are consistent with its placement in the Taphrinomycotina, suggesting that the ability to form ascomata arose early in the evolution of Ascomycota. Monophyly of the Taphrinomycotina is not strongly supported by current analyses, however, and it is possible that the genera in question arose independently, possibly during the early radiation of Ascomycota.

Saccharomycotina consist of organisms most biologists recognize as yeasts or "true yeasts" and is home to one of the best-known species of fungi, *Saccharomyces cerevisiae*, better known as the baker's yeast. Although most Saccharomycotina are primarily unicellular, numerous species do make abundant hyphae, but none produce ascoma (Barnett et al. 1990). Phylogenies within the Saccharomycotina are among the most developed in the fungi because the taxon sampling is very dense (Kurtzman and Robnett 2003).

Pezizomycotina contain well more than 90% of the members of Ascomycota. Most species exhibit a dominant hyphal growth form, with almost all of the sexually reproducing forms possessing ascomata. Members of Pezizomycotina fall into two major categories: ascohymenial, which form after the initial sexual fertilization event, and ascolocular, which form before the initial sexual fertilization event. Ascohymenial ascomata may be closed (cleistothecium), open by a narrow orifice (perithecium), or broadly open like a cup (apothecium; see fig. 12.3). They may be less than a millimeter in diameter in the case of perithecia and cleistothecia, or up to 10 cm in diameter in the case of some apothecia. The common names often used to denote groups possessing ascohymenial ascomata include "plectomycetes" for the cleistothecial species, "pyrenomycetes" for the "perithecial" species, and "discomycetes" for the apothecial species. The ascolocular ascomata are referred to as ascostromata, and the common name given to these fungi is the "loculoascomycetes." Most current phylogenetic hypotheses propose that the apothecium (discomycetes in fig. 12.5) is the most primitive ascomatal morphology within the Pezizomycotina (Gernandt et al. 2001, Eriksson et al. 2003) and that the remaining ascomatal morphologies are more derived, in some cases through numerous independent events of convergent and parallel evolution (fig. 12.5, Berbee and Taylor 1992, Spatafora and Blackwell 1994, Suh and Blackwell 1999, Lutzoni et al. 2001). Pezizomycotina contain species of all ecologies, including plant pathogens (e.g., *Pyrenophora tritici-repentis*), animal pathogens (e.g., *Cordyceps militaris*), mycorrhizae (e.g., *Tuber melanosporum*), endophytes (e.g., *Rhytisma acerinum*), and innumerable plant decay fungi. Importantly, Pezizomycotina include more than 98% of fungi that are lichenized. Lichenized fungi are an amazingly suc-

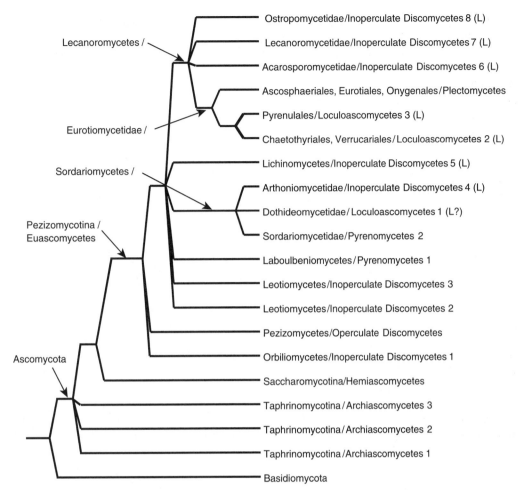

Figure 12.5. Depiction of the current understanding of relationships among members of Ascomycota, sister group to the Basidiomycota (adapted from Suh and Blackwell 1999, Bhattacharya et al. 2000, Platt and Spatafora 2000, Gernandt et al. 2001, Kirk et al. 2001, Lutzoni et al. 2001, McLaughlin et al. 2001, Kauff and Lutzoni 2002). Higher taxa of Eriksson et al. (2003) and "common names" are shown on the tree before and after "/," respectively. Taxa listed at the tips of terminal branches that include lichen-forming species are denoted "(L)." Note the phylogenetic uncertainty among several groups, including Taphrinomycotina (= Archiascomycetes) and within the Pezizomycotina (= Euascomycetes). Common groups such as the "inoperculate discomycetes" (e.g., Orbiliomycetes, Leotiomycetes, Lecanoromycetidae, and Ostropomycetidae) and "loculoascomycetes" (e.g., Chaetothyriales, Dothideomycetidae, Verrucariales, and Pyrenulales) do not denote monophyletic groupings. Most cleistothecial fungi ("plectomycetes") occur in a monophyletic group (Ascospheriales, Eurotiales, Onygenales; Geiser and LoBuglio 2001), whereas others are derived members of other groups such as the Sordariomycetes ("pyrenomycetes"). The vast majority of "pyrenomycetes" are members of Sordariomycetes, with a few unique and poorly known perithecial species among Laboulbeniomycetes (Weir and Blackwell 2001). The Lecanoromycetes, a recently established group of mostly lichen-forming species, include four major subgroups of Ascomycota: Acarosporomycetidae, Eurotiomycetidae, Lecanoromycetidae, and Ostropomycetidae.

cessful group, accounting for approximately 42% of all described species of Ascomycota and probably close to 50% of the known members of Pezizomycotina. Lichens are ecologically important organisms that cover as much as 8% of Earth's land surface, serve as important food sources for animals in harsh arctic environments, and function as pollution indicators in industrialized parts of the world. Lichens were widely believed to have arisen independently multiple times, accounting for the high diversity and mixed occurrence of lichenized and nonlichenized fungal species within Ascomycota (Gargas et al. 1995). A recent comparative phylogenetic study reported that lichens may have evolved earlier than previously believed within Pezizomycotina, and that independent gains of lichenization have occurred one to three times during Ascomycota evolution but have been followed by multiple independent losses of the lichen symbiosis

(Lutzoni et al. 2001). As a consequence, major Ascomycota groups of exclusively non-lichen-forming species, which include the medically important species *Exophiala* and *Penicillium* (e.g., Chaetothyriales and Plectomycetes), would have been derived from lichen-forming ancestors (fig. 12.5).

Although most of the recent molecular phylogenetic efforts have been directed at the Pezizomycotina, interrelationships of the major groups within Pezizomycotina are still poorly understood and not confidently resolved by phylogenetic analyses of the current data. Figure 12.5 presents the most current understanding of the relationships of the major groups within the Pezizomycotina; detailed discussion is available in Alexopoulos et al. (1996), Holst-Jensen et al. (1997), Berbee (1998), Liu et al. (1999), Eriksson et al. (2003), Gernandt et al. (2001), Lutzoni et al. (2001), and Miadlikowska and Lutzoni (in press), to name a few.

Basidiomycota

The Basidiomycota (Gr. *basidion*, small base or pedestal; *mykes*, fungi) contain roughly 22,000 described species, which is approximately 35% of the known species of fungi (Hawksworth et al. 1995, Kirk et al. 2001). Basidiomycetes include some of the most familiar and conspicuous of all fungi, namely, mushrooms and polypores, as well as yeasts (single-celled forms) and other relatively obscure taxa. Some basidiomycetes are economically important edible species, including button mushrooms (*Agaricus bisporus*), shiitake mushrooms (*Lentinula edodes*), and chanterelles (*Cantharellus cibarius*), whereas others are deadly poisonous (e.g., *Amanita phalloides*) or hallucinogenic (*Psilocybe* spp.). The latter play important roles in traditional shamanic cultures of Central America (Wasson 1980).

The overwhelming majority of basidiomycetes are terrestrial, but some species can be found in marine or freshwater habitats, including many basidiomycete yeasts (Fell et al. 2001). Some basidiomycetes have free-living, saprotrophic (decomposer) lifestyles, whereas others live in symbiotic associations with plants, animals, and other fungi. The oldest fossils of the group are hyphae with diagnostic clamp connections from the Pennsylvanian period [~290 million years ago (Mya)], but recent molecular clock estimates suggest that the common ancestor of all modern basidiomycetes lived at least 500 Mya, and maybe 1.0 billion years ago (Dennis 1970, Berbee and Taylor 2001, Heckman et al. 2001).

Tremendous progress has been made in basidiomycete phylogenetics through the use of molecular characters. Three major groups are now recognized, the Urediniomycetes, Ustilaginomycetes, and Hymenomycetes (Swann and Taylor 1995), and the major clades within these groups largely have been delimited (fig. 12.6). Nevertheless, many aspects of the relationships within and among the major groups remain poorly understood.

Characteristics and Life History

The dominant phase of the life cycle in most basidiomycetes is a heterokaryotic mycelium, which is a network of hyphae, in which each cell contains two different types of haploid nuclei resulting from the mating of two monokaryotic (haploid, uninucleate) mycelia (fig. 12.7). Historically, it has been very difficult to determine the longevity and spatial distribution of mycelia, but recently molecular markers have been used to study this phase of the life cycle—with astonishing results. In the "honey mushroom," *Armillaria* (Hymenomycetes), mycelia have been discovered that inhabit continuous patches of forest of many acres. One giant *Armillaria* mycelium in a Michigan forest was estimated to be about 1500 years old, with a mass of around 10,000 kg (Smith et al. 1992). *Armillaria* is a wood-decaying timber pathogen that forages along the forest floor using rootlike rhizomorphs. Most other basidiomycetes, especially those that colonize patchy, ephemeral resources (e.g., dung) or that lack rhizomorphs, probably have much more limited mycelia.

Sexually reproducing basidiomycetes produce cells called basidia (from which the group derives its name), in which the two haploid nuclei fuse, immediately undergo meiosis, and give rise to haploid spores (fig. 12.7). Thus, there is usually only a single diploid cell in the entire life cycle. In most species, the spores are discharged from the basidia by a forcible mechanism termed ballistospory that is unique to basidiomycetes. Ballistospory has been secondarily lost in puffballs and their relatives (which produce spores within enclosed fruit bodies), as well as in aquatic species and in most smut fungi. Basidia often are produced in elaborate, multicellular fruiting bodies (the basidioma; plural, basidiomata), although some species produce basidia directly from single-celled yeasts. Fruiting bodies are the most visible stage of the life cycle and encompass an amazing diversity of forms, including mushrooms, puffballs, bracket fungi, false truffles, jelly fungi, and others.

Numerous variations on the basic life cycle described above have evolved in basidiomycetes. In many groups, asexual spores are produced, from either monokaryotic or heterokaryotic hyphae, and some basidiomycetes have no known sexual stage at all (fig. 12.7). Some basidiomycetes are heteromorphic, alternating between a yeast phase and a filamentous phase. The most complex life cycles in basidiomycetes are those of the plant pathogens called rusts (Urediniomycetes), which have multiple spore-producing stages that may be formed on two, unrelated plant hosts.

Ecological Importance

Basidiomycetes play diverse ecological roles, but the decay of wood and other plant tissues may be the single most important process performed by the group. Although other fungi, particularly certain groups in the ascomycetes, can digest cellulose and lignin (the major components of plant

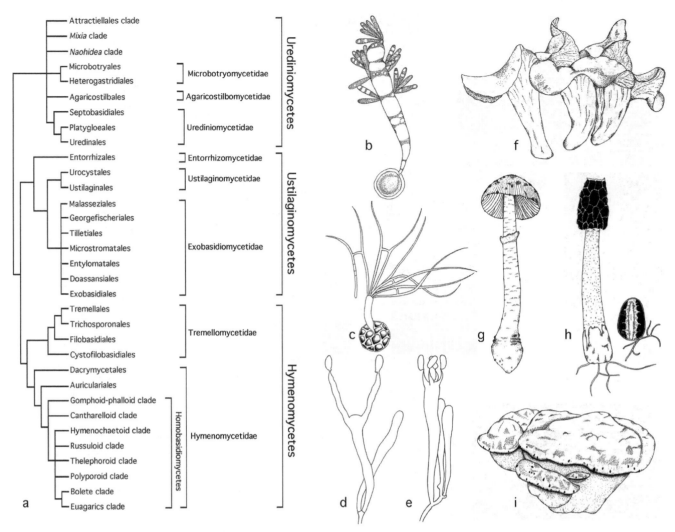

Figure 12.6. Phylogenetic relationships, basidia, and fruiting bodies of basidiomycetes. (A) Phylogenetic relationships of basidiomycetes, based on trees and classifications published by Swann et al. (2001: fig. 1); Swann and Taylor (1995: figs. 1–2); Bauer et al. (2001: figs. 33, 34); Hibbett and Thorn (2001: figs. 1F, 2); Fell et al. (2001: fig. 19B); and Wells and Bandoni (2001). Several minor clades of uncertain placement are not shown. (B–E). Diversity of basidia. (B) *Leucosporidium fellii* (Urediniomycetes; after Fell et al. 2001: fig. 3). (C) *Tilletia caries* (Ustilaginomycetes; after Oberwinkler 1977: fig. 24). (D) *Dacrymyces stillatus* (Hymenomycetes; after Wells and Bandoni 2001: fig. 13). (E) *Cantharellus cibarius* (Hymenomycetes; after Oberwinkler 1977: fig. 28). (F–I) Diversity of fruiting bodies in the Hymenomycetes. (F) *Phlogiotis helvelloides*. (G) *Amanita* species. (H) *Phallus* species (primordium on right). (I) *Inonotus dryadeus*. Drawings by Zheng Wang.

cell walls), this ability is best developed in the Hymenomycetes (Rayner and Boddy 1988, Hibbett and Thorn 2001, Hibbett and Donoghue 2001). With few exceptions, the major timber pathogens and saprotrophic wood decayers are basidiomycetes—this role makes their impact on forest systems substantial from both ecological and management perspectives (Edmonds et al. 2000, Rayner and Boddy 1988). Basidiomycetes use a diverse array of enzymes to digest wood and plant debris in leaf litter and soil (Cullen 1997, Reid 1995). Because of their enzymatic capabilities, basidiomycetes have come under scrutiny for possible applications in

bioremediation and biopulping (involved in paper production). A recent project to sequence the genome of the wood-decaying basidiomycete *Phanerochaete chrysosporium* (Hymenomycetes) was motivated, in part, by the potential of its enzymes for degrading recalcitrant substrates.

Ectomycorrhizal symbiosis (an association involving fungal hyphae and the roots of trees) is another major role that is well developed within the basidiomycetes. Ectomycorrhizal basidiomycetes have been shown to scavenge mineral nutrients directly from organic matter, thereby providing their host trees exclusive access to nutrient pools

Basidiomycota

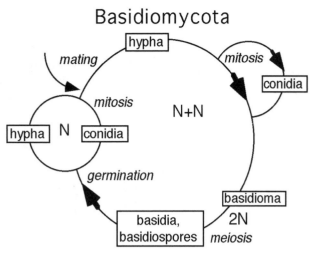

Figure 12.7. Basidiomycota life cycle. The haploid hyphal individual mates early in the life cycle and then persists as a dikaryon, so basidiomycetes found in nature are most often dikaryons. Both haploid and dikaryotic individuals are able to reproduce clonally via conidia in some species. Completion of the sexual cycle involves nuclear fusion in basidia, followed immediately by meiosis to produce basidiospores. Basidia and basidiospores in some groups are produced on basidioma made of dikaryotic hyphae, for example, mushrooms. Conidia and basidiospores germinate to produce hyphae.

that are unavailable to most plants (Haselwandter et al. 1990, Perez-Moreno and Read 2000). In return, ectomycorrhizal basidiomycetes receive sugars from their plant hosts. More than 6000 species of Hymenomycetes are known or suspected to be ectomycorrhizal, as well as a handful of ascomycetes and even zygomycetes (Molina and Trappe 1982, Smith and Read 1997). The plants that are involved in ectomycorrhizal symbioses include pines, oaks, poplars, chestnuts, birches, dipterocarps, eucalypts, and caesalpinoid legumes—that is, the dominant tree species in many temperate and some tropical forest ecosystems. There is strong evidence that ectomycorrhizal basidiomycetes have been derived multiple times from saprotrophic ancestors (Bruns et al. 1998, Gargas et al. 1995), and some analyses suggest that reversions to saprotrophy also have occurred (Hibbett et al. 2000).

Plant parasitism is phylogenetically the most widespread ecological niche within the basidiomycetes. The rusts (Urediniomycetes), with more than 7000 described species, are a particularly successful group. Wheat rusts, coffee rust, and fusiform and blister rust of pines are excellent examples of species that have a major economic impact on agriculture and forestry (Edmonds et al. 2000, Swann et al. 2001). Rusts use angiosperms, gymnosperms, lycopods, and pteridophytes as hosts, whereas closely related taxa parasitize mosses and scale insects. The smuts, which comprise a polyphyletic group composed of members of both the Ustilaginomycetes and Urediniomycetes (fig. 12.6), are important parasites that attack a huge diversity of angiosperms. *Ustilago* and *Tilletia* species (e.g., *U. hordei*, *U. tritici*, *U. maydis*, *T. caries*, and *T. controversa*) that occur on cereal crops cause large agricultural losses. In both the rusts and smuts there is widespread phylogenetic tracking of hosts, but jumps to unrelated hosts are well documented (Bauer et al. 2001, Sjamsuridzal et al. 1999, Vogler and Bruns 1998).

Saprotrophy, ectomycorrhizal symbiosis, and plant parasitism are by no means the only lifestyles represented in basidiomycetes. Basidiomycetes also parasitize other fungi and animals—an example is the human parasite *Filobasidiella neoformans*, causative agent of cryptococcosis. Basidiomycota form symbioses with insects, such as bark beetles and the leaf-cutter ants of the neotropics (Chapela et al. 1994). They also attack and digest bacteria and microscopic invertebrates, apparently as a means by which they acquire additional nitrogen (Barron 1988, Thorn and Barron 1984, Klironomos and Hart 2001). Basidiomycota also enter into lichenized symbioses with photosynthetic algae (Gargas et al. 1995, Lutzoni and Pagel 1997). These examples demonstrate some of the ecological diversity of basidiomycetes but hide the fact that we actually know very little about the basic ecology of the majority of species in this clade. For example, numerous basidiomycete yeasts can be isolated from soil and plant and animal substrates and grown on synthetic media, but little is known about how they function in nature (Fell et al. 2001). Even within the mushroom-forming basidiomycetes, our knowledge is limited usually to where they grow, if that, and the details about what they do and how they manage to successfully establish and compete often remain obscure.

Phylogeny

The traditional taxonomy of basidiomycetes was based largely on the morphology of fruiting bodies and basidia. Since the late 1980s, understanding of the phylogenetic relationships of basidiomycetes has been revolutionized through the use of molecular characters, especially sequences of ribosomal genes (rDNA). Three major clades are recognized now: Urediniomycetes, Ustilaginomycetes, and Hymenomycetes (fig. 12.6; Swann and Taylor 1995). The branching order among these three groups is not well resolved by rDNA data; however, this is one area where additional data from genome studies may help add resolution.

The Urediniomycetes consist of roughly 7400 (34%) of the described species of basidiomycetes (Swann et al. 2001, Hawksworth et al. 1995, Kirk et al. 2001). Members of Urediniomycetes include yeasts and filamentous forms, which function as saprotrophs and pathogens of plants, animals, and fungi. When they occur, fruiting bodies in this group usually are small and inconspicuous (Swann et al. 2001). Monophyly of Urediniomycetes appears to be supported by biochemical features of cell wall composition (cell wall sugars; Prillinger et al. 1993), ultrastructural aspects of the hyphal septa, and other characters that are visible only with transmission electron microscopy (Swann et al. 1999, 2001).

The Urediniomycetes are divided into six major clades (fig. 12.6). Relationships among the clades, however, are poorly resolved by rDNA data. By far the largest clade in Urediniomycetes is the Urediniomycetidae, which includes more than 7000 species, most of which are the plant pathogenic rusts (Uredinales). One intriguing member of Urediniomycetidae is *Septobasidium*, which parasitizes colonies of living scale insects as they feed on plant sap. Some groups now recognized as Urediniomycetes were formally classified among distantly related groups of fungi. For example, the Microbotryomycetidae include anther smuts that were formerly placed along with true smuts in Ustilaginomycetes (fig. 12.6). Similarly, *Mixia osmundae*, a fern parasite, was once thought to be a member of the ascomycetes, but rDNA data clearly place it in the Urediniomycetes (Nishida et al. 1995). Recognition of the monophyletic Urediniomycetes is a triumph of fungal molecular systematics. Nevertheless, the lack of resolution among the major clades remains a barrier to understanding pathways of morphological and ecological evolution in this group.

The Ustilaginomycetes contain about 1300 (6%) of the described species of basidiomycetes (Bauer et al. 2001, Hawksworth et al. 1995, Kirk et al. 2001) and includes plant parasites, which often are dimorphic with a saprotrophic yeast phase. Smuts of corn, barley, and wheat are economically important members of this group. Corn smut (*Ustilago maydis*) produces a large gall on maize ears that is eaten in the traditional cuisine of Mexico, as cuitlacoche. Monophyly of Ustilaginomycetes has received strong support in analyses of nSSU rDNA sequences (Swann and Taylor 1993) but only moderate support in more densely sampled studies of nuclear large subunit rDNA sequences (Begerow et al. 1997). The composition of cell wall sugars and ultrastructural aspects of host–fungus interaction provide additional characters that support monophyly of the Ustilaginomycetes (Bauer et al. 2001).

Three major clades have been recognized within Ustilaginomycetes: Entorrhizomycetidae, Ustilaginomycetidae, and Exobasidiomycetidae (fig. 12.6). The Exobasidiomycetidae are not strongly supported as monophyletic by rDNA data, however, and the branching order among the three clades is not well resolved. Bauer et al. (2001) have developed a detailed classification of Ustilaginomycetes (fig. 12.6) and have inferred patterns of evolution of morphological characters and host associations.

The Hymenomycetes include about 13,500 (60%) of the described species of basidiomycetes (Swann and Taylor 1993, Hawksworth et al. 1995, Kirk et al. 2001). A unifying character for this group is the production of a "dolipore" septum between cells. Typically, the dolipore septum is flanked by a membrane bound structure termed a parenthesome, the configuration of which is useful for delimiting major groups within Hymenomycetes. Diverse fruiting bodies are formed in Hymenomycetes, including some of the most complex forms that have evolved within the fungi.

The Hymenomycetes consist of seven main clades; six of them (Tremellales, Trichosporonales, Filobasidiales, Cystofilobasidiales, Dacrymycetales, and Auriculariales) include many members of the heterobasidiomycetes *sensu* Wells and Bandoni (2001), and the seventh (homobasidiomycetes) includes the better known mushrooms, shelf fungi, and puffballs (fig. 12.6). The heterobasidiomycetes encompass a tremendous range of morphologies, including yeasts and filamentous forms, and a wide range of ecological modes, including saprotrophs and parasites of fungi and animals. Fruiting bodies of heterobasidiomycetes are typically gelatinous and translucent, giving rise to the common name "jelly fungi." Familiar examples include "witches butter" (*Tremella mesenterica*) and the edible wood-ear (*Auricularia auricula-judae*), which is cultivated in Asia.

The homobasidiomycetes include more than 90% of the species in Hymenomycetes, suggesting that this group has undergone an increase in diversification rate relative to heterobasidiomycetes. Homobasidiomycetes include the mushroom-forming fungi, which display an incredible diversity of fruiting body forms. Yeast phases are generally absent from this group. Traditionally, taxonomy of homobasidiomycetes depended on morphological and anatomical characters of fruiting bodies. This group has been sampled intensively by fungal systematists (Bruns et al. 1998, Moncalvo et al. 2002, Hibbett et al. 2000). Although many aspects of morphology-based classifications have been upheld, there have also been major rearrangements, especially concerning the placement of the taxonomically enigmatic gasteromycetes, such as puffballs, false truffles, earthstars, and stinkhorns (Hibbett et al. 1997). Hibbett and Thorn (2001) proposed a classification of the homobasidiomycetes that includes eight major clades (fig. 12.6). Relationships among the clades are generally not well resolved, however, and recent analyses suggest that there are also some additional minor clades of homobasidiomycetes (Hibbett and Binder 2002).

Conclusions

Taxonomy of basidiomycetes has progressed dramatically in recent years, but significant questions remain. Relationships within and among major clades are often unresolved, which limits understanding of the pathways of evolution in basidiomycetes, and their role in the evolution of ecosystems. One major class of questions concerns the causes of the different patterns of apparent species richness observed from clade to clade. For example, why are homobasidiomycetes and rusts so diverse? The diversity seems too great simply to be due to the ease with which large mushrooms are recognized or to the intense economic interest in rusts. Did these two groups diversify in response to some environmental change, such as the rise of angiosperms, or are there intrinsic properties of these groups that contributed to their success?

Zygomycota

Species of the Zygomycota (Gr. *zygos*, marriage pairing; *mykes*, fungi) are remarkable for their morphological and ecological diversity (Hawksworth et al. 1995, Kirk et al. 2001), even though they account for fewer than 2% of all described fungal species. This group includes fast-growing molds responsible for storage rots of fruits, such as peaches and strawberries. Other species can cause life-threatening infections in humans and other animals, especially in immunocompromised or artificially immunosuppressed patients and diabetics (Rinaldi 1989). Most of the approximately 1000 described members of Zygomycota, however, are not encountered by humans and lack common names because of their microscopic size coupled with the fact that approximately half of the species cannot be cultured axenically. Economically and ecologically, the most important zygomycetes are represented by Glomales, whose members are all asexual, obligate symbionts of the great majority of vascular plants (Sanders 1999, Redecker et al. 2000b, Schüßler et al. 2001). This specialized fungus–plant root symbiosis (mycorrhizae; Gr. *mykes*, fungi; *rhiza*, root) functions as an auxiliary root system that is critical for ecosystem function and plant diversity. The mycorrhizal symbiosis is vital for phosphate uptake by plants, especially in nutrient-poor soils. In addition, such fungi are hypothesized to have been instrumental in the colonization of land by the first terrestrial plants (Pirozynski and Malloch 1975, Simon et al. 1993). Molecular clock estimates indicate that Glomales diverged after the divergences among zoosporic fungi (Chytridiomycota), at least 600 Mya and possibly as much as 1.2–1.4 billion years ago (Heckman et al. 2001, Berbee and Taylor 2001). Extant glomalean species are remarkably similar to fossils from the Ordovician period 460 Mya (Redecker et al. 2000a).

Beneficial species within Mucorales are used in the production of the traditional east Asian soybean-based fermented foods sufu (i.e., Chinese cheese) and tempeh. Another species within the Murorales, *Phycomyces blakesleeanus*, is used as a model system for understanding the genetics of phototropism and sensory transduction, in part because it responds to light over the same range as the human eye (Eslava and Alvarez 1996). Species within the Entomophthorales (Gr. *entoma*, insect; *phthora*, destroyer) have enormous potential as natural biological control agents of pest insects.

Characteristics and Life Cycle

Although there are relatively few species of Zygomycota, compared with Ascomycota and Basidiomycota, they exhibit a remarkable diversity of life history strategies and ecological specializations. Zygomycota species function as ecto- and endomycorrhizal symbionts of vascular plants, obligate mycoparasites, entomopathogens, endocommensials of aquatic arthropods, terrestrial saprobes, and endo- or ectoparasites of protozoa, nematodes, and other invertebrates (Benjamin

1979). A generalized life cycle is presented in figure 12.8. Hyphal thalli typically consist of branched or unbranched tubular filaments (fig. 12.9A) that either are predominately nonseptate (i.e., coenocytic: Mucorales, Entomophthorales, Glomales, and some Zoopagales and Endogonales) or are regularly septate (Kickxellales, Dimargaritales, Harpellales, and some Zoopagales). Where known, thalli have cell walls composed of chitin plus chitosan or chitin plus β-glucan (Bartnicki-Garcia 1987). Septa or cross walls are simple partitions in hyphae, except in the Harpellales, Kickxellales, and Dimargaritales, where they are flared with a plugged central pore. Species-specific differences in the mating system determine whether thalli are self-fertile (i.e., homothallic) or self-sterile (i.e., heterothallic, requiring the union of thalli of different mating types). Sexual reproduction, where known, involves the fusion of differentiated (fig. 12.9B) or undifferentiated hyphae followed by the development of a variously enlarged unicellular zygosporangium (fig. 12.9C–E), within which is formed a single zygospore. The zygospore is the only diploid stage in the life cycle and the site of meiosis. Relatively few studies have documented meiosis and zygospore germination, in part because these thick-walled spores require a dormancy period before they germinate to give rise to a haploid mycelium. Although this group derives its name from the sexual stage, phylogenetic studies are needed to assess whether the zygospore is synapomorphic for this group. Zygomycota also are united by the production of asexual nonflagellated mitospores in uni- to multispored sporangia (fig. 12.9F–O). Asexual spores also can be produced as intercalary or terminal modifications of the vegetative mycelium, or very rarely as a yeastlike phase. Mitospores are passively released, except in Entomophthorales, where they frequently are ejected forcibly (fig. 12.9k), and in the coprophilic mucoralean genus *Pilobolus* (Gr. *pileos*, hat; *bolus*, to throw), where the entire sporangium is discharged as far as 2 m toward light.

Although members of the largest order, Mucorales, comprise only one-third of all described Zygomycota taxa, they represent the overwhelming majority of zygomycetous species in axenic culture because they all grow saprobically (O'Donnell 1979). Representatives of the other seven orders account for less than half of all members of Zygomycota in culture, in part because they include obligate parasites (Dimigaritales, Zoopagales, and many Entomophthorales), obligate arthropodphilous symbionts (Harpellales), and ecto- and endomycorrhizal species (Endogonales and Glomales, respectively). Except for one mycoparasitic species, all Kickxellales species can be cultivated axenically. Mycoparasitic species of Dimargaritales and Zoopagales typically are cultured on their mucoralean hosts, but some of these species can be grown axenically on specialized media (Benjamin 1979). Specific culture collections have been established for Entomophthorales (Humber and Hansen 2003) and Harpellales (Lichtwardt et al. 2001). In addition, several phylogenetically diverse collections of the

Zygomycota

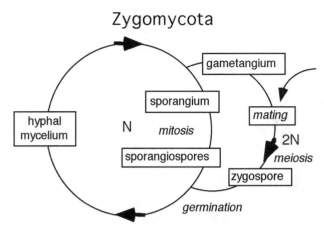

Figure 12.8. Generalized Zygomycota life cycle. Individuals in nature typically are hyphal and haploid. Vegetative hyphae can differentiate into reproductive structures for clonal (sporangia, sporangiospores) or sexual reproduction (gametangia). Sexual reproduction involves mating by gametangial fusion to produce a diploid zygote. In almost all cases, there is no fruiting body surrounding the zygospores. Both mature zygospores and conidia germinate to produce haploid hyphae. In the case of zygospores, the germinating hypha immediately differentiates to make a sporangium and sporangiospores.

obligately mycorrhizal Glomales are available (http://invam.caf.wvu.edu, http://res2.agr.ca/ecorc/ginco-can/ and http://www.ukc.ac.uk/bio/beg/). In these collections, Glomales species are maintained *in vivo* in host plants, stored as dried inoculum, or kept as cryogenically preserved material, or accessioned by all three methods.

Phylogenetic Relationships and Taxonomic Implications

Zygomycota appear to be non-monophyletic in most SSU rRNA and some β-tubulin gene analyses. However, the monophyly of this group has not been tested fully through analyses of the available molecular phylogenetic data. These analyses are based primarily on SSU rRNA (Bruns et al. 1992, Gehrig et al. 1996, James et al. 2000, Jensen et al. 1998, Nagahama et al. 1995, Schüßler et al. 2001, Tanabe et al. 2000), β-tubulin (Keeling et al. 2000) and several protein-coding genes within the mitochondrial genome (Forget et al. 2002, Lang 2001). Interestingly, Zygomycota may be monophyletic, if the putative long-branch taxon *Basidiobolus ranarum* (Entomophthorales), which clusters with Chytridiomycota in unconstrained SSU rRNA analyses, is excluded from the analysis [see James et al. (2000) for more information on *Basidiobolus*; see the section on Chytridiomycota below).

Relationships among orders of Zygomycota are poorly resolved by SSU rRNA phylogenies, except for a Harpellales + Kickxellales + *Spiromyces* clade (Gottlieb and Lichtwardt 2001, O'Donnell et al. 1998), with Zoopagales as a putative sister group (Tanabe et al. 2000). Overall, the available SSU data suggest that the orders as presently circumscribed, based

on morphological apomorphies, nutritional mode, and ecological specialization, are monophyletic except for Mortierellaceae, which may not form a monophyletic group with the Mucorales (Gehrig et al. 1996). Three orders of Zygomycota described recently (Cavalier-Smith 1998) are not accepted here, however, because Geosiphonales appears to be nested within Glomales, and too few data are available to assess the phylogenetic validity Mortierellales and Basidiobolales. Also, a new group, Glomeromycota, proposed to accommodate Glomales *sensu* Schwarzott et al. (2001), is based primarily on SSU rRNA data. It should be considered provisional until more robust molecular phylogenetic data become available.

Recent molecular phylogenies have advanced our knowledge of Zygomycota by providing novel hypotheses of evolutionary relationships within Glomales (Simon et al. 1993, Gehrig et al. 1996, Redecker et al. 2000b, Schüßler et al. 2001, Schwarzott et al. 2001), Harpellales and Kickxellales (Gottlieb and Lichtwardt 2001, O'Donnell et al. 1998), Entomophthorales (Jensen et al. 1998), Mucorales (O'Donnell et al. 2001), and Dimargaritales and Zoopagales (Tanabe et al. 2000). Two classes have been recognized in all recent taxonomic schemes for Zygomycota (Benny 2001, Benny et al. 2001): Trichomycetes (Gr. *thrix*, hair; *mykos*, fungi), represented by four arthropodophilous orders, Amoebidiales, Harpellales, Eccrinales and Ascellariales (Lichtwardt 1986); and Zygomycetes. However, polyphyletic Trichomycetes is not accepted here. Molecular phylogenetic analyses based on SSU rRNA indicate members of Amoebidiales are protists (Ustinova et al. 2000, Benny and O'Donnell 2000), as long suspected because their cell walls lack chitin and they produce amoeboid cells, which otherwise are unknown in Fungi (although some zoospores of Chytridiomycota can exhibit amoeboid movement). Phylogenetic evidence from SSU rRNA data also has identified Harpellales as a sister group to a *Spiromyces* + Kickxellales clade or to *Spiromyces* within Zygomycetes (Gottlieb and Lichtwardt 2001, James et al. 2000, O'Donnell et al. 1998). Lastly, Eccrinales and Asellariales are treated as *incertae sedis* until their phylogenetic relationships are resolved.

Chytridiomycota

Chytridiomycota are a relatively poorly known group at the base of the fungal tree, accounting for 1% or 2% of described fungal species. Chytridiomycetes, or chytrids, as they commonly are known, are microscopic and have a simple morphology. The distinguishing feature of the group is reproduction through a motile zoospore. The chytridiomycete zoospore typically possesses a single, smooth flagellum that is inserted on the cell posterior to the direction of motility. The chytridiomycetes have been variously classified through the years with other fungi and protists; as recently as 1990 Chytridiomycota were placed in Protoctista (Barr 1990). Because they produce zoospores, chytrids are generally thought to be aquatic fungi.

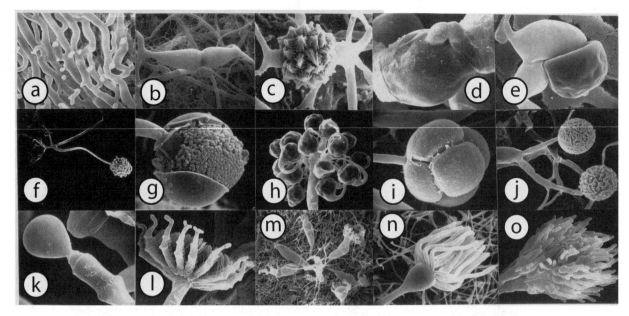

Figure 12.9. Scanning electron micrographs of Zygomycota. (A) Coenocytic mycelium with aerial hyphae beginning to form. (B–E) Sexual reproduction. (B) Gametangial fusion. (C–E) Zygosporangia. (F–O) Asexual reproduction. (F) Aerial, terminal multispored sporangium with basal rhizoids. (G) Multispored sporangium. (H and I) Few-spored sporangia. (J–L) Unispored sporangia. (M) Vesiculate mycoparasite growing on mucoraceous host. (N) Terminal fertile vesicle of mycoparasite. (O) Terminal fertile branch of a mycoparasite with two-spored sporangia.

This characterization is inaccurate, because they readily are isolated from soil. Originally described in the 19th century as curious "asterospheres" in living algae, these fungi have a strong habitat association as parasites and saprophytes on algae (Sparrow 1960). Chytrids, however, also play an important role in the decomposition of recalcitrant substrates, such as chitin, keratin, pollen, insect exuviae, plant debris, and so forth (Powell 1993). As a group, chytrids are ubiquitous in lakes, ponds, and soil. Many can be cultured, and the current study of chytrids generally involves observations of species in pure culture, whereas past descriptions focused on "gross culture" or their study on freshly collected substrates. Chytrids easily can be isolated from environmental samples by baiting with appropriate substrates, for example, pollen, cellophane, purified shrimp exoskeletons, and snake skin (Barr 1987).

The chytridiomycetes may be regarded as the economically least important major group of fungi, but there are several notable exceptions. Neocallimastigales are a clade of chytrids whose members are found in the rumen and hindgut of mammalian herbivores, where they aid in the digestion of plant fibers (Orpin 1988). Other economically important chytrids are the generalist plant pathogens *Synchytrium* and *Physoderma*. Species in both genera cause agricultural diseases in tropical climes, and *Synchytrium endobioticum* causes plant disease in the temperate zone. This parasite causes a malformation of potato tubers known as black wart. As recently as 2000, it was responsible for a one-year total quarantine on

the importation of potatoes from Prince Edward Island into the United States, resulting in a loss of at least $30 million to Canadian farmers. Finally, chytrids are parasites also on metazoans, primarily on soil invertebrates, such as nematodes and tardigrades. A notable exception is the vertebrate pathogen *Batrachochytrium dendrobatidis*, which infects frogs and has been associated with the recent global trend of amphibian declines (Berger et al. 1998, Longcore et al. 1999). If *Basidiobolus ranarum* truly is a chytrid (see below), then this amphibian and sometimes human pathogen would join *B. dendrobatidis* as a chytrid pathogen of vertebrates.

Taxonomy

Chytridiomycota consist of five orders, containing approximately 120 genera and 1000 species (Longcore 1996). Blastocladiales include *Allomyces macrogynus*, well known for studies on its cytology, genetics, and physiology, and *Coelomomyces stegomyiae*, a parasite of mosquito larvae. Fungi in this clade are distinguished by zoospores with a prominent "nuclear cap" of ribosomes. Monoblepharidales embrace only five genera; these aquatic chytrids are rarely seen but can be collected on decaying plant material such as fruits and twigs. Monoblepharids are distinguished by oogamous sexual reproduction (i.e., the female gamete is not motile and is larger than the uniflagellate male gamete) and vacuolate cells. Members of Spizellomycetales are ubiquitous in soil; one distinguishing feature is the amoeboid movement of zoospores during swimming (Barr 2001). Neocallimastigales are reserved for chytrids that inhabit anaerobic, rumen, and hindgut environments. These fungi either are uniflagellate or possess multiple flagella. The final and largest order, Chytridiales (~80 genera), contains a diversity of morphological forms. Most of the algal parasites are found in this clade.

Morphology

Chytridiomycete classification, traditionally, has been based on characteristics of vegetative growth and reproductive structures. The primary reproductive structure is the sporangium, a saclike structure whose contents are cleaved internally into zoospores (fig. 12.10A,B). Sporangia generally are subtended by a system of rhizoids that penetrate the substrate and facilitate anchoring and nutrient absorption. In some chytrids, the rhizoid system develops into an indeterminate, interconnected group of filaments, termed a rhizomycelium. Numerous sporangia can be produced from a rhizomycelium, which typically is coenocytic and lacks true septa. At maturity, zoospores are released from sporangia either through a small rounded opening (papillus) or a discharge tube. In some chytrids, the presence of a lidlike cover at the site of zoospore release can be seen clearly. This structure, the operculum, played an important role in previous classifications of chytrids (fig. 12.10B; Sparrow 1960, Karling 1977). A final, distinguishing character of many chytrids is the production of a resting spore. These thick-walled spores are desiccation resistant and can germinate into a sporangium after many years of dormancy. Although sexual reproduction generally results in the production of a resting spore, these spores also are produced asexually.

Life Cycle

Sexual reproduction has been observed in very few chytrids, but the variety of described mating systems is excitingly varied. Different modes of reproduction include the fusion of zoospores, gametangia, or rhizoids with subsequent transformation of the zygote into a resting spore (wherein meiosis is believed to occur; Doggett and Porter 1996). Oogamous reproduction occurs in Monoblepharidales, as mentioned above. In some species of Blastocladiales, an alternation of generations occurs between diploid sporophytes and haploid gametophytes. *Allomyces* species are hermaphoditic in that both male and female gametangia are produced on the same thallus. Sexual reproduction has been observed neither in Spizellomycetales nor in Neocallimastigales (Barr 2001). A representative Chytridiales life cycle is shown in figure 12.11.

Ultrastructure

Most chytrids have a simple and variable body plan that presents few characters on which to base a phylogentically meaningful taxonomy. Consequently, their ultrastructure as revealed by the transmission electron microscope is important in classification. Useful characters have been discovered in the zoospore (Lange and Olsen 1979); this special spore has proven to be exceptionally informative because of its internal complexity and conserved features (fig. 12.12). The zoospore is bounded by a membrane but lacks a cell wall. The zoospore of most chytrids contains a nucleus associated with an electron dense microbody and one to several lipid globules (fig. 12.12). The arrangement of these organelles is called the microbody–lipid globule complex and was used to group chytridiomycete zoospores

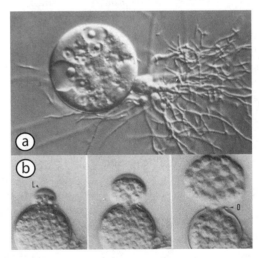

Figure 12.10. (A) Light micrograph of a developing sporangium with rhizoids of *Chytriomyces hyalinus*. (B) Light micrographs of zoospore discharge in *Chytriomyces hyalinus* showing an operculum (O) and a lenticular, expanding net of fibers (L) that constrains the zoospores for a brief period before they mature and swim away. From Taylor and Fuller (1981).

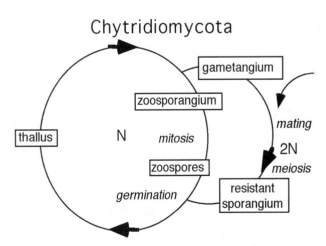

Figure 12.11. Generalized Chytridiomycota life cycle. The haploid thallus can differentiate to produce a zoosporangium with clonal zoospores, or to mate and produce a resistant sporangium. The resistant sporangium may germinate to release zoospores. Upon finding a suitable substrate, zoospores form cysts and the cysts germinate to produce a new thallus.

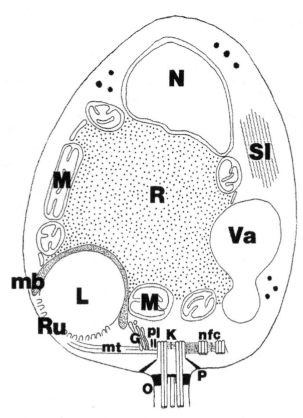

Figure 12.12. Ultrastructure of a typical Chytridiales zoospore as exemplified by *Podochytrium dentatum*. G, Golgi apparatus; K, functional kinetosome at the base of the flagellum; L, lipid globule; M, mitochondrion; mb, microbody; mt, microtubules; N, nucleus; nfc, second (nonfunctional) kinetosome; O, transition-zone plug; P, prop; pl, plates; R, ribosomes; Ru, rumposome; SI, striated inclusion; Va, vacuole. From Longcore (1992).

into broad taxonomic categories (Powell 1978). Another important feature of the zoospore is the rumposome, a fenestrated membrane located near the posterior portion of the zoospore adjacent to the spore membrane (Fuller and Reichle 1968). This organelle has been observed only in members of Chytridiales and Monoblepharidales. More recently, emphasis has been placed on the fine details of the flagellar apparatus (Barr 1990, 2001, James et al. 2000). Important characters include the connection of the non-flagellated centriole to the kinetosome (base of the flagellum) and the arrangement of microtubules and other kinetosomal roots. Zoospore ultrastructure currently is the only phenotypic means of accurately classifying chytrids into orders and even genera (Barr 1980, 2001).

Phylogenetic Relationships

Although the chytridiomycetes were recently classified in the Protoctista (Barr 1990), the link between Chytridiomycota and other members of Fungi already had been suggested by the

presence of chitinous cell walls, use of glycogen as a storage molecule, and presence of flattened mitochondrial cristae (Cavalier-Smith 1987, Powell 1993). Early phylogenies based on nSSU rDNA confirmed that Chytridiomycota are part of a monophyletic Fungi and are basal within Fungi (Förster et al. 1990, Dore and Stahl 1991, Bowman et al. 1992). The basal position of Chytridiomycota in Fungi suggests that the common ancestor of all fungi possessed motile zoospores. Therefore, the retention of a zoospore stage by the chytrids is considered a pleisiomorphy (ancestral character), which makes tenuous the unification and classification of chytrids based on the presence of a zoospore, because multiple independent losses of the flagellum may have occurred. For this reason, it is possible that Chytridiomycota is not a monophyletic group.

At present, few molecular phylogenetic data are available for the chytrids. Relationships of Chytridiomycota to other fungi have been examined, using primarily the SSU rRNA gene (Li and Heath 1992, Bruns et al. 1992, Nagahama et al. 1995, Jensen et al. 1998, James et al. 2000, Tanabe et al. 2000). These data are unclear as to whether the chytrids are monophyletic, because Blastocladiales typically groups with Zygomycota, rendering Chytridiomycota paraphyletic. In addition, placement of the putative zygomycete *Basidiobolus ranarum* within Chytridiomycota in SSU rRNA phylogenies has raised the possibility that some zygomycete orders may be chytrids that have experienced independent losses of the flagellum (Nagahama et al. 1995, Jensen et al. 1998). In support of the multiple independent losses of flagella is the observation that *Basidiobolus* species, which lack flagella, harbor an organelle resembling the centriole-like kinetosome found at the cellular end of flagella in Chytridioimycota; no such organelle is found in Zygomycota (McKerracher and Heath 1985). Confusing the picture is the placement of *B. ranarum* in Chytridiales by nSSU rDNA analyses but in Zygomycota by using β-tubulin analyses (Keeling et al. 2000). One possible explanation is that tubulin molecules evolve in similar ways when the constraint of flagellar function is lost, as might have occurred in *B. ranarum* and Zygomycota. The resolution of the possible non-monoplyly of Chytridiomycota awaits further sampling of genes and taxa.

Only one molecular phylogenetic study has heavily sampled taxa within Chytridiomycota (James et al. 2000). The authors of this study concluded that zoospore ultrastructure was concordant with the SSU rRNA phylogeny and that the five orders of chytrids seem to be monophyletic, with the exception of the largest order, Chytridiales. Within Chytridiales, well-supported clades were found, and these were consistent with groupings based on zoospore ultrastructure. However, relationships among clades of Chytridiales as well as among the orders were unresolved. Molecular phylogenies also confirmed the suspicion that chytrid gross morphology is of little use in classification. Indeed, pure culture studies have shown plasticity of developmental characters previously thought to be important in chytrid classification (Roane and Paterson 1974, Powell and Koch 1977). In contrast, zoospore ultrastructure has proven to be quite informative, and further investigation of these characters is warranted.

Studies of other gene regions also have shed some light on phylogenetic relationships of the chytridiomycetes. As mentioned above, analyses of β-tubulin gene sequences conflict with nSSU rDNA analyses over the placement of *Basidiobolus* (Keeling et al. 2000). Unfortunately, β-tubulin sequences show minimal variation among chytrids and provide little resolution of relationships among orders, making it imperative to examine other protein-coding genes to understand relationships of Chytridiomycota and Zygomycota. One promising development is the effort of the Fungal Mitochondrial Genome Project, which has sequenced the entire mitochondrial genome of several chytrids (Paquin et al. 1997, Forget et al. 2002, Bullerwell et al. 2003). Their analyses with concatenated mitochondrial proteins suggest a Spizellomycetales + Chytridiales clade, with Monoblepharidales as a sister group. These data also show a paraphyletic Chytridiomycota because *Allomyces* (Blastocladiales) again groups with the nonzoosporic fungi (including Zygomycota). Unfortunately, analysis of whole mitochondrial genomes must exclude the amitochondriate Neocallimastigales. In analyses of SSU rRNA, however, these fungi appear to be allied to Spizellomycetes, the order in which they previously were placed (Heath et al. 1983).

Based on current knowledge, it is possible to suggest a plausible phylogenetic hypothesis for Chytridiomycota for future testing (fig. 12.13). We may have been conservative in treating Chytridiomycota and Zygomycota as monophyletic groups and not as non-monophyletic groups, as shown in figures 12.2 and 12.13. However, until data from additional genes and taxa are available, we prefer to consider the treatment in figure 12.13 to be a hypothesis. In addition, more diversity continues to be uncovered as new chytrids are described and investigated with the electron microscope (Nyvall et al. 1999). Characterizing this diversity at the molecular level may result in the discovery of new major clades.

Fungi and Geologic Time

Our knowledge of the geologic history of Fungi is the subject of debate, mostly because of a lack of good fossils. The fossil record for fungi is based on very few specimens compared with that for plants and animals, probably because of a combination of factors: (1) fungi are mostly microscopic and are therefore easy to miss, (2) their tissues do not preserve very well, and (3) there are relatively few paleontologists looking for fungal fossils. Indeed, many of the best fossils are known only in association with a preserved plant or animal host. Some very well preserved fossils have been discov-

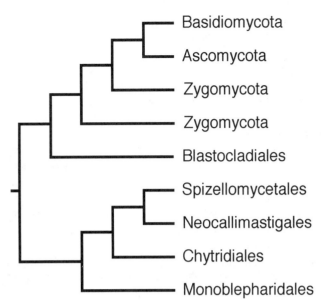

Figure 12.13. Phylogenetic relationships of Chytridiomycota orders to other fungi.

ered, but they provide only a few, hazy pictures of the long history of fungi. The oldest convincing fossils of Fungi were discovered in the Ordovician (~460 Mya) of Wisconsin, as hyphae and spores that strongly resemble modern structures in the genus *Glomus* (Redecker et al. 2000a). Otherwise, the vast majority of the oldest fungal fossils come from a single site, the lower Devonian (~400 Mya) Rhynie Chert of Scotland. A wide variety of fossils have been taken from this location, mostly members of Zygomycota and Chytridiomycota (Taylor and Taylor 1997). These fossils include zygomycete lichens associated with probable cyanobacterial photobionts (Taylor et al. 1995a, 1997), chytrid fungi resembling members of the modern genera *Allomyces* (Blastocladiales; Taylor et al. 1994, Remy et al. 1994a) and *Entophlyctis* (Chytridiales; Taylor et al. 1992), and glomalean fungi (Remy et al. 1994b, Taylor et al. 1995b). Most surprising, fossils morphologically very similar to extant members of Sordariomycetes (Ascomycota) were identified in the Rhynie Chert associated with the early land plant *Asteroxylon* (Taylor et al. 1999). The Rhynie Chert fossils indicate that a wide variety of fungi were present in the early Devonian period, including some resembling modern taxa thought to have evolved much more recently.

With few fossils available, analysis of DNA sequence is an attractive and powerful tool for inferring the times of origin for the major groups of Fungi. Different sets of molecular data have been used for these analyses and different analyses have used different calibration times for the divergence of animals and fungi; their results are summarized in table 12.1. Most approaches to date divergence times of organisms assume a molecular clock, where a rate of sequence evolution is identified for a particular gene region, and use a known calibration point, for example, the age of a known fossil or an independently estimated divergence time for fungi and animals. With these assumptions and data, divergence times between fungal divergences can be estimated. The first comprehensive attempts to date fungal divergences used nSSU rDNA and dated the origin of terrestrial fungi from the aquatic chytrids at approximately 550 Mya, in the Cambrian (Berbee and Taylor 1993). Using the knowledge that Fungi and Animalia probably share a common ancestor (Wainright et al. 1993), and a date of 965 Mya for that divergence (Doolittle et al. 1996), Berbee and Taylor (2001) revised their estimates based on nSSU rDNA and found that most inferred divergence times were pushed 50–100 million years earlier. Using the revision of Feng et al. (1997) for the divergence of animals and fungi, from 965 to 1200 Mya, would only have increased that effect. Berbee and Taylor (2001) used one gene for which sequences from many taxa were available, but more recent studies have used the ever-expanding DNA sequence databases to analyze more genes from fewer taxa. Wang et al. (1999) used amino acid sequences from 50 genes to explore the origin of animals, plants, and fungi. Although the majority of genes supported animals and fungi as closest ancestors, others supported animal and plant or plants and fungi as closest relatives, with an estimate of approximately 1576 Mya for the origin of these three kingdom-like clades. Using this and other molecular calibration points, Heckman et al. (2001) used amino acid sequences from 119 genes to estimate the divergence times of the major groups of fungi and inferred that most major groups evolved deep in the Precambrian, long before the points from which we have good fos-

sils. These authors note that nSSU rDNA data give a similar result, provided that a date of about 1576 Mya is used for the divergence of animals and fungi. This result leaves us to wonder what fungi were doing on Earth for a billion years before they were preserved as the fossils we know to exist. A point strongly in favor of the older estimate for the divergence of animals and fungi is the multiple gene estimate of ~670 Mya for the divergence of Sordariomycetes, which accommodates the discovery of a 400 Mya sordariomycete fossil from the lower Devonian. The age of this fossil is in conflict with the SSU estimate of ~310 Mya for the sordariomycete divergence, which is calibrated by a divergence of animals and fungi of 900 Mya (table 12.1).

In summary, both newly discovered fossils and molecular data have pushed back our estimates of the origins of the major fungal groups (Taylor et al. 1999, Redecker et al. 2000a, Berbee and Taylor 2001, Heckman et al. 2001). Ancient origins of fungi strongly suggest that fungi played an important role in the early colonization of land by plants and animals, both by changing the physical and chemical environment and by establishing mutualistic symbioses such as mycorrhizae and lichens (Selosse and Le Tecon 1998, Redecker et al. 2000b, Lutzoni et al. 2001, Heckman et al. 2001). The discrepancies between the history of fungi told by the fossil record and that by a molecular clock suggest that far more data are needed. Precambrian sources should be analyzed further for fungal fossils, and reports of Silurian fossils of Ascomycota (Sherwood-Pike and Gray 1985) deserve renewed attention. New methods of analysis that can

Table 12.1

Divergence Times within Major Fungal Groups.

Groups compared (reference group in parentheses)	rDNA[a] estimate (Mya)	rDNA[b] estimate (Mya)	119 protein gene[c] estimate (Mya)	Age of oldest known fossil in ref. group (Mya)
(Chytridiomycota) versus Zygomycota + Ascomycota + Basidiomycota	~550	~660	1458 ± 70	~400[d]
Chytridiomycota + (Zygomycota) versus Ascomycota + Basidiomycota	~490	~590	1107 ± 56[e]	~460[f]
(Ascomycota) versus Basidiomycota	~390	~560	1208 ± 108	~400[g]
(Hymenomycetes) versus Ustilaginomycetes	~380	~430	966 ± 86	~290[h]
(Taphrinomycotina) versus Saccharomycotina + Pezizomycotina	~320	~420	1144 ± 77	None
Saccharomycotina versus (Pezizomycotina)	~310	~370	1085 ± 81	~400[g]
Eurotiomycetes versus (Sordariomycetes)	~290	~310	670 ± 71	~400[g]

[a]Molecular clock calibrated using fungal fossils (Berbee and Taylor 1993).

[b]Molecular clock calibrated using fungal fossils and divergence time of fungi vs. animals estimated at 965 Ma (Doolittle et al. 1996, Berbee and Taylor 2001).

[c]Molecular clocks calibrated using divergence of plants, animals, and fungi estimated at 1576 Mya, divergence of nematodes and arthropods at 1177 Mya, and arthropods and chordates at 993 Mya, each of which was in turn based on a 75-gene molecular clock calibrated with the vertebrate fossil record (Heckman et al. 2001).

[d]Several different fossilized chytrids from Rhynie Chert (Taylor and Taylor 1997).

[e]No glomalean fungi were included in this study. The Glomales, which represent the oldest reliable fungi in the fossil record, are probably the most recently derived major clade of the Zygomycota.

[f]Fossilized glomalean spores and hyphae from the Ordovician period (Redecker et al. 2000a).

[g]Fossilized Pyrenomycete from Rhynie Chert (Taylor et al. 1999).

[h]Fossilized hyphae with clamp connections (Dennis 1970).

accommodate rate variation among lineages (e.g., Sanderson 2002) should be investigated and compared with other methods. Our current estimates of the timing of events in fungal evolution undoubtedly are crude and are sure to be improved as data and methods improve. However, it is essential that they be made, even knowing that they can be improved, because time is the common currency of evolutionary biologists, and only by making such estimates can events in the history of Fungi be compared with those in the other major kingdom-like groups. We hope that those who design museum displays will note these efforts and include fungi in their work.

Last Word

Looking back on a dozen years of fungal molecular phylogenetics, it is clear that no approach since microscopy has had such a profound influence on our understanding of fungal evolution. Owing to their microscopic size and ability to live in their food, fungi are cryptic in a way that no angiosperm or vertebrate can imitate. This fact has made the research of fungal molecular phylogenetics even more valuable, because it enables ecologists, finally, to add the fungi to their studies. Over the next decade, we look forward to the improved phylogenetic resolution that genomics and improved analytical methods promise, and to the application of microarray technology to ecological studies. The latter should automate fungal identification and make it possible to more accurately estimate fungal biodiversity. That information should provide some further surprises and it seems sure to close the gap between the 100,000 described fungi and the 1.5 million estimated to exist in nature.

Acknowledgments

We thank Joyce Longcore for advice on the manuscript and Zheng Wang for the illustrations used in figure 12.6. We acknowledge the support of the National Science Foundation (Research Coordination Networks in Biological Sciences: A Phylogeny for Kingdom Fungi; NSF 0090301).

Literature Cited

Agrios, G. N. 1997. Plant pathology. 4th ed. Academic Press, San Diego.

Alexopoulos, C. J., C. W. Mims, and M. Blackwell. 1996. Introductory mycology. John Wiley and Sons, New York.

Arnold, A. E., Z. Maynard, and G. S. Gilbert. 2001. Fungal endophytes in dicotyledonous neotropical trees: patterns of abundance and diversity. Mycol. Res. 105:1502–1507.

Barnett, J. A., R. W. Payne, and D. Yarrow. 1990. Yeasts: characteristics and identification. Cambridge University Press, Cambridge.

Barr, D. J. S. 1980. An outline for the reclassification of the Chytridiales, and for a new order, the Spizellomycetales. Can. J. Bot. 58:2380–2394.

Barr, D. J. S. 1987. Isolation, culture and identification of Chytridiales, Spizellomycetales, and Hyphochytriales. Pp. 118–120 in Zoosporic fungi in teaching and research (M. S. Fuller and A. Jaworski, eds.). Southeastern Publishing, Athens, GA.

Barr, D. J. S. 1990. Phylum Chytridiomycota. Pp. 454–466 in Handbook of Protoctista (L. Margulis, J. O. Corliss, M. Melkonian, and D. J. Chapman, eds.). Jones and Bartlett, Boston.

Barr, D. J. S. 2001. Chytridiomycota. Pp. 93–112 in The mycota VIIA, systematics and evolution (D. J. McLaughlin, E. G. McLaughlin, and P. A. Lemke, eds.). Springer-Verlag, Berlin.

Barron, G. L. 1988. Microcolonies of bacteria as a nutrient source for lignicolous and other fungi. Can. J. Bot. 66:2505–2510.

Bartnicki-Garcia, S. 1987. The cell wall: a crucial structure in fungal evolution. Pp. 389–403 in Evolutionary biology of the fungi (A. D. M. Rayner, C. M. Brasier, and D. Moore, eds.). Cambridge University Press, Cambridge.

Bauer, R., D. Begerow, F. Oberwinkler, M. Piepenbring, and M. L. Berbee. 2001. Ustilaginomycetes. Pp. 57–84 in The mycota VIIB, systematics and evolution (D. J. McLaughlin, E. G. McLaughlin, and P. A. Lemke, eds.). Springer-Verlag, Berlin.

Begerow, D., R. Bauer, and F. Oberwinkler. 1997. Phylogenetic studies on nuclear large subunit ribosomal DNA sequences of smut fungi and related taxa. Can. J. Bot. 75:2045–2056.

Benjamin, R. K. 1979. Zygomycetes and their spores. Pp.573–616 in The whole fungus. The sexual-asexual synthesis (B. Kendrick, ed.), vol. 2. Museums of Canada and the Kananaskis Foundation, Ottawa.

Benjamin, R. K., M. Blackwell, I. Chapella, R. A. Humber, K. G. Jones, K. A. Klepzig, R. W. Lichtwardt, D. Malloch, H. Noda, R. A. Roeper, J. W. Spatafora, and A. Weir. In press. The search for diversity of insects and other arthropod associated fungi. In Biodiversity of fungi: standard methods for inventory and monitoring (G. M. Mueller, G. F. Bills, and M. Foster, eds.). Academic Press, New York.

Benny, G. L. 2001. Zygomycota: Trichomycetes. Pp. 147–160 in The mycota VIIA, systematics and evolution (D. J. McLaughlin, E. G. McLaughlin, and P. A. Lemke, eds.). Springer-Verlag, Berlin.

Benny, G. L., R. A. Humber, and J. B. Morton. 2001. Zygomycota: Zygomycetes. Pp. 113–146 in The mycota VIIA, systematics and evolution (D. J. McLaughlin, E. G. McLaughlin, and P. A. Lemke, eds.). Springer-Verlag, Berlin.

Benny, G. L., and K. O'Donnell. 2000. *Amoebidium parasiticum* is a protozoan, not a Trichomycete. Mycologia 92:1133–1137.

Berbee, M. L. 1998. Loculoascomycete origins and evolution of filamentous ascomycete morphology based on 18S rRNA gene sequence data. Mol. Biol. Evol. 13:462–470.

Berbee, M. L., and J. W. Taylor. 1992. Convergence in ascospore discharge mechanism among Pyrenomycete fungi based on 18S ribosomal RNA gene sequence. Mol. Phylogenet. Evol. 1:59–71.

Berbee, M. L., and J. W. Taylor. 1993. Dating the evolutionary radiations of the true fungi. Can. J. Bot. 71:1114–1127.

Berbee, M. L., and J. W. Taylor. 2001. Fungal molecular evolution: gene trees and geologic time. Pp. 229–245 in The mycota VIIB, systematics and evolution (D. J. McLaughlin, E. G. McLaughlin, and P. A. Lemke, eds.). Springer-Verlag, Berlin.

Berger, L., R. Speare, P. Daszak, D. E. Green, A. A. Cunningham, C. L. Goggin, R. Slocombe, M. A. Ragan, A. D. Hyatt, K. R. McDonald, et al. 1998. Chytridiomycosis causes amphibian mortality associated with population declines in the rain forests of Australia and Central America. Proc. Natl. Acad. Sci. USA 95:9031–9036.

Bhattacharya, D., F. Lutzoni, V. Reeb, D. Simon, and F. Fernandez. 2000. Widespread occurrence of spliceosomal introns in the rDNA genes of Ascomycetes. Mol. Biol. Evol. 17:1971–1984.

Blackwell, M. 2000. Perspective: Evolution: Terrestrial life—fungal from the start? Science 289:1884–1885.

Bowman, B. H., J. W. Taylor, J. L. Brownlee, S.-D. Lu, and T. J. White. 1992. Molecular evolution of the fungi: relationship of the Basidiomycetes, Ascomycetes, and Chytridiomycetes. Mol. Biol. Evol. 9:258–296.

Brodo, I. M., S. Duran Sharnoff, and S. Sharnoff. 2001. Lichens of North America. Yale University Press, New Haven, CT.

Bruns, T. D., T. M. Szaro, M. Gardes, K. W. Cullings, J. Pan, D. L. Taylor, and Y. Li. 1998. A sequence database for the identification of ectomycorrhizal fungi by sequence analysis. Mol. Ecol. 7:257–272.

Bruns, T. D., R. Vilgalys, S. M. Barns, D. Gonzalez, D. S. Hibbett, D. J. Lane, L. Simon, S. Stickel, T. M. Szaro, W. G. Weisburg, and M. L. Sogin. 1992. Evolutionary relationships within the Fungi: analyses of nuclear small subunit rRNA sequences. Mol. Phylogenet. Evol. 1:231–241.

Bullerwell, C. E., L. Forget, and B. F. Lang. 2003. Evolution of monoblepharidalean fungi based on complete mitochondrial genome sequences. Nucleic Acids Res. 31:1614–1623.

Carroll, G. 1988. Fungal endophytes in stems and leaves: from latent pathogen to mutualistic symbiont. Ecology 69:2–9.

Carroll, G. 1995. Forest endophytes: pattern and process. Can. J. Bot. 73:S1316–S1324.

Carroll, G. C., and D. T. Wicklow, 1992. The fungal community: its organization and role in the ecosystem. Marcel Dekker, New York.

Cavalier-Smith, T. 1987. The origin of Fungi and pseudofungi. Pp. 339–353 in Evolutionary biology of the fungi (A. D. M. Rayner, C. M. Brasier, and D. Moore, eds.). Cambridge University Press, Cambridge.

Cavalier-Smith, T. 1998. A revised six-kingdom of life. Biol. Rev. 73:203–266.

Chapela, I., S. A. Rehner, T. R. Schultz, and U. G. Mueller. 1994. Evolutionary history of the symbiosis between fungus-gorwing ants and their fungi. Science 266:1691–1694.

Clay, K. 2001. Symbiosis and the regulation of communities. Am. Zool. 41:810–824.

Cole, G. T., and B. Kendrick. 1981. Biology of conidial fungi. Academic Press, New York.

Cullen, D. 1997. Recent advances on the molecular genetics of ligninolytic fungi. J. Biotechnol. 53:273–289.

Currie, C. R., B. Wong, A. E. Stuart, T. R. Schultz, S. A. Rehner, U. G. Mueller, G. H. Sung, J. W. Spatafora, and N. A. Straus. 2003. Ancient tripartite coevolution in the attine ant-microbe symbiosis. Science 299:386–388.

Dennis, R. L. 1970. A Middle Pennsylvanian basidiomycete mycelium with clamp connections. Mycologia 62:578–564.

Doggett, M. S., and D. Porter. 1996. Sexual reproduction in the fungal parasite, Zygorhizidium planktonicum. Mycologia 88:720–732.

Doolittle, R. F., D. F. Feng, S. Tsang, G. Cho, and E. Little. 1996. Determining divergence times of the major kingdoms of living organisms with a protein clock. Science 271:470–477.

Dore, J., and D. A. Stahl. 1991. Phylogeny of anaerobic rumen Chytridiomycetes inferred from small subunit ribosomal RNA sequence comparisons. Can. J. Bot. 69:1964–1971.

Edmonds, R. L., J. K. Agee, and R. I. Gara. 2000. Forest health and protection. McGraw-Hill Series in Forestry. McGraw-Hill, Boston.

Eriksson, O. E., H.-O. Baral, R. S. Currah, K. Hansen, C. P. Kurtzman, G. Rambold, and T. Laessøe. 2003. Pp. 1–89 in Outline of Ascomycota—2003. Myconet Umeå, Umeå University. Available: http://www.umu.se/myconet/Myconet.html. Last accessed 3 December 2003.

Eslava, A. P., and M. I. Alvarez. 1996. Genetics of Phycomyces. Pp. 385–406 in Fungal genetics: principles and practice (C. J. Bos, ed.). Marcel Dekker, New York.

Fell, J. W., T. Boekhout, A. Fonseca, and J. P. Sampaio. 2001. Basidiomycetous yeasts. Pp. 3–35 in The mycota VIIB, systematics and evolution (D. J. McLaughlin, E. G. McLaughlin, and P. A. Lemke, eds.). Springer-Verlag, Berlin.

Feng, D. F., G. Cho, and R. F. Doolittle. 1997. Determining divergence times with a protein clock: update and reevaluation. Proc. Natl. Acad. Sci. USA 94:13028–13033.

Forget, L., J. Ustinova, Z. Wang, V. A. R. Huss, and B. F. Lang. 2002. The "colorless green alga" Hyaloraphidium curvatum: a linear mitochondrial genome, tRNA editing, and an evolutionary link to lower fungi. Mol. Biol. Evol. 19:310–319.

Förster, H., M. D. Coffey, H. Elwood, and M. L. Sogin. 1990. Sequence analysis of the small subunit ribosomal RNAs of three zoosporic fungi and implications for fungal evolution. Mycologia 82:306–312.

Fuller, M. S., and R. E. Reichle. 1968. The fine structure of Monoblepharella sp. zoospores. Can. J. Bot. 46:279–283.

Gargas, A., P. T. DePriest, M. Grube, and A. Tehler. 1995. Multiple origins of lichen symbioses in fungi suggested by SSU rDNA phylogeny. Science 268:1492–1495.

Gehrig, H., A. Schüßler, and M. Kluge. 1996. Geosiphon pyriforme, a fungus forming endocytobiosis with Nostoc (Cyanobacteria), is an ancestral member of the Glomales: evidence by SSU rRNA analysis. J. Mol. Evol. 43:71–81.

Geiser, D. M., and K. F. LoBuglio. 2001. The monophyletic Plectomycetes: Ascosphaeriales, Onygenales, Eurotiales. Pp. 201–219 in The mycota VIIA, systematics and evolution (D. J. McLaughlin, E. G. McLaughlin, and P. A. Lemke, eds.). Springer-Verlag, New York.

Gernandt, D. S., J. L. Platt, J. K. Stone, J. W. Spatafora, A. Holst-Jensen, R. C. Hamelin, and L. M. Kohn. 2001. Phylogenetics of Helotiales and Rhytismatales based on partial small subunit nuclear ribosomal DNA sequences. Mycologia 93:915–933.

Glass, N. L., and I. A. J. Lorimer. 1991. Ascomycete mating types. Pp 193–216 in More gene manipulations in fungi (J. W. Bennett and L. L. Lasure, eds.). Academic Press, Orlando, FL.

Gottlieb, A. M., and R. W. Lichtwardt. 2001. Molecular variation within and among species of Harpellales. Mycologia 93:66–81

Griffin, D. H. 1994. Fungal physiology. 2nd ed. Wiley-Liss, New York.

Haselwandter, K., O. Bolbleter, and D. J. Read. 1990. Degradation of 14C-labelled lignin and dehydropolymer of coniferyl alchohol by ericoid and ectomycorrhizal fungi. Arch. Microbiol. 153:352–354.

Hawksworth, D. L. 1991. The fungal dimension of biodiversity: magnitude, significance and conservation. Mycol. Res. 95:641–655.

Hawksworth, D. L. 2001. The magnitude of fungal diversity: the 1.5 million species estimate revisited. Mycol. Res. 105:1422–1432.

Hawksworth, D. L., P. M. Kirk, B. C. Sutton, and D. N. Pegler. 1995. Ainsworth and Bisby's dictionary of the fungi. 8th ed. CAB International, Wallingford, UK.

Heath, I. B., T. Bauchop, and R. A. Skipp. 1983. Assignment of the rumen anaerobe Neocallimastix frontalis to the Spizellomycetales (Chytridiomycetes) on the basis of its polyflagellate ultrastructure. Can. J. Bot. 61:295–307.

Heckman, D. S., D. M. Geiser, B. R. Eidell, R. L. Stauffer, N. L. Kardos and S. B. Hedges. 2001. Molecular evidence for the early colonization of land by fungi and plants. Science 293:1129–1133.

Hibbett, D. S., and M. Binder. 2002. Evolution of complex fruiting body morphologies in homobasidiomycetes. Proc. R. Soc. Lond. B 269:1963–1969.

Hibbett, D. S., and M. J. Donoghue. 2001. Analysis of character correlations among wood decay mechanisms, mating systems, and substrate ranges in homobasidiomycetes. Syst. Biol. 50:215–242.

Hibbett, D. S., L. B. Gilbert, and M. J. Donoghue. 2000. Evolutionary instability of ectomycorrhizal symbioses in basidiomycetes. Nature 407:506–508.

Hibbett, D. S., E. M. Pine, E. Langer, G. Langer, and M. J. Donoghue. 1997. Evolution of gilled mushrooms and puffballs inferred from ribosomal DNA sequences. Proc. Natl. Acad. Sci. USA 94:12002–12006.

Hibbett, D. S., and R. G. Thorn. 2001. Basidiomycota: Homobasidiomycetes. Pp. 121–168 in The mycota VIIB, systematics and evolution (D. J. McLaughlin, E. G. McLaughlin, and P. A. Lemke, eds.). Springer-Verlag, Berlin.

Holst-Jensen, A., L. M. Kohn, and T. Schumacher. 1997. Nuclear rDNA phylogeny of the Sclerotiniaceae. Mycologia 89:885–899.

Humber, R. A., and K. S. Hansen. USDA-ARS Collection of Entomopathogenic Fungal Cultures (ARSEF). Available: http://www.ppru.cornell.edu/mycology/Insect_mycology.htm. Last accessed 3 December 2003.

Ingold, C. T. 1965. Spore liberation. Clarendon Press, Oxford.

James, T. Y., D. Porter, C. A. Leander, R. Vilgalys, and J. E. Longcore. 2000. Molecular phylogenetics of the Chytridiomycota supports the utility of ultrastructural data in chytrid systematics. Can. J. Bot. 78:336–350.

Jensen, A. B., A. Gargas, J. Eilenberg, and S. Rosendahl. 1998. Relationships of the insect-pathogenic order Entomophtorales (Zygomycota, Fungi) based on phylogenetic analyses of nuclear small subunit ribosomal DNA sequences (SSU rDNA). Fungal Genet. Biol. 24:325–334.

Karling, J. S. 1977. Chytridiomycetarum Iconographia. Lubrecht and Cramer, Monticello, NY.

Kauff, F., and F. Lutzoni. 2002. Phylogeny of the Gyalectales and Ostropales (Ascomycota, Fungi): among and within order relationships based on nuclear ribosomal RNA small and large subunits. Mol. Phylogenet. Evol. 25:138–156.

Keeling, P. J., M. A. Luker, and J. D. Palmer. 2000. Evidence from beta-tubulin phylogeny that microsporidia evolved from within the fungi. Mol. Biol. Evol. 17:23–31.

Kirk, P. M., P. F. Cannon, J. C. David, and J. A. Stalpers. 2001. Ainsworth and Bisby's dictionary of the fungi. 9th ed. CAB International, Wallingford, UK.

Klironomos, J. N., and M. M. Hart. 2001. Animal nitrogen swap for plant carbon. Nature 410:651–652.

Kohlmeyer, J., and E. Kohlmeyer. 1979. Marine mycology. The higher fungi. Academic Press, New York.

Kurtzman, C. P., and C. J. Robnett. 2003. Phylogenetic relationships among yeasts of the "Saccharomyces complex" determined from multigene sequence analyses. FEMS Yeast Res. 1554:1–16.

Kwon-Chung, K. J., and J. E. Bennett. 1992. Medical mycology. Lea and Febiger, Philadelphia.

Landvik, S. 1996. Neolecta, a fruit-body-producing genus of the basal ascomycetes, as shown by SSU and LSU rDNA sequences. Mycol. Res. 100:199–202.

Landvik, S., O. E. Eriksson, and M. L. Berbee. 2001. Neolecta—a fungal dinosaur? Evidence from β-tubulin amino acid sequences. Mycologia 93:1151–1163.

Lang, B. F. 2001. Fungal Mitochondrial Genome Project. Http://megasun.bch.umontreal.ca/People/lang/FMGP/.

Lang, B. F, C. O'Kelly, T. Nerad, M. W. Gray, and G. Burger. 2002. The closest unicellular relative of animals. Curr. Biol. 12:1773–1778.

Lange, L., and L. W. Olson. 1979. The uniflagellate phycomycete zoospore. Dansk Bot. Arkiv. 33:7–95.

Li, J., and I. B. Heath. 1992. The phylogenetic relationships of the anaerobic chytridiomycetous gut fungi (Neocallimasticaceae) and the Chytridiomycota. I. Cladistic analysis of rRNA sequences. Can. J. Bot. 70:1738–1746.

Lichtwardt, R. W. 1986. The trichomycetes, fungal associates of arthropods. Springer-Verlag, New York.

Lichtwardt, R. W., M. J. Cafaro, and M. M. White. 2001. The Trichomycetes: rev. ed. Available: http://www.nhm.ku.edu/%7Efungi/monograph/text/mono.htm. Last accessed 3 December 2003.

Liu, Y. J., S. Whelen, and B. D. Hall. 1999. Phylogenetic relationships among ascomycetes: evidence from an RNA polymerse II subunit. Mol. Biol. Evol. 16:1799–1808.

Longcore, J. E. 1992. Morphology, occurrence, and zoospore ultrastructure of Podochytrium dentatum sp. nov. (Chytridiales). Mycologia 84:183–192.

Longcore, J. E. 1996. Chytridiomycete taxonomy since 1960. Mycotaxon 60:149–174.

Longcore, J. E., A. P. Pessier, and D. K. Nichols. 1999. Batracho-

chytrium dendrobatidis gen. et sp. nov., a chytrid pathogenic to amphibians. Mycologia 91:219–227.

Loytynoja, A., and M. C. Milinkovitch. 2001. Molecular phylogenetic analyses of the mitochondrial ADP-ATP carriers: the Plantae/Fungi/Metazoa trichotomy revisited. Proc. Natl. Acad. Sci. USA 98:10202–10207.

Luttrell, E. S. 1951. Taxonomy of the Pyrenomycetes. Univ. Mo. Stud. 24:1–120.

Luttrell, E. S. 1955. The ascostromatic Ascomycetes. Mycologia 47:511–532.

Lutzoni, F., and M. Pagel. 1997. Accelerated evolution as a consequence of transitions to mutualism. Proc. Nat. Acad. Sci. USA 94:11422–11427.

Lutzoni, F. L., M. Pagel, and V. Reeb. 2001. Major fungal lineages are derived from lichen symbiotic ancestors. Nature 411:937–940.

Marsh, L. 1991. Signal transduction during pheromone response in yeast. Annu. Rev. Cell Biol. 7:699–728.

McKerracher, L. J., and I. B. Heath. 1985. The structure and cycle of the nucleus-associated organelle in two species of *Basidiobolus*. Mycologia 77:412–417.

McLaughlin, D. J., E. G. McLaughlin, and P. A. Lemke. 2001. The mycota VIIA: systematics and evolution. Springer-Verlag, New York.

Miadlikowska, J., and F. Lutzoni. In press. Phylogenetic classification of peltigeralean fungi (Peltigerales, Ascomycota) based on ribosomal RNA small and large subunits. Am. J. Bot.

Molina, R., and J. M. Trappe. 1982. Patterns of ectomycorrhizal host specificity and potential among pacific northwest conifers and fungi. Forest Sci. 28:423–458.

Moncalvo, J. M., R. Vilgalys, S. A. Redhead, J. E. Johnson, T. Y. James, M. C. Aime, V. Hofstetter, S. J. W. Verduin, E. Larsson, T. J. Baroni, et al. 2002. One hundred and seventeen clades of Euagarics. Mol. Phylogenet. Evol. 23:357–400.

Nagahama, T., H. Sato, M. Shimazu, and J. Sugiyama. 1995. Phylogenetic divergence of the entomophthoralean fungi: evidence from nuclear 18s ribosomal RNA gene sequence. Mycologia 87:203–209.

Nash, T. H. 1996, Lichen biology. Cambridge University Press, Cambridge.

Nishida, H., K. Ando, Y. Ando, A. Hirata, and J. Sugiyama. 1995. *Mixia osmundae*: transfer from the Ascomycota to the Basidiomycota based on evidence from molecules and morphology. Can. J. Bot. 73:S660–S666.

Nishida, H., and J. Sugiyama. 1994. Archiascomycetes: detection of a major new linage within the Ascomycota. Mycoscience 35:361–366.

Nyvall, P., M. Pedersén, and J. E. Longcore. 1999. *Thalassochytrium gracilariopsidis* (Chytridiomycota), gen. et sp. nov., endosymbiotic in *Gracilariopsis* sp. (Rhodophyceae). J. Phycol. 35:176–185.

Oberwinkler, F. 1977. Das neue System der Basidiomyceten. Pp. 59–105 *in* Biologie der niederen Pflanzen (H. Frey, H. Hurka, and F. Oberwinkler, eds.). G. Fischer, Stuttgart.

O'Donnell, K. 1979. Zygomycetes in culture. Department of Botany, University of Georgia, Athens.

O'Donnell, K., E. Cigelnik, and G. L. Benny. 1998. Phylogenetic relationships among the Harpellales and Kickxellales. Mycologia 90:624–639.

O'Donnell, K., F. M. Lutzoni, T. J. Ward, and G. L. Benny. 2001. Evolutionary relationships among mucoralean fungi (Zygomycota): evidence for family polyphyly on a large scale. Mycologia 93:286–296.

Orpin, C. G. 1988. Nutrition and biochemistry of anaerobic Chytridiomycetes. Biosystems 21:365–370.

Paquin, B., M.-J. Laforest, L. Forget, I. Roewer, Z. Wang, J. Longcore, and F. Lang. 1997. The fungal mitochondrial genome project: evolution of fungal mitochondrial genomes and their gene expression. Curr. Genet. 31:380–395.

Perez-Moreno, J., and D. J. Read. 2000. Mobilization and transfer of nutrients from litter to tree seedlings via the vegetative mycelium of ectomycorrhizal plants. New Phytol. 145:301–309.

Pirozynski, K. A., and D. W. Malloch. 1975. The origin of land plants: a matter of mycotropism. Biosystems 6:153–164.

Platt, J. L., and J. W. Spatafora. 2000. Evolutionary relationships of nonsexual lichenized fungi: molecular phylogenetic hypotheses for the genera *Siphula* and *Thamnolia* from SSU and LSU rDNA. Mycologia 92:475–487.

Powell, M. J. 1978. Phylogenetic implications of the microbody-lipid globule complex in zoosporic fungi. Biosystems 10:167–180.

Powell, M. J. 1993. Looking at mycology with a Janus face: a glimpse at Chytridiomycetes active in the environment. Mycologia 85:1–20.

Powell, M. J., and W. J. Koch. 1977. Morphological variations in a new species of *Entophlyctis*. II. Influence of growth conditions on morphology. Can. J. Bot. 55:1686–1695.

Prillinger, H., F. Oberwinkler, C. Umile, K. Tlachac, R. Bauer, C. Dörfler, and E. Taufratzhofer. 1993. Analysis of cell wall carbohydrates (neutral sugars) from ascomycetous and basidiomycetous yeasts with and without derivatization. J. Gen. Appl. Microbiol. 39:1–34.

Raju, N. B. 1992. Genetic control of the sexual cycle in *Neurospora*. Mycol. Res. 96:241–262.

Rayner, A. D. M., and L. Boddy. 1988. Fungal decomposition of wood: its biology and ecology. John Wiley, Chichester, UK.

Redecker, D., R. Kodner, and L. E. Graham. 2000a. Glomalean fungi from the Ordovician. Science 289:1920–1021.

Redecker, D., J. B. Morton, and T. D. Bruns. 2000b. Ancestral lineages of arbuscular mycorrhizal fungi (Glomales). Mol. Phylogenet. Evol. 14:276–284.

Reid, I. D. 1995. Biodegradation of lignin. Can. J. Bot. 73:S1011–S1018.

Remy, W., T. N. Taylor, and H. Hass. 1994a. Early Devonian fungi—a Blastocladalean fungus with sexual reproduction. Am. J. Bot. 81:690–702.

Remy, W., T. N. Taylor, H. Hass, and H. Kerp. 1994b. 4–Hundred-million-year-old vesicular-arbuscular mycorrhizae. Proc. Natl. Acad. Sci. USA 91:11841–11843.

Rinaldi, M. G. 1989. Zygomycosis. Infect. Dis. Clin. N. Am. 3:19–41.

Roane, M. K., and R. A. Paterson. 1974. Some aspects of morphology and development in the Chytridiales. Mycologia 66:147–164.

Sanders, I. R. 1999. No sex please, we're fungi. Nature 399:737–739.

Sanderson, M. J. 2002. Estimating absolute rates of molecular evolution and divergence times: a penalized likelihood approach. Mol. Biol. Evol. 19:101–109.

Schüßler, A., D. Schwarzott, and C. Walker. 2001. A new fungal phylum, the Glomeromycota: phylogeny and evolution. Mycol. Res. 105:1413–1421.

Schwarzott, D., C. Walker, and A. Schüßler. 2001. *Glomus*, the largest genus of the arbuscular mycorrhizal Fungi (Glomales), is nonmonophyletic. Mol. Phylogenet. Evol. 21:190–197.

Selosse, M. A., and F. Le Tacon. 1998. The land flora: a phototroph-fungus partnership? Trends Ecol. Evol. 13:15–20.

Sherwood-Pike, M. A., and J. Gray. 1985. Silurian fungal remains: probable records of the Class Ascomycetes. Lethaia 18:1–20.

Simon, L., J. Bousquet, R. C. Lévesque, and M. Lalonde. 1993. Origin and diversification of endomycorrhizal fungi and coincidence with vascular land plants. Nature 363:67–69.

Sjamsuridzal, W., H. Nishida, H. Ogawa, M. Kakishima, and J. Sugiyama. 1999. Phylogenetic positions of rust fungi parasitic on ferns: evidence from 18S rDNA sequence analysis. Mycoscience 40:21–27.

Smith, M. L., J. N. Bruhn, and J. B. Anderson. 1992. The fungus *Armillaria bulbosa* is among the largest and oldest living organisms. Nature 356:428–434.

Smith, S. E., and D. J. Read. 1997. Mycorrhizal symbiosis. Academic Press, San Diego.

Sparrow, F. K. 1960. Aquatic phycomycetes. 2nd rev. ed. University of Michigan Press, Ann Arbor.

Spatafora, J. W. 1995. Ascomal evolution among filamentous ascomycetes: evidence from molecular data. Can. J. Bot. S811–S815.

Spatafora, J. W., and M. Blackwell. 1994. The polyphyletic origins of ophiostomatoid fungi. Mycol. Res. 98:1–9.

Spatafora, J. W., B. Volkmann-Kohlmeyer, and J. Kohlmeyer. 1998. Independent terrestrial origins of the Halosphaeriales (marine Ascomycota). Am. J. Bot. 85:1569–1580.

Sugiyama, J. 1987. Pleomorphic fungi: the diversity and its taxonomic implications. Elsevier, Amsterdam.

Suh, S.-O., and M. Blackwell. 1999. Molecular phylogeny of the cleistothecial fungi placed in Cephalothecaceae and Pseudeurotiaceae. Mycologia 91:836–848.

Suh, S.-O., C. Marshall, J. V. McHugh, and M. Blackwell. 2003. Wood ingestion by passalid beetles in the presence of xylose-fermenting gut yeasts. Mol. Ecol. 12:3137–3145.

Swann, E. C., E. M. Frieders, and D. J. McLaughlin. 1999. *Microbotryum, Kriegeria* and the changing paradigm in basidiomycete classification. Mycologia 91:51–66.

Swann, E. C., E. M. Frieders, and D. J. McLaughlin. 2001. Urediniomycetes. Pp. 37–56 in The mycota VIIB, systematics and evolution (D. J. McLaughlin, E. G. McLaughlin, and P. A. Lemke, eds.). Springer-Verlag, Berlin.

Swann, E. C., and J. W. Taylor. 1993. Higher taxa of basidiomycetes: an 18S rRNA gene perspective. Mycologia 85:923–936.

Swann, E. C., and J. W. Taylor. 1995. Phylogenetic perspectives on basidiomycete systematics: evidence from the 18S rRNA gene. Can. J. Bot. 73:S862–S868.

Tanabe, Y., K. O'Donnell, M. Saikawa, and J. Sugiyama. 2000. Molecular phylogeny of parasitic Zygomycota (Dimargaritales, Zoopagales) based on nuclear small subunit ribosomal DNA sequences. Mol. Phylogenet. Evol. 16:253–262.

Tanabe, Y., Watanabe, M. M., and J. Sugiyama. 2002. Are Microsporidia really related to fungi? A reappriasal based on additional gene sequenc es from basal fungi. Mycol. Res. 106:1380–1391.

Taylor, J. W. 1995. Making the Deuteromycota redundant: a practical integration of mitosporic and meiosporic fungi. Can. J. Bot. 73:S754–S759.

Taylor, J. W., B. Bowman, M. L. Berbee, and T. J. White. 1993. Fungal model organisms: phylogenetics of *Saccharomyces, Aspergillus* and *Neurospora*. Syst. Biol. 42:440–457.

Taylor, J. W., and M. S. Fuller. 1981. The Golgi apparatus, zoosporogenesis, and development of the zoospore discharge apparatus of *Chytridium confervae*. Exp. Mycol. 5:35–59.

Taylor, T. N., H. Hass, and H. Kerp. 1997. A cyanolichen from the Lower Devonian Rhynie Chert. Am. J. Bot. 84:992–1004.

Taylor, T. N., T. Hass, and H. Kerp. 1999. The oldest fossil ascomycetes. Nature 399:648–648.

Taylor, T. N., H. Hass, W. Remy, and H. Kerp. 1995a. The oldest fossil lichen. Nature 378:244–244.

Taylor, T. N., W. Remy, and H. Hass. 1992. Fungi from the Lower Devonian Rhynie Chert—Chytridiomycetes. Am. J. Bot. 79:1233–1241.

Taylor, T. N., W. Remy, and H. Hass. 1994. Allomyces in the Devonian. Nature 367:601–601.

Taylor, T. N., W. Remy, H. Hass, and H. Kerp. 1995b. Fossil arbuscular mycorrhizae from the Early Devonian. Mycologia 87:560–573.

Taylor, T. N., and E. L. Taylor. 1997. The distribution and interactions of some Paleozoic fungi. Rev. Palaeobot. Palynol. 95:83–94.

Tehler, A., J. S. Farris, D. L. Lipscomb, and M. Källersjo. 2000. Phylogenetic analyses of the fungi based on large rDNA data sets. Mycologia 92:459–474.

Thorn, R. G., and G. L. Barron. 1984. Carnivorous mushrooms. Science 224:76–78.

Ustinova, I., L. Krienitz, and V. A. R. Huss. 2000. *Hyaloraphidium curvatum* is not a green alga, but a lower fungus; *Amoebidium parasiticum* is not a fungus, but a member of the DRIPs. Protist 151:253–262.

Varma, A., and B. Hock. 1999. Mycorrhiza: structure, function, molecular biology, biotechnology. Springer-Verlag, New York.

Vogler, D. R., and T. D. Bruns. 1998. Phylogenetic relationships among the pine stem rust fungi (*Cronartium* and *Peridermium* spp.). Mycologia 90:244–257.

Wainright, P. O., G. Hinkle, M. L. Sogin, and S. K. Stickel. 1993. Monophyletic origins of the Metazoa: "an evolutionary link with fungi." Science 260:340–342.

Wang, D. Y. C., S. Kumar, and S. B. Hedges. 1999. Divergence time estimates for the early history of animal phyla and the origin of plants, animals and fungi. Proc. R. Soc. Lond. B 266:163–171.

Wasson, G. 1980. The wondrous mushroom: mycolatry in Mesoamerica. McGraw-Hill, New York.

Weir, A., and M. Blackwell. 2001. Molecular data support the Laboulbeniales as a separate class of Ascomycota, Laboulbeniomycetes. Mycol. Res. 105:715–722.

Wells, K., and R. J. Bandoni. 2001. Heterobasidiomycetes. Pp. 85–120 in The mycota VIIB, systematics and evolution (D. J. McLaughlin, E. G. McLaughlin, and P. A. Lemke, eds.). Springer-Verlag, Berlin.

Wessels, J. G. H. 1994. Developmental regulation of fungal cell wall formation. Annu. Rev. Phytopathol. 32:413–437.

Wu, C. G., and J. W. Kimbrough. 1992. Ultrastructural studies of ascosporogenesis in *Ascobolus immersus*. Mycologia 84:459–466.

V

The Relationships of Animals: Overview

Douglas J. Eernisse
Kevin J. Peterson

The History of Animals

This is an exciting time for zoologists. A dramatic upsurge in interest in the interrelationships among animals has occurred across the biological subdisciplines; before the last decade, the topic of high-level animal relationships was one largely confined to zoological texts and older monographs. Revolutionary advances in the fields of phylogenetic analysis, paleontology, developmental biology, and microscopic anatomy, combined with a new wealth of relevant data such as DNA and protein sequences, have led to new insights into animal genealogy. These insights are crucial in this era of "omics": a deeper understanding of any process, including molecular processes, requires an understanding of the underlying pattern, particularly the phylogenetic topology of the systems under consideration.

One of the most significant changes to occur with our understanding of animal evolution is the recognition that animals should be arranged on a phylogenetic tree, and ancestors inferred from character states, rather than the ladder-like progression from protozoans to mammals with ancestors inferred from "archetypes." Despite this new appreciation for the necessity of phylogenetic patterns, it is important to emphasize that even if the topology were somehow precisely known, there would still be uncertainties concerning the appearance or life history attributes of many ancestral metazoan taxa, to say nothing of gene regulatory networks and molecular cascades.

What follows is our attempt to synthesize what is known about high-level (i.e., interphylum) animal relationships,

including the controversies that surround some of the crucial cladogenic events. We start from the base of the animal tree and proceed to the individual subclades of bilaterian metazoans, with the latter summarized only briefly because these topics are considered in much greater detail elsewhere in this book. Controversies still remain, but it is also true that agreement among zoologists has never been greater; the basic pattern of animal evolution has largely been resolved into a few major lineages. This congruence is shown in figure 13.1. Figure 13.1A summarizes where the field is with respect to animal interrelationships. This by necessity is a very conservative tree with many polytomies, yet compared with the state of the field just 15 years ago, we have made remarkable progress, and we expect that most of these polytomies will be resolved with the wealth of data being generated. Figure 13.1B is our total-evidence tree, where we combined our morphological data matrix (modified from Peterson and Eernisse 2001) with 335 small subunit (SSU) or 18S ribosomal DNA (rDNA) sequences, and 43 myosin heavy chain type II inferred amino acid sequences (details are provided in the appendix). The common names of many of these taxa are given in table 13.1, as is the number of SSU rDNA and myosin II sequences analyzed for each taxon, and the Bremer support index for selected nodes of interest. Although our data set is able to resolve all of the polytomies, many with high Bremer support (table 13.1), these should be viewed as tentative hypotheses rather than a consensus among workers in the field. We now discuss the interrelationships of the major animal groups; the reader should refer to figure 13.1 and table 13.1 throughout the remainder of the chapter to see the branching patterns discussed in each section and to compare the consensus nodes with those that are more equivocal.

Are Metazoans Monophyletic?

Until just recently, it seemed possible that sponges arose independently from unicellular ancestors different from those giving rise to all other animals. However, it is now clear from both morphological and molecular analyses that all multicellular animals, including sponges, are monophyletic. The morphological evidence for monophyly consists of many derived attributes that co-occur with the origin of multicellularity at the base of Metazoa ("Met" in fig. 13.1B), including the presence not only of multicellularity but also of the extracellular matrix (Morris 1993) and septate junctions (Nielsen 2001), as well as reproductive features such as eggs with polar bodies and spermatozoa. Furthermore, the molecular support extends beyond SSU rDNA (e.g., Wainright et al. 1993) to include combined SSU rDNA and large subunit (LSU, or 28S) rDNA (Medina et al. 2001), heat-shock protein HSP70 (Borchiellini et al. 1998, Snell et al. 2001), the largest subunit of RNA polymerase II (Stiller et al. 2001, Stiller and Hall 2002), and EF-2 and β-tubulin proteins (King and Carroll 2001). Because the monophyly of Metazoa is robust, multicellularity evolved just once within the animal lineage.

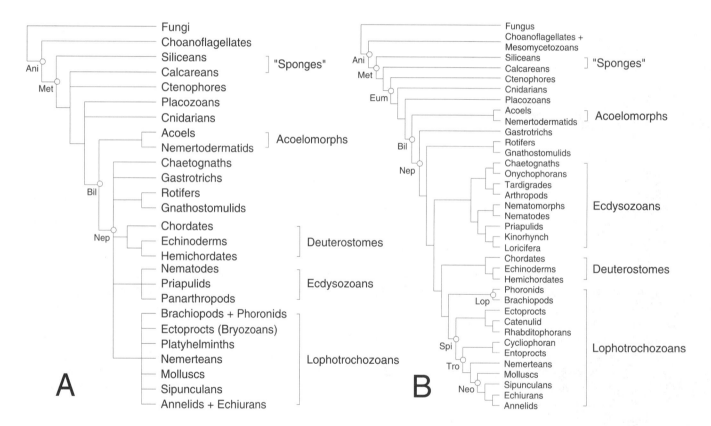

Figure 13.1. The interrelationships among major animal groups. (A) The consensus view from the literature. Although the general structure is apparent, there are several places where much controversy (and work) exists, including the base of Eumetazoa, and especially among the lophotrochozoan taxa. (B) Summary of our combined data set analysis of metazoans. This is the strict consensus summary of first 2000 most parsimonious trees (1115 parsimony-informative characters for 337 taxa, including two with only morphology data; branch length, L = 12,700). To simplify results, the resolution of some terminal taxa scored and analyzed separately are not depicted (see text for details). Bremer support indices and the number of taxa analyzed for SSU rDNA and myosin II are given in table 13.1. Some selected nodes have been labeled with a three-letter taxon abbreviation: Ani, Animalia; Bil, Bilateria; Eum, Eumetazoa; Lop, Lophophorata; Met, Metazoa; Neo = Neotrochozoa; Nep, Nephrrozoa; Spi, Spiralia; Tro, Trochozoa. Nexus format data matrices, search blocks, and full consensus tree descriptions as well as details of sequences analyzed are available from D.J.E.

Table 13.1

Bremer, Support Indices (BSI) for Terminal and Selected Higher Metazoan Taxa for Combined Analysis of Morphology, SSU rDNA, and Myosin II Data Sets.

Taxa	Common name	BSI
Terminal Taxa		
(No. SSU/myosin II)		
Silicea (10/0)	Siliceous sponges	2
Calcarea (4/0)	Calcareous sponges	4
Ctenophora (3)	Comb jellies	23
Cnidaria (27/3)	Cnidarians	8
Placozoa (2/0)	*Trichoplax*	22
Acoela (11/3)	Acoel flatworms	28
Nemertodermatida (2/1)	Nemertodermatid flatworms	37
Gastrotricha (2/0)	Gastrotrichs	12
Rotifera (6/1)	Rotifers	19
Gnathostomulida (3/0)	Gnathostomulids	13
Chaetognatha (3/0)	Arrow worms	15
Onychophora (2/0)	Velvet worms	27
Tardigrada (6/0)	Water bears	18
Arthropoda (47/9)	Arthropods	1
Nematomorpha (3/0)	Horsehair worms	14
Nematoda (17/3)	Round worms	20
Priapulida (6/1)	Priapulids	4
Kinorhyncha (1/0)	Kinorhynchs	—
Loricifera (0/0)	Loriciferans	—
Chordata (24/6)	Chordates	5
Echinodermata (6/0)	Echinoderms	12
Hemichordata (6/0)	Hemichordates	3
Phoronida (3/1)	Phoronids	9
Brachiopoda (20/1)	Brachiopods	10
Ectoprocta (2/0)	Bryozoans	4
Catenulida (1/0)	Catenulid flatworms	—
Rhabditophora (38/5)	Rhabditophoran flatworms	11
Cycliophora (1/0)	Cycliophorans	—
Entoprocta (2/0)	Entoprocts	15
Nemertea (4/1)	Ribbon worms	5
Mollusca (12/3)	Mollusks	1
Sipuncula (7/1)	Peanut worms	27
Echiura (3/1)	Spoon worms	18
Annelida (39/3)	Segmented worms	1
Selected higher taxa		
Metazoa	Multicellular animals	6
Eumetazoa	Eumetazoans	6
Bilateria	Bilaterians	36
Acoelomorpha	Acoelomorphs	1
Nephrozoa	Nephrozoans	6
Ecdysozoa	Ecdysozoans	4
Deuterostomia	Deuterostomes	6
Lophotrochozoa	Lophotrochozoans	1
Lophophorata	Brachiopods + phoronids	6
Spiralia	Spiralians	1
Trochozoa	Trochozoans	1
Neotrochozoa	Neotrochozoans	3

Although animal monophyly is firmly established, controversies still remain. One crucial issue relates to whether particular features shared by sponges and all other animals are truly derived for animals or whether they could be more primitive (i.e., found outside of Metazoa). A good example is the presence of receptor tyrosine kinases, a group of mol-ecules involved in cell–cell signaling and thought to be apomorphic for Metazoa (Suga et al. 1999). King and Carroll (2001) recently found a receptor tyrosine kinase in the choanoflagellate *Monosiga*, raising the possibility that many molecules (including those involved in such traditional multicellular activities as cell-to-cell communication and development) currently thought to exist only in animals (and known to be absent in fungi) might be present in choanoflagellates as well. This problem is not restricted to choanoflagellates: the absence of molecules that characterize higher level metazoan groups in "poriferans" is often the result of negative PCR experiments, and until we have a genome sequence from a sponge, all absences fall into the category of "absence of evidence" rather than the preferable "evidence of absence." As a point in fact, nerve cell genes such as *Pax* transcription factors have recently been isolated in sponges (Gröger et al. 2000), suggesting that they might be much more complex than usually presupposed (e.g., Müller 2001).

What Is the Sister Taxon of Metazoans?

Molecular data support the monophyly of a subclade of eukaryotes called Opisthokonta (Baldauf and Palmer 1993, Baldauf et al. 2000, Atkins et al. 2000, Zettler et al. 2001; see Loytynoja and Milinkovitch 2001), which includes metazoans, choanoflagellates, fungi, and several other poorly known unicellular eukaryotic taxa. Within Opisthokonta, metazoans and choanoflagellates appear quite closely related compared with the more distantly related fungi. The morphology of choanoflagellates has long suggested an affinity with animals, specifically sponges. The similarity between the feeding "collar" cells of sponges and those single-celled but frequently colonial choanoflagellates, first noticed more than a century ago (James-Clark 1866, 1868), is striking, and all morphological and molecular analyses conclude that this similarity is not due to convergence but instead was present in the last common ancestor of animals ("Ani" in fig. 13.1B: Animalia = Choanoflagellata + Metazoa; Nielsen 1995).

There is also another recently recognized group, the mesomycetozoans (alternatively known as ichthyosporeans), which are closely related to choanoflagellates and/or metazoans. Mesomycetozoans are parasites of various fish, birds, mammals, and snails (reviewed in Mendozoa et al. 2002; see also Hertel et al. 2002). In some analyses, Mesomycetozoa is resolved as the sister taxon of choanoflagellates, whereas in others it is the sister taxon of metazoans (Medina et al. 2001, Peterson and Eernisse 2001). King and Carroll (2001) argued that, even if mesomycetozoans comprise the sister taxon of metazoans, choanoflagellates are still the most appropriate metazoan outgroups to study because, as parasites, mesomycetozoans are more likely to have experienced general genomic simplification events. Nonetheless, it is prudent to include both choanoflagellates and mesomycetozoans as outgroups when estimating metazoan basal branching patterns. The diversity of choanoflagellates and mesomycetozoans is still poorly known, and it is possible that additional opisthokont taxa will be discovered (Moon-van der Staay et al. 2001).

Are Sponges Monophyletic?

Porifera is usually assumed to be monophyletic, and this notion is supported by their possession of the water-canal system, a unique arrangement of canals and pores not found in other metazoans. Nonetheless, recent analyses of SSU rDNA that have included an appropriate assortment of sponges, other animals such as cnidarians, and non-metazoan outgroups have instead found sponges to be paraphyletic (e.g., Borchiellini et al. 2001, Peterson and Eernisse 2001, Medina et al. 2001). In particular, those sponges whose skeleton is composed of calcareous spicules (Calcarea) have been supported as comprising the sister taxon of Eumetazoa ("Eum" in fig. 13.1B), the clade composed of all "nonsponge" metazoans, whereas the remaining sponges with a skeleton composed of siliceous spicules (Silicea) comprise the monophyletic sister taxon of the Calcarea + Eumetazoa clade. If the recent SSU rDNA analyses are accurate, then the name "Porifera" should be abandoned and replaced by Calcarea and Silicea. The controversy has important implications. Sponge paraphyly would simplify the optimization of ancestral conditions in ancient metazoans because then the last common ancestor of eumetazoans and calcareans would be more confidently spongelike, complete with a water-canal system. This is because the most proximal outgroup to the Calcarea + Eumetazoa clade, Silicea, also has a water-canal system indistinguishable from the calcarean water-canal system. Furthermore, sponge paraphyly would suggest that the last common ancestor of all animals had a water-canal system as well, and that the acquisition of a spongelike body plan occurred during the early evolution of metazoans and was lost early in the evolution of eumetazoans. Despite the prevailing textbook view of sponge monophyly, as well as our morphology-only analysis (Peterson and Eernisse 2001), sponge paraphyly is consistent with the presence of cross-striated rootlets in calcareous sponges and eumetazoans, but not in siliceous sponges or choanoflagellates (Nielsen 2001). Even if sponges are monophyletic, the near certain monophyly of metazoans and the placement of spongelike choanoflagellates as a near outgroup together imply that our ancient ancestors were "sponges." If living sponges represent a paraphyletic grade, not a clade, of basal metazoans, then the similarities between Silicea and Calcarea reflect only what they lack: the derived traits associated with the eumetazoan body plan.

What Are the Basal Relationships within Eumetazoa?

As for Metazoa, the monophyly of Eumetazoa is strongly supported by morphological evidence. Eumetazoans have clear body symmetry (either radial or bilateral), a mouth and gut, a nervous system, and tissues with characteristic organization, including a basement membrane layer as well as gap junctions and belt desmosomes, all of which are lacking in sponges (Nielsen 2001). Eumetazoa consists of four monophyletic groups whose interrelationships are still unresolved: Cnidaria (anemones and jellies), Ctenophora (comb jellies), Placozoa (a taxon of simple two-layered animals represented by the genus *Trichoplax*), and Bilateria (i.e., all remaining eumetazoans, which primitively have bilateral symmetry; also referred to as the triploblasts because of their three-layered bodies).

Although cnidarians, like sponges, have been popularly represented as models for our ancient ancestors, there is a fundamental difference: unlike sponges, there is substantial molecular evidence for cnidarian monophyly (Collins 2002). This is consistent with various morphological synapomorphies (Schuchert 1993), including their unique production of nematocysts, extracellular encapsulated structures that cnidarians produce in association with their predatory feeding (Tardent 1995). Also unequivocal is the close relationship between cnidarians and bilaterians to the exclusion of the sponges.

What is equivocal is how ctenophores and placozoans fit into the eumetazoan topology. SSU rDNA studies often find that ctenophores group either with the calcareous sponges (e.g., Wainright et al. 1993, Cavalier-Smith et al. 1996, Collins 1998, Kim et al. 1999, Medina et al. 2001, Podar et al. 2001) or basal to calcareous sponges and the remaining eumetazoan taxa (e.g., Peterson and Eernisse 2001), resulting in a paraphyletic Eumetazoa. In contrast, morphological studies have strongly supported ctenophores as comprising the sister taxon of bilaterians (Nielsen et al. 1996, Zrzavý et al. 1998, Peterson and Eernisse 2001). The almost insurmountable difficulty with clade Ctenophora + Calcarea is that complex systems like the nervous system, in addition to many other characters such as tissues, must have evolved twice, once in ctenophores and once in the remaining eumetazoans (or secondarily lost in calcareous sponges), a conclusion advocated by Cavalier-Smith et al. (1996). When a combined analysis of morphology and SSU rDNA sequence data is attempted, the multiple morphological synapomorphies for Eumetazoa, as well as the few supporting Ctenophora + Bilateria, cancel out the SSU rDNA synapomorphies such that neither cnidarians nor ctenophores are robustly supported as comprising a sister taxon of bilaterians (e.g., Peterson and Eernisse 2001). In fact, our new combined analysis (fig. 13.1B) finds a topology distinct from, but influenced by, both data sets: Eumetazoa is monophyletic, but ctenophores are basal to the remaining eumetazoans. This placement is also consistent with newly emerging data on *Hox* and *Parahox* genes, which appear to support a basal eumetazoan position because ctenophores seem to lack most, if not all, of these genes (Martindale et al. 2002). As above, we emphasize that this absence might not be primary because it is a possible secondary loss or merely absence due to methodological problems.

Placozoans are equally problematic. As discussed above, molecular results tend to suggest an affinity with either bilaterians or (more rarely) cnidarians, whereas morphologists and morphological cladistic analyses have favored a basal position among eumetazoans (Bonik et al. 1976, Grell and Ruthmann 1991, Nielsen et al. 1996, Collins 1998, Zrzavý et al. 1998, Peterson and Eernisse 2001). A position within Cnidaria, specifically within the Medusazoa (*sensu* Collins 2002; e.g., Bridge et al. 1995) is convincingly rejected by Ender and Schierwater (2003), who show that placozoans have a normal circular mitochondrial genome, not the derived linear version known exclusively from medusozoans. Contrary to morphology, analysis of SSU rDNA suggests a more apical position for placozoans, often as comprising the sister taxon of Bilateria, and the addition of morphology does not change this result (fig. 13.1B). Therefore, their simplicity might be better explained by reduction from a more complex body plan than by primitive simplicity relative to the other more complex eumetazoan taxa.

Resolving the interrelationships among eumetazoans is crucial because only by doing so will we elucidate which eumetazoan subgroup is the sister group of bilaterians. It appears that comparisons with cnidarians will remain most productive (Martindale et al. 2002) even should placozoans be found more proximal to bilaterians than are cnidarians. This is because of the similarities between cnidarians and bilaterians in developmental complexity and because the placozoan body plan is likely highly reduced.

Bilaterian Relationships

Of all the nodes found on the metazoan tree, none are more strongly supported than the monophyly of Bilateria ("Bil" in fig. 13.1B). Characters supporting the monophyly of Bilateria include (1) distinct anterior-posterior, dorsoventral, and left right axes [but see Martindale et al. (2002) for possible antecedents in cnidarians and ctenophores]; (2) mesoderm as a distinct germ layer giving rise to, for example, circular and longitudinal muscles; (3) nerves organized into distinct ganglia; (4) an expansion of the *Hox* complex to include at least seven genes; (5) the polar bodies positioned on the animal pole; and (6) the specification of one body axis during oogenesis (Peterson and Eernisse 2001). Two other characters, the presence of nephridia and a through-gut with mouth and anus, depend on the phylogenetic position of acoelomorph flatworms, as discussed below. Hence, all morphological studies find strong support for bilaterian monophyly (e.g., Nielsen et al. 1996, Zrzavý et al. 1998, Peterson and Eernisse 2001). SSU rDNA data are equally unequivocal (reviewed in Adoutte et al. 1999, 2000), as are myosin heavy-chain data (Ruiz-Trillo et al. 2002).

The traditional "textbook" approach to bilaterian phylogeny is to view the evolution of the coelom as a proxy for the evolution of bilaterians themselves. This view is traditionally ascribed to Hyman (1940; see also Hyman 1951), who in turn credits Schimkewitsch (1891). This is the familiar view that acoelomate flatworms are the most basal group; then come the "pseudocoelomates," including nematodes, priapulids, and most other "aschelminth" groups; and then finally the coelomates, including arthropods, mollusks, annelids, and chordates. Although Hyman (1940) clearly viewed this transition as a grade of increasing complexity, not always corresponding to phylogenetic pattern, she argued forcefully against the notion of acoelomate and pseudocoelomate conditions as secondarily derived. Nonetheless, the first morphological cladistic analyses based on explicit data matrices did not support the "Hyman" hypothesis of progressive acquisition of a coelomic condition. Schram (1991) found the "aschelminths" to be basal to both flatworms and coelomates, and Eernisse et al. (1992; see also for a reanalysis of the Schram data set) found nematodes grouping with the arthropods, and flatworms grouping with the spirally cleaving protostomes such as annelids and mollusks.

Nonetheless, it was not until SSU rDNA studies starting with Field et al. (1988) that a different view of bilaterian evolution began to emerge (Adoutte et al. 1999). Rather than viewing bilaterian evolution as a ladder of coelomic complexity, instead bilaterians can be divided into three major groups independent of the presence/absence of the coelom: (1) the deuterostomes, composed of echinoderms, hemichordates, and chordates; (2) the lophotrochozoans (Halanych et al. 1995), composed of lophophorates (brachiopods and phoronids), those taxa possessing a trochophore larva (e.g., annelids, mollusks), the catenulid and rhabidophoran flatworms, and many other minor groups, including rotifers, cycliophorans, and possibly gastrotrichs and gnathostomulids; and (3) the ecdysozoans (Aguinaldo et al. 1997), composed of panarthropods, nematodes, priapulids, and other minor aschelminth groups such as kinorhynchs and nematomorphs. Hence, Lophotrochozoa consists of conventional coelomate, pseudocoelomate, and acoelomate groups, and Ecdysozoa consists of "coelomate" groups such as arthropods and most of the pseudocoelomate taxa. This tripartite division removes "intermediate" taxa such that characters thought to apply only to coelomates now characterize all bilaterians (Adoutte et al. 1999). Thus, the story underlying bilaterian evolution seems to be one of an initial complexity followed by numerous simplifications within Ecdysozoa and Lophotrochozoa, as well as Deuterostomia (Takacs et al. 2002).

Although the monophyly of each of these groups is fairly well supported, the interrelationships among the three are not clear. Usually, a monophyletic Protostomia is assumed, and one character supporting this hypothesis is the presence of the UbdA signature peptide, a stretch of about 11 amino acids C-terminal of the homeodomains of the *Ubx*, *Abd-A*, *Lox-2*, and *Lox-4 Hox* genes (de Rosa et al. 1999, Saló et al. 2001). However, not a single SSU rDNA study has demonstrated any appreciable support for the monophyly of

Protostomia, nor has any other arrangement been strongly supported.

The Deuterostomes

Traditionally, deuterostomes consisted of six taxa: echinoderms, hemichordates, chordates, lophophorates, ectoprocts, and chaetognaths. However, both molecular and morphological analyses agree that lophophorates, ectoprocts, and chaetognaths are not deuterostomes. Deuterostomia *sensu stricto* consists of hemichordates and echinoderms (collectively called ambulacrarians), and the chordates, the monophyletic sister group of the ambulacrarians. For further discussion of deuterostome evolution, see Smith et al. (ch. 22 in this vol.).

The Lophotrochozoa

By far the most phylogenetically challenging group is Lophotrochozoa. Named by Halanych et al. (1995) to reflect its primary taxonomic constituents, the lophophorates (brachiopods and phoronids) and trochozoans (i.e., those protostome phyla having trochophore larva, e.g., annelids and mollusks), as well as groups such as ectoprocts that do not fit under either category, this is by far the largest group of higher level metazoan taxa, containing up to about 14 phyla. Furthermore, it is the least studied group with respect to molecular investigations, because none of its members are currently genetic model systems. In general, we can say very little about how lophotrochozoan phyla are related to one another. There are few morphological characters for resolving deep-level lophotrochozoan relationships, and there is virtually no resolution with SSU rDNA (for discussion and references, see Halanych 1998, Peterson and Eernisse 2001, Giribet 2002). Analyses of LSU (Mallat and Winchell 2002) and the myosin heavy chain (Ruiz-Trillo et al. 2002) have also failed to provide robust and biologically reasonable interrelationships among lophotrochozoans. Even the monophyly of some of the more conspicuous phyla, such as Annelida and Mollusca, is rarely recovered using molecular data.

Our best estimate of lophotrochozoan relationships divides this group into three subgroups: lophophorates [restricted in Peterson and Eernisse (2001) to brachiopods and phoronids], platyzoans (rotifers, gnathostomulids, platyhelminths, and possibly gastrotrichs; Cavalier-Smith 1998; but see Zrzavý et al. 2003 for gastotrichs), and the trochozoans (entoprocts, nemerteans, annelids, mollusks, echiurans, and sipunculans, modified from Ghiselin 1988; compare Beklemishev 1969). There is strong morphological support for the monophyly of lophophorates (e.g., Peterson and Eernisse 2001), but the monophyly of Lophophorata, as well as the monophyly of the remaining groups, is still under debate with respect to molecular data. Giribet and colleagues (Giribet et al. (2000, Giribet 2002) recovered a monophyletic Platyzoa, as did Peterson and Eernisse (2001) in their morphological analysis. With respect to trochozoans, all analyses agree that these taxa are more closely related to one another than to any platyzoan subgroup, but the interrelationships among these taxa are obscure at the moment, as is the taxonomic constituency of such taxa as Annelida (Halanych et al. 2002).

Morphology alone strongly suggests that lophophorates are basal lophotrochozoans, because they lack several important spiralian (Spiralia = Platyzoa + Trochozoa) and trochozoan characters such as spiral cleavage and a trochophore larval form, respectively (Peterson and Eernisse 2001). The difficulty is that most SSU rDNA analyses place the lophophorates within the trochozoans, often as the sister group to a mollusk or annelid subgroup, but usually with very little support. Nonetheless, this hypothesis is supported by the possession of annelid-like setae in brachiopods (Ghiselin 1989). The reason the position of the lophophorates is critical is that characters supporting the monophyly of Lophotrochozoa depend heavily on the relative position of lophophorates. If Lophophorata is nested within Trochozoa, then all of the traditional developmental characters, such as spiral cleavage and the possession of a prototroch, would constitute basal lophotrochozoan characters (with the interesting by-product of making Lophotrochozoa equivalent to Spiralia). As Giribet (2002) pointed out, Halanych et al. (1995) did not include any platyzoans in their original analysis when first diagnosing Lophotrochozoa, so the potential membership of platyzoans in Lophotrochozoa must depend on their position relative to lophophorates. If lophophorates are basal to Spiralia, then the only nonsequence characters presently supporting the monophyly of Lophotrochozoa are the possession of two *Abd-B Hox* genes, *post-1* and *post-2* (see Callaerts et al. 2002; note that this is known for only brachiopods, annelids, and mollusks), and the *Lox-5* signature peptide, a stretch of eight amino acids C-terminal of the homeodomain of the *Lox5* gene, known in platyhelminths, nemerteans, annelids, brachiopods, and mollusks (de Rosa et al. 1999, Saló et al. 2001, reviewed in Balavoine et al. 2002).

Although there are several other lophotrochozoan taxa, such as the ectoprocts, virtually nothing can be said about how they fit into the lophotrochozoan tree. One of the problems is that sequences for these taxa have been few and taxonomic sampling has been sparse. In some cases (e.g., ectoprocts), this can be easily remedied. In other cases (e.g., cycliophorans), there are relatively few extant species to sample, so multiple gene sequence comparisons are more apt to help.

The Ecdysozoa

Perhaps the most surprising result of SSU rDNA analyses was the formulation of Ecdysozoa by Aguinaldo et al. (1997). Instead of using long-branch nematode taxa like *Caenorhabditis elegans*, Aguinaldo et al. (1997) found shorter branched taxa that, when analyzed phylogenetically, grouped robustly with arthropods. This was unusual given that all previous analyses found nematodes to be basal bilaterians, supporting the traditional notion of a basal Pseudocoelomata (e.g., Winnepenninckx et al. 1995). Since Aguinaldo et al.'s (1997) analysis, numerous SSU rDNA studies (e.g., Giribet et al. 2000, Peterson and Eernisse 2001) have found strong support for a clade consisting of panarthropods, nematodes, nematomorphs, priapulids, kinorhynchs, and loriciferans (assumed, based on morphology alone, to be closely related to kinorhynchs and priapulids). Moreover, the monophyly of Ecdysozoa is further supported by phylogenetic analyses of LSU (Mallatt and Winchell 2002) and myosin heavy chain (fig. 13.1B; Ruiz-Trillo et al. 2002). In addition, a monophyletic Ecdysozoa is recovered using morphological data (Zrzavý et al. 1998, Peterson and Eernisse 2001); ecdysozoans share similarities in their cuticle and ecdysis pathways (Schmidt-Rhaesa et al. 1998), a terminal mouth, a distinct *Abd-B* gene (Van Auken et al. 2000), an internal triplication within the [-*thymosin* gene (Manuel et al. 2000), neural expression of horseradish peroxidase (HRP) immunoreactivity (Haase et al. 2001), the absence of cannabinoid receptors (McPartland et al. 2001), and the absence of the *Parahox* gene *Xlox* (Ferrier and Holland 2001)]. They might also share similarities in their circumpharyngeal brain (Eriksson and Budd 2000). Thus, the monophyly of Ecdysozoa is recovered using a variety of data sets (fig. 13.1).

Both morphological and molecular analyses agree on the monophyly of the three main Ecdysozoan groups: (1) Scalidophora (Lemburg 1995, Schmidt-Rhaesa et al. 1998, also referred to as Cephalorhyncha by some authors), consisting of priapulids, kinorhynchs and loriciferans; (2) Nematoida (Schmidt-Rhaesa 1996), consisting of nematodes and nematomorphs; and (3) Panarthropoda (Nielsen 1995), consisting of arthropods, onychophorans, and tardigrades. However, the interrelationships among these three groups are unclear.

The Chaetognath Problem

One of the more difficult groups to place phylogenetically is Chaetognatha. Chaetognaths show an odd mix of deuterostome and aschelminth-type characters (Hyman 1959), but because preference was usually given to embryological characters, chaetognaths were traditionally one of the six major deuterostome groups. Initial studies based on cladistic arguments found grouping with either deuterostomes (e.g., Brusca and Brusca 1990) or aschelminths (Schram 1991). Initial SSU rDNA analyses (Telford and Holland 1993, Turbeville et al. 1994, Wada and Satoh 1994; see also Giribet et al. 2000) did not support a placement within Deuterostomia but could not place them with any significant support elsewhere within Bilateria. Halanych (1996) argued that they were the sister group of the nematodes and argued that this was not due to long-branch attraction. More recent analyses seemed to confirm a placement within Ecdysozoa (e.g., Peterson and Eernisse 2001). Morphological analyses alone also suggest

that chaetognaths are basal ecdysozoans (Peterson and Eernisse 2001, Zrzavý et al. 2001), sharing with Ecdysozoa proper a terminal mouth, possibly a chitinous cuticle, absence of a ciliated epidermis, absence of an apical organ, and other larval structures, and they share with nematoidans the absence of circular muscles. A basal position to Ecdysozoa *sensu stricto* is also supported by the absence of HRP immunoreactivity in the chaetognath nervous system (Haase et al. 2001).

It has recently been shown that two characters usually given for a deuterostome affinity were misunderstood in chaetognaths. First, the presence of a trimeric arrangement of the coeloms is at best questionable in chaetognaths because the septum that divides the trunk into anterior and posterior compartments is not a primary septum but a secondary division derived from coelomic cells (Kapp 2000). Second, radial cleavage does not occur in chaetognaths. Instead, they have a tetrahedral four-cell embryo whose cleavage planes are similar to those of crustacean arthropods and nematodes (Shimotori and Goto 2001), and also comparable with the Precambrian embryos described by Xiao et al. (1998). The remaining deuterostome characters, for example, mouth not derived from blastopore, may represent bilaterian plesiomorphies (Peterson and Eernisse 2001). Thus, all available evidence points to an affinity with ecdysozoans, but where they fall within this group remains speculative at best. Because chaetognaths have the most strongly guanine + cytosine–biased sequences among all animal SSU rDNA sequences sampled to date (Peterson and Eernisse 2001), it would be desirable to test this hypothesis with amino acid comparisons instead of (or in addition to) the traditional SSU rDNA or LSU analyses.

The Acoelomorph Problem

One of the more interesting results to emerge from SSU rDNA analyses is the purported basal position of acoelomorph flatworms (Ruiz-Trillo et al. 1999, Jondelius et al. 2002), a placement that could shed much light on the plesiomorphic state of the early bilaterians (e.g., Ruiz-Trillo et al. 1999, 2002, Adoutte et al. 2000, Jondelius et al. 2002). Acoelomorphs (collectively the acoel and nemertodermatid flatworms) were conventionally considered basal platyhelminths because they possess neoblasts, a unique stem cell found only in flatworms (Ax 1996, Gschwentner et al. 2001, Ramachandra et al. 2002), and morphology-alone analyses confirm a flatworm affinity (e.g., Peterson and Eernisse 2001). Because of their possession of neoblasts, a basal position within Bilateria appeared suspicious, a suspicion that seemed justified given that acoels were also very long-branched taxa (Adoutte et al. 2000, Peterson and Eernisse 2001). Peterson and Eernisse (2001) tested this hypothesis and found that acoels strongly attract random DNA sequences and, to the extent that distant outgroups such as cnidarians might be behaving effectively as random sequences, their attraction to a basal position was considered to be potentially artifactual. In contrast, the internal branch between protostomes and deuterostomes was never attracted to random outgroups, yet that is where the root attached when acoelomorphs and selected other taxa subject to long-branch attraction were removed.

Nevertheless, Ruiz-Trillo et al. (2002) analyzed myosin heavy-chain type II sequences from a variety of bilaterians, including acoelomorphs, and similar to their SSU rDNA result, found acoelomorphs to be basal bilaterians. Consistent with these results, our total-evidence tree also finds a basal Acoelomorpha (fig. 13.1B). A basal position is only moderately less consistent with the morphological data: placing acoelomorphs basally adds only four steps to the analysis. Furthermore, Saló et al. (2001) reported that they were unable to find more than three *Hox/ParaHox* genes in the acoels *Paratomella* and *Convoluta*, and these observations are consistent with the basal bilaterian position supported for acoelomorphs based on available sequence data sets. Therefore, Jondelius et al. (2002) proposed the name Nephrozoa ("Nep" in fig. 13.1B; reflecting the evolution of nephridia) to include the last common ancestor of all bilaterians except acoelomorphs and all descendants of that last common ancestor living or extinct. Nephrozoa would also be characterized by the possession of a through-gut, complete with mouth and anus, which was most likely lost secondarily in platyhelminths (now restricted to exclude acoelomorphs).

The Biology of the Earliest Bilaterians

The implications for a basal position of Acoelomorpha (or "acoelomorph" grade) are striking. Baguñà et al. (2001) proposed that if their mode of development is primitive then it is likely that the earliest bilaterians were small, benthic, directly developing animals without a coelom, segments, a true brain, or nephridia. Of their conclusions, the proposed lack of a true brain in the earliest bilaterians might need reconsideration in light of the recently demonstrated brain primordium in the acoel *Neochildia*, as assessed by the expression of *POU* genes (Ramachandra et al. 2002). Jondelius et al. (2002) further proposed that acoelomorphs arose via progenesis from a planula-like larva. This is a very different scenario for early bilaterian evolution than that espoused, for example, by Davidson and colleagues (e.g., Davidson et al. 1995, Peterson et al. 2000), which postulated indirect development to be primitive and the earliest bilaterians to be small planktonic larval forms. It also differs from the morphology-biased prediction of Peterson and Eernisse (2001), that the last common ancestor of bilaterians (including acoelomorphs) was a large organism with deuterostome-like development (including possibly the possession of a "dipleurula-like" larva) and a tripartite arrangement of coeloms similar to modern hemichordates. However, trimery can no longer be considered primitive for Bilateria because neither phoronids (Bartolomaeus 2001) nor cha-

etognaths (Kapp 2000) are trimeric, which reduces trimery to a novel synapomorphy for Ambulacraria (see Smith et al., ch. 22 in this vol.). Furthermore, this result suggests that there is no reason to postulate that a coelom is primitive for either Bilateria or Nephrozoa (*contra* Budd and Jensen 2000).

We find it intriguing that if acoelomorphs are basal to other bilaterians, this strengthens the inference that the earliest bilaterians were small, interstitial, or meiofaunal animals. Within the remaining bilaterians, small body size is widespread, so it is at least feasible that the last common ancestor of the most familiar animals (e.g., vertebrates, insects, mollusks) was likewise small and benthic. The results (not shown) of SSU rDNA plus morphology alone still support acoelomorphs as basal bilaterians but differ from the total-evidence tree (fig. 13.1B) in that gastrotrichs, gnathostomulids, and rotifers are basal lophotrochozoans. We also found the more conventional split between protostomes (ecdysozoans + lophotrochozoas) and deuterostomes exclusive of Acoelomorpha. If this topology is further supported, then the case for a small, creeping, and direct-developing last common ancestor of not only Nephrozoa but also Protostomia is strongly supported, because the outgroup(s) (acoelomorphs) and basal lineages of at least Lophotrochozoa are small bodied. This could explain why trace fossils are absent during the earliest phase of bilaterian evolution dating from about 600 million years ago (K. J. Peterson, J. B. Lyons, K. S. Nowak, C. M. Takacs, M. J. Wargo, and M. A. McPeek, unpubl. obs.) to 555 million years ago, when traces make their first appearance in the rock record (Martin et al. 2000). The story underlying bilaterian evolution may be one of initial genetic complexity not manifested until the Cambrian explosion.

Conclusions

What continually strikes us is that, aside from a few minor controversies, disparate data sets lead to a remarkably similar topology of the major animal groups. But equally as important (and interesting) is that no single data set is entirely accurate. For example, morphology alone might be "incorrect" (albeit relatively weak) in supporting a monophyletic Porifera, a sister grouping between ctenophores and bilaterians, and placing acoelomorphs within Platyhelminthes. On the other hand, morphology, but not SSU rDNA, can potentially resolve the interrelationships among trochozoans. Along the same vein as our earlier works (e.g., Eernisse 1997, Peterson and Eernisse 2001), we continue to advocate a total-evidence approach with several different types of data derived from numerous taxa. The ever continual advancement in phylogenetic software, molecular tools, and scientific perspective can only lead to a better understanding of the interrelationships among the major animal lineages and, of course, to animal evolution itself.

Appendix: Materials and Methods

The morphology matrix is a revised version of the "morphology" analysis presented in Peterson and Eernisse (2001). Our new matrix consists of 168 characters; it is not exclusively morphological because it also includes coding of developmental or biochemical variation, as well as coding of some molecular aspects such as inferred *Hox* gene duplication events and genetic code differences. The results of this analysis are only slightly different from our previous study and largely agree with those derived from sequence data despite a general perception that molecular results differ fundamentally from what might be inferred from morphology. The modified matrix is available from either author.

We also analyzed two different molecular data sets: 43 myosin heavy-chain type II inferred amino acid sequences, and a data set of 335 selected and manually aligned SSU rDNA sequences (the full matrix is available upon request from D. J. E.). The myosin heavy-chain data set, recently assembled by Ruiz-Trillo et al. (2002), is the newest non-rDNA data set available for a broad range of metazoan taxa and is probably the most promising current alternative to the widely studied SSU rDNA data set [see Giribet (2002) for a review of the others]. In order to combine these data sets, we matched myosin heavy-chain sequences with sequences from the same or related species whose SSU rDNA sequences we analyzed, and then treated each combined sequence as a single taxon. This is similar to the method employed by Ruiz-Trillo et al. (2002) except that, whereas they limited their analysis to only those taxa represented by myosin heavy-chain sequences, we kept the nearly 300 SSU rDNA sequences not matched by particular myosin heavy-chain sequences in the combined analysis, coding the myosin heavy-chain portion for those sequences as missing data. Also unlike those authors, we also combined these molecular data with our morphology matrix. As in Peterson and Eernisse (2001), we did not attempt to code corresponding morphology scores for each of the 335 taxa whose SSU rDNA sequences we analyzed. Instead, for our morphology analysis we gave equivalent morphology scores to each of the sequenced species within each of our terminal taxa. This will create bias in the combined data set favoring the monophyly of these terminal taxa; usually this was not a problem because most of these taxa were already found to be monophyletic in the molecular analyses. The few exceptions, such as annelids and mollusks, that were monophyletic in the combined but not the SSU rDNA analysis could be monophyletic merely because of the groupwide morphology scores they were given.

Methods used for sequence alignment, exclusion of those sites with ambiguous alignment, data set combination, and two-step heuristic search strategy in PAUP* (ver. 4b10; Swofford 2002), are very similar to those employed in Peterson and Eernisse (2001; see also Eernisse and Kluge 1992, Eernisse

1997). We did not include one of the redundant rodent myosin heavy-chain sequences in the combined analysis. Our SSU rDNA data set consisted of 278 of the 302 SSU rDNA sequences analyzed in Peterson and Eernisse (2001), plus 57 additional SSU rDNA sequences beyond those analyzed previously, added to bolster previously underrepresented taxa. We also varied the taxon composition of the SSU rDNA and myosin heavy-chain sequence data sets, and analyzed a number of these different taxon combinations plus our reported 335 taxon SSU rDNA data set with different algorithms, specifically using minimum evolution heuristic searches (HKY85 and LogDet distances as implemented in PAUP*) and Bayesian inference searches using Mr. Bayes software (ver. 2.01; Huelsenbeck and Ronquist 2001). All of these results were consistent with the general pattern resulting from the reported analyses, with the most substantial differences typically involving where particular "long-branch" sequences (e.g., chaetognaths, nemertodermatids, gnathostomulids, onychophorans) happened to be resolved within Bilateria. For example, the nemertodermatid and gnathostomulid sequences were observed to group together or apart anywhere from basally within Bilateria, to within chordates, to within the panarthropods as sister group to onychophorans, and such movement was characteristic of all algorithms employed in the case of the SSU rDNA analyses.

Literature Cited

Adoutte, A., G. Balavoine, N. Lartillot, and R. de Rosa. 1999. Animal evolution: the end of intermediate taxa? Trends Genet. 15:104–108.

Adoutte, A., G. Balavoine, N. Lartillot, O. Lespinet, B. Prud'homme, and B. de Rosa. 2000. The new animal phylogeny: reliability and implications. Proc. Natl. Acad. Sci. USA 97:4453–4456.

Aguinaldo, A. M. A., J. M. Turbeville, L. S. Linford, M. C. Rivera, J. R. Garey, R. A. Raff, and J. A. Lake. 1997. Evidence for a clade of nematodes, arthropods and other molting animals. Nature 387:489–493.

Ahlrichs, W. 1995. Ultrastruktur und Phylogenie von *Seison nebaliae* (Grube 1859) und *Seison annulatus* (Claus 1876). Hypothesen zu phylogenetischen Verwandtschaftsverhältnissen innerhalb der Bilateria. Cuvillier Verlag, Göttingen.

Atkins, M. S., A. G. McArthur, and A. P. Teske. 2000. Ancyromonadida: a new phylogenetic lineage among the Protozoa closely related to the common ancestor of Metazoans, Fungi, and Choanoflagellates (Opisthokonta). J. Mol. Evol. 51:278–285.

Ax, P. 1996. Multicellular animals: a new approach to the phylogenetic order in nature, vol. 1. Springer, Berlin.

Baguñà, J., I. Ruiz-Trillo, J. Paps, M. Loukota, C. Ribera, U. Jondelius, and M. Riutort. 2001. The first bilaterian organisms: simple or complex? New molecular evidence. Int. J. Dev. Biol. 45:S133–S134.

Balavoine, G., R. de Rosa, and A. Adoutte. 2002. Hox clusters and bilaterian phylogeny. Mol. Phylogenet. Evol. 24:366–373.

Baldauf, S. L., and J. D. Palmer. 1993. Animals and fungi are each other's closest relatives: congruent evidence from multiple proteins. Proc. Natl. Acad. Sci. USA 90:11558–11562.

Baldauf, S. L., A. J. Roger, I. Wenk-Siefert, and W. F. Doolittle. 2000. A kingdom-level phylogeny of eukaryotes based on combined protein data. Science 290:972–977.

Bartolomaeus, T. 2001. Ultrastructure and formation of the body cavity lining in *Phoronis muelleri* (Phoronida, Lophophorata). Zoomorphology 120:135–148.

Beklemishev, V. N. 1969. Principles of comparative anatomy of invertebrates (J. M., MacLennan, trans.; Z. Kabata, ed.). University of Chicago Press, Chicago.

Bonik, K., M. Grasshoff, and W. F. Gutmann. 1976. Die Evolution der Tierkonstruktionen I. Problemlage und Prämissen. Vielzeller und die Evolution der Gallertoide. Nat. Mus. 106:129–143.

Borchiellini, C., N. Boury-Esnault, J. Vacelet, and Y. Le Parco. 1998. Phylogenetic analysis of the Hsp70 sequences reveals the monophyly of Metazoa and specific phylogenetic relationships between animals and fungi. Mol. Biol. Evol. 15:647–655.

Borchiellini, C., M. Manuel, E. Alivon, N. Boury-Esnault, J. Vacelet, and Y. Le Parco. 2001. Sponge paraphyly and the origin of Metazoa. J. Evol. Biol. 14:171–179.

Bridge, D., C. W. Cunningham, R. Desalle, and L. W. Buss. 1995. Class-level relationships in the phylum Cnidaria: molecular and morphological evidence. Mol. Biol. Evol. 12:679–689.

Brusca, R. C., and G. J. Brusca. 1990. Invertebrates. Sinauer, Sunderland, MA.

Budd, G. E., and S. Jensen. 2000. A critical reappraisal of the fossil record of the bilaterian phyla. Biol. Rev. Camb. Philos. Soc. 75:253–295.

Callaerts, P., P. N. Lee, B. Hartmann, C. Farfan, D. W. Y. Choy, K. Ikeo, K.-F. Fischback, W. J. Gehring, and H. Gert de Couet. 2002. *HOX* genes in the sepiolid squid *Euprymna scolopes*: implications for the evolution of complex body plans. Proc. Natl. Acad. Sci. USA 99:2088–2093.

Cavalier-Smith, T. 1998. A revised six-kingdom system of life. Biol. Rev. 73:203–266.

Cavalier-Smith, T., M. T. E. P. Allsopp, E. E. Chao, N. Boury-Esnault, and J. Vacelet. 1996. Sponge phylogeny, animal monophyly, and the origin of the nervous system: 18S rRNA evidence. Can. J. Zool. 74:2031–2045.

Collins, A. G. 1998. Evaluating multiple alternative hypotheses for the origin of Bilateria: an analysis of 18S rRNA molecular evidence. Proc. Natl. Acad. Sci. USA 95:15458–15463.

Collins, A. G. 2002. Phylogeny of Medusozoa and the evolution of cnidarian life cycles. J. Evol. Biol. 15:418–432.

Davidson, E. H., K. J. Peterson, and R. A. Cameron. 1995. Origin of adult bilaterian body plans: evolution of developmental regulatory mechanisms. Science 270:1319–1325.

de Rosa, R., J. K. Grenier, T. Andreeva, C. E. Cook, A. Adoutte, M. Akam, S. B. Carroll, and G. Balavoine. 1999. Hox genes in brachiopods and priapulids and protostome evolution. Nature 399:772–776.

Eernisse, D. J. 1997. Arthropod and annelid relationships re-examined. Pp. 43–56 *in* Arthropod relationships (R. A. Fortey, and R. H. Thomas, eds.). Systematics Association Special Volume Series 55. Chapman and Hall, London.

Eernisse, D. J., J. S. Albert, and F. E. Anderson. 1992. Annelida and Arthropoda are not sister taxa: a phylogenetic analysis of spiralian metazoan morphology. Syst. Biol. 41:305–330.

Eernisse, D. J., and A. Kluge. 1993. Taxomonic congruence versus total evidence, and amniote phylogeny inferred from fossils, molecules, and morphology. Mol. Biol. Evol. 10:1170–1195.

Ender, A., and B. Schierwater. 2003. Placozoa are not derived cnidarians: evidence from molecular morphology. Mol. Biol. Evol. 20:130–134.

Eriksson, B. J., and G. E. Budd. 2000. Onychophoran cephalic nerves and their bearing on our understanding of head segmentation and stem-group evolution of Arthropoda. Arthrop. Struct. Dev. 29:197–209.

Ferrier, D. E. K., and P. W. H. Holland. 2001. Sipunculan ParaHox genes. Evol. Dev. 3:263–270.

Field, K. G., G. J. Olsen, D. J. Lane, S. J. Giovannoni, M. T. Ghiselin, E. C. Raff, N. R. Pace, and R. A. Raff. 1988. Molecular phylogeny of the animal kingdom. Science 239:748–753.

Ghiselin, M. T. 1988. The origin of molluscs in the light of molecular evidence. Oxford Surv. Evol. Biol. 5:66–95.

Ghiselin, M. T. 1989. Summary of our present knowledge of metazoan phylogeny. Pp. 262–272 *in* The hierarchy of life (B. Fernholm, K. Bremer, and H. Jörnvall, eds.). Elsevier Science Publishers, Amsterdam.

Giribet, G. 2002. Current advances in the phylogenetic reconstruction of metazoan evolution: a new paradigm for the Cambrian explosion? Mol. Phylogenet. Evol. 24:345–357.

Giribet, G., D. L. Distel, M. Polz, W. Sterrer, and W. C. Wheeler. 2000. Triploblastic relationships with emphasis on the acoelomates and the position of Gnathostomulida, Cycliophora, Plathelminthes, and Chaetognatha: a combined approach of 18S rNDA sequences and morphology. Syst. Biol. 49:539–562.

Grell, K. G., and A. Ruthmann. 1991. Placozoa. Pp. 13–28 *in* Microscopic anatomy of invertebrates (F. W. Harrison and J. A. Westfall, eds.), vol. 2. Wiley-Liss, New York.

Gröger, H., P. Callaerts, W. J. Gehring, and V. Schmid. 2000. Characterization and expression analysis of an ancestor-type *Pax* gene in the hydrozoan jellyfish *Podocoryne carnea*. Mech. Dev. 94:157–169.

Gschwentner, R., P. Ladurner, K. Nimeth, and R. Rieger. 2001. Stem cells in a basal bilaterian: S-phase and mitotic cells in *Convolutriloba longifissura* (Acoela, Platyhelminthes). Cell Tissue Res. 304:401–408.

Haase, A., M. Stern, K. Wächtler, and G. Bicker. 2001. A tissue-specific marker of Ecdysozoa. Dev. Genes Evol. 211:428–433.

Halanych, K. M. 1996. Testing hypotheses of chaetognath origins: long branches revealed by 18S ribosomal DNA. Syst. Biol. 45:223–246.

Halanych, K. M. 1998. Consideration for reconstructing metazoan history: signal, resolution, and hypothesis testing. Am. Zool. 38:929–941.

Halanych, K. M., J. D. Bacheller, A. M. A. Aguinaldo, S. M. Liva,

D. M. Hillis, and J. A. Lake. 1995. Evidence from 18S ribosomal DNA that the lophophorates are protostome animals. Science 267:1641–1643.

Halanych, K. M., T. G. Dahlgren, and D. McHugh. 2002. Unsegmented annelids? Possible origins of four lophotrochozoan worm taxa. Integ. Comp. Biol. 42:678–684.

Hertel L. A., C. J. Bayne, and E. S. Loker. 2002. The symbiont *Capsaspora owczarzaki*, nov. gen. nov. sp., isolated from three strains of the pulmonate snail *Biomphalaria glabrata* is related to members of the Mesomycetozoea. Int. J. Parasitol. 32:1183–1191.

Hoffman, P. F., A. J. Kaufman, G. P. Halverson, and D. P. Schrag. 1998. A Neoproterozoic snowball Earth. Science 281:1342–1346.

Huelsenbeck, J. P., and F. Ronquist. 2001. MRBAYES: Bayesian inference of phylogeny. Bioinformatics L. H. 17:754–755.

Hyman, L. H. 1940. The invertebrates, Vol. 1: Protozoa through Ctenophora. McGraw-Hill, New York.

Hyman, L. H. 1951. The invertebrates, Vol. 2: Platyhelminthes and Rhynchocoela. McGraw-Hill, New York.

Hyman, L. H. 1959. The invertebrates, Vol. 5: Smaller Coelomate groups. McGraw Hill, New York.

James-Clark, H. 1866. Note on the infusoria flagellata and the spongiae ciliatae. Am. J. Sci. 1:113–114.

James-Clark, H. 1868. On the spongiae ciliatae as infusoria flagellata; or observations on the structure, animality and relationship of *Leucosolenia botryoides*, Bowerbank. Ann. Mag. Nat. Hist. 1:133–142.

Jondelius, U., I. Ruiz-Trillo, J Baguñà, and M. Riutort. 2002. The Nemertodermatida are basal bilaterians and not members of the Platyhelminthes. Zool. Scr. 31:201–215.

Kapp, H. 2000. The unique embryology of Chaetognatha. Zool. Anz. 239:263–266.

Kim, J., W. Kim, and C. W. Cunningham. 1999. A new perspective on lower metazoan relationships from 18S rDNA sequences. Mol. Biol. Evol. 16:423–427.

King, N., and S. B. Carroll. 2001. A receptor tyrosine kinase from choanoflagellates: molecular insights into early animal evolution. Proc. Natl. Acad. Sci. USA 98:15032–15037.

Lemburg, C. 1995. Ultrastructure of the introvert and associated structures of the larvae of *Halicryptus spinulosus* (Priapulida). Zoomorphology 115:11–29.

Loytynoja, A., and M. C. Milinkovitch. 2001. Molecular phylogenetic analyses of the mitochondrial ADP-ATP carriers: the Plantae/Fungi/Metazoa trichotomy revisited. Proc. Natl. Acad. Sci. USA 98:10202–10207.

Mallatt, J., and C. J. Winchell. 2002. Testing the new animal phylogeny: first use of combined large-subunit and small-subunit rRNA gene sequences to classify the protostomes. Mol. Biol. Evol. 19:289–301.

Manuel, M., M. Kruse, W. E. G. Müller, and Y. Le Parco. 2000. The comparison of -thymosin homologues among Metazoa supports and arthropod-nematode clade. J. Mol. Evol. 51:378–381.

Martin, M. W., D. V. Grazhdankin, S. A. Bowring, D. A. D. Evans, M. A. Fedonkin, and J. L. Kirschvink. 2000. Age of Neoproterozoic bilaterian body and trace fossils, White Sea, Russia: implications for metazoan evolution. Science 288:841–845.

Martindale, M. Q., J. R. Finnerty, and J. Q. Henry. 2002. The Radiata and the evolutionary origins of the bilaterian body plan. Mol. Phylogenet. Evol. 24:358–365.

McPartland, J., V. Di Marzo, L. De Petrocellis, A. Mercer, and M. Glass. 2001. Cannabinoid receptors are absent in insects. J. Comp. Neurol. 436:423–429.

Medina, M., A. G. Collins, J. D. Silberman, and M. L. Sogin. 2001. Evaluating hypotheses of basal animal phylogeny using complete sequences of large and small subunit rRNA. Proc. Natl. Acad. Sci. USA 98:9707–9712.

Mendozoa, L., J. W. Taylor, and L. Ajello. 2002. The class Mesomycetozoea: a heterogeneous group of microorganisms at the animal-fungal boundary. Annu. Rev. Microsc. 56:315–344.

Moon-van der Staay, S. Y., R. De Wachter, and D. Vaulot. 2001. Oceanic 18S rDNA sequences from picoplankton reveal unsuspected eukaryotic diversity. Nature 409:607–610.

Morris, P. J. 1993. The developmental role of the extracellular matrix suggests a monophyletic origin of the kingdom Animalia. Evolution 47:152–165.

Müller, W. E. G. 2001. Review: how was metazoan threshold crossed? The hypothetical Urmetazoa. Comp. Biochem. Physiol. A 129:433–460.

Nielsen, C. 1995. Animal evolution: interrelationships of the living phyla. Oxford University Press, Oxford.

Nielsen, C. 2001. Animal evolution: interrelationships of the living phyla. 2nd ed. Oxford University Press, Oxford.

Nielsen, C., N. Scharff, and D. Eibye-Jacobsen. 1996. Cladistic analysis of the animal kingdom. Biol. J. Linn. Soc. 57:385–410.

Peterson, K. J., R. A. Cameron, and E. H. Davidson. 2000. Bilaterian origins: significance of new experimental observations. Dev. Biol. 219:1–17.

Peterson, K. J., and D. J. Eernisse. 2001. Animal phylogeny and the ancestry of bilaterians: inferences from morphology and 18S rDNA gene sequences. Evol. Dev. 3:170–205.

Podar, M., S. H. D. Haddock, M. L. Sogin, and G. R. Harbison. 2001. A molecular phylogenetic framework for the phylum Ctenophora using 18S rRNA genes. Mol. Phylogenet. Evol. 21:218–230.

Ramachandra, N. B., R. D. Gates, P. Ladurner, D. K. Jacobs, and V. Hartenstein. 2002. Embryonic development in the primitive bilaterian *Neochildia fusca*: normal morphogenesis and isolation of POU genes *Brn-1* and *Brn-3*. Dev. Genes Evol. 212:55–69.

Ruiz-Trillo, I., J. Paps, M. Loukota, C. Ribera, U. Jondelius, J. Baguña, and M. Riutort. 2002. A phylogenetic analysis of myosin heavy chain type II sequences corroborates that Acoela and Nemertodermatida are basal bilaterians. Proc. Natl. Acad. Sci. USA 99:11246–11251.

Ruiz-Trillo, I., M. Riutort, D. T. J. Littlewood, E. A. Herniou, and J. Baguñà. 1999. Acoel flatworms: earliest extant bilaterian metazoans, not members of Platyhelminthes. Science 283:1919–1923.

Saló, E., J. Tauler, E. Jimenez, J. R. Bayascas, J. Gonzalez-Linares, J. Garcia-Ferdandez, and J. Baguñà. 2001. Hox and ParaHox genes in flatworms: characterization and expression. Am. Zool. 41:652–663.

Schimkewitsch, W. 1891. Versuch einer Klassifikation des Tierreichs. Biol. Zent. Bl. 11:291–295.

Schmidt-Rhaesa, A. 1996. The nervous system of *Nectonema*

munidae and *Gordius aquaticus*, with implications for the ground pattern of the Nematomorpha. Zoomorphology 116:133–142.

Schmidt-Rhaesa, A., T. Bartolomaeus C. Lemburg, U. Ehlers, and J. Garey. 1998. The position of the Arthropoda in the phylogenetic system. J. Morphol. 238:263–285.

Schram, F. R. 1991. Cladistic analysis of metazoan phyla and the placement of fossil problematica. Pp. 35–46 *in* The early evolution of Metazoa and the significance of problematic taxa (A. M. Simonetta and S. Conway Morris, eds.). Cambridge University Press, Cambridge.

Schuchert, P. 1993. Phylogenetic analysis of the Cnidaria. Z. Zool. Syst. Evol. 31:161–173.

Shimotori, T., and T. Goto. 2001. Developmental fates of the first four blastomeres of the chaetognath *Paraspadella gotoi*: relationship to protostomes. Dev. Growth Differ. 43:371–382.

Snell, E. A., R. F. Furlong, and P. W. H. Holland. 2001. Hsp70 sequences indicate that choanoflagellates are closely related to animals. Curr. Biol. 11:967–970.

Stiller, J. W., and B. D. Hall. 2002. Evolution of the RNA polymerase II C-terminal domain. Proc. Natl. Acad. Sci. USA 99:6091–6096.

Stiller, J. W., J. Riley, and B. D. Hall. 2001. Are red algae plants? A critical evaluation of three key molecular data sets. J. Mol. Evol. 52:527–539.

Suga, H., M. Koyanagi, D. Hoshiyama, K. Ono, N. Iwabe, K.-I Kuma, and T. Miyata. 1999. Extensive gene duplication in the early evolution of animals before the parazoan-eumetazoan split demonstrated by G proteins and protein tyrosine kinases from sponge and hydra. J. Mol. Evol. 48:646–653.

Swofford, D. L. 2002. PAUP* phylogenetic analysis using parsimony (* and other methods), ver. 4.0b10 for Macintosh. Sinauer Associates, Sunderland, MA.

Takacs, C. M., V. N. Moy, and K. J. Peterson. 2002. Testing putative hemichordate homologues of the chordates dorsal nervous system and endostyle: expression of *NK2.1* (*TTF-1*) in the acorn worm *Ptychodera flava* (Hemichordata, Ptychoderidae). Evol. Dev. 4:405–417.

Tardent, P. 1995. The cnidarian cnidocyte, a high-tech cellular weaponry. Bioessays 17:351–362.

Telford, M. J., and P. W. H. Holland. 1993. The phylogenetic affinities of the chaetognaths: a molecular analysis. Mol. Biol. Evol. 10:660–676.

Turbeville, J. M., J. R. Schultz, and R. A. Raff. 1994. Deuterostome phylogeny and the sister group of the chordates: evidence from molecules and morphology. Mol. Biol. Evol. 11:648–655.

Van Auken, K., D. C. Weaver L. G. Edgar, and W. B. Wood. 2000. *Caenorhabditis elegans* embryonic axial patterning requires two recently discovered posterior-group Hox genes. Proc. Natl. Acad. Sci. USA 97:4499–4503.

Wada, H., and N. Satoh. 1994. Details of the evolutionary history from invertebrates to vertebrates, as deduced from the sequences of 18S rDNA. Proc. Natl. Acad. Sci. USA 91:1801–1804.

Wainright, P. O., G. Hinkle, M. L. Sogin, and S. K. Stickel. 1993. Monophyletic origins of the Metazoa: an evolutionary link with Fungi. Science 260:340–342.

Winnepenninckx, B., T. Backeljau, L. Y. Mackey, J. M. Brooks, R. De Wachter, S. Kumar, and J. R. Garey. 1995. 18S rRNA data indicate that Aschelminthes are polyphyletic in origin and consist of at least three distinct clades. Mol. Biol. Evol. 12:1132–1137.

Xiao, S., Y. Zhang, and A. H. Knoll. 1998. Three-dimensional preservation of algae and animal embryos in a Neoproterozoic phosphorite. Nature 391:553–558.

Zettler, L. A. A., C. J. O'Kelly, T. A. Nerad, and M. L. Sogin. 2001. The nucleariid amoebae: more protists at the animal-fungal boundary. J. Eukaryot. Microbiol. 48:293–297.

Zrzavý, J. 2003. Gastrotricha and metazoan phylogeny. Zool. Scripta 32:61–81.

Zrzavý, J., V. Hypsa, and D. F. Tietz. 2001. Myzostomida are not annelids: molecular and morphological support for a clade of animals with anterior sperm flagella. Cladistics 17:170–198.

Zrzavý, J., S. Mihulka, P. Kepka, A. Bezdek, and D. Tietz. 1998. Phylogeny of the Metazoa based on morphological and 18S ribosomal DNA evidence. Cladistics 14:249–285.

D. Timothy J. Littlewood

Maximilian J. Telford

Rodney A. Bray

Protostomes and Platyhelminthes

The Worm's Turn

The simplest partitioning of the bilaterally symmetrical animals (Bilateria) is the split between the Deuterostomia and Protostomia, divisions founded primarily on very different modes of embryonic development. The protostomes (Gr. "mouth first") include those animals in which, after gastrulation, the mouth is formed at or near the blastopore opening rather than being a secondary opening (deuterostomy). Here we introduce some of the major protostome groups not treated elsewhere in this volume, with a particular emphasis on the flatworms (phylum Platyhelminthes) and their allies. We cover 15 phyla, including a number of important but enigmatic groups that have either flirted with shared ancestry with the flatworms or that are difficult to place among the protostomes. Early phylogenetic scenarios often placed the flatworms as basal bilaterian groups (even as ancestral archetypes) from which a range of more complex protostomes arose. As with any phylogenetic tree, the placement of the most basal group has important consequences for our understanding of an evolutionary radiation. Consequently, identifying the basal bilaterian is pivotal for understanding the evolutionary radiation of the major animal phyla.

The Protostomia are currently split into Lophotrochozoa, with members characterized by spiral embryonic cleavage patterns, and Ecdysozoa, characterized by animals that molt an exoskeleton as they grow and develop (see Eernisse and Peterson, ch. 13 in this vol.; see also Gilbert 2000). The taxa we cover (shown in boldface in fig. 14.1) are to be found in both groups, or have yet to be placed convincingly in the tree.

Untangling the inter- and intraphyletic relationships of the various taxa covered here has been driven variously by purely systematic goals and evolutionary questions but also, importantly, within some phyla, by a need to understand parasites and parasitism. Some of the most medically and economically important parasites are found among the Platyhelminthes, Nematoda and Acanthocephala. Additionally, some protostome species have been model organisms for the latest developments in genome research; the nematode *Caenorhabditis elegans*, for example, was the first multicellular animal to have its entire genome sequenced and remains a favored organism for understanding gene function. These applied aspects of biology have often provided both the need to resolve wider patterns of evolutionary radiation and the sources of characters with which phylogenies can be estimated.

Our starting point is the tree shown in figure 14.1. The lack of resolution indicated by the collapsed nodes, and the tentative placement of taxa with dashed lines, indicates conflict and uncertainty over the interrelationships of the protostomes. The tree is an updated version of what Adoutte et al. (2000) termed "the new animal phylogeny." Based largely on ribosomal RNA (rRNA) gene sequences, and other molecular data, it is overall poorly resolved but represents major groupings that are well supported.

Almost without exception, the groups we consider here have little or no fossil record. As soft-bodied animals, their fossil record is, at best, restricted to traces, which provide few reliable characters for phylogenetic analysis. This is in

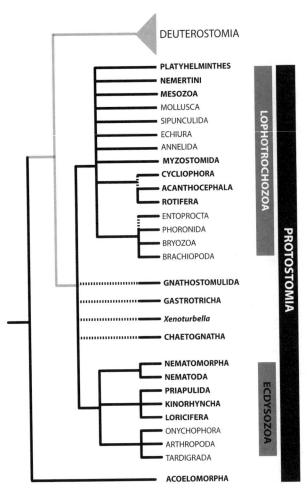

Figure 14.1. Interrelationships of the major protostome groups. Groups covered in this chapter are shown in boldface. Broken lines indicate possible affinities of various groups. Acoelomorpha, once considered members of Platyhelminthes, are now convincingly placed at the base of Bilateria (Deuterostomia + Protostomia).

contrast to other protostome groups such as Arthropoda and Mollusca, which are addressed elsewhere in this volume. Wherever possible, we provide detailed interrelationships within each phylum considered.

Basal Bilaterians

Which was the first bilaterally symmetrical animal to evolve, and what did it look like? As members of Bilateria ourselves, this question generates more than an intrinsic academic interest. The fossil record has not been able to help us because it seems most likely that the first bilaterian was a soft-bodied and possibly microscopic organism that has left few, if any, clues as to its identity. Identifying the earliest branching taxon within Bilateria has been difficult, and different lines of evidence have not converged on a single satisfactory solution.

Because we can work only with extant organisms, the best we can do is to reveal the earliest divergent living group while necessarily ignoring the extinct groups that have left little or no trace. Notwithstanding the usual conflict in opinion over character homology when comparing deeply branching taxa, an additional problem has been that confusion reigns when we inadvertently use para- or polyphyletic taxa as monophyletic groups for coding purposes or when subsequently interpreting a tree. A case in point concerns the acoelomorph flatworms, which, combined with the perceived "primitiveness" of all the flatworms, are responsible for pulling the Platyhelminthes to the base of Bilateria or Protostomia in works dating before the 1990s.

Acoelomorpha

Acoelomorph flatworms include two groups, Acoela (19 families, 120 genera) and the far less species-rich Nemertodermatida (two families, six genera; fig. 14.2). Long established as the most basal members of the phylum Platyhelminthes, along with Catenulida (Ehlers 1985a), the apparent simplicity of members of Acoelomorpha may have contributed to the view that the whole phylum is an early offshoot of the bilateral Metazoa, in combination of course with the phylum's lack of anus or coelom. Their simplicity in form makes the acoelomorphs attractive candidates from which more complex forms may be postulated to have evolved, and although there are no derived characters that unite acoelomorphs with other flatworm groups (Tyler 2001), they have long since been considered members of Platyhelminthes. Insidiously, this simple body plan shared by acoelomorphs and all other flatworms has led to a deal of conflict and a range of scenarios, with flatworms being variously placed as sister group to all protostomes or nestled within Lophotrochozoa. Smith and Tyler (1985) and Smith et al. (1986) were the first to question the monophyly of Platyhelminthes, and Haszprunar (1996b), suggested that acoels alone should be considered basal, with the catenulid and rhabditophoran flatworms as more closely related offshoots of a para- if not polyphyletic Platyhelminthes. Molecular data from small subunit (SSU) ribosomal DNA (rDNA) began a renewed debate that carries on to this day. Ruiz-Trillo et al. (1999) presented evidence that acoels are the most basal bilaterian group and are not monophyletic with the other flatworms, which were indeed Lophotrochozoa. Efforts to avoid long-branch attraction, where divergent taxa spuriously appear at the base of a rooted tree, failed to convince some authors that the basal placement of Acoela was anything but artifact (e.g., Adoutte et al. 1999), and appears to have plagued others (Peterson and Eernisse 2001). The polyphyly of Acoelomorpha in Ruiz-Trillo et al.'s (1999) study has also not helped the case, because strong morphological characters unite the two constituent groups (see below; see also Ehlers 1992, Littlewood et al. 1999b).

Jondelius et al. (2002) have since shown that the sequence attributed to the nemertodermatid *Nemertinoides elongatus* was probably that of a rhabditophoran flatworm and have subsequently provided a denser sampling of SSU rDNA, and a recent study of complete large subunit (LSU) rDNA by Telford et al.(2003) shows that both Acoela and Nemertodermatida appear as basal bilaterians. Although in these studies Acoelomorpha remains weakly paraphyletic, its basal position is robust, setting this group apart from both catenulid and rhabditophoran platyhelminths, which appear convincingly among the Lophotrochozoa. Evidence from a further gene, coding for myosin II, strongly corroborates the basal position of Acoelomorpha (Ruiz-Trillo et al. 2002). Although not giving evidence of a basal position for Acoela, developmental studies demonstrated unique duet spiral cleavage (Henry et al. 2000). Members of Acoela apparently lack ectomesoderm and all musculature, and peripheral parenchyma is of entomesodermal origin, suggesting a possible link with Ctenophora (see Henry et al. 2000, Martindale and Henry 1999a, 1999b). Evidence from neuronal cytochemistry continues to support the uniqueness of Acoelomorpha when contrasted with other flatworm groups (Reuter et al. 2001a, 2001b). The interrelationships of Acoela have been explored phylogenetically most recently with SSU by Hooge et al. (2002), and the relatively species-poor Nemertodermatida has been tackled thoroughly by a morphological analysis (Lundin 2000).

Apomorphies of Acoelomorpha (Ehlers 1985a)
Epidermal cilia with shelflike termination
Rostral rootlet of epidermal cilia with kneelike bend
Posterior rootlet of epidermal cilia with two fiber
    bundles
Reduction of protonephridia

Additional molecular data from the gene that encodes EF-1α protein (Littlewood et al. 2001b) and surveys of mitochondrial genetic code assignment throughout the flatworms (Telford et al. 2000) have conclusively demonstrated the separation of acoelomorphs and rhabditophoran flatworms, and strong evidence from three genes places acoelomorphs as the most basal Bilateria (fig. 14.1).

Gnathostomulida: The Jaw Worms

Gnathostomulida (fig. 14.3), a group consisting of about 100 species of nonsegmented microscopic marine worms, is considered by some to be the sister group to Platyhelminthes; synapomorphies include hermaphroditism, direct transfer of sperm and internal fertilization of egg cells, threadlike sperm, and no mitosis in somatic cells (Ax 1996). In turn, this clade, Plathelminthomorpha, was postulated to be the sister group to all other members of Bilateria (Ax 1985). Recent studies

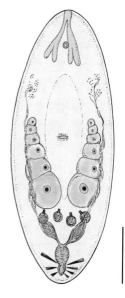 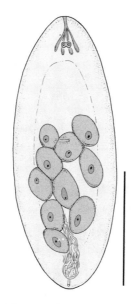

Convoluta norwegica
Convolutidae
ACOELA

Nemertoderma bathycola
Nemertodermatidae
NEMERTODERMATIDA

Figure 14.2. Representatives of each of the constituent acoelomorph groups; courtesy of Queensland Museum, from Cannon (1986), with permission. Scale bars, 200 μm.

including both molecular and combined analyses have prompted a bewildering array of possibilities: based on SSU rDNA, they have been placed among the Ecdysozoa (Littlewood et al. 1998); based on SSU rDNA and morphology, with Platyhelminthes in a clade, Platyzoa, that includes Cycliophora, Syndermata (Acanthocephala + Rotifera), and Gastrotricha (Giribet et al. 2000); based on a combined analysis of SSU rDNA and morphology, in a clade that unites Gnathostomulida and Gastrotricha affiliated with Ecdysozoa (Zrzavý et al. 1998); and based on morphology, in a clade with Rotifera and Acanthocephala related to Lophotrochozoa (Sørensen et al. 2000).

Although gnathostomulids are not strong contenders for the title of most basal bilaterian, their position among the Metazoa is unstable, based on both molecular and morphological studies, and so additional evidence is needed to secure their true position. Recent morphological studies on jaw ultrastructure suggest an affiliation with Rotifera and Micrognathozoa (Sørensen and Sterrer 2002).

Apomorphy of Gnathostomulida (Ax 1985)
Pharynx with jaws and basal plate

Gastrotricha

Gastrotrichs are microscopic, cryptic animals that usually live between grains of sand and silt in both freshwater and saltwater.

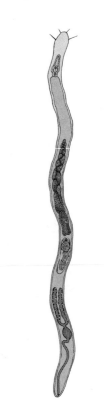

Gnathostomula lutheri
Bursovaginoidea
GNATHOSTOMULIDA

Figure 14.3. Member of the enigmatic Gnathostomulida;
redrawn from Ax (1996).

They are nonsegmented, have a through-gut with a pharynx, and are generally microscopic (50–1000 μm; fig. 14.4). The two constituent orders are very different from one another, but the group has long been considered monophyletic from a morphological perspective. As with other members of the meiofauna, gastrotrichs are relatively poorly studied but constitute an important and ubiquitous component of limnetic and marine sediments and detritus. Ciliated, hermaphroditic, and often bottle-shaped, gastrotrichs are usually flattened ventrally with the posterior end sometimes split into a fork.

Apomorphies of Gastrotricha
Unique, cuticle-covered duo-gland adhesive organ
Multilayered epicuticle
Cuticle-covered locomotory and sensory cilia
Possibly unique left and right helicoidal muscle

The order Macrodasyida, with six recognized families, includes exclusively marine or brackish, interstitial creatures, whereas the more species-rich Chaetonotida, with seven families, includes freshwater and epibenthic animals. Each order is defined primarily on the fine structure of the pharynx. Gastrotricha is another phylum that has vied for the position of most basal bilaterian, with apparent affinities to

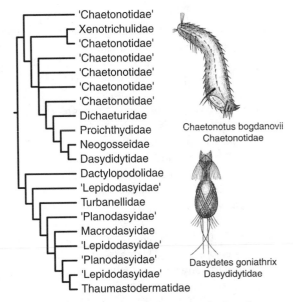

Figure 14.4. Interrelationships of Gastrotricha based on a morphological analysis (Hochberg and Litvaitis 2000), with line drawings of representatives.

Gnathostomulida, according to some (e.g., Boaden 1985), or to the ecdysozoan phyla, most notably Nematoda, by others (Ruppert 1991, Wallace et al. 1996). Many of the early cladistic analyses appear to have suffered from choosing characters unique to chaetonotids, rather than ones apomorphic for the phylum (e.g., see discussion in Hochberg and Litvaitis 2000). The position is even more confused from molecular estimates, largely because of poor sampling of both genes and taxa.

Early molecular studies have variously placed the gastrotrichs as a sister group to Acanthocephala or Nematomorpha (Carranza et al. 1997), Gnathostomulida (Littlewood et al. 1998), or Platyhelminthes (Winnepenninckx et al. 1995). In each case, rarely more than a single gastrotrich sequence was used. Subsequent denser sampling of taxa produced unsatisfactory results because only a limited sampling and range of other metazoan taxa were employed (Wirz et al. 1999), with the result that the group's affinities are not well supported by molecular or by morphological data. Combined molecular and morphological analyses have placed the group as sister group to Gnathostomulida (Zrzavý et al. 1998) or Platyhelminthes (Giribet et al. 2000) and within Ecdysozoa (Peterson and Eernisse 2001).

The gastrotrichs are at least resolved as a monophyletic phylum, although this has yet to be confirmed with molecular data, and recent morphological analyses lend some resolution to the interrelationships of constituent families (see fig. 14.4; see also Hochberg and Litvaitis 2000), although the authors urge caution and suggest the main use of such a phylogeny is for hypothesis testing and appropriate future sampling for molecular studies. Clearly, much has to be done with this neglected but fascinating phylum.

Xenoturbellida

One species alone, *Xenoturbella bockii*, forms the taxon Xenoturbellida (fig. 14.5). As with the Acoelomorpha, *Xenoturbella* is very simple, and perhaps as a consequence it has been placed as a sister group to both Acoelomorpha (Franzén and Afzelius 1987, Hyman 1951) and Bilateria (Ehlers and Sopott-Ehlers 1997a, 1997b, Franzén and Afzelius 1987). The worm shares with members of Acoelomorpha the ability to resorb worn or damaged ciliated epidermal cells (Lundin 2001) and a shelf-like termination of the epidermal cilia (Ax 1996). Affinities with Mollusca, based on molecular data (Norén and Jondelius 1997), appear to results from contamination in the study, because sequences are almost identical to those from the species of protobranch mollusks that *Xenoturbella* feeds on, and affinities based on morphology are likely misinterpretations (Israelsson 1999). Ultrastructural evidence argues strongly against molluscan affinities (Lundin 1998, Lundin and Schander 1999, Raikova et al. 2000), and in the absence of crucial corroborative molecular data, we are left with no concrete idea of the position of *Xenoturbella*. However, new sequence data in the form of SSU rDNA and two mitochondrial genes have recently demonstrated that *Xenoturbella* falls among Deuterostomia perhaps as sister group to Ambulacraria (Bourlat et al. 2003).

Platyhelminthes: The Flatworms

Platyhelminthes includes Acoelomorpha, Catenulida, and Rhabditophora, but as described above, there is no synapomorphy uniting these taxa, and the acoelomorphs appear to be sufficiently different to consider them apart from the other two. Recent analyses of full LSU and SSU rDNA place Catenulida and Rhabditophora as sister groups, and it is these two groups we believe constitute Platyhelminthes to the exclusion of acoelomorphs, although again, there are no

morphological synapomorphies for this grouping. The monophyly of the two constituent groups is not in doubt, and the resolution of rhabditophoran relationships has progressed considerably, although some groupings remain contentious (see fig. 14.6). An excellent online taxonomic database for the free-living flatworms is Tyler (2003). Although a total evidence estimate of the interrelationships of members of Platyhelminthes has been attempted, combining morphological and molecular data (Littlewood et al. 1999a), there are still many problems in resolving a stable phylogeny because we are limited by numbers of morphological characters and by problems in their coding and, in many cases, establishing homology.

Catenulida

Although little systematic effort has been expended in resolving the interrelationships of members of Catenulida, it is clear that as sister group to all other (rhabditophoran) flatworms, it deserves greater attention from both morphological and molecular perspectives. With five families, 11 genera, and more than 100 species, it is surprising that only two species have been sequenced for various molecular estimates of phylogeny. All catenulids are free-living, primarily in freshwater but some in marine environments (see fig. 14.7). When scored for various platyhelminth features, they are notably lacking in many systems that define the majority of other groups, such as a duo-gland adhesive system, or show a great deal of variability in the presence or absence of other features between the families.

> *Apomorphies of Catenulida (Ehlers 1985a)*
> Unpaired protonephridium
> Unique organization of the cyrtocyte
> Dorsally located male genital porus
> Aciliary spermatozoa

Rhabditophora

The majority of Platyhelminthes are members of Rhabditophora, and the group is very readily recognized as monophyletic. With the exception of members of Catenulida, the rhabditophorans encapsulate the full diversity of the phylum. The clade is split into a number of distinct groups that have variously been ascribed class, ordinal, and family level status.

> *Apomorphies of Rhabditophora (Ehlers 1985a,*
> *Telford et al. 2000)*
> Lamellated rhabdites
> Duo-gland adhesive system
> Duo-cell weir of the protonephridia
> Multiciliary terminal cells of the protonephridia
> Unusual codon usage in mitochondrial genes: AAA =
>   Asn not Lys, AUA = Ile not Met

Xenoturbella bocki
XENOTURBELLIDA

Figure 14.5. *Xenoturbella bocki*, Xenoturbellida; redrawn from Ax (1996).

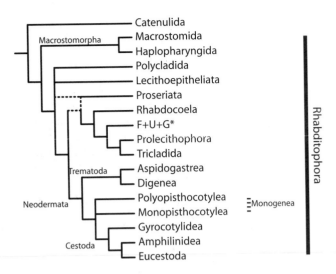

*F+U+G = Fecampiida+Urastomidae+Genostomatidae

Figure 14.6. Interrelationships of Platyhelminthes based on various sources.

We take each of the major groups in turn. Figure 14.6 illustrates the best estimate of their interrelationships based on both morphological and molecular evidence.

Macrostomorpha

Macrostomorphs encompass Haplopharyngida and Macrostomida, and in almost all phylogenetic analyses they appear as the sister group to the remaining Rhabditophora groups. Haplopharyngida are represented by just three species, all marine, whereas the macrostomids, with three families, 23 genera, and many hundreds of species, is much more diverse with representatives found in marine, brackish, and freshwater. Early competing hypotheses as to the interrelationships of the constituent groups are founded on phylogenies prepared on assessments of the adhesive system (Tyler 1976, Tyler and Rieger 1977) or the construction of the pharynx (Doe 1981). Most recently, Rieger (2001) using these features plus the rhammites and the female canal system, offered two alternative phylogenies, the consensus of which suggests that complementary molecular sequencing will serve well in resolving the interrelationships of this, perhaps most basal rhabditophoran group. If Rhabditophora do indeed represent the majority of flatworms (or at least the monophyletic Platyhelminthes), this group is pivotal within the phylum (fig. 14.7).

Apomorphies of Macrostomorpha (Doe 1986, Rieger 2001)
Duo-gland adhesive organs emerge in one collar of modified microvilli
Pharynx simplex coronatus
Aciliary spermatozoa

Polycladida

Although some representatives of the 37 families of polyclads live in fresh or brackish water, these generally large worms are predominantly marine, free-living flatworms. Many are associated with other organisms as symbionts, and tropical representatives found on reefs include some of the most spectacularly colorful invertebrates. Split into Cotylea, whose members have a pseudosucker posterior to the female genital pore, and Acotylea, whose members do not, polyclads are often resolved as relatively deep-branching platyhelminths. A combined morphological and molecular assessment places the group as sister taxon to Macrostomorpha at the base of Rhabditophora (Littlewood et al. 1999b), although molecular data alone fail to adequately resolve monophyly of a Polycladida + Macrostomorpha clade (Littlewood and Olson 2001). The internal relationships have yet to be tackled, but such a phylogeny will be invaluable because the group includes many unique larval forms, and the evolution of their development will prove fascinating (fig. 14.7).

Apomorphies of Polycladida (Ehlers 1985a, Littlewood et al. 1996b)
Extensive intestinal branching
Resorption of certain blastomeres during development
Characteristic plicatus-type pharynx

Figure 14.7. Basal flatworm groups; images courtesy of Queensland Museum, from Cannon (1986), with permission. Scale bars, 200 μm.

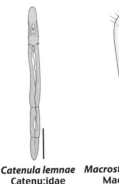

Catenula lemnae
Catenu:idae
CATENUL:DA

Macrostomum curvituba
Macrostomidae
MACROSTOM:DA

Discocelis australis
Discocoe:idae
POLYCLAD:DA

Gnosonesima borealis
Gnosonesimidae
LEC:THOEP:THEL:ATA

Lecithoepitheliata

Of the major flatworm groups this is probably the least studied as regarding internal phylogeny, at least in terms of modern phylogenetic systematic methods. The worms are free-living and are found in freshwater, marine, and terrestrial environments. The only morphological assessment places them as the sister group to Prolecithophora + Rhabdocoela (Littlewood et al. 1999b), but there are no explicit autapomorphies for the group. Composed of two families, Prorhynchidae and Gnosonesimidae, according to Timoshkin (1991), the group has no well-defined homology to unite it and may not even be monophyletic. SSU rDNA data have been collected only for one genus of Prorhynchidae, but the analyses including the most densely sampled flatworms places the lecithoepitheliates as a basal group united with macrostomorphs (fig. 14.7).

Proseriata

Proseriates are marine worms, predominantly interstitial but occupying a variety of trophic levels. Seven families are recognized and include more than 250 species. Although a recent combined molecular assessment of Proseriata cast doubt as to whether the group is truly monophyletic (Littlewood et al. 2000), additional evidence based on complete SSU and LSU rDNA has since demonstrated monophyly (Lockyer et al. 2003). Three synapomorphies were erected to describe the group, but each of these has been found to be present in non-proseriates, and Curini-Galletti (2001) considers this sufficient reason to focus on the constituent clades, which each have strong autapomorphies. The two proseriate groups are Unguiphora and Lithophora, and molecular estimates using both complete SSU and LSU rDNA show each to be monophyletic.

Apomorphies for Unguiphora (Curini-Galletti 2001)
"Multiple" ovaries
Claw-shaped stylet
Cocoons with up to nine openings

Apomorphies for Lithophora (Curini-Galletti 2001)
One pair of compact ovaries
Sclerotized copulatory structures never claw shaped
Cocoons with one opening

Figure 14.8 depicts the interrelationships of four of the six lithophoran families according to recent analysis (Curini-Galletti 2001, and see discussion therein). Of the remaining families, it seems likely that Monotoplanidae are not monophyletic and probably fall within the Monocelididae.

Tricladida

There are more than 100 genera of triclads and many hundreds of species. Originally considered to be sister group to Proseriata in a clade called Seriata, Tricladida is not placed anywhere near the proseriates by SSU rDNA. Instead, these ubiquitous, free-living worms inhabit freshwater, brackish, marine, and terrestrial environments and appear quite robustly in a clade that includes Prolecithophora and a small but remarkable group of non-neodermatan parasitic flatworms, Fecampiida + Urastomidae + Genostomatidae (see below and fig. 14.6). In the most densely sampled SSU rDNA analysis, Tricladida is sister group to Prolecithophora (Littlewood and Olson 2001), but there are no obvious morphological synapomorphies for this grouping. The internal relationships of Tricladida have been estimated using a variety of gene fragments and, although requiring additional evidence, are shown in figure 14.9 (Baguñà et al. 2001, Carranza et al. 1998). Triclads, or more commonly planarians, include some of the best-known free-living flatworms. Many have the ability to regenerate after being cut in two or more pieces and are therefore excellent candidates for studies on developmental genetics (Baguñà 1998). Additionally, some, such as Dugesiidae, appear to be potential indicators of terrestrial biodiversity (Sluys 1999), so their phylogeny has been investigated in some detail (Sluys 2001).

Apomorphies of Tricladida (Baguñà et al. 2001, Carranza et al. 1998, Ehlers 1985a, Littlewood et al. 1999b)
Three-branched intestine
Two germaria located at anterior end of germo-vitelloducts
Formation of transitory embryonic pharynx
Crossing over of pharynx muscles
Cerebral position of female gonads

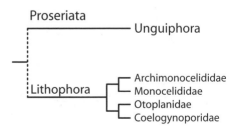

Nematoplana ciliovesiculae
Nematoplanidae
Unguiphora

Promonotus orthocirrus
Monocelididae
Lithophora

Figure 14.8. Interrelationships of Proseriata; images courtesy of Queensland Museum, from Cannon (1986), with permission. Scale bars, 200 μm.

Tricladida

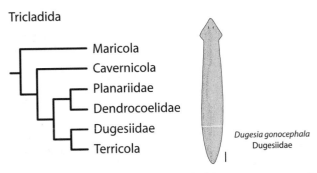

Figure 14.9. Interrelationships of Tricladida; image courtesy of Queensland Museum, from Cannon (1986), with permission. Scale bar, 200 μm.

Serial arrangement of many nephridiopores
Marginal adhesive zone

Prolecithophora

There are approximately 150 species of this group classified into 11 or so families. The most recent assessment of prolecithophoran interrelationships combines a predominantly molecular approach with an assessment of sperm characters (Jondelius et al. 2001). Not all the families have been sampled for molecular analysis, but SSU rDNA resolves most of the interfamilial phylogeny quite well. A combination of results from two separate molecular studies is shown in figure 14.10 (see Jondelius et al. 2001, Littlewood and Olson 2001, and D. T. J. Littlewood, unpubl. obs.).

   Apomorphy of Prolecithophora (Ehlers 1988)
   Abundantly folded membrane derivatives in the
      aflagellar sperm cells

Rhabdocoela

This group originally included all remaining flatworms, a clade composed of Dalyelliida, Temnocephalida, Kalyptorhynchia, Typhloplanida, and Neodermata. With a single putative morphological autapomorphy, "unspecialized phar-

ynx bulbosus," molecular data consistently fail to resolve the group as a whole, and Rhabdocoela is now restricted to the original constituent taxa, excluding Neodermata. Consequently, there is no apomorphy for the group, although it seems well supported at least from SSU rDNA analyses (Littlewood and Olson 2001, Littlewood et al. 1999b). Temnocephalida, whose members are characterized by an epidermis made of multiple syncytial plates (Joffe and Cannon 1998), and Dalyelliida and Typhloplanida likely form a clade according to SSU rDNA, but their interrelationships need further investigation; kalyptorhynchs appear consistently as the sister group to these three taxa (Littlewood and Olson 2001, Littlewood et al. 1999b; fig. 14.11).

Little effort has been made to elucidate the interrelationships of the constituent groups of Rhabdocoela except among polcystid Kalyptorhynchia (Artois and Schockaert 1998) and the Temnocephalida (Cannon and Joffe 2001). The temnocephalids are all ectosymbiotic and have developed a distinct posterior sucker. A recent analysis of interrelationships by Cannon and Joffe (2001), which also includes a list of apomorphies for the group, is shown in figure 14.11. Watson (2001) has provided additional apomorphies from studies on sperm and spermiogenesis for Temnocephalida and Kalyptorhynchia.

Fecampiida, Urastomidae, Genostomatidae, and a Note on the Revertospermata

Three enigmatic groups of flatworms that have been allied historically with the free-living taxa, but are all found in close association with vertebrate or invertebrate hosts, were recently thought to be members of a clade including the obligate parasites, the Neodermata. The clade, termed Revertospermata [so named because of a peculiar migration of the sperm nucleus relative to sperm tail seen in neodermatans and these taxa (Kornakova and Joffe 1999)], is not supported by molecular data. However, Fecampiida, Urastomidae, and Genostomatida do form a convincing clade, and molecular data place the clade as sister to Tricladida + Prolecithophora. A revertospermatan clade is compelling from a parasitological perspective, uniting most of the flatworms with a close

Figure 14.10. Interrelationships of Prolecithophora; images courtesy of Queensland Museum, from Cannon (1986), with permission. Scale bars, 200 μm.

Prolecithophora

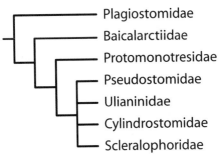

Allostoma pallidum
Cylindrostomidae

Baicalarctia gulo
Baicalarctiidae

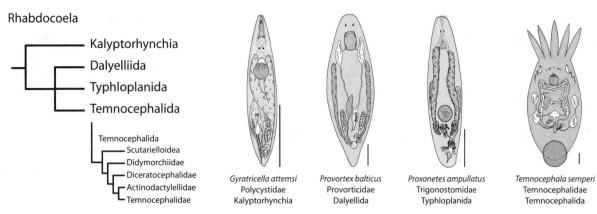

Rhabdocoela

Kalyptorhynchia
Dalyelliida
Typhloplanida
Temnocephalida

Temnocephalida
Scutarielloidea
Didymorchiidae
Diceratocephalidae
Actinodactylellidae
Temnocephalidae

Gyratricella attemsi
Polycystidae
Kalyptorhynchia

Provortex balticus
Provorticidae
Dalyellida

Proxonetes ampullatus
Trigonostomidae
Typhloplanida

Temnocephala semperi
Temnocephalidae
Temnocephalida

Figure 14.11. Interrelationships of Rhabdocoela; images courtesy of Queensland Museum, from Cannon (1986), with permission. Scale bars, 200 μm.

association with invertebrate and/or vertebrate hosts, but remains controversial.

Neodermata

The major obligate parasite groups of the Platyhelminthes (i.e., the Cestoda, Trematoda, and Monogenea) have often been thought distinct enough not to be closely related. Bychowsky (1937), for example, postulated that Trematoda and Cestoda/Monogenea or Cercomeromorphae were each derived independently from Rhabdocoela. Other authors have been in favor of even more disparate origins, with Digenea not even being flatworms (Sinitsin 1911) or having a common ancestor with the Mesozoa (Wright 1971) and Cestoda being derived from poriferan-like forms (Ubelaker 1983)!

The first detailed cladistic treatments of the phylum Platyhelminthes by Ehlers (1984, 1985a, 1985b) and Brooks et al. (1985b) produced strong evidence for the monophyly of these parasites, for which Ehlers (1984) coined the name Neodermata, referring to the replacement of the epidermis during ontogeny. Later work, particularly molecular phylogenies (e.g., Baverstock et al. 1991, Blair 1993, Littlewood et al. 1999b, Rohde et al. 1993) have provided further evidence of this monophyly, although some studies (Joffe and Kornakova 2001, Rohde 2001) have thrown doubt on some of the original apomorphies. Littlewood et al. (1999b) considered the monophyly of the Neodermata "beyond doubt," and Joffe and Kornakova (2001) considered this problem "finally solved," citing as evidence three unique insertions in the SSU rDNA sequence.

Apomorphies of Neodermata (Brooks et al. 1985b, Ehlers 1984, 1985a, 1985b, Littlewood et al. 1999b)
Multiciliated ectoderm is limited to "larval" stages and is shed later and replaced by syncytial neodermis with subepidermal perikarya each separately connected to surface layer

Protonephridia with a two-cell weir
Epidermal locomotory cilia with single, cranial rootlet
Epithelial sensory receptors with electron-dense collars
Complete incorporation of both axonemes is sperm body
Two long and one short insertions in SSU rDNA sequence (Joffe and Kornakova 2001)

The relationships of the major groups within the Neodermata are becoming well accepted, although new molecular data may add some confusion. Neodermata consist of two sister groups, Cercomeromorphae (i.e., Cestoda + Monogenea) and Trematoda (fig. 14.12a). The separateness and main constituents of these taxa were recognized as early as Baer (1931) and Bychowsky (1937), although recognition of their joint monophyly awaited cladistic study (see above). Although almost all recent morphological (Brooks et al. 1985b, Ehlers 1985b, Zamparo et al. 2001) and molecular (Baverstock et al. 1991, Littlewood and Olson 2001, Littlewood et al. 1999b) analyses agree with this dichotomy, a recent study using LSU rDNA sequences indicated the relationship (Trematoda, Cestoda) Monogenea (Lockyer at al. 2003; fig. 14.12b,c)]. This unusual result relies solely on data from one gene, and until corroborated, Cercomeromorphae continue to be recognized. The monophyly of the Monogenea is discussed below.

Cercomeromorphae

Janicki (1930), on the basis that the digenean cercarial tail, the monogenean opisthaptor, and the cestode cercomer are homologues, erected the taxon Cercomeromorphae. The name is now used for the taxon including Monogenea and Cestoda, with the posterior hooklets as the major innovation.

Apomorphy of Cercomeromorphae (Littlewood et al. 1999b)
Posterior hook of larva and adults (ancestrally probably 16 hooks)

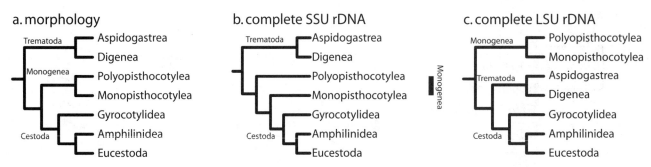

Figure 14.12. Interrelationships of the Neodermata: competing hypotheses (a–c) based on morphology and molecular data.

Monogenea = Monogenoidea

Members of Monogenea, occasionally referred to as Monogenoidea, are as diverse as any of the obligate flatworm parasites despite using only single hosts in their life cycle. Predominantly ectoparasites of marine and freshwater teleost fishes, usually clinging to species-specific regions of the outer surfaces of gills and body, some groups have successfully exploited a wide range of aquatic vertebrates, including elasmobranchs, dipnoi, teleosts, amphibians, and even the hippopotamus. The monophyly of the group has been challenged by molecular data (Justine 1998, Littlewood et al. 1999a, Mollaret et al. 1997) and some sperm morphology (Justine 1993), but neither challenge has been conclusive. Indeed, the rates of evolution of both SSU and LSU rDNA, both of which suggest paraphyly, are so different between the major constituent groups of Monogenea that additional molecular evidence is required to solve the problem (Olson and Littlewood 2002). It seems likely that, whether monophyletic or paraphyletic, members of Monogenea radiated very rapidly from their ancestral stock. Morphology alone suggests monophyly (Littlewood et al. 2001a; fig. 14.12a), complete SSU suggests paraphyly but monophyly of the Cercomeromorphae (Littlewood et al. 2001a; fig. 14.12b), partial LSU suggests paraphyly and non-monophyly of the Cercomeromorphae (Mollaret et al. 1997), and complete LSU suggests monophyly of Monogenea and non-monophyly of Cercomeromorphae (Lockyer et al. 2003; fig. 14.12c).

Apomorphies of Monogenea (Boeger and Kritsky 2001)
Larva with three ciliated zones
Larva and adult with two pairs of pigmented eyes
One pair of ventral anchors
One egg filament

As with other parasitic flatworms, members of Monogenea possess attachment organs, and as appropriate for ectoparasites, these can be quite elaborate with various arrangements of suckers, hooks, and anchors. Monogenea have anterior and posterior structures, and it is predominantly the structure of the posterior organ that delineates

the two major constituent groups: Polyonchoinea and Heteronchoinea. The naming of these two groups is as hotly debated as their interrelationship, but at least each group is recognized as being monophyletic. Polyonchoinea, most commonly referred to as Monopisthocotylea, are supported by eight synapomorphies, and Heteronchoinea, composed of the Polystomatidae, Sphyranuridae, and Polyopisthocotylea, are supported by six (Boeger and Kritsky 2001; see fig. 14.13). The interrelationships of families are based on a multitude of adult features and, to a lesser extent, the unique larval form called the oncomiracidium. Boeger and Kritsky are responsible for much of the modern morphologically based phylogenetic systematic (i.e., cladistic) work on Monogenea, and their publications provide a review of characters, hypotheses on interrelationships, and interpretations based on host associations (Boeger and Kritsky 1993, 1997, 2001); their most recent estimate of interrelationships is shown in figure 14.13. A recent analysis of interrelationships based on molecular evidence is given by Olson and Littlewood (2002).

Cestoda

This taxon includes all gutless tapewormlike groups, including those in which no serial repetition of the genitalia occurs. Xylander (2001) enumerated eight autapomorphies as evidence of the monophyly of Cestoda. Some workers have doubted the monophyly of the group, including Ubelaker (1983), who found it "tempting" to propose an early poriferan-like ancestor and went as far as considering "Cestoidea" a phylum, which did not include Gyrocotylidea or Amphilinidea. More recent morphological and molecular evidence, however, supports the monophyly of these groups (Brooks et al. 1985b, Ehlers 1985a, Littlewood et al. 1999b, Zamparo et al. 2001).

Apomorphies of Cestoda (Xylander 2001)
All stages without intestine
Neodermis with distinct type of microvilli (microtriches, microthrix)

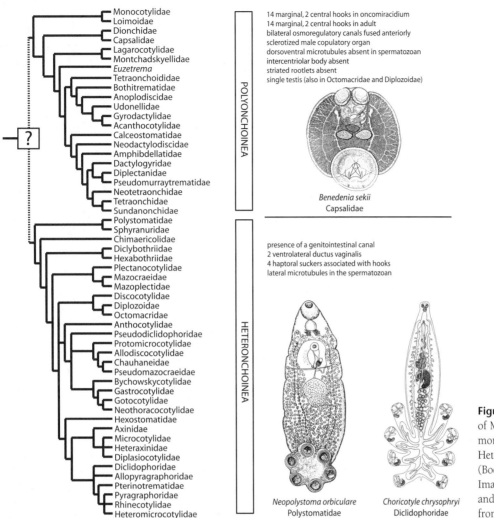

14 marginal, 2 central hooks in oncomiracidium
14 marginal, 2 central hooks in adult
bilateral osmoregulatory canals fused anteriorly
sclerotized male copulatory organ
dorsoventral microtubules absent in spermatozoan
intercentriolar body absent
striated rootlets absent
single testis (also in Octomacridae and Diplozoidae)

POLYONCHOINEA

Benedenia sekii
Capsalidae

presence of a genitointestinal canal
2 ventrolateral ductus vaginalis
4 haptoral suckers associated with hooks
lateral microtubules in the spermatozoan

HETERONCHOINEA

Neopolystoma orbiculare
Polystomatidae

Choricotyle chrysophryi
Diclidophoridae

Figure 14.13. Interrelationships of Monogenea with autapomorphies for Polyonchoinea and Heteronchoinea, redrawn from (Boeger and Kritsky, 2001). Images courtesy of John Wiley and Sons, Inc., with permission from Yamaguti (1963).

First canal cell of protonephridium lacks cell gap and desmosome

Reticulate protonephridial system in postlarvae

Cell bodies of protonephridial canal cells under basal lamina

Larval epidermis is syncytial, neodermal tissue does not reach body surface

10 larval hooks (Littlewood et al. 1999b)

Large body dimensions

Apical pit forms when in first host

Male copulatory organ a cirrus

Vertebrate host in life cycle

Gyrocotylidea

Gyrocotylideans are a small group (~10 species in one genus) of monozoic worms found exclusively in the stomach of holocephalan fishes. They are large worms, which normally occur in pairs, attach by a posterior rosette organ, which has complex folds that mesh with the folds of the stomach wall (Bandoni and Brooks 1987b). The life cycle is not known, but larval forms are found embedded in the parenchyma of adults. Some early workers have considered gyrocotylideans to be monogeneans, claiming that the rosette is the homologue of the monogenean opisthaptor (Williams et al. 1987). The consensus opinion on the position is now that it is a basal cestode, based on morphological apomorphies (Bandoni and Brooks 1987b) and molecular results.

Apomorphies of Gyrocotylidea (Xylander 2001)
Lycophore (larva) epidermis without nuclei
Parasite of Holocephali
Parenchymatic postlarvae
Neodermal spine shape
No intraepithelial multiciliary sensory structures
Caudal rosette organ
Apical proboscis

Nephroposticophora (Amphilinidea + Eucestoda)

This relatively recently recognized group contains two superficially dissimilar groups, the amphilinids and the "true" tapeworms. It was originally recognized on morphological grounds (Bandoni and Brooks 1987a, Ehlers 1985a) and has been confirmed by several molecular studies (Littlewood et al. 1999a, Littlewood and Olson 2001).

> *Apomorphies of Nephroposticophora (Xylander 2001)*
> Unpaired excretory pore at postlarval posterior end
> Larger nephridioduct unciliated

Amphilinidea

This small group (about eight genera, 20 species) consists of leaflike monozoic forms found in the body cavity of chondrosteans, teleosts, and freshwater chelonians (Bandoni and Brooks 1987a). Crustaceans are used as intermediate hosts. Some early workers (see Gibson et al. 1987, Janicki 1930) considered amphilideans to be paedomorphic eucestode plerocercoids, but later morphological (Brooks et al. 1985a, Ehlers 1985a, Zamparo et al. 2001) and molecular (e.g., Baverstock et al. 1991, Littlewood et al. 1999a, 1999b, Littlewood and Olson 2001) evidence has shown that they are very likely to be the sister taxon to Eucestoda.

> *Apomorphies of Amphilinidea (Xylander 2001)*
> All stages coelomic parasites
> Neodermal microvilli short and stubby
> Uterus tripartite
> Uterine pore at anterior end
> Leaflike shape
> Characteristic apical organ

Eucestoda (Cestoidea)

The Eucestoda, or "true" tapeworms, are a large group of 750 genera and up to 5000 species and includes all the forms with proglottidization and segmentation as well as the monozoic Caryophyllidea. Hoberg et al. (1999) reckoned that true cestodes arose in basal teleosts and subsequently have spread to elasmobranchs (where numerous groups are found) and tetrapods. Numerous life cycles are known (Beveridge 2001), giving evidence that arthropods are the primitive intermediate host, which has been lost in some terrestrial cestodes. With the exception of Caryophyllidea and Spathebothriidea, the eucestodes are segmented. They may be tiny with few segments or huge with thousands of segments (in whales some tapeworms grow to many tens of meters in length). The anterior attachment organ, the scolex, has many forms and may use suckers, hooks, proboscides, muscular pads, or folded ridges to adhere to the intestinal wall of the host. Each segment con-

tains a full set of hermaphroditic sexual organs such that vast numbers of eggs may be produced in the lifetime of the worm (the human tapeworm *Diphyllobothrium latum* is said to shed up to one million eggs per day). The serial repetition of sexual organs (proglottidization) may not always be reflected in surface segmentation, but it usually is.

The phylogeny of Eucestoda is probably better developed than any of the other equivalent platyhelminth groups, and the major internal taxa tend to be more satisfactorily delimited. Brooks et al. (1991) and Hoberg et al. (1999, 1997) have presented morphological phylogenies, Mariaux (1998) and Olson and Caira (1999) produced molecular phylogenies (summarized by Mariaux and Olson 2001) and Hoberg et al. (2001) and Olson et al. (2001) used combined evidence. A consensus is appearing on several significant features of cestode phylogeny, including the possibility that some apparently well-established groups (e.g., Tetraphyllidea, Pseudophyllidea) are not monophyletic (fig. 14.14). The monozoic Caryophyllidea are now recognized as the basal eucestodes and it is likely that internal proglottidization developed before external segmentation (Olson et al. 2001).

> *Apomorphies of Eucestoda (Xylander 2001)*
> Neodermis with typical microtriches
> Spermatozoa without mitochondria
> First larval stage without sensory structures and
>   cerebrum
> Reduction of several tissues and organs in the primary
>   larval stage (coracidium)
> Cercomer shed during larval development
> Six caudal hooks

Note that proglottidization and segmentation are not apomorphies, because they are lacking in the Caryophyllidea.

Trematoda

This taxon, erected in 1808, was recognized as containing both the ectoparasitic and the endoparasitic flukes, until Baer (1931) and Bychowsky (1937) proposed that the ectoparasitic Monogenea was closer to Cestoda. Trematoda is now recognized to contain two sister taxa, Aspidogastrea and Digenea. Early workers have postulated that Aspidogastrea were derived from within Digenea (Cable 1974, Poche 1926), but morphological (Brooks et al. 1985b, Ehlers 1985a, Gibson 1987, Pearson 1992), molecular (Blair 1993, Littlewood and Olson 2001), and combined (Cribb et al. 2001, Littlewood et al. 1999b) evidence strongly indicates the sister-group relationship of the taxa (fig. 14.15).

> *Apomorphies of Trematoda (Littlewood et al. 1999b)*
> Ciliated epidermal cells of larva separated by cyto-
>   plasm of neodermis
> Male copulatory organ a cirrus
> Molluscan first host

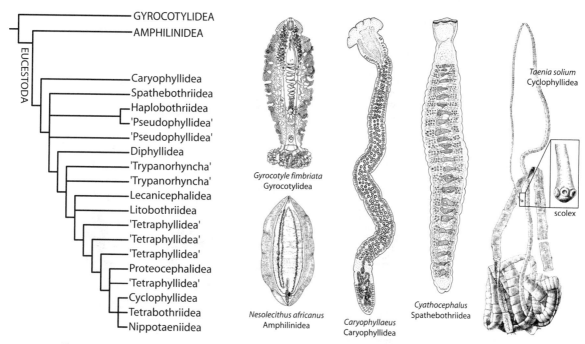

GYROCOTYLIDEA
AMPHILINIDEA
EUCESTODA
Caryophyllidea
Spathebothriidea
Haplobothriidea
'Pseudophyllidea'
'Pseudophyllidea'
Diphyllidea
'Trypanorhyncha'
'Trypanorhyncha'
Lecanicephalidea
Litobothriidea
'Tetraphyllidea'
'Tetraphyllidea'
'Tetraphyllidea'
Proteocephalidea
'Tetraphyllidea'
Cyclophyllidea
Tetrabothriidea
Nippotaeniidea

Taenia solium
Cyclophyllidea

scolex

Gyrocotyle fimbriata
Gyrocotylidea

Nesolecithus africanus
Amphilinidea

Caryophyllaeus
Caryophyllidea

Cyathocephalus
Spathebothriidea

Figure 14.14. Interrelationships of the major cestode groups redrawn from Olson et al. (2001); images courtesy of Willi Xylander from Westheide and Rieger (1996) and courtesy of Taylor and Francis from Williams and Jones (1994) with permission.

Aspidogastrea

This small group (~12 genera, 80 species) is generally considered uncontroversially monophyletic, although the constituent genera are morphologically diverse (Rohde 2001). Despite the relatively small number of species, they are found as adults in lamellibranchs, gastropods, holocephalans, elasmobranchs, teleosts, and chelonians. A mollusk-inhabiting stage is present in all known life cycles, and the presence of aspidogastrean adults in mollusks is considered facultative by Rohde (2001). The series of alveoli on the adhesive disk or the rows of suckers on the ventral surface are considered evidence of "pseudosegmentation" by Rohde (2001). Molecular evidence confirms the basal status of Rugogastridae (fig. 14.15).

Apomorphies of Aspidogastrea (*Littlewood et al. 1999b*)
Larva (cotylocidium) with ventrocaudal sucker, becoming alveolated adhesive organ in adults
Few ciliated cells in larvae
Neodermis with characteristic microvilli (= microtubercles)

Digenea

This is the largest group of flatworms, with some 18,000 nominal species and more than 2700 genera. Adults are found in all types of jawed vertebrates, although they are less common in elasmobranchs (Bray and Cribb 2003). The life cycle is complex, usually with three hosts in sequence. The first host is a mollusk, in which the parasite reproduces asexually, producing numerous motile free-living cercariae (Cribb et al. 2001). Infection of the final host is usually by way of ingestion of a second intermediate host, which may be an invertebrate or a vertebrate. The asexual reproduction in the mollusk ensures the large number of offspring necessary for such a precarious lifestyle. The relationship with the mollusk, which they share with the aspidogastreans, has been thought to be the primitive parasitological association in the group, the vertebrate host having been acquired later. The fact that all neodermatan groups parasitize vertebrates, and that Neodermata is now convincingly demonstrated as monophyletic, suggests, on the other hand, that the association with the vertebrate is more primitive.

Digenea are considered to be clearly monophyletic, with several convincing apomorphies. Relationships within the taxon are less defined and are still controversial (e.g., Brooks et al. 1985a, Pearson 1992). Molecular results and combined molecular and morphological analyses (Cribb et al. 2001) do not resolve the basal digeneans unequivocally. Early suggestions that Heronimidae consist of basal digeneans (Brooks et al. 1985a) have been criticized (Gibson 1987, Pearson 1992) and are at odds with most molecular results (Barker et al. 1993, Cribb et al. 2001; but see Campos et al. 1998). Cribb et al. (2001), using morphological, molecular, and combined evidence approaches found different topologies. In the molecular and combined evidence results (fig. 14.15),

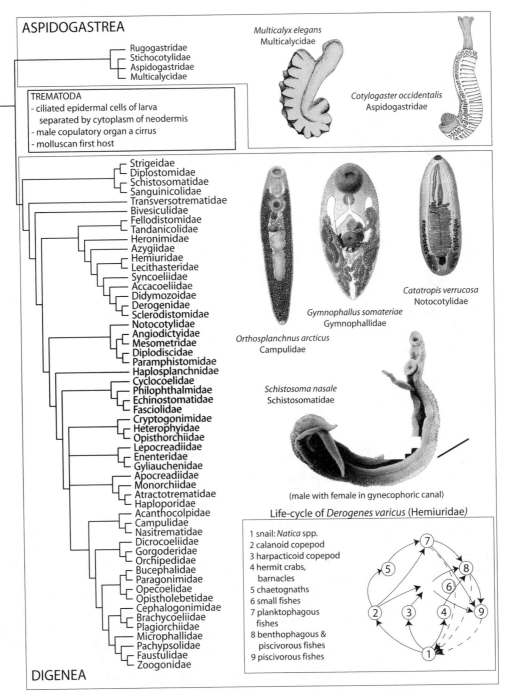

Figure 14.15. Interrelationships of Trematoda, including Aspidogastrea, redrawn from Rohde (2001), and Digenea, redrawn from Cribb et al. (2001). Inset shows life cycle of the most common digenean, *Derogenes varicus*. Schistosomes, causative agents of schistosomiasis, are some of the few flatworms with separate sexes, here shown *in copula*; image courtesy of Vaughan Southgate and Kluwers, from Southgate et al. (1990) with permission.

the clade (Diplostomoidea + Schistosomatoidea) was found basal, whereas various morphological analyses, based on different premises, resolved Transversotrematidae and Bivesiculidae as basal. A resolution of this point is crucial to our understanding of the evolution of Digenea because mem-

bers of Transversotrematidae are ectoparasites and those of Bivesiculidae lack suckers. Included in the group are the most medically important flatworms, the schistosomes. Members of Schistosomatidae infect one species of crocodile and many species of birds and mammals, and among humans, five species cause various forms of schistosomiasis, a debilitating disease affecting more than 200 million people worldwide but predominantly in the tropics.

> *Apomorphies of Digenea (Cribb et al. 2001, Littlewood et al. 1999b)*
> Series of asexual generations in first intermediate (mollusk) host
> Ciliated epidermal cells of miracidium arranged in regular transverse rows
> Jawed vertebrates in complex life cycle
> Cercaria
> Miracidium and mother sporocyst without digestive system

Nemertea: The Ribbonworms

There are currently about 1000 recognized species of nemerteans. Their eversible (inside-out) proboscis used in food capture and, in some cases locomotion, is perhaps the most obvious unique character of this phylum; the proboscis is separate from the gut and is connected to a rhynchocoel, which serves to evert it rapidly through hydrostatic pressure (Senz 1995). These predominantly marine worms (some are found in freshwater and even in damp terrestrial environments) are often brightly colored and can be extremely long and thin, up to 30 m in the case of the appropriately named *Lineus longissimus*. Their great length perhaps could not be achieved without a circulatory system to distribute nutrients throughout the body, and this is their second key innovation: a closed blood system. Uniquely for invertebrates, their blood vessels are lined with a cellular epithelium rather than the more usual nonepithelial basement membrane (Ruppert and Carle 1983). Their embryonic development is similar to that of classic spiralians such as annelids and mollusks, and some groups have a trochophore-like larval stage known as a pilidium (Nielsen 2001).

Nemertea have traditionally been considered acoelomate animals and have consequently been associated with the acoelomate platyhelminths as an early branch within Bilateria (Nielsen 2001). Ultrastructural analyses, however, have convincingly homologized their closed circulatory system and rhynchocoel with the coelomic cavities of other invertebrates, showing they are not in fact acoelomate (Turbeville and Ruppert 1985). Molecular studies (both of SSU and LSU rRNA and *Hox* genes) strongly support this contention and link the nemerteans with other coelomate protostomes with spiral cleavage and trochophore-type larvae such as the annelids, mollusks, sipunculids, and echiurans as well as with Platyhelminthes (Turbeville et al. 1992).

Within the phylum, two classes have been recognized on the basis of morphology: Anopla, whose members have a post oral brain and an unarmored proboscis, and Enopla, whose members have a postoral brain and often have a proboscis armored with stylets. Both classes are further divided into two orders: Palaeonemertea and Heteronemertea within Anopla, and Hoplonemertea and Bdellonemertea within Enopla (Meglitsch and Schram 1991).

Molecular phylogenetic analysis supports some aspects of these traditional divisions. In an analysis of SSU rRNA sequences from 15 species representing all four classes, Enopla, Hoplonemertea, and Bdellonemertea were robustly grouped. This analysis suggested, however, that Anopla is paraphyletic (Sundberg and Saur 1998, Sundberg et al. 1998, 2001; see fig. 14.16).

Figure 14.16. Interrelationships of Nemertea.

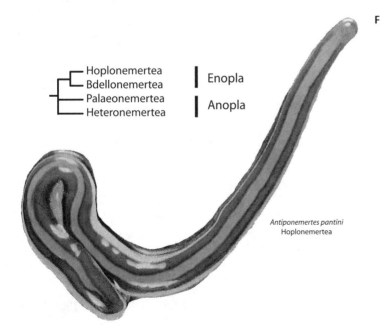

Hoplonemertea
Bdellonemertea
Palaeonemertea
Heteronemertea

Enopla

Anopla

Antiponemertes pantini
Hoplonemertea

Apomorphies of Nemertea
Reversible proboscis and rhynchocoel
Closed blood system lined with epithelium
Great powers of regeneration

Rotifera and Acanthocephala (Syndermata): Rotifers and Thorny-Headed Worms

Syndermata is the name given to the taxon that includes the two phyla Rotifera (the rotifers) and Acanthocephala (the thorny-headed worms; fig. 14.17). Almost 2000 species have been described within each group. Rotifers, at one time known as "wheel-animalcules," are common members of the microscopic fauna in freshwater. They are extremely tiny, at one time considered the "smallest of all Metazoa" (Borradaile et al. 1963). On the other hand, acanthocephalans are robust worms, up to 1 m long, and are obligate parasites of vertebrates.

Apomorphies of Syndermata (Wallace et al. 1996, Zrzavý 2001)
Syncytial integument with intrasyncytial skeletal lamina (includes Micrognathozoa)
Anteriorly directed sperm flagella (shared with Myzostomida)
Sperm has acrosome

Primordial germ cells invaginated separately before gastrulation
Loss of cilia in protonephridial canals
Jaw characters (shared with Gnathostomulida and Micrognathozoa, lost in Acanthocephala)

Zrzavý (2001) listed five morphological characters used as evidence for the close relationship (monophyly) of Acanthocephala and Rotifera, jointly forming Syndermata (= Trochata). Gene trees based on SSU rDNA sequences (García-Varela et al. 2000, Near et al. 1998) provide further evidence for this conclusion. García-Varela et al. (2000) found statistically significant evidence for Acanthocephala as a sister group to Eurotatoria (Bdelloidea + Monogononta). Rotifera, however, consists of Eurotatoria and Seisonida (Garey et al. 1998, Melone et al. 1998), and evidence from a heat-shock protein gene (Hsp82; Mark Welch 2000) indicates that, whereas Acanthocephala are the sister group to Eurotatoria, Seisonidea may be the sister group to Acanthocephala + Eurotatoria, meaning that acanthocephalans are rotifers. This is in conflict with earlier evidence, based on relatively few SSU rDNA sequences, that Acanthocephala are the sister group of Bdelloidea (Garey et al. 1996), forming Lemniscea. The morphological evidence for the monophyly of Lemniscea is also disputed (Ricci 1998). Zrzavý (2001) reckoned that monophyly of Seisonida + Acanthocephala is supported by morphological data and possibly

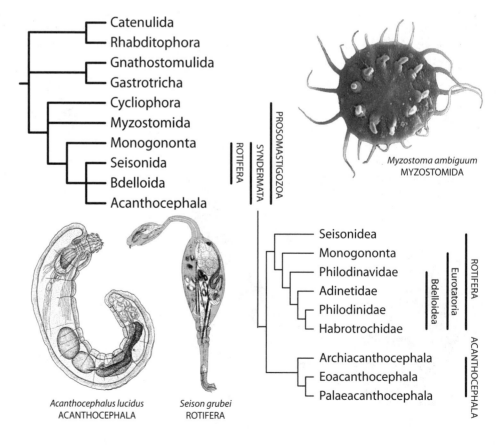

Figure 14.17. Interrelationships of Syndermata and their proposed relationships among Platyzoa. Image of *Myzostoma* courtesy of Igor Eeckhaut, reproduced with permission.

Myzostoma ambiguum
MYZOSTOMIDA

Acanthocephalus lucidus
ACANTHOCEPHALA

Seison grubei
ROTIFERA

by SSU rDNA, forming Pararotatoria, a sister group to Eurotatoria in Syndermata.

The internal phylogeny of Acanthocephala appears quite well resolved. Molecular evidence (García-Varela et al. 2000, Near et al. 1998) indicates the relationship Archiacanthocephala (Eoacanthocephala, Palaeacanthocephala) for its major subgroups. The detailed morphological analysis of Monks (2001) also finds Eoacanthocephala and Palaeacanthocephala monophyletic and sister taxa but finds Archiacanthocephala paraphyletic and with its constituent species basal to the other groups. Herlyn (2001) provided further evidence for the monophyly of Eoacanthocephala in recognizing the apomorphic status of the epidermis cone in this group.

Cycliophora

Composed of a single species (*Symbion pandora*, an ectocommensal of the Norway lobster *Nephrops norvegicus* measuring only 350 μm), this recently discovered phylum has been shown to have syndermatan affinities using both SSU rDNA (Giribet et al. 2000, Winnepenninckx et al. 1998) and morphology (Funch and Kristensen 1995, Sørensen et al. 2000). Cycliophora may be the sister group to Entoprocta, sharing the presence of mushroom-shaped extensions into the epidermis, originating from the basal lamina, according to Sørensen et al. (2000), or from a combined molecular and morphological analysis the sister group of Syndermata in Platyzoa (Giribet et al. 2000; fig. 14.17). Unique features include an anterior feeding region termed the buccal funnel and a complicated life cycle with distinct sexual and asexual phases and chordoid and pandora larval forms (Funch 1996).

Myzostomida

Myzostomida are composed of about 150 species, and all are symbionts on echinoderms, predominantly Crinoida. Through this lifestyle, they have become adapted so uniquely that it is difficult to place the group among the Metazoa. The animals are incompletely segmented, acoelomate, have five pairs of parapodia with chaetae, and exhibit a trochophore larva (see Eeckhaut et al. 2000). Morphology has dictated an annelid affiliation, although sperm morphology and cladistic analyses have variously placed them in a clade including the Sipuncula, Echiura, and Annelida (Haszprunar 1996a), within the polychaete annelids (Rouse and Fauchald 1997), and in a clade including the Echiura, Pogonophora, and Annelida (Zrzavý et al. 1998). A recent but relatively sparsely sampled molecular study using the gene for EF-1α protein suggests that Myzostomida are sister group to Platyhelminthes (Eeckhaut et al. 2000), although some doubt has been cast on this (Littlewood et al. 2001b), not least because no syndermatan taxa were included. The latest study combining SSU rDNA, morphology, life cycle, and developmental data places them as sister group to Cycliophora closely related to the Rotifera + Acanthocephala (Syndermata) clade (Zrzavý et al. 2001). This latter study appears to be the most exhaustive to date and is summarized in figure 14.17.

Chaetognatha: The Arrow Worms

The chaetognaths, or arrow worms, comprise a small, extremely homogenous phylum (150–200 species) of strictly marine worms. The majority are planktonic, and they occur in huge numbers in the entire world's oceans, where they are important predators, eating large numbers of copepods and fish fry. They have a torpedo-shaped body with one or two fins laterally and a dorsoventrally flattened tail fin with which they can propel themselves rapidly. They have large numbers of sensory bristles on their bodies and anterodorsal eyes that combine to enable them to find their prey, which they grab with their impressive chitinous jaws.

Charles Darwin described the chaetognaths as being "remarkable . . . from the obscurity of their affinities," and this remains true today (Darwin 1844). Because of similarities in embryology (radial cleavage, deuterostomous mouth formation, and formation of the mesoderm and coeloms by outpocketing of the archenterons), they were long considered relatives of the deuterostomes. Molecular studies have rejected this possibility, but because of the fast rate of evolution of their rRNA genes relative to most other animals, a more accurate placement has not been possible (Telford and Holland 1993). Halanych (1996) study of SSU linked them to the nematodes, but this may be due to long-branch attraction. Littlewood et al. (1998) likewise grouped them with the nematodes and gnathostomulids within the ecdysozoan clade. By contrast, other authors have linked the chaetognaths to other phyla with similar jaws (e.g., rotifers and gnathostomulids within Lophotrochozoa, Nielsen 2001).

Within the phylum a single extant class is recognized: Sagittoidea (Bieri 1991, Casanova 1985, Tokioka 1965a, 1965b). The main division within Sagittoidea is among three orders: Monophragmophora, Biphragmophora, and Aphragmophora, the first two of which have a transverse sheet of muscle (phragma) crossing the body (Bieri 1991, Casanova 1985, Tokioka 1965a, 1965b). A single molecular study of a rapidly evolving portion of the LSU from 26 species lends support to the division between Phragmophora and the other two orders, at least for the species sampled, but was unable to determine reliably further divisions with the phylum (Telford and Holland 1997). This partial LSU study suggested that all extant chaetognaths derive from a relatively recent radiation, and the close grouping of SSU sequences from members of the Aphragmophora and Biphragmophora supports this view (Halanych 1996; fig. 14.18).

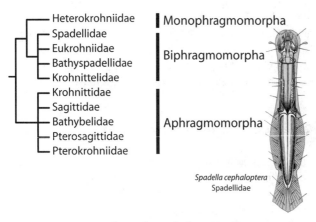

Figure 14.18. Interrelationships of Chaetognatha.

Apomorphies of Chaetognatha
Mesoderm and coelom formation by enterocoely
Chitinous retractable jaws
Multilayered epithelium on body
Cephalic hood
Retrocerebral organ and ciliary loop of unknown
    function

Ecdysozoa

The segmented, coelomate arthropods were long thought by most zoologists to be most closely related to the segmented, coelomate annelids. Analyses of SSU, most notably that of Aguinaldo et al. (1997), have radically revised this view. It now seems that the closest relatives of the arthropods is an assortment of pseudocoelomate worms: Nematoda and Nematomorpha (probably grouped together as Nematoida), and Priapulida, Kinorhyncha, and Loricifera (probably related within Cephalorhyncha). This entire assemblage (including the arthropods) has been termed Ecdysozoa because of the common character of ecdysis or periodic molting (Aguinaldo et al. 1997). It follows that the annelid and arthropod shared characters are either primitive within the Metazoa or convergently derived in these two groups.

There is corroboration of the SSU results from a recent study of the LSU molecule (Mallatt and Winchell 2002). In addition, the nematodes share with the arthropods an unusual triplicated β-thymosin molecule found nowhere else in the Metazoa (Manuel et al. 2000). Common characteristics of *Hox* gene amino acid sequences also support the ecdysozoan clade (de Rosa et al. 1999, Telford 2000), as does examination of the binding of an anti-HRP (horseradish peroxidase) antibody that stains the nervous system only in ecdysozoans (Haase et al. 2001). Potentially contradictory, however, is the discovery of a fusion of prolyl and glutamyl transfer RNA synthetase genes common to the arthropod *Drosophila melanogaster* and vertebrates but unfused in the nematode *Caenorhabditis elegans* and in outgroups. This observation needs investigating in other potential ecdysozoan phyla (Berthonneau and Mirande 2000). Other contradictory evidence comes from Blair et al.'s (2002) study of 100 genes. Although in some sense less reliable because of the small number of taxa sampled, these authors reject the idea of a monophyletic ecdysozoan clade.

Apomorphies of Ecdysozoa
Lack of primary ciliated trochophore type larva
Chitinous cuticle molted under influence of
    ecdysteroid hormones
Lack of locomotory cilia
Radial cleavage (may be primitive within Metazoa)?

Relationships within Ecdysozoa

There is no consensus regarding the relationships of the ecdysozoan phyla. The nematodes and nematomorphs have long been considered related, and some have even suggested that the nematomorphs are derived from within the nematodes and are sister group of the mermithoids, which are also parasites of arthropods. This is not supported by analyses of SSU rDNA, which do, however, support the monophyly of Nematoida (M. J. Telford, unpubl. obs.). SSU rDNA also supports the link between kinorhynchs and priapulids (loriciferans have not been sampled; e.g., Littlewood et al. 1998). The grouping of kinorhynch, priapulid, and loriciferans has been named Cephalorhyncha (or Scalidophora after the scalids or spines around their introverts). Some form of introvert is shared by Nematoida and may be homologous with the mouth cone of the tardigrades, which are likely basal arthropods. Members of Cephalorhyncha share chitinous cuticle, rings of scalids on their introvert, flosculi (sensory pits of unique morphology), and characteristic musculature for retracting the introvert (Nielsen 2001). The relationships of all of these groups to the arthropods are unclear.

Nematoda: The Roundworms and Thread Worms

Treated only very briefly here, roundworms are both ubiquitous and numerous. Whether free-living or parasitic, they have been found in almost every environment, and they range in size from the microscopic (100 mm) to the enormous (~9 m, parasite of a sperm whale); estimated numbers of species range from 40,000 to 10 million. With thin tapering, unsegmented, cylindrical bodies and a muscular suctorial pharynx/esophagus, it is perhaps their cuticle and cuticular structures that have afforded them such success in so many habitats (see chart in figure 14.19). Possession of a cuticle places them in the molting clade Ecdysozoa, although this placement was first recognized on the basis of SSU rDNA

(Aguinaldo et al. 1997). There seems little doubt that Nematoda are a monophyletic phylum. According to SSU rDNA, Nematoda do not separate into sister taxa Adenophorea and Secernentea, a long-established split based principally on trophic ecology and habitat, but instead five major clades are identified, with Chromadorida paraphyletic (Blaxter et al. 1998, Dorris et al. 1999; fig. 14.19). The latest phylogenetic estimates, based on SSU rDNA, support a monophyletic Secernentea and resolve a paraphyletic Adenophorea but have yet to be supported by additional gene sequencing (see also Kampfer et al. 1998). Nevertheless, the solution has prompted many reevaluations of morphology and biology (Schierenberg 2000), which appear to lend support to the new scheme. Parasitism evolved several times in the group (Blaxter 2001, Schierenberg 2000), with many of the major clades including novel associations of animal-parasitic, plant- and fungus-parasitic, and free-living groups. A number of molecular studies, using LSU rDNA and mitochondrial gene fragments, have been undertaken to elucidate further the interrelationships of nematode groups, but there are no data sets rivaling the SSU rDNA to estimate overall nematode phylogeny.

Apomorphies of Nematoda
6 + 6 + 4 cephalic sensillae and amphids
Lateral epidermal cords with the perikarya

Nematomorpha: Horsehair Worms

The nematomorphs or horsehair worms are a phylum of nematodelike worms all of which parasitize arthropods. Their body is an extremely long and slender cylinder, in some species more than 1 m long yet only 1 mm in diameter (Bresciani 1991, Nielsen 2001). Their similarities to other ecdysozoan worms (especially nematodes) are perhaps seen most clearly in their larvae, which have a retractable (although not invert-ible) proboscis on an anterior introvert, which has backward-pointing cuticular spines. Roughly 325 extant species have been described in two orders. The marine order Nectone-matoidea has a single genus, *Nectonema*, with just four species. *Nectonema* larvae parasitize marine decapods, and the adults have bristles on the body that enable them to swim; they have dorsal and ventral nerve cords and an unpaired gonad. Species in the order Gordioidea are terrestrial, and their larval stages parasitize insects; they have only a ventral nerve cord and paired gonads (Schmidt-Rhaesa 1998).

Apomorphies of Nematomorpha
Parasites of arthropods during larval stage
Extremely long and thin
Periodic molting of collagenous cuticle
Reduced or no mouth; nutrient absorption via cuticle
Nonfeeding adults; adults without guts

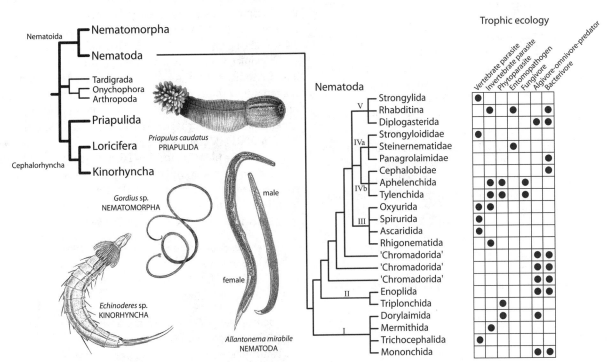

Figure 14.19. Interrelationships of the wormlike ecdysozoan groups, with a phylogeny of Nematoda, taken from Blaxter et al. (1998), indicating multiple origins of parasitism and feeding habits.

Apomorphies of Nematoida (Nematoda and Nematomorpha; Nielsen 2001, Schmidt-Rhaesa 1998)
Cuticle with layers of crossing collagenous (not chitinous) fibrils
Reduction of circular body muscles
Epidermal longitudinal nerve cords
Cloaca in both sexes
Spermatozoa without a flagellum

Priapulida

The priapulids are bottom-living marine worms ranging from 0.5 mm to >20 cm in length (Storch 1991). There are fewer than 20 species currently recognized, but their characteristic body plan can be recognized in numerous fossils from the Cambrian onward. They have a cylindrical body with a significant anterior introvert. The introvert can be everted by contraction of the trunk muscles, with the fluid-filled body cavity acting as a hydroskeleton and inverted through contraction of two rings of retractor muscles (Nielsen 2001). Eversion and inversion allow the animals to burrow through the sands and muds where they live and are also used for feeding. The posterior end has one or two caudal appendages that are most probably for gas exchange. The body is covered in a cuticle that contains chitin and is periodically molted during growth. The embryology is poorly known, but radial cleavage has been seen in *Priapulus* and *Halicryptus*.

Recent morphology-based phylogenies (Wills 1998) support classification of the extant genera in three families. The most speciose, Priapulidae, contains four living genera—*Acanthopriapulus*, *Priapulus*, *Priapulopsis*, and *Halicryptus*—and the Carboniferous fossil *Priapulites*. Maccabeidae have just one genus, *Maccabeus*. Tubiluchidae have two genera, *Tubiluchus* and *Meiopriapulus*. There are five fossil families: Ottoidae, Selkirkiidae, Miskoiidae, Ancalagonidae, and Fieldidae. Morphological cladistic analyses of the relationships between these families group Priapulidae and Maccabeidae and furthermore suggest that all families still extant are monophyletic with respect to the Cambrian fossils (Wills 1998). We are not aware of any molecular analyses of priapulid intraphyletic relationships.

Apomorphies of Priapulida
Large, spiny, retractable presoma (introvert)
Terminal caudal appendage in most species
Large body cavity with amoebocytes and erythrocytes

Loricifera

Loricifera are a recently discovered (1983) phylum of microscopic interstitial or infaunal marine animals. Very little has been published on this phylum, and few members have been thoroughly described, although more than 100 species have been found (Nielsen 2001). Kristensen has placed the described species in a single order, Nanaloricida, with two families at present, Nanaloricidae and Pliciloricidae (Kristensen 1991). Their body consists of a trunk covered with a chitinous exoskeleton called a lorica (girdle) consisting of 6–30 longitudinal cuticular plates and an anterior introvert surrounded by several hundred complex cuticular appendages or scalids in two to seven rows. These scalids are of differing morphology and presumably function (sensory, locomotory). The cuticle is molted repeatedly during growth of the larva (known as a Higgins larva), which is similar in morphology to the adult but has toes that serve to propel it in Nanaloricidae and to act as adhesive pads in Pliciloricidae.

Apomorphies of Loricifera (Kristensen 1991)
Higgins larva
Chitinous lorica on trunk
Scalids with muscles

Kinorhyncha

The kinorhynchs are a very uniform phylum of approximately 150 species. All are small (<1 mm long), marine, and benthic, living in coastal bottom mud (Nielsen 2001). All have a body consisting of 13 segments (with segmentation of muscles and nervous system as well as external cuticle), the anteriormost of which is an introvert with up to seven rings of spines or scalids (sensory and locomotory), followed by a neck and 11 trunk segments (Kristensen and Higgins 1991). The newly hatched larvae have just 11 segments (nine in the trunk), with the two additional adult segments added after periodic molts of the chitinous cuticle. Despite the homogeneity of their morphology, they are classified in two orders: Cyclorhagida (which contain four families and seven genera) and Homalorhagida (two families and four genera; Nielsen 2001). Members of Cyclorhagida (e.g., *Pycnophyes*) have a circular pharynx and 14–16 cuticular plates (placids) on their neck segment, and their body is round or oval in cross section; members of Homalorhagida (e.g., *Echinoderes*) have a triradiate pharynx and two to eight placids on their neck, and their body is flattened ventrally and arched dorsally.

Apomorphies of Kinorhyncha (Kristensen 1991)
Truly segmented (including muscle, nervous system and cuticle); 11 segments in larvae and 13 segments in adults

Apomorphies of Cephalorhyncha/Scalidophora (Kinorhyncha + Loricifera + Priapulida)
Neuropileous nerve ring in a terminal position
Introvert with scalids
Reversible foregut
Tanycytes (tonofibril-containing ectodermal cells in brain; Nebelsick 1993)

Summary

Although we cover metazoan taxa not mentioned elsewhere in this volume, there are few features that unite them. Indeed, it is this very problem that has prevented a phylogenetic resolution for the protostome phyla based on morphology alone. Of course, many phyla do share common features, and matrices have been constructed in order to best estimate relationships. However, molecular data have played a significant role in generating independent estimates or as supplements to morphology. The call for more genes and additional molecular markers is as loud as ever. The so-called lesser phyla, which are often poorly studied because they are few in number, microscopic, cryptic, or mistakenly appear "simple," are in fact critical if we are to understand the interrelationships of the Metazoa and their radiation. Simple does not necessarily equate with primitive, and a common lack of characters has suggested affiliation where little or none exists. As a result, morphological matrices are arguably best used currently as a source of mappable characters in order to establish or confirm homology *a posteriori*. The distribution of phyla on the tree enables the mapping of unique and shared characters alike. Although a total evidence approach may be possible or even preferred by some, reciprocal illumination between independent data sets enlightens our understanding of both morphological and molecular characters as we learn how each has evolved. Acoelomorph flatworms appear to be basal members of Bilateria, with some other taxa still vying for the position and worthy of closer attention. True flatworms (Platyhelminthes, composed of Catenulida and Rhabditophora) appear to be derived bilaterians occupying a position within Lophotrochozoa. The split in Protostomia between Lophotrochozoa and Ecdysozoa is still not as simple as rDNA would have us believe. Taxa such as Acoelomorpha, Chaetognatha, Gastrotricha, and Xenoturbellida suggest the need for other branches and fewer polytomies. Far greater attention is required among the commercially and medically unimportant, yet richly diverse and ecologically important groups that comprise the lophotrochozoans. Meanwhile, some stability is appearing among the major ecdysozoan groups, helped by a healthy interaction between morphologists, molecular systematists, and evolutionary developmental geneticists. As molecular trees promote the reevaluation of morphological characters and highly unexpected topologies sometimes question the utility of types of molecular data, affiliations throughout the protostomes at all taxonomic levels within the tree will evolve by consensus and be resolved only by a sustained effort with all taxa included neither prejudged as lesser nor minor.

Acknowledgments

We are grateful to the following for permission to reproduce figures: Lester Cannon and the Queensland Museum, Igor Eeckhaut, Vaughan Southgate, Willi Xylander, and the following publishers: Gustav Fischer Verlag, Kluwer Academic, Taylor and Francis, and John Wiley and Sons. D.T.J.L. and M.J.T. are funded individually through Wellcome Trust fellowships, for which we are most grateful (D.T.J.L., 043965; M.J.T., 060503).

Literature Cited

Adoutte, A., G. Balavoine, N. Lartillot, and R. de Rosa. 1999. Animal evolution—the end of the intermediate taxa? Trends Genet. 15:104–108.

Adoutte, A., G. Balavoine, N. Lartillot, O. Lespinet, B. Prud'homme, and R. de Rosa. 2000. The new animal phylogeny: reliability and implications. Proc. Natl. Acad. Sci. USA 97:4453–4456.

Aguinaldo, A. M. A., J. M. Turbeville, L. S. Linford, M. C. Rivera, J. R. Garey, R. A. Raff, and J. A. Lake. 1997. Evidence for a clade of nematodes, arthropods and other moulting animals. Nature 387:489–493.

Artois, T. J., and E. R. Schockaert. 1998. A cladistic reassessment of the *Polycystis* species complex (Polycystididae, Eukalyptorhynchia). Hydrobiologia 383:97–102.

Ax, P. 1985. The position of the Gnathostomulida and Platyhelminthes in the phylogenetic system of the Bilateria. Pp. 168–180 *in* The origins and relationships of lower invertebrates (S. Conway Morris, J. D. George, R. Gibson, and H. M. Platt, eds.). Oxford University Press, Oxford.

Ax, P. 1996. Multicellular animals: a new approach to the phylogenetic order in nature. Springer, Berlin.

Baer, J.-G. 1931. Étude monographique du groupe des temnocéphales. Bull. Biol. Fr. Belg. 65:1–59.

Baguñà, J. 1998. Planarians. Pp. 135–165 *in* Cellular and molecular basis of regeneration: from invertebrates to humans (P. Ferretti and J. Géraudie, eds.). John Wiley and Sons, Chichester.

Baguñà, J., S. Carranza, J. Paps, I. Ruiz-Trillo, and M. Riutort. 2001. Molecular taxonomy and phylogeny of the Tricladida. Pp. 49–56 *in* Interrelationships of the Platyhelminthes (D. T. J. Littlewood and R. A. Bray, eds.). Taylor and Francis, London.

Bandoni, S. M., and D. R. Brooks. 1987a. Revision and phylogenetic analysis of the Amphilinidea Poche, 1922 (Platyhelminthes: Cercomeria: Cercomeromorpha). Can. J. Zool. 65:1110–1128.

Bandoni, S. M., and D. R. Brooks. 1987b. Revision and phylogenetic analysis of the Gyrocotylidea Poche, 1926 (Platyhelminthes: Cercomeria: Cercomeromorpha). Can. J. Zool. 65:2369–2389.

Barker, S. C., D. Blair, A. R. Garrett, and T. H. Cribb. 1993. Utility of the D1 domain of nuclear 28S rRNA for phylogenetic inference in the Digenea. Syst. Parasitol. 26:181–188.

Baverstock, P. R., R. Fielke, A. M. Johnson, R. A. Bray, and I. Beveridge. 1991. Conflicting phylogenetic hypotheses for the parasitic platyhelminths tested by partial sequencing of 18S ribosomal RNA. Int. J. Parasitol. 21:329–339.

Berthonneau, E., and M. Mirande. 2000. A gene fusion event in the evolution of amino-acyl-tRNA synthetases. FEBS Lett. 470:300–304.

Beveridge, I. 2001. The use of life-cycle characters in studies of the evolution of cestodes. Pp. 250–256 *in* Interrelationships of the Platyhelminthes (D. T. J. Littlewood and R. A. Bray, eds.). Taylor and Francis, London.

Bieri, R. 1991. Systematics of the Chaetognatha. Pp. 122–136 *in* The biology of Chaetognaths (Q. Bone, H. Kapp and A. C. Pierrot-Bults, eds.). Oxford University Press, Oxford.

Blair, D. 1993. The phylogenetic position of the Aspidobothrea within the parasitic flatworms inferred from ribosomal RNA sequence data. Int. J. Parasitol. 23:169–178.

Blair, J. E., K. Ikeo, T. Gojobori, and S. B. Hedges. 2002. The evolutionary position of nematodes. BMC Evol. Biol. 2:7.

Blaxter, M. L. 2001. Molecular analysis of nematode evolution. Pp. 1–24 *in* Parasitic nematodes—molecular biology, biochemistry and immunology (M. W. Kennedy and W. Harnett, eds.). CABI Publishing, Wallingford, UK.

Blaxter, M. L., P. de Ley, J. R. Garey, L. X. Liu, P. Scheldeman, A. Vierstraete, J. R. Vanfleteren, L. Y. Mackey, M. Dorris, L. M. Frisse, J. T. Vida, and W. K. Thomas. 1998. A molecular evolutionary framework for the phylum Nematoda. Nature 392:71–75.

Boaden, P. J. S. 1985. Why is a gastrotrich? Pp. 248–260 *in* The origins and relationships of the lower invertebrates (S. Conway Morris, J. D. George, R. Gibson, and H. M. Platt, eds.). Clarendon Press, Oxford.

Boeger, W. A., and D. C. Kritsky. 1993. Phylogeny and a revised classification of the Monogenoidea Bychowsky, 1937 (Platyhelminthes). Syst. Parasitol. 26:1–32.

Boeger, W. A., and D. C. Kritsky. 1997. Coevolution of the Monogenoidea (Platyhelminthes) based on a revised hypothesis of parasite phylogeny. Int. J. Parasitol. 27:1495–1511.

Boeger, W. A., and D. C. Kritsky. 2001. Phylogenetic relationships of the Monogenoidea. Pp. 92–102 *in* Interrelationships of the Platyhelminthes (D. T. J. Littlewood and R. A. Bray, eds.). Taylor and Francis, London.

Borradaile, L. A., F. A. Potts, L. E. S. Eastham, J. T. Saunders, and G. A. Kerkut. 1963. The invertebrata. 4th ed. Cambridge University Press, Cambridge.

Bourlat, S. J., C. Nielsen, A. E. Lockyer, D. T. J. Littlewood, and M. J. Telford. 2003. *Xenoturbella* is a deuterostome that eats molluscs. Nature 424:925–928.

Bray, R. A., and T. H. Cribb. 2003. The digeneans of elasmobranchs—distribution and evolutionary significance. Pp. 67–96 *in* Taxonomie, écologie et évolution de metazoaires parasites (C. Combes and J. Jourdane, eds.). Presses Universitaire de Perpignan, Perpignan, France.

Bresciani, J. 1991. Nematomorpha. Pp. 197–218 *in* Microscopic anatomy of invertebrates (F. W. Harrison and E. E. Ruppert, eds.), vol. 4. Wiley-Liss, New York.

Brooks, D. R., E. P. Hoberg, and P. J. Weekes. 1991. Preliminary phylogenetic systematic analysis of the major lineages of the Eucestoda (Platyhelminthes, Cercomeria). Proc. Biol. Soc. Wash. 104:651–668.

Brooks, D. R., R. T. O'Grady, and D. R. Glen. 1985a. Phylogenetic analysis of the Digenea (Platyhelminthes: Cercomeria) with comments on their adaptive radiation. Can. J. Zool. 63:411–443.

Brooks, D. R., R. T. O'Grady, and D. R. Glen. 1985b. The phylogeny of the Cercomeria Brooks, 1982 (Platyhelminthes). Proc. Helminthol. Soc. Wash. 52:1–20.

Bychowsky, B. E. 1937. [Ontogenesis and phylogenetic interrelationships of parasitic flatworms] (in Russian). Izv. Akad. Nauk SSSR, Ser. Biol. 4:1354–1383 (Engl. trans., Virginia Institute of Marine Science, Translation Series 26, 1981).

Cable, R. M. 1974. Phylogeny and taxonomy of trematodes with reference to marine species. Pp. 173–193 *in* Symbiosis in the sea (W. B. Weinberg, ed.). University of South Carolina Press, Columbia.

Campos, A., M. P. Cummings, J. L. Reyes, and J. P. Laclette. 1998. Phylogenetic relationships of Platyhelminthes based on 18S ribosomal gene sequences. Mol. Phylogenet. Evol. 10:1–10.

Cannon, L. R. G. 1986. Turbellaria of the world: a guide to families & genera. Queensland Museum, Brisbane.

Cannon, L. R. G., and B. Joffe. 2001. The Temnocephalida. Pp. 83–91 *in* Interrelationships of the Platyhelminthes (D. T. J. Littlewood and R. A. Bray, eds.). Taylor and Francis, London.

Carranza, S., J. Baguñà, and M. Riutort. 1997. Are the Platyhelminthes a monophyletic primitive group? An assessment using 18S rDNA sequences. Mol. Biol. Evol. 14:485–497.

Carranza, S., D. T. J. Littlewood, K. A. Clough, I. Ruiz-Trillo, J. Baguñà, and M. Riutort. 1998. A robust molecular phylogeny of the Tricladida (Platyhelminthes: Seriata) with a discussion on morphological synapomorphies. Proc. R. Soc. Lond. B 265:631–640.

Casanova, J.-P. 1985. Description de l'appareil genital primitif du genre *Heterokrohnia* et nouvelle classification des Chaetognathes. C. R. Acad. Sci. III 301:397–402.

Cribb, T. H., R. A. Bray, D. T. J. Littlewood, S. Pichelin, and E. A. Herniou. 2001. The Digenea. Pp. 168–185 *in* Interrelationships of the Platyhelminthes (D. T. J. Littlewood and R. A. Bray, eds.). Taylor and Francis, London.

Curini-Galletti, M. 2001. The Proseriata. Pp. 41–48 *in* Interrelationships of the Platyhelminthes (D. T. J. Littlewood and R. A. Bray, eds.). Taylor and Francis, London.

Darwin, C. 1844. Observations on the structure and propagation of the genus *Sagitta*. Ann. Mag. Nat. Hist. 13:1–6.

de Rosa, R., J. K. Grenier, T. Andreeva, C. E. Cook, A. Adoutte, M. Akam, S. B. Carroll, and G. Balavoine. 1999. Hox genes in brachiopods and priapulids and protostome evolution. Nature 399:772–776.

Doe, D. A. 1981. Comparative ultrastructure of the pharynx simplex in Turbellaria. Zoomorphology 97:133–192.

Doe, D. A. 1986. Ultrastructure of the copulatory organ of *Haplopharynx quadristimulus* and its phylogenetic significance (Plathelminthes, Haplopharyngida). Zoomorphology 106:163–173.

Dorris, M., P. de Ley, and M. L. Blaxter. 1999. Molecular analysis of nematode diversity and the evolution of parasitism. Parasitol. Today 15:188–193.

Eeckhaut, I., D. McHugh, P. Mardulyn, R. Tiedemann, D. Monteyne, M. Jangoux, and M. C. Milinkovitch. 2000. Myzostomida: a link between trochozoans and flatworms. Proc. R. Soc. Lond. B. 267:1383–1392.

Ehlers, U. 1984. Phylogenetisches System der Plathelminthes. Verhandlungen der Naturwissenschaftlichen Vereins In Hamburg (NF) 27:291–294.

Ehlers, U. 1985a. Das phylogenetische System der Plathelminthes. Gustav Fischer, Stuttgart.

Ehlers, U. 1985b. Phylogenetic relationships within the Platyhelminthes. Pp. 143–158 in The origins and relationships of lower invertebrates (S. Conway Morris, J. D. George, R. Gibson, and H. M. Platt, eds.). Clarendon Press, Oxford.

Ehlers, U. 1988. The Prolecithophora—a monophyletic taxon of the Platyhelminthes. Fortschr. Zool. 36:359–365.

Ehlers, U. 1992. Frontal glandular and sensory structures in Nemertoderma (Nemertodermatida) and Paratomella (Acoela): ultrastructure and phylogenetic implications for the monophyly of the Euplathelminthes (Plathelminthes). Zoomorphology 112:227–236.

Ehlers, U., and B. Sopott-Ehlers. 1997a. Ultrastructure of the subepidermal musculature of Xenoturbella bocki, the adelphotaxon of the Bilateria. Zoomorphology 117:71–79.

Ehlers, U., and B. Sopott-Ehlers. 1997b. Xenoturbella bocki: organization and phylogenetic position as sistertaxon of the Bilateria. Verhandlungen der Deutschen Zoologischen Gesellschaft 90:68.

Franzén, A., and B. A. Afzelius. 1987. The ciliated epidermis of Xenoturbella bocki (Platyhelminthes, Xenoturbellida) with some phylogenetic considerations. Zool. Scr. 16:9–17.

Funch, P. 1996. The chordoid larva of Symbion pandora (Cycliophora) is a modified trochophore. J. Morphol. 230:231–263.

Funch, P., and R. M. Kristensen. 1995. Cycliophora is a new phylum with affinities to Entoprocta and Ectoprocta. Nature 378:711–714.

García-Varela, M., G. Pérez-Ponce de León, P. de la Torre, M. P. Cummings, S. S. S. Sarma, and J. P. Laclette. 2000. Phylogenetic relationships of Acanthocephala based on analysis of 18S ribosomal RNA gene sequences. J. Mol. Evol. 50:532–540.

Garey, J. R., T. J. Near, M. R. Nonnemacher, and S. A. Nadler. 1996. Molecular evidence for Acanthocephala as a subtaxon of Rotifera. J. Mol. Evol. 43:287–292.

Garey, J. R., A. Schmidt-Rhaesa, T. J. Near, and S. A. Nadler. 1998. The evolutionary relationships of rotifers and acanthocephalans. Hydrobiologia 388:83–91.

Gibson, D. I. 1987. Questions in digenean systematics and evolution. Parasitology 95:429–460.

Gibson, D. I., R. A. Bray, and C. B. Powell. 1987. Aspects of the life history and origins of Nesolecithus africanus (Cestoda: Amphilinidea). J. Nat. Hist. 21:785–794.

Gilbert, S. 2000. Developmental biology. 6th ed. Sinauer Associates, Sunderland, MA.

Giribet, G., D. L. Distel, M. Polz, W. Sterrer, and W. C. Wheeler. 2000. Triploblastic relationships with emphasis on the acoelomates and the position of Gnathostomulida, Cycliophora, Plathelminthes, and Chaetognatha: a combined approach of 18S rDNA sequences and morphology. Syst. Biol. 49:539–562.

Haase, A., M. Stern, K. Wächtler, and G. Bicker. 2001. A tissue-specific marker of Ecdysozoa. Dev. Genes Evol. 211:428–433.

Halanych, K. M. 1996. Testing hypotheses of chaetognath origins—long branches revealed by 18S ribosomal DNA. Syst. Biol. 45:223–246.

Haszprunar, G. 1996a. The Mollusca: coelomate turbellarians or mesenchymate annelids? Pp. 1–28 in Origin and evolutionary radiation of the Mollusca (J. D. Taylor, ed.). Oxford University Press, Oxford.

Haszprunar, G. 1996b. Plathelminthes and Plathelminthomorpha —paraphyletic taxa. J. Zool. Syst. Evol. Res. 34:41–48.

Henry, J. Q., M. Q. Martindale, and B. C. Boyer. 2000. The unique developmental program of the acoel flatworm, Neochildia fusca. Dev. Biol. 220:285–295.

Herlyn, H. 2001. First description of an apical epidermis cone in Paratenuisentis ambiguus (Acanthocephala: Eoacanthocephala) and its phylogenetic implications. Parasitol. Res. 87:306–310.

Hoberg, E. P., S. L. Gardner, and R. A. Campbell. 1999. Systematics of the Eucestoda: advances toward a new phylogenetic paradigm, and observations on the early diversification of tapeworms and vertebrates. Syst. Parasitol. 42:1–12.

Hoberg, E. P., J. Mariaux, and D. R. Brooks. 2001. Phylogeny among orders of the Eucestoda (Cercomeromorphae): Integrating morphology, molecules and total evidence. Pp. 112–126 in Interrelationships of the Platyhelminthes (D. T. J. Littlewood and R. A. Bray, eds.). Taylor and Francis, London.

Hoberg, E. P., J. Mariaux, J.-L. Justine, D. R. Brooks, and P. J. Weekes. 1997. Phylogeny of the orders of the Eucestoda (Cercomeromorphae) based on comparative morphology: historical perspectives and a new working hypothesis. J. Parasitol. 83:1128–1147.

Hochberg, R., and M. K. Litvaitis. 2000. Phylogeny of the Gastrotricha: a morphology-based framework of gastrotrich relationships. Biol. Bull. 198:299–305.

Hooge, M. D., P. A. Haye, S. Tyler, M. K. Litvaitis, and I. Kornfield. 2002. Molecular systematics of the Acoela (Acoelomorpha, Platyhelminthes) and its concordance with morphology. Mol. Phylogenet. Evol. 24:333–342.

Hyman, L. H. 1951. The invertebrates: Platyhelminthes and Rhynchocoela. The acoelomate Bilateria. Vol. 2. McGraw-Hill, New York.

Israelsson, O. 1999. New light on the enigmatic Xenoturbella (phylum uncertain): ontogeny and phylogeny. Proc. R. Soc. Lond. B 266:835–841.

Janicki, C. 1930. Über die jüngsten Zustände von Amphilina foliacea in der Fischleibeshöhle, sowie Generelles zur Auffassung des Genus Amphilina G. Wagen. Zool. Anz. 90:190–205.

Joffe, B. I., and L. R. G. Cannon. 1998. The organisation and evolution of the mosaic of the epidermal syncytia in the Temnocephalida (Plathelminthes: Neodermata). Zool. Anz. 237:1–14.

Joffe, B. I., and E. E. Kornakova. 2001. Flatworm phylogeneticist: between molecular hammer and morphological anvil. Pp. 279–291 in Interrelationships of the Platyhelminthes (D. T. J. Littlewood and R. A. Bray, eds.). Taylor and Francis, London.

Jondelius, U., M. Norén, and J. Hendelberg. 2001. The Prolecithophora. Pp. 74–80 in Interrelationships of the Platyhelminthes (D. T. J. Littlewood and R. A. Bray, eds.). Taylor and Francis, London.

Jondelius, U., I. Ruiz-Trillo, J. Baguñà, and M. Riutort. 2002. The Nemertodermatida are basal bilaterians and not members of the Platyhelminthes. Zool. Scr. 31:201–215.

Justine, J. L. 1993. Phylogeny of the Monogenea based upon a parsimony analysis of characters of spermiogenesis and spermatozoon ultrastructure including recent results. Bull. Fr. Pêche Pisc. 328:137–155.

Justine, J. L. 1998. Non-monophyly of the monogeneans? Int. J. Parasitol. 28:1653–1657.

Kampfer, S., C. Sturmbauer, and J. Ott. 1998. Phylogenetic analysis of rDNA sequences from adenophorean nematodes and implications for the Adenophorea-Secernentea controversy. Invert. Biol. 117:29–36.

Kornakova, E. E., and B. I. Joffe. 1999. A new variant of the neodermatan-type spermiogenesis in a parasitic "turbellarian," *Notentera ivanovi* (Platyhelminthes) and the origin of the Neodermata. Acta Zool. 80:135–151.

Kristensen, R. M. 1991. Loricifera. Pp. 351–375 *in* Microscopic anatomy of invertebrates (F. W. Harrison and E. E. Ruppert, eds.), vol. 4. Wiley-Liss, New York.

Kristensen, R. M., and R. P. Higgins. 1991. Kinorhyncha. Pp. 377–404 *in* Microscopic anatomy of invertebrates (F. W. Harrison and E. E. Ruppert, eds.), vol. 4. Wiley-Liss, New York.

Littlewood, D. T. J., T. H. Cribb, P. D. Olson, and R. A. Bray. 2001a. Platyhelminth phylogenetics—a key to understanding parasitism? Belg. J. Zool. 131:35–46.

Littlewood, D. T. J., M. Curini-Galletti, and E. A. Herniou. 2000. The interrelationships of Proseriata (Platyhelminthes : Seriata) tested with molecules and morphology. Mol. Phylogenet. Evol. 16:449–466.

Littlewood, D. T. J., and P. D. Olson. 2001. Small subunit rDNA and the Platyhelminthes: signal, noise, conflict and compromise. Pp. 262–278 *in* Interrelationships of the Platyhelminthes (D. T. J. Littlewood and R. A. Bray, eds.). Taylor and Francis, London.

Littlewood, D. T. J., P. D. Olson, M. J. Telford, E. A. Herniou, and M. Riutort. 2001b. Elongation factor 1-alpha sequences alone do not assist in resolving the position of the Acoela among the Metazoa. Mol. Biol. Evol. 18:437–442.

Littlewood, D. T. J., K. Rohde, R. A. Bray, and E. A. Herniou. 1999a. Phylogeny of the Platyhelminthes and the evolution of parasitism. Biol. J. Linn. Soc. 68:257–287.

Littlewood, D. T. J., K. Rohde, and K. A. Clough. 1999b. The interrelationships of all major groups of Platyhelminthes: phylogenetic evidence from morphology and molecules. Biol. J. Linn. Soc. 66:75–114.

Littlewood, D. T. J., M. J. Telford, K. A. Clough, and K. Rohde. 1998. Gnathostomulida—an enigmatic metazoan phylum from both morphological and molecular perspectives. Mol. Phylogenet. Evol. 9:72–79.

Lockyer, A. E., P. D. Olson, and D. T. J. Littlewood. 2003. Utility of complete 28S rRNA genes in resolving the phylogeny of the Platyhelminthes and a review of the cercomer theory. Biol. J. Linn. Soc. 78:155–173.

Lundin, K. 1998. The epidermal ciliary rootlets of *Xenoturbella bocki* (Xenoturbellida) revisited: new support for a possible kinship with the Acoelomorpha (Platyhelminthes). Zool. Scr. 27:263–270.

Lundin, K. 2000. Phylogeny of the Nemertodermatida (Acoelomorpha, Platyhelminthes). A cladistic analysis. Zool. Scr. 29:65–74.

Lundin, K. 2001. Degenerating epidermal cells in *Xenoturbella bocki* (phylum uncertain), Nemertodermatida and Acoela (Platyhelminthes). Belg. J. Zool. 131:153–157.

Lundin, K., and C. Schander. 1999. Ultrastructure of gill cilia and ciliary rootlets of *Chaetoderma nitidulum* Loven 1844 (Mollusca, Chaetodermomorpha). Acta Zool. 80:185–191.

Mallatt, J., and C. J. Winchell. 2002. Testing the new animal phylogeny: first use of combined large-subunit and small-subunit rRNA gene sequences to classify the protostomes. Mol. Biol. Evol. 19:289–301.

Manuel, M., M. Kruse, W. E. G. Müller, and Y. Le Parco. 2000. The comparison of β-thymosin homologues among Metazoa supports and arthropod-nematode clade. J. mol. Evol. 51:378–381.

Mariaux, J. 1998. A molecular phylogeny of the Eucestoda. J. Parasitol. 84:114–124.

Mariaux, J., and P. D. Olson. 2001. Cestode systematics in the molecular era. Pp. 127–134 *in* Interrelationships of the Platyhelminthes (D. T. J. Littlewood and R. A. Bray, eds.). Taylor and Francis, London.

Mark Welch, D. B. 2000. Evidence from a protein-coding gene that acanthocephalans are rotifers. Invert. Biol. 119:17–26.

Martindale, M. Q., and J. Q. Henry. 1999a. Intracellular fate mapping in a basal metazoan, the ctenophore *Mnemiopsis leidyi*, reveals the origins of mesoderm and the existence of indeterminate cell lineages. Dev. Biol. 214:243–257.

Martindale, M. Q., and J. J. Q. Henry. 1999b. The origins of mesoderm. A cell lineage analysis in basal metazoans. Dev. Biol. 210:159.

Meglitsch, P. A., and F. R. Schram. 1991. Invertebrate zoology. 3rd ed. Oxford University Press, Oxford.

Melone, G., C. Ricci, H. Segers, and R. L. Wallace. 1998. Phylogenetic relationships of phylum Rotifera with emphasis on the families of Bdelloidea. Hydrobiologia 387/388:101–107.

Mollaret, I., B. G. M. Jamieson, R. D. Adlard, A. Hugall, G. Lecointre, C. Chombard, and J.-L. Justine. 1997. Phylogenetic analysis of the Monogenea and their relationships with Digenea and Eucestoda inferred from 28S rDNA sequences. Mol. Biochem. Parasitol. 90:433–438.

Monks, S. 2001. Phylogeny of the Acanthocephala based on morphological characters. Syst. Parasitol. 48:81–116.

Near, T. J., J. R. Garey, and S. A. Nadler. 1998. Phylogenetic relationships of the Acanthocephala inferred from 18S ribosomal DNA sequences. Mol. Phylogenet. Evol. 10:287–298.

Nebelsick, M. 1993. Introvert, mouth cone, and nervous system of *Echinoderes capitatus* (Kinorhyncha, Cyclorhagida) and implications for the phylogenetic relationships of Kinorhyncha. Zoomorphology 113:211–232.

Nielsen, C. 2001. Animal evolution: interrelationships of the living phyla. 2nd ed. Oxford University Press, Oxford.

Norén, M., and U. Jondelius. 1997. *Xenoturbella*'s molluscan relatives. Nature 390:31–32.

Olson, P. D., and J. N. Caira. 1999. Evolution of the major lineages of tapeworms (Platyhelminthes: Cestoidea) inferred from 18S ribosomal DNA and elongation factor-1a. J. Parasitol. 85:1134–1159.

Olson, P. D., and D. T. J. Littlewood. 2002. Phylogenetics of the Monogenea—evidence from a medley of molecules. Int. J. Parasitol. 32:233–244.

Olson, P. D., D. T. J. Littlewood, R. A. Bray, and J. Mariaux. 2001. Interrelationships and evolution of the tapeworms (Platyhelminthes: Cestoda). Mol. Phylogenet. Evol. 19:443–467.

Pearson, J. C. 1992. On the position of the digenean family Heronimidae: an inquiry into a cladistic classification of the Digenea. Syst. Parasitol. 21:81–166.

Peterson, K. J., and D. J. Eernisse. 2001. Animal phylogeny and the ancestry of bilaterians: inferences from morphology and 18S rDNA gene sequences. Evol. Dev. 3:170–205.

Poche, F. 1926. Das System der Platodaria. Arch Naturgesch 91:1–459.

Raikova, O. I., M. Reuter, U. Jondelius, and M. K. S. Gustaffson. 2000. An immunocytochemical ultrastructural study of the nervous and muscular systems of *Xenoturbella westbladi* (Bilateria inc. sed.). Zoomorphology 120:107–118.

Reuter, M., O. I. Raikova, and M. K. S. Gustafsson. 2001a. Patterns in the nervous and muscle systems in lower flatworms. Belg. J. Zool. 131:47–53.

Reuter, M., O. I. Raikova, U. Jondelius, M. K. S. Gustafsson, A. G. Maule, and D. W. Halton. 2001b. Organisation of the nervous system in the Acoela: an immunocytochemical study. Tissue Cell 33:119–128.

Ricci, C. 1998. Are lemnisci and proboscis present in the Bdelloidea? Hydrobiologia 387/388:93–96.

Rieger, R. M. 2001. Phylogenetic systematics of the Macrostomorpha. Pp. 28–38 in Interrelationships of the Platyhelminthes (D. T. J. Littlewood and R. A. Bray, eds.). Taylor and Francis, London.

Rohde, K. 2001. The Aspidogastrea: an archaic group of Platyhelminthes. Pp. 159–167 in Interrelationships of the Platyhelminthes (D. T. J. Littlewood and R. A. Bray, eds.). Taylor and Francis, London.

Rohde, K., C. Hefford, J. T. Ellis, P. R. Baverstock, A. M. Johnson, N. A. Watson, and S. Dittmann. 1993. Contributions to the phylogeny of Platyhelminthes based on partial sequencing of 18S ribosomal DNA. Int. J. Parasitol. 23:705–724.

Rouse, G. W., and K. Fauchald. 1997. Cladistics and polychaetes. Zool. Scr. 26:139–204.

Ruiz-Trillo, I., J. Paps, M. Loukota, C. Ribera, U. Jondelius, J. Baguñà, and M. Riutort. 2002. A phylogenetic analysis of myosin heavy chain type II sequences corroborates that Acoela and Nemertodermatida are basal bilaterians. Proc. Natl. Acad. Sci. USA 99:11246–11251.

Ruiz-Trillo, I., M. Riutort, D. T. J. Littlewood, E. A. Herniou, and J. Baguñà. 1999. Acoel flatworms: earliest extant bilaterian metazoans, not members of Platyhelminthes. Science 283:1919–1923.

Ruppert, E. E. 1991. Gastrotricha. Pp. 41–109 in Microscopic anatomy of invertebrates, Vol. 4: Aschelminthes (F. W. Harrison and E. E. Ruppert, eds.). Wiley-Liss, New York.

Ruppert, E. E., and K. J. Carle. 1983. Morphology of metazoan circulatory systems. Zoomorphology 103:193–208.

Schierenberg, E. 2000. New approaches to a better understanding of nematode phylogeny: molecular and developmental studies. J. Zool. Syst. Evol. Res. 38:129–132.

Schmidt-Rhaesa, A. 1998. Phylogenetic relationships of the nematomorpha—a discussion of current hypotheses. Zool. Anz. 236:203–216.

Senz, W. 1995. The "Zentralraum": an essential character of nemertean organisation. Zool. Anz. 47:53–62.

Sinitsin, D. F. 1911. [Parthenogenetic generation of trematodes and its progeny in molluscs of the Black Sea] (in Russian). Zap. Imp. Akad. Nauk 30:1–127 (Engl. trans. by A. M. Bagusin).

Sluys, R. 1999. Global diversity of land planarians (Platyhelminthes, Tricladida, Terricola): a new indicator-taxon in biodiversity and conservation studies. Biodivers. Conserv. 8:1663–1681.

Sluys, R. 2001. Towards a phylogenetic classification and characterization of dugesiid genera (Platyhelminthes, Tricladida, Dugesiidae): a morphological perspective. Pp. 57–73 in Interrelationships of the Platyhelminthes (D. T. J. Littlewood and R. A. Bray, eds.). Taylor and Francis, London.

Smith, J., and S. Tyler. 1985. The acoel turbellarians: kingpins of metazoan evolution or a specialized offshoot? Pp. 123–142 in The origins and relationships of lower invertebrates (S. Conway Morris, J. D. George, R. Gibson, and H. M. Platt, eds.). Oxford University Press, Oxford.

Smith, J. P. S., S. Tyler, and R. M. Rieger. 1986. Is the Turbellaria polyphyletic? Hydrobiologia 132:13–21.

Sørensen, M. V., P. Funch, E. Willerslev, A. J. Hansen, and J. Olesen. 2000. On the phylogeny of the Metazoa in the light of the Cycliophora and Micrognathozoa. Zool. Anz. 239:297–318.

Sørensen, M. V., and W. Sterrer. 2002. New characters in the gnathostomulid mouth parts revealed by scanning electron microscopy. J. Morphol. 253:310–334.

Southgate, V. R., D. Rollinson, J. Debont, J. Vercruysse, D. Vanaken, and J. Spratt. 1990. Surface topography of the tegument of adult *Schistosoma nasale* Rao, 1933 from Sri Lanka. Syst. Parasitol. 16:139–147.

Storch, V. 1991. Priapulida. Pp. 333–350 in Microscopic anatomy of invertebrates (F. W. Harrison and E. E. Ruppert, eds.), vol. 4. Wiley-Liss, New York.

Sundberg, P., and M. Saur. 1998. Molecular phylogeny of some European heteronemertean (Nemertea) species and the monophyletic status of *Riseriellus*, *Lineus*, and *Micrura*. Mol. Phylogenet. Evol. 10:271–280.

Sundberg, P., J. M. Turbeville, and M. S. Harlin. 1998. There is no support for Jensen's hypothesis of nemerteans as ancestors to the vertebrates. Hydrobiologia 365:47–54.

Sundberg, P., J. M. Turbeville, and S. Lindh. 2001. Phylogenetic relationships among higher nemertean (Nemertea) taxa inferred from 18S rDNA sequences. Mol. Phylogenet. Evol. 20:327–334.

Telford, M. J. 2000. Turning Hox "signatures" into synapomorphies. Evol. Dev. 2:360–364.

Telford, M. J., E. A. Herniou, R. B. Russell, and D. T. J. Littlewood. 2000. Changes in mitochondrial genetic codes as phylogenetic characters: two examples from the flatworms. Proc. Natl. Acad. Sci. USA 97:11359–11364.

Telford, M. J., and P. W. H. Holland. 1993. The phylogenetic affinities of the chaetognaths—a molecular analysis. Mol. Biol. Evol. 10:660–676.

Telford, M. J., and P. W. H. Holland. 1997. Evolution of 28S ribosomal DNA in chaetognaths: duplicate genes and molecular phylogeny. J. Mol. Evol. 44:135–144.

Telford, M. J., A. E. Lockyer, C. Cartwright-Finch, and D. T. J. Littlewood. 2003. Combined large and small subunit ribosomal RNA phylogenies support a basal position of the acoelomorph flatworms. Proc. R. Soc. Lond. B 270:1077–1083.

Timoshkin, O. A. 1991. Turbellaria Lecithoepitheliata: morphology, systematics, phylogeny. Hydrobiologia 227:323–332.

Tokioka, T. 1965a. Supplementary notes on the systematics of Chaetognatha. Publ. Seto Mar. Biol. Lab. 13:231–242.

Tokioka, T. 1965b. The taxonomical outline of Chaetognatha. Publ. Seto Mar. Biol. Lab. 12:335–357.

Turbeville, J. M., K. G. Field, and R. A. Raff. 1992. Phylogenetic position of phylum Nemertini, inferred from 18S rRNA sequences: molecular data as a test of morphological character homology. Mol. Biol. Evol. 9:235–249.

Turbeville, J. M., and E. E. Ruppert. 1985. Comparative ultrastructure and the evolution of nemertines. Am. Zool. 25:53–71.

Tyler, S. 1976. Comparative ultrastructure of adhesive systems in the Turbellaria. Zoomorphologie 84:1–76.

Tyler, S. 2001. The early worm—origins and relationships of the lower flatworms. Pp. 3–12 in Interrelationships of the Platyhelminthes (D. T. J. Littlewood and R. A. Bray, eds.). Taylor and Francis, London.

Tyler, S., and R. M. Rieger. 1977. Ultrastructural evidence for the systematic position of the Nemertodermatida (Turbellaria). Acta Zool. Fenn. 154:193–207.

Ubelaker, J. E. 1983. The morphology, development and evolution of tapeworm larvae. Pp. 235–296 in Biology of the Eucestoda (C. Arme and P. W. Pappas, eds.). Academic Press, London.

Wallace, R. L., C. Ricci, and G. Melone. 1996. A cladistic analysis of pseudocoelomate (aschelminth) morphology. Invert. Biol. 115:104–112.

Watson, N. A. 2001. Insights from comparative spermatology in the "urbellarian" Rhabdocoela. Pp. 217–230 in Interrelationships of the Platyhelminthes (D. T. J. Littlewood and R. A. Bray, eds.). Taylor and Francis, London.

Westheide, W., and R. M. Rieger. 1996. Spezielle Zoologie. G. Fischer Verlag, Stuttgart.

Williams, H., and A. Jones. 1994. Parasitic worms of fish. Taylor and Francis, London.

Williams, H. H., J. A. Colin, and O. Halvorsen. 1987. Biology of gyrocotylideans with emphasis on reproduction, population ecology and phylogeny. Parasitology 95:173–207.

Wills, M. A. 1998. Cambrian and recent disparity: the picture from priapulids. Paleobiology 24:177–199.

Winnepenninckx, B., T. Backeljau, and R. De Wachter. 1995. Phylogeny of protostome worms derived from 18S ribosomal RNA sequences. Mol. Biol. Evol. 12:641–649.

Winnepenninckx, B. M. H., T. Backeljau, and R. M. Kristensen. 1998. Relations of the new phylum Cycliophora. Nature 393:636–638.

Wirz, A., S. Pucciarelli, C. Miceli, P. Tongiorgi, and M. Balsamo. 1999. Novelty in phylogeny of Gastrotricha: evidence from 18S rRNA gene. Mol. Phylogenet. Evol. 13:314–318.

Wright, C. A. 1971. Flukes and snails. George Allen and Unwin, London.

Xylander, W. E. R. 2001. The Gyrocotylidae, Amphilinidea and the early evolution of the Cestoda. Pp. 103–111 in Interrelationships of the Platyhelminthes (D. T. J. Littlewood and R. A. Bray, eds.). Taylor and Francis, London.

Yamaguti, S. 1963. Systema helminthum, Vol. 4: Monogenea and Aspidocotylea. John Wiley and Sons, New York.

Zamparo, D., D. R. Brooks, E. P. Hoberg, and D. A. McLennan. 2001. Phylogenetic analysis of the Rhabdocoela (Platyhelminthes) with emphasis on the Neodermata and relatives. Zool. Scr. 30:59–77.

Zrzavý, J. 2001. The interrelationships of metazoan parasites: a review of phylum and higher-level hypotheses from recent morphological and molecular phylogenetic analyses. Folia Parasitol. 48:81–103.

Zrzavý, J., V. Hypsa, and D. Tietz. 2001. Myzostomida are not annelids: molecular and morphological support for a clade of animals with anterior sperm flagella. Cladistics 17:170–198.

Zrzavý, J., S. Mihulka, P. Kepka, A. Bezdek, and D. Tietz. 1998. Phylogeny of the Metazoa based on morphological and 18S ribosomal DNA evidence. Cladistics 14:249–285.

VI

The Relationships of Animals: Lophotrochozoans

Mark E. Siddall
Elizabeth Borda
Gregory W. Rouse

Toward a Tree of Life for Annelida

The basic characteristics of Annelida, the quintessential "worms," are immediately recognizable to most people, if only from having seen countless earthworms creeping over grass or braving the streets after a hard summer rain. The most recognizable feature of annelids, besides their shape and propensity for exuding mucus when disturbed, is the segmented nature of their bodies. This segmentation, or "somatic metamerism," has been central in the history of ideas about their relationships, although it is thought now to have been somewhat misleading. From the iceworms living deep in the Gulf of Mexico to the Pompeii worm that can withstand water temperatures that approach boiling, it is clear that annelids are a remarkably diverse group with a range of morphologies, life history strategies, and habitat preferences that rivals any other group of organisms considered in this volume. Although clearly the oligochaetes (which includes the earthworms) would probably be the most readily recognized as belonging to this group, there are also the much more numerous, principally marine, bristleworms (or polychaetes) and, of course, the much-maligned leeches (fig. 15.1). Additional, less known groups belong to the phylum Annelida, and whether or not others have evolved from annelid ancestors remains a matter of debate and intense scientific scrutiny.

The importance of annelids to ecology received a considerable boost in the 1800s with Charles Darwin's (1881) detailed demonstration that earthworms are responsible for recycling and aerating soils. Since that time, and particularly in the last century, annelid species have been central in assessments of water quality both in freshwater and in marine ecosystems as indicators of oxygen content, salinity, organic chemical pollutants, and heavy metal concentrations (Lauristen et al. 1985, Uzunov et al. 1988, Metcalfe et al. 1988, Verdonschot 1989, McNicol et al. 1997). The ubiquitous use of worms as bait by sport fishermen is testament to the direct role worms play in global food webs, where they may constitute more than one-third of the benthic animal diversity associated with coral reefs or intertidal shore life (Grassle 1973).

But, too, there is a darker ecological side to annelids as it relates to their parasitological role. Although, generally speaking, leeches are painless thieves of scant quantities of blood from unsuspecting hosts, a few transmit deadly blood flagellates to their victims. Small tubificid oligochaetes serve as the intermediate hosts for myxosporeans that cause "whirling disease" in salmon by infecting the brain and other neurological tissues of the fish (Kent et al. 2001). Even a few of the marine polychaetes have been found to wreak havoc on important mollusk species by boring into their shells and thus threatening millions of dollars of fishery resources and aquaculture operations (Fitzhugh and Rouse 1999, Kuris and Culver 1999, Lafferty and Kuris 1996).

As with most of the major branches on the eukaryotic tree, our understanding of the anatomical and ecological complexities of annelids would be greatly enhanced with a solid accounting of the evolutionary history of the group. For example, if we knew where leeches came from (or, specifically, with which group they share a recent ancestry), we

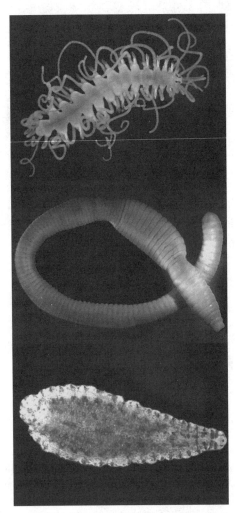

Figure 15.1. Three among many of the principal groups of annelids are polychaetes, oligochaetes, and leeches, represented here by a syllid polychaete (top), a glossoscolescid earthworm (middle), and a glossiphoniid leech (bottom). Photos by G. Rouse (top) and M. Siddall.

seems, by Linnaeus himself. Granted, there was the superficial similarity among wormy animals in that they lacked prominent appendages and were longer than they were wide, but there was little else (save convenience) to suggest this potpourri of animal life should be held together. Soon Linnaeus began the deconstruction of Vermes, first by removing snakes to a more sensible location with other vertebrates. Similarly, near the beginning of his career, Lamarck (1802) recognized the segmented nature (fig. 15.2) of a large collection of the remaining worms, creating the taxon Annélides (= Annelida) for them but leaving the remainder in Vermes.

Almost immediately, differences in opinion arose regarding the closest relatives to annelids. Lamarck (1809) clearly had them grouped with mollusks in a derivation separate from the insects and crustaceans. However, Lamarck's chief detractor, Georges Cuvier, placed annelid worms with the arthropods together as one of the major "embranchments" of life, creating what we would today regard as the superphylum Articulata (Cuvier 1812). The principal rational for this amalgamation of worms possessing a hydrostatic skeleton with crustaceans and insects possessing an exoskeleton was the recognition that each exhibits a longitudinal repetition of portions of the body in which the segments are separated by walls or septa. The influential Haeckel (1866) agreed that this axial mesodermal somatic metamerism justified the

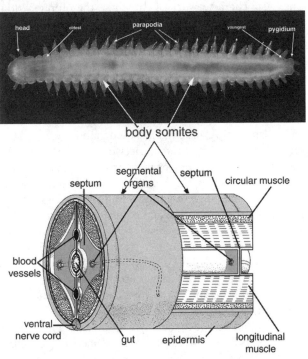

Figure 15.2. An obvious feature of annelids, yet one that historically has led to come confusion regarding relationships, is their segmentation. The name "Annelida" is derived from the Latin word for "ring." Each body ring, or somite, is separated from the next by a septum, and each has a series of structures that repeats in successive somites through the body. Photo by G. Rouse; drawing modified from Rouse and Pleijel (2001).

might be afforded important clues regarding the origins of the very powerful salivary compounds they harbor that prevent blood from clotting, which in turn might open new avenues for research into treating those prone to strokes or heart disease. Thankfully, there has been good progress in this direction in the last decade. We now have a more complete picture of what annelids are related to, what groups of worms should rightly be included in the phylum, and in certain instances a very good idea of how portions of the annelidan tree have branched and diversified. However, important gaps remain in our knowledge in each of these three contexts. It is our hope that this chapter will stimulate greater interest in solving those concerns once and for all.

The Sister Search

The enormous subkingdom of life "Vermes" created by Linnaeus was not taken seriously for very long—not even, it grouping and drew Articulata as one of the largest limbs emerging from his stylized tree of life (depicted in the introduction, fig. I.2). After the Darwinian revolution, this affiliation of annelids and arthropods carried more weight in that there was an easy suggestion of a "transitional form" between annelids and arthropods embodied by the limbed onychophoran velvet worms (e.g., Snodgrass 1938, Meglitsch and Schram 1991). A few systematists continued to wonder whether or not Lamarck was right in grouping annelids with mollusks (e.g., Pelseneer 1899, Naef 1913), but this hypothesis did not receive serious consideration until the advent of molecular phylogenetics in the 1980s.

The availability of universal primers for PCR (polymerase chain reaction) amplification and sequencing of the ribosomal DNA (rDNA) encoding the small subunit (SSU, 18S) of ribosomal RNA (rRNA) provided a means for testing many notions about the evolutionary history of groups of organisms (Medlin et al. 1988). One of the first groupings to come into doubt in light of these new data was Cuvier's Articulata. Contrary to the broadly held and widely taught belief in the primacy of somatic metamerism, 18S rDNA suggested a monophyletic group comprising onychophorans and arthropods, quite separate from another that included mollusks and annelids (Field et al. 1988, Ghiselin 1988). It was quickly recognized that, although this would require independent evolution of metamerism, the latter group was characterized by the presence of pelagic trochophore larvae. Those molecular results were quickly corroborated by additional DNA data (Lake 1990) and by an analysis of morphological characters (Eernisse et al. 1992), but they then came into doubt again in the face of contradictory analyses both of molecular and morphological data sets (Wheeler et al. 1993, Rouse and Fauchald 1995). Eventually the weight of evidence continued to mount against Cuvier's Articulata. Since 1995, reanalyses of rRNA genes and morphological data, whether separately (Conway Morris and Peel 1995, Ax 1996, Halanych et al. 1995, Winnepenninckx et al. 1995, Aguinaldo et al. 1997) or in combination (Zrzavý et al. 1998, Peterson and Eernisse 2001), or of mitochondrial gene sequences (Garcia-Machado et al. 1999) and even mitochondrial gene order (Boore and Brown 2000), all indicate that Annelida has a more recent common ancestry with Mollusca and other groups in Lophotrochozoa than with Arthropoda and what are now known as the molting Ecdysozoa.

What Is a Worm and What Is It Not?

Commensurate with the difficulties in determining the differences between the Articulata hypothesis and the Trochozoa hypothesis have been those associated with the specific composition of Annelida itself. Many early phylogenetic analyses of the problem suffered from presuming that various groups were monophyletic, such as by including a single taxon "Annelida" or only a few representatives of the group (e.g., Eernisse et al. 1992, Wheeler et al. 1993). As such,

higher level determinations that tested whether or not annelids and arthropods had a recent common ancestry did not necessarily settle the question of just what is an annelid. Polychaetes and oligochaetes have hairlike chaetae (or setae) projecting from each of their body somites (indeed, their names effectively mean very hairy and a little hairy, respectively), but then so do other animals, such as brachiopods, echiurans, and beard-worms (pogonophorans). Besides, leeches have no hairs at all, and no one doubted that leeches are related to oligochaetes. These latter two groups comprise the larger Clitellata by virtue of each having the saddlelike clitellum about one-third of the way down from the head. On close examination in a modern phylogenetic context, Rouse and Fauchald (1995) noted that, with the possible exception of the presence of a "nuchal organ," there was no reason to suppose even Polychaeta to be monophyletic, much less Annelida, if various groups such as pogonophorans were excluded.

The pogonophorans (which includes deep-sea hydrothermal vent Vestimentifera) are marine tube-forming worms that have an occluded gut and do not exhibit metamerism in the same way that annelids do. The varied and complex taxonomy of the group represents one of the more fascinating tales in animal systematics (see Rouse 2001). The fact that they tend to be found in deep-sea sediments resulted in the first member of this group, *Siboglinum weberi*, not being described until 1914. The anatomy of the worms was variously interpreted during the 20th century such that some were described in a way that was upside down and the larvae were back to front. Complete specimens of the worms were not even found until the 1960s. There are now more than 100 nominal species described, most from abyssal regions. Some, such as *Riftia*, are large and spectacular members of hydrothermal-vent communities (Jones 1981), whereas others are smaller and found in association with reducing sediments, methane seeps, rotting whale carcasses, or with sunken terrestrial-plant debris. The nutritional requirements for these worms are met through their symbiotic relationship with chemoautotrophic bacteria that occupy cells in the expanded gut wall (Southward 1993). *Riftia pachyptila* has the fastest growth rate of a marine invertebrate: it can colonize a new hydrothermal vent site, grow to sexual maturity, and have tubes of 1.5 m in length, all in less than two years (Lutz et al. 1994). This rapid growth would appear to be essential because their habitat is ephemeral and lasts for only a few years or decades. In contrast, *Lamellibrachia* that live in cold seeps on the Louisiana slope (Gulf of Mexico) grow very slowly, reaching more than 2 m in tube length but taking more than 100 years to do so (Fisher et al. 1997).

Shortly after their discovery, there was some suggestion that pogonophorans may be related to the polychaetes (Uschakov 1933, Hartman 1954), although others considered them to be more similar to the hemichordate acorn worms. The spiralian nature of pogonophorans was eventually conceded, but most invertebrate systematists continued to hold them to be in a separate phylum (e.g., Nørrevang

1970, Ivanov 1988). Rouse and Fauchald's (1995) work indicated that morphological data were unable to separate pogonophorans and vestimentiferans from the polychaetes and predicted that these aberrant worms would eventually group with the sabellid polychaetes (which also form protective tubes). Shortly thereafter, this hypothesis was corroborated in the context of morphological assessments of polychaetes (Bartolomaeus 1995, Rouse and Fauchald 1997), and these odd worms are now included among polychaetes in the family Siboglinidae (fig. 15.3).

Initial attempts to confirm these results using elongation factor gene sequences (McHugh 1997) offered some corroboration of the polychaete ancestry for these extraordinary deep-sea worms but also suffered from the use of too few taxa or too small a portion of the gene (Siddall et al. 1998). Eventually, the combined use of histone gene sequences and ribosomal gene sequences (Brown et al. 1999) lent strong support to the morphological results previously obtained (fig. 15.3). Even mitochondrial gene order corroborates the annelidan origins for the Siboglinidae (Boore and Brown 2000). Each of those analyses, in addition to demonstrating that Polychaeta logically had to include the pogonophorans, also indicated that the clitellate annelids [Oligochaeta and Hirudinida (leeches)] arose from within the polychaetes, and that possibly so too did the spoon-worm echiurans.

Regarding the latter, several analyses place Echiura either within Annelida (McHugh 1997, Brown et al. 1999), sister to Annelida (Brown et al. 1999), or perhaps closer to Mollusca (Siddall et al. 1998). The body of echiurans is unsegmented with an extrusible proboscis anteriorly and with hooks posteriorly. Their trochophore larval stages are similar to certain polychaetes. Although common in intertidal zones around the world, there are few more than a hundred species described (more from lack of interest than lack of diversity). The Californian "innkeeper worm," *Urechis caupo*, lives in a U-shaped burrow providing a safe home to several species of crabs, polychaetes, and even small fish (Arp et al. 1992).

The position of Echiura remains problematic, and molecular data have also recently necessitated the removal of myzostomids (a strange group of ectosymbiotic worms once thought to be annelids) from Annelida in light of their closer relationship to rotifers and acanthocephalans (Zrzavý et al. 2001). Meanwhile, there is vanishing support for the notion that Polychaeta constitute a natural group; rather, they are expected to be found to be synonymous with Annelida as a whole (Rouse and Fauchald 1998).

Clitellata: From the Leaves to the Trunk

Although the preceding efforts progressed in terms of delineating the limits of Annelida from the bottom of the spiralian tree upward, several researchers have been engaged in ascertaining the relative relatedness of subsets of annelids such as the leeches, the tubificid oligochaetes, and other groups, all

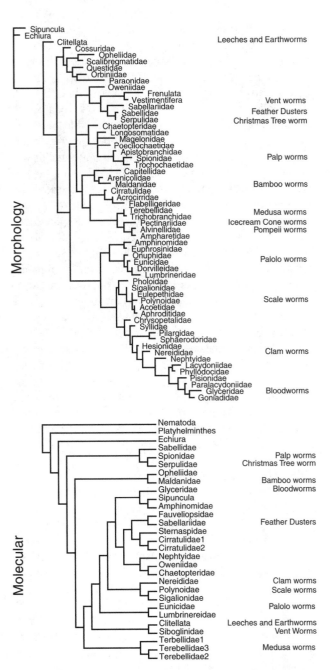

Figure 15.3. Morphological data (top, slightly modified from Rouse and Fauchald, 1997, their fig. 58) and molecular data (bottom, tree redrawn from Brown et al. 1999). Both provide support for the hypothesis that pogonophorans and vestimentiferans (Siboglinidae) evolved from within Polychaeta. The latter result suggests that oligochaetes and leeches (Clitellata) also are derived from polychaetes. If correct, Polychaeta would be synonymous with Annelida. The position of echiurans remains unclear.

with the expressed intention of eventually combining their data in a larger analysis of clitellate annelids. This top-down approach has proven successful in demonstrating that Oligochaeta are destined for a fate similar to that suggested above for the paraphyletic Polychaeta, principally because leeches and their allies group inside of oligochaetes.

Like earthworms, leeches are clitellates but with special adaptations to blood-feeding. They have a muscular caudal sucker made up of the last seven somites of the segmented body that is critical for maintaining position on a host and is used as a swimming fluke by the medicinal leeches (Hirudinidae). The anterior six somites likewise are modified into a region with a ventral sucker surrounding a mouth pore. Leeches are subdivided into two basic groups based on anatomical variations in blood-feeding mechanisms. The large, wormlike members of Arhynchobdellida, of which *Hirudo medicinalis* is typical (fig. 15.4A), have three muscular jaws each with a row of teeth for cutting through skin into capillary-rich tissues. In contrast, members of Rhynchobdellida, as the name implies, have a muscular proboscis to effect blood-feeding from vascularized deeper tissues.

Blood-feeding arhynchobdellids include the aquatic Hirudinidae ("medicinal leeches") and the terrestrial Haemadipsidae ("jungle leeches"). The European medicinal leech has for centuries been used in phlebotomy (blood-letting) in a variety of regions, including China (in Wang Chung's *Lun Hêng*, circa 30 A.D.), India (in Kunja Lal Sharma's *Su'sruta Samhitá*, circa 200 A.D.), ancient Rome (in Pliny's *Natural History*, circa 50 A.D.), and throughout Europe (Shipley 1927). Use in Europe, however, reached its peak in the 19th century after the ascendancy of Napoleon's army surgeon Broussais and his student Broussard, known together as the "Grand Sangeurs." Leeching was a dubious cure considered for everything from simple headaches and insomnia to ulcers and obesity. Nonetheless, harvesting leeches from European lakes and ponds continued intensively, with importations in the 1830s to France exceeding 50 million annually (and that notwithstanding a duty of 1 franc per thousand). France was hardly alone in this endeavor—Russia and Hungary each imposed hefty export duties and fines for trafficking in the worms, and more than seven million leeches per year were used in London hospitals as late as 1863 (see Sawyer 1981, Elliott and Tullett 1984, 1992).

The consequences of this demand for *Hirudo medicinalis* have been profound. As early as 1823, the Hanover government acted to restrict trade in light of declining numbers, forbidding all exports. Sardinia followed suit in 1828 and eventually Moldavia, Wallachia, Spain, Portugal, Bohemia, and Italy had either exhausted populations or had banned their export so as to conserve what was left (Sawyer 1981). By the 1990s, *Hirudo medicinalis* was declared either threatened or endangered in more than 15 countries, had been included in the *IUCN Invertebrate Red Data Book* (1983), and was listed as Appendix II in CITES (Wells et al. 1983, Elliott and Tullett 1992).

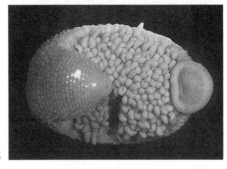

Figure 15.4. (A) *Hirudo medicinalis*, the European medicinal leech. Photo by M. Siddall. (B) Glossiphoniid leeches such as *Placobdelloides jaegerskeoldi* exhibit a strong degree of parental care by brooding their young. Photo by J. Oosthuizen (deceased).

In tropical wet forests, haemadipsids are more frequently encountered than are hirudinids (fig. 15.6). Both of these groups are equipped with a parabolic arc of 10 eyespots that permit the detection of contrasting movement in three dimensions. Haemadipsids have an unusual biogeographic distribution, being found only on the Indian subcontinent and in Southeast Asia, Wallacea, Australia, Melanesia, Madagascar, and the Seychelles, but not in Africa or in South America. All other leech families have a global distribution. No other group of leeches has inspired such passionate accounts by travelers or naturalists. Even North America's most prolific hirudinologist was particularly awestruck by this family where, in the "dank tropical jungles, the misty ravines and the showery, forested mountain-sides of this extensive region they are among the most dominant and self-assertive elements" (Moore 1927: 224). "Leeches swarmed with incredible profusion . . . they got into my hair, hung from my eyelids and crawled up my back" [*Himalayan Journals* (Hooker 1854)]. They were "so close together that your eyes had to be focused at your feet to find a place where you could step . . . I finally compromised with the leeches . . . letting them get their fill . . . so long as they kept away from my face and the fly of my trousers" [*Burma Surgeon Returns* (Seagrave 1946)].

Terrestrial leeches have the additional adaptation of respiratory auricles near their caudal sucker, allowing for gas exchange without excessive loss of fluid. Moreover, they have well-developed sensory systems probably for detecting vibrations, carbon dioxide, and heat. The terrestrial habits and the nature of the global distribution of the haemadipsids have been cause for speculation regarding their evolutionary history. Considerably distantly related terrestrial blood-feeders such as *Mesobdella gemmata* in Chile (Blanchard 1893), *Malagobdella* species in Madagascar (Blanchard 1917), and the Seychellian *Idiobdella* species (Harding 1913) naturally caused some consternation for Autrum (1939) in his attempt to explain the world's distribution of this group.

The two groups of proboscis-bearing Rhynchobdellida have pairs of centrally arranged eyespots that sense at least two-dimensional movement. The small fish leeches, or Piscicolidae, exhibit a form of parental care that promotes their offspring achieving an early blood meal. Rather than abandoning a secreted "cocoon" on shore, as the arhynchobdellids do, the piscicolids cement dozens of egg cases to the surface of shrimp or crabs. When that crustacean is eaten by a fish, juvenile leeches jump off, attaching to the buccal surfaces or migrating to the gills in order to acquire a blood meal. The Glossiphoniidae, such as *Haementeria ghilianii*, are broad and flattened, normally feeding on turtles or amphibians. Glossiphoniids secrete a membranous bag to hold their eggs on their underside. Covering their eggs (fig. 15.4B), adults will fan the brood until they hatch. The brood then will turn and attach to the venter of their parent, and when the parent finds its next blood meal, they are carried to their first.

Leeches have gained importance not only in terms of their use in microsurgery but also in relation to the isolation of bioactive compounds from their saliva. Vertebrate blood has a plethora of coagulation factors, and a leech ill-equipped for preventing the activation of this system would surely perish. Most leeches need to feed for 20–40 minutes, but blood can clot in much less time. Should the ingested blood-meal coagulate in their gut, this would render mating, avoidance of predators, or seeking another meal quite impossible. Leeches not only have dramatically circumvented the end points of the mammalian coagulation cascade (cross-linkage of platelets, thrombin's production of a fibrin matrix, and the cross-linking of that matrix into a hard clot) but also have interfered with no fewer than seven points in the mammalian clotting system. Hirudin, a potent thrombin inhibitor, was the first anticoagulant to be isolated from a leech. Most other leech-derived anticoagulants also are protease inhibitors (of killikrein, fibrinogen, or factors Xa and XIIIa; Chopin et al. 2000). Calin blocks von Willebrandt's factor and platelet aggregation. Platelet aggregation inhibitors from North American species of *Macrobdella* and *Placobdella* (decorsin and ornatin, respectively) block the IIb/IIIa site (Seymour et al. 1990, Mazur et al. 1991). Yet, the most frequently discovered anticoagulants are protease inhibitors that block factor Xa, thus preventing conversion of prothrombin to thrombin and that also seem to have an ability to prevent tumor metastasis (Brankamp et al. 1990, Blakenship et al. 1990). Beyond simply stopping the formation of clots, the giant Amazonian leech *Haementeria ghilianii* has also evolved

ways to break them down (Budzynski et al. 1981, Malinconico et al. 1984). There even are known anti-inflammatory agents such as eglin, bdellin, and cytin that have been isolated from leeches.

Many leeches do not feed on blood at all. Glossiphoniids, such as species of *Helobdella* and *Glossiphonia*, feed on aquatic oligochaetes and snails. The jawless Erpobdellidae members feed on chironomid larvae, and the jawed members of Haempidae consume whole earthworms, shredding them over jaws with two rows of large teeth. In addition, there are rarely encountered families such as the South American Americobdellidae and Cylicobdellidae that are terrestrial earthworm hunters and of uncertain phylogenetic affinities. Typically, it has been assumed that non-blood-feeding varieties are more primitive than those with the "advanced" behavior of blood-feeding.

In addition to Oligochaeta and Hirudinea, two other groups of annelids possess a clitellum and are included in Clitellata: the orders Branchiobdellida and Acanthobdellida. Branchiobdellidans, commonly known as crayfish worms, as the name implies, are ectoparasitic of astacoid crayfish (Crustacea: Astacidae) and are endemic to the Holarctic (Eurasia and North America) region. They are subdivided into five families consisting of 21 genera and approximately 150 species and have a constant number of 15 body segments (somites). The first four constitute the head region, with the first somite forming an adhesive oral surface around the mouth. The last segment forms a posterior disk-shaped attachment organ (Gelder et al. 1988). Branchiobdellidans possess a dorsal and ventral denticulate jaw (Odier 1823) and, like leeches, lack hairlike chaetae. The second group, monotypic with *Acanthobdella peledina* Grube 1851, is specifically parasitic on salmon and also endemic to the Holarctic. *Acanthobdella* is characterized by a constant number of 29 somites, an anterior sucker composed of the first five somites, with hooklike chaetae limited to this region, and a posterior sucker.

Resolution of the evolutionary lineages and relationships among subgroups within Clitellata has been a topic of debate deliberated for more than a century (Odier 1823, Vejdowsky 1884, Livanow 1906, 1931, Sawyer 1986, Brinkhurst and Gelder 1989, Siddall and Burreson 1996, Brinkhurst 1999, Siddall et al. 2002). A close relationship between branchiobdellidans and leeches, with *Acanthobdella* as their sister taxon and with the lumbriculids as a linkage between these and the rest of Oligochaeta, has long been suspected (Odier 1823, Livanow 1906, 1931, Sawyer 1986). Before the advent of molecular phylogenetics, these studies used morphology to discern relationships among the groups, but because of subjective interpretations of clitellate anatomy, agreement and resolution of the classification have been problematic.

In particular, the taxonomic position of branchiobdellidans and *Acanthobdella* within Clitellata has been problematic because of their possession of combinations, or "transitional" (Holt 1965, Purschke et al. 1993) forms, of hirudinean (leech) and/or oligochaete characters. Odier (1823) and Livanow

(1906) hypothesized that a common ancestor existed for these worms and leeches based on their possession of "leech-like" characters: an attachment organ, loss of chaetae, constant number of body segments, and an ectocommensal life history strategy. Michaelsen (1919) was first to counter this view, arguing that because *Acanthobdella* possessed cephalic (head) chaetae and an oligochaete-type seminal funnel, it should fall within Oligochaeta. He therefore attributed the leechlike characters to convergence, or independent evolution, because of the adoption of an ectocommensalistic lifestyle. Livanow (1931) later reiterated his contention that *Acanthobdella* and branchiobdellidans are more closely related to leeches. Contrary to Holt (1965), who denied that Branchiobdellida and *Acanthobdella* are phylogenetically associated with leeches, Sawyer (1986) proposed four subclasses grouping all of the ectocommensal clitellates as subclasses of Hirudinea, with the inclusion of agriodrilidans (carnivorous lumbriculids proposed to be ancestral to leeches). Holt (1989) countered this, again affirming that the only common characteristic was their possession of a clitellum and that the remaining similarities must be convergences due to ectocommensalism.

The reinvestigation of the systematic position and synapomorphies (shared derived characters) of various annelids with leeches continued. Several studies dismissed the obvious similarities (Holt 1989, Brinkhurst and Gelder 1989, Purschke et al. 1993, Brinkhurst 1994), despite phylogenetic results corroborating various synapomorphies. Purschke et al. (1993) and Brinkhurst (1994), for example, reexamined the morphology of *Acanthobdella* and branchiobdellidans by reconstructing cladograms that showed their monophyly with leeches and a lumbriculid sister group. In each case they rejected their own findings. Brinkhurst and Gelder (1989) argued that the variability in the number of somites (hirudinids, branchiobdellidans, and *Acanthobdella* have 34, 15, and 27, respectively) was evidence of nonhomology of having a fixed number of segments, unlike the variable number in oligochaetes. Additionally, the presence or absence of chaetae is not consistent, being absent both in leeches and in branchiobdellidans but limited to the cephalic (head) region in *Acanthobdella*. In comparison to lumbriculids, the coelom (fluid-filled body cavity) in branchiobdellidans is reduced in the extremities where muscles are well developed, whereas in leeches and *Acanthobdella* it is completely reduced, with only the latter retaining septa (coelomic tissue walls between somites). A muscular posterior sucker, absent in oligochaetes but present in leeches and *Acanthobdella*, has been referred to as a non-muscular "attachment disk" with supposedly nonhomologous adhesive secretions or a "duo-adhesive" organ in branchiobdellidans (Weigl 1994, Gelder and Rowe 1988), suggesting the latter is not a sucker per se. Based on the lack of precise correspondence of morphology—the basis of monophyly among branchiobdellidans, *Acanthobdella*, leeches, and therefore lumbriculid oligochaetes—the hypothesis of convergent evolution still remained (Brinkhurst 1999).

Inasmuch as overall morphological similarities appeared to be inconclusive, sperm ultrastructure had also been used for phylogenetic analysis (Franzén 1991, Ferraguti and Erséus 1999). Although this offered a different perspective and broadened the basis in assessing relationships, it did not provide conclusive resolution. Ferraguti and Erséus (1999) presented synapomorphies in sperm structure corroborating the sister-group relationship of leeches and *Acanthobdella*, but they found no evidence in support of an exact position for Branchiobdella within Clitellata.

Conversely, a reconstruction of leech phylogeny based on morphology (Siddall and Burreson 1995) seemed to be in agreement, proposing several speculative evolutionary relationships. Because *Acanthobdella* does not directly feed on blood from the host, feeding mostly on dermal tissue, they hypothesized that the common ancestor of leeches was in fact not a blood-feeder and, as Sawyer (1986) proposed, that blood-feeding was acquired independently in rhynchobdellids and arynchobdellids. Avoiding the discrepancies caused by conflicting interpretations of morphology and in response to the broad convergence argued by Brinkhurst (1994) and Purschke et al. (1993), Siddall and Burreson (1996) took a different approach by examining the evolution of life history strategies of leeches in contrast to oligochaete plesiotypic (ancestral) conditions. In all cases, *Acanthobdella* and Branchiobdellida retained "oligochaete" conditions with these states being inherited by the hirudinids and later modified into conditions more typical of leeches, which Siddall and Burreson (1996) took as affirmation of the inclusion of these three groups within Oligochaeta.

Since the mid-1990s, the collection and addition of molecular data to known annelid morphology, ecology, and life histories (within and among various groups) began to shed light on resolving higher level relationships of leeches down to family-level phylogenies. Siddall and Burreson (1998) investigated the molecular phylogenetic relationships of leeches for the first time, using mitochondrial cytochrome c oxidase subunit I (mtCOI). This preliminary study confirmed previously suspected internal relationships but also suggested the existence of a sister-group relationship between the piscicolids (fish leeches) and Arhynchobdellida. Additionally, Oligochaeta seemed to be paraphyletic, with a split of lumbriculids from the rest of the oligochaetes, followed by a divergence of subsequent clitellate taxa (i.e., Acanthobdellida, Branchiobdellida, and Hirudinida, respectively). Since then, the use of a combination of ribosomal and mitochondrial gene sequences with morphological data has successfully been employed (fig. 15.5) to resolve family, genus, and higher level taxa in leeches (Apakupakul et al. 1999, Light and Siddall 1999, Siddall 2002, Siddall and Borda 2002).

In the same way that interpretations of morphology created a platform for debates, conflicting results were also noted using molecular data because of low or uneven taxon sampling and different methods of data analysis. Martin et al. (2000) examined the phylogenetic relationships of Clitellata

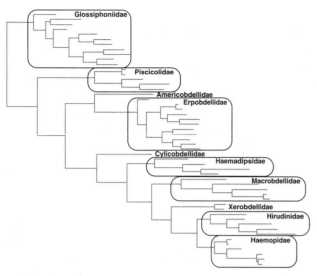

Figure 15.5. Phylogenetic relationships of the principal families of leeches based on morphological data, 18S rDNA, and 28S rDNA, as well as mtCOI and mitochondrial 12S rDNA.

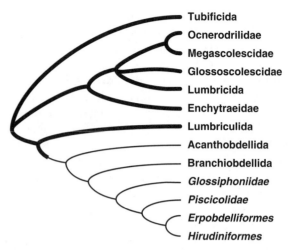

Figure 15.6. Phylogeny of the Clitellata based on a coordinated approach from several labs using nuclear and mitochondrial gene sequences. Oligochaetous lineages are represented by thicker lines. Leech taxa are italicized. Based on combined information from Siddall et al. (2001), Erséus et al. (2000), and B. Jameison (unpubl. obs.).

with maximum likelihood using 18S rRNA and mtCOI, in separate and combined analysis. They reported that, although their data suggested that leeches and leechlike worms do in fact fall within a paraphyletic Oligochaeta, different sequencing alignment methods gave conflicting results, and resolution of Clitellata was deemed to be confounded by faster evolving lineages.

At the 1994 International Meeting of Aquatic Oligochaete Biology, Siddall, Burreson, Coates, Erseus, and Gelder agreed on which genes would be pursued in order to finally solve the question of clitellate relationships: mtCOI and 18S rDNA. Commensurate with these data being gathered for leeches (Apakupakul et al. 1999), substantial members of aquatic oligochaetes had been similarly analyzed (Nylander et al. 1998, Erséus et al. 1999), with the attendant discovery that Naididae and Tubificidae are in dire need of revision. Once these data were complete for Branchiobdellida (see Gelder and Siddall 2001), it was possible to combine all in a broad assessment of clitellate relationships some eight years after the authors had agreed to do so. Nuclear 18S rDNA and mtCOI data for a total of 101 annelids were analyzed (Siddall et al. 2002), excluding morphological data so as to eliminate the criticism that results would be influenced by morphological convergence. The results of this cooperative phylogenetic work was the unambiguous validation of Livanow's (1906, 1931) assertions that branchiobdellidans and *Acanthobdella* share a recent common ancestor with leeches, which together form the sister lineage to the lumbriculid oligochaetes (fig. 15.6).

Although results so far are compelling, there is still considerable work to be accomplished among clitellate lineages. Most notable is our relative lack of megadrile oligochaetes such as the earthworm and allied taxa. Incorporating these families will require considerable fieldwork acquiring fresh specimens, particularly from South America, Africa, and Asia.

Primacy for Polychaetes

Polychaetes are generally small and cryptic. However, if one deliberately seeks them, for example, in a grab of marine sediment hauled up from a few hundred meters' depth, the number and variety of polychaetes can be overwhelming, and it may take weeks of work to identify them. Apart from the impact of polychaete diversity on specialists, there are a number of ways in which polychaetes do impinge on general human awareness.

One of the few annelids regularly eaten by people is the palolo worm (*Palola viridis*). *Palola viridis* is a eunicid polychaete with robust jaws that it uses to burrow through coral, where they form large galleries. Periodically, and usually at night, the posterior ends of these worms, about 20 cm long and filled with eggs or sperm, detach and swim toward the sea surface. There, people gather the worms, greatly regarded as a delicacy. The name "palolo" is Samoan, and in Samoa there are two breeding events, during the third quarter of the moon in both October and November. There are a number of *Palola* species around the world, including the Mediterranean and off California, that are also known to swarm (Fauchald 1992). Samoans and other South Pacific peoples for centuries have known of a relationship between the emergence of the worms, the "palolo risings," and the phase of the moon, now regarded as a classic example of lunar periodicity in animals (Caspers 1984, Fauchald 1992). The anterior end of the worm survives the spawning event and grows a new posterior to spawn again.

Swarming of annelids occurs in other parts of the world, and a number of different kinds of polychaete engage in this behavior. The phenomenon is broadly known as epitoky. Those with schizogamous epitoky, such as the palolo worm, detach their gamete-filled posteriors and live to breed another day. Others with epigamous epitoky, mostly in the Nereididae, transform their bodies entirely to allow them to swim up to the surface (e.g., by producing enlarged eyes, special paddle chaetae, and major muscle development). After spawning, the worms cannot possibly return to their life on the bottom and so die. Other annelids have epigamous epitoky but survive to breed again. The most famous of these is the syllid *Odontosyllis enopla*, also known as the "Bermudian fireworm" because their swarming is associated with a bright green luminescence. These 1–cm-long worms swarm in vast numbers in the evenings just after the full moons of June and July and create luminescent displays thought to help them attract mates near the surface of the water. After spawning, the worms descend to the bottom again and resume their lives (Fischer and Fischer 1995). It has been suggested that the light Christopher Columbus described the evening before his landfall in the Caribbean in October 1492 may have been the glow of *Odontosyllis* swarms (Crawshay 1935).

Annelids have direct economic importance to human society through their ecological function in the creation and maintenance of marine and terrestrial soils and sediments. Some people also make their livelihood from worms, supplying them as bait for recreational fishing. Marine worms in groups such as Arenicolidae, Glyceridae, Eunicidae, Nephtyidae, Nereididae, and Onuphidae are used as bait, whether caught in the wild or farmed in aquaculture systems. For instance, the glycerid *Glycera dibranchiata* and nereidid *Nereis virens* are manually harvested from mud flats of Maine with a wholesale value of several million U.S. dollars (Olive 1994). In Europe and Asia there are several commercial worm farms that supply tons of worms to the fishing industry (Olive 1994). At present this does not compare with the amount harvested from the wild, with all its attendant potential degradation of habitat.

Two polychaete groups one must be careful of are Amphinomida and Glyceridae. Amphinomids, commonly referred to as fireworms, induce a burning pain on anyone foolish enough to pick them up. Commonly found under rubble in coral reef environments, large (15–20 cm) amphinomids such as *Eurythoe* and *Hermodice* have elongate pink or green bodies with tufts of white chaetae emerging dorsally. These chaetae are unusually brittle and thin and may break off in the skin, producing an intense itchy or burning sensation that may last for days (Kem 1988). Members of Glyceridae can reach 40 cm in length and have four jaws at the end of their eversible proboscis, each armed with a venom gland. They inject this venom into their prey (crustaceans and other annelids), inducing paralysis (Kem 1988). People who have been bitten by these worms have reported intense pain and swelling, although there have apparently been no deaths to date.

Alvinellidae ("Pompeii worms" and "Palm worms") are a relatively recently discovered annelid group known only from sites associated with deep-sea hydrothermal vents in the Pacific Ocean. Given this recent discovery, they are surprisingly well studied, particularly *Alvinella pompejana* (Desbruyères and Laubier 1980, Desbruyères et al. 1998). Tolerating some of the most extreme living conditions of any animal, they are called Pompeii worms because they live in tubes on the sulfide chimney walls of active hydrothermal vents. As such, they are continuously in the presence of an unrelenting downpour of mineral particles that result from fluctuating thermal and chemical reactions of the hydrothermal fluid and surrounding seawater. Worms have been recorded crawling at temperatures exceeding 100°C! Only the crushing pressure of 250 atmospheres keeps the surrounding water from boiling. Desbruyères and Toulmond (1998) recently described an extraordinary new hesionid polychaete *Sirsoe methanicola* (as *Hesiocaeca*; see Pleijel 1998) living in large numbers on frozen methane hydrate mounds associated with cold methane "seeps" in the Gulf of Mexico (Fisher at al. 2000). This animal is also known as the "iceworm," but thus far little is known about its biology.

The broad-level systematics of polychaetes, after a period of relative stability, is undergoing major reassessment. The most recent comprehensive systematization of polychaetes was proposed by Rouse and Fauchald (1997) based on a series of morphological cladistic analyses. Allowing for the likely errors in the placement of many taxa, and the fact that there were conflicting results included in the original analyses by Rouse and Fauchald (1997), the most fundamental problem inherent in their systematization may be that of the placement of the root for any tree of Annelida. This has major implications for the taxon Clitellata (which is now synonymous with Oligochaeta) and the name Polychaeta itself, which may become synonymous with Annelida. Rouse and Fauchald (1997) assessed the monophyly of Polychaeta and relationships among the taxa usually included in the group and those traditionally excluded. Polychaete "families" and groups such as Sipuncula, Echiura, Clitellata, Pogonophora, and Vestimentifera were used as terminal taxa, largely because this allowed the most heuristic assessment of relationships based on present knowledge. It also permitted many of the current problems in the systematics of polychaetes to be highlighted. They found that the traditionally formulated Annelida were monophyletic and comprised two clades, Clitellata and Polychaeta, although the monophyly of the latter was not well supported at all, which is not that surprising, given the tremendous diversity of the group (fig. 15.7). There was no obvious sister group for Clitellata within Polychaeta that could be identified on current morphological evidence. Rouse and Fauchald (1997) then presented a new classification of polychaetes based on one of the analyses.

Rouse and Fauchald (1997), Pleijel and Dahlgren (1998) and most previous influential systematizations of polychaetes (e.g., Fauchald 1977) recognize a taxon Phyllodocida, explicitly or implicitly accepting that this is a clade. Basal

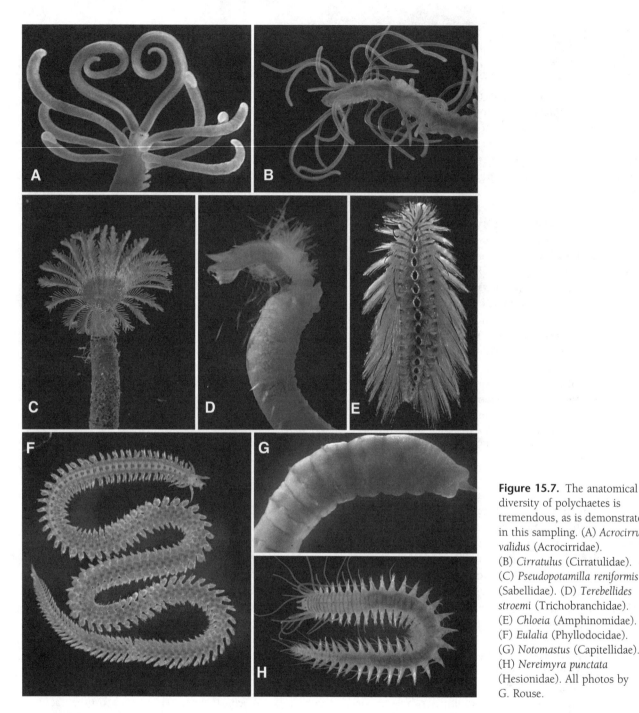

Figure 15.7. The anatomical diversity of polychaetes is tremendous, as is demonstrated in this sampling. (A) *Acrocirrus validus* (Acrocirridae). (B) *Cirratulus* (Cirratulidae). (C) *Pseudopotamilla reniformis* (Sabellidae). (D) *Terebellides stroemi* (Trichobranchidae). (E) *Chloeia* (Amphinomidae). (F) *Eulalia* (Phyllodocidae). (G) *Notomastus* (Capitellidae). (H) *Nereimyra punctata* (Hesionidae). All photos by G. Rouse.

annelids, according to Rouse and Fauchald (1997), are taxa such as Clitellata and simple-bodied polychaete groups like *Questa* and Paraonidae. This rooting of Annelida was based on outgroup choices such as Mollusca and Sipuncula and may well be misleading. There currently is little evidence that is not ad hoc to justify other ways of rooting this tree with morphological data. However, several of the alternative hypotheses (e.g., Westheide 1997, Conway Morris and Peel 1995) are similar in that they suggest that the root for the annelid tree should be placed within Phyllodocida or Aciculata (Phyllodocida plus Eunicida).

In addition to the rooting problem, the phenomenon of paraphyletic taxa in polychaete systematics may be a common situation for several reasons. Most polychaete taxa have been named without reference to any tree topology. Classifications based only on similarity will inevitably lead to paraphyly. In their review of those polychaete taxa with a rank of family, Fauchald and Rouse (1997) found that of the 80 families that they accepted as "valid," they could provide no evidence of monophyly for 21, including such well-known taxa as Eunicidae and Polynoidae. It should be noted that even where Fauchald and Rouse (1997) suggested features that provided evidence of monophyly for the remaining 59 families, this must be regarded as provisional. Until comprehensive detailed cladistic analyses are performed across relevant sets of taxa such assumptions of monophyly for these groups probably are unfounded. For example, Fauchald and Rouse (1997) provided apomorphies supporting the monophyly of Spionidae, of Longosomatidae, of Poecilochaetidae, of Trochochaetidae, and of Uncispionidae. Subsequently, a cladistic analysis by Blake and Arnofsky (1997) showed that Spionidae was rendered paraphyletic relative to the other four, which should now be regarded as junior synonyms.

Within the numerous polychaete taxa, there have also been few detailed systematic studies. Rouse and Pleijel (2001) found that there have been cladistic analyses only of the following polychaete taxa: Opheliidae, Orbiniidae, *Questa*, Eunicida, Dorvilleidae, Onuphidae, Chrysopetalidae, Hesionidae, Namanereidinae (in Nereididae), Pilargidae, Syllidae, Phyllodocidae, *Notophyllum* (in Phyllodocidae), *Phyllodoce* (in Phyllodocidae), Glyceriformia, Sabellidae, Serpulidae, Siboglinidae, Terebelliformia, Terebellinae (Terebellidae), and Spionidae. Clearly, there is much work to be done toward our basic understanding of the relationships among polychaetes.

At an even more fundamental level, it is certain that there are many more polychaetes to be described and that they represent an important component of the diversity of marine animals. This is exemplified by studies on the variety of polychaetes in a small area. In a well-known example, Grassle (1973) found 1441 polychaetes in a single chunk of coral weighing a few kilograms. He placed these polychaetes into 103 nominal species and noted that they represented two-thirds of the macrofauna collected. More recent surveys on diversity of deep-sea polychaetes have shown a similar pattern: dominance in terms of individuals and taxa (e.g., Grassle and Maciolek 1992). What is more striking about these surveys is the number of undescribed polychaetes that were found (e.g., 64% by Grassle and Maciolek 1992). Arguably, we will not arrive at a comprehensive understanding of annelid origins and phylogeny until more of extant polychaete diversity is found and described.

Quo Vadimus?

Certainly there has been no lack of effort regarding the morphological characterization of annelidan groups on a broad scale (e.g., Rouse and Fauchald 1997, Siddall and Burreson 1995, Purschke et al. 1993, Brinkhurst 1994, 1999). Homologizing those characters and states among disparate subsets of worms has proven more difficult and often an intractable task for lack of independent corroboration of sister-group relationships. Although the use of molecular sequence data provides an opportunity to achieve those aims, there has yet to be either a full accounting of which loci are available across the phylum or, more importantly, what information those data together might provide regarding support for group membership. Currently, there are about 800 gene sequences, divided into roughly one-third from polychaetes and two-thirds from clitellates (of which more than half are from leeches alone). Sampling has yet to be coordinated among various laboratories, but it can be and should start with the complete amalgamation of sequences in a data set of approximately 365 taxa and about 4000 sites newly aligned and analyzed. Our expectation for wholly sensible results from that are, however, rather low. We estimate that more than two-thirds of the preliminary matrix will be missing for lack of overlap in data across taxa. Still, that work would create a springboard from which several labs cooperating internationally (Australia, France, and the United States of America) might focus sequencing efforts on existing DNA isolates or samples in a way that would most efficiently ameliorate topological instability. This first phase might take less than two years to bring to completion. A more full accounting of annelid phylogeny will need another complementary approach and considerably more time.

The main questions that need answers include the following:

Where does the root for Annelida lie?
What is the sister group to Clitellata?
Do other major taxa, such as Brachiopoda, Echiura, and Sipunculida, to name a few, belong within Annelida or are they sister to it?

These broad questions all are interlinked and, once satisfactorily resolved, will allow for a multitude of more detailed analyses among less inclusive annelid groups. How would one best approach these questions, given the equivocal results to date? The answer is, of course, more data, and lots of it. This first means an extensive array of gene sequence data for many terminals. The genes to be sequenced would comprise parts of both nuclear and mitochondrial genomes. To make the most of the data available already, these arguably would be four nuclear regions—SSU rDNA (18S), large subunit rDNA (28S), histone H3, elongation factor EF-1α—plus the mtCOI and mitochondrial 16S regions. Additionally, the sequenced specimens should be studied with a range of morphological techniques. This would then allow for a fuller development of the morphological data set presented in Rouse and Fauchald (1997). Much of the data used in that study was based on observations more than a century old, and there are many gaps in our knowledge for many taxa. Using light and electron microscopy of both internal and external features, as well as larval development, a comprehensive suite of anatomical characters could then be added to the molecular data set. A sound tree at this level will provide the basis for resolving many other problems in annelid systematics. The homology of many body regions in annelids is unresolved, and this is reflected in the multitude of names for the "same" parts. Simplifying terminology will make the taxonomy of the various groups easier, allowing many more people to study annelid systematics as a whole.

Moreover, the full scope of diversification of life-history roles and the phylum's expansion across the planet in space and time could then be understood. Our understanding of fundamental questions such as the evolution of reproductive mechanisms, feeding strategies, and physiology can only be enhanced with a better understanding of annelid evolution. In the next five years we predict it will truly be the worms' turn.

Literature Cited

Aguinaldo, A. M. A., J. M. Turbeville, L. S. Linford, M. C. Rivera, J. R. Garey, R. A. Raff, and J. A. Lake. 1997. Evidence for a clade of nematodes, arthropods and other moulting animals. Nature 387:489–493.

Apakupakul, K., M. E. Siddall, and E. M. Burreson. 1999. Higher level relationships of leeches (Annelida: Clitellata: Euhirudinea) based on morphology and gene sequences. Mol. Phylogenet. Evol. 12:350–359.

Arp, A. J., B. M. Hansen, and D. Julian. 1992. Burrow environment and coelomic fluid characteristics of the echiuran worm *Urechis caupo* from populations at three sites in northern California. Mar. Biol. 113:613–623

Autrum, H. 1939. Hirudineen. Geographische Verbreitung. Pp. 32–639 in Klassen und Ordnungen des Tierreichs (H. S. Bronns, ed.), vol. 4, sec. 3, bk. 4, no. 2. Akademische Verlagsgesellschaft, Leipzig.

Ax, P. 1996. Multicellular animals. A new approach to the phylogenetic order in nature. Springer Verlag, Berlin

Bartolomaeus, T. 1995. Structure and formation of the uncini in *Pectinaria koreni*, *Pectinaria auricoma* (Terebellida) and *Spirorbis spirorbis* (Sabellida): implications for annelid phylogeny and the position of the Pogonophora. Zoomorphology 115:161–177.

Blake, J. A., and P. L. Arnofsky. 1997. Reproduction and larval development of the spioniform polychaeta with application to systematics and phylogeny. In Developments in Hydrobiology. Proceeding of the Meeting on Reproductive Strategies in Marine Invertebrates. Osnabrück, Germany.

Blakenship, D. T., R. G. Brankamp, G. D. Manley, and A. D. Cardin. 1990 Amino acid sequence of ghilanten: anticoagulant antimetastatic principle of the South American leech, *Haementeria ghilianii*. Biochem. Biophys. Res. Commun. 166:1384–1389

Blanchard, R. 1893. Courtes notices sur les Hirudinées. VIII. Sur l'*Hirudo brevis* Grube, 1871. Bull. Soc. Zool. 18:26–29.

Blanchard, R. 1917. Monographie des Haemadipsines (Sangsues Terrestres). Bull. Soc. Pathol. Exot. 10:640–675.

Boore, J. L., and W. M. Brown 2000 Mitochondrial genomes of *Galathealinum*, *Helobdella*, and *Platynereis*: sequence and gene arrangement comparisons indicate that Pogonophora is not a phylum and Annelida and Arthropoda are not sister taxa. Mol. Biol. Evol. 17(1):87–106.

Brankamp, R. G., D. T. Blankenship, P. S. Sunkara, and A. D. Cardin. 1990. Ghilantens: anticoagulant antimetastatic proteins from the South American leech, *Haementeria ghilianii*. J. Lab. Clin. Med. 115:89–97.

Brinkhurst, R. O. 1994. Evolutionary relationships within the Clitellata: an update. Megadrilogica 5:109–112.

Brinkhurst, R. O. 1999. Lumbriculids, branchiobdellidans and leeches: an overview of recent progress in phylogenetic research on clitellates. Hydrobiologia 406:281–290.

Brinkhurst, R. O., and S. R. Gelder. 1989. Did the lumbriculids provide the ancestors of the branchiobdellidans, acanthobdellidans and leeches? Hydrobiologia 180:7–15.

Brown, S., G. Rouse, P. Hutchings, and D. Colgan. 1999. Assessing the usefulness of histone H3, U2 snRNA and 28S rDNA in analyses of polychaete relationships. Australian Journal of Zoology 47(5):499–516.

Budzynski, A. Z., S. A. Olexa, B. S. Brizuela, R. T. Sawyer, and G. S. Stent. 1981. Anticoagulant and fibrinolytic properties of salivary proteins from the leech *Haementeria ghilianii*. Proc. Soc. Exp. Biol. Med. 168:261–275.

Caspers, H. 1984. Spawning periodicity and habitat of the palolo worm *Eunice viridis* (Polychaeta: Eunicidae) in the Samoan Islands. Mar. Biol. 79:229–236.

Chopin, V., M. Salzet, L. Baert, F. Vandenbulcke, P. E. Sautiere, J. P. Kerckaert, and J. Malecha. 2000. Therostasin, a novel clotting factor Xa inhibitor from the rhynchobdellid leech, *Theromyzon tessulatum*. J. Biol. Chem. 275:32701–32707.

Conway Morris, S., and J. S. Peel. 1995. Articulated Halkieriids from the lower Cambrian of north Greenland and their role in early protostome evolution. Philos. Trans. R. Soc. Lond. B 347:305–358.

Crawshay, L. R. 1935. Possible bearing of a luminous syllid on the question of the landfall of Columbus. Nature 136:559–560.

Cuvier, G. 1812. Sur un nouveau rapprochment à établir entre les classes qui composant le Règne Animal. Ann. Mus. Natl. Hist. Nat. Paris 19:73–84.

Darwin, C. 1881. The formation of vegetable mould, through the action of worms with observations on their habits. J. Murray, London.

Desbruyères, D., P. Chevaldonne, A. M. Alayse, D. Jollivet, F. H. Lallier, C. Jouin-Toulmond, F. Zal, P. M. Sarradin, R. Cosson, J. C. Caprais, et al. 1998. Biology and ecology of the Pompeii worm (*Alvinella pompejana* Desbruyères and Laubier), a normal dweller of an extreme deep-sea environment: a synthesis of current knowledge and recent developments. Deep Sea Res. II Top. Stud. Oceanogr. 45:383–422.

Desbruyères, D., and L. Laubier. 1980. *Alvinella pompejana* gen. sp. nov., Ampharetidae aberrant des sources hydrothermales de la ride Est-Pacifique. Oceanol. Acta 3:267–274.

Desbruyères, D., and A. Toulmond. 1998. A new species of hesionid worm, *Hesiocaeca methanicola* sp. nov. (Polychaeta: Hesionidae), living in ice-like methane hydrates in the deep Gulf of Mexico. Cahiers Biol. Mar. 39:93–98.

Eernisse, D. J., J. S. Albert, and F. E. Anderson. 1992. Annelida and Arthropoda are not sister taxa: a phylogenetic analysis of spiralean metazoan morphology. Syst. Biol. 41:305–330.

Elliott, J. M., and P. A. Tullett. 1984. The status of the medicinal leech Hirudo medicinalis in Europe and especially in the British Isles. Biol. Conserv. 29:15–26.

Elliott, J. M., and P. A. Tullett. 1992. The medicinal leech. Biologist 39:153–158.

Erséus, C., T. Prestegaard, and M. Källersjö. 2000. Phylogenetic analysis of Tubificidae (Annelida, Clitellata) based on 18S rDNA sequences. Mol. Phylogenet. Evol. 15:381–389.

Fauchald, K. 1977. The Polychaete worms. Definitions and keys to the orders, families and genera. Nat. His. Mus. Los Angeles County Science Ser. 28:1–188.

Fauchald, K. 1992. Review of the types of *Palola* (Eunicidae: Polychaeta). J. Nat. Hist. 26:1177–1225.

Fauchald, K., and G. Rouse. 1997. Polychaete systematics: past and present. Zoologica Scripta 26:71–138.

Ferraguti, M., and C. Erséus. 1999. Sperm types and their use for a phylogenetic analysis of aquatic clitellates. Hydrobiologia 402:225–237.

Field, K. G., G. J. Olsen, D. J. Lane, S. J. Giovannoni, M. T. Ghiselin, E. C. Raff, N. R. Pace, and R. A. Raff. 1988. Molecular phylogeny of the animal kingdom. Science 239:748–753.

Fischer, A., and U. Fischer. 1995. On the life-style and life-cycle of the luminescent polychaete *Odontosyllis enopla* (Annelida, Polychaeta). Invert. Biol. 114:236–247.

Fisher, C. R., I. R. MacDonald, R. Sassen, C. M. Young, S. A. Macko, S. Hourdez, R. S. Carney, S. Joye, and E. McMullin. 2000. Methane ice worms: *Hesiocaeca methanicola* colonizing fossil fuel reserves. Naturwissenschaften 87:184–187.

Fisher, C. R., I. A. Urcuyo, M. A. Simpkins, and E. Nix. 1997. Life in the slow lane: growth and longevity of cold-seep vestimentiferans. Mar. Ecol. 18:83–94.

Fitzhugh, K., and G. W. Rouse. 1999. A remarkable new genus and species of fan worm (Polychaeta: Sabellidae: Sabellinae) associated with some marine gastropods. Invert. Biol. 118:357–390.

Franzén, Å. 1991. Spermiogenesis and sperm ultrastructure in *Acanthobdella peledina* (Hirudinea) with some phylogenetic considerations. Invert. Reprod. Dev. 19:245–256.

Garcia-Machado, E., M. Pempera, N. Dennebouy, M. Oliva-Suarez, J. C. Mounolou, and M. Monnerot. 1999. Mitochondrial genes collectively suggest the paraphyly of Crustacea with respect to Insecta. J. Mol. Evol. 49:142–149.

Gelder, S. R., and J. P. Rowe. 1988. Light microscopical and cytochemical study on the adhesive and epidermal gland cell secretions of the branchiobdellid *Cambarincola fallax* (Annelida: Clitellata). Can. J. Zool. 66:2057–2064.

Gelder, S. R., and M. E. Siddall. 2001. Phylogenetic assessment of the Branchiobdellidae (Annelida: Clitellata) using 18S rDNA and mitochondrial cytochrome *c* oxidase subunit I characters. Zool. Scr. 30:215–222.

Ghiselin, M. T. 1988. The origin of molluscs in the light of molecular evidence. Oxf. Surv. Evol. Biol. 5:66–95.

Grassle, J. F. 1973. Variety in coral reef communities. Pp. 247–270 *in* Biology and geology of coral reefs (O. A. Jones, and R. Endean, eds.). Academic Press, New York.

Grassle, J. F., and N. J. Maciolek. 1992. Deep-sea species richness: regional and local diversity estimates from quantitative bottom samples. Amer. Nat. 139:313–341.

Haeckel, E. H. P. 1866. Generelle Morphologie Der Organismen: Allgemeine Grundzüge Der Organischen Formen-Wissenschaft, mechanisch begründet durch die von Charles Darwin reformirte Descendenz-Theorie. G. Reimer, Berlin.

Halanych, K. M., J. D. Bacheller, A. M. A. Aguinaldo, S. M. Liva, D. M. Hillis, and J. A. Lake. 1995. Evidence from 18S ribosomal DNA that the lophophorates are protostome animals. Science 267:1641–1643.

Harding, W. A. 1913. On a new land leech from the Seychelles. Trans. Linn. Soc. Lond. 16:39–43.

Hartman, O. 1954. Pogonophora Johansson, 1938. Syst. Zool. 3:183–185.

Holt, P. C. 1965. The systematic position of the Branchiobdellidae (Annelida, Clitellata). Syst. Zool. 14: 25–32.

Holt, P. C. 1989. Comments on the classification of the Clitellata. Hydrobiologia 180:1–5.

Hooker, J. D., Sir. 1854. Himalayan journals: notes of a naturalist in Bengal, the Sikkim and Nepal Himalayas, the Khasia Mountains. J. Murray, London

Ivanov, A. V. 1988. Analysis of the embryonic development of Pogonophora in connection with the problems of phylogenetics. Z. Zool. Syst. Evolutionsforsch. 26:161.

Jones, M. L. 1981. *Riftia pachyptila* Jones: observations on the vestimentiferan worm from the Galápagos Rift. Science 213:333–336.

Kem, W. R. 1988. Worm toxins. Pp. 353–378 *in* Handbook of natural toxins (A. T. Tu, ed.). Marcel Dekker, New York.

Kent, M. L., K. P. Andree, J. P. Bartholomew, M. El-Matbouli, S. S. Desser, R. H. Devlin, R. P. Hedrick, R. W. Hoffmann, J. W. Khattra, S. L. Hallett, et al. 2001. Recent advances in our knowledge of the Myxozoa. J. Eukary. Microbiol. 48:395–413.

Kuris, A. M., and C. S. Culver. 1999. An introduced sabellid polychaete pest infesting cultured abalones and its potential spread to other California gastropods. Invert. Biol. 118: 391–403.

Lafferty, K. D., and A. M. Kuris. 1996. Biological control of marine pests. Ecology 77(7):1989–2000.

Lake, J. A. 1990. Origin of the Metazoa. Proc. Natl. Acad. Sci. USA 87:763–766.

Lamarck, J.-B. D. 1802. La nouvelle classes des Annélides. Bulletin du Muséum d'Histoire Naturelle, Paris An X: Disc. d'ouverture, 27 Floréal (reprint: 1907, Bull. Biol. Fr. Belg. 60:56).

Lamarck, J.-B. D. 1809. Philosophie zoologique, ou exposition des considérations relative à l'histoire naturelle des animaux. Dentu et L'Auteur, Paris.

Lauristen, D. D., S. C. Mozley, and D. S. White. 1985. Distribution of oligochaetes in Lake Michigan and comments on their use as indices of pollution. J. Great Lakes Res. 11:67–76.

Light, J. E., and M. E. Siddall. 1999. Phylogeny of the leech family Glossiphoniidae based on mitochondrial gene sequences and morphological data. J. Parasitol. 85:815–823.

Livanow, N. 1906. *Acanthobdela peledina* Grube, 1851. Zool. Jahrb. Anat. 22:637–866.

Livanow, N. 1931. Die organisation der Hirudineen und die Beziehungen dieser Gruppe zu den Oligochäten. Ergeb. Fortschr. Zool. 7:378–484.

Lutz, R. A., T. M. Shank, D. J. Fornari, R. M. Haymon, M. D. Lilley, K. L. Von Damm, and D. Desbruyères. 1994. Rapid growth at deep-sea vents. Nature 371:663–664.

Malinconico, S. M., J. B. Katz, and A. Z. Budzynski. 1984. Fibrinogen degradation by hementin, a fibrinogenolytic anticoagulant from the salivary glands of the leech *Haementeria ghilianii*. J. Lab. Clin. Med. 104:842–854.

Martin, P., I. Kaygorodova, D. Y. Sherbakov, and E. Verheyen.

2000. Rapidly evolving lineages impede the resolution of phylogenetic relationships among Clitellata (Annelida). Mol. Phylogenet. Evol. 15:355–68

Mazur, P., W. J. Henzel, J. L. Seymour, and R. A. Lazarus. 1991. Ornatins: potent glycoprotein IIb IIIa antagonists and platelet aggregation inhibitors from the leech *Placobdella ornata*. Eur. J. Biochem. 202:1073–1082

McHugh, D. 1997. Molecular evidence that echiurans and pogonophorans are derived annelids. Proc. Natl. Acad. Sci. USA 94:8006–8009.

McNicol, D. K., M. L. Mallory, G. Mierle, A. M. Scheuhammer, and A. H. K. Wong. 1997. Leeches as indicators of dietary mercury exposure in non-piscivorous waterfowl in central Ontario, Canada. Environ. Pollut. 95:177–181.

Medlin, L., H. J. Elwood, S. Stickel, and M. L. Sogin. 1988. The characterization of enzymatically amplified eukaryotic 16S-like rRNA coding regions. Gene 71:491–499.

Meglitsch, P. A., and Schram, F. R. 1991. Invertebrate zoology. Oxford University Press, Oxford.

Metcalfe, J. L., M. E. Fox, and J. H. Carey. 1988. Freshwater leeches (Hirudinea) as a screening tool for detecting organic contaminants in the environment. Environ. Monit. Assess. 11:147–169.

Michaelsen, W. 1919. Über die Beziehungen der Hirudineen zu den Oligochäten. Mitt. Zool. Mus. Hamburg 36:131–153.

Moore, J. P. 1927. Arynchobdellae. Pp. 97–295 *in* The fauna of British India: Hirudinea (W. A. Harding, W. A., and J. P. Moore, eds.). Taylor and Francis, London.

Naef, A. 1913. Studien zur generaellen Morphologie der Mollusken. 1. Teil: über Torsion und Asymmetrie der Gastropoden. Ergeb. Fortschr. Zool. 3:73–164.

Nørrevang, A. 1970. The position of Pogonophora in the phylogenetic system. Z. Zool. Syst. Evolutionsforsch. 8:161–172.

Nylander, J. A. A., C. Erséus, and M. Kallersjo. 1999. A test of monophyly of the gutless Phallodrilinae (Oligochaeta, Tubificidae) and the use of a 573-bp region of the mitochondrial cytochrome oxidase I gene in analysis of annelid phylogeny. Zool. Scr. 28:305–313.

Odier, A. 1823. Memoire sur le Branchiobdelle nouveau genre d'Annelides de la famille des Hirudiner. Mem. Soc. Hist. Nat. Paris 1:69–78.

Olive, P. J. W. 1994. Polychaeta as a world resource: a review of patterns of exploitation as sea angling baits, and potential for aquaculture based production. Mem. Mus. Natl. Hist. Nat. 162:603–610.

Pelseneer, P. 1899. Recherches morphologiques et phylogénétiques sur les mollusques Archaiques. Mem. Acad. R. Sci. Belg. 57:1–112.

Peterson, K. J., and D. J. Eernisse. 2001. Animal phylogeny and the ancestry of bilaterians: inferences from morphology and 18S rDNA gene sequences. Evol. Dev. 3:170–205.

Pleijel, F. 1998. Phylogeny and classification of Hesionidae (Polychaeta). Zool. Scr. 27:89–163.

Pleijel, F., and T. G. Dahlgren. 1998. Phylogeny of Phyllodocida and Nereidiformia (Polychaeta, Annelida). Cladistics 14:129–150.

Purschke, G., W. Westheide, D. Rohde, and R. O. Brinkhurst. 1993. Morphological reinvestigation and phylogenetic

relationship of *Acanthobdella peledina* (Annelida, Clitellata). Zoomorphology 113:91–101.

Rouse, G. W. 2001. A cladistic analysis of Siboglinidae Caullery, 1914 (Polychaeta, Annelida): formerly the phyla Pogonophora and Vestimentifera. Zool. J. Linn. Soc. 132:55–80.

Rouse, G. W., and K. Fauchald. 1995. The articulation of annelids. Zool. Scr. 24:269–301.

Rouse, G. W., and K. Fauchald. 1997. Cladistics and polychaetes. Zool. Scr. 26:139–204.

Rouse, G. W., and K. Fauchald. 1998. Recent views on the status, delineation and classification of the Annelida. Am. Zool. 38:953–964.

Sawyer, R. T. 1981. Why we need to save the medicinal leech. Oryx 16:165–168.

Sawyer, R. T. 1986. Leech biology and behavior. Clarendon Press, Oxford.

Seagrave, G. S. 1946. Burma surgeon returns. V. Gollancz, London.

Seymour, J. L., W. J. Henzel, B. Nevins, J. T. Stults, and R. A. Lazarus. 1990. Decorsin. A potent glycoprotein IIb IIIa antagonist and platelet aggregation inhibitor from the leech *Macrobdella decora*. J. Biol. Chem. 265:10143–10147.

Shipley, A. E. 1927. Historical preface. Pp. v–xxxii *in* The fauna of British India: Hirudinea (W. A. Harding and J. P. Moore, eds.). Taylor and Francis, London.

Siddall, M. E. 2002. Phylogeny and revision of the leech family Erpobdellidae (Hirudinida: Oligochaeta). Invert. Taxon. 16:1–6.

Siddall, M. E., K. Apakupakul, E. M. Burreson, K. A. Coates, C. Erséus, S. R. Gelder, M. Källersjö, and H. Trapido-Rosenthal. 2001. Validating Livanow: molecular data agree that leeches, branchiobdellidans and *Acanthobdella peledina* form a monophyletic group of Oligochaetes. Mol. Phylogenet. Evol. 21:346–351.

Siddall, M. E., and E. Borda. 2002. Phylogeny of the leech genus *Helobdella* (Glossiphoniidae) based on mitochondrial gene sequences and morphological data. Zool. Scr.32:23–33.

Siddall, M. E., and E. M. Burreson. 1995. Phylogeny of the Euhirudinea: independent evolution of blood feeding by leeches? Can. J. Zool. 73:1048–1064

Siddall, M. E., and E. M. Burreson. 1996. Leeches (Oligochaeta?: Euhirudinea), their phylogeny and the evolution of life-history strategies. Hydrobiologia 334:277–285.

Siddall, M. E., and E. M. Burreson. 1998. Phylogeny of leeches (Hirudinea) based on mitochondrial cytochrome *c* oxidase subunit I. Mol. Phylogenet. Evol. 9:156–162.

Siddall, M. E., K. Fitzhugh, and K. Coates. 1998. Problems determining the phylogenetic position of echiurans and pogonophorans with limited data. Cladistics 14:401–410.

Snodgrass, R. E. 1938. Evolution of the Annelida, Onychophora and Arthropoda. Smithson. Misc. Collect. 97:1–159.

Southward, E. C. 1993. Pogonophora. Pp. 327–369 *in* Microscopic anatomy of invertebrates, Vol. 12: Onychophora, Chilopoda and lesser Protostomata (F. W. Harrison and M. E. Rice, eds.). Wiley-Liss, New York.

Uschakov, P. V. 1933. Eine neue Form aus der Familie Sabellidae (Polychaeta). Zool. Anz. 104:205–208.

Uzunov, J., V. Kosel, and V. Sladecek. 1988. Indicator value of freshwater Oligochaeta. Acta –Hydrochim. Hydrobiol. 16:173–186.

Vejdovsky, F. 1884. System und Morphologie des Oligochaeten. Prague.

Verdonschot, P. F. M. 1989. The role of oligochaetes in the management of waters. Hydrobiologia 180:213–227.

Weigl, A. M. 1994. Ultrastructure of the adhesive organs in branchiobdellids (Annelida: Clitellata). Trans. Am. Microsc. Soc. 113:276–301.

Wells, S. M., R. M. Pyle, and N. M. Collins. 1983. The IUCN invertebrate red data book. IUCN, Gland, Switzerland.

Westheide, W. 1997. The direction of evolution within the Polychaeta. J. Nat. Hist. 31:1–15.

Wheeler, W. C., P. Cartwright, and C. Y. Hayashi. 1993. Arthropod phylogeny: a combined approach. Cladistics 9:1–39.

Winnepenninckx, B., T. Backeljau, and R. de Wachter. 1995. Phylogeny of protostome worms derived from 18S rRNA sequences. Mol Bioland Evol 12:641–649.

Zrzavý, J., J. Hypa, and D. F. Tietz. 2001. Myzostomida are not annelids: molecular and morphological support for a clade of animals with anterior sperm flagella. Cladistics 17:170–198.

Zrzavý, J., S. Mihulka, P. Kepka, A. Bezdek, and D. Tietz. 1998. Phylogeny of the Metazoa based on morphological and 18S ribosomal DNA evidence. Cladistics 14:249–285.

David R. Lindberg
Winston F. Ponder
Gerhard Haszprunar

The Mollusca: Relationships and Patterns from Their First Half-Billion Years

Mollusks are bilaterally symmetrical eumetazoans that are diverse in body form and size, ranging from giant squids more than 20 m in length to adult body sizes of about 500 μm. They are often considered to be the second largest phylum next to Arthropoda, with about 200,000 living species, of which about 75,000 living and 35,000 fossil have been named, making them one of the better known invertebrate groups. They also exhibit a great range of physiological, behavioral, and ecological adaptations. Mollusks have an excellent fossil record extending back some 560 million years to the early Cambrian, and perhaps into the Precambrian as well. Three major classes, Gastropoda (snails, slugs, limpets), Bivalvia (scallops, clams, oysters, mussels) and Cephalopoda (squid, cuttlefish, octopuses, nautilus), are recognized, as well as four or five minor living classes [Aplacophora (spicule worms)—which are often divided into two separate classes, Polyplacophora (chitons), Scaphopoda (tusk shells), and Monoplacophora (a small group of deep sea limpets with a long fossil history)]. A few extinct groups often treated as classes are also recognized.

The majority of mollusks are marine, but large numbers also occupy freshwater and terrestrial habitats. They are extremely diverse in their food habits, ranging from grazers and browsers on many different biotic substrates to suspension feeders, predators, and parasites. Many are economically important as food, cultural objects, hosts for human parasites, or pests. Many nonmarine taxa are also in jeopardy as a result of human activities. Despite only a small fraction of

the world's nonmarine molluscan faunas being adequately assessed, there are more recorded extinctions of these mollusks than of birds and mammals combined (Ponder 1997, Killeen et al. 1998, Seddon 1998). In addition, alien species are resulting in the homogenization of many previously unique biotas, especially on islands (Cowie 2002).

Some common morphological features enable Mollusca to be characterized as a monophyletic group. These include having the body, which typically has a head, foot, and visceral mass, covered with a pallium or mantle that typically secretes the shell (or, more rarely, spicules), although this is secondarily lost in some groups (e.g., slugs, octopuses). Typically, there are one or more pairs of gills (ctenidia), which lie in a posterior pallial (i.e., mantle) cavity or in a postero-lateral groove surrounding the foot, into which the kidneys, gonads, and anus open and which also contains a pair of sensory osphradia. The buccal cavity contains a radula—a ribbon of teeth supported by a muscular odontophore (lost in bivalves). There is a ventral foot used in locomotion using muscular waves and/or cilia in combination with mucus. They are coelomate, although the coelom is small and represented by the kidneys, gonads, and pericardium, the main body cavity being a haemocoel. They lack segmentation and have spiral cleavage. Trochophore and/or veliger larvae are found in many aquatic taxa, but direct development is also common.

The earliest undoubted mollusks are found in the early Cambrian (~560 million years ago), when several major groups (gastropods, bivalves, monoplacophorans, and rostroconchs) appear. Cephalopods are found from the Middle Cambrian, polyplacophorans from the Late Cambrian, and scaphopods from the Middle Ordovician. Studies on molluscan evolution are able to use this rich fossil diversity and can be particularly illuminating when combined with morphological, ultrastructural, embryological, and molecular studies on taxa from the Recent period. Studies on the genetics, diversity, phylogeny, and ecology of mollusks have provided important insights into evolutionary biology, biogeography, and ecology in general.

Phylogenetic Scenarios and Hypotheses

There have been two traditions for placing Mollusca on the Tree of Life—one paleontological (using fossils) and the other neontological (using living taxa). These traditions extend to varying degrees into the subclades that make up Mollusca. Every so often workers unify these traditions with varying degrees of success. An early example was Dall's (1893) noting of the symmetry of the adductor scars of Paleozoic monoplacophoran fossils and that they "paralleled in some particulars the organization of some of the Chitons of that ancient time." It was 45 years before the same suggestion was made by Wenz (1938–1944), and another 19 years before the discovery of living monoplacophorans (Lemche and Wingstrand 1959) confirmed Dall's insight into the nontorted state of these animals. Like Dall, Knight (1952) used observations on living gastropods and applied them to fossil gastropod morphologies, creating new evolutionary scenarios and generating a renaissance in thinking about gastropod evolution.

However, by the late 1960s, interest in systematics was waning, and a new generation of paleontologists, including S. Gould, D. Raup, S. Stanley, J. Valentine, and G. Vermeij, moved the field to a more theoretical position from which to evaluate patterns and processes of taxic evolution. For many of these workers, Mollusca was the taxon of choice because of its diversity and record from deep time. Systematics continued, especially on Paleozoic taxa, where E. Yochelson, J. Pojecta, B. Runnegar, S. Bengsten, J. Peel, and their colleagues were discovering new major lineages and setting the stage for reinterpreting previous findings (see Runnegar 1996). New evolutionary scenarios for patterns seen in the fossil record were proposed, and molluscan groups were often used to test many of these new theories, including patterns of heterochrony and punctuated equilibrium, theoretical morphospaces, and community and phyletic patterns of ecological interactions. Such an integrated approach quickly brought molluscan evolutionary biology into a much more paleontological framework. A notable exception during this period was the work of L. Salvini-Plawen, who continued to study molluscan origins from an almost exclusively neontological position (Salvini-Plawen 1972, 1980). Molecular data have

recently joined these two more traditional molluscan data sets and—as would be predicted under Murphy's Law—currently falsifies neither the paleontological nor the neontological views.

To ultimately render robust hypotheses of molluscan origins and relationships, all of these data sets need to be compared, combined, parsed, and analyzed. It is likely that too much time has passed since the divergences and/or the time span is too short to preserve that perfect phylogenetic marker. This problem has been recognized and examined in paleontological studies (e.g., Wagner 2001), in morphological studies (e.g., Lindberg and Ponder 1996, Ponder and Lindberg 1996, 1997), and more recently in molecular studies (e.g., Giribet 2002).

What Makes a Spiralian Taxon a Mollusk?

Currently there is no consensus as to the identity of the sister taxon of Mollusca. Contenders include Brachiopoda, suggested by the 28S data set (Mallat and Winchell 2001). Haszprunar (1996; fig. 16.1) has suggested the kamptozoans based on developmental data (body wall cuticle, blood sinuses) and larval characters (cuticle, ciliary gliding sole with pedal gland). However, confirmation of these details is needed because only one description of a kamptozoan larva has appeared in the literature (Nielsen 1971). Sipuncula has been suggested by Scheltema (1993, 1996) based on developmental and larval characters. Traditionally, Annelida have been considered the sister taxon of Mollusca by most workers and in some text books (Brusca and Brusca 2002). The mollusks and annelids share several characters, including the trochophore larvae, anteriorly positioned ferrous oxide structures as teeth and jaws, and a cross configuration of micromeres during early development. However, the Arthropoda–Annelida–Mollusca triad, which dominated invertebrate classification for more than 75 years, was ultimately overturned by molecular and other data, revealing that the supposed relationship of these three taxa (based on the supposed shared "similarity" of body segmentation) was actually convergent.

Ghiselin (1988) and Winnepenninckx et al. (1994, 1995) provide some of the earliest analyses of small subunit (18S) ribosomal DNA (rDNA), and for many years this served as the basis for many molluscan outgroup comparisons. These and other studies suggested that mollusks reside among the lophotrochozoan taxa (mollusks, annelids, brachiopods, bryozoans, and phoronids; Halanych et al. 1995; fig. 16.2). However, the relative branching of these taxa is not clearly delineated by 18S data (Medina and Collins 2003). Zrzavý et al. (1998), using a combined analysis of 18S data and morphology, suggested that the sipunculids were the sister taxon of the mollusks. However, Boore and Staton (2002), using partial mitochondrial DNA (mtDNA) gene order data, suggested the sipunculids are actually more closely related to annelids rather than to mollusks. In addition, Mallat and

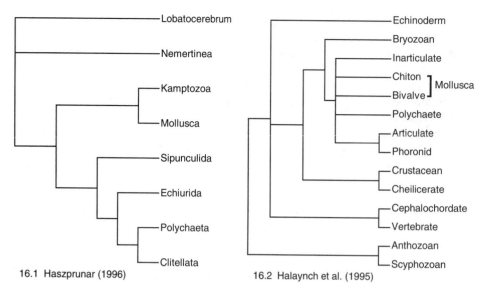

16.1 Haszprunar (1996)

16.2 Halaynch et al. (1995)

Figures 16.1 and 16.2. Phylogenetic relationships of putative molluscan outgroups. 16.1. Morphological data (Haszprunar 1996). 16.2. Molecular data (18S rDNA; Halanych et al. 1995).

Winchell (2001) suggested that brachiopods and/or phoronids may be the molluscan sister group based on their analyses of complete 28S sequences. Surprisingly, there is little molecular evidence to test the hypothesis of Annelida as the sister taxa of Mollusca, although morphological and developmental evidence of this relationship has been long-standing (Ghiselin 1988). mtDNA gene order data may be important in understanding the position of Mollusca on the Tree of Life (Medina and Collins 2003) because, unlike many other phyla, all the molluscan mtDNA genomes examined so far show major rearrangements (Boore and Brown 1994, Boore 1999). However, as a cautionary note, Adoutte et al. (2000) have suggested that the inability to clearly identify a sister taxon of Mollusca may result from the burst of rapid speciation in the Cambrian within the three major bilaterian lineages.

Any of the outgroups discussed above would suggest a worm *bauplan* for the last common ancestor of the molluscan taxa. Whether or not the worm was covered with a cuticle, spicules, or shell cannot be determined because hardening of the ectoderm is present in several outgroups, including the brachiopods (both calcium carbonate and calcium phosphate shells), annelids (fibrous cuticle, secondary calcium carbonate tubes), and members of Kamptozoa (chitinous cuticle). A crossed lamella-like microstructure in the molluscan shell appears to be plesiomorphic by outgroup comparisons (hyoliths); foliated structures are present in both mollusks and brachiopods and, along with nacre, have been independently derived in bivalve and gastropod mollusks (Hedegaard and D. R. Lindberg, unpubl. obs.).

Molluscan Characters, Plesiomorphy, Apomorphy, and Homoplasy

The presence of a pericardium—a coleomic cavity that encloses the heart and performs ultrafiltration in several taxa

(Andrews 1988, Meyhoefer and Morse 1996)—is a synapomorphy of Mollusca. Addition of repeated structures from posterior to anterior and a radula and a tripartite mantle edge divisible into outer, middle, and inner folds are also molluscan synapomorphies [see Haszprunar (1996) for additional ultrastructure characters].

Most mollusks have a space between the mantle and the side of the foot that forms the pallial (or mantle) groove. Typically, the groove deepens posteriorly and forms a cavity that contains a pair of gills or ctenidia, as well as openings of the rectum, paired renal organs, and gonads from the dorsal visceral mass. Although the molluscan pallial cavity has long been considered a single defining system, character transformations of many of the individual components that make up the pallial cavity system can be problematic (Lindberg and Ponder 2001). For example, a single pair of ctenidia is common in hypothetical ancestors of the major clades, but its distribution on the tree is not informative, and its current function in many groups is likely autapomorphic. Members of Mollusca, like other lophotrochozoans, have gills that have both respiration and ventilation functions. In several taxa (within and outside mollusks), filter feeding is a third part of the repertoire of gills, and they also play a role in brooding larvae in several taxa.

Lindberg and Ponder (2001) argued that phyletic size increase in Gastropoda increased selective pressure for increased efficiency of the gills and the separation of ventilation and respiration functions. Suggestions of the same conflict are present in the other molluscan taxa and well illustrate the nested sets of parallel evolution present throughout the molluscan tree.

For example, the Polyplacophora increase both respiratory and ventilation surfaces simultaneously by adding gills in serial repetition from posterior to anterior as phyletic size increases (Lindberg 1985). In Monoplacophora, ventilation currents appear to be generated by the ctenidia (added in serial repetition from posterior to anterior), and the pallial groove serves as the respiratory surface (Lindberg and Ponder 1996, Haszprunar and Schaefer 1997a). In Bivalvia, the hypothetical ancestral states are inferred from the deposit-feeding protobranchs where the paired gills are used as ventilators and respirators alone within a spacious pallial cavity. These structures are probably reliable analogues of the likely progenitors of the larger, more complex gills of other bivalves that are highly modified for suspension feeding. In Cephalopoda, *Nautilus* alone has two pairs of ctenidia; the remainder, one pair. Ventilation currents are produced by muscular contractions of the mantle or funnel (Ghiretti 1966), and the gills are used solely in respiration. The circulatory system is closed with the ctenidia, in many living cephalopods, having auxiliary hearts that increase the rate of blood passing through the gills in these large, very active animals. Scaphopods lack gills, but the elongate pallial cavity is large, and strong bands of cilia drive water circulation along with regular muscular contractions. Lastly, the plesiomorphic state of gastropods was paired gills with a small shallow pallial cavity (Lindberg and Ponder 2001), although this configuration is highly modified in most taxa.

In the chaetodermomorphs, paired gills are present in a small posterior pallial cavity; in the nonburrowing Neomenimorpha, only gill folds are present around a rudimentary posterior pallial cavity. Thus, whether members of the aplacophoran (grade or clade) represent the clade Aculifera, or are the stem taxa of Mollusca, they do not assist in polarizing the outgroup node for the plesiomorphic character states of the conchiferian ctenidium (primary gill). The inability to polarize gill character states continues within Conchifera. Thus, the only character states for the gill of the molluscan common ancestor that can be strongly argued are filament shape and ventilation (table 16.1). Although there are certainly majority rule candidates among the other gill characters (e.g., paired ctenidia, ctenidia + pallial cavity respiration), none of the remaining character states are supported by the duplet rule (Maddison et al. 1984) at any node in previously reported phylogenies (figs. 16.3–16.6). There are other majority rule characters that are often cited as molluscan ancestral states, including the presence of a head region (lacking tentacles and eyes), a ventral muscular foot, a dorsal visceral mass, and an enveloping mantle (= pallium) that secretes spicules and/or the shell, but these characters, like the gill characters, cannot be unequivocally confirmed by outgroup analysis. This inability to estimate character polarity is a common outcome throughout the molluscan tree (for Gastropoda, see Ponder and Lindberg 1997).

The digestive system of mollusks follows a common pattern, although in some aplacophoran and conchiferan groups (cephalopods, bivalves, and some gastropods) it is highly modified. The molluscan digestive system is autapomorphic to potential outgroups and consists of numerous glands and sacs associated with the buccal chamber. The mouth opens to a buccal cavity that typically contains paired jaws and a

muscular odontophore that typically bears the radula and a pair of salivary glands. All of these structures, other than the mouth, are lost in bivalves. An esophagus, sometimes with glandular pouches, opens to a typically complex stomach where a large pair of digestive glands also open. Ciliary tracts sort food particles from the waste material in the stomach, and digestion occurs in the digestive gland. Waste is moved to the intestinal part of the stomach that typically starts as a style sac in which the waste string is rotated and bound with mucus before being passed into the intestine proper. In most bivalves and some gastropods, a crystalline style, a rotating rod of muco-protein that releases digestive enzymes, lies in the style sac. The hindgut or intestine is often long and looped or coiled. Fecal material is released through the anus that typically lies within the pallial cavity.

All mollusks other than cephalopods (as noted above) have an open circulatory system with blood sinuses, a heart, blood vessels, and respiratory pigment, usually hemocyanin. The heart is enclosed within the pericardium and has multiple (usually two, one in many gastropods) auricles and a single ventricle. Cephalopods have a closed system with arteries and veins. Gas exchange is via gills, lungs, or the body surface. Excretion takes place by means of kidneys (nephridia) that excrete waste into the pallial cavity. The excretory system is paired and connected to the pericardium as well as the gonads in some taxa. The gonads are also paired but can be fused into a single structure (Polyplacophora) or reduced to a single organ (Gastropoda and Scaphopoda). Separate gonoducts are present in some taxa, and in other taxa the gonads empty into the kidneys. These connected, mesodermal structures (pericardium, kidneys, and gonads) likely represent the coelom of Mollusca.

Most mollusks are dioecious (separate sexes); some, monoecious (hermaphroditic). Some groups have internal fertilization and produce various forms of jelly or capsule-covered eggs that contain the embryo for at least part of its development; others release their gametes into the water column and their development is entirely pelagic, passing through both trochophore and veliger stages. Some planktonic larvae feed on the plankton and other suspended particles (planktotrophic); others feed on nutrients stored in the egg (lecithotrophic). Some species have direct development, with juveniles emerging from the egg capsule or from a brood pouch within the parent. Internal fertilizing taxa may transfer sperm during copulation involving a penis or, as in cephalopods and some gastropods, by transferring spermatophores—packets of sperm.

The nervous system consists of four main paired centers—cerebral, visceral, pedal, and pleural ganglia. They are connected by commissures; in the plesiomorphic condition the paired pedal nerve cords extend ladderlike through the foot. Sensory and nervous systems are concentrated in the head region, especially in gastropods and cephalopods. Highly specialized sense organs are on the head (eyes, tactile organs such as tentacles), as well as statocysts for balance

Table 16.1

Assumed Plesiomorphic Character States for Respiratory Structures in the Molluscan Pallial Cavity.

Character	Polyplacophora	Neomeniomorpha	Chaetodermomorpha	Monoplacophora	Bivalvia	Scaphopoda	Cephalopoda	Gastropoda
Pallial groove	Long, narrow groove around foot	Absent	Shallow posterior embayment	Long, narrow groove around foot	Large, surrounds entire animal	Large, elongate, extends length of animal inside shell	Large ventral embayment	Shallow (deep in advanced taxa), anterior embayment
Ctenidia	5–60 pairs	Absent	1 pair	3–6 pairs	1 pair	Absent	*Nautilus*, 2 pairs; all others, 1 pair	1 pair (reduced to one ctenidium or lost in most gastropods)
Skeletal rods	Absent	NA	Absent	Absent	Present (efferent)	NA	Present (afferent)	Absent
Filament shape	Semicircular	NA	Semicircular	Semicircular	Semicircular	NA	Semicircular	Semicircular (triangular to elongate in most gastropods)
Ventilation	Ctenidia	NA	Ctenidia	Ctenidia	Ctenidia	Ciliary bands, musculature	Musculature	Ctenidia
Respiration	Ctenidia + pallial cavity?	Subcutaneous	Ctenidia + pallial cavity?	Pallial cavity	Ctenidia + Pallial cavity	Pallial cavity	Ctenidia	Ctenidia + pallial cavity

From Haszprunar (1988) and Lindberg and Ponder (2001).

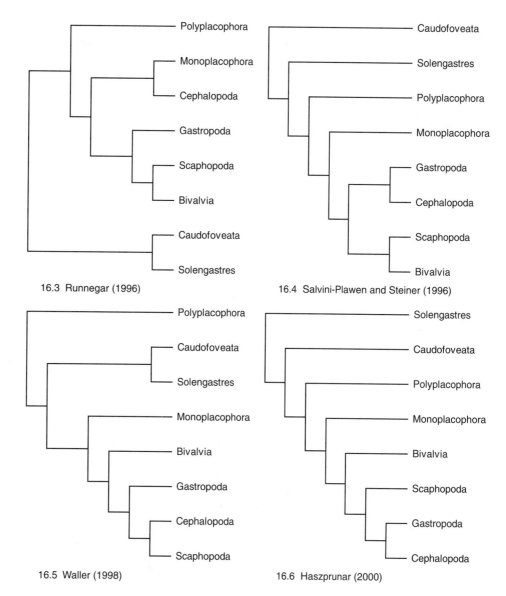

Figures 16.3–16.6. Phylogenetic relationships of living molluscan classes based on morphological data. 16.3. Runnegar (1996), with extinct taxa removed. 16.4. Salvini-Plawen and Steiner (1996). 16.5. Waller (1998). 16.6. Haszprunar (2000).

and chemosensory osphradia, a pair of specialized patches in the pallial cavity. Light receptors are found on the dorsal surface of some mollusks (e.g., chitons) and on the mantle edge, particularly in some bivalves (where they may be structurally complex and eyelike). Many gastropods have small cephalic eyes, which are rather complex in some groups. Most living cephalopods have large, complex eyes that parallel those of vertebrates.

During development, mollusks are one of several invertebrate phyla that undergo spiral cleavage. Embryological studies show that they have true coelomic cavities formed by the splitting of embryonic mesodermal masses (schizocoely) and that they have protostomous development (mouth develops before the anus); these characteristics are shared with several other phyla that are grouped as Eutrochozoa within Spiralia.

Many mollusks pass through free-swimming larval stages called trochophore and veliger larvae. The trochophore larva,

characterized by its apical tuft of cilia and ciliated bands, is found in primitive gastropods and many bivalves, as well as aplacophorans, scaphopods, and chitons. Similar larvae are also found in other marine invertebrate phyla, including Annelida, Sipuncula, and Entroprocta. Veliger larvae are characteristic of gastropods and bivalves and have a bilobed, ciliated swimming organ known as the velum that, in feeding larvae, also collects food particles from the water.

The molluscan body plan has been substantially modified, both among and within groups (table 16.2). Diversification appears to have occurred early in the history of Mollusca, but there has been surprisingly little change in some groups. For example, the shells of some Late Cambrian monoplacophorans are almost identical to those of living taxa despite 450 million years of evolution. Other examples of little change to molluscan body plan include protobranch bivalves, nautiloids, and scaphopods.

Table 16.2

Morphological Diversity of Living Adult Members of the Major Molluscan Clades.

Taxon	Anopedal flexure	Wormlike	Shell absent	Number of shells	Coiled	Slug	Limpets	Fishlike
Polyplacophora				8				
Neomeniomorpha		**	**	NA				
Chaetodermomorpha		**	**	NA				
Monoplacophora				1			**	
Bivalvia		*		2				
Scaphopoda	*			1				
Cephalopoda	*		*	1	(*)			**
Gastropoda	*	(*)	*	1 (2)	*	*	*	(*)
Patellogastropoda	*			1			**	
Cocculinida	*			1				
Vetigastropoda	*		(*)	1	**		*	
Neritopsina	*		(*)	1	**	(*)	(*)	
Caenogastropoda	*	(*)	(*)	1	**	(*)	(*)	(*)
Heterobranchia	*	(*)	*	1 (2)	**	**	(*)	(*)

** = predominant, * = well represented, (*) = rare.
Data compiled by D. R. Lindberg and W. F. Ponder.

Fossil History

Mollusca include some of the oldest metazoans known. Late Precambrian rocks of southern Australia and the White Sea region in northern Russia contain bilaterally symmetrical, benthic animals with a univalved shell (*Kimberella*) that resembles those of mollusks in some respects. The earliest unequivocal mollusks are helcionelloid mollusks that date from Late Vendian rocks (Gubanov and Peel 2000). In the Early Cambrian the Coeloscleritophora are also present. Most of the familiar groups, including gastropods, bivalves, monoplacophorans, and rostroconchs, all date from the Early Cambrian, whereas cephalopods are first found in the Middle Cambrian, polyplacophorans in the Late Cambrian, and Scaphopoda in the Middle Ordovician (Wen 1990). Most of these taxa tend to be small (<10 mm in length; Runnegar 1983). The Late Vendian–Early Cambrian taxa bear little resemblance to the Cambrian–Ordovician lineages (most of which remain extant today). After their initial appearances, taxonomic diversity tends to remain low until the Ordovician, when gastropods, bivalves, and cephalopods show strong increases in diversity. For bivalves and gastropods, this diversification increases throughout the Phanerozoic, with relatively small losses at the end-Permian and end-Cretaceous extinction events. Cephalopod diversity is much more variable through the Phanerozoic, whereas the remaining groups (monoplacophorans, rostroconchs, polyplacophorans, and scaphopods) maintain low diversity over the entire Phanerozoic or became extinct (Sepkoski and Hulver 1985).

There is a diversity of views on whether many of the Cambrian univalved mollusks should be interpreted as either gastropods or untorted taxa, and substantially divergent phylogenetic scenarios can result. In the most recent scheme,

Parkhaev (2002) has proposed a new gastropod subclass (Archaeobranchia) to contain taxa he considered to be torted and, therefore, gastropods. His action is based in part by the allocation of Helcionellacea to Gastropoda by Knight and Yochelson (1958). Most of the taxa allocated to the Archaeobranchia have also been treated as monoplacophorans (e.g., Runnegar 1983, Runnegar and Pojeta 1985) or as a separate class (Helcionelloida; Peel and Yockelson 1987, Peel 1991). However, when these controversial extinct taxa are removed from paleontological analysis of molluscan relationships, the resulting trees are often remarkably similar to current phylogenetic schemes based on living taxa (Runnegar 1996; fig. 16.3).

Habitats and Habits

Mollusks occur in almost every habitat found on Earth, where they are often the more conspicuous organisms and sometimes predominant (table 16.3). Although most are found in the marine environment, where they extend from the supralittoral to the deepest oceans, several major gastropod clades predominantly live in freshwater or terrestrial habitats. Marine diversity is highest nearshore and becomes reduced as depth increases beyond the shelf slope. Like many other organisms, marine mollusks reach their highest diversity in the tropical western Pacific and decrease in diversity toward the poles. Only one comprehensive study on molluscan diversity has been carried out in the tropical western Pacific, where around 3000 species have been found within a single site in coral reef habitat in New Caledonia (Bouchet et al. 2002). In terrestrial communities, gastropods can achieve reasonably high diversity and abundance: as many as 95 species may coexist in a single square kilometer of Cameroon rainforest (de Winter and Gittenberger 1998), and abundance in leaf litter can exceed more than 500 individuals in four liters of litter. Abundance and diversity for some groups can also be higher in temperate communities than in tropical settings. In freshwater communities, where both gastropods and bivalves co-occur, species diversity can also be high. Historically, in rivers of the southeastern United States, more than 100 species of mollusks (97 bivalves and a minimum of 12 different species of gastropods) were found on a single mussel shoal (P. J. Johnson, pers. comm.), and abundances of native freshwater unionid bivalves can approach 300 clams/m^2 in this same region (Johnson and Brown 2000). But these numbers pale compared with the introduced zebra mussel (*Dreissena polymorpha*), which can exceed more than 30,000 individuals/m^2 in North America (Dermott and Munawar 1993).

Marine mollusks occur on a large variety of substrates, including rocky shores, coral reefs, mud flats, and sandy beaches. Gastropods and chitons are characteristic of these hard substrates, and bivalves are commonly associated with softer substrates, where they burrow into the sediment. However, there are many exceptions: the largest living bivalve, *Tridacna gigas*, nestles on coral reefs, many bivalves (e.g., mussels, oysters) are attached to hard substrates; microscopic gastropods live interstitially between sand grains, and some are stygobionts.

The adoption of different feeding habits appears to have had a profound influence on molluscan diversification (table 16.4). The change from grazing to other forms of food acquisition is one of the major features in the adaptive radiation of the group (Ponder and Lindberg 1997, Vermeij and Lindberg 2000). Based on our current understanding of relationships (figs. 16.3–16.6), the earliest mollusks were carnivores or grazed on encrusting animals and detritus. Such feeding may have been selective or indiscriminate and will have encompassed algal, diatom, or cyanobacterial films and mats, or encrusting colonial animals. Truly herbivorous grazers are relatively rare and are limited to some polyplacophorans and a few gastropod groups (Vermeij and Lindberg 2000). Most chaetodermomorph aplacophorans, monoplacophorans, and scaphopods feed on protists and/or bacteria, whereas neomeniomorph aplacophorans graze on cnidarians. Cephalopods are mainly active predators as are some gastropods, whereas a few chitons and septibranch bivalves capture microcrustaceans. Most bivalves are either suspension or deposit feeders that indiscriminately take in particles but then elaborately sort them based on size and weight.

Cephalopods are typically active carnivores specialized on mobile prey such as fish, crustaceans, and other cephalopods. Because they are so abundant in pelagic systems, cephalopods are often important food sources for larger fishes, marine mammals, and seabirds. In the gastropods, members of Janthinidae are planktic pelagic carnivores feeding on cnidarians, whereas the heteropods (Caenogastropoda) and the gymnosomes (Opisthobranchia), like the cephalopods, are active swimmers in search of prey. These taxa spend their entire lives in the water column feeding on other mollusks (including small cephalopods), crustaceans, and even fishes. In addition to these more typical trophic strategies and interactions, some are endo- or ectoparastic, and the glochidium larvae of freshwater unionid bivalves parasitize fish and amphibians, although the adults are free living (see below).

Molluscan groups are ubiquitous and diverse in marine habitats, but only the bivalves and gastropods have invaded freshwater habitats, and only gastropods have invaded terrestrial ones. In nonmarine habitats, gastropods can be found in the wettest environments of tropical rainforests and in the

Table 16.3

Habitats Occupied by Living Adult Members of the Major Molluscan Clades.

| Taxon | Marine benthic | | Water column | Estuarine | Freshwater | Terrestrial | |
	Shallow	Deep				Damp	Arid
Polyplacophora	**	(*)					
Neomeniomorpha	(*)	**					
Chaetodermomorpha	*	**					
Monoplacophora		**					
Bivalvia	**	*	(*)	*	*		
Scaphopoda	**	*					
Cephalopoda	*	*	**	(*)			
Gastropoda	**	*	(*)	*	*	*	*
Patellogastropoda	**	(*)		(*)			
Cocculinida		**					
Vetigastropoda	**	**		*			
Neritopsina	**	*		*	*	*	
Caenogastropoda	**	*	*	*	*	*	(*)
Heterobranchia	**	*	*	*	*	**	*

** = predominant, * = well represented, (*) = rare.
Data compiled by D. R. Lindberg and W. F. Ponder.

Table 16.4

Feeding Types in the Major Molluscan Clades.

Taxon	Detritivory	Macroherbivory	Grazing carnivory	Microcarnivory	Hunting	Parasitic	Suspension
Polyplacophora		*	**		(*)		
Neomeniomorpha			**				
Chaetodermomorpha				**			
Monoplacophora	**						
Bivalvia	*			(*)			**
Scaphopoda				**			
Cephalopoda					**		?
Gastropoda	**	*	**	*	*	(*)	(*)
Patellogastropoda	(*)	**					
Cocculinida	**						
Vetigastropoda	**	*	**				(*)
Neritopsina	**	**					
Caenogastropoda	**	*	*	*	*	(*)	(*)
Heterobranchia	*	*	**	*	**	(*)	(*)

** = predominant, * = well represented, (*) = rare.
Data from D. R. Lindberg and W. F. Ponder.

driest deserts, where their annual activity patterns may be measured in hours. Some live below ground in the lightless world of aquifers and caves, and others interstitially in groundwater (stygobionts). The major terrestrial clade is the pulmonate gastropods, which originated at least by the Carboniferous period (Solem and Yochelson 1979), but other taxa that have nonmarine groups such as the neritopsines and caenogastropods are likely Devonian/Silurian in origin (Frýda 2001, Frýda and Blodgett 2001, Wagner 2001). Often the terrestrial groups are among the most basal of the extant taxa in the clade. For example, in both Neritopsina and Caenogastropoda, nonmarine taxa are thought to be more basal than marine members of these groups (Ponder and Lindberg 1997). These patterns could result from competition among sister taxa and the relegation of one taxon to a unique habitat while the other diversified in the ancestral setting.

Shell morphology is often thought to be correlated with lifestyle and habitat, and some substantial changes in body form are clearly associated with major adaptive changes. Frequently, however, morphology is not readily correlated with habitat, and similar shell morphologies do not necessarily indicate similar habits or habitats. For example, limpet taxa occur on wave-swept platforms, on various substrates in the deep sea, at hot vents, in fast-flowing rivers, in quiet lakes and ponds, and as parasites on oysters and starfish. It is often suggested that strong wave action selects for limpet morphology, but it is obvious from their known habitat distributions that mollusks with limpet-shaped shells do very well in a wide range of habitats (Ponder and Lindberg 1997).

Suspension feeding is characteristic of most bivalves but has also evolved in some gastropods such as the vetigastropod *Umbonium* and several caenogastropods (e.g., turritellids and calyptraeids) and in the pelagic heterobranch group Thecosomata. Some groups with carnivorous diets have undergone what appear to be true, explosive adaptive radiations (e.g., the Neogastropoda). Others that are food specialists such as the neomeniomorph aplacophorans and scaphopods have low diversity and abundance.

Several groups of bivalves, including Lucinidae and Solemyidae, have developed symbiotic relationships with bacteria that live in their modified gills and reduce or even eliminate the need for the uptake of alternative food supplies. The giant clams (or tridacnids), a number of other bivalves, and a few opisthobranch gastropods have symbiotic relationships with zooxanthellae embedded in their tissues.

Large concentrations of gastropods and bivalves are found at hydrothermal vents in the deep sea. Living in these or other dysoxic habitats appears to be a plesiomorphic condition for Mollusca and several outgroups. For example, the fauna of Paleozoic hydrothermal vent communities includes the molluscan groups Bivalvia, Monoplacophora, and Gastropoda as well as the outgroups Brachiopoda and Annelida (Little et al. 1997).

Outline of Major Groups

How important is the molluscan branch on the Tree of Life? Molluscan history is filled with incredible diversifications. Numerical abundance and diversification of living species have been previously referred to, but the total number of living species likely represents less than 5% of the total molluscan diversity that has ever lived. Many of the major lineages of the gastropods and bivalves survived the great extinctions. Some other major groups of mollusks did not, such as the ammonites, which did not survive the Cretaceous–Tertiary extinction. Taxa with high taxonomic diversity are often thought of as evolutionarily successful and therefore important in evolutionary studies. However, more than just numerical dominance should be considered in laying out an evolutionary research program. For example, although beetles, amphibians, and mollusks are numerically and ecologically diverse, the first two groups are rare in the fossil record compared with Mollusca. Although patterns of current diversity are intriguing, the degree of resolution of these patterns and the ability to deduce and test potential processes responsible for them through time are of great importance in diversity studies. Mollusca is one of the few groups that provides adequate data in this historical context.

As this volume attests, the state of our knowledge of metazoan phylogeny and taxa (including the Mollusca), and the wealth of new data that are now appearing from molecular, developmental, morphological, and paleontological work, will cause any classification proposed here to become rapidly outdated. Several traditional classifications are available in the references cited below. However, few are based on hypotheses of relationships, but are instead based on overall similarity and ad hoc scenarios of evolution.

Classifications based solely on morphology have been especially problematic, and much of this confusion has resulted from problematic taxa such as the aplacophorans, scaphopods, and bivalves, where possible reduction and loss of organs or other secondary simplification have produced morphologies that may be argued as either primitive or highly derived. Many of classification have also focused exclusively on the morphology of living taxa and have ignored potential, fossil members of Mollusca. If extinct fossil taxa are included in evolutionary scenarios, they are typically limited to distinctive clades such as Rostroconchia and Bellerophonta. Other more problematic extinct taxa (e.g., hyoliths) are systematically ignored, arbitrarily excluded from Mollusca without analysis, or shoe-horned into extant groups.

In some classifications (figs. 16.3, 16.4), the higher taxa have been treated as classes and arranged into several groupings, for example, the Conchifera (Gastropoda + Monoplacophora + Bivalvia + Scaphopoda + Cephalopoda), the Visceroconcha (Gastropoda + Cephalopoda) and the Diasoma (= Loboconcha; Bivalvia + Scaphopoda). In these classifications, the sister taxa of Mollusca have included Annelida, Lophophorates, and Kamptozoa (figs. 16.1, 16.2), and within Mollusca, both Polyplacophora and aplacophoran taxa have been argued as the most primitive taxa and therefore the outgroup to all Conchifera (figs. 16.3–16.6).

Most classifications have also assumed a single cladogenetic event in the origin of the Conchifera from the supposedly more primitive placophoran groups. Alternative hypothesis have derived the conchiferans in an unresolved polytomy from a hypothetical ancestral mollusk, or HAM. Some workers have interpreted the Cambrian Burgess Shale taxon *Wiwaxia* and other less complete halkieriid-like fossils as molluscan (e.g., Conway Morris and Peel 1990), whereas others have argued *Wiwaxia* to have annelid worm affinities (e.g., Butterfield

1990). However, the discovery of an articulated halkieriid from the lower Cambrian and the existence of these and other multi-shelled placophorans necessitate the reexamination of long-held assumptions of molluscan ancestry and monophyly. The rapidly increasing knowledge of coeloscleritophoran diversity suggests that we should not rule out the possibility that they shelter independent ancestors for extant molluscan groups (Lindberg and Ponder 1996).

Early molecular phylogenies for Mollusca using nuclear and mtDNA sequences initially had limited success in resolving a monophyletic molluscan clade or even producing robust or reasonable groupings within Mollusca (e.g., the bivalves and gastropods). These problems most likely result because of the deep, Paleozoic divergence of many of the molluscan taxa and the variable rates of change in genomes across taxa. We are now witnessing a new period in molluscan molecular studies with the addition of new genes, secondary structures, in situ hybridizations, and more. These data are currently providing analyses that are converging on a relatively small subset of polytomies within some molluscan groups (for a review, see Lydeard and Lindberg 2003).

The major groups of living mollusks are clearly dissimilar from one another and have long been recognized as distinct taxa. However, not all were originally recognized as belonging to Mollusca. For example, the wormlike bodies of the aplacophorans were perplexing to early biologists and required study of their internal anatomy to ultimately recognize their affinities with the other molluscan groups. This problem becomes especially acute with fossil taxa; the extinct groups (indicated below with a †) may or may not be mollusks in our current delimitation of the taxon based on living representatives. However, it is probable that with some more inclusive grouping, these fossil taxa share common ancestors with living molluscan groups.

The converse problem relates to living taxa. For example, although it is possible to relate living taxa to one another using both morphologic and molecular characters, there exists the real possibility that the living taxa do not share a single most recent common ancestor, but may have had multiple, independent derivations from distantly related mollusks or mollusk-like taxa that are now extinct (see below). These and other alternative hypotheses require that both fossils and living taxa be studied and incorporated into evolutionary scenarios and hypotheses of molluscan relationships, especially when the fossil record provides such a wealth of fossils and putative relatives.

Possible Mollusks

† Coeloscleritophora—represented worldwide as small, hollow, calcareous sclerites in the Precambrian and Cambrian. Insights into these enigmatic fossils have been obtained from articulated specimens (Conway Morris and Peel 1990, Bengtson 1992). Nevertheless, their relationship to Mollusca

remains uncertain, although at least some members of this possibly polyphyletic group may share common ancestry with mollusks, annelids, or brachiopods.

† Hyolitha—sometimes treated as a separate extinct phylum. The hyoliths have bilaterally symmetrical closed tubes with the aperture closed with an operculum. They first appear in the Early Cambrian and were extinct by the end of the Paleozoic (Runnegar 1980).

† Stenothecoida—bivalved Early to Middle Cambrian fossils in five or six genera that are sometimes regarded as mollusks (Pojeta and Runnegar 1976, Yochelson 2000). Waller (1998) considered Stenothecoida to represent the sister taxon of the Rostroconchia + Bivalvia.

Higher Molluscan Taxa

Polyplacophora (Chitons, Amphineura)

Morphology and Biology

Chitons (fig. 16.7) are flattened and elongate-oval, with eight overlapping dorsal shell plates or valves, bordered by a thick girdle that may be covered with spines, scales, or hairs and is formed from the mantle. The pallial cavity containing multiple pairs of small gills surrounds the foot, with which the animal typically clings to hard surfaces. The plates are greatly reduced or even internal in a few species, these sometimes having an elongate, somewhat wormlike body. Most are small (0.5–5 cm), but one species reaches more than 30 cm in length.

Chitons possess a heart and an open blood system, a pair of kidneys that open to the pallial cavity, a simple nervous system with two pairs of nerve cords, and many special minute sensory organs (aesthetes) that pass through the shell valves. Some of these are specialized as light receptors, having a minute lens and retinalike structure. The mouth is surrounded by a simple fold, and the head lacks tentacles or eyes. They feed on encrusting organisms such as sponges and bryozoans and nonselectively on diatoms and algae that are scraped from the substrate with their radula, which is hardened by the incorporation of metallic ions. One group captures small crustaceans by trapping them under the anterior part of their body (McLean 1962).

Chitons are generally dioecious, with sperm released by males into the water. In most chitons, fertilized eggs are shed singly or in gelatinous strings, and once fertilized in the water column, these develop into trochophore larvae that soon elongate and then directly develop into juvenile chitons; there is no veliger stage. In brooding species the eggs remain in the pallial cavity of the female, where they are fertilized by sperm moving through with the respiratory currents. Upon hatching from the brooded eggs, the offspring may remain in the pallial cavity until they crawl away as young chitons or exit the pallial cavity as trochophores for a short pelagic phase before settling.

Habitat

All chitons are marine, and the group has a worldwide distribution. Most live in the rocky intertidal zone or shallow

Figures 16.7–16.11. The lesser molluscan classes. 16.7. Polyplacophora (chitons; redrawn from Gray 1850). 16.8. Caudofoveata (or Chaetodermomorpha; redrawn from Beesley et al. 1998). 16.9. Solenogastres (or Neomeniomorpha; redrawn from Beesley et al. 1998). 16.10. Monoplacophora (or Tryblidia; redrawn from Lemche 1957). 16.11. Scaphopoda (tusk shells). All drawings by C. Huffard.

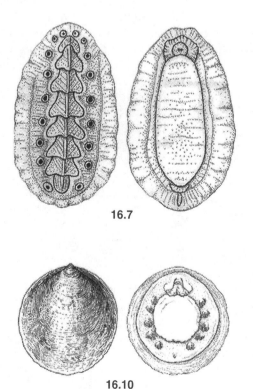

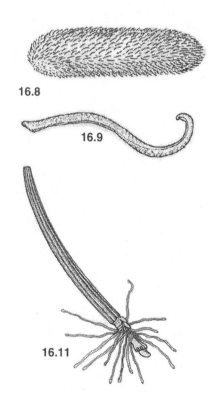

sublittoral, but some live in deep water to more than 7000 m. A few species are associated with algae and marine plants, and in the deep sea water-logged wood is a common habitat for one group.

Diversity and Fossil History

This relatively small group has been estimated to be between 650 and 800 recent species. The group first appears in the Late Cambrian (*Mattheva*).

Major Groups

Two groups (Paleoloricata and Neoloricata) are currently recognized, one of which are extinct. All living chitons are included in Neoloricata.

State of Knowledge

Our understanding of the species-level diversity of polyplacophorans has been greatly enhanced by the systematic work of Kaas and van Belle (1987–1994); Paleozoic taxa have been recently treated by Hoare (2000). However, given chiton diversity and abundances along rocky shores, and their importance in rooting analyses of other putative molluscan classes, it is surprising that a modern phylogenetic treatment of the group remains to be done.

Aplacophora (Caudofoveata and Solenogastres or Chaetodermomorpha and Neomeniomorpha, Spicule Worms)

Morphology and Biology

These wormlike mollusks (figs. 16.8, 16.9) lack shells but instead have calcareous scales or spicules in their integument, and they range in size from 1 mm to 30 cm. Caudofoveates are burrowers that feed on bottom-dwelling microorganisms such as formanifera, whereas most soleonogasters feed on cnidarians. Both groups have a radula and lack true nephridia.

Overall, the aplacophoran body plan is similar to that of the chitons. Aplacophorans and polyplacophorans differ from the monoplacophorans by having a dorsal gonad rather than a posterior gonad. The pericardium is similar in all three groups, as are many of the other organ systems and positions. Major differences are found in the type of spicules secreted by the dorsal mantle epidermis.

The calcareous spicules that cover the bodies of most aplacophorans give the animals a striking sheen. These spicules are secreted by the mantle epidermis and are the probable homologue of the shell of other molluscan groups. Spicule morphology varies over the body of the aplacophoran, and in some taxa spicules are modified into scales.

It is the internal anatomy that provides evidence of the molluscan identity of the aplacophorans. In both groups, the anterior end of the alimentary system includes a radula and odontophore. In Chaetodermomorpha, the radula and odontophore are strongly developed, and the alimentary system is more differentiated than in Neomeniomorpha. Both groups

have a dorsal gonad that opens into the pericardium, which contains the heart. From the posterior portion of the pericardium, there extends a coelomoduct that loops or bends and ultimately opens into the pallial cavity. In Neomeniomorpha, the posterior portion of the coelomoducts is modified for reproductive functions such as sperm storage or brooding young. The nervous system is ladderlike, with a well-developed cerebral ganglion. Radular configurations are quite variable and show a wide range of tooth development and modifications that include jawlike structures, denticles with cones, and sweepers. This is second only to the range of radular variation found in gastropods and is in marked contrast to the lack of variation found in Monoplacophora, Polyplacophora, and Scaphopoda.

Development includes trochophores or a test cell larval stage in which the three tissue types (mesoderm, ectoderm, endoderm) align and differentiate within an exterior cell layer constructed of large test cells. Aplacophoran eggs are relatively large and free-spawned in Chaetodermomorpha and fertilized internally in Neomeniomorpha; some Neomeniomorpha members brood their young to various stages of development. After the formation in the test cell larva of an apical tuft and prototroch, the posterior development of the differentiating larva quickly outgrows the exterior test and develops directly into the juvenile aplacophoran.

Habitat

All are marine and many live in the deep sea (to 6000 m or more).

Diversity and Fossil History

Around 320 species are known. There are no undoubted aplacophoran fossils, although some fossil organisms have been incorrectly attributed to them (e.g., Sutton et al. 2001).

Major Groups

Aplacophora is probably paraphyletic (Haszprunar 2000, Salvini-Plawen and Steiner 1996), although Scheltema (1996) regards this taxon as monophyletic and considers it to be equivalent in rank to the other classes.

Caudofoveata (or Chaetodermomorpha; fig. 16.8). Contains about one third of the known aplacophoran species, all of which are footless and vermiform and live in sediments. They have a circumoral sensory cuticular shield, the midgut separated into a stomach and glandular digestive diverticulum, and a pair of ctenidia in the small pallial cavity and are dioecious. They lack a foot and pedal groove and serial sets of lateroventral muscle bands.

Solenogastres (or Neomeniomorpha; fig. 16.9). Contains about two-thirds of the known aplacophoran species, which typically live in association with cnidarians such as hydroids and alcyonaceans. They have a narrow foot in a ventral groove with which they can creep, no oral shield, a sensory supraoral vestibule, a simple midgut (combined stomach and digestive gland), and serial sets of lateroventral muscle bands and are simultaneous hermaphrodites. They lack ctenidia in the rudimentary pallial cavity.

State of Knowledge

Recent studies and interpretations of aplacophoran phylogeny (Haszprunar 2000, Waller 1998) have focused attention on this small group of mollusks. Primarily because of the detailed studies (and contrasting interpretations) of Salvini-Plawen and Scheltema [see references in Haszprunar (2000) and Waller (1998)], the morphology of aplacophorans are relatively well known for a numerically and physically small-sized group of organisms. This knowledge base is even more remarkable when you consider that this is primarily a deep-water taxon, but species-level diversity is undoubtedly still severely understudied in this poorly collected group. Molecular phylogenetic studies of this taxon are lacking, and its placement on the molluscan tree remains problematic.

Monoplacophora (Tryblidia, Helcionelloidea, and Tergomya)

Morphology and Biology

Extant monoplacophorans are small and limpet-like, having a single, cap-like shell (fig. 16.10). Some organs (kidneys, heart, gills) are repeated serially, giving rise to the now falsified hypothesis that they have a close relationship with segmented organisms such as annelids and arthropods (Wingstrand 1985, Haszprunar and Schaefer 1997).

In recent and fossil patelliform monoplacophoran shells, the apex is typically positioned at the anterior end of the shell, and in some species it actually overhangs the anterior edge of the shell. Aperture shapes vary from almost circular to pear shaped. Shell height is also variable and ranges from relatively flat to tall. The monoplacophoran animal has a poorly defined head with an elaborate mouth structure on the ventral surface. The mouth is typically surround by a V-shaped, thickened anterior lip and postoral tentacles in a variety of morphologies and configurations. Behind the head lies the circular foot. In the pallial groove, between the lateral sides of the foot and the ventral mantle edge, are found five or six pairs of gills (fewer in minute taxa).

Internally, the monoplacophoran is organized with a long, looped alimentary system, one to three pairs of gonads, and multiple paired excretory organs (some of which also serve as gonoducts). A bilobed ventricle lies on either side of the rectum and is connected via a long aorta to a complex plumbing of multiple paired atria. The nervous system is cordlike and has weakly developed anterior ganglia; paired muscle bundles surround the visceral mass. Large dorsal paired cavities are extensions of glands associated with the esophagus. The monoplacophoran radula is docoglossate, each row having a central tooth, three pairs of lateral teeth, and two pairs of marginal teeth. There are no developmental studies of monoplacophorans.

Recent monoplacophorans form a clade (Wingstrand 1985), and their similarities and differences with the other extant molluscan groups are easily recognized. There is little question that some Paleozoic taxa are also members of this clade. However, the characters that distinguish some Paleozoic monoplacophorans from the torted gastropods and vice versa are open to alternative interpretations, and the relationships of several major groups of early-shelled mollusks have therefore been the subject of much debate (see above).

Habitat

Monoplacophorans are found both on soft bottoms and on hard substrates on the continental shelf and seamounts. Paleozoic taxa are associated with relatively shallow water faunas (<100 m).

Diversity and Fossil History

Monoplacophorans are the first undoubted mollusks, being found from the earliest Cambrian. Although diverse in the Paleozoic, the first living member of this exclusively marine taxon was not discovered until 1952 (Lemche 1957). About 25 living species of monoplacophorans have been discovered worldwide, living at depths between 174 and 6500 m.

Major Groups

Two groups, Helcionelloidea and Tergomya, are often treated as separate classes or subclasses. Recent monoplacophorans belong to Tergomya, whereas the youngest known helcionelloideans are from the earliest Ordovician.

State of Knowledge

Our knowledge of living members of Monoplacophora comes from the original anatomical description of *Neopilina galathaea* by Lemche and Wingstrand (1959). Wingstrand (1985) added additional observations and interpretations; Haszprunar and Schaefer (1997a) and Schaefer and Haszprunar (1997) provide additional anatomy of two Antarctic species. All of this work has been reviewed by Haszprunar and Schaefer (1997b).

Paleozoic members of Monoplacophora are still the subject of much conjecture. Pojeta and Runnegar (1976) and Peel (1991) consider almost all Cambrian cap-shaped taxa as well as the coiled Helcionelloida and some, if not all, of the bellerophontiform taxa to be untorted monoplacophorans, whereas others, including Knight and Yochelson (1958), Golikov and Starobogatov (1988), and Parkhaev (2002), limit the diagnosis of Monoplacophora to cap-shaped taxa and consider the remaining Helcionelloida and bellerophontiform taxa to be torted gastropods. Because these positions are based on the interpretations of a small suite of muscle insertion characters and cartoonlike reconstructions of possible water flow patterns, it is difficult to test either position.

Scaphopoda (Tusk Shells)

Morphology and Biology

Scaphopods are benthic, infaunal animals with slender, tubular shells open at both ends (fig. 16.11). The pallial cavity is large and surrounds much of the body, and there is a very simple head and well-developed burrowing foot located at the ventral (wider) end of the shell. Clublike feeding tentacles extend from the head, which lacks eyes, and a radula is present. Paired kidneys are present, but there is no heart (a reduced pericardium may be present) or gills. Foot morphology is variable and has been used as a taxonomic character. Water passing through the pallial cavity enters and exits through the dorsal aperture.

The scaphopod shell is a calcium carbonate tube with equal or unequal apertures; the tube may be either inflated or bowed. The shell microstructure includes prismatic and crossed-lamellar components; the latter is similar in structure to elements seen in members of Bivalvia.

Unlike the previously discussed groups, scaphopods have a U-shaped gut rather than an anterior–posterior configuration of the mouth and anus. The stomach and digestive gland are in juxtaposition, and the intestine loops before passing through the excretory organ and opening into the pallial cavity. The posterior portion of the digestive gland overlies the gonad that connects with the pallial cavity via the excretory organ. The radula consists of a central plate, a single lateral tooth, and a lateral plate.

The ontogeny of several species has been documented (Moor 1983, Wanninger and Haszprunar 2001). The trochophore larva has an apical tuft and prototroch. The foot rudiment appears early followed by differentiation of the mantle. The mantle and the protoconch fuse ventrally producing a characteristic median ventral fusion line on the embryonic shell. During metamorphosis, the prototroch is shed and the protoconch stops growing. The adult shell begins to form, as do the trilobate foot, cephalic captacula, and the buccal apparatus. Animals are able to feed a few days after metamorphosis.

Scaphopods have an intriguing set of molluscan characters that have been allied to several scenarios of molluscan evolution and relationships. Shell structure and earlier observations of their development suggest bivalve affinities, but scaphopods also have a radula. The gross morphology of the scaphopod gut is U-shaped, like that of gastropods and cephalopods, rather than linear as in monoplacophorans, polyplacophorans, and aplacophorans, and recent molecular studies of shell formation suggest affinities with the gastropods and cephalopods, as well (Wanninger and Haszprunar 2001).

It has been suggested that scaphopods are descended from ribeirid rostroconchs (Pojeta and Runnegar 1976), therefore grouping them with Bivalvia. Although there is little doubt that scaphopods share some characters with Bivalvia, the direct derivation of scaphopods from a ribeirid rostroconch is contradicted by the U-shaped gut present in scaphopods because rostroconchs are thought to have had a linear gut based on reconstructions of shell morphology and musculature.

Habitat

Scaphopods are infaunal organisms and feed on foraminiferans and other interstitial organisms. They occur from the intertidal zone to depths in excess of 7000 m and are present in all the major oceans.

Diversity and Fossil History

There are approximately 600 recent species. Members of the class first appear in the Early Paleozoic, and the taxon has maintained a slow but steady rate of increase in morphological diversification since then.

Major Groups

Two orders, the Dentalida and Gadilida, are recognized.

State of Knowledge

Morphological cladistic analyses of the Scaphopoda have been performed by Steiner (1992) and Reynolds and Okusu (1999). A molecular study was conducted by Reynolds and Peters (1998). However, the relationships within the taxon are still some way from resolution (Reynolds 1997, 2002). Several recent morphological analyses (figs. 16.5, 16.6), as well as unpublished molecular studies (e.g., Steiner and Dreyer 2002), are resolving Scaphopoda with Cephalopoda and Gastropoda rather than their more traditional association with Bivalvia.

Bivalvia (Bivalves, Clams, Lamellibranchs, Pelecypoda)

Morphology and Biology

Bivalves, including the oysters, mussels, and clams (figs. 16.12–16.14), are the second largest group of mollusks. They have the shell composed of a pair of laterally compressed hinged valves, and the pallial cavity surrounds the whole body (fig. 16.12).

The bivalve shell consists of two valves that are hinged dorsally, usually with shelly interlocking teeth (the hinge), and always with a horny ligament that connects the two valves along their dorsal surfaces and acts to force the valves apart. The interior of the valves contains scars of the various muscles attached to it, in particular the (usually two, sometimes one) adductor muscles that, on contraction, close the valves. Another scar, the pallial line, represents the line of attachment of the mantle to the shell, and a posterior embayment in this line (the pallial sinus) is related to siphonal length in some bivalves. The shell can be internal and reduced (or even absent), and the bivalve animal can be wormlike, such as in "shipworms" (*Teredo*; fig. 16.14). Bivalve shells are

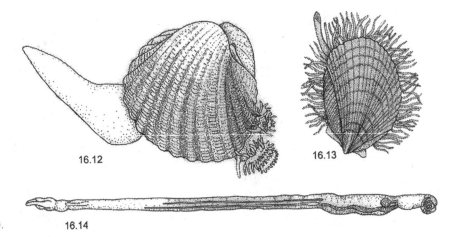

Figures 16.12–16.14. Bivalvia. 16.12. *Cardium* (cockle). 16.13. Pectinidae (scallop). 16.14. *Teredo* (shipworm). All redrawn by C. Huffard from Gray (1857).

constructed of different shell fabrics, including crossed lamellar, nacreous, and foliated microstructures. Most of the variability in shell structure sorts along higher taxon divisions. For example, nacreous structures are present primarily in the basal members of the group (Protobranchia, Pteriomorpha, Unionida), whereas crown taxa have primarily crossed lamellar shells (Heterodonta).

Bivalves typically display bilateral symmetry both in shell and anatomy, but there are significant departures from this theme in such taxa as scallops and oysters.

Bivalves lack a buccal apparatus, radula and jaws. Although the plesiomorphic feeding state for bivalves is probably deposit feeding using long labial palps, the ctenidia provide an effective filter-feeding mechanism in most taxa, with numerous levels or grades of organization. In most bivalves, the pallial cavity contains a pair of very large gills that are used to capture food particles suspended in the inhalant water current. The food is bound in mucus in strings that are carried by cilia, along food grooves on the edges of the gills, to the mouth region. Here particles are sorted on the ciliated labial palps before they enter the mouth. The bivalve stomach is large and complex with sophisticated ciliary sorting mechanisms and, usually, a rotating hyaline rod, the crystalline style, which liberates enzymes into the stomach. Digestion is carried out in the large paired digestive diverticula.

The visceral mass is primarily situated above the pallial cavity and continues ventrally into the foot. The intestine is irregularly looped and opens dorsally into the exhalant area. Also opening into this region are the paired kidneys and, when separate from the kidneys, the gonopores of the paired gonads. The heart typically lies below the center of the valves and consists of two auricles and a single ventricle that supplies both anterior and posterior aorta. The nervous system is made up of three pairs of ganglia. These innervate the oral apparatus, musculature, mantle, viscera, ctenidia, and siphons. They receive sensory input from oral lappets, statocysts, osphradium, various siphonal sensory structures, and photoreceptors along the mantle margin.

The bivalve foot is modified as a powerful digging tool in many groups, but in those that live a permanently attached life (e.g., oysters) it is very reduced. In many bivalve larvae or juveniles, a special gland, the byssal gland, can produce organic threads used for temporary attachment. In some groups, such as mussels, byssal threads permanently anchor the adults. A few groups of bivalves, such as oysters, are cemented permanently to the substrate.

The mantle edge in some primitive forms is open around the entire edge of the shell, but in most bivalves the mantle is fused to a greater or lesser extent, with openings for the foot (anterior and ventral) and posteriorly, the exhalant opening through which the water is expelled from the pallial cavity and which also carries waste products and gametes. The inhalant opening, through which water is carried into the pallial cavity, is also posteriorly located in most bivalves, lying just below the exhalant opening. In burrowing bivalves, the mantle edge around the inhalant and exhalant apertures is extended as separate or fused siphons that can be longer than the shell length. The mantle edge is also where contact is made with the external world and is, consequently, where most sense organs are located. These are usually simple sensory cells, but in some there are pallial eyes and/or sensory tentacles.

Bivalves are hermaphrodite or have separate sexes. Eggs of the protobranchs are large and yolky, whereas those of the remaining taxa are typically small and not very yolk-rich. Fertilization is usually external but in brooding species occurs in the pallial cavity. Cleavage patterns are spiral, and both polar lobes and unequal cleavage patterns are present throughout the group. Those embryos developing in the water column go through both trochophore and veliger ("spat") larval stages. Although morphologically similar to the gastropod veliger stage, phylogenetic analyses (Ponder and Lindberg 1997, Waller 1998) suggest that the veliger stage is homoplastic rather than homologous. The initial uncalcified shell grows laterally in two distinct lobes to envelop the body. Larval bivalves have a byssal gland that may assist with flotation while planktic but later attaches the juvenile to the substrate. Many bivalves retain their eggs in the pallial cavity and suck in sperm with the inhalant water current. In these brooding bivalves, the larvae develop in special pouches in the gills in some taxa, whereas in others they simply lie in the pallial cavity. Many brooding bivalves release their young as swimming veliger larvae, whereas others retain them longer and release them as juveniles. Freshwater mussels (Unionoidea) have glochidial larvae that attach to fish as ectoparasites.

Habitat

Most bivalves are marine, but there are also substantial radiations in brackish and freshwater habitats. They may be infaunal or epifaunal, and epifaunal taxa may be either sessile (cemented or byssally attached) or motile (fig. 16.13).

Diversity and Fossil History

The bivalves are an extremely diverse group with about 20,000 living species that range in adult size from 0.5 mm to giant clams that reach 1.5 m. Although the first occurrences of Bivalvia are found in Lower Cambrian deposits (Pojeta 2000), it is not until the Lower Ordovician that bivalve diversification, both taxonomic and ecological, explodes in the fossil record. This diversification continues unabated through the Phanerozoic, with relatively small losses at the end-Permian and end-Cretaceous extinction events. Two other extinct Cambrian bivalved groups, Stenothecoida and Siphonoconcha (Parkhaev 1998), may also nest within Mollusca, but the absence of bilateral symmetry, enigmatic hinge structures, and shell composition place them outside of Bivalvia as currently diagnosed.

Major Groups

Five major groups, usually given the rank of subclass, are recognized.

Protobranchia are mostly small sized with the hinge typically composed of many similar, small teeth (taxodont condition) and include the so-called nut shells (Nuculidae). They differ from other bivalves in that their large labial palps are used in deposit feeding and the gills are used only for respiration. This group is entirely marine, and the interior of the shell is nacreous in some families. All are shallow burrowers. One group, Solemyidae, farm symbiotic bacteria in their gills (Kraus 1995) and have a reduced gut. There are about 10, mostly small-sized families in all. Many are only found in deep water.

Pteriomorphia are an important, entirely marine group that includes many of the familiar bivalves—scallops (Pectinidae), oysters (Ostraeidae), pearl oysters (Pteriidae), mussels (Mytilidae), and arcs (Arcidae)—as well as about 18 other families. The hinge is taxodont or has a few reduced teeth, or the teeth are absent. A number of families have lost one of the adductor muscles (the monomyarian condition), and some have a nacreous shell interior. Many pteriomorphs are free-living epifaunal animals, are byssally attached, or are cemented and have a reduced foot. Others are shallow burrowers.

Palaeoheterodonta include the broach shells (Trigoniidae) and the freshwater mussels arranged in two superfamilies—Unionoidea (Unioniidae, Hyriidae, Margaritiferidae) and Muteloidea (Mutelidae, Mycetopodidae, and Etheri-

dae). The shell interior is often nacreous, and the hinge is composed of a few, often large teeth. All are shallow burrowers. The freshwater mussels have glochidial larvae that parasitize fish.

Heterodonta are a large group that includes the majority of familiar burrowing bivalves—the so-called clams, with more than 40 families including the very large family Veneridae, the cockles (Cardiidae), a family that now includes the giant clams (Tridacnidae), mactrids or trough shells (Mactridae), and the tellins (Tellinidae). Although most of the above groups are shallow burrowers, the heterodonts also include the deep-burrowing soft-shelled clams (Myiidae), the shipworms (Teredinidae), and rock borers (Pholadidae). One family (Chamidae) is cemented, and some members of the very diverse, mostly small-sized Galeommatoidea are commensals with a wide range of invertebrates. The shells of heterodonts have a complex hinge composed of relatively small numbers of different types of teeth, and the shell is never nacreous. Some members of this group are found in freshwater (notably Corbiculidae and Sphaeriidae), and the lucinoids farm symbiotic bacteria in their gills that provide most of their food requirements.

Anomalodesmata are a rather diverse group includes the watering pot shells (Clavagellidae) and about a dozen other small families, some of which are found only in rather deep water. Members of a few taxa are cemented to the substrate, but most are shallow burrowers and all are marine. One group of mostly deep-water families (collectively known as septibranchs) have the gills modified as pumping septa and feed on small crustaceans. The shells of some anomalodesmatans are nacreous, and most have a simple hinge.

State of Knowledge

A Hennigian analysis of bivalve morphology by Waller (1998) provides an overview of bivalve phylogeny and the relationships of the bivalves to the other molluscan classes. Waller's treatment is also somewhat unique in that it combines both fossil and living taxa in the analysis. Combined molecular and morphological studies of bivalve phylogeny have recently taken a substantial step forward with the high-level analysis of Giribet and Wheeler (2002). Strictly molecular analyses of bivalve relationships include Steiner and Muller (1996), Adamkewicz et al. (1997), and Canapa et al. (1999).

†Rostroconchia

The rostrochonchs look like bivalves but have a single larval shell that is transformed into a nonhinged, gaping bivalve shell as the animal grew. They are thought to have evolved from helcionelloidean monoplacophorans in the Early Cambrian and underwent an extensive Late Cambrian and Early Ordovician radiation; they survived until the Permian (Pojeta and Runnegar 1976). Rostroconchs are thought to share common ancestry with Bivalvia (Pojeta and Runnegar 1976, Waller 1998).

State of Knowledge

The seminal treatment of Rostroconchia is Pojeta and Runnegar (1976). Waller (1998) provides apomorphies and discussion of character states for Rostroconchia along with those for Stenothecidae and Bivalvia.

Gastropoda (Univalves, Limpets, Snails, Slugs)

Gastropods (literally "stomach/foot"; figs. 16.15–16.19) have figured prominently in paleobiological and biological studies and have served as study organisms in numerous evolutionary, biomechanical, ecological, physiological, and behavioral investigations.

Morphology and Biology

Gastropods are characterized by the possession of a single (often coiled) shell (figs. 16.15–16.18), although this is lost in some slug groups (fig. 16.19), and a body that has undergone torsion (see below) so that the pallial cavity faces forward. They have a well-developed head that bears eyes and a pair of cephalic tentacles and a muscular foot used for "creeping" in most species, while in some it is modified for swimming or burrowing. The foot typically bears an operculum that seals the shell opening (aperture) when the head-foot is retracted into the shell. Although this structure is present in all gastropod veliger larvae, it is absent in the embryos of some direct-developing taxa and in the juveniles and adults of many members of Heterobranchia. The nervous and circulatory systems are well developed, with the concentration of nerve ganglia being a common evolutionary trend.

Externally, gastropods appear to be bilaterally symmetrical; however, they are one of the most successful clades of asymmetric organisms known. The ancestral state of this group is clearly bilateral symmetry (e.g., chitons, cephalopods, bivalves; see above), but during development their organ systems can be twisted into figure eights, they can differentially develop or lose organs on either side of their midline, or they can generate shells that coil to the right or left. The best-documented source of gastropod asymmetry is the developmental process known as torsion. Like other mollusks, gastropods pass through a trochophore stage and then form a characteristic stage of development known as the veliger. During the veliger stage a 180° rotation of the pallial cavity from posterior to anterior places the anus and renal openings over the head and twists organ systems that pass through the snail's "waist" (the area between the foot and visceral mass) into a figure eight. This rotation is accomplished by a combination of differential growth and muscular contraction. In some taxa the contribution of each process is about 50:50, but in other taxa the entire rotation is accomplished by differential growth. Although the results of torsion are the best-known asymmetries in gastropods, numerous other asymmetries appear independent of the torsion process (Lindberg and Ponder 1996). Anopedal flexure (differential growth that places the mouth and anus in juxtaposition), which sometimes is considered a feature of torsion, is widely distributed in Mollusca and is present in the extinct hyoliths as well as in Scaphopoda and Cephalopoda (and to a lesser extent in the Bivalvia; Lindberg 1985).

Externally the animal has a well-developed head bearing a pair of cephalic tentacles and eyes that are primitively situated near the outer bases of the tentacles. In some taxa, the eyes are located on short to long eyestalks. The mantle edge in some taxa is extended anteriorly to form an inhalant si-

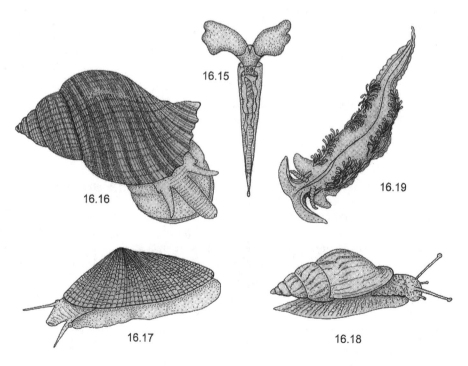

Figures 16.15–16.19. Gastropoda. 16.15. Pteropoda (Caenogastropoda). 16.16. Buccinidae (Caenogastropoda). 16.17. Patellogastropoda (limpet). 16.18. Pulmonate land snail (Heterobranchia). 16.19. Nudibranchia (Heterobranchia). All redrawn by C. Huffard from Gray (1842; figs. 16.15–16.18) and Gray (1850; fig. 16.19).

phon, and this is sometimes associated with an elongation of the aperture of the shell. The foot is usually rather large and is typically used for crawling. It can be modified for burrowing, leaping (as in conchs—Strombidae), swimming, or clamping (as in limpets; fig. 16.17).

They are extremely diverse in size, body, and shell morphology and in habits and occupy the widest range of ecological niches of all mollusks, being the only group to have invaded the land. Gastropod feeding habits are extremely varied, although most species make use of a radula in some aspect of their feeding behavior. Gastropods include grazers, browsers, suspension feeders, scavengers, detritivores, and carnivores. Carnivory in some taxa may simply involve grazing on colonial animals, whereas others engage in hunting their prey. Some gastropod carnivores drill holes in their shelled prey, this method of entry having being acquired independently in several groups (e.g., Muricidae and Naticidae). Some gastropods feed suctorially and have lost the radula.

Most aquatic gastropods are benthic and mainly epifaunal, but some are planktonic—a few, such as the violet snails (Janthinidae) and some nudibranchs (*Glaucus*), drift on the surface of the ocean, where they feed on floating siphonophores, whereas others (heteropods and Gymnosomata) are active predators swimming in the plankton (fig. 16.15). Some snails (e.g., the whelk *Syrinx aruanus*) reach about 600 mm in length, but there is also a very large (and poorly known) fauna of microgastropods that live in marine, freshwater, and terrestrial environments. It is among these tiny snails (0.5–4 mm) that many of the undescribed species lie.

Most gastropods have separate sexes, but some groups (mainly the Heterobranchia) are hermaphroditic, although most hermaphroditic forms do not normally engage in self-fertilization. The basal gastropods release their gametes into the water column, where they undergo development, but others use a penis to copulate or exchange spermatophores and produce eggs surrounded by protective capsules or jelly. The first gastropod larval stage is typically a trochophore that transforms into a veliger and then settles and undergoes metamorphosis to form a juvenile snail. Although many marine species undergo larval development, there are also numerous marine taxa that have direct development, this mode being the norm in freshwater and terrestrial taxa. Brooding of developing embryos is widely distributed throughout the gastropods, as are sporadic occurrences of hermaphrodism in the non-heterobranch taxa. The basal groups have nonfeeding larvae, whereas veligers of many neritopsines, caenogastropods, and heterobranchs are planktotrophic. Egg size is reflected in the initial size of the juvenile shell or protoconch, and this feature has been useful in distinguishing feeding and nonfeeding larvae in both recent and fossil taxa.

Phylogenetic patterns in gastropod evolution often feature a reduction in the complexity of many characters (Haszprunar 1988, Ponder and Lindberg 1997). These in-

clude reduction of the number of radular teeth, simplification (thought to be due to shell coiling) of the renopericardial system (loss of right auricle and renal organ), reduction of ctenidia (loss of the right gill), and associated circulatory and nervous system changes. There is also a reduction of diversity of shell microstructures, simplification of the buccal cartilages and muscles, reduced coiling of the hindgut, and simplification of the stomach. Other characters show an increase in complexity, such as life-history characters (e.g., internal fertilization with penis and spermatophores and associated reproductive organs). This increase in complexity is correlated with the ability to produce egg capsules and the evolution of planktotrophic larvae and direct development. There is also a phyletic increase in chromosome number, and greater complexity of sensory structures (e.g., eyes, osphradium; Haszprunar 1988). In the pulmonates (land snails; fig. 16.18), the pallial cavity is modified into a pulmonary cavity or lung, whereas the opisthobranchs (sea slugs) have secondary gills and elaborate neurosecretory structures.

Habitat

Gastropods occupy all marine habitats ranging from the deepest ocean basins to the supralittoral, as well as freshwater habitats and other inland aquatic habitats, including salt lakes. They are also terrestrial, being found in virtually all habitats ranging from high mountains, to deserts, to rainforests and from the tropics to high latitudes.

Diversity and Fossil History

Gastropods are one of the most diverse groups of animals, in form, habit, and habitat. They are by far the largest group of mollusks, with more than 62,000 described living species, and comprise about 80% of living mollusks. Estimates of total extant species range from 40,000 to more than 100,000 but may number as high as 150,000, with about 13,000 named genera for both recent and fossil species (Bieler 1992). They have a long and rich fossil record from the Early Cambrian that shows periodic extinctions of subclades followed by diversification of new groups (Erwin and Signor 1991).

Major Groups

The traditional classification of the gastropods was to divide it into three subclasses, Prosobranchia, Opisthobranchia, and Pulmonata. Prosobranchia (= Patellogastropoda + Vetigastropoda + Cocculinida + Neritopsina + Caenogastropoda and some members of Heterobranchia in the classification below) is paraphyletic, whereas Opisthobranchia + Pulmonata (Euthyneura) is now known to be but a major clade within a wider monophyletic group, Heterobranchia. Prosobranchia were often further divided into Archaeogastropoda, Mesogastropoda, and Neogastropoda; Archaeogastropoda and Mesogastropoda are both paraphyletic (Hickman 1988, Haszprunar 1988, Ponder and Lindberg 1997). There is as yet no general agreement regarding the ranks applied to the major groups within the gastropods that have now been confirmed from several morphological and molecular studies. The two main clades (Eogastropoda and Orthogastropoda) have been used as subclasses, but some authors prefer to assign subclass rank to the next highest category (Patellogastropoda, Vetigastropoda, etc.).

Eogastropoda. Patellogastropoda (= Docoglossa) include the true limpets (Patellidae, Acmaeidae, Lottiidae, Nacellidae, and Lepetidae). All are marine and limpet-shaped, and many live in the intertidal zone. This group was previously included within "Archaeogastropoda." The shell is foliated in some taxa, and the operculum is absent in adults. Their radula has several teeth in each row, some of which are strengthened by the incorporation of metallic ions such as iron.

Orthogastropoda. Vetigastropoda contain the keyhole and slit limpets (Fissurellidae), abalones (Haliotiidae), slit shells (Pleurotomariidae), top shells (trochids), and about 10 other families. All are marine and have coiled to limpet-shaped shells. This group was previously included within "Archaeogastropoda." The shell is nacreous in many of these taxa, and an operculum is usually present. The radula has many teeth in each row (rhipidoglossate). Many of the hydrothermal vent taxa are members of this group, including the neomphalids.

Neritopsina (or Neritimorpha) contain the nerites (Neritidae), which have marine, freshwater, and terrestrial members, and a few other small terrestrial and marine families. They have coiled to limpet-shaped shells, with only one species (family Titiscaniidae) being a slug. This group was previously included within "Archaeogastropoda." The shell is never nacreous, and an operculum is present in adults. The radula has many teeth in each row (rhipidoglossate).

Cocculinida contain a group of small white limpets that occur on waterlogged wood and other organic substrates in the deep sea. The operculum is absent in adults, and the radula has many teeth in each row, similar to the vetigastropods and nerites.

Caenogastropoda are a very large, diverse group containing about 100, mostly marine families, including littorines (Littorinidae), cowries (Cypraeidae), creepers (Cerithiidae, Batellariidae, and Potamididae), worm snails (Vermetidae), moon snails (Naticidae), frog shells (Ranellidae and Bursidae), apple snails (Ampullariidae), and a large, almost entirely marine group of about 20 families that are all carnivores and belong to Neogastropoda. These include whelks (Buccinidae), muricids (Muricidae), volutes (Volutidae), harps (Harpidae), cones (Conidae), and augers (Terebridae). Caenogastropod shells are typically coiled, a few being limpetlike (e.g., the slipper limpets, Calyptraeidae), and one family (Vermetidae) has shells resembling worm tubes. Although most caenogastropods possess a shell that encloses the animal, it is reduced in some and has become a small internal remnant in the sluglike Lamellariidae. Eulimidae are all parasitic on echinoderms; most are shelled ectoparasites, but some have become shell-less, wormlike internal parasites. Some groups have invaded freshwater, the most important being Viviparidae,

Ampullariidae, and Thiaridae (and several closely related families), and smaller sized snails belong to the diverse families Hydrobiidae, Bithyniidae, and Pomatiopsidae. There are a few terrestrial taxa, the cyclophorids being the most significant family.

Caenogastropods previously consisted of the monophyletic Neogastropoda and the paraphyletic Mesogastropoda. The shell is never nacreous, and an operculum is typically present in adults. Apart from members of Neogastropoda, the radula usually has only seven teeth in each row (taenioglossate). The radula of neogastropods has one to five teeth in each row (stenoglossate); the radula is absent in some.

Heterobranchia are a very large group composed of several marine and one freshwater group (Valvatidae) that were previously included in "Mesogastropoda" and two very large groups previously given subclass status—Opisthobranchia and Pulmonata (collectively Euthyneura). The more basal members consist of about a dozen families that are mostly small sized, mainly rather poorly known operculate groups, including the sundial shells (Architectonicidae) and a huge group of small-sized ectoparasites, Pyramidellidae. The opisthobranchs consist of about 25 families and 4000 species of bubble shells (Cephalaspidea) and seaslugs (Nudibranchia), as well as the seahares (Anaspidea). Virtually all opisthobranchs are marine, with most showing shell reduction or shell loss and only some of the "primitive" shell-bearing taxa having an operculum as adults. The pulmonates comprise the majority of land snails and slugs—a very diverse group consisting of many families and about 20,000 species. A few marine pulmonates (including the limpet-shaped Siphonariidae) comprise groups that mostly inhabit estuaries. A basal group of mainly estuarine air-breathing slugs (Onchidiidae) also has terrestrial relatives (Veronicellidae, Rathouisiidae). Some important groups of freshwater snails are also included here—the Lymnaeidae, Planorbidae, Physidae, and Ancylidae. The operculum is absent in all pulmonates except the estuarine Amphibolidae and the freshwater Glacidorbidae. The shells of heterobranchs are never nacreous.

State of Knowledge

Although both Haszprunar (1988) and Ponder and Lindberg (1997) present detailed phylogenetic analysis of Gastropoda, some of the ordering of the stem-based gastropod groups on the Tree of Life remains poorly understood, but there is mounting evidence that Patellogastropoda represents the sister taxon of all other gastropods. The base of Eogastropoda remains a polytomy of Cocculinidae, Vetigastropoda, and Neritopsina in recent analyses. In addition, branching patterns within relatively well-known groups such as Caenogastropoda, Vetigastropoda, and Euthyneura can vary markedly between analyses and data sets. Within Heterobranchia, there have been recent morphological and molecular analyses of Nudibranchia by Wagele and Willan (2000) and Wollscheid-Lengeling et al. (2001), and a recent molecular analysis of Pulmonata by Wade et al. (2001).

Cephalopoda (Octopuses, Squids, Cuttlefish, Chambered Nautilus)

Morphology and Biology

Cephalopods (literally "head-foot") are dorsiventrally elongated (figs. 16.20, 16.21), have well-developed sense organs and large brains, and are thought to be the most intelligent of all invertebrates. Nearly all are predatory, and most very active swimmers. A few taxa are benthic, drifters or medusalike, and some are detritus feeders. All are active carnivores in marine benthic and pelagic habitats from nearshore to abyssal depths. Giant squid (*Architeuthis*) are the largest invertebrates, and the cephalopods include the largest living as well as largest extinct mollusks: ammonite shells extend to more than 2 m across, and body sizes of living squid extend up to 8 m, with tentacles exceeding 21 m in length. The smallest cephalopods are around 2 cm in length.

Cephalopods are the most complex and motile of the nonvertebrate metazoans and show numerous modifications of the general molluscan body plan. The chambered nautilus has an external shell, but all other living cephalopods have either reduced and internalized the shell or have lost it completely. The calcareous shell of *Sepia* or cuttlefish (the cuttlebone) is internal, as is that of the ram's horn squid (*Spirula*), but other squid have the shell reduced to a horny pen, and octopuses lack a shell. The shells of cephalopods (other than the reduced gladius or pen in squids) have gas-filled chambers that assist with buoyancy.

Cephalopods have an amazing ability to rapidly change color (using numerous chromatophores in the skin), body shape, and texture, all of which is under nervous control. Their highly developed, efficient circulatory system differs from that of other mollusks in being closed and including a pair of accessory hearts (except in *Nautilus*). Most cephalopods can swim using jet propulsion, the pulses generated by the muscular walls of the pallial cavity. Some also use undulating movements of paired fins at the distal end of the mantle for swimming. Many can expel a cloud of ink to create a "smoke screen" to assist escape. Tentacles (cephalic in origin) surround the mouth on the head and capture prey. They often bear suckers, sometimes hooks, and a pair of retractile tentacles (arms) is found in some groups. They have powerful, modified jaws (beaks) and a small radula. The gut is dominated by muscle and enzymes and uses extracellular digestion. The large salivary glands in some squids and octopuses can produce highly toxic venoms, and there is a large digestive gland. The muscular stomach mixes the enzymes and food and passes the semidigested contents to a large caecum, where ciliated leaflets sort the particles.

The nervous system is highly advanced, with three major ganglia concentrated to form a large, efficient brain that is further enhanced by the formation of lobes. Coleoid cephalopods also have two large stellate ganglia on the mantle that control both respiratory and locomotory functions of the mantle. Experiments on cephalopods have been shown that they can learn and have good memories and excellent powers of discrimination (Hanlon and Messenger 1996). Their eyes are by far the most advanced in the invertebrates, are strongly convergent on vertebrate eyes, and are capable of resolving brightness, shape, size, and orientation. Additional sensory structures include statocysts and olfactory organs.

Cephalopods have a single gonad and separate sexes, with males transferring spermatophores to females after typically complex courtship. The spermatophore is transferred by the male using a penis (some squid, vampire squids, and cirrate octopuses) or (in nearly all others) a modified arm (hectocotylus). *Nautilus* uses four modified arms. Some taxa are highly sexually dimorphic. Fertilization is internal, with egg capsules being laid, and development is direct. Eggs are large and yolk-rich. There is no larval stage, just direct development into juveniles, although, as in some benthic taxa, these may have a pelagic phase. Both the eggs and young may be brooded, benthic, or pelagic. The shell of the paper argonaut (*Argonauta*) is the egg case, not a true shell.

Cephalopods are thought to have evolved from monoplacophoran-like ancestors (Pojeta and Runnegar 1976). Septa formed at the apex as the animal grew and withdrew into a newly formed body chamber. The old chambers are gas filled and provide buoyancy for the organism. The foot was modified into a funnel that provided jet propulsion for movement.

Habitat

Cephalopods are found worldwide, all are marine, and only a few can tolerate brackish water. All are found in benthic and pelagic habitats from nearshore to abyssal depths.

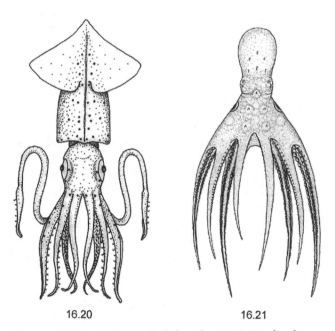

16.20   **16.21**

Figures 16.20 and 16.21. Cephalopoda. 16.20. Decabrachia (squid). 16.21. Octobrachia (octopus). All drawn by C. Huffard.

Diversity and Fossil History

Cephalopods were once one of the dominant marine animals, but there are only about 700 living species. More than 20,000 species are known as fossils.

Cephalopods are much more variable in their diversity through time than are other molluscan groups. They have experienced numerous extinctions (e.g., terminal Permian, Triassic, Cretaceous events) but typically showed rapid replacement (and subsequent radiation) by the survivors.

Major Groups

Three major clades (usually treated as subclasses) are recognized: Nautiloidea, Coleoidea, and Ammonoidea.

Nautiloidea include the pearly or chambered nautilus and its many fossil relatives. They first appeared in the Late Cambrian and underwent a rapid diversification in the Ordovician. All have a spiral nacreous shell with interconnected internal chambers. The head is covered with a hood and has numerous short, suckerless tentacles; there are two pairs of gills and no ink sac.

Coleoidea have 8–10 suckered or hooked tentacles and a single pair of gills, and an ink sac is often present. There are two main groups: *Octobrachia* and *Decabrachia*. *Octobrachia* (= Octopodiformes; fig. 16.20) includes octopuses, paper argonauts, the pelagic cirrate octopods (Octopoda), and vampire squid (Vampyromorpha). These all have four pairs of tentacles and no internal shell. *Decabrachia* (fig. 16.21) contains the ram's horn squid (Spirulida), the cuttlefish and dumpling squid (Sepioidea), and the squid (Teuthoidea). These all have four pairs of nonretractable arms and one pair of retractable arms (tentacles), and most have an internal shell (reduced to a chitinous pen in squids). The extinct Belemnoidea also belongs to this group.

Ammonoidea are a large, diverse clade of extinct shelled cephalopods that appeared in the Devonian and died out at the end of the Mesozoic. Impressions of animals suggest that they had 8–10 tentacles.

State of Knowledge

The last few years have witnessed a substantial increase in morphological, molecular, and combined analyses of cephalopod groups. Early morphological analyses include Young and Vecchione (1996; coleoid cephalopods) and Anderson (1996; loliginid squids). Concurrent molecular analyses include Bonnaud et al. (1996; also coleoid cephalopods) and Boucher-Rodoni and Bonnaud (1996) and Bonnaud et al. (1997), who examined higher level cephalopod relationships. More recently, Vecchione et al. (2000) investigated the relationships of neocoleoid cephalopods using molecular characters, and Anderson (2000a, 2000b) first used mtDNA sequences and then combined data sets to further examine relationships among the loliginid squids. Molecular and morphological data sets have also been compared in the analysis of Octopoda by Carlini et al. (2001). Phylogenetic analyses have recently extended back into deep time with morphological analyses of Neoammonoidea (Engeser and Keupp 2002) and the hamitid ammonites (Monks 2002).

The Future: Significant Problems Remaining, New Developments, and Targets

As discussed above, there remains a lack of resolution of the sister taxon to Mollusca. Convincing resolution of this problem will require new molecular data, acquisition of additional detailed morphological (including ultrastructure and immunocytochemistry) data for adults and larvae, and developmental information, for basal molluscan taxa and putative outgroups.

There is an urgent need for more sequence data in all groups, especially from a larger set of genes, both coding and noncoding. In addition, data sets using secondary structure (Lydeard et al. 2000, 2002) and mtDNA gene order (Boore and Brown 1994, Ueshima and Nishizaki 1994) have already proved to have great potential utility. There is a need to resolve not only the deep branches (the relationships of the "classes") but also the relationships within the major monophyletic groups, virtually all of which have Paleozoic roots. For example, within the gastropods, the placement of neritopsines and various groups of limpets on the tree is still problematic, in part because of long-branch attraction (Colgan et al. 2000). The development of methodologies to overcome long-branch problems would greatly benefit such studies.

Many long branches cannot be easily resolved (e.g., by adding additional taxa) because of the extinction of major clades. Incorporation of these extinct taxa in phylogenetic reconstructions may be difficult with mollusks because most of the characters used (anatomical, cytological ultrastructural, molecular) are not preserved. However, shell characters have, when properly used, been shown to be as useful as other characters at various levels of phylogenetic reconstruction (Wagner 1996, Schander and Sundberg 2001), especially if preservation is adequate to enable the incorporation of the fine structure of larval shells or shell microstructure. Such findings give more hope that the relationships of Paleozoic and Mesozoic taxa will ultimately be successfully resolved, and as argued above, there exists the real possibility that the recognized groupings of living taxa do not share a single common ancestor but may have had multiple, independent derivations from distantly related mollusks or mollusk-like taxa that are now extinct.

Phylogenetic resolution within non-gastropod clades is also fragmentary, poorly resolved, or lacking. There is a continuing need of better, parallel anatomical data for many groups and more comprehensive, phylogenetically based, comparative studies of organ systems (Ponder and Lindberg 1996), ideally incorporating histological studies.

Several ultrastructural data sets have contributed considerably to our understanding of molluscan, and especially gastropod, phylogeny. In particular, data on the osphradium

(Haszprunar 1985a) and sperm (Healy 1998, Buckland-Nicks 1995) have made major contributions, whereas smaller data sets such those on cephalic tentacles in gastropods (e.g., Künz and Haszprunar 2001) and details of the central nervous system (Huber 1993) have added important markers. Additional data on other systems, such as the work of Lundin and Schander (2001a, 2001b, 2001c) on cilial ultrastructure, is needed to expand coverage and provide additional characters. The recent compilation by Harrison and Kohn (1994, 1997) and their colleagues provides a comprehensive overview of the state of our knowledge in molluscan ultrastructure.

The use of developmental data has been extremely important in delineating spiralian taxa but has only infrequently been used in studies on molluscan phylogeny. Only three of the 117 characters used in Ponder and Lindberg's (1997) data set were developmental, mainly because of the lack of data for many of the critical taxa. Freeman and Lundelius (1992), van den Biggelaar and Haszprunar (1996), and Guralnick and Lindberg (2001) have shown that cleavage patterns and cell lineages can be successfully employed in reconstructing gastropod phylogeny. Studies on organogenesis have provided many valuable insights but are currently unfashionable, although the use of transmission electron micrography has been shown to be a valuable tool to provide much improved interpretation (e.g., Page 1998). Other imaging techniques such as confocal microscopy have been used to examine the development of musculature and other organ systems in chitons (Wanninger and Haszprunar 2002a), scaphopods (Ruthensteiner et al. 2001, Wanninger and Haszprunar 2002b), and gastropods (Wanninger et al. 1999), resulting in the resolution of several long-standing controversies. And although we may not agree with the phyletic placement of the spiculate animals described by Sutton et al. (2001), the

imaging techniques used to resurrect these creatures from solid rock will likely provide researchers with a wealth of new, detailed morphological data from deep time.

Although the literature has many detailed descriptions of larval development for higher gastropods and bivalves, there are relatively few for basal taxa. Comparative studies on trochophore and veliger larvae to address phylogenetic questions within mollusks are a potentially valuable field of study. For example, it has been suggested that planktotrophy may have arisen as many as three times within gastropods based on supposed larval differences (e.g., Ponder 1991, Ponder and Lindberg 1997), but no study has yet made a detailed comparison of the larvae from all three feeding clades (Neritopsina, Caenogastropoda, and Heterobranchia), and the most parsimonious scenario remains a single origination (Lindberg and Guralnick 2003).

Some of the issues identified above result from the unequal coverage and treatment of molluscan groups. For example, the number of papers with "Gastropoda" appearing as a key word in the BIOSIS literature database (available at http://www.biosis.org/) is in excess of 15,000 papers over the last 8 years, whereas "Monoplacophora" papers number only 30 (fig. 16.22: solid bars). However, these numbers can be misleading relative to the biodiversity of these groups, and a more accurate metric might be the ratio of "species" to papers (fig. 16.22: open bars). Using this ratio, the Monoplacophora, Bivalvia, and Cephalopoda are actually pretty well represented by research publications (the latter two taxa most likely because of their commercial importance, the former because of its status as a supposed "living fossil"). Although the relatively understudied status of Aplacophora, Polyplacophora, and Scaphopoda is not surprising (see also Lindberg 1985), this status for Gastropoda may come as a surprise to many given the seeming overabundance of gastropod work-

Figure 16.22. Research effort on major living molluscan taxa. Data from BIOSIS key word searches of papers published from 1995 through 2002. Solid bars, number of taxon papers; open bars, number of "species" estimated in each taxon, divided by the number of taxon papers.

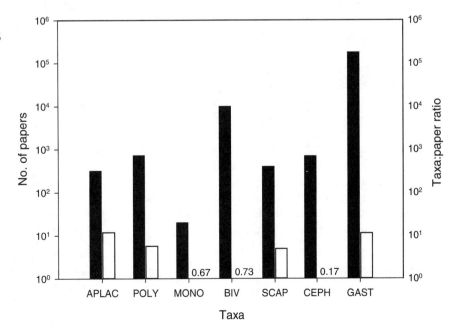

ers and publications relative to other molluscan groups. However, the sheer diversity of this group simply overwhelms even this relatively large number of workers.

Conclusions

Although great progress has been made over the last 15 years to resolve molluscan relationships, their relationships to other spiralian taxa, and thus their precise placement on the Tree of Life, remain unresolved. Within mollusks, different data sets are used in phylogenetic studies and in developing evolutionary scenarios. These include fossils (shell morphology), anatomy and histology, larval characters, ultrastructure, and molecular data. More recently, there have been some attempts to combine some or all of these kinds of data. However, robust hypotheses of molluscan origins and finer level relationships still appear to be some way off. This is unfortunate because the lack of such hypotheses (and the resultant stable classifications) may contribute to the lack of a modern (post-1960) treatment of Mollusca in many textbooks (e.g., Brusca and Brusca 2002) and to the continued use of paraphyletic taxa, falsified evolutionary scenarios, and just-so stories in teaching and the popular literature.

In a more positive light, phylogenetic studies of molluscan groups have produced many new insights into molluscan evolution, especially in Gastropoda, and many of these patterns are also present in other molluscan groups, and at the level of Mollusca as well. These include pronounced asymmetries in diversity, morphology, and ecology; evolutionary patterns in respiration and ventilation; phyletic changes in early developmental timing; and stunning examples of morphological and biological convergence. Evaluation of these and other character distributions, as well as testing of alternative hypotheses of molluscan evolution, requires a rigorous phylogenetic analysis of the data and continuing evaluation of the alternative theories and interpretations. New approaches such as gene expression and mtDNA gene order are beginning to be employed to resolve phylogenetic questions, but there is also a great need for additional data in more traditional areas on critical taxa (e.g., detailed anatomy, histology, ultrastructure, developmental data, and standard sequencing). With their diversity, abundance, and excellent fossil record, mollusks are an excellent group for exploring a wide range of evolutionary hypotheses. Well-resolved phylogenies will undoubtedly reduce the variance in all investigations and markedly enhance the already rich literature on the genetics, diversity, and ecology of mollusks that have provided important insights into evolutionary biology, biogeography, and ecology in general.

Acknowledgments

We thank J. Cracraft and M. Donoghue for the opportunity to participate in the Assembling the Tree of Life symposium, and P. D. Johnson and C. Lydeard for sharing their knowledge of freshwater molluscan faunas with us. The manuscript was improved by the comments of an anonymous reviewer and the artistic talents of C. Huffard.

Literature Cited

Adamkewicz, S. L., M. G. Harasewych, J. Blake, D. Saudek, and C. J. Bull. 1997. A molecular phylogeny of the bivalve mollusks. Mol. Biol. Evol. 14:619–629.

Adoutte, A., G. Balavoine, N. Lartillot, O. Lespinet, B. Prudhomme, and R. de Rosa. 2000. The new animal phylogeny: reliability and implications. Proc. Natl. Acad. Sci. USA 97:4453–4456.

Anderson, F. E. 1996. Preliminary cladistic analyses of relationships among loliginid squids (Cephalopoda: Myopsida) based on morphological data. Am. Malacol. Bull. 12:113–128.

Anderson, F. E. 2000a. Phylogenetic relationships among loliginid squids (Cephalopoda: Myopsida) based on analyses of multiple data sets. Zool. J. Linn. Soc. 130:603–633.

Anderson. F. E. 2000b. Phylogeny and historical biogeography of the loliginid squids (Mollusca: Cephalopoda) based on mitochondrial DNA sequence data. Mol. Phylogenet. Evol. 15:191–214.

Andrews, E. B. 1988. Excretory systems of molluscs. Pp. 381–448 in The Mollusca. Form and function (E. R. Trueman and M. R. Clarke, eds.), vol. 11. Academic Press, San Diego.

Beesley, P. L., G. J. B. Ross, and A. Wells (eds.). 1998. Mollusca: the southern synthesis: Pts. A and B, Fauna of Australia, vol. 5. CSIRO Publishing, Melbourne.

Bengtson, S. 1992. The cap-shaped Cambrian fossil Maikhanella and the relationship between coeloscleritophorans and molluscs. Lethaia 25:401–420.

Bieler, R. 1992. Gastropod phylogeny and systematics. Annu. Rev. Ecol. Syst. 23:311–338.

Bonnaud, L., R. Boucher-Rodoni, and M. Monnerot. 1996. Relationship of some coleoid cephalopods established by 3' end of the 16S rDNA and cytochrome oxidase III gene sequence comparison. Am. Malacol. Bull. 12:87–90.

Bonnaud, L., R. Boucher-Rodoni, and M. Monnerot. 1997. Phylogeny of cephalopods inferred from mitochondrial DNA sequences. Mol. Phylogenet. Evol. 7:44–54.

Boore, J. L. 1999. Animal mitochondrial genomes. Nucleic Acids Res. 27:1767–1780.

Boore, J. L., and W. M. Brown. 1994. Mitochondrial genomes and the phylogeny of mollusks. Nautilus 108(suppl. 2):61–78.

Boore, J. L., and J. L. Staton. 2002. The mitochondrial genome of the sipunculid Phascolopsis gouldii supports its association with Annelida rather than Mollusca. Mol. Biol. Evol. 19:127–137.

Boucher-Rodoni, R., and L. Bonnaud. 1996. Biochemical and molecular approach to cephalopod phylogeny. Am. Malacol. Bull. 12:79–85.

Bouchet, P., P. Lozouet, P. Maestrati, and V. Heros. 2002. Assessing the magnitude of species richness in tropical marine environments: exceptionally high numbers of molluscs at a New Caledonian site. Biol. J. Linn. Soc. 75:421–436.

Brusca, R. C., and G. J. Brusca. 2002. Invertebrates, 2nd ed. Sinauer Associates, Sunderland, MA.

Buckland-Nicks, J. 1995. Ultrastructure of sperm and sperm-egg interaction in Aculifera: implications for molluscan phylogeny. Mem. Mus. Natl. Hist. Nat. 166:129–153.

Butterfield, N. J. 1990. A reassessment of the enigmatic Burgess Shale, British Columbia, Canada fossil *Wiwaxia corrugata* Matthew and its relationship to the polychaete *Canadia spinosa* Walcott. Paleobiology 16:287–303.

Canapa, A., I. Marota, F. Rollo, and E. Olmo. 1999. The small-subunit rRNA gene sequences of venerids and the phylogeny of Bivalvia. J. Mol. Evol. 48:463–468.

Carlini, D. B., R. E. Young, and M. Vecchione. 2001. A molecular phylogeny of the Octopoda (Mollusca: Cephalopoda) evaluated in light of morphological evidence. Mol. Phylogenet. Evol. 21:388–397.

Colgan, D. J., W. F. Ponder, and P. E. Eggler. 2000. Gastropod evolutionary rates and phylogenetic relationships assessed using partial 28S rDNA and histone H3 sequences. Zool. Scr. 29:29–63.

Conway Morris, S., and J. S. Peel 1990. Articulated halkieriids from the Lower Cambrian of North Greenland, Arctic Ocean. Nature 345:802–805.

Cowie, R. H. 2002. Invertebrate invasions on Pacific islands and the replacement of unique native faunas: a synthesis of the land and freshwater snails. Biol. Invas. 3:119–136.

Dall, W. H. 1893. The phylogeny of the Docoglossa. Proc. Acad. Natl. Sci. Phila. 1893:285–287.

Dermott, R., and M. Munawar. 1993. Invasion of Lake Erie offshore sediments by *Dreissena*, and its ecological implications. Can. J. Fish. Aquat. Sci. 50:2298–2304.

de Winter, A. J., and E. Gittenberger. 1998. The land snail fauna of a square kilometer patch of rainforest in southwestern Cameroon: high species richness, low abundance and seasonal fluctuation. Malacologia 40:231–250.

Engeser, T., and H. Keupp. 2002. Phylogeny of the aptychi-possessing Neoammonoidea (Aptychophora nov., Cephalopoda). Lethaia 35:79–96.

Erwin, D. H., and P. W. Signor. 1991. Extinction in an extinction-resistant clade: the evolutionary history of the Gastropoda. Pp. 152–160 in The unity of evolutionary biology (E. C. Dudley, ed.). Dioscorides Press, Portland, OR.

Freeman, G., and J. W. Lundelius. 1992. Evolutionary implications of the mode of D quadrant specification in coelomates with spiral cleavage. J. Evol. Biol. 5:205–247.

Frýda, J. 2001. Discovery of the larval shell in Middle Paleozoic subulitoidean gastropods with description of two new species from the early Devonian of Bohemia. Bull. Czech Geol. Surv. 76:29–37.

Frýda, J., and R. B. Blodgett. 2001. The oldest known heterobranch gastropod, *Kuskokwimia* gen. nov., from the Early Devonian of west-central Alaska, with notes on theeh early phylogeny of higher gastropods. Bull. Czech Geol. Surv. 76:39–53.

Ghiretti, F. 1966. Respiration. Pp. 175–208 in Physiology of Mollusca (K. M. Wilbur and C. M. Yonge, eds.), vol. 2. Academic Press, New York.

Ghiselin, M. T. 1988. The origin of molluscs in the light of molecular evidence. Oxford Sur. Evol. Biol. 5:66–95.

Giribet, G. 2002. Current advances in the phylogenetic reconstruction of metazoan evolution. A new paradigm for the Cambrian explosion? Mol. Phylogenet. Evol. 24:345–357.

Giribet, G., and W. Wheeler. 2002. On bivalve phylogeny: a high-level analysis of the Bivalvia (Mollusca) based on combined morphology and DNA sequence data. Invert. Biol. 121:271–324.

Golikov, A. N., and Y. I. Starobogatov. 1988. Problems of phylogeny and system of the prosobranchiate gastropods. Proc. Zool. Inst. USSR 187:4–77.

Gray, M.-E. 1842. Figures of molluscous animals, vol. 1. Longman and Co. and J. Bailliere, London.

Gray, M.-E. 1850. Figures of molluscous animals, vol. 2. Longman, Brown, Green, and Longmans, London.

Gray, M.-E. 1857. Figures of molluscous animals, vol. 5. Longman, Brown, Green, Longmans, and Roberts, London.

Gubanov, A. P., and J. S. Peel. 2000. Cambrian monoplacophoran molluscs (Class Helcionelloida). Am. Malacol. Bull. 15:139–145.

Guralnick, R. P., and D. R. Lindberg. 2001. Reconnecting cell and animal lineages: what do cell lineages tell us about the evolution and development of Spiralia? Evolution 55:1501–1519.

Halanych, K. M., J. D. Bacheller, A. M. A. Aguinaldo, S. M. Liva, D. M. Hills, and J. A. Lake. 1995. Evidence from 18S ribosomal DNA that the lophophorates are protostome animals. Science 267:1641–1643.

Hanlon, R. T., and J. B. Messenger. 1996. Cephalopod behaviour. Cambridge University Press, Cambridge.

Harrison, F. W., and A. J. Kohn (eds.). 1994. Mollusca I, microscopical anatomy of invertebrates, vol. 5. Wiley-Liss, New York.

Harrison, F. W., and A. J. Kohn (eds.). 1997. Mollusca IIA and B, microscopical anatomy of invertebrates, vol. 5. Wiley-Liss, New York.

Haszprunar, G. 1985a. The fine morphology of the osphradial sense organs of the Mollusca. I. Gastropoda, Prosobranchia. Philos. Trans. R. Soc. Lond. B 307:457–496.

Haszprunar, G. 1985b. The fine morphology of the osphradial sense organs of the Mollusca. II. Allogastropoda (Architectonicidae, Pyramidellidae). Philos. Trans. R. Soc. Lond. B 307:497–505.

Haszprunar, G. 1988. On the orgin and evolution of major gastropod groups, with special reference to the Streptoneura (Mollusca). J. Mollusc. Stud. 54:367–441.

Haszprunar, G. 1996. The Mollusca: coelomate turbellarians or mesenchymate annelids? Pp. 1–28 in Origin and evolutionary radiation of the Mollusca (J. D. Taylor, ed.). Oxford University Press, Oxford.

Haszprunar, G. 2000. Is the Aplacophora monophyletic? A cladistic point of view. Am. Malacol. Bull. 15:115–130.

Haszprunar, G., and K. Schaefer. 1997a. Anatomy and phylogenetic significance of *Micropilina arntzi* (Mollusca, Monoplacophora, Micropilinidae Fam. Nov.). Acta Zool. 77:315–334.

Haszprunar, G., and K. Schaefer 1997b. Monoplacophora. Pp. 415–457 in Microscopic anatomy of invertebrates, Vol. 6B: Mollusca II (F. Harrison and A. J. Kohn, eds.). Wiley-Liss, New York.

Healy, J. M. 1998. The Mollusca. Pp. 122–173 *in* Invertebrate zoology (D. T. Anderson, ed.). Oxford University Press, Melbourne.

Hickman, C. S. 1988. Archaeogastropoda evolution, phylogeny and systematics: a re-evaluation. Pp. 17–34 *in* Prosobranch phylogeny (W. F. Ponder, ed.). Malacol. Rev. suppl. 4. Ann Arbor, MI.

Hoare, R. D. 2000. Considerations on Paleozoic Polyplacophora including the description of *Plasiochiton curiosus* n. gen. and sp. Am. Malacol. Bull. 15:131–137.

Huber, G. 1993. On the cerebral nervous system of marine Heterobranchia (Gastropoda). J. Mollusc. Stud. 59:381–420.

Johnson, P. D., and K. M. Brown, 2000. The importance of microhabitat factors and habitat stability to the threatened Louisiana pearl shell, *Margaritifera hembeli* (Conrad). Can. J. Zool. 78:271–277.

Kaas, P., and R. A. van Belle. 1987–1994. Monograph of living chitons: (Mollusca—Polyplacophora). E. J. Brill, Leiden.

Killeen, I. J., M. B. Seddon, and A. M. Holmes (eds.). 1998. Molluscan conservation: a strategy for the 21st century. J. Conchol. Spec. Publ. 2.

Knight, J. B. 1952. Primitive fossil gastropods and their bearing on gastropod classification. Smith. Misc. Coll. 117:1–56.

Knight, J. B., and E. L. Yochelson. 1958. A reconsideration of the relationships of the Monoplacophora and the primitive Gastropoda. Proc. Malacol. Soc. Lond. 33:37–48.

Kraus, D. W. 1995. Heme proteins in sulfide-oxidizing bacteria/mollusc symbioses. Am. Zool. 35:112–120.

Künz, E., and G. Haszprunar. 2001. Comparative ultrastructure of gastropod cephalic tentacles: Patellogastropoda, Neritaemorphi and Vetigastropoda. Zool. Anz. 240:137–165.

Lemche, H. 1957. A new living deep-sea mollusc of the Cambro-Devonian class Monoplacophora. Nature 179:413–416.

Lemche, H., and K. G. Wingstrand. 1959. The anatomy of *Neopilina galatheae* Lemche, 1957. Galathea Rep. 3:9–71.

Lindberg, D. R. 1985. Aplacophorans, monoplacophorans, polyplacophorans and scaphopods: the lesser classes. Pp. 230–247 *in* Mollusks. Notes for a short course (T. W. Broadhead, ed.). Department of Geological Science Studies in Geology 13. University of Tennessee, Knoxville, TN.

Lindberg, D. R., and R. P. Guralnick. 2003. Phyletic patterns of early development in gastropod molluscs. Evol. Dev. 5:494–507.

Lindberg, D. R., and W. F. Ponder. 1996. An evolutionary tree for the Mollusca: branches or roots? Pp. 67–75 *in* Origin and evolutionary radiation of the Mollusca (J. D. Taylor, ed.). Oxford University Press, Oxford.

Lindberg, D. R., and W. F. Ponder. 2001. The influence of classification on the evolutionary interpretation of structure: a re-evaluation of the evolution of the pallial cavity of gastropod molluscs. Org. Divers. Evol. 1:273–299.

Little, C. T. S., R. J. Herrington, V. V. Maslennikov, N. J. Morris, and V. Zaykov. 1997. Silurian hydrothermal-vent community from the southern Urals, Russia. Nature 385:146–148.

Lundin, K., and C. Schander. 2001a. Ciliary ultrastructure of neomeniomorphs (Mollusca, Neomeniomorpha = Solenogastres). Invert. Biol. 120:342–349.

Lundin, K., and C. Schander. 2001b. Ciliary ultrastructure of polyplacophorans (Mollusca, Amphineura, Polyplacophora). J. Submicrosc. Cytol. Pathol. 33:93–98.

Lundin, K., and C. Schander. 2001c. Ciliary ultrastructure of protobranchs (Mollusca, Bivalvia). Invert. Biol. 120:350–357.

Lydeard, C., W. E. Holznagel, M. N. Schnare, and R. R. Gutell. 2000. Phylogenetic analysis of molluscan mitochondrial LSU rDNA sequences and secondary structures. Mol. Phylogenet. Evol. 15:83–102.

Lydeard, C., W. E. Holznagel, R. Ueshima, and A. Kurabayashi. 2002. Systematic implications of extreme loss or reduction of mitochondrial LSU rRNA helical-loop structures in gastropods. Malacologia 44:349–352.

Lydeard, C., and D. R. Lindberg. 2003. Molecular systematics and phylogeography of mollusks: past, present, and future research opportunities. Pp. 1–13 *in* Molecular systematics and phylogeography of mollusks (C. Lydeard and D. R. Lindberg, eds.). Smithsonian Institution Press, Washington, DC.

Maddison, W. P., M. J. Donoghue, and D. R. Maddison. 1984. Outgroup analysis and parsimony. Syst. Zool. 33:83–103.

Mallatt, J. M., and C. J. Winchell. 2001. Use of combined large-subunit and small-subunit ribosomal RNA sequences to classify the protostomes and deuterostomes. Am. Zool. 41:1512–1513.

McLean, J. H. 1962. Feeding behavior of the chiton *Placiphorella*. Proc. Malacol. Soc. Lond. 35:23–26.

Medina, M., and A. G. Collins. 2003. The role of molecules in understanding molluscan evolution. Pp. 14–44 *in* Molecular systematics and phylogeography of mollusks (C. Lydeard and D. R. Lindberg, eds.). Smithsonian Institution Press, Washington, DC.

Meyhoefer, E., and M. P. Morse. 1996. Characterization of the bivalve ultrafiltration system in *Mytilus edulis, Chlamys hastata*, and *Mercenaria mercenaria*. Invert. Biol. 115:20–29.

Monks, N. 2002. Cladistic analysis of a problematic ammonite group: the Hamitidae (Cretaceous, Albian-Turonian) and proposals for new cladistic terms. Palaeontology 45:689–707.

Moor, B. 1983. Organogenesis. Pp. 123–177 *in* The Mollusca, Vol. 3: Development (N. H. Verdonk, J. A. M. van den Biggelaar, and A. S. Tompa, eds.). Academic Press, San Diego.

Nielsen, C. 1971. Entoproct life-cycles and the entoproct/ectoproct relationship. Ophelia 9:209–341.

Page, L. R. 1998. Sequential developmental programmes for retractor muscles of a caenogastropod: reappraisal of evolutionary homologues. Proc. R. Soc. Lond. B 265:2243–2250.

Parkhaev, P. Y. 1998. Siphonoconcha—a new class of early Cambrian bivalved organisms. Paleon. Zhur. 32:1–15.

Parkhaev, P. Y. 2002. Phylogenesis and the system of the Cambrian univalved mollusks. Paleon. Zhur. 36:27–39.

Peel, J. S. 1991. Functional morphology, evolution and systematics of early Palaeozoic univalved molluscs. Grønl. Geol. Under. Bull. 161:1–116.

Peel, J. S., and E. L. Yochelson. 1987. New information on *Oelandia* Mollusca from the Middle Cambrian of Sweden. Bull. Geol. Soc. Denm. 36:263–274.

Pojeta, J., Jr. 2000. Cambrian Pelecypoda (Mollusca). Am. Malacol. Bull. 15:157–166

Pojeta, J., Jr., and B. Runnegar. 1976. The paleontology of rostoconch mollusks and the early history of the phylum Mollusca. USGS Prof. Pap. 968:1–88.

Ponder, W. F. 1991. Marine valvatoidean gastropods—implications for early heterobranch phylogeny. J. Mollusc. Stud. 57:21–32.

Ponder, W. F. 1997. Conservation status, threats and habitat requirements of Australian terrestrial and freshwater mollusca. Mem. Mus. Victoria 56:421–430.

Ponder, W. F., and D. R. Lindberg. 1996. Gastropod phylogeny—challenges for the '90s. Pp. 135–154 in Origin and evolutionary radiation of the Mollusca (J. D. Taylor ed.). Oxford University Press, Oxford.

Ponder, W. F., and D. R. Lindberg. 1997. Towards a phylogeny of gastropod molluscs—a preliminary anaylsis using morphological characters. Zool. J. Linn. Soc. 119:83–265.

Reynolds, P. D. 1997. The phylogeny and classification of Scaphopoda (Mollusca): an assessment of current resolution and cladistic reanalysis. Zool. Scr. 26:13–21.

Reynolds, P. D. 2002. The Scaphopoda. Adv. Marine Biol. 42:137–236.

Reynolds, P. D., and A. Okusu. 1999. Phylogenetic relationships among families of the Scaphopoda (Mollusca). Zool. J. Linn. Soc. 126:131–154.

Reynolds, P. D., and J. Peters. 1998. Molecular and morphological phylogenetic systematics of the Scaphopoda (Mollusca). Am. Zool. 37:54A.

Runnegar, B. 1980. Hyolitha, status of the phylum. Lethaia. 13:21–25.

Runnegar, B. 1983. Molluscan phylogeny revisted. Mem. Assoc. Austral. Palaeont. 1:121–144.

Runnegar, B. 1996. Early evolution of the Mollusca: the fossil record. Pp. 77–87 in Origin and evolutionary radiation of the Mollusca (J. D. Taylor, ed.). Oxford University Press, Oxford.

Runnegar, B., and J. Pojeta, Jr. 1985. Origin and diversification of the Mollusca. Pp. 1–57 in The Mollusca, Vol. 10: Evolution (E. R. Truman and M. R. Clarke, eds.). Academic Press, San Diego.

Ruthensteiner, B., A. Wanninger, and G. Haszprunar. 2001. The protonephridial system of the tusk shell, Antalis entalis (Mollusca, Scaphopoda). Zoomorphology 121:19–26.

Salvini-Plawen, L. V. 1972. Zur Morphologie und Phylogenie der Mollusken: Die Beziehungen der Caudofoveata und der Solenogastres als Aculifera, als Mollusca und als Spiralia. Zeit. Wiss. Zool. 184:205–394.

Salvini-Plawen, L. V. 1980. A reconsideration of systematics in the Mollusca (phylogeny and higher classification). Malacologia 19:249–278.

Salvini-Plawen, L. V., and G. Steiner. 1996. Synapomorphies and pleistomorphies in higher classification of Mollusca. Pp. 29–51 in Origin and evolutionary radiation of the Mollusca (J. D. Taylor, ed.). Oxford University Press, Oxford.

Schaefer, K., and G. Haszprunar. 1997. Anatomy of Laevipilina antarctica, a monoplacophoran limpet (Mollusca) from Antarctic waters. Acta Zool. 77:295–314.

Schander, C., and P. Sundberg. 2001. Useful characters in gastropod phylogeny: soft information or hard facts? Syst. Biol. 50:136–141.

Scheltema, A. H. 1993. Aplacophora as progenetic aculiferans and the coelomate origin of mollusks as the sister taxon of Sipuncula. Biol. Bull. 184:57–78.

Scheltema, A. H. 1996 Phylogenetic position of Sipuncula, Mollusca and the progeentic Aplacophora. Pp. 53–58 in Origin and evolutionary radiation of the Mollusca (J. D. Taylor, ed.). Oxford University Press, Oxford.

Seddon, M. B. 1998. Red listing for molluscs: a tool for conservation? J. Conchol. Spec. Publ. 2:27–44.

Sepkoski, J. J., Jr., and M. L. Hulver. 1985. An atlas of Phanerozoic clade diversity diagrams. Pp. 11–39 in Phanerozoic diversity patterns (J. W. Valentine, ed.). Princeton University Press, Princeton, NJ.

Solem, A., and E. L. Yochelson. 1979. North American Paleozoic land snails, with a summary of other Paleozoic nonmarine snails. USGS Prof. Pap. 1072:1–42.

Steiner, G. 1992. Phylogeny and classification of Scaphopoda. J. Mollusc. Stud. 58:385–400.

Steiner, G., and H. Dreyer. 2002. Cephalopoda and Scaphopoda are sister taxa: an evolutionary scenario. Zoology 105(suppl. 5):95.

Steiner, G., and M. Muller. 1996. What can 18S rDNA do for bivalve phylogeny? J. Mol. Evol. 43:58–70.

Sutton, M. D., D. E. G. Briggs, and D. J. Siveter. 2001. An exceptionally preserved vermiform mollusc from the Silurian of England. Nature 410:461–463.

Ueshima, R., and N. Nishizaki. 1994. Evolutionary divergence in genomic structures of gastropodian mitochondrial DNA. Zool. Sci. 11(suppl.):36.

van den Biggelaar, J. A. M., and Haszprunar, G. 1996. Cleavage patterns and mesentoblast formation in the Gastropoda: an evolutionary perspective. Evolution 50:1520–1540.

Vecchione, M., R. E. Young, and D. B. Carlini. 2000. Reconstruction of ancestral character states in neocoleoid cephalopods based on parsimony. Am. Malacol. Bull. 15:179–193.

Vermeij, G., and D. R. Lindberg. 2000. Delayed herbivory and the assembly of marine benthic ecosystems. Paleobiology 26:419–430.

Wade, C. M., P. B. Mordan, and B. Clarke. 2001. A phylogeny of the land snails (Gastropoda: Pulmonata). Proc. R. Soc. Lond, B 268:413–422.

Wagele, H., and R. C. Willan. 2000. Phylogeny of the Nudibranchia. Zool. J. Linn. Soc.130:83–181.

Wagner, P. J. 1996. Contrasting the underlying patterns of active trends in morphologic evolution. Evolution 50:990–1007.

Wagner, P. J. 2001. Gastropod phylogenetics: progress, problems, and implications. J. Paleontol. 75:1128–1140.

Waller, T. R. 1998. Origin of the molluscan class Bivalvia and a phylogeny of the major groups. Pp. 1–45 in Bivalves: an eon of evolution (P. A. Johnston and J. W. Haggard, eds.). University of Calgary Press, Alberta, Canada.

Wanninger, A., and G. Haszprunar. 2001. The expression of an engrailed protein during embryonic shell formation of the tusk-shell, Antalis entalis (Mollusca, Scaphopoda). Evol. Dev. 3:312–321.

Wanninger, A., and G. Haszprunar. 2002a. Chiton myogenesis: perspectives for the development and evolution of larval and adult muscle systems in molluscs. J. Morphol. 251:103–113.

Wanninger, A., and G. Haszprunar. 2002b. Muscle development in *Antalis entalis* (Mollusca, Scaphopoda) and its significance for scaphopod relationships. J. Morphol. 254:53–64.

Wanninger, A., B. Ruthensteiner, W. J. A. G. Dictus, and G. Haszprunar. 1999. The development of the musculature in the limpet *Patella* with implications on its role in the process of ontogenetic torsion. Invert. Reprod. Dev. 36:211–215.

Wen, Y. 1990. The first radiation of shelled molluscs. Palaeont. Cathayana 5:139–170.

Wenz, W. 1938–1944. Gastropoda, Teil 1: Allgemeiner Teil und Prosobranchia. Pp. 1–1639 *in* Handbuch der Paläozoologie (O. H. Schindewolf, ed.), vol. 6. Gebrüder Bornträger, Berlin.

Wingstrand, K. G. 1985. On the anatomy and relationships of recent Monoplacophora. Galathea Rept. 16:7–94.

Winnepenninckx, B., T. Backeljau, and R. De Wachter. 1994.

Small ribosomal subunit RNA and the phylogeny of Mollusca. Nautilus 108(suppl. 2):98–110.

Winnepenninckx, B., T. Backeljau, and R. De Wachter. 1995. Phylogeny of protostome worms derived from 18S rRNA sequences. Mol. Biol. Evol. 12:641–649.

Wollscheid-Lengeling, E., J. L. Boore, W. Brown, and H. Waegele. 2001. The phylogeny of Nudibranchia (Opisthobranchia, Gastropoda, Mollusca) reconstructed by three molecular markers. Org. Divers. Evol. 1:241–256.

Yochelson, E. L. 2000. Concerning the concept of extinct classes of Mollusca: or what may/may not be a class of mollusks. Am. Malacol. Bull. 15:195–202.

Young, R. E., and M. Vecchione. 1996. Analysis of morphology to determine primary sister-taxon relationships within coleoid cephalopods. Am. Malacol. Bull. 12:91–112.

Zrzavý, J., S. Mihulka, P. Kepka, A. Bezdek, and D. Tietz. 1998. Phylogeny of the Metazoa based on morphological and 18S ribosomal DNA evidence. Cladistics 14:249–285.

The Relationships of Animals: Ecdysozoans

Ward C. Wheeler

Gonzalo Giribet

Gregory D. Edgecombe

Arthropod Systematics

The Comparative Study of Genomic, Anatomical,
and Paleontological Information

Arthropods are perhaps the most diverse creatures on Earth, with the number of known species approaching one million, and perhaps 10 times as many left to discover. Comprised today of Hexapoda (insects and relatives), Myriapoda (centipedes, millipedes, and allies), Crustacea (shrimps, crabs, lobsters, crayfish, barnacles, etc.), and Chelicerata (arachnids, horseshoe crabs, and sea spiders), the arthropods vary over four orders of magnitude in size (from <1 mm mites and parasitic wasps to >4 m spider crabs), are herbivores and carnivores, free-living and parasitic (endo and ecto), and solitary and social, and constitute the great majority of animal biomass. Arthropods are ubiquitous. They are found on all continents, the deepest oceans, and highest mountains. Extinct groups include trilobites, marrellomorphs, anomalocaridids, and euthycarcinoids, some of which may well be equal in taxonomic status to those we know today.

As members of the triploblastic Metazoa, arthropods are characterized by a segmented, hardened, chitinous cuticular exoskeleton and paired, jointed appendages. This exoskeleton is composed of a series of dorsal, ventral, and lateral plates that undergoes molting (ecdysis), sometimes periodically. Primitively, arthropods share a compound eye with a subunit structure that is unique within the animal kingdom.

The geological history of arthropods extends back over 520 million years (to the Lower Cambrian) with extinct lineages of great diversity (e.g., trilobites). This history has undergone several dramatic rounds of extinction and diversification, most prominently in the Paleozoic Era near the end of the Ordovician Period and at the Permian-Triassic boundary. The Cambrian and Ordovician body fossil record of arthropods is exclusively marine, but terrestrial forms (including arachnids, millipedes, and centipedes) appear from the Upper Silurian, more than 400 million years ago.

Relatives

The closest relatives of the arthropods are the enigmatic water bears (Tardigrada) and velvet worms (Onychophora). All of these animals share paired appendages and a chitinous cuticle. There are approximately 800 species of tardigrades that live in marine, freshwater, and terrestrial habitats. Marine tardigrades are an important component of the meiofauna, crawling between sand grains. Terrestrial tardigrades are mostly found on mosses and bryophytes and may occur in huge densities (hundreds of thousands to millions per square meter). Tardigrades are small (between 150 and 1000 µm); have a round mouth and four pairs of legs, the last one being terminal; and, like arthropods and a few other phyla, grow by molting. Terrestrial tardigrades can live in extreme environments, surviving desiccation or freezing by entering into cryptobiosis. The cryptobiotic stage has been recorded to last more than 100 years, and in this stage they can be dispersed by wind. The Onychophora are a group of exclusively terrestrial, predatory creatures that live in humid temperate (mostly southern hemisphere) and tropical forests of

America, Southern Africa, Australia, and New Zealand. The velvet worms are characterized by a soft body with pairs of "lobopod" walking limbs, a pair of annulated antennae, jaws, and oral ("slime") papillae. About 150 extant species have been named, but there were many more types including marine "armored" or plated lobopods in the Early Paleozoic. Onychophorans and arthropods share a dorsal heart with segmental openings (ostia) and a unique structure of the nephridia, the excretory organs. Lack of these organs in tardigrades may be due to miniaturization. It is thought that Tardigrada is the sister taxon of Arthropoda and Onychophora, the next closest relative (Giribet et al. 1996, 2001).

It has been long thought that there was an evolutionary progression from wormlike creatures, to lobopodous forms like Onychophora, to modern arthropods. This was expressed in the "Articulata" hypothesis that linked annelid worms (polychaetes and oligochaetes, including leeches) to Onychophora and Arthropoda. Recent work, especially from DNA sequences, has largely replaced this view, instead allying arthropods, tardigrades, and onychophorans with other molting creatures such as the nematodes, kinorhynchs, and priapulids in Ecdysozoa (after ecdysis or molting; Aguinaldo et al. 1997, Giribet and Ribera 1998, Schmidt-Rhaesa et al. 1998), and uniting the annelids with mollusks, nemerteans, sipunculans, and entoprocts in Trochozoa (Eernisse et al. 1992, Halanych et al. 1995, Giribet et al. 2000).

Extant Groups

The major extant arthropod groups are discussed in separate chapters and so are only briefly discussed here.

Hexapoda

The insects are by far the most diverse known arthropod group (but mites might come close), with hundreds of thousands of species known to science. Hexapods are characterized by possession of three body tagma (head, thorax, abdomen), the second of which possesses three limb-bearing segments. Insecta comprise most of the diversity within Hexapoda, insects being those hexapods with an antenna developed as a flagellum without muscles between segments. The hexapod head (like that of crustaceans and myriapods) has a large, generally robust mandible used for food maceration, a single pair of sensory antennae, and both compound and simple eyes. There are 30 commonly recognized hexapod "orders" further organized into several higher groups: Entognatha (those with internal mouthparts)—Protura, Diplura, and Collembola (springtails); Archaeognatha (bristletails); Zygentoma (silverfish); Ephemerida (mayflies), Odonata (damselflies and dragonflies); orthopteroids—Plecoptera (stoneflies), Embiidina (web spinners), Dermaptera (earwigs), Grylloblattaria (ice insects), Phasmida (walking sticks), Orthoptera (crickets, grasshoppers), Zoraptera, Isoptera (termites), Man-todea (praying mantises), Blattaria (roaches), Mantophasmatodea; hemipteroids—Hemiptera (true bugs and hoppers), Thysanoptera (thrips), Psocoptera, Pthiraptera (lice); and the Holometabola—Coleoptera (beetles), Neuroptera (lacewings, dobsonflies, snakeflies), Hymenoptera (bees, ants, wasps), Trichoptera, Lepidoptera (moths and butterflies), Siphonaptera (fleas), Mecoptera (snow fleas), Strepsiptera, and Diptera (flies). Basal hexapods (Protura, Collembola, Diplura, Archaeognatha, and Zygentoma) are wingless, whereas the more derived insect orders generally possess two pairs of wings. Members of Neoptera (Pterygota—winged insects except for the "paleopteran" ephemerids and odonates) possess wing hinge structures that allow folding their wings back over their abdomen. Those insects with complex development, Holometabola, are the most diverse, with beetles leading the way with more than 300,000 recognized species. Insects are found over the world in terrestrial and freshwater habitats, and many have economic importance as pests or medical interest for causing or carrying disease. An extensive fossil record of hexapods commences with the Devonian collembolan *Rhyniella* (Whalley and Jarzembowski 1981), through other Paleozoic and Mesozoic deposits, to the dramatic and beautiful amber-preserved insects from Lebanon, the Baltic, and the Dominican Republic (Carpenter 1992, Grimaldi 2001).

Myriapoda

The centipedes, millipedes, symphylans, and pauropods are multilegged, mostly soil-adapted creatures. Generally without compound eyes (except for scutigeromorph centipedes) but possessing a single pair of sensory antennae, the myriapods are most easily recognized by their large numbers of legs and the trunk not being differentiated into distinct tagmata. Almost all postcephalic segments bear a single (centipedes, pauropods, symphylans) or double (millipedes) pair of legs, numbering into the hundreds in some taxa. These arthropods are generally small (<5–10 cm), but there are several dramatically larger examples (*Scolopendra gigantea* at 30 cm). There are four main lineages of myriapods: Diplopoda (millipedes), Chilopoda (centipedes), Pauropoda, and Symphyla. The basic division among myriapods lies between Chilopoda, whose members have the genital opening at the posterior end of the body, and the other three lineages, grouped as Progoneata on the basis of the genital opening being located anteriorly on the trunk, behind the second pair of legs (Dohle 1998). The millipedes are by far the most diverse group, with approximately 11,000 described species. The chilopods are the other diverse group (~2,800 known species). Pauropods and symphylans are less speciose, with a few hundred described taxa. In general, myriapods are soil creatures feeding on detritus, with the centipedes exclusively predatory and possessing a modified fang and the ability to deliver toxins to their prey. It is probable, but far from universally agreed, that the myriapods share a single common ancestor (Edgecombe and Giribet 2002). The movement and connections of the head endoskeleton (the tentorium), structure and musculature of the mandible, and most DNA sequence evidence support the single origin of Myriapoda, but several hypotheses place myriapod lineages with hexapods (Kraus 1998). There are few well preserved myriapod fossils, but the extant chilopod order Scutigeromorpha and the diplopod group Chilognatha both have fossil representatives from the Late Silurian (Almond 1985, Shear et al. 1998). The extinct group Arthropleurida, thought to be members of Diplopoda (Wilson and Shear 2000), may have reached 2 m in length.

Crustacea

Crustaceans are perhaps the most morphologically diverse group of arthropods (>30,000 species known), with huge variation in numbers and morphology of appendages, body organization (tagmosis), mode of development, and size (<1 mm to >4 m). These creatures are generally characterized by having two pairs of antennae (first and second), biramous (branched) appendages, and a specialized swimming larval stage (nauplius). They usually possess both simple ("naupliar") and compound eyes (the latter frequently stalked). Like myriapods and hexapods, crustaceans possess strongly sclerotized mandibles that are distinguished by frequently having a segmented palp. The Crustacea are generally marine, with several freshwater and terrestrial groups (e.g., some isopods, the woodlice). Crustacean phylogeny is an area of active debate with the status of some long-recognized groups under discussion (see Schram and Koenemann, ch. 19 in this vol.). Currently, several higher groups are recognized (Martin and Davis 2001) with their interrelationships (and even interdigitiation) unclear: Remipedia (12 species; *Speleonectes*, *Lasionectes*, and three other genera), Cephalocarida (few species; *Hutchinsoniella* and three other genera), Branchiopoda (1000 species; fairy shrimp, water fleas, tadpole shrimp, clam shrimp), Maxillopoda (10,000 species; copepods, barnacles, ostracods, fish lice), and Malacostraca (20,000 species; mantis shrimp, crayfish, lobsters, crabs, isopods, amphipods). Many of the debates on crustacean relationships center on the position of the recently discovered remipedes as either the most basal lineage resembling, in some respects, the first Crustacea, or a more derived position having little to do with crustacean origins. The fossil group Phosphatocopina is probably the earliest Crustacea or the closest relative of the extant Crustacea (Walossek 1999), first occurring in the Lower Cambrian in England and being known from fine preservational quality, notably in the three-dimensional Orsten Cambrian fauna (Müller 1979).

Chelicerata

The sea spiders, horseshoe crabs, and arachnids are characterized by division of body segments into two tagmata: pro-soma and opisthosoma (generally), and the first leg-bearing head segment being modified into chelifores or chelicerae. With the exception of horseshoe crabs (the American *Limulus* and the Asian *Carcinoscorpius* and *Tachypleus*), extant chelicerates do not possess compound eyes, and none have antennae. Horseshoe crabs and arachnids have one pair of median eyes, whereas sea spiders have a second pair. Of the three main divisions of chelicerates [Pycnogonida—sea spiders (1000 species), Xiphosura—horseshoe crabs (four species), and Arachnida—spiders, scorpions, etc. (92,000 species)], the sea spiders and horseshoe crabs are marine and arachnids are terrestrial, with the exception of some groups of mites. Many groups of Acari (mites and ticks) are parasites of plants and animals, both vertebrates and invertebrates, and being ecto- and endoparasitic, mostly of respiratory organs. The arachnids are the most diverse component of the Chelicerata, with the Acari and Araneae (spiders) constituting the vast majority of taxa. Other arachnid groups include Opiliones (harvestmen, daddy longlegs), Scorpiones (scorpions), Solifugae (sun, camel, or wind spiders), Pseudoscorpiones ("false" scorpions), Ricinulei, Palpigradi (micro-whip scorpions), Amblypygi (tailless whip scorpions or whip spiders), Uropygi (vinegaroons), and Schizomida. The Paleozoic eurypterids are an aquatic (mostly brackish water) group, generally considered to be the closest relatives of Arachnida, although some workers consider them especially related to scorpions (see Dunlop and Braddy 2001 for a discussion of the evidence). The largest eurypterids are 1.8 m long, among the largest arthropods ever. The sea spiders graze on corals, anemones, or seaweeds and vary in size from quite small (<1 cm) to almost a meter in leg span. Horseshoe crabs and arachnids are almost entirely predatory, with spiders the dominant arthropod predators in many environments. Horseshoe crabs scavenge and prey on small animals in seaweeds, and like the Opiliones, they digest their food internally. Most arachnids, however, digest food extraorally, ingesting their prey in the form of digested fluids.

Fossil History and Extinct Lineages

No doubt there are more extinct lineages of arthropods than extant. More likely than not, most will remain unknown to science, but several major groups we do know about have a great effect on our notions of higher level relationships among the arthropods (living and extinct). Trilobites are among the best-known group of extinct arthropods. First known from the Lower Cambrian, trilobites had huge radiations in the Paleozoic. Trilobites were an exclusively marine group (10,000 species described) characterized by two longitudinal furrows dividing the body into three lobes (hence the name). The body segments are organized into three tagmata (cephalon, thorax, pygidium). Trilobites possessed compound eyes and a single pair of antennae and had biramous appendages. All post-antennal appendages in trilobites are basically similar in structure (Whittington 1975). The imbricated lamellar setae in the exopods suggest that trilobites are closely related to the Chelicerata (being similar to the book gills of Xiphosura and Eurypterida), together with numerous other extinct lineages constituting the group Arachnata. Anomalocaridids or Dinocarida: Radiodonta are a group of large (up to 2 m), predatory Cambrian arthropod relatives. With unmineralized but sclerotized cuticle, they were known initially only by their raptorial feeding/grasping appendages that were anterior to a circular mouth that was surrounded by a ring of plates (Collins 1996). Their phylogenetic affinities are uncertain, but most recent work places them in the stem group of Arthropoda (Budd 2002), probably more closely related to extant arthropods than are tardigrades (Dewel et al. 1999). Marrellomorphs comprise a clade known from the Burgess Shale (Middle Cambrian, Canada) and Hunsrück Slate (Lower Devonian, Germany) that possess two pairs of antenniform limbs and two pairs of long spines that curve back over the body. *Marrella* is the most abundant arthropod in the Burgess Shale fauna (Whittington 1971). Euthycarcinoids are an enigmatic group that ranges from the Ordovician or Lower Silurian to the Middle Triassic, having potential affinities with myriapods or crustaceans (Edgecombe and Morgan 1999). They possessed a single pair of antennae and numerous pairs of uniramous legs. A diversity of lobopodian taxa has recently come to light via soft-part-preserved specimens, mainly from the Lower Cambrian of China. The marine lobopodians are thought to be related to living terrestrial Onychophora or Tardigrada, or some may be positioned higher on the arthropod stem group. Several of the Cambrian lobopodians possessed elaborate spines and armored plates (Ramsköld and Chen 1998). The "Orsten" fauna of Sweden contains amazingly well-preserved, three-dimensional Upper Cambrian fossils, most importantly of basal crustacean-like taxa (Walossek and Müller 1998). Several of these forms (e.g., *Martinssonia*) are important to understanding the origins and relationships of Crustacea. Among the most productive Paleozoic fossil deposits are the Burgess Shale, Chengjiang and Orsten (Cambrian), Rhynie Chert and Gilboa (Devonian), and Mazon Creek (Carboniferous) deposits.

The Relationships of the Arthropod "Classes"

The question of arthropod relationships has been and is still unsettled, despite the large effort invested by researchers. Excellent literature sources and reviews on many issues about arthropod relationships can be found in the recent volumes edited by Edgecombe (1998), Fortey and Thomas (1998), and Melic et al. (1999). These volumes complement the classical treatises by Snodgrass (1938), Boudreaux (1979), and Gupta (1979).

Of the living taxa (Chelicerata, Crustacea, Myriapoda, Hexapoda), it seems clear that those groups that possess

mandibles (robust, sclerotized, chewing mouthparts), the clade Mandibulata: Crustacea, Myriapoda, and Hexapoda, share a unique common ancestor (fig. 17.1). The biting edge of mandibles is formed by the same segment, the coxa, of the same limb (third limb-bearing segment in Crustacea), with a distinctive expression pattern of the *Distal-less* gene (Popadić et al. 1998, Scholtz et al. 1998). Within this group, things become less clear. There are two main competing hypotheses: Tracheata or Atelocerata (myriapods and insects) versus Tetraconata or Pancrustacea (crustaceans and insects). The Tracheata hypothesis is supported by some anatomical evidence, notably the similar tentorial head endoskeleton, an absence of limbs on the head segment (intercalary segment) innervated by the third brain ganglia, and similar respiratory and excretory organs (Klass and Kristensen 2001). Molecular sequence data and an alternative set of anatomical features, notably ommatidium structure, the optic neuropils, and neurogenesis, support the Tetraconata hypothesis (Dohle 2001).

This is a somewhat simplistic view of arthropod relationships that assumes that the four main classes are each monophyletic. However, pycnogonids may challenge this premise, and recent studies have shown them as the putative sister group to all remaining arthropods (Zrzavý et al. 1998, Giribet et al. 2001), in part supported by the presence of a terminal mouth as in many other non-arthropod ecdysozoans (Schmidt-Rhaesa et al. 1998) and absence of arthropod-type nephridia and intersegmental tendons. Fossil pycnogonids demonstrate their presence as far back as the Cambrian (Waloszek and Dunlop 2002). Also, many proponents of the Tracheata hypothesis supported myriapod paraphyly (Snodgrass 1938, Tiegs 1947, Dohle 1965). Paraphyly or polyphyly of crustaceans has also been proposed (Moura and Christoffersen 1996).

Mandibulata is supported by most molecular and total evidence analyses (Wheeler et al. 1993, Giribet and Ribera 1998, Wheeler 1998a, 1998b, Zrzavý et al. 1998, Edgecombe et al. 2000, Giribet et al. 2001). Alternatives to the clade Mandibulata have also appeared based on molecular sequence data analyses (Turbeville et al. 1991, Friedrich and Tautz 1995, Giribet et al. 1996, Hwang et al. 2001), although this seems to be an artifact of deficient taxonomic sampling because most other molecular analyses support Mandibulata (Regier and Shultz 1997, 1998). A second molecular alternative places Chelicerata as sister to Tetraconata (Regier and Shultz 2001, Shultz and Regier 1999), but again this result seems to be a bias toward particular genes.

Although relationships within Mandibulata are debated, molecular data from all sources tend to agree that crustaceans and insects form a monophyletic group, with the exception of some total evidence analyses (Wheeler et al. 1993, Wheeler 1998b, Edgecombe et al. 2000), but not from the most recent one including eight genes and morphology (Giribet et al. 2001).

The addition of fossil arthropods to the phylogenetic mix has rendered a strikingly different view from that of mor-

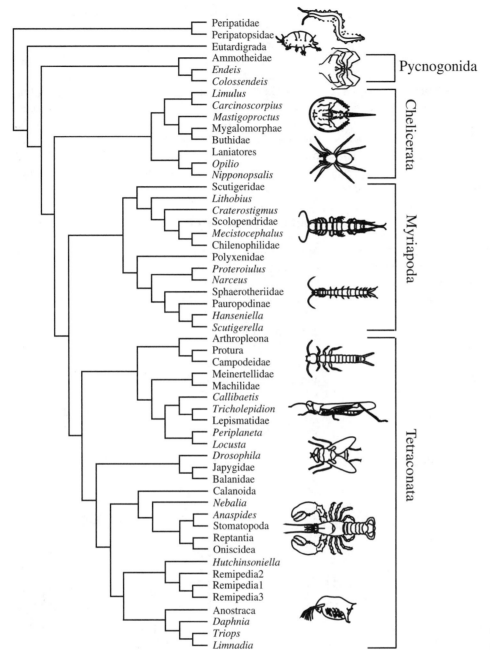

Figure 17.1. Cladogram of extant arthropod relationships, after Giribet et al. (2001).

phologists and molecular biologists, notably a hypothesis uniting all arthropods with biramous appendages in a clade named Schizoramia (Cisne 1974, Briggs et al. 1992, Budd 1996, Wills et al. 1998). Schizoramia contains the extant crustaceans and chelicerates, as well as many extinct lineages, including trilobites.

Monophyly versus Polyphyly

Arthropods were considered to be monophyletic since the 19th century (Siebold and Stannius 1848, Haeckel 1866) and were treated as such by most zoologists until the mid 20th century (Snodgrass 1938). A diphyletic current then ap-

peared, grouping the myriapods and hexapods together with the velvet worms to form Uniramia, versus Trilobita, Crustacea, and Chelicerata (Tiegs 1947, Tiegs and Manton 1958; named TCC by Cisne 1974). The diphyletic theory relied upon functional morphology arguments, based on the idea that the synapomorphies defining arthropods, such as the presence of a chitinous exoskeleton with jointed appendages and the presence of compound eyes, were convergences due to a similar mode of life.

The diphyletic theory further evolved into a polyphyletic theory in which the only previous taxon to be maintained was Uniramia. This was proposed by Manton (1964, 1973, 1977, 1979) and Anderson (1973, 1979). Manton proposed that the mandibles of crustaceans were not homologous to those of insects and myriapods, although she did not indicate an explicit relationship for the crustaceans or chelicerates. Anderson (1979) used embryonic fate maps to suggest a close relationship among annelids, onychophorans, and atelocerates (insects and myriapods). Subsequently, Schram (1978) joined the polyphyletists and used fate maps to endorse a relationship between pycnogonids and chelicerates.

The arguments in defense of arthropod polyphyly were not based on phylogenetic thinking or identifying alternative sister groups to different arthropod clades and were refuted by morphological (e.g., Weygoldt 1986, Kukalová-Peck 1992, 1998, Shear 1992, Wägele 1993), developmental (e.g., Weygoldt 1979, Panganiban et al. 1995, Popadić et al. 1996, 1998, Scholtz et al. 1998, Abzhanov and Kaufman 1999), and molecular (e.g., Wheeler et al. 1993, Edgecombe et al. 2000, Giribet et al. 2001) evidence. Also recently, homeobox genes have suggested homology between the chelicerae and the antennae of myriapods and insects and the first antennae of crustaceans (Damen et al. 1998, Telford and Thomas 1998, Abzhanov et al. 1999, Mittmann and Scholtz 2001). The only recent defenses of arthropod polyphyly (Fryer 1996, 1998) have resorted to imaginary worms rather than real taxa to force arthropod non-monophyly.

Schizoramia versus Mandibulata

With the issue of arthropod monophyly settled, arguments about the relationships among the main arthropod lineages grew, especially in relation to Schizoramia versus Mandi-

bulata. The TCC (Tiegs 1947, Cisne 1974) concept groups extinct trilobites and allied "trilobitomorophs" with extant chelicerates and crustaceans based on the primitive biramous nature of their appendages (Hessler and Newman 1975, Briggs and Fortey 1989, Bergström 1992, Briggs et al. 1992, Wills et al. 1995, 1998). This hypothesis, however, does not find support in molecular analyses, but this is not unexpected because TCC is based on the combinations of character states found in the extinct fauna. The Schizoramia concept obviously conflicts with Mandibulata (fig. 17.2), which finds support in morphological and molecular analyses (see discussion above).

Tracheata versus Tetraconata

Another major issue in arthropod systematics is the relative position of the mandibulate taxa. Classically, myriapods and insects were grouped together in Tracheata (or Atelocerata; Snodgrass 1938, 1950, 1951, Wägele 1993, Kraus and Kraus 1994, 1996, Kraus 1998, 2001, Wheeler 1998a, 1998b) based on morphological evidence (see discussion above). The addition of molecular data to study arthropod relationships, however, suggested an alternate relationship of crustaceans and hexapods (Boore et al. 1995, 1998, Friedrich and Tautz 1995, Giribet et al. 1996, 2001, Regier and Shultz 1997, 1998, Giribet and Ribera 1998), originally named Pancrustacea (Zrzavý et al. 1998) and later on formalized as Tetraconata (Dohle 2001) in reference to the ommatidium structure (four-part crystalline cone) shared by crustaceans and insects.

Figure 17.2. Signal synapomorphies for Mandibulata (mandible, shown for the chilopod *Ethmostigmus*) versus Schizoramia (biramous appendages, shown for the cephalocarid *Hutchinsoniella*).

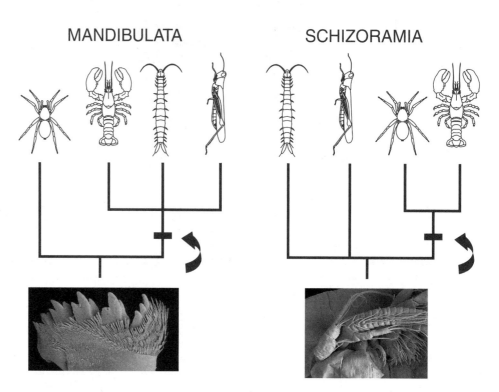

Other aspects of heated argumentation about arthropod evolution are the monophyly of Crustacea (see Schram and Koenemann, ch. 19 in this vol.) and the monophyly of Myriapoda (see Edgecombe and Giribet 2002).

Current Status and the Role of Fossils

In summary, arthropod systematists recognize the monophyly of the group, with Euarthropoda closely related to velvet worms (Onychophora) and water bears (Tardigrada). The arthropods can be divided into four main lineages, Chelicerata, Myriapoda, Crustacea, and Hexapoda, and a minor lineage of more uncertain affinities, Pycnogonida. Agreement about the monophyly of Mandibulata and Tetraconata seems to emerge from combined analyses of morphology and molecules (e.g., Giribet et al. 2001; fig. 17.1), but these groupings are not recognized universally, especially not so when the extinct diversity is brought into the picture. With regard to the sea spiders, emerging evidence suggests that they could be the sister group to the remaining arthropods, although a relationship to chelicerates cannot be rejected.

To evaluate these and other hypotheses, we attempted an analysis including almost 250 arthropods, living and extinct, and other related animals, together with information on more than 800 morphological characters and more than 2 kb (kilobases) of molecular sequence data. The aim of this study was to bring together the vast array of information known for extant arthropods and begin the integration of extinct taxa.

New Analysis

Taxa

The analysis of Giribet et al. (2001) contained 54 well-sampled, extant taxa but did not attempt any examination of extinct lineages. Here we have enlarged the sample of living taxa from 54 to 247, including seven Paleozoic taxa. These extinct lineages were Trilobita, coded largely from Whittington (1975: *Olenoides*); *Emeraldella* (from Bruton and Whittington 1983); *Sidneyia* (from Bruton 1981); Eurypterida, coded largely from Selden (1981); the Devonian pycnogonid *Palaeoisopus* (from Bergström et al. 1980); and the putative stem group crustacean *Martinssonia* (from Müller and Walossek 1986). Anomalocaridids are coded from *Parapeytoia* (Hou et al. 1995), but the coding precedes the reinterpretation (Budd 2002) of the grasping appendage as pre-antennal (with respect to crown group euarthropods). These morphological data were coded for 128 lineages, and the specific molecular taxa were treated as exemplars, with each member of the morphologically defined lineage (if there are several) receiving the same character coding (see supporting materials, see Wheeler 2003).

Of the 247 total taxa, 227 were sampled for molecular data [227 taxa for 18S ribosomal DNA (rDNA) and 135 taxa for 28S rDNA]. The remaining 20 taxa were sampled only for morphological data, seven because they are extinct, and the remainder due to the unavailability of sequence data.

Characters

Three sources of data were used in this study: morphological, small subunit rDNA (18S), and large subunit (28S) rDNA. The morphological characters include information from external and internal anatomy, behavior, ultrastructure, gene order, and development (see Wheeler 2003 for data). Overall, the morphological data had 13 additive multistate and 795 nonadditive characters. The small- and large-subunit sequence data are the same fragments used in Giribet et al. (2001). There were 10.7% missing and 14.5% inapplicable anatomical cells, 8.10% missing 18S rDNA sequences, and 45.3% missing 28S rDNA sequences (including extinct lineages).

Analysis

Morphological and molecular data were analyzed under parsimony using the program POY (vers. 2.7; Gladstein and Wheeler 1997–2002) on a 560 CPU PIII Linux cluster at the American Museum of Natural History and morphological analyses verified with NONA (vers. 2.0; Goloboff 1998). Cladogram costs were calculated for unequal length sequences using direct optimization (Wheeler 1996). A sensitivity analysis (Wheeler 1995) was performed using a variety of indel:transversion cost ratios (1:1, 2:1, 4:1, 8:1, and 16:1) and transversion:transition costs (1:1, 2:1, 4:1, and 8:1). This diversity of analyses was performed to assess the effects of analytical assumptions on phylogenetic conclusions.

Results

Analysis of the living taxa data set via NONA produced 100 equally parsimonious cladograms of length 1669, consistency index (CI) 0.60, and retention index (RI) 0.87, the strict consensus of which is shown in figure 17.3A. The inclusion of the seven extinct lineages resulted in 110 equally parsimonious cladograms of length 1720 (CI, 0.58; RI, 0.87), the strict consensus of which is shown in figure 17.3B. The two analyses jibe nearly completely with each other except for three areas: pycnogonids, remipedes/cephalocarids, and tracheates.

The living-taxa-only analysis shows a rather standard extant taxon hierarchy with the sea spiders as sister group to a clade of Xiphosura (horseshoe crabs) + arachnids. This is consistent with Snodgrass (1938), Wheeler et al. (1993), and the basal placement of pycnogonids by Giribet et al. (2001). The total taxon analysis (extinct + extant), however, inverts this relationship, placing Pycnogonida as sister to

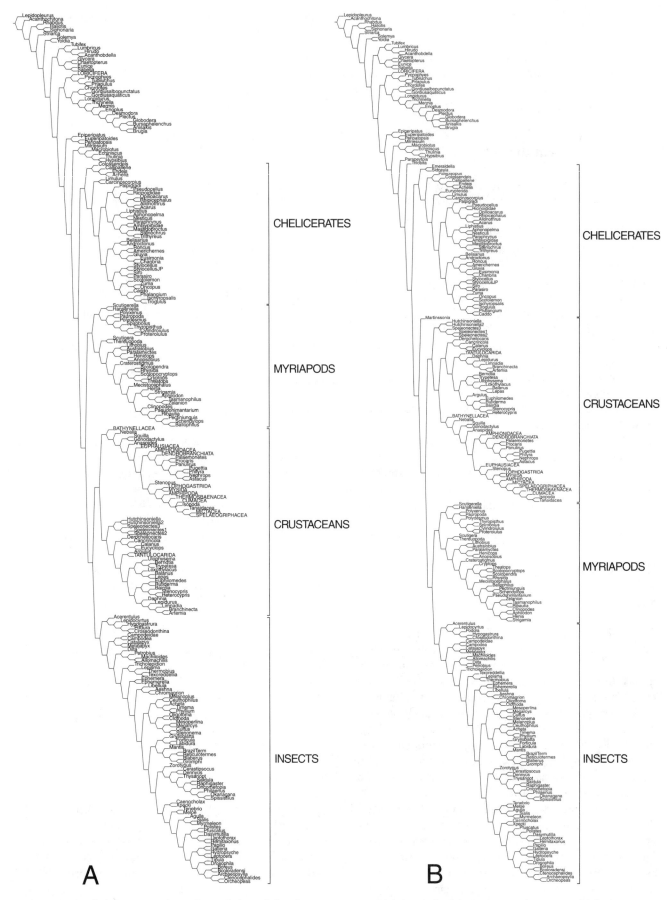

Figure 17.3. Phylogenetic analysis of morphological data for major groups of arthropods. (A) Extant taxa data set, and (B) extant + extinct data sets. Cladogram realized using WINCLADA (ver. 1.0; Nixon 2002).

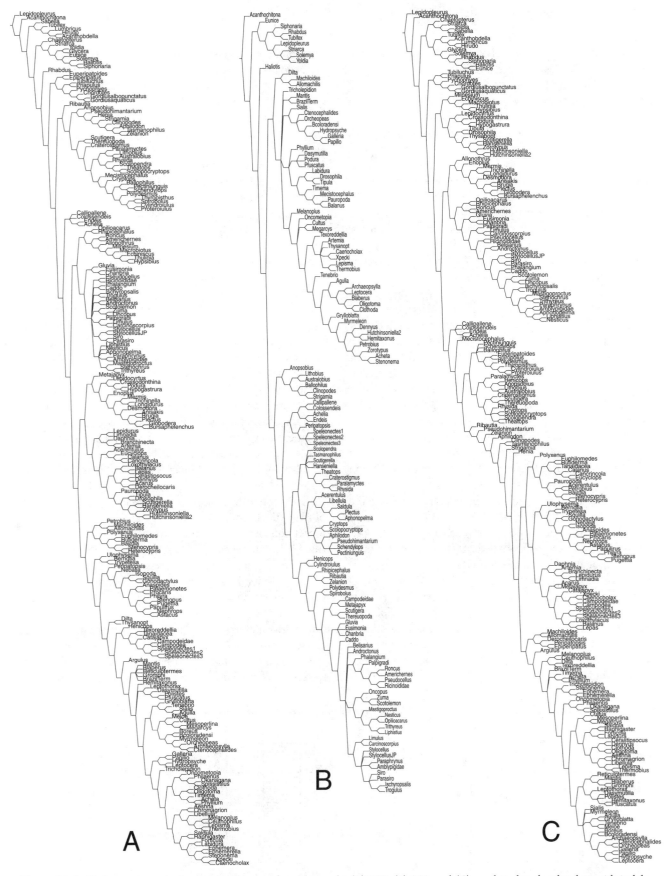

Figure 17.4. Phylogenetic analysis of molecular data for arthropods. (A) 18S, (B) 28S, and (C) combined molecular data with indels costing 8; transversions, 1; and transitions, 1; and morphological transformations costing 8. Cladogram realized using WINCLADA (ver. 1.0; Nixon 2002).

Arachnida, with the eurypterids, Xiphosura, trilobites, and *Emeraldella* + *Sidneyia* as successive sister groups. The inclusion of extinct lineages inverts the pattern based on living taxa. This is in part because of the additional scorable states in the pycnogonid opisthosoma due to *Palaeoisopus*, and the biramous limbs of the trilobites and other basal arachnates.

A second difference comes in the basal lineages of Crustacea. Both analyses support a major division between the malacostracan and maxillopodan + branchiopodan lineages. The placement of the remipedes and cephalocarids differs. In the more restrictive analysis (extant taxa only), these two putatively basal taxa group with Malacostraca, whereas in the complete taxon analysis the remipedes are the sister group to the remaining crustaceans, with *Hutchinsoniella* grouping with the non-malacostracan lineages.

The highest-level disagreement between these analyses is in the relative placement of Crustacea, Myriapoda, and Hexapoda. The extant taxa analysis supports Crustacea + Hexapoda (= Tetraconata), whereas the total-taxon analysis supports Hexapoda + Myriapoda (= Tracheata). The interactions here are complex. Certainly the role of the crustacean-like *Martinssonia* as a basal mandibulate (Wägele 1993, Moura and Christoffersen 1996) is central. The extinct lineages have altered the basal relationships of both the crustaceans and the chelicerates, and therefore their basalmost character states. Uniramy, as an example, has gone from the primitive condition in arthropods to a derived condition uniting tracheates on one side and arachnids + pycnogonids on the other. This is reinforced by both *Martinssonia* and the status of the anomalocarids (i.e., *Parapeytoia*) as sister group to crown group Euarthropoda (Dewel et al. 1999).

Molecular analyses show a diversity of patterns depending on the analytical parameters used to derive cladograms. There is a general pattern, however, of linking and even intermixing the crustacean and hexapod taxa (fig. 17.4). This pattern has been seen in molecular analyses of arthropod data for some time (e.g., Wheeler et al. 1993, Regier and Shultz 1997, Zrzavý et al. 1998, Giribet et al. 2001). The four pycnogonid representatives group together and separate from the arachnid lineages.

Combined analyses show an interesting distinction between extant and total-taxon analysis. As far as the relationships among the "classes," the extant taxa analyses are completely robust (fig. 17.5, left panel). In each of the 20 cases examined (e.g., fig. 17.6A), the crustaceans and hexapods form a clade. This is not terribly surprising in that both the morphological analysis of living taxa and the molecular data show this pattern. The Tetraconata (Dohle 2001) ["Pancrustacea" of Zrzavý et al. (1998) is based on crustacean paraphyly] is ubiquitous. When the extinct taxa are included, however, the pattern becomes less clear. At lower indel costs, Tetraconata is favored, whereas at higher indel costs (>2:1 over base substitutions), Tracheata is most parsimonious (figs. 17.5, right panel, and 17.6B). The "TCC" grouping was never found. Several patterns are common to the analyses. In both cases, the major groups (Crustacea, Chelicerata, Myriapoda, and Hexapoda) are monophyletic. Furthermore, the pycnogonids are brought to the base of chelicerates (sister group to Xiphosura + Arachnida), with *Emeraldella* + *Sidneyia* as stem-group chelicerates in the total-taxa analysis. Both analyses also support Remipedia + Cephalocarida (found in Giribet et al. 2001), which is not supported by either morphological taxon set. However, this clade is sister to the remaining crustaceans when the extinct lineages are included. Another noteworthy difference concerns the status of the entomostracan crustaceans, monophyletic based on the extant taxa (see Walossek and Müller 1998) but paraphyletic with respect to Malacostraca when fossils are included.

Inclusion of the molecular data affects the position of some of the extinct groups. Morphology alone resolves Trilobita in a frequently endorsed position in an arachnate clade (fig. 17.3B), in the chelicerate stem group (Wills et al. 1995, 1998, among many others). Analysis with the molecular data, however, shifts the trilobites outside Arachnata (fig. 17.6B), perhaps in part caused by character conflict when pycnogonids are placed as sister group of euchelicerates. This latter resolution, with trilobites as sister group to other euarthropods, allows that the lack of differentiation of post-antennal appendages in trilobites could be a primitive condition, rather than the reversal forced by their deep nesting in Arachnata.

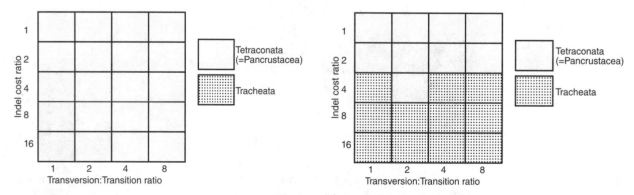

Figure 17.5. Sensitivity plots for (left panel) extant and (right panel) extant + extinct taxa showing the support for Tetraconata and Tracheata over varied analytical parameter assumptions.

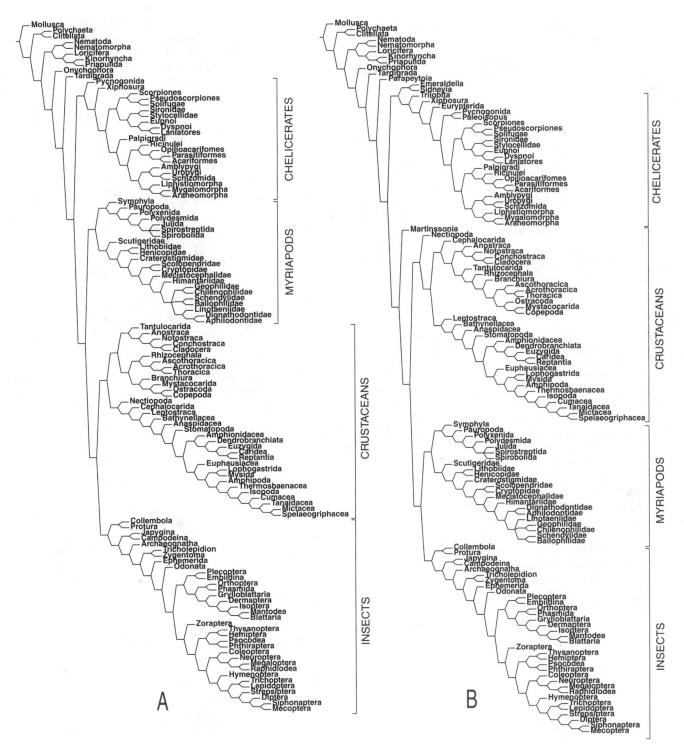

Figure 17.6. Combined (all data) analysis for (A) extant and (B) extant + extinct taxa with indels costing 8 transversions 1 and transitions 1 and morphological transformations 8. Cladogram realized using WINCLADA (ver. 1.0; Nixon 2002).

Discussion

The most striking result of this analysis and summary of current data on arthropod relationships is the importance of extinct lineages. Although we are able to examine a great deal of extant arthropod anatomy and molecular biology, the patterns of diversification and extinction in these groups make sampling limited to living taxa insufficient. Furthermore, even though this initial attempt at uniting these lineages resulted in unavoidably large levels of missing data in both molecular and morphological analysis, the effects of including even a few extinct taxa were profound.

At this point, several overall patterns in arthropod relationships can be identified as having support: monophyly of each of the major groups, Crustacea, Myriapoda, Hexapoda, and Chelicerata (with the possible exception of the Pycnogonida); monophyly of Mandibulata (crustaceans, hexapods, and myriapods); and outgroup status of Tardigrada and Onychophora. Several other important questions remain, including the position of the pycnogonids, the basalmost lineages of Crustacea and the sister group to Hexapoda. As we have shown here, these problems are sensitive to the inclusion of extinct lineages and are unlikely to be resolved with any great confidence until a broader sample of extinct diversity is incorporated into this analysis. Our results changed radically when we had 3% extinct lineages; what will happen when we have 99%?

Literature Cited

Abzhanov, A., and T. C. Kaufman. 1999. Homeotic genes and the arthropod head: expression patterns of the *labial*, *proboscipedia*, and *deformed* genes in crustaceans and insects. Proc. Natl. Acad. Sci. USA 96:10224–10229.

Abzhanov, A., A. Popadić, and T. C. Kaufman. 1999. Chelicerate Hox genes and the homology of arthropod segments. Evol. Dev. 1:77–89.

Aguinaldo, A. M. A., J. M. Turbeville, L. S. Lindford, M. C. Rivera, J. R. Garey, R. A. Raff, and J. A. Lake. 1997. Evidence for a clade of nematodes, arthropods and other moulting animals. Nature 387:489–493.

Almond, J. E. 1985. The Silurian and Devonian fossil record of the Myriapoda. Philos. Trans. R. Soc. Lond. B. 309:227–237.

Anderson, D. T. 1973. Embryology and phylogeny in annelids and arthropods. Pergamon, Oxford.

Anderson, D. T. 1979. Embryos, fate maps, and the phylogeny of arthropods. Pp. 59–106 *in* Arthropod phylogeny (A. P. Gupta, ed.). Van Nostrand, New York.

Bergström, J. 1992. The oldest Arthropoda and the origin of the Crustacea. Acta Zool. 73:287–291.

Bergström, J., W. Sturmer, and G. Winter. 1980. *Palaeoisopus, Palaeopantopus* and *Palaeothea*, pycnogonid arthropods from the Lower Devonian Hunsrück Slate, West Germany. Paläont. Z. 54:7–54.

Boore, J. L., T. M. Collins, D. Stanton, L. L. Daehler, and W. M. Brown. 1995. Deducing the pattern of arthropod phylogeny from mitochondrial DNA rearrangements. Nature 376:163–165.

Boore, J. L., D. V. Lavrow, and W. M. Brown. 1998. Gene translocation links insects and crustaceans. Nature 392:667–668.

Boudreaux, H. B. 1979. Arthropod phylogeny with special reference to insects. John Wiley and Sons, New York.

Briggs, D. E. G., and R. A. Fortey. 1989. The early radiation and relationships of the major Arthropod groups. Science 246:241–243.

Briggs, D. E. G., R. A. Fortey, and M. A. Wills. 1992. Morphological disparity in the Cambrian. Science 256:1670–1673.

Bruton, D. L. 1981. The arthropod *Sidneyia inexpectans*, Middle Cambrian, Burgess Shale, British Columbia. Philos. Trans. R. Soc. Lond. B. 295:619–656.

Bruton, D. L., and H. B. Whittington. 1983. *Emeraldella* and *Leanchoilia*, two arthropods from the Burgess Shale, Middle Cambrian, British Columbia. Philos. Trans. R. Soc. Lond. B. 300:553–585.

Budd, G. E. 1996. The morphology of *Opabinia regalis* and the reconstruction of the arthropod stem-group. Lethaia 29:1–14.

Budd, G. E. 2002. A palaeontological solution to the arthropod head problem. Nature 417:271–275.

Carpenter, F. M. 1992. Hexapoda. Treatise on invertebrate paleontology: Pt. R, Arthropoda 4. Geological Society of America, University of Kansas Press, Lawrence.

Cisne, J. L. 1974. Trilobites and the origin of arthropods. Science 186:13–18.

Collins, D. 1996. The "evolution" of *Anomalocaris* and its classification in the arthropod class Dinocarida (nov.) and order Radiodonta (nov.). J. Paleontol. 70:280–293.

Damen, W. G. M., M. Hausdorf, E. A. Seyfarth, and D. Tautz. 1998. A conserved mode of head segmentation in arthropods revealed by the expression pattern of hox genes in a spider. Proc. Natl. Acad. Sci. USA 95:10665–10670.

Dewel, R. A., G. E. Budd, D. F. Castano, and W. C. Dewel. 1999. The organization of the subesophageal nervous system in tardigrades: insights into the evolution of the arthropod hypostome and tritocerebrum. Zool. Anz. 238:191–203.

Dohle, W. 1965. Über die Stellung der Diplopoden im System. Zool. Anz. 28(suppl.):597–606.

Dohle, W. 1998. Myriapod-insect relationships as opposed to an insect-crustacean sister group relationship. Pp. 305–315 *in* Arthropod relationships (R. A. Fortey and R. H. Thomas, eds.). Systematics Association spec. ser. vol. 55. Chapman and Hall, London.

Dohle, W. 2001. Are the insects terrestrial crustaceans? A discussion of some new facts and arguments and the proposal of the proper name Tetraconata for the monophyletic unit Crustacea + Hexapoda. Ann. Soc. Entomol. Fr. (n.s.) 37:85–103.

Dunlop, J. A., and Braddy, S. J. 2001. Scorpions and their sister-group relationships. Pp. 1–24 *in* Scorpions 2001. In Memoriam Gary A. Polis (V. Fet and P. A. Selden, eds.). British Arachnological Society, Burham Beeches, Buckinghamshire, UK.

Edgecombe, G. D. 1998. Arthropod fossils and phylogeny. Columbia University Press, New York.

Edgecombe, G. D., and G. Giribet. 2002. Myriapod phylogeny and the relationships of Chilopoda. Pp. 143–168 *in* Biodiversidad, taxonomía y biogeografía de artrópodos de México: hacia una síntesis de su conocimiento (J. E. Llorente Bousquets and J. J. Morrone, eds.). Prensas de Ciencias, Universidad Nacional Autónoma de México, Mexico DF.

Edgecombe, G. D., and H. Morgan. 1999. *Synaustrus* and the euthycarcinoid puzzle. Alcheringa 23:193–213.

Edgecombe, G. D., G. D. F. Wilson, D. J. Colgan, M. R. Gray, and G. Cassis. 2000. Arthropod cladistics: combined analysis of Histone H3 and U2 snRNA sequences and morphology. Cladistics 16:155–203.

Eernisse, D. J., J. S. Albert, and F. E. Anderson. 1992. Annelida and Arthropoda are not sister taxa: a phylogenetic analysis of spiralian metazoan morphology. Syst. Biol. 41:305–330.

Fortey, R. A., and R. H. Thomas. 1998. Arthropod relationships. Chapman and Hall, London.

Friedrich, M., and D. Tautz. 1995. Ribosomal DNA phylogeny of the major extant arthropod classes and the evolution of myriapods. Nature 376:165–167.

Fryer, G. 1996. Reflections on arthropod evolution. Biol. J. Linn. Soc. 58:1–55.

Fryer, G. 1998. A defence of arthropod polyphyly. Pp. 23–33 *in* Arthropod relationships (R. A. Fortey and R. H. Thomas, eds.). Chapman and Hall, London.

Giribet, G., S. Carranza, J. Baguñà, M. Riutort, and C. Ribera. 1996. First molecular evidence for the existence of a Tardigrada + Arthropoda clade. Mol. Biol. Evol. 13:76–84.

Giribet, G., D. L. Distel, M. Polz, W. Sterrer, and W. C. Wheeler. 2000. Triploblastic relationships with emphasis on the acoelomates and the position of Gnathostomulida, Cycliophora, Plathelminthes, and Chaetognatha: a combined approach of 18S rDNA sequences and morphology. Syst. Biol. 49:539–562.

Giribet, G., G. D. Edgecombe, and W. C. Wheeler. 2001. Arthropod phylogeny based on eight molecular loci and morphology. Nature 413:157–161.

Giribet, G., and C. Ribera. 1998. The position of arthropods in the animal kingdom: a search for a reliable outgroup for internal arthropod phylogeny. Mol. Phylogenet. Evol. 9:481–488.

Gladstein, D. S., and W. C. Wheeler. 1997–2002. POY: the optimization of alignment characters, ver. 2.7 [program and documentation]. New York. Available: ftp.amnh.org/pub/molecular. Last accessed 15 December 2003.

Goloboff, P. A. 1998. NONA, ver. 2.0 [program and documentation]. Available: www.cladistics.com. Last accessed 15 December 2003.

Grimaldi, D. 2001. Insect evolutionary history from Handlirsch to Hennig, and beyond. J. Paleontol. 75:1152–1160.

Gupta, A. P. 1979. Arthropod phylogeny. Van Nostrand Reinhold, New York.

Haeckel, E. 1866. Generelle Morphologie der Organismen. Georg Reimer, Berlin.

Halanych, K. M., J. D. Bacheller, A. M. Aguinaldo, S. M. Liva, D. M. Hillis, and J. A. Lake. 1995. Evidence from 18S ribosomal DNA that the lophophorates are protostome animals. Science 267:1641–1643.

Hessler, R. R., and W. A. Newman. 1975. A trilobitomorph origin for the Crustacea. Fossils and Strata 4:437–459.

Hou, X.-G., J. Bergström, and P. Ahlberg. 1995. *Anomalocaris* and other large animals in the Lower cambrian Chengjiang fauna of southwest China. Geologisk. Föreningens Förhandl. 117:163–183.

Hwang, U. W., M. Friedrich, D. Tautz, C. J. Park, and W. Kim. 2001. Mitochondrial protein phylogeny joins myriapods with chelicerates. Nature 413:154–157.

Klass, K.-D., and N. P. Kristensen. 2001. The ground plan and affinities of hexapods: recent progress and open problems. Ann. Soc. Entomol. Fr. (n.s.) 37:265–298.

Kraus, O. 1998. Phylogenetic relationships between higher taxa of tracheate arthropods. Pp. 295–303 *in* Arthropod relationships (R. A. Fortey and R. H. Thomas, eds.). Systematics Association spec. ser. vol. 55. Chapman and Hall, London.

Kraus, O. 2001. "Myriapoda" and the ancestry of the Hexapoda. Ann. Soc. Entomol. Fr. (n.s.) 37:105–127.

Kraus, O., and M. Kraus. 1994. Phylogenetic system of the Tracheata (Mandibulata): on "Myriapoda": Insecta interrelationships, phylogenetic age and primary ecological niches. Verh. Naturwiss. Ver. Hamburg 34:5–31.

Kraus, O., and M. Kraus. 1996. On myriapod/insect interrelationships. Mem. Mus. Natl. Hist. Nat. 169:283–290.

Kukalová-Peck, J. 1992. The "Uniramia" do not exist: the ground plan of the Pterygota as revealed by Permian Diaphanopterodea from Russia (Insecta: Paleodictyopteroidea). Can. J. Zool. 70:236–255.

Kukalová-Peck, J. 1998. Arthropod phylogeny and 'basal' morphological structures. Pp. 249–268 *in* Arthropod relationships (R. A. Fortey and R. H. Thomas, eds.). Chapman and Hall, London.

Manton, S. M. 1964. Mandibular mechanisms and the evolution of Arthropods. Philos. Trans. R. Soc. Lond. B 247:1–183.

Manton, S. M. 1973. Arthropod phylogeny—a modern synthesis. J. Zool. Lond. 171:11–130.

Manton, S. M. 1977. The Arthropoda: habits, functional morphology, and evolution. Clarendon Press, Oxford.

Manton, S. M. 1979. Functional morphology and the evolution of the hexapod classes. Pp. 387–466 *in* Arthropod phylogeny (A. P. Gupta, ed.). Van Nostrand Reinhold, New York.

Martin, J. W., and G. E. Davis. 2001. An updated classification of the Recent Crustacea. Nat. Hist. Mus. LA County Contrib. Sci. 39:1–124.

Melic, A., J. J. de Haro M. Méndez, and I. Ribera (eds.). 1999. Evolución y filogenia de Arthropoda [monograph]. Bol. Soc. Entomol. Aragonesa no. 26.

Mittmann, B., and G. Scholtz. 2001. Distal-less expression in embryos of *Limulus polyphemus* (Chelicerata, Xiphosura) and *Lepisma saccharina* (Insecta, Zygentoma) suggests a role in the development of mechanoreceptors, chemoreceptors, and the CNS. Dev. Gen. Evol. 211:232–243.

Moura, G., and M. Christoffersen. 1996. The system of the mandibulate arthropods; Tracheata and Remipedia as sister groups, "Crustacea" non-monophyletic. J. Comp. Biol. 1:95–113.

Müller, K. J. 1979. Phosphatocopine ostracodes with preserved appendages from the Upper Cambrian of Sweden. Lethaia 12:1–27.

Müller, K. J., and D. Walossek. 1986. *Martinssonia elongata* gen. et sp. n.: a crustacean-like euarthropod from the Upper Cambrian of Sweden. Zool. Scr. 15:73–92.

Nixon, K. G. 2002. WINCLADA ver. 1.0. Available: http://cladistics.com. Last accessed 15 December 2003.

Panganiban, G., A. Sebring, L. Nagy, and S. Carroll. 1995. The development of crustacean limbs and the evolution of arthropods. Science 270:1363–1366.

Popadić, A., G. Panganiban, D. Rusch, W. A. Shear, and T. C. Kaufman. 1998. Molecular evidence for the gnathobasic derivation of arthropod mandibles and for the appendicular origin of the labrum and other structures. Dev. Gen. Evol. 208:142–150.

Popadić, A., D. Rusch, M. Peterson, B. T. Rogers, and T. C. Kaufman. 1996. Origin of the arthropod mandible. Nature 380: 395.

Ramsköld, L., and J.-Y. Chen. 1998. Cambrian lobopodians: morphology and phylogeny. Pp. 107–150 in Arthropod fossils and phylogeny (G. D. Edgecombe, ed.). Columbia University Press, New York.

Regier, J. C., and J. W. Shultz. 1997. Molecular phylogeny of the major arthropod groups indicates polyphyly of crustaceans and a new hypothesis for the origin of hexapods. Mol. Biol. Evol. 14:902–913.

Regier, J. C., and J. W. Shultz. 1998. Molecular phylogeny of arthropods and the significance of the Cambrian "explosion" for molecular systematics. Am. Zool. 38:918–928.

Regier, J. C., and J. W. Shultz. 2001. Elongation factor-2: a useful gene for arthropod phylogenetics. Mol. Phylogenet. Evol. 20:136–148.

Schmidt-Rhaesa, A., T. Bartolomaeus, C. Lemburg, U. Ehlers, and J. R. Garey. 1998. The position of the Arthropoda in the phylogenetic system. J. Morphol. 238:263–285.

Scholtz, G., B. Mittmann, and M. Gerberding. 1998. The pattern of Distal-less expression in the mouthparts of crustaceans, myriapods and insects: new evidence for a gnathobasic mandible and the common origin of Mandibulata. Int. J. Dev. Biol. 42:801–810.

Schram, F. R. 1978. Arthropods: a convergent phenomenon. Fieldiana Geol. 39:61–108.

Selden, P. A. 1981. Functional morphology of the prosoma of *Baltoeurypterus tetragnophthalmus* (Fischer) (Chelicerata: Eurypterida). Trans. R. Soc. Edinb. Earth Sci. 72:9–48.

Shear, W. A. 1992. End of the "Uniramia" taxon. Nature 359:477–478.

Shear, W. A., A. J. Jeram, and P. A. Selden. 1998. Centipede legs (Arthropoda, Chilopoda, Scutigeromorpha) from the Silurian and Devonian of Britain and the Devonian of North America. Am. Mus. Nov. 3231:1–16.

Shultz, J. W., and J. C. Regier. 2000. Phylogenetic analysis of arthropods using two nuclear protein-encoding genes supports a crustacean + hexapod clade. Proc. R. Soc. Lond. B 267:1011–1019.

Siebold, C. T. W. V., and H. Stannius. 1848. Lehrbuch der vergliechenden Anatomie der Wirbellosen Tiere. Veit, Berlin.

Snodgrass, R. E. 1938. Evolution of the Annelida, Onychophora and Arthropoda. Smithson. Misc. Coll. 97:1–159.

Snodgrass, R. E. 1950. Comparative studies on the jaws of mandibulate arthropods. Smithson. Misc. Coll. 116:1–85.

Snodgrass, R. E. 1951. Comparative studies on the head of mandibulate arthropods. Comstock Publishing, Ithaca, NY.

Telford, M. J., and R. H. Thomas. 1998. Expression of homeobox genes shows chelicerate arthropods retain their deutocerebral segment. Proc. Natl. Acad. Sci. USA 95:10671–10675.

Tiegs, O. W. 1947. The development and affinities of the Pauropoda, based on a study of Pauropus silvaticus. Q. J. Microsc. Sci. 88:275–336.

Tiegs, O. W., and S. M. Manton. 1958. The evolution of the Arthropoda. Biol. Rev. 33:255–337.

Turbeville, J. M., D. M. Pfeifer, K. G. Field, and R. A. Raff. 1991. The phylogenetic status of arthropods, as inferred from 18S rRNA sequences. Mol. Biol. Evol. 8:669–686.

Wägele, J. W. 1993. Rejection of the "Uniramia" hypothesis and implications of the Mandibulata concept. Zool. Jb. Syst. 120:253–288.

Walossek, D. 1999. On the Cambrian diversity of Crustacea. Pp. 3–27 in Crustaceans and the biodiversity crisis (F. R. Schram and J. C. von Vaupel Klein, eds.). Brill, Leiden.

Walossek, D., and K. J. Muller. 1998. Early arthropod phylogeny in light of the Cambrian "Orsten" fossils. Pp. 107–150 in Arthropod fossils and phylogeny (G. D. Edgecombe, ed.). Columbia University Press, New York.

Waloszek, D., and J. A. Dunlop. 2002. A larval sea spider (Arthropoda: Pycnogonida) from the Upper Cambrian "Orsten" of Sweden, and the phylogenetic position of pycnogonids. Palaeontology 45:421–446.

Weygoldt, P. 1979. Significance of later embryonic stages and head development in arthropod phylogeny. Pp. 107–136 in Arthropod phylogeny (A. P. Gupta, ed.). Van Nostrand, New York.

Weygoldt, P. 1986. Arthropod interrelationships—the phylogenetic–systematic approach. Z. Zool. Syst. Evol. 24:19–35.

Whalley, P. E. S., and E. A. Jarzembowski. 1981. A new assessment of Rhyniella, the earliest known insect, from the Devonian of Rhynie, Scotland. Nature 291:317.

Wheeler, W. C. 1995. Sequence alignment, parameter sensitivity, and the phylogenetic analysis of molecular data. Syst. Biol. 44:321–331.

Wheeler, W. C. 1996. Optimization alignment: the end of multiple sequence alignment in phylogenetics? Cladistics 12:1–9.

Wheeler, W. C. 1998a. Molecular systematics and arthropods. Pp. 9–32 in Arthropod fossils and phylogeny (G. D. Edgecombe, ed.). Columbia University Press, New York.

Wheeler, W. C. 1998b. Sampling, groundplans, total evidence and the systematics of arthropods. Pp. 87–96 in Arthropod relationships (R. A. Fortey and R. H. Thomas, eds.). Chapman and Hall, London.

Wheeler, W. C. 2003. Supporting data. Available: ftp.amnh.org/pub/molecular/data/tol-appendices.doc.

Wheeler, W. C., P. Cartwright, and C. Y. Hayashi. 1993. Arthropod phylogeny: a combined approach. Cladistics 9:1–39.

Whittington, H. B. 1971. Redescription of Marrella splendens (Trilobitoidea) from the Burgess Shale, Middle Cambrian, British Columbia. Bull. Geol. Soc. Can. 209:1–24.

Whittington, H. B. 1975. Trilobites with appendages from the Middle Cambrian Burgess Shale, British Columbia. Fossils Strata 4:97–136.

Wills, M. A., D. E. G. Briggs, R. A. Fortey, and M. Wilkinson. 1995. The significance of fossils in understanding arthropod evolution. Verh. Deutsch. Zool. Ges. 88:203–215.

Wills, M. A., D. E. G. Briggs, R. A. Fortey, M. Wilkinson, and P. H. A. Sneath. 1998. An arthropod phylogeny based on fossil and recent taxa. Pp. 33–105 *in* Arthropod fossils and phylogeny (G. D. Edgecombe, ed.). Columbia University Press, New York.

Wilson, H. M., and W. A. Shear. 2000. Microdecemplicida, a new order of minute arthropleurideans (Arthropoda;

Myriapoda) from the Devonian of New York State, U.S.A. Trans. R. Soc. Edinburgh Earth Sci. 90:351–375.

Zrzavý, J., V. Hypsa, and M. Vlaskova. 1998. Arthropod phylogeny: taxonomic congruence, total evidence and conditional combination approaches to morphological and molecular data sets. Pp. 97–107 *in* Arthropod relationships (R. A. Fortey and R. H. Thomas, eds.). Chapman and Hall, London.

Jonathan A. Coddington

Gonzalo Giribet

Mark S. Harvey

Lorenzo Prendini

David E. Walter

Arachnida

Although the earliest arachnids were apparently marine, arachnid diversity has been dominated by terrestrial forms from at least the Devonian. Even though arachnid fossils are scarce (perhaps only 100 pre-Cenozoic taxa), representatives of all major arachnid clades are known or cladistically implied from the Devonian or earlier, suggesting very early origins (Selden and Dunlop 1998). The more recent great radiation of insects, in contrast, seems to be Permian (Kukalová-Peck 1991, Labandeira 1999). Taxonomically, arachnids today are composed of approximately 640 families, 9000 genera, and 93,000 described species (table 18.1), but untold hundreds of thousands of new mites and spiders, and several thousand species in the remaining orders, are still undescribed. Arachnida include 11 classically recognized recent clades, ranked as "orders," although some acarologists regard Acari as a subclass with three superorders. Acari (ticks and mites) are by far the most diverse, with Araneae (spiders) second, and the remaining orders much less diverse. Discounting secondarily freshwater and marine mites, and a few semiaquatic spiders and one palpigrade, all extant arachnid taxa are terrestrial. Arachnids evidently arose in the marine habitat (Dunlop and Selden 1998, Selden and Dunlop 1998, Dunlop and Webster 1999), invaded land independently of other terrestrial arthropod groups such as myriapods, crustaceans, and hexapods (Labandeira 1999), and solved the problems of terrestrialization (skeleton, respiration, nitrogenous waste, locomotion, reproduction, etc.) in different ways.

Arachnids and Chelicerata

The monophyly of extant Euchelicerata—the arachnids and their marine sister group, the horseshoe crabs or merostomes—is consistently indicated by both morphology and molecular data (Snodgrass 1938, Wheeler 1998, Zrzavý et al. 1998, Giribet and Ribera 2000, Giribet et al. 2001, Shultz 2001). However, their relationship to the "sea spiders" (Pycnogonida), an enigmatic and morphologically highly specialized group of marine predators, remains controversial. Pycnogonids are variously seen as sister to euchelicerates (Weygoldt and Paulus 1979, Weygoldt 1998, Giribet and Ribera 2000, Shultz and Regier 2000, Regier and Shultz 2001, Waloszek and Dunlop 2002) or as sister to euchelicerates and all remaining arthropods (Zrzavý et al. 1998, Giribet et al. 2001).

Phylogeny of Arachnida

Arachnid monophyly is supported by at least 11 synapomorphies, among which extraintestinal digestion (although some mites and all members of Opiliones are particulate feeders), slit sense sensilla (absent in palpigrades), a single medial genital opening, and an anteroventrally directed mouth are particularly convincing (Weygoldt and Paulus 1979, Shultz 1990, 2001). If fossils are considered, arachnid monophyly is less certain mainly because of the character conflict

Table 18.1

Arachnid Diversity at the Family, Genus, and Species (Described and Estimated) Levels.

	Families	Genera	Species Described	Species Estimated
Arachnida	650	9500	100,000	~1 million
Acari	~430	~3300–4000	~50,000	0.5–1 million
Araneae	109	3471	37,596	76,000–170,000
Opiliones	43	1500	5000	7500–10,000
Pseudoscorpiones	24	425	3261	3500–5000
Scorpiones	17	163	1340	4,000
Solifugae	12	141	1084	1,115
Amblypygi	5	17	142	?
Schizomida	2	39	237	?
Palpigradi	2	6	78	100
Uropygi	1	16	101	?
Ricinulei	1	3	55	85

From Adis and Harvey (2000), Harvey (2003), Platnick (2002), Fet et al. (2000).

created by marine scorpions and eurypterids. Paleontologists consider some fossil scorpions to have been marine (Jeram 1998, Dunlop 1998, Dunlop and Webster 1999, Dunlop and Selden 1998), which, if true, implies either that terrestrial scorpions invaded land independently, or that they returned to the seas secondarily. If the former, the similar arachnid innovations for terrestrial life may be convergent rather than homologous (Jeram 1998, Dunlop and Selden 1998, Dunlop and Webster 1999). Some paleontologists have argued that scorpions are derived merostomes (Dunlop 1999, Dunlop and Selden 1998, Jeram 1998, Dunlop and Braddy 2001), but the paucity of informative characters and the poor or incomplete preservation of the (very) few fossils that exist make conclusions ambiguous and tentative. Paleontologists now recognize three extinct arachnid orders: the clearly tetrapulmonate Trigonotarbida (50 species, including Anthracomarta; Dunlop 1996b), Haptopoda (one species), and Phalangiotarbida (26 species), the latter two orders of uncertain affinities (Selden and Dunlop 1998, Dunlop 1996b, 1999). The paleontological arguments tend to emphasize a few characters (e.g., absence of respiratory structures on the genital somite and subdivision of the abdomen into a proximal broader section and a distal tail) while discounting contrary evidence, especially that not preserved in fossils. Cladistic analyses based on morphological data for extant taxa place scorpions deep inside the recent arachnid clade, possibly related to Opiliones, pseudoscorpions, and solifuges (Shultz 1990, 2000, Wheeler and Hayashi 1998, Giribet et al. 2002), but this clade becomes ambiguously resolved when fossil scorpions and eurypterids are coded, possibly because of the large amount of conflicting character states, because of the aquatic habitat and missing data imposed by the fossils (Giribet et al. 2002). The extinct eurypterids are also chelicerates and are apparently closer to arachnids than to

xiphosurans (Weygoldt and Paulus 1979). Molecular data sometimes place scorpions as true arachnids (Wheeler et al. 1993, Giribet et al. 2001, 2002) but can nest horseshoe crabs within "true" arachnids as well (Wheeler 1998, Wheeler and Hayashi 1998, Edgecombe et al. 2000, Giribet et al. 2002).

The phylogeny of Arachnida itself is contentious, but not as contentious as a perusal of the recent literature might suggest. Specialists may disagree on analytical methodology and interpretation of fossil morphology but largely agree that more data are needed before incongruence should be taken seriously. Classical morphological analysis more or less strongly suggests various clades: Acaromorpha (= ricinuleids–mites), Haplocnemata (= pseudoscorpions–solifuges), Camarostomata (= whip scorpions–schizomids), and Tetrapulmonata (four-lunged arachnids: Araneae, Uropygi, Schizomida, Amblypygi). Besides the controversy over scorpions mentioned above, the positions of Palpigradi, Opiliones, Ricinulei, and Acari are unsettled (Weygoldt and Paulus 1979, Weygoldt 1998, Shultz 1990, 1998, Wheeler et al. 1998, Giribet et al. 2002). Weygoldt and Paulus's early analysis was the first explicit phylogenetic treatment of arachnid relationships, selecting characters that they considered to be of phylogenetic importance while dismissing contradictory evidence as convergence or secondary loss without regard to parsimony. Later authors analyzed morphology and/or molecular evidence cladistically (or using other numerical analytical methods). Parsimony analysis of morphological data from extant groups by different researchers generally agrees with the topology presented in figure 18.1. However, most of the morphological phylogenetic analyses of Arachnida published so far are based on groundplan codings for each order instead of using multiple representatives of each order showing the particular combinations of character states in those terminals. This alternative way of coding terminals has been recently discussed by Prendini (2001a), and it is

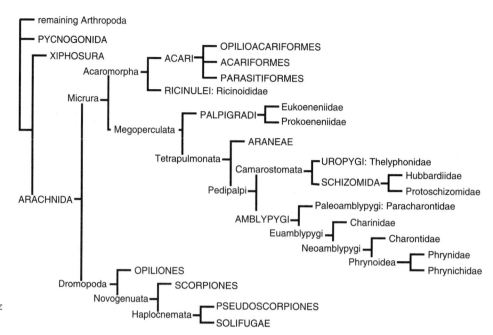

Figure 18.1. Phylogeny of arachnid orders based on the morphological analysis of Shultz (1990).

clearly superior at least in the sense that it allows testing for monophyly of the arachnid orders. Such an exemplar coding has been recently attempted (although with some groundplan codings remaining) in the context of arachnid phylogeny by Giribet et al. (2002).

Recent analyses based on molecular data neither confirm much of the tree based on morphology nor agree on an alternative. Two nuclear loci, 18S and 28S ribosomal RNA are usually employed at the interordinal level (Wheeler and Hayashi 1998, Giribet and Ribera 2000, Giribet et al. 2001, 2002), on the grounds that rates of change in these loci seem appropriate for reconstructing divergences this old. Elongation factor-1α (EF-1α), EF-2, and RNA polymerase II have also been studied at the level of arthropod relationships (Regier and Shultz 1997, 1998, 2001, Shultz and Regier 2000), but few data are available for the interordinal chelicerate relationships. The Uropygi–Schizomida doublet is always corroborated, but the molecular data either deny Acari–Ricinulei (Wheeler and Hayashi 1998, Giribet et al. 2002) or include them in a trichotomy with sea spiders (Wheeler 1998). The monophyly of Tetrapulmonata is strongly supported by morphology, contradicted by some molecular-only analyses (Wheeler and Hayashi 1998, Giribet et al. 2001) and confirmed by others (Giribet et al. 2002). But even the latter found a novel internal topology for Tetrapulmonata (Amblypygi (Araneae (Uropygi, Schizomida))). If viewed as an unrooted network, its spider subclade was correct, but morphology clearly roots the subclade differently (see below). Wheeler and Hayashi (1998) did recover Opiliones–Acari (but excluding Ricinulei). However, this clade was sister to horseshoe crabs, requiring another hypothesis of secondary marine invasion.

In general, the molecular results to date tend to agree with morphology on fairly low-level relationships (monophyly of

harvestmen, haplocnemates, camarostomes, scorpions, spiders, etc.) but to disagree with some morphologically based deeper nodes. Besides nesting exclusively marine groups inside terrestrial arachnids, examples include scorpions as sister to Camarostomata, Acari falling outside a group including mollusks, myriapods, and chelicerates (Wheeler and Hayashi 1998), scorpions as sister to spiders (Giribet et al. 2002), a diphyletic Acari (Giribet et al. 2002; although monophyly of Acari is, of course, not universally agreed upon even among acarologists), amblypygids and pseudoscorpions as sister to the remaining chelicerates, palpigrades nested within spiders (Wheeler 1998), scorpions as sister to ricinuleids, or spiders as sister to uropygids exclusive of schizomids (Giribet and Ribera 2000). The lack of consistency in molecular results at the ordinal level from one study to the next casts doubt on the robustness and accuracy of the molecular data gathered to date. On the other hand, molecular data have tested the monophyly of arachnid orders more strictly than has morphology by including multiple exemplars within each order. Furthermore, very few molecular analyses specifically address arachnid interrelationships, and the same loci (18S and 28S rRNA) have been used consistently. Studies of metazoan or arthropod phylogeny tend to include only a few chelicerates, and the topological incongruities seen are probably due at least in part to sparse taxon sampling.

When the currently available molecular data are combined with morphology (Wheeler 1998, Wheeler and Hayashi 1998, Giribet et al. 2001, 2002), the latter tend to dominate at the deepest nodes. The ordinal topology of the combined analysis by Wheeler and Hayashi (1998: fig. 7) agrees almost perfectly with the morphology based analysis of Shultz (1990) and differs strongly from the molecules-only tree. This is not as true of the largest analysis to date by Giribet et al. (2002).

However, given the conflict in molecules alone, it seems wiser to recommend the morphological cladogram of Shultz (fig. 18.1) as a working hypothesis for arachnid phylogeny.

Although this review focuses more on the controversies than the consensus, some nodes in figure 18.1 are well supported. The tetrapulmonates share the subchelate condition of the mouthparts, the unique 9 + 3 axoneme sperm morphology, the narrow or petiolar connection between cephalothorax and abdomen, the reduction to four prosomal endosternal components, and the complex coxo-trochanteral joint. According to a recent anatomical study of the musculoskeletal system, Pedipalpi share 31 morphological synapomorphies (Shultz 1999), although many of these characters are not independent, and the extent of homoplasy in other arachnids is unclear. Camarostomata is also strongly supported by at least six synapomorphies. Haplocnemata (= Pseudoscorpiones-Solifugae) also has substantial morphological support. Dromopoda (= Scorpiones-Pseudoscorpiones-Solifugae-Opiliones) and Micrura (= Tetrapulmonata-Ricinulei-Acari) have been considered the weakest nodes morphologically (Weygoldt 1998).

Mites and Ticks (Acari or Acarina)

Mites are the "go anywhere, do anything" arachnids (Walter and Proctor 1999). They occur on every continent, including Antarctica, where they dominate the endemic terrestrial fauna (Pugh 1993). On land, they form a minute, scurrying plankton that coats the vegetation, from the canopies of the tallest rainforests down into the soil, at least as deep as roots can penetrate (Walter 1996, Walter and Behan-Pelletier 1999). Every bird, mammal, reptile, and social insect species plays host to symbiotic mites, as do many amphibians, slugs, spiders, scorpions, opilionids, myriapods, and nonsocial insects. Animal- and plant-associated mites are commonly commensals that scavenge a living on their hosts'

surfaces, and sometimes provide beneficial services, but all too often are parasites capable of damaging or killing their hosts. Although originating on land, mites have reinvaded and radiated into both freshwater (around six invasions, >5000 described species) and marine systems (around three invasions, hundreds of known species) from the intertidal to the deepest marine trenches (Walter and Proctor 1999).

More than 50,000 species of the "subclass" Acari have been described and distributed across three superorders, six orders, more than two dozen suborders and "cohorts" (~infrasuborders), >400 families, and 3000–4000 genera (see Table 18.2). Roughly 90 fossil species have been described (Selden 1993a). Like the artificial assemblage that we call reptiles, mites are easily recognized as such, but the monophyly of Acari is open to question. Mites have long been studied in isolation from other arachnids, and characters that once appeared to unite the Acari are now known to be more general. For example, the hexapod larva and the headlike capitulum (gnathosoma) were once thought unique to mites, but both are also found in ricinuleids (Lindquist 1984). Other supposedly unique characters, such as the ventral fusion of the palpal coxae, occur in many arachnids (e.g., ricinuleids, schizomids, pseudoscorpions) and may even have evolved twice within mites (Walter and Proctor 1999). Modern phylogenetic methods, especially using molecules, have only recently been applied to Acari, but most of these studies have been restricted to economically important parasites (Navajas and Fenton 2000).

Although Acari are not clearly monophyletic (van der Hammen 1989), each of the three acarine superorders probably is (Grandjean 1936). Opilioacarans are fairly large (2–3 mm) tracheate mites, superficially resembling small opilionids, which retain a number of plesiomorphic characters. Like early derivative acariform mites and most opilionids, opilioacarans ingest solid food, using large, three-segmented chelicerae to grasp small arthropods or fungi, and

---

Table 18.2

Systematic Synopsis and Distribution of Major Mite Lineages.

Class Arachnida, Acari (Acarina): mites and ticks
    Superorder Opilioacariformes: Order Opilioacarida—1 family, 9 genera, ~20 species
    Superorder Acariformes: mitelike mites
  Order Sarcoptiformes: Endeostigmata, "Oribatida," Astigmata——~230 families, >15,000 described species, including the paraphyletic oribatid mites (~1100 genera in >150 families); stored product mites; house dust, feather, and fur mites; and scabies and their relatives
  Order Trombidiformes: Sphaerolichida, Prostigmata——125 families, >22,000 described species, including spider mites and their relatives (Tetranychoidea); earth mites and their relatives (Eupodoidea); gall and rust mites (Eriophyoidea); soil predators and fungivores; hair, skin, and follicle mites (Cheyletoidea); straw itch mites (Pyemotidae); chiggers, velvet mites, water mites, and their relatives (Parasitengona)
Superorder Parasitiformes: ticks and ticklike mites
  Order Ixodida (Metastigmata)—ticks—3 families, <900 described species
  Order Holothyrida: holothyrans—3 families, <35 described species
  Order Mesostigmata (Gamasida): Monogynaspida + Trigynaspida *sensu lato* (often treated as 3–4 separate suborders)——~70 families, <12,000 described species, including poultry mites, nasal mites, bird mites, and rat mites (Dermanyssoidea); major soil predators; biocontrol agents (Phytoseiidae); tortoise mites (Uropodoidea)

serrated hypertrophied palpal coxal setae (rutella) on either side of the buccal opening to saw the food into bite-sized chunks that can be swallowed. Fossil opilioacarans are unknown, although Dunlop (1995) speculates that they may be related to the curious Carboniferous Phalangiotarbida. Opilioacariformes may be a sister group to Parasitiformes, but convincing synapomorphies have yet to be demonstrated. A sister-group relationship of Opilioacariformes and Parasitiformes has been recently proposed based on molecular data (Giribet et al. 2002).

Acariformes are supported by several synapomorphies unique within Arachnida, including prodorsal trichobothria, the loss of all primary respiratory structures or remnants (e.g., the ventral sacs in Palpigradi), the fusion of the tritosternum to the palpal coxal endites to form a subcapitulum, and genital papillae (osmoregulatory structures). Acariformes share the nonfeeding, hexapod prelarval stage, the rutella, and particulate feeding with Opilioacariformes. Particulate feeding also occurs in Opiliones (see above) and in horseshoe crabs.

Acariformes consist of two orders, Sarcoptiformes and Trombidiformes, both corroborated by a morphological cladistic analysis (OConnor 1984) that established the relationship between the suborders Sphaerolichida (two families previously attributed to the basal suborder Endeostigmata) and Prostigmata. Although no comprehensive analysis of Prostigmata has been published, five cohorts (fig. 18.2) are well supported by morphological characters. Of these, only Heterostigmata have received a thorough morphological cladistic analysis (Lindquist 1986), but parts of Parasitengona

are currently under molecular and morphological review (e.g., Soeller et al. 2001).

Prostigmatans display anterior dorsal stigmatal openings and feed only on fluids. Almost half of all known mite species belong to Prostigmata, including major radiations of mites parasitic on vertebrates, invertebrates, and plants (Table 18.2). All of the major acarine plant parasites belong here, including the smallest known terrestrial animals, gall mites (Eupodina: Eriophyoidea) as small as 0.07 mm in length as adults (Walter and Proctor 1999). In contrast, Parasitengona contains more than 7000 described species of terrestrial and aquatic mites, including some of the largest known (16 mm long).

Some traditional subdivisions of the Sarcoptiformes are obviously paraphyletic, but few cladistic analyses, even of a preliminary nature, have been published. Astigmata, often given subordinal rank, is monophyletic (Norton 1998) but derived from within the traditional suborder Oribatida (also Oribatei, Cryptostigmata), thereby rendering Oribatida paraphyletic. Oribatida consist of the beetle mites that form a dominant part of the soil fauna. Sarcoptiformans were among the earliest terrestrial animals and probably invaded land directly from the ocean by way of interstices in moist beach sand as minute animals that exchanged gases across their cuticles (Walter and Proctor 1999). By the Early Devonian (380–400 million years ago), sarcoptiformans were diverse members of the soil fauna, and 11 species are known from the Gilboa shales and Rhynie Chert (Norton et al. 1988, Kethley et al. 1989). Based on extensive fecal remains, it appears that sarcoptiform mites were major components of

Figure 18.2. Phylogeny of Acari.

the detritivore system in Palaeozoic coal swamps (Labandeira et al. 1997). A later radiation in association with animals (Astigmata: Psoroptida) has produced a dazzling diversity of nest, feather, fur and skin inhabitants and a source of some interesting host–symbiont analyses (e.g., Klompen 1992, Dabert et al. 2001).

Parasitiformes are supported by a number of unique character states, including a plate above or behind leg IV bearing a stigmatal opening and peritreme, a biflagellate tritosternum, a sclerotized ring formed by fusion of the palps around the chelicerae (possibly representing a fusion of a ricinuleid-like cucullus to the palpal coxae), horn-shaped corniculi (possibly homologous with the rutella) that support the salivary stylets, a recessed sensory array on leg I (called Haller's organ in ticks), and by the use of the chelicerae to transfer sperm. Additional characters supporting Parasitiformes include suppression of the prelarval stage and widespread fluid feeding (the general condition among arachnids).

The internal relationships of Parasitiformes are the best studied of any of the three acarine superorders, but this is faint praise indeed. Relationships among ticks (Ixodida) and between ticks and other suborders are the best resolved (e.g., Klompen et al. 1996). However, some exemplary morphological and molecular analyses of parts of Mesostigmata are starting to appear (e.g., Naskrecki and Colwell 1998, Cruickshank and Thomas 1999). The monophyly of ticks, perhaps the most familiar of all mites because of their large size and bloodthirsty habits, is supported by several modifications of the chelicerae and hypostome for blood-feeding. Molecular evidence suggests that holothyrans, large (2–7 mm long), reddish to purplish armored mites, are close relatives of ticks. Holothyrans are rare, known only from Gondwanan continents and Indo-Pacific Islands, where they scavenge on fluids from dead arthropods (Walter and Proctor 1998). A uniquely formed all-encompassing dorsal shield and lateral peritrematal plate support the monophyly of Holothyrida.

The group consisting of (Holothyrida + Ixodida) is the sister to Mesostigmata. Characters supporting the monophyly of the latter are mostly developmental, for example, suppression of the tritonymphal stage and of the genital opening until the adult, and the appearance of sclerotized plates on the opisthosoma in nymphs. Mesostigmata can be split into two suborders, each with five cohorts based on variation in the female genital shield. In Monogynaspida (Gamasina), the plesiomorphic condition of four genital shields (found in Holothyrida and Cercomegistina) is reduced to a single genital shield by fusion of the laterals (latigynials) to the median genital shield and the loss of the anterior genital shield. Trigynaspida *sensu lato* shows a general trend toward fusion of the latigynials with other shields, and is only weakly supported. Trigynaspines often have restricted distributions but are prominent members of tropical forest faunas, as are members of Uropodina. A group comprising Uropodina, Sejina, and Microgyniina is supported by the development of a heteromorphic deutonymph (i.e., a differently formed

phoretic stage) that disperses on insects via an anal attachment organ.

Within Monogynaspida, the cohort Dermanyssina is clearly separated by the presence of a secondary insemination and sperm-storage system in the female and an inseminatory sperm finger on the male chelicera. Dermanyssines occur on all continents, including Antarctica. About half of the described species are free-living predators in soil litter, rotting wood, compost, herbivore dung, carrion, nests, house dust, or similar detritus-based systems. These predators are usually abundant and voracious enough to regulate the populations of other small invertebrates and are often used in biocontrol. A few mesostigmatans have switched from external digestion of prey to ingesting fungal spores and hyphae. Others feed on pollen, nectar, and other plant fluids. Pollen feeding is common in the Phytoseiidae, a family that has successfully colonized the leaf-surface habitat and accounts for about 15% of described species of Mesostigmata. Many Ascidae (Naskrecki and Colwell 1998) and Ameroseiidae have become venereal diseases of plants, that is, pollen- and nectar-feeding flower mites vectored by insect or bird pollinators. The Dermanyssoidea contain several massive radiations of vertebrate and invertebrate parasites, including such well-known pests as the bird and rat mites and the varroa mite of bees.

Ricinuleids (Ricinulei)

Ricinulei are an enigmatic group of curious, slow-moving arachnids that possess a series of unique modifications, including a hinged plate, the cucullus, at the front of the prosoma, which acts as a hood covering the mouthparts; a locking mechanism between the prosoma and the opisthosoma (shared with the fossil trigonotarbids) that can be uncoupled during mating and egg-laying; and a highly modified male third leg that is used for sperm transfer during mating. This leg structure is analogous to the modified pedipalp of male spiders, and provides a series of species-specific character states helpful in delimiting taxa.

Ricinulei are probably the sister group of mites (Lindquist 1984, Weygoldt and Paulus 1979, Shultz 1990, Wheeler and Hayashi 1998, Giribet et al. 2002). Savory (1977) proposed a relationship to Opiliones, even suggesting paraphyly of Opiliones by including Ricinulei, but that hypothesis remains quite dubious. More recently, addition of the extinct order Trigonotarbida as well as molecular data suggested a possible relationship to tetrapulmonates (Dunlop 1996a, Giribet et al. 2002). Internal relationships of extant Ricinulei have been explored by Platnick (1980).

Hansen and Sørensen (1904) provided the first comprehensive taxonomic account of this order, in which they recognized a single family, Cyptostemmatoidae, with eight species grouped in the genera *Cryptostemma* and *Cryptocellus*. The order, as currently defined, contains just a single recent family, Ricinoididae, with three genera (Harvey 2003). *Ricinoides* (10 species) occurs in the rainforests of western and central Africa. *Cryptocellus* (27 species) and *Pseudocellus* (18 species) occur in forest and cave ecosystems of Central America as far north as Texas and as far south as Peru. Selden (1992) proposed a classification for the order that divided it into two suborders, Palaeoricinulei for the two families of Carboniferous ricinuleids (15 species total) and the Neoricinulei for Ricinoididae.

Palpigrades (Palpigradi)

Palpigrades or micro-whip scorpions are one of the most enigmatic arachnid orders, with just 78 species in six genera and two families (Harvey 2003), and an unresolved phylogenetic position because of doubts regarding the many reductional apomorphies these small animals possess. Only one fossil species is known (Selden and Dunlop 1998). Their phylogenetic placement based on molecular data is similarly equivocal (Giribet et al. 2002). Palpigrades bear a long, multi-segmented flagellum, three-segmented chelicerae, sub-segented pedipalpal and pedal tarsi, and a host of other modifications, including lack of slit sensillae, a dorsal hinged joint between the trochanter and femur on the walking legs (Shultz 1989), and a pair of anteromedial sensory organs (Shultz 1990). Palpigrades occur primarily in endogean habitats—soil, litter, under rocks, in caves and other subterranean voids—but the remarkable genus *Leptokoenenia* occurs in littoral deposits of Saudi Arabia and Congo.

Until recently only a single family, Eukoeneniidae, was recognized, but Condé (1996) transferred *Prokoenenia* and *Triadokoenenia* to a separate family, Prokoeneniidae. These two families can be distinguished by the presence (Prokoeneniidae) or absence (Eukoeneniidae) of abdominal ventral sacs on sternites IV–VI. This arrangement has not been tested cladistically, nor has the monophyly of each of the six genera. The genera are disproportionately sized: *Eukoenenia* consists of 60 named species, and the remaining five genera possess a total of just 18 species. Although the differences between families and genera are well understood (Condé 1996), their interrelationships have never been examined cladistically.

Spiders (Araneae)

Spiders currently consist of 110 families, about 3500 genera, and more than 38,000 species (Platnick 2002). Roughly 600 fossil species have been described (Selden 1996, Selden and Dunlop 1998). Strong synapomorphies support the clade: cheliceral venom glands, male pedipalpi modified for sperm transfer, abdominal spinnerets and silk glands, and lack of the trochanter-femur depressor muscle (Coddington and Levi 1991). The advent of the scanning electron microscope in the 1970s rejuvenated spider systematics: micro-

structures on the cuticle (sensory tarsal organs, the kinds and distributions of silk spigots on spinnerets) are now fundamental to phylogenetic research. Roughly 67 quantitative cladistic analyses of spiders have been published to date, covering about 905 genera (about 25% of the known total), on the basis of approximately 3200 morphological characters. Nine of these studies focus on interfamilial relationships (Coddington 1990a, 1990b, Platnick et al. 1991, Goloboff 1993, Griswold 1993, Griswold et al. 1998, 1999, Bosselaers and Jocqué 2002, Silva Davila 2003). Many of the others that focus on single families, however, include multiple outgroups that overlap from one study to another (Coddington 1986a, 1986b, Jocqué 1991, Rodrigo and Jackson 1992, Hormiga 1994, 2000, Davies 1995, 1998, 1999, Harvey 1995, Hormiga et al. 1995, Gray 1995, Ramírez 1995a, 1995b, 1997, Pérez-Miles et al. 1996, Ramírez and Grismado 1997, Scharff and Coddington 1997, Sierwald 1998, Huber 2000, 2001, Platnick 1990, 2000, Davies and Lambkin 2000, 2001, Griswold 2001, Griswold and Ledford 2001, Wang 2002, Schütt 2003). The trend has been to address unknown parts of the spider tree, thus yielding a first-draft, higher level phylogeny for the order, rather than repeating or intensifying lower level analyses. On the one hand, overlap and congruence have been fortuitously sufficient to permit "adding" results together manually; on the other, they are so sparse that many details in figure 18.3 are certain to change with more data and more detailed taxon sampling. Molecular work, at least above the species level, is still almost nonexistent (but see Huber et al. 1993, Hausdorf 1999, Piel and Nutt 1997, Hedin and Maddison 2001). Some molecular results are strongly contradicted by morphology, such as rooting the spider clade among arachnids on an araneomorph rather than a mesothele (Wheeler and Hayashi 1998).

The comparative data for the most inclusive groupings of spiders have been known for more than a century, but the data were not rigorously analyzed from a phylogenetic point of view until the mid-1970s (Platnick and Gertsch 1976). This analysis clearly showed a fundamental division between two suborders: the plesiomorphic mesotheles (one family, Liphistiidae; two genera; about 85 species) and the derived opisthotheles. Although mesotheles show substantial traces of segmentation, for example, in the abdomen and nervous system, the opisthothele abdomen is usually smooth and the ventral ganglia fused. Opisthotheles is composed of two major lineages: the baboon spiders (or tarantulas) and their allies (Mygalomorphae, 15 families, about 300 genera, 2500 species) and the so-called "true" spiders (Araneomorphae, 94 families, 3200 genera, 36,000 species) (Platnick 2003).

Mygalomorphs resemble mesotheles. They tend to be fairly large, often hirsute animals with large, powerful chelicerae that live in burrows and, apparently, rely little on silk for prey capture, at least compared with many araneomorph spiders. Within mygalomorphs, the atypoid tarantulas are probably sister to the remaining lineages (Raven 1985, Goloboff 1993), although some evidence supports the mono-

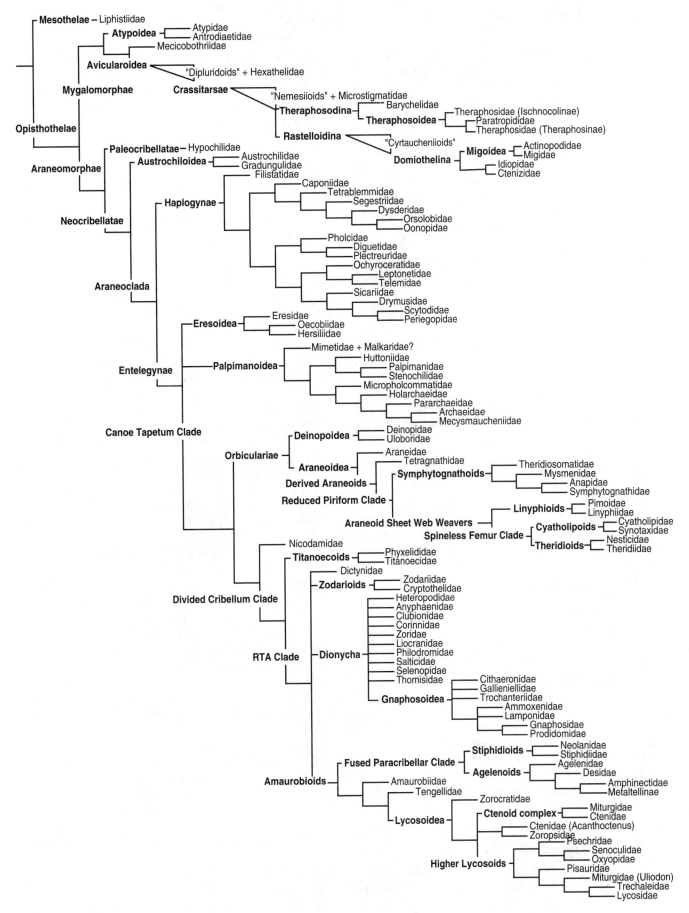

Figure 18.3. Phylogeny of Araneae.

phyly of Mecicobothriidae and the atypoids. The atypoid sister group is Avicularioidea, of which the basal taxon, Dipluridae, seems to be a paraphyletic assemblage. One of the larger problems in mygalomorph taxonomy concerns Nemesiidae, currently 38 genera and 325 species (Goloboff 1993, 1995). The group is conspicuously paraphyletic. The remaining mygalomorph families are relatively derived and more closely related to each other than to the preceding. Two seemingly distinct groups are the theraphosodines [baboon spiders or true "tarantulas" and their allies (Pérez-Miles et al. 1996), typically vagabond] and the rastelloidines (typically trap door spiders). Because of the evident paraphyly of several large mygalomorph "families" (Dipluridae, Nemesiidae, Cyrtaucheniidae), the number of mygalomorph family-level lineages will probably increase dramatically with additional research.

Araneomorphs include more than 90% of known spider species; they are derived in numerous ways and are quite different from mesotheles or mygalomorphs. Although repeatedly lost, a strong synapomorphy of this clade is the fusion and specialization of the anterior median spinnerets into a flat spinning plate (cribellum) with hundreds to thousands of spigots that produce a dry yet extremely adhesive silk (cribellate silk). Many araneomorph lineages independently abandoned the sedentary web-spinning lifestyle to become vagabond hunters, but the plesiomorphic foraging mode seems to be a web equipped with dry adhesive silk (austrochiloids, Filistatidae among the haplogynes, oecobiids and eresids among eresoids, many entelegyne groups). Within Araneomorphae, the relictually distributed Hypochilidae (two genera, 11 species) are sister to the remaining families (Platnick et al. 1991). Some austrochiloid genera have lost webs, and most haplogynes are also vagabonds. These haplogyne taxa tend to live in leaf litter or other soil habitats (Caponiidae, Tetrablemmidae, Orsolobidae, Oonopidae, Telemidae, Leptonetidae, Ochyroceratidae, etc.; Platnick et al. 1991). The haplogyne cellar spiders (Pholcidae) are exceptional for their relatively elaborate, large webs. Some of the most common and ubiquitous commensal spider species are pholcids.

The entelegyne "node" in spiders is supported by several synapomorphies (Griswold et al. 1999). Among other things, the copulatory apparatus fundamentally changed in both males and females. One theory is that the change was driven by cryptic female choice: the tendency of females to choose males on the basis of their effectiveness in genitalic stimulation during copulation (Eberhard 1985). Females evolved a complex antechamber to their gonopore and acquired a second opening of the reproductive system to the exterior coupled with an unusual "flow-through" sperm management system in which deposited sperm are stored in separate chambers for later use in fertilizing eggs. Females also evolved a special sort of silk used only in egg sacs, which is almost universally present among entelegynes although its function

is unknown. Male genitalia became hydraulically rather than muscularly activated and more elaborate; the interaction with the equally complicated female genitalia became more complex. This "hydraulic bulb" of the male genitalia is so flexible during its operation that males have evolved various levers and hooks that seem to serve mainly to stabilize and orient their own genitalia during copulation. One of these, the "retrolateral tibial apophysis" has given its name to a fairly large clade of entelegyne families (the "RTA clade"; Coddington and Levi 1991, Griswold 1993, Sierwald 1998). Non-entelegynes, in contrast, have relatively simple male and female genitalia in which the female anatomy is one or two pairs or an array of blind receptacula, and the male intromittent organ is a smooth and simple hypodermiclike structure operated by tarsal muscles.

Among entelegynes the "eresoid" families seem basal. No clear synapomorphies define this group; in various analyses, eresoids may be paraphyletic (Coddington 1990a, Griswold et al. 1999). Perhaps the hottest current controversy in entelegyne systematics concerns the Palpimanoidea (10 families, 54 genera). Before their relimitation as a monophyletic group (Forster and Platnick 1984), palpimanoid families were dispersed throughout entelegyne classification: mimetids, archaeids, and micropholcommatids in particular were considered to be araneoids. The two classic features defining Palpimanoidea are setae shortened and thickened to function as cheliceral teeth (very rare in spiders) and the concentration of cheliceral glands on a raised mound. However, these two features are homoplasious within palpimanoids, and evidence is building that some palpimanoid taxa are araneoids after all (Schütt 2000).

One of the larger entelegyne lineages is the Orbiculariae. It unites two robustly monophyletic superfamilies (Araneoidea, 12 families, 980 genera; and Deinopoidea, 2 families, 23 genera) mainly but not entirely on the basis of web architecture and morphology associated with web spinning (Coddington 1986b and references therein). Both groups spin orb webs. Ethological research on orb weavers shows that orbs are constructed in fundamentally similar ways, although the deinopoid orb uses the plesiomorphic cribellate silk, whereas araneoids use the derived viscid silk (Griswold et al. 1998). Araneoidea are by far the larger taxon and includes many ecologically dominant web-weaving species. Interestingly, derived araneoids (the "araneoid sheet web weavers," six families, 685 genera) no longer spin orbs (some may not even spin webs) but rather sheets, tangles, and cobwebs (Griswold et al. 1998). There is a strong trend among araneoids to reduce and stylize the spinning apparatus (Hormiga 1994, 2000).

The sister taxon of Orbiculariae remains a mystery, although the most recent research suggests that most other entelegyne lineages are more closely related to each other than any is to the orb weavers (Griswold et al. 1999). Thus, the orbicularian sister group at present seems likely to be a very large, hitherto unrecognized lineage consisting of amauro-bioids (Davies 1995, 1998, 1999, Davies and Lambkin 2000, 2001), "wolf" spiders [Lycosoidea (Griswold 1993)], two-clawed hunters (Dionycha; Platnick 1990, 2000), and other, smaller groups (Jocqué 1991). Many of these lineages are relictual austral groups whose diversity is very poorly understood.

The phylogenetic structure among non-orbicularian entelegynes, therefore, is highly provisional at this point. Because of a long-standing emphasis on symplesiomorphy, many of the classical entelegyne families (most seriously Agelenidae, Amaurobiidae, Clubionidae, Ctenidae, and Pisauridae) were paraphyletic. Dismembering these assemblages into monophyletic units has been difficult because the monophyly of related families is also often doubtful (e.g., Amphinectidae, Corinnidae, Desidae, Liocranidae, Miturgidae, Tengellidae, Stiphidiidae, Titanoecidae). Therefore neither the RTA clade, nor the two-clawed hunting spider families (Dionycha) may be strictly monophyletic, although in each is certainly a large cluster of closely related lineages. Dionychan relationships are quite unknown, although some headway has been made in the vicinity of Gnaphosidae (Platnick 2000). In contrast, Lycosoidea was supposedly based on a clear apomorphy in eye structure, but recent results suggest that this feature evolved more than once or, less likely, has been repeatedly lost (Griswold et al. 1998). The nominal families Liocranidae and Corinnidae are massively polyphyletic (Bosselaers and Jocqué 2002). The nodes surrounding Entelegynae will certainly change in the future.

In sum, phylogenetic understanding of spiders has advanced remarkably since the early 1980s. We are on the cusp of having at least a provisional, quantitatively derived hypothesis at the level of families, but on the other hand, the density and consistency of the data for subsidiary taxa will remain soft for some years to come.

Whip Spiders (Amblypygi)

Whip spiders, also known as tailless whip scorpions, are a conspicuous group of mostly medium to large, dorsoventrally flattened arachnids distributed throughout the humid tropics and subtropics with a few species occurring in the arid regions of southern Africa. Although most species are epigean, several troglobite species are known.

Monophyly of Amblypygi is supported by several features, including the morphology and orientation of the pedipalps, the enormously elongated antenna-like first legs that act as tactile organs, and the presence of a cleaning organ on the palpal tarsus. The order belongs to Pedipalpi as the sister to Camarostomata (Uropygi + Schizomida) (Shultz 1990, 1999, Giribet et al. 2002), although some treatments place them as the sister to Araneae (e.g., Platnick and Gertsch 1976, Weygoldt and Paulus 1979, Wheeler and Hayashi 1998).

Current understanding of the internal phylogeny and classification of Amblypygi is almost entirely the work of Weygoldt (1996, 2000), who recognized five families, placed in two suborders, Paleoamblypygi and Euamblypygi. Paleoamblypygi contain a single West African species, *Paracharon caecus* (Paracharontidae), as well as five Carboniferous species that remain unplaced in a family. Paleoamblypygi differ in various features, including an anteriorly produced carapace and reduced pedipalpal spination. The Euamblypygi consist of the remaining whip spiders, including the circumtropical Charinidae, which contains three genera and 43 species. Charinidae may not be monophyletic (Weygoldt 2000). The remaining three families comprise Neoamblypygi, which is in turn divided into the Charontidae and Phrynoidea; the latter includes the Phrynidae and Phrynichidae. The Charontidae consist of two genera and 11 species from Southeast Asia and Australasia. The Phrynidae contain four genera and 55 species from the Americas, with a single outlying species from Indonesia (Harvey 2002a). The Phrynichidae contain 31 species in seven genera from Africa, Asia, and South America.

Whip Scorpions (Uropygi)

Whip scorpions are large, heavily sclerotized arachnids that have changed little since the Carboniferous. They primarily inhabit tropical rainforests but some, such as the well-known North American *Mastigoproctus giganteus*, occupy arid environments. Like other members of the Pedipalpi, tarsus I is subsegmented and is used as a tactile organ. They possess a number of distinctive features, including palpal chelae with the movable finger supplied with internal musculature (Barrows 1925), a long, multisegmented flagellum, raptorial pedipalps, and a long rectangular carapace. The abdomen bears a pair of glands that discharge at the base of the flagellum and are used to direct a spray of acetic acid (vinegar) at potential predators (Eisner et al. 1961; Haupt et al. 1988). On account of this unusual ability, whip scorpions are known as vinegaroons (or vinegarones) in the southern United States.

Uropygi are consistently placed as sister to Schizomida, and the gross morphology of its members suggests monophyly. Dunlop and Horrocks (1996) suggested that the Carboniferous uropygid *Proschizomus* may represent the sister to Schizomida, rendering Uropygi paraphyletic. The sole family Thelyphonidae is divided into four subfamilies: Hypoctoninae (4 genera, 25 species: Southeast Asia, South America, west Africa), Mastigoproctinae (4 genera, 18 species: Americas, Southeast Asia), Typopeltinae (1 genus, 10 species), and Thelyphoninae (7 genera, 48 species: Southeast Asia and Pacific) (Rowland and Cooke 1973, Harvey 2003). Eight fossil species have been described (Selden and Dunlop 1998, Harvey 2003). Only Typopeltinae and

Thelyphoninae are well supported by apomorphic character states; Hypoctoninae and Mastigoproctinae appear to be solely defined by plesiomorphies (M. Harvey, unpubl. obs.).

Schizomids (Schizomida)

Schizomids are small (<1 cm), weakly sclerotized arachnids that can be recognized by the presence of a short abdominal flagellum that generally in females consists of three or four segments and in males is single segmented. The shape and setation of the male flagellum are species specific (e.g., Rowland and Reddell 1979, Harvey 1992b, Reddell and Cokendolpher 1995), probably reflecting its use during courtship and mating, in which it is gripped in the mouthparts of the female (Sturm 1958).

The order contains two families, the Central American Protoschizomidae and the widely distributed Hubbardiidae. Three fossil species have been described (Selden and Dunlop 1998). Protoschizomidae are represented by two genera and 11 species from Mexico or Texas, many from caves (Rowland and Reddell 1979, Reddell and Cokendolpher 1995). The Hubbardiidae consist of two subfamilies. Megaschizominae are represented by two species of *Megaschizomus* from Mozambique and South Africa. The widespread Hubbardiinae consists of 205 species in 35 genera (Harvey 2003), the vast proportion of which have been named in the last 40 years because of an increased awareness of previously overlooked character systems such as female genitalia. Cokendolpher and Reddell (1992) presented a cladistic analysis of the basal clades of Schizomida but refrained from including individual hubbardiine genera, whose systematics are still in a state of flux.

Harvestmen (Opiliones)

Commonly known as "daddy longlegs," harvestmen, shepherd spiders, or harvest spiders (among other names), the Opiliones were well known to North Temperate farmers and shepherds because of their abundance at harvest time. These are the only nonacarine arachnids known to ingest vegetable matter, but generally they prey on insects, other arachnids, snails, and worms. They can ingest particulate food, unlike most arachnids, which are liquid, external digesters. The order is reasonably well studied, although many of the Southern Hemisphere families are still poorly understood taxonomically.

Opiliones contain 43 families, about 1500 genera, and about 5000 species, but many more species await discovery and description. Most members of Opiliones are small to medium in size (<1 mm to almost 2.5 cm in the European species *Trogulus torosus*) and inhabit moist to wet habitats on all continents except Antarctica. Laniatores include large (>2 cm), colorful, well-armored Opiliones, most diverse in tropical regions of the Southern Hemisphere, but many

laniatorids are also very small. Eupnoi and Dyspnoi are more widely distributed and are especially abundant in the Northern Hemisphere. Members of Cyphophthalmi are distributed worldwide but are among the smallest (down to 1 mm) and most obscure members of the Opiliones.

Opilionids are typical arachnids with two basic body regions, and their junction is not constricted, giving them the appearance of "waistless" spiders. The cephalothorax generally has a pair of median simple eyes surmounting the ocular tubercle. Cyphophthalmi either lack eyes entirely or have a pair of eyes (some stylocellids), possibly lateral eyes. The anterior rim of the cephalothorax bears the large openings of a pair of secretory organs, known as repugnatorial glands. These differ in position and type among different groups within Opiliones, being most obvious in the suborder Cyphophthalmi, whose members take the shape of cones, named ozophores. The cephalothorax bears one pair of chelate three-segmented chelicerae for manipulating the food particles, one pair of pedipalps of either tactile or prehensile function, and four pairs of walking legs. The legs can be enormously long (>15 cm) in some Eupnoi and Laniatores species. Laniatorid palps are usually large and equipped with parallel rows of ventral spines that act as a grasping organ. The second pair of walking legs is sometimes modified for a tactile or sensory function.

The abdomen is clearly segmented in most species, although some segments may appear fused to different degrees. One pair of trachea for respiration opens ventrally on the sternite of the first abdominal segment. The genital aperture and its associated structures (operculum) open on the same segment. The anal region is very often modified; certain Cyphophthalmi males have anal glands, secondary sexual characters that are probably secretory. Females may have a long ovipositor with sensory organs on the tip that is used to check the soil quality for egg deposition. Males have a muscular or hydraulically operated penis, or copulatory organ. Some mites have vaguely similar structures, but otherwise ovipositors and penises are unique to Opiliones. Fertilization is thus internal and direct.

The monophyly of Opiliones is strongly supported by the presence of five unambiguous synapomorphies: (1) the presence of repugnatorial glands, (2) the special vertical bicondylar joint between the trochanter and femur of the walking legs, (3) the paired tracheal stigmata on the genital segment, (4) the male penis, and (5) the female ovipositor (Shultz 1990, Giribet et al. 2002). Opilionid taxonomy supposes a basic division between Cyphophthalmi (no common name, six families) and the remaining harvestmen ("Phalangida"), consisting of Eupnoi (six families), Dyspnoi (seven families), and Laniatores (24 families). Eupnoi and Dyspnoi have been traditionally grouped in Palpatores (fig. 18.4).

Cyphophthalmids are small (1–6 mm), hard-bodied, soil animals that superficially resemble mites. Six families are recognized (Shear 1980, 1993, Giribet 2000), although some

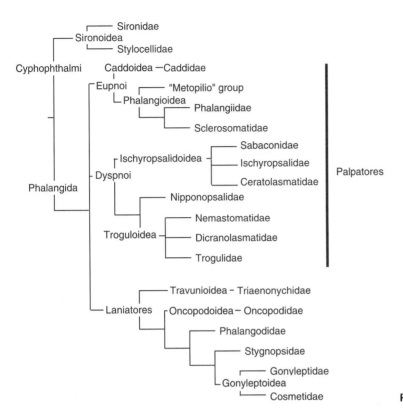

Figure 18.4. Phylogeny of Opiliones.

do not withstand cladistic tests (Giribet and Boyer 2002). "Palpatores" are diverse and heterogeneous; their monophyly is disputed. The component Eupnoi and Dyspnoi, however, are well-supported monophyletic clades, each with two superfamilies. Eupnoi includes Caddoidea (one family) and Phalangioidea (five families), and Dyspnoi includes Ischyropsalidoidea (three families) and Troguloidea (four families). The caddoids and especially the phalangioids include the typical "daddy long legs" of the Holarctic region, although Gondwanan families of both groups also exist. Ischyropsalidoids and troguloids are diverse but more poorly known. Laniatores, in contrast, are heavily sclerotized, usually short-legged, often fantastically armored animals with diversity concentrated in the Southern Hemisphere.

Only recently have workers focused on the internal phylogenetic structure of Opiliones. Five modern quantitative cladistic studies have been published to date, covering about 50 genera and directed mainly at interfamilial relationships (Shultz 1998, Giribet et al. 1999, 2002, Giribet and Wheeler 1999, Shultz and Regier 2001). In contrast to the situation in spiders, molecular data are strongly represented and largely agree with morphology. Despite the relatively small size of the group, no phylogeny to date has included all families. Martens and coworkers (Martens 1976, 1980, 1986, Martens et al. 1981) and Shear (1986) provided an early overview of aspects of opilionid phylogeny and emphasized the phylogenetic value of the male genital organs. Martens rejected the division between Cyphophthalmi and Phalangida,

instead suggesting the taxon "Cyphopalpatores," consisting of Cyphophthalmi nested within a paraphyletic Palpatores. The idea depended largely on penis morphology, but because a penis among arachnids is unique to Opiliones (convergent in some mites), the character transformation was polarized and ordered by evolutionary speculations rather than outgroups. If the features are left unordered, Cyphopalpatores disappear under parsimony (Shultz 1998, Giribet et al. 2002). All later work has decisively rejected the Cyphopalpatores hypothesis and agrees that Phalangida are monophyletic.

Opinions diverge on groups within Phalangida. Three monophyletic groups clearly exist: Eupnoi, Dyspnoi, and Laniatores, as recognized by Hansen and Sørensen (1904), but the monophyly of Palpatores is still disputed. Molecular data (18S rRNA and 28S rRNA) separately and combined with morphology suggest Dyspnoi as sister to Laniatores, thus rendering Palpatores paraphyletic (Giribet 1997, Giribet et al. 1999). The morphological codings employed in these studies were later criticized by Shultz and Regier (2001), who presented new molecular data to support Palpatores monophyly but dismissed the morphological evidence. A more inclusive analysis of morphology and molecular data including 35 genera of Opiliones recently reaffirmed Palpatores paraphyly, a result stable under a wide variety of analytical parameters (Giribet et al. 2002). This result also accords with a study of internal Cyphophthalmi relationships (Giribet and Boyer 2002). The studies of Shultz and Regier (2001) and of

Giribet et al. (2002) disagree on the internal resolution of Troguloidea and Ischyopsalidoidea, possibly because of the sparser taxon sampling in Shultz and Regier's analysis or differences in information content between the genes used.

Phylogeny of Laniatores is still in its infancy. No analysis has yet included a large sample with the exception of a study on Gonyleptoidea (Kury 1993) and the more recent molecular (Shultz and Regier 2001) and total evidence (Giribet et al. 1999, 2002) analyses considering Opiliones as a whole. The Laniatores are a well-supported monophyletic group originally divided into two groups, Oncopodomorphi and Gonyleptomorphi, by Šilhavý (1961). Martens (1976) later divided Laniatores into the three superfamilies Travunioidea, Oncopodoidea, and Gonyleptoidea, although it has been suggested (A. B. Kury, unpubl. obs.) suggests that Gonyleptoidea could be paraphyletic with respect to Oncopodoidea, constituting a clade informally named "Grassatores." The tripartite relationship proposed by Martens (1976) for Laniatores was also corroborated by total evidence analyses (Giribet et al. 1999, 2002), but many laniatorean families remain untested and their phylogenetic affinities unexplored.

Fossil members of Opiliones are rare, and their fossil record is currently restricted to a few Paleozoic and Mesozoic examples plus a more diverse Tertiary record based principally on the Florissant Formation and on Baltic and Dominican ambers (for reviews, see Cokendolpher and Cokendolpher 1982, Selden 1993b). The majority of known fossil harvestmen strongly resemble members of Eupnoi and Dyspnoi. Laniatores is currently only known from Tertiary ambers, and all the Dominican amber harvestmen described so far are Laniatores (Cokendolpher and Poinar 1998). A single fossil of the suborder Cyphophthalmi is known from Bitterfeld amber, Sachsen-Anhalt, Germany (Dunlop and Giribet in press).

Scorpions (Scorpiones)

Although their placement in Arachnida remains controversial (Weygoldt and Paulus 1979, Shultz 1990, 2000, Sissom 1990, Starobogatov 1990, Wheeler et al. 1993, Dunlop 1998, Dunlop and Selden 1998, Jeram 1998, Weygoldt 1998, Wheeler and Hayashi 1998, Dunlop and Webster 1999, Dunlop and Braddy 2001, Giribet et al. 2002), scorpions are unquestionably monophyletic. The clade is supported by 11 synapomorphies, including pectines (ventral abdominal sensory appendages), chelate pedipalps, and a five-segmented postabdomen (metasoma) terminating with a modified telson, including a pair of venom glands internally and a sharp aculeus distally, which functions as a stinging apparatus for offense and defense (Shultz 1990, Wheeler et al. 1993, Wheeler and Hayashi 1998, Giribet et al. 2002).

The approximately 1340 extant (Recent) scorpion species in 163 genera and 17 families (Fet et al. 2000, Lourenço 2000, Prendini 2000, Fet and Selden 2001, Soleglad and Sissom 2001, Kovařík 2001, 2002) constitute a monophyletic crown group with a post-Carboniferous common ancestor (Jeram 1994a, 1998). Fossil representatives comprise 92 species assigned to 71 genera and 42 families (Fet et al. 2000), of which only six species can be placed in two extant families. All Paleozoic scorpions form the stem group of this clade, with *Palaeopisthacanthus* the most crownward stem taxon (Jeram 1994b, 1998), sister to recent scorpions (Soleglad and Fet 2001). Paleozoic scorpions were far more diverse than present forms and are pivotal to resolving the phylogenetic placement of the order (Jeram 1998, Dunlop and Braddy 2001), but their phylogeny and classification are controversial and largely decoupled from that of Recent scorpions. Some classifications (Kjellesvig-Waering 1986, Starobogatov 1990) were typological and overly detailed (Sissom 1990, Fet et al. 2000). Kjellesvig-Waering (1986) placed Paleozoic scorpions into two suborders, five infraorders, 21 superfamilies, and 48 families; only Palaeopisthacanthidae was placed with the suborder containing Recent period scorpions. Starobogatov (1990) treated scorpions and eurypterids as two superorders and recognized two orders and seven suborders of scorpions. Other classifications, although based on phylogenetic analysis (Stockwell 1989, Selden 1993a, Jeram 1994a, 1994b, 1998), were hampered by the limited quantity and quality of data obtainable from fragmentary fossils. These treat scorpions as a class Scorpionida, with two extinct and one Recent order, the latter containing several suborders and infraorders, of which, again, only one contains all living representatives. In the latest classification of Paleozoic scorpions (Jeram 1998), hierarchical ranks are not established because the rank of the crown group is uncertain and there is no point of reference for the stem group clades.

Stockwell (1989) conducted the first quantitative phylogenetic analysis of Recent scorpions, excluding Buthidae, and proposed a new higher classification. Stockwell retrieved four major clades of Recent scorpions, ranked as superfamilies: Buthoidea (Buthidae and Chaerilidae), Chactoidea (Chactidae, Euscorpiidae, and Scorpiopidae), Scorpionoidea (Bothriuridae, Diplocentridae, Ischnuridae, Scorpionidae, and Urodacidae), and Vaejovoidea (Iuridae, Superstitioniidae, and Vaejovidae). However, Stockwell used groundplans derived from often paraphyletic genera as terminals (Prendini 2001b), casting doubt on his cladistic findings and resulting classification. Further, only his proposed revisions to the suprageneric classification of North American Chactoidea and Vaejovoidea were actually published (Stockwell 1992), although others, notably Lourenço (1998a, 1998b, 2000), have since implemented some of his other unpublished revisions.

Only two significant family-level morphological analyses appeared since Stockwell (1989). One treats Scorpionoidea using exemplar species (Prendini 2000). The other treats the chactoid family Euscorpiidae using genera as terminals (Soleglad and Sissom 2001). Soleglad and Fet (2001) recently attempted to illuminate basal relationships among extant scorpions (placement of the enigmatic Chaerilidae and monotypic Pseudochactidae), in an analysis based solely on trichobothrial characters, and Fet et al. (2003) presented an analysis of 17 buthid exemplar species based on 400–450 bp of 165 rDNA. A molecular analysis of the entire order, based on nuclear and mitochondrial DNA loci, to be combined with available morphological data, is underway (L. Prendini and W. Wheeler, unpubl. obs.).

Stockwell's (1989) unpublished cladogram remains the only comprehensive hypothesis for nonbuthid families and genera. Addressing the internal relationships of Buthidae (~50% and 43% of all generic and species diversity, respectively) is a major goal of future research. Although it will certainly change, the most reasonable working hypothesis of scorpion phylogeny is basically Stockwell's (1989) cladogram for nonbuthids as emended by Prendini (2000), Soleglad and Sissom (2001), and Soleglad and Fet (2001) and including the little that is known about buthid phylogeny (fig. 18.5). Most of Lourenço's (1996, 1998b, 1998c, 1999, 2000) proposed familial and superfamilial emendations cannot be justified phylogenetically (Prendini 2001b, 2003a, 2003b, Soleglad and Sissom 2001, Volschenk 2002) but are included here because they represent the most recent published opinion.

Most authorities agree that the basal dichotomy among Recent scorpions separates buthids (Buthoidea) from nonbuthids, a hypothesis supported by morphological, embryological, toxicological, and DNA sequence data (Lamoral 1980, Stockwell 1989, Sissom 1990, Fet and Lowe 2000, Soleglad and Fet 2001, Fet et al. 2003, L. Prendini and W. Wheeler, unpubl. obs.). The divergence predates the breakup of Pangaea. Similarly, it is clear that the buthoid clade is monophyletic, although the monogeneric Microcharmidae (Lourenço 1996, 1998c, 2000) renders Buthidae paraphyletic (Volschenk 2002). Within Buthidae *sensu lato*, a basal dichotomy between

New and Old World genera has also been retrieved with toxicological and DNA sequence data (Froy et al. 1999, Tytgat et al. 2000, L. Prendini and W. Wheeler, unpubl. obs.).

The Buthidae are the largest and most widely distributed scorpion family (81 genera, 570 species). Buthids are characterized by eight chelal carinae, the type A trichobothrial pattern, and flagelliform hemispermatophore, whereas most also display a triangular sternum (Vachon 1973, Stockwell 1989, Sissom 1990, Prendini 2000). Buthidae include the majority of species known to be highly venomous to humans. Buthid scorpion toxins block sodium and potassium channels, preventing transmission of action potentials across synapses (Tytgat et al. 2000). At the clinical level, this results in severe systemic symptoms and signs of neurotoxicosis (extreme pain extending beyond the site of envenomation, disorientation, salivation, convulsions, paralysis, asphyxia, and often death). Toxins affecting sodium channels are better known and divided into two major classes, alpha and beta, according to physiological effects and binding properties (Froy et al. 1999). Alpha toxins occur among Old and New World buthids, whereas beta toxins occur only among New World buthids.

Examining the phylogenetic placements of the enigmatic Chaerilidae (one genus and 19 species, Khatoon 1999, Kovařík 2000) from tropical South and Southeast Asia, and recently described monotypic Pseudochactidae (Gromov 1998), known only from Central Asia, is critical for resolving basal relationships of scorpions. Both display autapomorphic trichobothrial patterns, dubbed type B (Vachon 1973) and type D (Soleglad and Fet 2001), respectively, along with a peculiar mix of buthid and nonbuthid character states. Chaerilidae additionally exhibit an autapomorphic, fusiform hemispermatophore (Stockwell 1989, Prendini 2000). Although Stockwell (1989) placed Chaerilidae as sister taxon of Buthidae, mounting evidence confirms earlier opinions

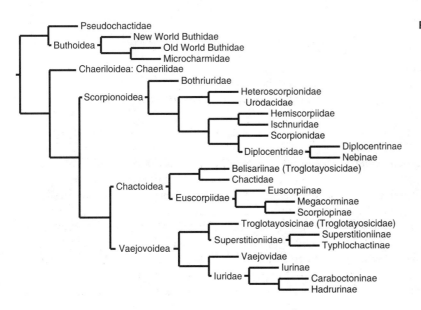

Figure 18.5. Phylogeny of Scorpiones.

that they are sister group of nonbuthids (Lamoral 1980, Lourenço 1985, Prendini 2000, Soleglad and Fet 2001), whereas Pseudochactidae may, instead, be sister group of buthids (Fet 2000, Soleglad and Fet 2001, Fet et al. 2003). Neither hypothesis based on evidence supports Lourenço's (2000) proposal to place Pseudochactidae with Chaerilidae in a unique subfamily, Chaeriloidea.

All remaining scorpions are characterized by the type C trichobothrial pattern and the lamelliform hemispermatophore, whereas most display 10 chelal carinae and a pentagonal sternum (Vachon 1973, Stockwell 1989, Sissom 1990, Prendini 2000). According to morphological and molecular evidence (Stockwell 1989, L. Prendini and W. Wheeler, unpubl. obs.), the type C scorpions comprise two distinct clades, corresponding to Stockwell's (1989) superfamilies Scorpionoidea and (Chactoidea + Vaejovoidea).

Relationships in the scorpionoid clade (37 genera and 380 species, or 23% and 29% of generic and species diversity) are better understood. All scorpionoid families are monophyletic according to morphological and molecular evidence (Stockwell 1989, Prendini 2000, L. Prendini and W. Wheeler, unpubl. obs.). Placement of Bothriuridae, a Gondwanan group with species in South America, Africa, India, and Australia, remains contentious. Bothriuridae was placed as sister to the chactoid-vaejovoid clade in some reconstructions (Lamoral 1980, Lourenço 1985) and, more recently (Lourenço 2000), in a unique superfamily Bothriuroidea. However, quantitative analyses (Stockwell 1989, Prendini 2000) place it as sister to the remaining scorpionoid families, monophyly of which is, in turn, well supported by embryological and reproductive characters, the most important being katoikogenic development. Embryos develop in ovariuterine diverticula and obtain nutrition through specialized connections with digestive caeca, rather than developing in the lumen of the ovariuterus (apoikogenic development) as in other scorpions. Katoikogenic scorpions occur mostly in the Old World and include some of the largest and most impressive scorpions. Relationships among the katoikogenic scorpionoid families, portrayed in figure 18.5, are well supported, except for the sister group relationship of Malagasy Heteroscorpionidae and Australian Urodacidae, which warrants additional testing (Prendini 2000).

Monophyly of the chactoid-vaejovoid clade (42 genera and 360 species, or 26% and 27% of generic and species diversity) appears well supported by morphological and molecular data (Stockwell 1989, L. Prendini and W. Wheeler, unpubl. obs.), but relationships among its component families, and monophyly thereof, are uncertain. Chactoidea and Vaejovoidea, as conceptualized by Stockwell (1989), may not withstand further analysis.

The chactoid-vaejovoid lineage includes the traditional and severely paraphyletic families Chactidae, Iuridae, and Vaejovidae. In an attempt to achieve monophyly, Stockwell (1989, 1992) removed Scorpiopidae from Vaejovidae, and Superstitioniidae and Euscorpiidae (to which he transferred the chactid subfamily Megacorminae) from Chactidae. Soleglad and Sissom (2001) further altered these families by placing the scorpiopid genera into Euscorpiidae and transferring Chactopsis from Chactidae to Euscorpiidae. Chactid monophyly, particularly inclusion of the North American Nullibrotheas in an otherwise exclusively neotropical group, is untested. Euscorpiidae, comprising species from Europe, Asia, and the Americas, and Vaejovidae, including most North American species, now appears to be monophyletic. This cannot be said for Superstitioniidae, a family consisting almost entirely of eyeless, depigmented troglobites from Mexico that, in Stockwell's (1989, 1992) view, included two additional troglobites: Troglotayosicus from Ecuador and Belisarius from the Pyrenees (France and Spain). Sissom (2000) questioned their inclusion in Superstitioniidae. Lourenço (1998b) placed them in a new family, Troglotayosicidae, because of their eyeless, troglobite habitus. Notwithstanding that eyelessness may have evolved convergently in the caves of Ecuador and the Pyrenees, morphological and molecular evidence (Soleglad and Sissom 2001, L. Prendini and W. Wheeler, unpubl. obs.) indicates that Belisarius is more closely related to Euscorpiidae than to Troglotayosicus, which probably is a superstitioniid.

Iuridae, also in the chactoid-vaejovoid clade, include six genera from North America, South America, and southwestern Eurasia, formerly distributed among two families and four subfamilies. This heterogeneous group is united by a single synapomorphy—a large, ventral tooth on the cheliceral movable finger (Francke and Soleglad 1981, Stockwell 1989). However, mounting morphological and molecular evidence (L. Prendini and W. Wheeler, unpubl. obs.) suggests that it is paraphyletic. Few agree on placement of the monotypic North American Anuroctonus in the chactoid-vaejovoid clade at large, although it might be related to Hadrurus, also from North America (Stockwell 1989, 1992). The South American Caraboctonus and Hadruroides form a monophyletic group, as do the Eurasian Calchas and Iurus, but the Eurasian genera display significant trichobothrial and pedipalp carinal differences, suggesting that their putative relationship to the other genera is spurious.

Pseudoscorpions (Pseudoscorpiones)

Pseudoscorpions, false scorpions, or book scorpions are a cosmopolitan group that consists of 24 families, 425 genera, and 3261 species (Harvey 1991, 2002b, M. S. Harvey, unpubl. obs.). They represent a monophyletic clade strongly supported by several features, but only one, the presence of a silk producing apparatus discharging through the movable cheliceral finger, is deemed to be autapomorphic. Other important features include the presence of chelate pedipalps, loss of the median eyes, median claw absent from all legs but replaced by an arolium, and two-segmented chelicerae. They represent the sister group of Solifugae, together comprising the

Haplocnemata (Shultz 1990, Wheeler et al. 1993, Wheeler and Hayashi 1998, Giribet et al. 2002).

Chamberlin (1931) provided the first modern classification of the order, recognizing the groups Heterosphyronida and Homosphyronida. The former consisted solely of the Chthonioidea, whereas the latter consisted of two suborders, Diplosphyronida (Neobisioidea and Garypoidea) and Monosphyronida (Feaelloidea, Cheiridioidea, and Cheliferoidea). Beier (1932a, 1932b) adopted this classification but changed the subordinal names to Chthoniinea, Neobisiinea, and Cheliferiinea. These complementary classifications remained in place, with various new families being added or synonymized, until Harvey (1992a) presented a cladistic analysis of the group based upon 200 morphological and behavioral characters. Harvey's analysis (fig. 18.6) hypothesized a different arrangement, with the suborder Epiocheirata, composed of the superfamilies Chthonioidea and Feaelloidea, representing the sister to the remaining Iocheirata. Epiocheiratans lack a venom apparatus in the chelal fingers, and adults and later nymphal instars always possess a small unique diploid trichobothrium on the distal end of the fixed chelal finger. Chthonioidea are dominated by the cosmopolitan Chthoniidae (30 genera, 612 species). Tridenchthoniidae (15 genera, 70 species) is largely tropical, whereas Lechytiidae (*Lechytia*, 22 species) is sporadically distributed. Whereas the superfamily Chthonioidea and the families Tridenchthoniidae and Lechytiidae are each clearly monophyletic, Chthoniidae probably are not. The Pseudotyrannochthoniinae and some other apparently basal taxa such as *Sathrochthonius* may warrant removal from the family. Feaelloidea are curiously distributed with Pseudogarypidae (seven species, two genera) in North America and Tasmania, and Feaellidae (11 species, one genus), on continents bordering the Indian Ocean. These distributions are undoubtedly

vicariant (Harvey 1996). The group was once more widely distributed, because three species of *Pseudogarypus* are known from Oligocene Baltic amber deposits.

The larger suborder, Iocheirata, is characterized by the presence of a venom apparatus in the chelal fingers (later lost in one finger in several lineages) and absence of the diploid trichobothrium. Iocheirata contains Hemictenata (Neobisioidea) and Panctenata (Olpioidea, Garypoidea, Sternophoroidea, and Cheliferoidea).

Neobisioidea are a basal clade containing Bochicidae (10 genera, 38 species) and Ideoroncidae (9 genera, 54 species), successively followed by the Hyidae (three genera, nine species), Gymnobisiidae (four genera, 11 species), Neobisiidae (33 genera, 499 species), Syarinidae (16 genera, 96 species), and Parahyidae (one genus, one species). Olpioidea contain two families, Olpiidae (52 genera, 324 species) and Menthidae (four genera, eight species), but there is little support for the monophyly of the former. Garypoidea consist of the basal Geogarypidae (three genera, 59 species), the Holarctic Larcidae (two genera, 12 species), the Garypidae (10 genera, 75 species), and two families previously placed in Cheiridioidea: Cheiridiidae (six genera, 71 species) and Pseudochiridiidae (two genera, 12 species). Cheiridioidea were recently reinstated as a separate superfamily by Judson (2000) but without a full reanalysis of the character set provided by Harvey (1992a).

The remaining taxa are placed in Elassommatina—consisting of the monofamilial Sternophoroidea (three genera, 20 species), a group of pallid, flattened, corticolous species distributed in various disparate regions of the world (Harvey 1991)—and Cheliferoidea. The perceived relationship of Sternophoridae with Cheliferoidea is only tentatively supported (Harvey 1992a), and knowledge of the mating be-

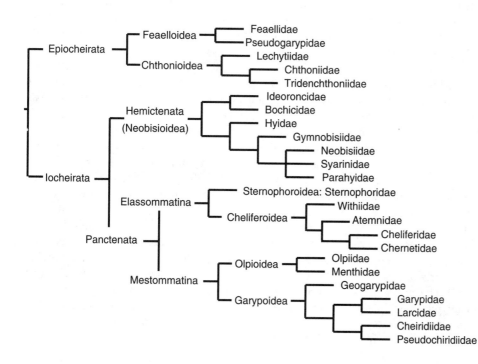

Figure 18.6. Phylogeny of Pseudoscorpiones.

havior of sternophorids may assist in determining their phylogenetic status. Cheliferoidea consist of Withiidae (34 genera, 153 species), Cheliferidae (59 genera, 274 species), and Chernetidae (111 genera, 646 species). The resolution of this clade depends on mating behavior and spermatophore morphology (Proctor 1993). Cheliferoids are the only pseudoscorpions with sperm storage receptacula (spermathecae) in females.

The fossil fauna consists of 35 named species, most of which were found as inclusions in Tertiary ambers. Cretaceous pseudoscorpions are known (Schawaller 1991), but the earliest known taxon is *Dracochela deprehendor* from Devonian shales in New York (Schawaller et al. 1991).

Harvey (1992a) confirmed the monophyly of most families, but the original analysis is currently being extended to include more taxa to test further the monophyly and internal phylogeny of various clades.

Solifuges, Camel Spiders (Solifugae)

Solifuges or solpugids are a bizarre group of specialized, mostly nocturnal, errant hunting arachnids notable for their huge powerful chelicerae and voracious appetite (Punzo 1998). Besides their large powerful chelicerae, solifuges are unique in having sensory malleoli (or racket organs) on the fourth coxae and trochanters, and many other peculiar features (prosomal stigmata, male cheliceral flagellae, palpal coxal gland orifices, adhesive palpal organs, a monocondylar walking leg joint between the femur and patella).

The Solifugae contain 1,084 species in 141 genera and 12 families (Harvey 2003): Ammotrechidae (22 genera, 81 species), Ceromidae (three genera, 20 species), Daesiidae (28 genera, 189 species), Eremobatidae (eight genera, 183 species), Galeodidae (eight genera 199 species), Gylippidae (five genera, 26 species), Hexisopodidae (two genera, 23 species), Karschiidae (four genera, 40 species), Melanoblossiidae (six genera, 16 species), Mummuciidae (10 genera, 18 species), Rhagodidae (27 genera, 98 species), and Solpugidae (17 genera, 191 species). Only three fossil species are known (Selden and Dunlop 1998). They primarily occur in Old and New World semi-arid to hyperarid ecosystems but are absent from Australia and Madagascar. The Southeast Asian melanoblossiid *Dinorhax rostrumpsittaci* is unusual in residing in rainforest, whereas the peculiar mole solifuges (Hexisopodidae) from the deserts of southern Africa are highly modified for burrowing through soil (Lamoral 1972, 1973).

Relationships within the order are very poorly understood, largely because of the chaotic familial and generic classification promulgated by Roewer (1932, 1933, 1934) and continued with many reservations by later workers (e.g., Muma 1976, Panouse 1961, Turk 1960). The current classification is a flat structure devoid of any phylogenetic signal (Harvey 2002b, 2003). There has been no detailed phylogenetic work on any solifuge group, let alone a synopsis, and no monophyly arguments exist for any family, although some (e.g., Hexisopodidae) seem to be defined by obvious autapomorphies. The group urgently needs higher level cladistic analysis.

Conclusions

The last decade has seen substantial progress in research on major arachnid clades. Considering family rank as indicating "major" lineages, at least preliminary hypotheses are available for five of the 13 "orders" (Araneae, Amblypygi, Opiliones, Scorpiones, and Pseudoscorpiones), but an additional four (Ricinulei, Palpigradi, Uropygi, and Schizomida) have only one or two clades ranked as families, so relationships at that level are trivial. Solifugae (12 families, 141 genera) and Acari (~400 families, ~4000 genera) remain as substantial lineages without explicit family-level phylogenies. Although solifuge taxonomy is so completely artificial that it is difficult to know how to begin, the main reason is lack of workers: only two or three solifuge specialists exist worldwide. Mites similarly suffer from a lack of taxonomists, but the few acarologists must deal with a much greater taxonomic tangle. There are so many autapomorphic mite lineages and so much diversity that relationships are obscured, resulting in an overly split higher classification. The very small size of mites makes molecular work difficult, although not impossible (e.g., Dabert et al. 2001), and they are so morphologically diverse (and often highly simplified) that morphological work is no easier.

The current conflict between molecules and morphology at the ordinal level in arachnid phylogeny is intriguing but probably temporary. Deeper nodes in arachnid phylogeny are hard to recover consistently with 18S and 28S rRNA sequence data. Curiously, the same loci do provide robust signal on still deeper nodes (e.g., arthropods; see Wheeler et al., ch. 17 in this vol.), as well as shallower nodes such as Opiliones (Giribet et al. 2002) and Scorpiones (L. Prendini and W. Wheeler, unpubl. obs.). The problem, therefore, seems to be, on the one hand, exploratory—loci robustly informative for these presumably Lower Palaeozoic divergences are as yet unknown—and on the other, technical, because the few loci that seem to have worked in other taxa at comparable levels have not been studied in arachnids. Edgecombe et al. (2000) also point out that the "anomalous" nodes in molecular results are usually weakly supported. The sheer quantity of molecular data make a single, most parsimonious tree almost inevitable, but that obscures the often very tenuous support for some nodes. Because fewer comparisons are usually possible, morphological data are more likely to produce multiple most parsimonious trees so that dubious nodes disappear in the strict consensus tree. No doubt as more genes are analyzed and taxon sampling improves, the discrepancies will decrease and the congruence of the total evidence will improve.

Acknowledgments

We thank Heather Proctor, Jeff Shultz, Jeremy Miller, and Greg Edgecombe for comments on the manuscript, and the National Science Foundation (EAR-0228699, DEB-9712353, and DEB-9707744 to J.A.C.) and the Smithsonian Neotropical Lowlands and Biodiversity Programs for funding.

Literature Cited

Adis, J., and M. S. Harvey. 2000. How many Arachnida and Myriapoda are there worldwide and in Amazonia? Studies on Neotrop. Fauna and Env. 35:139–141.

Barrows, W. M. 1925. Modification and development of the arachnid palpal claw, with especial reference to spiders. Annals of the Entomological Society of America 18:483–516.

Beier, M. 1932a. Pseudoscorpionidea I. Subord. Chthoniinea et Neobisiinea. Tierreich 57:i–xx, 1–258.

Beier, M. 1932b. Pseudoscorpionidea II. Subord. C. Cheliferinea. Tierreich 58:i–xxi, 1–294.

Bosselaers, J., and R. Jocqué. 2002. Studies in Corinnidae: cladistic analysis of 38 corinnid and liocranid genera, and transfer of Phrurolithinae. Zool. Scr. 31:241–270.

Chamberlin, J. C. 1931. The arachnid order Chelonethida. Biol. Sci. (Stanford Univ. Publ.) 7:1–284.

Coddington, J. A. 1986a. The genera of the spider family Theridiosomatidae. Smithson. Contrib. Zool. 422:1–96.

Coddington, J. A. 1986b. The monophyletic origin of the orb web. Pp. 319–363 in Spiders: webs, behavior, and evolution (W. A. Shear, ed.). Stanford University Press, Palo Alto, CA.

Coddington, J. A. 1990a. Cladistics and spider classification: araneomorph phylogeny and the monophyly of orbweavers (Araneae: Araneomorphae, Orbiculariae). Acta Zool. Fennica 190:75–87.

Coddington, J. A. 1990b. Ontogeny and homology in the male palpus of orb-weaving spiders and their relatives, with comments on phylogeny (Araneoclada: Araneoidea, Deinopoidea). Smithson. Contrib. Zool. 496:1–52.

Coddington, J. A., and H. W. Levi. 1991. Systematics and evolution of spiders (Araneae). Annu. Rev. Ecol. Syst. 22:565–592.

Cokendolpher, J. C., and J. E. Cokendolpher. 1982. Reexamination of the Tertiary harvestmen from the Florissant Formation, Colorado (Arachnida: Opiliones: Palpatores). J. Paleontol. 56:1213–1217.

Cokendolpher, J. C., and G. O. Poinar, Jr. 1998. A new fossil harvestman from Dominican Republic amber (Opiliones, Samoidae, *Hummelinckiolus*). J. Arachnol. 26:9–13.

Cokendolpher, J. C., and J. R. Reddell. 1992. Revision of the Protoschizomidae (Arachnida: Schizomida) with notes on the phylogeny of the order. Tex. Mem. Mus. Speleol. Monogr. 3:31–74.

Condé, B. 1996. Les Palpigrades, 1885–1995: acquisitions et lacunes. Rev. Suisse Zool. hors ser. 1:87–106.

Cruickshank, R. H., and R. H. Thomas. 1999 Evolution of haplodiploidy in dermanyssine mites (Acari: Mesostigmata). Evolution 53:1796–1803

Dabert, J., M. Dabert, S. V. Mironov, J. D. Holloway, M. J. Scoble, and C. Lofstedt. 2001. Phylogeny of feather mite subfamily Avenzoariinae (Acari: Analgoidea: Avenzoariidae) inferred from combined analyses of molecular and morphological data. Mol. Phylogenet. Evol. 20:124–135.

Davies, V. T. 1995. A new spider genus (Araneae: Amaurobioidea: Amphinectidae) from the wet tropics of Australia. Mem. Queensl. Mus. 38:463–469.

Davies, V. T. 1998. A revision of the Australian metaltellines (Araneae: Amaurobioidea: Amphinectidae: Metaltellinae). Invertebr. Taxon. 12:212–243.

Davies, V. T. 1999. *Carbinea*, a new spider genus from north Queensland, Australia (Araneae, Amaurobioidea, Kababininae). J. Arachnol. 27:25–36.

Davies, V. T., and C. Lambkin. 2000. *Wabua*, a new spider genus (Araneae: Amaurobioidea: Kababininae) from north Queensland, Australia. Mem. Queensl. Mus. 46:129–147.

Davies, V. T., and C. Lambkin. 2001. A revision of *Procambridgea* Forster & Wilton, (Araneae: Amaurobioidea: Stiphidiidae). Mem. Queensl. Mus. 46:443–459.

Dunlop, J. 1995. Are the fossil phalangiotarbids just big opilioacarid mites? Newsl. Br. Arachnol. Soc. 74:8–9.

Dunlop, J. A. 1996a. Evidence for a sister group relationship between Ricinulei and Trigonotarbida. Bull. Br. Arachnol. Soc. 10:193–204.

Dunlop, J. A. 1996b. Systematics of the fossil Arachnida. Rev. Suisse Zool. h.s. 1:173–184.

Dunlop, J. A. 1998. The origins of tetrapulmonate book lungs and their significance for chelicerate phylogeny. Pp. 9–16 in Proceedings of the 17th European College of Arachnology, Edinburgh 1997 (P. A. Selden, ed.). British Arachnological Society, Burnham Beeches, Buckinghamshire, UK.

Dunlop, J. A. 1999. A redescription of the Carboniferous arachnid *Plesiosiro madeleyi* Pocock 1911 (Arachnida: Haptopoda). Trans. R. Soc. Edinb. Earth Sci. 90:29–47.

Dunlop, J. A., and S. J. Braddy. 2001. Scorpions and their sister-group relationships. Pp. 1–24 in Scorpions 2001. In Memoriam Gary A. Polis (V. Fet, and P. A. Selden, eds.). British Arachnological Society, Burnham Beeches, Buckinghamshire, UK.

Dunlop, J. A., and G. Giribet. In press. The first fossil cyphophthalmid (Arachnida: Opiliones), from Bitterfeld amber, Germany. J. Arachnol.

Dunlop, J. A., and C. A. Horrocks. 1996. A new Upper Carboniferous whip scorpion (Arachnida: Uropygi: Thelyphonida) with a revision of the British Carboniferous Uropygi. Zool. Anzeiger 234:293–306.

Dunlop, J. A., and P. A. Selden. 1998. The early history and phylogeny of the chelicerates. Pp. 221–235 in Arthropod relationships (R. A. Fortey and R. H. Thomas, eds.). Chapman and Hall, London.

Dunlop, J. A., and M. Webster. 1999. Fossil evidence, terrestrialization and arachnid phylogeny. J. Arachnol. 27:86–93.

Eberhard, W. G. 1985. Sexual Selection and Animal Genitalia. Harvard University Press, Cambridge, MA.

Edgecombe, G. D., G. D. F. Wilson, D. J. Colgan, M. R. Gray, and G. Cassis. 2000. Arthropod cladistics: combined analysis of histone H3 and U2 snRNA sequences and morphology. Cladistics 16:155–203.

Eisner, T., J. Meinwald, A. Monro, and R. Ghent. 1961. Defense mechanisms of arthropods–I. The composition and function of the spray of the whipscorpion, Mastigoproctus giganteus (Lucas) (Arachnida: Pedipalpida). J. Insect Physiol. 6:272–298.

Fet, V. 2000. Family Pseudochactidae Gromov, 1998. P. 426 in Catalog of the scorpions of the world (1758–1998) (V. Fet, W. D. Sissom, G. Lowe, and M. E. Braunwalder, eds.). New York Entomological Society, New York.

Fet, V., B. Gantenbein, A. V. Gromov, G. Lowe, and W. R. Lourenço. 2003. The first molecular phylogeny of buthidae (Scorpiones). Euscorpius 4:1–10.

Fet, V., and G. Lowe. 2000. Family Buthidae C. L. Koch, 1837. Pp. 54–286 in Catalog of the scorpions of the world (1758–1998) (V. Fet, W. D. Sissom, G. Lowe, and M. E. Braunwalder, eds.). New York Entomological Society, New York.

Fet, V., and P. A. Selden (eds.). 2001. Scorpions 2001. In Memoriam Gary A. Polis. British Arachnological Society, Burnham Beeches, Buckinghamshire, UK.

Fet, V., W. D. Sissom, G. Lowe, and M. E. Braunwalder (eds.). 2000. Catalog of the scorpions of the world (1758–1998). New York Entomological Society, New York.

Forster, R. R., and N. I. Platnick. 1984. A review of the archaeid spiders and their relatives, with notes on the limits of the superfamily Palpimanoidea (Arachnida, Araneae). Bull. Am. Mus. Nat. Hist. 178:1–106.

Francke, O. F., and M. E. Soleglad. 1981. The family Iuridae Thorell (Arachnida, Scorpiones). J. Arachnol. 9:233–258.

Froy, O., T. Sagiv, M. Pore, D. Urbach, N. Zilberberg, and M. Gurevitz. 1999. Dynamic diversification from a putative common ancestor of scorpion toxins affecting sodium, potassium and chloride channels. J. Mol. Evol. 48:187–196.

Giribet, G. 1997. Filogenia molecular de Artrópodos basada en la secuencia de genes ribosomales. Ph.D. thesis, Universitat de Barcelona, Barcelona.

Giribet, G. 2000. Catalogue of the Cyphophthalmi of the world (Arachnida, Opiliones). Rev. Iber. Aracnol. 2:49–76.

Giribet, G., and S. Boyer. 2002. A cladistic analysis of the cyphophthalmid genera (Opiliones, Cyphophthalmi). J. Arachnol. 30:110–128.

Giribet, G., G. D. Edgecombe, and W. C. Wheeler. 2001. Arthropod phylogeny based on eight molecular loci and morphology. Nature 413:157–161.

Giribet, G., G. D. Edgecombe, W. C. Wheeler, and C. Babbitt. 2002. Phylogeny and systematic position of Opiliones: a combined analysis of chelicerate relationships using morphological and molecular data. Cladistics 18:5–70.

Giribet, G., M. Rambla, S. Carranza, M. Riutort, J. Baguñà, and C. Ribera. 1999. Phylogeny of the arachnid order Opiliones (Arthropoda) inferred from a combined approach of complete 18S, partial 28S ribosomal DNA sequences and morphology. Mol. Phylogenet. Evol. 11:296–307.

Giribet, G., and C. Ribera 2000. A review of arthropod phylogeny: new data based on ribosomal DNA sequences and direct character optimization. Cladistics 16:204–231.

Giribet, G., and W. C. Wheeler. 1999. On gaps. Mol. Phylogenet. Evol. 13:132–143.

Goloboff, P. A. 1993. A reanalysis of mygalomorph spider families. Am. Mus. Nov. 3056:1–32.

Goloboff, P. A. 1995. A revision of the South American spiders of the family Nemesiidae (Araneae, Mygalomorphae). Part 1: Species from Peru, Chile, Argentina, and Uruguay. Bull. Am. Mus. Nat. Hist. 224:1–189.

Grandjean, F. 1936. Un acarien synthétique: Opilioacarus segmentatus With. Bull. Soc. Hist. Nat. Afr. Nord Alger 27:413–444.

Gray, M. R. 1995. Morphology and relationships within the spider family Filistatidae (Araneae: Araneomorphae). Rec. West. Aus. Mus. 52(suppl.):79–89.

Griswold, C. E. 1993. Investigations into the phylogeny of the lycosoid spiders and their kin (Arachnida, Araneae, Lycosoidea). Smithson. Contrib. Zool. 539:1–39

Griswold, C. E. 2001. A monograph of the living world genera and Afrotropical species of cyatholipid spiders (Araneae, Orbiculariae, Araneoidea, Cyatholipidae). Mem. Calif. Acad. Sci. 26:1–251.

Griswold, C., and J. Ledford. 2001. A monograph of the migid trap-door spiders of Madagascar, with a phylogeny of world genera (Araneae, Mygalomorphae, Migidae). Occ. Pap. Calif. Acad. Sci. 151:1–120.

Griswold, C. E., J. A. Coddington, G. Hormiga, and N. Scharff. 1998. Phylogeny of the orb-web building spiders (Araneae, Orbiculariae: Deinopoidea, Araneoidea). Zool. J. Linn. Soc. 123:1–99.

Griswold, C. E., J. A. Coddington, N. I. Platnick, and R. R. Forster. 1999. Towards a phylogeny of entelegyne spiders (Araneae, Entelegynae). J. Arachnol. 27:53–63.

Gromov, A. V. 1998. [A new family, genus and species of scorpion (Arachnida, Scorpiones) from southern Central Asia] (in Russian). Zool. Zh. 77:1003–1008. (Engl. summ.).

Hansen, H. L., and W. Sorensen. 1904. On two orders of Arachnida. Cambridge University Press, Cambridge.

Harvey, M. S. 1991. Catalogue of the Pseudoscorpionida. Manchester University Press, Manchester.

Harvey, M. S. 1992a. The phylogeny and systematics of the Pseudoscorpionida (Chelicerata: Arachnida). Invertebr. Taxonomy 6:1373–1435.

Harvey, M. S. 1992b. The Schizomida (Chelicerata) of Australia. Invertebr. Taxonomy 6:77–129.

Harvey, M. S. 1995. The systematics of the spider family Nicodamidae (Araneae: Amaurobioidea). Invertebr. Taxonomy 9:279–386.

Harvey, M. S. 1996. The biogeography of Gondwanan pseudo-scorpions (Arachnida). Rev. Suisse Zool. hors ser. 1:255–264.

Harvey, M. S. 2002a. The first Old World species of Phrynidae (Amblypygi): Phrynus exsul from Indonesia. J. Arachnol. 30:470–474.

Harvey, M. S. 2002b. The neglected cousins: what do we know about the smaller arachnid orders? J. Arachnol. 30:373–382.

Harvey, M. S. 2003. Catalogue of the smaller arachnid orders of the world: Amblypygi, Uropygi, Schizomida, Palpigradi, Ricinulei and Solifugae. CSIRO Publishing, Melbourne.

Haupt, J., G. Hohne, H. Schwartz, B. Chen, W. Zhao, and Y. Zhang. 1988. Chinese whip scorpion using 2–ketones in defense secretion (Arachnida: Uropygi). J. Comp. Physiol. B 157:883–885.

Hausdorf, B. 1999. Molecular phylogeny of araneomorph spiders. J. Evol. Biol. 12:980–985.

Hedin, M. C., and W. P. Maddison. 2001. A combined molecular approach to phylogeny of the jumping spider subfamily Dendryphantinae (Araneae: Salticidae). Mol. Phylogenet. Evol. 18:386–403.

Hormiga, G. 1994. Cladistics and the comparative morphology of linyphiid spiders and their relatives (Araneae, Araneoidea, Linyphiidae). Zool. J. Linn. Soc. 111:1–71.

Hormiga, G. 2000. Higher level phylogenetics of erigonine spiders (Araneae, Linyphiidae, Erigoninae). Smithson. Contrib. Zool. 600:1–160.

Hormiga, G., W. G. Eberhard, and J. A. Coddington. 1995. Web construction behavior in Australian *Phonognatha* and the phylogeny of nephiline and tetragnathid spiders (Araneae, Tetragnathidae). Austr. J. Zool. 43:313–343.

Huber, B. A. 2000. New World pholcid spiders (Araneae: Pholcidae): a revision at generic level. Bull. Am. Mus. Nat. Hist. 254:1–348.

Huber, B. A. 2001. The pholcids of Australia (Araneae; Pholcidae): taxonomy, biogeography, and relationships. Bull. Am. Mus. Nat. Hist. 260:1–144.

Huber, K. C., T. S. Haider, M. W. Muller, B. A. Huber, R. J. Schweyen, and F. G. Barth. 1993. DNA sequence data indicates the polyphyly of the family Ctenidae (Araneae). J. Arachnol. 21:194–201.

Jeram, A. J. 1994a. Carboniferous Orthosterni and their relationship to living scorpions. Palaeontology 37:513–550.

Jeram, A. J. 1994b. Scorpions from the Viséan of East Kirkton, West Lothian, Scotland, with a revision of the infraorder Mesoscorpionina. Trans. R. Soc. Edinb. Earth Sci. 84:283–299.

Jeram, A. J. 1998. Phylogeny, classification and evolution of Silurian and Devonian scorpions. Pp. 17–31 in Proceedings of the 17th European Colloquium of Arachnology, Edinburgh 1997 (P. A. Selden, ed.). British Arachnological Society, Burnham Beeches, Buckinghamshire, UK.

Jocqué, R. 1991. A generic revision of the spider family Zodariidae (Araneae). Bull. Am. Mus. Nat. Hist. 201:1–160.

Judson, M. L. I. 2000. *Electrobisium acutum* Cockerell, a cheiridiid pseudoscorpion from Burmese amber, with remarks on the validity of the Cheiridioidea (Arachnida, Chelonethi). Bull. Nat. Hist. Mus. Geol. 56:79–83.

Kethley, J. B., R. A. Norton, P. M. Bonamo, and W. A. Shear. 1989. A terrestrial alicorhagiid mite (Acari: Acariformes) from the Devonian of New York. Micropaleontology 35:367–373.

Khatoon, S. 1999. Scorpions of Pakistan (Arachnida: Scorpionida). Proc. Pak. Congr. Zool. 19:207–225.

Kjellesvig-Waering, E. N. 1986. A restudy of the fossil Scorpionida of the world. (Palaeontogr. Am. 55:1–287). Organized for publication by A. S. Caster and K. E. Caster. Paleontological Research Institution. 287 pp. Ithaca, NY.

Klompen, J. S. H. 1992. Phylogenetic relationships in the mite family Sarcoptidae (Acari: Astigmata). Misc. Publ. Mus. Zool. University Mich.180:1–154

Klompen, J. S. H., W. C. Black, IV, J. E. Keirans, and J. H. Oliver, Jr. 1996. Evolution of ticks. Annu. Rev. Entomol. 41:141–161.

Kovařík, F. 2000. Revision of family Chaerilidae (Scorpiones), with descriptions of three new species. Serket 7:38–77.

Kovařík, F. 2001. Catalog of the Scorpions of the World (1758–1998) by V. Fet, W. D. Sissom, G. Lowe, and M. Braunwalder (New York Entomological Society, 2000:690 pp.): discussion and supplement for 1999 and part of 2000. Serket 7:78–93.

Kovařík, F. 2002. Co nového u štíru v roce 2000 (in Czech). Akv. Ter. 45:55–61.

Kukalová-Peck, J. 1991. Fossil history and the evolution of hexapod structures. Pp. 141–179 in The insects of Australia (I. D. Nauman, ed.). Cornell University Press, Ithaca, NY.

Kury, A. 1993. Análise filogenética de Gonyleptoidea (Arachnida, Opiliones, Laniatores). Ph.D. thesis, Universidade de São Paulo, São Paulo.

Labandeira, C. C. 1999. Insects and other hexapods. Pp. 604–624 in Encyclopedia of paleontology (R. Singer, ed.). Fitzroy and Dearborn, Chicago.

Labandeira, C. C., T. L. Phillips, and R. A. Norton. 1997. Oribatid mites and the decomposition of plant tissues in Paleozoic coal-swamp forests. Palaios 12:319–353.

Lamoral, B. H. 1972. New and little known scorpions and solifuges from the Namib Desert, South West Africa. Madoqua 1:117–131.

Lamoral, B. H. 1973. The arachnid fauna of the Kalahari Gemsbok National Park, Pt 1: A revision of the "mole solifuges" of the genus *Chelypus* Purcell, 1901 (Family Hexisopodidae). Koedoe 16:83–102.

Lamoral, B. H. 1980. A reappraisal of the suprageneric classification of recent scorpions and their zoogeography. Pp. 439–444 in Verhandlungen. 8. Internationaler Arachnologen—Kongress abgehalten ander Universität für Bodenkultur Wien, 7–12 Juli, 1980 (J. Gruber, ed.). H. Egermann, Vienna.

Lindquist, E. E. 1984. Current theories on the evolution of major groups of Acari and on their relationships with other groups of Arachnida, with consequent implications for their classification. Pp. 28–62 in Acarology VI (D. A. Griffiths and C. E. Bowman, eds.), vol. 1. Ellis Horwood Ltd., Chichester, UK.

Lindquist, E. E. 1986. The world genera of Tarsonemidae (Acari: Heterostigmata): a morphological, phylogenetic, and systematic revision, with a reclassification of the family-group taxa in the Heterostigmata. Mem. Entomol. Soc. Can. 136:1–517

Lourenço, W. R. 1985. Essai d'Interprétation de la distribution du genre *Opisthacanthus* (Arachnida, Scorpiones, Ischnuridae) dans les régions néotropicales et afrotropicale. Étude taxinomique, biogéographique, évolutive et écologique. Thése de Doctorat d'État, Université Paris VI.

Lourenço, W. R. 1996. Faune de Madagascar. 87. Scorpions (Chelicerata, Scorpiones). Muséum National d'Histoire Naturelle, Paris.

Lourenço, W. R. 1998a. Designation of the scorpion subfamily Scorpiopsinae Kraepelin, 1905 as family Scorpiopsidae Kraepelin, 1905: its generical composition and a description of a new species of *Scorpiops* from Pakistan (Scorpiones, Scorpiopsidae). Entomol. Mitt. Zool. Mus. Hamb. 12:245–254.

Lourenço, W. R. 1998b. Panbiogeographie, les distribution disjointes et le concept de famille relictuelle chez les scorpions. Biogeographica 74:133–144.

Lourenço, W. R. 1998c. Une nouvelle famille est nécessaire

pour des microscorpions humicoles de Madagascar et d'Afrique. C. R. Acad. Sci. III, Sci. Vie 321:845–848.

Lourenço, W. R. 1999. Considérations taxonomiques sur le genre *Hadogenes* Kraepelin, 1894; création de la sous-famille des Hadogeninae n. subfam., et description d'une espèce nouvelle pour l'Angola (Scorpiones, Scorpionidae, Hadogeninae). Rev. Suisse Zool. 106:929–938.

Lourenço, W. R. 2000. Panbiogéographie, les familles des scorpions et leur répartition géographique. Biogeographica 76:21–39.

Lourenço, W. R. 2003. The first molecular phylogeny of Buthidae (Scorpiones). Euscorpius 4:1–10.

Martens, J. 1976. Genitalmorphologie, System und Phylogenie der Weberknechte (Arachnida: Opiliones). Entomol. Germ. 3:51–68.

Martens, J. 1980. Versuch eines phylogenetischen Systems der Opiliones. Pp. 355–360 in Verhandlungen. 8. Internationaler Arachnologen—Kongress abgehalten ander Universität für Bodenkultur Wien, 7–12 Juli, 1980 (J. Gruber, ed.). H. Egermann, Vienna.

Martens, J. 1986. Die Grossgliederung der Opiliones und die Evolution der Ordnung (Arachnida). Actas X Congr. Int. Arachnol. Esp. (J. A. Barrientos, ed.) 1:289–310.

Martens, J., U. Hoheisel, and M. Gotze. 1981. Vergleichende Anatomie der Legeröhren der Opilions als Beitrag zur Phylogenie der Ordnung (Arachnida). Zool. Jb. Anat. 105:13–76.

Muma, M. H. 1976. A review of solpugid families with an annotated list of western hemisphere solpugids. Publication of the Office of Research, vol. 2. Western New Mexico University, Silver City.

Naskrecki, P., and R. K. Colwell. 1998. Systematics and host plant affiliations of hummingbird flower mites of the genera *Tropicoseius* Baker and Yunker and *Rhinoseius* Baker and Yunker (Acari: Mesostigmata: Ascidae). Thomas Say Foundation Monographs. Entomological Society of America Monographs. 128 pp.

Navajas, M., and B. Fenton. 2000. The application of molecular markers in the study of diversity in acarology: a review. Exp. Appl. Acarol. 24:751–774.

Norton, R. A. 1998. Morphological evidence for the evolutionary origin of Astigmata (Acari: Acariformes) Exp. Appl. Acarol. 22:559–594.

Norton, R. A., P. M. Bonamo, J. D. Grierson, and W. A. Shear. 1988. Oribatid mite fossils from a terrestrial Devonian deposit near Gilboa, New York. J. Paleontol. 62:259–269.

OConnor, B. M. 1984. Phylogenetic relationships among higher taxa in the Acariformes, with particular reference to the Astigmata. Pp. 19–27 in Acarology VI (D. A. Griffiths and C. E. Bowman, eds), vol. 1. Ellis Horwood Ltd, Chichester.

Panouse, J. B. 1961. Note complémentaire sur la variation des caractères utilisés dans la taxonomie des Solifuges. Bull. Soc. Sci. Nat. Phys. Maroc. 40:121–129.

Pérez-Miles, F., S. M. Lucas, P. I. Da Silva, Jr., and R. Bertani. 1996. Systematic revision and cladistic analysis of Theraphosinae (Araneae: Theraphosidae). Mygalomorph 1:33–68.

Piel, W. H., and K. J. Nutt. 1997. *Kaira* is a likely sister group to *Metepeira*, and *Zygiella* is an araneid (Araneae, Araneidae): evidence from mitochondrial DNA. J. Arachnol. 25:262–268.

Platnick, N. I. 1980. On the phylogeny of Ricinulei. Presented at 8th Internationaler Arachnologen-Kongress abgehalten ander Universitat fur Bodenkultur Wien, 7–12 Juli 1980, Vienna.

Platnick, N. I. 1990. Spinneret morphology and the phylogeny of ground spiders (Araneae, Gnaphosoidea). Am. Mus. Nov. 2978:1–42.

Platnick, N. I. 2000. A relimitation and revision of the Australasian ground spider family Lamponidae (Araneae: Gnaphosoidea). Bull. Am. Mus. Nat. Hist. 245:1–330.

Platnick, N. I. 2003. The world spider catalog, ver. 4.5. American Museum of Natural History, New York, NY. Available: http://research.amnh.org/entomology/spiders/catalog/index.html. Last accessed 14 December 2003.

Platnick, N. I., J. A. Coddington, R. R. Forster, and C. E. Griswold. 1991. Spinneret evidence and the higher classification of the haplogyne spiders (Araneae, Araneomorphae). Am. Mus. Nov. 3016:1–73.

Platnick, N. I., and W. J. Gertsch. 1976. The suborders of spiders: a cladistic analysis. Am. Mus. Nov. 2607:1–15.

Prendini, L. 2000. Phylogeny and classification of the Superfamily Scorpionoidea Latreille 1802 (Chelicerata, Scorpiones): an exemplar approach. Cladistics 16:1–78.

Prendini, L. 2001a. Species or supraspecific taxa as terminals in cladistic analysis? Groundplans versus exemplars revisited. Syst. Biol. 50:290–300.

Prendini, L. 2001b. Two new species of *Hadogenes* (Scorpiones, Ischnuridae) from South Africa, with a redescription of *Hadogenes bicolor* and a discussion on the phylogenetic position of *Hadogenes*. J. Arachnol. 29:146–172.

Prendini, L. 2003a. A new genus and species of bothriurid scorpion from the Brandberg Massif, Namibia, with a reanalysis of bothriurid phylogeny and a discussion on the phylogenetic position of *Lisposoma* Lawrence. Syst. Entomol. 28:1–24.

Prendini, L. 2003b. Revision of the genus *Lisposoma* Lawrence, 1928 (Scorpiones: Bothriuridae). Insect Syst. Evol. 34:241–264.

Proctor, H. C. 1993. Mating biology resolves trichotomy for cheliferoid pseudoscorpions (Pseudoscorpionida, Cheliferoidea). J. Arachnol. 21:156–158.

Pugh, P. J. A. 1993. A synonymic catalogue of the Acari from Antarctica, the Sub-Antarctic Islands and the Southern Ocean. J. Nat. Hist. 27:323–421.

Punzo. F. 1998. The biology of camel spiders (Arachnida Solifugae). Kluwer, New York.

Ramírez, M. J. 1995a. A phylogenetic analysis of the subfamilies of Anyphaenidae (Arachnida, Araneae). Ent. Scand. 26:361–384.

Ramírez, M. J. 1995b. Revisión y filogénia del género *Monapia*, con notas sobre otras Amaurobiodinae (Araneae: Anyphaenidae). Bol. Soc. Concepción Chile 66:71–102.

Ramírez, M. J. 1997. Revisión y filogénia de los géneros *Ferrieria*, y *Acanthoceto* (Araneae: Anyphaenidae, Amaurobioidinae). Iheringia Sér. Zool. Porto Alegre 82:173–203.

Ramírez, M. J., and C. J. Grismado. 1997. A review of the spider family Filistatidae in Argentina (Arachnida, Araneae), with a cladistic reanalysis of filistatid genera. Ent. Scand. 28:319–349.

Raven, R. J. 1985. The spider infraorder Mygalomorphae: cladistics and systematics. Bull. Am. Mus. Nat. Hist. 182:1–180.

Reddell, J. R., and J. C. Cokendolpher. 1995. Catalogue, bibliography, and generic revision of the order Schizomida (Arachnida). Tex. Mem. Mus. Speleol. Monogr. 4:1–170.

Regier, J. C., and J. W. Shultz. 1997. Molecular phylogeny of the major arthropod groups indicates polyphyly of crustaceans and a new hypothesis for the origin of hexapods. Mol. Biol. Evol. 14:902–913.

Regier, J. C., and J. W. Shultz. 1998. Molecular phylogeny of arthropods and the significance of the Cambrian "explosion" for molecular systematics. Am. Zool. 38:918–928.

Regier, J. C., and J. W. Shultz. 2001. Elongation factor-2: a useful gene for arthropod phylogenetics. Mol. Phylogenet. Evol. 20:136–148.

Rodrigo, A. G., and R. R. Jackson. 1992. Four jumping spider genera of the *Cocalodes*-group are monophyletic with genera of the Spartaeinae (Araneae: Salticidae). N. Z. Nat. Sci. 19:61–67.

Roewer, C. F. 1932. Solifugae, Palpigradi. Klassen und Ordnungen des Tierreichs. 5: Arthropoda. IV: Arachnoidea (H. G. Bronns, ed.), Vol. 5(IV)(4)(1):1–160. Akademische Verlagsgesellschaft M. B. H., Leipzig.

Roewer, C. F. 1933. Solifugae, Palpigradi. Klassen und Ordnungen des Tierreichs. 5: Arthropoda. IV: Arachnoidea (H. G. Bronns, ed.), Vol. 5(IV)(4)(2–3):161–480. Akademische Verlagsgesellschaft M. B. H., Leipzig.

Roewer, C. F. 1934. Solifugae, Palpigradi. Klassen und Ordnungen des Tierreichs. 5: Arthropoda. IV: Arachnoidea (H. G. Bronns, ed.). Vol. 5(IV)(4)(4–5):481–723. Akademische Verlagsgesellschaft M. B. H., Leipzig.

Rowland, J. M., and J. A. L. Cooke. 1973. Systematics of the arachnid order Uropygida (= Thelyphonida). J. Arachnol. 1:55–71.

Rowland, J. M., and J. R. Reddell. 1979. The order Schizomida (Arachnida) in the New World. I. Protoschizomidae and dumitrescoae group (Schizomidae: *Schizomus*). J. Arachnol. 6:161–196.

Savory, T. H. 1977. Arachnida. Academic Press, New York.

Scharff, N., and J. A. Coddington. 1997. A phylogenetic analysis of the orb-weaving spider family Araneidae (Arachnida, Araneae). Zool. J. Linn. Soc. 120:355–434.

Schawaller, W. 1991. The first Mesozoic pseudoscorpion, from Cretaceous Canadian amber. Paleontology 34:971–976.

Schawaller, W., W. A. Shear, and P. M. Bonamo. 1991. The first Paleozoic pseudoscorpions (Arachnida, Pseudoscorpionida). Am. Mus. Nov. 3009:1–24.

Schütt, K. 2000. The limits of the Araneoidea (Arachnida: Araneae). Austr. J. Zool. 48:135–153.

Schütt, K. 2003. Phylogeny of Symphytognathidae s.l. (Araneae, Araneoidea). Zool. Scripta 32:129–151.

Selden, P. A. 1992. Revision of the fossil ricinuleids. Trans. R. Soc. Edinb. Earth Sci. 83:595–634.

Selden, P. A. 1993a. Arthropoda (Aglaspidida, Pycnogonida and Chelicerata). Pp. 297–320 in The fossil record 2 (M. J. Benton, ed.). Chapman and Hall, New York.

Selden, P. A. 1993b. Fossil arachnids—recent advances and future prospects. Mem. Queensl. Mus. 33:389–400.

Selden, P. A. 1996. Fossil mesothele spiders. Nature 379:498–499.

Selden, P. A., and J. A. Dunlop. 1998. Fossil taxa and relationships of chelicerates. Pp. 303–331 in Arthropod fossils and phylogeny (G. D. Edgecombe, ed.). Cambridge University Press, New York.

Shear, W. A. 1980. A review of the Cyphophthalmi of the United States and Mexico, with a proposed reclassification of the Suborder (Arachnida, Opiliones). Am. Mus. Nov. 2705:1–34.

Shear, W. A. 1986. A cladistic analysis of the opilionid superfamily Ischyropsalidoidea, with descriptions of the new family Ceratolasmatidae, the new genus *Acuclavella*, and four new species. Am. Mus. Nov. 2844:1–29.

Shear, W. A. 1993. The genus *Troglosiro* and the new family Troglosironidae (Opiliones, Cyphophthalmi). J. Arachnol. 21:81–90.

Shultz, J. W. 1989. Morphology of locomotor appendages in Arachnida: evolutionary trends and phylogenetic implications. Zool. J. Linn. Soc. 97:1–56.

Shultz, J. W. 1990. Evolutionary morphology and phylogeny of Arachnida. Cladistics 6:1–38.

Shultz, J. W. 1998. Phylogeny of Opiliones (Arachnida): an assessment of the "Cyphopalpatores" concept. J. Arachnol. 26:257–272.

Shultz, J. W. 1999. Muscular anatomy of a whipspider, *Phrynus longipes* (Pocock) (Arachnida: Amblypygi), and its evolutionary significance. Zool. J. Linn. Soc. 126:81–116.

Shultz, J. W. 2000. Skeletomuscular anatomy of the harvestman *Leiobunun aldrichi* (Weed, 1893) (Arachnida: Amblypygi) and its evolutionary significance. Zool. J. Linn. Soc. 128:401–438.

Shultz, J. W. 2001. Gross muscular anatomy of *Limulus polyphemus* (Xiphosura, Chelicerata) and its bearing on evolution in the Arachnida. J. Arachnol. 29:283–303.

Shultz, J. W., and J. C. Regier. 2000. Phylogenetic analysis of arthropods using two nuclear protein-encoding genes supports a crustacean + hexapod clade. Proc. R. Soc. Lond. B 267:1011–1019.

Shultz, J. W., and J. C. Regier. 2001. Phylogenetic analysis of Phalangida (Arachnida: Opiliones) using two nuclear protein-encoding genes supports monophyly of Palpatores. J. Arachnol. 29:189–200

Sierwald, P. 1998. Phylogenetic analysis of pisaurine nursery web spiders, with revisions of *Tetragonopthalma* and *Perenethis* (Araneae, Lycosoidea, Pisauridae). J. Arachnol. 25:361–407.

Šilhavý, V. 1961. Die Gundsätze der modernen Wberknechttaxonomie und Revisions des bisherigen Systems der Opilioniden. Verh Int. Kongr. Entomol. 1:262–267.

Silva Davila, D. 2003. Higher-level relationships of the spider family Ctenidae (Araneae, Ctenoidea). Bull. Am. Mus. Nat. Hist. 274:1–85.

Sissom, W. D. 1990. Systematics, biogeography and paleontology. Pp. 64–160 in The biology of scorpions (G. A. Polis, ed.). Stanford University Press, Stanford, CA.

Sissom, W. D. 2000. Family Superstitioniidae Stahnke, 1940. Pp. 496–500 in Catalog of the scorpions of the world (1758–1998) (V. Fet, W. D. Sissom, G. Lowe, and M. E. Braunwalder, eds.). New York Entomological Society, New York.

Snodgrass, R. E. 1938. Evolution of the Annelida, Onychophora and Arthropoda. Smithson. Misc. Collect. 97:1–159.

Soeller, R., A. Wohltmann, H. Witte, and D. Blohm. 2001. Phylogenetic relationships within terrestrial mites (Acari: Prostigmata, Parasitengona) inferred from comparative DNA sequence analysis of the mitochondrial cytochrome oxidase subunit I gene. Mol. Phylogenet. Evol. 18:47–53.

Soleglad, M. E., and V. Fet. 2001. Evolution of scorpion orthobothriotaxy: a cladistic approach. Euscorpius 1:1–38.

Soleglad, M. E., and W. D. Sissom. 2001. Phylogeny of the family Euscorpiidae Laurie, 1896: a major revision. Pp. 25–111 in Scorpions 2001. In Memoriam Gary A. Polis (V. Fet, and P. A. Selden. eds.). British Arachnological Society, Burnham Beeches, Buckinghamshire, UK.

Starobogatov, Ya. I. 1990. The systematics and phylogeny of the lower chelicerates (a morphological analysis of the Paleozoic groups). Paleont. J. 1:2–16.

Stockwell, S. A. 1989. Revision of the phylogeny and higher classification of scorpions (Chelicerata). Ph.D. thesis, University of California, Berkeley.

Stockwell, S. A. 1992. Systematic observations on North American Scorpionida with a key and checklist of the families and genera. J. Med. Entomol. 29:407–422.

Sturm, H. 1958. Indirekte Spermatophorenübertragung bei dem Geisselskorpion Trithyreus sturmi Kraus (Schizomidae, Pedipalpi). Naturwissenschaften 45:142–143.

Turk, F. A. 1960. On some sundry species of solifugids in the collection of the Hebrew University of Jerusalem. Proc. Zool. Soc. Lond. 135:105–124.

Tytgat, J., K. G. Chandy, M. L. Garcia, G. A. Gutman, M. F. Martin-Eauclaire, J. Van der Walt, and L S. Possani. 2000. A unified nomenclature for short-chain peptides isolated from scorpion venoms: alpha-KTx molecular subfamilies. Trends Pharmacol. Sci. 20:444–447.

Vachon, M. 1973. Étude des caractères utilisés pour classer les familles et les genres de scorpions (Arachnides). 1. La trichobothriotaxie en arachnologie. Sigles trichobothriaux et types de trichobothriotaxie chez les scorpions. Bull. Mus. Natl. Hist. Nat. Paris 140:857–958.

Van der Hammen, L. 1989. An introduction to comparative arachnology. SPB Academic Publishing, The Hague.

Volschenk, E. S. 2002. Systematic revision of the Australian scorpion genera in the family Buthidae. Ph.D. thesis, Curtin University of Technology, Perth.

Waloszek, D., and J. A. Dunlop. 2002. A larval sea spider (Arthropoda: Pycnogonida) from the Upper Cambrian "Orsten" of Sweden, and the phylogenetic position of pycnogonids. Palaeontology 45:421–446.

Walter, D. E. 1996. Living on leaves. Mites, tomenta, and leaf domatia. Annu. Rev. Entomol. 41:101–114.

Walter, D. E., and V. Behan-Pelletier. 1999. Mites in forest canopies: filling the size distribution shortfall? Annu. Rev. Entomol. 44:1–19.

Walter, D. E., and H. C. Proctor. 1998. Feeding behaviour and phylogeny: observations on early derivative Acari. Exp. Appl. Acarol. 22:39–50.

Walter, D. E., and H. C Proctor. 1999. Mites: ecology, evolution and behaviour. University of NSW Press, Sydney, and CABI, Wallingford.

Wang, X.-P. 2002. A generic-level revision of the spider subfamily Coelotinae (Araneae, Amaurobiidae). Bull. Am. Mus. Nat. Hist. 269:1–150.

Weygoldt, P. 1996. Evolutionary morphology of whip spiders: towards a phylogenetic system (Chelicerata: Arachnida: Amblypygi). J. Zool. Syst. Evol. Res. 34:185–202.

Weygoldt, P. 1998. Evolution and systematics of the Chelicerata. Exp. Appl. Acarol. 22:63–79.

Weygoldt, P. 2000. Whip spiders. Their biology, morphology and systematics. Apollo Books, Stenstrup, Denmark.

Weygoldt, P., and H. F. Paulus. 1979. Untersuchungen zur Morphologie, Taxonomie und Phylogenie der Chelicerata. I. Morphologische Untersuchungen. II. Cladogramme und die Enfaltung der Chelicerata. Z. Zool. Syst. Evolut. Forsch. 17:85–116, 177–200.

Wheeler, W. C. 1998. Sampling, groundplans, total evidence, and the systematics of arthropods. Pp. 87–96 in Arthropod relationships (R. A. Fortey and R. H. Thomas, eds.). Chapman and Hall, London.

Wheeler, W. C., and C. Y. Hayashi. 1998. The phylogeny of the extant chelicerate orders. Cladistics 14:173–192.

Wheeler, W. C., P. Cartwright, P., and C. Hayashi. 1993. Arthropod phylogeny: a combined approach. Cladistics 9:1–39.

Zrzavý, J., V. Hypsa, and M. Blaskova. 1998. Arthropod phylogeny: taxonomic congruence, total evidence, and conditional combination approaches to morphological and molecular data sets. Pp. 97–107 in Arthropod relationships (R. A. Fortey and R. H. Thomas, eds.). London, Chapman and Hall.

Frederick R. Schram
Stefan Koenemann

19

Are the Crustaceans Monophyletic?

Grasses of the Distant Past

In these times of rampant Tree of Life cultivation, it is perhaps hard to realize that up until 20 years ago the "phylogenetic lawn" of figure 19.1 entailed everything we as carcinologists knew about crustacean relationships. This poorly resolved "twig" was, and to some people still is, a widely excepted consensus that dates back to the 1962 conference on the *Phylogeny and Evolution of Crustacea* held at the Museum of Comparative Zoology at Harvard (Dahl 1963). Moreover, this was considered state of the art for the field during that period!

The situation within the major groups of crustaceans was little better. Like the deer frozen in the middle of a roadway by oncoming headlights on its way to becoming a venison pancake, crustacean workers were overawed by the multiplicity of body plans all too evident in their animals. Somehow, they thought, one could not make sense of such diversity. Other arthropod groups exhibited greater uniformity of body plans. An insect is an insect: a head, three sets of legs, two sets of wings, and gonopores at the end of the abdomen. Each one of the millions of insect species conforms in principle to this basic body plan. So, too, with other arthropod groups: an arachnid is an arachnid, a sea spider is a sea spider. Even myriapods, although they vary in length, conform to similar body plans. However, such uniformity of body plan is not true for crustaceans. Some crustaceans have long bodies, whereas others are short. Some have gonopores

in the front of the body, most someplace in the middle (but by no means in the same position from group to group), and one or two even have the gonopores at the posterior end of the body. Some bear body shields, or carapaces; others do not. Some possess legs on every segment, but many do not. "Entomologists may have more species," trumpeted carcinologists, "but we have greater diversity of body plans." However, there is little reflection on just what this greater diversity really meant.

Attempts to Get "True" Trees

Morphology

Beginning in mid-1980s, long after cladistic methods of analysis had been taken up in other arthropod groups, more rigorous approaches to crustacean tree building began to emerge. Sieg (1983) published the first true cladograms for any crustacean group, namely, the Tanaidacea, albeit using the paper-and-pencil method of Hennig. The next year, Schram (1984) published the first numerical cladistic analysis of eumalacostracan crustaceans, and a few years later (Schram 1986) employed computer based cladistic analyses for the crustaceans as a whole in his book-length overview of the subphylum.

Furthermore, based on the results of these analyses, Schram (1986) produced a higher taxonomy of Crustacea

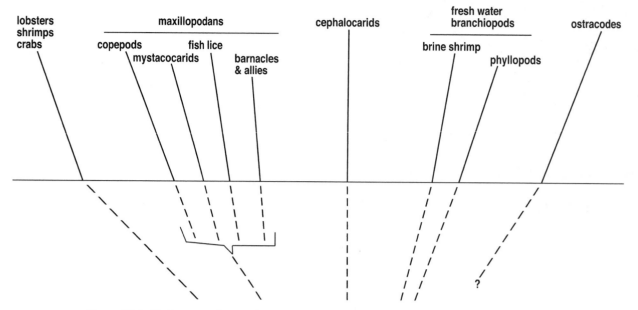

Figure 19.1. Phylogenetic "lawn" of crustacean relationships. This view, with unconnected lines, prevailed for more than 20 years from the early 1960s to the 1980s. Diverse body plans could not be reconciled, so carcinologists made little effort toward "growing" a tree. Modified from Dahl (1963).

that grew out of the cladistic analyses. Essentially, four classes were recognized and defined on the base of apomorphic features: Remipedia, Malacostraca, Maxillopoda, and Phyllopoda (fig 19.2A). Several authorities criticized this arrangement, viewing the disappearance or relocation of favorite higher taxa with alarm. None of the 1986 trees were arrived at a priori; that is, they emerged as the simplest patterns of relationships derived from character matrixes. But some workers took exception to them purely on the basis that the patterns of relationships did not conform with what people had thought before that time. These critics preferred to argue from an evolutionary systematic viewpoint, using a few characters to a priori judge the affinities within the crustaceans and arrange the higher taxa accordingly. Nevertheless, Schram (1986) had a positive effect on the field of carcinology because in ensuing years cladistic analyses of smaller groups within the crustaceans began to appear.

By 1990, a special conference held at Kristineberg, Sweden, was deemed necessary to address the origin of crustaceans and their evolution, within which an attempt was made at generating a new computer-based cladistic analysis of morphological features (Wilson 1992). One is tempted to assume that because the matrix upon which this analysis was based was the product of the deliberations of all the participants at the conference, it would contain few errors. This proved not to be the case (Schram 1993) and had serious consequences for the trees derived from that analysis.

To address this last problem, Schram and Hof (1997) undertook a more comprehensive analysis of fossil and recent crustaceans (fig 19.2B). This paralleled an independently

conceived and carried out attempt directed at the same time by Wills (1997). Wills achieved a "cleaner" result in that he obtained a tree in which the four classes of Schram were more-or-less clearly evident (fig 19.3), albeit in a slightly different branching sequence, but his study was essentially an ingroup analysis rooting the tree to the long-bodied remipedes. Schram and Hof (1997) took a broader approach using insects and myriapods as outgroups (fig 19.2B). In addition, they tested various alternative partitions of the database; for instance, they wanted to determine what happened with and without fossils included (and with fossils alone), or with and without soft anatomy (so often impossible to assess for fossil forms). The shifting patterns of relationship they observed (for details, see Schram and Hof 1997) indicated great instability to the underlying data. Interestingly, the trees of Schram and Hof (1997) did not confirm the results uncovered by Wilson (1992), nor did it convincingly support the four classes derived by Schram (1986).

Molecules

Meanwhile, it was inevitable that the analysis of molecule sequences would enter into the picture. The sequencing of various molecules to elucidate the phylogeny of many crustacean groups proceeded apace through the 1990s. In any analysis of this sort, it is critical to perform comprehensive analyses with a wide array of crustaceans and molecules. Nevertheless, these type studies focus instead on model types or selected species and use only one or two molecules. Perhaps then it is not too surprising that just about every con-

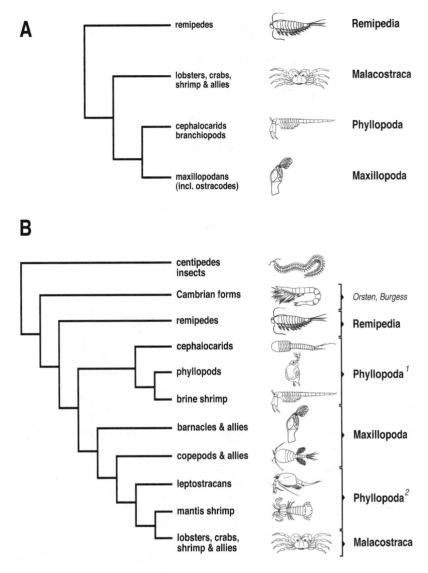

Figure 19.2. (A) A phylogenetic tree for Crustacea derived from a computer-analyzed morphological matrix showing the hypothesized relationships amongst four recognized classes (modified from Schram 1986). (B) A more comprehensive analyses of relationships of fossil and recent crustaceans (simplified from Schram and Hof 1998). This analysis revealed that the Schram (1986) clade of branchiopods and cephalocarids (and allies) is probably polyphyletic (some of those taxa occur near the base, whereas others are high in the tree), and that the maxillopodans are paraphyletic (two separate clades side by side).

ceivable result one could hope for was in fact obtained by these methods (fig 19.4).

Unsurprisingly, one of the first molecules sequenced was 18S ribosomal DNA (rDNA), and studies using this molecule remain the most comprehensive to date. They culminated in the results (fig 19.4A) presented in Spears and Abele (1997). These 18S rDNA data yielded a pattern of a *polyphyletic* Crustacea, that is, crustaceans interspersed among other groups of arthropods. Because Spears and Abele made a conscious attempt to sample a wide array of taxa and several species from each major group, these results have to be considered seriously.

Another laboratory undertook a separate but not quite so taxonomically comprehensive an analysis (fig 19.4B) employing elongation factor-1α (EF-1α) and later augmented this data with RNA polymerase II (Pol II) (Regier and Shultz 1998, Shultz and Rieger 2000). The EF-1α (both alone and with Pol II) yielded trees with at least *paraphyletic* crustaceans vis-á-vis an insect/hexapod clade, but these analyses suffered from a limited array of taxa sampled.

A final set of analyses tried to address the issue of comprehensiveness of the character set (Edgecombe et al. 2000, Giribet et al. 2001). As a result, these studies used total evidence approaches, combining morphology together with molecular data, but still examined only a limited number of taxa. The results (fig 19.4C) stand in contrast to the above studies in that virtual *monophyly* of the limited number of crustaceans (as well as other arthropod groups) emerged.

All of these molecular data sets exhibited the phenomenon of long-branch attraction, especially prominent in Spears and Abele (1997) and Giribet et al. (2001), with the terminal taxa of long branches tending to emerge in single clades. We can summarize the results derived from these molecular studies as follow: (1) The phylogenies do not agree with each other. (2) The results are quite at odds with the analysis based on morphology alone. (3) The varying patterns of relationship within their own data depending on different methods of analysis. Like the differences noted between morphological analyses, the inconsistent results

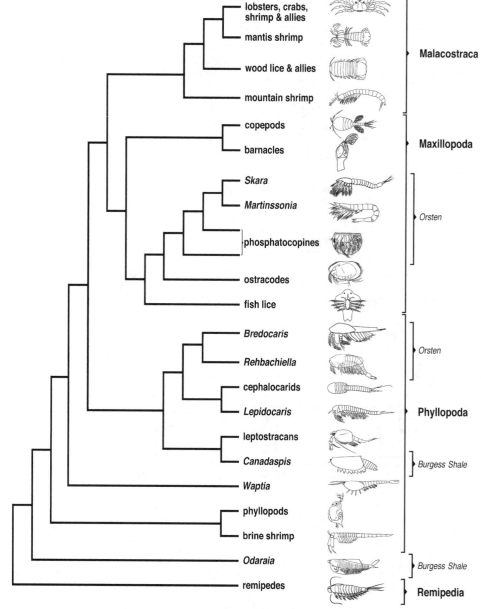

Figure 19.3. Comprehensive analyses of relationships of fossil and recent crustaceans (simplified from Wills 1997). Note that Wills uncovered a pattern of relationships similar to that shown in figure 19.2A, except that the clade of branchiopods and cephalocarids (and allies) is shifted toward the base of the tree and is paraphyletic. Orsten and Burgess fossils appear in different places in this tree.

obtained from molecule sequencing are vexing. Again, this may indicate some underlying problem with the assumptions concerning either the nature of, or membership in, what we have come to call Crustacea—reflecting maybe the great disparity of body plans.

Challenge of the Cambrian

Naturally, when one discusses phylogeny, one cannot avoid deliberating on fossils; the crustaceans are no exception. We could categorize crustacean fossils into two types. On the one hand, are taxa that extend throughout the fossil record and essentially amplify aspects of the deep history of extant forms.

However, this record is uneven. For example, the background history concerning "lobsters" is relatively well known, whereas the situation within a group such as the copepods is opaque (for details, see Schram 1986). On the other hand, some fossils do not easily fit into living groups. These are largely Paleozoic, especially Cambrian, in age, and consideration of these species in a phylogenetic context presents real challenges. Yet these taxa are crucial to understanding the history of crustaceans and crustacean-like creatures because these fossils come from a time when the basic body plans took origin. We can delineate two groups of these fossils: the short-bodied, micro-arthropods of the Cambrian Orsten and the long-bodied arthropods from faunas such as the Burgess Shale and Chengjiang (both described below). All these fos-

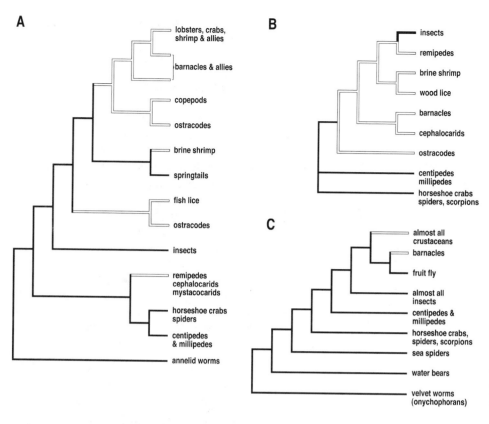

Figure 19.4. Disparity of phylogenetic results from molecular sequencing: trees simplified from the published versions, with crustacean clades highlighted with open lines. (A) Tree derived from 18S rDNA (simplified from Spears and Abele 1997); note polyphyletic crustaceans, and the long-branch attraction of a clade of remipedes, cephalocarids, and mystacocarids. (B) Tree derived from EF-1α and Pol II (simplified from Shultz and Regier 2000); note paraphyletic crustaceans. (C) Total evidence tree derived from eight molecular loci and morphology (simplified from Giribet et al. (2001); note a more-or-less monophyletic Crustacea (given that long branch attractions are undoubtedly operant in producing the barnacle/fruit fly clade).

sils fortunately are characterized by high quality of preservation, such that the monographs describing these animals often resemble in their detail those for living forms.

The Orsten

Few fossil discoveries in the last century have so consistently yielded amazing insights into the minutest aspects of Cambrian life as have the micro-arthropods of the Orsten, or "stink stones." The first, and still most productive, sites come from the bituminous limestones of Sweden, but similar localities occur around the world. The literature is voluminous, prolific, and detailed. The often unusual preservation of these fossils (fig 19.5) allows insights into not only the fine details of features such as appendage setation (see Müller and Waloßek 1986b, Waloßek and Müller 1998, Waloßek 1999), but also knowledge of larval forms (Müller and Waloßek 1986a; Waloßek and Müller 1989). In many instances these larvae can be related to adult and subadult forms into coherent life cycles. Two basic forms of larvae are noted by Waloßek and Müller (1997): those that have four sets of limbs [antennae + three additional limb pairs], and those that have three sets of limbs [antennae + two additional locomotory/feeding limb pairs = nauplius]. The insights these fossils offer into the unfolding process of "cephalization"; that is, the formation of a distinct head region is new.

Furthermore, another critical feature of the Orsten fossils, which we believe has not been fully absorbed by most authorities, is the generally short-bodied nature of their structural plans. Orsten fossils not only are characterized by small body size, in the range of a few millimeters but also exhibit a relatively small number of body segments. In this, the Orsten animals differ from the long-bodied living remipedes and also from many of the elongate Cambrian species from the Burgess Shale and Chengjiang faunas (fig 19.6).

Consideration of the Orsten fossils has provided us with many hypotheses about the evolution of arthropods, and especially crustacean body plans. Some of these, for example, the differences between an "abdomen" and a "pleon" (Waloßek 1999), have received unexpected confirmation from developmental genetics (Schram and Koenemann 2003). Other models of morphological evolution, for example, the suggested significance of "proximal endites" (see Waloßek and Müller 1997), may have to contend with alternative hypotheses (Schram and Koenemann 2001). Nevertheless, Orsten arthropods play a key role in reconstructing the phylogeny of the crustaceans.

The Burgess and Chengjiang Faunas

The fossils of the Burgess Shale of British Columbia, Canada, have long been known for their unusual preservation and unique array of arthropod body forms (see Gould 1989). In

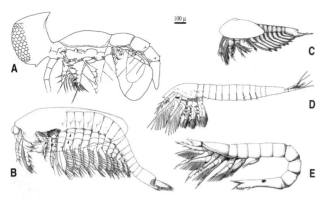

Figure 19.5. Some representative Cambrian Orsten "crustaceomorphs" all drawn to the same scale. (A) *Henningsmoenicaris scutula* (modified from Waloßek 1999). (B) *Rehbachiellakinnekulensis* (modified from Waloßek 1993). (C) *Bredocaris aadmirabilis* (modified from Müller and Waloßek 1988). (D) *Skara anulata* (modified from Müller and Waloßek 1985). (E) *Martinssonia elongata* (modified from Müller and Waloßek 1986b).

recent years, the Burgess fossils have been joined by an equally well-preserved series of fossils from the Cambrian beds of Chengjiang, China (Hou and Bergström 1997, Bergström and Hou 1998). These assemblages largely consist of macrofossils with body sizes up to 120 mm. They present a contrasting group of species in apposition to the micro-arthropods of the Orsten. Unlike the short-bodied Orsten taxa, the trunks of many (although not all) of the Canadian and Chinese fossils have a large number of homonomous segments each bearing a pair of limbs identical to each other (fig 19.6A,B). In this they resemble many modern forms, for example, the long-bodied myriapods and

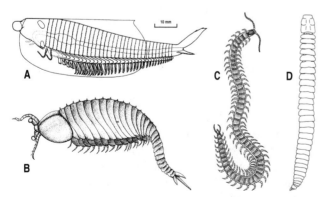

Figure 19.6. Some typical long-bodied arthropods. (A). *Odaraia alata*, from the Cambrian Burgess Shale, Canada (from Briggs 1981). (B) *Fuxianhuia protensa*, from the Cambrian Chengjiang, China. (C) A living geophilomorph centipede (from Meglitsch and Schram 1991). (D) Body outline of *Speleonectes lucayensis*, a remipede crustacean (from Schram et al. 1986).

the remipede crustaceans (fig 19.6C and 19.6D) that also have long homonomous trunks.

Clues from Developmental Genetics

The most exciting new source of information in recent years that can be applied to issues of phylogeny is emerging from the field of developmental genetics. This area of research is leading to fundamentally new insights into relationships of arthropods and our understanding of comparative anatomy. As an example, Averof and Akam (1993, 1995) and Akam et al. (1994) suggested that the patterns of expression of the *Hox* gene complexes indicate that crustaceans and hexapods share a common body plan. However, the *Hox* condition in myriapods, chelicerates, or the near-arthropods such as tardigrades and onychophorans was then, and still is largely now, not known. To resolve relationships between two groups, one needs to assess their position in reference to a third, potential outgroup. We now know that *Hox* genes are shared by *all* higher metazoans. Therefore, *Hox* genes in broad aspect are plesiomorphic features and thus tell us little directly about phylogenetic relationships within the groups.

For example, Cartwright et al. (1993) uncovered multiple copies of the *Hox* gene complex in the horseshoe crab, *Limulus*. These resemble the *Hox* B sequence in the mouse. The multiple *Hox* clusters in horseshoe crabs could be considered an autapomorphy of the limulids, probably. Even though this is the only known occurrence of multiple *Hox* gene complexes outside of the Chordata, no one is about to suggest on this basis a return to the old arachnid theory for vertebrate origins. We have to be careful with conclusions on arthropod relationships based on *Hox* genes alone.

Nevertheless, individual aspects of *Hox* gene complexes can be used for assessing homologies. A stunning example concerns the examination of *Hox* expression in chelicerates (Damen et al. 1998, Telford and Thomas 1998). There appears clear evidence to indicate that the old ideas about the lack of a deutocerebrum in chelicerate brains were wrong. Rather, *Hox* gene patterns indicate that the relevant "deutocerebral segment" is in fact present in chelicerates. As a result, the chelicerae, those diagnostic limbs for horseshoe crabs, true spiders, and sea spiders, are seen as homologous to the hexapod antennae and, by extension, the crustacean antennules (first antennae).

Schram and Koenemann (2001) examined modes of limb formation among arthropods and compared the resulting limb patterns with those found in fossils. In this instance, the purported close affiliation of two well-known fossil species, *Lepidocaris rhyniensis* from the Devonian and *Rehbachiella kinnekullensis* from the Cambrian Orsten, to the living brine shrimp could not be supported. The development of the legs seen in the fossils is entirely different from that of the living brine shrimp and its relatives, precluding close relationship.

Straitjackets of the Past

Finally, we need to highlight one other factor that strongly influences discussions of phylogenetic relationships. It is orthodoxy, the dogma of long-standing assumptions that constrain thought and have become the heavy shackles of tradition. "Analyzing phylogeny" is an exercise closely akin to scriptural exegesis. The weight of authority sits heavy. Although a book could be written on the subject, discussion of two issues will illustrate the point.

We alluded above to the generally short-bodied nature of Orsten arthropod body plans and how the potential significance of this has not been fully absorbed by the most authorities. The classic theories about the direction of arthropod evolution have featured long, equally segmented bodies as ancestral forms that gave rise to shorter-bodied organisms. These descendants were believed to have segments specialized into regions (like a thorax, or abdomen, or pleon). It is difficult to identify the ultimate source for this idea. The long-bodied ancestor theory for arthropods is the result of a consensus arrived at when it was first suggested that annelid worms were the precursors of arthropods in the "great chain of being," or "ladder of life." Snodgrass (1935, 1952) is only one of the more recent advocates of this view (fig 19.7).

The effect of this fully emerges when examining morphology-based phylogenies of arthropods such as illustrated by

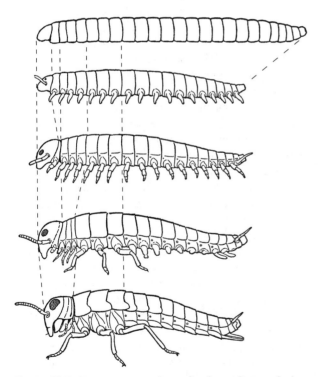

Figure 19.7. Diagrammatic schema for the evolution of a long-bodied articulate (at the top), through successive degrees of body regionalization and specialization, into an arthropod (at the bottom). Modified from Snodgrass (1935).

Schram and Hof (1997) or Wills (1997). The latter is particularly relevant here because Wills (1997), as mentioned above, performed an ingroup analysis, rooting his tree to the Remipedia, because he believed this class of crustaceans had the most primitive body plan. Schram and Hof (1997) achieved the same result without explicitly making the assumption of "long-bodied = primitive." The effect was the same because their chosen outgroups induced polarities determined by the use of longer-bodied forms such as centipedes.

Another example of historical constraint arises out of the classic definition of Crustacea, namely, that they possess a "second antenna." When some crustaceans really do bear a second set of antennae on the segment immediately posterior to the "first antennae," we then deal with a true apomorphy. However, we cannot score any limb on that segment as a second antenna. In actual fact, the "first post-antennal limb" is by far one of the most variable appendages amongst all the arthropods (fig 19.8). In addition to real antennal specializations of that limb (fig 19.8D), we also see at that position purely locomotory limbs (fig 19.8C), limbs that function in feeding and locomotion (fig 19.8A,B), limbs that serve as attachment structures, limbs that assist in copulation (fig 19.8E), and reduced limbs of uncertain function (fig 19.8F)—and these variants are only the alternatives seen among crustaceomorphs. To this we might add limbs modified as a "labrum" that apparently occur in insects and myriapods. By assigning "traditional" simple labels in character analyses, for example, presence or absence of a labrum or a second antenna, phylogenetically important information is potentially disregarded. By ignoring the rich variety of anatomy expressed on that limb we mask some possible apomorphies that could help sort phylogenetic relationships among crustaceomorph arthropods.

A New Analysis

In light of the above commentary, we have conducted a new, preliminary analysis of fossil and recent crustaceomorphs. (We prefer this term, which serves to denote all the traditional Crustacea *sensu stricto* and the various fossil stem forms.) In this we paid particular attention to incorporating as much useful information as possible from the Cambrian and other relevant fossil forms. In combination with morphology of modern forms, we attempted to be as "theory free" as we could in scoring characters. In addition, constructed a matrix using features that have not figured prominently in the past, if at all, for example, body plans, gonopore locations, *Hox* expressions, and limb ontogeny. We polarized our data with a hypothetical ancestor that represents a short-bodied form. A total of 42 characters (half of them multistate) were used for 31 end groups. A series of alternative analyses were performed, but in the end we preferred a partially weighted and unordered data set for which we obtained 45 trees. We took the strict consensus of those trees and ex-

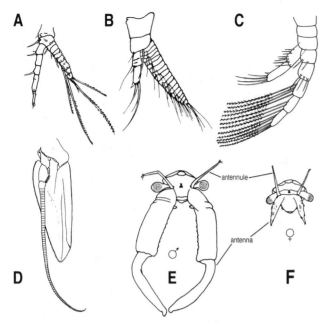

Figure 19.8. Variations of morphology and function found on the so-called second antenna of crustaceans. (A) The mystacocarid, *Derocheilocaris typicus*, a locomotory/feeding limb. (B) The cephalocarid, *Hutchinsoniella macracantha*, a locomotry/possibly feeding limb. (C) Devonian fossil species, *Lepidocaris rhyniensis*, probably a feeding limb. (D) The decapod, *Penaeus setiferous*, a true "second" antenna. (E) Male brine shrimp, *Branchinecta campestris*, a copulatory clasper. (F) Female *Branchinecta,* with rudimentary limb stub of uncertain function. All modified from Schram (1986).

plored alternative branchings for polychotomies to find resolved branching patterns of minimum tree length.

Our working hypothesis is presented in figure 19.9. We have highlighted the traditionally recognized crustacean taxa. These are groups that can for the most part be defined by concise sets of characters related to gonopore location and other aspects of the body plan. The tree renders three separate clades that until now we might have placed within the Crustacea *sensu stricto*. These clades often contain fossil taxa, but we did not automatically include fossils within the highlighted clades if doing so would obscure the definition of the clade. One of these clades, which might be termed "Eucrustacea," contains the Malacostraca (lobsters, crabs, and allies), a reduced array of maxillopodans (including copepods, barnacles, and closely related forms), remipedes, and cephalocarids. We obtained a sister clade to this assemblage, which we might term the "Pancrustacea." This includes Branchiopoda, insects, and some Devonian and Cambrian Orsten fossil species often included within branchiopods. We cannot, however, include those fossils within the branchiopods because they appear to be separated from the brine shrimp and allies by the insects. A third crustacean clade can be seen deeper in the tree and is composed of the fish lice (branchiurans) and the mystacocarids allied with *Skara*, an Orsten genus.

The tree clearly reveals a series of individual stem taxa and stem clades leading toward the three clades with living crustacean forms. However, it appears that at least some of the crustacean-containing clades also contain within them other stem forms, for example, Pancrustacea can be interpreted as a crown clade with *Lepidocaris* and *Rehbachiella* as stem forms to it. The juxtaposition of insects with branchiopods is not as startling as it might appear. Galent and Carroll (2002) and Ronshaugen et al. (2002) have recently shown that slight changes in the *Hox* gene *Ultrabithorax* can achieve an astounding shift of morphology in the fruit fly, *Drosophila*, and the brine shrimp, *Artemia*. Finally, the large clade near the base of the present tree in figure 19.9 contains an array of long-bodied taxa, However, it is unclear whether this clade will persist as we expand the database to include some additional taxa and characters, as we move toward a more definitive analysis.

Are Crustacea Monophyletic?

What is a crustacean? The classic definition (see Schram 1986) says that crustaceans are arthropods that have (1) a five-segmented head including two pairs of antennae, mandibles, and two pairs of maxillae; (2) a tendency to fuse the head segments to form a cephalic shield and to develop from the back of the head a posterior out-growth (carapace); (3) a tendency to regionalize the trunk into an anterior thorax and a posterior abdomen, or pleon; and (4) anamorphic development that typically begins with a unique larva or ontogenetic stage termed the nauplius, or egg nauplius. This definition has essentially served since the time of Cuvier and Latreille in the early 1800s. However, Waloßek (1999), in light of his work on Orsten arthropods, proposed some modification of this definition: (1) the antenna (2nd antenna) with a coxa proximal to the basis; (2) the maxillules (1st maxilla) specialized for food transport; (3) the labrum, atrium oris, paragnaths, sternum, and feeding areas of limbs covered with fine hairs (setae) (even in the nauplius); (4) the trunk terminating in a conical segment that bears the anus terminally and a pair of caudal rami; and (5) the first larval or developmental stage as a nauplius. Under either definition, Crustacea remain as a monophyletic crown taxon.

Therefore, results of our new analysis indicating polyphyly are rather significant. It would appear that the Crustacea *sensu stricto* do not constitute a monophyletic group. Waloßek (1999, and elsewhere) has argued forcefully and convincingly for a stem group/crown group understanding of crustacean history. In our present analysis, we do obtain stem taxa arrays. However, we have a multiplicity of crown groups, each of which has their fossil stem taxa. All these clades in turn form a transition series from the root of a "crustaceomorph" clade near the Orsten genus *Cambrocaris* extending to diverse terminal groups that we have traditionally identified as "Crustacea."

Is their any validity to the concept of a monophyletic taxon Crustacea? We could of course *redefine* what it is to be a crus-

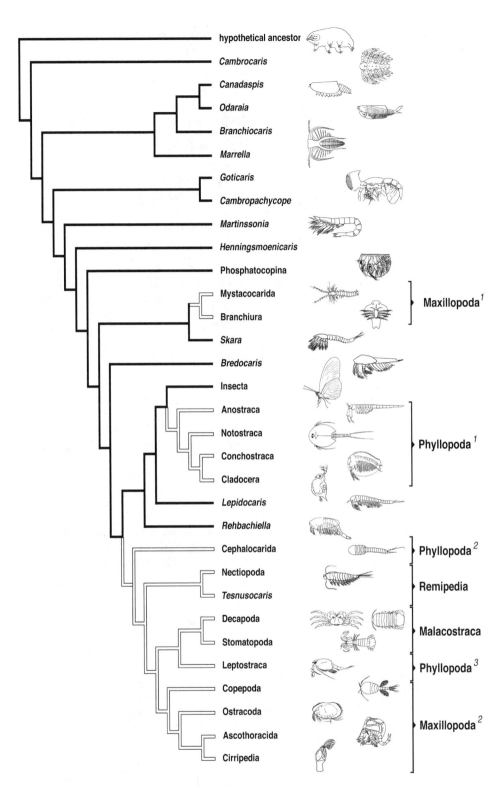

Figure 19.9. Preliminary cladistic analysis of "crustaceomorph" arthropods. Note the three distinct clades of "crustaceans" highlighted with the open lines. There is a short-thorax, mystacocarid/fish-lice (Branchiura) clade, typically placed in the past within Maxillopoda. Brachiopoda sit in a larger clade with insects (Pancrustacea) and other fossil forms, which in Schram (1986) was a part of the class Phyllopoda. Eucrustacea, with the remaining crustaceans that have gonopores located midbody on the 6th through 8th thoracic segments, which includes what Schram (1986) classified as Remipedia, Malacostraca, most of the Maxillopoda, and miscellaneous phyllopods.

tacean, that is, seek to identify the last common ancestor of all the currently recognized groups we refer to as "Crustacea" and extract the ground pattern for *that* "ancestor." These features then could serve to define that monophylum. However, in doing that we would (1) dilute our understanding of what it is to be a crustacean (and we might ask, in so doing would such a watered-down definition really have any meaning?) and (2) simply leave untouched the issue of unreconciled disparity of body plans.

We prefer to accept the results of the analysis at face value and acknowledge that Crustacea *sensu stricto* are not a monophyletic group. This has two advantages. First, it maintains the primacy of Baupläne, body plans, as a raison d'être for recognizing major groups of arthropods. Second, it allows us to better reconcile the disparity of results between morphological and molecular analyses. The message of many of the molecular studies has been that there is a problem with maintaining monophyly for crustaceans. By reassessing the assumptions upon which the morphological studies have been based, we now can begin to see the way clear toward a grand synthesis of cladistic phylogeny that can more effectively integrate crustaceans into the Tree of Life.

Many times morphologists summarily reject the results of molecular phylogenies because these trees do not agree frequently with those derived from morphology alone. The skepticism of morphologists may be justified, because molecular databases are often less comprehensive regarding taxon sampling. However, morphologists ignore at their peril the message of the molecules, which should at least compel a reexamination of the assumptions implicit in morphological character surveys. Conversely, molecular systematists look askance at the morphologists and derided them for their archaic and "subjective" approaches to phylogeny development. However, it is hubris on the part of molecular systematists to persist in ignoring the conflicting results of different molecule databases and not to acknowledge the financial and technical limits of achieving real comprehensiveness of taxon and character sampling in a molecular framework. Obviously, a truce is needed for real cooperation to occur and promise the hope of a coordinated consensus on this vexing issue of crustaceomorph relationships with other arthropods on the Tree of Life.

Acknowledgments

We want to express special thanks to Jan van Arkel, who prepared the graphic figures, and Joris van der Ham, who assisted us with cladistic analysis of the Cambrian Orsten forms. We also want to pay tribute to the many workers on Cambrian fossil arthropods, whose careful and detailed studies published over the years have allowed us to better "see" the deep history of the arthropods. In addition, special mention should go to Dieter Waloßek, with whom we have had numerous and intense discussions over the years, and, although we do not always agree, these exchanges have always served as a learning process. Finally, Trisha Spears has been a guiding light for us on matters concerning molecular phylogenies.

Literature Cited

Akam, M., M. Averof, J. Castelli-Gair, R. Dawes, F. Falciani, and D. Ferrier. 1994. The evolving role of *Hox* genes in arthropods. Development 1994 (suppl.):209–215.

Averof, M., and M. Akam. 1993. HOM/*Hox* genes of *Artemia*: implications for the origin of insect and crustacean body plans. Curr. Biol. 3:73–78.

Averof, M., and M. Akam. 1995. *Hox* genes and the diversification of insect and crustacean body plans. Nature 376:420–423.

Bergström, J., and X. Hou. 1998. Chengjiang arthropods and their bearing on early arthropod evolution. Pp. 151–184 in Arthropod fossils and phylogeny (G. Edgecombe, ed.). Columbia University Press, New York.

Briggs, D. E. G. 1981. The arthropod *Odaraia alata* Walcott, Middle Cambrian, Burgess Shale, British Columbia. Phil. Trans. R. Soc. Lond. B 291:541–584.

Cartwright, P., M. Dick, and L. W. Buss. 1993. *HOM/Hox* type homeoboxes in the chelicerate *Limulus polyphemus*. Mol. Phylogenet. Evol. 2:185–192.

Dahl, E. 1963. Main evolutionary lines among recent Crustacea. Pp. 1–15 in Phylogeny and evolution of Crustacea (H. Whittington and W. D. I. Rolfe, eds.). Museum of Comparative Zoology, Cambridge.

Damen, W. G. M., M. Hausdorf, E.-A. Seyfarth, and D. Tautz. 1998. A conserved mode of head segmentation in arthropods revealed by the expression pattern of *Hox* genes in a spider. Proc. Natl. Acad. Sci. USA 95:10665–10670.

Edgecombe, G. D., G. D. F. Wilson, D. J. Colgan, M. R. Gray, and G. Casis. 2000. Arthropod cladistics: combined analysis of histone H3 and U2 sequences and morphology. Cladistics 16:155–203.

Galent, R., and S. B. Carroll. 2002. Evolution of a transcriptional repression domain in an insect Hox protein. Nature 415:910–913.

Giribet, G., G. D. Edgecombe, and W. C. Wheeler. 2001. Arthropod phylogeny based on eight molecular loci and morphology. Nature 413:157–161.

Gould, S. J. 1989. Wonderful life. Norton, New York.

Hou, X., and J. Bergström. 1997. Arthropods of the Lower Cambrian Chengjiang fauna, southwest China. Fossils Strata 45:1–116.

Meglitsch, P. A., and F. R. Schram. 1991. Invertebrate zoology 3rd ed. Oxford University Press, New York.

Müller, K. J., and D. Waloßek. 1985. Skaracarida, a new order of Crustacea from the Upper Cambrian of Västergötland, Sweden. Fossils Strata 17:1–65.

Müller, K. J., and D. Waloßek. 1986a. Arthropod larvae from the Upper Cambrian of Sweden. Trans. R. Soc. Edinb. Earth Sci. 77:157–179.

Müller, K. J., and D. Waloßek. 1986b. *Martinssonia elongata* gen. et sp. n., a crustacean-like euarthropod from the Upper Cambrian "Orsten" of Sweden. Zool. Scr. 15:73–92.

Müller, K. J., and D. Waloßek. 1988. External morphology and larval development of the Upper Cambrian maxillopod *Bredocaris admirabilis*. Fossils Strata 23:1–70.

Regier, J. C., and J. W. Shultz. 1998. Molecular phylogeny of arthropods and the significance of the Cambrian "explosion" for molecular systematics. Am. Zool. 38:918–928.

Ronshaugen, M., N. McGinnis, and W. McGinnis. 2002. Hox protein mutation and macroevolution of the insect body plan. Nature 415:914–917.

Schram, F. R. 1984. Relationships within eumalacostracan crustaceans. Trans. San Diego Soc. Natl. Hist. 20:301–312.

Schram, F. R. 1986. Crustacea. Oxford University Press, New York.

Schram, F. R. 1993. Review of: Boxshall, G. A., J.-O. Strömberg, and E. Dahl. 1992. The Crustacea: origin and evolution. Acta Zool. 73:271–392. J. Crust. Biol. 13: 820–822.

Schram, F. R., and C. H. J. Hof. 1997. Fossils and the interrelationships of major crustacean groups. Pp. 233–302 *in* Arthropod fossils and phylogeny (G. Edgecombe, ed.). Columbia University Press, New York.

Schram, F. R., and S. Koenemann. 2001. Developmental genetics and arthropod evolution: part I, on legs. Evol. Dev. 3:343–354.

Schram, F. R., and S. Koenemann. 2003. Developmental genetics and arthropod evolution: on body regions of Crustacea. Crust. Issues 15:75–92.

Schram, F. R., J. Yager, and M. J. Emerson. 1986. Remipedia, Pt. 1: Systematics. San Diego Soc. Nat. Hist. Mem. 15:1–60.

Shultz, J. W., and J. C. Regier. 2000. Phylogenetic analysis of arthropods using two nuclear protein-encoding genes supports a crustacean + hexapod clade. Proc. R. Soc. Lond. B 267:1011–1019.

Sieg, J. 1983. Evolution of Tanaidacea. Crust. Issues 1:229–256.

Snodgrass, R. E. 1935. Principles of insect morphology. McGraw-Hill, New York.

Snodgrass, R. E. 1952. A textbook of arthropod anatomy. Comstock, Ithaca, NY.

Spears, T., and L. G. Abele. 1997. Crustacean phylogeny inferred from 18S rDNA. Pp. 169–187 *in* Arthropod relationships (R. Fortey and R. Thomas, eds.). Chapman and Hall, New York.

Telford, M. J., and R. H. Thomas. 1998. Expression of homeobox genes shows chelicerate arthropods retain their deutocerebral segment. Proc. Natl. Acad. Sci. USA 95:10671–10675.

Waloßek, D. 1993. The Upper Cambrian *Rehbachiella* and the phylogeny of Branchiopoda and Crustacea. Fossils Strata 32:1–202.

Waloßek, D. 1999. On the Cambrian diversity of Crustacea. Pp. 3–27 *in* Crustaceans and the biodiversity crisis (F. R. Schram and J. C. von Vaupel Klein, eds.). Koninklijke Brill, Leiden.

Waloßek, D., and K. J. Müller. 1989. A second type A-nauplius from the Upper Cambrian "Orsten" of Sweden. Lethaia 22:301–306.

Waloßek, D., and K. J. Müller. 1990. Upper Cambrian stem-lineage crustaceans and their bearing upon the monophyletic origin of Crustacea and the position of *Agnostus*. Lethaia 23:409–427.

Waloßek, D., and K. J. Müller. 1997. Cambrian "Orsten"-type arthropods and the phylogeny of Crustacea. Pp. 139–153 *in* Arthropod relationships (R. Fortey and R. Thomas, eds.). Chapman and Hall, New York.

Waloßek, D., and K. J. Müller. 1998. Early arthropod phylogeny in light of the Cambrian "Orsten" fossils. Pp. 185–231 *in* Arthropod fossils and phylogeny (G. Edgecombe, ed.). Columbia University Press, New York.

Wills, M. A. 1997. A phylogeny of recent Crustacea derived from morphological characters. Pp. 189–209 *in* Arthropod relationships (R. Fortey and R. Thomas, eds.). Chapman and Hall, New York.

Wilson, G. D. F. 1992. Computerized analysis of crustacean relationships. Acta Zool. 73:383–389.

Rainer Willmann

Phylogenetic Relationships and Evolution of Insects

More than 1.2 million recent insect species have been described, but recent estimates suggest several million additional species are to be expected (see the early discussion in Weber 1933). About 25,000 fossil species are known, but more than one billion insect species must have existed in the past, of which only a small minority have left any trace in sediments, and again, of these, only a small fraction will ever be found (Willmann 2002). Systematically, insects are one of the most studied groups. Modern biosystematics was developed with insects as one of its main targets, because Willi Hennig (1913–1976), the founder of phylogenetic systematics, was an entomologist dealing mainly with Diptera. However, the branching sequence of the insect tree is difficult to reconstruct for several reasons. First, insects are a fast-evolving, enormously diverse group, and synapomorphies of subordinate groups may have become veiled by more recent evolutionary changes. Second, some taxa have preserved ancient characters, and it appears that structures once lost may reappear from time to time, leading to confusion among phylogenetists. Third, the amount of homoplasy has been underestimated. Fourth, although a huge number of fossil insects have been assembled and described, the gaps in the fossil record are considerable. Fossils have repeatedly shown that some phylogenetic conclusions based on extant taxa are untenable. Last but not least, there are far too few entomological morphologists, and therefore, there is an enormous lack of knowledge about the details of their structural disparity, constructional morphology, relationships of taxa of

all hierarchical levels, and ground patterns of recognized monophyla.

Many authors prefer the name "Hexapoda" over "Insecta." Snodgrass (1952), for example, pointed out that "Insecta" has been used in very different ways, and indeed Linné (1758) included with them crustaceans, myriapods, and chelicerates. Some recent authors such as Jamieson et al. (1999), Kristensen (e.g., 1975, 1991), Ross et al. (1982), and Wheeler et al. (2001) considered "Insecta" a synonym of "Ectognatha," whereas Whiting et al. (1997: fig. 5) equated "Insecta" with "Pterygota" (but did not use the latter term). This recalls Mayer (1876), who had argued that Thysanura and Collembola are primarily wingless and also used the term Insecta to refer only to winged and secondarily wingless kinds. Thirty years later, Handlirsch (1908), uncertain of the phylogenetic relationships of Collembola, Diplura, Archaeognatha, and Zygentoma (Protura had only been introduced in 1907), distinguished four hexapod classes, calling Pterygota "Insecta s. str." In this chapter, "Insecta" is used as synonym for "Hexapoda," which is in accordance with the works of Hennig (e.g., 1953, 1969, 1981), many widespread textbooks both old and recent (e.g., Naumann 1991, Kaestner 1973, Richards and Davies 1977, Arnett 2000), and common use (e.g., Snodgrass 1935).

In the following discussion, I offer insights into the arguments for and against different relationships. Many assumed nodes are supported by very few characters. For example, Wheeler et al. (2001) have, on average, seven morphological characters per node (not counting autapomorphies of terminal taxa) and two to six characters in about 50% of the nodes; Beutel and Gorb (2001) have even fewer (111 characters for 35 nodes). Exclusion or inclusion of just a few characters would easily produce new phylogenetic hypotheses. This emphasizes the need for more morphological data as well as for more thorough morphological studies.

Many of the taxa treated in the following have been given categorical ranks, and most of the widely known taxa appear as "orders" in traditional classifications. In the following, no use is made of categorical ranks, because different authors differ in their opinion as to the rank of a particular taxon. Many systematists state that sister groups must be assigned the same categorical rank, which implies that usually a particular rank (e.g., order) can only be used twice along a particular evolutionary lineage. For example, if Lepidoptera (moths and butterflies) is ranked as an order, then their sister group (Trichoptera, caddis flies) would be an order as well, whereas the immediately superordinate taxon of the two (Amphiesmenoptera) would deserve a higher rank. Even higher ranks would be attributed to Mecopteria (which includes Amphiesmenoptera), Holometabola (which includes, e.g., Mecopteria), Eumetabola (which includes Holometabola and Acercaria), Neoptera (which includes Eumetabola) and Polyneoptera (if this is a monophylum). Odonates (damselflies and dragonflies) as the sister group of Neoptera would require the same categorical rank as the latter, which is certainly not the rank "order". Thus, although cladograms can be matched up with systematizations using Linnaean categorical ranks because both are hierarchies, categorical formal ranks are impractical in a phylogenetic system. However, the main issue is that Linnaean categorical formal ranks like family, order, class, and so forth, were not introduced to indicate sister group relationships. They were not coined in a phylogenetic context but to serve classifications on the basis of Aristotelian logic instead (Griffiths 1974, 1976, Willmann 1987, 1997), a major step in systematics being the transformation of a purely logical attempt into a phylogenetical science (Burckhardt 1903). As Artois (2001:10) put it, "[T]he nested hierarchical structures of the Linnean system and the nested hierarchy found by phylogenetic analysis are based on completely different premises and only superficially resemble one another." The debate on the issue is ongoing in various directions (e.g., Artois 2001, Nixon and Carpenter 2000, Papavero et al. 2001).

Origin and Sister Group of the Insects

Insects are primarily wingless. Along with the possession of three tagmata (head, thorax, and abdomen), two corneagene cells of the ocelli developed as primary pigment cells (Paulus 1979), posterior tentorial apodemes that are elements of the head skeleton (Koch 2000), six leg segments (Kristensen 1981, Willmann 1998; possibly a plesiomorphy, because the

number is also present in Symphyla and Pauropoda, but homology of the podomeres among Tracheata is not clear), 14 body segments (possibly a plesiomorphy shared with progoneates), and lack of appendages of any kind on abdominal segment 10. Hexapody is considered to be one of the apomorphies of insects. Contrasting with this view, Manton (1977) believed that hexapody has developed within Hexapoda independently five times, in Diplura, Protura, Collembola, Thysanura, and Pterygota, because leg mechanics are different in the groups and because the mechanism in one of them cannot be ancestral to that in one of the others. What Manton had overlooked, however, is the fact that the different mechanisms may have had their origin in a common ancestor with a leg mechanism not found in any of the recent taxa. Although insect monophyly is doubted by some (see Dohle 1998 and discussion in Klass and Kristensen 2001), most molecular sequence studies have supported the view that insects are a natural taxon (Wheeler et al. 2001), which is accepted here.

It has been suggested that insects are the sister group of or derive directly from crustaceans [Crustacea + Insecta = Tetraconata (Dohle 2001), because they possess a crystal cone consisting of four pieces in the ommatidia of their compound eyes). But according to most morphological data, the Tetraconata hypothesis is in all probability not true, because insects share an impressive number of derived features with Chilopoda (centipedes) and Progoneata (millipedes and relatives), together constituting Tracheata. Possible synapomorphies are the loss of the first post-antennal head segment (intercalary segment), loss of the mandibular palp, loss of the pretarsal levator muscle with only one muscle remaining, possession of ectodermal Malpighian tubules, and tracheae (all discussed by Snodgrass 1938, but also many subsequent authors), as well as the organ of Tömösvary (which is a receptor on the head described under various names in Chilopoda, Diplopoda, Symphyla, Pauropoda, Collembola, and Protura), possession of anterior tentorial arms (Kristensen 1989), addition to the head of a sixth segment bearing the second maxillae, and the centriole adjunct in the sperm (Jamieson et al. 1999). [Hilken's (1998) conclusion as to the multiple origins of tracheae within Tracheata is contradicted here.] According to this phylogenetic hypothesis, the crystal cone in the compound eye is assumed to be lost in centipedes and progoneates. Moreover, the progoneates (Symphyla + Dignatha) share a number of derived characters only with hexapods.

The progoneates are characterized by several derived characters, for example, the position of the genital opening that is in the anterior part of the body, the development of the midgut within the yolk (the midgut lumen is therefore free from the yolk), the formation of a fat body out of vittelophages, trichobothria with basal bulb, and loss of the palps of the first maxilla (Dohle 1998, Kraus 2001). Therefore, progoneates may well be monophyletic and the sister group of the insects. The group Insecta + Progoneata has been called Labiata (Snodgrass 1938) because insects and one progoneate taxon, the symphylans, have their posterior mouth parts fused into the so-called labium that is often considered to be a derived labiatan ground pattern character (but see below).

Some derived characters are possessed by insects and symphylans only. These include styli on the underside of the body, vesicles in a ventral position (vs. their presence at the basal podomere in some diplopodans), and appendages at the labium (glossa and paraglossa). Furthermore, the labium, a mouth part consisting of the united second maxillae that has been used as an argument for the Labiata hypothesis (see above), is in fact present only in Symphyla and Insecta: it cannot be traced in Dignatha because it has the second maxillae reduced. Therefore, it is difficult to decide between the Progoneata + Insecta hypothesis and the Symphyla + Insecta hypothesis (see Willmann 2003 for details). The central problem with the latter grouping is the anterior genital openings possessed by the progoneates only. Indeed, some authors claim that the anterior genital opening is not a synapomorphy of the progoneates but that the positions of the genital openings in insects somewhere in the posterior body area derive from a progoneate situation. A fourth hypothesis suggests that insects constitute the sister group to the myriapods (Myriapoda = Chilopoda + Progoneata). This hypothesis is not discussed here at length, because the assumption that myriapods are monophyletic is not supported by morphological evidence (Dohle 1998), although Baccetti (1979), Jamieson (1987), and Jamieson et al. (1999) point to a derived similarity in sperm structure in chilopods and progoneates. A striated cylinder surrounding the 9 + 2 axoneme is considered to be an autapomorphy of Myriapoda, although it has been demonstrated only in chilopods and pauropods. The cylinder is possibly the homologue of the coarse fibers (intersinglet or intertubular material) of the insectan sperm. A phylogenetic tree based on DNA data for eight loci and morphological characters produced by Giribet et al. (2001) show myriapods as monophyletic and as the nearest relatives of Pancrustacea (= Crustacea + Insecta). These results are difficult to interpret, however, because insects appear scattered among Crustacea, results that are "confusing," in the words of the authors (e.g., Diplura–Campodeidae as sister group to Protura/Hexapoda, Diplura–Japygidae as sister group to barnacles/crustaceans, whereas Giribet et al. accept that japygids are basal hexapods).

To summarize: (1) current morphological knowledge offers no clear support for the existence of a sister-group relationship between insects and a nontracheate group, and (2) under the tracheate hypothesis it is not clear whether Progoneata or Symphyla is the nearest relative of the insects. Only a few molecular studies support this latter view, such as the 18S ribosomal DNA (rDNA) tree published by Wheeler et al. (2001). It must be noted in this context, however, that very few molecular studies have been undertaken to address this question.

Insect Phylogeny and Evolution

Insect evolution began more than 400 Myr (million years) ago, as deduced from fossil springtails of Lower Devonian Age, 395 Myr old (*Rhyniella praecursor*), and the discovery of possible ectognathan mandibles from the same locality and of the same age (*Rhyniognatha hirsti*). Springtails, which include about 6000 described species, have unique derived characters such as specialized appendages at the end of their abdomen that are united to form a furca or spring used for jumping. When at rest, the spring is held under the abdomen and fixed by the retinaculum, another organ produced by abdominal legs. The springtails also have a so-called ventral tube, developed from fused abdominal vesicles, and many other derived characters. These structures are known from the Devonian springtails as well, and because it is unlikely that they evolved in a short period of time, it is probable that insects had a long history before that epoch.

Today, no insect has abdominal legs. Yet, because the jumping organ of the springtails consists of three segments, the last ancestor of insects must have had abdominal legs with at least three podomeres. Legs of that kind must have been lost independently in the two basal lineages of insects: in Ellipura, on the one hand, and in Euentomata (= Diplura + Ectognatha), on the other. A few fossils seem to fill either the gap between the insects and their multilegged sister group or the gap between the last stem species of Insecta and the first split within Euentomata (for figures see Haas in Bechly 2001, Willmann 2003).

The springtails and Protura (telsontails, including about 500 described species) are subgroups of Ellipura. Their close relationship is well substantiated: they have no abdominal tracheal stigmata and no styli, they possess cranial folds covering the mouth parts in a unique way, and they possess a longitudinal fold on the underside of the head and neck not found in any other insect, the so-called linea ventralis. On the anterior part of the abdomen (first segment in Collembola, segments 1–3 in the Protura) are appendages consisting of an eversible vesicula that are paired in telsontails but fused in springtails.

Diplura (800 species) poses a major problem in insect phylogeny. "Diplurans" means doubletails, and the name refers to their cerci, which may either be long and multisegmented or short and used for grasping. The latter character state is derived. Until a few years ago almost all entomologists agreed that the diplurans are most closely related to the ellipurans (fig. 20.1), because their mouth parts are also hidden by cranial folds (Diplura + Ellipura = Entognatha). Other possible autapomorphies of Entognatha are the reduction of the Malpighian tubules and the absence of a centriole adjunct in the sperm. However, a structure superficially resembling an adjunct is present in telsontails (*Acerentulus*; Jamieson et al. 1999), and according to Koch (1997, 2000, 2001) it is probable that the cranial folds have developed independently. Now it seems likely that the dip-

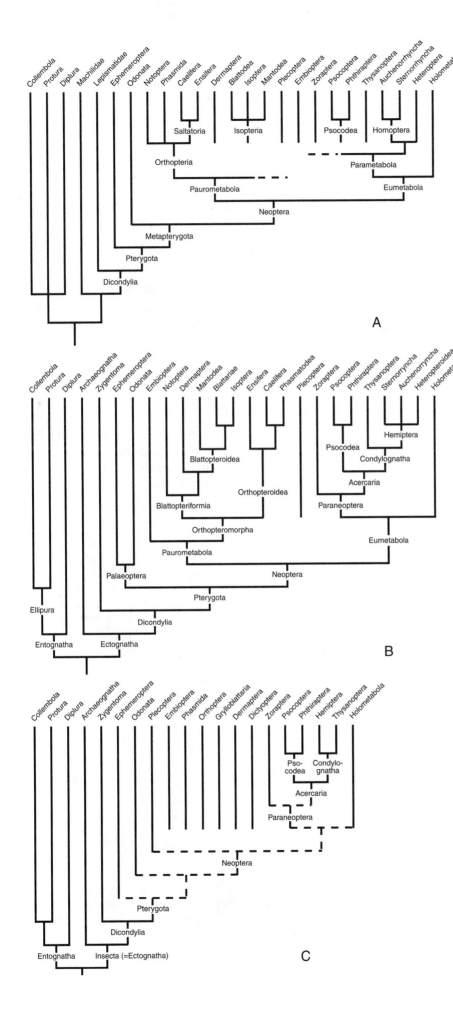

Figure 20.1. Previous hypotheses of relationships among insects based on morphology, illustrating advancements in insect phylogenetics over 30 years, beginning with the first "Stammbaumentwurf" of Hennig (for Holometabola, see Whiting 2002). (A) Hennig (1953). (B) Hennig (1969). (C) Kristensen (1981; all redrawn from sources). The taxon names are those used by the respective author. Dashed lines indicate uncertainty in relationships. Hennig (1969) inadvertently united Protura and Diplura in his figure as "Ellipura," which is corrected here. It should be noted that these authors have favored a particular view, but they have always discussed alternative ideas.

lurans are the sister group of the rest of the insects, whose mouth parts are externally visible. These are called Ectognatha. Few characters support the hypothesis that Diplura are the sister group of Ellipura, whereas a number of characters appear to be synapomorphies of the Diplurans and the ectognathous insects: for example, the lack of Tömösvary's organs on the head, the lack of abdominal legs, a new mode of molting, the possession of long filamentous appendages of the 11th abdominal segment (cerci), the structure of the tail of the sperm, superficial cleavage (character state uncertain as the type of cleavage is unknown in Protura), and epimorphosis (the young hatch with the full number of abdominal segments, whereas the plesiomorphous state is a hatchling that adds several abdominal segments after having left the egg). Some authors have suggested that a movable appendage of the mandible is further evidence for the monophyly of the clade Diplura + Ectognatha (Richter et al. 2002).

Ectognatha

Ectognatha consists of Archaeognatha (bristletails, about 390 described species), Zygentoma (silverfish, 400 species), and Pterygota (winged insects). Monophyly of Ectognatha has never been doubted, because its members have a large number of derived characters in common (figs. 20.1, 20.2, 20.3).

The more obvious ones are an antenna with a long flagellum that lacks muscles and possesses Johnston's organ in its second segment, and females with an ovipositor whose elements are contributed by ventral sclerites of the 8th and 9th abdominal segments.

The characters used for reconstructing insect phylogeny are sometimes very complex and no doubt determined by a large number of genes. This makes morphological structures a powerful tool in reconstructing phylogeny. To give an example: the monophyly of the ectognathous insects is also supported by the structure of blood vessels. First, members of Tracheata have vessels extending from the head into the antennae. Primitively, the antennal blood vessels are connected to the large dorsal vessel, but in the ectognathous insects these vessels are separate and thus there are several circulatory systems. And second, in Pterygota (winged insects), each antennal vessel has a pulsatile ampulla that functions as a pump or as an "antennal heart" (Pass 1998).

Dicondylia

The first two branching events in the phylogeny of Ectognatha were the central focus of a classic controversy in systematics. Because bristletails and silverfish are superficially similar, they were often united in one group. It has long been known, how-

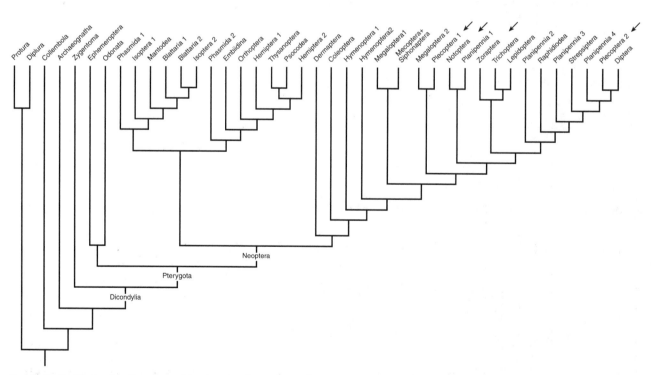

Figure 20.2. Cladogram from a combined molecular analysis of insects minimizing character incongruence between molecular data sets (18S rDNA and 28S rDNA data); redrawn and condensed from Wheeler et al. (2001). Each insertion:deletion event was weighted 1, as were transitions and transversions. Arrows point to non-holometabolan taxa within the clade next to Dermaptera, which consists mainly of Holometabola. Holometabola are generally accepted as monophyletic. Numbers identify taxa that are split into several units and so appear at different places in the cladogram.

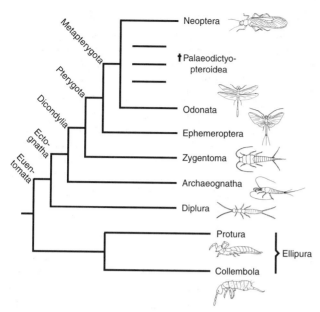

Figure 20.3. Basal phylogenetic relationships of insects as favored in this chapter (but see text for alternative hypotheses). For the higher winged Insecta (Neoptera), with the exclusion of Holometabola, see figures 20.4 and 20.5.

ever, that bristletails are more plesiomorphous than are silverfish and that silverfish are more closely related to the winged insects than to bristletails. In 1953 Hennig introduced the name "Dicondylia" for silverfish and winged insects, reflecting their phylogenetic relationships (the two taxa share, e.g., a mandible with two articulations, or condyli, with the head capsule). Yet it was only 25 years ago that Hennig's systematic framework of basal insects was generally taken into textbooks, and in fact, there are still a number of textbooks that present the old classification [e.g., Ross et al. (1982: 284), which unites Archaeognatha and Thysanura into Apterygota; Richards and Davies (1977), which subsumes under Apterygota all primarily winged insects; Borroret al. (1992), which also considered Archaeognatha and Thysanura as "apterygote insects"].

It may be, however, that things are not that simple. In California, the species *Tricholepidion gertschi* (Wygodzinsky 1961), the bristlefish, is the only surviving representative of a taxon originally described from Baltic amber, Lepidotrichidae. This species is usually regarded as belonging to Zygentoma (e.g., Boudreaux 1979, Kristensen 1998, Wygodzinsky 1961), but it may also be the sister group of Zygentoma + Pterygota (Kristensen 1991, Klass 1998, Staniczek 2000; for a summary of the evidence, see Willmann 2003).

Pterygota

The oldest known winged insects come from the uppermost Mississippian, or middle Carboniferous (*Delitzschala bitterfeldensis*, *Ampeliptera limburgica*, *Stygne roemeri*, *Brodioptera* *stricklani*; age between 317 and a little more than 320 Myr). They were already advanced because they were fully winged and capable of flight, and it is unknown what the first pterygotes looked like. It is also unknown what the function of the first winglike structures was. They were certainly very small, and they cannot have served as flight organs but may have supported thermoregulation.

For decades it has been debated whether mayflies or odonates are the first side branch of the Pterygota or whether the two combined form a clade of their own (Palaeoptera, fig. 20.1B). Although the Palaeoptera hypothesis persists (Hovmöller et al. 2002), several characters used to support it are hardly tenable [aquatic larvae, possibly a convergence; fusion of galea and lacinia, but the fused parts of the maxilla in Odonata may not represent galea and lacinia (Staniczek 2000); short antennal flagellum, apparently a convergence (Soldán 1997)], whereas the character state of other structures (wing: anterior media fused to the radial sector, intercalary veins) is uncertain (Willmann 1999). More convincing is another hypothesis, favored by Kristensen (e.g., 1981, 1989, 1991; see also Hennig 1953, 1986; fig. 20.1A,C). Based on evidence from head morphology, the mayflies are the sister group of the remaining pterygotes (Staniczek 2000). Another indication that this may be so is the subimago, a flying stage followed by the final flying stage, the reproducing imago. The subimaginal stage is considered to be an ancient character, and it is not retained in any other recent insect group. Grimaldi (2001) has stressed that loss of an imaginal molt in odonates and Neoptera cannot simply be attributed to convergence (I know of no evidence that some Paleozoic neopterans molted as flying life stages; see Kristensen, 1989, 1991). Again, odonates, Neoptera, and also Palaeodictyopteroidea have lost their paracercus (the median terminal filament).

With respect to their copulatory apparatus, mayflies resemble derived winged insects, the Neoptera. The males have long styli on their 9th abdominal segment that serve to grasp the female's abdomen. It is difficult to tell whether or not this is a synapomorphy, because the grasping organ in mayflies and neopterans might be a convergent similarity. As an argument supporting this view, one could point to Odonata and Palaeodictyopteroidea. Odonata have an indirect mode of sperm transfer, and it is unlikely that it derives from a gonopore-to-gonopore transfer (Bechly et al. 2000). Palaeodictyopteroidea were minute to huge insects (wing span up to 55 cm) of impressive diversity and species richness, and they are known from the Carboniferous to the Triassic. Some of them had very small copulatory organs, which may not have served to hold the female (they may have instead been tactile organs), and palaeodictyopteroids are usually considered to be more closely related to neopterans than are mayflies and odonates. In fact, however, it is unclear where they belong. It is generally stated that they are monophyletic because they have elongate mouthparts forming a beak, but mouthparts are well known in only a few specimens and do not always form a distinctive proboscis (Novokshonov and

Willmann 1999). Therefore, palaeodictyopteroids may not be a natural group. That the transfer of sperm from gonopore to gonopore may have evolved independently in mayflies and Neoptera has already been discussed by Kristensen (1981).

Odonates and the other winged insects have mandibles and mandibular muscles quite different from those in mayflies and primarily wingless insects. For this reason, Börner (1908) united Odonata and Neoptera (higher winged insects) under the name Metapterygota. Odonata consist of Zygoptera (damselflies) and Epiprocta (= Epiproctophora), which are in turn composed of Epiophlebioptera and Anisoptera (dragonflies). Zygoptera, often considered to be a paraphyletic group, are certainly monophyletic, as evidenced by numerous autapomorphies, among them the distinctly stalked wings, extremely broadened hammer-shaped head capsule with widely separated eyes, extreme obliqueness of the pterothorax, and an ovipositor pouch formed by the enlarged outer gonapophyses (valvulae 3) of the 9th abdominal segment of the female (Bechly 1996, Lohmann 1996). Members of Epiprocta have enlarged eyes. The wing nodus lies almost in the middle of the fore margin. The larvae have rectal folds containing tracheal gills in a rectal gill chamber. Today, Epiophlebioptera consists only of *Epiophlebia*. The widely used name "Anisozygoptera" is no longer in use among phylogenetists because it denoted a paraphyletic group. Lohmann (1996) has attempted to reconstruct in detail the phylogeny of Anisoptera, but the relationships within the group are disputed (see Bechly 1996) and monophyly of many odonatan taxa (e.g., the zygopteran "families") has yet to be demonstrated (Jarzembowski et al. 1998).

Neoptera

The wings of the Neoptera—all recent pterygotes not belonging in the mayflies or odonates—are probably more advanced than those of mayflies and odonates. In particular, neopterans have sclerites at the wing base, thus allowing the wings to be folded back over the abdomen (Martynov 1925, Hennig 1969, Hörnschemeyer 1998, 2002).

Basal relationships among neopteran groups have been under dispute because of the uncertain position of Plecoptera (stoneflies). This is a cosmopolitan group of common insects with about 2300 species. Their nymphs are virtually ubiquitous in rivers and brooks. The phylogenetic relationships within the group were the topic of one of the classical studies in phylogenetic systematics (Zwick 1973), and the results it revealed are still considered valid (Zwick 2002). In Plecoptera, two sister taxa with very different distributional patterns have been recognized. Arctoperlaria (Systellognatha + Euholognatha, >1500 species) occurs mainly in the Northern Hemisphere, whereas Notonemouridae and representatives of Perlidae live in the Southern Hemisphere. Antarctoperlaria (Eusthenioidea + Gripopterygoidea, ~300

species) are strictly confined to Australia, South America, and New Zealand (Zwick 1973, 1980). Eusthenioids are commonly colored, which is unusual in Plecoptera, because most other species are grayish brown.

Because plecopterans appear to be more plesiomorphous than other neopterans in the segmental arrangement of their testes, and because they have a transversal muscle in the stipes, otherwise known only from Archaeognatha and Zygentoma, Zwick (1980) hypothesized that they represent the sister group of the remaining neopterans. Beutel and Gorb (2001) believed the aquatic larvae of plecopterans to be another plesiomorphy compared with the terrestrial larvae found in other neopterans, but this is unlikely because neither Archaeognatha nor Zygentoma has an aquatic early life stage, and the larvae of Palaeodictyopteroidea were, as far as known, terrestrial as well. Stys and Bilinski (1990) and Büning (1998) assume that Plecoptera are the sister taxon to a monophyletic group consisting of Dermaptera and Eumetabola. Disturbingly, according to the most recent molecular analyses (Wheeler et al. 2001), Plecoptera appear in Holometabola in a combined molecular analysis (18S rDNA and 28S rDNA) that minimized character incongruence between the molecular data sets. 18S rDNA analyses put Plecoptera as a sister group to Psocodeans + Zoraptera + Thysanoptera, whereas 28S rDNA data place them as the sister group to thrips. These, as well as other acercarians, form the sister group to Hymenoptera. In a total evidence cladogram including morphological data (Wheeler et al. 2001), Plecoptera appeared as a sister group to Embiida. In this case, the trees resulting from molecular data were modified according to one interpretation of morphological data, but characters are interpretations. Thus, structural similarities between Embiida and Plecoptera were considered to be plesiomorphies by Rähle (1970) and Zwick (1980).

Based on evidence from wing structures, it appears likely that Plecoptera are part of a species-rich group named Polyneoptera by Martynov (1925). The remaining species possibly fall into two other neopteran groups, Acercaria and Holometabola.

Polyneoptera

Polyneopterans (fig. 20.4) are characterized by a number of probably derived hind wing structures. Two veins, the second cubitus and the first anal vein, are almost straight and run parallel to one another. The remaining anal veins form a fan, and the second anal vein splits into two or more branches, whereas the others do not. Some polyneopterans, such as the rock crawlers (Notoptera, Grylloblattaria), are wingless, whereas others have small wings with reduced venation like the Embiida (web spinners), but their male genitalia and molecular data support the view that they belong in the group. However, as in many other cases, the aforementioned hind wing structures have been accorded differing significance, being convergences in Kristensen's (1991) opin-

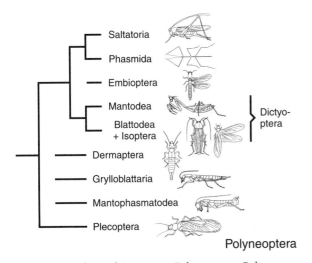

Figure 20.4. Relationships among Polyneoptera. Polyneopteran monophyly is not generally accepted, and the positions of most taxa, especially Embioptera, Dermaptera, and Grylloblattaria, are controversial.

ion. Dermapteran hind wings resemble those of the (other) polyneopterans only superficially.

The relationships within Polyneoptera are only partially clear. It has long been suggested that the praying mantids, roaches, and termites (Mantodea, Blattodea, Isoptera) form a systematic unity called Dictyoptera, and morphological work on the gut structure and female genitalia has supported this view (Klass 1995, 1998: fig. 4), in accordance with several (Wheeler et al. 2001; fig. 20.2) but not all molecular analyses. One major step in the evolution of Dictyoptera was the development of sociality, when some cockroaches became termites. According to this evolutionary scenario, which was proposed by Wheeler (1904, 1928) and Handlirsch (1908), termites are a highly evolved subgroup of the roaches. It appears that the roach *Cryptocercus* is the closest relative of Isoptera. Isopterans and *Cryptocercus* share a rich diversity of hindgut symbionts belonging to Oxymonadida/Metamonada and Hypermastigida/Parabasalia. Klass (2001b) believes that it is unlikely that the associations are due to lateral transfer, as suggested by Grandcolas and Deleporte (1996). Grandcolas (1994, 1996, 1997) assumed that xylophagy and intestinal symbiosis of *Cryptocercus* and Isoptera is a matter of convergence because he was of the opinion that *Cryptocercus* has a subordinate position within the Polyphaginae/ Blattaria. Klass (2000, 2001b, and previous publications) presented evidence that *Cryptocercus* is only distantly related to Polyphaginae. He showed that most of the autapomorphies indicated in the cladogram of Grandcolas had to be rejected as supporting the respective clades largely because of erroneous homologies, but he also stated (Klass 2001b:263) that blattarian phylogeny itself is not finally settled. Lo et al. (2000) found strong support for the clade *Cryptocercus* + Isoptera based on the combined analysis of several gene sequences. The oldest known termites come from Cretaceous

sediments, and today they are important modifiers of soil structure in tropical environments, with thousands of billions of individuals. The queen of *Bellicositermes natalensis* lays one egg every two seconds, which gives a total of 43,000 eggs per day.

The closest allies of the Dictyoptera are possibly the dermapterans (earwigs, about 1900 species), whose cerci are usually transformed into a forceps. It is sometimes believed that the very short ovipositor is a synapomorphy of Dictyoptera and Dermaptera, but Jurassic earwigs with a long ovipositor show that this is not correct. Indicators of a close relationship between Dermaptera and Dictyoptera are the pterothoracic musculature and similarities in wing venation (Klass 1998, Willmann 2003; but see below).

Grimaldi (2001) lists four apomorphies in favor of a Dermaptera + (Zoraptera + Embiida) relationship, but none of them (three-segmented tarsi, ovipositor highly reduced, loss of ocelli, cerci reduced to a one- or two-segmented appendage) were developed in Jurassic earwigs (Vishniakova 1980) and thus do not pertain to the dermapteran ground plan (Willmann 1990, 2003). The view that Embioptera are most closely related to Dermaptera receives weak support from a spermatozoal similarity (shared oblique implantation fossa), but this is in conflict with a spermatozoal apomorphy shared by Phasmatodea and Dermaptera (double anterior axonemal cylinder; Jamieson et al. 1999).

The situation with regard to the Dermaptera is even more complicated than indicated above. The hind wing similarities between Dermaptera (including its stem-group representatives that are usually united under the term "Protelytroptera") and other polyneopterans are only superficial because of the apomorphic structure of the former. This is certainly not evidence of a position outside Polyneoptera, but Büning (1998) assumes that the earwigs are the sister group of Eumetabola (Dermaptera + Eumetabola = Meroista) based on similarities in the ovarioles. Interestingly, Mesozoic male dermapterans had well-developed gonobases and gonostyli (Vishniakova 1980) that do not occur in other Polyneoptera. This demonstrates that reduction of the structures has occurred independently.

Dermaptera include one taxon, *Hemimerus* (~10 species), that has no forceps but segmented cerci instead. Popham (1985) believed the cercal structure of *Hemimerus* to be plesiomorphous, and the earwigs were therefore subdivided into two subordinate taxa, Hemimerina and Forficulina. As some Jurassic Dermaptera had unsegmented cerci (e.g., *Turanoderma*) but were plesiomophous in many other respects, and because *Hemimerus* shares several apomorphies only with recent earwigs, Willmann (1990) concluded that *Hemimerus* has secondarily segmented cerci due to pedomorphosis and that subdividing Dermaptera into Hemimerina and Forficulina is unfounded. This corresponds to the view of Giles (1974), who regarded the forceps as an autapomorphy of Dermaptera, later lost in *Hemimerus*. This view has gained strong support from detailed morphological studies (Klass 2001a).

Stick insects and leaf insects (phasmids) include one species-poor taxon (*Timema*, in California), and the higher phasmids or Euphasmatodea, composed of about 3000 species. Their classification has been typological and was based on work by Günther (1953) until Bradler (1999) began a phylogenetic analysis using morphological data, soon followed by molecular sequence studies. Within euphasmatodeans, wingless *Agathemera* (10 species in South America) appears to be the sister group of Neophasmatidae, which includes Phyllinae, Heteropteryginae, Eurycanthinae, Lanceocercata (200 species; Australia, southern Asia, Madagascar), and various taxa commonly called stick insects (Bradler 2000, 2002).

Phasmida (or Phasmida + Embiida) are probably the sister group of Orthoptera or Saltatoria, that is, grasshoppers, crickets, and allies (fig. 20.4). Earliest saltatorians are known from the Pennsylvanian or upper Carboniferous period. The earliest certain fossil stick and leaf insects are known from the Mesozoic, but the group must be as old as the saltatorians, if they are their closest relatives. Saltatoria and Phasmatodea share a large precostal area in the wing that is derived but is lost in all extant and some of the Mesozoic phasmatodeans (Sharov 1968, Willmann 2003). Among Recent phasmids, *Heteropteryx* exhibits the most plesiomorphic wing structure. The forewings are elongated, the longitudinal veins radius, radial sector and media are parallel to one another, and the cubitus consists of two branches. The venation is very similar to that of the Cretaceous *Coniphasma*, differing only in the fusions in *Heteropteryx* and the shortage of wing in *Coniphasma* (Willmann 2003). This is in conflict with the results of Whiting et al. (2003) based on DNA sequence data, where *Agathemera* + *Heteropteryx* + *Haaniella* appear as one of the most derived phasmatodean subgroups.

Saltatoria are composed of more than 20,000 species, belonging to two monophyletic groups, Ensifera and Caelifera. The hypothesis of saltatorian monophyly is founded on the enlarged hind femora containing the extensor muscle of the tibia that enables the animals to jump, the presence of prothoracic cryptopleury (which means that the saddle-shaped pronotum covers the prothoracic sides), the fusion of the 1st and 2nd tarsal segments, and other characters interpretable as autapomorphies of the group. The ensiferans are plesiomorphous with respect to their long antennae, but they have lost their arolium, which is an adhesive structure of the tarsus, and exhibit numerous derived wing characters. Caeliferans, by contrast, have short antennae that are not longer than the combined head and prothorax (a derived state). In their digestive tract, caeliferans have lost the proventriculus (which means they have to use their mouth parts intensively), and their tarsi consist of only three tarsomeres at most. Both Ensifera and Caelifera came out as monophyletic in an analysis of molecular sequences by Rowell and Flook (1998), who also proposed a division of Caelifera into subunits based on an investigation of about 150 species.

The position of Notoptera (rock crawlers, 16 species), which are confined to East Asia and North America, is unclear. Almost every group within Polyneoptera has been contemplated as their sister group, which in turn implies that the sister group of any of the polyneopteran taxa is uncertain as well. Rowell and Flook (1998) grouped them along with Dermaptera and Plecoptera in one clade based on analysis of genome sequences. In 2002, a new insect taxon was described, Mantophasmatodea, a name suggesting relationship to praying mantises and phasmids, but which has no close affinities to the former, whereas its proventriculus and midgut structure is similar to that in Notoptera (Klass et al. 2002a, 2002b). Like notopterans, mantophasmatodeans are wingless and live on other arthropods. Members of the group had already been described from Baltic amber five years before (Arillo et al. 1997), although without assignment to any insect taxon of higher rank.

The phylogenetic position of the Embiida (web spinners, >1500 species, with many remaining undescribed; Ross 2000) is unclear (figs. 20.1, 20.2, 20.4). Engel and Grimaldi (2000) and Grimaldi (2001) regard them as the closest relatives of Zoraptera. The two groups have in common the reduction of the cerci (two-segmented in the ground pattern of Zoraptera), the enlargement of the hind femora, the presence of at least some wingless morphs, the shedding of wings along a basal fraction zone, brood care, and the reduction in the number of tarsomeres (three in Embiida, two in Zoraptera). Some of these similarities are not convincingly interpretable as synapomorphies (Rasnitsyn 1998), and Rähle (1970) has pointed to several derived similarities shared by Embiida and Phasmida, among them a gula or gulalike structure (although a gula does not pertain to the ground pattern of the phasmids; Bradler 1999, Kristensen 1975), the structure of the propleura, and the possession of both a ventral and a dorsal flexor of the paraglossae and two furca-furcasternal muscles inserting at the profurcal sternite. Molecular sequence studies (Rowell and Flook 1998) have also supported a close relationship between phasmids and web spinners, and these two combined constitute the sister group to Saltatoria.

Acercaria

The second species-rich branch of neopteran insects is Acercaria (fig. 20.5), so-called because this group lacks cerci. Additional derived characters include a reduction in the number of the Malpighian tubules (four at most), loss of the first abdominal sternum, possession of a single ganglionic complex in the abdomen, and the loss of the perforatorium of the sperm. Within Acercaria, two main branches are distinguishable. The first branch is Hemiptera, which consist of Heteropterida (Heteroptera, bugs; and Coleorrhyncha, a group of fewer than 30 described species occurring in the

Southern Hemisphere), Auchenorrhyncha (cicadas), and Sternorrhyncha (plant lice), one of the most successful lineages of insects. The relationships among the three have not been worked out. The assumption of the monophyly of Auchenorrhyncha, based for example on a pair of sound producing organs in the first abdominal segment, is sometimes doubted (Mahner 1993) but has gained support from examinations of the forewing base (Yoshizawa and Saigusa 2001). The second acercarian clade is Micracercaria, or small Acercaria. Their wings have an easily recognizable area formed by the first cubitus, the areola postica, and their tarsus consists of only three segments. One group belonging to them is Psocodea, which contain the wingless sucking and biting lice (the certainly monophyletic Phthiraptera with >3000 species; Königsmann 1960 and many subsequent authors) and book lice (Psocoptera, >3000 species). They share, for example, a unique sclerotization of the esophagus and therefore possess a so-called cibarial sclerite, and they are equipped with a modification of the basal part of their antennal flagellomeres to facilitate rupture, which is interpreted as an escape device (Königsmann 1960, Seeger 1975). Monophyly of Psocoptera has been doubted, but Seeger (1979) found embryological and egg structural evidence that it is a natural taxon. Lyal (1985) pointed to similarities of Phthiraptera and Liposcelidae/Psocoptera that might indicate a sister-group relationship between the two but concluded that they are most probably convergences. The other micracercarian group is Thysanoptera (thrips, >4500 species). They range in body length from less than 1 mm to 15 mm; their name refers to their fringed wings, which, however, also occur in small members of other insect taxa. The thrips have usually been considered to be the closest allies of Hemiptera, but according to the fossil record they are linked to Posocodea instead. In the Mesozoic and the Permian there were the psocodean-like lophioneurids, which share two striking apomorphies with thrips: a tarsus with only two segments and a bladderlike structure at its tip (Vishniakova 1981). The Jurassic *Karataothrips* is already similar to recent thrips, but its venation is more primitive. The view that thrips are the nearest relatives of Psocodea is also supported by the total evidence cladogram of Wheeler et al. (2001).

Many have accepted the view that Hemiptera and Thysanoptera constitute a taxon called Condylognatha, which Börner (1904) had erected based on a study of head structures. However, the interpretation of decisive similarities as possible synapomorphies has been doubted by several authors, among them Königsmann (1960). There appears to be no spermatozal apomorphy supporting the monophyly of the Condylognatha (Jamieson et al. 1999), but Yoshizawa and Saigusa (2001) have found two possible synapomorphies of Thysanoptera and Hemiptera in the sclerites of the forewing base (fusion of basisubcostale and second axillary sclerite; distal median plate placed next to the second axillary sclerite).

Zoraptera

The Zoraptera (fig. 20.5) are a little-known insect group, for which no popular name exists. In German they are called Bodenläuse (i.e., groundlice). They are up to 3 mm long, and fewer than 30 species have been described. Their systematic position is unclear. In the literature, they appear as the sister group of Isoptera (which is untenable because isopterans share derived internal head sclerite structures with cockroaches and mantises that zorapterans do not), or as the sister group of Dictyoptera, Embiida (see above), Dermaptera + Dictyoptera, Dermaptera, Acercaria, Holometabola, and others. Similarities with some groups are due to reductions or losses (e.g., the gonostyli, appendages of the male genital apparatus, are lacking). A sister group relationship with Acercaria, for example, has been postulated because of a reduction in number of the Malpighian tubules, an abdominal ventral nerve cord that consists of two ganglia only (reduced to one in Acercaria; Hennig 1969, 1986, Kristensen 1981, Königsmann 1960, Seeger 1979), and the shared presence in the wings of some groups (the micracercarians) of a so-called areola postica formed by the first cubitus that is one

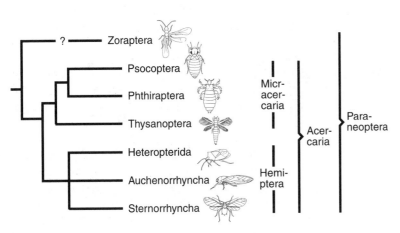

Figure 20.5. Phylogenetic relationships of Acercaria. Monophyly of Paraneoptera is doubtful because of the uncertain position of Zoraptera, which may be closely related to polyneopterans.

of the posterior veins (significance unclear). Kristensen (1991), however, feels that zorapterans, generally simplified because of their minute size, might well have had their origin among the polyneopterans.

Eumetabola (= Acercaria + Holometabola)

The Acercaria are possibly the closest ally of Holometabola, as evidenced by the development of the male genital structures (fig. 20.6). So far, however, none of the cladograms based on molecular sequence data alone supports the Eumetabola hypothesis (Whiting et al. 1997, Wheeler et al. 2001). In fact, acercarians appear scattered within Polyneoptera and Holometabola in the consensus cladogram for the 18S rDNA data (Wheeler et al. 2001) in which hemipterans are the nearest relatives to a group consisting mainly of Holometabola, but also of *Metajapyx* (Diplura) and *Grylloblatta* (Notoptera), whereas thrips and psocodeans are grouped with Zoraptera among some of the polyneopterans. According to 28S rDNA data, Acercaria seems to be part of Holometabola, which also includes the stoneflies [(((Hemiptera + Psocodea) + (Thysanoptera + Plecoptera)) + Hymenoptera); Wheeler et al. 2001].

It has been estimated that more than 75% of all organisms belong in the insects, and of these, more than 75% belong in Holometabola. The insects discussed to this point have young that gradually become more and more similar to

the adult, but holometabolans have a larval stage that is very different from the adult and a pupal stage between the larva and adult. Sometimes, the pupa is described as a stage of rest, and in fact it is almost motionless and usually does not take up food. But it is actually that life stage during which the most fundamental changes in ontogeny occur, because the larval body is entirely restructured to become equipped with adult characters. In the last five or so decades, holometabolan monophyly has not been doubted by morphologists (*contra* numerous earlier publications), but none of the more detailed molecular sequence studies has produced a cladogram with a monophyletic Holometabola (Chalwatzis et al. 1996, Whiting et al. 1997, Wheeler et al. 2001). (For more detail about the phylogenetic relationships with the Holometabola, see Whiting, ch. 21 in this vol.).

What Is Really Known?

It may appear that nothing in insect phylogeny and systematics is well established, and indeed morphological characters considered to be useful for phylogeny reconstruction have consistently been interpreted in different ways. However, the significance of many structures has been clarified, and a major reason for this is that phylogenetic thinking has contributed much to an entirely different approach to analytic examination of characters. Although some authors in the middle of the 20th century held the view that insect wings may have developed independently twice, because there are two different types this assertion is no longer considered to be tenable, because similarities in wing structure outweigh the probability of convergence. The same applies to many other structures, but in many cases—and this has been underestimated by morphologists—even apparently complex body parts seem to have evolved in different evolutionary lineages. This dilemma has not been solved yet. It is certain that in many cases, structures appear to be superficially similar until more detailed investigations often unveil differences (and nonhomology). Sometimes, a name appears to be all that structures share (e.g., "sperm pump" in Mecoptera and Diptera). This has also practical aspects: not only is a new generation of skilled morphologists needed, but such studies are also time-consuming. Yet, the reward of years of hard comparative work is deep insight not only into structural complexity as well as constructional morphology, functions, ecology and behavior; most important, a deeper understanding of the organism and its evolutionary context will ultimately emerge.

Different possible interpretations of similarities limit the value of any cladogram, and in fact, phylogeneticists used to discuss the meaning and significance of every single structure that appeared to relate different taxa. Consequently, computer-generated cladograms of all of Insecta based on morphological evidence, or combined molecular sequence and morphological data, have not, with rare exceptions, led to entirely new and convincing hypotheses of relationship because it is not char-

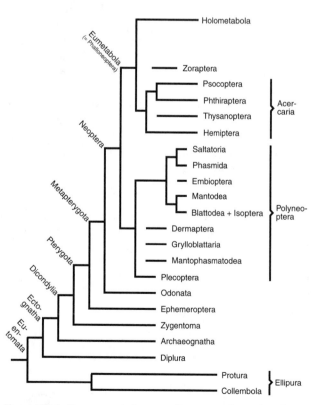

Figure 20.6. Summary cladogram of insects as favored in this chapter.

acters that are being coded, but rather character interpretations. Unveiling relationships of groups of closely related insect species seems to be much less problematic.

So, what do we know? Insects are probably monophyletic, as supported by most molecular studies. Almost all easily distinguishable major taxa are monophyletic, namely, Collembola, Protura, Diplura, Archaeognatha, Ephemeroptera, Odonata, Plecoptera, Notoptera, Mantophasmatodea, Dermaptera, Embioptera, Saltatoria, Phasmida, Mantodea, Isoptera, Zoraptera, Phthiraptera, Psocoptera, Thysanoptera, Heteroptera, Coleorhyncha, Auchenorrhyncha, and Sternorrhyncha (see fig. 20.6); and among Holometabola, the Coleoptera, Planipennia, Raphidiodea, Megaloptera, Strepsiptera, Hymenoptera, Lepidoptera, Trichoptera, Diptera, and Siphonaptera are also monophyletic. However, Blattodea are probably paraphyletic in terms of Isoptera, serious doubts as to the monophyly of Mecoptera exist, and Zygentoma may be paraphyletic. Until recently, the monophyly of several more taxa had been uncertain, for example, Diplura, Dermaptera, and Megaloptera. Collembola, Protura, and Diplura are basal insect lineages and do not belong in the entity composed of Archaeognatha, Zygentoma, and pterygotes. Archaeognatha are the sister taxon to Dicondylia, which are composed of Zygentoma (monophyly not certain) and Pterygota. Odonata and Ephemeroptera are closely related (but possibly not sister taxa), and most probably Neoptera forms a clade (fig. 20.6). The Holometabola appear to be a natural taxon, and probably Acercaria (Hemiptera, Thysanoptera, Psocodea) are also monophyletic, being the sister group to holometabolans. The Zoraptera are often thought to be the nearest relatives of Acercaria (Zoraptera + Acercaria = Paraneoptera; fig. 20.5), but this needs confirmation. The positions of the remaining groups are also uncertain. They may constitute a natural group ("Polyneoptera," figs. 20.4, 20.6) or form a series of taxa between the root of Neoptera and acercarian- (or paraneopteran-) holometabolan node. Among them are Mantodea and Blattodea (inclusive of termites), which have long been known to be a natural unit (Dictyoptera). Almost certainly, Phasmida and Saltatoria are more closely related to each other than either of them is to any other neopteran group, with the possible exception of Embioptera.

Acknowledgments

I thank the organizers of the Tree of Life Symposium for having invited me to speak. The comments of an anonymous referee are greatly appreciated. This work was in part supported by grants from the Deutsche Forschungsgemeinschaft.

Literature Cited

Arillo, A., V. M. Ortuño, and A. Nel. 1997. Description of an enigmatic insect from Baltic amber. Bull. Soc. Entomol. Fr. 102:11–14.

Arnett, R. 2000. American insects. A handbook of the insects of America north of Mexico. 2nd ed. CRC Press, Boca Raton, FL.

Artois, T. 2001. Cladistic analysis of the Polycystididae (Platyhelminthes Kalyptorhynchia), with application of phylogenetic nomenclature. Proefschrift voorgelegd tot het behalen van de graad van Doctor in de Wetenschappen, richting Biologie, te verdedigen door. Limburgs Universitair Centrum, Diepenbeek, Belgium.

Baccetti, B. 1979. Ultrastructure of sperm and its bearing on arthropod phylogeny. Pp. 609–644 in Arthropod phylogeny (A. P. Gupta, ed.). Van Norstrand Peinhold, New York.

Bechly, G. 1996. Morphologische Untersuchungen am Flügelgeäder der rezenten Libellen und deren Stammgruppenvertreter (Insecta; Pterygota; Odonata, unter besonderer Berücksichtigung der Phylogenetischen Systematik und des Grundplanes der *Odonaten. Petalura 2(spec vol):1–402.

Bechly, G. 2001. Ur-Geziefer. Die faszinierende Evolution der Insekten. Stutt. Beitr. Nat. C 49:1–94.

Bechly, G., C. Brauckmann, W. Zessin, and E. Gröning. 2000. New results concerning the morphology of the most ancient dragonflies (Insecta: Odonatoptera) from the Namurian of Hagen-Vorhalle (Germany). J. Zool. Syst. Evol. Res. 39:209–226.

Beutel, R., and S. Gorb. 2001. Ultrastructure of attachment specialisations of hexapods (Arthropoda): evolutionary patterns inferred from a revised ordinal phylogeny. J. Zool. Syst. Evol. Res. 39:177–207.

Börner, C. 1904. Zur Systematik der Hexapoden. Zoolog. Anz. 27:511–533.

Börner, C. 1908. Neue Homologien zwischen Crustaceen und Hexapoden. Die Beißmandibel der Insekten und ihre phylogenetische Bedeutung. Archi. Metapterygota. Zool. Anz. 34:100–125.

Borror, D., C. Triplehorn, and N. Johnson. 1992. An introduction to the study of insects. 6th ed. Saunders College Publishing, San Diego.

Boudreaux, H. B. 1979. Arthropod phylogeny with special reference to insects. John Wiley and Sons, New York.

Bradler, S. 1999. The vomer of Timema Scudder, 1865 (Insecta: Phasmatodea) and its significance for phasmatodean phylogeny. Cour. Forsch. Inst. Senckenb. 215:43–47.

Bradler, S. 2000. On the systematic position of Agathemera Stål, 1875 within the Phasmatodea (Insecta). Zoology 103(suppl. 3):99.

Bradler, S. 2002. Phasmatodea, Gespenstschrecken. Pp. 251–261 in Lehrbuch der Speziellen Zoologie I: Insecta (H. H. Dathe, ed.). G. Fischer, Heidelberg, Germany.

Büning, J. 1998. The ovariole: structure, type and phylogeny. Pp. 897–932 in Microscopic anatomy of invertebrates 11C (Insecta) (M. Locke and F. W. Harrison, eds.). John Wiley and Sons, New York.

Burckhardt, R. 1903. Zur Geschichte der biologischen Systematik. Verh. Naturforsch. Ges. Basel. 16:388–440.

Chalwatzis, N., J. Hauf, Y. van de Peer, R. Kinzelbach, and F. K. Zimmermann. 1996. 18S ribosomal RNA genes of insects: primary structure of the genes and molecular phylogeny of the Holometabola. Ann. Entomol. Soc. Am. 89:788–803.

Dohle, W. 1998. Myriapod-insect relationships as opposed to

an insect-crustacean sister group Relationship. Pp. 305–315 *in* Arthropod relationships (R. Fortey and R. Thomas, eds.). Chapman and Hall, London.

Dohle, W. 2001. Are the insects terrestrial crustaceans? A discussion of some new facts and arguments and the proposal of the proper name "Tetraconata" for the monophyletic unit Crustacea + Hexapoda. Ann. Soc. Entomol. Fr. (n.s.) 37:85–103.

Engel, M., and D. Grimaldi. 2000. A winged *Zorotypus* in Miocene amber from the Dominican Republic, with discussion on relationships of and within the order. Acta Geol. Hisp. 35:149–164.

Giles, E. T. 1974. Relationships between the Hemimerina and other Dermaptera—a case for reinstating the Hemimerina in the Dermaptera based on numerical procedure. Trans. R. Entomol. Soc. Lond. 126:189–206.

Giribet, G., G. D. Edgecombe, and W. C. Wheeler. 2001. Arthropod phylogeny based on eight molecular loci and morphology. Nature 413:157–161.

Grandcolas, P. 1994. Phylogenetic systematics of the subfamily Polyphaginae, with the assignment of *Cryptocercus* Scudder, 1862 to this taxon (Blattaria, Blaberoidea, Polyphagidae). Syst. Entomol. 19:145–158.

Grandcolas, P. 1996. The phylogeny of cockroach families: a cladistic appraisal of morpho-anatomical data. Can. J. Zool. 74:508–527.

Grandcolas, P. 1997. What did the ancestors of the woodroach *Cryptocercus* look like? A phylogenetic study of the origin of subsociality in the subfamily Polyphaginae (Dictyoptera, Blattaria). Mem. Mus. Natl. Hist. Nat. 173:231–252.

Grandcolas, P., and P. Deleporte. 1996. The origin of protistan symbionts in termites and cockroaches: a phylogenetic perspective. Cladistics 12:93–98.

Griffiths, G. 1974. On the foundations of biological systematics. Acta Biotheor. 23:85–131.

Griffiths, G. 1976. The future of Linnean nomenclature. Syst. Zool. 25:168–173.

Grimaldi, D. 2001. Insect evolutionary history from Handlirsch to Hennig, and beyond. J. Paleont. 75:1152–1160.

Günther, K. 1953. Über die taxonomische Gliederung und die geographische Verbreitung der Insektenordnung der Phasmatodea. Beitr. Entomol. 3:541–563.

Handlirsch, A. (1906–1908). Die fossilen Insekten und die Phylogenie der rezenten Formen. Verlag von Wilhelm Engelmann, Leipzig.

Hennig, W. 1953. Kritische Bemerkungen zum phylogenetischen System der Insekten. Beitr. Entomol. 3:1–85.

Hennig, W. 1969. Die Stammesgeschichte der Insekten. W. Kramer Verlag, Frankfurt am Main.

Hennig, W. 1981. Insect phylogeny. Academic Press, New York.

Hennig, W. 1986. Taschenbuch der speziellen Zoologie, Wirbellose 2: Gliedertiere. Harri Deutsch, Frankfurt.

Hilken, G. 1998. Vergleich von Tracheensystemen unter phylogenetischem Aspekt. Verh. Naturwiss. Verh. Hamb. 37:5–94.

Hörnschemeyer, T. 1998. Morphologie und Evolution des Flügelgelenks der Coleoptera und Neuropterida. Bonner Zool. Monogr. 43:1–126.

Hörnschemeyer, T. 2002. Phylogenetic significance of the wing-base of the Holometabola (Insecta). Zool. Scr. 31:17–30.

Hovmöller, R., T. Pape, and M. Källersjö. 2002. The Palaeoptera problem: basal pterygote phylogeny inferred from 18S and 28R DNA sequences. Cladistics 18:313–323.

Jamieson, B. 1987. The ultrastructure and phylogeny of insect spermatozoa. Cambridge University Press, Cambridge.

Jamieson, B. G. M., R. Dallai, and B. A. Afzelius. 1999. Insects. Their spermatozoa and phylogeny. Science Publishers, Enfield, NH.

Jarzembowski, E. A., X. Martínez-Delclòs, G. Bechly, A. Nel, R. Coram, and F. Escuillié. 1998. The Mesozoic non-calopterygoid Zygoptera: description of new genera and species from the Lower Cretaceous of England and Brazil and their phylogenetic significance (Odonata, Zygoptera, Coenagrionoidea, Hemiphlebioidea, Lestoidea). Cretaceous Res. 19:403–444.

Kaestner, A. 1973. Lehrbuch der Speziellen Zoologie, Bd. I, 3. Insecta: B. Spezieller Teil. Gustav Fischer Verlag, Jena.

Klass, K.-D. 1995. Die Phylogenie der Dictyoptera. Cuvillier Verlag, Göttingen.

Klass, K.-D. 1998. The proventriculus of the Dicondylia, with comments on evolution and phylogeny in Dictyoptera and Odonata (Insecta). Zool. Anz. 237:15–42.

Klass, K.-D. 2000. The male abdomen of the relic termite *Mastotermes darwiniensis* (Insecta: Isoptera: Masto-termitidae). Zool. Anz. 239:231–262.

Klass, K.-D. 2001a. The female abdomen of the viviparous earwig *Hemimerus vosseleri* (Insecta: Dermaptera: Hemimeridae), with a discussion of the postgenital abdomen of Insecta. Zool. J. Linn. Soc. 131:251–307.

Klass, K.-D. 2001b. Morphological evidence on blattarian phylogeny: "phylogenetic histories and stories" (Insecta, Dictyoptera). Mitt. Mus. Nat. Berl. Dtsch. Entomol. Z. 48:223–265.

Klass, K.-D., and N. P. Kristensen. 2001. The ground plan and affinities of hexapods: recent progress and open problems. Ann. Soc. Entomol. Fr. (n.s.) 37:265–298.

Klass, K.-D., O. Zompro, and J. Adis. 2002b. Ordnung Mantophasmatodea. Pp. 161–166 *in* Lehrbuch der Speziellen Zoologie I: Insecta (H. H. Dathe, ed.). G. Fischer, Heidelberg, Germany.

Klass, K.-D., O. Zompro, N. P. Kristensen, and J. Adis. 2002a. Mantophasmatodea: a new insect order with extant members in the Afrotropics. Science 296:1456–1459.

Koch, M. 1997. Monophyly and phylogenetic position of the Diplura (Hexapoda). Pedobiologia 41:9–12.

Koch, M. 2000. The cuticular cephalic endoskeleton of primarily wingless hexapods: ancestral state and evolutionary changes. Pedobiologia 44:374–385.

Koch, M. 2001. Mandibular mechanisms and the evolution of hexapods. Ann. Soc. Entomol. Fr. (n.s.) 37:129–174.

Königsmann, E. 1960. Zur Phylogenie der Parametabola unter besonderer Berücksichtigung der Phtiraptera. Beitr. Entomol. 10:705–744.

Kraus, O. 2001. Myriapoda" and the ancestry of the Hexapoda. Ann. Soc. Entomol. Fr. (n.s.) 37:105–127.

Kristensen, N. P. 1975. The phylogeny of hexapod "orders." A critical review of recent accounts. Z. Zool. Syst. Evolut. Forsch. 13:1–44.

Kristensen, N. P. 1981. Phylogeny of insect orders. Annu. Rev. Entomol. 26:135–157.

Kristensen, N. P. 1989. Insect phylogeny based on morphological evidence. Pp. 295–306 *in* The hierarchy of life. Molecules and morphology in phylogenetic analysis (B. Fernholm, K. Bremer, L. Brundin, H. Jörnvall, L. Rutberg, and H.-E. Wanntorp, eds.). Excerpta Medica, Amsterdam.

Kristensen, N. P. 1991. Phylogeny of extant hexapods. Pp. 125–140 *in* The insects of Australia (I. Naumann, ed.), vol. 1. Cornell University Press, Ithaca, NY.

Kristensen, N. P. 1998. The groundplan and basal diversification of the hexapods. Pp. 280–293 *in* Arthropod relationships (R. Fortey and R. Thomas, eds.). Chapman and Hall, London.

Linné, C. 1758. Systema naturae. Editio Decima. Holmiae.

Lo, N., G. Tokuda, H. Watanabe, H. Rose, M. Slaytor, K. Maekawa, C. Bandi, and H. Noda. 2000. Evidence from multiple gene sequence indicates that termites evolved from wood-feeding cockroaches. Curr. Biol. 10:801–804.

Lohmann, H. 1996. Das phylogenetische System der Anisoptera (Odonata). Entomol. Z. 106:209–296.

Lyal, C. 1985. Phylogeny and classification of the Psocodea, with particular reference to the lice (Psocodea: Phthiraptera). Syst. Entomol. 10:145–165.

Mahner, M. 1993. Systema Cryptoceratorum Phylogeneticum (Insecta, Heteroptera). Zoologica 48:1–302.

Manton, S. 1977. The Arthropoda. Habits, functional morphology and evolution. Clarendon Press, Oxford.

Martynov, A. B. 1925. Über zwei Grundtypen der Flügel bei den Insecten und ihre Evolution. Z. Morphol. Okol. Tiere 4:465–501.

Mayer, P. 1876. Ueber Ontogenie und Phylogenie der Insekten. Jen. Z. Naturwiss. 10:125–221.

Naumann, I. 1991. The insects of Australia. 2 vols. Cornell University Press, Ithaca, NY.

Nixon, K. C., and J. M. Carpenter. 2000. On the other "phylogenetic systematics." Cladistics 16:298–318.

Novokshonov, V. G., and R. Willmann. 1999. On the morphology of *Asthenohymen uralicum* (Insecta; Diaphanopterida: Asthenohymenidae) from the lower Permian of the Urals. Paleontol. J. 33:539–545.

Papavero, N., J. Llorente-Bousquets, and J. Abe. 2001. Proposal of a new system of nomenclature for phylogenetic systematics. Arq. Zool. 36:1–145.

Pass, G. 1998. Accessory pulsatile organs. Microsc. Anat. Invert. 11B:621–640.

Paulus, H. F. 1979. Eye structure and the monophyly of the Arthropoda. Pp. 299–383 *in* Arthropod phylogeny (A. P. Gupta, ed.). Van Nostrand Reinhold, New York.

Popham, E. 1985. The mutual affinities of the major earwing taxa (Insecta, Dermaptera). Z. Zool. Syst. Evol. Forsch. 23:199–214.

Rähle, W. 1970. Untersuchungen an Kopf und Prothorax von *Embia ramburi* Rimsky-Korsakoff 1906 (Embioptera, Embiidae). Zool. Jb. Anat. 87:248–330.

Rasnitsyn, A. 1998. On the taxonomic position of the insect order Zorotypida = Zoraptera. Zool. Anz. 237:185–194.

Richards, O. W., and R. G. Davies. 1977. Imm's general textbook of entomology, Vol. 2: Classification and biology. Chapman and Hall, London.

Richter, S., G. D. Edgecombe, and G. D. F. Wilson. 2002. The

lacinia mobilis and similar structures—a valuable character in arthropod phylogenetics? Zool. Anz. 241:339–361.

Ross, E. S. 2000. Contributions to the biosystematics of the insect order Embiidina. 1. Origin, relationships and integumental anatomy. Occ. Pap. Calif. Acad. Sci. 149:1–53.

Ross, H. H., C. A. Ross, and J. R. P. Ross. 1982. A textbook of entomology. 4th ed. John Wiley and Sons, New York.

Rowell, C. H. F., and P. K. Flook. 1998. Phylogeny of the Caelifera and the Orthoptera as derived from ribosomal gene sequences. J. Orthopt. Res. 7:147–156.

Seeger, W. 1975. Funktionsmorphologie an Spezialbildungen der Fühlergeißel von Psocoptera und anderen Paraneoptera (Insecta); Psocodea als monophyletische Gruppe. Z. Morphol. Tiere 81:137–159.

Seeger, W. 1979. Spezialmerkmale an Eihüllen und Embryonen von Psocoptera im Vergleich zu anderen Paraneoptera (Insecta); Psocoptera als monophyletische Gruppe. Stutt. Beitr. Nat. 329:1–57.

Sharov, A. G. 1968. Phylogeny of the orthopteroidea. Trudy Paleontol. Inst. Akad. Nauk. SSSR 118 (repr.: 1971, Keter Press, Jerusalem).

Snodgrass, R. E. 1935. Principles of insect morphology. Cornell University Press, Ithaca, NY.

Snodgrass, R. 1938. Evolution of the Annelida, Onychophora, and Arthropoda. Smithsonian Misc. Coll. 97(6):1–159.

Snodgrass, R. E. 1952. A textbook of arthropod anatomy. Comstock Publ. Assoc., Ithaca, NY.

Soldán, T. 1997. The Ephemeroptera: whose sister-group are they? Pp. 514–519 *in* Ephemeroptera and Plecoptera: biology–ecology–systematics (P. Landolt and M. Sartori, eds.). Mauron + Tinguely and Lachat SA, Fribourg, Switzerland.

Staniczek, A. H. 2000. The mandible of silverfish (Insecta: Zygentoma) and mayflies (Ephemeroptera): its morphology and phylogenetic significance. Zool. Anz. 239:147–178.

Stys, P., and S. Bilinski. 1990. Ovariole types and the phylogeny of hexapods. Biol. Rev. 65:401–429.

Vishniakova, V. 1980. Earwigs (Insecta, Forficulida) from the Upper Jurassic of the Karatau Range. Paleont. J. (no vol.):63–79.

Vishniakova, V. 1981. Novye paleozoyskie i mezozoyskie lophioneuridy (Thripida, Lophioneuridae) (in Russian). [New paleozoic and mesozoic lophioneurids (Thripida, Lophioneuridae).] Trudy Paleontol. Inst. Akad. Nauk, SSSR 183:43–63.

Weber, H. 1933. Lehrbuch der Entomologie. G. Fischer Verlag, Jena.

Wheeler, W. C., M. Whiting, Q. D. Wheeler, and J. M. Carpenter. 2001. the phylogeny of the extant hexapod orders. Cladistics 17:113–169.

Wheeler, W. M. 1904. The phylogeny of termites. Biol. Bull. 8:29–37.

Wheeler, W. M. 1928. The social insects. Their origin and evolution. Paul, Trench, Trubner & Co., London, and Harcourt, Brace and Company, New York.

Whiting, M. F., S. Bradler, and T. Maxwell. 2003. Loss and recovery of wings in stick insects. Nature 421:264–267.

Whiting, M. F., J. C. Carpenter, Q. D. Wheeler, and W. C. Wheeler. 1997. The Strepsiptera problem: phylogeny of the holometabolous insect orders inferred from 18S and 28S ribosomal DNA sequences and morphology. Syst. Biol. 46:1–69.

Willmann, R. 1987. Phylogenetic systematics, classification and the plesion concept. Verh. Naturwiss. Ver. Hamb. 29:221–233.

Willmann, R. 1990. Die Bedeutung paläontologischer Daten für die zoologische Systematik. Verh. Dtsch. Zool. Ges. 83:277–289.

Willmann, R. 1997. Phylogeny and the consequences of phylogenetic systematics. Pp. 499–510 in Ephemeroptera and Plecoptera: biology–ecology–systematics (P. Landolt and M. Sartori, eds.). Mauran and Tinguely and Lachat SA, Fribourg, Switzerland.

Willmann, R. 1998. Advances and problems in insect phylogeny. Pp. 269–279 in Arthropod relationships (R. Fortey and R. Thomas, eds.). Chapman and Hall, London.

Willmann, R. 1999. The upper carboniferous *Lithoneura lameerei* (Insecta, Ephemeroptera ?). Palaontol. Z. 73:289–302.

Willmann, R. 2002. Phylogenese und System der Insecta. Pp. 1–64 in Lehrbuch der Speziellen Zoologie: Insecta (H. H. Dathe, ed.). G. Fischer, Heidelberg, Germany.

Willmann, R. 2003. Die phylogenetischen Beziehungen der Insekten: Offene Fragen und Probleme. Verh. Westdeutsch. Entomol. Tag. 2003:1–64.

Wygodzinsky, P. 1961. On a surving representative of the Lepidotrichidae (Thysanura). Ann. Entomol. Soc. Am. 54:621–627.

Yoshizawa, K., and T. Saigusa. 2001. Phylogenetic analysis of paraneopteran orders (Insecta: Neoptera) based on forewing base structure, with comments on monophyly of Auchenorrhyncha (Hemiptera). Syst. Entomol. 26:1–13.

Zwick, P. 1973. Insecta: Plecoptera. Phylogenetisches System und Katalog. Das Tier. 94:1–465.

Zwick, P. 1980. Plecoptera (Sternfliegen) [sic]. Handb. Zool. IV Liefg. 26:1–121. Eine Naturgeschichte der Stämme des Tierreichs. Walter de Gruyter, Berlin.

Zwick, P. 2002. Plecoptera. Steinfliegen. Pp. 144–154 in Lehrbuch der speziellen Zoologie I: Insecta (H. H. Dathe, ed.). G. Fischer, Heidelberg, Germany.

Michael F. Whiting

Phylogeny of the Holometabolous Insects

The Most Successful Group of Terrestrial Organisms

The radiation and diversification of the holometabolous insects stand as two of the grandest events in all of evolutionary history, representing an unprecedented explosion in species coupled with extensive anatomical and physiological specialization. The defining characteristic for Holometabola is complete metamorphosis: every insect in this group, with rare exception, passes through an egg, larval, pupal, and adult stage. This is in contrast to the non-holometabolous insect groups in which juveniles have more or less the same form as the adult, live in the same environment, and exploit similar resources. Although it has never been thoroughly tested, it is thought that the evolution of complete metamorphosis was the key innovation allowing these insects to partition habitats between adults and juveniles, resulting in a wider range of niches that could be occupied by the nascent species. And occupied they have. Holometabola includes well more than one million species representing roughly 80% of all described insect species and just more than half of the total number of described species on Earth today (Kristensen 1999, Wilson 1988). The immense size of this group and their unique morphological specializations present a serious challenge to phylogenetic systematics. However, current research is providing new insight into the evolution and diversification of this, the most successful group of terrestrial organisms, and in the past few years researchers have finally begun to unravel the Tree of Life for holometabolous insects.

Holometabola appear to be a true evolutionary group in the sense that all members of Holometabola can trace their evolutionary history back to a single ancestor (i.e., Holometabola are monophyletic). This is evidenced by the fact that all members of Holometabola undergo complete metamorphosis, and that they have some other distinct morphological characteristics shared by no other insect groups (Kristensen 1999, Whiting 1998a). For instance, holometabolans are the only insects in which the larval eyes disintegrate and the adult eyes develop de novo during the last immature stage. The developing wings in the larvae of holometabolous insects are kept inside the body until the larval-pupal molt, whereas in other insect groups the developing wing appears on the outside of the body in early nymphal stages. In fact, the group Holometabola is often called Endopterygota (internal-winged) because of this feature. Likewise, external genitalia do not appear until the penultimate (larval-pupal) molt. In addition, phylogenetic analysis of DNA sequence data consistently supports the monophyly of Holometabola. With the possible exception of the group Neoptera (winged insects), there is no other major group of insects whose monophyly is more strongly supported than that of Holometabola.

Holometabola are composed of 11 major living lineages, each of which is also a monophyletic group (with one exception, described below). Entomologists have given each of these lineages the taxonomic ranking of an order, but the number of species within each of these orders is drastically unequal, reflecting both the morphological specialization and the differential success of particular groups (table 21.1). The majority of holometabolous insect species are placed

Table 21.1

Holometabolous Insect Orders and Common Names.

Order	Common name
Coleoptera	Beetles
Neuroptera	Lacewings, antlions, owlflies
Megaloptera	Alderflies, fishflies, dobsonflies
Raphidioptera	Snakeflies
Hymenoptera	Bees, wasps, ants
Trichoptera	Caddisflies
Lepidoptera	Butterflies, moths, skippers
Mecoptera	Scorpionflies
Siphonaptera	Fleas
Strepsiptera	Twisted-winged parasites
Diptera	Flies
Nannomecoptera	Nannochoristid scorpionflies
Neomecoptera	Snow fleas (Boreidae)

within four megadiverse orders: approximately 500,000 species of beetles (order Coleoptera), 160,000 species of bees, wasps, and ants (order Hymenoptera), 150,000 species of flies (order Diptera), and 150,000 species of butterflies and moths (order Lepidoptera). Additional species are added to each of these orders on almost a daily basis, and it is clear that we have only scratched the surface of species diversity within these groups. The remaining seven orders are less diverse, although they include some of the most peculiar and specialized forms. These include caddisflies (order Trichoptera) with roughly 7000 species, lacewings (order Neuroptera) with 6000 species, fleas (order Siphonaptera) with ~2400 species, twisted-winged parasites (order Strepsiptera) with 532 species, scorpionflies (order Mecoptera) with 500 species, dobsonflies and alderflies (order Megaloptera) with 270 species, and snakeflies (order Raphidioptera) with 205 species. There are good morphological characters to support the monophyly of most of these groups, and for well more than a century any newly described insect with complete metamorphosis could be easily assigned to one of these living lineages.

What we do not know, however, is the exact pattern of phylogenetic relationships among each of the 11 holometabolous insect orders. A child can tell a beetle from a wasp from a butterfly, but even the entomologically erudite is left pondering which two insects are most closely related. A few hypotheses of interordinal phylogenetic relationships will be presented below, but there many unanswered questions still remain. Likewise, relationships within each of the holometabolous insect orders are often obscure, although major insights are being made each year. This chapter focuses on what we think we know about holometabolan phylogeny, what relationships are more dubious, and pinpointing major gaps in our knowledge of holometabolan phylogeny.

Interordinal Phylogeny

Many hypotheses have been presented for phylogenetic relationships among the holometabolous insect orders over the past century; these reflect the general difficulty of reconstructing the evolutionary history of this important insect group and the variety of opinions on the matter. Summaries of the most influential and current hypotheses are presented in figure 21.1. Boudreaux (1979; fig. 21.1A) and Hennig (1981; fig. 21.1B) presented phylogenies based on different interpretations of morphological characters. Both of these workers compiled and discussed evidence for insect phylogeny based on morphological (anatomical) data, but because neither presented any formal analyses of these data, it remained unclear how well a particular phylogenetic tree was supported by the underlying data. Boudreaux placed Strepsiptera + Coleoptera as the most primitive holometabolan lineage and then argued for the placement of Hymenoptera at the base of the remaining orders. However, the questionable morphological data he presented coupled with the particular twist he put on the interpretation of these data (e.g., arguing that the most common morphological feature must be the most primitive feature), leave his conclusions unsatisfying. Hennig was influential in the development of phylogenetic theory and is widely considered the father of modern phylogenetics, although he was also challenged by his attempts to provide a complete view of insect ordinal relationships. Hennig was uncertain as to the placement of Hymenoptera and Siphonaptera but argued for a sister-group relationship between Strepsiptera and Coleoptera, and associated Trichoptera + Lepidoptera with Diptera + Mecoptera. Kristensen is the most influential morphological worker in recent memory, and his summaries of insect ordinal phylogeny (Kristensen 1975, 1981, 1991, 1995, 1999) provide excellent commentary on the wide variety of morphological evidence that has been garnered to support different phylogenetic hypotheses. In his most recent summary (Kristensen 1999; fig. 21.1C), Holometabola are divided into two main divisions. The Coleoptera + Neuropterid lineages (Neuroptera, Megaloptera, and Raphidioptera) form one division, and the remaining orders are placed in a second division (Hymenoptera + Mecopterida), with uncertainty as to the position of the enigmatic Strepsiptera (more on this below). Recently, Beutel and Gorb (2001) added a suite of morphological characters associated with the tarsi of insects and proposed a phylogeny that agrees with Kristensen (1999) except for the position of Strepsiptera as sister group to Coleoptera.

Although a few attempts had been made from a molecular standpoint to decipher holometabolan phylogeny (Carmean et al. 1992, Chalwatzis et al. 1996, Pashley et al. 1993), Whiting et al. (1997) was the first the presentation of a formal analysis of morphological data in combination with extensive DNA sequence data for Holometabola. These

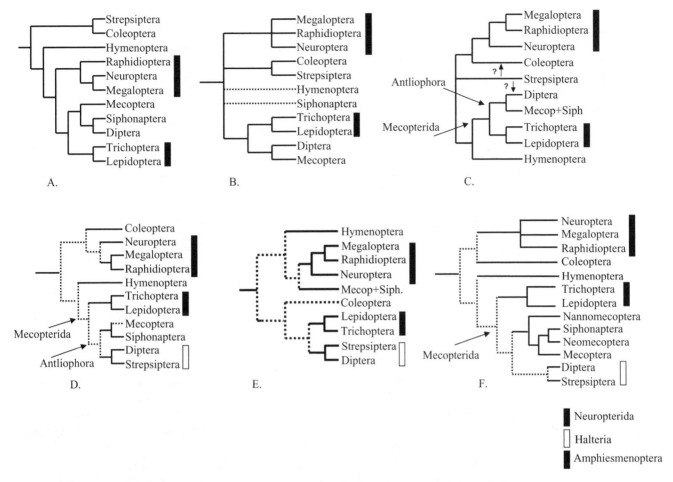

Figure 21.1. Previous phylogenetic hypotheses of relationships among holometabolous insect orders. (A) Boudreaux (1979), based on morphology. (B) Hennig (1981), based on morphology. (C) Kristensen (1999), based on morphology. (D) Whiting et al. (1997) and Wheeler et al. (2001), based on morphology and DNA. (E) Whiting (2002c), based on extensive sample of DNA sequences. (F) Summary tree representing current state of knowledge. Dashed lines represent uncertain relationships.

data consisted of 176 morphological characters coded across Holometabola and outgroups, and portions of the 18S ribosomal DNA (rDNA) molecule (~1000 nucleotides) and 28S rDNA (~400 nucleotides). Wheeler et al. (2001) expanded this study to include all hexapod orders and used a new analytical tool that obviates the need to generate a multiple alignment of the DNA sequence data before phylogenetic reconstruction (i.e., optimization alignment). Both studies largely concurred in their view of holometabolan phylogeny (fig. 21.1D). These results were surprising in three ways: (1) they suggested a sister-group relationship between the enigmatic Strepsiptera and Diptera; (2) they demonstrated a close association of fleas with a family placed within the scorpionflies (Mecoptera); and (3) although their topology is largely congruent with those trees presented by Kristensen, their results indicate that many holometabolan interordinal relationships are not particularly well supported. Whiting

(2002b, 2002c) performed more extensive molecular analyses based on the entire 18S rDNA gene for roughly three times more holometabolan species than in earlier studies. Although this increased species sampling helped resolve some relationships (e.g., better support for Neuropterida), the general pattern of relationships provided by this single molecule is in some cases different than those found with morphology (fig. 21.1E).

So what do these studies tell us? All workers agree that there are two well-supported relationships among the holometabolous insect orders (table 21.2). The first is a sister-group relationship between Lepidoptera and Trichoptera to form a group called Amphiesmenoptera. This relationship is supported by more than 15 morphological characters, including the female heterogamy (essentially, females possess the XY chromosome) and the presence of scales or hairs on the wing surface between veins (Hennig 1981, Kristensen

1997, Whiting et al. 1997). This group has been found in every DNA phylogenetic analysis to date (Chalwatzis et al. 1996, Wheeler et al. 2001, Whiting 2002c, Whiting et al. 1997) and is considered the best-supported sister-group relationship in all of insect ordinal phylogeny. Second, all hypotheses agree that the orders Neuroptera, Raphidioptera, and Megaloptera should be placed in a single group called Neuropterida. The monophyletic grouping of the neuropterids is supported by a series of specializations associated with the female ovipositor (Mickoleit 1973), and this group is also consistently recovered in phylogenetic analyses based on DNA sequence data (Wheeler et al. 2001, Whiting 2002b, 2002c). Molecular data consistently support a sister-group relationship between Megaloptera and Raphidioptera, which agrees with some morphological evidence (Wheeler et al. 2001, Whiting 2002c). An alternative hypothesis is that Megaloptera and Neuroptera are sister groups based on the presence of aquatic larvae, found in all Megaloptera and one primitive family of Neuroptera (Nevrorthidae), although the vast majority of neuropterans are terrestrial, with the exception of the more derived spongillaflies (Aspöck et al. 2001).

Beyond Neuropterida and Amphiesmenoptera, the picture becomes murky and the hypotheses more controversial. This is largely because most of the holometabolous insect orders are so highly specialized that it becomes difficult to unravel the morphological clues required to determine phylogenetic affinity. Very often the morphological evidence presented to support hypothesized relationships consists of only one or two characteristics that are not universally shared by members of those groups, and the homology among these characters is questionable. Moreover, different specialists have different interpretations of morphology leading to dramatically different estimates of phylogeny.

Current morphological analyses suggest that Holometabola may be divided into two major groups: Coleoptera + Neuropterida and Hymenoptera + Mecopterida (= Trichoptera + Lepidoptera + Mecoptera + Siphonaptera + Diptera). The position of the enigmatic Strepsiptera is discussed below. The sister-group relationship between the lacewings and the beetles is supported by specific modifications of the ovipositor (Kristensen 1991) and characters associated with the

base of the hind wing in these insects (Hörnschemeyer 2002; fig. 21.1C). The monophyly of Mecopterida is supported by the presence of a muscle that is attached between the thorax wall (i.e., pleuron) and a hardened structure at the base of the wing (i.e., first axillary sclerite; Kristensen 1999), although this character is not present in the wingless fleas. Within Mecopterida, Lepidoptera and Trichoptera form the group Amphiesmenoptera (as discussed above), and Diptera + Mecoptera + Siphonaptera form another group. Morphological data combined with molecular data suggest that fleas actually are an offshoot of one scorpionfly lineage. Boudreaux (1979) placed Hymenoptera as one of the most basal members of Holometabola (fig. 21.1A) but did not provide a convincing argument to support this position. Kristensen (1991, 1999) argues that Hymenoptera should be placed as sister group to Mecopterida, based on two characters associated with the form of the larvae and one based on a particular modification of the sucking pump in the adult insect (Kristensen 1999).

DNA sequences are presently being generated to try and provide independent estimates of ordinal phylogeny, and although these data have provided new insight into some of the more nebulous questions, the overall view of ordinal phylogeny is still under construction. From a molecular standpoint, the problem has been that the few DNA markers that are commonly used in insect ordinal phylogeny are not informative for all portions of the phylogeny, so additional gene regions need to be investigated to provide a more robust estimate of the holometabolan branches of the Tree of Life. The hope is that these additional data will provide new insights in the patterns of diversification across Holometabola. Although the picture is not yet clear, the current DNA data have pointed to some very interesting relationships.

For instance, data from four independent genes suggest that the fleas are sister group to the snow scorpionflies (Boreidae), a family of scorpionflies that live on the snow and are closely associated with moss (Whiting 2002a). Once the molecular data suggested this relationship, a reevaluation of morphology demonstrated that this is a plausible hypothesis. Morphological features supporting this relationship include the presence of unusual spines in the gut (proventriculous; Schlein 1980), multiple sex chromosomes (Bayreuther and Brauning 1971), a series of specializations associated with the female ovaries (Bilinski et al. 1998), and the ability to jump via a similar mechanism. These data suggest that fleas did not evolve from a group of flies, as has been proposed (Byers 1996), but rather were living on the snow and then shifted to mammal burrows where they became obligate, external parasites. An additional mecopteran lineage of small and obscure insects (Nannochoristidae) is the most primitive group of Mecoptera, based on both molecular (Whiting 2002a) and morphological data (Simiczyjew 2002, Willmann 1987). These findings indicate that Mecoptera are not monophyletic and that if the Siphonaptera are to be retained as a recognized order, it must be subdivided

Table 21.2
Superordinal Groups in Insect Phylogeny.

Superordinal name	Groups included
Neuropterida	Neuroptera + Megaloptera + Raphidioptera
Mecopterida	Lepidoptera + Trichoptera + Siphonaptera + Diptera + Strepsiptera
Amphiesmenoptera	Lepidoptera + Trichoptera
Antliophora	Mecoptera + Siphonaptera + Diptera + Strepsiptera
Halteria	Strepsiptera + Diptera

into additional insect orders. Given that current classification does not allow non-monophyletic groups to be formally named, it is necessary to recognize the additional orders Nannomecoptera (for Nannochoristidae) and Neomecoptera (snow scorpionflies; fig. 21.1F). Hinton (1958) was the first to present a series of morphological characters to elevate snow scorpionflies to their own order, Neomecoptera.

The most perplexing question in holometabolan phylogeny, and the one that has received the most attention in recent years, has been the controversy surrounding the placement of Strepsiptera. This is an unusual group of insects, members of which spend most of their lives as obligate internal parasites of other insects. From a morphological standpoint, the adult females are so highly reduced and larvalike that they leave no clues as to their phylogenetic position. The males are highly derived with unusual eyes, mouthparts, and other structures and are so specialized that it has been very difficult to assign them to any particular phylogenetic group. This perplexing amalgamation of morphological reduction in females and extreme modification in males, combined with unusual biology and larval characteristics, has challenged systematic placement of this group for more than two centuries. Strepsiptera were associated with Coleoptera, either as a member of Coleoptera (Crowson 1960) or as sister group to Coleoptera, based on wing morphology and function (Kathirithamby 1989, Kristensen 1981, 1991, Kukalova-Peck and Lawrence 1993). Detailed examination of this evidence, however, suggests that these characters are based on mistaken descriptions of strepsipteran wing morphology and function (Beutel and Haas 2000, Kinzelbach 1990, Pix et al. 1993, Whiting 1998b). Current DNA sequence data strongly support a sister-group relationship between Strepsiptera and Diptera to form a group called Halteria (Wheeler et al. 2001, Whiting 2002c, Whiting et al. 1997, Whiting and Wheeler 1994). This result has been challenged as a methodological artifact of a particular mode of data analysis (Huelsenbeck 1997), although, as has been argued elsewhere (Sidall and Whiting 1999, Whiting 1998a) that these criticisms are off the mark. If Strepsiptera are sister group to Diptera, then the similarities in the form and function of their modified wings might be attributed to evolution via shifts in development, providing new insights into how organisms can evolve in leaps and bounds across evolutionary time. Nonetheless, Diptera + Strepsiptera is still controversial, and additional data are needed before this relationship is universally accepted.

In summary, current DNA sequence data support the monophyly of most of the holometabolous insect orders, in agreement with morphology. DNA also supports the superordinal groups Amphiesmenoptera, Neuropterida, and Halteria and the relationship among Mecoptera and Siphonaptera as described above. DNA has not, however, been successful at confirming the relationships hypothesized by morphology, such as Mecopterida, Hymenoptera + Mecopterida, or Coleoptera + Neuropterida. A tree sum-

marizing the current state of affairs in holometabolan phylogeny (fig. 21.1F) indicates that further work is needed to elucidate the more ancient patterns of holometabolan evolution and diversification.

Coleoptera (Beetles)

Beetles are widely considered the most successful group of organisms, with estimated numbers of species ranging from 500,000 to several million (Hammond 1992). Coleoptera appears to be a well-supported monophyletic group characterized by the presence of front wings that are rigid, hardened, and typically cover the entire abdomen (elytra), as well as 20 morphological features unique to this group (Beutel and Haas 2000). Ironically, all molecular studies to date suggest that beetles do not form a natural grouping of species (Caterino et al. 2002, Wheeler et al. 2001, Whiting 2002b, Whiting et al. 1997), but this is probably more indicative of the inadequacy of the current DNA evidence rather than substantial evidence of coleopteran paraphyly.

Coleoptera are divided into four major lineages that are treated as suborders: Archostemata, Myxophaga, Adephaga, and Polyphaga (fig. 21.2). Except for the basal placement of Archostemata, relationships among the other three suborders are controversial. Morphological evidence places Adephaga as sister group to Myxophaga + Polyphaga (Beutel and Haas 2000), but recent molecular analyses suggest that Adephaga are sister group to Polyphaga, with Myxophaga placed at their base (Caterino et al. 2002). Archostemata include four small, living families, although this group was more extensive formerly, as shown by the fossil record. Archostematan larvae are wood borers, and the monophyly of this suborder is supported by some discrete adult and larval characteristics. Myxophaga also include four families of small to minute semiaquatic beetles, and overall this group appears to be well supported based on a series of morphological features (Beutel and Haas 2000). Myxophaga and Archostemata account for less than 1% of the living beetle diversity.

Adephaga include ~30,000 species in a dozen families and comprises ~10% of beetle diversity. This group includes tiger beetles, ground beetles, whirligigs, predaceous diving beetles, wrinkled bark beetles, and others. The monophyly of this suborder also appears to be well supported, although relationships among the constituent families are more controversial and are focused on whether the aquatic taxa (Hydradephaga, six families) and terrestrial taxa (Geadephaga, six families) form two distinct lineages within this suborder. A recent molecular analysis suggests that the aquatic taxa are monophyletic and proposes a phylogeny for the 12 families (Shull et al. 2001).

The suborder Polyphaga includes the vast majority of beetle diversity, with at least 300,000 described species from more than 100 families. In polyphagan beetles, the lateral side of the prothorax (pleuron) is not externally visible, making the pro-

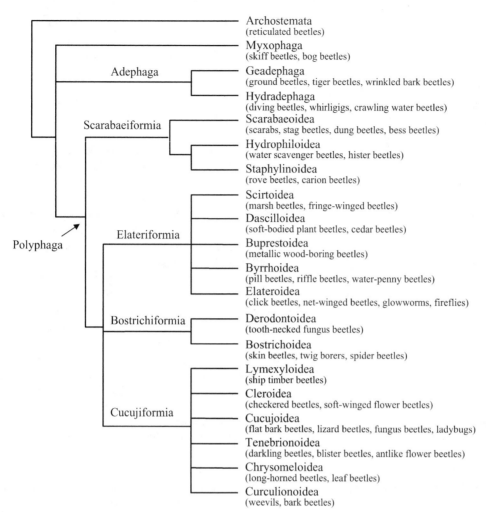

Figure 21.2. Summary phylogeny of beetles (Coleoptera).

thorax appear as a single dorsal plate that wraps around the lateral sides of the prothorax. It appears likely that adoption of a plant feeding lifestyle in these beetles early in angiosperm evolution correlates with the great number of species in some of the major beetle lineages (Farrell 1998). Detailed phylogenetic relationships among most families are unknown, and this is large part because of the overwhelming diversity of anatomical features in this group and the enormous number of species the systematist must deal with. The monophyly of some families is in doubt, but work by a number of beetle specialists has provided a glimpse of polyphagan phylogeny (Crowson 1960, Lawrence and Newton 1982, 1995, Lawrence et al. 1995). Polyphaga are divided into four major lineages, Scarabaeiformia, Elateriformia, Bostrichiformia, and Cucujiformia, although relationships among these lineages are largely unknown. Scarabaeiformia include three superfamilies: Scarabaeoidea (13 families, including scarabs, stag beetles, dung beetles, bess beetles), Hydrophiloidea (four families, including water scavenger beetles and hister beetles), and Staphylinoidea (seven families, including carrion beetles and the extremely large family of rove beetles). Elateriformia include five superfamilies, phylogenetic relationships among which are largely unknown. This group includes Scirtoidea (four families, including marsh beetles and

fringe-winged beetles), Dascilloidea (two families, including soft-bodied plant beetles and cedar beetles), Buprestoidea (one family, the metallic wood-boring beetles), Byrrhoidea (12 families, including pill beetles, riffle beetles, water-penny beetles), and Elateroidea (16 families, including click beetles, net-winged beetles glowworms, fireflies, soldier beetles, etc.). Bostrichiformia are composed of two superfamiles: Derodontoidea (one family, tooth-necked fungus beetles) and Bostrichoidea (six families, including skin beetles, twig borers, and spider beetles). Cucujiformia are the largest and most diverse beetle lineage, including the vast majority of plant-eating beetles. The monophyly of this group is supported by a specialized type of malpighian tubule (essentially, the insect kidney) and is composed of six superfamilies. Lymexeloidea (one family, ship timber beetles), Cleroidea (seven families, including checker beetles and soft-winged flower beetles), Cucujoidea (31 families, including flat bark beetles, lizard beetles, pleasing fungus beetles, ladybugs, etc.), Tenebrionoidea (26 families, including darkling beetles, blister beetles, antlike flower beetles, tumbling flower beetles, etc.), Chrysomeloidea (four families, including long-horn beetles and leaf beetles), and Curculionoidea (nine families, including weevils and bark beetles). Given the enormous size of Coleoptera, it may take half a century to construct a phylogeny as detailed as those currently available for most vertebrate groups.

Neuropterida (Lacewings, Snakeflies, Alderflies, Dobsonflies)

Neuropterida are composed of three closely related orders: Neuroptera (17 families), Megaloptera (two families), and Raphidioptera (two families). Adults have large, separated eyes, mandibulate mouthparts, and multisegmented antennae. Collectively, this group includes individuals that exhibit a broad range of morphological and biological diversity, and the living species are remnants of what were once more diverse lineages, as evidenced by their rich fossil record (Aspöck et al. 2001). As larvae, many neuropterans are voracious predators of other insects, especially the brown and green lacewings and the antlions. Other families have become more specialized, including the spider egg-sac predation in the mantis lacewings (Mantispidae) and the freshwater sponge-feeding spongillaflies (Sisyridae).

The monophyly of Neuroptera is supported chiefly by the larvae possessing piercing, sucking tubes modified from the primitive chewing mouthparts. In addition, the anterior intestinal tract is not connected to the posterior intestinal tract in the larvae, such that they are unable to pass solid waste until the insect becomes an adult and the gut is fully connected (Aspöck et al. 2001). The monophyly of Megaloptera is supported by the presence of lateral, segmented tracheal gills in larvae that allows the larval insect to respire underwater. The monophyly of Raphidioptera is supported by an elongated neck and a pronotum that wraps around the lateral (pleural) regions of the thorax (Wheeler et al. 2001). There has been a suggestion that the megalopteran alderflies (Sialidae) may be sister group to the snakeflies (Raphidioptera), rendering the Megaloptera paraphyletic (Stys and Bilinksy 1990), but this interpretation is not widely accepted (Aspöck et al. 2001). As discussed above, there is a debate as to the phylogenetic relationships among these orders, with the molecular data strongly arguing for Megaloptera + Raphidioptera, as well as some morphological characters (Whiting 2002b, 2002c), versus some revised morphological characters arguing for Megaloptera + Neuroptera (Aspöck et al. 2001).

Relationships among neuropteran families have been historically controversial and have most recently been investigated quantitatively by Aspöck et al. (2001) and Aspöck (2002). According to Aspöck, Neuroptera are divided into three main lineages: antlion-like lacewings (Myrmeleontiformia), lacewing-like (Hemerobiiformia), and Nevrorthiformia, including one obscure family (Nevrorthidae; fig. 21.3). The Myrmeleontiformia include antlions (Myrmeleontidae), owlflies (Ascalaphidae), spoon-winged lacewings (Nemopteridae), and two additional, rather obscure families. This group is supported by wing and larval characteristics and is one of only two well-supported relationship across neuropteran phylogeny. There is debate as to the relationships within Myrmeleontiformia,

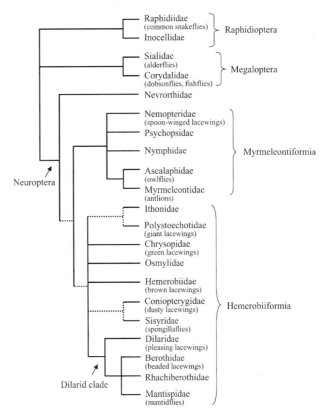

Figure 21.3. Summary phylogeny of Neuropterida, including Megaloptera (alderflies and dobsonflies), Raphidioptera (snakeflies), and Neuroptera (lacewings, antlions, owlflies, etc.). Dashed lines represent uncertain relationships.

particularly regarding the position of Psychopsidae and Nymphidae.

Hemerobiiformia consist of 11 families, including brown and green lacewings (Hemerobiidae and Chrysopidae), dusty wings (Coniopterygidae), mantidflies (Mantispidae), spongillaflies (Sisyridae), and other groups. The monophyly of this group is questionable, although the "dilarid clade," including Dilaridae, Mantispidae, Rhachiberothidae, and Berothidae, is well supported by characteristics associated with the larval head capsule. With the exception of the dilarid clade, relationships among the constituent families within this group are also questionable. The Nevrorthiformia include an obscure group of lacewings with aquatic larvae that have been placed as the most primitive group within Neuroptera, although this is certainly open to further investigation.

One of the more interesting questions in neuropteran evolution has been the suggestion that Neuroptera were derived from an aquatic ancestor. This hypothesis is based on a phylogenetic topology where the entirely aquatic Megaloptera are sister group to Neuroptera, and the aquatic Nevrorthidae are the most basal neuropteran lineage (Aspöck et al. 2001). If it turns out that Megaloptera and Raphidioptera are indeed sister groups, as indicated by current molecular data, or that Nevrorthidae are not the most basal lineage, then the aquatic origin hypothesis will be left without much merit. Clearly, there is a need to further investigate phylogenetic relationships among these interesting insects.

Hymenoptera (Sawflies, Bees, Wasps, Ants)

Hymenoptera are currently composed of ~150,000 described species, but when all the undescribed species are added, the group may be twice this size (Kristensen 1999), putting it on par with Coleoptera. Hymenopterans are found within most terrestrial ecosystems and play a vital role in pollination of flowering plants and as predators and parasites of other insects, with ants alone forming a major component of tropical ecosystems. Hymenopterans range in size from microscopic parasites of insect eggs to very large bees and wasps. This group is characterized by the presence of specialized hooks that join the hind wings to the forewings (hamuli), absence of notal coxal muscles, and the presence of a unique reproductive mode known as haplodiploidy.

Hymenoptera have been traditionally divided into two groups: Symphyta (sawflies and allies) and Apocrita (bees, wasps, and ants; fig. 21.4). In Symphyta, the thorax is three segmented and broadly joined to the abdomen, and the wing venation is relatively complete. Most of the members of this group are external feeders on foliage and have an ovipositor that is somewhat sawlike, hence the common name "sawflies." Comparative morphological work suggests that Symphyta as a whole are not monophyletic, but Tenthredinoidea (five sawfly families) and Megalodontoidea (two families, web-spinning sawflies) are monophyletic (Ronquist et al. 1999, Schulmeister et al. 2002, Vilhelmsen 1997). The xyelid sawflies are considered the most primitive of all

Hymenoptera, and morphological data suggest that the parasitic wood wasps (Orussidae) form a sister group to Apocrita (Ronquist 1999), although molecular data suggest other alternatives (Dowton and Austin 1999).

The monophyletic Apocrita contain the vast majority of hymenopteran species diversity. In contrast to Symphyta, in Apocrita the first abdominal segment (propodeum) is fused to the thorax to form a mesosoma, and the second abdominal segment (and sometimes the third) is constricted to form a petiole, the threadlike waist seen in wasps, bees, and ants. Traditionally, Apocrita are divided into the parasitic and aculeate wasps (Rasnitsyn 1988), and although Aculeata are clearly monophyletic, Parasitica include a large number of lineages whose phylogenetic relationships are largely unknown. Within the paraphyletic "Parasitica," Evaniomorpha are composed of a diverse number of lineages, including stephanid wasps, ceraphronid wasps, and ensign wasps, and this group is probably not monophyletic. There are, however, some well-established groupings within Parasitica, some of which have undergone formal phylogenetic investigation, including Cynipoidea, Chalcidoidea, Platygastroidea, and Ichneumonoidea (Rasnitsyn 1988, Ronquist et al. 1999). Chalcidoidea include 20 families of very small wasps (0.5–3 mm) that are primarily the parasites of other insects, attacking chiefly the egg or larval stage of the host. Cynipoidea are composed of five families of mostly minute wasps that are primarily gall makers. Ichneumonoidea include three families of relatively large wasps that are parasitoids of other insects. All of these groups have a large number of species, and phylogenetic relationships among most of the constituent species remain virtually unknown.

Aculeatans are hymenopterans in which the ovipositor has been modified into a stinger. Aculeata consists of three major lineages: Chrysidoidea, Vespoidea, and Sphecidae + Apoidea. Chrysidoidea (cuckoo wasps and allies) include seven families, and the basic phylogenetic relationships among these groups are moderately well understood (Carpenter 1999). Vespoidea (ants, vespid wasps, sphecid wasps, spider wasps, velvet ants, etc.) are a diverse assemblage of lineages composed of roughly 10 families. Phylogenetic analysis suggest, among other things, that ants are sister group to vespid and scoliid wasps and that bees (Apoidea) evidently arose from a single lineage of sphecid wasps (Brothers 1999, Brothers and Carpenter 1993). Given the minute size of many hymenopterans, and the vast diversity of this group as a whole, completing the hymenopteran branch of the Tree of Life will take many years.

Lepidoptera (Butterflies, Moths, Skippers)

Lepidoptera are a large group of primarily terrestrial insects characterized by having wings with a dense covering of setae in the more primitive groups and scales in the more advanced groups. Although the current estimate of described lepi-

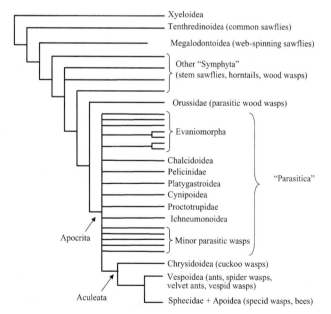

Figure 21.4. Summary phylogeny of bees, wasps, and ants (Hymenoptera).

dopteran species is approximately 150,000, the total number of extant species may be as high as 500,000, making Lepidoptera the largest lineage of primarily herbivorous animals (Kristensen and Skalski 1999). When most people think of Lepidoptera, they think of two groups: butterflies and moths. Although the butterflies are certainly the most popular and well-known lepidopterans, which do indeed form a monophyletic group, they are only a splash in the bucket of lepidopteran diversity. The vast majority of lepidopteran species represent an almost infinite variety of small, drab moths from multiple evolutionary lineages, and the key to unraveling the story of lepidopteran evolution lies in deciphering the phylogeny of moths. Over the last 30 years, extensive morphological studies of the more primitive Lepidoptera, and some more recent molecular studies, have led to a relatively well-established hypothesis of phylogenetic relationships among the more primitive moth groups (Davis 1986, Krenn and Kristensen 2000, Kristensen and Skalski 1999, Wiegmann et al. 2002). However, phylogenetic relationships among the more advanced Lepidoptera, and the more detailed relationships at the family level that include some of the major species radiations, are still unresolved and in need of further phylogenetic investigation.

Kristensen (1999) recognized 46 lepidopteran superfamilies and presented a phylogeny based on a compilation of morphological data. Although the monophyly of most of these superfamilies is relatively well established, superfamilial relationships, particularly among the more derived groups, are very tentative. Lepidopteran phylogeny can be envisioned as a comb (fig. 21.5), where a succession of morphological modifications across a few small groups eventually gave rise to a body type that allowed the organisms to radiate in bursts of speciation events. The first three basal lineages (Micropterigoidea, Agathiphagoidea, Heterobathmioidea) comprise very primitive moths that have retained mandibles and associated muscles for chewing, along with an unmodified, inner pair of lobes (glossa) on the labium, or insect "lower lip." These insects are detritivores, feeding primarily on plant debris in the soil, or are miners, boring into the seeds or leaves of gymnosperms. The mandibulate moths probably reflect very closely the morphology of the trichopteran-lepidopteran ancestor and lack many of the modifications of the more advanced lepidopterans.

The first major evolutionary innovation in lepidopteran morphology was the reduction and loss of the chewing mandibles in the adult insect, which were replaced by extension and fusion of the inner lobes of the labium to form a coilable, sucking proboscis typical of most Lepidoptera. This morphological shift rendered the adults of all higher lepidopterans dependent exclusively on fluid nutrients, which opened a new niche that these insects were uniquely suited to exploit. Hence, a shift from a gymnosperm feeding, mandibulate moth to that of an angiosperm nectaring, proboscis-bearing moth allowed higher lepidopterans to diversify concomitantly with their angiosperm hosts (Kristensen 1997) and is

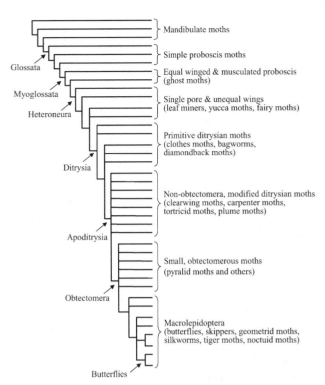

Figure 21.5. Summary phylogeny of butterflies and moths (Lepidoptera).

largely the reason why this is such a diverse and successful group. More than 99.9% of all lepidopteran species possess these sucking tubes and collectively are placed in the group Glossata, named after their possession of the glossa modified into the all-important proboscis. A proboscis that is adapted for nectar feeding should be long and flexible and should have particular sensory equipment allowing for control of probing movements and the detection of concealed nectar in elongated corollae (Krenn 1998). The development of the proboscis did not occur as a single evolutionary event, however, but a succession of gradual transformations leading to the refinement in sensory equipment and muscle control occurred as lepidopterans diversified. The most primitive glossatans (Eriocranioidea, Acanthopteroctetoidea, and Lophocoronoidae) have a relatively simple proboscis with limited movement due to a lack of true intrinsic musculature (Nielsen and Kristensen 1996). The group Myoglossata possesses true intrinsic musculature of the proboscis as well as advanced sensory organs for the more efficient detection of nectar in flowering plants.

Two other evolutionary changes in morphology have played a key role in the evolution and diversification of Lepidoptera. The first was a shift from the forewings and hind wings being approximately the same size with a similar pattern of venation ("homoneuran" condition), to a condition in which the hind wing is smaller than the forewing, and has certain veins fused together. This latter group is termed Heteroneura, meaning "different veined." The myoglossatan, "homoneuran" groups include ghost moths and their allies (Neopseustoidea, Hepialoidea, and Mnesarchaeoidea). The second major evolutionary change was a shift from a single genital pore to a double genital pore in Lepidoptera females. Primitive Lepidoptera females exhibit the typical insect condition of having a single genital orifice that is used for copulation and egg deposition. In the more advanced lepidopterans (group Ditrysia), there is one orifice for copulation (on the eighth ventral abdominal segment) and a separate orifice for egg laying (abdominal segment 9–10), with an internal communication between sperm receiving and oviduct systems. The heteroneuran, non-ditrysian groups consist of four major lineages (Nepticuloidea, Incurvaroidea, Palaephatoidea, and Tischerioidea), including leaf miners, yucca moths, and fairy moths, but these groups are sparse in species numbers relative to Ditrysia. Roughly 98% of all lepidopterans belong to Ditrysia, and there are no major species radiations before the development of this unique reproductive system (Kristensen and Skalski 1999).

Phylogenetic relationships among the ditrysian lineages are more difficult to ascertain, in large part because of the extensive modifications in morphology and the explosion of species numbers. The primitive Ditrysia consist of four lineages (Tineoidea, Gracillarioidea, Yponomeutoidea, and Gelechioidea), including clothes moths, bagworms, and diamondback moths. The more advanced ditrysians (Apoditrysia) are characterized by the presence of specific modifications of the endoskeletal structure of the second abdominal segment (Kristensen and Skalski 1999). Within Apoditrysia is the group Obtectomera, which is characterized by the abdominal segments 1–4 being immovable and the wings being appressed next to the body while in the pupal stage. The non-obtectomeran, apoditrysian moths consist of eight lineages, including clearwing moths, carpenter moths, plume moths, and totrticid moths. Phylogenetic relationships among these lineages, some of which are very large with more than 10,000 described species, are almost entirely unknown. The obtectomeran moths can be divided roughly into two groups: "Microlepidoptera" and Macrolepidoptera. The obtectomeran microlepidoptera consist of six lineages, the largest of which includes the pyralids or snout moths, and relationships among these lineages are unknown, although it is likely that as a whole these microlepidopterans are not monophyletic. Macrolepidoptera, as the name indicates, include the large moths and butterflies that have broad wings and a unique elongation on a portion of the wing base associated with the hinge (first axillary sclerite). This group includes the most spectacular lepidopteran species, including silkworm moths, tiger moths, geometrid moths, noctuids, skippers, and butterflies. Within Macrolepidoptera, there are three major radiations (noctuid moths, geometrid moths, and butterflies with more than 20,000 species each), one moderate-sized radiation (silkworm lineage and allies), and four relatively minor lineages. One group, Noctuoidea, has more than 30,000 described species and represents by far the largest radiation of any lepidopteran

group, and getting a handle on even the basic diversity of this group is a daunting task. So, although a basic skeletal structure of lepidopteran phylogeny exists, the real challenge in lepidopteran systematics for the next century will be to flesh out the phylogenetic relationships of these diverse groups in more detail.

Trichoptera (Caddisflies)

Trichoptera are a large group of semi-aquatic insects whose larvae are found in lakes, streams, and rivers around the world and form a major component of most freshwater ecosystems. Trichopteran adults have a mothlike appearance but with hair rather than scales on the wings, three- to five-segmented maxillary palps, and three-segmented labial palps. As discussed above, a sister-group relationship between Trichoptera and Lepidoptera is well established, but trichopterans lack the sucking, tubelike mouthparts characteristic of Lepidoptera. Like lepidopterans, caddisflies are capable of spinning silk from specially modified salivary glands, and the diversity of ways this silk is used probably accounts for the success of the order as a whole (Mackay and Wiggins 1978). Trichoptera includes approximately 10,000 species placed within 45 recognized families, and the group is quite diverse in terms of the aquatic microhabitats and trophic niches occupied by the species (Morse 1997a).

Phylogenetic relationships within Trichoptera are somewhat controversial, although ongoing research is providing new insights on the evolution of this group. Current classifications recognize three major suborders that are largely characterized by different ways in which silk is used by the larvae (fig. 21.6). Annulipalpia (retreat-makers) include nine families, and these caddisflies make fixed retreats or capture

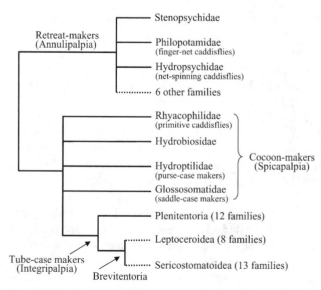

Figure 21.6. Summary phylogeny of caddisflies (Trichoptera). Dashed lines represent uncertain relationships.

nets under rocks, logs, and other objects in streams, rivers, lakes, and ponds. All retreat makers possess a ringlike (annulated) last segment of the maxillary palp. Integripalpia are the largest group of caddisflies (33 families), and this group includes species that make mobile, tubelike cases. These tube-making caddisflies use silk to attach small rocks, sticks, and other material to form a case that they carry around with them as they move, and can retract their heads and thorax inside the case for protection as needed.

Spicipalpia (cocoon-makers) are composed of four families, including free-living and predaceous caddisflies (Rhyacophilidae and Hydrobiosidae), caddisflies that make a small purselike case (Hydroptilidae), and the tortoise-case and saddle-case caddisflies (Glossosomatidae). Although the monophyly of both the retreat making group and the tube making group appears well supported by morphological (Morse 1997b) and molecular data (Kjer et al. 2002), the monophyly of the diverse cocoon makers is still debatable. Previous phylogenetic hypotheses have included all possible ways of arranging these three groups (Ross 1967, Weaver 1984, Wiggins and Wichard 1989), but the most recent data suggest that the retreat maker group is the most basal suborder, with the remaining caddisflies (Spicipalpia and Integripalpia) forming a monophyletic group (Kjer et al. 2002).

Relationships within retreat-makers are still unclear. Kjer et al. (2002) recognize four distinct lineages (Stenopsychidae, Philopotamidae, Hydropsychidae, and the remaining families), although relationships among these lineages and even the monophyly of each of these lineages is in need of additional investigation. As mentioned above, the cocoon-makers may be paraphyletic, but each of the four families composing this group is probably monophyletic. There appears to be two distinct lineages within the tube-case makers: Plenitentoria (12 families) and Brevitentoria (21 families). Specific familial relationships within Plenitentoria have been suggested by Gall (1994), but current molecular data have not been robust enough to examine this hypothesis in detail. Brevitentoria may consist of two lineages (Leptoceroidea and Sericostomatoidea), but again the monophyly of these groups and relationships within them still require further investigation (Kjer et al. 2002, Scott 1993, Weaver and Morse 1986).

Mecoptera (Scorpionflies, Hangingflies)

Mecoptera (in the broad, classical sense) are a small but morphologically diverse insect order with approximately 600 extant described species placed in nine families and 32 genera (Penny 1997, Penny and Byers 1979). The common name for this group is derived from the fact that the male 9th abdominal segment of one family (Panorpidae) is enlarged and bulbous and curves anterodorsally, resembling the stinger of a scorpion. This group is not monophyletic because fleas are sister group to snow scorpionflies (Boreidae), and the nan-

nochoristid scorpionflies are probably the most basal lineage. As discussed above, both of these groups are deserving of ordinal status (fig. 21.7).

Mecoptera include seven families, two of which—Panorpidae (true scorpionflies) and Bittacidae (hangingflies)—contain 90% of mecopteran species. The remaining five families are much less diverse, but they include groups that exhibit a wide degree of morphological specialization from the wingless Apteropanorpidae, to the earwig flies (Meropeidae), to the fossil-like eomeropid scorpionflies. Mecoptera have a very well documented fossil history and are among the most conspicuous part of the insect fauna of the Lower Permian. The monophyly of each mecopteran family is well established by morphological and molecular data (Byers 1991, Kaltenbach 1978, Whiting 2002a, Willmann 1987).

A number of phylogenetic hypotheses have been presented for relationships, and each has resulted in somewhat different conclusions. Kaltenbach (1978) presented Mecoptera subdivided into three suborders, Protomecoptera (Meropeidae + Eomeropidae), Neomecoptera (Boreidae), and Eumecoptera (remaining families), but did not present a specific phylogeny for these taxa. In a comprehensive analysis of mecopteran morphology from extinct and extant taxa, Willmann (1987, 1989) presented a phylogeny in which Nannochoristidae are the basalmost taxon, with Panorpidae + Panorpodidae forming the most apical clade. This phylogeny was not the result of a formal quantitative analysis of a coded character matrix,

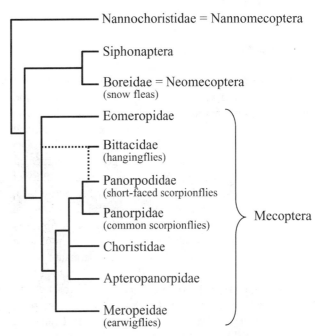

Figure 21.7. Summary phylogeny of scorpionflies (Mecoptera) showing the relative positions of fleas (Siphonaptera). The snow scorpionflies (Boreidae) and nannochoristid scorpionflies are not members of the true scorpionfly lineage (Mecoptera) but are given their own ordinal status. Hangingflies (Bittacidae) are either the sister group to Panorpodidae or at the base of Mecoptera.

but Willmann did provide an explicit explanation of the characters supporting each node of the phylogeny. Whiting (2002a) sequenced four genes across multiple representatives of Mecoptera and performed a preliminary analysis in which Bittacidae appeared as sister group to Panorpodidae. However, inclusion of additional data suggests a more basal placement of Bittacidae and a sister-group relationship between Panorpidae and Panorpodidae, more in line with the phylogeny presented by Willmann. The phylogeny of Mecoptera stands as probably the best-known phylogeny within Holometabola.

Siphonaptera (Fleas)

Fleas are laterally compressed, wingless insects that possess mouthparts modified for piercing and sucking. They have highly modified combs and setae on their body and legs to help stay attached to their vertebrate hosts, and their hind legs are modified for jumping. There are approximately 2400 described flea species placed in 15 families and 238 genera (Lewis and Lewis 1985). Fleas are entirely ectoparasitic, with ~100 species as parasites of birds and the remaining species as parasites of mammals (Holland 1964). Flea distribution extends to all continents, including Antarctica, and fleas inhabit a range of habitats and hosts from equatorial deserts, through tropical rainforests, to the arctic tundra. Fleas are of tremendous economic importance as vectors of several diseases important to human health, including bubonic plague, murine typhus, and tularemia (Dunnet and Mardon 1991).

From a phylogenetic standpoint, Siphonaptera are perhaps the most neglected of holometabolous insect orders. Although we have a reasonable knowledge of flea taxonomy at the species and subspecific level, and a relatively good record of their biology and role in disease transmission, phylogenetic relationships among fleas at any level have remained virtually unexplored. Classically, the major obstacle in flea phylogenetics has been their extreme morphological specializations associated with ectoparasitism, and the inability of systematists to adequately homologize characters across taxa. The majority of characters used for species diagnoses are based on the shape and structure of their extraordinarily complex genitalia, or the presence and distribution of setae and spines. Although these characters are adequate for species diagnoses, they are of limited utility for phylogenetic reconstruction. There is no generally accepted higher classification for Siphonaptera, and several classifications published in recent years have significantly conflicting treatments of superfamilial relationships (Dunnet and Mardon 1991, Lewis and Lewis 1985, Mardon 1978, Smit 1979, Traub and Starcke 1980, Traub et al. 1983).

Molecular data are beginning to provide a more complete view of flea phylogeny (Whiting 2002a) and Whiting (unpubl. obs.; fig. 21.8). These data support the monophyly of the fami-

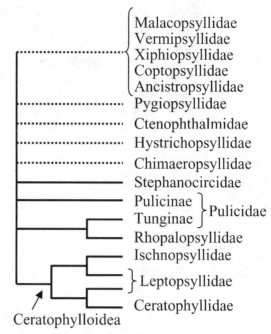

Figure 21.8. Summary phylogeny of fleas (Siphonaptera). Dashed lines represent uncertain relationships.

lies Certaophyllidae, Ischnopsyllidae (bat fleas), Rhopalopsyllidae, and Stephanocircidae. The Leptopsyllidae are paraphyletic, but the superfamilial group Ceratophylloidea is monophyletic. Pulicidae are paraphyletic, but the subfamilies that comprise this family (Pulicinae and Tunginae) are each monophyletic. These data suggest that about half of the families are paraphyletic (e.g., Chimaeropsyllidae, Hystrichopsyllidae, Pygiopsyllidae, Leptopsyllidae, Pygiopsyllidae, and Ctenophthalmidae), although 5 out of 20 subfamilies that could be assessed with these data are monophyletic. Collectively, these data suggest that many of the flea families are artificial assemblages of species, and certain families that have been used as a catchall for a wide range of divergent taxa (e.g., Ctenophthalmidae) are almost certainly paraphyletic groups, suggesting that family-level revision of this group is warranted. However, at the subfamily level, the current groupings more closely reflect phylogenetic relationships. It is still unclear which flea group is most primitive, and further data are required to refine current phylogenetic estimates.

Diptera (Flies)

Diptera are a major order of insects with approximately 125,000 species currently described, but the actual number of extant species is probably at least twice this number. Dipterans are easily distinguished from other insects by the modification of the hind wings into organs (halteres) used for balance during flight. Mouthparts range from lapping to biting and sucking, and flies have had a tremendous impact on humans owing to their transmission of deadly diseases such as malaria. Higher level phylogenetic relationships within Diptera have probably received more attention than those of any other holometabolous insect order, and yet relationships among the major constituent groups continue to elude entomologists. The current state of dipteran phylogeny is outlined in an outstanding recent review by Yeates and Wiegmann (1999).

Diptera have traditionally been divided into two major groups (fig. 21.9): long horned (Nematocera, flies with long antennae) and short horned (Brachycera). Recent research demonstrates that although the short-horned flies form a monophyletic group, the long-horned flies are a large assemblage of ancient lineages, which as a whole are probably not monophyletic (Yeates and Wiegmann 1999). The long-horned flies are generally divided into six major groups, but phylogenetic relationships among these groups are not well resolved. Ptychopteromorpha contains two families (Tanyderidae and Ptychopteridae), including primitive and phantom craneflies. The Culicomorpha are composed of 8 families and contains all of the blood-sucking primitive flies, including mosquitoes, black flies, biting midges, and midges. This is a well-supported monophyletic group based on features associated with the modified larval mouthparts used for filter feeding. Blephariceromorpha include three families, and all of these midges have specially modified prolegs in larvae for attaching to the substrate in fast flowing streams. Bibionomorpha

are composed of five families, including march flies, fungus gnats, and gall midges, but the monophyly of this group based on morphological characters is questionable. Tipulomorpha are a large group containing two cranefly families, and Psychodomorpha contain five families, including moth flies, sand flies, and wood gnats. The monophyly of both of these two groups is also questionable.

Brachycera, the short-horned flies, are a well-supported monophyletic group based on reduction in antenna size, modifications of the larval head capsule, and specific mouthpart specializations. This group is composed of four infra-orders, Stratiomyomorpha (soldier and xylomyid flies), Tabanomorpha (horse flies, snipe flies, and athericid flies), Xylophagomorpha (xylophagid flies), and Muscomorpha, which includes the vast majority of fly families. A recent comprehensive morphological analysis suggests that Tabanomorpha are sister group to Xylophagomorpha, with Stratiomyomorpha at its base, and that this group is in turn sister to Muscomorpha (Yeates 2002). Nemestrinoidea (small headed and tangle-vein flies) are thought to contain the most basal members of Muscomorpha, although there is some evidence that they should be placed within the Tabanomorpha. Asiloidea are composed of six families, including robber flies, flower-loving flies, mydas flies, stiletto flies, and bee flies, and the monophyly of this group is supported by a particular configuration of spiracles in the larvae. The group Empidoidea,

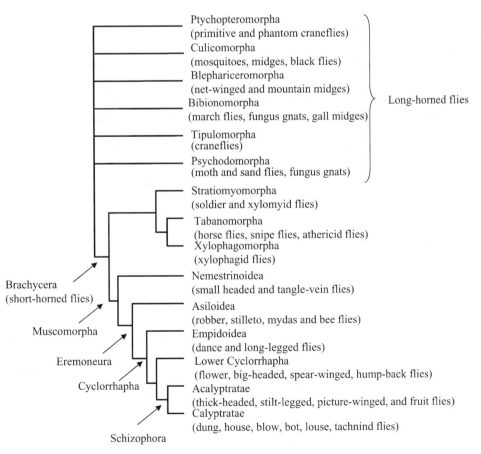

Figure 21.9. Summary phylogeny of flies (Diptera).

dance flies and long-legged flies, is sister to Cyclorrhapha, a large lineage of flies that have a reduced larval head capsule and feeding structures, and pupation occurs within a specially formed puparium. Cyclorrhapous Diptera were traditionally divided into two groups: Schizophora and Aschiza; however, the latter is not monophyletic but rather a compilation of at least 10 distinct families assigned to the "lower Cyclorrhapha." These include flower flies, big-headed flies, humpback flies, flat-footed flies, spear-winged flies, and phylogenetic relationships among these groups are controversial. Schizophora contain at least 75 families and comprises the majority of family-level diversity within Diptera. Schizophoran flies emerge from the puparium by the inflation of a membranous head sac, the ptilinum. Schizophora are traditionally divided into two groups: Acalypteratae and Calypteratae. Acalypteratae include a wide variety of families, including thick-headed flies, stilt-legged flies, fruit flies, picture-winged flies, leaf miner flies, and many others, and the monophyly of this group is not well established. Calypterate flies, on the other hand, are a very well-supported monophyletic group, and these flies have the lower lobe of the front wing (calypter) well developed. Calypteratae are composed of three superfamiles: Hippoboscoidea (primarily ectoparasitic flies that are blood feeders), Muscoidea (house flies, dung flies, and others), and Oestroidea (flesh flies, bot flies, house flies, tachinid flies). The monophyly of each of these subgroups appears relatively well supported, but relationships within each of these subgroups deserve further scrutiny. In short, there is an obvious need for further investigation into the relationships of long-horned flies, primitive short-horned flies, lower Cyclorhappha, and acalypterate flies.

Strepsiptera (Twisted-Winged Parasites)

Strepsiptera (twisted-winged parasites) are a cosmopolitan order of small insects (males, 1–7 mm; females, 2–30 mm) that are obligate insect endoparasites. The order is composed of ~550 species placed within eight extant and one extinct family (Kathirithamby 1989). Strepsiptera derive their common name from the male front wing, which is haltere-like, and early workers considered it to be twisted in appearance

when dried specimens were examined. All members of this group spend the majority of their life cycle as internal parasites of other insects and, consequently, have a highly specialized morphology, extreme sexual dimorphism, and a unique biology. The adult male strepsipteran is free-living and winged, whereas the adult female is entirely parasitic within the host, with the exception of one family (Mengenillidae) where the female last larval instar leaves the host to pupate externally. Strepsiptera parasitize species from seven insect orders: Zygentoma, Orthoptera, Blattaria, Mantodea, Hemiptera, Hymenoptera, and Diptera. In one family (Myrmecolacidae), the males are known to parasitize ant hosts whereas the females are parasites of Orthoptera. The life cycle of most strepsipteran species is unknown, and only a few species have been studied in detail.

The difficulty of placing this group among the other insect orders was described above. Investigation of phylogenetic relationships among strepsipteran families has not received the same attention as the ordinal placement of this group. Kinzelbach (1971, 1990) used adult morphological features to investigate this group, but he did not perform a formal quantitative analysis of these data. Recently, Pohl (2002) used characteristics of the first instar larvae and standard analytical techniques to infer phylogenetic relationships. The phylogeny he produced is somewhat different from that presented by Kinzelbach, but the overall pattern is the same.

Strepsiptera are divided into two main lineages: the primitive Mengenillidia and the more advanced Stylopidia (fig. 21.10). The former lineage includes one extinct and one living family and is characterized by presence of robust mandibles, a single genital tube in the female, specific characteristics associated with a vein in the hind wing (MA1 broad), and a primitive type of larvae (Pohl 2002). In this group, the female leaves the host to pupate, in contrast to Stylopidia, where the female remains within the body of the host during the pupal and adult stage. Stylopidia can be further distinguished by the females possessing multiple genital openings and the hind wing in males with only a remnant of the MA1 vein. Relationships within the Stylopidia are less known. Current data suggest that Corioxenidae is the most primitive family in this group, but further investigation is necessary to fully resolve relationships among the members of this unusual insect order.

Future Prospects

Entomologists have long been humbled by the immense size of Holometabola, and understanding the pattern of diversification among its constituent lineages has largely eluded scientific investigation for well more than two centuries. A clear view of the Holometabola branch of the Tree of Life is just beginning to emerge. Entomologists are a long way from exhausting the usefulness of morphological data for reconstructing holometabolan phylogeny, and for many groups

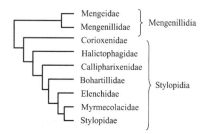

Figure 21.10. Summary phylogeny of twisted-winged parasites (Strepsiptera).

further investigation of anatomical similarities is bound to reveal a treasure trove of useful information. The advent of molecular systematics in the past decade brought with it not only a new set of tools with which to infer phylogeny, but also the ability to take a broad-stroke look at Holometabola in a new way, by selecting a few exemplars from a large range of diverse groups for molecular screening. Even the best current efforts in insect molecular systematics will seem primitive by tomorrow's standards, and it is clear that, like morphology, molecular systematics has not yet reached the pinnacle of usefulness in insects.

Many challenges still remain in unraveling the evolutionary history these insects: the challenge to catalog the immense number of species that are members of this group; the challenge to train a new generation of entomologists in insect morphology and systematics; the challenge to find novel genetic markers that better track the phylogeny of these lineages; and the challenge to overcome the computational limitations of organizing and analyzing the mountains of data emerging on insect phylogeny. But for the first time we are beginning to see a surge of researchers zeroing in on unraveling the complete phylogenetic structure of Holometabola, tossing their whole arsenal of tools into the fray and providing exciting new insights into the most wondrous event in evolution: the diversification of insects and the evolution of their most successful group, Holometabola.

Acknowledgments

I thank J. Cracraft and M. Donoghue for the invitation to speak at the Tree of Life symposium, and M. Terry, H. Ogden, K. Jarvis, J. Cherry, J. Robertson, and A. Whiting for assistance with the manuscript. This work was supported by National Science Foundation grants DEB-9806349 and DEB-9983195.

Literature Cited

Aspöck, U. 2002. Phylogeny of the Neuropterida (Insecta: Holometabola). Zool. Scr. 31:51–56.

Aspöck, U., J. D. Plant, and H. L. Nemeschkal. 2001. Cladistic analysis of Neuroptera and their systematic position within Neuropterida (Insecta: Holometabola: Neuropterida: Neuroptera). Syst. Entomol. 26:73–86.

Bayreuther, K., and S. Brauning. 1971. Die Cytogenetik der Flohe (Aphaniptera). Chromosoma (Berl.) 33:19–29.

Beutel, R. G., and S. N. Gorb. 2001. Ultrastructure of attachment specializations of Hexapods (Arthropoda): evolutionary patterns inferred from a revised ordinal phylogeny. J. Zool Syst. Evol. Res. 39:177–207.

Beutel, R. G., and F. Haas. 2000. Phylogenetic relationships of the suborders of Coleoptera (Insecta). Cladistics 16:103–142.

Bilinski, S., J. Bünnig, and B. Simiczyjew. 1998. The ovaries of Mecoptera: basic similarities and one exception to the rule. Fol. Histochem. Cytobiol. 36:189–195.

Boudreaux, H. B. 1979. Arthropod phylogeny with special reference to insects. John Wiley and Sons, New York.

Brothers, D. J. 1999. Phylogeny and evolution of wasps, ants, and bees (Hymenoptera, Chrysidoidea, Vespoidea, and Apoidea). Zool. Scr. 28:233–249.

Brothers, D. J., and J. M. Carpenter. 1993. Phylogeny of Aculeata: Chrysidoidea and Vespoidea (Hymenoptera). J. Hymenopt. Res. 2:227–304.

Byers, G. W. 1991. Mecoptera. Pp. 696–704 in Mecoptera (I. D. Naumann, P. B. Carne, J. F. Lawrence, E. S. Nielsen, J. P. Spradberry, R. W. Taylor, M. J. Whitten, and M. J. Littlejohn, eds.). CSIRO, Melbourne University Press, Melbourne.

Byers, G. W. 1996. More on the origin of Siphonaptera. J. Kans. Entomol. Soc. 69:274–277.

Carmean, D., L. S. Kimsey, and M. L. Berbee. 1992. 18S rDNA sequences and holometabolous insects. Mol. Phylogenet. Evol. 1:270–278.

Carpenter, J. M. 1999. What do we know about chrysidoid (Hymenoptera) relationships? Zool. Scr. 28:215–231.

Caterino, M. S., V. L. Shull, P. M. Hammond, and A. P. Vogler. 2002. Basal relationships of Coleoptera inferred from 18S rDNA sequences. Zool. Scr. 31:41–49.

Chalwatzis, N., J. Hauf, Y. V. Peer, R. Kinzelbach, and F. K. Zimmerman. 1996. 18S ribosomal RNA genes of insects: primary structure of the genes and molecular phylogeny of the Holometabola. Ann. Entomol. Soc. 89:788–803.

Crowson, R. A. 1960. The phylogeny of Coleoptera. Annu. Rev. Entomol. 5:111–134.

Davis, D. R. 1986. A new family of montrysian moths from austral South America (Lepidoptera: Palaephatidae), with a phylogenetic review of the Monotrysia. Smithson. Contrib. Zool. 434:1–202.

Dowton, M., and A. D. Austin. 1999. Models of analysis for molecular datasets for the reconstruction of basal hymnopteran relationships. Zool. Scr. 28:69–74.

Dunnet, G. M., and D. K. Mardon. 1991. Siphonaptera. Pp. 125–140 in Siphonaptera (P. B. C. I. D. Naumann, J. F. Lawrence, E. S. Nielsen, J. P. Spradberry, R. W. Taylor, M. J. Whitten, and M. J. Littlejohn, eds.). CSIRO and Melbourne University Press, Melbourne.

Farrell, B. D. 1998. "Inordinate fondness" explained: why are there so many beetles? Science 281:555–558.

Gall, W. K. 1994. Phylogenetic studies in the Limnephiloidea, with a revision of the world genera of Goeridae (Trichoptera). Phylogenetic studies in the Limnephiloidea, with a revision of the world genera of Goeridae (Trichoptera). University of Toronto, Toronto.

Hammond, P. 1992. Species inventory. Pp. 21–25 in Species inventory (B. Groombridge, ed.). Chapman and Hall, London.

Hennig, W. 1981. Insect phylogeny. Academic Press, New York.

Hinton, H. E. 1958. The phylogeny of the panorpoid orders. Annu. Rev. Entomol. 3:181–206.

Holland, G. P. 1964. Evolution, classification, and host relationships of Siphonaptera. Annu. Rev. Entomol. 9:123–146.

Hörnschemeyer, T. 2002. Phylogenetic signficance of the wing-base of the Holometabola (Insecta). Zool. Scr. 31:17–30.

Huelsenbeck, J. P. 1997. Is the Felsenstein zone a fly trap? Syst. Biol. 46:69–74.

Kaltenbach, A. 1978. Mecoptera (Schnabelhafte, Schnabelfliegen). Mecoptera (Schnabelhafte, Schnabelfliegen). Walter de Gruyter, Berlin, New York.

Kathirithamby, J. 1989. Review of the order Strepsiptera. Syst. Entomol. 14:41–92.

Kinzelbach, R. K. 1971. Morphologische Befunde and Facherfluglern und ihre phylogenetische bedeutung (Insecta: Strepsiptera). Zool. 119:1–256.

Kinzelbach, R. K. 1990. The systematic position of Strepsiptera (Insecta). Am. Entomol. 36:292–303.

Kjer, K., R. J. Blahnik, and R. W. Holzenthal. 2002. Phylogeny of caddisflies (Insecta, Trichoptera). Zool. Scr. 31:83–91.

Krenn, H. W. 1998. Proboscis sensilla in Vanessa cardui (Nymphalidae: Lepidoptera): functional morphology and significance in flower-probing. Zoomorphology 118:23–30.

Krenn, H. W., and N. P. Kristensen. 2000. Early evolution of the proboscis of Lepidoptera (Insecta): external morphology of the galea in basal glossatan moths lineages, with remarks on the origin of pilifers. Zool. Anz. 239:179–196.

Kristensen, N. P. 1975. The phylogeny of hexapod "orders." A critical review of recent accounts. Z. Zool. Syst. Evol. Forsch. 13:1–44.

Kristensen, N. P. 1981. Phylogeny of insect orders. Annu. Rev. Entomol. 26:135–157.

Kristensen, N. P. 1991. Phylogeny of extant hexapods. Pp. 125–140 in Phylogeny of extant hexapods (P. B. C. I. D. Naumann, J. F. Lawrence, E. S. Nielsen, J. P. Spradberry, R. W. Taylor, M. J. Whitten, and M. J. Littlejohn, eds.). CSIRO, Melbourne University Press, Melbourne.

Kristensen, N. P. 1995. Fourty [sic] years' insect phylogenetic systematics. Zool. Beitr. N. F. 36:83–124.

Kristensen, N. P. 1997. Early evolution of the Lepidoptera + Trichoptera lineage: phylogeny and the ecological scenario. Mem. Mus. Natl. Hist. Nat. 173:253–271.

Kristensen, N. P. 1999. Phylogeny of endopterygote insects, the most successful lineage of living organisms. Eur. J. Entomol. 96:237–253.

Kristensen, N. P., and A. W. Skalski. 1999. Phylogeny and paleontology. Pp. 7–25 in Phylogeny and paleontology (N. P. Kristensen, ed.). De Gruyter, New York.

Kukalova-Peck, J., and J. F. Lawrence. 1993. Evolution of the hind wing in Coleoptera. Can. Entomol. 125:181–258.

Lawrence, J. F., and A. F. J. Newton. 1982. Evolution and classification of beetles. Annu. Rev. Ecol. Syst. 13:261–290.

Lawrence, J. F., and A. F. J. Newton. 1995. Families and subfamilies of Coleoptera (with selected genera, notes, references and data on family-group names). Pp. 779–1006 in Families and subfamilies of Coleoptera (with selected genera, notes, references and data on family-group names) (J. Pakaluk and S. A. Slipinski, eds.). Muzeum i Instytut Zoologii, PAN, Warszawa.

Lawrence, J. F., S. A. Slipinski, and J. Pakaluk. 1995. From Latreille to Crowson: a history of the higher-level classification of beetles. Pp. 87–154 in From Latreille to Crowson: a history of the higher-level classification of beetles (J. Pakaluk and S. A. Slipinski, eds.). Muzeum i Instytut Zoologii, PAN, Warszawa.

Lewis, R. E., and J. H. Lewis. 1985. Notes on the geographical distribution and host preferences in the order Siphonaptera. J. Med. Entomol. 22:134–152.

Mackay, R. J., and G. B. Wiggins. 1978. Ecological diversity in the Trichoptera. Annu. Rev. Entomol. 24:185–208.

Mardon, D. K. 1978. On the relationships, classification, aedeagal morphology and zoogeography of the genera of Pygiopsyllidae (Insecta : Siphonaptera). Aust. J. Zool. Suppl. Ser. 64:1–69.

Mickoleit, G. 1973. Uber den ovipositor der Neuropteroidea und Coleoptera und seine phylogenetische Bedeutung (Insecta, Holometabola). Z. Morphol. Tiere 74:37–64.

Morse, J. C. 1997a. Checklist of the world Trichoptera. Pp. 339–342 in Checklist of the world Trichoptera (R. W. Holzenthal and O. S. Flint, Jr., eds.). Ohio Biological Survey, Columbus.

Morse, J. C. 1997b. Phylogeny of Trichoptera. Annu. Rev. Entomol. 42:427–450.

Nielsen, E. S., and N. P. Kristensen. 1996. The Australian moth family Lophocoronidae and the basal phylogeny of the Lepidoptera–Glossata. Invert. Taxon. 10:1199–302.

Pashley, D. P., B. A. McPheron, and E. A. Zimmer. 1993. Systematics of holometabolous insect orders based on 18S ribosomal RNA. Mol. Phylogenet. Evol. 2:132–142.

Penny, N. D. 1997. World checklist of extant Mecoptera species. World checklist of extant Mecoptera species. California Academy of Sciences, San Francisco, CA.

Penny, N. D., and G. W. Byers. 1979. A check-list of the Mecoptera of the World. Acta Amazon. 9:365–388.

Pix, W., G. Nalbach, and J. Zeil. 1993. Strepsipteran forewings are haltere-like organs of equilibrium. Naturwissenschaften 80:371–374.

Pohl, H. 2002. Phylogeny of the Strepsiptera based on morphological data of the first instar. Zool. Scr. 31:123–134.

Rasnitsyn, A. P. 1988. An outline of evolution of the hymenopterous insects. Orient. Insects 22:115–145.

Ronquist, F. 1999. Phylogeny of the Hymenoptera (Insecta): the state of the art. Zool. Scr. 28:3–11.

Ronquist, F., A. P. Rasnitsyn, A. Roy, K. Eriksson, and M. Lindgren. 1999. Phylogeny of the Hymenoptera: a cladistic reanalysis of Rasnitsyn's (1998) data. Zool. Scr. 28:139–164.

Ross, H. H. 1967. The evolution and past dispersal of the Trichoptera. Annu. Rev. Entomol. 12:169–206.

Schlein, Y. 1980. Morphological similarities between the skeletal structures of Siphonaptera and Mecoptera. Pp. 359–367 in Proceedings of the International Conference on Fleas (R. Traub and H. Starcke, eds.). A. A. Balkema, Rotterdam.

Schulmeister, S., W. C. Wheeler, and J. M. Carpenter. 2002. Simulatenous analysis of the basal lineages of Hymenoptera (Insecta) using sensitivity analysis. Cladistics 18:455–484.

Scott, K. M. F. 1993. Three recently erected Trichoptera families from South Africa, the Hydrosalpingidae, Petrothrincidae and Barbarochthonidae (Integripalpia: Serocostomatoidea), with a cladistic analysis of character states in the twelve families here considered as belonging to the Sericostomatoidea, by F.C. de Moor. Ann. Cape Prov. Mus. 18:293–354.

Shull, V. L., A. P. Vogler, M. D. Baker, D. R. Maddison, and P. M. Hammond. 2001. Sequence alignment of 18S ribosomal RNA and the basal relationships of adephagan beetles: evidence for the monophyly of aquatic families and the placement of Trachypachidae. Syst. Biol. 50:945–969.

Sidall, M. E., and M. F. Whiting. 1999. Long-branch abstractions. Cladistics 15:9–24.

Simiczyjew, B. 2002. Structure of the ovary in *Nannochorista neotropica* Navas (Insecta: Mecoptera: Nannochorsittdae) with remarks on mecopteran phylogeny. Acta Zool. 83:61–66.

Smit, F. G. A. M. 1979. The fleas of New Zealand (Siphonaptera). J. R. Soc. N. Z. 9:143–232.

Stys, P., and S. Bilinksy. 1990. Ovariole types and the phylogeny of hexapods. Biol. Rev. Cambr. Philos. Soc. 65:401–429.

Traub, R., and H. Starcke, eds. 1980. Fleas: Proceedings of the International Conference on Fleas. A. A. Balkema, Rotterdam.

Traub, R. M., M. Rothschild, and J. Haddow. 1983. The Ceratophyllidae, key to the genera and host relationships. Academic Press, New York.

Vilhelmsen, L. 1997. The phylogeny of lower Hymenoptera (Insecta), with a summary of the early evolutionary history of the order. J. Zool. Syst. Evol. Res. 35:49–70.

Weaver, J. S., III. 1984. The evolution and classification of Trichoptera, Part 1: the groundplan of Trichoptera. Pp. 413–419 *in* The evolution and classification of Trichoptera, Pt 1: The groundplan of Trichoptera (J. C. Morse, ed.). Dr. W. Junk, The Hague.

Weaver, J. S., III, and J. C. Morse. 1986. Evolution of feeding and case-making behavior in Trichoptera. J. N. A. Benthol. Soc. 5:150–158.

Wheeler, W. C., M. F. Whiting, Q. D. Wheeler, and J. M. Carpenter. 2001. The phylogeny of the extant hexapod orders. Cladistics 17:113–169.

Whiting, M. F. 1998a. Long-branch distraction and the Strepsiptera. Syst. Biol. 47:134–138.

Whiting, M. F. 1998b. Phylogenetic position of the Strepsiptera: review of molecular and morphological evidence. Int. J. Morphol. Embryol. 27:53–60.

Whiting, M. F. 2002a. Mecoptera is paraphyletic: multiple genes and phylogeny of Mecoptera and Siphonaptera. Zool. Scr. 31:93–104.

Whiting, M. F. 2002b. Phylogeny of the holometabolous insect orders based on 18S ribosomal data: when bad things happen to good data. Pp. 69–84 *in* Phylogeny of the holometabolous insect orders based on 18S ribosomal data: when bad things happen to good data (R. DeSalle, W. C. Wheeler and G. Giribet, eds.). Birkhauser, Basel.

Whiting, M. F. 2002c. Phylogeny of the holometabolous insect orders: molecular evidence. Zool. Scr. 31:3–16.

Whiting, M. F., J. C. Carpenter, Q. D. Wheeler, and W. C. Wheeler. 1997. The Strepsiptera problem: phylogeny of the Holometabolous insect orders inferred from 18S and 28S ribosomal DNA sequences and morphology. Syst. Biol. 46:1–68.

Whiting, M. F., and W. C. Wheeler. 1994. Insect homeotic transformation. Nature 368:696.

Wiegmann, B. W., J. C. Regier, and C. Mitter. 2002. Combined molecular and morphological evidence on the phylogeny of the earliest lepidopteran lineages. Zool. Scr. 31:67–81.

Wiggins, G. B., and W. Wichard. 1989. Phylogeny of pupation in Trichoptera, with proposals on the origin and higher classification of the order. J. N. A. Benthol. Soc. 8:260–276.

Willmann, R. 1987. The phylogenetic system of the Mecoptera. Syst. Entomol. 12:519–524.

Willmann, R. 1989. Evolution und Phylogenetisches System der Mecoptera (Insecta: Holometabola). Abh. Senckenb. Naturforsch. Ges. 544:1–153.

Wilson, E. O. 1988. The current state of biological diversity. Pp. 3–18 *in* The current state of biological diversity (E. O. Wilson, ed.). National Academy Press, Washington, DC.

Yeates, D. K. 2002. Relationships of extant lower Brachycera (Diptera): a quantitative synthesis of morphological characters. Zool. Scr. 31:105–121.

Yeates, D. K., and B. M. Wiegmann. 1999. Congruence and controversy: toward a higher-level phylogeny of the Diptera. Annu. Rev. Entomol. 44:397–428.

VIII

The Relationships of Animals: Deuterostomes

Andrew B. Smith

Kevin J. Peterson

Gregory Wray

D. T. J. Littlewood

From Bilateral Symmetry to Pentaradiality

The Phylogeny of Hemichordates and Echinoderms

Nested within the clade of bilaterally symmetrical animals, variously called the triploblasts or the bilaterians, lies a most unusual group. Although most bilaterians have a bilaterally symmetric body plan, with a clear anterior–posterior axis and in most cases a differentiated head region, the echinoderm adult is constructed on a pentaradiate plan and lacks an obvious anterior–posterior axis (e.g., figs. 22.1, 22.6, and 22.8). Yet echinoderms clearly start out life as bilateral organisms, and their peculiar body plan is a secondary modification that arises during the metamorphosis that transforms them from larva to adult. It is because echinoderms are so very different in appearance from their closest relatives, the hemichordates, that they provide a fascinating and important group for evolutionary and developmental studies.

Based on their pattern of development, both echinoderms and their bilateral relatives the hemichordates clearly fall among the deuterostomes. Until comparatively recently, five major groups (Echinodermata, Hemichordata, Chordata, Lophophorata, and Chaetognatha) were considered to be deuterostomes. However, molecular evidence now overwhelmingly suggests that only the echinoderms, hemichordates, and chordates belong together (Adoutte et al. 2000, Cameron et al. 2000, Giribet et al. 2000, Peterson and Eernisse 2001, Winchell et al. 2002). TheLophophorata are now recognized to be members of the protostome clade, specifically part of "Lophotrochozoa," which includes the lophophorates and the classically spirally cleaving taxa such as annelids and mollusks (Halanych et al. 1995). The phy-

logenetic affinity of chaetognaths (arrow worms) has been more difficult to resolve, but the first studies to address their affinity based on 18S ribosomal DNA (rDNA) data showed that they were not deuterostomes (Telford and Holland 1993, Wada and Satoh 1994). The bulk of evidence that has since accumulated suggests that chaetognaths are ecdysozoans (Halanych 1996, Peterson and Eernisse 2001), although their precise position within that group remains uncertain (e.g., Giribet et al. 2000, Zrzavý et al. 1998, Littlewood et al. 1998).

Deuterostome Relationships

There are sound reasons for hypothesizing that echinoderms, hemichordates, and chordates are all closely related: unequivocal synapomorphies for the clade Deuterostomia include the shared presence of endogenous sialic acids (Warren 1963, Segler et al. 1978) and gill slits (although these are present in stem-group echinoderms only). Furthermore, De Rosa et al. (1999) suggested that deuterostomes also share two (presumably) independent Hox gene duplications, one involving the generation of Hox6, Hox7, and Hox8, and the other involving the generation of the apomorphic Abd-B or 9–13 complex. However, we find the evidence for the central class duplication being a synapomorphy for Deuterostomia far from convincing (K. J. Peterson et al., unpubl. obs.), as did Telford (2000).

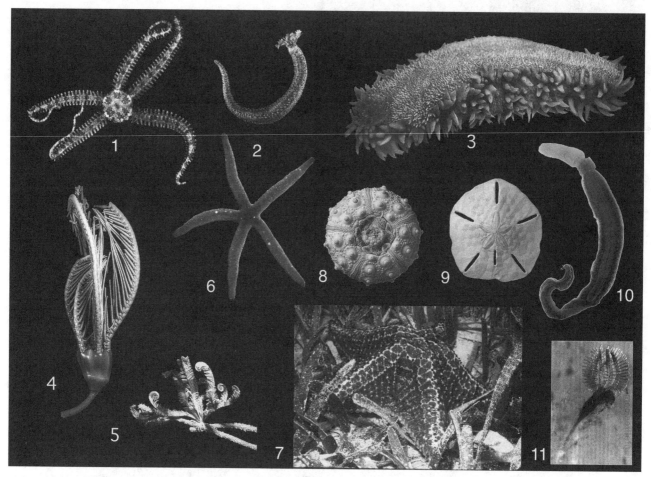

Figure 22.1. Representative ambulacrarian taxa. (1–9) Echinoderms: (1) brittlestar (Ophiuroidea, *Ophiactis*); (2 and 3) sea cucumbers (Holothuroidea, *Holothuria* and *Thelenota*); (4) sea lily (Crinoidea, *Anachalypsicrinus*); (5) feather star (Crinoidea, *Oligometra*); (6 and 7) starfishes (*Linckia*, *Oreaster*); (8) regular sea urchin (Echinoidea, *Eucidaris*); (9) sand dollar (Echinoidea, *Leodia*). (10 and 11) Hemichordates: (10) acorn worm (Enteropneusta, *Saccoglossus*); (11) colonial hemichordate (Pterobranchia, *Cephalodiscus*). From Rigby (1993).

Resolving the relationships of these three deuterostome groups has proved controversial. One reason for this is that hemichordates have an echinodermlike larva but a chordate-like adult. As a consequence, depending upon whether adult or larval characters have been emphasized, either an echinoderm or chordate affinity has been proposed. Thus, Metschnikoff in 1881 emphasized larval similarity when arguing that hemichordates and echinoderms are more closely related, and it was he who proposed uniting them in the taxon Ambulacraria. Others, starting with Bateson in 1885, have emphasized the adult similarities and thus come to regard hemichordates as more closely related to chordates than to echinoderms [see Hyman (1959) for all historical references]. The first cladistic analyses of morphological characters seemed to confirm Bateson's hypothesis. Schaeffer (1987), Gans (1989), Brusca and Brusca (1990), Cripps (1991), Schram (1991, 1997), Nielsen (1995, 2001, Nielsen et al. 1996), and Peterson (1995) all found hemichordates to be the sister group

of the chordates, not of the echinoderms. In some of these analyses hemichordates were either paraphyletic (Cripps 1991, Peterson 1995) or polyphyletic (Schram 1991, Nielsen 1995, 2001), with enteropneusts the sister group of Chordata. Characters shared between echinoderms and hemichordates (e.g., dipleurula larva, trimery) were seen as either deuterostome plesiomorphies or were not considered.

However, starting with the analyses of Turbeville et al. (1994) and Wada and Satoh (1994), virtually all 18S rDNA analyses have found significant support for the monophyly of Ambulacraria (reviewed in Adoutte et al. 2000; see also Bromham and Degnan 1999, Cameron et al. 2000, Giribet et al. 2000, Peterson and Eernisse 2001, Winchell et al. 2002, Furlong and Holland 2002). There are now also several molecular markers supporting the monophyly of Ambulacraria. For example, the mitochondrial genetic code for the transfer RNA lys-1 protein gene in both echinoderms and hemichordates carries the anticodon CTT rather than TTT as found in most other metazoans, whereas ATA encodes for isoleucine rather than methionine, a reversal to the primitive condition (Castresana et al. 1998a, 1998b). Finally, the *Hox11/13a* and *Hox11/13b* genes of echinoderms (Long and Byrne 2001) have orthologues in hemichordates (specifically the ptychoderid *Ptychodera flava*) but are unknown from other taxa (K. J. Peterson et al., unpubl. obs.).

Morphological characters also lend support to the monophyly of Ambulacraria (Peterson and Eernisse 2001). The close similarity between the larva of enteropneust hemichordates and asteroid echinoderms is striking, and indeed the former was long thought to be the larva of an unknown asteroid. Both have a preoral feeding band that creates an upstream feeding current using monociliated cells and a perioral ciliated band that manipulates food into the esophagus. Their basic tricoelomate body organization is also very similar, both possessing a protocoel and paired mesocoels and metacoels (called axocoels, hydrocoels, and somatocoels, respectively, in echinoderms). Peterson and Eernisse (2001) considered trimery a possible bilaterian plesiomorphy because they believed both phoronids and potentially chaetognaths also had a trimeric body plan. However, Bartolomaeus (2001) has recently shown that phoronids are not trimeric—the "protocoel" is actually an enlarged subepidermal extracellular matrix. Hence, neither phoronids nor brachiopods possess a distinct protocoel. The situation in chaetognaths is equally dubious because Kapp (2000) noted that the transverse septum dividing the female part of the trunk from the male part of the trunk is associated only with the development of the gonads and forms from coelomic cells. Therefore, it appears that true trimery is a synapomorphy uniting the ambulacrarians.

In terms of adult morphology, the most conspicuous derived character uniting hemichordates and echinoderms is the axial complex (Ruppert and Balser 1986, Balser and Ruppert 1990). This is the metanephridium ("kidney") of the adult in which fluid from the blood vascular system is pressure filtered by contractions of the madreporic vesicle (echinoderms) or the heart vesicle (hemichordates) across a layer of podocytes in the axial gland (echinoderms) or glomerulus (hemichordates) into the axocoel (echinoderms) or protocoel (hemichordates). This coelom contains a pore (hydropore) through which the filtrate is expelled into the external environment. The extensive development of the mesocoel/hydrocoel to form a tubular network of tentacles used in feeding is a second obvious similarity between pterobranchs and echinoderms but, as shown below, is probably not homologous.

Traditionally, hemichordates have usually been considered closer to chordates than echinoderms because both have pharyngeal openings (gills). There is striking morphological similarity between the gill anatomy of enteropneusts and chordates and the similarities extend to the molecular level, because both taxa express the same transcription factor in the gills (Ogasawara et al. 1999). There is therefore little doubt that the structures are indeed homologous. As

pharyngeal slits are absent from crown-group echinoderms, this has been taken as evidence that echinoderms are primitive and sister group to the clade chordates plus hemichordates. However, because echinoderms and hemichordates are sister taxa, the evidence only implies that the possession of pharyngeal slits is plesiomorphic for deuterostomes as a whole, and so their loss is an apomorphy of crown-group echinoderms (fig. 22.2). When precisely echinoderms lost these structures is something that paleontological data can shed light on. Evidence that stem group echinoderms may have had gill slits comes from the careful work of Jefferies and students (e.g., Dominguez et al. 2002). They have shown that structures comparable with pharyngeal slits are widely developed amongst a subgroup of the pre-pentameral stem-group Echinodermata loosely termed carpoids. Not all, however, agree that these structures represent gill slits, and some recent analyses place carpoids within crown group Echinodermata (Sumrall 1997, David et al. 2000).

Most of the other traditional deuterostome characters can be shown to be either bilaterian plesiomorphies (e.g., radial cleavage, enterocoely, posterior fate of the blastopore) or restricted to just the ambulacrarians (e.g., trimery, "dipleurula" larva). In fact, Peterson and Eernisse (2001) suggested that, because lophophorates (phoronids and brachiopods, *sensu* Peterson and Eernisse 2001) were basal lophotrochozoans, and chaetognaths were basal ecdysozoans, many of the traditional characters ascribed to deuterostomes are in fact bilaterian plesiomorphies. Thus the latest common ancestor of bilaterians may have been very deuterostome-like.

In summary, a substantial body of corroborative evidence now exists, from comparative anatomy of both larval and adult form, from molecular data and from the fossil record, that echinoderms and hemichordates are sister group to the exclusion of chordates (fig. 22.2).

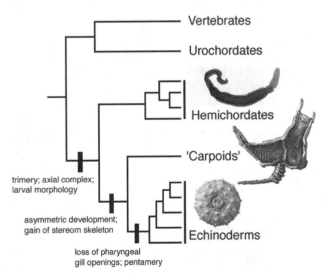

Figure 22.2. Deuterostome relationships showing principal morphological characters of Ambulacraria.

Hemichordates

The phylum Hemichordata has traditionally been partitioned into two groups, the enteropneusts, or acorn worms, and the pterobranchs. There are approximately 75 species of acorn worm grouped into 11 genera, and about 20 species of pterobranchs grouped in only two valid genera *Cephalodiscus* and *Rhabdopleura* [see Benito (1982) for classification]. A third group, Planctosphaeroidea, are known only as fairly large and distinctive larvae and are assumed to be the larval form of an unknown enteropneust (Benito and Pardos 1997). All hemichordates are benthic marine animals as adults, and those with indirect development pass through a planktonic larval stage called a tornaria. Their body is constructed around five coeloms bilaterally arranged, a single anterior protocoel, and paired mesocoels and metacoels. The anterior part of the body associated with the protocoel is the shield (pterobranchs) or proboscis (enteropneusts). The mesocoel region forms the collar, and a long trunk contains the metacoels. There is either one or a pair of protocoel pores, a pair of mesocoelic ducts and one or more pairs of gill pores in the anterior part of the metacoel together with genital openings [see Benito and Pardos (1997) for a detailed description].

Enteropneusts

Enteropneusts are wormlike creatures (fig. 22.1.10), with an anterior proboscis (protosome), a short collar (mesosome) and a long cylindrical trunk (metasome). The mouth opens between the proboscis and collar, and the anus is terminal at the end of the trunk. There is a series of gill pores on left and right of the anterior part of the trunk, and unlike pterobranchs, the paired mesocoelomic ducts open into the first pair of gill slits. Enteropneusts also differ from pterobranchs in having no feeding tentacles developed from the collar. Enteropneusts are solitary and are common in the intertidal zones where they usually live buried in soft sediment, although a few are known from depths of up to 400 m, with one (*Saxipendium coronatum*) associated with the Galapagos geothermal vent community. They vary in size from a few centimeters long (*Saccoglossus pygmaeus* of the North Sea) to 2 m or more in length (*Balanoglossus gigas* of Brazil).

Three families of enteropneusts have traditionally been recognized, Ptychoderidae, Spengelidae, and Harrimaniidae (Benito 1982). Ptychoderidae is usually considered the most complicated and "advanced" family united by several synapomorphies, including the possession of well-developed genital ridges with lateral septa in the trunk, a pygocord, and externally visible hepatic sacculations. They also possess synapticules, but as argued below, this may be a plesiomorphy. Spengelidae are considered intermediate between the ptychoderids and the harrimaniids. Spengelids are characterized by having an appendix on the anterior end of the stomochord or buccal diverticulum. All known ptychoderids and spengelids pass through a tornaria larval stage and hence are indirect developers. The most basic or "primitive" family is Harrimaniidae. Harrimaniids have proboscis skeleton crura, which create dorsolateral grooves in the stomochord, and well-developed proboscis musculature. Development is of the direct type and is best known in the genus *Saccoglossus*. A fourth monotypic family, Protoglossidae, has been proposed, but most hemichordate workers consider *Protoglossus* a member of Harrimaniidae (e.g., Giray and King 1996). Woodwick and Sensenbaugh (1985) erected a new family, Saxipendiidae, for the vent worm *Saxipendium* because it does not clearly belong to any of the three traditional enteropneust families.

As nonskeletonized animals, enteropneusts have a scanty fossil record. The earliest definitive occurrence is from the Pennsylvanian Mazon Creek fauna (Bardack 1997), with a second occurrence from the Lower Jurassic or northern Italy (Arduini et al. 1981). A distinctive trace fossil from the Lower Triassic of northern Italy has been assigned to Enteropneusta (Twitchett 1996), but surely many fossilized burrows and traces reflect the activities of enteropneusts. In fact, Jensen et al. (2000) suggest that an enteropneust may have been the maker of the trace fossil *Treptichnus pedum*, the fossil that defines the base of the Cambrian system in the stratotype section in Newfoundland. *Yunnanozoon*, an enigmatic form from the famous Early Cambrian Chengjiang Lagerstätte of China, has been described as a chordate (Chen et al. 1995), an enteropneust hemichordate (Shu et al. 1996), or a stem-group deuterostome (Budd and Jensen 2000). In our view, *Yunnanozoon* shows two chordate apomorphies, a notochord and segmented muscles, and resembles hemichordates only in shared primitive characters such as pharyngeal slits. Hence, we agree with Chen and Li (1997) that *Yunnanozoon* is best considered a member of the phylum Chordata.

Pterobranchs

Pterobranchs have the same tripartite body plan as enteropneusts (fig. 22.1.11). There is a platelike anterior shield (protosome), a narrow U-shaped collar (mesosome) from which a paired series of feeding tentacles arise, and a bipartite trunk (metasome) from which an extensible stalk with a terminal sucker arises. There are paired mesocoelic ducts and pores and, in *Cephalodiscus*, a pair of gill pores that penetrate the pharynx (*Rhabdopleura* lacks gill pores, although traces marking their position remain). Pterobranchs are much less common than are enteropneusts and are small (generally > 1 cm). All are colonial and attached to the seafloor, either aggregating (*Cephalodiscus*) or colonial (*Rhabdopleura*), and both inhabit a horny tube (coenecium). Although they can move out of their tube, they generally remain attached by their sucker. Reproduction is direct and asexual budding occurs, with new individuals arising from the stalk. They are ubiquitous and range in depth from 5 to 5000 m.

Pterobranchs are fairly common fossils with both rhabdopleurid and cephalodiscid-like fossils known from as early as the Middle Cambrian (Chapman et al. 1995). Of course, what is found is just the collagenous tube built by the animal (the coenecium). The most important hemichordate fossil group are the graptolites, which thrived from the Middle Cambrian until the Late Carboniferous and are especially important for biostratigraphy from the Early Ordovician through the Early Devonian. The graptolite coenecium is very similar to modern, and fossil pterobranchs both in terms of structure (Crowther 1981) and composition (Armstrong et al. 1984). However, many graptolites possessed a structure on the coenecium called a nema that was not known to be part of any pterobranch coenecium, and to some this absence precluded a pterobranch affinity for graptolites (Rigby 1993). Fortunately, Dilly (1993) described a new species of *Cephalodiscus*, *C. graptolitoides*, collected in deep water off the coast of New Caledonia that possesses a spine virtually indistinguishable from the graptolite nema. The demonstration of a nema on a recent pterobranch effectively removed the last barrier to ascribing a pterobranch affinity for graptolites (Rigby 1993).

Hemichordate Phylogeny and Classification

A clear account of the history of hemichordate classification is provided by Hyman (1959). In early cladistic analyses of deuterostomes the monophyly of Hemichordata was assumed, with hemichordates treated as a terminal taxon. However, Cripps (1991), Schram (1991, 1997), Nielsen (1995, 2001, Nielsen et al. 1996) and Peterson (1995) coded for Pterobranchia, and Enteropneusta separately and all found Hemichordata to be either paraphyletic (Cripps, Peterson) or polyphyletic (Schram, Nielsen). Cripps (1991) even found Pterobranchia to be paraphyletic, with *Cephalodiscus* more closely related to enteropneusts, echinoderms, and chordates than to *Rhabdopleura*.

In their recent analysis of metazoan taxa, Peterson and Eernisse (2001) found support for a monophyletic Hemichordata. They identified two hemichordate synapomorphies: (1) the stomochord, a unique extension of the dorsal wall of the pharynx into the protosome, and (2) the mesocoelomic ducts, which connect the mesocoel directly to the exterior (see also Ruppert 1997).

The monophyly of Pterobranchia, although not tested by Peterson and Eernisse (2001), seems clear. Pterobranch synapomorphies include the presence of tentacular arms, the U-shaped gut, the coenecium secreted by the protosome or cephalic shield, and mesocoelomic ducts that communicate through pores not connected with the gill slits (as they are in "enteropneusts"). On the other hand, Enteropneusta were shown to be paraphyletic, with Harrimaniidae identified as sister taxon to Pterobranchia, both possessing a ventral postanal stalk. Some harrimaniids also possess two hydropores like pterobranchs, raising the possibility that harmaniids themselves are paraphyletic. However, molecular data (see below) suggest that this is unlikely, at least for the genera

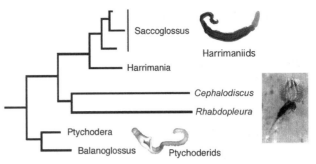

Figure 22.3. Phylogenetic relationships of hemichordates based on 18S rRNA data (from Cameron et al. 2000).

Harrimania and *Saccoglossus*. Finally, Peterson and Eernisse (2001) also found support for the monophyly of Ptychoderidae + Spengelidae, both, for example, having metacoelomic peribuccal spaces in the collar (Benito 1982).

Molecular data are consistent with the morphological data reviewed above. Studies involving 18S rDNA by Halanych (1996), Cameron et al. (2000), and Peterson and Eernisse (2001) support the two major conclusions derived solely from the morphological analysis, namely, the monophyly of Hemichordata, and that harrimaniids are the sister group of the pterobranchs. Furthermore, Cameron et al. (2000) show with 18S rDNA data that Ptychoderidae, Harrimaniidae, and Pterobranchia are each monophyletic. 28S rDNA data, on the other hand, suggest that pterobranchs are the sister taxon of enteropneusts, and hence Enteropneusta is monophyletic, although this is not supported in the combined 18S + 28S analysis (Winchell et al. 2002).

Combining available molecular and morphological data (fig. 22.3; for data, see Smith 2003b) leads to the following conclusions: (1) Hemichordata is a monophyletic taxon, and (2) Enteropneusts are a paraphyletic grade, with Harrimaniidae as more closely related to pterobranchs than to the other enteropneust families.

If this is a correct phylogeny, then it implies that pterobranchs may have undergone some secondary simplification associated with miniaturization. *Cephalodiscus*, rather than having a complicated gill skeleton, has just two relatively simple gill pores, and gill slits are entirely wanting in *Rhabdopleura*. Furthermore, pterobranchs have a simple neuronal ganglion in the collar region, whereas enteropneusts have a dorsal nerve cord whose development in at least saccoglossids is reminiscent of chordates (Bateson 1885). Finally, it also implies that the water vascular system of echinoderms and the tentacles of pterobranchs must have been independently acquired.

Echinoderms

Echinodermata are a well-characterized group of exclusively marine invertebrates that includes the familiar starfishes and sea urchins. They are solitary and almost exclusively benthic as adults. The group first appears near the base of the Cambrian and has expanded to colonize a wide range of marine habitats from intertidal to abyssal trench depths. There are about 6000 species alive today, and several groups have left an extensive fossil record.

Echinoderm Autapomorphies

Echinoderms are unique within Bilateria in having an adult body plan that is pentaradiate in construction, although their larvae are clearly bilaterally symmetrical. In addition to their obvious pentaradiate body plan, echinoderms share four other important morphological traits that identify them as a monophyletic clade: (1) In the transition from larval rudiment to adult, there is a striking asymmetry in the fate of coelomic compartments. Although there is variation in detail within echinoderm classes (e.g., Janies and McEdward 1993), in all the right hydrocoel is reduced in size and plays no part in adult structures, whereas the remaining coeloms ultimately become vertically stacked, with the right somatocoel aboral to the left somatocoel and the left somatocoel aboral to the left hydrocoel (see Hyman 1955, Peterson et al. 2000). (2) The left hydrocoel gives rise to a system of tentacles, as in hemichordates, but in living forms these are not free extensions, because they remain embedded within the body-wall and associated with somatocoel components even when prolonged into a filtration fan. (3) There are no gill pores, at least among extant representatives. (4) There is a mesodermal skeleton of calcite that takes the form of a distinctive meshwork termed stereom. This is present in all groups, although in holothurians it is typically reduced to microscopic spicules, and may occasionally be wanting altogether.

Molecular data are equally unambiguous as to the monophyly of echinoderms. Phylogenetic analysis of ribosomal RNA sequence data (Field et al. 1988, Littlewood et al. 1997, Janies 2001, Peterson and Eernisse 2001) all identify echinoderm exemplars as forming a monophyletic clade with strong bootstrap and Bremer support.

Echinoderm Body Plan Organization

One question has long puzzled echinodermologists: What is the relationship of the adult pentaradiate body plan of an echinoderm to the bilateral symmetrical plan of a chordate or hemichordate? In contrast to other deuterostomes, an adult echinoderm has no obvious anteroposterior, dorsoventral or left–right axes (fig. 22.4A). Echinoderm researchers have tended to avoid the whole question of body axis homologies by referring echinoderm orientation not to an anterior–posterior axis but to an oral–aboral axis. But recent work on the developmental molecular genetics has finally provided an answer.

One possibility is that each of the five ambulacra in an echinoderm represents a serially duplicated anterior–poste-

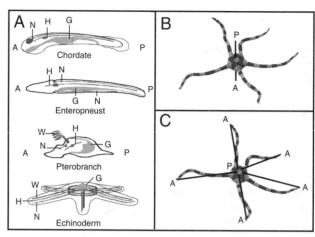

Figure 22.4. Schematic representation of body axes in echinoderms and other deuterostomes. (A) The body outlines show the arrangement of the nervous system (N), hemal system (H), digestive system (G), and hydrocoel system (W) in chordates, enteropneusts, pterobranchs, and echinoderms. A, anterior; P, posterior. (B and C) Two alternative interpretations of anterior–posterior body axis in a brittlestar.

rior axis in echinoderms. So a starfish would have five anterior–posterior axes, with each arm tip being the equivalent of a bilaterian anterior (fig. 22.4C). This idea found initial support from developmental genetics, when it was shown that the regulatory gene *orthodenticle*, which in arthropods and vertebrates is a "specifier" of anterior structures, is expressed distally in the arms of developing ophiuroids and starfish. Another developmental regulatory gene, *engrailed*, is active along the anterior–posterior axis of the central nervous system of several bilaterally symmetrical metazoan phyla and is also expressed along the developing arms of echinoderms (Lowe and Wray 1997). However, because developmental regulatory genes can readily be co-opted into different roles, such evidence is weak (Wray and Lowe 2000).

More convincing evidence has come from following the fate of the bilaterally symmetrical coeloms from larva to adult Peterson et al. (2000) pointed out that because "posterior" *Hox* genes are expressed colinearly in the posterior coeloms (the somatocoels; see also Arenas-Mena et al. 2000), this must be the primitive locus of expression. If true, then this means that the primitive adult anterior–posterior axis can be seen in the larval mesoderm, specifically the paired coelomic sacs. The development of the adult body plan involves a rotation of the coeloms such that the right somatocoel comes to lie underneath the left somatocoel, with both coeloms giving rise to extraxial skeletal structures at the aboral end of the animal. Because the primitive axis is mesodermal, this means that the modified anterior–posterior axis runs from the oral surface through the left hydrocoel, then the left somatocoel, and finally the right somatocoel at the aboral end of the animal (fig. 22.4B). Furthermore, their pentamery is an expression of secondary lateral outgrowth, not a duplication of primary body axes as suggested by Raff (1996).

The Five Classes and Their Relationships

There are five extant classes of echinoderms: the crinoids (sea lilies and feather stars), asteroids (starfishes), ophiuroids (brittlestars), echinoids (sea urchins), and holothurians (sea cucumbers). These five classes are well characterized from both morphological and molecular perspectives. A sixth class, Concentricycloidea (sea daisies), composed of one genus with two deep-sea species, has been proposed (Baker et al. 1986), but recent molecular work (Janies and Mooi 1999) has shown that this taxon nests well inside Asteroidea.

The crinoids stand clearly apart from the other four classes. They are primitively stalked and sessile (fig. 22.1.4), although in one important but derived subclade, the comatulids (fig. 22.1.5), the stalk is lacking and they are able to swim. In crinoids, the mouth faces away from the seafloor and the anus opens in close proximity on the same anatomical surface. A system of branched arms, which carry extensions of the somatocoel and water vascular system, form a filtration fan for food capture. The plates that make up the arms and that bear the radial water vessels have traditionally been thought of as ambulacral in origin and thus homologous to the ambulacral plates in other echinoderms. However, the presence of somatocoel and somatocoel-related structures (e.g., gonads) in the arms is evidence for there being part of the aboral plating system (extraxial plating of David and Mooi 1999, Mooi and David 1997) rather than ambulacral (axial) plates. Extraxial plating thus is much more extensively de-

veloped than axial plating. In addition the nervous system of crinoids is very different from that in other echinoderms, being dominated by the subepithelial component rather than the epithelial component that dominates in other echinoderms and hemichordates (Heinzeller and Welsch 2001). Crinoids have a long fossil record going back to the start of the Ordovician, although extant crinoids all belong to a clade whose origins are much more recent, at about 250 Mya (million years ago; Simms 1999).

The four other echinoderm classes are free-living and have been grouped together under the name Eleutherozoa. They live mouth downward and have a nervous system dominated by the ectoneural component. The starfish (Asteroidea) are stellate forms whose body projects as five or more arms from a central region (fig. 22.1.6–7). Major body organs such as the gonads and stomach extend into the arms. Aboral (extraxial) and ambulacral (axial) surfaces are approximately equally developed in almost all taxa, and the ossicles around the mouth are relatively unspecialized and do not form a jaw apparatus. Finally, the radial nerve lies externally within the epithelial layer (fig. 22.5).

Brittlestars (Ophiuroidea) resemble starfish in shape but have a much more clearly demarked boundary between the central disk and the narrow, whiplike arms (fig. 22.1.1). The arms differ fundamentally from those of starfishes in having a cylindrical core of ossicles (vertebrae) that are modified ambulacral plates. Aboral (extraxial) and oral (axial) plating systems are again equally developed. In a few taxa the gonads extend into the arms, and this was probably much more common in primitive, extinct representatives. During development, the radial nerve and radial water vessel become

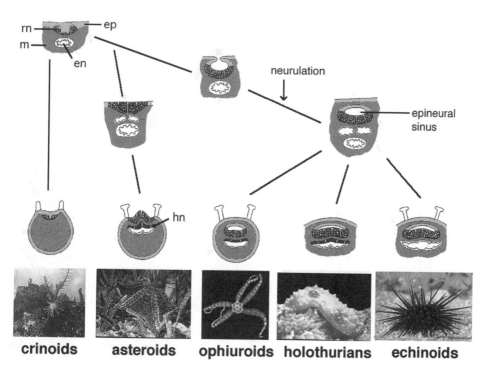

Figure 22.5. Schematic cross sections through the body wall to show radial nerve arrangement in echinoderms. en = ectoneural plexus; ep, epithelial tissue; hn, hyponeural plexus; m, mesoderm; rn, ectoneural plexus. Phylogenetic relationships are indicated by lines. From Heinzeller and Welsch (2001).

crinoids   asteroids   ophiuroids   holothurians   echinoids enveloped by epithelial flaps and a secondary cavity, the epineural sinus, is created (fig. 22.5). All brittlestars and most starfishes have a blind gut and lack an anus.

Sea urchins (Echinoidea) are primitively globular forms but have over geological time evolved into a wide range of shapes (figs. 22.1.8–9). Irrespective of shape, most of their body skeleton is formed of axial components and thus homologous to the oral surface of starfish and brittlestars. Aboral (extraxial) components in sea urchins are confined to the 10 plates of the apical disk and the periproctal system they enclose. Sea urchins also primitively have a complex internal jaw apparatus, known as the Aristotle's Lantern, composed at least in part of modified ambulacral plates. The lantern is secondarily lost in some irregular echinoids.

Sea cucumbers (Holothuroidea) are mostly sausage or worm-shaped animals (figs. 22.1.2–3) whose skeleton is reduced to microscopic spicules embedded in their thick collagenous skin. Their mouth and anus are situated at opposite poles, as in echinoids, with the mouth encircled by a ring of large feeding tentacles. The only substantial skeletal structure is an internal ring of 10 ossicles that surrounds the buccal cavity. Interestingly, holothurians are the only group of echinoderms that pass through metamorphosis with little torsion (Smiley 1988).

The relationships among these four eleutherozoan groups has been much disputed and remain far from settled. Traditionally, they have been subdivided into two groups, Asterozoa for the stellate starfishes and brittlestars, and Echinozoa for the globular to cylindrical sea urchins and sea cucumbers (e.g., Fell 1967). However, one or other body form is presumably the primitive condition for Eleutherozoa as a whole. The transformation between the two body plans requires only a modest change in the relative production of aboral and oral (extraxial and axial) plating systems. Simply by retarding the production of aboral plating, starfishes such as *Podosphaeraster* take on an echinoid-like form (see Blake 1984).

Smith (1984) has argued that Asterozoa are a paraphyletic grouping, with brittlestars more closely related to Echinozoa (i.e., Echinoida + Holothuroida) than to starfishes. This was based on the similarity of larval form, jaw apparatus construction, internal coelom arrangement, and the enclosure of the radial nerve and water vessel in ophiuroids and echinoids. A cladistic analysis of a large morphological data matrix supported this view (Littlewood et al. 1997), as did a more detailed analysis of the nervous system of echinoderms (Heinzeller and Welsch 2001). A revised and emended morphological character matrix compiled by Janies (2001) also supported the same topology.

An alternative view (Sumrall 1997, Mooi and David 1997, 2000, David and Mooi 1996, 1999) is that Asterozoa are monophyletic. Strongest support for this grouping initially came from mitochondrial genome order (but see below). Just two morphological synapomorphies support this group; the presence of a saccate gut and the presence of a system of adambulacral ossicles.

The relationship between sea urchins and sea cucumbers is more difficult to establish on morphological grounds. This is in part because of the extreme skeletal reduction in holothurians, making detailed comparisons difficult. In addition, there are also major uncertainties over the homologies of certain structures, such as the calcareous ring and radial water vessel. Holothurians, other than apodids, have five radial water vessels that run along the length of the body and give rise to tube-feet. These lie within the mesoderm and have an overlying epineural sinus exactly as in echinoids (Heinzeller and Welsch 2001; see fig. 22.5), suggesting secondary enclosure. However, Mooi and David (1997) point out that the tube-feet are added irregularly along the length rather than terminally, and that they arise secondarily after the oral tentacles have formed. Under their model only the buccal tentacles are homologous to the radial water vessels in echinoids. They homologize only the oral region of holothurians with the body of echinoids and believe the trunk is a novel structure that has been derived from the extraxial portion of larval tissue.

The fossil record provides crucial evidence linking the echinoids and holothurians, because of the unusual character combination found in the extinct and probably paraphyletic Ophiocistioida. Ophiocistioids have a complex lantern that is homologous in almost every detail to the lantern of echinoids. Furthermore, they have an arrangement of plates similar to that seen in the most primitive of echinoids, in which there is a central uniserial series of plates in each ambulacral zone (Smith and Savill 2002). Yet advanced members reduce their skeleton to wheel-shaped spicules and platelets that are almost indistinguishable from those of holothurians (Gilliland 1993). This combination of holothurian and echinoid traits implies sister-group relationship between the two living groups.

Molecular Evidence for Echinoderm Class Relationships

Molecular evidence, derived principally from nuclear and mitochondrial ribosomal RNA (rRNA) genes, provides clear support for the monophyly of each of the five classes. Exemplars of each class always group together, confirming the long-standing picture from the fossil record that crown-group diversification within each class is relatively recent compared with the time at which the classes diverged from one another. However, the relationships of the five classes are much more controversial.

Ribosomal Sequence Data

The pioneering analysis of Raff et al. (1988), based on partial 18S rRNA sequences, identified Asterozoa as paraphyletic, with asteroids as sister group to Echinozoa. Littlewood et al. (1997) undertook a more comprehensive analysis for both complete 18S and partial 28S sequences. They found that, although both Eleutherozoa and Echinozoa were well supported, other groups were only poorly supported. The most parsimonious solution had ophiuroids as sister group to Echinozoa, but the two other possible solutions (asteroids as sister group to Echinozoa, and asteroids and ophiuroids as sister group) were only one step longer. In the analysis of Littlewood et al. (1997) regions of ambiguous alignment were removed before analysis, and consensus sequences were constructed for each class based on the sequences then available. Janies (2001) added a considerable number of asterozoan 18S rRNA sequences to the database and carried out both separate and combined analyses of molecular and morphological data. Unlike Littlewood et al. (1997), Janies used the full sequence data aligned using CLUSTAL (Thompson and Jeanmougin 2001) with various weightings. This identified asteroids as sister group to Echinozoa when indels, transversions, and transitions were all equally weighted, and ophiuroids as sister group when indels and transversions were given a weight of 2.

Janies (2001) then applied a dynamic analysis of the combined morphological and molecular data (POY; Wheeler and Gladstein 2000) whereby alignment and tree building occur together so as to co-optimize all available data. Once again, there was strong support for Eleutherozoa, Echinozoa, and each of the five classes. His best total evidence tree identified Asterozoa as a clade, but with very weak Bremer support. Just suboptimal is a tree that has asteroids as sister group to Echinozoa. Significantly, although Echinozoa, Eleutherozoa, and all five classes can be recovered under a wide range of parameters (indicating that there is strong support for these groups), the grouping Asterozoa was only recovered under a small subset of conditions, and the ophiuroid-echinoid-holothurian clade was hardly ever recovered.

We have reanalyzed the now quite extensive rRNA sequences in various ways (aligned sequences are provided at in Smith 2003b) both under parsimony (Paup 4*; Swofford 2001) and Bayesian inference (MrBayes 2.01; Huelsenbeck and Ronquist 2001). Individually, both 18S and 28S rRNA sequences rooted on hemichordates identify the same topology, namely (crinoids(asteroids(ophiuroids(echinoids, holothurians)))), but with weakest support for the ophiuroid-echinoid-holothurian pairing. The same topology resulted from a combined sequence analysis irrespective of whether only exemplars common to both data sets are used or whether taxa whose 28S rRNA sequences are currently unknown were included (fig. 22.6A).

For the combined morphological and molecular analysis, instead of using gene sequences from exemplars, we have constructed consensus gene sequences for each class. For each variable position, the consensus sequence replaces two or more alternate bases with the international nucleotide code encompassing the uncertainty. The logic behind this approach is that it removes the variation within each class that has arisen since the crown group started to diverge. The sequences for each order were then aligned, and fast-evolving regions where alignment was ambiguous (usually because of the presence of long strings of N values) were removed. The results are shown in figure 22.6B. High Bremer support was found for most branches, with the most parsimonious solution placed ophiuroids as sister group to Echinozoa.

Mitochondrial Gene Order

When the first complete mitochondrial genomes of echinoderms became available, it was quickly realized that the order in which genes were arranged around the circle differed significantly between asteroids and echinoids (Smith et al. 1989). A 4.6-kilobase section of the genome, incorporating four protein coding genes, was inverted. Subsequently, when holothurian and ophiuroid mitochodrial genome order became known, it was shown that echinoids and holothurians had one arrangement and asteroids and ophiuroids another (Smith et al. 1993). Comparison with vertebrates as outgroup showed that it was the asteroid-ophiuroid arrangement that was inverted, suggesting that the inversion was a synapomorphy for Asterozoa. However, it is becoming clear that the order of genes may not be so reliable a marker (e.g., Mindell et al. 1998) even though there has been reasonable stability of the mitochondrial gene order among echinoid groups that last shared a common ancestor some 170 Mya (Giorgi et al. 1996).

The recent publication of the complete crinoid mitochondrial gene sequence (Scouras and Smith 2001) has confirmed this view. Crinoids, as the immediate outgroup to Eleutherozoa, should provide the most appropriate sequence for determining which mitochondrial genome arrangement is primitive. However, the crinoid arrangement is significantly different from both the asterozoan and echinozoan arrangements. Specifically the crucial 4.6–kilobase section is partly inverted as in Asterozoa and partly normal as in echinozoans (fig. 22.7). Thus, the initially strong evidence for an asterozoan clade is now much more problematic to interpret. Considerable gene rearrangement is required to transform the outgroup vertebrate mitochondrial gene sequence to any of the three echinoderm arrangements, although the echinozoan sequence requires slightly fewer steps. Both crinoids and asterozoans have an inverted portion of the genome compared with either vertebrates or echinoids. It appears, therefore, that there has been a complicated pattern of rearrangement of the mitochondrial genomes in the lines leading up to the recent echinoderm classes. Again, there is no clear solution: either an inversion has occurred before crown group separation and then been reversed in Echinozoa, or crinoids and Asterozoa have independently inverted part of their genome sequence.

Other Molecular Data

Scouras and Smith (2001) used amino acid and sequence data of the cytochrome oxidase gene complex to explore echinoderm relationships. The ophiuroid sequence was unfortunately very strongly divergent, and although they were able to demonstrate the monophyly of Eleutherozoa, they were unable to resolve interclass relationships with any statistical confidence.

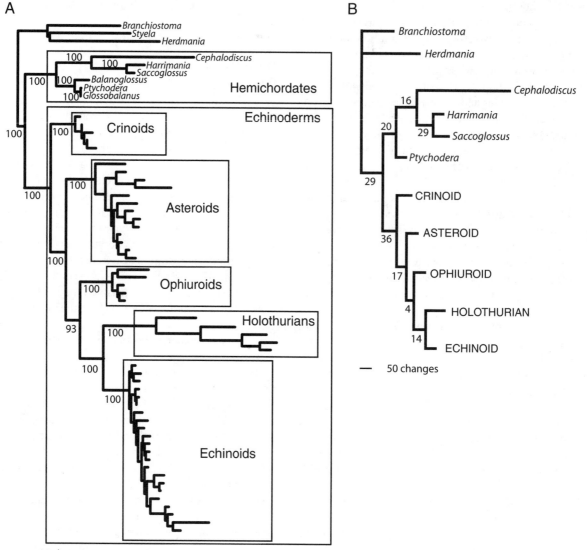

Figure 22.6. Phylogenetic relationships of the major clades of Ambulacraria. (A) Tree derived from Bayesian inference of complete large subunit (LSU) ribosomal rRNA and partial small subunit (SSU) ribosomal rRNA sequences of the 59 taxa whose complete large subunit (LSU) ribosomal rRNA sequences are known (the matrices can be found at http://puffin.nhm.ac.uk:81/ iw-mount/default/main/Internet/WORKAREA/palaeontology/Web-Site/palaeontology/I&p/abs/ abs.html). Bayesian inference analysis used the following parameters: nst = 6, rates = invgamma, ncat = 4, shape = estimate, inferrates = yes, basefreq = empirical, which corresponds to the GTR + I + G model. Posterior probabilities were approximated using more than 200,000 generations via four simultaneous Markov chain Monte Carlo chains with every 100th tree saved. Nodal support is shown, estimated as posterior probabilities (Huelsenbeck et al. 2001). (B) Tree derived from parsimony analysis of the combined complete LSU ribosomal rRNA and partial SSU ribosomal rRNA sequences and morphological data (all data can be found at the web site noted above noted above). Consensus sequences were constructed for each echinoderm class with positions that vary in base composition within each class scored with the international nucleotide code to reflect this uncertainty. Bremer support values are given for each node.

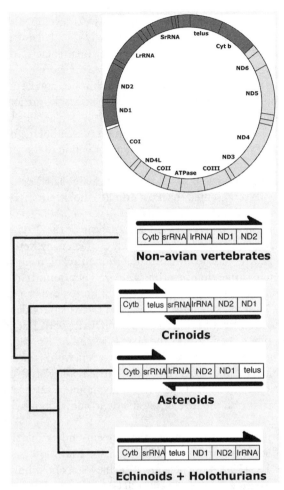

Figure 22.7. Mitochondrial gene order in echinoderms. The gray zones in the mitochondrial gene represent the variable region. Arrows indicate transcription polarity (after Scouras and Smith 2001).

In conclusion, there is strong support from both morphological and molecular data for the monophyly of Echinodermata and for a basal crinoid-eleutherozoan split. Within Eleutherozoa, all molecular data support a pairing of Echinoida and Holothuroida, and there is also some morphological data to support the monophyly of Echinozoa, as well, depending upon how one interprets certain structures. The ophiuroid-asteroid-echinozoan trichotomy remains the most difficult to resolve, but both morphology and molecular data point to an ophiuroid-echinozoan sister group (Cryptosyringida), albeit with a reduced level of statistical support.

Relationships within Echinoderm Classes

There are marked differences as to how well we currently understand relationships of the families and higher taxa within each of the five classes. This only partially reflects the amount of work that has been carried out, because there is also variation in how well morphological and molecular es-

timates agree. Both morphological and molecular phylogenies are well advanced and show a high degree of congruence in echinoids, for example, whereas relationships of the major clades of asteroids remain highly problematic, with different data sets giving highly conflicting results.

Crinoids

The basic taxonomy of crinoids that we have today is founded on the monographic efforts of A. H. Clark and A. M. Clark (Clark 1915–1950, Clark and Clark 1967). The group is relatively small with approximately 560 extant species. Of these, more than 500 belong to the free-living Comatulida, the remainder being stalked crinoids that are rarely encountered and because they are entirely deep-water creatures today. Although workers continue to add to our understanding of the species-level taxonomy, surprisingly little progress has been made in unraveling the relationships of the major crinoid lineages. The principal cladistic analysis for the group remains that of Simms (1988, 1999; fig. 22.8). According to Simms, crown-group diversification started in the Early Mesozoic (~250 Mya). He recognized two major groups, Millericrinida and Isocrinida. These two groups differ from their Paleozoic antecedents in having the axial nerves buried within the skeleton of the cup and in having pinnulate arms. Of the two groups, the obligate deep-sea millericrinids are less common today. Isocrinida include both the stemmed deep-sea isocrinids and the very much more diverse shallow-water commatulids. Commatulids are stemless as adults and are primarily reef dwellers, having undergone a major radiation since the Mesozoic. Isocrinidans are characterized by having synarthrial columnal articulations and, except for the deep-water bourguetticrinids, all also possess fingerlike cirri for gripping the seafloor.

Crinoids were a major constituent of benthic faunas in the Paleozoic, appearing first in the earliest Ordovician and remaining diverse through to the Permian. Four major groups existed throughout this period, one of which (the Cladida) gave rise to modern crinoids. An excellent summary of crinoid biology and palaeontology is given in Hess et al. (1999).

There are no detailed molecular studies of crinoid relationships available as yet, although one is currently being undertaken (M. Ruse, pers. comm.).

Asteroids

This is the second largest of the echinoderm classes, composed of some 1400 species. There is little consensus at present about the phylogenetic framework for asteroid orders. The morphological analyses of Gale (1987) and Blake (1987) used data from both extant and extinct asteroids but disagreed about key character polarities and character definitions (fig. 22.9). A more extensive reappraisal of the morphological data that takes into account the rival views of character scoring is urgently needed.

The molecular analyses all suffer to a greater or lesser extent from limited taxonomic sampling and long-branch

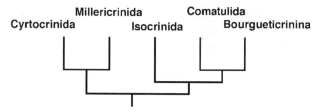

Figure 22.8. Cladogram for crinoids based on morphological analysis (after Simms 1999).

problems. Lafay et al. (1995) used partial 28S rRNA gene sequence data for nine asteroids and found almost no information about ordinal relationships. Wada et al. (1996) used 12S and 16S rDNA in combination to investigate phylogenetic relationships. Collapsing branches with less than 50% bootstrap support in the Wada et al. topology produces a topology congruent with that of Lafay et al. (1995) except for the placement of *Crossaster*. Smith (1997) reanalyzed the data for the two genes separately and combined with 28S rRNA data. Knott and Wray (2000) sequenced a large number of species for two mitochondrial genes (*tRNA* and *COI*) and analyzed these both separately and combined with previous data. Finally, Janies (2001) has provided a number of new asteroid 18S rDNA sequences carried out new methods of analysis. It is difficult to see any common thread emerging from this work. Forcipulatids appear to be monophyletic, but most other major traditional groupings were not recovered in the analyses of Janies or Knott and Wray (fig. 22.9). Furthermore, different methods of analysis give very different groupings. It would appear, therefore, that there is very little signal in the molecular data currently available with which to resolve asteroid relationships. Even the question of whether Paxillosida is basal or not remains ambiguous based on molecular data. Asteroids have a rather poor and patchy fossil record. The earliest asteroids come from the basal Ordovician (Smith 1988). However, it is clear that the modern crown group asteroids arose in the early part of the Mesozoic and that, like other groups, the major orders had become established by the Middle Jurassic.

Holothurians

Until very recently holothurians remained the most poorly known of the echinoderm classes. Twenty-five families in six orders are currently distinguished based on body form spiculation. Apodidans are slender wormlike forms that lack tube-feet and respiratory trees. The body wall is thin, and its spicules are wheel-shaped ossicles that are present throughout life (Chirodotidae and Myriotrochidae) or in larvae only (Synaptidae). Similar ossicles are found in the extinct ophiocistioids. Elasipodans are entirely deep-water forms and include the only holopelagic (swimming) echinoderm. They often have highly modified dorsal tube-feet that are fused to form curtainlike structures. Aspidochirotidans have shieldlike tentacles with internal ampullae and

usually creep along the ground on a well-developed sole. Dendrochirotids include the most heavily plated of holothurians and have branched, dendritic feeding tentacles without ampullae that can be retracted into an oral introvert. This group is split into two orders, Dactylochirotida and Dendrochirotida, differing in how branched the tentacles are and in the structure of the calcareous ring. Finally, Molpadiida have 10 or 15 simple tentacles and the posterior end of the body is narrowed into a "tail." A useful introduction to the group can be found in Kerr (2003).

The accepted view was that the heavily plated dendrochirotids represented the most primitive holothurians (e.g., Pawson 1966). However, this view has recently been overturned, and the recent cladistic analysis of the 25 extant families based on 47 morphological characters by Kerr and Kim (2001) has now placed holothurian relationships on a much firmer footing (fig. 22.10). They found strong support for the monophyly of four of the six orders (Apodida, Elasipoda, Aspidochirotida, and Dactylochirotida) but found Dendrochirotida to be paraphyletic, with Dactylochirotida nested inside. The class is rooted on Apodida. A second, more detailed analysis of the genera within the three families of Apodida has also been carried out (Kerr 2001). This again found that the current taxonomic classification consisted of a mixture of paraphyletic and monophyletic groups.

Few molecular sequence data are currently available to test this phylogeny (Smith 1997, Kerr and Kim 1999). Complete 18S rRNA sequences are available for six holothurians (representing four of the six orders), and these generate a phylogeny fully congruent with the morphology-based tree (fig. 22.10). The basal position of Apodida in phylogenies is particularly robust based on molecular data.

Although the fossil record of holothurians is poor compared with that of other echinoderm groups, isolated body wall spicules recovered from sedimentary samples are frequently encountered and can be used to deduce much about the timing of appearance of holothurian groups in the fossil record Gilliland (1993). Kerr and Kim (2001) found a good match between their phylogeny and the stratigraphic record based on spicules.

One of the most interesting outcomes of this work is that the holothurian crown group appears to be considerably older than the crown group of any other echinoderm class. Holothurians are the only class in which undisputed crown-group clades appear well before the end of the Paleozoic, and the dichotomy between Apodida and other holothurians has a Bremer support more than twice the value at which the clades within other classes collapse to a polytomy.

Ophiuroids

Ophiuroids are the most diverse of extant classes, with around 2000 extant species. Despite this diversity, many workers follow Mortensen (1927) in recognizing just two orders, Euryalina for forms with arm ossicle articulations that are hour-

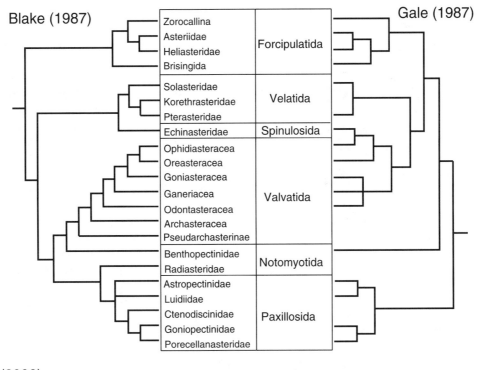

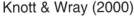

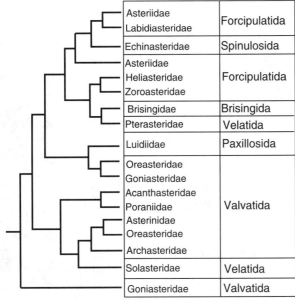

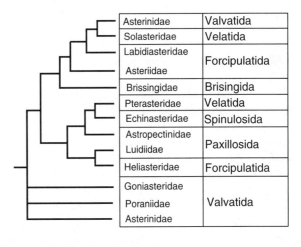

Figure 22.9. Alternative phylogenetic hypotheses for asteroids.

glass-shaped (streptospondyline), and Ophiurina for forms with a peg-and-socket-type articulation between arm ossicles (zygospondyline). The former group includes both simple-armed forms and the basket stars with branched arms and has long been considered primitive with respect to Ophiurina.

Smith et al. (1995b) undertook a cladistic analysis of the 27 extant families that confirmed the paraphyletic nature of Euryalina. This suggested that, although the multiarmed basket stars (Gorgonocephalidae and Euryalidae) from a clade together with certain simple-armed forms, Ophiomyxidae were a more derived clade and sister group to Ophiurina, whereas *Ophiocanops* might be sister group to all other extant ophiuroids.

Smith et al. (1995b) also used partial 28S rRNA sequence data from 10 representative taxa to test the morphological hypothesis. Unfortunately, no simple-armed euryalinans were included, and the resultant trees had most internal nodes rather poorly supported. Subsequently, both *Ophiomyxa* and *Ophio-*

Morphology   **18S rRNA**

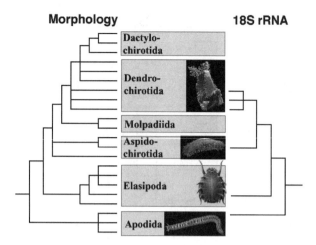

Figure 22.10. Morphological and molecular phylogenies for holothurians (after Kerr and Kim 1999, 2001).

canops have had their 18S gene sequenced, and a partial 28S gene sequence is available for *Ophiocanops*. Analysis of total molecular data confirms that *Ophiocanops* is the sister taxon to the Ophiurina, but 18S rRNA data alone place *Ophiocanops* and *Ophiomyxa* as sister taxa nested within Ophiurina (but with low bootstrap support). Better sampling of both taxa and genes is required to generate a more robust phylogeny for the class.

Ophiuroids first appear in the fossil record near the start of the Ordovician, about 490 Mya, but the modern orders all appear to stem back to a major crown-group radiation of the class that occurred in the Late Triassic or Early Jurassic.

Echinoids

Of all echinoderm classes, the echinoids have the most detailed and well-established phylogeny. There are about 900 extant species equally divided between regular forms whose anus opens in the aboral plated surface and that live epifaunally, and irregular forms whose anus is displaced out from the aboral plates into the posterior interambulacral zone and that live predominantly infaunally. The most basal group is Cidaroida, which differs from all other echinoids in having lantern muscle attachments that are interradial in position (apophyses) and simple ambulacral plating. Other major regular echinoid groups have lantern muscle supports that are radial in position (auricles), and all but Echinothurioida have soft-tissue extensions of the internal coelom called buccal expansion sacs. Echinothurioids, which are deep sea forms, differ further in having an entirely flexible skeleton. The remaining regular echinoids are divided on their tooth and lantern structure, and on whether tubercles are perforate or imperforate. Diadematoida and Pedinoida have simple U-shaped teeth in cross section, like cidaroids and echinothurioids, whereas Camarodonta and Stirodonta have teeth that are T-shaped in cross section. Camarodonta are the more derived of the two because they also have a fused brace in their lantern.

There are two major extant groups of irregular echinoid alive today. One group are the heart urchins, which have

secondary bilateral symmetry and have completely lost their lantern. Heart urchins (orders Spatangoida and Holasteroida) are exclusively deposit feeders. The other group consists also of deposit feeders but ones that have retained a much more obvious pentameral symmetry. Traditionally, two orders have been distinguished, Clypeasteroida and Cassiduloida, but the latter is paraphyletic and requires reclassifying (Smith 2001). Clypeasteroida includes the well-known sand dollars and have the distinct synapomorphy of having large numbers of tube-feet to each ambulacral plate (all other echinoids have just a single tube-foot to each plate). Irregular echinoids first appeared in the Early Jurassic and diversified rapidly as deposit feeders. Clypeasteroids are the most recent group to have arisen, first appearing about 50 Mya. A general introduction to sea urchin morphology, biology and systematics can be found in Smith (2003a).

In recent years, many groups of echinoid have begun to be analyzed cladistically, and in some cases with both morphological and molecular data (Smith, 1988, 2001, Smith et al. 1995a, Harold and Telford 1990, Mooi and David 1996, Jeffery et al. 2003). The primary framework for ordinal relationships is well established through the work of Littlewood and Smith (1995). They used a combined morphological and molecular approach (18S and 28S rRNA gene sequences) from a wide range of taxa to construct a phylogenetic hypothesis. Both approaches proved closely comparable topologies, although with some differences among the camarodont taxa (fig. 22.11). Echinoids first appeared in the Middle Ordovician but were never particularly diverse during the Paleozoic. Just two lineages passed through into the Mesozoic, one of which gave rise to modern cidaroids, and the other, to all other extant echinoids. Most of the higher taxa were established during the Late Triassic to Middle Jurassic.

The Importance of Ambulacraria in Metazoan Phylogeny

The Ambulacraria hold an important position within the Metazoa for several reasons.

(1) As the immediate sister group to chordates, Ambulacraria provides the closest outgroup from which to establish basal character polarities in early chordate evolution. Significant difficulties in reconstructing the evolutionary history of deuterostome body plans remain, yet the fact that phylogenetic relationships among the deuterostome phyla are now clear means that inferences about body plan changes are on a more secure footing. For instance, it is no longer necessary to derive the chordate body plan from precursors with trimerous coeloms and a hydropore (see above). Likewise, anatomical similarities in the larva shared between living enteropneust hemichordates and eleutherozoan echinoderms (Strathmann 1988) can no longer be taken as ancestral features that were modified or lost during the origin of chordates (Garstang 1928). This is not to say that we can

Morphology

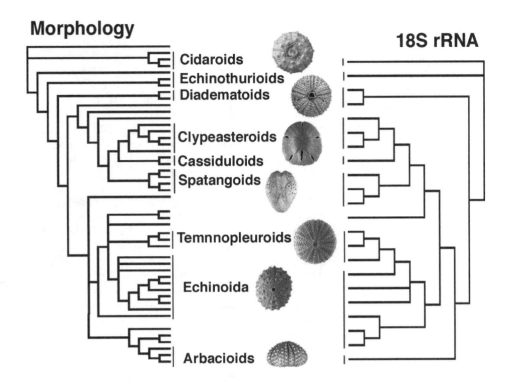

18S rRNA

Figure 22.11. Morphological and molecular phylogenies for orders of echinoids (after Littlewood and Smith 1995).

confidently rule out trimery or any of the other features uniquely shared by hemichordates and echinoderms as also being a plesiomorphic condition within the stem lineage leading to the urochordate + chordate clade. It simply requires positive evidence for the possession of the trait, either from fossils or living taxa.

(2) Ambulacrarians are turning out to be crucial in developing our understanding of the genetic basis of the evolution of body-plans. Despite the tremendous progress of developmental genetics during the past two decades, most of what we know about body plan patterning still comes from two phyla: arthropods and chordates. Echinoderms (and, increasingly, hemichordates) have emerged as a crucial group for studying the evolution of the developmental mechanisms that establish animal body plans (Wray and Lowe 2000, Davidson 2001, Tagawa et al. 2001).

It is clear that the basic genetic mechanisms that govern body patterning among bilaterians were already established in the latest common ancestor of Bilateria (Gerhart and Kirschner 1997, Peterson and Davidson 2000, Carroll et al. 2001). Furthermore, we now have a good working understanding of the way in which the regulatory molecules that carry out these functions operate. The transcription factors that regulate gene expression and the signaling systems that define the morphogenetic fields that establish the bilaterally symmetrical body plan are reasonably well understood in forms as distant as insects, nematodes, and mouse (Gellon and McGinnis 1998, Carroll et al. 2001).

These genetic controls and mechanisms are also present in the echinoderms (Davidson 2001), but the body plan that results is drastically different. Echinoderms, with their radial

body organization, are thus likely to provide crucial evidence as to what sort of modifications in ancient regulatory genes are required to generate such a large shift in basic body organization (Wray and Lowe 2000). Long and Byrne (2001) reviewed the *Hox* gene clusters in the five classes of echinoderm and identified orthologues for most of the chordate *Hox* genes, and orthologues of many other crucial regulatory genes have been identified as well. Thus, the evolutionary modifications in developmental mechanisms that resulted in the echinoderm body plan must have included co-option and modification of roles and expression domains of preexisting bilaterian regulatory genes (Wray and Lowe 2000). Nonetheless, echinoderms do show some autapomorphic uses of regulatory genes (Lowe and Wray 1997, Wray and Lowe 2000), including the absence of *Hox* gene function in the sea urchin embryo (Arenas-Mena et al. 2000). Therefore, echinoderms provide a unique opportunity to investigate the genetic basis of pattern formation and morphogenesis in the generation of novel evolutionary structures.

(3) Echinoderms, like many animal phyla, are composed largely of species that develop indirectly, by means of a larva that is ecologically and anatomically distinct from the adult. Because evolutionary changes in larval ecology occur commonly in the echinoderm crown group, including multiple transitions from planktotrophy to lecithotrophy and from lecithotrophy to brooding, the group has become one of the best studied in terms of understanding diverse aspects of larval ecology (Hart et al. 1997, McEdward and Miner 2001). Comparisons of larval and life-history diversity have taken advantage of the growing understanding of phylogenetic relationships within echinoderms to formulate spe-

cific hypotheses about evolutionary history (Wray 1992, 1996).

(4) Echinoderms are the dominant component of the macrobenthos in the deep sea, forming more than 90% of the biomass in abyssal settings, the largest single ecosystem in the world (Kerr and Kim 2001). Many echinoderms have a complex endoskeleton and an excellent fossil record, making them ideal subjects for investigating patterns and processes of evolution within a rigorous phylogenetic framework.

Literature Cited

Adoutte, A., G. Balavoine, N. Lartillot, O. Lespinet, B. Prud'homme, and B. de Rosa. 2000. The new animal phylogeny: reliability and implications. Proc. Natl. Acad. Sci. USA 97:4453–4456.

Arduini, P., G. Pinna, and G. Teruzzi. 1981. *Megaderaion sinemuriense* n.g. n.sp., a new fossil enteropneust of the Sinemurian of Osteno in Lombardy. Atti Soc. Ital. Sci. Nat. Mus. Civ. Stor. Nat. Milano 122:104–108.

Arenas-Mena, C., A. R. Cameron, and E. H. Davidson. 2000. Spatial expression of *Hox* cluster genes in the ontogeny of a sea urchin. Development 127:4631–4643.

Armstrong, W. G., P. N. Dilly, and A. Urbanek. 1984. Collagen in the pterobranch coenecium and the problem of graptolite affinities. Lethaia 17:145–152.

Baker, A. N., F. W. E. Rowe, and H. E. S. Clark.1986. A new class of Echinodermata from New Zealand. Nature 321:862–864.

Balser, E. J., and E. E. Ruppert. 1990. Structure, ultrastructure and function of the preoral heart-kidney in *Saccoglossus kowalevskii* (Hemichordata, Enteropneusta) including new data on the stomochord. Acta Zool. 71:235–249.

Bardack, D. 1997. Wormlike animals: Enteropneusta (acorn worms). Pp. 89–92 in Richardson's guide to the fossil fauna of Mazon Creek (D. W. Shabica and A. A. Hay, eds.). Northeastern Illinois University, Chicago.

Bartolomaeus, T. 2001. Ultrastructure and formation of the body cavity lining in *Phoronis muelleri* (Phoronida, Lophophoroata). Zoomorphology 120:135–148.

Bateson, W. 1885. The later stages in the development of *Balonoglossus kowalevskii*, with a suggestion as to the affinities of the Enteropneusta. Quart. J. Microscopical Science 26:535–571.

Benito, J. 1982. Hemichordata. Pp. 819–820 in Synopsis and classification of living organisms (S. P. Parker, ed.), vol. 2. McGraw-Hill, New York.

Benito, J., and Pardos, F. 1997. Hemichordata. Pp. 15–102 in Microscopic anatomy of invertebrates, Vol. 15: Hemichordata, Chaetognatha, and the invertebrate chordates (F. W. Harrison and E. E. Ruppert, eds.). Wiley-Liss, New York.

Blake, D. B. 1984. Constructional morphology and life habits of the Jurassic sea star *Sphaeraster* Quenstedt. N. Jb. Geol. Palaontol. Abh. 169:74–101.

Blake, D. B. 1987. A classification and phylogeny of post-Palaeozoic sea stars (Asteroidea: Echinodermata). J. Nat. Hist. 21:481–528.

Bromham, L. D., and B. M. Degnan. 1999. Hemichordates and

deuterostome evolution: robust molecular phylogenetic support for a hemichordate plus echinoderm clade. Evol. Dev. 1:166–171.

Brusca, R. C., and G. J. Brusca, 1990. Invertebrates. Sinauer Associates, Sunderland, MA.

Budd, G. E., and S. Jensen. 2000. A critical reappraisal of the fossil record of bilaterian phyla. Biol. Rev. 75:253–295.

Cameron, C. B., J. R. Garey, and B. J. Swalla. 2000. Evolution of the chordate body plan: new insights from phylogenetic analyses of deuterostome phyla. Proc. Natl. Acad. Sci. USA 97:4469–4474.

Carroll, S. B., J. K. Grenier, and S. D. Weatherbee. 2001. From DNA to diversity: molecular genetics and the evolution of animal design. Blackwell Science, Oxford.

Castresana, J., G. Feldmaier-Fuchs, and S. Pääbo. 1998a. Codon reassignment and amino acid composition in hemichordate mitochondria. Proc. Natl. Acad. Sci. USA 95:3703–3707.

Castresana, J., G. Feldmaier-Fuchs, S. Yokobori, N. Satoh, and S. Pääbo. 1998b. The mitochondrial genome of the hemichordate *Balanoglossus carnosus* and the evolution of deuterostome mitochondria. Genetics 150:1115–1123.

Chapman, A. J., P. N. Durman, and R. B. Rickards. 1995. Rhabdopleuran hemichordates: new fossil forms and review. Proc. Geol. Assoc. 106:293–303.

Chen, J., and C. Li. 1997. Early Cambrian chordate from Chenjiang, China. Pp. 257–273 in The Cambrian explosion and the fossil record (J. Chen, Y. N. Cheng, and H. V. Iten, eds.), vol. 10. National Museum of Natural Science, Taichung.

Chen, J. Y., J. Dzik, G. D. Edgecombe, L. Ramskold, and G. Q. Zhou. 1995. A possible early Cambrian chordate. Nature 377:720–722.

Clark, A. H. 1915–1950. A monograph of the existing crinoids. U.S. Natl. Mus. Bull. 82 Pt. 1 (1915), 406 pp., Pt. 2 (1921), 795 pp., Pt. 3 (1931), 816 pp., Pt. 4a (1941) 603 pp., Pt. 4b (1947), 473 pp., Pt. 4c (1950), 383 pp.

Clark, A. H., and A. M. Clark. 1967. A monograph of the existing crinoids, Pt. 5. U.S. Natl. Mus. Bull. 82:1–860.

Cripps, A. P. 1991. A cladistic analysis of the cornutes, stem chordates. Zool. J. Linn. Soc. 100:333–366.

Crowther, P. R. 1981. The fine structure of graptolite periderm. Spec. Pap. Palaeontol. 26:1–119.

David, B., B. Lefevre, R. Mooi, and R. Parsley. 2000. Are homalozoans echinoderms? An answer from the extraxial-axial theory. Paleobiology 26:529–555.

David, B., and R. Mooi. 1996. Embryology supports a new theory of skeletal homologies for the phylum Echinodermata. C. R. Acad. Sci. Paris 319:577–584.

David, B., and R. Mooi. 1999. Comprendre les echinodermes: la contribution du modele extraxial-axial. Bull. Soc. Geol. Fr. 170:91–101.

Davidson, E. H. 2001. Genomic regulatory systems: development and evolution. Academic Press, San Diego.

De Rosa, R, J. K. Grenier, T. Andreeva, C. E. Cook, and A. Adoutte. 1999. Hox genes in brachiopods and priapulids and protostome evolution. Nature 399:772–776.

Dilly, P. N. 1993. *Cephalodiscus graptolitoides* sp. nov. a probable extant graptolite. J. Zool. 229:69–78.

Dominguez, P., A. G. Jacobson, and R. P. J. Jefferies. 2002. Paired gill slits in a fossil with a calcite skeleton. Nature 417:841–844.

Fell, H. B. 1967. Echinoderm ontogeny. Pp. S60–S85 *in* Treatise on invertebrate paleontology, Pt. S: Echinodermata 1 (R. C. Moore, ed.). Geological Society of America and University of Kansas Press, Boulder, CO.

Furlong, R. F., and P. W. H. Holland. 2002. Bayesian phylogenetic analysis supports monophyly of ambulacraria and of cyclostomes. Zool. Sci. 19:593–599.

Field, K. G., M. T. Ghislen, D. J. Lane, G. J. Olsen, N. R. Pace, E. C. Raff, and R. A. Raff. 1988. Molecular phylogeny of the animal kingdom. Science 239:748–753.

Gale, A. S. 1987. Phylogeny and classification of the Asteroidea (Echinodermata). Zool. J. Linn. Soc. 89:107–132.

Gans, C. 1989. Stages in the origins of vertebrates: analysis by means of scenarios. Biol. Rev. 64:221–268.

Garstang, W. 1928. The morphology of the Tuinicata and its bearing on the phylogeny of the Chordata. Q. J. Microsc. Sci. 72:51–187.

Gellon, G., and W. McGinnis. 1998. Shaping animal body plans in development and evolution by modulation of Hox expression patterns. Bioessays 20:116–125.

Gerhart, J., and M. Kirschner. 1997. Cells, embryos, and evolution: towards a cellular and developmental understanding of phenotypic variation and evolutionary adaptability. Blackwell Science, Malden, MA.

Gilliland, P. M. 1993. The skeletal morphology, systematics and evolutionary history of holothurians. Spec. Pap. Palaeontol. 47:1–147.

Giorgi, C. de, A. Martiradonna, C. Lanave, and C. Saccone. 1996. Complete sequence of the mitochondrial DNA in the sea urchin *Arbacia lixula*: conserved features of the echinoid mitochondrial genome. Mol. Phylogenet. Evol. 5:323–332.

Giray, C., and G. M. King. 1996. Protoglossus graveolens, a new hemichordate (Hemichordata: Enteropneusta: Harrimanidae) from the northwest Atlantic. Proc. Biol. Soc. Wash. 109:430–445.

Giribet, G., D. L. Distel, M. Polz, W. Sterrer, and W. C. Wheeler. 2000. Triploblastic relationships with emphasis on the acoelomates and the position of Gnathostomulida, Cycliophora, Plathelminthes and Chaetognatha: a combined approach of 18S rDNA sequences and morphology. Syst. Biol. 49:539–562.

Halanych, K. 1996. Testing hypotheses of chaetognath origins: long branches revealed by 18S ribosomal DNA. Syst. Biol. 45:223–246.

Halanych, K. M., J. D. Bacheller, A. M. A. Aguinaldo, S. M. Liva, D. M. Hillis, and J. A. Lake. 1995. Evidence from 18S ribosomal DNA that the lophophorates are protostome animals. Science 267:1641–1643.

Harold, A. S., and M. Telford. 1990. Systematics, phylogeny and biogeography of the genus *Mellita* (Echinoidea: Clypeasteroida). J. Nat. Hist. 24:987–1026.

Hart, M. W., M. Byrne, and M. J. Smith. 1997. Molecular phylogenetic analysis of life-history evolution in asterinid starfish. Evolution 51:1848–1861.

Heinzeller, T., and U. Welsch. 2001. The echinoderm nervous system and its phylogenetic interpretation. Pp. 41–75 *in* Brain evolution and cognition (G. Roth and M. F. Wullimann, eds.). John Wiley and Sons, New York.

Hess, H., W. I. Ausich, C. E. Brett, and M. J. Simms. 1999. Fossil crinoids. Cambridge University Press, Cambridge.

Huelsenbeck, J. P., and F. Ronquist. 2001. MrBayes: Bayesian inference of phylogeny. Computer program. Available: http://morphbank.ebc.uu.se/mrbayes. Last accessed 17 December 2003.

Huelsenbeck, J. P., F. Ronquist, R. Nielsen, and J. P. Bollback. 2001. Bayesian inference of phylogeny and its impact on evolutionary biology. Science 294:2310–2314.

Hyman, L. 1955. The invertebrates: Echinodermata, vol. 4. McGraw Hill, New York.

Hyman, L. 1959. Phylum Hemichordata. Pp. 72–207 *in* The invertebrates, vol. 5. McGraw Hill, New York.

Janies, D. 2001. Phylogenetic relationships of extant echinoderm classes. Can. J. Zool. 79:1232–1250.

Janies, D., and R. Mooi, 1999. *Xyloplax* is an asteroid. Pp. 311–16 *in* Echinoderm research 1998 (M. D. Candia Carnevali and B. Francesco, eds.). A. A. Balkema, Rotterdam.

Janies, D. A., and L. R. McEdward. 1993. Highly derived coelomic and water-vascular morphogenesis in a starfish with pelagic direct development. Biol. Bull. 185:56–76.

Jeffery, C. H., R. B. Emlet, and D. T. L. Littlewood. 2003. Phylogeny and evolution of development in temnopleurid echinoids. Mol. Phylogen. Evol. 28:99–118.

Jensen, S., B. Z. Saylor, J. G. Gehling, and G. J. B. Germs. 2000. Complex trace fossils from the terminal Proterozoic of Namibia. Geology 28:143–146.

Kapp, H. 2000. The unique embryology of Chaetognatha. Zool. Anz. 239:263–266.

Kerr, A. M. 2001. Phylogeny of the apodan holothurians (Echinodermata) inferred from morphology. Zool. J. Linn. Soc. 133:53–62.

Kerr, A. M. 2003. Holothuroidea. Available: http://tolweb.org/tree?group=Holothuroidea&contgroup=Echinodermata. Last accessed November 2003.

Kerr, A. M., and J. H. Kim. 1999. Bi-penta-bi-decaradial symmetry: a review of evolutionary and developmental trends in Holothuroidea (Echinodermata). J. Exp. Zool. (Mol. Dev. Evol.) 285:93–103.

Kerr, A. M., and J. H. Kim. 2001. Phylogeny of Holothuroidea (Echinodermata) inferred from morphology. Zool. J. Linn. Soc. 133:63–81.

Knott, K. E., and G. A. Wray. 2000. Controversy and consensus in asteroid systematics: new insights to ordinal and familial relationhips. Am. Zool. 40:382–392.

Lafay, B., A. B. Smith, and R. Christen. 1995. A combined morphological and molecular approach to the phylogeny of asteroids (Asteroidea: Echinodermata). Syst. Biol. 44:190–208.

Littlewood, D. T. J., and A. B. Smith. 1995. A combined morphological and molecular phylogeny for echinoids. Philos. Trans. R. Soc. Lond. B 347:213–234.

Littlewood, D. T. J., A. B. Smith, K. A. Clough, and R. H. Ensom. 1997. The interrelationships of the echinoderm classes: morphological and molecular evidence. Biol. J. Linn. Soc. 61:409–438.

Littlewood, D. T. J., M. J. Telford, K. A. Clough, and K. Rohde. 1998. Gnathostomulida—an enigmatic metazoan phylum from both morphological and molecular perspectives. Mol. Phylogenet. Evol. 9:72–79.

Long, S., and M. Byrne. 2001. Evolution of the echinoderm Hox gene cluster. Evol. Dev. 3:302–311.

Lowe, C. J., and G. A. Wray. 1997. Radical alterations in the roles of homeobox genes during echinoderm evolution. Nature 389:718–722.

McEdward, L. R., and B. G. Miner. 2001. Larval and life-cycle patterns in echinoderms. Can. J. Zool. Rev. Can. Zool. 79:1125–1170.

Mindell, D. P., M. D. Sorenson, and D. E. Dimcheff. 1998. Multiple independent origins of mitochondrial gene order in birds. Proc. Natl. Acad. Sci. USA 79:7195–7199.

Mooi, R., and B. David. 1996. Phylogenetic analysis of extreme morphologies: deep-sea holasteroid echinoids. J. Nat. Hist. 30:913–953.

Mooi, R., and B. David. 1997. Skeletal homologies of echinoderms. Paleont. Soc. Pap. 3:305–336.

Mooi, R., and B. David. 2000. What a new model of skeletal homologies tells us about asteroir evolution. Am. Zool. 40:326–339.

Mortensen, T. 1927. Handbook of the echinoderms of the British Isles. Oxford University Press, London.

Nielsen, C. 1995. Animal evolution. Interrelationships of the living phyla. Oxford University Press, Oxford.

Nielsen, C. 2001. Animal evolution. Interrelationships of the living phyla. 2nd ed. Oxford University Press, Oxford.

Nielsen, C., N. Scharff, and D. Eibye-Jacobsen. 1996. Cladistic analyses of the animal kingdom. Biol. J. Linn. Soc. 57:385–410.

Ogasawara, M., H. Wada, H. Peters, and N. Satoh. (1999). Developmental expression of *Pax1/9* genes in urochordate and hemichordate gills: insight into function and evolution of the pharyngeal epithelium. Development 126:2539–2550.

Pawson, D. L. 1966. Phylogeny and evolution of holothuroids. Pp. U641–U646 *in* Treatise on invertebrate paleontology. (U) Echinodermata 3 (R. C. Moore, ed.). University of Kansas and Geological Society of America, Boulder, CO.

Peterson, K. J. 1995. A phylogenetic test of the Calcichordate scenario. Lethaia 28:25–38

Peterson, K. J., C. Arenas-Mena, and E. H. Davidson. 2000. The A/P axis in echinoderm ontogeny and evolution: evidence from fossils and molecules. Evol. Dev. 2:93–101.

Peterson, K. J., and E. H. Davidson. 2000. Regulatory evoilution and the origin of the bilaterians. Proc. Natl. Acad. Sci. USA 97:4430–4433.

Peterson, K. J., and D. J. Eernisse. 2001. Animal phylogeny and the ancestry of bilaterians: inferences from morphology and 18S rDNA gene sequences. Evol. Dev. 3:170–205.

Raff, R. A. 1996. The shape of life. genes, development, and the evolution of animal form. University of Chicago Press, Chicago.

Raff, R. A., K. G. Field, M. T. Ghiselin, D. J. Lane, G. L. Olsen, N. R. Pace, A. L. Parks, B. A. Parr, and E. C. Raff. 1988. Molecular analysis of distant phylogenetic relationships in echinoderms. Pp. 29–41 *in* Echinoderm phylogeny and evolutionary biology (C. R. C. Paul, and A. B. Smith, eds.). Oxford University Press, Oxford.

Rigby, S. 1993. Graptolites come to life. Nature 362:209–210.

Ruppert, E. E. 1990. Structure, ultrastructure and function of the neural gland complex of *Ascidia interrupta* (Chordata, Ascidiacea): clarification of hypotheses regarding the evolution of the vertebrate pituitary. Acta Zool. 71:135–149.

Ruppert, E. E. 1997. Introduction: microscopic anatomy of the notochord, heterochrony, and chordate evolution. Pp. 1–13 *in* Microscopic anatoimy of invertebrates, Vol. 15: Hemichordata (F. W. Harrison, and E. E. Ruppert, eds.). Wiley-Liss, New York.

Ruppert, E. E., and E. J. Balser. 1986. Nephridia in the larvae of hemichordates and echinoderms. Biol. Bull. 171:188–196

Schaeffer, B. 1987. Deuterostome monophyly and phylogeny. Evol. Biol. 21:179–235.

Schram, F. R. 1991. Cladistic analysis of metazoan phyla and the placement of fossil problematica. Pp. 35–46 *in* The early evolution of metazoa and the significance of problematic taxa (A. M. Simonetta, and S. Conway Morris, eds.). Cambridge University Press, Cambridge.

Schram, F. R. 1997. Of cavities—and kings. Contrib. Zool. 67:143–150.

Scouras, A., and M. J. Smith. 2001. A novel mitochondrial gene order in the crinoid echinoderm *Florometra serratissima*. Mol. Biol. Evol. 18:61–73.

Segler, K., H. Rahmann, and H. Rösner. 1978. Chemotaxonomical investigations of the occurrence of sialic acids in Protostomia and Deuterostomia. Biochem. Syst. Ecol. 6:87–93.

Shu, D. G., X. Zhang, and L. Chen. 1996. Reinterpretation of *Yunnanozoon* as the earliest known hemichordate. Nature 380:428–430.

Simms, M. J. 1988. The phylogeny of post-Palaeozoic crinoids. Pp. 269–286 *in* Echinoderm phylogeny and evolutionary biology (C. R. C. Paul, and A. B. Smith, eds.). Oxford University Press, Oxford.

Simms, M. J. 1999. Systematics, phylogeny and evolutionary history. Pp. 31–40 *in* Fossil crinoids (H. Hess, W. I. Ausich, C. E. Brett, and M. J. Simms, eds.). Cambridge University Press, Cambridge.

Smiley, S. 1988. The phylogenetic relationships of holothurians: a cladistic analysis of the extant echinoderm classes. Pp. 69–84 *in* Echinoderm phylogeny and evolutionary biology (C. R. C. Paul, and A. B. Smith, eds.). Oxford University Press, Oxford.

Smith, A. B. 1984. Echinoid palaeobiology. George Allen and Unwin, London.

Smith, A. B. 1988. Phylogenetic relationships, divergence times, and rates of molecular evolution for camarodont sea urchins. Mol. Biol. Evol. 5:345–365.

Smith, A. B. 1997. Echinoderm phylogeny: how congruent are morphological and molecular estimates? Paleont. Soc. Pap. 3:337–355.

Smith, A. B. 2001. Probing the cassiduloid origins of clypeasteroid echinoids using stratigraphically restricted parsimony analysis. Paleobiology 27:392–404.

Smith, A. B. 2003a. The echinoid directory. Available: http://www.nhm.ac.uk/paleontology/echinoids/index.html. Last accessed 17 December 2003.

Smith, A. B. 2003b. Supporting matrices. Available: http://puffin.nhm.ac.uk:81/iw-mount/default/main/Internet/WORKAREA/palaeontology/I&p/abs/abs.html.

Smith, A. B., D. T. J. Littlewood, and G. A. Wray. 1995a. Comparing patterns of evolution: larval and adult life-history stages and small ribosomal RNA of post-Palaeozoic echinoids. Philos. Trans. R. Soc. Lond. B 349:11–18.

Smith, A. B., G. L. Patterson, and B. Lafay. 1995b. Ophiuroid phylogeny and higher taxonomy: morphological, molecular and palaeontological perspectives. Zool. J. Linn. Soc. 114:213–243.

Smith, A. B., and J. Savill. 2002. *Bromidechinus*, a new Middle Ordovician echinozoan (Echinodermata), and its bearing on the early history of echinoids. Trans. R. Soc. Edinb. Earth Sci. 92:137–147.

Smith, M. J., A. Arndt, S. Gorski, and E. Fajber. 1993. The phylogeny of echinoderm classes based on mitochondrial gene arrangements. J. Mol. Evol. 36:545–554.

Smith, M. J., D. K. Banfield, K. Doteval, S. Gorski, and D. J. Kowbel. 1989. Gene arrangement in sea star mitochondrial DNA demonstrates a major inversion event during echinoderm evolution. Gene 76:181–185.

Strathmann, R. R. 1988. Larvae, phylogeny, and von Baer's Law. Pp. 53–68 in Echinoderm phylogeny and evolutionary biology (C. R. C. Paul and A. B. Smith, eds.). Clarendon Press, Oxford.

Sumrall, C. D. 1997. The role of fossils in the phylogenetic reconstruction of Echinodermata. Paleont. Soc. Pap. 3:267–288.

Swofford, D. 2001. PAUP* 4.0. Sinauer Associates, Sunderland, MA.

Tagawa, K., N. Satoh, and T. Humphreys. 2001. Molecular studies of hemichordate development: a key understanding the evolution of bilateral animals and chordates. Evol. Dev. 3:443–454.

Telford, M. J. 2000. Evidence for the derivation of the *Drosophila fushi tarazu* gene from a Hox gene orthologous to lophotrochozoan *Lox5*. Curr. Biol. 10:349–352.

Telford, M. J., and P. W. H. Holland. 1993. The phylogenetic affinities of the chaetognaths: a molecular analysis. Mol. Biol. Evol. 10:660–676.

Thompson, J., and F. Jeanmougin. 2001. Clustal W: multiple sequence alignment program (ver. 1.8). Available: www.ebi.ac.uk/clustalW.

Turbeville, J. M., J. R. Schulz, and R. A. Raff, 1994. Deuterostome phylogeny and the sister group of chordates: evidence from molecules and morphology. Mol. Biol. Evol. 11:648–655.

Twitchett, R. J. 1996. The resting trace of an acorn-worm (Class: Enteropneusta) from the Lower Triassic. J. Paleontol. 70:128–131.

Wada , H., M. Komatsu, and N. Satoh. 1996. Mitochondrial rDNA phylogeny of the Asteroidea suggests the primitiveness of the Paxillosida. Mol. Phylogenet. Evol. 6:97–106.

Wada, H., and N. Satoh. 1994. Details of the evolutionary history from invertebrates to vertebrates, as deduced from the sequences of 18S rDNA. Proc. Natl. Acad. Sci. USA 91:1801–1804.

Warren, L. 1963. The distribution of sialic acids in nature. Comp. Biochem. Physiol. 10:153–171.

Wheeler, W., and D. Gladstein. 2000. POY. Computer program. Available: ftp.amnh.org/pub/molecular/poy/. Last accessed 17 December 2003.

Winchell, C., J. Sullivan, C. Cameron, B. Swalla, and J. Mallatt. 2002. Evaluating hypotheses of deuterostome phylogeny and chordate evolution with new LSU and SSU ribosomal DNA data. Mol. Biol. Evol. 19:762–776.

Woodwick, K. H., and T. Sensenbaugh. 1985. *Saxipendium coronatum*, new genus, new species (Hemichordata: Enteropneusta): the unusual spaghetti worms of the Galapagos rift hydrothermal vents. Proc. Biol. Soc. Wash. 98:351–365.

Wray, G. A. 1992. The evolution of larval morphology during the post-Paleozoic radiation of echinoids. Paleobiology 18:258–287.

Wray, G. A. 1996. Parallel evolution of nonfeeding larvae in echinoids. Syst. Biol. 45:308–322.

Wray, G. A., and C. J. Lowe. 2000. Developmental regulatory genes and echinoderm evolution. Syst. Biol. 49:28–51.

Zrzavý, J., S. Mihulka, P. Kepka, A. Bezdek, and D. Tietz. 1998. Phylogeny of the Metazoa based on morphological and 18S ribosomal DNA evidence. Cladistics 14:249–285.

Timothy Rowe

Chordate Phylogeny and Development

Chordata, our own lineage (fig. 23.1), belongs to the successively more inclusive clades Deuterostomata, Bilateria, Metazoa, and so forth. The organization of chordates is distinctively different from that of its metazoan relatives, and much of this distinction is conferred by unique mechanisms of development (Slack 1983, Schaeffer 1987). Throughout chordate history, modulation and elaboration of developmental systems are persistent themes underlying diversification. Only by understanding how ontogeny itself evolved can we fully apprehend chordate history, diversity, and our own unique place in the Tree of Life. My goal here is to present a contemporary overview of chordate history by summarizing current views on relationships among the major chordate clades in light of a blossoming understanding of molecular, genetic, and developmental evolution, and a wave of exciting new discoveries from deep in the fossil record.

Chordates comprise a clade of approximately 56,000 named living species that includes humans and other animals with a notochord—the embryological precursor of the vertebral column. Chordate history can now be traced across at least a half billion years of geological time, and twice that by some estimates (Wray et al. 1996, Ayala et al. 1998, Bromham et al. 1998, Kumar and Hedges 1998, Hedges 2001). Chordates are exceptional among multicellular animals in diversifying across eight orders of size magnitudes and inhabiting virtually every terrestrial and aquatic environment (McMahon and Bonner 1985). New living chordate species are still being discovered both through traditional explorations and as molecular analyses discover cryptic taxa in lineages whose diversities were thought to be thoroughly mapped. But it is unknown whether the pace of discovery is now keeping up with the pace of extinction, which is accelerating across most major chordate clades in the wake of human population growth (Dingus and Rowe 1998).

Many chordate clades have long been recognized by characteristic adult features, for instance, birds by their feathers, mammals by their hair, or turtles by their shells. But owing in large part to such distinctiveness, few adult morphological features have been discovered that decisively resolve the relationships among the chordate clades, and even after 300 years of study broad segments of chordate phylogeny remain *terra incognita*.

Much of the hypothesized hierarchy of higher level chordate relationships has been deduced from paleontology and developmental biology (Russell 1916). Thanks to the advent of phylogenetic systematics, both fields are expressing resurgent interest and progress on the question of chordate phylogeny. And, as they are becoming integrated with molecular systematic analyses, a fundamental new understanding of chordate evolution and development is emerging.

In most other metazoans, the adult fate of embryonic cells is determined very early in ontogeny. However as chordate ontogeny unfolds, the fates of embryonic cells are plastic for a longer duration. Chordate cells differentiate as signals pass

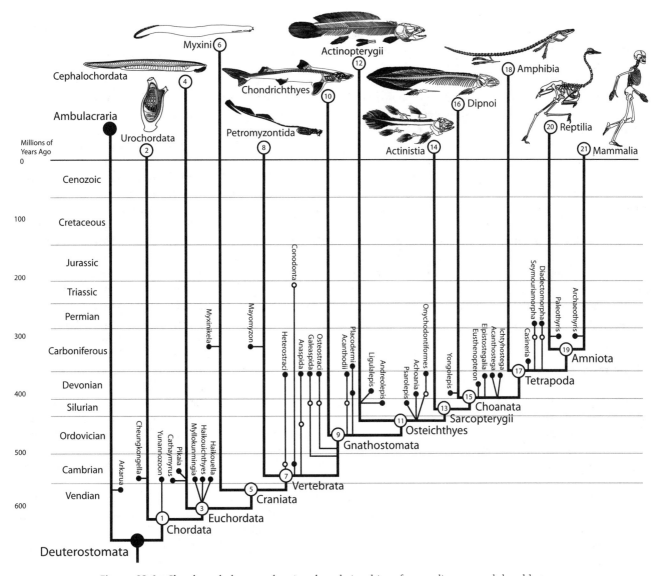

Figure 23.1. Chordate phylogeny, showing the relationships of extant lineages and the oldest fossils, superimposed on a geological time column. Nodal numbers are keyed to text headings.

between adjacent cells and tissues during the integration of developing cell lineages into functioning tissues, organs, and organ systems. Seemingly subtle modulations in early ontogeny by this information exchange system have occurred many times over chordate history to yield cascades of subsequent developmental effects that underlie chordate diversity (Hall 1992). Molecular and developmental genetic studies are now revealing the intricate details of this unique, hierarchical system of information transfer as genes are expressed in cells and tissues in early ontogeny. These analyses, moreover, generate data that possess a recoverable phylogenetic signal and are yielding fundamental insights into the evolution of development.

An important conclusion already evident is that major innovations in chordate design were generally derived from preexisting genetic and developmental pathways, whose alteration transformed ancestral structures into distinctive new features with entirely different adult functions (Shubin et al. 1997). Increase in numbers of genes was a primary mediator of this change, and the inductive nature of chordate development amplified that change via epigenesis, which occurs as familiar physical forces and dynamic processes interact with the cells and tissues of a developing organism. These include gravity, adhesion, diffusion, mechanical loading, electrical potentials, phase separations, differential growth among tissues and organs, and many others (Rowe 1996a, 1996b). Morphogenic and patterning effects are the developmental outcomes of these recognized physical phenomena, because they affect interactions among virtually all developing cells, tissues, and organs (Newman and Comper 1990). In the inductive environment of chordate ontogeny, epigenesis has been especially influential, triggering its own cascades of rapid and nonlinear developmental change. Understanding how epigenesis mediates the genetic blueprint of ontogeny is fundamental to understanding how such diverse chordates as sea squirts, coelacanths, and humans emerged from their unique common ancestor.

Recognizing that most biologists reading this volume study living organisms, the focus below is on extant taxa. However, extinct taxa are discussed as well, and their inclusion helps to emphasize the timing of origins of the major extant chordate clades and to acknowledge the diversity and antiquity of the lineages of which they are a part. Moreover, the framework of chordate relationships presented below came from the simultaneous consideration of all available evidence. In resolving several parts of the chordate tree discussed below, evidence afforded by fossils proved more important than that derived from living species (Gauthier et al. 1988a, 1989, Donoghue et al. 1989).

Taxonomic Names, Ancestry, and Fossils

Older views of chordate relationships make reference to groups united on general similarity or common gestalt. In contrast, the names used below designate lineages whose members appear to be united by common ancestry (de Queiroz and Gauthier 1992). To avoid ambiguity, the meanings of these names are defined in terms of particular ancestors of two or more living taxa (i.e., node-based or crown clade names). I follow an arbitrary but useful narrative convention in specifying the crown clade names used below in terms of their most recent common ancestry with humans. For example, the name Chordata refers to the clade stemming from the last common ancestor that humans share with living tunicates and lancelets; the name Vertebrata designates the clade stemming from the last common ancestor that humans share with lampreys; and so on (fig. 23.1). This is arbitrary in the sense that many other possible living specifiers among amniotes (viz., birds, turtles, crocodilians, lizards) in place of humans would designate the same clades.

Stem-based names are used in reference to a node or terminal taxon, plus all extinct taxa that are more closely related to it than to some other node or terminal taxon. In the interests of simplifying the complex taxonomy that evolved under the Linnaean system, I follow a convention now gaining popularity that employs the prefix "Pan-" to designate stem + crown lineages (Gauthier and de Queiroz 2001). For example, Pan-Mammalia refers to the clade Mammalia, plus all extinct species closer to Mammalia than to its extant sister taxon Reptilia. The clade Pan-Vertebrata includes Vertebrata plus all extinct taxa closer to Vertebrata than to hagfishes, and so forth.

Chordate Relationships

Node 1. The Chordates (Chordata)

Chordata (fig. 23.1) comprise the lineage arising from the last common ancestor that humans share with tunicates and lancelets. Tunicates are widely regarded as the sister taxon to all other chordates (Gegenbaur 1878, Schaeffer 1987, Cameron et al. 2000), and tunicate larvae are commonly viewed as manifesting the organization of the adult ancestral chordate (e.g., Meinertzhagen and Okamura 2001). But some systematists contend that lancelets are the more distant outgroup (Løvtrup 1977, Jeffries 1979, 1980, 1986, Jeffries and Lewis 1978). The controversy stems in part from the fact that living adult tunicates are small and built from a small number of cells. Even their larvae appear highly divergent from other living chordate larvae. It now seems likely that they were secondarily simplified in having lost half or more of the *Hox* genes from the single cluster that was probably present in deuterostomes ancestrally (Holland and Garcia-Fernàndez 1996), hence, too, the loss of adult structures governed by these genes. As adults, tunicates are derived in losing the coelom and hindgut (Holland and Chen 2001) and are speculated to be pedomorphic in having lost segmentation (Holland and Garcia-Fernàndez 1996). One character shared by tunicates and craniates, to the exclusion of lancelets, is expression of the *Pax 2/5/8* gene in a region of the developing brain known as the isthmocerebellar-midbrain-hindbrain boundary. The lack of *Pax 2/5/8* expression in lancelets implies either secondary loss, or independent expression in tunicates and craniates (Butler 2000), or that tunicates share closer common ancestry with other chordates than do lancelets. Having separated from other chordates by at least a half-billion years ago (Wray et al. 1996, Bromham et al. 1998, Kumar and Hedges 1998, Hedges 2001), and without a useful fossil record (below), relationships among these chordates must be viewed as tenuous (Gauthier et al. 1988a, Donoghue et al. 1989). More for narrative convenience than conviction, I follow current convention in treating tunicates as sister lineage to all other chordates.

Chordate Characters

The notochord.   The namesake feature of chordates is a premiere example of embryonic induction and patterning, in which differentiation of the embryo along a dorsoventral axis launches a cascade of subsequent developmental events (Slack 1983, Schaeffer 1987). "Dorsalization" is controlled by the *Hedgehog* gene and signaling by bone morphogenesis protein, or BMP (Shimeld and Holland 2000). As in other bilaterians, chordates develop from three primary embryonic layers. These are the outer ectoderm, the inner endoderm, and the mesoderm, which arises from cells that migrate between the inner and outer layers. Chordate mesoderm develops in the upper hemisphere of the embryonic gastrula, its identity being induced partly as its cells stream across the dorsal lip of the primordial opening (blastopore) into the inner cavity (archenteron) of the embryo, and partly by signaling from endoderm at the equator of the embryo (Hall 1992). Mesoderm cells reaching the dorsal midline condense into a strip of cells known as chordamesoderm, which later differentiates to become the notochord. The notochord in turn induces overlying ectoderm to form the dorsal neural plate, triggering another morphogenic chain of events as the chordate central nervous system (CNS) differentiates and begins to grow. In most chordates, the mesoderm immediately adjacent to the notochord takes on special properties, as does the ectoderm immediately adjacent to the neural plate. Elaboration of these dorsal structures is tied closely to evolution of the organs of information acquisition and integration, as well as to locomotion.

The chordate central nervous system.   Induction of a dorsal neural plate is directed by the underlying chordamesoderm (above). This is the first step of neurulation, in which the nervous system arises, becomes organized, and helps direct the integration of other parts of the developing embryo. During neurulation, longitudinal neural folds arise along the edges of the neural plate, perhaps under the direction of the adjacent mesoderm (Jacobson 2001), and meet on the midline to enclose a space that initially lay entirely outside of the embryo. This "hollow" comprises the adult ventricular system of the brain and central canal of the spinal cord. It is lined with ciliated ependymal cells and its lumen fills with cerebrospinal fluid. This original "periventricular" layer becomes the primary region from which subsequent neural cells arise in the brain (Butler and Hodos 1996).

Molecular signaling during neurulation also produces anteroposterior regionalization in chordate embryos. The rostral end of the central nerve cord swells to form the brain, which differentiates into three regions that express distinct gene families and which have distinct adult fates. The rostral-most (diencephalic) domain of the neural tube expresses the *Otx* gene family and is connected to specialized light-sensitive cells. Behind this is a caudal (hindbrain–spinal cord) division, in which *Hox* genes are active and which receives nonvisual sensory inputs. Between the two lies an intermediate region marked by expression of the *Pax 2/5/8* patterning gene that is more problematically compared with a region known as the isthmocerebellar-midbrain-hindbrain boundary and involves the ear (Meinertzhagen and Okamura 2001, Butler 2000, Shimeld and Holland 2000). *Pax 2/5/8* is expressed in tunicates and craniates, but not lancelets (below).

Other bilaterians have a longitudinal nerve cord and brain but it is ventrally positioned; hence, biologists long maintained that the chordate dorsal nerve cord arose independently. However, both brains express orthologous homeobox genes in similar spatial patterns. For instance, the fruit fly has a regionalized neural tube with similarities in rostrocaudal and mediolateral specification to chordates (Arendt and Nübler-Jung 1999, Nielsen 1999, Butler 2000; for alternative view, see Gerhart 2000). Its rostral brain is specified by

the regulatory gene *Orthodenticle*, a homologue to the chordate *Otx* family genes, and it receives input from paired eyes. This suggests a common blueprint. Biologists long found it difficult to accept the two nerve cords as homologous owing to their different positions relative to the mouth, but it now appears that the deuterostome mouth is a new structure and not homologous to the mouth in protostomes (Nielsen 1999).

Special sensory organs of the head.   An eye and ear of unique design were probably present in chordates ancestrally. The master control gene *Pax6* is expressed during early development in paired neural photoreceptors—eyes—in chordates and many other bilaterians. Paired eyes and ears, however rudimentary, were almost certainly present in chordates ancestrally (Gehring 1998). However, *Pax6* expression in chordates is manifested in eye morphogenesis that follows a unique hierarchy of pathways and inductive signals, and in which considerable diversity evolved among the different chordates lineages. Living tunicates, lancelets, and hagfish each appear uniquely derived, leaving equivocal exactly what type of eye was present in chordates ancestrally. In tunicates, the larval eye forms a small vesicle that contains a sunken, pigmented mass. Internal to the pigment lies a layer of cells that are directed radially toward it, and overlying the pigment are two hemispherical refractive layers (Gegenbaur 1878). These same relationships occur in all other chordates. However, in tunicates an optic vesicle is present only in larvae and is generally unpaired. Nevertheless, it is an outgrowth of the *Otx*-expressing region of the forebrain and it expresses *Pax6*, as do the paired eyes of vertebrates and unlike the median pineal eye (Meinertzhagen and Okamura 2001). In lancelets there is a single, median frontal eye, which also expresses *Pax6*, and like the bilateral eyes of vertebrates it is linked with cells in the primary motor center (Lacalli 1996a, 1996b, Butler 2000). In the case of lancelets, the forward extension of the notochord may be implicated in secondary fusion of the single eye. Hagfish have paired eyes, but they are poorly developed compared with most vertebrates.

The chordate ear or otic system eventually differentiated into the organs of both balance and hearing in vertebrates. Adult tunicates have sensory hair cells that support a pigmented otolith and are grouped into gelatinous copular organs located in the atrium of the adult. These cells express members of the *Pax 2/5/8* gene family, as do the otic placodes in craniates (but not lancelets), and in early development they are topographically similar to craniate otic placodes. However, placodes themselves are not yet present. Similar gene expression, cellular organization, and topography point to the probable homology of the otic organ in all chordates (Shimeld and Holland 2000, Jeffries 2001, Meinertzhagen and Okamura 2001).

Hormonal glands.   Two hormonal glands arose in chordates ancestrally to exert novel control over growth and metabolism. The pituitary is a compound structure that forms via the interaction between neurectoderm, which descends from the developing brain toward the roof of the pharynx, and oral ectoderm that folds inward to line the inside of the mouth. Ectoderm forms Rathke's pouch and becomes the glandular part of the pituitary, whereas neural tissue from the floor of the diencephalon becomes its infundibular portion. The infundibulum is present in lancelets and craniates, but its homologue in tunicates is unclear. However, in tunicates the homologue of the glandular portion, known as the neural gland, lies in the same position with respect to both brain and pharyngeal roof (Barrington 1963, 1968, Maisey 1986).

The second hormonal gland, the endostyle, develops in a groove in the floor of the larval pharynx in tunicates, lancelets, and in larval lampreys. Its cells form thyroid follicles that secrete iodine-binding hormones. Its homologue in gnathostomes is probably the thyroid gland, which also develops in a median out-pocketing in the floor of the pharynx, and also forms thyroid follicles that secrete iodine-binding hormones (Schaeffer 1987). Thyroid hormone production is controlled in large measure by the pituitary gland and affects growth, maintenance of general tissue metabolism, reproductive phenomena, and in some taxa metamorphosis.

Tadpole-shaped larva. Unlike the ciliated egg-shaped larvae of hemichordates and echinoderms, the chordate larva is tadpole shaped, with a swollen rostral end and a muscular tail. The rostral end houses the brain, beneath which lie the rostral end of the notochord, and the pharynx and gut tube. Behind the pharynx is a tail equipped with muscle deriving from caudal mesoderm (Maisey 1986, Schaeffer 1987). Although lacking tails as adults, the larvae of many species have tails of comparatively simple construction with muscle that form bilateral bands, in contrast to the segmental muscle blocks found in euchordates (below). A recent study of tailed and tailless tunicate larvae (Swalla and Jeffery 1996) found that the *Manx* gene is expressed in the cells of the tailed form but it is down-regulated in the tail-less species, and that complete loss of the tail can be attributed to disrupted expression of the single gene. Whether *Manx* was central to the origin of the tail in chordates is unknown, but this study highlights the potential genetic simplicity underlying complex adult structures.

Pan-Chordata

Although an extensive fossil record is known for many clades lying within Chordata, no fossils are known at present that lie with any certainty on its stem.

Node 2. The Tunicates or Sea Squirts (Urochordata)

Chordate species all can be distributed between the tunicates and euchordates, its two principal sister clades (fig. 23.1). The tunicates comprise a diverse marine clade that includes roughly 1300 extant species distributed among the sessile ascidians, and the pelagic salps and larvaceans (Jamieson 1991). Tunicate monophyly is well supported (Gegenbaur

1878, Maisey 1986). As adults, the tunicate body is enclosed within the tunic, an acellular membrane made of cellulose-like tunicin. It is derived from ectoderm, and in tunicates it may contain both amorphous and crystalline calcium carbonate spicules (Aizenberg et al. 2002). Echinoderms possess crystalline calcium in ectodermal structures, raising the question of whether biomineralization was present in deuterostomes ancestrally (see below). The tunic presents an outwardly simple body, but it cloaks a much more complex and derived organism. The pharynx is perforated by two pairs of slits and is enormously enlarged for suspension feeding. The pharynx size obliterates the coelom, a cavity inside the body walls that surrounds the gut in tunicate larvae and most adult chordates. Unique incurrent and excurrent pores supply a stream of water through the huge pharynx, which in some species serves in locomotion. All tunicates are mobile as larvae, but not all species have larval tails. The pelagic salps and larvaceans are thought to be more basal and to reflect the primitive adult lifestyle.

Pan-Urochordata

The fossil record of tunicates is sparse and tentative, but potentially long. The oldest putative tunicate, *Cheungkongella ancestralis*, from the Early Cambrian of China (Shu et al. 2001a) is known from a single specimen. It evidently preserves a two-fold division of the body into an enlarged pharyngeal region with pharyngeal openings, a large oral siphon surrounded by short tentacles, and a smaller excurrent siphon. The body appears wholly enclosed in a tuniclike outer covering. It has short tail-like attachment structure, a derived feature placing *Cheungkongella* among crown tunicates. This fossil, if properly interpreted, marks the Early Cambrian as the minimum age of divergence of tunicates from other chordates and implies a Precambrian origin for Chordata.

A possible stem tunicate fossil was brought to light through a reinterpretation of *Jaekelocarpus oklahomensis*, a Carboniferous "mitrate" (Dominguez et al. 2002). High-resolution X-ray computed tomography (e.g., Rowe et al. 1995, 1997, 1999, Digital Morphology 2003) provided new details of internal anatomy and revealed the presence of paired tunicate-like gill skeletons. *Jaekelocarpus* and a number of similar, tiny Paleozoic fossils have a calcite exoskeleton over their head and pharynx and are generally thought to lie as stem members of echinoderms or various basal chordate clades (Jeffries 1986, Dominguez et al. 2002). The mitrates may prove to be paraphyletic, and its members assignable to different deuterostome clades. The eventual placement of all of these fossils will have bearing on our interpretation of basal chordate relationships, and on the structure and history of mineralized tissues.

Node 3. Chordates with a Brain (Euchordata)

Euchordata comprise the last common ancestor that humans share with lancelets (but see caveats above), and all of its descendants (fig. 23.1). Apart from the tunicates and a single ancient fossil of uncertain affinities (below), all other chordates are members of Euchordata. Expanding on the innovations that arose in chordates ancestrally, euchordates manifest more complex genetic control over development. This was accompanied by further elaboration of the CNS and special sense organs, and a fundamental reorganization of the trunk musculature and locomotor system.

Euchordate Characters

Increased genetic complexity I. Euchordates express *Msx*, *HNF-3*, and *Netrin* genes, whereas only *Hedgehog* is expressed in tunicates. This evident increase in homeobox expression corresponds to elaborated dorsoventral patterning in the CNS. Additional genes are also expressed in more elaborate anteroposterior regionalization, including *BF1* and *Islet* genes (Holland and Chen 2001). Tunicates express only one to five *Hox* genes, whereas lancelets express 10 *Hox* genes in one cluster, affecting broader regions of the brain and nerve cord. Although poorly sampled, at least one hemichordate (*Saccoglossus*) expresses nine *Hox* genes in its single cluster. Tunicates therefore may have lost genes that were present in deuterostomes ancestrally (Holland and Garcia-Fernàndez 1996).

Elaboration of the brain I. Lancelets were long thought to have virtually no brain at all, but recent structural studies reveal an elaborate brain and several unique resemblances to the brain in craniates (Lacalli 1996a, 1996b, Butler 2000). Reticulospinal neurons differentiate in the hindbrain, where they are involved in undulatory swimming and movements associated with the startle reflex. Also present in lancelets are homologues of trigeminal motor neurons, which are involved in pharyngeal movement, and possibly other cranial nerves (Fritzsch 1996, Butler 2000). Additionally, the neural tube is differentiated into an inner ependymal cell layer (gray matter) and synaptic outer fibrous layer (white matter; Maisey 1986) and is innervated by intermyotomal dorsal nerve roots that carry sensory and motor fibers (Schaeffer 1987). Several of these features lie partly or wholly within the expression domain of *Hox* genes.

Elaboration of the special senses I. An olfactory organ occurs in lancelets, in the form of the corpuscles of de Quatrefages. These are a specialized group of anterior ectodermal cells that send axonal projections to the CNS via the rostral nerves. They are marked by expression of the homeobox gene *AmphiMsx*, which is also expressed in craniate ectodermal thickenings known as placodes (below), but no true placodes have been observed in lancelets or tunicates (Shimeld and Holland 2000). The olfactory organ is highly developed in nearly all other euchordates.

Segmentation. Segmentation arises when mesoderm along either side of the notochord subdivides to form somites. These are hollow spheres of mesoderm that mature into muscle blocks known as myomeres, which are separated by sheets of connective tissue (myocomata). Only the mesoderm lying close to the notochord becomes segmented, whereas more laterally the mesoderm produces a sheet of muscle that surrounds the coelomic cavity. The segmented muscles enable powerful locomotion, producing waves of contraction that pass backward and propel the body ahead. Segmentation is accompanied by *Fringe* (or its homologue) expression and signaling by the Notch protein, features shared with other segmented bilaterians. These regulate the timing and synchronization of cell-to-cell communication required of segmental patterning and the formation of tissue boundaries (Evrard et al. 1998, Jiang et al. 2000).

Other features. Also arising from mesoderm is a blood circulatory system of stereotyped arterial design, with a dorsal and ventral aorta linked by branchial vessels, and a complementary venous system (Maisey 1986). Other transformations traceable to the ancestral euchordate yielded a larva that is essentially a miniature, bilateral adult. As adults, a median fin ridge increases thrust area while helping to stabilize movement through the water (Schaeffer 1987).

Pan-Euchordata

The oldest stem euchordate fossil may be the Early Cambrian *Yunnanozoon* from the Chengjiang lagerstätte of southern China (Chen et al. 1995, Shu et al. 2001b, Holland and Chen 2001). It is known from a single specimen that shows evidence of segmental muscle blocks, an endostyle, a notochord, and a nonmineralized pharyngeal skeleton. Little more than a flattened smear, the chordate affinities of this problematic fossil are debatable.

Node 4. The Lancelets (Cephalochordata)

The lancelets, sometimes known as amphioxus, form an ancient lineage that today consists of only 30 species (Gans and Bell 2001). *Branchiostoma* consists of 23 species and *Epigonichthyes* includes seven (Poss and Boschung 1996, Gans et al. 1996). Lancelets are suspension feeders distributed widely in tropical and warm-temperate seas. The larvae are pelagic, and one possibly pedomorphic species remains pelagic as an adult. Adults of the other species burrow into sandy substrate, protruding their heads into the water column to feed.

Adult lancelets lack an enlarged head. They are unique in the extent of both the notochord and cranial somites, which extend to the very front of the body. A single median eye also distinguishes them, which, based on *AmphiOtx* expression, may be homologous to the paired eyes of other chordates and bilaterians (Lacalli 1996a, Butler 2000). Their feeding apparatus involves a unique ciliated wheel organ surrounding the mouth, and a membranous antrum that surrounds the pharynx (Maisey 1986, Holland and Chen 2001).

Pan-Cephalochordata

A single fossil from the Early Cambrian of China, known as *Cathaymyrus* (Shu et al. 1996), may be a stem cephalochordate and the oldest representative of the clade. *Pikaia gracilens* from the Middle Cambrian Burgess Shale is known from numerous specimens and is popularly embraced as a cephalochordate (Gould 1989), but this is now questionable (Holland and Chen 2001). A mitrate known as *Lagynocystis pyramidalis*, from the lower Ordovician of Bohemia, may also be a stem cephalochordate (Jeffries 1986). In all cases, more specimens and more detailed anatomical preservation are needed to have any confidence in these assignments.

Node 5. Chordates with a Head (Craniata)

Craniata contain the last common ancestor that humans share with hagfish, and all its descendants (fig. 23.1). Even contemporary literature often confuses this clade name with the designation Vertebrata. However, Vertebrata are properly regarded as a clade lying within Craniata (Janvier 1996). Compared with their euchordate ancestors, craniates have increased genetic complexity, a larger brain, and more elaborate paired sense organs. Larvae probably persisted as suspension feeders (Mallatt 1985), but adults shifted to active predation with higher metabolic levels, more powerful locomotion, and a sensory system perceptive to multiple modes of environmental signal (Jollie 1982, Northcutt and Gans 1983;, but see Mallatt 1984, 1985). A rigid skull protects and supports the brain, special sense organs, and feeding apparatus. Most important, the neural crest blooms in early development as a unique population of motile cells that induce new structures and assist the many parts of the increasingly complex head and pharynx to integrate as a functional whole.

Craniate Characters

Increased genetic complexity II. Craniates have at least two *Hox* gene clusters, and perhaps three or four clusters were present ancestrally (Holland and Garcia-Fernàndez 1996). This increase in number is correlated with further elaboration of the neurosensory system over that of lancelets and tunicates. Several additional gene families increased in number, including those encoding transcription factors (*ParaHox*, *En*, *Otx*, *Msx*, *Pax*, *Dlx*, *HNF3*, *bHLH*), signaling molecules (hh, IGF, BMP), and others (Shimeld and Holland 2000). The mechanism of duplication is uncertain.

Elaborated brain and sensory organs II. The craniate brain includes new cell types and neuronal groups. It now integrates input from elaborated special sensory organs that develop from paired ectodermal thickenings known as placodes, with the assistance of cells of the neural crest (Northcutt and Gans 1983, Webb and Noden 1993, Butler 2000, Shimeld and Holland 2000). Placodes are typically induced by the underlying mesoderm, and they develop into organs and structures that contribute sensory input to the brain. Although there is evidence for olfactory, optic, and otic organs earlier in chordate history, the integration of placodes with neural crest cells marks a first blossoming of acute, highly complex special sense organs. At least two placode types can now be distinguished. Sensory placodes are involved in the olfactory sacs, lens, ear vesicles, and lateral line system, whereas neurogenic placodes contribute sensory neurons to cranial ganglia. Both categories include some rather different structures, and the different placodes probably had separate histories (Northcutt 1992, Webb and Noden 1993).

The craniate brain is also fully segmented in early ontogeny and differentiates into discrete adult regions associated with special cranial nerves that have specific sensory functions, motor components, or both. Up to 22 cranial nerves are know in some craniates (Butler 2000). The fore- and midbrain regions are expanded and compartmentalized to degrees not seen in other chordates. The forebrain differentiates from segmented prosomeres into an anterior telencephalon that receives input from highly developed olfactory nerves, and the diencephalon to which project the paired eyes (Butler and Hodos 1996). The pineal eye was probably also a part of this system ancestrally. Adult hagfish lack a pineal eye, evidently an ontogenetic loss as the entire visual system degenerates (Hardisty 1979, Forey 1984b). The midbrain arises from segmental mesomeres (Butler and Hodos 1996). The hindbrain develops from segmental rhombomeres controlled by *Hox* genes via *Krox-20* and *Kreisler* expression (Shimeld and Holland 2000). Also elaborated is the otic system, which functions in both vestibular and acoustic reception. Two semicircular canals were present ancestrally (Maisey 2001, Mazan et al. 2000). A lateral line system also arises from head and body placodes (Northcutt 1992). Its functions in electroreception (Bodsnick and Northcutt 1981), and also in mechanoreception by sensing water currents and turbulence, aiding locomotion and hunting (Pohlmann et al. 2001). Also, an autonomic nervous system helps control the endocrine system and other internal functions, and the spinal cord is equipped with dorsal root ganglia.

The internal skeleton. The cartilaginous precursor of an internal skeleton was present in the head, and along the notochord as paired neural and hemal arches. These elements develop via induction between the mesodermal sclerotome and the adjacent notochord and/or spinal chord (Maisey 1986, 1988), but only later in chordate history do they become mineralized or ossified (below). Although lacking jaws and teeth, the ancestral craniate probably had specialized hard mouthparts built from noncollagenous enamel proteins that formed mineralized denticles along the pharyngeal arches at the borders of the gill clefts. These are sites where endoderm and ectoderm interact, and neural crest may also contribute to their mineralization (Smith and Hall 1990). Even in hagfish, high molecular weight amelogens are associated with pharyngeal tissues (Slavkin et al. 1983, Delgado et al. 2001) and the calcium regulatory hormone calcitonin is present (Schaeffer 1987, Maisey 1988).

The neural crest. Origin of the neural crest was perhaps the most remarkable morphogenic event in deuterostome history, owing to the diverse structures that these cells induce or contribute to directly, and help to integrate (Northcutt and Gans 1983, Schaeffer 1987). Neural crest cells are themselves induced by mesoderm along the edges of the overlying neural plate. They migrate to new locations throughout the head, where they produce the cartilaginous neurocranium, a unique structure housing the expanded brain and providing a rigid armature that suspends the special sense organs. Neural crest cells also form a cartilaginous branchial arch system. Neural crest cells also arise from the developing spinal cord to form spinal ganglia, the sympathetic nervous system, pigment cells, and adrenalin glands.

Neural crest cells do not differentiate nor are the structures that they build present in tunicates or lancelets. However, several neural crest cell–inducing genes occur in lancelets. These include the *Msx*, *Slug/Snail*, and *Distalless* gene families, which are expressed in lateral neural plate, and *Pax-3/7*, which is expressed in immediately adjacent ectoderm (Butler 2000, Shimeld and Holland 2000). *Hox* regulatory elements have also been identified in lancelets that in craniates drive spatially localized expression of neural crest cells in the derivatives of placodes and the branchial arches (Manzanares et al. 2000). Thus, well before the emergence of the ancestral craniate, the relative spatial expression patterns of several genes involved in neural crest induction were present.

Pharyngeal arch elaboration. In lancelets, there is a more or less stiff framework of several pairs of collagenous arches. Between adjacent arches are branchial clefts that function primarily in suspension feeding (Mallatt 1984, 1985). In contrast, craniate pharyngeal arches are major structural elements, composed of segmented cartilage or bone that suspend heavily vascularized gills within the clefts. The arches are muscular, and under CNS control they power a pump involved in both respiration and feeding. In craniates, for the first time, the pharyngeal clefts may properly be called gill slits (Schaeffer 1987, Maisey 1988). Each arch develops from an outer covering of ectoderm, an inner covering of endoderm, and a mesenchymal core derived from neural crest and mesoderm (Graham and Smith 2001). The majority of the neural crest cells forming the arches arise adjacent to the hindbrain rhombomeres, each arch with a neural crest population tied to a specific group of rhombomeres. This ensures the faithful transfer of segmental patterning information from the CNS to the arches, establishing a correspondence between innervations and effector muscles. The neural crest segregates into discrete arch populations partly through apoptosis, or preprogrammed cell death, in a process similar to that which sculpts the discrete digits in the tetrapod hands and feet (below). In both instances, key components in the cell death program are the genes encoding Msx2 and BMP4 (Graham and Smith 2001, Zhou and Niswander 1996).

Elaborated muscular system. Muscle ontogeny follows a unique pathway in craniates. First, mesodermal somitomeres appear in strict rostral to caudal order during gastrulation, as segmental arrays of paraxial mesenchymal cells condense along the length of the embryo (Jacobson 1988, 2001). Cranial somitomeres then disperse to form the striated muscles

of the head, including extrinsic muscles of the eye (except in hagfish, which may have lost them secondarily), and branchial musculature. In the trunk, the somitomeres gradually condense to form somites. Lateral to the developing somites the mesoderm differentiates into three separate populations of cells. These are the sclerotome, which later forms part of the cranium and much of the vertebral column, the dermatome, which forms the connective tissues of the dorsal trunk, and the myotome, which forms the striated muscles of the trunk. The adult trunk musculature consists of sequential chevron-shaped myomeres. Finally, the unsegmented lateral plate splits and the coelomic cavity forms between its two layers. The gut, which is no longer ciliated internally, becomes invested by a layer of smooth muscle that provides peristaltic contractions for the movement of ingested food (Schaeffer 1987, Maisey 1986).

Powerful heart and circulatory system. A powerful two-chambered heart is present in craniates along with red blood cells, hemoglobin, and vasoreceptors that monitor pressure and gas levels of the blood passing through the heart. Associated with the elaborated circulatory system is a highly innervated kidney (Schaeffer 1987, Maisey 1988).

Additional endodermal derivatives. The liver and pancreas arise from endoderm through new inductive signals from mesoderm. Also deriving form this source are elaborate endocrine glands including the parathyroids, which control calcium and phosphate metabolism with the plasma calcium-regulatory hormone (calcitonin), and the adrenal glands, all of which are controlled to varying degrees by the autonomic nervous system. The larval endostyle metamorphoses into the adult thyroid gland, becoming a true endocrine gland, directing its secretions into the circulatory rather than digestive system (Schaeffer 1987).

Paired and median fin folds. Primordia of the paired lateral and median appendages arise in craniates via mesodermal-epithelial induction, whereas the dorsal fin arises via interaction between the epidermis and trunk neural crest. A median fin fold is present in lancelets, but it develops without the neural crest interaction.

High metabolic capacity. Craniates possess a well-developed capacity for anaerobic metabolism, resulting in the formation of lactic acid. This probably evolved in association with burst activity that is unobtainable by relying solely on aerobic metabolism (Ruben and Bennett 1980).

Pan-Craniata

The oldest putative pancraniate is *Haikouella lanceolata*, known by more than 300 specimens from the Chengjiang lagerstätte of southern China (Chen et al. 1995, 1999, Shu et al. 2001a, 2001b, Holland and Chen 2001). It has a three-part brain and paired eyes. Its mouth has 12 oral tentacles, and the pharynx has six nonmineralized pharyngeal arches bearing gill filaments that lie in separate visceral clefts. A pair of grooves in its floor suggests an endostyle. There may be several mineralized denticles on the third arch, but preservation leaves this uncertain. About two dozen paired straight myomeres are separated by myosepta behind the 5th visceral arch. Stains are preserved that may represent a heart with ventral and dorsal aorta, and anterior branchial artery. The notochord extends about 85% the length of the body, stopping short of the rostrum, and slight banding can be seen resembling the immature vertebral elements of lampreys (Holland and Chen 2001). It also has dorsal, caudal, and ventral midline fins. *Haikouella* has also been hypothesized to lie on the lamprey stem (Chen et al. 1999), but support is weak (Janvier 1999). From the same deposits, possibly lying on the craniate stem, are *Haikouichthyes* and *Myllokunmingia*, each known from a single fusiform fossil (Shu et al. 1999a, 1999b). The rostral two-thirds of their bodies comprises the pharyngeal region, with Z-shaped myomeres making up the rest. A median dorsal fin shows faint striations that may be fin rays. There are also paired lateral structures, but it is doubtful whether they are homologous with the fins of gnathostomes (below). In *Haikouichthyes* are nine pharyngeal arches and a complex skull, probably built of cartilage, suggesting the presence of neural crest cells. Neither specimen shows evidence of mineralization (Shimeld and Holland 2000, Holland and Chen 2001).

Node 6. The Hagfish (Myxini)

Hagfish comprise a poorly known chordate lineage that includes 58 living species (Froese and Pauly 2001). Throughout their life cycles, hagfish generally occupy deep marine habitats in temperate seas, ranging from 25 to 5000 m in depth (Moyle and Cech 2000). They scavenge large carcasses, burrow into soft substrate for invertebrates, and pursue small prey through the water column. But they are difficult to observe and little is known of their development.

The monophyly of Myxini is well supported. They have three pairs of unique tactile barbels around the nostril and mouth, and a single median nostril of distinct structure. Many other features distinguish them from other craniates, but some may reflect secondary loss, including absence of the epiphysis and pineal organ, reduction of the eyes, presence of only a single adult semicircular canal, and a vestigial lateral line system confined to the head (Hardisty 1979, Maisey 1986, 2001).

Pan-Myxini

Only three fossil species have been allied to the hagfish. The least equivocal is *Myxinikela siroka*, from the Carboniferous Mazon Creek deposits of Illinois (Bardack 1991). A second specimen from these same beds, *Pipiscus zangerli* (Bardack and Richardson 1977), is more problematically a hagfish and has also been allied to lampreys (below). *Xidazoon stephanus*, known by three specimens from the Lower Cambrian of China, has been compared with *Pipiscius* (Shu et al. 1999a, 1999b). Its mouth is defined by a circlet of about 25 plates, and it may have a dilated pharynx and segmented tail. But

other assignments are equally warranted by the vague anatomy it preserves, and whether it is even a chordate remains questionable.

Node 7. Chordates with a Backbone (Vertebrata)

Vertebrata comprise the last common ancestor that humans share with lampreys, and all its descendants. The relationship of hagfish and lampreys to other craniates is long debated. Hagfish and lampreys were once united either as Cyclostomata or Agnatha, jawless fishes grouped by what its members lacked instead of by shared unique similarities, and they were considered ancestral to gnathostomes (e.g., Romer 1966, Carroll 1988). This grouping was largely abandoned as diverse anatomical data showed lampreys to share more unique resemblances with gnathostomes than with hagfish (Stensiö 1968, Løvtrup 1977, Hardisty 1979, 1982, Forey 1984b, Janvier 1996). But controversy persists, and recent studies of the feeding apparatus have resurrected a monophyletic Cyclostomata (Yalden 1985, Mallatt 1997a, 1997b). Cyclostome monophyly is also supported by ribosomal DNA (rDNA; Turbeville et al. 1994, Lipscomb et al. 1998, Mallatt and Sullivan 1998, Mallatt et al. 2001), vasotocin complementary DNA (cDNA; Suzuki et al. 1995), and globin cDNA (Lanfranchi et al. 1994). However, the results from small subunits of rDNA were overturned when larger ribosomal sequences were used, and morphological analyses that sample many different systems also refute cyclostome monophyly (Philippe et al. 1994, Donoghue et al. 2000). The question may not be settled, but I follow current convention and treat lampreys and hagfish as successive sister taxa to gnathostomes.

Vertebrate Characters

Increased genetic complexity III. A tandem duplication of *Hox*-linked *Dlx* genes occurred in vertebrates ancestrally, encoding transcription factors expressed in several developing tissues and structures. They are expressed in an expanded forebrain, cranial neural crest cells, placodes, pharyngeal arches, and the dorsal fin fold. An additional duplication evidently occurred independently in lampreys and gnathostomes (Amores et al. 1998, Niedert et al. 2001, Holland and Garcia-Fernàndez 1996).

Elaboration of the brain and special senses III. In vertebrates, exchange of products between blood and cerebrospinal fluid occurs via the choroid plexus, a highly vascularized tissue developing in the two thinnest parts of the ventricular roof of the brain. Vertebrate eyes are also enhanced by a retinal macula, a small spot of most acute vision at the center of the optic axis of the eye, and by synaptic ribbons that improve retinal signal processing. Extrinsic musculature originating from the rigid orbital wall provides mobility to the bilateral eyeballs. The pineal body is also photosensory, and in some vertebrates differentiates into a well-developed pineal eye with retina and lens. In addition, the lateral line system extends along the sides of the trunk (Maisey 1986).

Correspondingly, an extensive cartilaginous braincase that includes embryonic trabecular cartilages arises beneath the forebrain, and an elaborate semirigid armature supports the brain and its special sensory organs.

Locomotor and circulatory systems. Vertebrates have dorsal, anal, and caudal fins that are stiffened by fin rays, increasing thrust and steering ability. The circulatory and muscular systems were also bolstered. The heart comes under nervous regulation and a stereotyped vascular architecture carries blood to and from the gills. Myoglobin stores oxygen in the muscles, augmenting scope and magnitude in bursts of activity. The kidney is also elaborated for more sensitive osmoregulation and more rapid and thorough filtration of the blood (Maisey 1986).

Pan-Vertebrata

The oldest putative stem vertebrates are the heterostracans, an extinct lineage extending from Late Cambrian (*Anatolepis*) to the Late Devonian (Maisey 1986, 1988, Gagnier 1989, Janvier 1996). Their skeleton consists of plates of acellular membranous bone. Precise relationships of this clade are controversial, but if correct the position of heterostracans as the sister taxon to Vertebrata may suggest that lampreys may have secondarily lost a bony external skeleton. However, in the absence of direct evidence that lampreys ever possessed bone, heterostracan fossils and the characteristics of bone are treated below (see Pan-Gnathostomata, below).

Node 8. The Lampreys (Petromyzontida)

There are approximately 35 living lamprey species, all but three of which inhabit the northern hemisphere (Froese and Pauly 2001). In most, larvae hatch and live as suspension feeders in freshwaters for several years, then migrate to the oceans as metamorphosed adults, where they become predatory and parasitic. Nonparasitic freshwater species are known (Beamish 1985) and in some cases the metamorphosed adults are nonpredatory and do not feed during their short adult lives (Moyle and Cech 2000).

Lamprey monophyly is diagnosed by a unique feeding apparatus. It consists of an annular cartilage that supports a circular, suction-cup mouth lined with toothlike keratinized denticles. A mobile, rasping tongue is supported by a unique piston cartilage and covered by denticles whose precise pattern diagnoses many of the different species. Lampreys attach to a host, rasp a hole in its skin, and feed on its body fluids. Lampreys also eat small invertebrates. The structure of the branchial skeleton (Mallatt 1984, Maisey 1986) and the single median nasohypophysial opening (Janvier 1997) are unique. Lampreys have a distinctive suite of olfactory receptor genes that serves in the detection of odorants such as bile acids (Dryer 2000). There is also evidence that lampreys are apomorphic in having undergone duplication of a tandem pair of *Dlx* genes, followed by loss of several genes, independent of a comparable du-

plication and subsequent loss that occurred in gnathostomes (Niedert et al. 2001)

Pan-Petromyzontida

Haikouichthyes ercaicunensis (Shu et al. 1999b) from the Early Cambrian of China is the oldest fossil lamprey reported, but the data for its placement are tenuous (Janvier 1999). *Mayomyzon pieckoensis*, known by several specimens from the Late Carboniferous Mazon Creek beds of Illinois (Bardack and Zangerl 1968), is the oldest unequivocal lamprey, preserving unique lamprey feeding structures, including the annular and piston cartilages. *Hardistiella montanensis* (Janvier and Lund 1983) from the Lower Carboniferous of Montana preserves less detail, and it is not clear whether either lies within or outside of (crown) Petromyzontida. *Pipiscus zangerli* (Bardack and Richardson 1977) from the same Mazon Creek beds as *Mayomyzon* is sometimes also tied to lampreys, as well as hagfish, but it preserves little relevant evidence.

Node 9. Chordates with Jaws (Gnathostomata)

Gnathostomata comprise the last common ancestor that humans share with Chondrichthyes, and all of its descendants (fig. 23.1). Its origin was marked by additional increases in complexity of the genome, which mediated several landmark innovations, including jaws, paired appendages, several types of bone, and the adaptive immune system. Although the positions of certain basal fossils are debated, there is little doubt regarding gnathostome monophyly.

Gnathostome Characters

Increased genetic complexity IV. Gnathostomes have at least four *Hox* gene clusters, and some have as many as seven. In addition to specifying the fate of cell lineages along the anteroposterior axis, these gene clusters mediate limb development and other outgrowths from the body wall. It is questionable whether as many as four *Hox* clusters arose earlier, either in vertebrates or craniates ancestrally (Holland and Garcia-Fernàndez 1996), but in gnathostomes their expression nevertheless manifests more complex morphology. There was also duplication of *Hox*-linked *Dlx* genes and several enhancer elements, leading to elaboration of cranial neural crest in the pharyngeal arches, placodes, and the dorsal fin fold (Niedert et al. 2001). Immunoglobin and recombinase activating genes also arose in gnathostomes, marking the origin of the adaptive immune system.

Brain and sensory receptor enhancement IV. The gnathostome forebrain is enlarged, primarily reflecting enhancement of the olfactory and optic systems. The extrinsic muscles of the eyeball are rearranged and an additional muscle (the obliquus inferior) is added to the suite present in vertebrates ancestrally (Edgeworth 1935). In the ear, a third (horizontal) semicircular canal arises, lying in nearly the same plane as the synaptic ribbons of the eye, and correlates with *Otx1* expression (Maisey 2001, Mazan et al. 2000). In addition, the lateral line system is elaborated over much of the head and trunk. On the trunk, it is developmentally linked to the horizontal septum and becomes enclosed by mineralized tissues that insulate and tune directional electroreception by the lateral line system (Northcutt and Gans 1983). The gnathostome lateral line system derives from neural crest and lateral plate mesoderm induction, heralding a new stage in developmental complexity. Myelination of many nerve fibers improves impulse transmission through much of the body (Maisey 1986, 1988).

Mineralized, bony skeleton.   Many bilaterians produce mineralized tissues, and both echinoderms and tunicates generate amorphous and crystalline calcium carbonate spicules (Aizenberg et al. 2002). Biomineralization is thus an ancient property, although its erratic expression outside of Craniata affords only equivocal interpretations of its history in this part of the tree. Certain other components required for bone mineralization, such as calcitonin, were already present but did not lead to bone production. However, in gnathostomes, different types of bone form in the head and body (Maisey 1988). Bone development requires the differentiation of specialized cell types, including fibroblasts, ameloblasts, odontoblasts, and osteoblasts, which are derived from the ectoderm and cephalic neural crest. In the formation of membranous bone, fibroblasts first lay down a fibrous collagen framework around which the other cells deposit calcium phosphate as crystalline hydroxyapatite. Another type of bone development typically involves preformation by cartilage, followed by deposition of hydroxyapatite crystals around the cartilage (perichondral ossification), or within and completely replacing it (endochondral ossification). Chondral ossification occurred first in the head in the oldest extinct gnathostomes (see Pan-Ganthostomata, below), and it later spread to the axial skeleton and shoulder girdle. Ossification in the shoulder girdle is of interest because it is the first such transformation of the embryonic lateral plate mesoderm and because it signals the initiation of neural crest activity in the trunk (Maisey 1988). In the shark lineage, the internal skeleton consists of cartilage that is sheathed in a layer of crystalline apatite, but fossil evidence suggests that this is a derived condition (below).

Elaborated skull.   Cartilage and/or chondral bone surround the brain and cranial nerves, providing a semirigid armature for the special sensory organs. At the back of the head, the cephalicmost vertebral segment is "captured" during ontogeny by the skull to form a back wall of the braincase. Thereby, it confines several cranial nerves and vessels to a new passage through the base of the skull, known in embryos as the metotic fissure. Cellular membranous bone was also present, covering the top and contributing to other parts of the skull (Maisey 1986, 1988).

Jaws.   The namesake characteristic of gnathostomes arises in ontogeny from the first pharyngeal arch, known now as the mandibular arch. Its upper half is the palatoquadrate cartilage, which is attached to the braincase primitively by ligaments, whereas the lower half of the arch, Meckel's cartilage, forms the lower jaw and hinges to the palatoquadrate at the back of the head. Teeth and denticles develop on inner surfaces of these cartilages through an induction of ectoderm and endoderm. Neural crest cells populating the mandibular arch derive from the mesomeres and from hindbrain rhombomeres 1 and 2, whereas the second pharyngeal arch, the hyoid arch, derives its neural crest from rhombomere 4 (Graham and Smith 2001).

Paired appendages.   Other bilaterians have multiple sets of paired appendages that serve a broad spectrum of functions. It was long believed that their evolution was entirely independent of the paired appendages in gnathostomes, but this appears only partly true today. Common *Hox* patterning genes were likely present in the last common ancestor of chordates and arthropods, if not a more inclusive group. The *SonicHedgehog* gene specifies patterning along anteroposterior, dorsoventral, and proximodistal axes of the developing limb, via BMP2 signaling proteins (Shubin et al. 1997). In gnathostomes, independent expression of orthologous genes occurs in the elaboration of fins, feet, hands, and wings. As expressed in gnathostomes, the distal limb elements are the most variable elements. In basal gnathostomes they comprise different kinds of stiffening rays, whereas in tetrapods they are expressed as fingers and toes (Shubin et al. 1997). Moreover, somite development transformed to provide for muscularization of the limbs, as certain somite cells became motile and moved into the growing limb buds (Galis 2001). Thus, although the *Hox* genes have a more ancient history of expression, in gnathostomes they are expressed across a unique developmental cascade.

The adaptive immune system.   One of the most remarkable gnathostome innovations is the adaptive immune system (Litman et al. 1999, Laird et al. 2000). It responds adaptively to foreign invaders or antigens such as microbes, parasites, and genetically altered cells. Other animals have immune mechanisms, but unique to gnathostomes is a system that is specific, selective, remembered, and regulated. Its fundamental mediators are immunoglobin and recombinase activation genes, which are present throughout gnathostomes but absent in lampreys and hagfish. The immune system is expressed in a diverse assemblage of immunoreceptor-bearing lymphocytes that circulate throughout the body in search of antigens. Gnathostome lymphocytes present an estimated 10^{16} different antigen receptors, which arose seemingly instantaneously as an "immunological big bang" (Schluter et al. 1999) in gnathostomes ancestrally.

New endodermal derivatives.   In gnathostomes, the endoderm elaborates to form the pancreas, spleen, stomach, and a spiral intestine (Maisey 1986).

Pan-Gnathostomata

Several extinct lineages lie along the gnathostome stem. Their relationships remain problematic, and most have been allied with virtually every living chordate branch (Forey 1984a,

Maisey 1986, 1988, Donoghue et al. 2000). All preserve mineralized and bony tissues of some kind, and the phylogenetic debate revolves in large degree around interpreting the history of tissue diversification. The most ancient, if problematic extinct pangnathostome lineage is Conodonta. Known to paleontologists for decades only from isolated, enigmatic mineralized structures, conodonts range in the fossil record from Late Cambrian to Late Triassic. The recent discovery of several complete body-fossils demonstrated that these objects are toothlike structures aligned along the pharyngeal arches and bordering the gill clefts. They are built of dentine, calcified cartilage, and possibly more than one form of hypermineralized enamel (Sansom et al. 1992). Microwear features indicate that they performed as teeth, occluding directly with no intervening soft tissues. They formed along the same zones of endoderm-ectoderm induction as the pharyngeal teeth in more derived vertebrates. The mineralized oropharyngeal skeleton and dentition arose at the base of the gnathostome stem, Cambrian conodont fossils providing its oldest known expression (Donoghue et al. 2000).

Branching from or possibly below the gnathostome stem are the heterostracans (see Pan-Vertebrata, above), whose skeleton consists of external plates of acellular membranous bone. In heterostracans, bones formed around the head, and the cranial elements seemingly grew continually throughout life. Their bone is formed of a basal lamina, a middle layer of spongy arrays of enameloid, and an outer covering of enameloid and dentine. Heterostracan fossils suggest that bone was acellular at first.

The next most problematic taxon is Anaspida, which range from Middle Silurian to Late Devonian (Forey 1984a, Maisey 1986, 1988, Donoghue et al. 2000). Anaspids are diagnosed by the presence of branchial and postbranchial scales, pectoral plates, and continuous bilateral fin folds. Perichondral ossification occurred in neural and hemal arches, and the appendicular skeleton, whereas endochondral ossification occurred in fin radials and dermal fin rays in the tail. The anaspid trunk squamation pattern suggests the presence of the horizontal septum, a critical feature in the trunk-powered locomotion that is also tied developmentally to the lateral line system. Anaspid lateral fin folds may prove to be precursors of the paired appendages of crown gnathostomes.

Lying closer to the gnathostome crown clade is Galeaspida, which range through the Silurian and Devonian. Its members are distinguished by a large median dorsal opening that communicates with the oral cavity and pharyngeal chamber. Galeaspids also have 15 or more pharyngeal pouches. Their chondral skeleton appears mineralized around the brain and cranial nerves, however the bone is primitive in being acellular (Maisey 1988). Lying closer to the gnathostome crown is Osteostraci, a lineage with a similar character and temporal range as galeaspids. Osteostracans have a dorsal head shield with large dorsal and lateral sensory fields. They share with crown gnathostomes cellular calcified tissues and perichondral ossification of the headshield, which encloses the brain and cranial nerve roots. Ossification surrounds the orbital wall, otic capsules, and calcified parachordal cartilages, structures developing in extant gnathostomes via inductions between the CNS, notochord, and the ectomesenchyme. Perichondral mineralization of the otic capsule implies interaction between mesenchyme and the otic placode (Maisey 1988). Also present are lobed, paired pectoral fins that are widely viewed as homologous to the pectoral appendages in crown Gnathostomata (Forey 1984a, Maisey 1986, 1988, Shubin et al. 1997, Donoghue et al. 2000). Supportive of this view is the ontogenetic sequence in most extant gnathostomes, in which pectoral appendages arise before pelvic.

Node 10. Sharks and Rays (Chondrichthyes)

Chondrichthyes includes sharks, skates, rays, and chimaeras (fig. 23.1). The chimaeras (Holocephali) include roughly 30 living species, and there are about 820 living species of skates and rays (Batoidea) plus sharks (Moyle and Cech 2000). Morphology suggests that the species commonly known as sharks do not by themselves constitute a monophyletic lineage, and that some are more closely related to the batoids than to other "sharks" (Maisey 1986).

Earlier authors argued that these different groups evolved independently from more primitive chordates, and that Chondrichthyes was a grade that also included several cartilaginous actinopterygians (below). Cartilage is an embryonic tissue in all craniates, and it persists throughout life in sharks and rays (and a few other chordates), but the perception that "cartilaginous fishes" are primitive is mistaken. In its more restricted reference to sharks, rays, and chimaeras, the name Chondrichthyes designates a monophyletic lineage. Histological examination reveals bone at the bases of the teeth, dermal denticles, and some fin spines. This suggests that this restricted distribution of bone is a derived condition in chondrichthyans (Maisey 1984, 1986, 1988).

Other apomorphic characters include the presence of micromeric prismatically calcified tissue in dermal elements and surrounding the cartilaginous endoskeleton. Chondrichthyans also possess a specialized labial cartilage adjacent to the mandibles, the males possess pelvic claspers, and the gill structure is unique. The denticles (scales) possess distinctive neck canals (but these may not be unique to chondrichthyans), and the teeth have specialized nutrient foramina in their bases with a unique replacement pattern in which replacing teeth attach to the inner surface of the jaws as dental arcades (Maisey 1984, 1986). Fin structure also presents a number of unique modifications (Maisey 1986). Relationships among chondrichthyans have received a great deal of attention (Compagno 1977, Schaeffer and Williams 1977, Maisey 1984, 1986, Shirai 1996, de Carvalho 1996).

Pan-Chondrichthyes

The extinct relatives of chondrichthyans have a long, rich fossil record. The oldest putative fossils are scales with neck canals from the Late Ordovician Harding Sandstone of Colorado (Sansom et al. 1996). Although present in extant sharks and chimeroids, most well-known Paleozoic sharks lack them. From the Silurian onward, chondrichthyan teeth are abundantly preserved, although in most cases their identification rests on solely phenetic grounds, and they provide little useful information on higher level phylogeny. The oldest anatomically complete fossils are the Late Devonian Symmoriidae and *Cladoselache*, which are known from numerous skeletons that in some case preserve body outlines and other evidence of soft tissues. Both are stem chondrichthyans.

Node 11. Chordates with Lungs (Osteichthyes)

Osteichthyes (fig. 23.1) comprise the lineage stemming from the last common ancestor that humans share with actinopterygians. The name means "bony fishes" and was coined in pre-Darwinian times in exclusive reference to the fishlike members of this clade. In the phylogenetic system (de Queiroz and Gauthier 1992), the name now refers to all members of the clade, roughly half of which are the chordate species adapted to life on land.

Osteichthyan Characters

An extensive composite bony skeleton.   All conclusions about skeletal evolution at this node are weak, because chondrichthyans lack an ossified internal bony skeleton that can be compared directly with that in osteichthyans. Nonetheless, the fossil record offers assistance and suggests that a bony skeleton likely arose in early pangnathostomes, and that it was further elaborated in Osteichthyes. The membranous skeleton of the head forms laminae that descend from the braincase and offer attachment to muscles of the jaws and pharyngeal skeleton. The jaws themselves are invested in a layer of membranous bone, with teeth attached to their margins (Rosen et al. 1981, Maisey 1986). Around the pharyngeal chamber is an extensive series of dermal gular and opercular bones, which improve pharyngeal function as a suction chamber in both respiration and feeding. The pectoral girdle became ossified, primitively more through perichondral than endochondral processes. Lastly, in the fins are stiffening rays known as lepidotrichia, which represent rows of slender scales that replace the primitive covering of body scales (Maisey 1986, 1988).

Lungs.   Lungs develop as ventral outgrowths from the rostral end of the gut tube and are often associated with skeletal structures of mesenchymal origin. Over the course of osteichthyan history, these diverticula become modified for radically different functions that range from respiration, to buoyancy regulation, to communication. In most terrestrial members of the clade, lungs completely replace gills. They are secondarily lost in some small living amphibian species, where cutaneous respiration takes over. Lungs develop as branching tubular networks constructed of sheetlike cellular epithelia. There can be hundreds to millions of branches

in the network, yet they must also have a regular patterning and structure to ensure proper function. A signaling pathway mediated by fibroblast growth factor (FGF) occurs in development of the branched lungs in the mouse, as well as in the branched respiratory tracheae in the fruit fly, raising the question of whether their common ancestor had a branched respiratory structure. But because the tracheal system lungs in insects are ectodermal and the osteichthyan lung is endodermal, this seems unlikely. Moreover, FGF is implicated in other branched structures and has probably been co-opted throughout metazoan history to produce different kinds of structures. The patterning mechanism is ancient, but its expression in the osteichthyan lung is unique (Metzger and Krasnow 1999).

Pan-Osteichthyes

Two problematic extinct lineages, Acanthodii and Placodermi, arguably lie along the osteichthyan stem, but the evidence is equivocal and a wide spectrum of other possibilities have been proposed. Although some gnathostomes went on to lose one or both sets of limbs, acanthodians are the only clade to exceed the primitive number of two pairs. An anterior spine stiffens each fin. Acanthodian fossils are known from the Late Silurian to the Late Devonian. Placoderms comprise a much more diverse clade whose fossil record extends from Early Devonian to Early Carboniferous. Placoderms are heavily armored, with a distinctive pattern of membranous bones forming a head shield that hinges to a membranous thoracic shield in a pair of ball-in-socket joints. Acanthodians and placoderms share with Osteichthyes the presence of the clavicle and interclavicles and other membranous elements in the pectoral girdle. Placoderms lie closer to Osteichthyes based on descending laminae of membranous bone in the neurocranium, lepidotrichia in the fins, and other features (Gardiner 1984). Difficulties in comparing skeletal features in these fossils with chondrichthyans, which largely lack a bony skeleton, complicate understanding the relationships of these extinct lineages (Maisey 1986).

Node 12. The Ray-Finned Fishes (Actinopterygii)

The ray-finned fishes (fig. 23.1) include nearly 23,000 living species and comprise nearly half of extant chordate diversity (Lauder and Liem 1983). The most basal divergence among extant actinopterygians is represented by the bichirs and reedfish (Polypteriformes), which commonly (but not unanimously) are regarded as sister taxon to all others. Next most basal was the divergence between the sturgeons and paddlefishes (Acipenseriformes), followed by gars (Ginglymodi) and bowfins (Halecomorpha). Among these basal clades alone are nearly 300 extinct genera named for fossils. However, this part of the actinopterygian tree remains a frontier, in large part because the fossil morphology is known only superficially (Grande and Bemis 1996). The rest of extant actinopterygian diversity resides among the teleosts (De Pinna

1996). Today actinopterygians occupy virtually every freshwater and marine environment. Their economic importance underlies the base of a huge global market, and actinopterygian conservation increasingly is involved in conflicts with development and use of the world's water resources. One member of this clade, the zebrafish, is growing in importance for biomedicine as an important model organism. Actinopterygian history and diversity are reviewed by Stiassny et al. (Stiassny et al., ch. 24 in this vol.).

Pan-Actinopterygii

The fossil record of stem actinopterygians extends tentatively into the Late Silurian (Long 1995, Arratia and Cloutier 1996). The Late Silurian *Andreolepis* and Early Devonian *Ligulalepis* are the oldest purported panactinopterygians fossils. They are known only from scales, which overlap in a seemingly distinctive tongue-in-groove arrangement often considered diagnostic of actinopterygians. However, an ossified Early Devonian braincase, possibly referable to *Ligulalepis* (Basden et al. 2000) closely resembles the braincase in the extinct Early Devonian shark *Pucapampella* (Maisey and Anderson 2001), although it is ossified. Hence, we may expect continued reassessment of character distributions and view as tentative the phylogenetic assignment of extinct taxa at this deep part of the tree. By the Early Devonian, actinopterygian fossils are found worldwide, but diversity is low. In the Middle and Late Devonian, only 12 species and seven genera are recognized. The best known is *Cheirolepis*, whose skeleton is known in detail (Arratia and Cloutier 1996). From Middle and Late Devonian rocks, abundant fossils of *Mimia* and *Moythomasia* have been recovered, representing the oldest members of crown clade Actinopterygii (Grande and Beamis 1996).

Node 13. Chordates with Lobe Fins (Sarcopterygii)

Sarcopterygians include the last common ancestor that humans share with coelacanths, and lungfishes and all its descendants (fig. 23.1). Just less than half of chordate diversity lies within this clade (Cloutier and Ahlberg 1996). Its early members were all aquatic, but from the Carboniferous onward most sarcopterygians have been terrestrial (Gauthier et al. 1989). Today only eight living species retain the ancestral life style. Two are coelacanths and the other six are lungfish, whereas the remainder of sarcopterygian diversity resides among the tetrapods.

Sarcopterygian monophyly is strongly supported, but relationships within are far from settled, especially when fossils are concerned. Leaving fossils aside for the moment, morphological, and molecular analyses continue to provide conflicting results (Marshall and Schultze 1992, Schultze 1994, Meyer 1995, Zhu and Schultze 1997). Older studies placed coelacanths outside of tetrapods + actinopterygians (von Wahlert 1968, Wiley 1979), and even with chondrichthyans (Løvtrup 1977, Lagios 1979). Parvalbumin sequences

also support the placement of the *Latimeria* outside of Osteichthyes (Goodwin et al. 1987). Morphology consistently places Actinistia closer to tetrapods than to actinopterygians (Romer 1966, Rosen et al. 1981, Maisey 1986, Nelson 1989, Chang 1991), a position also supported by 28S rDNA (Hillis and Dixon 1989). But whether lungfish or coelacanths are closer to tetrapods, or whether lungfish and coelacanths together form a clade independent of tetrapods is still debated. A larger 28S sequence (Zardoya and Meyer 1996) found coelacanths and lungfishes to be the sister lineage to tetrapods. A genomic DNA analysis (Venkatesh et al. 1999, 2001) and morphology (Rosen et al. 1981, Maisey 1986, Cloutier and Ahlberg 1996) favor lungfishes and coelacanths as successive outgroups to tetrapods, the position that is followed here.

Sarcopterygian Characters

Lobe fins. The sarcopterygian pectoral and pelvic appendages form muscular lobes that protrude from the lateral body wall with a distinct skeletal architecture. In gnathostomes ancestrally there were multiple basal elements in each limb, but in sarcopterygians there is a single proximal element, followed distally by a pair of radial cartilages. This arrangement enables the insertion of muscles between the radials, giving the fin flexibility along its axis (Clack 2000). Fundamental similarities in branching occur within the embryonic digital arch in lungfishes and tetrapods, producing the familiar pattern of a single proximal element (humerus or femur), followed by a pair of elements (radius/ulna or tibia/fibula), followed by the more complex pattern of wrist and ankle bones. This branching sequence is known as the metapterygial axis, and it reflects further influence by *SonicHedgehog* (via BMP2 signaling proteins), which specifies patterning along anteroposterior, dorsoventral, and proximodistal axes of the developing limb. Expressed from the beginnings of gnathostome history in the development of fins, modified expression of orthologous genes lead to the elaboration of lobefins, feet, hands, and wings in sarcopterygians (Shubin and Alberch 1986, Shubin et al. 1997, Cloutier and Ahlberg 1996).

Enamel. A thin layer of enamel covers the teeth in sarcopterygians, and at their bases the enamel is intricately infolded into the dentine, in a pattern known as labyrinthodonty. Infolded enamel enhances tooth strength as well as the strength of attachment to the jaw (Long 1995).

Pan-Sarcopterygii

The acceptance by earlier researchers of paraphyletic groups such as the crossopterygians (e.g., Romer 1966) and the search for direct ancestors of tetrapods in these "amphibian-like fishes" left controversial the relationship among the extinct Paleozoic sarcopterygians (Rosen et al. 1981, Maisey 1986). However, most of these extinct taxa are now assignable as stem lungfish (Pan-Dipnoi) or stem tetrapods (Pan-Tetrapoda). However, two recent fossil discoveries lie on the sarcopterygian stem and provide the oldest evidence of the clade. These are *Psarolepis romeri*, from the Late Silurian and Early Devonian of Asia (Ahlberg 1999, Zhu et al. 1999) and *Achoania jarvikii* (Zhu et al. 2001) from the Early Devonian of China. Lying at the base of either the sarcopterygian stem (Long 1995, Clack 2000) or the choanate stem is Onychodontiformes, a poorly known Devonian lineage whose members reached 2 m in length and are characterized by daggerlike tooth whorls. It is possible that *Psarolepis* lies within this clade.

Node 14. The Coelacanths (Actinistia)

Coelacanth history is at least 400 million years (Myr) long (Forey 1998), but only two species survive today. *Latimeria chalumnae* inhabits coastal waters along southeastern Africa, and a second population was recently discovered in the waters off Sulawesi (Erdmann et al. 1998). Divergent DNA sequences reportedly diagnose *Latimeria menadoensis* (Pouyaud et al. 1999), but it shows little morphological distinction. However, sequences from parts of two mitochondrial genes also diagnose the Sulawesi species, and molecular clock estimates suggest that it diverged from its common ancestor with the African species 5.5 Mya (Holder et al. 1999). Monophyly of the lineage has never been seriously questioned, and it is diagnosed by such features as the absence of the maxilla, absence of the surangular, absence of the branchiostegal rays, presence of a rostral electric organ, presence of numerous supraorbital bones, and a distinctive tassle on the tail.

Pan-Actinistia

The coelacanth fossil record ranges back to the Middle Devonian but it ends in the Late Cretaceous, or more tenuously the Paleocene (Cloutier and Ahlberg 1996). Approximately 125 extinct coelacanth species have been named (Cloutier 1991a, 1991b, Cloutier and Ahlberg 1996, Forey 1998). Although often described as a living fossil (Forey 1984b), a phylogenetic analysis of *Latimeria chalumnae* and its extinct relatives showed that the living species differ by many dozens of apomorphies from their Paleozoic relatives (Cloutier 1991a). Some of these characters represent losses of elements in the cheek and opercular region, leading to suggestions that coelacanth history was characterized by pedomorphosis (Lund and Lund 1985, Forey 1984b). However, there are also elaborations in complexity of skeletal elements, which indicate that the history of actinistians involved more than a single developmental trend and that living coelacanths are not "living fossils" (Cloutier 1991a).

Node 15. The Breathing Chordates (Choanata)

Choanata comprise the last common ancestor that humans share with lungfishes (fig. 23.1), and all its descendants (= Rhipidistia of Cloutier and Ahlberg 1996). Choanata monophyly is supported by genomic DNA (Venkatesh et al. 2001, Hyodo et al. 1997) and morphology (Rosen et al. 1981,

Maisey 1986, Cloutier and Ahlberg 1996), although it remains among the more controversial nodes within Chordata (above).

Choanata Characters

The choanate nose and respiratory system. Its namesake feature is a palatal opening called the choana that communicates externally via paired external nostrils to the lungs and pharynx. The interpretation of this region is controversial in both Paleozoic fossils and Recent taxa, and whether the choana was actually present ancestrally is in dispute (Rosen et al. 1981, Maisey 1986, Carroll 2001). Despite debate over this feature, other transformations of nasal architecture and function were underway. A nasolacrimal canal is present, connecting the orbit with the narial passageway (Maisey 1986). The snout in front of the orbits is elongated in association with these passageways. These facial changes appear related to modifications in the internal structure of the lung tied to increase in efficiency of air breathing with the addition of pulmonary circulation and augmentation of the heart with two auricles (Johansen 1970, Rosen et al. 1981).

Simplification of the pharyngeal skeleton. The opercular elements that enclosed the pharynx in osteichthyans ancestrally are reduced and the pharyngeal arches are simplified with the loss of their dorsal (pharyngobranchial) and ventral (interhyal) elements (Rosen et al. 1981). The upper division of the second arch, the hyomandibula, is reduced and freed from its primitive role as a support between the cranium and jaws. This may signal the beginning of its function in sound transduction.

Tetrapodous locomotion. Well-developed pectoral and pelvic skeletons with two primary joints are present, signaling the beginnings of stereotyped locomotor patterns (Rosen et al. 1981). In the forelimb, the humerus articulates with the shoulder girdle in a ball-in-socket joint. Distal to that is the radius and ulna, which articulate to the humerus in a synovial elbow joint. The presence of these elements represents the unfolding of fundamental patterning at a cellular level (Oster et al. 1988) that persists through most members of the clade. The pelvis is also strengthened by ventral fusion of its right and left halves to form a single girdle. In addition, the musculature that powers the limbs is segmented, paving the way for a blossoming of limb diversification.

Pan-Choanata

Lying along the stem of either Choanata or Sarcopterygii lies a poorly known lineage known as Onychodontida (Cloutier and Ahlberg 1996). If this placement proves correct, its Early Devonian fossils would be the oldest crown sarcopterygians yet discovered.

Node 16. The Lungfishes (Dipnoi)

The lungfishes (fig. 23.1) have a 400 Myr history but today include only six living species. Four live in freshwaters of tropical Africa (*Protopterus dolloi, P. annectens, P. aethiopicus*, and *P. amphibious*), one in South America (*Lepidosiren paradoxa*), and one in Australia (*Neoceratodus forsteri*). The monophyly of dipnoans has never been challenged. Their most distinctive features involve the feeding apparatus (Schultze 1987, 1992, Cloutier and Ahlberg 1996). Lungfish may have teeth along the margins of their jaws as juveniles, but they are lost in adults. The adult dentition consists of tooth plates that line the roof and floor of the mouth. The plates grow by the continual addition of new teeth and dentine, which consolidate into dental plates that are not shed (Reisz and Smith 2001).

Pan-Dipnoi (= Dipnomorpha)

Approximately 280 extinct species are known, their record extending back to the Early Devonian. The earliest dipnomorphs retain marginal teeth but also have palatal tooth plates. The earliest members of the lineage are from the Early Devonian and occupied marine waters, but by the mid-Devonian skeletal structures associated with air breathing had appeared and soon thereafter members of the lineage had moved to the freshwaters that all living species inhabit (Cloutier and Ahlberg 1996). *Yongolepis* and Porolepiformes are extinct lineages known from Devonian rocks that lie at the base of the stem of the lungfish lineage (Clack 2000).

Node 17. Chordates with Hands and Feet (Tetrapoda)

Tetrapoda (fig. 23.1) comrpise the last common ancestor that humans share with amphibians, and all its descendants. The sister relationship between amphibians and amniotes (below) is supported by molecular (Hedges et al. 1993) and morphological data (Schultze 1970, 1987, Rosen et al. 1981, Cloutier and Ahlberg 1996). Historically, the name Tetrapoda designated all sarcopterygians possessing limbs with digits rather than fin rays, such as the Devonian *Ichthyostega* and *Acanthostega*. Although it is true that a wide morphological gap separates the fingers and toes of *Ichthyostega*, from more basal sarcopterygians that lack discrete digits such as *Eusthenopteron*, the limbs of *Ichthyostega* are quite different from those inferred to have been present in the last common ancestor of living tetrapod species. It was once believed that some of the extant tetrapod lineages arose independently from fishlike sarcopterygians, (Jarvik 1996), but recent phylogenetic analyses conclude that extant amphibians and amniotes share a more recent common ancestor that is not also shared with *Ichthyostega* or *Acanthostega*. The history of Tetrapoda was long considered to extend back to the Late Devonian, but under this more restrictive definition of the name, the oldest known tetrapods are Carboniferous fossils (Paton et al. 1999).

Tetrapod Characters

The tetrapod limb. In crown tetrapods, the shoulder girdle has a prominent scapular blade and a posterior corocoidal region, and the humerus has a discrete shaft. There are fully differentiated proximal and distal carpals in the wrist and phalanges in the hand. The ankle also has separate proximal and distal tarsals and phalanges (Gauthier et al. 1988b). The evolution of fingers and toes is associated with changes in the timing and position of expression of the more ancient *Hox* genes that regulate development of the body axis and appendages (Shubin et al. 1997, Carroll 2001). In sampled actinopterygians, the *Hoxd-9* to *Hoxd-13* genes are expressed in an overlapping sequence from the proximal to distal ends of the posterior surface of the fin. In tetrapods the most distal gene, *Hoxd-13*, is expressed over a more anterior portion of the distal end of the limb, directing distal expansion of the limb and the formation of fingers and toes. Key components in the development of separate digits are cell death (apoptosis) programs directed by the genes encoding *Msx2* and BMP4 (Graham and Smith 2001). These were first expressed in the development of separate pharyngeal arches. In tetrapods they are co-opted to direct apoptosis in the tissues that lie between the digits, to produce discrete fingers and toes (Zhou and Niswander 1996). Lost from the tetrapod limb are the ectodermal lepidotrichia, along with axial elements tied to axial locomotion through water, including the caudal fin rays.

Tetrapod skull. Reduction occurred in the dermal bones tied to aquatic feeding and respiration, including loss of the last opercular elements (subopercular, preopercular) and anterior tectal and internasal (Gauthier et al. 1988b). The braincase is further enclosed, as the metotic fissure becomes floored by the basioccipital and basisphenoid, and ossified lateral "wings" of the parasphenoid expand beneath the otic capsules. An elongated parasphenoidal cultriform process extends forward below much of the brain. Tetrapods also develop an ossified occiput and craniovertebral joint, heralding independence and mobility of the head on the neck. Also, the lateral line system of the skull lies almost entirely in open canals.

Vomeronasal organ. The vomeronasal organ is a paired structure located in the floor of the nasal chamber, on either side of the nasal septum. It is a chemoreceptor similar in general function to the olfactory epithelium and olfactory nerves and bulb. But unlike olfactory epithelium, its lining is nonciliated and it has separate innervation by the vomeronasal nerve, which projects to an accessory olfactory bulb, rather than to the main bulb as do the olfactory nerves. Its function is largely in reception of pheromones and other molecular mediators of social interaction. There is great elaboration of the vomeronasal organ in squamates, in which it takes on more general environmental functions. The vomeronasal organ was once thought to be absent in primates. But it is present in early development in nearly all mammals, and may be present in humans (Margolis and Getchell 1988, Butler and Hodos 1996, Keverne 1999).

Pan-Tetrapoda

The fossil record of stem tetrapods extends from the Middle Devonian through the Permian and is represented in many parts of the world. However, the fossil record of its sister taxon (Pan-Dipnoi) suggests that the tetrapod stem extends to the Early Devonian or Late Silurian (Clack 2000). At the base of Pan-Tetrapoda lies Osteolepiformes, a diverse group that ranged from the Middle Devonian to Early Permian. One especially well-known member is *Eusthenopteron* (Jarvik 1996), long thought to be ancestral to tetrapods, now seen as a distant cousin. Monophyly of Osteolepiformes is not strongly defended, and some of its members may eventually find other positions near the base of this part of the chordate tree. Also near the base of Pan-Tetrapoda is Rhizontida, which ranged through much of the Devonian and Carboniferous. Its monophyly is well supported by pectoral fin morphology and scale composition (Cloutier and Ahlberg 1996). Some of its members were predators that grew to great size.

Still closer to the tetrapod crown is Elpistostegalia, which include only *Elpistostege* and *Panderichthyes*, from the Late Devonian of North America and Eastern Europe. These taxa are similar to Tetrapoda in having a cranial roofing pattern consisting of paired frontals that lie anterior to the parietals, and in the flattened shape of the head. They also have a straight tail lacking dorsal and ventral lobes, and the dorsal and anal fins are lost. All of these may indicate a shallow-water lifestyle (Clack 2000).

The Devonian taxa *Ichthyostega* (Jarvik 1996) and *Acanthostega* (Clack 1998) are still closer to the tetrapod crown and were long considered to be the basalmost tetrapods because they have hands and feet with discrete digits. However, their hands and feet were very different from those of extant tetrapods, as well as from the condition that was present in their last common ancestor (Gauthier et al. 1989). They have up to eight toes and retain primitive features such as a well-developed gill arch skeleton and lepidotrichia along the tail (lost in Tetrapoda), suggesting that they remained primarily aquatic (Coates and Clack 1990, Cloutier and Ahlberg 1996). One important feature *Ichthyostega* and *Acanthostega* share with crown tetrapods is the fenestra vestibuli, an opening through which the stapes communicates to the inner ear, signaling the beginnings of an airborne-impedance-matching ear.

Node 18. The Amphibians (Amphibia)

Extant amphibians (= Lissamphibia) comprise 4700 extant species that are all distributed among the distinctive frog, salamander (fig. 23.1), and limbless caecilian lineages. All are small and insectivorous and have wet skins that in many cases convey oxygen and other exogenous materials into the body. Hence, they are important as sensitive barometers of freshwater and riparian environments, and many species are facing decline. Their skeletons are pedomorphic in many respects, for example, in the maintenance of extensive cartilage in the adult skeleton, and in the absence of many membranous roofing bones (Djorović and Kalezić 2000). However, they are also highly derived in other respects, and none of the

extant species closely resembles its Paleozoic ancestors. Both molecular and morphological data suggest that frogs and salamanders are more closely related than either is to caecilians (Zardoya and Meyer 1996).

Pan-Amphibia

Relationships at the base of Pan-Amphibia are especially problematic, and more than 100 extinct species have been named for Permo-Carboniferous fossils alone. The problematic aïstopods (Carroll 1998, Anderson et al. 2003), nectrideans, and microsaurs are often regarded as basal members of Pan-Amphibia. However, all are highly derived and their positions uncertain. The most basal divergence among panamphibians was that of the extinct Paleozoic loxammatids (Beaumont and Smithson 1998, Milner and Lindsay 1998). Temnospondyles are generally regarded to include all other panamphibians. Temnospondyles include large extinct *Edops*, *Eryops*, and mastodonsaurids (Damiani 2001) in addition to extant amphibians and a host of other fossils. These basal taxa include large and fully aquatic or amphibious carnivores, some exceeding 2 m in total length. They are distinguished by the opening of large fenestrae in the roof of the palate. However, the extinct lepospondyles have also been regarded as closer relatives of extant amphibians than temnospondyles (Laurin 1998a, 1998b), and the debate remains active. In either case, amphibians and amniotes had diverged from the ancestral tetrapod by the early Carboniferous.

By the Late Triassic, frogs, salamanders, and caecilians had diverged, and left a fairly detailed fossil record. One of the most exciting discoveries occurred in Late Jurassic sediments of northern China, where 500 exceptionally well-preserved salamander specimens were recently recovered. The new finds implicate Asia as the place of salamander diversification (Gao and Shubin 2001). Amphibian history is reviewed in detail by Cannatella and Hillis (ch. 25 in this vol.).

Node 19. Terrestrial Chordates (Amniota)

Amniota (fig. 23.1) comprise the last common ancestor that humans share with Reptilia, and all its descendants (Gauthier et al. 1988a). Although some members became secondarily aquatic, the origin of amniotes heralded the first fully terrestrial chordates. Its monophyly is strongly supported, and its membership is noncontroversial (with the exception of certain Paleozoic fossils). However, relationships among the major living amniote clades are debated. Of principle concern is whether mammals are closest to birds (Gardiner 1982, Løvtrup 1985) or are the sister taxon to other amniotes (Gauthier et al. 1988a, Laurin and Reisz 1995). Arguments linking birds and mammals are based on analyses confined to extant taxa alone, or they treat extant taxa primarily and then secondarily fit selected fossils to that tree. However, when all evidence is analyzed simultaneously, mammals are the sister taxon to other amniotes (Gauthier et al. 1988a, Laurin and Reisz 1995).

Amniote Characters

Amniote egg. The amniote egg and attendant equipment for internal fertilization present a complex of ontogenetic innovations affording reproductive independence from the water. Incubation of the amniote embryo is a more protracted process than before, because the larval stage and metamorphosis are lost, and instead a fully formed young emerges from the egg. Amniote eggs are larger than those of most nonamniotes, with larger volumes of yolk. As the embryo grows, its size produces special problems with respect to metabolic intensity, the exchange of respiratory gases, structural support, and the mobilization and transport of nutrients (Packard and Seymour 1997, Stewart 1997). The outer eggshell takes on an important role in mediating metabolism. It is made of semipermeable collagen fibers and varying proportions of crystalline calcite, which permits respiration while preventing desiccation. The eggshell also provides a calcium repository for the developing skeleton. The embryo is also equipped with several novel extra-embryonic membranes. The amnion encloses a fluid filled cavity in which the embryo develops. The allantois stores nitrogenous wastes, and the chorion is a respiratory membrane. A single penis with erectile tissue is also apomorphic of Amniota (Gauthier et al. 1988a).

The amniote skeleton and dentition. Amniotes have a ball-in-socket craniovertebral joint, which increases the mobility and stability of the head on the neck. They also have two coracoid ossifications in the shoulder girdle, an ossified astragalus in the ankle joint, and they lose fishlike bony scales from the dorsal surface of the body. Teeth are present on the pterygoid transverse process, but there is no infolding of enamel anywhere in the dentition. Also present is an enlarged caniniform maxillary tooth. These changes reflect fully terrestrial feeding and locomotor patterns (Gauthier et al. 1988b, Laurin and Reisz 1995, Sumida 1997).

Loss of lateral line system. The lateral line placodes fail to appear in amniotes (Northcutt 1992), and with their loss is the complete absence of a lateral line system. This is consistent with the view that amniote origins represent increasingly terrestrial habits.

Pan-Amniota

The amniote stem is represented by fossils that extend to the Early Carboniferous (Gauthier et al. 1988b), the oldest being *Casineria* (Paton et al. 1999). The best-known members of the amniote stem include the Carboniferous-Permian anthracosauroids, seymouriamorphs, and diadectomorphs, and a handful of other extinct taxa (Gauthier et al. 1988a, Sumida 1997). Many of the osteological transformations occurring among stem amniotes involved modifications of the dentition and palate, and specialization of the atlanto-axial joint between the head a neck. These modifications reflect an increased role of the mouth in capturing and manipulating terrestrial prey items. Also, there was increased

strengthening of the vertebral column via swelling of the neural arches, the girdles were expanded, the pelvis has an expanded attachment to the sacrum, and the limbs are elongated. Loss of the lateral line system was marked by the disappearance of the canals that it etches into the skull roofing bones. Collectively these features indicate that increasingly terrestrial patterns of locomotion, predation, and prey manipulation preceded the origin of Amniota.

Node 20. The Turtles, Lizards, Crocodilians, and Birds (Reptilia)

The Reptilia are the lineage stemming from the last common ancestor of birds and turtles (fig. 23.1). Reptilians comprise nearly 17,000 living species and enjoy a long and rich fossil record (Gauthier et al. 1988b, Laurin and Reisz 1995, Dingus and Rowe 1998). The name Reptilia was long used in reference to a paraphyletic assemblage of ectothermic amniotes, including turtles, lizards and snakes, crocodilians, and a host of extinct forms. Although long considered to have evolved from reptiles, mammals and birds were excluded from actual membership within it. More recently, the name Reptilia was brought into the phylogenetic system by defining its meaning in reference to the last common ancestor of turtles and birds, and by including birds within it. The name Reptilia has also been used to encompass the extinct relatives of mammals, once known as "mammal-like reptiles." But in the phylogenetic system, these taxa are now referred to under the term Pan-Mammalia (= Synapsida), and the name is rendered monophyletic by including mammals, plus all extinct taxa closer to mammals than to reptiles, within it. Reptile phylogeny is discussed elsewhere in this volume (Lee et al., ch. 26, and Cracraft et al., ch. 27).

Pan-Reptilia

The fossil record of Pan-Reptilia extends into the Late Carboniferous (Gauthier et al. 1988a). *Archaeothyris*, from the Joggins fauna of Nova Scotia, is the oldest panreptile that is known in some detail. In the Early Permian a diversity of poorly known forms are allied as Parareptilia (Gauthier et al. 1988a, Laurin and Reisz 1995, Berman et al. 2000), a tentatively monophyletic clade of extinct taxa that all differ considerably from one another. Their relationships to one another, and to extant turtles and diapsids remains unstable. Included among parareptiles are the Carboniferous-Permian mesosaurs, which seemed highly derived and adapted to a fully aquatic existence. Also often included are the small terrestrial bolosaurids, milleretids, and possibly also the procolophonids and pareiasaurs. The latter two are considered as possible extinct relatives of turtles, and the pareiasaurs are the only members of this basal part of the tree that grew to large adult weights (1000 kg). Pan-Mammalia (see below) dominates the early fossil record of amniotes, because many of its members expressed an early trend toward size increase. Panreptiles, with the exception of pareiasaurs, remained small. By the end of the Triassic, however, these roles reversed, and from then on, the panreptiles dominate the fossil record and extant reptiles are far more numerous and diverse than mammals.

Node 21. Chordates with Hair (Mammalia)

Mammalia comprise the last common ancestor that humans (fig. 23.1) share with living monotremes, plus all its descendants (Rowe 1987, 1988, 1993, Gauthier et al. 1988a, Rowe and Gauthier 1992). It includes approximately 5000 living species and a long fossil record. The mammalian crown extends to the Middle or Early Jurassic, whereas the base of the mammalian stem (Pan-Mammalia or Synapsida) traces to the Late Carboniferous. Mesozoic mammals and their closest extinct relatives were tiny animals, and their fossils are notoriously difficult to collect. Most Mesozoic taxa are named from isolated dentitions or broken jaws, and the early history of mammals was long shrouded by incompleteness. But a host of exciting new discoveries from Asia and South America have yielded relatively complete ancient skeletons. Some were announced together with detailed phylogenetic analyses that are rapidly revising and detailing the early phylogeny of mammals (Hu et al. 1997, Luo et al. 2001a, 2001b, 2002, Rougier et al. 1998). Mammalia is apomorphic in the brain and special senses, body covering, musculature, skeleton, circulatory system, respiratory system, digestive system, reproductive system, metabolism, molecular structure, and behavior (see Rowe 1988, 1993, 1996a, 1996b, Gauthier et al. 1988a: appx. B). Only a few of these are discussed below.

Mammalian Characters

The neocortex.   Compared with even their closest extinct relatives, mammals have large brains. The additional volume marks an episode of heterochrony (peramorphosis) in which the brain began to grow further into ontogeny and more rapidly than in their extinct relatives, marked by the origin of the mammalian neocortex. Its two hemispheres each have a columnar organization of six radial layers, generated in ontogeny by waves of migrating cells that originate from the ventricular zone and move radially outward to their adult positions. This inside-out pattern of neural growth produces a huge cortical volume in mammals. The developing mammalian forebrain hypertrophies into inflated lobes that swell backward over the midbrain and forward around the bases of the olfactory bulbs, which themselves are inflated. The cerebellum is also expanded and deeply folded. The neocortex supports heightened olfactory and auditory senses, and coincident, overlapping sensory and motor maps of the entire body surface. The enlarged cerebellum is related to acquisition and discrimination of sensory information, and the adaptive coordination of movement through a complex three-dimensional environment. These changes may reflect invasion of a nocturnal and/or arboreal niche and have been implicated in the evolution of endothermy (Rowe 1996a, 1996b).

The mammalian middle ear.   In adults, the middle ear skeleton lies suspended beneath the cranium and behind the jaw. It is an impedance matching lever system that contains a chain of tiny ossicles connecting an outer tympanum to the fluid-filled neurosensory inner ear. Its parallel histories in ontogeny and phylogeny are among the most famous in comparative biology. The middle ear arose in premammalian history as an integral component of the mandible. Over a 100 Myr span of premammalian history, its bones were gradually reduced to tiny ossicles, reflecting specialization for increasingly high-frequency hearing, whereas the dentary undertook a greater role in the mandible. Hearing and feeding were structurally linked in premammalian history, but in mammals these functions became decoupled as the auditory chain was detached from the mandible and repositioned behind it, and a new craniomandibular joint arose between the dentary and squamosal bones. Separation of the ossicles from the mandible occurs in all adult mammals and was widely regarded as the definitive mammalian character under Linnaean taxonomy (Rowe 1987, 1988). In ontogeny the auditory chain differentiates and begins growth attached to the mandible. But the connective tissues joining them are torn as the brain grows, and the entire auditory chain (stapes, incus, malleus, ectotympanic) is carried backward during the next few weeks to its adult position behind the jaw. Transposition of the auditory chain is a consequence of its differential growth with respect to the brain. The tiny ear bones quickly reach adult size, whereas the brain continues to grow for many weeks thereafter. As the developing brain balloons, it loads and remodels the rear part of the skull, detaching the ear ossicles from the developing mandible. Many other features of the skull were altered by this dynamic epigenetic relationship between the rapidly growing brain and the tissues around it (Rowe 1996a, 1996b).

Enhanced olfactory system.   The mammalian olfactory system is unique in the breadth of its discriminatory power. Approximately 1000 genes encode odorant receptors in the mammalian nose, making this the largest family in the entire genome (Ressler et al. 1994). Each gene encodes a different type of odorant receptor, and the individual receptor types are distributed in topographically distinct patterns in the olfactory epithelium of the nose. Their discriminatory power is multiplied by increased surface area provided by elaborate scrolling of the bony ethmoid turbinals. This rigid framework enhances olfactory discrimination by facilitating the detection of spatial and temporal information as odorant molecules disperse within the nasal cavity. Each odorant receptor transmits signals directly to a single glomerulus in the olfactory bulb without any intervening synapses; hence, the topographic distribution of odorant receptors over the ethmoid turbinals is mapped in the spatial organization of the olfactory bulb. Ossified turbinals occur only in mammals (and independently in a few birds), although there is ample evidence of unossified turbinals among their extinct relatives. Bone is fundamentally structural, and turbinal ossification may have arisen in response to tighter scrolling, increased surface area, and an increase in the number of olfactory odorant receptors in mammals compared with their closest extinct relatives. The ossified ethmoid turbinal complex may thus be viewed as the skeleton of the olfactory system, arising as an integral component of its distinctive forebrain.

Pan-Mammalia

The mammalian stem lineage, also known as Synapsida, contains mammals plus all extinct species closer to mammals than to Reptilia. Panmammalian fossils range back to the Late Carboniferous, and an exceptionally complete sequence of fossils links extant mammals to the base of their stem. Before phylogenetic systematics, the focus of study was to elucidate the reptile-to-mammal transition. The premammalian segment of this history was believed marked by rampant convergence in the evolution of mammal-like sensory, masticatory, and locomotor systems, and Mammalia itself was held to be a grade rather than a clade. The major debate involved rationalizing which character should mark the boundary between reptilian and mammalian grades. Few claims of homoplasy were substantiated when the characters were subjected to rigorous parsimony analyses, and as synapsids were placed in a taxonomy based on common ancestry (Rowe and Gauthier 1992).

Pan-Mammalia are diagnosed by the lower temporal fenestra and a forward-sloping occiput. Its early history saw enhancement of the locomotor system for fast, agile movement, and elaboration of the feeding system for macro predation. The primitive armature of a tympanic impedance matching ear also appeared early on (Kemp 1983).

A major node on the mammalian stem is Therapsida, whose fossils date back to the Late Permian. The temporal fenestra is larger than before, and there is a deeply incised reflected lamina of the angular (the homologue of the mammalian ectotympanic), and a deep external auditory meatus. These denote an ear more sensitive to a broader range of frequencies. Limb structure indicates a somewhat more erect posture and narrow-tracked gait, possibly facilitating breathing while running and a higher metabolic rate (Kemp 1983).

Cynodontia comprise a node within Therapsida whose monophyly is supported by numerous characters that where passed on to living mammals. The overriding feature of basal cynodonts is that their brain had expanded to completely fill the endocranial cavity, impressing its outer surficial features into the inner walls of the braincase (Rowe 1996a, 1996b, Rowe et al. 1995). Osteological synapomorphies include a broad alisphenoid (epipterygoid) forming the lateral wall to the braincase, and a double occipital condyle that permitted wide ranges of stable excursion of the head about the craniovertebral joint (Kemp 1983). The dentition is differentiated into simple incisiform teeth, a long canine, and postcanine teeth with multiple cusps aligned into a longitudinal row. The dentary was elongated over the postdentary elements, which are reduced and more sensitive to higher frequencies.

Among nonmammalian cynodonts, those closest to crown Mammalia were tiny animals. Miniaturization involved elaborate repackaging of the brain and special sense organs, remodeling of the masticatory system, an accelerated rate of evolution in a complex occlusal dentition. The vertebral column became more strongly regionalized, and the limbs and girdles were modified for scansorial movement. Several episodes of inflation in the size of the brain occurred before the origin of mammals. The recent discovery of *Hadrocodon* (Luo et al. 2001b), from the Early Jurassic of China, may indicate that the neocortex and middle ear transformation originated just outside the mammalian crown, but it is questionable whether *Hadrocodon* lies outside or within the crown. In either event, inflation of the neocortex and detachment of the middle ear appear to coincide.

Discussion

Many of the innovations in chordates design described above arose as unique expressive pathways or as elaborations of preexisting genetic and developmental mechanisms. For example, in all chordates, molecular signaling during neurulation produces anteroposterior regionalization of the embryo, and a brain that divides into rostral, middle, and caudal divisions, each with its own region of unique genetic expression. The genes themselves are more ancient, being expressed in the same tripartite anteroposterior regionalization of the brain in arthropods and other bilaterians. But the inductive pathway of expression in chordates is unique, and it produces a nervous system radically different from that in arthropods, or in what was likely to have been present in bilaterians ancestrally.

Another pattern of morphogenesis and diversification corresponds to successive increases in the numbers of genes. The first episode occurred in either Chordata or Euchordata ancestrally, and in either case was associated with elaboration of brain and sensory organs, as well as with the appearance of mesodermal segmentation. The second occurred in craniates ancestrally and was accompanied by segmentation of the brain into prosomeres, mesomeres, and rhombomeres in early development, as well as enhancement of the adult brain and sensory organs. The third increase occurred in Vertebrata, and the fourth in Gnathostomata ancestrally, each in association with further elaboration of the brain and special senses. Mammalian origins also coincided with an unprecedented increase in the number of olfactory genes. Mammalian olfaction is the most sensitive of any chordate, and with up to 1000 genes coding for different odorant molecule receptors, olfactory genes comprise the largest single mammalian gene family. We can expect many similar examples of this pattern of gene increase and structural elaboration to be mapped in the near future.

The inductive nature of chordate ontogeny provided an especially rich substrate for evolutionary change. The most spectacular example is the neural crest, whose motile cells are induced by the underlying mesoderm and in turn induce many tissues and structures. The neural crest arose in craniates ancestrally, building the embryonic cartilaginous cranium, providing a rigid armature for the brain and special senses, and the skeleton of the pharynx, and providing a novel substrate for the tremendous range of evolutionary variation.

Epigenesis further multiplied these agents of morphogenesis. Origin of the mammalian middle ear may have been one such episode, in which early changes in the timing of development and rate of growth of the brain altered the adjacent connective tissues and the adult structures forming within them. In the wake of the ballooning brain, the rear of the developing mammalian skull is remodeled, and the middle ear ossicles and eardrum were detached and displaced backward from their embryonic attachment to the mandible. The differentiation of neurectoderm is one of the earliest events in ontogeny, and virtually anything that affects its pattern of development will set into motion a new dynamic in the surrounding connective tissues, potentially altering the adult structures that form within them. Just how much adult chordate morphology is epigenetically produced remains to be determined. These examples illustrate that mapping and understanding the relationship between molecules and morphology, as it unfolds in the course of ontogeny, is fundamental to chordate systematics and comparative biology, and understanding our place in the Tree of Life.

Literature Cited

Ahlberg, P. E. 1999. Something fishy in the family tree. Nature 397:564–565.

Aizenberg, J., G. Lambert, S. Weiner, and L. Addadi. 2002. Factors involved in the formation of amorphous and crystalline calcium carbonate: a study of an ascidian skeleton. J. Am. Chem. Soc. 124:32–39.

Amores, A., A. Force, Y. L. Yan, L. Joly, C Amemiya, A. Fritz, R. K. Ho, J. Langeland, V. Prince, Y. L. Wang, et al. 1998. Zebrafish hox clusters and vertebrate genome evolution. Science 282:1711–1714.

Anderson, J. S., R. L. Carroll, and T. Rowe. 2003. New information on *Lethiscus stocki* (Tetrapoda, Lepospondyli, Aïstopoda) from high-resolution x-ray computed tomography and a phylogenetic analysis of Aïstopoda. Can. J. Earth Sci. 40:1071–1083.

Arendt, D., and K. Nübler-Jung. 1996. Common ground plans in early brain development in mice and flies. Bioessays 18:255–259.

Arendt, D., and K. Nübler-Jung. 1999. Comparison of early nerve cord development in insects and vertebrates. Development 126:2309–2325.

Arratia, G., and M. Cloutier. 1996. Reassessment of the morphology of *Cheirolepis canadensis* (Actinopterygii). Pp. 165–197 *in* Devonian Fishes and Plants of Miguasha, Quebec, Canada (H.-P. Schultze and R. Cloutier, eds.). Verlag Dr. Friedrich Pfeil, Munich.

Ayala, F. J., A. Rzhetsky, and F. J. Ayala. 1998. Origin of the metazoan phyla: molecular clocks confirm paleontological estimates. Proc. Natl. Acad. Sci. 95:606–611.

Bardack, D. 1991. First fossil hagfish (Myxinoidea): a record from the Pennsylvanian of Illinois. Science 254:701–703.

Bardack, D., and E. S. Richardson. 1977. New agnathous fishes from the Pennsylvanian of Illinois. Fieldiana Geol. 33:489–510.

Bardack, D., and R. Zangerl. 1968. First fossil lamprey: a record from the Pennsylvanian of Illinois. Science 162:1265–1267.

Barrington, E. J. W. 1963. An introduction to general and comparative endocrinology. Clarendon Press, Oxford

Barrington, E. J. W. 1968. Phylogenetic perspectives in vertebrate endocrinology. Pp. 1–46 *in* Perspectives in Endocrinology (E. J. W. Barrington and C. B. Jorgensen, eds.). Academic Press, New York.

Basden, A. M., G. C. Young, M. I. Coates, and A. Ritchie. 2000. The most primitive osteichthyan braincase? Nature 403:185–188.

Beamish, R. J. 1985. Freshwater parasitic lamprey on Vancouver Island and a theory of the evolution of the freshwater parasitic and nonparasitic life history types. Pp. 123–140 *in* Evolutionary biology of primitive fishes (R. E. Foreman, A. Gorbman, M. J. Dodd, and R. Olsson, eds.). Plenum, New York.

Beaumont, E. H., and T. R. Smithson. 1998. The cranial morphology and relationships of the aberrant Carboniferous amphibian *Spathicephalus mirus* Watson. Zool. J. Linn. Soc. 122:187–209.

Berman, D. S., R. R. Reisz, D. Scott, A. C. Henrici, S. Sumida, and T. Martens. 2000. Early Permian bipedal reptile. Science 290:969–972.

Bodsnick, D., and R. G. Northcutt. 1981. Elecroreception in lampreys: evidence that the earliest vertebrates were electroreceptive. Science 212:465–467.

Bromham, L., A. Rambaut, R. Forey, A. Cooper, and D. Penny. 1998. Testing the Cambrian explosion hypothesis by using a molecular dating technique. Proc. Natl. Acad. Sci. USA 95:12386–12389.

Butler, A. B. 2000. Chordate evolution and the origin of craniates: an old brain in a new head. Anat. Rec. (New Anat.) 261:111–125.

Butler, A. B., and W. Hodos. 1996. Comparative vertebrate neuroanatomy. Wiley-Liss, New York.

Cameron, C. B., J. R. Garey, and B. J. Swalla. 2000. Evolution of the chordate body plan: new insights from phylogenetic analyses of deuterostome phyla. Proc. Natl. Acad. Sci. USA 97:4469–4474.

Carroll, R. L. 1988. Vertebrate paleontology and evolution. W. H. Freeman, New York.

Carroll, R. L. 1998. Cranial anatomy of ophiderpetontid aïstopods: paleozoic limbless amphibians. Zool. J. Linn. Soc. 122:143–166.

Carroll, R. L. 2001. The origin and early radiation of terrestrial vertebrates. J. Paleontol. 75:1202–1213.

Chang, M.-M. 1991. "Rhipidistians," dipnoans, and tetrapods. Pp. 3–28 *in* Origins of the higher groups of tetrapods (H. P. Schultze and L. Treub, eds.). Comstock, Ithaca, NY.

Chen, J.-Y., J. Dzik, G. D. Edgecombe, L. Ramsköld, and G.-Q. Zhou. 1995. A possible Early Cambrian chordate. Nature 377:720–722.

Chen, J.-Y., D.-Y. Huang, and C.-W. Li. 1999. An early Cambrian craniate-like chordate. Nature 402:518–522.

Clack, J. A. 1998. The neurocranium of *Acanthostega gunnari* Jarvik and the evolution of the otic region in tetrapods. Zool. J. Linn. Soc. 122:61–97.

Clack, J. A. 2000. The origin of tetrapods. Pp. 979–1029 *in* Amphibia Biology, Vol. 4: Paleontology, the evolutionary history of amphibians (H. Heatwole and R. L. Carroll, eds.). Surrey Beatty and Sons, London.

Cloutier, R. 1991a. Diversity of extinct and living actinistian fishes (Sarcopterygii). Environ. Biol. Fishes 32:59–74.

Cloutier, R. 1991b. Patterns, trends, and rates of evolution within the Actinistia. Environ. Biol. Fishes 32:23–58.

Cloutier, R., and P. E. Ahlberg. 1996. Morphology, characters, and the interrelationships of basal sarcopterygians. Pp. 445–479 *in* Interrelationships of fishes (M. L. J. Stiassny, L. R. Parenti, and G. D. Johnson, eds.). Academic Press, San Diego.

Coates, M. I., and J. A. Clack, 1990. Polydactyly in the earliest known tetrapod limbs. Nature 347:66–69.

Compagno, L. J. V. 1977. Phyletic relationships of living sharks and rays. Am. Zool. 17:303–322.

Damiani, R. J. 2001. A systematic revision and phylogenetic analysis of Triassic mastodonsaurids (Temnospondyli: Stereospondyli). Zool. J. Linn. Soc. 133:379–482.

de Carvalho, M. R. 1996. Higher-level elasmobranch phylogeny, basal squaleans, and paraphyly. Pp. 35–62 *in* Interrelationships of fishes (M. L. J. Stiassny, L. R. Parenti, and G. D. Johnson, eds.). Academic Press, San Diego.

Delgado, S., D. Casane, L. Bonnaud, M. Laurin, J.-Y. Sire, and M. Girondot. 2001. Molecular evidence for Precambrian origin of amelogenin, the major protein of vertebrate enamel. Mol. Biol. Evol. 18:2146–2153.

De Pinna, M. C. C. 1996. Teleostean monophyly. Pp. 147–162 *in* Interrelationships of fishes (M. L. J. Stiassney, L. R. Parenti, and G. D. Johnson, eds.). Academic Press, San Diego.

de Queiroz, K., and J. A. Gauthier. 1992. Phylogenetic taxonomy. Annu. Rev. Ecol. Syst. 23:449–480.

Digital Morphology. 2003. Available: http://www.digimorph. org. Last accessed 25 December 2003.

Dingus, L., and T. Rowe. 1998. The mistaken extinction: dinosaur evolution and the origin of birds. W. H. Freeman, New York.

Djorović, A., and M. L. Kalezivć. 2000. Paedogenesis in European newts (Triturus: Salamandridae): cranial morphology during ontogeny. J. Morphol. 243:127–139.

Dominguez, P., A. G. Jacobson, and R. P. F. Jeffries. 2002. Paired gill slits in a fossil with a calcite skeleton. Nature 417:841–844.

Donoghue, M. J., J. Doyle, J. A. Gauthier, A. G. Kluge, and T. Rowe 1989. Importance of fossils in phylogeny reconstruction. Annu. Rev. Ecol. Syst. 20:431–460.

Donoghue, P. C. J., P. L. Forey, and R. J. Aldridge. 2000. Conodont affinity and chordate phylogeny. Biol. Rev. 75:191–251.

Dryer, L. 2000. Evolution of odorant receptors. Bioessays 22:803–810.

Edgeworth, F. H. 1935. The cranial muscles of vertebrates. Cambridge University Press, Cambridge.

Erdmann, M. V., R. L. Caldwell, S. L. Jewett, and M. K. Moosa. 1998. Indonesian "king of the sea" discovered. Nature 395:335.

Evrard, Y. A., Y. Lun, A. Aulehla, L. Gan, and R. L. Johnson. 1998. *Lunatic fringe* is an essential mediator of somite segmentation and patterning. Nature 394:377–381.

Forey, P. L. 1984a. The coelacanth as a living fossil. Pp. 166–169 *in* Living fossils (N. Eldredge and S. M. Stanley, eds.). Springer Verlag, New York.

Forey, P. L. 1984b. Yet more reflections on agnathan-gnathostome relationships. J. Vert. Paleontol. 4:330–343.

Forey, P. L. 1998. History of the coelacanth fishes. Chapman and Hall, London.

Fritzsch, B. 1996. Similarities and differences in lancelet and craniate nervous systems. Isr. J. Zool. 42:S147–S160.

Froese, R., and D. Pauly. 2001. Fishbase. Database. Available: http://www.fishbase.org. Last accessed 25 December 2003.

Gagnier, P.–Y. 1989. The oldest vertebrate: a 470 million years old jawless fish, *Sacabambaspis janvieri*, from Ordovician of Bolivia. Natl. Geogr. Res. 5:250–253.

Galis, F. 2001. Evolutionary history of vertebrate appendicular musculature. Bioessays 23:383–387.

Gans, C., and C. J. Bell. 2001. Vertebrates, overview. Pp. 755–766 *in* Encyclopedia of biodiversity (S. A. Levin, ed.). Academic Press, San Diego.

Gans, C., N. Kemp, and S. Poss (eds.). 1996. The lancelets (Cephalochordata): a new look at some old beasts. Isr. J. Zool. 42(suppl.).

Gao, K.-Q., and N. H. Shubin. 2001. Late Jurassic salamanders from northern China. Nature 410:574–577.

Gardiner, B. 1982. Tetrapod classification. Zool. J. Linn. Soc. 74:207–232.

Gardiner, B. G. 1984. The relationships of placoderms. J. Vert. Paleontol. 4:379–395.

Gauthier, J. A., D. Cannatella, K. de Queiroz, A. G. Kluge, and T. Rowe. 1989. Tetrapod phylogeny. Pp. 337–353 *in* The hierarchy of life. (B. Fernholm, H. Bremer, H. Jörnvall, eds.). Nobel Symposium 70. Excerpta Medica, Amsterdam.

Gauthier, J. A., and K. de Queiroz. 2001. Feathered dinosaurs, flying dinosaurs, crown dinosaurs, and the name "Aves." Pp. 7–41 *in* New perspectives on the origin and early evolution of birds (J. Gauthier and L. Gall, eds.). Yale University Press, New Haven, CT.

Gauthier, J., A. G. Kluge, and T. Rowe. 1988a. Amniote phylogeny and the importance of fossils. Cladistics 4:105–209.

Gauthier, J., A. G. Kluge, and T. Rowe. 1988b. The early evolution of the Amniota. Pp. 103–155 *in* The phylogeny and classification of the tetrapods, Vol. 1: Amphibians, reptiles and birds (M. Benton, ed.). Syst. Assoc. Spec. Vol. 35a. Clarendon Press, Oxford.

Gegenbaur, C. 1878. Elements of comparative anatomy. Macmillan, London.

Gehring, W. J. 1998. Master control genes in development and evolution. Yale University Press, New Haven, CT.

Gerhart, J. 2000. Inversion of the chordate body axis: are there alternatives? Proc. Natl. Acad. Sci. USA 97:4445–4448.

Goodwin, M., M. M. Miyamoto, and J. Czelusniak. 1987.

Pattern and process in vertebrate phylogeny revealed by coevolution of molecules and morphology. Pp. 141–176 *in* Molecules and morphology in evolution: conflict or compromise? (Colin Patterson, ed.). Cambridge University Press, Cambridge.

Gould, S. J. 1989. Wonderful life. Norton, New York.

Graham, A., and A. Smith. 2001. Patterning the pharyngeal arches. Bioessays 23:54–61.

Grande, L., and W. E. Bemis. 1996. Interrelationships of Acipenseriformes, with comments on "Chondrostei." Pp. 85–116 *in* Interrelationships of fishes (M. L. J. Stiassney, L. R. Parenti, and G. D. Johnson, eds.). Academic Press, San Diego.

Hall, B. K. 1992. Evolutionary developmental biology. Chapman and Hall, London.

Hardisty, M. W. 1979. Biology of the cyclostomes. Chapman and Hall, London.

Hardisty, M. W. 1982. Lampreys and hagfishes: analysis of cyclostome relationships. Pp. 165–260 *in* The biology of lampreys (M. W. Hardisty and I. C. Potter, eds.), vol. 4B. Academic Press, New York.

Hedges, S. B. 2001. Molecular evidence for the early history of living vertebrates. Pp. 119–134 *in* Major events in early vertebrate evolution (P. Ahlberg, ed.). Taylor and Francis, London and New York.

Hedges, S. B., C. A. Hass, and L. R. Maxson. 1993. Relations of fish and tetrapods. Nature 363:501–502.

Hillis, D. M., and M. T. Dixon 1989. Vertebrate phylogeny: evidence from 28S ribosomal DNA sequences. Pp. 355–367 *in* The hierarchy of life (B. Fernholm, K. Bremer, and H. Jörnvall, eds.). Elsevier, New York.

Holder, M. T., M. V. Erdmann, T. P. Wilcox, R. L. Caldwell, and D. M. Hillis. 1999. Two living species of coelacanths? Proc. Natl. Acad. Sci. USA 96:12616–12620.

Holland, N. D., and J. Chen. 2001. Origin and early evolution of the vertebrates: new insights from advances in molecular biology, anatomy, and palaeontology. Bioessays 23:142–151.

Holland, P. W. H., and J. Garcia-Fernàndez. 1996. Hox genes and chordate evolution. Dev. Biol. 173:382–394.

Hu, Y., Y. Wang, Z. Luo, and C. Li. 1997. A new symmetrodont mammal from China and its implications for mammalian evolution. Nature 390:137–142.

Hyodo, S., S. Ishii, and J. M. P. Joss. 1997. Australian lungfish neurohypophysial hormone genes encode vasotocin and [Phe²] mesotocin precursors homologous to tetrapod-type precursors. Proc. Natl. Acad. of Sci. USA 94:13339–13344.

Jacobson, A. G. 1988. Somitomeres: mesodermal segments of vertebrate embryos. Development 104(suppl.):209–220.

Jacobson, A. G. 2001. Somites and head mesoderm arise from somitomeres. Pp. 16–29 *in* The origin and fate of somites (E. J. Sanders, ed.). IOS Press, Amsterdam.

Jamieson, B. G. M. 1991. Fish evolution and systematics: evidence from spermatozoa. Cambridge University Press, Cambridge.

Janvier, P. 1996. Early vertebrates. Clarendon Press, Oxford.

Janvier, P. 1997. Hyperoarteia *in* Tree of Life (D. Maddison, ed.). Available: http://www.tolweb.org. Last accessed 25 December 2003.

Janvier, P. 1999. Catching the first fish. Nature 402:21–22.

Janvier, P., and R. Lund. 1983. *Hardistiella montanensis* n. gen. et sp. (Petromyzontida) from the Lower Carboniferous of Montana, with remarks on the affinities of lampreys. J. Vert. Paleontol. 2:407–413.

Jarvik, E. 1996. The evolutionary importance of *Eusthenopteron foordi* (Osteolepiformes). Pp. 285–315 *in* Devonian fishes and plants of Miguasha, Quebec, Canada (H.-P. Schultze and R. Cloutier, eds.). Verlag Dr. Friedrich Pfeil, Munich.

Jeffries, R. P. S. 1979. The origin of chordates—a methodological essay. Pp. 443–477 *in* The origins of major invertebrate groups (M. R. House, ed.). Syst. Assoc. Spec. Vol. 12. Academic Press, London.

Jeffries, R. P. S., 1980. Zur Fossilgeschichte des Ursprungs der Chordaten und der Echinodermaten. Zool. Jahrb. Jena Abt. Anat. 103:285–353.

Jeffries, R. P. S. 1986. Ancestry of the vertebrates. Cambridge University Press, New York.

Jeffries, R. P. S. 2001. The origin and early fossil record of the acustico-lateralis system, with remarks on the reality of the echinoderm-hemichordate clade. Pp. 40–66 *in* Major events in early vertebrate evolution (P. Ahlberg, ed.). Taylor and Francis, London.

Jeffries, R. P. S., and D. N. Lewis. 1978. The English Silurian fossil *Placocystites forbesianus* and the ancestry of the vertebrates. Philos. Trans. R. Soc. Lond. 282:205–323.

Jiang, Y. J., B. L. Aerne, L. Smithers, C. Haddon, D. Ish-Horowicz, and J. Lewis. 2000. Notch signaling and the synchronization of the somite segmentation clock. Nature 408:475–479.

Johansen, K. 1970. Air breathing in fishes. Pp. 361–413 *in* Fish phylogeny (W. S. Hoar and D. J. Randall, eds.), vol. 4. Academic Press, London.

Jollie, M. 1982. What are the "Calcichordata"? and the larger question of the origin of chordates. Zool. J. Linn. Soc. 75:167–188.

Kemp, T. S. 1983. Mammal-like reptiles and the origin of mammals. Academic Press, London.

Keverne, E. B. 1999. The vomeronasal organ. Science 286:716–720.

Kumar, S., and S. B. Hedges. 1998. A molecular time scale for vertebrate evolution. Nature 392:917–920.

Lacalli, T. C. 1996a. Frontal eye circuitry, rostral sensory pathways and brain organization in amphioxus larvae: evidence from 3D reconstructions. Philos. Trans. R. Soc. Lond. B 351:243–263.

Lacalli, T. C. 1996b. Landmarks and subdomains in the larval brain of Branchiostoma: vertebrate homologs and invertebrate antecedents. Isr. J. Zool. 42:S131–S146.

Lagios, M. D. 1979. The Coelacanth and Chondrichthyes as sister groups: a review of shared apomorphous characters and a cladistic analysis and reinterpretation. Pp. 25–44 *in* The biology and physiology of the living coelacanth (J. E. McCosker and M. D. Lagios, eds.). Calif. Acad. Sci. Occasional Papers 134.

Laird, D. J., A. W. De Tomaso, M. D. Cooper, and I. L. Weissman. 2000. 50 million years of chordate evolution: seeking the origins of adaptive immunity. Proc. Natl. Acad. Sci. USA 97:6924–6926.

Lanfranchi, G., A. Pallavicini, P. Laveder, and G. Valle. 1994. Ancestral hemoglobin switching in lampreys. Dev. Biol. 164:402–408.

Lauder, G. V., and K. F. Liem. 1983. The evolution and interrelationships of the actinopterygian fishes. Bull. Mus. Comp. Zool. 150:95–197.

Laurin, M. 1998a. The importance of global parsimony and historical bias in understanding tetrapod evolution. Part I—systematics, middle ear evolution, and jaw suspension. Ann. Sci. Nat. Zool. Paris 19:1–42.

Laurin, M. 1998b. The importance of global parsimony and historical bias in understanding tetrapod evolution. Part II—vertebral centrum, costal ventilation, and paedomorphosis. Ann. Sci. Nat. Zool. Paris 19:99–114.

Laurin, M., and Reisz, R. R. 1995. A reevaluation of early amniote phylogeny. Zool. J. Linn. Soc. 113:165–223.

Lipscomb, D. L., J. S. Farris, M. Källersjo, and A. Tehler. 1998. Support, ribosomal sequences and the phylogeny of eukaryotes. Cladistics 14:303–338.

Litman, G. W., M. K. Anderson, and J. P. Rast. 1999. Evolution of antigen binding receptors. Annu. Rev. Immunol. 17:109–147.

Long, J. A. 1995. The rise of fishes. Johns Hopkins University Press, Baltimore, MD.

Løvtrup, S. 1977. The phylogeny of Vertebrata. John Wiley and Sons, London.

Løvtrup, S. 1985. On the classification of the taxon Tetrapoda. Syst. Zool. 34:463–470.

Lund, R., and W. L. Lund. 1985. Coelacanths from the Bear Gulch Limestone (Namurian) of Montana and the evolution of the Coelacanthiformes. Bull. Carnegie Mus. Nat. Hist. 25:1–74.

Luo, Z.-X., R. L. Cifelli, and Z. Kielan-Jaworowska. 2001a. Dual origin of tribosphenic mammals. Nature 409:53–57.

Luo, Z.-X., A. W. Crompton, and A.-L. Sun. 2001b. A new mammaliaform from the Early Jurassic and evolution of mammalian characteristics. Science 292:1535–1540.

Luo, Z.-X., Z. Kielan-Jaworowska, and R. L. Cifelli. 2002. In quest for a phylogeny of Mesozoic mammals. Acta Palaeontol. Pol. 47:1–78.

Maisey, J. 1984. Chondrichthyan phylogeny: a look at the evidence. J. Vert. Paleontol. 4:359–371.

Maisey, J. 1986. Heads and tails: a chordate phylogeny. Cladistics 2:201–256.

Maisey, J. 1988. Phylogeny of early vertebrate skeletal induction and ossification patterns. Pp. 1–36 in Evolutionary biology (M. K. Hecht, B. Wallace, and G. T. Prance, eds.), vol. 22. Plenum, New York.

Maisey, J. G. 2001. Remarks on the inner ear of elasmobranches and its interpretation from skeletal labyrinth morphology. J. Morphol. 250:236–264.

Maisey, J. G., and E. M. Anderson. 2001. A primitive chondrichthyan braincase from the Early Devonian of South Africa. J. Vert. Paleontol. 21:702–713.

Mallatt, J. 1984. Feeding ecology of the earliest vertebrates. Zool. J. Linn. Soc. 82:261–272.

Mallatt, J. 1985. Reconstructing the life cycle and the feeding of ancestral vertebrates. Pp. 59–68 in Evolutionary biology of primitive fishes (R. E. Foreman, A. Gorbman, J. M. Dodd, and R. Olsson, eds.). Plenum, New York.

Mallatt, J. 1997a. Crossing a major morphological boundary: the origin of jaws in vertebrates. Zoology 100:128–140.

Mallatt, J. 1997b. Shark pharyngeal muscles and early vertebrate evolution. Acta Zool. Stockh. 78:279–294.

Mallatt, J., and J. Sullivan. 1998. 28S and 18S rDNA sequences support the monophyly of lampreys and hagfishes. Mol. Biol. Evol. 15:1706–1718.

Mallatt, J., J. Sullivan, and C. J. Winchell. 2001. The relationship of lampreys to hagfishes: a spectral analysis of ribosomal DNA sequences. Pp. 106–118 in Major events in early vertebrate evolution (P. Alberch, ed.). Taylor and Francis, London.

Manzanares, M., H. Wada, N. Itasaki, P. R. Trainor, R. Krumlauf, and P. W. H. Holland. 2000. Conservation and elaboration of Hox gene regulation during evolution of the vertebrate head. Nature 408:854–857.

Margolis, F. L., and T. V. Getchell (eds.). 1988. Molecular neurobiology of the olfactory system. Plenum, New York.

Marshall, C., and H.-P. Schultze. 1992. Relative importance of molecular, neontological and paleontological data in understanding the biology of the vertebrate invasion of the land. J. Mol. Evol. 35:93–101.

Mazan, S., D. Jaillard, B. Baratte, and P. Janvier. 2000. OTX1 gene-controlled morphogenesis of the horizontal semicircular canal and the origin of the ganthostome characteristics. Evol. Dev. 2:186–193.

McMahon, T., and J. T. Bonner. 1985. On size and life. Scientific American Library, New York.

Meinertzhagen, I. A., and Y. Okamura 2001. The larval ascidian nervous system: the chordate brain from its small beginnings. Trends Neurosci. 24:401–410.

Metzger, R. J., and M. A. Krasnow. 1999. Genetic control of branching morphogenesis. Science 284:1635–1639.

Meyer, A. 1995. Molecular evidence on the origin of tetrapods and the relationships of the coelacanth. Trends Ecol. Evol. 10:111–116.

Milner, A. R., and A. C. Lindsay. 1998. Postcranial remains of Baphetes and their bearing on the relationships of the Baphetidae (= Loxommatidae). Zool. J. Linn. Soc. 1998:211–235.

Moyle, P. B., and J. J. Cech, Jr. 2000. Fishes, an introduction to ichthyology. 4th ed. Prentice Hall, New York.

Nelson, G. 1989. Phylogeny of major fish groups. Pp. 325–336 in The hierarchy of life (B. Fernholm, K. Bremer, and H. Jörnvall, eds.). Nobel Symposium 70. Excerpta Medica, Amsterdam.

Newman, S. A., and W. D. Comper. 1990. "Generic" physical mechanisms of morphogenesis and pattern formation. Development 110:1–18.

Niedert, A. H., V. Virupannavar, G. W. Hooker, and J. A. Langeland. 2001. Lamprey Dlx genes and early vertebrate evolution. Proc. Natl. Acad. Sci. USA 96:1665–1670.

Nielsen, C. 1999. Origin of the chordate central nervous system and the origin of chordates. Dev. Genes Evol. 209:198–205.

Northcutt, R. G. 1992. The phylogeny of the octavolateralis ontogenies: a reaffirmation of Garstang's phylogenetic hypothesis. Pp. 21–47 in The evolutionary biology of hearing (D. B. Webster, R. R. Fay, and A. N. Popper, eds.). Springer Verlag, New York.

Northcutt, R. G., and C. Gans. 1983. The genesis of neural crest and epidermal placodes: a reinterpretation of vertebrate origins. Q. Rev. Biol. 58:1–28.

Oster, G. F., N. Shubin, J. D. Murray, and P. Alberch. 1988. Evolution and morphogenetic rules: the shape of the vertebrate limb in ontogeny and phylogeny. Evolution 42:862–884.

Packard, M. J., and R. S. Seymour. 1997. Evolution of the amniote egg. Pp. 265–290 in Amniote origins (S. S. Sumida and K. L. M. Martin, eds.). Academic Press, San Diego.

Paton, R. L., T. R. Smithson, and J. A. Clack. 1999. An amniote-like skeleton from the Early Carboniferous of Scotland. Nature 1999:508–513.

Philippe, H., A. Chenuil, and A. Adoutte. 1994. Can the Cambrian explosion be inferred through molecular phylogeny? Development (suppl.):15–25.

Pohlmann, K., F. W. Grasso, and T. Breithaupt. 2001. Tracking wakes: the nocturnal predatory strategy of piscivorous catfish. Proc. Natl. Acad. Sci. USA 98:7371–7374.

Poss, S. G., and H. T. Boschung. 1996. Lancelets (Cephalochordata: branchiostomidae): How many species are valid? Isr. J. Zool. 42:S13–S66.

Pouyaud, L., S. Wirjoatmodjo, I. Rachmatika, A. Tjakrawidjaja, R. Hadiaty, and W. Hadie. 1999. Un nouvelle espèce de cœlacanthe: preuves génétiques et morphologiques. Compt. Rend. Acad. Sci. 322:261–267.

Reisz, R. R., and M. M. Smith. 2001. Lungfish dental pattern conserved for 360 million years. Nature 411:548.

Ressler, K. J., S. L. Sullivan, and L. B. Buck. 1994. Information coding in the olfactory system: evidence for a stereotyped and highly organized epitope map in the olfactory bulb. Cell 79:1245–1255.

Romer, A. S. 1966. Vertebrate paleontology. University of Chicago Press, Chicago.

Rosen, D. E., P. L. Forey, B. G. Gardiner, and C. Patterson. 1981. Lungfishes, tetrapods, paleontology, and plesiomorphy. Bull. Am. Mus. Nat. Hist. 167:162–275.

Rougier, G. W., J. R. Wible, and M. J. Novacek. 1998. Implications of Deltatheridium specimens for early marsupial history. Nature 396:459–463.

Rowe, T. 1987. Definition and diagnosis in the phylogenetic system. Syst. Zool. 36:208–211.

Rowe, T. 1988. Definition, diagnosis and origin of Mammalia. J. Vert. Paleontol. 8:241–264.

Rowe, T. 1993. Phylogenetic systematics and the early history of mammals. Pp. 129–145 in Mammalian phylogeny (F. S. Szalay, M. J. Novacek, and M. C. McKenna, eds.). Springer-Verlag, New York.

Rowe, T. 1996a. Brain heterochrony and evolution of the mammalian middle ear. Pp. 71–96 in New perspectives on the history of life (M. Ghiselin and G. Pinna, eds.). Calif. Acad. Sci. Memoir 20.

Rowe, T. 1996b. Coevolution of the mammalian middle ear and neocortex. Science 273:651–654.

Rowe, T., C. A. Brochu, and K. Kishi (eds.). 1999. Cranial morphology of Alligator and phylogeny of Alligatoroidae. J. Vert. Paleontol. 19(2 suppl.), mem. 6.

Rowe, T., W. Carlson, and W. Bottorff. 1995. Thrinaxodon: digital atlas of the skull (CD-ROM). 2nd ed. (MS Windows and Macintosh). University of Texas Press, Austin.

Rowe, T., and J. Gauthier. 1992. Ancestry, paleontology, and definition of the name Mammalia. Syst. Biol. 41:372–378.

Rowe, T., J. Kappelman, W. D. Carlson, R. A. Ketcham, and C. Denison. 1997. High-resolution computed tomography: a breakthrough technology for Earth scientists. Geotimes 42:23–27.

Ruben, J. A., and A. F. Bennett. 1980. Antiquity of the vertebrate pattern of activity metabolism and its possible relation to vertebrate origins. Nature 286:886–888.

Russell, E. S. 1916. Form and function. John Murray, London.

Sansom, I. J., M. P. Smith, H. A. Armstrong, and M. M. Smith. 1992. Presence of earliest vertebrate hard tissues in conodonts. Science 256:1308–1311.

Sansom, I. J., M. M. Smith, and M. P. Smith. 1996. Scales of thelodont and shark-like fishes from the Ordovician of Colorado. Nature 379:628–630.

Schaeffer, B. 1987. Deuterostome monophyly and phylogeny. Pp. 179–235 in Evolutionary biology (M. K. Hecht, B. Wallace, and G. T. Prance, eds.), vol. 21. Plenum, New York.

Schaeffer, B., and M. Williams. 1977. Relationships of fossil and living elasmobranches. Am. Zool. 17:293–302.

Schluter, S. F., R. M. Bernstein, H. Bernstein, and J. J. Marchalonis. 1999. "Big bang" emergence of the combinatorial immune system. Dev. Comp. Immunol. 23:107–111.

Schultze, H.-P. 1970. Folded teeth and the monophyletic origin of tetrapods. Am. Mus. Nov. 2408:1–10.

Schultze, H.-P. 1987. Dipnoans as sarcopterygians. J. Morphol. 1(suppl.):39–74.

Schultze, H.-P. 1992. Dipnoi. Pp. 1–464 in Fossilium catalogus I: Animalia (F. Westphal, ed.), pt. 131. Kugler Publications, Amsterdam.

Schultze, H.-P. 1994. Comparisons of hypotheses on the relationships of sarcopterygians. Syst. Biol. 43:155–173.

Shimeld, S. M., and P. W. H. Holland. 2000. Vertebrate innovations. Proc. Natl. Acad. Sci. USA 97:4449–4452.

Shirai, S. 1996. Phylogenetic interrelationships of neoselachians (Chondrichthyes: Euselachii). Pp. 9–34 in Interrelationships of fishes (M. L. J. Stiassny, L. R. Parenti, and G. D. Johnson, eds.). Academic Press, San Diego.

Shu, D. G., L. Chen, J. Han, and X.-L. Zhang. 2001a. An Early Cambrian tunicate from China. Nature 411:472–473.

Shu, D. G., S. Conway Morris, J. Han, L. Chen, X. L. Zhang, H. Q. Liu, Y. Li, and J. N. Liu. 2001b. Primitive deuterostomes from the Chengjiang Lagerstätte (Lower Cambrian, China). Nature 414:419–424.

Shu, D. G., S. Conway Morris, and X. L. Zhang. 1996. A Pikaia-like chordate from the Lower Cambrian of China. Nature 384:157–158.

Shu, D. G., S. Conway Morris, X.-L. Zhang, L. Chen, Y. Li, and J. Han. 1999a. A pipiscid-like fossil from the Lower Cambrian of south China. Nature 400:746–749.

Shu, D. G., H. L. Luo, S. Conway Morris, X. L. Zhang, S. X. Hu, L. Chen, J. Han, M. Zhu, Y. Li, and L. Z. Chen. 1999b. Lower Cambrian vertebrates from south China. Nature 402:42–46.

Shubin, N. H., and P. Alberch. 1986. A morphogenetic approach to the origin and basic organization of the tetrapod limb. Pp. 319–387 in Evolutionary biology (M. K. Hecht, B. Wallace, and G. T. Prance, eds.), vol. 20. Plenum, New York.

Shubin, N. H., C. Tabin, and S. Carroll. 1997. Fossils, genes and the evolution of animal limbs. Nature 388:639–647.

Slack, J. M. W. 1983. From egg to embryo: determinative events in early developmental history. Cambridge University Press, London.

Slavkin, H. C., E. E. Graham, M. Zeichner-David, and W. Hildemann. 1983. Enamel-like antigens in hagfish: possible evolutionary significance. Evolution 37:404–412.

Smith, M. M., and B. K. Hall. 1990. Development and evolutionary origins of vertebrate skeletogenic and odontogenic tissues. Biol. Rev. 65:277–373.

Stensiö, E. A. 1968. They cyclostomes with special reference to the diphyletic origin of the Petromyzontida and Myxinoidea. Pp. 13–71 in Current problems in lower veterbrate phylogeny (T. Ørvig, ed.). Nobel Symposium 4. Excerpta Medica, Amsterdam.

Stewart, J. 1997. Morphology and evolution of the egg of oviparous amniotes. Pp. 291–326 in Amniote origins (S. S. Sumida and K. L. M. Martin, eds.). Academic Press, San Diego.

Sumida, S. S. 1997. Locomotor features spanning the origin of amniotes. Pp. 353–398 in Amniote origins (S. S. Sumida and K. L. M. Martin, eds.). Academic Press, San Diego.

Suzuki, M., K. Kubokawa, H. Nagasawa, and A. Urano. 1995. Sequence analysis of vasotocin cDNAs of the lamprey Lampetra japonica, and the hagfish Eptatretus burgeri: evolution of cyclostome vasotocin precursors. J. Mol. Endocrinol. 14:67–77.

Swalla, B. J., and W. R. Jeffery. 1996. Requirement of the Manx gene for expression of chordate features in a tailless ascidian larva. Science 274:1205–1208.

Turbeville, J. M., J. R. Schulz, and R. A. Raff. 1994. Deuterostome phylogeny and the sister group of the chordates—evidence from molecules and morphology. Mol. Biol. Evol. 11:648–655.

Venkatesh, B., M. V. Erdmann, and S. Brenner. 2001. Molecular synapomorphies resolve evolutionary relationships of extant jawed vertebrates. Proc. Natl. Acad. Sci. USA 98:11382–11387.

Venkatesh, B., Y. Ning, and S. Brenner. 1999. Late changes in spliceosomal introns define clades in vertebrate evolution. Proc. Natl. Acad. Sci. USA 96:10267–10271.

von Wahlert, G. 1968. Latimeria und die Geschichte der Wirbeltiere: eine evolutionsbiologische Untersuchung. Gustav Fischer Verlag, Stuttgart.

Webb, J. F., and D. M. Noden. 1993. Ectodermal placodes: contributions to the development of the vertebrate head. Am. Zool. 33:434–447.

Wiley, E. O. 1979. Ventral gill arch muscles and the phylogenetic relationships of Latimeria. Pp. 56–67 in The biology and physiology of the living coelacanth (J. E. McCosker and M. D. Lagios, eds.). Calif. Acad. Sci. Occasional Papers 134.

Wray, G. A., J. S. Levinton, and L. H. Shapiro. 1996. Molecular evidence for deep Precambrian divergences among metazoan phyla. Science 274:568–573.

Yalden, D. W. 1985. Feeding mechanisms as evidence of cyclostome monophyly. Zool. J. Linn Soc. 84:291–300.

Zardoya, R., and A. Meyer. 1996. Evolutionary relationships of the coelacanth, lungfishes, and tetrapods based on the 28S ribosomal RNA gene. Proc. Natl. Acad. Sci. USA 93:5449–5454.

Zhou, H., and L. Niswander. 1996. Requirement for BMP signaling in interdigital apoptosis and scale formation. Science 272:738–741.

Zhu, M., and H.–P. Schultze. 1997. The oldest sarcopterygian fish. Lethaia 30:293–304.

Zhu, M., X. Yu, and P. E. Ahlberg. 2001. A primitive fossil sarcopterygian fish with an eyestalk. Nature 410:81–84.

Zhu, M., X. Yu, and P. Janvier. 1999. A primitive fossil fish sheds light on the origin of bony fishes. Nature 397:607–610.

M. L. J. Stiassny

E. O. Wiley

G. D. Johnson

M. R. de Carvalho

Gnathostome Fishes

Gnathostomata are a species-rich assemblage that, with the exclusion of the Petromyzontiformes (lampreys, 45 spp.), represents all living members of Vertebrata. Gnathostomes are most notably characterized by the possession of endoskeletal jaws primitively formed of dorsal palatoquadrate and ventral Meckelian cartilages articulating at a mandibular joint. Our task here is to provide a review of a large (paraphyletic) subset of gnathostome diversity—an artificial grouping often referred to as the "jawed fishes": chondrichthyans, "piscine sarcopterygians," and actinopterygians. We treat all living jawed vertebrates with the exclusion of most Sarcopterygii—the tetrapods—since they are discussed in other chapters. After a review of the chondrichthyans or cartilaginous fishes and a brief summary of the so-called "piscine sarcopterygians," we focus our contribution on the largest and most diverse of the three groups, the Actinopterygii, or rayfin fishes.

As a guide to the chapter, figure 24.1 presents, in broad summary, our understanding of the interrelationships among extant gnathostome lineages and indicates their past and present numbers (with counts of nominal families indicated by column width through time). Much of the stratigraphic information for osteichthyans is from Patterson (1993, 1994), and that for chondrichthyans is mostly from Cappetta et al. (1993).

Chondrichthyes (Cartilaginous Fishes)

Chondrichthyans (sharks, rays, and chimaeras) include approximately 1000 living species (Compagno 1999), several dozen of which remain undescribed. Recent sharks and rays are further united in the subclass Elasmobranchii (975+ spp.), whereas the chimaeras form the subclass Holocephali (35+ spp.). All chimaeras are marine; as are most sharks and rays, but about 15 living elasmobranch species are euryhaline, and some 30 are permanently restricted to freshwater (Compagno and Cook 1995).

Chondrichthyans are characterized by perichondral prismatic calcification; the prisms form a honeycomb-like mosaic that covers most of the cartilaginous endoskeleton (Schaeffer 1981, Janvier 1996). Paired male intromittent organs derived from pelvic radials (claspers) are probably another chondrichthyan synapomorphy, although they are unknown in some early fossil forms (e.g., the Devonian *Cladoselache* and Carboniferous *Caseodus*), but all recent chondrichthyans and most articulated fossil taxa have them (Zangerl 1981). Earlier notions that sharks, rays, and chimaeras evolved independently from placoderm ancestors (Stensiö 1925, Holmgren 1942, Ørvig 1960, 1962; Patterson 1965), culminating in the Elasmobranchiomorphi (placoderms + chondrichthyans) of Stensiö (e.g., 1958, 1963, 1969) and Jarvik (e.g., 1960, 1977, 1980), have not survived close inspection (e.g., Compagno 1973, Miles and Young 1977); chondrichthyan monophyly is no longer seriously challenged (Schaeffer 1981, Maisey 1984).

Sharks, rays, and chimaeras form an ancient lineage. The earliest putative remains are dermal denticles from the Late Ordovician of Colorado [some 450 million years ago (Mya)]; the first braincase is from the Early Devonian of South Af-

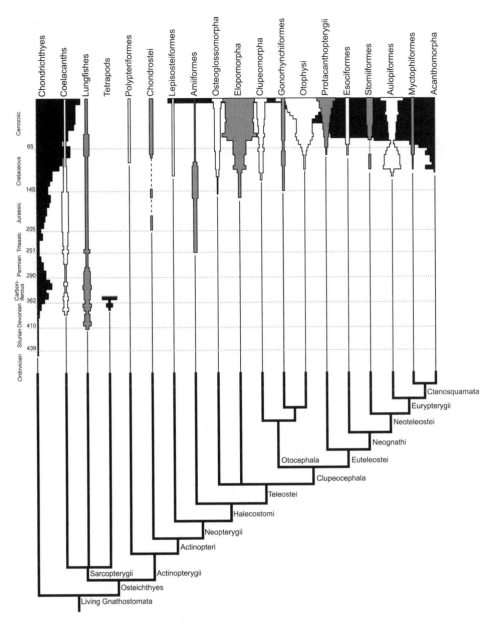

Figure 24.1. Current estimate of relationships among extant gnathostome lineages. Past and present counts of nominal families are indicated by column width through time (tetrapod diversity truncated, chondrichthyan diversity truncated to the left, and acanthomorph diversity truncated to the right). Stratigraphic information for Osteichthyes is taken from Patterson (1993, 1994) but with new data for Polypteriformes from Dutheil (1999) and for Otophysi from Filleul and Maisey (in press). Data for Chondrichthyes are from Cappatta et al. (1993), with complementary information from Janvier (1996) and other sources. For practical reasons, familial diversity is charted and this does not necessarily reflect known species diversity.

rica some 60 million years later (Maisey and Anderson 2001). The divergence between elasmobranchs and holocephalans is also relatively old, because isolated holocephalan tooth plates are known from the Late Devonian (Zangerl 1981, Stahl 1999), and articulated specimens from the Early Carboniferous (320 Mya; Lund 1990, Janvier 1996). A few of the earliest known fossil sharks may be basal to the elasmobranch–holocephalan dichotomy, such as *Pucapampella* from the Devonian of Bolivia (Maisey 2001), but much work remains to be done in early chondrichthyan phylogeny (Coates and Sequeira 1998). Sharks were remarkably diverse morphologically and ecologically during much of the Paleozoic, considerably more so than early bony fishes. Some 32 families existed during the Carboniferous, but many of these went extinct before the end of the Permian (Cappetta et al. 1993; fig. 24.1).

The entrenched notion that sharks are primitive or ancestral vertebrates because of their antiquity, "generalized design," and lack of endochondral (cellular) bone (e.g., Dean 1895, Woodward 1898) is contradicted by the theory that bone may have been lost in sharks, because it is widely distributed among stem gnathostomes (Stensiö 1925, Maisey 1986). Furthermore, acellular bone is present in the dorsal spine-brush complex of an early shark (*Stethacanthus*; Coates et al. 1999) and also in the teeth, denticles, and vertebrae of extant chondrichthyans (Kemp and Westrin 1979, Hall 1982, Janvier 1996), supporting the assertion that sharks evolved from bony ancestors. Highly complex, derived attributes of elasmobranchs, such as their semicircular canal arrangement (Schaeffer 1981), internal fertilization, and formation of maternal–fetal connections ("placentas" of some living forms; Hamlett and Koob 1999), reveal, in fact, that sharks are much more "advanced" than previously thought.

Elasmobranchs (Sharks and Rays)

Modern sharks and rays share with certain Mesozoic fossils (e.g., *Palaeospinax, Synechodus*) calcified vertebrae and specialized enameloid in their teeth (both secondarily lost in some living forms) and are united with them in Neoselachii (Schaeffer 1967, 1981, Schaeffer and Williams 1977, Maisey 1984). Most of modern elasmobranch diversity originated in the Late Cretaceous to Early Tertiary (some 55–90 Mya), but several extant lineages have fossil members, usually represented by isolated teeth, dating back to the Early Jurassic (some 200 Mya).

Recent phylogenetic studies have recognized two major lineages of living elasmobranchs, Galeomorphi (galeomorph sharks) and Squalomorphi (squalomorphs or squaleans; Shirai 1992, 1996, Carvalho 1996; fig. 24.2). These studies, however, differ in the composition of Hexanchiformes and Squaliformes, and in relation to the coding and interpretation of many features; the tree adopted here (fig. 24.2) is modified from Carvalho (1996).

The phylogeny in figure 24.2 is the most supported by morphological characters, but an alternative scheme has been proposed on the basis of the nuclear *RAG-1* gene (J. G. Maisey, pers. comm.), in which modern sharks are monophyletic without the rays (an "all-shark" hypothesis). Stratigraphic data are slightly at odds with both hypotheses, but more so with the morphological one, because there are no Early Jurassic squaloids, pristiophoroids, or squatinoids. But lack of stratigraphic harmony will persist unless these taxa

are demonstrated to comprise a crown group within a monophyletic "all-shark" collective (i.e., with galeomorphs basal to them). Nonetheless, dozens of well-substantiated morphological characters successively link various shark and all batoid groups in Squalomorphi, many of which would have to be overturned if sharks are to be considered monophyletic to the exclusion of rays.

Historically, some of the difficulties in discerning relationships among elasmobranchs have been due to the highly derived design of certain taxa (e.g., angelsharks, sawsharks, batoids, electric rays), which has led several workers (e.g., Regan 1906, Compagno 1973, 1977) to isolate them in their own lineages, ignoring their homologous features shared with other elasmobranch groups (Carvalho 1996). Elevated levels of homoplasy (Shirai 1992, Carvalho 1996, McEachran and Dunn 1998), coupled with the lack of dermal ossifications (a plentiful source of systematically useful characters in bony fishes), hinders the recovery of phylogenetic patterns within elasmobranchs. Moreover, the (erroneous) notion that there is nothing left to accomplish in chondrichthyan systematics is unfortunately common. In fact, the situation is quite the contrary, because many taxa are only "phenetically" defined and require rigorous phylogenetic treatment (e.g., within Carcharhiniformes and Myliobatiformes). However, many morphological complexes still require more in-depth descriptive and comparative study (in the style of Miyake 1988, Miyake et al. 1992) before they can be confidently used in phylogenetic analyses.

The general morphology, physiology, and reproduction of extant sharks and rays are comprehensively reviewed in Hamlett (1999). Fossil forms are discussed in Cappetta (1987) and Janvier (1996). Below is a brief account of ex-

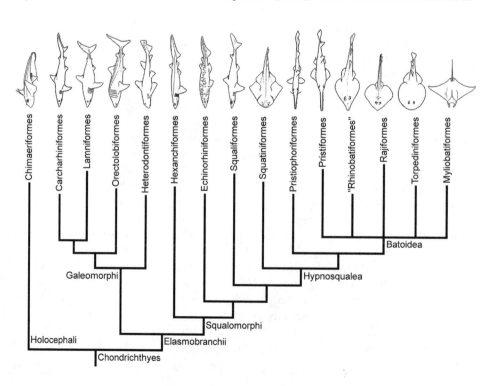

Figure 24.2. Intrarelationships of extant chondrichthyan lineages based mostly on Carvalho (1996). Relationships among rays (Batoidea) are left unresolved, with guitarfishes (Rhinobatiformes) in quotation marks because the group is probably not monophyletic (see McEachran et al. 1996).

tant elasmobranch orders; their monophyly ranges from the relatively well established (Orectolobiformes) to the poorly defined (Squaliformes; Compagno 1973, 1977, Shirai 1996, Carvalho 1996).

Galeomorph sharks encompass four orders (fig. 24.2): Heterodontiformes (bullhead sharks), Orectolobiformes (carpet sharks), Lamniformes (mackerel sharks), and Carcharhiniformes (ground sharks). Galeomorphs have various specializations (Compagno 1973, 1977), such as the proximity between the hyomandibular fossa and the orbit on the neurocranium, and are the dominant sharks of shallow and epipelagic waters worldwide (Compagno 1984b, 1988, 2001).

The two most basal galeomorph orders are primarily benthic, inshore sharks. Bullheads (*Heterodontus*, eight spp.) are distributed in tropical and warm-temperate seas of the western and eastern Pacific Ocean and western Indian Ocean (Compagno 2001). *Heterodontus* has a unique dentition, composed of both clutching and grinding teeth, and is oviparous. It was once believed to be closely related to more primitive Mesozoic hybodont sharks (which also had dorsal fin spines) and therefore regarded as a living relic (e.g., Woodward 1889, Smith 1942), but its ancestry with modern (galeomorph) sharks is strongly corroborated (Maisey 1982). Orectolobiforms (14 genera, 32+ spp.) are among the most colorful elasmobranchs, occurring in tropical to warm-temperate shallow waters; they are most diverse in the Indo-West Pacific region but occur worldwide. Species are aplacentally viviparous or oviparous. One orectolobiform, the planktophagous whale shark (*Rhincodon typus*), is the largest known fish species, reaching 15 m in length. Derived characters of carpet sharks include their complete oronasal grooves and arrangement of cranial muscles (Dingerkus 1986, Goto 2001). Their taxonomy is reviewed in Compagno (2001), and their intrarelationships in Dingerkus (1986) and Goto (2001). An alternative view recognizes bullheads and carpet sharks as sister groups (Compagno 1973; fig. 24.2).

From a systematic perspective, Lamniformes (10 genera, 15 spp.) contain some of the best-known sharks, characterized by their "lamniform tooth pattern" (Compagno 1990, 2001). Although their low modern-day diversity pales compared with the numerous Cretaceous and Tertiary species described from isolated teeth (Cappetta 1987), this order contains some of the most notorious sharks, such as the great white (Klimley and Ainley 1997), its gigantic fossil cousin *Carcharodon megalodon* (Gottfried et al. 1996), the megamouth (now known from some 15 occurrences worldwide; Yano et al. 1997), and the filter-feeding basking shark. Lamniforms are yolk-sac viviparous, and adelphophagy (embryos consuming each other in utero) and oophagy (embryos eating uterine eggs) have been documented in some species (Gilmore 1993). Molecular data sets (Naylor et al. 1997, Morrissey et al. 1997) are at odds with morphological ones (and with each other), indicating that the jury is still out in relation to the evolutionary history of lamniform genera.

Carcharhiniformes (48 genera, 216+ spp.) are by far the largest order of sharks, containing more than half of all living species, and about half of all shark genera (Compagno 1984b). Carcharhiniforms have specialized secondary lower eyelids (nictitating eyelids), as well as unique clasper skeletons (Compagno 1988). Species are oviparous (Scyliorhinidae) or viviparous, with or without the development of a yolk-sac placenta (Hamlett and Koob 1999). Ground sharks range from sluggish, bottom-dwelling catsharks (Scyliorhinidae, the largest shark family) to epipelagic, streamlined, and active requiem sharks (Carcharhinidae), which includes some of the most common and economically important species (e.g., blue and tiger sharks, *Carcharhinus* spp.). Hammerhead sharks (Sphyrnidae) are morphologically very distinctive (Nakaya 1995) and capable of complex behavioral patterns (e.g., Myrberg and Gruber 1974). Some ground sharks may be restricted to freshwater (*Glyphis* spp.), and the bull shark, *Carcharhinus leucas*, penetrates more than 4000 km up the Amazon River, reaching Peru. New species have been described in recent years, particularly of catsharks (e.g., Nakaya and Séret 1999, Last 1999), and additional new species await formal description (Last and Stevens 1994). Phylogenetic relationships among ground sharks requires further study (Naylor 1992), which may eventually result in the merging of several currently monotypic genera and some of the families. Compagno (1988) presents a comprehensive review of the classification and morphology of Carcharhiniformes.

Squalomorphs (equivalent to the Squalea of Shirai 1992) are a very diverse and morphologically heterogeneous group that includes the six- and seven-gill sharks (Hexanchiformes), bramble sharks (Echinorhiniformes), dogfishes and allies (Squaliformes), angelsharks (Squatiniformes), sawsharks (Pristiophoriformes), and rays (Batoidea; fig. 24.2). These taxa share complete precaudal hemal arches in the tail region, among many other features (Shirai, 1992, 1996, Carvalho 1996). Many previous authors defended similar arrangements for the squalomorphs, but usually excluded one group or another (e.g., Woodward 1889, White 1937, Glickman 1967, Maisey 1980). The most dramatic evolutionary transition among elasmobranchs has taken place within the squalomorphs—the evolution of rays from sharklike ancestors, which probably took place in the Early Jurassic (some 200 Mya). *Protospinax*, from the Late Jurassic (150 Mya) Solnhofen limestones of Germany, is an early descendent of the shark–ray transition because it is the most basal hypnosqualean (fig. 24.2), sister group to the node uniting angelsharks, sawsharks, and batoids, and has features intermediate between sharks and rays (Carvalho and Maisey 1996).

Basal squalomorph lineages are relatively depauperate; hexanchiforms (four genera, five spp.) and bramble sharks (*Echinorhinus*, two spp.) are mostly deep-water inhabitants of the continental slopes but occasionally venture into shallow water. All species are aplacentally viviparous. Hexanchiforms have a remarkable longevity; fossil skeletons date from the Late Jurassic. They are united by several derived characters, such as an extra gill arch and pectoral propterygium separated from its corresponding radials (Compagno 1977, Carvalho 1996; compare Shirai 1992, 1996, which do not support hexanchiform monophyly). The frilled shark, *Chlamydoselachus anguineus*, is one of the strangest living sharks, with an enormous gape, triple-cusped teeth, and eel-like body. Some researchers even thought it was a relic of Paleozoic "cladodont" sharks (reviewed in Gudger and Smith 1933). *Echinorhinus* has traditionally been classified with the Squaliformes (Bigelow and Schroeder 1948, Compagno 1984a) but was given ordinal status by Shirai (1992, 1996, Carvalho 1996); studies of its dentition further support this conclusion (Pfeil 1983, Herman et al. 1989).

Squaliformes (20 genera, 121+ spp.), Squatiniformes (*Squatina*, 15+ spp.), and Pristiophoriformes (two genera, five or more spp.) form successive sister groups to the rays (Batoidea, 73+ genera, 555+ spp.). The squaliform dogfishes are mesopelagic, demersal, and deep-water species that vary greatly in size (from 25 cm *Euprotomicrus* to 6 m *Somniosus*). Many species are economically important, and new species continue to be described (Last et al. 2002). They are aplacentally viviparous, and some have the longest gestation periods of all vertebrates (*Squalus*, some 24 months). Shirai (1992, 1996) and Carvalho (1996) disagree in relation to the composition of this order, which is recognized as monophyletic by Carvalho, but broken into several lineages by Shirai. Squatiniforms (angelsharks) are morphologically unique, benthic sharks that resemble rays in being dorsoventrally flattened with expanded pectoral fins. They are distributed worldwide, but most species are geographically restricted (Compagno 1984a). Pristiophoriforms (sawsharks) are poorly known benthic inhabitants of the outer continental shelves (Compagno 1984a). They first appear in the fossil record during the Late Cretaceous of Lebanon (some 90 Mya) and have an elongated rostral blade ("saw") with acute lateral rostral spines that are replaced continuously through life; the saw is used to stun and kill fishes by slashing it from side to side. Similar to angelsharks, sawsharks are yolk-sac viviparous.

Rays (batoids), once thought to represent a gargantuan evolutionary leap from sharklike ancestors (e.g., Regan 1906), are best understood as having evolved through stepwise anatomical transformations from within squalomorphs. Sawsharks are their sister group, sharing with rays various characters (Shirai 1992), such as enlarged supraneurals extending forward to the abdominal area. But at least one feature traditionally considered unique to rays (the antorbital cartilage) can be traced down the tree to basal squalomorphs, in the form of the ectethmoid process (Carvalho and Maisey 1996) of hexanchiforms, *Echinorhinus*, and squaliforms, or as an unchondrified "antorbital" in pristiophoriforms (Holmgren 1941, Carvalho 1996). Even though "advanced" rays are very modified (e.g., *Manta*), basal rays retain various sharklike traits such as elongated, muscular tails with dorsal fins.

In precladistic days, Batoidea were traditionally divided into five orders (e.g., Compagno 1977): Pristiformes (sawfishes, two genera, five or more spp.), "Rhinobatiformes" (guitarfishes, nine genera, 50+ spp.), Rajiformes (skates, 28 genera, 260+ spp.), Torpediniformes (electric rays, 10 genera, 55+ spp.), and Myliobatiformes (stingrays, 24 genera, 185+ spp.). Phylogenetic analyses have revealed that Rhinobatiformes is not monophyletic (Nishida 1990, McEachran et al. 1996), but all other groups are morphologically well defined (Compagno 1977, McEachran et al. 1996). There is conflict as to which batoid order is the most basal, whether it is sawfishes (Compagno 1973, Heemstra and Smith 1980, Nishida 1990, Shirai 1996) or electric rays (Compagno 1977, McEachran et al. 1996). The most comprehensive phylogenetic study to date is that of McEachran et al. (1996); molecular analyses have hitherto contributed very little to the resolution of this problem (e.g., Chang et al. 1995). Rays are clearly monophyletic, with ventral gill openings, synarcual cartilages, and an anteriorly expanded propterygium, among other characters (e.g., Compagno 1973, 1977). There is as much morphological distinctiveness among the different groups of rays as there is among the orders of sharks. The oldest ray skeletons are from the Late Jurassic of Europe and are morphologically reminiscent of modern guitarfishes (Saint-Seine 1949, Cavin et al. 1995), but their relationships require further study (see Carvalho, in press).

Sawfishes are large batoids (up to 6 m long), present in inshore seas and bays, but also in freshwaters. The precise number of species is difficult to determine because of the paucity of specimens but is between four and seven; some are critically endangered because of overfishing and habitat degradation (Compagno and Cook 1995). They differ from sawsharks in the arrangement of canals for vessels and nerves within the rostral saw and in the mode of attachment of rostral spines. Guitarfishes are widespread in tropical and warm temperate waters, and are economically important. Much work is needed on their species level taxonomy; the last comprehensive revision was by Norman (1926). Characters supporting their monophyly are known, but they are undoubtedly a heterogeneous assemblage that requires subdivision (as in McEachran et al. 1996); for simplicity they are treated as a single taxon in figure 24.2. Electric rays are notorious for their electrogenic abilities. Although known since antiquity, they have been neglected taxonomically until very recently (e.g., Carvalho 1999, 2001). Their electric organs are derived from pectoral muscles and can produce strong shocks that are actively used to hunt prey (Bigelow and Schroeder 1953, Lowe et al. 1994). All electric ray species are marine, in tropical to temperate waters, and some occur in deep water. Skates are oviparous (all other rays are viviparous), marine, mostly deep water and more abundant in temperate areas. They also produce weak discharges from caudal electric organs (Jacob et al. 1994). Even though skates are the most species-rich chondrichthyan assemblage, they are rather conservative morphologically. Rajiform intrarelationships have been studied by McEachran (1984), McEachran and Miyake (1990), and McEachran and Dunn (1998). Many new species still await description (J. D. McEachran, pers. comm.). Stingrays are also highly diverse (Last and Stevens 1994) and are found in both marine and freshwaters (the 20+ species of South American potamotrygonid stingrays are the only supraspecific chondrichthyan group restricted to freshwater). Stingrays can be very colorful and range from 15 cm (*Urotrygon microphthalmum*) to 5 m (*Manta*) across the disk. Stingray intrarelationships have been recently investigated by Nishida (1990), Lovejoy (1996), and McEachran et al. (1996). Stingray embryos are nourished in utero by milk-like secretions from trophonemata (Hamlett and Koob 1999); there are at least 10 undescribed species.

Holocephalans (Chimaeras)

Living holocephalans represent only a fraction of their previous (mostly Carboniferous) diversity. As a result, fossil holocephalans (summarized in Stahl 1999) have received more attention from systematists than have extant forms. The single surviving holocephalan order (Chimaeriformes) contains three extant families: Chimaeridae (2 genera, 24+ spp.), Callorhynchidae (*Callorhinchus*, three spp.), and Rhinochimaeridae (three genera, eight spp.). Chimaeras are easily distinguished from elasmobranchs, with opercular gill covers, open lateral-line canals, three pairs of crushing tooth plates with hypermineralized pads (tritors), and frontal tenacula on their foreheads (Didier 1995). Most species are poorly known, deep-water forms of relatively little economic significance. All chimaeras are oviparous, and some of their egg capsules are highly sculptured (Dean 1906). Relationships among living holocephalans is summarized by Didier (1995). New species are still being described (e.g., Didier and Séret 2002), but relationships among chimaeriform species are unknown.

Osteichthyes (Bony Fishes)

Before the advent of *Phylogenetic Systematics* (Hennig 1950, 1966, and numerous subsequent authors), Osteichthyes constituted only bony fishes; tetrapod vertebrates were classified apart as coordinate groups (usually ranked as classes). With the recognition that vertebrate classifications should strictly reflect evolutionary relationships, it has become apparent that Osteichthyes cannot include only the bony fishes, but must also include the tetrapods. Thus, there are two great osteichthyan groups of approximately equal size: Sarcopterygii (lobefins and tetrapods) and Actinopterygii (rayfins). Here, we briefly review the so-called "piscine sarcopterygians," or lobefins, before considering the largest, and most diverse radiation of the jawed fishes, the actinopterygians or rayfins.

Sarcopterygii (Lobefin Fishes and Tetrapods)

The lobefin fishes and tetrapods comprise some 24,000+ living species of fishes, amphibians, and amniote vertebrates (mammals; birds, crocodiles; turtles; snakes, lizards, and kin) with a fossil record extending to the Upper Silurian. All sarcopterygians are characterized by the evolutionary innovation of having the pectoral fins articulating with the shoulder girdle by a single element, known as the humerus in tetrapods. In contrast, actinopterygian fishes retain a primitive condition similar to that seen in sharks, in which numerous elements connect the fin with the girdle. A rich record of fossil lobefin fishes provides numerous "transitional forms" leading to Tetrapoda (Cloutier and Ahlberg 1996, Zhu and Schultze 1997, Zhu et al. 1999, Clack 2002). Two living groups survive, lungfishes and coelacanths.

Lungfishes

There are six living species of lungfishes, one in Australia (*Neoceratodus forsteri*), one in South America (*Lepidosiren paradoxa*), and four in Africa (*Protopterus* spp.). All are freshwater, but there are more than 60 described fossil genera dating back to the Devonian, almost all of which were marine. Of the living lungfishes all except the Australian species share an ability to survive desiccation by aestivating in burrows. This lifestyle is ancient; Permian lungfishes are commonly found preserved in their burrows. Considerable controversy surrounds the interrelationship of lungfishes. Most recent studies place them at (Zhu and Schultze 1997) or near (Cloutier and Ahlberg 1996) the base of the sarcopterygian tree, although some ichthyologists have claimed that they are the closest relatives of Tetrapoda (Rosen et al. 1981), a view recently supported with molecular evidence by Venkatesh et al. (2001).

Coelacanths

Coelacanths were once thought to have become extinct in the Cretaceous. The discovery of a living coelacanth off the coast of South Africa in 1938 caused a sensation in the zoological community [Weinberg (2000) presents a very readable history; see also Forey (1998)]. Between the 1950s and the 1990s, extant coelacanths were thought to be endemic to the Comoro Islands. But in 1997 Arnaz and Mark Erdmann photographed a specimen in a fish market in Indonesia (Sulawesi) and eventually obtained a specimen through local fishermen (Erdmann 1999). Since that time, coelacanths have been discovered off South Africa, Kenya, and Madagascar [see Third Wave Media Inc. (2003) for accounts of these discoveries and other coelacanth news]. Like lungfishes, the phylogenetic position of coelacanths has been subject to some dispute. Cloutier and Ahlberg (1996) placed them at the base of Sarcopterygii; Zhu and Schultze (1997) placed them near the clade containing Tetrapoda.

Actinopterygii (Rayfin Fishes)

The actinopterygian fossil record is rich, but unlike that of most other vertebrate groups, there are far more living forms than known fossils. The exact number of rayfin fishes remains to be determined, but most authors agree that the group minimally consists of some 23,600–26,500 living species, with approximately 200–250 new species being described each year (Eschmeyer 1998). Early actinopterygian fishes are characterized by several evolutionary innovations (synapomorphies) still found in extant relatives (Schultze and Cumbaa 2001). These include several technical features of the skull and paired fins, and the composition and morphology of the scales [see Janvier (1996) for an excellent overview of actinopterygian anatomy]. The earliest well-preserved actinopterygian, *Dialipina*, from the lower Devonian of Canada and Siberia, retains several primitive features of their osteichthyian and gnathostome ancestors, such as two dorsal fins (Schultze and Cumbaa 2001).

Living actinopterygian diversity resides mostly in the crown group Teleostei (see below), but between the species-rich teleosts and the base of Actinopterygii are a number of small but interesting living groups allied with a much more diverse but extinct fauna. For example, an actinopterygian thought to represent the closest living relative of teleost fishes is the North American bowfin, *Amia calva* (Patterson 1973, Wiley 1976, Grande and Bemis 1998). The bowfin is the last remaining survivor of a much larger group of fishes (the Halecomorphi) that radiated extensively in the Mesozoic and whose fossil representatives have been found in marine and freshwater sediments worldwide. As another example, between and below the branches leading to the living bichirs and the living sturgeons and paddlefishes are a whole series of Paleozoic fishes generally termed "palaeoniscoids." They display a dazzling array of morphologies, many paralleling the body forms now observed among teleost fishes and probably reflecting similar life styles. A review of this fossil diversity is beyond the scope of this chapter, but the reader can refer to Grande (1998) and Gardiner and Schaeffer (1989). However, fossil diversity has important consequences for our study of the evolution of characters. When we only consider living groups on the Tree of Life, we might get the impression that the appearance of some groups was accompanied by massive morphological change. This is usually not the case. When the fossils are included, we gain a very different impression: most of the evolutionary innovations we associate with major groups are gained over many speciation events, and the distinctive nature of the living members of the group is largely due to the extinction of its more basal members. Thus, it is true that the living teleost fishes are distinguished from their closest relatives by a large number of evolutionary innovations (DePinna 1996). Yet, when we include all the fossil diversity, this impressive number is, according to Arratia (1999), significantly reduced. Of course,

this is to be expected; evolution by large saltatory steps is more the exception than the rule, because derived characters were acquired gradually. Another example is that gnathostomes, today remarkably diverse and divergent in anatomy, appear to have been very similar to each other shortly after their initial separation, because many features were primitively retained in now extinct stem gnathostome lineages (Basden et al. 2000, Maisey and Anderson 2001, Zhu et al. 2001, Zhu and Schultze 2001).

Living Actinopterygian Diversity and Basal Relationships

Wiley (1998) and Stiassny (2002) provide nontechnical overviews of basal actinopterygian diversity, and the review of Lauder and Liem (1983) remains a valuable and highly readable summary of actinopterygian relationships. The most basal of living actinopterygians are the bichirs (Polypteridae), a small group (11 spp.) of African fishes previously thought to be related to the lobefin fishes (sarcopterygians), or to form a third group. Despite past controversy, two recent molecular studies provide additional support for the birchirs as the basal living actinopterygian lineage (Venkatesh et al. 2001, Inoue et al. 2002), and this placement now seems well established. Compared with other rayfin fishes, birchirs are distinctive in having a rather broad fin base (even giving the external appearance of a lobe fin), a dorsal fin composed of a series of finlets running atop an elongate body, and only four gill arches. Although the analysis by Schultze and Cumbaa (2001) places them one branch above the basal *Dialipina*, their fossil record only just extends to the Lower Cretaceous (Dutheil 1999), a geologic enigma, but such a disparity between the phylogenetic age of a taxon and its first known fossil occurrence is not uncommon among rayfin fishes (fig. 24.1).

The living chondrostean fishes include the sturgeons of the Holarctic and the North American and Chinese paddlefishes. The comprehensive morphological analyses of Grande and Bemis (1991, 1996) have established a hypothesis of relationships among the living and fossils members of this group, which originated in the Paleozoic. The diversification of the living chondrosteans may go back to the Jurassic (Zhu 1992), when paddlefishes and sturgeons were already diversified. Paddlefishes and sturgeons retain many primitive characters, such as a strongly heterocercal tail that led some 19th century ichthyologists to believe that they are related to sharks. Sturgeons are among the most endangered, sought after, and largest of freshwater fishes. The Asian beluga *Huso huso* reaches at least 4 m in length, and a large female may yield 180 kg of highly prized caviar. Paddlefish caviar is also prized, and the highly endangered Chinese paddlefish grows to twice the size of its American cousins, reaching 3 m.

The remaining rayfin fishes belong to the clade Neopterygii. Garfishes (Lepisosteidae) are considered by most to be the basal group (Patterson 1973, Wiley 1976). They form an exception among rayfin fishes in that there are as many living gars (a mere seven species) as fossil forms. Although fossils are known from many regions of the world and their record extends to the Lower Cretaceous, living gars are now confined to North and Middle America and Cuba.

Amia calva, the North American bowfin, is the sole living representative of Halecomorphi, a group that radiated in the Mesozoic. It shares a number of evolutionary innovations with teleost fishes (first detailed by Patterson 1973) but also displays a number of teleost characters that are now considered convergent, such as having cycloid rather than ganoid scales. Although most workers have followed Patterson (1973) in the recognition of *Amia* as the closest living relative of the Teleostei, there remains some controversy about their systematic position (Patterson 1994); alternative schemes of basal neopterygian relationships and the proximate relatives of the Teleostei are reviewed in Arratia (2001).

Teleostei

Among vertebrates, without doubt, Teleostei dominate the waters of the planet. The earliest representatives of living teleost lineages (the Teleocephala of DePinna 1996) date to the Late Jurassic some 150 Mya, but as noted by Arratia (2001), if definitions of the group are to include related fossil lineages, this date is pushed back into the Late Triassic–Early Jurassic (~200–210 Mya). Regardless of how fossil lineages are incorporated into definitions of the group, today's teleosts occupy almost every conceivable aquatic habitat from high-elevation mountain springs more than 5000 m above sea level to the ocean abyss almost 8500 m below. Estimates of the number of living species vary, but most authors agree that a figure of around 26,000 is reasonable. Although discovery rates are more or less constant at around 200–250 new species a year, for some groups, particularly those in little explored or inaccessible habitats, new species are being described in extraordinary numbers, for example, 30 new snailfishes from deep water off Australia (Stein et al. 2001) with some 70 more to be described from polar seas, or an estimated 200 new rock-dwelling cichlids from Lake Victoria, Africa (Seehausen 1996). There are more teleost species than all other vertebrates combined, and their number contrasts starkly with the low species diversity in their immediate amiiform relatives, or indeed of all basal actinopterygian lineages. Among actinopterygians the extraordinary species richness of the teleostean lineage is noteworthy, and although "adaptationist" explanations are not readily testable, it seems probable that much of their success may be attributed to the evolution of the teleost caudal skeleton, permitting increased efficiency and flexibility in movement (Lauder 2000), and to the evolution of powerful suction feeding capabilities that have facilitated a wide range of feeding adaptations (Liem 1990).

Teleostean Basal Relationships

Systematic ichthyology has a rich history, and the past three centuries have seen waves of progress and revision. But in the modern era, perhaps one of the most important contributions on teleost relationships was that of Greenwood et al. (1966; fig. 24.3). In that paper, the authors presented a tentative scheme of relationships among three main lineages, Elopomorpha (tarpons and eels), Osteoglossomorpha (elephantfishes and kin), and what are now known as the Euteleostei (all "higher teleosts," including such groups as cods and basses). Greenwood et al. (1966) found placement of Clupeomorpha (herrings and allies) problematic, but most subsequent workers have placed them as the basal euteleosts. Recently, however, this alignment has been challenged (see below). As Patterson (1994) later noted, it was as if the distinction between monotremes, marsupials, and placental mammals was not recognized until the mid 1960s.

By 1989, Gareth Nelson summarized the previous 20 years of ichthyological endeavor with the by now much quoted observation that "recent work has resolved the bush at the bottom but that the bush at the top persists." He presented a summary tree that showed a fully resolved scheme of major teleostean lineages as a comb leading to the spiny rayed Acanthomorpha that contains the percomorph "bush at the top."

The outstanding problem of Percomorpha is discussed below, but it is perhaps also worth noting that some recent studies have begun to challenge the notion of a fully resolved teleostean tree and to question the monophyly of some lineages (e.g., Lê et al. 1993, Johnson and Patterson 1996, Arratia 1997, 1999, 2000, 2001, Filleul and Lavoué 2001, Inoue et al. 2001, Miya et al. 2001, 2003). This is perhaps not surprising given that Nelson (1989) was somewhat guarded in his optimism and noted that although the interrelationships of major groups of fishes were resolved no group was defined by more than a few characters. Results of more refined matrix-based analyses that incorporate broader taxon sampling than the previously more standard "exemplar " approaches, the inclusion of new high quality fossil data, and the beginnings of more sophisticated multigene molecular studies indicate that character support for many teleost nodes is weak, ambiguous, or entirely wanting. Some of these changes or uncertainties are reflected in figure 24.1, in which basal teleostean relationships are represented as unresolved. For example, in a highly influential paper, Patterson and Rosen (1977) hypothesized that osteoglossomorphs are the sister group of elopomorphs and other living teleosts, whereas Shen (1996) and Arratia (e.g., 1997, 1999) have proposed that elopomorphs occupy that basal position.

We turn now to a brief review of diversity within extant non-acanthomorph teleost groups. Osteoglossomorpha consist of two freshwater orders: the North American Hiodonti-

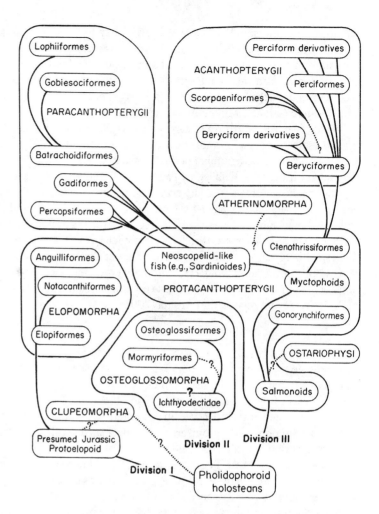

Figure 24.3. Diagram of teleostean relationships from Greenwood et al. (1966). This remarkably prescient, precladistic study delineated for the first time the major groups of teleostean fishes and thereby laid an important foundation for the "modern era" of teleostean systematics that was to follow.

formes (mooneyes; two spp., one family) and mostly Old World Osteoglossiformes (bony tongues, knifefishes, and elephantfishes; 220+ spp., five families). Osteoglossomorpha are an ancient group with a long fossil history dating to the Jurassic (Patterson 1993, 1994, Li and Wilson 1996) and displaying a number of primitive characters as well as two evolutionary innovations; a complex tongue-bite mechanism and a gut that uniquely coils to the left of the stomach. The most speciose and perhaps the most interesting members of this group are the elephantfishes (Mormyridae), which create an electric field with muscles of the caudal region and use it to find prey and avoid obstacles in their turbid water habitats. Relationships among mormyrids and the evolution of their electric organs have recently been elucidated with molecular data by Sullivan et al. (2000) and Lavoué et al. (2000). Other osteoglossiforms include the large (to 2.5 m) bonytongues of South America, Asia, and Africa. Li and Wilson (1996) analyzed phylogenetic relationships and discussed evolutionary innovations of osteoglossomorphs, and a recent molecular analysis (Kumazawa and Nishida 2000) corroborates osteoglossomorph monophyly but differs in its assessment of osteoglossiform interrelationships.

Elopomorpha are a heterogeneous group united by the unique, leaflike, transparent leptocephalus larval stage, once considered a distinct taxonomic group, and by the possession of derived sperm morphology (Mattei 1991, Jamieson 1991). All are marine, although some venture into brackish waters. Elopomorph intrarelationships are poorly understood; however, most studies agree in placing Elopiformes (tarpons and ladyfishes; eight spp., two families) as the basal order. Albuliformes (bonefishes, two spp., one family) are a small group highly prized by fishermen. Notacanthiformes (halosaurs and spiny eels, 25 spp., two families) are marine, deep-water fishes. The bulk of elopomorph diversity lies in the Anguilliformes (true eels, 750+ spp., 15 families), which includes morays (200 spp.), snake eels (250 spp.), conger eels (150 spp.), and the anadromous freshwater eels (15 spp.). Saccopharyngiformes (deep-water gulper eels, 25 spp., three families) contains among the most bizarre of living vertebrates, with luminescent organs and huge mouths capable of swallowing prey several times their body size. Forey et al. (1996) accepted elopomorph monophyly and presented a detailed study of their intrarelationships, using both morphological and molecular characters. However, two recent studies (Filleul and Lavoué 2001, Obermiller and Pfeiler 2003) have challenged elopomorph monophyly, and Filleul and Lavoué (2001) place the four orders as *incertae sedis* among basal teleosts.

Until 1996, the remaining teleost fishes were grouped into two putative lineages, Clupeomorpha (herrings and allies, 360+ spp., five families) and Euteleostei. Euteleostei have proven to be a problematic group, persistently defying unambiguous diagnosis (Fink 1984).

Following the molecular work of Lê et al. (1993), Lecointre (1995) and Lecointre and Nelson (1996) suggested, based on both morphological and molecular characters, that ostariophysans (minnows, catfishes, and allies) are not euteleosts but instead are the sister group of clupeomorphs. Further evidence is emerging, both molecular (Filleul and Lavoué 2001, G. Orti pers. comm.) and morphological (Arratia 1997, 1999, M. DePinna pers. comm.) to support this hypothesis, which removes one of the stumbling blocks to understanding the evolution of euteleosts, but its validity and implications are not yet fully understood. For example, Ishiguro et al. (2003) find mitogenomic support for an Ostariophysan-clupeomorph clade, but one that also includes the alepocephaloids (slickheads, see below) nested within it.

With the ostariophysans removed, Johnson and Patterson (1996) argued that four unique evolutionary innovations characterize the "new" Euteleostei and recognized two major lineages. The first, Protacanthopterygii, is a refinement of the group first proposed by Greenwood et al. (1966). The second (Neognathi) placed the small order Esociformes (the freshwater Holarctic pikes and mudminnows; about 10 spp., two families) as the sister group of the remaining teleosts (Neoteleostei). The relationships of the pikes and mudminnows remain problematic, but they share two unique evolutionary innovations with neoteleosts (Johnson and Patterson 1996).

The reconstituted Protacanthopterygii consists of two orders, Salmoniformes and Argentiniformes, each with two suborders. Salmoniformes includes the whitefishes, Holarctic salmons and trouts, Salmonoidei (65+ spp., one family) and the northern smelts, noodlefishes, southern smelts and allies, and Osmeroidei (75+ spp., three families). The Argentiniformes include the marine herring smelts and allies (Argentinoidei; 60+ spp., four families), most of which occur in deep water, and the deep-sea slickheads and allies (Alepocephaloidea; 100+ spp., three families).

Morphological character support for a monophyletic Neoteleostei and the monophyly and sequential relationships of the three major neoteleost groups leading to Acanthomorpha, depicted in figure 24.1, appears strong (Johnson 1992, Johnson and Patterson 1993, Stiassny 1986, 1996), and it is perhaps at this level on the teleostean tree that most confidence can currently be placed. Stomiiformes (320+ spp., four families) are a group of luminescent, deep-sea fishes with exotic names such as bristlemouths and dragonfishes that complement their morphological diversity (fig. 24.4). Two genera of midwater bristlemouths (*Cyclothone* and *Gonostoma*) have the greatest abundance of individuals of any vertebrate genus on Earth (Marshall 1979). Harold and Weitzman (1996) provide the most recent analysis of stomiiform intrarelationships. Aulopiformes (220+ spp., 15 families) are a diverse group of nearshore and mostly deep-sea species, including the abyssal plain tripod fishes, the familiar tropical and temperate lizardfishes, and midwater predators such as the sabertooths and lancetfishes (for the most recent analyses of their intrarelationships, see Johnson et al. 1996, Baldwin and Johnson 1996, Sato and Nakabo 2002). Members of Myctophiformes—lanternfishes and allies (240+ spp., two families)—are also ubiquitous midwater fishes, most with luminescent organs. They are a major food source for economically important midwater feeders, from tunas to whales, and many undertake vertical migrations into surface waters at night to feed, returning to depths during the day, thereby contributing significantly to biological nutrient cycling in the deep ocean. Stiassny (1996) and Yamaguchi (2000) provide recent analyses of their intrarelationships.

Acanthomorpha and the "Bush at the Top"

The spiny-rayed fishes, Acanthomorpha, are the crown group of Teleostei. With more than 300 families and approximately 16,000 species, they comprise more than 60% of extant teleosts and about one-third of all living vertebrates. This immense group of fishes exhibits staggering diversity in adult and larval body form, skeletal and soft anatomy, size (8 mm to 15 m), habitat, physiology, and behavior. Acanthomorphs first appear in the fossil record at the base of the Late Cretaceous (Cenomanian) represented by more than 20 genera assignable to four or five extant taxa (fig. 24.1). By the late Paleocene the fauna is somewhat more diverse, but at the Middle Eocene, as seen in the Monte Bolca Fauna, an explosive radiation seems to have occurred, wherein the majority of higher acanthomorph diversity is laid out (Patterson 1994, Bellwood 1996). To date, because of the uncertainty of structure and relationships of many of the earlier fossils and the rapid appearance of most extant families, fossils have offered little to our understanding of acanthomorph relationships.

Acanthomorpha originated with Rosen's (1973) seminal paper on interrelationships of higher euteleosts and was based on five ambiguously distributed characters. In an attempt to define the largest and most diverse acanthomorph assemblage, Percomorpha, Johnson and Patterson (1993) proposed a morphology-based hypothesis of acanthomorph relationships. In so doing, they reviewed and evaluated support for previous hypotheses, including acanthomorph monophyly, for which they identified eight evolutionary innovations. Perhaps the most convincing of these are the presence in the dorsal and anal fins of true fin spines, as well as a single median chondrified rostral cartilage associated with specific rostral ligaments (Hartel and Stiassny 1986, Stiassny 1986) that permit the jaws to be greatly protruded while feeding. Johnson and Patterson (1993) proposed a phylogeny for six basal acanthomorph groups leading sequentially to a newly defined Percomorpha. Below, we briefly discuss acanthomorph diversity in this proposed phylogenetic order (fig. 24.5).

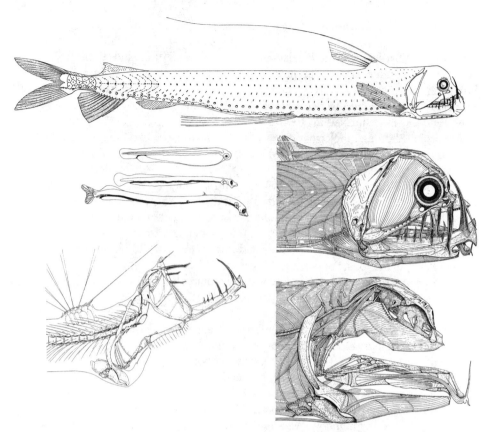

Figure 24.4. The viperfish, *Chauliodus sloani*; anatomical detail from Tchernavin (1953). Larvae redrawn after Kawaguchi and Moser (1984). Teleostean fishes are biomechanically complex; the head alone is controlled by some 50 muscles operating more than 30 movable skeletal parts. Such anatomical complexity, plus a wide range of ontogenetic variation, ensures a continued pivotal role for anatomical input into systematic study.

Interestingly, Lampridiformes (opahs and allies) were once placed among the perciform fishes at the top of the tree. They are a small (20 spp., seven families) but diverse group, characterized by a uniquely configured, highly protrusible upper jaw mechanism. Except for the most primitive family, the velifers, which occur in near shore-waters, the remaining families are meso- and epipelagic. In body shape they range from the deep-bodied opahs to extremely elongate forms such as the oarfish (*Regalecus glesne*), which is the longest known bony fish, reported to reach 15 m. The position of lampridiforms as a basal acanthomorph group has been supported by both morphological (Olney et al. 1993) and molecular data sets (Wiley et al. 2000, Miya et al. 2001, 2003, Chen et al. 2003).

Polymixiiformes (beard fishes; 10 spp., one family) are characterized by two chin barbels supported by the first branchiostegals and occur on the continental shelf and upper slope. The fossil record for this group is considerably more diverse than its living representation. Recent molecular studies have confirmed a basal position for these fishes, but some suggest a placement within a large clade consisting otherwise of paracanthopterygian and zeoid lineages (e.g., Miya et al. 2001, 2003, Chen et al. 2003).

Paracanthopterygii (1,200+ spp., 37 families) are an odd and almost certainly unnatural assemblage of freshwater and marine fishes first proposed by Greenwood et al. (1966) and refined to its present form by Patterson and Rosen (1989). Most of the hypothesized evolutionary innovations proposed by these authors are suspect (Gill 1996), and molecular studies by Wiley et al. (2000) and Miya et al. (2001) suggest that although the freshwater Percopsiformes (troutperches; six spp., three families) and Gadiformes (cods; 500+ spp., nine families) are basal acanthomorphs, the other groups may be scattered through the higher acanthomorph lineages. These orders include Ophidiiformes (cuskeels; 380+ spp., 18 families), Batrachoidiformes (toadfishes; 70 spp., three families), and Lophiiformes (anglerfishes; 300+ spp., 18 families). Most species belonging to these orders are marine. The dismemberment of all or part of Paracanthopterygii will have significant implications for acanthomorph relationships, perhaps particularly those within the perciforms.

Between the paracanthopterygians and the immense diversity of Percomorpha are three small, but phylogenetically critical, marine lineages. Stephanoberyciformes (90 spp., nine families) is a monophyletic group of marine benthic and deep-water fishes commonly called pricklefishes and whalefishes. Johnson and Patterson (1996) separated this group from Beryciformes, but molecular data suggest that at least some members of the group might rejoin Beryciformes (Wiley et al. 2000, Colgan et al. 2000, Chen et al. 2003). Zeiformes (45 spp., five families) includes the dories, a marine group of deep-bodied fishes that includes the much-valued John

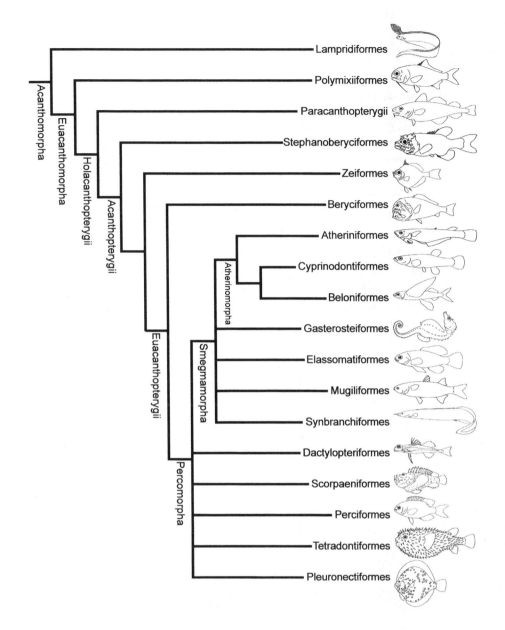

Figure 24.5. Intrarelationships among acanthomorph lineages after Johnson and Patterson (1993).

Dory of the Atlantic. Recent molecular studies suggest a relationship between the dories and the codfishes and/or beardfishes (Wiley et al. 2000, Miya et al. 2001, Chen et al. 2003), but this conclusion might be due to the relatively low numbers of species included in these studies. Beryciformes (140+ spp., seven families) includes some of the most familiar reefdwelling fishes, the squirrelfishes. Beryciforms are entirely marine and occur worldwide from shallow depths, where they are nocturnal, to the deep sea. External bacterial luminescent organs characterize the pinecone fishes and flashlight fishes, the latter having a complex mechanism for rapidly occluding the large subocular light organ by rotating it downward or covering it with a lidlike shutter. Two genera of the closely related roughies (Trachichthyidae) have internal luminescent organs, and the orange roughy (*Hoplostethus atlanticus*) is an overexploited food fish.

Percomorpha, the Bush at the Top

Percomorph (14,000+ spp., 244 families) are the crown group of the spiny-rayed fishes and best represent what Nelson (1989) called the "bush at the top." The name Percomorpha originated with Rosen (1973) and was essentially the equivalent of Greenwood et al.'s (1966) Acanthopterygii, which consisted of beryciforms, perciforms, and groups placed between and beyond those two, such as lampridiforms, zeiforms, gasterosteiforms, scorpaeniforms, pleuronectiforms, and tetraodontiforms. Rosen presented no characters in support of his Percomorpha, nor have any been supported subsequently (but see Stiassny 1990, 1993, Stiassny and Moore 1992, Roberts 1993). The major goal of Johnson and Patterson's (1993) analysis was to sort out basal lineages of acanthomorphs and revise the composition of Percomorpha to represent a monophyletic group diagnosed by derived characters. In the process, they erected a new, putatively monophyletic assemblage, Smegmamorpha, which, together with "the perciforms and their immediate relatives," constituted the newly defined Percomorpha. They identified eight evolutionary innovations of the Percomorpha, all of which are homoplasious. Although monophyly of Johnson and Patterson's Percomorpha has not been challenged subsequently with morphological analyses, it is considered tenuous, particularly in view of our ignorance of the composition and intrarelationships of Perciformes and allies (below) and strong doubts about paracanthopterygian monophyly. To date, no molecular analyses have captured a monophyletic Percomorpha without the inclusion of certain "paracanthopterygian" lineages.

Smegmamorpha (1,700+ spp., 37 families) of Johnson and Patterson (1993) are a diverse group consisting of spiny and swamp eels (Synbranchiformes; 90 spp., three families), gray mullets (Mugiliformes; 80 spp., one family), pygmy sunfishes (Elassomatiformes; six spp., one family), sticklebacks, pipefishes and allies (Gasterosteiformes; 275 spp., 11 families), and the speciose silversides, flyingfishes, killifishes, and allies (Atherinomorpha; 1225+ spp., 21 families, four orders). The recognition of this group was greeted with some skepticism because swamp and spiny eels had traditionally been allied with the perciforms whereas pygmy sunfishes had been considered centrarchids (sunfish and basses), a family deeply embedded in one suborder of Perciformes. Smegmamorpha is united by a single evolutionary innovation, a specialized attachment of the first intermuscular bone (epineural) at the tip of a prominent transverse process on the first vertebra, but several additional specializations are shared by most smegmamorphs. There have been no comprehensive morphological analyses to challenge smegmamorph monophyly; however, Parenti (1993) suggested that atherinomorphs might be the sister group of paracanthopterygians, and Parenti and Song (1996) identified a pattern of innervation of the pelvic fin in mullets and pygmy sunfishes that is shared with more derived perciforms. Molecular analyses have failed to capture monophyly of smegmamorphs, although major components of the group are recognized (e.g., Wiley et al. 2000, Miya et al. 2003, Chen et al. 2003). Although relationships among smegmamorphs remain unknown, Stiassny (1993) suggested grey mullets (Mugilidae) may be most closely related to atherinomorphs, and Johnson and Springer (1997) presented evidence suggesting a possible relationship between pygmy sunfishes and sticklebacks.

The remaining groups comprise some 12,000+ species in more than 207 families. In their cladogram of percomorph relationships (fig. 24.4), Johnson and Patterson (1993) placed Perciformes (perches and allies) in an unresolved polytomy with Smegmamorpha and four remaining groups traditionally classified as orders: the scorpionfishes and allies (Scorpaeniformes), flying gurnards (Dactylopteriformes), flatfishes (Pleuronectiformes), and triggerfishes, pufferfishes, and allies (Tetraodontiformes). However, they saw no rea-

son to exclude these last four orders from the traditional Perciformes and believed it likely that they are nested within it. Subsequently, Mooi and Gill (1995) classified Scorpaeniformes within Perciformes. To date, no morphological or molecular synapomorphies support a monophyletic Perciformes in either the restricted or expanded sense that would include any or all of the orders Johnson and Patterson (1993) placed in their terminal polytomy. Many questions remain about monophyly and interrelationships of a number of the approximately 25 suborders and more than 200 families included in that polytomy. Certainly the possibility that affinities of some members lie with other acanthomorphs, or vice versa, cannot be dismissed. With these observations in mind, we review the remaining orders.

Perciformes (9800+ spp., 163 families) are the largest and most diverse vertebrate order. Perciforms range in size from the smallest vertebrate, the 8 mm *Trimmatom nanus* (for which an estimated 3674 individuals would be needed to make up one quarter-pound gobyburger), to the 4.5 m black marlin (*Makaira indica*). Although there are a number of freshwater perciforms (mostly contained within the large cichlid clade), most species are marine, and they represent the dominant component of coral reef and inshore fish faunas. In a taxonomic sense, Perciformes is a catchall assemblage of families and suborders whose relationships have not been convincingly shown to lie elsewhere. Although there is reasonably good support for monophyly of about half of the suborders, others remain poorly defined, most notably the largest suborder, Percoidei (3,500+ spp., 70 families), another catch-all or "wastebasket group," for which not a single diagnostic character has been proposed. Percoids are usually referred to as perchlike fishes, and although this general physiognomy characterizes many families, such as freshwater perches (Percidae), sunfishes (Centrarchidae), sea basses (Serranidae), and others, percoids encompass a wide range of body forms, from the deep-bodied moonfishes (Menidae), butterflyfishes (Chaetodontidae), and more, to very elongate, eel-like forms such as bandfishes (Cepolidae) and bearded snakeblennies (Notograptidae). For lists and discussions of perciform suborders and percoid families, see Johnson (1993), Nelson (1994), and Johnson and Gill (1998), each of which, not surprisingly, differ somewhat in definition and composition of the two groups.

Scorpaeniformes (lionfishes and allies; 1,200+ spp., 26 families) were included within Perciformes by Mooi and Gill (1997) based on a specific pattern of epaxial musculature shared with some perciforms. It is a large, primarily marine group characterized by the presence of a bony stay of questionable homology that extends from the third infraorbital across the cheek to the preopercle. Monophyly, group composition, and relationships remain controversial, but most recent work supports two main lineages, scorpaenoids and cottoids (e.g., Imamura and Shinohara 1998), and preliminary molecular studies suggest a close relationship between zoarcoids and the cottoid lineage (Miya et al. 2003, Smith 2002, Chen et al. 2003). Whether the scorpaenoid and cottoid lineages are sister groups is open to question, and clarification of scorpaeniform relationships is an important component of the "percomorph problem."

Dactylopteriformes (flying gurnards; seven spp., one family) are a small, clearly monophyletic, group of inshore bottom-dwelling marine fishes characterized by a thick, bony, "armored" head with an elongate preopercular spine and colorful, greatly enlarged, fanlike pectoral fins. Their relationships are obscure (Imamura 2000), and they have been variously placed with, among other groups, the scorpaeniforms and gasterosteiforms. Molecular studies to date have shed little light on placement, with weak support for an alignment with flatfishes (Miya et al. 2001), gobioids (Miya et al. 2003), or syngnathoids (Chen et al. 2003).

Pleuronectiformes (flatfishes; 540+ spp., seven families) are widely distributed, bottom-dwelling fishes containing a number of commercially important species. These are characterized by a unique, complex evolutionary innovation in which one eye migrates ontogenetically to the opposite side of the head, so that the transformed juveniles and adults are asymmetrical and lie, eyeless side down, on the substrate. Their relationships as shown by morphological analysis have most recently been reviewed by Chapleau (1993) and Cooper and Chapleau (1998). A molecular analysis of mitochondrial ribosomal sequences by Berendzen and Dimmick (2002) suggests an alternative hypothesis of relationship. Interestingly, a recent mitogenomic study provides quite strong nodal support for a relationship with the jacks (Carangidae), but taxon sampling in this region of the tree is quite sparse (Miya et al. 2003).

Tetraodontiformes (triggerfishes, puffers, and allies; 350+ spp., 10 families) are a highly specialized and diverse order of primarily marine fishes, ranging in size from the 2 cm diamond leatherjacket (*Rudarius excelsus*) to the 3.3 m (>1000 kg) ocean sunfish (*Mola mola*). They are characterized by small mouths with few teeth or teeth incorporated into beaklike jaws, and scales that are either spine like or, more often, enlarged as plates or shields covering the body as in the boxfishes (Ostraciidae). Members of three families have modified stomachs that allow extreme inflation of the body with water as a defensive mechanism. Relationships of tetraodontiforms have been treated in large monographs dealing with comparative myology (Winterbottom 1974) and osteology (Tyler 1980). Although tetraodontiforms have been considered as highly derived percomorphs, Rosen (1984) proposed that they are more closely related to caproids and the apparently more basal zeiforms. Johnson and Patterson (1993) rejected that hypothesis, as do ongoing molecular studies (Holcroft 2002, N. I. Holcroft pers. comm.). However, it is defended in a recent morphological analysis (Tyler et al. 2003).

Concluding Remarks

Systematic ichthyologists were early to adopt Hennig's methods and have made great progress toward understanding the evolutionary diversification of fishes. Much of the new phylogenetic structure is underpinned by morphological character data, most of it from the skeleton and much of it gathered anew or reexamined and refined during the last 35 years. Another seminal innovation appeared fortuitously on the cusp of the cladistic revolution—the use of trypsin digestion in cleared and stained preparations, followed by the ability to stain cartilage as well as bone. These techniques revolutionized fish osteology and greatly facilitated detailed study of skeletal development adding significantly to our understanding of character transformation and homology. However, there is still much to do. Our understanding of the composition and relationships of Percomorpha, with more than half the diversity of all bonyfishes, remains chaotic—a state of affairs proportionally equivalent to not knowing the slightest thing about the relationships among amniote vertebrates.

Fishes are a tremendously diverse group of anatomically complex organisms (e.g., fig. 24.4) and undoubtedly morphology will continue to play a central role in systematic ichthyology. However, as in other groups of organisms, molecular analyses are increasingly beginning to make significant contributions, especially for fish groups with confusing patterns of convergent evolution. The combination of molecular and morphological data sets, and the reciprocal illumination they shed, augurs an exciting new phase in systematic ichthyology. We are, perhaps, at the halfway point of our journey.

Acknowledgments

We gratefully acknowledge the numerous colleagues whose studies of fish phylogenetics have helped to elucidate the present state of the art for the piscine limb of the Tree of Life, and extend our apologies to those we may have omitted or inadvertently misrepresented in our efforts to keep this chapter to a manageable length. Thanks also to Scott Schaefer and Leo Smith (AMNH) for some helpful comments on an early draft of the manuscript, and additional thanks to Leo for his artful help with the figures that accompany the chapter. Part of this work was funded through grant DEB-9317881 from the National Science Foundation to E.O.W. and G.D.J. and through the Scholarly Research Fund of the University of Kansas to E.O.W. We thank both institutions. Ongoing support from the Axelrod Research curatorship to M.L.J.S. is also gratefully acknowledged. Finally our thanks to Joel Cracraft and Mike Donoghue for so successfully having taken on the formidable task of organizing the Tree of Life symposium and without whose constant nudging this chapter would never have seen the light of day.

Literature Cited

Arratia, G. 1997. Basal teleosts and teleostean phylogeny. Palaeo Ichthyologica 7. Pfeil, Munich.

Arratia, G. 1999. The monophyly of Teleostei and stem-group teleosts. Consensus and disagreements. Pp. 265–334 *in*

Mesozoic fishes 2: Systematics and fossil record (G. Arratia and H.-P. Schultze, eds). Pfeil, Munchen.

Arratia, G. 2000. Phylogenetic relationships of Teleostei. Past and present. Estud. Oceanol. 19:19–51.

Arratia, G. 2001. The sister-group of Teleostei: consensus and disagreements. J. Vert. Paleontol. 21(4):767–773.

Baldwin, C. C., and G. D. Johnson. 1996. Interrelationships of Aulopiformes. Pp. 355–404 in Interrelationships of fishes (M. L. J. Stiassny, L. Parenti, and G. D. Johnson, eds.). Academic Press, San Diego.

Basden, A. M., G. C. Young, Coates, M. I., and A. Ritchie. 2000. The most primitive osteichthyan braincase? Nature 403:185–188.

Bellwood, D. R. 1996. The Eocene fishes of Monte Bolca: the earliest coral reef fish assemblage. Coral Reefs 15:11–19.

Berendzen, P. B., and W. W. Dimmick. 2002. Phylogenetic relationships of Pleuronectiformes based on molecular evidence. Copeia 2002(3):642–652

Bigelow, H. B., and W. C. Schroeder. 1948. Fishes of the northwestern Atlantic. Part I. Lancelets, cyclostomes, and sharks. Memoirs of the Sears Foundation for Marine Research 1. Yale University, New Haven, CT.

Bigelow, H. B., and W. C. Schroeder. 1953. Fishes of the northwestern Atlantic. Part II. Sawfishes, guitarfishes, skates, rays and chimaeroids. Memoirs of the Sears Foundation for Marine Research 2. Yale University, New Haven, CT.

Cappetta, H. 1987. Chondrichthyes II. Mesozoic and Cenozoic Elasmobranchii. Pp. 1–193 in Handbook of paleoichthyology (H.-P. Schultze, ed.), vol. 3B. Gustav Fisher Verlag, Stuttgart.

Cappetta, H., C. Duffin, and J. Zidek. 1993. Chondrichthyes. Pp 593–609 in The fossil record (M. J. Benton, ed.), vol. 2. Chapman and Hall, London.

Carvalho, M. R. de. 1996. Higher-level elasmobranch phylogeny, basal squaleans and paraphyly. Pp. 35–62 in Interrelationships of fishes (M. L. J. Stiassny, L. R. Parenti, and G. D. Johnson, eds.). Academic Press, San Diego.

Carvalho, M. R. de. 1999. A synopsis of the deep-sea genus Benthobatis Alcock, with a redescription of the type-species Benthobatis moresbyi Alcock, 1898 (Chondrichthyes, Torpediniformes, Narcinidae). Pp. 231–255 in Proceedings of the 5th Indo-Pacific Fishes Conference (Nouméa, 3–8 November, 1997) (B. Séret and J.-Y. Sire, eds.). Société Francaise d'Ichtyologie and Institut de Recherche pour le Développement, Paris.

Carvalho, M. R. de. 2001. A new species of electric ray, Narcine leoparda, from the tropical eastern Pacific Ocean (Chondrichthyes: Torpediniformes: Narcinidae). Proc. Biol. Soc. Wash. 114 (3):561–573.

Carvalho, M. R. de. In press. A Late Cretaceous thornback ray from southern Italy, with a phylogenetic reappraisal of the Platyrhinidae (Chondrichthyes: Batoidea). In Mesozoic fishes 3 (G. Arratia and A. Tintori, eds.). Verlag Dr. F. Pfeil, Munich.

Carvalho, M. R. de, and J. G. Maisey. 1996. Phylogenetic relationships of the Late Jurassic shark Protospinax Woodward, 1919 (Chondrichthyes: Elasmobranchii). Pp. 9–46 In Mesozoic fishes, systematics and paleoecology (G. Arratia and G. Viohl, eds.). Verlag Dr. F. Pfeil, Munich.

Cavin, L., Cappetta, H., and B. Séret. 1995. Révision de Belemnobatis morinicus (Sauvage, 1873) du Portlandien du Boulonnais (Pas-de-Calais, France). Comparaison avec quelques Rhinobatidae Jurassiques. Geol. Palaeontol. 29:245–267.

Chang, H.-Y., T.-K. Sang, K.-Y. Jan, and C.-T. Chen. 1995. Cellular DNA contents and cell volumes of batoids. Copeia 3:571–576.

Chapleau, F. 1993. Pleuronectiform relationships: a cladistic reassessment. Bull. Mar. Sci. 52:516–540.

Chen, W.-J., C. Bonillo, and G. Lecointre. 2003. Repeatability of clades as a criterion of reliability: a case study for molecular phylogeny of Acanthomorpha (Teleostei) with larger number of taxa. Mol. Phylogenet. Evol. 26:262–288.

Clack, J. E. 2002. Gaining ground, the origins and evolution of tetrapods. Indiana University Press, Bloomington.

Cloutier, R., and P. E. Ahlberg. 1996. Morphology, characters, and the interrelationships of basal sarcopterygians. Pp. 425–426 in Interrelationships of fishes (M. L. J. Stiassny, L. R. Parenti, and G. D. Johnson, eds.). Academic Press, San Diego.

Coates, M. I., and S. E. K. Sequeira. 1998. The braincase of a primitive shark. Trans. R. Soc. Edinb. Earth Sci. 89:63–85.

Coates, M. I., S. E. K. Sequeira, I. J. Sansom, and M. M. Smith. 1999. Spines and tissues of ancient sharks. Nature 396:729–730.

Colgan, D. J., C.-D. Zhang, and J. R. Paxton. 2000. Phylogenetic investigations of the Stephanoberyciformes and Beryciformes, particularly whalefishes (Euteleostei: Cetomimidae) based on partial 12S rDNA and 16S rDNA sequences. Mol. Phylogenet. Evol. 17(1):15–25.

Compagno, L. J. V. 1973. Interrelationships of living elasmobranch fishes. Pp. 15–61 in Interrelationships of fishes (P. H. Greenwood, C. Patterson, and R. Miles, eds.). Academic Press, London.

Compagno, L. J. V. 1977. Phyletic relationships of living sharks and rays. Am. Zool. 17(2):303–322.

Compagno, L. J. V. 1984a. FAO species catalogue, vol. 4, pt. I. Sharks of the world. An annotated and illustrated catalogue of shark species known to date. Food and Agriculture Organization of the United Nations, Rome.

Compagno, L. J. V. 1984b. FAO species catalogue, vol. 4, pt. II. Sharks of the world. An annotated and illustrated catalogue of shark species known to date. Food and Agriculture Organization of the United Nations, Rome.

Compagno, L. J. V. 1988. Sharks of the order Carcharhiniformes. Princeton University Press, Princeton, NJ.

Compagno, L. J. V. 1990. Relationships of the megamouth shark, Megachasma pelagios (Lamniformes: Megachasmidae), with comments on its feeding habits. Pp. 357–379 in Elasmobranchs as living resources (H. L. Pratt, Jr., S. H. Gruber, and T. Taniuchi, eds.). Advances in the Biology, Ecology, Systematics, and the Status of the Fisheries. NOAA Technical Reports no. 90. U.S. Department of Commerce, Washington, DC.

Compagno, L. J. V. 1999. Checklist of living elasmobranches. Pp. 471–498 in Sharks, skates, and rays, the biology of elasmobranch fishes (W. C. Hamlett, ed.). Johns Hopkins University Press, Baltimore.

Compagno, L. J. V. 2001. Sharks of the world. An annotated

and illustrated catalogue of shark species known to date, Vol. 2: Bullhead, mackerel and carpet sharks (Heterodontiformes, Lamniformes and Orectolobiformes). Food and Agriculture Organization of the United Nations, Rome.

Compagno, L. J. V., and S. F. Cook. 1995. The exploitation and conservation of freshwater elasmobranchs: status of taxa and prospects for the future. Pp. 62–90 in The biology of freshwater elasmobranchs, a symposium to honor Thomas B. Thorson (M. I. Oetinger and G. D. Zorzi, eds.). J. Aquar. Aquat. Sci. 7. The Written Word, Parkville, MO.

Cooper, J. E., and F. Chapleau. 1998. Monophyly and intra-relationships of the family Pleuronectidae (Pleuronectiformes), with a revised classification. Fish. Bull. 96:686–726.

Dean, B. 1895. Fishes, living and fossil. Macmillan, New York.

Dean, B. 1906. Chimaeroid fishes and their development. Publ. no. 32. Carnegie Institute, Washington, DC.

DePinna, M. C. C. 1996. Teleostean monophyly. Pp. 147–162 in Interrelationships of fishes (M. L. J. Stiassny, L. R. Parenti, and G. D. Johnson, eds.). Academic Press, San Diego.

Didier, D. 1995. Phylogenetic systematics of extant chimaeroid fishes (Holocephali, Chimaeroidei). Am. Mus. Nov. 3119.

Didier, D., and B. Séret. 2002. Chimaeroid fishes of New Caledonia with description of a new species of Hydrolagus (Chondrichthyes, Holocephali). Cybium 26(3):225–233.

Dingerkus, G. 1986. Interrelationships of orectolobiform sharks (Chondrichthyes: Selachii). Pp. 227–245 in Indo-Pacific fish biology, proceedings of the second International Conference on Indo-Pacific Fishes (T. Uyeno, R. Arai, T. Taniuchi, and K. Matsuura, eds.). Ichthyological Society of Japan, Tokyo.

Dutheil, D. B. 1999. An overview of the freshwater fish fauna from the Kem Kem beds (Late Cretaceous: Cenomanian) of southeastern Morocco. Pp. 553–563 in Mesozoic fishes 2: Systematics and fossil record (G. Arratia and H.-P. Schultze, eds.). Pfeil, Münich.

Erdmann, M. V. 1999. An account of the first living coelacanth known to scientists from Indonesian waters. Environ. Biol. Fish. 54:439–443.

Eschmeyer, W. N., ed. 1998. Catalog of fishes. Special publication. California Academy of Sciences, San Francisco.

Filleul, A., and S. Lavoué. 2001. Basal teleosts and the question of elopomorph monophyly. Morphological and molecular approaches. C.R. Acad. Sci. Paris 324:393–399.

Filleul, A., and J. Maisey. In press. Redescription of Santanichthys diasii (Otophysi, Characiformes) from the Albian of the Santana Formation and comments on its implications for Otophysan relationships. Am. Mus. Novitates.

Fink, W. L. 1984. Stomiiforms: relationships. Pp. 181–184 in Ontogeny and systematics of fishes (H. G. Moser, W. J. Richards, D. M. Cohen, M. P. Fahay, A. W. Kendall, and S. L. Richardson, eds.). Spec. publ. 1. American Society of Ichthyologists and Herpetologists, Lawrence, KS.

Forey, P. L. 1998. History of the coelacanth fishes. Chapman and Hall, New York.

Forey, P. I., D. T. J. Littlewood, P. Riche, and A. Meyer. 1996. Interrelationships of elopomorph fishes. Pp. 175–191 in Interrelationships of fishes (M. L. J. Staissny, L. R. Parenti, and G. D. Johnson, eds.). Academic Press, San Diego.

Gardiner, B. G., J. G. Maisey, and D. T. J. Littlewood. 1996. Interrelationships of basal neopterygians. Pp. 117–146 in Interrelationshps of fishes (M. L. J. Stiassny, L. R. Parenti, and G. D. Johnson, eds.). Academic Press, San Diego.

Gardiner, B. G., and B. Schaeffer. 1989. Interrelationships of lower actinopterygian fishes. Zool. J. Linn. Soc. Lond. 97:135–187.

Gill, A. C. 1996. Comments on an intercalar path for the glossopharygeal (cranial IX) nerve as a synapomorphy of the Paracanthopterygii, and on the phylogenetic position of the Gobiesocidae (Teleostei: Acanthomorpha). Copeia (1996):1022–1029.

Gilmore, R. G. 1993. Reproductive biology of lamnoid sharks. Environ. Biol. Fishes 38:95–114.

Glickman, L. S. 1967. Subclass Elasmobranchii (sharks). Pp. 292–352 in Fundamentals of paleontology (D. V. Obruchev, ed.), vol. 2. Israel Program for Scientific Translations, Jerusalem.

Goto, T. 2001. Comparative anatomy, phylogeny and cladistic classificationof the order Orectolobiformes (Chondrichthyes, Elasmobranchii). Mem. Grad. Sch. Fish. Sci. 48(1):1–100.

Gottfried, M. D., L. J. V. Compagno, and S. C. Bowman. 1996. Size and skeletal anatomy of the giant "megatooth" shark Carcharodon megalodon. Pp. 55–66 in Great white sharks, the biology of Carcharodon carcharias (A. P. Klimley and D. G. Ainley, eds.). Academic Press: San Diego.

Grande, L. 1998. Fishes through the ages. Pp. 27–31 in Encyclopedia of fishes. A comprehensive guide by international experts, 2nd ed. (J. R. Paxton and W. N. Eschmeyer, eds.). Academic Press, San Diego.

Grande, L., and W. E. Bemis. 1991. Osteology and phylogenetic relationships of fossil and Recent paddlefishes (Polyodontidae) with comments on the interrelationships of Acipenseriformes. J. Vert. Paleontol. 11(suppl. 1):1–121.

Grande, L., and W. E. Bemis. 1996. Interrelationships of Acipenseriformes, with comments on "Chondrostei." Pp. 85–115 in Interrelationships of fishes (M. L. J. Stiassny, L. R. Parenti, and G. D. Johnson, eds.). Academic Press, San Diego.

Grande, L., and W. E. Bemis. 1998. A comprehensive phylogenetic study of amiid fishes (Amiidae) based on comparative skeletal anatomy. An empirical search for interconnected patterns of natural history. J. Vert. Paleontol. 18(suppl. 4):1–690.

Greenwood, P. H., D. E Rosen, S. H. Weitzman, and G. S. Myers. 1966. Phyletic studies of teleostean fishes, with a provisional classification of living forms. Bull. Am. Mus. Nat. Hist. 131:339–456.

Gudger, E. W., and B. G. Smith. 1933. The natural history of the frilled shark, Chlamydoselachus anguineus. Pp. 245–319 in Bashford Dean Memorial Volume: Archaic fishes (E. W. Gudger, ed.), V (1933). American Museum of Natural History, New York.

Hall, B. K. 1982. Bone in the cartilaginous fishes. Nature 298:324.

Hamlett, W. C. 1999. Sharks, skates, and rays. The biology of elasmobranch fishes. Johns Hopkins University Press, Baltimore.

Hamlett, W. C., and T. J. Koob. 1999. Female reproductive system. Pp. 398–443 in Sharks, skates, and rays. The biology of elasmobranch fishes (W. C. Hamlett, ed.). Johns Hopkins University Press, Baltimore.

Harold, A. C., and S. H. Weitzman. 1996. Interrelationships of stomiiform fishes. Pp. 333–353 *in* Interrelationships of fishes (M. L. J. Stiassny, L. R. Parenti, and G. D. Johnson, eds.). Academic Press, San Diego.

Hartel, K. E., and M. L. J. Stiassny. 1986. The identification of larval *Parasudis* (Teleostei, Chlorophthalmidae); with notes on the anatomy and relationships of aulopiform fishes. Breviora 487:1–23.

Heemstra, P. C., and M. M. Smith. 1980. Hexatrygonidae, a new family of stingrays (Myliobatiformes: Batoidea) from South Africa, with comments on the classification of batoid fishes. Ichthyol. Bull. J.L.B. Smith Inst. 43:1–17

Hennig, W. 1950. Grundzüge einer Theorie der Phylogenetischen Systematik. Deutscher Zentralverlag, Berlin.

Hennig, W. 1966. Phylogenetic systematics. University of Illinois Press, Urbana.

Herman, J., Hovestadt-Euler, M., and D. C. Hovestadt. 1989. Contributions to the study of the comparative morphology of teeth and other relevant ichthyodorulites in living supraspecific taxa of chondrichthyan fishes. Part A: Selachii. No. 3. Order: Squaliformes. Families: Echinorhinidae, Oxynotidae and Squalidae. Bull. Inst. R. Sci. Nat. Belg. Biol. 59:101–157.

Holcroft, N. I. 2002. A molecular study of the phylogenetic relationships of the tetraodontiform fishes and the relationships to other teleost fishes. P. 172 *in* Program book and abstracts, Joint Meeting of Amer. Soc. Ichthyologists and Herpetologists, Kansas City, KS.

Holmgren, N. 1941. Studies on the head in fishes. Part II. Comparative anatomy of the adult selachian skull, with remarks on the dorsal fins in sharks. Acta Zool. 22:1–100.

Holmgren, N. 1942. Studies on the head in fishes. Part III. The phylogeny of elasmobranch fishes. Acta Zool. 23:129–262.

Imamura, H. 2000. An alternative hypothesis for the position of the family Dactylopteridae (Pisces: Teleostei), with a proposed new classification. Ichthyol. Res. 47:203–222.

Imamura, H., and G. Shinohara. 1998. Scorpaeniform fish phylogeny: an overview. Bull. Nat. Sci. Mus. Tokyo Ser. A. 24:185–212.

Inoue, J. G., M. Miya, K. Tsukamoto, and M. Nishida. 2001. A mitogenomic perspective on the basal teleostean phylogeny: resolving higher-level relationships with longer DNA sequences. Mol. Phylogenet. Evol. 20(1):275–285.

Inoue, J. G., M. Miya, K. Tsukamoto, and M. Nishida. 2002. Basal actinopterygian relationships: a mitogenomic perspective on the phylogeny of the "ancient fish." Mol. Phylogenet. Evol. 26: 110–120.

Ishiguro, N. B., M. Miya, and M. Nishida. 2003. Basal euteleostean relationships: a mitogenomic perspective on the phylogenetic reality of the "Protacanthopterygii." Mol. Phylogenet. Evol. 27:476–488.

Jacob, B. A., J. D. McEachran, and P. L. Lyons. 1994. Electric organs in skates: variation and phylogenetic significance (Chondrichthyes: Rajoidei). J. Morphol. 221:45–63.

Jamieson, B. G. M. 1991. Fish evolution and systematics: evidence from spermatozoa. Cambridge University Press, Cambridge.

Janvier, P. 1996. Early vertebrates. Clarendon Press, Oxford.

Jarvik, E. 1960. Théories de l'évolution des vertebres, reconsidérés a la lumière des récentes decouvertes sur les vertébrés inférieurs. Monographies Scientifiques, Masson and Cie., Paris.

Jarvik, E. 1977. The systematic position of acanthodian fishes. Linn. Soc. Symp. Ser. 4:199–225.

Jarvik, E. 1980. Basic structure and evolution of vertebrates, 2 vols. Academic Press, London.

Johnson, G. D. 1992. Monophyly of the euteleostean clades— Neoteleostei, Eurypterygii, and Ctenosquamata. Copeia 1992:8–25.

Johnson, G. D. 1993. Percomorph phylogeny: progress and problems. Bull. Mar. Sci. 52:3–28.

Johnson, G. D., C. Baldwin, M. Okiyama, and Y. Tominaga. 1996. Osteology and relationship of *Pseudotrichonotus altivelis* (Teleostei: Aulopiformes: Pseudotrichonotidae). Ichthyol. Res. 43:17–45.

Johnson, G. D., and A. C. Gill. 1998. Perches and their allies. Pp. 181–194 *in* Encyclopedia of fishes. A comprehensive guide by international experts, 2nd ed. (J. R. Paxton and W. N. Eschmeyer, eds.). Academic Press, San Diego.

Johnson, G. D., and C. Patterson. 1993. Percomorph phylogeny: a survey of acanthomorphs and a new proposal. Bull. Mar. Sci. 52(1):554–626.

Johnson, G. D., and C. Patterson. 1996. Relationships of lower euteleostean fishes. Pp. 251–332 *in* Interrelationships of fishes (M. L. J. Stiassny, L. R. Parenti, and G. D. Johnson, eds.). Academic Press, San Diego.

Johnson, G. D., and V. G. Springer. 1997. *Elassoma*: another look. P. 176 *in* Amer. Soc. Ichthyologists and Herpetologists program and abstracts for 1997 annual meetings. Seattle, WA.

Kawaguchi, K., and H. G. Moser. 1984. Stomiatoidea: development. Pp. 169–181 *in* Ontogeny and systematics of fishes (H. G. Moser, W. J. Richards, D. M. Cohen, M. P. Fahay, A. W. Kendall, and S. L. Richardson, eds.). Spec. publ. 1. American Society of Ichthyologists and Herpetologists, Lawrence, KS.

Kemp, N. E., and S. K. Westrin, 1979. Ultrastructure of calcified cartilage in the endoskeletal tesserae of sharks. J. Morphol. 160:75–102.

Klimley, A. P., and D. G. Ainley. 1996. Great white sharks, the biology *of Carcharodon carcharias*. Academic Press, San Diego.

Kumazawa, Y., and M. Nishida. 2000. Molecular phylogeny of osteoglossoids: a new model for Gondwanian origin and plate tectonic transportation of the Asian Arowana. Mol. Biol. Evol. 17(12):1869–1878.

Last, P. R. 1999. Australian catsharks of the genus *Asymbolus* (Carcharhiniformes: Scyliorhinidae). CSIRO Marine Labs. Rep. 239:1–35.

Last, P. R., G. H. Burgess, and B. Séret. 2002. Description of six new species of lantern-sharks of the genus *Etmopterus* (Squaloidea: Etmopteridae) from the Australasian region. Cybium 26(3):203–223.

Last, P. R., and J. D. Stevens. 1994. Sharks and Rays of Australia. CSIRO, Melbourne.

Lauder, G. V. 2000. Function of the caudal fin during locomotion in fishes: kinematics, flow visualization, and evolutionary patterns. Am. Zool. 40:101–122.

Lauder, G. V., and K. F. Liem. 1983. The evolution and relationships of the actinopterygian fishes. Bull. Mus. Comp. Zool. 150(3):95–197.

Lavoué, S., R. Bigorne, G. Lecointre, and J.-F. Agnèse. 2000.

Phylogenetic relationships of mormyrid electric fishes (Mormyridae; Teleostei) inferred from Cytochrome b sequences. Mol. Phylogenet. Evol. 14(1):1–10.

Lê, H. L. V., G. Lecointre, and R. Perasso. 1993. A 28S rRNA-based phylogeny of the gnathostomes: first steps in the analysis of conflict and congruence with morphologically based cladograms. Mol. Phylogenet. Evol. 2(1):31–51.

Lecointre, G. 1995. Molecular and morphological evidence for a Clupeomorpha-Ostariophysi sister-group relationship (Teleostei). Geobios. Mem. Spez. 19:204–210.

Lecointre, G., and G. Nelson. 1996. Clupeomorpha, sister-group of Ostariophysi. Pp. 193–207 in Interrelationships of fishes (M. L. J. Stiassny, L. R. Parenti, and G. D. Johnson, eds.). Academic Press, San Diego.

Li, G.-Q., and M. V. H. Wilson. 1996. Phylogeny of Osteoglossomorpha. Pp. 163–174 in Interrelationships of fishes (M. L. J. Stiassny, L. R. Parenti, and G. D. Johnson, eds.). Academic Press, San Diego.

Liem, K. F. 1990. Aquatic versus terrestrial feeding modes: possible impacts on the trophic ecology of vertebrates. Am. Zool. 30:209–221.

Lovejoy, N. R. 1996. Systematics of myliobatoid elasmobranchs: with emphasis on the phylogeny and historical biogeography of neotropical freshwater stingrays (Potamotrygonidae: Rajiformes). Zool. J. Linn. Soc. 117:207–257.

Lowe, C. G., R. N. Bray, and D. Nelson. 1994. Feeding and associated behavior of the Pacific electric ray *Torpedo californica* in the field. Mar. Biol. 120:161–169.

Lund, R. 1990. Chondrichthyan life history styles as revealed by the 320 million years old Mississippian of Montana. Environ. Biol. Fishes 27:1–19.

Maisey, J. G. 1980. An evaluation of jaw suspension in sharks. Am. Mus. Nov. 2706:1–17.

Maisey, J. G. 1982. The anatomy and interrelationships of Mesozoic hybodont sharks. Am. Mus. Nov. 2724:1–48.

Maisey, J. G. 1984. Chondrichthyan phylogeny: a look at the evidence. J. Vert. Paleontol. 4:359–371.

Maisey, J. G. 1986. Heads and tails: a chordate phylogeny. Cladistics 1986(2):201–256.

Maisey, J. G. 2001. A primitive chondrichthyan braincase from the middle Devonian of Bolivia. Pp. 263–288 in Major events in early vertebrate evolution. Palaeontology, phylogeny, genetics and development (P. E. Ahlberg, ed.). Taylor and Francis, London.

Maisey, J. G., and M. E. Anderson. 2001. A primitive chondrichthyan braincase from the early Devonian of South Africa. J. Vert. Paleontol. 21(4):702–713.

Marshall, N. B. 1979. Developments in deep-sea biology. Blandford Press, Poole, Dorset, UK.

Mattei, X. 1991. Spermatozoon ultrastructure and its systematic implications in fishes. Can. J. Zool. 69:3038–3055.

McEachran, J. D. 1984. Anatomical investigations of the New Zealand skates, *Bathyraja asperula* and *B. spinifera*, with an evaluation of their classification within Rajoidei (Chondrichthyes, Rajiformes). Copeia 1984(1):45–58.

McEachran, J. D., and K. A. Dunn. 1998. Phylogenetic analysis of skates, a morphologically conservative group of elasmobranchs. Copeia 1998(3):271–293.

McEachran, J. D., Dunn, K., and T. Miyake. 1996. Interrelationships of batoid fishes. Pp. 63–84 in Interrelationships of fishes (M. L. J. Stiassny, G. D. Johnson, and L. Parenti, eds.). Academic Press, San Diego.

McEachran, J. D., and T. Miyake. 1990. Phylogenetic interrelationships of skates: a working hypothesis (Chondrichthyes, Rajoidei). Pp. 285–304 in Elasmobranchs as living resources: advances in the biology, ecology, systematics, and the status of the fisheries (H. L. Pratt, S. H. Gruber, and T. Taniuchi, eds.). NOAA Tech. Rep. 90. U.S. Department of Commerce, Washington, DC.

Miles, R. S., and G. C. Young. 1977. Palcoderm interrelationships reconsidered in the light of ner ptyctodontids from Gogo, western Australia. Linn. Soc. Symp. Ser. 4:123–198.

Miya, M., A. Kawaguchi, and M. Nishida. 2001. Mitogenomic exploration of higher teleostean phylogenies: a case study for moderate-scale evolutionary genomics with 38 newly determined complete mitochondrial DNA sequences. Mol. Biol. Evol. 18(11):1993–2009.

Miya, M., H. Takeshima, H. Endo, N. B Ishiguro, J. G. Inoue, T. Mukai, T. P. Satoh, M. Yamaguchi, A. Kawaguchi, T. Mabuchi, S. Shirai, and M. Nishida. 2003. Major patterns of higher teleostean phylogenies: a new perspective based on 100 complete mitochondrial DNA sequences. Mol. Phylo. Evol. 26(1):121–138.

Miyake, T., J. D. McEachran, P. J. Walton, and B. K. Hall. 1992. Development and morphology of rostral cartilages in batoid fishes (Chondrichthyes: Batoidea), with comments on homology within vertebrates. Biol. J. Linn. Soc. 46:259–298.

Miyake, Y. 1988. The systematics of the stingray genus *Urotrygon* with comments on the interrelationships within Urolophidae (Chondrichthyes, Myliobatiformes). Ph.D. thesis, Texas A & M University, College Station.

Mooi, R. D., and A. C. Gill. 1995. Association of epaxial musculature with dorsal-fin pterygiophores in acanthomorph fishes, and its phylogeneticsignificance. Bull. Nat. Hist. Mus. Lond. (Zool.) 61:121–137.

Morrissey, J. F., K. A. Dunn, and F. Mulé. 1997. The phylogenetic position of *Megachasma pelagios* inferred from mtDNA sequence data. Pp. 33–36 in Biology of the megamouth shark (K. Yano, J. F. Morrissey, Y. Yabumoto, and K. Nakaya, eds.). Tokai University Press, Tokyo.

Myrberg, A. A., and S. H. Gruber. 1974. The behavior of the bonnethead shark, *Sphyrna tiburo*. Copeia 1974(2):358–374.

Nakaya, K. 1995. Hydrodynamic funstion of the head in the hammerhead sharks (Elasmobranchii: Sphyrnidae). Copeia 1995(2):330–336.

Nakaya, K., and B. Séret. 1999. A new species of deepwater catshark, *Apristurus albisoma* n. sp. From New Caledonia (Chondrichthyes: Carcharhiniformes: Scyliorhinidae). Cybium 23(3):297–310.

Naylor, G. J. P. 1992. The phylogenetic relationships among requiem and hammerhead sharks: inferring phylogeny when thousands of equally most parsimonious trees result. Cladistics 8:295–318.

Naylor, G. J. P., A. P. Martin, E. G. Mattison, and W. M. Brown. 1997. Interrelationships of lamniform sharks: testing phylogenetic hypotheses with sequence data. Pp. 199–218 in Molecular systematics of fishes (T. D. Kocher and C. A. Stepien, eds.). Academic Press, San Diego.

Nelson, G. 1989. Phylogeny of major fish groups. Pp. 325–336 *in* The hierarchy of life (B. Fernholm, K. Bremer, L. Brundin, H. Jörnvall, L. Rutberg, and H.-E. Wanntorp, eds.). Elsevier Science, Amsterdam.

Nelson, J. S. 1994. Fishes of the world. 3rd ed. Wiley, New York.

Nishida, K. 1990. Phylogeny of the suborder Myliobatidoidei. Mem. Fac. Fish. Hokkaido Univ. 37(1/2):1–108.

Norman, J. R. 1926. A synopsis of the rays of the family Rhinobatidae, with a revision of the genus *Rhinobatus*. Proc. Zool. Soc. Lond. 62(4):941–982.

Obermiller, L. E., and E. Pfeiler. 2003. Phylogenetic relationships of elopomorph fishes inferred from mitochondrial ribosomal DNA sequences. Mol. Phylogenet. Evol. 26:202–214.

Olney, J. E., G. D. Johnson, and C. C. Baldwin. 1993. Phylogeny of lampridiform fishes. Bull. Mar. Sci. 52(1):137–169.

Ørvig, T. 1960. New finds of acanthodians, arthrodires, crossopterygians, ganoids and dipnoans in the upper middle Devonian Calcareous Flags (Oberer Plattenkalk) of the Bergisch-Paffrath Trough (Part I). Palaont. Z. 34:295–335.

Ørvig, T. 1962. Y a-t-il une relation directe entre les arthrodires ptyctodontides et les holocéphales? Colloq. Int. Cent. Nat. Rech. Sci. 104:49–61.

Parenti, L. R. 1993. Relationships of atherinomorph fishes (Teleostei). Bull. Marine Sci. 52:170–196.

Parenti, L. R., and J. Song. 1996. Phylogenetic significance of the pectoral-pelvic fin association in acanthomorph fishes: a reassessment using comparative neuroanatomy. Pp. 427–444 *in* Interrelationships of fishes (M. L. J. Stiassny, L. R. Parenti, and G. D. Johnson, eds.). Academic Press, San Diego.

Patterson, C. 1965. The phylogeny of the chimaeroids. Philos. Trans. R. Soc. Lond. B 249:101–219.

Patterson, C. 1973. Interrelationship of holosteans. Pp. 233–305 *in* Interrelationships of fishes (P. H. Greenwood, R. S. Miles, and C. Patterson, eds.). Academic Press, London.

Patterson, C. 1993. Osteichthyes: Teleostei. Pp. 621–656 *in* The fossil record 2 (M. J. Benton, ed.). Chapman and Hall, London.

Patterson, C. 1994. Bony fishes. Pp. 57–84 *in* Major features of vertebrate evolution (D. R. Prothero and R. M. Schoch, eds.). The Paleontological Society Short Course in Paleontology no. 7.

Patterson, C., and D. E. Rosen. 1989. The Paracanthopterygii revisited: order and disorder. Pp. 5–36 *in* Papers on the systematics of gadiform fishes (D. M. Cohen, ed.). Science Series No. 32, Nat. Hist. Mus. Los Angeles Co.

Pfeil, F. H. 1983. Zahnmorphologische Untersuchungen in rezenten und fossilen Haien der Ordnungen Chlamydoselachiformes und Echinorhiniformes. Palaeoichthyologica 1:1–315.

Regan, C. T. 1906. A classification of selachian fishes. Proc. Zool. Soc. Lond. 1906:722–758.

Roberts, C. 1993. The comparative morphology of spined scales and their phylogenetic significance in the Teleostei. Bull. Mar. Sci. 52:60–113.

Rosen, D. E. 1973. Interrelationships of higher euteleostean fishes. Pp. 397–513 *in* Interrelationships of fishes (P. H. Greenwood, R. S. Miles, and C. Patterson, eds.). Academic Press, London.

Rosen, D. E. 1984. Zeiforms as primitive plectognath fishes. Amer. Mus. Nov. 2782:1–45.

Rosen, D. E., P. L. Forey, B. G. Gardiner, and C. Patterson. 1981. Lungfishes, tetrapods, paleontology, and plesiomorphy. Bull. Am. Mus. Nat. Hist. 167:159–276.

Rosen, D. E., and C. Patterson. 1969. The structure and relationships of the paracanthopterygian fishes. Bull. Am. Mus. Nat. Hist. 141(3):361–474.

Saint-Seine, P. 1949. Les poissons des calcaires lithographiques de Cérin (Ain). Nouv. Arch. Mus. Hist. Nat. Lyon 2:1–357.

Sato, T., and T. Nakabo. 2002. Paraulopidae and *Paraulopus*, a new family and genus of aulopiform fishes with revised relationship within the order. Ichthyol. Res. 49:25–46.

Schaeffer, B. 1967. Comments on elasmobranch evolution. Pp. 3–35 *in* Sharks, skates and rays (P. W. Gilbert, R. F. Matthewson, and D. P. Rall, eds.). The Johns Hopkins Press, Baltimore.

Schaeffer, B. 1981. The xenacanth shark neurocranium, with notes on elasmobranch monophyly. Bull. Am. Mus. Nat. Hist. 169(1):1–66.

Schaeffer, B., and M. Williams. 1977. Relationships of fossil and living elasmobranchs. Am. Zool. 17(2):293–302.

Schultze, H.-P., and S. L. Cumbaa. 2001. *Dialipina* and the characters of basal actinopterygians. Pp. 315–332 *in* Major events in early vertebrate evolution. Paleontology, phylogeny, genetics and development (P. E. Ahlberg, ed.). Talyor and Francis, London.

Seehausen, O. 1996, Lake Victoria rock cichlids. Taxonomy, ecology and distribution. Verduyn Cichlids, Zevenhuizen, Netherlands.

Shen, M. 1996. Fossil "osteoglossomorphs" in East Asia and their implications in teleostean phylogeny. Pp. 261–272 *in* Mesozoic fishes 2: Systematics and fossil record (G. Arratia and H.-P. Schultze, eds.). Pfeil, Münich.

Shirai, S. 1992. Squalean phylogeny: a new framework of "squaloid" sharks and related taxa. Hokkaido University Press, Sapporo.

Shirai, S. 1996. Phylogenetic interrelationships of neoselachians (Chondrichthyes: Euselachii). Pp. 63–84 *in* Interrelationships of fishes (M. L. J. Stiassny, L. Parenti, and G. D. Johnson, eds.). Academic Press, San Diego.

Smith, B. G. 1942. The heterodontid sharks: their natural history, and the external development of *Heterodontus japonicus* based on notes and drawings by Bashford Dean. Pp. 649–770, pls. 1–7 *in* Bashford Dean memorial volume: archaic fishes (E. W. Gudger, ed.), vol. 8. American Museum of Natural History, New York.

Smith, W. L. 2001. Is *Normanichthys crockeri* a scorpaeniform? P. 279 *in* Program book and abstracts, Amer. Soc. Ichthyologists and Herpetologists. Kansas City, MO.

Stahl, B. J. 1999. Chondrichthyes III. Holocephali. Pp. 1–164 *in* Handbook of paleoichthyology (H.-P. Schultze, ed.), vol. 4. Pfeil, Munich.

Stein, D. L., V, Chernova-Natalia, and P. Andriashev-Anatoly. 2001. Snailfishes (Pisces: Liparidae) of Australia, including descriptions of thirty new species. Rec. Aust. Mus. 53(3):341–406.

Stensiö, E. A. 1925. On the head of macropetalichthyids with certain remarks on the head of other arthrodires. Publ. Field Mus. Nat. Hist. Geol. 4:89–198.

Stensiö, E. A. 1958. Les cyclostomes fossiles ou Ostracodermes. Pp. 173–425 *in* Traité de zoologie (P. P. Grassé, ed.), vol. 13. Masson, Paris.

Stensiö, E. A. 1963. Anatomical studoes on the arthrodiran head, Pt. 1: Preface, geological and geographical distribution. The organization of the arthrodires. The anatomy of the head in the Dolichothoraci, Coccosteomorphi and Pachyosteomorphi. K. Svenska Vetensk. Akad. Handl. 9:1–419.

Stensiö, E. A. 1969. Elasmobranchiomorphi, Placodermata, Arthrodires. Pp. 71–642 *in* Traité de paléontologie (J. Piveteau, ed.), vol. 4(2). Masson, Paris.

Stiassny, M. L. J. 1986. The limits and relationships of the acanthomorph teleosts. J. Zool. B 1:411–460.

Stiassny, M. L. J. 1990. Notes on the anatomy and relationships of the bedotiid fishes of Madagascar, with a taxonomic revision of the genus *Rheocles* (Atherinomorpha: Bedotiidae). Am. Mus. Nov. 2979:1–33.

Stiassny, M. L. J. 1993. What are grey mullets? Bull. Mar. Sci. 52(2):197–219.

Stiassny, M. L. J. 1996. Basal ctenosquamate relationships and the interrelationships of the myctophiform (Scopelomorph) fishes. Pp. 405–436 *in* Interrelationships of fishes (M. L. J. Stiassny, L. R. Parenti, and G. D. Johnson, eds.). Academic Press, San Diego.

Stiassny, M. L. J. 2002. Bony fishes. Pp. 192–197 *in* Life on Earth: an encyclopedia of biodiversity, ecology, and evolution (N. Eldredge, ed.). ABI-CLIO Press, Santa Barbara, CA.

Stiassny, M. L. J., and J. S. Moore. 1992. A review of the pelvic girdle of acanthomorph fishes, with comments on hypotheses of acanthomorph intrarelationships. Zool. J. Linn. Soc. 104:209–242.

Sullivan, J. P., S. Lavoué, and C. D. Hopkins. 2000. Molecular systematics of the African electric fishes (Mormyroidea: Teleostei) and a model for the evolution of their electric organs. J. Exp. Biol. 203:665–683.

Tchernavin, V. V. 1953. The feeding mechanism of a deep sea fish. *Chauliodus sloani* Schneider. British Museum (Natural History), London. 99 pp.

Tyler, J. C. 1980. Osteology, phylogeny, and higher classification of the fishes of the order Plectognathi (Tetraodontiformes). NOAA Tech. Rept. NMFS Circ. 434:1–422.

Tyler, J. C., B. O'Toole, and R. Winterbottom. 2003. Phylogeny of the genera and families of zeiform fishes, with comments on their relationships to tetraodontiforms and caproids. Smithson. Contrib. Zool. 618:1–110.

Venkatesh, B., M. V. Erdmann, and S. Brenner. 2001. Molecular synapomorphies resolve evolutionary relationships of extant jawed vertebrates. Proc. Natl. Acad. Sci. USA 98:11382–11387.

Weinberg, S. 2000. A fish caught in time: the search for the coelacanth. HarperCollins, New York.

White, E. G. 1937. Interrelationships of the elasmobranchs with a key to the order Galea. Bull. Am. Mus. Nat. Hist. 74(2):25–138.

Wiley, E. O. 1976. The phylogeny and biogeography of fossil and recent gars (Actinopterygii: Lepisosteidae). Misc. Publ. Univ. Ks. Mus. Nat. Hist. 64:1–111.

Wiley, E. O. 1998. Birchirs and allies. Pp. 75–79 *in* Encyclopedia of fishes. A comprehensive guide by international experts. 2nd ed. (J. R. Paxton and W. N. Eschmeyer, eds.). Academic Press, San Diego.

Wiley, E. O., G. D. Johnson and W. W. Dimmick. 1998. The phylogenetic relationships of lampridiform fishes (Lampridiformes, Acanthomorpha), based on a total evidence analysis of morphological and molecular data. Mol. Phylogenet. Evol. 10:471–425.

Wiley, E. O., G. D. Johnson, and W. W. Dimmick. 2000. The interrelationships of acanthomorph fishes: a total evidence approach using molecular and morphological data. Biochem. Syst. Ecol. 28(2000):319–350.

Winterbottom, R. 1974. The familial phylogeny of the Tetraodontiformes (Acanthopterygii: Pisces) as evidenced by their comparative myology. Smithson. Contrib. Zool. 155:1–201.

Woodward, A. S. 1889. Catalog of the fossil fishes in the British Museum, pt. I. British Museum, London.

Woodward, A. S. 1898. Outlines of vertebrate paleontology for students of zoology. Cambridge University Press, Cambridge.

Yamaguchi, M. 2000. Phylogenetic analyses of myctophid fishes using morphological characters: progress, problems, and future prospects (in Japanese). Jap. J. Ichthyol. 47(2):87–107 (Engl. abstr.).

Yano, K., J. F. Morrissey, Y. Yabumoto, and K. Nakaya. 1997. Biology of the megamouth shark. Tokai University Press, Tokyo.

Zangerl, R. 1981. Chondrichthyes I. Paleozoic Elasmobranchii. Pp. 1–115 *in* Handbook of paleoichthyology (H.-P. Schultze, ed.), vol. 3A. Fischer, New York.

Zhu, Z. 1992. Review on *Peipiaosteus* based on new material of *P. pani* (in Chinese). Vertebr. Palasiat. 30:85–101 (Engl. summ.).

Zhu, M., and H.-P. Schultze. 1997. The oldest sarcopterygian fish. Lethaia 30:192–206.

Zhu, M., and H.-P. Schultze. 2001. Interrelationships of basal osteichthyans. Pp. 289–314 *in* Major events in early vertebrate evolution. palaeontology, phylogeny, genetics and development (P. E. Ahlberg, ed.). Taylor and Francis, London.

Zhu, M., Yu, X., and P. E. Ahlberg. 2001. A primitive sarcopterygian fish with an eyestalk. Nature 410:81–84.

Zhu, M., X. Yu, and P. Janvier. 1999. A primitive fossil fish sheds light on the origin of bony fishes. Nature 397:607–610.

David Cannatella
David M. Hillis

Amphibians

Leading a Life of Slime

Amphibians have generally not been regarded with much favor. An often-cited paraphrasing of Linnaeus's *Systema Naturae* suggests that they are such loathsome, slimy creatures that the Creator saw fit not to make many of them. In fact, the number of living amphibians, about 5300, exceeds that of our own inclusive lineage, Mammalia (Glaw and Köhler 1998). The rate of discovery of new species exceeds that of any other vertebrate group. Since the publication of *Amphibian Species of the World* (Frost 1985), the number of recognized amphibians has increased by 36%. More than 100 undescribed frog species have been reported in Sri Lanka (Meegaskumbura et al. 2002). Yet, the decline and extinction of amphibian populations are visible signals of environmental degradation (Hanken 1999).

Amphibians are named for their two-phased life history: larva and adult. Typically, the larva is aquatic and metamorphoses into a terrestrial adult. In a loose, descriptive sense, amphibians bridge the gap between fishes, which are fully aquatic, and amniotes, which have completely escaped a watery environment and have abandoned metamorphosis. However, amphibians are not in any sense trapped in an evolutionary cul-de-sac, because they exhibit a far greater diversity of life history modes than do amniotes.

Each type of living amphibian—frog, salamander, and caecilian—is highly distinctive. Frogs are squat, four-legged creatures with generally large mouths and eyes and elongate hind limbs used for jumping. There is no tail (the meaning of Anura), because the caudal vertebrae have coalesced into a bony strut. About 90% of the living amphibian species are frogs; they rely mostly on visual and auditory cues. Salamanders are more typical-looking tetrapods, all with a tail (hence, Caudata) and most with four legs. Some are elongate and have reduced the limbs and girdles; these are usually completely aquatic or fossorial species. In general, they rely more on olfactory cues. Living caecilians are all limbless and elongate. Grooved rings encircle the body, evoking the image of an earthworm; most caecilians are fossorial, but some are aquatic. All have reduced eyes, although the root *caecus*—Latin for "blind"—is a misnomer. Near the eye or the nostril is a unique protrusible tentacle used for olfaction. The tail is essentially absent.

Modern Amphibians

By modern amphibians, we mean the lineage minimally circumscribed by living taxa; this is known as the crown clade Amphibia. In the language of phylogenetic taxonomy (discussed below), Amphibia are a node-based name defined as the most recent common ancestor of frogs, salamanders, caecilians, and all the descendants of that ancestor (Cannatella and Hillis 1993). Frost (1985) and Duellman (1993) summarized the species of amphibians. Up-to-date Internet resources include Frost (2002) and D. B. Wake (2003). The distribution of modern amphibians is treated in Duellman (1999). Aspects of modern amphibian biology can be found in two recent textbooks (Pough et al. 2001, Zug et al. 2001) and in a treatise (Laurent 1986). The most comprehensive treatment is that of Duellman and Trueb (1986).

Modern amphibians are at times called lissamphibians to distinguish them from the Paleozoic forms referred to as "amphibians." Modern amphibians include frogs, salamanders, and caecilians, and their Mesozoic [245–65 million years ago (Mya)] and Cenozoic (65 Mya to present) extinct relatives (including albanerpetontids), all of which are readily identifiable as belonging to this group. In contrast, their Paleozoic relatives include the traditional groups termed the Labyrinthodontia and Lepospondyli. Labyrinthodonts, including the earliest four-legged vertebrates, ranged from the Upper Devonian (375 Mya) through the Permian (290 Mya), with numbers declining into the Triassic and one small lineage persisting into the Cretaceous. Lepospondyls range from the Lower Carboniferous (240 Mya) to the base of the Upper Permian (250 Mya). Labyrinthodonts are a paraphyletic group and also gave rise to amniotes. Lepospondyls are a heterogeneous group but have a characteristic vertebral morphology (Carroll et al. 1999); their monophyly is unclear.

Several features set modern amphibians apart from other vertebrates. Some of these characters support monophyly of the group compared with both fossil and living taxa. The significance of other characters, such as soft tissue features (Trueb and Cloutier 1991a), is less certain because they cannot be assessed in extinct forms. But these characters do support amphibian monophyly relative to amniotes and fishes.

Most adult amphibians have teeth that are pedicellate and bicuspid, or modified from this condition. Pedicellate teeth have a zone of reduced mineralization between the crown and the base (pedicel). In fossils the crowns are often broken off, leaving a cylindrical base with an open top. Pedicellate teeth are also found in a few temnospondyl labyrinthodonts believed to be closely related to modern amphibians (Bolt 1969).

Living amphibians also share the absence or reduction of several skull bones. On the dorsal skull, the jugals, postorbitals, postparietals, supratemporals, intertemporals, and tabulars are absent. On the palate, the pterygoid, ectopterygoid, and palatines are reduced or absent so as to produce a large space, the interpterygoid vacuity, below the eye sockets (Reiss 1996). The reduction/loss of many skull bones in modern amphibians is a result of pedomorphosis (Alberch et al. 1979). Pedomorphosis is a pattern derived from a change the timing of development; specifically, a species becomes sexually mature (adult) at an earlier stage of development than its immediate ancestor. As a result, the adult of amphibians resembles the juvenile (or larval) stage of Paleozoic relatives. A secondary result of pedomorphosis is miniaturization (Hanken 1985); because living amphibians mature at an earlier age, they are typically much smaller than the Paleozoic forms (Bolt 1977, Schoch 1995).

Amphibians employ a buccal force-pump mechanism for breathing (Brainerd et al. 1993, Gans et al. 1969). Air is forced back into the lungs by positive pressure from the mouth cavity. In contrast, amniotes use aspiration to fill the lungs, in which the rib cage and/or diaphragm creates negative pressure in the thorax. Amphibians have distinctive short ribs that do not form a complete rib cage as in amniotes, so aspiration is not possible.

In addition to the stapes-basilar papilla sensory system of tetrapods, living amphibians have a second acoustic pathway, the opercular-amphibian papilla system. This system is more sensitive to lower frequency vibrations than is the stapes-basilar papilla pathway. In frogs and salamanders, the operculum (a bone of the posterior aspect of the braincase, not in any way similar to the homonymous bone of fishes) is also connected to the shoulder girdle by way of a modified levator scapulae muscle, the opercularis. This muscle transmits vibrations from the ground through the forelimb and shoulder girdle to the inner ear.

The skin is a significant respiratory organ; it is supplied by cutaneous branches of the ductus arteriosus (the presence of these is not clear in caecilians). The skin has a stratum corneum (outer layer) like that of other tetrapods, although it is thinner than that of amniotes. However, living amphibians retain the primitive feature of mucous glands and granular glands. Granular glands secrete poisons of varying toxicity, some lethal. Mucous glands keep the skin moist, which allows the dissipation of heat, as well as the loss of water through the skin. Many caecilians have dermal scales, similar to those of teleost fishes, embedded in the skin.

The Name "Amphibia"

In the *Systema Naturae* of Carolus Linnaeus, the Amphibia were one of six major groups of animals, the others being mammals, birds, fish, insects, and mollusks. The group included not only frogs, salamanders, and caecilians but also reptiles and some fish that lacked dermal scales. Later, as early fossil tetrapods were uncovered, these were also relegated to "Amphibia" because of their presumed ancestral position to other tetrapods. In 1866 the great German biologist Ernst Haeckel divided Amphibia into Lissamphibia (salamanders and frogs) and Phractamphibia (caecilians and fossil labyrinthodonts; Haeckel 1866). "Liss-" refers to the naked skin of frogs and salamanders, and "phract-" means helmet, in reference to the armor of dermal skull bones and scales found in early tetrapods and, in a reduced form, in caecilians. Gadow (1901) transferred the caecilians from Phractamphibia to Lissamphibia.

For most of the 20th century, the name Amphibia was used for tetrapods that were not reptiles, birds, or mammals. Thus, the earliest tetrapods (labyrinthodonts from the Devonian) were included in Amphibia, as were the Lepospondyli. This rendition of Amphibia appeared in most comparative anatomy and paleontology texts, largely because of the influence of the paleontologist Alfred Romer. Modern amphibians were believed to be polyphyletic and derived from different "amphibian" lineages; frogs from Labyrinthodontia, and salamanders and caecilians from Lepospondyli. Parsons and Williams adduced evidence for the monophyly of modern amphibians and resurrected Gadow's Lissamphibia for living amphibians (Parsons and Williams 1962, 1963). However, the term Lissamphibia is used mainly among specialists to distinguish the modern groups from extinct Paleozoic forms. Most biologists and most textbooks refer to frogs, salamanders, and caecilians simply as amphibians.

Use of Amphibia in the Romerian sense of a paraphyletic taxon has been largely abandoned and the name has been redefined as a monophyletic group in two contrasting ways (fig. 25.1). First, the name Amphibia is applied to the node that is the last (most recent) ancestor of living frogs, salamanders, and caecilians (de Queiroz and Gauthier 1992). Amphibia includes this ancestor and all its descendants, which are the modern forms, including albanerpetontids. Second, Amphibia is defined as the stem or branch that contains living frogs, caecilians, salamanders, and all other taxa more closely related to these than to amniotes (e.g., Gauthier et al. 1989, Laurin 1998a). In other words, the stem-based name Amphibia includes all taxa along the stem leading to modern amphibians; this includes either the temnospondyls, the lepospondyls, or both, depending on which phylogeny one accepts. Under a stem-based definition, the content of Amphibia, in terms of fossil taxa, may change dramatically. Laurin (1998a) proposed such changes based on his application of principles of priority and synonymy to phylogenetic taxonomy. He argued that the definition of Amphibia as a stem-based name by Gauthier et al. (1989) must be accorded priority over the node-based definition of Amphibia of de Queiroz and Gauthier (1992). One result of accepting the stem-based definition is that the content of Amphibia under Laurin's phylogeny (Laurin and

Reisz 1997) is very different compared with the content under other definitions of Amphibia.

Node- and stem-based names have their respective advantages in communicating taxonomy. However, a stem-based definition of Amphibia, a name in general parlance, has an undesirable effect, because generalizations about the biology of modern amphibians can be wrongly extended to extinct temnospondyls and/or lepospondyls (de Queiroz and Gauthier 1992). These groups bear little resemblance to the living forms, and their biology was presumably very different. Under a stem-based definition of Amphibia, the common statement "all amphibians have mucous glands" would be interpreted to mean that lepospondyls had mucous glands, an inference for which there is no evidence. In contrast, under the node-based definition of Amphibia, one can reasonably infer that extinct frogs, salamanders, and caecilians have mucous glands, but the inference does not extend inappropriately to extinct temnospondyls and lepospondyls. Although some neontologists and most paleontologists appreciate the semantic distinction between Amphibia and Lissamphibia, most biologists use Amphibia to mean frogs, salamanders, and caecilians.

Amphibians and the Origin of Tetrapods

The exact relationships of modern amphibians to extinct Paleozoic forms is not clear. Heatwole and Carroll (2000) provided a summary of the phylogeny of various fossil groups. The favored family of hypotheses (fig. 25.2A,B) posits that the group of frogs, salamanders, and caecilians is monophyletic and that this clade is nested within dissorophoid temnospondyls (Bolt 1977, 1991, Milner 1988, 1993, Trueb and Cloutier 1991a). (Temnospondyls are labyrinthodonts that include Edopoidea, Trimerorhachoidea, Eryopoidea, Stereospondyli, and Dissorophoidea.) The most thorough and data-rich analysis, in terms of characters and taxa (Ruta et al. 2003; fig. 25.2B), also reached this conclusion.

A recent variant of the monophyly hypothesis (fig. 25.2C) is that modern amphibians are nested within the lepospondyls (e.g., Anderson 2001), particularly within the Microsauria (Laurin 1998a, 1998b, Laurin et al. 2000a, 2000b, Laurin and Reisz 1997; but see Coates et al. 2000, Ruta et al. 2003). Because temnospondyls are distantly related to amphibians under this second hypothesis, the derived similarities between them and dissorophoid temnospondyls are interpreted as convergent.

A very different hypothesis claims polyphyly of the modern groups (fig. 25.2D), with caecilians derived from goniorhynchid microsaurs (Carroll 2000b, Carroll and Currie 1975), and salamanders and frogs from temnospondyls. The polyphyly hypothesis gained some strength with the discovery of the fossil *Eocaecilia* (see below), which possessed characters seemingly intermediate between goniorhynchid

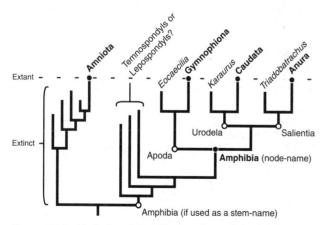

Figure 25.1. Node-based (boldface) and stem-based definitions of Amphibia.

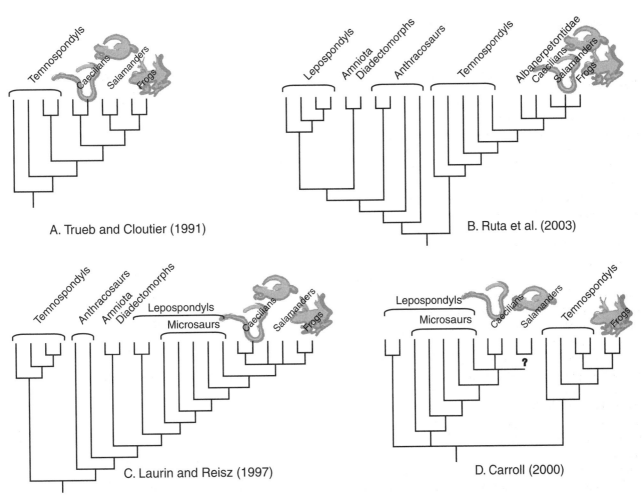

Figure 25.2. (A–D) Alternative relationships among modern amphibians (caecilians, frogs, and salamanders) and Paleozoic groups (temnospondyls, microsaurs, and lepospondyls).

microsaurs and living caecilians (Carroll 2000a)—this interpretation remains controversial.

Interrelationships of Modern Amphibians

Two general alternative hypotheses have been considered for relationships among the groups of modern amphibians. One tree, based primarily, but not exclusively, on non-molecular data, allies frogs and salamanders, with caecilians as the odd group out (fig. 25.2A,B). The name Batrachia, formerly synonymous with Amphibia, has been applied to this clade. In the second hypothesis, the earliest analyses of DNA sequence data slightly favored salamanders and caecilians, a group named Procera, as closest relatives (Feller and Hedges 1998, Hedges and Maxson 1993, Hedges et al. 1990), as in figure 25.2D. However, Zardoya and Meyer (2001) analyzed complete mitochondrial sequences of one species each of a frog, salamander, and caecilian and found the frog and salamander to be sister groups. Although their level of taxon sampling was shallow, the results suggest sig-

nificant uses for character-rich data sets such as mitochondrial genomes.

A fourth group of amphibians is Albanerpetontidae, known only from fossils from the Jurassic to the Miocene (Milner 2000); the name Allocaudata has been used infrequently for these, because it is redundant with Albanerpetontidae. This group closely resembles salamanders in skull shape and in the primitive tetrapod features of a generalized body shape, four limbs and a tail. Albanerpetontids lack most of the same dorsal skull bones as do living amphibians but do not have pedicellate teeth. They have been considered to be nested within salamanders, or the sister group of Batrachia (McGowan and Evans 1995); the most recent and extensive analysis (Gardner 2001) placed them in the latter position. Ruta et al. (2003; fig. 25.2B) placed them in a basal polytomy with the modern forms.

Both nucleotide sequence data and "soft" anatomy ally frogs, salamanders, and caecilians as a clade relative to living amniotes and fishes. Because fossils do not so easily yield information about nucleotides or soft tissue characters, these data sets provide no direct evidence for the monophyly of

Amphibia with respect to Paleozoic tetrapods (Trueb and Cloutier 1991a, 1991b).

Caecilians

The node-based name for modern caecilians is Gymnophiona, meaning "naked snake." Caecilians include 165 extant species, restricted to tropical America, Africa, and Asia. They are grouped into five or six families (fig. 25.3, table 25.1).

Because of their habits, caecilians are rarely seen in the wild. A dedicated herpetologist might find them by digging, and occasionally individuals are found on the surface of the ground after a heavy tropical rains Most caecilians are 0.3–0.5 m long, although one species is as large as 1.5 m and one as small as 0.1 m. All caecilians are elongate, but some are more elongate than others; the number of vertebrae ranges from 86 to 285. Caecilians are almost unique among amphibians (two species of frogs are the exception) in having a male intromittent organ, the phallodeum, and internal fertilization occurs during copulation.

Living caecilians have reduced eyes with small orbits, and scolecomorphids and some caeciliids have eyes covered by the skull bones. Compared with other amphibians, the skulls of caecilians are highly ossified and many bones are fused. The resulting wedge-shaped cranium is used for digging and compacting the soil. Most caecilians are oviparous with free-living larvae. Viviparous species occur in a few families; in some of these the embryos derive nutrition from the lining of the oviduct, so far as is known. They have a species-specific "fetal dentition" that apparently is used to help ingest the nutritive secretions. Most caecilians are fossorial, but the

Typhlonectidae are aquatic and most have laterally compressed bodies, especially posteriorly, and a slight dorsal "fin," presumably for swimming.

Fossil caecilian vertebrae are known from the Upper Cretaceous, Tertiary, and Quaternary of Africa north of the Sahara and Mexico to Bolivia and Brazil (summarized in Wake et al. 1999). Although living caecilians are limbless, and nearly or completely tailless, the earliest putative caecilian had legs and a tail! *Eocaecilia micropodia* from the Jurassic has a somewhat elongate body and small but well-developed limbs (Carroll 2000a, Jenkins and Walsh 1993, Wake 1998). *Eocaecilia* has pedicellate teeth and a groove in the edge of the eye socket is interpreted to be for passage of the tentacle; thus *Eocaecilia* is inferred to have a feature otherwise unique to living caecilians. The evidence suggests it is the sister group of all other caecilians. The stem-based name for the clade containing *Eocaecilia* + Gymnophiona is Apoda (Cannatella and Hillis 1993).

Gymnophiona are the least understood of all vertebrate lineages, given its size. Caecilians are restricted to tropical regions of America, Africa (excluding Madagascar), the Seychelles Islands, and much of Southeast Asia. In general, phylogenetic relationships among caecilian families have not generated as much controversy as have those among salamanders or frogs, but little work has been done and sampling of species is poor. Taylor (1968) presented a monographic revision of the systematics of caecilians that stimulated work for the next 30 years, including considerable molecular and morphological research. Lescure et al. (1986) presented a radically different classification of caecilians based on sparse new data. Nussbaum and Wilkinson (1989) reviewed this unorthodox classification in a larger context; they argued for maintaining the current generic and familial relationships pending further research.

Hedges et al. (1993) analyzed sequence data for the 12S and 16S ribosomal RNA (rRNA) genes for 13 species in 10 genera; and M. Wilkinson et al. (2002) examined relationships among Indian species. Although molecular data have added substantially to caecilian phylogenetics, new morphological characters have contributed as well. Wake (1993, 1994) found that neuroanatomical characters in isolation are not a robust character base, but are useful within a larger morphological set; Wilkinson (1997) confirmed the "eccentricity" of the neuroanatomical set. The description of a bizarre new typhlonectid used 141 morphological characters and resulted in a new analysis of Typhlonectidae (Wilkinson and Nussbaum 1999). Similarly, phylogenetic analysis of Uraeotyphlidae has made use of new anatomical features (Wilkinson and Nussbaum 1996). Only recently has the osteology of the entire group been surveyed (M. H. Wake 2003).

Rhinatrematidae are almost universally considered to be the sister taxon of other living gymnophiones (fig. 25.3) based on both morphological and molecular data (Hedges et al. 1993, Nussbaum 1977). These caecilians retain a very short tail behind the cloaca, as do the Ichthyophiidae, in contrast to other

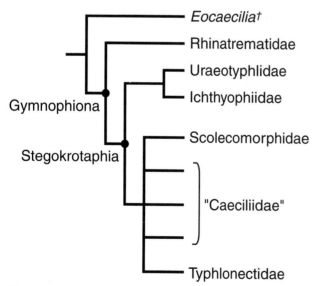

Figure 25.3. A generally accepted phylogenetic hypothesis of relationships among caecilians. "Caeciliidae" indicates a group that is paraphyletic with respect to Scolecomorphidae and Typhlonectidae. The dagger indicates extinction.

Table 25.1

Geographical Distribution of the Major Extant Groups of Amphibia.

Taxon	Distribution
Gymnophiona	
Rhinatrematidae	Northern South America
Ichthyophiidae	India, Sri Lanka, Southeast Asia
Uraeotyphlidae	South India
Scolecomorphidae	Africa
"Caeciliidae"	Mexico, Central and South America, Africa, Seychelles, India, Southeast Asia
Typhlonectidae	South America
Caudata	
Hynobiidae	Continental Asia to Japan
Sirenidae	Eastern United States and adjacent Mexico
Cryptobranchidae	China, Japan, eastern United States
Ambystomatidae	North America
Rhyacotritonidae	Northwest United States
Dicamptodontidae	Western United States and adjacent Canada
Salamandridae	Eastern and western North America, Europe and adjacent western Asia, northwest Africa, eastern Asia
Proteidae	Eastern United States and Canada, Adriatic coast of Europe
Amphiumidae	Southeast United States
Plethodontidae	North and Central America, northern South America, Italy and adjacent France, Sardinia
Anura	
Ascaphus	Northwest United States and adjacent Canada
Leiopelma	New Zealand
Bombinatoridae	Europe and eastern Asia, Borneo and nearby Philippine Islands
Discoglossidae	Europe, northern Africa
Pipidae	South America and adjacent Panama, sub-Saharan Africa
Rhinophrynidae	Central America, Mexico, and south Texas
Pelobatidae	North America, Europe, western Asia
Pelodytidae	Western Europe, western Asia
Megophryidae	Southern Asia to Southeast Asia
Heleophryne	Southern Africa
Myobatrachinae	Australia, New Guinea
Limnodynastinae	Australia, New Guinea
"Leptodactylidae"	South America, Central America, Mexico, southern United States
Bufonidae	All continents (including Southeast Asia) except Australia and Antarctica
Centrolenidae	Mexico, Central and South America
Dendrobatidae	Northern South America, Southeast Brazil, Central America
Sooglossidae	Seychelles
Hylidae	The Americas, Europe and adjacent Asia, northern Africa, eastern Asia, Japan, New Guinea, Australia
Pseudidae	South America
Rhinoderma	Southern South America
Allophryne	Northern South America
Brachycephalidae	Atlantic forests of southeastern Brazil
Microhylidae	Southern United States, Mexico, Central America, South America, sub-Saharan Africa, Madagascar, southern Asia, Southeast Asia, New Guinea, northeastern Australia
"Ranidae" (including Mantellinae)	All continents (northern South America only northeastern Australia only)
Arthroleptidae	Sub-Saharan Africa
Hyperoliidae	Sub-Saharan Africa, Madagascar, Seychelles
Hemisus	Sub-Saharan Africa
Rhacophoridae	Sub-Saharan Africa, Madagascar, southern Asia, Southeast Asia, Japan

family groups. Ichthyophiidae are a group of semi-fossorial forms from southern and Southeast Asia. Uraeotyphlidae, generally considered the sister taxon of Ichthyophiidae, are also from southern Asia; these are tailless.

Most taxonomic uncertainty resides in the geographically and biologically diverse taxon "Caeciliidae," which is prob-

ably paraphyletic with respect to Scolecomorphidae and Typhlonectidae. Caeciliids occur pantropically, and include a great diversity of taxa—including the smallest and largest species—and many reproductive modes, such as egg-layers with free-living larvae, direct developers, and viviparous forms, and several kinds of maternal care.

Scolecomorphidae are an African group with some bizarre features; in some taxa the eye is completely covered by a layer of bone, and in at least one species the eye can be protruded beyond the skull because of its attachment to the base of the tentacle (O'Reilly et al. 1996). The last group of caecilians, Typhlonectidae, is semi-aquatic to aquatic with attendant modifications, such as slight lateral compression of the posterior part of the body. Hedges et al. (1993) found the one species of Typhlonectidae analyzed to be nested among neotropical caeciliids. Accordingly, they synonymized the Typhlonectidae within Caeciliidae. Wilkinson and Nussbaum (1996, 1999) rejected that conclusion because of poor taxon sampling, preferring to wait until the relationships of the Caeciliidae, *sensu lato*, were fully explored.

Salamanders

The node-based name for living salamanders is Caudata. The 502 species of living salamanders are arranged into 10 families (fig. 25.4, table 25.1). Historically, salamanders are a primarily Holarctic group of the north temperate regions; one clade, the Bolitoglossini, has diversified in the Neotropics. The largest salamanders are the Cryptobranchidae; adult *Andrias* can reach 1.5m in total length. The smallest are *Thorius* (Plethodontidae), which may have an adult length as small as 30 mm.

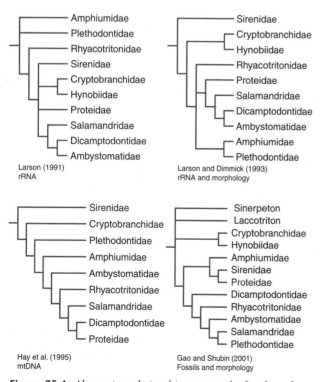

Figure 25.4. Alternative relationships among the families of salamanders.

Several salamanders are elongate and have reduced limbs. Some are larger, aquatic, neotenic forms, such as Sirenidae, Proteidae, and Amphiumidae. Fully aquatic salamanders typically retain gill slits, and some have external gills resembling crimson tufts of feathers. Elongate terrestrial salamanders typically have reduced limbs and digits, and occupy a semifossorial niche in leaf litter or burrows. At another extreme are arboreal forms with palmate hands and feet and reduced digits resulting from heterochrony.

Most of the major groups of salamanders have internal fertilization accomplished by way of a spermatophore, typically a mushroom-shaped mass of spermatozoa and mucous secretions. The male deposits a spermatophore either in water or on land, depending on the group. The female retrieves it with her cloaca during courtship. The sperm may be retained live in a cloacal pocket, the spermatheca, for months or even years. Fertilized eggs are deposited and develop either directly, in which case a small salamander hatches, or indirectly, in which a larval salamander emerges, and later metamorphoses.

Relationships among Salamanders

Karaurus sharovi, the oldest salamander, is a fully articulated Middle Jurassic fossil from Kazakhstan. The stem-based name for the clade of *Karaurus* + Caudata is Urodela ("with a tail"), so *Karaurus* is a urodele but not part of Caudata. Although the fossil *Karaurus* firmly established salamanders in the Jurassic, the fossil record of salamanders has not contributed to resolution of relationships among extant taxa until recently. However, crown-group salamanders belonging to the Cryptobranchidae are now known from the Middle Jurassic (Gao and Shubin 2003). Also, Gao and Shubin's (2001) analysis of Jurassic urodeles (fig. 25.4) placed these at the base of the extant salamander tree with Hynobiidae and Cryptobranchidae (Cryptobranchoidea). The Sirenidae formed a clade with two other neotenic taxa (Proteidae and Amphiumidae). In contrast, Duellman and Trueb (1986) placed Sirenidae as the sister of all other salamanders, followed by Cryptobranchoidea as sister to remaining salamanders. A possible explanation for this discordance is that salamanders are notorious for the amount of homoplasy in pedomorphic features (Wake 1991). Of course, this alone does not explain incongruence in nuclear and mitochondrial rRNA data (mt-rRNA; see below).

Larson and Wilson (1989) and Larson (1991) presented a tree (fig. 25.4) based on nuclear-encoded rRNA, which differed dramatically in placing Plethodontidae and Amphiumidae at the base of the tree. Larson and Dimmick (1993) combined these molecular data with morphological data from Duellman and Trueb (1986). The resulting tree effectively rerooted the Larson (1991) tree to place Sirenidae and Cryptobranchoidea at its base. Analyses of 12S and 16S mitochondrial DNA (mtDNA; Hay et al. 1995, Hedges and Maxson 1993) also placed Sirenidae at the base (fig. 25.4), but with different relationships among other taxa.

Compared with caecilians and frogs, the placement of family-level groups of salamanders remains in an extreme state of flux, with very different topologies resulting from different data sets (sequences, morphology, and fossils) and combinations of those data sets. In contrast, there is almost no disagreement about the content of the Linnaean families. Ten families of living salamanders are generally recognized; all clearly are monophyletic. Four are species-rich and extensively sampled using molecular techniques. Substantial progress has been made in generating phylogenetic hypotheses at the species level, in contrast to frogs and caecilians. In several families nearly all species have been examined.

Sirenidae include two genera of non-metamorphosing, elongate neotenic forms that retain external gills as adults. In contrast to most elongate salamanders, the front limbs are present and robustly developed, whereas the hind limbs and pelvic girdle is absent. Cryptobranchoidea are a clade generally acknowledged to be among the most plesiomorphic of living salamanders. The included families are Cryptobranchidae and Hynobiidae. Cryptobranchidae include the largest salamanders; adult *Andrias* may reach 1.5 m in length. Recently described Jurassic cryptobranchid fossils (Gao and Shubin 2003) represent the oldest crown-group salamanders, i.e., members of Caudata. All Hynobiidae but 2 of the 42 species have been studied using mtDNA (A. Larson and R. Macey, unpubl. obs.).

The Dicamptodontidae and Rhyacotritonidae each include one living genus. *Dicamptodon* and *Rhyacotriton* have been considered closely related and were united in the Dicamptodontidae, but recent analyses (Good and Wake 1992, Larson and Dimmick 1993) place them as separate but adjacent lineages.

Amphiumidae include only *Amphiuma*. This elongate neotenic form lacks external gills and has limbs reduced to spindly projections with remnants of the digits. Proteidae include species both in North America and Europe. *Necturus*, the beloved mudpuppy of comparative anatomy labs, is a large pedomorphic salamander with large fluffy external gills. *Proteus*, a very elongate and aquatic cave-dweller in SE Europe, also retains external gills.

Ambystomatidae include 30 extant species of *Ambystoma*. Nearly all have using mtDNA sequences and allozymes (Shaffer 1984a, 1984b, Shaffer et al. 1991). Some species are facultatively neotenic and retain the ability to metamorphose; others are obligately trapped in the larval morphology, spending their entire lives in lakes. Most have a larval period that is always followed by metamorphosis to the adult condition.

Many species of Salamandridae are aposematic (having a bright warning coloration) and have highly effective cutaneous poison glands to deter predators. At least two species are viviparous, a rare occurrence among salamanders. Salamandridae are also diverse in morphology and life history, although not as speciose as Plethodontidae (see below). Relationships among salamandrids have been examined using morphological data (Özeti and Wake 1969, Wake and

Özeti 1969), although not with current phylogenetic algorithms. All 62 species of this widely distributed family have been studied using molecular markers (Titus and Larson 1995, D. Weisrock and A. Larson, unpubl. obs.).

Plethodontidae are by far the largest family taxon, with 27 genera and 348 species (from a total of 502 species of salamanders). Plethodontids are lungless and use primarily cutaneous respiration. The release of the hyoid musculoskeleton from the constraints of buccal force-pump breathing has apparently permitted the diversification of mechanisms prey capture by tongue protrusion. In addition to being the most diverse in morphology and life history—there are highly arboreal, aquatic, terrestrial, saxicolous, and fossorial forms—this is the only clade with a neotropical radiation. Four major groups of Plethodontidae are recognized: Desmognathinae, Plethodontini, Bolitoglossini, and Hemidactyliini; work has concentrated on relationships within each clade, and relationships among the four are not clear.

All species of Desmognathinae have been studied with mtDNA (Titus and Larson 1995, 1996). Detailed studies of many species of Plethodontini have been published by Mahoney (2001), and studies of all species are in progress (M. Mahoney, D. Weisrock, and D. Wake et al., unpubl. obs.). The Bolitoglossini have been sampled broadly. About 40% of all salamanders are in the mainly Middle American clade *Bolitoglossa*, and all genera and about 80% of its species have some sequence data (García-París et al. 2000a, 2000b, García-París and Wake 2000, Parra-Olea et al. 1999, 2001, Parra-Olea and Wake 2001). Data from three mtDNA genes have been collected for almost all tropical species in the lab of D. Wake (pers. comm.). Jackman et al. (1997) examined relationships of bolitoglossines based on a combination of morphological and molecular data sets. Work is also underway on the mostly aquatic plethodontids, the Hemidactyliini, using ribosomal mtDNA and recombination activating protein 1 (RAG-1) (P. Chippindale and J. Wiens, unpubl. obs.).

Frogs

Living frogs include about 4837 species arranged in 25–30 families (fig. 25.4, table 25.1). The earliest forms considered as proper frogs are *Notobatrachus* and *Vieraella* from the Middle Jurassic of Argentina. *Prosalirus vitis* from Lower Jurassic of Arizona (Jenkins and Shubin 1998, Shubin and Jenkins 1995) is fragmentary, but clearly a frog. All of these have skeletal features that indicate that the distinctive saltatory locomotion of frogs had evolved by this time.

The sister group of frogs proper is *Triadobatrachus massinoti*, known from a single fossil from the Lower Triassic of Madagascar. It has been called a proanuran and retains many plesiomorphic features, such as 14 presacral vertebrae (living frogs have nine or fewer) and lack of fusion of the radius and ulna and also of the tibia and fibula (living frogs have fused elements, the radioulna and tibiofibula; Rage and Rocek

1989, Rocek and Rage 2000). The clade containing *Triado-batrachus* and all frogs is named Salientia.

Frogs have a dazzling array of evolutionary novelties associated with reproduction. Their diverse vocal signals of the males are used for mate advertisement and territorial displays. Parental care is highly developed in many lineages, including brooding of developing larvae on a bare back, in pouches on the back of females, in the vocal sacs of males, and in the stomach of females. Some females in some unrelated lineages of Hylidae and Dendrobatidae raise their tadpoles in the watery confines of a bromeliad axil and supply their own unfertilized eggs as food. Whereas amniotes escaped from the watery environment once with their evolution of the amniote egg, frogs have done so many times; direct development, with terrestrial eggs in which the tadpole stage is bypassed in favor of development to a froglet, has evolved at least 20 times.

Although some frogs have escaped an aquatic existence, many have embraced it, taking the biphasic life to an extreme. In contrast to caecilian and salamander larvae, frog tadpoles are highly morphological specialized to exploit their transitory and often unpredictable larval niche. The tadpole is mostly a feeding apparatus in the head and locomotor mechanism in the tail. The feeding apparatus is a highly efficient pump that filters miniscule organic particles from the water. Tadpoles do not reproduce; there are no neotenic forms. They live their lives eating until it is time to make a quick and awkward metamorphosis to a froglet.

Anura and Salientia

The names of higher frog taxa are used here following Ford and Cannatella (1993). Their general rationale was (1) to recognize only monophyletic groups except when it was not feasible to reduce the nonmonophyletic group to smaller clades (as in the case of "Leptodactylidae" and "Ranidae"), (2) to identify nonmonophyletic groups as such, and (3) to avoid the use of family names that are redundant with the single included genus.

Ford and Cannatella (1993) defined Anura as the ancestor of living frogs and all its descendants. The use of living taxa as reference points or anchors for the definition follows the rationale of de Queiroz and Gauthier (1990, 1992), who convincingly argued that this stabilizes a definition. In contrast, the incompleteness of fossil taxa and the discovery of new fossils renders definitions based on extinct reference taxa less stable.

The taxonomy of frogs illustrates this issue of taxonomic practice. The Jurassic fossil *Notobatrachus* was considered by Estes and Reig (1973) to be closely related to, and in the same family as, the living taxa *Ascaphus* and *Leiopelma*. Thus, *Notobatrachus* would be included in Anura according to Ford and Cannatella's definition. In contrast, the analyses by Cannatella (1985) and Báez and Basso (1996) placed *Notobatrachus* as

the sister group to the clade containing *Ascaphus*, *Leiopelma*, and other living frogs. The latter placement means that *Notobatrachus* is not part of Anura, because Anura is defined as the last common ancestor of living frogs and all its descendants. Some herpetologists or paleontologists may be rankled by the proposition that the very froglike *Notobatrachus* is not part of Anura. But this concern is based on a typological notion that the definition of a taxon name is tied to a combination of characters, rather than to a branch of the Tree of Life.

We can ask, Are there characters that make a frog a frog? The *eidos* of a frog requires a big head, long legs, no tail, and a short vertebral column. But how short? Most living frogs have eight presacral vertebrae. *Notobatrachus* and two of the most "primitive" frogs, *Ascaphus* and *Leiopelma*, have nine. Another Jurassic fossil, *Vieraella herbsti*, has 10 vertebrae (Báez and Basso 1996). All of the aforementioned look like proper "frogs." The Triassic fossil *Triadobatrachus* has 14 vertebrae (Rage and Rocek 1989). It has several unambiguous synapomorphies that place it as the sister group of frogs. It is considered froglike, but not quite a frog. In summary, it seems the consensus of published work is that 10 or fewer presacral vertebrae make a frog a frog.

When fossil X with 11 presacral vertebrae is discovered, will the boundary of "frogness" move one node lower in the tree, so as to include fossil X? This question highlights the problem: when a taxon name is defined by a diagnostic character, each new fossil with an intermediate condition will stretch the definition of the name (Rowe and Gauthier 1992). However, it is less likely that the discovery of a new living frog species will stretch our concept of frogness. Therefore, attaching the taxon name Anura to a node circumscribed by living taxa will yield a more stable definition. Because Anura is defined as the ancestor of living frogs and all its descendants, the discovery of a new fossil just below this node, no matter how froglike, will not require a change in the meaning of Anura. And, we can still argue about which characters make a frog a frog.

"Salientia" is the stem-based name for the taxon including Anura and taxa (all fossils) more closely related to Anura than to other living amphibians. Salientia include *Triadobatrachus*, *Vieraella*, *Notobatrachus* (Báez and Basso 1996), *Czatkoba-trachus* (Evans and Borsuk-Bialynicka 1998), and *Prosalirus* (Shubin and Jenkins 1995). Because the name is tied to a stem, the discovery of new fossils on this stem will not destabilize the name. The use of Salientia for *Triadobatrachus* plus all other frogs is widespread and not controversial.

Our understanding of frog phylogeny rests primarily on morphological data (Griffiths 1963, Inger 1967, Kluge and Farris 1969, Lynch 1973, Noble 1922, Trueb 1973), summarized by Duellman and Trueb (1986) and Ford and Cannatella (1993; fig. 25.5). In general, morphological characters resolved the plesiomorphic basal branches known as archaeobatrachians (Cannatella 1985, Duellman and Trueb 1986, Haas 1997). The family-level relationships within Neo-

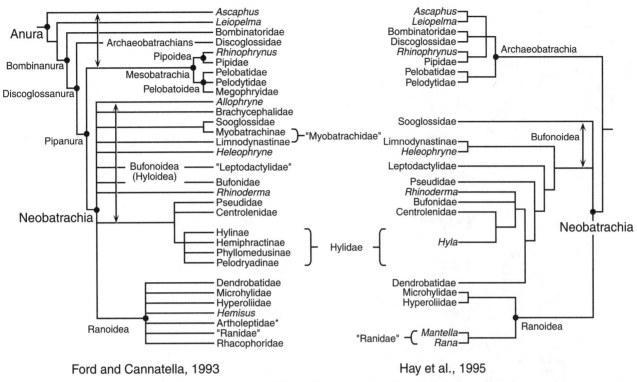

Figure 25.5. Alternative phylogenies of frogs. The tree on the left is labeled with taxon names (see text).

batrachia, a large clade with more than 95% of frog species, are mostly unresolved (Ford and Cannatella 1993) by morphological data, although Ranoidea is strongly supported. Most remaining neobatrachians are known as hyloids or bufonoids, but no morphological evidence for their monophyly has been proposed (with the possible exception of sperm morphology; Lee and Jamieson 1992). Ranoidea is primarily Old World; hyloids are mostly New World.

The distinctive and diverse morphology of tadpoles has been a source of characters to elucidate frog phylogeny. At one time it was thought that the larval morphology of the pipoid frogs argued for their position as the most primitive (early-branching in this context), but highly specialized, group (Starrett 1968, 1973). However, other interpretations (Cannatella 1999, Haas 1997, Sokol 1975, 1977) indicate that although pipoids are highly specialized, the discoglossoids are the earliest-branching frog lineages (see below). However, the most comprehensive analysis of larval morphology (Haas 2003) found *Ascaphus* to be the most basal frog and pipoids to be the next adjacent clade (fig. 25.6), rather than other discoglossoids. Maglia et al. (2001) reported Pipoidea to be the sister taxon of all other frogs, a hypothesis reminiscent of Starrett (1968, 1973).

The fossil record of frogs was thoroughly reviewed by Sanchiz (1998). Báez and Basso (1996) included Jurassic fossils in a phylogenetic analysis of early frogs. Gao and Wang (2001) analyzed data for a combined treatment of fossil and living archaeobatrachians and pre-archaeobatrachians, but

they reached very different conclusions than did Ford and Cannatella (1993); a full analysis of this is beyond the scope of this chapter.

A range of morphological phylogenetic studies treats relationships within particular family-level groups: Pelobatoidea (Maglia 1998); Hyperoliidae (Drewes 1984); Rhacophoridae and Hyperoliidae (Liem 1970); Myobatrachidae *sensu lato*, including Myobatrachinae and Limnodynastinae (Heyer and Liem 1976); Leptodactylidae (Heyer 1975); Hylinae (da Silva 1998); Microhylidae (Wu 1994); Hemiphractinae (Mendelson et al. 2000); and Pipidae (Cannatella and Trueb 1988a).

Sequences from both nuclear and mt-rRNA genes provided new data (Emerson et al. 2000, Graybeal 1997, Hay et al. 1995, Hedges and Maxson 1993, Hedges et al. 1990, Hillis et al. 1993, Ruvinsky and Maxson 1996, Vences et al. 2000). Several alternative hypotheses emerged from these works, including (1) monophyly of "Archaeobatrachia," (2) weak monophyly of the bufonoids (= Hyloidea), (3) dendrobatids excluded from Ranoidea, and (4) extensive paraphyly of the large families Hylidae and Leptodactylidae.

The "Basal" Frogs—Discoglossoids

A group of plesiomorphic lineages includes *Ascaphus*, *Leiopelma*, Bombinatoridae, and Discoglossidae (Ford and Cannatella 1993); this group has been called discoglossoids and is paraphyletic with respect to other frogs, the Pipanura.

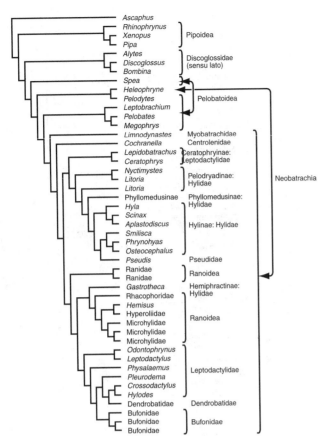

Figure 25.6. A phylogeny of frogs based mostly on larval morphology, simplified from Haas (2003: fig. 3).

Ascaphus and *Leiopelma* are plesiomorphic, now-narrowly distributed relicts of a once more widely distributed Mesozoic frog fauna (Green and Cannatella 1993). The family name Ascaphidae is redundant with *Ascaphus*. The family name Leiopelmatidae is redundant with the single genus *Leiopelma*. Formerly, The name Leiopelmatidae (*sensu lato*) has also been used to include *Ascaphus* and *Leiopelma*, a group that is probably paraphyletic.

"Bombinanura" is the node-based name for the last common ancestor of Bombinatoridae + Discoglossanura; Bombinatoridae is the node name for the ancestor of *Bombina* and *Barbourula* and all of its descendants (Ford and Cannatella 1993); this node is well supported (Cannatella 1985; but see Haas 2003).

The names Discoglossoidei (Sokol 1977) and Discoglossoidea (e.g., Duellman 1975, Lynch 1971) were used for the group containing *Ascaphus*, *Leiopelma*, *Bombina*, *Barbourula*, *Alytes*, and *Discoglossus*. The Discoglossoidei of Sokol (1977) and Duellman and Trueb (1986) were a clade; however, other morphological analyses strongly reject this conclusion. As an informal term, the name discoglossoids is a useful catchall for plesiomorphic anurans that are not part of Pipanura. One general primitive feature of this group is the rather rounded,

disklike tongue; hence the name. *Alytes* and *Discoglossus* are included in the Discoglossidae, although the two are fairly divergent and evidence of monophyly is not overwhelming. Some evidence indicates that Discoglossidae are more closely related to other frogs than to Bombinatoridae, *Ascaphus*, or *Leiopelma* (Ford and Cannatella 1993).

Pipanura

Pipanura consists of Pipoidea, Pelobatoidea, and Neobatrachia, that is, living frogs minus discoglossoids. Specifically, it is the node name for the last ancestor of Mesobatrachia + Neobatrachia, and all of its descendants (Ford and Cannatella 1993). Pipoidea and Pelobatoidea are regarded as intermediate lineages between discoglossoids and Neobatrachia. Mesobatrachia is the node name applied to the last ancestor of Pelobatoidea + Pipoidea. Support for this clade is not strong (Cannatella 1985). Pelobatoids and pipoids are represented by a large number of Cretaceous and Tertiary fossils (Rocek 2000, Sanchiz 1998).

The node name Pelobatoidea was defined by Ford and Cannatella (1993) as the (last) common ancestor of living Megophryidae, Pelobatidae, and *Pelodytes*, and all its descendants. The content of Pelobatoidea is not controversial. Historically, Pelobatidae has included Megophryidae as a subfamily (e.g., Duellman and Trueb 1986), although recent summaries recognize Megophryidae (e.g., Zug et al. 2001. This follows Ford and Cannatella (1993), who defined Pelobatidae as the node name for the last common ancestor of *Pelobates*, *Scaphiopus*, and *Spea*, and all its descendants. This definition was based on a sister-group relation between the European (*Pelobates*) and American spadefoots (*Scaphiopus* + *Spea*), which were united by synapomorphies related to their habitus as fossorial species (Cannatella 1985, Maglia 1998).

In contrast, García-París et al. (2003) reexamined relationships among all pelobatoids using mtDNA and found *Scaphiopus* + *Spea* to be the sister group of other pelobatoids (*Pelobates*, Pelodytidae, and Megophryidae). Because *Scaphiopus* + *Spea*, which they termed Scaphiopodidae, were no longer related to *Pelobates*, they inferred the fossorial habitus of the two groups to be convergent. The taxonomic implication of this finding is that Pelobatidae as defined by Ford and Cannatella (1993) applies to the same node as Pelobatoidea. One solution would be to redefine Pelobatidae as a stem name so as to include the fossil taxa that are thought to be closely related, such as *Macropelobates*. But the issue remains unresolved.

The node name Megophryidae was used by Ford and Cannatella (1993) for the group of taxa referred to as megophryines, previously been considered to be a subfamily (Megophryinae) of Pelobatidae. Although preliminary work exists (Lathrop 1997), relationships among the Megophryidae have not been assessed in detail; however, the content is uncontroversial. In contrast to most of the family-level names, Pelodytidae was defined as a stem name by Ford and Cannatella (1993) because its use as a node name for the clade of living taxa would make it redundant with *Pelodytes*. Also, use of a stem name retains the several taxa of fossil pelodytids within Pelodytidae, a placement that is well supported (Henrici 1994).

Pipoidea was implicitly defined as the node name for the most recent common ancestor of Pipidae and Rhinophrynidae, and all its descendants. By this definition, the fossil family Palaeobatrachidae are included within Pipoidea, as has generally been the case (but see Spinar 1972). Relationships among pipoids have been examined by Cannatella and Trueb (Báez 1981, Báez and Trueb 1997, Cannatella and de Sá 1993, Cannatella and Trueb 1988a, 1988b, de Sá and Hillis 1990)

As pointed out by Ford and Cannatella (1993), the phylogenetic definition of the name Pipidae excluded several fossils previously and currently included in Pipidae (Báez 1996). The stem name Pipimorpha was proposed to accommodate these. Because the name applies to those taxa that are more closely related to (living) Pipidae than to Rhinophrynidae, it is a useful descriptor for the increasingly specialized taxa on the stem leading to the Pipidae. Báez and Trueb (1997) defined Pipidae slightly differently; their tree is unresolved at the crucial point. The single species of highly fossorial frog *Rhinophrynus dorsalis* is regarded to be the sister group of Pipidae, among living forms. Like Pelodytidae, the name Rhinophrynidae was defined as a stem name by Ford and Cannatella (1993).

Neobatrachia

Neobatrachia consist of the "advanced" frogs and includes 95% of living species. Except for the Late Tertiary, they are not well represented in the fossil record. Neobatrachia is well supported by both morphological and molecular data (Ford and Cannatella 1993, Ruvinsky and Maxson 1996, but see Haas 2003). Two groups of Neobatrachia have been generally recognized: Bufonoidea (Hyloidea has priority; see below) for arciferal neobatrachians, and Ranoidea for the firmisternal neobatrachians. These correspond roughly to the classic Procoela and Diplasiocoela of Nicholls (1916) and Noble (1922), respectively. Hyloidea are primarily a New World clade, and Ranoidea an Old World group, although the hyloids have significant radiations in the Australopapuan region as do Ranidae and Microhylidae in the New World.

Hyloidea (formerly Bufonoidea) include Bufonidae, Hylidae, "Leptodactylidae," Centrolenidae, Pseudidae, Brachycephalidae, *Rhinoderma*, and *Allophryne*. Ford and Cannatella (1993) noted that Hyloidea and Bufonoidea apply to a nonmonophyletic group, that is, neobatrachians that were not ranoids. Ranoidea (see below) consist of ranids (including arthroleptids and mantellines), hyperoliids, rhacophorids, *Hemisus*, and microhylids. Some authors have placed

microhylids in the superfamily Microhyloidea to reflect the distinctiveness of the microhylid larva (e.g., Starrett 1973). But agreement is universal that microhylids are more closely related to ranoids than to hyloids.

Lynch (1973) considered Pelobatoidea an explicitly paraphyletic group transitional between "archaic frogs" and the "advanced frogs." He included here Pelobatidae, Pelodytidae, Heleophrynidae, the myobatrachids, and Sooglossidae. His dendrogram (Lynch 1973: fig. 3-6) showed Bufonoidea and Ranoidea as independently derived from the paraphyletic Pelobatoidea. Duellman (1975) used Reig's (1958) Neobatrachia to include Lynch's Bufonoidea and Ranoidea. Subsequent morphological and molecular analyses have supported monophyly of Neobatrachia (Cannatella 1985, Hay et al. 1995, Ruvinsky and Maxson 1996). However, supposed basal neobatrachians such as myobatrachids, sooglossids, and *Heleophryne* are of uncertain position.

Until recently, Limnodynastinae and Myobatrachinae were included as subfamilies of "Myobatrachidae" (e.g., Heyer and Liem 1976). Ford and Cannatella (1993) could find no synapomorphies for "Myobatrachidae." However, Lee and Jamieson (1992) provided some characters from spermatozoan ultrastructure that support myobatrachid monophyly. Some textbooks (Zug et al. 2001) have recognized each group as a distinct family [which was not Ford and Cannatella's (1993) intention]. Ruvinsky and Maxson (1996) placed Myobatrachinae, Limnodynastinae, and *Heleophryne* (Heleophrynidae) in a clade of at the base of Hyloidea. Some recent phylogenies have placed Sooglossidae as the sister group of all other Hyloidea (Ruvinsky and Maxson 1996), sister group to Ranoidea (Emerson et al. 2000), basal to both (Hay et al. 1995), or as the sister of Myobatrachidae (Duellman and Trueb 1986) or Myobatrachinae (Ford and Cannatella 1993).

Hyloidea

Hyoidea has been used to refer to neobatrachians with an arciferal pectoral girdle, in contrast to those with a firmisternal girdle, the ranoids. The name has Linnaean priority over Bufonoidea (Dubois 1986), although it has not been used often. Ford and Cannatella found no published data to support its monophyly. Hay et al. (1995) were the first to use character data to support the monophyly of Hyloidea (as Bufonoidea). This lineage included Myobatrachidae, Heleophrynidae, and Dendrobatidae, Centrolenidae, Hylidae, Bufonidae, Rhinodermatidae, Pseudidae, and Leptodactylidae. They also identified the Sooglossidae as a "distinct major lineage" of Neobatrachia apart from Hyloidea and Ranoidea. Ruvinsky and Maxson (1996), using mostly the same data as Hay et al. (1995), concluded that Sooglossidae was included within Hyloidea.

Darst and Cannatella (in press) identified a well-supported clade (fig. 25.7) for which they defined the name Hyloidea in a phylogenetic context. They excluded from the definition taxa such as Dendrobatidae whose phylogenetic position might make the content of this taxon unstable. Also, they excluded certain neobatrachian groups whose placement is more relatively basal and also less well resolved, such as Myobatrachinae, Limnodynastinae, and Sooglossidae.

"Leptodactylidae" are a hodgepodge of hyloids that lack distinctive apomorphies. Historically, the derived features of the other hyloid families separated them from Leptodactylidae, suggesting it was paraphyletic. Hylidae have cartilaginous intercalary elements between the ultimate and penultimate phalanges of the hands and feet; Centrolenidae has the two elongate ankle bones (tibiale and fibulare) fused into a single element; Pseudidae have bony intercalary elements, in contrast to the generally cartilaginous ones found in hylids; has a Bidder's organ present in males; this is a portion of embryonic gonad that retains an ovarian character. Rhinodermatidae have rearing of larvae in the vocal sac of the male; Brachycephalidae lack a well-developed sternum.

Phylogenetic relationships among the genera of "Leptodactylidae" were analyzed using morphology by Heyer (1975). Basso and Cannatella (2001) analyzed relationships among leptodactyloid frogs from 12S and 16S mtDNA and found "Leptodactylidae" to be polyphyletic. Darst and Cannatella (2003) also found the same, based on a smaller sample of leptodactylid taxa (fig. 25.7).

Pseudidae, Centrolenidae, Brachycephalidae, and Dendrobatidae are node names whose content is not controversial. Recent work has clarified the relationships of some of these groups. Darst and Cannatella (in press) found Dendrobatidae to be nested clearly within Hyloidea and were able to reject the alternate hypothesis that dendrobatids are within Ranoidea (Ford 1989, Ford and Cannatella 1993). Duellman (2001) reduced Pseudidae to a subfamily. However, this action stopped short of what would be demanded by the Linnaean system. If Pseudidae is not acceptable as a family within Hylidae, then Pseudinae cannot be accepted as a subfamily within the subfamily Hylinae. Darst and Cannatella (in press) also found Pseudidae to be nested within hylines, specifically the sister group to *Scarthyla ostinodactyla*. Assuming an adherence to Linnaean taxonomy coupled with a desire to recognize only monophyletic groups, then there is no basis for recognition of the group at a subfamily or even tribe level.

Darst and Cannatella also found Brachycephalidae to be within eleutherodactylines ("Leptodactylidae"); the taxonomic changes necessitated by these new findings are in progress. *Allophryne ruthveni* is an enigmatic hyloid (Fabrezi and Langone 2000) that has been placed in a monotypic (and redundant) family Allophrynidae; it is probably the sister group of Centrolenidae (Austin et al. 2002). The two species of *Rhinoderma* have been placed in Rhinodermatidae. Were it not for the apomorphic life history of the two species, in which the males brood the developing larvae in their vocal sacs, *Rhinoderma* would be included in the "Leptodactylidae." Ford and Cannatella (1993) provided phylogenetic names for these taxa.

Hylidae is the node name for the most recent common ancestor of Hemiphractine, Phyllomedusinae, Pelodryadinae, and Hylinae, and all of its descendants. These latter four names have not been formally defined in a phylogenetic manner, but the composition of each is well established. Some workers elevated Pelodryadinae to family level (Dubois 1984, Savage 1973). Morphology-based phylogenies of Hylinae and Hemiphractinae exist (da Silva 1998, Mendelson et al. 2000). According to Darst and Cannatella (in press), Hylidae is polyphyletic; however, their sample of hemiphractines, which are the troublesome species, was small.

Bufonidae is also a node name. Recent work (Gluesenkamp 2001, Graybeal 1997, Graybeal and Cannatella 1995) found no basis for the subfamilies or tribes recognized by Dubois (1984). Relationships among the higher groups of Bufonidae are unresolved.

Ranoidea

Ford and Cannatella (1993) defined Ranoidea as the node-based name for the clade anchored by the last common ancestor of hyperoliids, rhacophorids, ranids, dendrobatids, *Hemisus*, arthroleptids, and microhylids. With the possible exception of the controversial dendrobatids, the content of this group includes the classic "firmisternal" frogs, Firmisternia. Wu (1994) treated the Ranoidea and Microhyloidea as the two components of Firmisternia. The resurrection of this arrangement has merit in recognizing the two major clades of firmnisternal frogs, as in the past where the groups were Microhyloidea and Ranoidea. Duellman (1975), for example, recognized distinct superfamilies Microhyloidea and Ranoidea.

Growing evidence suggests that Microhylidae (or at least a large clade of those) is the sister group to Hyperoliidae or Hyperoliidae + arthroleptines within the Ranoidea (Darst and Cannatella in press, Emerson et al. 2000, Hay et al. 1995) rather than the sister group of all other ranoids. Thus, inclusion of Microhylidae within Ranoidea is appropriate in one sense. However, one could argue equally that Microhyloidea could include Microhylidae (minimally the type-genus) and whatever else is more closely related to these than to Ranidae. Microhyloidea and Ranoidea would be sister taxa in Firmisternia. For example, Darst and Cannatella (in press) and Emerson et al. (2000) each recovered two major clades of ranoids, one including hyperoliids, arthroleptids, microhylids (including brevicipitines), and *Hemisus*, and the other containing rhacophorids, mantellines, and the remaining "ranids." However, Blommers-Schlösser (1993) recognized Microhyloidea as consisting of Microhylidae, Sooglossidae, Dendrobatidae, and Hemisotidae. We have not followed this unusual rearrangement pending a broader synthesis of morphological and molecular data of ranoids. For the moment, we continue the use of Ranoidea for all these firmisternal frogs because of its recent common use.

Perhaps the most controversial group within Neobatrachia has been Dendrobatidae. Hay et al. (1995) and Ruvinsky

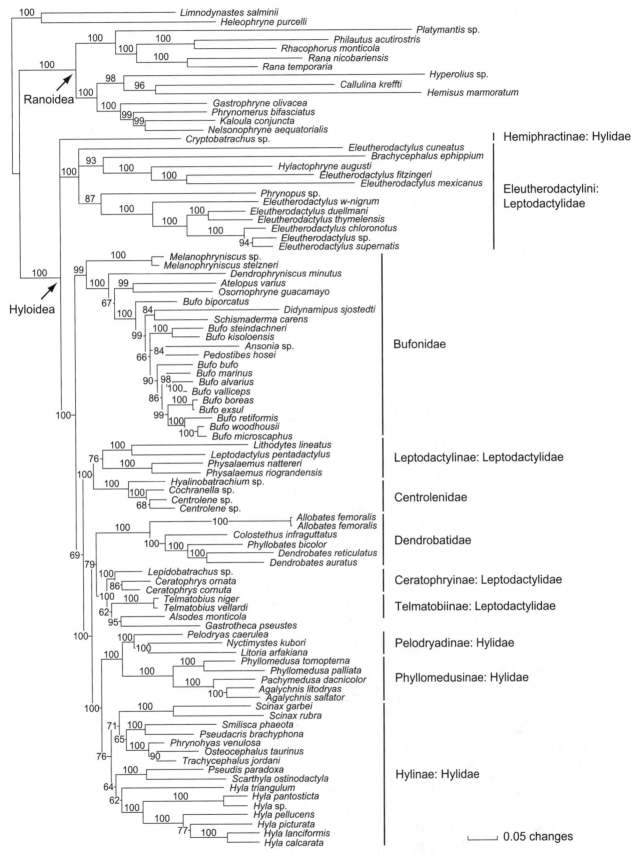

Figure 25.7. Phylogeny of Hyloidea based on a Bayesian analysis, after Darst and Cannatella (in press). The numbers on the branches are posterior probabilities.

and Maxson (1996) found dendrobatids to be nested within hyloids (bufonoids), and this was corroborated with a broader taxon sample by Darst and Cannatella (in press) and with mostly larval data by Haas (2003). In retrospect, Ford and Cannatella's (1993) inclusion of a potentially unstable taxon (Dendrobatidae) as a specifier taxon for Ranoidea was not wise. Accepting the new evidence for the position of Dendrobatidae, Ranoidea as they defined it has now the same content as Hyloidea + Ranoidea; this is a drastic departure from its usual content. Rather than redefine Ranoidea here (because of work in progress), for the moment we consider statements about Ranoidea to exclude Dendrobatidae.

Microhylidae were found to be nested within Ranoidea by Ford and Cannatella (1993), Hay et al. (1995), Ruvinsky and Maxson (1996), and (at least close to some) Haas (2003). Ford and Cannatella (1993) used larval evidence from Wassersug (1984, 1989) to formally recognize Scoptanura as a large clade within Microhylidae. This was corroborated by Haas (2003).

Wu (1994) produced the most comprehensive survey of microhylid osteology, examining 188 characters in 105 species in 56 of the 64 named genera. He adopted a rankless taxonomy, placing Microhyloidea and Ranoidea as sister groups in the Firmisternia. His Ranoidea included Hyperoliidae, Mantellidae, Ranidae, and Rhacophoridae, and his Microhyloidea consisted of two families, Brevicipitidae and Microhylidae. Wu's Brevicipitidae was unusual in that it included a clade Hemisotinae composed of *Hemisus* and *Rhinophrynus*. The latter has never been placed within Neobatrachia and shares many molecular and morphological synapomorphies with Pipidae (Cannatella 1985, Hay et al. 1995).

Relationships of ranoid frogs (microhylids aside) are in a kinetic state, and the taxonomy we follow is certainly arbitrary. For years an accepted arrangement was Ranidae, Hyperoliidae, and Rhacophoridae, the latter two families being treefrog morphs independently derived from within Ranidae. It was generally appreciated that the mantelline ranids (Mantellinae or Matellidae) shared some derived features with Rhacophoridae (e.g., Duellman and Trueb 1986). Ford and Cannatella (1993) embellished "Ranidae" with quotes to indicate its status as a nonmonophyletic group.

Most recent attempts to establish a classification of Ranidae have been based on a hypothesis of phylogeny (but see Dubois 1992, Inger 1996). Phylogenetic analyses of both sequence data and morphological characters exist for Hyperoliidae and Rhacophoridae (Channing 1989, Drewes 1984, Liem 1970, Richards and Moore 1996, 1998, J. Wilkinson et al. 2002). Although Rhacophoridae have generally been thought to be monophyletic, accumulating evidence suggests that the Malagasy rhacophorids are not the closest relatives of the Asian rhacophorids (J. Wilkinson et al. 2002) and may be more closely related to other Malagasy lineages, such as mantellines.

The most comprehensive analysis of ranoids (Emerson et al. 2000), which used mostly published molecular data and

10 morphological characters, found familiar results: the close relationship of Microhylidae and Hyperoliidae (Hay et al. 1995); the placement of Sooglossidae outside of Ranoidea (Hay et al. 1995); and mantelline ranids most closely related to, or nested within, rhacophorids (Channing 1989, Ford 1989). Relationships among a small sample of Indian ranoids were examined by Bossuyt and Milinkovitch (2001).

As had historically happened with hyloids, the recent taxonomic tendency for ranoids has been to elevate some loosely defined subfamily groups to family status; for example, the recognition of Arthroleptidae by Dubois (1984). These have usually been considered to be a subfamily of Ranidae, and its elevation to family level was more because of taxonomic tinkering than any new knowledge of relationships. Ford and Cannatella (1993) considered it a metataxon.

Blommers-Schlösser (1993) recognized a clade Ranoidea comprising Arthroleptidae, Hyperoliidae, and Ranidae, the last including Mantellinae and Rhacophorinae. Emerson et al. (2000: table 1) listed subfamilies of Ranidae as Raninae, Mantellinae, and Rhacophorinae, reportedly from Blommers-Schlösser (1993). Actually, Blommers-Schlösser (1993) included these three, plus Cacosterninae, Nyctibatrachinae, Petropedetinae, and Indiraninae, for a total of seven subfamilies of Ranidae. Of these, the petropedetines have been arbitrarily elevated to familial rank by some. *Hemisus*, one of the few frogs known to burrow headfirst, has usually been placed in the redundant family Hemisotidae. It was considered to be derived from some group of African ranids, but recent molecular analysis suggests closer relationships to brevicipitine microhylids (Darst and Cannatella in press; fig. 25.7), as did a morphological analysis (Blommers-Schlösser 1993).

Prospects for the Future

Rather than address the future of the systematics of Amphibia, we offer some general comments are possibly applicable to all groups. Information age technology has changed the nature of systematics. The flood of data from molecular systematics continue to rise as new technologies facilitate its collection. The program solicitation for the National Science Foundation's Assembling the Tree of Life competition (National Science Foundation 2003) indicated the need for "scaling up" the level of activity of data collection. But scaling up in nature is rarely isometric; a change in size demands a change in shape. Put another away, we will not reach the goal of the Tree of Life (or the Tree of Amphibia) without doing systematics differently. We suggest that some of the core practices of systematics pose a severe impediment to completing the Tree of Life. Methods and theory of tree construction have "gone to warp speed" relative to the practices of taxonomy, nomenclature, and biodiversity studies.

Our facility at reconstructing phylogeny now exceeds our ability to describe new species in a reasonable amount of time.

Classically trained systematists, even those with active programs in molecular systematics, must still linger over species descriptions. Descriptions of new species are not much different than those published more than a century ago. Some systematists have bemoaned the dearth of jobs for classically trained taxonomists. But even if positions were available, would there be systematists interested in filling them? Proposals for automation of species descriptions have not received rave reviews. Is the practice of taxonomy really a different enterprise than phylogenetic analysis (Donoghue 2001)? Perhaps it is time to redefine the mode and meaning of "describing a new species."

We are not advocating a reductionist, barcode approach (Blaxter 2003) in which a sequence of one gene is sole diagnosis of a species. However, DNA sequences are a powerful source of data for species discovery and description, and we welcome a fusion between traditional activities of species description and the opportunities offered by information technology. The nature of this compromise is not clear, but it is evident that our mandate will not succeed without consideration of this issue.

Related to the description of new species is nomenclature, the rules for bestowing and keeping track of names. Although the term "Phyloinformatics" has entered the language of systematics, it lacks a meaningful definition. We do not attempt one here, but certainly any concept of phyloinformatics must include storage and retrieval systems for taxonomy and nomenclature. Like others, we suggest that the Linnaean system needs informatics-based reengineering; it is a square peg in the world of information technology.

Last, the increasing difficulty of on-site biodiversity studies must be addressed. Legitimate concerns over the loss of natural resources and opportunities through bioprospecting and biopiracy have grown in the same regions that harbor the greatest proportion of biodiversity. If natural history collections and related information are as precious as we claim, then we must invest in the countries of origin to enable the development of those resources on-site. The alternative, the removal of collections to another country largely for reasons of convenience, meets with increasing and justifiable resistance. This investment must be genuine and durable, so that local researchers are enabled to do long-term research. Only this type of investment will ensure the survival of the biodiversity that we all value.

Acknowledgments

We thank Joel Cracraft and Michael Donoghue for the opportunity to participate in the symposium. David Wake and Marvalee Wake were coauthors on the symposium presentation and offered much useful criticism on this manuscript; however, the opinions expressed herein are our own.

Literature Cited

Alberch, P., S. J. Gould, Oster, G., and D. B. Wake. 1979. Size and shape in ontogeny and phylogeny. Paleobiology 5:296–317.

Anderson, J. S. 2001. The phylogenetic trunk: maximum inclusion of taxa with missing data in an analysis of the Lepospondyli (Vertebrata, Tetrapoda). Syst. Biol. 50:170–193.

Austin, J. D., S. C. Lougheed, K. Tanner, A. A. Chek, J. P. Bogart, and P. T. Boag. 2002. A molecular perspective on the evolutionary affinities of an enigmatic neotropical frog, *Allophryne ruthveni*. Zool. J. Linn. Soc. 134:335–346.

Báez, A. M. 1981. Redescription and relationships of *Saltenia ibanezi*, a Late Cretaceous pipid frog from northwestern Argentina. Ameghiniana 18:127–154.

Báez, A. M. 1996. The fossil record of the Pipidae. Pp. 329–347 in The biology of *Xenopus* (R. C. Tinsley and H. R. Kobel, eds.). Clarendon Press, Oxford.

Báez, A. M., and N. G. Basso. 1996. The earliest known frogs of the Jurassic of South America: review and cladistic appraisal of their relationships. Münch. Geowiss. Abh. 30A:131–158.

Báez, A. M., and L. Trueb. 1997. Redescription of the Paleogene *Shelania pascuali* from Patagonia and its bearing on the relationships of fossil and recent pipoid frogs. Sci. Pap. Mus. Nat. Hist. Univ. Kans. 4:1–41.

Basso, N., and D. C. Cannatella. 2001. The phylogeny of leptodactylid frogs based on 12S and 16S mtDNA. *In* Abstracts of the Soc. Study Amphibians and Reptiles-Herpetologists League (SSAR-HL) meetings, Indianapolis, Indiana, July 2001.

Blaxter, M. 2003. Counting angels with DNA. Nature 421:122–124.

Blommers-Schlösser, R. M. A. 1993. Systematic relationships of the Mantellinae Laurent 1946 (Anura Ranoidea). Ethol. Ecol. Evol. 5:199–218.

Bolt, J. R. 1969. Lissamphibian origins: possible protolissamphibian from the lower Permian of Oklahoma. Science 166:888–891.

Bolt, J. R. 1977. Dissorophoid relationships and ontogeny, and the origin of the Lissamphibia. J. Paleontol. 51:235–249.

Bolt, J. R. 1991. Lissamphibian origins. Pp. 194–222 in Origins of the higher groups of tetrapods (H.-P. Schultze and L. Trueb, eds.). Cornell University Press, Ithaca, NY.

Bossuyt, F., and M. C. Milinkovitch. 2001. Amphibians as indicators of Early Tertiary "Out of India" dispersal of vertebrates. Science 292:93–95.

Brainerd, E. L., J. S. Ditelberg, and D. M. Bramble. 1993. Lung ventilation in salamanders and the evolution of vertebrate air-breathing mechanisms. Biol. J. Linn. Soc. 49:163–183.

Cannatella, D. C. 1985. A phylogeny of primitive frogs (archaeobatrachians). Ph.D. thesis, University of Kansas, Lawrence.

Cannatella, D. C. 1999. Architecture: Cranial and axial musculoskeleton. Pp. 52–91 in Tadpoles. The biology of anuran larvae (R. W. McDiarmid and R. Altig, eds.). University of Chicago Press.

Cannatella, D. C., and de R. O. Sá. 1993. *Xenopus laevis* as a model organism. Syst. Biol. 42:476–507.

Cannatella, D. C., and D. M. Hillis. 1993. Amphibian phylogeny: phylogenetic analysis of morphology and molecules. Herpetol. Monogr. 7:1–7.

Cannatella, D. C., and L. Trueb. 1988a. Evolution of pipoid frogs: intergeneric relationships of the aquatic frog family Pipidae (Anura). Zool. J. Linn. Soc. 94:1–38.

Cannatella, D. C., and L. Trueb. 1988b. Evolution of pipoid frogs: morphology and phylogenetic relationships of *Pseudhymenochirus*. J. Herpetol. 22:439–456.

Carroll, R. L. 2000a. *Eocaecilia* and the origin of caecilians. Pp. 1402–1411 *in* Amphibian biology (H. Heatwole and R. L. Carroll, eds.), vol. 4. Surrey Beatty and Sons, Chipping Norton, New South Wales, Australia.

Carroll, R. L. 2000b. The lissamphibian enigma. Pp. 1270–1273 *in* Amphibian biology (H. Heatwole and R. L. Carroll, eds.), vol. 4. Surrey Beatty and Sons, Chipping Norton, New South Wales, Australia.

Carroll, R. L., and P. J. Currie. 1975. Microsaurs as possible apodan ancestors. Zool. J. Linn. Soc. 57:229–247.

Carroll, R. L., A. Kuntz, and K. Albright. 1999. Vertebral development and amphibian evolution. Evol. Dev. 1:36–48.

Channing, A. 1989. A re-evaluation of the phylogeny of Old World treefrogs. S. Afr. J. Sci. 24:116–131.

Coates, M. I., M. Ruta, and A. R. Milner. 2000. Early tetrapod evolution. TREE 15:327–328.

Darst, C. R., and D. C. Cannatella. In press. Novel relationships among hyloid frogs inferred from 12S and 16S mitochondrial DNA sequences. Mol. Phylogenet. Evol.

da Silva, H. R. 1998. Phylogenetic relationships of the family Hylidae with emphasis on the relationships within the subfamily Hylinae (Amphibia: Anura). Ph.D. thesis, University of Kansas, Lawrence.

de Queiroz, K., and J. Gauthier. 1990. Phylogeny as a central principle in taxonomy: phylogenetic definitions of taxon names. Syst. Zool. 39:307–322.

de Queiroz, K., and J. Gauthier. 1992. Phylogenetic taxonomy. Annu. Rev. Ecol. Syst. 23:449–480.

de Sá, R. O., and D. M. Hillis. 1990. Phylogenetic relationships of the pipid frogs *Xenopus* and *Silurana*: an integration of ribosomal DNA and morphology. Mol. Biol. Evol. 7:365–376.

Donoghue, M. J. 2001. A wish list for systematic biology. Syst. Biol. 50:755–757.

Drewes, R. C. 1984. A phylogenetic analysis of the Hyperoliidae (Anura): treefrogs of Africa, Madagascar, and the Seychelles Islands. Occas. Pap. Calif. Acad. Sci. 139:1–70.

Dubois, A. 1984. La nomenclature supragénérique des amphibiens anoures. Mem. Mus. Natl. Hist. Nat. Ser. A Z 131:1–64.

Dubois, A. 1986. Miscellanea taxinomica batrachologica (I). Alytes 5:7–95.

Dubois, A. 1992. Notes sur la classification des Ranidae (Amphibiens Anoures). Bull. Men. Soc. Linn. Lyon 61:305–352.

Duellman, W. E. 1975. On the classification of frogs. Occas. Pap. Mus. Nat. Hist. Univ. Kans. 42:1–14.

Duellman, W. E. 1993. Amphibian species of the world: additions and corrections. Univ. Kans. Mus. Nat. Hist. Spec. Publ. 21:1–372.

Duellman, W. E. 1999. Patterns of distribution of amphibians. A global perspective. Johns Hopkins University Press, Baltimore.

Duellman, W. E. 2001. Hylid frogs of Middle America, vol. 2. Soc. Study Amphib. Reptiles, New York.

Duellman, W. E., and L. Trueb. 1986. Biology of amphibians. McGraw-Hill, New York.

Emerson, S. B., C. Richards, R. C. Drewes, and K. M. Kjer. 2000. On the relationships among ranoid frogs: a review of the evidence. Herpetologica 56:209–230.

Estes, R., and O. A. Reig. 1973. The early fossil record of frogs: a review of the evidence. Pp. 11–63 *in* Evolutionary biology of the anurans (J. L. Vial, ed.). University of Missouri Press, Columbia.

Evans, S., and M. Borsuk-Bialynicka. 1998. A stem-group frog from the Early Triassic of Poland. Acta Palaeontol. Pol. 43:573–580.

Fabrezi, M., and J. A. Langone. 2000. Los caracteres morfológicos del controvertido Neobatrachia arboricola *Allophryne ruthveni* Gaige, 1926. Cuad. Herpetol. 14:47–59.

Feller, A. E., and S. B. Hedges. 1998. Molecular evidence for the early history of living amphibians. Mol. Phylogenet. Evol. 9:509–516.

Ford, L. S. 1989. The phylogenetic position of poison-dart frogs (Dendrobatidae): reassessment of the neobatrachian phylogeny with commentary on complex character systems. Ph.D. thesis, University of Kansas, Lawrence.

Ford, L. S., and D. C. Cannatella. 1993. The major clades of frogs. Herpetol. Monogr. 7:94–117.

Frost, D. 2002. Amphibian species of the world: an online reference, vers. 2.21. American Museum of Natural History, New York, NY. Available: http://research.amnh.org/herpetology/amphibia/index.html. Last accessed July 15, 2002.

Frost, D. R., ed. 1985. Amphibian species of the world: a taxonomic and geographic reference. Allen Press and the Association of Systematics Collections, Lawrence, KS.

Gadow, H. 1901. Amphibia and reptiles. Macmillan, London.

Gans, C., H. J. Dejongh, and J. Farber. 1969. Bullfrog (*Rana catesbeiana*) ventilation: how does the frog breathe? Science 163:1223–1225.

Gao, K.-Q., and N. H. Shubin. 2001. Late Jurassic salamanders from northern China. Nature 410:574–577.

Gao, K.-Q., and N. H. Shubin. 2003. Earliest known crown-group salamanders. Nature 422:424–428.

Gao, K.-Q., and Y. Wang. 2001. Mesozoic anurans from Liaoning province, China, and phylogenetic relationships of archaeobatrachian anuran clades. J. Vert. Paleontol. 21:460–476.

García-París, M., D. R. Buchholz, and G. Parra-Olea. 2003. Phylogenetic relationships of Pelobatoidea re-examined using mtDNA. Mol. Phylogenet. Evol. 28:12–23.

García-París, M., D. A. Good, G. Parra-Olea, and D. B. Wake. 2000a. Biodiversity of Costa Rican salamanders: implications of high levels of genetic differentiation and phylogeographic structure for species formation. Proc. Natl. Acad. Sci. USA 97:1640–1647.

García-París, M., G. Parra-Olea, and D. B. Wake. 2000b. Phylogenetic relationships within the lowland tropical salamanders of the *Bolitoglossa mexicana* complex (Amphibia: Plethodontidae). Pp. 199–214 *in* The biology of plethodontid salamanders (R. C. Bruce, R. G. Jaeger, and L. D. Houck, eds.). Kluwer Academic/Plenum Publishers, New York.

García-París, M., and D. B. Wake. 2000. Molecular phylogenetic analysis of relationships of the tropical salamander genera *Oedipina* and *Nototriton*, with descriptions of a new genus and three new species. Copeia 2000:42–70.

Gardner, J. D. 2001. Monophyly and affinities of albanerpetontid amphibians (Temnospondyli; Lissamphibia). Zool. J. Linn. Soc. 131:309–352.

Gauthier, J., D. Cannatella, K. de Queiroz, A. G. Kluge, and T. Rowe. 1989. Tetrapod phylogeny. Pp. 337–353 *in* The hierarchy of life (K. Bremer, H. Jörnvall, and B. Fernholm, eds.). Elsevier, Amsterdam.

Glaw, F., and J. Köhler. 1998. Amphibian species diversity exceeds that of mammals. Herpetol. Rev. 29:11–12.

Gluesenkamp, A. G. 2001. Developmental mode and adult morphology in bufonid frogs: a comparative analysis of correlated traits. Ph.D. thesis, University of Texas, Austin.

Good, D. A., and D. Wake. 1992. Geographic variation and speciation in the torrent salamanders of the genus *Rhyacotriton* (Caudata: Rhyacotritonidae). Publ. Zool. Univ. Calif. 126:1–91.

Graybeal, A. 1997. Phylogenetic relationships of bufonid frogs and tests of alternate macroevolutionary hypotheses characterizing their radiation. Zool. J. Linn. Soc. 119:297–338.

Graybeal, A., and D. C. Cannatella. 1995. A new taxon of Bufonidae from Peru, with descriptions of two new species and a review of the phylogenetic status of supraspecific bufonid taxa. Herpetologica 51:105–131.

Green, D. M., and D. C. Cannatella. 1993. Phylogenetic significance of the amphicoelous frogs, Ascaphidae and Leiopelmatidae. Ecol. Ethol. Evol. 5:233–245.

Griffiths, I. 1963. The phylogeny of the salientia. Biol. Rev. 38:241–292.

Haas, A. 1997. The larval hyobranchial apparatus of discoglossoid frogs: its structure and bearing on the systematics of the Anura (Amphibia: Anura). J. Zool. Syst. Evol. Res. 35:179–197.

Haas, A. 2003. Phylogeny of frogs as inferred from primarily larval characters (Amphibia: Anura). Cladistics 19:23–89.

Haeckel, E. 1866. Generelle Morphologie der Organismen. Berlin.

Hanken, J. 1985. Morphological novelty in the limb skeleton accompanies miniaturization in salamanders. Science 229:871–874.

Hanken, J. 1999. Why are there so many new amphibian species when amphibians are declining? TREE 14:7–8.

Hay, J. M., I. Ruvinsky, S. B. Hedges, and L. R. Maxson. 1995. Phylogenetic relationships of amphibian families inferred from DNA sequences of mitochondrial 12S and 16S ribosomal RNA genes. Mol. Biol. Evol. 12:928–937.

Heatwole, H., and R. L. Carroll (eds.). 2000. Paleontology. The evolutionary history of amphibians. Surrey Beatty and Sons, Chipping Norton, New South Wales, Australia.

Hedges, S. B., and L. R. Maxson. 1993. A molecular perspective on lissamphibian phylogeny. Herpetol. Monogr. 7:27–42.

Hedges, S. B., K. D. Moberg, and L. R. Maxson. 1990. Tetrapod phylogeny inferred from 18S and 28S ribosomal RNA sequences and a review of the evidence for amniote relationships. Mol. Biol. Evol. 7:607–633.

Hedges, S. B., R. A. Nussbaum, and L. R. Maxson. 1993.

Caecilian phylogeny and biogeography inferred from mitochondrial DNA sequences of the 12S rRNA and 16S rRNA genes (Amphibia: Gymnophiona). Herpetol. Monogr. 7:64–76.

Henrici, A. 1994. *Tephrodytes brassicarvalis*, new genus and species (Anura: Pelodytidae), from the Arikareean Cabbage Patch beds of Montana, USA, and pelodytid-pelobatid relationships. Ann. Carn. Mus. 63:155–183.

Heyer, W. R. 1975. A preliminary analysis of the intergeneric relationships of the frog family Leptodactylidae. Smithson. Contrib. Zool. 199:1–55.

Heyer, W. R., and D. S. Liem. 1976. Analysis of the intergeneric relationships of the Australian frog family Myobatrachidae. Smithson. Contrib. Zool. 233:1–29.

Hillis, D. M., L. K. Ammerman, M. T. Dixon, and R. O. de Sá. 1993. Ribosomal DNA and the phylogeny of frogs. Herpetol. Monogr. 7:118–131.

Inger, R. F. 1967. The development of a phylogeny of frogs. Evolution 21:369–384.

Inger, R. F. 1996. Commentary on a proposed classification of the family Ranidae. Herpetologica 52:241–246.

Jackman, T. R., G. Applebaum, and D. B. Wake. 1997. Phylogenetic relationships of bolitoglossine salamanders: a demonstration of the effects of combining morphological and molecular data sets. Mol. Biol. Evol. 14:883–891.

Jenkins, F. A., and N. H. Shubin. 1998. *Prosalirus bitis* and the anuran caudopelvic mechanism. J. Vert. Paleontol. 18:495–510.

Jenkins, F. A. J., and D. M. Walsh. 1993. An Early Jurassic caecilian with limbs. Nature 365:246–250.

Kluge, A. G., and J. S. Farris. 1969. Quantitative phyletics and the evolution of anurans. Syst. Zool. 18:1–32.

Larson, A. 1991. A molecular perspective on the evolutionary rela-tionships of the salamander families. Evol. Biol. 25:211–277.

Larson, A., and W. W. Dimmick. 1993. Phylogenetic relationships of the salamander families: an analysis of congruence among morphological and molecular characters. Herpetol. Monogr. 7:77–93.

Larson, A., and A. C. Wilson. 1989. Patterns of ribosomal RNA evolution in salamanders. Mol. Biol. Evol. 6:131–154.

Lathrop, A. 1997. Taxonomic review of the megophryid frogs (Anura: Pelobatoidea). Asiatic Herpetol. Res. 7:68–79.

Laurent, R. F. 1986. Sous classe des lissamphibiens. Systématique. Pp. 594–797 *in* Traité de zoologie (P.-P. Grassé and M. Delsol, eds.). Masson, Paris.

Laurin, M. 1998a. The importance of global parsimony and historical bias in understanding tetrapod evolution, Pt. I: Systematics, middle ear evolution and jaw suspension. Ann. Sc. Nat. 1:1–42.

Laurin, M. 1998b. The importance of global parsimony and historical bias in understanding tetrapod evolution, Pt. II: Vertebral centrum, costal ventilation, and paedomorphosis. Ann. Sc. Nat. 2:99–114.

Laurin, M., M. Girondot, and A. de Ricqlès. 2000a. Early tetrapod evolution. TREE 15:118–123.

Laurin, M., M. Girondot, and A. de Ricqlès. 2000b. Reply from M. Laurin, M. Girondot, and A. de Ricqlès. TREE 15:328.

Laurin, M., and Reisz, R. R. 1997. A new perspective on tetrapod phylogeny. Pp. 9–59 *in* Amniote origins. Completing the transition to land (S. S. Sumida and K. L. M. Martin, eds.). Academic Press, San Diego.

Lee, M. S. Y., and B. G. M. Jamieson. 1992. The ultrastruture of the spermatozoa of three species of myobatrachid frogs (Anura, Amphibia) with phylogenetic considerations. Acta Zool. (Stockh.) 73:213–222.

Lescure, J., S. Renous, and J.-P. Gasc. 1986. Proposition d'une nouvelle classification des amphibiens gymnophiones. Mem. Soc. Zool. Fr. (43):145–177.

Liem, S. S. 1970. The morphology, systematics, and evolution of the Old World Treefrogs (Rhacophoridae and Hyperoliidae). Field. Zool. 57:1–145.

Lynch, J. D. 1971. Evolutionary relationships, osteology, and zoogeography of leptodactyloid frogs. Misc. Publ. Mus. Nat. Hist. Univ. Kans. 53:531–238.

Lynch, J. D. 1973. The transition from archaic to advanced frogs. Pp. 133–182 in Evolutionary biology of the anurans (J. L. Vial, ed.). University of Missouri Press, Columbia.

Maglia, A. M. 1998. Phylogenetic relationships of the extant pelobatoid frogs (Anura: Pelobatoidea): evidence from adult morphology. Sci. Pap. Nat. Hist. Mus. Univ. Kans. 10:1–19.

Maglia, A. M., L. A. Pugener, and L. Trueb. 2001. Comparative development of anurans: using phylogeny to understand ontogeny. Am. Zool. 41:538–551.

Mahoney, M. J. 2001. Molecular systematics of Plethodon and Aneides (Caudata: Plethdoontidae: Plethodontini): phylogenetic analysis of an old and rapid radiation. Mol. Phylogenet. Evol. 18:174–188.

McGowan, G., and S. E. Evans. 1995. Albanerpetontid amphibians from the Cretaceous of Spain. Nature 373:143–145.

Meegaskumbura, M., F. Bossuyt, R. Pethiyagoda, K. Manamendra-Arachchi, M. C. Milinkovitch, and C. J. Schneider. 2002. Sri Lanka: a new amphibian hotspot. Science 298:379.

Mendelson, I., R. Joseph, H. R. da Silva, and A. M. Maglia. 2000. Phylogenetic relationships among marsupial frog genera (Anura: Hylidae: Hemiphractinae) based on evidence from morphology and natural history. Zool. J. Linn. Soc. 128:125–148.

Milner, A. R. 1988. The relationships and origin of living amphibians. Pp. 59–102 in The phylogeny and classification of the tetrapods, 1: Amphibians, reptiles, birds (M. J. Benton, ed.). Oxford University Press, Oxford.

Milner, A. R. 1993. The Paleozoic relatives of lissamphibians. Herpetol. Monogr. 7:8–27.

Milner, A. R. 2000. Mesozoic and Tertiary Caudata and Albanerpetontidae. Pp. 1412–1444 in Amphibian biology (H. Heatwole and R. L. Carroll, eds.), vol. 4. Surrey Beatty and Sons, Chipping Norton, New South Wales, Australia.

National Science Foundation. 2003. Assembling the Tree of Life. Progr. Solic. 03–536. National Science Foundation, Arlington, VA.

Nicholls, G. C. 1916. The structure of the vertebral column in the Anura Phaneroglossa and its importance as a basis of classification. Proc. Linn. Soc. Lond. Zool. 128:80–92.

Noble, G. K. 1922. The phylogeny of the Salientia. I. The osteology and the thigh musculature; their bearing on classification and phylogeny. Bull. Am. Mus. Nat. Hist. 46:1–87.

Nussbaum, R. A. 1977. Rhinatrematidae: a new family of caecilians (Amphibia: Gymnophiona). Occ. Pap. Mus. Zool. Univ. Mich. 682:1–30.

Nussbaum, R. A., and M. Wilkinson. 1989. On the classification and phylogeny of caecilians (Amphibia: Gymnophiona), a critical review. Herpetol. Monogr.:1–42.

O'Reilly, J. C., R. A. Nussbaum, and D. Boone. 1996. Vertebrate with protrusible eyes. Nature 382:33.

Özeti, N., and D. B. Wake. 1969. The morphology and evolution of the tongue and associated structures in salamanders and newts (family Salamandridae). Copeia 1969:205–215.

Parra-Olea, G., M. García-París, and D. B. Wake. 1999. Status of some populations of Mexican salamanders (Amphibia: Plethodontidae). Rev. Biol. Trop. 47:217–223.

Parra-Olea, G., T. J. Papenfuss, and D. B. Wake. 2001. New species of lungless salamanders of the genus Pseudoeurycea (Amphibia: Caudata: Plethodontidae) from Veracuz, Mexico. Sci. Pap. Nat. Hist. Mus. Univ. Kans. 20:1–9.

Parra-Olea, G., and D. B. Wake. 2001. Extreme morphological and ecological homoplasy in tropical salamanders. Proc. Natl. Acad. Sci. USA 98:7888–7891.

Parsons, T. S., and E. E. Williams. 1962. The teeth of Amphibia and their relation to amphibian phylogeny. J. Morphol. 110:375–383.

Parsons, T. S., and E. E. Williams. 1963. The relationships of the modern Amphibia: a re-examination. Q. Rev. Biol. 38:26–53.

Pough, F. H., R. M. Andrews, J. E. Cadle, M. L. Crump, A. A. Savitzky, and K. D. Wells. 2001. Herpetology. 2nd ed. Prentice Hall, Upper Saddle River, NJ.

Rage, J.-C., and Z. Rocek. 1989. Redescription of Triadobatrachus massinoti (Piveteau, 1936) an anuran amphibian from the early Triassic. Palaeontogr. Abt. A 206:1–16.

Reig, O. A. 1958. Proposiciones para una nueva macrosistemática de los anuros. Nota preliminar. Physis 21:109–118.

Reiss, J. O. 1996. Palatal metamorphosis in basal caecilians (Amphibia: Gymnophiona) as evidence for lissamphibian monophyly. J. Herpetol. 30:27–39.

Richards, C., and W. S. Moore. 1996. A phylogeny for the African treefrog family Hyperoliidae based on mitochondrial DNA. Mol. Phylogenet. Evol. 5:522–532.

Richards, C. M., and W. S. Moore. 1998. A molecular phylogenetic study of the Old World treefrog family Rhacophoridae. Herpetol. J. 8:41–46.

Rocek, Z. 2000. Mesozoic anurans. Pp. 1295–1331 in Amphibian biology (H. Heatwole and R. L. Carroll, eds.), Vol. 4. Surrey Beatty and Sons, Chipping Norton, New South Wales, Australia.

Rocek, Z., and J.-C. Rage. 2000. Proanuran stages (Triadobatrachus, Czatkobatrachus). Pp. 1283–1294 in Amphibian biology (H. Heatwole and R. L. Carroll, eds.), vol. 4. Surrey Beatty and Sons, Chipping Norton, New South Wales, Australia.

Rowe, T., and J. Gauthier. 1992. Ancestry, paleontology, and definition of the name Mammalia. Syst. Biol. 41:372–378.

Ruta, M., M. I. Coates, and D. L. J. Quicke. 2003. Early tetrapod relationships revisited. Biol. Rev. 78:251–345.

Ruvinsky, I., and L. Maxson. 1996. Phylogenetic relationships among bufonoid frogs (Anura:Neobatrachia) inferred from mitochondrial DNA sequences. Mol. Phylogenet. Evol. 5:533–547.

Sanchiz, B. 1998. Salientia. Pfeil, Münich.

Savage, J. M. 1973. The geographic distribution of frogs: patterns and predictions. Pp. 351–445 in Evolutionary biology of the anurans: contemporary research on major problems (J. L. Vial, ed.). University of Missouri Press, Columbia.

Schoch, R. R. 1995. Heterochrony in the development of the amphibian head. Pp. 107–124 in Evolutionary change and heterochrony (K. J. McNamara, ed.). John Wiley and Sons, New York.

Shaffer, H. B. 1984a. Evolution in a paedomorphic lineage. I. An electrophoretic analysis of the Mexican ambystomatid salamanders. Evolution 38:1194–1206.

Shaffer, H. B. 1984b. Evolution in a paedomorphic lineage. II. Allometry and form in the Mexican ambystomatid salamanders. Evolution 38:1207–1218.

Shaffer, H. B., J. M. Clark, and F. Kraus. 1991. When molecules and morphology clash: a phylogenetic analysis of the North American ambystomatid salamanders Caudata: Ambystomatidae). Syst. Zool. 40:284–303.

Shubin, N. H., and F. A. Jenkins. 1995. An Early Jurassic jumping frog. Nature 377:49–52.

Sokol, O. M. 1975. The phylogeny of anuran larvae: a new look. Copeia 1975:1–24.

Sokol, O. M. 1977. A subordinal classification of frogs (Amphibia: Anura). J. Zool. Lond. 182:505–508.

Spinar, Z. V. 1972. Tertiary frogs from central Europe. W. Junk, The Hague, the Netherlands.

Starrett, P. H. 1968. The phylogenetic significance of the jaw musculature in anuran amphibians. Ph.D. thesis, University of Michigan, Ann Arbor.

Starrett, P. H. 1973. Evolutionary patterns in larval morphology. Pp. 251–271 in Evolutionary biology of the anurans: contemporary research on major problems (J. L. Vial, ed.). University of Missouri Press, Columbia.

Taylor, E. H. 1968. The caecilians of the world. University of Kansas Press, Lawrence.

Titus, T. A., and A. Larson. 1995. A molecular phylogenetic perspective on the evolutionary radiation of the salamander family Salamandridae. Syst. Biol. 44:125–151.

Titus, T. A., and A. Larson. 1996. Molecular phylogenetics of desmognathine salamanders (Caudata: Plethodontidae): a reevaluation of evolution in ecology, life history, and morphology. Syst. Biol. 45:451–472.

Trueb, L. 1973. Bones, frogs, and evolution. Pp. 65–132 in Evolutionary biology of the anurans: contemporary research on major problems (J. L. Vial, ed.). University of Missouri Press, Columbia.

Trueb, L., and R. Cloutier. 1991a. A phylogenetic investigation of the inter- and intrarelationships of the Lissamphibia (Amphibia: Temnospondyli). Pp. 233–313 in Origins of the higher groups of tetrapods: controversy and consensus (H.-P. Schultze and L. Trueb, eds.). Cornell University Press, Ithaca, NY.

Trueb, L., and R. Cloutier. 1991b. Toward an understanding of the amphibians: two centuries of systematic history. Pp. 175–193 in Origins of the higher groups of tetrapods: controversy and consensus (H.-P. Schultze and L. Trueb, eds.). Cornell University Press, Ithaca, NY.

Vences, M., J. Kosuch, S. Lötters, A. Widmer, K.-H. Jungfer, J. Köhler, and M. Veith. 2000. Phylogeny and classification of poison frogs (Amphibia: Dendrobatidae), based on mitochondrial 16S and 12S ribosomal RNA gene sequences. Mol. Phylogenet. Evol. 15:34–40.

Wake, D. B. 1991. Homoplasy: the result of natural selection, or evidence of design limitations? Am. Nat. 138:543–567.

Wake, D. B. 2003. AmphibiaWeb. Available: http://elib.cs. berkeley.edu/aw/index.html.

Wake, D. B., and N. Özeti. 1969. Evolutionary relationships in the family Salamandridae. Copeia 1969:124–137.

Wake, M. H. 1993. Non-traditional characters in the assessment of caecilian phylogenetic relationships. Herpetol. Monogr. 7:42–55.

Wake, M. H. 1994. The use of unconventional morphological characters in the analysis of systematic patterns and evolutionary processes. Pp. 173–200 in Interpreting the hierarchy of nature—from systematic patterns to evolutionary process theories (L. Grande and O. Rieppel, eds.). Academic Press, New York.

Wake, M. H. 1998. Amphibian locomotion in evolutionary time. Zoology 100:141–151.

Wake, M. H. 2003. Chapter 6. The osteology of caecilians. In Amphibian biology (H. Heatwole and M. Davies, eds.). Surrey Beatty and Sons, Chipping Norton, New South Wales, Australia.

Wake, T. A., M. H. Wake, and R. Lesure. 1999. A Mexican archaeological site includes the first Quaternary fossil of caecilians. Quat. Res. 52:138–140.

Wassersug, R. J. 1984. The Pseudohemisus tadpole: a morphological link between microhylid (Orton type 2) and ranoid (Orton type 4) larvae. Herpetologica 40:138–149.

Wassersug, R. J. 1989. What, if anything, is a microhylid (Orton type II) tadpole? Pp. 534–538 in Trends in vertebrate morphology (H. Splechtna and H. Helge Hilgers, eds.). Gustav Fischer, Stuttgart, Germany.

Wilkinson, J., R. C. Drewes, and O. L. Tatum. 2002. A molecular phylogenetic analysis of the family Rhacophoridae with an emphasis on the Asian and African genera. Mol. Phylogenet. Evol. 24:265–273.

Wilkinson, M. 1997. Characters, congruence, and quality. A study of neuroanatomical and traditional data in caecilian phylogeny. Biol. Rev. 72:423–470.

Wilkinson, M., and R. A. Nussbaum. 1996. On the phylogenetic position of the Uraeotyphlidae (Amphibia: Gymnophiona). Copeia 1996:550–562.

Wilkinson, M., and R. A. Nussbaum. 1999. Evolutionary relationships of the lungless caecilian Atretochoana eiselti (Amphibia: Gymnophiona: Typhlonectidae). Zool. J. Linn. Soc. 126:191–223.

Wilkinson, M., J. A. Sheps, O. V. Oommen, and B. L. Cohen. 2002. Phylogenetic relationships of Indian caecilians

(Amphibia: Gymnophiona) inferred from mitochondrial rRNA sequences. Mol. Phylogenet. Evol. 23:401–407.

Wu, S.-H. 1994. Phylogenetic relationships, higher classification, and historical biogeography of the microhyloid frogs (Lissamphibia: Anura: Brevicipitidae and Microhylidae). Ph.D. thesis, University of Michigan, Ann Arbor.

Zardoya, R., and A. Meyer. 2001. On the origin of and phylogenetic relationships among living amphibians. Proc. Natl. Acad. Sci. USA 98:7380–7383.

Zug, G. R., L. J. Vitt, and J. P. Caldwell. 2001. Herpetology. An introductory biology of amphibians and reptiles. 2nd ed. Academic Press, San Diego.

Michael S. Y. Lee
Tod W. Reeder
Joseph B. Slowinski
Robin Lawson

26

Resolving Reptile Relationships

Molecular and Morphological Markers

What, If Anything, Is a Reptile?

Although the origin of tetrapods is often synonymized with the radiation of vertebrates into terrestrial habitats, most early tetrapods and many extant representatives ("amphibians") remained partly aquatic. They possessed permeable skin and (primitively) a breeding biology requiring free water, with external fertilization and aquatic eggs hatching into gilled larvae. Many tetrapod lineages (including some living amphibians) partly circumvented this dependence on water by acquiring internal fertilization and direct development. However, only one lineage, Amniota, evolved additional adaptations permitting full terrestriality, including a waterproof epidermis and the amniotic egg (Sumida and Martin 1997). The amniotic egg is one of the most significant vertebrate innovations, consisting of a tough eggshell, outer and inner protective membranes (chorion and amnion), a yolk sac for nourishing the developing embryo, and an allantois for storage of waste products and respiration. It allows the embryo to develop terrestrially in its own private "pond," bypassing the aquatic larval stage and hatching into a fully formed neonate. Amphibian-grade tetrapods breathe through their permeable skin, supplemented by rather inefficient buccal (throat-based) lung ventilation. The evolution of highly efficient costal (rib-based) lung ventilation has been proposed to be another critical amniote innovation, permitting them to abandon cutaneous respiration and thus waterproof their skin (Janis and Keller 2001).

Reptiles (Reptilia) are a subgroup of amniotes. However, exactly which amniotes have been termed "reptiles" has been in a state of flux. Historically (e.g., Romer 1966), Amniota has been divided "horizontally," by separating two advanced clades (birds and mammals) possessing endothermy and fluffy, insulatory body covering (feathers or hair). The leftovers, mostly ectothermic and scaly skinned, were termed "reptiles." This old definition of Reptilia included living forms such as turtles, tuataras, squamates (lizards and snakes), and crocodiles, as well as extinct forms such as plesiosaurs, "mammal-like reptiles" (pelycosaurs, therapsids), dinosaurs, and pterosaurs. Thus, as defined, reptiles excluded birds (even though these are closely related to crocodiles and dinosaurs), but included "mammal-like reptiles" (even though these are more closely related to mammals than to other reptiles). Furthermore, it has recently been discovered that many extinct groups traditionally included in reptiles, such as pterosaurs, advanced therapsids, and theropod dinosaurs, possessed insulatory integuments and (probably) high metabolic rates (similar to mammals and birds), which makes their inclusion in the traditionally defined Reptilia problematic. Thus, the old concept of Reptilia grouped together a heterogeneous assortment of primitive amniotes that were neither closely related nor even very similar to each other.

With the advent of modern systematic practices advocating classification according to phylogenetic relationships rather than vague notions of evolutionary "advancement" (e.g., Hennig 1966), this arrangement was increasingly seen as unsatisfactory. Therefore, the term Reptilia has recently been redefined by biologists to refer to a cohesive, monophyletic group (clade) of amniotes (e.g., Gauthier et al. 1988). The redefined Reptilia now include birds but excludes the "mammal-like reptiles," which have been transferred to Synapsida, the clade consisting of mammals and their extinct relatives (fig. 26.1). This rearrangement means that Amniota is now divided according to ancestry into its two principal lineages, Synapsida (mammals and their fossil relatives) and the newly reconstituted Reptilia (turtles, tuataras, squamates, crocodilians, birds, and their fossil relatives). The earliest amniotes can already be assigned to either the synapsid or reptile branch, indicating that this dichotomy occurred during the earliest phases of amniote evolution (Reisz 1997).

This newer interpretation of Reptilia is increasingly being adopted by the general community, partly because of the recent evidence that birds are directly descended from dinosaurian reptiles, and is the one used here. Thus, as presently understood, reptiles consist of three major living lineages (figs. 26.1, 26.2): lepidosaurs (lizards, snakes, and tuataras), archosaurs (crocodilians and birds), and testudines (turtles). Reptiles also have an excellent stratigraphic record, with many important groups known exclusively from fossils (fig. 26.1). In addition to the terrestrial adaptations found in all amniotes (discussed above), reptiles possess high levels of skin keratin, the ability to conserve water by excreting uric acid, and novel eye structures (Gauthier et al. 1988).

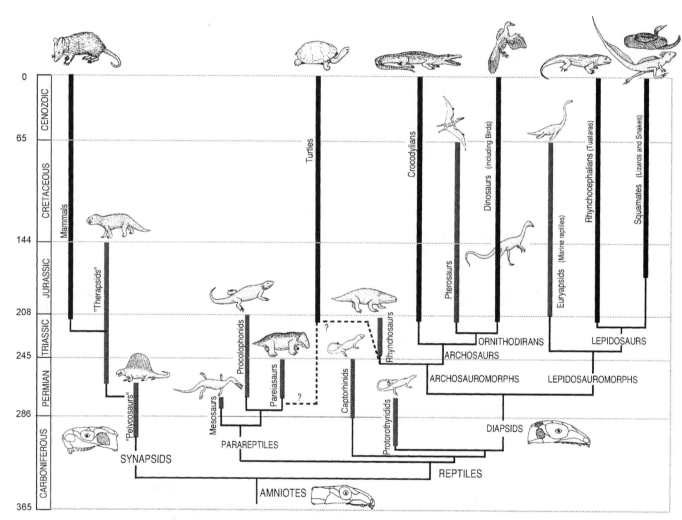

Figure 26.1. Relationships and temporal duration of the major groups of amniote vertebrates. The thick lines depict the known fossil duration for each group, excluding contentious finds (e.g., the Triassic "bird" *Protoavis* and the Cenozoic "therapsid" *Chronoperates*); black lines denote surviving groups; gray lines denote totally extinct groups. Dashed lines indicate uncertain relationships. Examples from each lineage are illustrated. The skull diagrams show the three major skull types found in amniotes: synapsid (found in synapsids), diapsid (found in diapsid reptiles), and anapsid (found in turtles, parareptiles, captorhinids and protorothyridids). Note that synapsid and diapsid skulls each characterize discrete lineages but the anapsid skull does not.

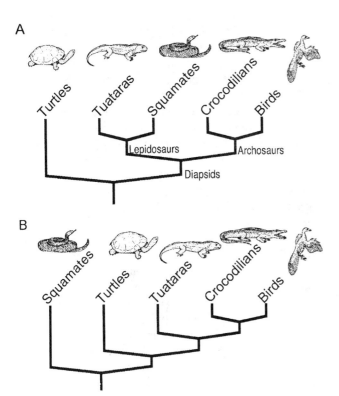

Figure 26.2. Relationships between extant reptiles based on anatomical traits (A; e.g., Gauthier et al. 1988) and well-sampled genes known for tuataras (B; e.g., Hedges and Poling 1999, Raxworthy et al. 2003). Note that although molecular data have often been suggested to require a reinterpretation of turtle affinities, it is actually squamates that shift position between the two trees. The relationships between turtles, tuataras, crocodiles, and birds remain constant.

Parareptiles and Other Primitive Reptiles

Most early reptiles possessed "anapsid" skulls with a solid temporal (or cheek) region (fig. 26.1; Williston 1917), the primitive condition inherited from their amphibian-grade ancestors. Many, but not all, of these anapsid-skulled reptiles belong to a lineage termed the Parareptilia (Laurin and Reisz 1995, Lee 2001). Examples include mesosaurs, procolophonids and pareiasaurs (fig. 26.1). Mesosaurs have long been enigmatic, but have recently been shown to have parareptilian affinities (Modesto 1999). They were small, aquatic forms with long necks, webbed feet, and narrow snouts bearing needle-like teeth. They were weak swimmers presumably incapable of transoceanic crossings, and the discovery of two closely related species on opposite sides of the present Atlantic Ocean was early evidence for continental drift. Procolophonids were the most diverse and longest surviving parareptiles (unless one considers turtles), and superficially resembled stout lizards. The latest forms possessed spiny skulls and molar-like teeth for crushing hard invertebrates. Pareiasaurs were large (up to 3 m), slow-moving herbivores with leaf-shaped teeth, heavy and highly

ornamented skulls, and armor plating over their back and sides.

A few early, anapsid-skulled reptiles do not belong within the parareptile clade (fig. 26.2). Protorothyridids were tiny, slender, long-limbed insectivores, whereas captorhinids were similar, but larger and more robust. Protorothyridids are among the earliest known reptiles (being found inside petrified tree hollows that are more than 300 million years old), and partly on this basis were long assumed to be ancestral to all other reptiles. However, recent cladistic analyses (Laurin and Reisz 1995) suggest that protorothyridids are not ancestral (basal) to all other reptiles, but like captorhinids are close relatives of the diapsid radiation (lepidosaurs and archosaurs).

Turtles

Turtles (Testudines or Chelonians; ~300 living species) are among the most distinct vertebrates, exhibiting striking morphological specializations that involve not just the shell but also associated modifications of the vertebrae, limbs, and skull. Although the skull in all turtles is technically anapsid, with a solid cheek region, the arrangement of bones in this area is rather different from that of other anapsid-skulled reptiles. This is consistent with the suggestion that the turtle skull might be a secondarily "defenestrated" diapsid skull (see below). Although no turtles have true cheek fenestrae, extensive emarginations along the posterior and ventral cheek margins have evolved repeatedly (Gaffney et al. 1991). All teeth on the jaw margins are lost and replaced by a keratinous beak (rhamphotheca). The orbits are positioned anteriorly, resulting in a short facial region and long cheek region.

The turtle shell is a boxlike structure consisting of a dorsal carapace and a ventral plastron, joined laterally by the "bridge." It is open anteriorly for the head and forelimbs, and posteriorly for the tail and hind limbs. The shell is unique among tetrapods in incorporating both dermal armor and internal skeletal elements (e.g., ribs and clavicles), a union that results when the lateral edges of the developing carapace ensnare the developing ribs (Gilbert et al. 2001). The shell is secondarily reduced in certain forms, especially aquatic taxa such as sea turtles and soft-shelled turtles. The dorsal vertebrae and ribs of turtles are immobile, being completely fused to the inside of the carapace, and the body and tail are shortened to fit within the confines of the shell. The limb girdles of turtles lie within (rather than outside) the ribcage, inside the protective shell, and project horizontally through the anterior and posterior shell openings, resulting in a low sprawling stance and broad trackway. Except in sea turtles, the limbs can be retracted into the shell.

Most anatomical studies place turtles within a plexus of primitive reptiles with anapsid skulls, and thus outside of other living reptiles (which possess diapsid skulls; see fig. 26.2A). In particular, turtles are often placed with pareiasaurs based on features such as a consolidated braincase, a shortened vertebral column, and the presence of dermal armor (Lee 1995, 2001; fig. 26.1). However, some other characters, principally those of the appendicular skeleton, link turtles with lepidosaurian diapsids (deBraga and Rieppel 1997, Rieppel and Reisz 1999). The phylogenetic relationships of turtles remain labile because, whereas many primitive cranial features suggesting a basal position among living reptiles, almost as many derived appendicular traits align them with lepidosaurs. Disconcertingly, recent analyses of mitochondrial and nuclear genes contradict both morphological hypotheses, instead consistently suggesting that turtles are related to archosaurian diapsids (e.g., Kumazawa and Nishida 1999, Hedges and Poling 1999, Janke et al. 2001, Rest et al. 2003), an arrangement with no anatomical support (Rieppel 2000). If so, the apparently primitive anapsid skull of turtles would represent an evolutionary reversal. Thus, anatomical and molecular trees cannot be reconciled, and at least one must be wrong.

Widespread adaptive convergence has been invoked to explain why the anatomical evidence might be misleading (Hedges and Poling 1999, Janke et al. 2001), and indeed, the morphological data contain much internal conflict. However, less consideration has been given to the problems of the molecular data sets. Most studies are plagued by poor taxon sampling and many also encounter additional problems (Zardoya and Meyer 2001) such as base composition bias [e.g., 18S and 28S ribosomal RNA (rRNA)], short sequences (e.g., nuclear amino acid residues), inappropriately fast substitution rates (e.g., mitochondrial genes), and potential paralogues (nuclear DNA sequences) or pseudogenes (mitochondrial DNA sequences). The molecular data also contain internal conflicts (C. J. Raxworthy, A. L. Clarke, S. Hauswaldt, J. B. Pramuk, L. A. Pugener, and C. A. Sheil, unpubl. ms.), although the trend that turtles cluster with, or within, archosaurs is sufficiently strong to warrant consideration as true phylogenetic signal (Hedges and Poling 1999, Kumazawa and Nishida 1999, Zardoya and Meyer 2001, Rest et al. 2003). However, there are also reasons why multiple genes could give a (relatively) concordant, but misleading picture. A recent combined analysis of all available molecular data (C. J. Raxworthy, A. L. Clarke, S. Hauswaldt, J. B. Pramuk, L. A. Pugener, and C. A. Sheil, unpubl. ms.), and an earlier one that only included well-sampled genes (Hedges and Poling 1999) both resulted in a tree (fig. 26.2B) that differs from the traditional tree (fig. 26.2A) only in the basal position of squamates. This shift pushes turtles up the tree as the sister group of tuataras and archosaurs (or archosaurs alone, if tuataras are not sampled). So, instead of asking why turtles are emerging high on the molecular tree, the question could be rephrased, Why are squamates emerging as basal? When the question is rephrased as such, an alternative answer emerges. Recent studies have shown that nuclear genetic evolution occurs much faster in squamates than in other reptiles (Hughes and Mouchiroud 2001). Mitochondrial genetic evolution also appears to have acceler-

ated in certain squamates such as agamids, chameleons, and snakes (Kumazawa and Nishida 1999, Rest et al. 2003, T. Reeder and T. Townsend, unpubl. obs.). Although rates in mammals have not (to our knowledge) been comprehensively compared with those in reptiles, mammalian rates do not appear to be any slower than those of typical reptiles (e.g., see Kumazawa and Nishida 1999, Janke et al. 2001), and the long period between the mammal–reptile divergence and the radiation of living mammals means that the synapsid clade will always be on a long branch. The rapid divergence between squamates and other reptiles, and the long temporal gap at the base of the mammal clade, means that the longest branches are those leading to squamates and to the outgroup (mammals). Long branch attraction could thus artificially force squamates toward the base of the reptile tree (Lee 2001). The elevated evolutionary rates throughout the nuclear genome of most squamates, and the mitochondrial genome of at least some, could therefore cause multiple genetic data sets to converge on the same but spurious tree.

The morphological–molecular conflict on turtle origins (or, more accurately, higher level reptile phylogeny in general) thus remains unresolved. Combined analyses (Eernisse and Kluge 1993, Lee 2001, C. J. Raxworthy, A. L. Clarke, S. Hauswaldt, J. B. Pramuk, L. A. Pugener, and C. A. Sheil, unpubl. ms.) still place turtles in the traditional position outside diapsid reptiles (fig. 26.2A). Nevertheless, if turtles are assumed to be related to archosaurs (as suggested by some molecular studies), it would be interesting to determine what fossil reptiles might be the nearest relatives of turtles. This can be ascertained by performing an analysis of all reptiles such that living turtles are "forced" to cluster with living archosaurs to the exclusion of other living reptiles, but all fossil forms are allowed to "float." Turtles then group with extinct herbivorous archosaur relatives called rhynchosaurs, based on shared features such as toothless, beaklike jaws and squat bodies (Lee 2001).

Relationships among turtles have been investigated using morphology alone (Gaffney et al. 1991) or combined with the mitochondrial gene cyt-b and 12S mitochondrial rRNA (Shaffer et al. 1997). The combined data set has been reanalyzed here, and the results are summarized in figure 26.3. The striking concordance between the morphological and molecular data sets (Shaffer et al. 1997) is upheld. Most clades have positive partitioned branch supports from both morphology and molecules, indicating concordant support (see Baker and DeSalle 1997, Gatesy et al. 1999). The most primitive turtles are Proganochelys from the Upper Triassic (Gaffney 1990) and the australochelids from the Upper Triassic and Lower Jurassic (Rougier et al. 1995). They are large, terrestrial herbivores with robust legs and extremely short digits, superficially similar to large modern land tortoises. Unlike living turtles, they could not retract their heads into the shell. Instead, the vulnerable neck region was protected by loose armor plates in Proganochelys and by an anterior expansion of the carapace in australochelids (Rougier et al.

1995). Both groups are more primitive than all other turtles ("casichelydians") in retaining lacrimal and supratemporal bones in the skull, a median opening in the palate (interpterygoid vacuity), separate rather than fused external nostrils, and a very weakly developed anterior process on the shoulder girdle. The remaining turtles (which include all living forms) have the derived condition in all these features and fall into two large clades, pleurodires and cryptodires (each diagnosed by a different method of retracting their head).

Pleurodires (side-necked turtles; ~75 species) retract their heads by folding their neck laterally. They also have a unique arrangement of jaw muscles (Gaffney 1975), where the main jaw closing muscle (adductor mandibulae) passes over a trochlear (pulley) formed by a bone in the roof of the mouth (the pterygoid). Fusion of the pelvis with the shell was formerly thought to be diagnostic of pleurodires, but this feature might be more widespread (Rougier et al. 1995). All living pleurodires are "terrapin-like" in morphology and fall into two lineages, the chelids (47 species) and the pelomedusoids (26 species). Both are now restricted to freshwater habitats of the Southern Hemisphere.

Cryptodires (~225 species) retract their heads by folding the neck in the vertical plane. As in pleurodires, the jaw muscles pass over a trochlear; however, in cryptodires this is formed by a lateral expansion of the braincase (Gaffney 1975). Living cryptodires fall into five major groups (fig. 26.3): trionychoids, chelydrids, chelonioids, kinosternoids, and testudinoids. The trionychoids (26 species) are unusual in that the last dorsal vertebra has been freed from the shell. They include soft-shelled and pig-nosed turtles, and are all highly

aquatic, predatory freshwater forms. These are fast swimmers and rely primarily on speed to escape predators. The shell is reduced and highly streamlined, being very flat and covered in smooth skin. Chelydrids (snapping turtles; two species) are highly sedentary freshwater scavengers and ambush predators; one species lures prey using a wormlike tongue. The chelonioids (sea turtles and leatherbacks; seven species) are all specialized marine forms characterized by limbs modified into flippers. The paddlelike forelimbs are enlarged and used in underwater flight. The buoyancy afforded by water has allowed some sea turtles to reach gigantic proportions. Unlike typical turtles, they partly rely on speed to escape predators and have reduced the shell and lost the ability to retract the skull and limbs. Kinosternoids (mud, musk, and tabasco turtles; 27 species) are unusual in having a shell with a ventral hinge that can close firmly to protect the animal. Finally, the testudinoids (~162 species) are a highly diverse group that includes most remaining living turtles, including familiar forms such as emydids (semi-aquatic to aquatic freshwater sliders) and testudinids (terrestrial tortoises with robust domed shells and elephantine limbs). Testudinoids are united mainly by specializations of the shell (Gaffney and Meylan 1988).

Diapsids (Lepidosaurs and Archosaurs)

Lepidosaurs, archosaurs, and their relatives all have skulls with two large fenestrae (holes) in each cheek, a condition termed "diapsid" (fig. 26.1; Osborn 1903). These fenestrae lighten the skull, and their rims provide insertion areas for the jaw-closing muscles. In addition, these forms possess a

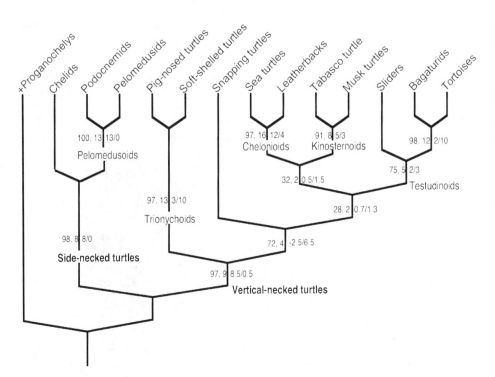

Figure 26.3. Relationships between the major groups of turtles, based on a combined analysis of morphological and molecular data (see text and appendix). The two numbers to the left of each branch show bootstrapping frequency and branch (Bremer) support, respectively; the two numbers to the right denote partitioned branch support (morphology/mitochondrial genes). + denotes totally extinct taxon.

pair of suborbital fenestrae in the roof of the mouth. These novel cranial features, and other traits, unite most diapsid-skulled reptiles as a distinct lineage (the Diapsida), to the exclusion of anapsid-skulled reptiles (fig. 26.1; see Gauthier et al. 1988, Laurin and Reisz 1995, Lee 2001). One possible independent evolution of the diapsid skull occurs in araeoscelids, a group of very primitive reptiles. Some (but not all) araeoscelids have diapsid skulls; a recent study suggests that they are distantly related to other diapsids, implying convergent evolution of the diapsid condition (C. J. Raxworthy, A. L. Clarke, S. Hauswaldt, J. B. Pramuk, L. A. Pugener, and C. A. Sheil, unpubl. ms.). There might also be at least one striking loss of the diapsid skull condition: if turtles are truly related to archosaurs (figs. 26.1, 26.2B), their cheeks are presumably secondarily closed (but see above). Diapsida split quite early in its history into two diverse lineages, one (the lepidosauromorphs) leading to living lepidosaurs, and the other (the archosauromorphs) leading to living archosaurs (fig. 26.1; see Gauthier et al. 1988). Most diapsid reptiles, except some early primitive forms, can be assigned confidently to one of these two clades.

Lepidosaurs (Tuataras, Lizards, and Snakes)

Lepidosaurs (Lepidosauria) include living forms such as *Sphenodon* (tuataras) and squamates (lizards and snakes). Their monophyly is supported by a transversely (rather than longitudinally) oriented cloacal slit, a separate ("sexual") segment in the kidney, novel features in the eye, and skin containing a unique type of keratin and that is shed in large pieces (Gauthier et al. 1988). There is also strong support for lepidosaur monophyly from well-sampled mitochondrial genes (e.g., Zardoya and Meyer 2001, Rest et al. 2003). Although the relatively few nuclear genes so far sequenced for both squamates and *Sphenodon* suggest lepidosaur paraphyly, with squamates basal to all other living reptiles (e.g., Hedges and Poling 1999), this arrangement might be an artifact of elevated substitution rates in squamates coupled with inadequate taxon sampling (see above).

Fossil relatives of living lepidosaurs include the euryapsids, which are marine reptiles such as the armored placodonts, long-necked plesiosaurs, and short-necked pliosaurs (fig. 26.1; Rieppel and Reisz 1999, Mazin 2001). Euryapsids are characterized by a diapsid skull with an extremely wide cheek region lacking the lower strut of bone, a condition termed "euryapsid" (Colbert 1945). The ichthyosaurs, a diverse radiation of fishlike reptiles, might also be related to lepidosaurs, although this is debated (Sander 2000). Among living lepidosaurs, the tuataras (*Sphenodon*) are the most primitive (or basal). They superficially resemble slow-moving, stout iguanas and have unusually slow metabolisms and life cycles, perhaps adaptations to their harsh cold habitat. They are famous "living fossils" and today consist of only two very similar species (only recently distinguished genetically;

Daugherty et al. 1990) restricted to small, rat-free islands off New Zealand. However, in the past the tuatara clade (rhynchocephalians) was much more diverse and included a variety of terrestrial forms as well as elongate marine forms (Wilkinson and Benton 1996).

Squamates (Lizards and Snakes)

Squamata are a diverse and successful radiation of more than 7000 species of lizards, amphisbaenians, and snakes (Vitt et al. 2003). Like most ectothermic tetrapods, they are most diverse and abundant in warmer regions. All squamates share numerous distinctive evolutionary novelties (Estes and Pregill 1988) such as a reduced cheek region with mobility of the quadrate bone that suspends the lower jaw (streptostyly), and a distinct type of vertebral joint (procoely; lost in some geckos). Male squamates have paired copulatory organs called hemipenes. Each hemipenis is generally a forked structure often covered in small spines for anchorage; they are usually ensheathed within the tail and are normally only everted during copulation. Squamates are the only reptiles to exhibit live birth (viviparity). This trait has evolved convergently up to 100 times within squamates, often in the context of cold climates (Shine 1989), and, when acquired, is rarely if ever lost (Lee and Shine 1998).

Several major clades of limbed squamates have long been recognized (e.g., Camp 1923, Estes and Pregill 1988). However, interrelationships between these clades, and the affinities of three highly modified limb-reduced groups (snakes, amphisbaenians, and dibamids), remain contentious. As a result, a phylogenetic analysis of squamates was undertaken combining a large anatomical and behavioral data set (399 characters, see Appendix) with sequences from four genes (mitochondrial 12S and 16S rRNA, nuclear *c-mos* and *c-myc*; see Appendix). The results are summarized in figure 26.4. The combined analysis corroborates the monophyly of many previously recognized groups, such as the lizard "families," as well as larger groupings such as Iguania, Iguanidae *sensu lato* (= Pleurodonta), Acrodonta, Scleroglossa, Gekkota, Pygopodidae + Diplodactylinae, Scincoidea, Lacertoidea, Teiioidea, Anguimorpha, and Varanoidea. Snakes are placed within anguimorphs. Although many traditional groups are supported, the basal divergences within Scleroglossa, and the position of dibamids and amphisbaenians, remain as enigmatic as ever.

Squamata encompasses two major basal clades: Iguania (1000 living species) and Scleroglossa (~6000 species). Iguanian lizards are divided into two groups that can be diagnosed by type of tooth implantation: pleurodont iguanians (traditionally known as iguanids; ~470 species) and the acrodont iguanians (consisting of the agamines, leiolepidines and chamaeleonids; ~535 species). As a group, iguanians are difficult to diagnose, but they generally have a fleshy dewlap in the chin region and often have other crests and ornaments over their skulls and bodies. They also have the ability for rapid and profound color change, a feature linked to male

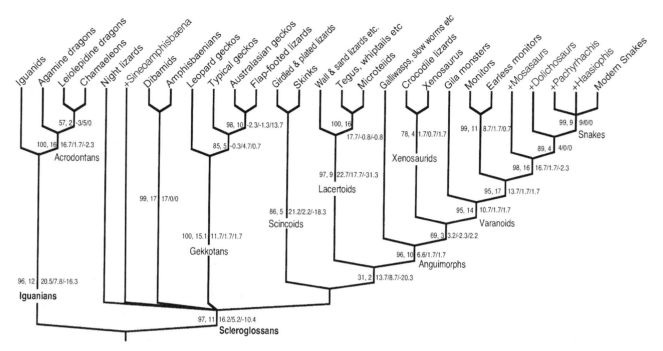

Figure 26.4. Relationships between the major groups of squamates (lizards, amphisbaenians, and snakes), based on a combined analysis of morphological and molecular data (see main text and appendix). The numbers to the left of each branch show bootstrapping frequency and branch (Bremer) support, respectively; the numbers to the right denote partitioned branch support (morphology/mitochondrial genes/nuclear genes). + denotes totally extinct taxon.

territoriality and visual displays, which are more highly developed in iguanians than in other lizards. Iguanians have lost one of the body muscles (the intercostalis ventralis); this simplified trunk musculature might have been a constraint preventing them from evolving a flexible snakelike morphology. Most of the (relatively few) herbivorous lizards are iguanians, a diet perhaps facilitated by their generally large size. Chameleons are among the most famous and bizarre lizards, and many of their unusual features are related to their sit-and-wait predation strategy: the rapid and extensive color changes (camouflage), grasping digits and prehensile tail (facilitating a permanent tight grip on branches), independently movable eyes on turrets, and long projectile tongue (enabling visual sweeps and prey capture without head movement).

The remaining (non-iguanian) squamates form a group named Scleroglossa (fig. 26.4), which is corroborated by distinct morphological novelties (Estes and Pregill 1988), but has not been supported by molecular data (e.g., Rest et al. 2003). Scleroglossans mainly use their teeth for capturing prey, rather than the tongue (as in iguanians), freeing the tongue for chemoreception ("tasting" the air). As a result, the tongue contains many scent-detecting cells, and the chemosensory Jacobson's organ in the palate is elaborated. Scleroglossans also have a flexible hinge in the skull roof, between the frontals and parietals. The hinge appears to be correlated with a shift of the pineal organ and foramen posteriorly away from the mobile frontoparietal

boundary (Schwenk 2000). It is notable that the only scleroglossans with a pineal apparatus on this boundary are certain mosasauroids, which have secondarily consolidated this joint.

Gekkotan lizards (geckos and flap-footed lizards; 1050 living species) appear to be a another relatively basal group of scleroglossans (fig. 26.4). They are usually nocturnal and accordingly have large and distinctive eyes with slitlike vertical pupils. In most, the eyelids are fused into a transparent "spectacle" that is cleaned by licks from a specialized pad on the tongue (Schwenk 2000). Unlike the vast majority of squamates, they have a reduced clutch size (usually fixed at two or one eggs). Most members have enlarged toe pads that enable them to scale smooth vertical surfaces. All gekkotans also lack many skull bones found in other squamates, and many lack well-formed vertebral joints, all probably due to early cessation of ossification (pedomorphosis). Vocal communication is highly developed, with some members having elaborate repertoires similar to those of many frogs. Accordingly, gekkotans have well-developed larynxes ("voiceboxes") and highly sensitive auditory structures. One lineage of gekkotans, the pygopodids (flap-footed lizards), has become very snakelike. However, their phylogenetic position within Gekkota as close relatives of diplodactylines (Australasian geckos) is strongly supported by both morphology (Kluge 1987) and mitochondrial and nuclear genes (fig. 26.4; see also Donnellan et al. 1999).

Scincomorph lizards are the most diverse and "typical" group of lizards, consisting mainly of small-bodied, generalized, insectivorous forms such as scincids (skinks; ~1260 species), cordylids (girdled lizards and plated lizards; ~85 species), lacertids (wall lizards, sand lizards, etc.; ~275 species), teiids (tegus, whiptails, etc.; ~117 species), gymnophthalminds (microteiids; ~190 species), and xantusiids (night lizards; 16 species). The evidence for scincomorph monophyly has always been very weak, with most of the features shared by scincomorphs also being generalized traits widespread in other lizards. In this analysis (fig. 26.4), there is strong evidence for three major lineages of scincomorphs, the scincoids (skinks, cordylids), lacertoids (lacertids, teiids, gymnophthalmids), and xantusiids. There is no evidence that these three lineages are each other's closest relatives, but no alternative arrangement is strongly supported. Most scincomorphs are agile, secretive, smallish forms that shelter beneath leaf litter or loose rocks. This intimate association with the substrate is most marked in skinks and gymnophthalmids and has probably facilitated the frequent (>30 times) evolution within these groups of burrowing habits, limb reduction, and body elongation (e.g., Greer 1989, Pellegrino et al. 2001).

Anguimorph lizards (180 living species, not counting snakes) are generally medium to large predators and include anguids (e.g., galliwasps, glass lizards, slow "worms," alligator lizards), *Heloderma* (the venomous Gila monsters and beaded lizards), xenosaurids (e.g., crocodile lizards), *Varanus* (typical monitors such as the komodo dragon), and *Lanthanotus* (earless monitors) All anguimorphs possess a specialized secretory gland on the lower jaw (gland of Gabe) and a distinctive pattern of tooth replacement. Many also have sharp recurved teeth and a distinct zone of flexibility in each lower jaw (the intramandibular joint; Estes and Pregill 1988). Anguimorphs also have a retractile, deeply forked tongue that is used to pick up airborne molecules ("scents") of prey and other objects (independently evolved in teiid lizards) that are then transmitted to the vomeronasal organ in the roof of the mouth. Differences in the intensity of the scent between the two prongs of the forked tongue allow the direction of the source to be determined. Although most scleroglossan lizards use this system, it is most strongly developed in anguimorphs (Schwenk 2000). All of these traits are related to feeding on large prey and are also found in snakes, which are most likely part of the anguimorph radiation. In this analysis (fig. 26.4), snakes cluster closely with extinct marine varanoids (mosasaurs and dolichosaurs).

Amphisbaenians (160 living species) are a highly aberrant group of long-bodied, limb-reduced squamates that superficially resemble large fat earthworms. They are highly specialized and efficient burrowers, with extremely solid skulls for ramming their way through the substrate, and scales and muscles arranged in rings around the body for gripping the sides of burrows. They have a novel median bone (the orbitosphenoid) surrounding the anterior brain-case and have reduced their right lung (other elongate squamates, including snakes, have reduced the left lung). Their eyes are among the most degenerate in vertebrates, and they rely largely on chemical and vibrational cues to locate prey. Their precise position within Squamata remains unclear, but the suggestion that they might be linked to the fossil *Sineoamphisbaena* is not supported in this study. Morphological data (Lee 2001) place amphisbaenians with dibamids (another highly modified limb-reduced group), but the possibility of pervasive adaptive convergence means this hypothesis of relationship requires independent corroboration. The current molecular data neither support nor contradict this grouping (fig. 26.4).

Snakes

Serpentes (2900 living species) are one of the many lineages of squamates that has undergone body elongation and limb reduction. Snakes range from tiny wormlike blindsnakes to giant constrictors such as boas and pythons, and deadly mambas, cobras, and sea snakes. Characteristic external features include eyelids fused into a transparent "spectacle," absence of the external eardrum, retractile forked tongue, and long, limb-reduced bodies. Each of these traits, however, has evolved independently in certain other squamates ("lizards"), and the key diagnostic features of snakes are internal (Underwood 1967, Estes and Pregill 1988, Greene 1997, Lee and Scanlon 2002). There are usually between 140 and 600 trunk vertebrae (more than in even the most elongate lizards), and the trunk muscles are highly elaborate, permitting both great flexibility and precise local control of body movement. The forelimb and pectoral girdle are totally lost (vestiges remain in even the most limb-reduced lizards). Snakes are characterized by extremely loose skulls with highly flexible upper and lower jaws loosely suspended from a central bony braincase. The tooth-bearing bones of the upper jaw are all mobile. The lateral element (the maxilla) is used to capture prey during the initial strike; later, the palatal elements (palatines and pterygoid) ratchet the prey into the esophagus during the swallowing phase. In many snakes, including most advanced forms, the left and right lower jaws are connected anteriorly by elastic ligaments and thus can separate to engulf of huge prey. This mechanism for increasing gape circumvents the problem that snakes have small heads relative to body size but swallow large prey whole (Greene 1983).

Even the earliest snakes had extensive adaptations for predation, and this constraint appears to have prevented snakes from evolving into omnivores or herbivores. All primitive snakes (and indeed 80% of all snakes) are aglyphous, lacking fangs and venom glands. Aglyphous snakes that take larger prey kill by constriction and continuous bites. However, several groups of advanced snakes have independently evolved fangs (enlarged teeth with grooves or canals for injecting venom) and venom glands (modified salivary glands). These venomous forms often do not constrict but adopt a strike-and-release strategy to avoid injury by large struggling prey. Opisthoglyphous snakes have fixed fangs at the back of the jaws. This arrangement has evolved repeatedly among colubrids (e.g., boomslangs). *Proteroglyphous* snakes have fixed fangs at the front of the jaws. This arrangement characterizes elapid snakes (e.g., cobras, sea snakes, coral snakes). Solenoglyphous snakes have mobile fangs that are only erected while striking. Because the fangs can be folded away when not in use, they can be very large. Vipers (e.g., rattlesnakes and adders) and some enigmatic colubroids (atractaspidids) have this arrangement.

Although there is widespread agreement that snakes evolved from lizards, the more precise details remain contentious. Most recent morphological analyses group snakes with either small fossorial amphisbaenians and dibamids (e.g., Rieppel and Zaher 2000), or large predatory anguimorph lizards (e.g., Lee 2003). The first arrangement is consistent with the hypothesis that snakes evolved from a lineage of burrowing lizards, which is further supported by the close association of burrowing habits with limb reduction in living lizards, and highly divergent eye structure suggesting that the eyes of snakes became reduced and then re-elaborated. The second idea links snakes to marine anguimorphs (mosasaurs and dolichosaurs) based on features such as a unique pattern of tooth eruption and increased flexibility of the jaw

joints, and would suggest that snakes evolved in a marine habitat for eel-like swimming. The combined morphological and molecular analysis of squamates favors this hypothesis (fig. 26.4).

The phylogeny of snakes summarized in figure 26.5 is based on a combined analysis of 263 anatomical and behavioral traits (Lee and Scanlon 2002) and sequences from four genes: mitochondrial 12S rRNA, 16S rRNA (Heise et al. 1995), *cyt-b*, and nuclear *c-mos* (Slowinski and Lawson 2002). The morphological and molecular data, separately and combined, support some traditionally recognized clades, namely, blindsnakes, alethinophidians, and colubroids. However, as discussed below, there are major disagreements regarding the position of dwarf boas and sunbeam snakes, leading to extensive character conflict as revealed by some large negative partitioned branch support (PBS) values.

The limbed marine snakes *Pachyrhachis* and *Haasiophis* emerge as the most basal snakes (fig. 26.5), supporting the view that their legs, low vertebral count, and cranial similarities to anguimorph lizards are retained primitive features (Lee and Scanlon 2002) rather than atavistic reversals (Tchernov et al. 2000, Rieppel and Zaher 2000). Their marine habits are thus relevant to the idea of a marine origin of snakes. The most primitive terrestrial snakes are large superficially "boalike" forms, *Dinilysia* and madtsoiids. These are

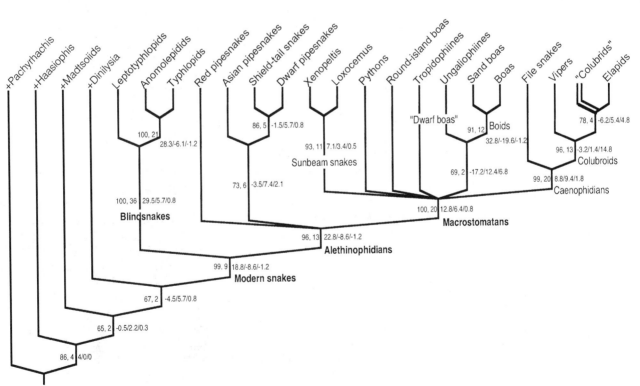

Figure 26.5. Relationships between the major groups of snakes, based on a combined analysis of morphological and molecular data (see main text and appendix). The numbers to the left of each branch show bootstrapping frequency and branch (Bremer) support, respectively; the numbers to the right denote partitioned branch support (morphology/mitochondrial genes/nuclear genes). + denotes totally extinct taxon.

too massive to burrow actively, an observation inconsistent with the suggested fossorial origin of snakes. The large ("macrostomatan") feeding apparatus of these fossil snakes has been interpreted as indicating affinities with higher snakes (e.g., Rieppel and Zaher 2000); however, the recent molecular studies that place some macrostomatan snakes as very basal among living snakes (see below) raise the possibility that the macrostomatan condition was primitive for snakes as a whole. If so, the presence of such gape adaptations in early and apparently basal fossil snakes is no longer problematic.

Among living snakes, the most basal forms are scolecophidians (blindsnakes): leptotyphlopids (~91 species), typhlopids (~225 species), and anomalepidids (15 species). However, they are not primitive by any means, but share a suite of unique specializations indicating their monophyly, such as bizarre consolidated skulls with spherical snouts (Lee and Scanlon 2002). This arrangement is also supported by molecular data (fig. 26.5). These generally small snakes are totally fossorial and accordingly have reduced eyes, cylindrical wormlike bodies, and glossy, dirt-resistant scales. They gorge themselves on ants and termites using rapid oscillations of their small, highly modified jaws (Kley and Brainerd 1999).

The remaining snakes, called alethinophidians (fig. 26.5), are characterized by evolutionary innovations such as a pair of bones (laterosphenoids) surrounding the anterior braincase, a median bony wall between the olfactory lobes of the brain, and the ability to subdue prey by constriction (lost in some advanced venomous forms). They are usually larger, have longer jaws, and have more developed eyes than scolecophidians. The most primitive alethinophidians are *Anilius* (red pipesnake; one species), *Cylindrophis* (Asian pipesnakes; seven species), *Anomochilus* (dwarf pipesnakes; two species) and uropeltids (shield-tail snakes; ~44 species). These are partly fossorial but also frequent surface or aquatic habitats. They lack the elaborate gape adaptations of more advanced snakes and therefore feed mainly on elongate prey with small cross sections, such as eels, caecilians, and earthworms (Greene 1983).

More derived alethinophidians, termed macrostomatans, have further evolutionary innovations to increase gape and permit a greater range of prey. These include a chin ligament that allows the left and right jaw rami to separate, longer jaw elements suspended from enlarged supratemporals, and looser palatal bones (Cundall and Greene 2000). These innovations, and molecular data, support their monophyly (fig. 26.5). They are active above ground for large parts or all of their lives and possess a row of transversely enlarged belly scales for more efficient terrestrial locomotion (lost in some sea snakes). Macrostomatans include most "familiar" snakes, such as boas and pythons, colubrids, and all venomous forms.

Xenopeltis and *Loxocemus*, called sunbeam snakes because of their iridescent scales, form Xenopeltidae (three species). They share many features of the snout and scale microstructure that indicate close relationship, an arrangement supported by molecular data (fig. 26.5). Morphological analyses place them as basal to all other macrostomatans (Lee and Scanlon 2002), and accordingly, they possess relatively weak development of macrostomatan feeding adaptations (Cundall and Greene 2000, Slowinski and Lawson 2002). However, molecular evidence places sunbeam snakes deep within "true" macrostomatans, as relatives of pythons, implying secondary reduction of their gape adaptations (e.g., Slowinski and Lawson 2002, Wilcox et al. 2002, Vidal and Hedges 2002).

Boas (35 species) and pythons (31 species) are typically large and include the largest living snakes. Many are arboreal, and can swallow very large, warm-blooded prey (mammals and birds). Accordingly, many boas and pythons have heat-sensitive lip organs to detect prey and well-developed powers of constriction. Erycines (sand boas; 13 species) are a group of fossorial boas that are generally smaller than typical boas, with most possessing highly bizarre fused tail vertebrae that they use as an antipredator defense. Dwarf boas (tropidophiines, ~20 species; ungaliophiines, three species) are small, boalike snakes that feed principally on reptiles and amphibians. Although traditionally classified as a single group, the two groups of dwarf boas are not close relatives. Morphological studies still place both tropidophiines and ungaliophiines high within snakes, although not as sister groups (Zaher 1994, Lee and Scanlon 2002), but multiple genes suggest a much more radical position for tropidophiines as basal alethinophidians (Slowinski and Lawson 2002, Vidal and Hedges 2002, Wilcox et al. 2002). Given that all other basal alethinophidians are fossorial and gape-limited, the occurrence of above-ground, macrostomatan forms in this part of the tree would imply extensive homoplasy of these traits in early snakes.

Bolyeriines (Round Island boas; two species) are remarkable in that each upper jaw element (maxilla) is divided into two moveable halves, an adaptation for gripping slippery prey such as skinks. One species (*Bolyeria*) has recently become extinct; the other (*Casarea*) is endangered. Morphological and molecular data agree that these groups are all basal macrostomatans but disagree about their precise interrelationships. The phylogeny presented here (fig. 26.5) results from the combined evidence. The morphological data alone place sunbeam snakes as the most basal macrostomatans, followed by a python-boa-erycine clade, with Round Island and dwarf boas being aligned with advanced snakes (Lee and Scanlon 2002). However, the molecular data alone group sunbeam snakes with pythons, whereas sand boas, true boas, and ungaliophiine dwarf boas form another clade (Slowinski and Lawson 2002).

File snakes (acrochordids; three species) are highly aquatic snakes with granular skin and sluggish, limp bodies. They have huge jaws and can swallow extremely large fish prey. However, they feed very infrequently and have very slow metabolisms, perhaps reproducing only once every decade (Shine and Houston 1993). Because of their bizarre morphology, and retention of a few apparently primitive features of the inner ear and lower jaw, they have sometimes been interpreted as the most basal living snakes, perhaps even more primitive than blindsnakes. However, these traits are reversals, because other morphological characters, such as a unique structure of the snout joint, and loss of the coronoid bone in the lower jaw, link acrochordids with the most advanced snakes (colubroids). This grouping (caenophidians) is also supported by molecular data (fig. 26.5).

Colubroidea (colubroids, ~2300 spp.) are the most rapidly diversifying and species-rich group of snakes, and have the dominant snakes on all continents. They are so diverse that their internal phylogenetic relationships are uncertain, and it is difficult to make generalizations about their morphology and biology. They usually possess an extremely mobile upper jaw, specialized dentitions, and elaborate palatal mechanisms for ratcheting prey down the throat (Cundall and Greene 2000). They also share unique elaborations of the trunk musculature and associated rib cartilages. These might be related to their ability for more rapid and precise movement than more primitive snakes, which in turn is correlated with their tendency to use more open habitats. Two groups of highly derived, venomous colubroids have long been recognized: vipers and elapids.

Vipers (Viperidae; ~245 species) are characterized by solenoglyphy (mobile front fangs). They are generally stout-bodied, sit-and-wait predators, but some arboreal forms are more slender. The venom is usually hemotoxic, damaging the blood circulatory system, muscles, and other tissues and often producing hideous wounds. Typical forms include rattlesnakes (*Crotalus*), adders (*Vipera*), and copperheads (*Agkistrodon*). Elapids (Elapidae; ~250 species) are characterized by proteroglyphy (fixed front fangs). Most are more slender and active than vipers, but again, many exceptions exist. The venom is usually neurotoxic, interfering with the nervous system. Elapids include the most deadly snakes, and are the dominant snakes in Australasia. Typical forms include cobras (*Naja*), coral snakes (*Micrurus*), mambas (*Dendroaspis*), and taipans (*Oxyuranus*). Living sea snakes represent two independent marine invasions by elapids (Slowinski and Keogh 2000, Scanlon and Lee in press): sea kraits (*Laticauda*) and true sea snakes (hydrophiines). All sea snakes accordingly have fixed front fangs that inject potent neurotoxins. They have laterally compressed bodies and paddlelike tails to facilitate swimming, and valves in the nostrils to exclude water. *Laticauda* periodically returns to shore to deposit eggs, whereas hydrophiines are totally marine, bearing live young underwater.

The remaining colubroids are often lumped into a wastebasket group, the "Colubridae" (~1800 species). Typical "colubrids" include ratsnakes (*Elaphe*), racers and whipsnakes (*Coluber*), grass snakes (*Natrix*), and boomslangs (*Dispholidus*). They are mainly aglyphous (lacking fangs and venom systems), although a sizable proportion are opisthoglyphous (having fixed rear fangs). The position of the fangs in the back of the

mouth might make it more difficult for them envenomate large victims (including humans). However, some opisthoglyphous colubrids (e.g., boomslangs) have caused many fatalities. The relationships of "colubrids" with each other and other colubroids (vipers and elapids) have long been problematic because of the species diversity of the group. However, they have recently been partly clarified based on molecular sequences (Kraus and Brown 1998, Slowinski and Lawson in press). Vipers are the most basal colubroids, as has been proposed previously based on anatomical data (Underwood 1967), with "colubrids" and elapids forming a clade. Elapids are nested within "colubrids," being related to certain African forms such as psammophiines (e.g., sandsnakes), boodontines (e.g., housesnakes), and atractaspidids (e.g., stiletto snakes). Such a relationship suggests an African origin for elapids. The "Colubridae" as currently construed is thus not a true evolutionary lineage. One solution might be to also include elapids within Colubridae, thereby restoring colubrid monophyly. However, given the medical importance of Elapidae, subsuming them into the (largely harmless) Colubridae might cause confusion, and an alternative would be to restrict Colubridae to a apply to a small monophyletic group.

Archosaurs (Crocodiles, Pterosaurs, Dinosaurs, and Birds)

The archosaurs (Archosauria) include some of the most spectacular reptiles, such as crocodilians, pterosaurs, dinosaurs, and birds (fig. 26.6; Brochu 2001b). They are characterized by numerous anatomical traits (Gauthier et al. 1988) such as a fully divided ventricle in the heart, special stomach chamber (gizzard) housing swallowed stones (gastroliths) used to pulverize food, novel pair of bones (the laterosphenoids) forming the front of the braincase, system of air sacs within the skull, and fenestrae in the snout and lower jaw (these snout fenestrae are secondarily closed in living crocodilians). Living archosaurs (crocodilians and birds) share behavioral traits such as nest building, parental care, and vocalizations (chirping) by nestlings. These habits are difficult to confirm in fossil archosaurs, but smoothly worn stomach stones have been found within complete dinosaur skeletons, and fossilized dinosaurs have recently been found brooding nests of eggs (Clark et al. 1999). Molecular studies reveal that the DNA of crocodiles and birds is very similar (e.g., Zardoya and Meyer 2001, C. J. Raxworthy, A. L. Clarke, S. Hauswaldt, J. B. Pramuk, L. A. Pugener, and C. A. Sheil, unpubl. ms.). The large number of advanced morphological, behavioral and genetic features shared by birds, crocodilians and (where known) fossil archosaurs reflect their close evolutionary relationship and justify the current practice of classifying birds with archosaurian reptiles, rather than the older approach of separating birds off from all reptiles as separate groups. The latter approach is further complicated by recent discoveries of numerous feathered, birdlike dinosaurs

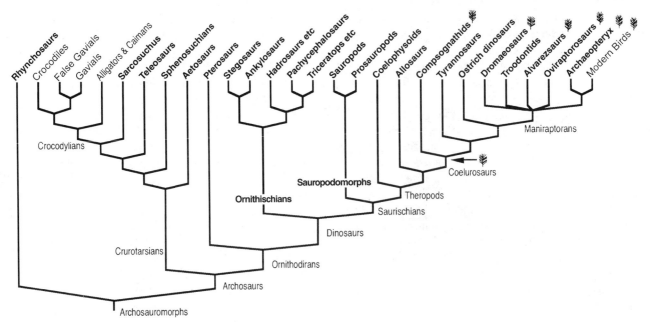

Figure 26.6. Relationships between the major groups of fossil and living archosauromorphs (crocodiles, birds, dinosaurs, pterosaurs and their relatives). Relationships depicted are based on Gauthier (1986), Brochu (1997), Sereno (1999b) and Gatesy et al. (2002). Taxa names with living representatives are shown in black; totally extinct taxa are shown in boldface type. Taxa known to possess feathers are indicated by symbol.

that blur the distinction between birds and nonavian reptiles (see below).

The monophyly of living archosaurs (crocodilians and birds), to the exclusion of other living reptiles, is strongly supported by both morphological traits (fig. 26.2A; Gauthier et al. 1988) and molecular sequences (fig. 26.2B; Janke et al. 2001, C. J. Raxworthy, A. L. Clarke, S. Hauswaldt, J. B. Pramuk, L. A. Pugener, and C. A. Sheil, unpubl. ms.). Relationships among extinct archosaurs are also well established (fig. 26.6). Fossil forms can be assigned to two major lineages, Crurotarsi, which leads to living crocodiles, and Ornithodira, leading to living birds (e.g., Gauthier et al. 1988, Sereno 1999b). However, one important fossil group, the rhynchosaurs, falls outside both living lineages of archosaurs. Rhynchosaurs were the dominant herbivores during the Triassic and had stout bodies, wide, short skulls, and crushing beaks instead of toothed jaws. If turtles are indeed related to archosaurs, as has been proposed by some molecular workers, then they might have affinities with rhynchosaurs (Lee 2001).

The lineage leading to living crocodilians (crurotarsans) includes heavily armored herbivorous forms such as aetosaurs, cursorial long-legged forms such as sphenosuchians that actively chased terrestrial prey, giants amphibious forms such as *Sarcosuchus* that were larger than the largest carnivorous dinosaurs, as well as the ocean-going teleosaurs with flippers and caudal fins (fig. 26.6; Gauthier et al. 1988, Brochu 2001b, Sereno et al. 2001).

Living crocodilians (Crocodylia; 24 living species) are all large, semi-aquatic predators. They are all morphologically

quite uniform, with long snouts, conical piercing teeth, longish bodies, short but robust limbs, laterally compressed tails, and leathery skin containing bony plates. There are two major living lineages, the alligatorids (alligators and caimans) and crocodylids (crocodiles and the "false gavial"). The relationships of true gavials have been contentious, with anatomical evidence suggesting that it represents an independent lineage lying outside of both alligatorids and crocodylids (Brochu 2001a). However, mitochondrial and nuclear sequences, some morphological characters such as narrow elongate jaws, and the combined sequence and morphological data place true gavials within crocodylids, next to the "false gavial" (fig. 26.6; Gatesy et al. 2002). All living crocodilians are ambush predators that (as adults) take sizable vertebrate prey, such as fish, amphibians, birds, and mammals captured either near or under water.

The lineage leading to living birds (ornithodirans) includes pterosaurs, dinosaurs, and some other less known groups (fig. 26.6; Gauthier 1986, Brochu 2001b). Pterosaurs were the first vertebrates to evolve powered flight. Their bones were extremely hollow and light (like those of birds), and their membranous wings were suspended by a greatly elongated fourth finger and stiff internal fibers. The shape of their wings has long been debated, but fossils preserving soft tissue have revealed that (at least in some taxa) the wing membrane was wide and stretched between the forelimbs and hind limbs, resulting in sprawling, clumsy gait. These fossils have also revealed that pterosaurs were covered in fine, hair-like structures (Unwin and Bakhurina 1994), and thus might have evolved endothermy ("warm-bloodedness") in response to the high metabolic demands of flapping flight.

Dinosaurs (including birds) are the most diverse and important archosaur lineage. Unlike all other reptiles, dinosaurs possess modifications of the hips and limbs for an upright (rather than sprawling) gait. This permits breathing while running and thus greater activity levels (Carrier and Farmer 2000). Dinosaurs were primitively bipedal, but facultative or obligate quadrapedality evolved repeatedly within the group. Very early in their evolution, dinosaurs split into two great lineages that each radiated extensively (fig. 26.6; Gauthier 1986, Sereno 1999b). Members of Ornithischia (bird-hipped dinosaurs) possess a (convergently) birdlike pelvis with a backward-pointing pubis, a new bone (predentary) at the tip of the snout, and distinct leaf-shaped teeth. They are all herbivores and include stegosaurs, ankylosaurs, ornithopods, ceratopsians, and pachycephalosaurs. Saurischia (lizard-hipped dinosaurs) are usually characterized by a reptilelike pelvis with a forward-pointing pubis, but this has reverted to an ornithischian-like arrangement in birds and some of their closest theropod relatives. Saurischians also possess elongated birdlike neck vertebrae. They consist of the herbivorous sauropods and prosauropods, as well as the carnivorous theropods. Birds are descended (or ascended) from theropod dinosaurs and are thus part of Saurischia, not Ornithischia.

The theropod–bird transition has recently become one of the most richly documented examples of macroevolution (e.g., Ostrom 1969, Gauthier and Gall 2001, Padian and Horner 2002). Many of the "key" features of birds, such as the wishbone (fused clavicles), enlarged shoulder girdle, and wrist structure permitting wing beat movements, appear in small, lightly built theropods such as dromaeosaurs (e.g., *Velociraptor*, *Deinonychus*). Even birdlike egg structure and brooding behavior have now been confirmed in theropods (Clarke et al. 1999). Perhaps most compelling featherlike integumentary structures have been observed in a range of theropods from exceptional deposits in China (e.g., Xu et al. 1999, 2001, 2003, Ji et al. 2001), and increasing complexity of such structures can be traced along the theropod lineage leading to birds (Prum and Brush 2002). The occurrence of proto-feathers in even quite basal theropods such as compsognathids implies that they were widely distributed throughout the group and arose at the base of Coelurosauria or even earlier. This means that feathers can most parsimoniously be inferred to have been present even in rather unbirdlike forms such as *Tyrannosaurus*. The possession of efficient insulation might have permitted theropods to thermoregulate at smaller body size. This might explain why theropods are the only group of dinosaurs showing a consistent trend toward size reduction; the evolution of small body size, in turn, might have facilitated the origin of flight.

Despite the overwhelming evidence that birds are nested within theropods, major questions remain. First, most theropods show no unequivocal adaptations for climbing, implying that flight probably evolved "from the ground up" via

cursorial theropods (but see Xu et al. 2003). However, this scenario has been argued to be biomechanically less plausible than the alternative view that flight evolved "from the trees down" via a gliding intermediate. The speculation that flight evolved via theropods leaping at prey from high vantage points might reconcile both viewpoints (Garner et al. 1999) but will be difficult to confirm. Also, the homologies of the avian digits remain contentious. There is clear phylogenetic evidence that the functional digits in theropod manus are 1, 2, and 3; digits 4 and 5 gradually diminish and disappear within the clade. However, developmental data suggest that the digits in birds are 2, 3, and 4. This conflict can be reconciled by assuming a homeotic frameshift occurred in the bird manus (Wagner and Gauthier 1999), but this explanation remains controversial (Galis et al. 2002). Finally, the precise position of many transitional taxa (maniraptorans; fig. 26.6) remains debated; for instance, the small, lightly built alvarezaurids and oviraptosaurs might be very birdlike nonavian dinosaurs, or secondarily flightless birds (Sereno 2001, Xu et al. 2002, Maryanska et al. 2002). The plethora of intermediates connecting dinosaurs and birds has shifted the question from whether birds are descended from dinosaurs, to where we draw should the line between dinosaurs and birds. There is now a strong consensus that birds are integral part of the dinosaurian radiation and must be classified as a subgroup of dinosaurs, in much the same way as humans must be considered a subgroup of primate mammals. This taxonomic arrangement correctly reveals that not all dinosaurs became extinct at the end of the Cretaceous; rather, one lineage (Aves) survived to diversify into more than 9000 living species.

Reptiles as a Barometer for Systematics

Phylogenetic studies of reptiles have not only furthered our knowledge of the biodiversity and evolution of this important and conspicuous group but also have generated some of the most important philosophical and methodological advances in systematics. For instance, the old concept of Reptilia represented a classic example of a paraphyletic assemblage (grade), and the shift toward redefining Reptilia as a discrete monophyletic group has reflected the trend toward delimiting taxa based on phylogenetic relationships, rather than vague impressions of similarity or evolutionary advancement. Many workers elaborating this approach (as "phylogenetic taxonomy"; de Queiroz and Gauthier 1992, Cantino and de Queiroz 2000), along with some strong opponents of this system, are reptile systematists. These ideas were thus initially used and debated heavily in the context of reptile studies (e.g., Gauthier 1986, de Queiroz and Gauthier 1992, Laurin and Reisz 1995, Lee 1995, 1998, Dilkes 1998, Sereno 1999a, Padian et al. 1999, Benton 2000). Thus, reptiles have been the empirical exemplar for some of the important advances in taxonomy, and this will continue in the years to come.

Key early papers advocating the importance of considering as many taxa as possible in recovering phylogenetic relationships dealt with reptiles, with these studies demonstrating that incomplete fossil taxa can be critical. For instance, if only living taxa are considered, birds and mammals group together, as the "Haematothermia" (e.g., Gardiner 1993). However, most of their similarities are not present in their putative fossil relatives (e.g., dinosaurs, therapsids). The inclusion of fossil stem taxa reveals that the apparent derived similarities uniting birds and mammals are convergences, thus separating these two taxa to opposite sides of the amniote tree (Gauthier et al. 1988). The wider implication is that partially known taxa of any kind (e.g., those with partial sequence data) can only be ignored at one's peril. Similarly, the earliest papers strongly advocating the "total evidence" or "simultaneous analysis" approach of using as many sources of data as possible in a single analysis to infer phylogenetic relationships were reptile studies (e.g., Kluge 1989, Eernisse and Kluge 1993), and as a result, combined morphological and molecular studies are more common in reptiles than in most other organisms (see Bromham et al. 2002). Systematists now have a wealth of disparate sources of information at their disposal (e.g., morphology, behavior, allozymes, DNA and amino acid sequences, microsatellites, genetic "language," SINEs). The problems and insights of integrating multiple data sets with (potentially) different histories and evolutionary dynamics represent some of the most promising and exciting areas of systematic biology. Some of the most important early contributions in these areas dealt with reptiles, and empirical studies on reptiles will continue to be fertile ground for the growth of phylogenetic methodology. Although this overview has perhaps focused on areas of conflict between morphology and molecules, it should be stressed that, by and large, they agree more often than they disagree. For instance, most of the major groups of reptiles (e.g., crocodiles, birds, turtles, squamates, snakes, amphisbaenians, most lizard and snake "families") were recognized long ago on the basis of morphological data and have since been corroborated by molecular data. However, molecular data corroborating "obvious" groupings are usually considered rather uninteresting, and usually hardly rate a mention in the literature. In contrast, the few areas of strong conflict (and thus novel molecular findings) often receive wider attention, being discussed at length in each study and furthermore encouraging publication in a higher profile journal (e.g., Hedges and Poling 1999, Gatesy et al. 2002). It is difficult to quantify the extent of this "systematic" bias, which is analogous to the greater probability of publication of experimental results rejecting the null hypothesis. However, such a bias is likely, and would have fostered the (erroneous) impression that morphology and molecules are widely or even generally in conflict, thereby encouraging the equally dubious assumption that morphology is not very useful for inferring phylogenetic relationships.

Appendix: Details of Analyses

The turtle data set was that of Shaffer et al. (1997), obtained from the senior author, and reanalyzed unmodified. The complete squamate and snake matrices are available in TreeBASE (2003). The squamate data set consists of the morphological characters of Lee (2000) and partial sequences of four genes: 12S rRNA, 16S rRNA, c-mos, and c-myc (Saint et al. 1998, T. Reeder, unpubl. obs.). The snake data set consisted of the morphological characters of Lee and Scanlon (2002), partial sequences of 12S rRNA and 16S rRNA from Heise et al. (1995), and complete cyt-b and partial c-mos sequences from Slowinski and Lawson (2002). Morphological characters were ordered as discussed in the original studies. Protein-coding genes (cyt-b, c-mos, c-myc) were aligned by eye using SEAL. RNA genes were aligned using Clustal (Gibson et al. 1997), using parameters listed in the data files; sensitivity of results to different alignment costs will be explored in more detail elsewhere. However, the caveat should be added that these are works in progress and the full analyses to follow will almost certainly contain a few alterations to the morphological data, as well as more thorough exploration of alignments, and additional taxa for sequenced for certain genes. Data entry and analyses were undertaken with MacClade (Maddison and Maddison 2000) and PAUP* (Swofford 2000). Analyses included all taxa in the data matrices (certain taxa subsequently pruned from the figured trees) and employed parsimony with all character transformations assigned unit weight. Gaps were treated as a fifth base; this approach was feasible because most parsimony-informative gapped regions were relatively short (the few long gaps were usually either autapomorphic or present throughout the ingroup). Alternative tree-building methods, character weightings, and gap treatments will be explored elsewhere. The overall support for each clade was assessed using branch support (Bremer 1988) and bootstrapping (Felsenstein 1985). Partitioned branch support (Baker and DeSalle 1997), as calculated by TreeRot (Sorenson 1999), was used to evaluate support from each data set for each clade; this was calculated manually from the PAUP log generated by TreeRot. The nonzero molecular PBS values for some basal clades of snakes are not errors but result from rearrangements among extant taxa that occur when calculating PBS.

Acknowledgments

M.S.Y.L. thanks the symposium organizers for the invitation and funding to attend, the Australian Research Council for ongoing research support, Brad Shaffer for providing the turtle data set, and Chris Raxworthy and John Gatesy for permitting citation of manuscripts in review. T.W.R. acknowledges the National Science Foundation for support, and William McJilton for collection of nuclear gene data. R.L. and J.B.S. thank the California Academy of Sciences and the National Science foundation for support.

Literature Cited

Baker, R. H., and R. DeSalle. 1997. Multiple sources of character information and the phylogeny of Hawaiian drosophilids. Syst. Biol. 46:654–673.

Benton, M. J. 2000. Stems, nodes, crown clades, and rank-free lists: is Linnaeus dead? Biol. Rev. 75:633–648.

Bremer, K. 1988. The limits of amino acid sequence data in angiosperm phylogenetic reconstruction. Evolution 42:795–803.

Brochu, C. A. 1997. Morphology, fossils, divergence timing, and the phylogenetic relationships of *Gavialis*. Syst. Biol. 46:479–522.

Brochu, C. A. 2001a. Crocodylian snouts in space and time: phylogenetic approaches toward adaptive radiation. Am. Zool. 41:564–585.

Brochu, C. A. 2001b. Progress and future directions in archosaur phylogenetics. J. Paleont. 75:1185–1201.

Bromham, L., M. Woolfit, M. S. Y. Lee, and A. Rambaut. 2002. Testing the relationship between morphological and molecular rates of change along phylogenies. Evolution.

Camp, C. L. 1923. Classification of the lizards. Bull. Am. Mus. Nat. Hist. 48:289–481.

Cantino, P. D., and K. de Queiroz. 2000. Phylocode: a phylogenetic code of biological nomenclature. Available: http://www.ohiou.edu/phylocode. Last accessed 28 November 2003.

Carrier, D., and C. G. Farmer. 2000. The evolution of pelvic aspiration in archosaurs. Paleobiology 26:271–293.

Clark, J. M., M. A. Norell, and L. M. Chiappe. 1999. An oviraptorid skeleton from the Late Cretaceous of Ukhaa Tolgod, Mongolia, preserved in an avianlike brooding position over an oviraptorid nest. Am. Mus. Nov. 3265:1–36.

Colbert, E. H. 1945. The dinosaur book: ruling reptiles and their relatives. American Museum of Natural History, Handb. Ser. 14. American Museum of Natural History, New York.

Cundall, D., and H. W. Greene. 2000. Feeding in snakes. Pp. 293–333 in Feeding: form, function and evolution in tetrapod vertebrates (K. Schwenk, ed.). Academic Press, San Diego.

Daugherty, C. H., A. Cree, J. M. Hay, and M. B. Thompson. 1990. Neglected taxonomy and continuing extinctions of tuatara (*Sphenodon*). Nature 347:177–179.

DeBraga, M., and O. Rieppel. 1997. Reptile phylogeny and the interrelationships of turtles. Zool. J. Linn. Soc. 120:281–354.

De Queiroz, K., and J. Gauthier. 1992. Phylogenetic taxonomy. Annu. Rev. Ecol. Syst. 23:449–480.

Dilkes, D. W. 1998. The early Triassic rhynchosaur *Mesosuchus browni* and the interrelationships of basal archosauromorph reptiles. Philos. Trans. R. Soc. Lond. B 353:501–541.

Donnellan, S. C., M. N. Hutchinson, and K. M. Saint. 1999. Molecular evidence for the phylogeny of Australian gekkonoid lizards. Biol. J. Linn. Soc. 67:97–118.

Eernisse, D. J., and A. G. Kluge. 1993. Taxonomic congruence versus total evidence, and amniote phylogeny inferred from fossils, molecules and morphology. Mol. Biol. Evol. 10:1170–1195.

Estes, R., and G. Pregill (eds.). 1988. Phylogenetic relationships of the lizard families. Stanford University Press, Stanford, CA.

Felsenstein, J. 1985. Confidence limits on phylogenies: an approach using the bootstrap. Evolution 39:783–791.

Gaffney, E. S. 1975. A phylogeny and classification of the higher categories of turtles. Bull. Am. Mus. Nat. Hist. 155:387–436.

Gaffney, E. S. 1990. The comparative osteology of the Triassic turtle *Proganochelys*. Bull. Am. Mus. Nat. Hist. 194:1–263.

Gaffney, E. S., and P. A. Meylan. 1988. A phylogeny of turtles. In The Phylogeny and classification of tetrapods, vol. 1 (M. J. Benton, ed.). Clrendon Press, Oxford.

Gaffney, E. S., P. A. Meylan, and A. R. Wyss. 1991. A computer assisted analysis of the relationships of the higher categories of turtles. Cladistics 7:313–335.

Galis, F, M. Kundrat, and B. Sinervo. 2002. An old controversy solved: bird embryos have five fingers. Trends Ecol. Evol. 18:7–9.

Gardiner, B. G. 1993. Haematothermia: warm-blooded amniotes. Cladistics 9:369–395.

Garner, J. P., G. K. Taylor, and A. L. R. Thomas. 1999. On the origins of birds: the sequence of character acquisition in the evolution of avian flight. Proc. R. Soc. Lond. B 266:1259–1266.

Gatesy, J., G. Amato, M. Norell, R. DeSalle, and C. Hayashi. 2003. Fossil and molecular evidence for extreme taxic atavism in gavialoid crocodylians. Syst. Biol. 52:403–422.

Gatesy, J., P. O'Grady, and R. H. Baker. 1999. Corroboration among data sets in simultaneous analysis: hidden support for phylogenetic relationships among higher-level artiodactyl taxa. Cladistics 15:271–313.

Gauthier, J. 1986. Saurischian monophyly and the origin of birds. Mem. Calif. Acad. Sci. 8:1–55.

Gauthier, J., and L. F. Gall (eds.). 2001. New perspectives on the origin and early evolution of birds. Yale University Press, New Haven, CT.

Gauthier, J., A. G. Kluge, and T. Rowe. 1988. Amniote phylogeny and the importance of fossils. Cladistics 4:105–209.

Gibson, T., D. Higgins, J. Thompson, and F. Jeanmougin. 1997. Clustal X (computer alignment program). EMBL, Heidelberg, Germany. Available: http://igbmc.ustrasbg.fr/BioInfo/ClustalX/Top.html. Last accessed 29 November 2003.

Gilbert, S. F., G. A. Loredo, A. Brukman, and A. C. Burke. 2001. Morphogenesis of the turtle shell: the development of a novel structure in tetrapod evolution. Evol. Dev. 3:47–58.

Greene, H. W. 1983. Dietary correlates of the origin and radiation of snakes. Am. Zool. 23:431–441.

Greene, H. W. 1997. Snakes—the evolution of mystery in nature. University of California Press, Berkeley.

Greer, A. E. 1989. The biology and evolution of Australian lizards. Surrey Beatty, Chipping Norton, Sydney.

Hedges, S. B., and L. L. Poling. 1999. A molecular phylogeny of reptiles. Science 283:998–1001.

Heise, P. J., L. R. Maxson, H. G. Dowling, and S. B. Hedges. 1995. Higher-level snake phylogeny inferred from mitochondrial DNA sequences of 12S rRNA and 16S rRNA genes. Mol. Biol. Evol. 12:259–265.

Hennig, W. 1966. Phylogenetic systematics. University of Illinois Press, Urbana.

Hughes, S., and D. Mouchiroud. 2001. High evolutionary rates in nuclear genes of squamates. J. Mol. Evol. 53:70–76.

Janis, C. M., and J. C. Keller. 2001. Modes of ventilation in early tetrapods: costal aspiration as a key feature of amniotes. Acta Palaeont. Pol. 46:137–170.

Janke, A., D. Erpenbeck, M. Nilsson, and U. Arnason. 2001. The mitochondrial genomes of the iguana (*Iguana iguana*) and the caiman (*Caiman crocodylus*): implications for amniote phylogeny. Proc. R. Soc. Lond. B 268:623–631.

Ji, Q., M. A. Norell, K.-Q. Gao, S.-A. Ji, and D. Ren. 2001. The distribution of integumentary structures in a feathered dinosaur. Nature 410:1084–1088.

Kley, N., and E. L. Brainerd. 1999. Feeding by mandibular raking in a snake. Nature 402:369–370.

Kluge, A. G. 1987. Cladistic relationships in the Gekkonoidea (Squamata, Sauria). Misc. Publ. Mus. Zool. Univ. Mich. 173:1–54.

Kluge, A. G. 1989. A concern for evidence and a phylogenetic hypothesis of relationships among *Epicrates* (Boidae, Serpentes). Syst. Zool. 38:7–25.

Kraus, F., and W. M. Brown. 1998. Phylogenetic relationships of colubroid snakes based on mitochondrial DNA sequences. Zool. J. Linn. Soc. 122:455–487.

Kumazawa, Y., and Nishida, M. 1999. Complete mitochondrial DNA sequences of the green turtle and blue-tailed mole skink: statistical evidence for archosarian affinity of turtles. Mol. Biol. Evol. 16: 784–792.

Laurin, M., and R. R. Reisz. 1995. A reevaluation of early amniote phylogeny. Zool. J. Linn. Soc. 113:165–223.

Lee, M. S. Y. 1995. Historical burden in systematics and interrelationships of "parareptiles." Biol. Rev. 70:459–547.

Lee, M. S. Y. 1998. Convergent evolution and character correlation in burrowing reptiles: towards a resolution of squamate phylogeny. Biol. J. Linn. Soc. 65:369–453.

Lee, M. S. Y. 2000. Soft anatomy, diffuse homoplasy, and the relationships of lizards and snakes. Zool. Scr. 29:101–130.

Lee, M. S. Y. 2001. Molecules, morphology and the monophyly of diapsid reptiles. Contrib. Zool. 70:1–22.

Lee, M. S. Y. 2003. Taxon sampling, data congruence, and squamate phylogeny. Zool. Scr.

Lee, M. S. Y., and J. D. Scanlon. 2002. Snake phylogeny based on osteology, soft anatomy, and behaviour. Biol. Rev. 77:333–401.

Lee, M. S. Y., and R. Shine. 1998. Reptilian viviparity and Dollo's law. Evolution 52:1441–1450.

Maddison, D. R., and W. P. Maddison. 2000. MacClade 4 (computer program and manual). Sinauer Associates, Sunderland, MA.

Maryanska, T., H. Osmolska, and M. Wolsan. 2002. Avialan status for Oviraptorosauria. Acta Palaeont. Pol. 47:97–116.

Mazin, J.-M. 2001. Mesozoic marine reptiles: an overview. Pp. 95–117 *in* Secondary adaptation of tetrapods to life in water (J.-M. Mazin and V. de Buffrenil, eds.). Pfeil, Munich.

Modesto, S. P. 1999. Observations on the structure of the early Permian reptile *Stereosternum tumidum* Cope. Palaeont. Africana 35:7–19.

Osborn, H. F. 1903. The reptilian subclasses Diapsida and Synapsida. Mem. Am. Mus. Nat. Hist. 1:449–507.

Ostrom, J. H. 1969. Osteology of *Deinonychus antirrhopus*, an unusual theropod from the Lower Cretaceous of Montana. Bull. Peabody Mus. Nat. Hist. 30:1–165.

Padian, K., and J. R. Horner. 2002. Typology versus transformation in the origin of birds. Trends Ecol. Evol. 17:120–124.

Padian, K., J. R. Hutchison, and T. R. Holtz. 1999. Phylogenetic definitions and nomenclature of the major taxonomic categories of the carnivorous Dinosauria (Theropoda). J. Vert. Paleontol. 19:69–80.

Pellegrino, K. C. M., M. T. Rodrigues, Y. Yonenaga-Yassuda, and J. W. Sites, Jr. 2001. A molecular perspective on the evolution of microteiid lizards (Squamata, Gymnophthalmidae) and a new classification of the family. Biol. J. Linn. Soc. 74:315–338.

Prum, R. O., and A. H. Brush. 2002. The evolutionary origin and diversification of feathers. Q. Rev. Biol. 77:261–295.

Reisz, R. R. 1997. The origin and early evolutionary history of amniotes. Trends Ecol. Evol. 12:218–222.

Rest, J. S., J. C. Ast, C. C. Austin, P. J. Waddell, E. A. Tibbetts, J. M. Hay, and D. P. Mindell. 2003. Molecular systematics of primary reptilian lineages and the tuatara mitochondrial genome. Mol. Phlyogen. Evol. 29:289–297.

Rieppel, O. 2000. Turtles as diapsid reptiles. Zool. Scr. 29:199–212.

Rieppel, O., and R. R. Reisz, 1999. The origin and early evolution of turtles. Annu. Rev. Ecol. Syst. 30:1–22.

Rieppel, O., and H. Zaher. 2000. The intramandibular joint in squamates, and the phylogenetic relationships of the fossil snake *Pachyrhachis problematicus*. Field. Geol. 1507:1–69.

Romer, A. S. 1966. Vertebrate paleontology. University of Chicago Press, Chicago.

Rougier, G. W., M. S. de la Fuente, and A. B. Arcucci. 1995. Late Triassic turtles from South America. Science 268:855–858.

Saint, K. M., C. C. Austin, S. C. Donnellan, and M. N. Hutchinson. 1998. *C-mos*, a nuclear, marker useful for squamata phylogenetic analysis. Mol. Phylogen. Evol. 10:259–263.

Sander, P. M. 2000. Ichthyosauria: their diversity, distribution and phylogeny. Palaeontol. Z. 74:1–35.

Scanlon, J. D., and M. S. Y. Lee. In press. Phylogeny of Australasian venomous snakes (Colubridae; Elapidae; Hydrophiidae) based on phenotypic and molecular evidence. Zool. Scr.

Schwenk, K. 2000. Feeding in Lepidosaurs. Pp. 175–291 *in* Feeding: form, function and evolution in tetrapod vertebrates (K. Schwenk, ed.). Academic Press, San Diego.

Sereno, P. C. 1999a. Definitions in phylogenetic taxonomy: critique and rationale. Syst. Biol. 48:329–351.

Sereno, P. C. 1999b. The evolution of dinosaurs. Science 284:2137–2147.

Sereno, P. C. 2001. Alvarezsaurids: birds or ornithomimisaurs? Pp. 69–97 *in* New perspectives on the origin and early evolution of birds (J. Gauthier and L. F. Gall, eds.). Yale University Press, New Haven, CT.

Sereno, P. C., H. C. E. Larsson, C. A. Sidor, and B. Gado. 2001. The giant crocodyliform *Sarcosuchus* from the Cretaceous of Africa. Science 294:1516–1519.

Shaffer, H. B., P. Meylan, and M. L. McKnight. 1997. Tests of turtle phylogeny: molecular, morphological, and paleontological approaches. Syst. Biol. 46:235–268.

Shine, R. 1989. Ecological influences on the evolution of vertebrate viviparity. Pp. 263–278 *in* Complex organismal functions: integration and evolution in vertebrates (D. B. Wake and G. Roth, eds.). John Wiley and Sons, Chichester.

Shine, R., and D. Houston. 1993. Acrochordidae. Pp. 322–324 *.in* Fauna and Flora of Australia (D. Walton, ed.), vol. 2. Australian Government Publishing Service, Canberra.

Slowinski, J. B., and J. S. Keogh. 2000. Phylogenetic relationships of elapid snakes based on cytochrome *b* mtDNA sequences. Mol. Phylogenet. Evol. 15:157–164.

Slowinski, J. B., and R. Lawson. 2002. Higher-level snake phylogeny based on mitochondrial and nuclear genes. Mol. Phylogenet. Evol. 24:194–202.

Slowinski, J. B., and R. Lawson. In press. Elapid relationships. *In* Ecology and evolution in the tropics: a herpetological perspective (M. A. Donnelly, B. I. Crother, C. Guyer, M. H. Wake, and M. E. White, eds.). University of Chicago Press, Chicago.

Sorenson, M. R. 1999. TreeRot, ver. 2 (computer program). M. R. Sorenson, Boston University, Boston.

Sumida, S. S., and K. L. M. Martin (eds.). 1997. Amniote origins: completing the transition to land. Academic Press, San Diego.

Swofford, D. L. 2000. PAUP*. Phylogenetic analysis using parsimony (*and other methods), ver. 4b8–10 (computer program). Sinauer Associates, Sunderland, MA.

Tchernov, E., O. Rieppel, H. Zaher, M. J. Polcyn, and L. L. Jacobs. 2000. A fossil snake with limbs. Science 287:2010–2012.

TreeBASE. 2003. TreeBASE: a database of phylogenetic knowledge. Available: www.treebase.org/treebase/. Last accessed 18 December 2003.

Underwood, G. 1967. A contribution to the classification of snakes. Br. Mus. Nat. Hist. Publ. 653:1–179.

Unwin, D. M., and N. N. Bakhurina. 1994. *Sordes pilosus* and the nature of the pterosaur flight apparatus. Nature 371:62–64.

Vidal, N., and S. B. Hedges. 2002. Higher-level relationships of snakes inferred from four nuclear and mitochondrial genes. C. R. Biol. 325:977–985.

Vitt, L. J., E. R. Pianka, W. E. Cooper, Jr., and K. Schwenk. 2003. History and global ecology of squamata reptiles. Amer. Nat. 162:44–60.

Wagner, G. P., and J. A. Gauthier. 1999. 1, 2, 3 = 2, 3, 4: a solution to the problem of the homology of the digits in the avian hand. Proc. Natl. Acad. Sci. USA 6:5111–5116.

Wilcox, T. P., D. J. Zwickl, T. A. Heath, and D. M. Hillis. 2002. Phylogenetic relationships of the dwarf boas and a comparison of Bayesian and bootstrap measures of phylogenetic support. Mol. Phylogenet. Evol. 25:361–371.

Wilkinson, M., and M. J. Benton. 1996. Sphenodontid phylogeny and the problems of multiple trees. Philos. Trans. R. Soc. Lond. B 351:1–16.

Williston, S. W. 1917. The phylogeny and classification of reptiles. J. Geol. 25:411–421.

Xu, X., M. A. Norell, X.-L. Wang, P. J. Makovicky, and X.-C. Wu. 2002. A basal troodontid from the Early Cretaceous of China. Nature 415:780–784.

Xu, X., X.-L. Wang, and X.-C. Wu. 1999. A dromaeosaurid dinosaur with a filamentous integument from the Yixian formation of China. Nature 401:262–266.

Xu, X., Z.-H. Zhou, and R. O. Prum. 2001. Branched integumental structures in *Sinornithosaurus* and the origin of feathers. Nature 410:200–204.

Xu, X., Z.-H. Zhou, X.-L. Wang, X.-W. Fuang, F.-C. Zhang, and X.-K. Du. 2003. Four-winged dinosaurs from China. Nature 421:335–340.

Zaher, H. 1994. Les tropidopheoidea (Serpentes; Alethinophidia) sont-ils réellement monophylétiques? Arguments en faveur de leur polyphylétisme. C. R. Acad. Sci. Paris 317:471–478.

Zardoya, R., and A. Meyer. 2001. The evolutionary position of turtles revised. Naturwissenschaften 88:193–200.

Joel Cracraft

F. Keith Barker

Michael Braun

John Harshman

Gareth J. Dyke

Julie Feinstein

Scott Stanley

Alice Cibois

Peter Schikler

Pamela Beresford

Jaime García-Moreno

Michael D. Sorenson

Tamaki Yuri

David P. Mindell

Phylogenetic Relationships among Modern Birds (Neornithes)

Toward an Avian Tree of Life

Modern perceptions of the monophyly of avian higher taxa (modern birds, Neornithes) and their interrelationships are the legacy of systematic work undertaken in the 19th century. Before the introduction of an evolutionary worldview by Charles Darwin in 1859, taxonomists clustered taxa into groups using similarities that reflected a vision of how God might have organized the world at the time of Creation. Such was the case with the Quinerian system of avian classification devised by Macleay (1819–1821) in which groups and subgroups of five were recognized, or of Strickland (1841) or Wallace (1856) in which affinities were graphed as unrooted networks (see O'Hara 1988).

After Darwin, this worldview changed. For those comparative biologists struggling to make sense of Earth's biotic diversity in naturalistic terms, Darwinism provided a framework for organizing similarities and differences hierarchically, as a pattern of ancestry and descent. The search for the Tree of Life was launched, and it did not take long for the structure of avian relationships to be addressed. The first to do so was no less a figure than Thomas Henry Huxley (1867), who produced an important and influential paper on avian classification that was explicitly evolutionary. It was also Huxley who provided the first strong argument that birds were related to dinosaurs (Huxley 1868).

Huxley was particularly influential in England and was read widely across Europe, but the "father of phylogenetics" and phylogenetic "tree-thinking" was clearly Ernst Haeckel. Darwin's conceptual framework had galvanized Haeckel,

and within a few short years after *Origin* and a year before Huxley's seminal paper, he produced the monumental *Generelle Morphologie der Organismen*—the first comprehensive depiction of the Tree of Life (Haeckel 1866). Haeckel's interests were primarily with invertebrates, but one of his students was to have a singular impact on systematic ornithology that lasted more than 125 years.

In 1888 Max Fürbringer published his massive (1751 pages, 30 plates) two-volume tome on the morphology and systematics of birds. Showing his classical training with Haeckel and the comparative anatomist Carl Gegenbaur, Fürbringer meticulously built the first avian Tree of Life—including front and hind views of the tree and cross sections at different levels in time. The vastness of his morphological descriptions and comparisons, and the scope of his vision, established his conception of relationships as the dominant viewpoint within systematic ornithology. All classifications that followed can fairly be said to be variations on Fürbringer's theme. Such was the magnitude of his insights. Indeed, as Stresemann (1959: 270) noted:

On the whole all the avian systems presented in the standard works in this century are similar to each other, since they are all based on Fürbringer and Gadow [who followed Fürbringer's scheme closely and, being fluent in German, was able to read the 1888 tome]. My system of 1934 [Stresemann 1927–1934] does not differ in essence from those which

Wetmore (1951) and Mayr and Amadon (1951) have recommended.

Fürbringer (1888) thus established the framework for virtually all the major higher level taxa in use today, and the fact that subsequent classifications, with relatively minor alterations, adopted his groups entrenched them within ornithology so pervasively that his classificatory scheme has influenced how ornithologists have sampled taxa in systematic studies to the present day.

Despite his monumental achievement in establishing the first comprehensive view of the avian branch of the Tree of Life, avian phylogeny soon became of only passing interest to systematists. Phylogenetic hypotheses—in the sense of taxa being placed on a branching diagram—were largely abandoned until the last several decades of the 20th century. For more than 80 years after Fürbringer the pursuit of an avian Tree of Life was replaced by an interest in tweaking classifications, the most important being those of Wetmore (1930, 1934, 1940, 1951, 1960), Stresemann (1927–1934), Mayr and Amadon (1951), and Storer (1960). Aside from reflecting relationships in terms of overall similarity, these classifications also shaped contemporary views of avian phylogenetics by applying the philosophy of evolutionary classification (Simpson 1961, Mayr 1969), which ranked groups according to how distinct they were morphologically.

What happened to "tree thinking" in systematic ornithology between 1890 and 1970? The first answer to this question was that phylogeny became characterized as the unknown and unknowable. Relationships were considered impossible to recover without fossils and resided solely in the eye of the beholder inasmuch as there was no objective method for determining them. Thus, Stresemann (1959: 270, 277) remarked,

> The construction of phylogenetic trees has opened the door to a wave of uninhibited speculation. Everybody may form his own opinion . . . because, as far as birds are concerned, there is virtually no paleontological documentation. . . . Only lucky discoveries of fossils can help us. . . .

A second answer is that phylogeny was eclipsed by a redefinition of systematics, which became more aligned with "population thinking." This view was ushered in by the rise of the so-called "New Systematics" and the notion that "the population . . . has become the basic taxonomic unit" (Mayr 1942: 7). The functions of the systematist thus became identification, classification ("speculation and theorizing"), and the study of species formation (Mayr 1942: 8–11). Phylogeny became passé [see also Wheeler (1995) for a similar interpretation]. Thus,

> The study of phylogenetic trees, of orthogenetic series, and of evolutionary trends comprise a field which was the happy hunting ground of the speculative-minded taxonomist of bygone days. The development of the

"new systematics" has opened up a field which is far more accessible to accurate research and which is more apt to produce tangible and immediate results. (Mayr 1942: 291)

A final answer was that, if phylogeny were essentially unknowable, it would inevitably be decoupled from classification, and the latter would be seen as subjective. The architects of the synthesis clearly understood the power of basing classifications on phylogeny (e.g., Mayr 1942: 280) but in addition to lack of knowledge, "the only intrinsic difficulty of the phylogenetic system consists in the impossibility of representing a 'phylogenetic tree' in linear sequence."

Twenty-seven years later, Stresemann summarized classificatory history to that date in starkly harsh terms:

> In view of the continuing absence of trustworthy information on the relationships of the highest categories [taxa] of birds to each other it becomes strictly a matter of convention how to group them into orders. Science ends where comparative morphology, comparative physiology, comparative ethology have failed us after nearly 200 years of effort. The rest is silence. (Stresemann 1959: 277–278)

The silence did not last. A mere four years after this indictment of avian phylogenetics, Wilhelm Meise, whose office was next to that of the founder of phylogenetic systematics, Willi Hennig, published the first explicitly cladistic phylogenetic tree in ornithology, using behavioral characters to group the ratite birds (Meise 1963). Avian systematics, like all of systematics, soon became transformed by three events. The first was the introduction of phylogenetic (cladistic) thinking (Hennig 1966) and a quantitative methodology for building trees using those principles (Kluge and Farris 1969; the first quantitative cladistic analysis for birds was included in Payne and Risley 1976). At the same time, the rise of cladistics logically led to an interest in having classifications represent phylogenetic relationships more explicitly, and that too became a subject of discussion within ornithology (e.g., Cracraft 1972, 1974, 1981). This desire for classifications to reflect phylogeny had its most comprehensive expression in the classification based on DNA–DNA hybridization, a methodology, however, that was largely phenetic (Sibley et al. 1988, Sibley and Ahlquist 1990, Sibley and Monroe 1990).

The second contribution that changed avian systematics was increased use of molecular data of various types. Techniques such as starch-gel electrophoresis, isoelectric-focusing electrophoresis, immunological comparisons of proteins, mitochondrial DNA (mtDNA) RFLP (restriction fragment length polymorphism) analysis, DNA hybridization, and especially mtDNA and nuclear gene sequencing have all been used to infer relationships, from the species-level to that of families and orders. Today, with few exceptions, investigators of avian higher level relationships use DNA sequencing, mostly of mtDNA, but nuclear gene sequences are now becoming increasingly important.

Finally, not to be forgotten were the continuous innovations in computational and bioinformatic hardware and software over the last three decades that have enabled investigators to collect, store, and analyze increasing amounts of data.

This chapter attempts to summarize what we think we know, and don't know, about avian higher level relationships at this point in time. In the spirit of this volume, the chapter represents a collaboration of independent laboratories actively engaged in understanding higher level relationships, but it by no means involves all those pursuing this problem. Indeed, there is important unpublished morphological and molecular work ongoing that is not included here. Nevertheless, it will be apparent from this synthesis that significant advances are being made, and we can expect the next five years of research to advance measurably our understanding of avian relationships.

Birds Are Dinosaurs

Considerable debate has taken place in recent years over whether birds are phylogenetically linked to maniraptorian dinosaurs, and a small minority of workers have contested this relationship (e.g., Tarsitano and Hecht 1980, Martin 1983, Feduccia 1999, 2002, Olson 2002). In contrast, all researchers who have considered this problem over the last 30 years from a cladistic perspective have supported a theropod relationship for modern birds (Ostrom 1976, Cracraft 1977, 1986, Gauthier 1986, Padian and Chiappe 1998, Chiappe 1995, 2001, Chiappe et al. 1999, Sereno 1999, Norell et al. 2001, Holtz 1994, 2001, Prum 2002, Chiappe and Dyke 2002, Xu et al. 2002), and that hypothesis appears as well corroborated as any in systematics (fig. 27.1).

Having said this, droves of fossils—advanced theropods as well as birds—are being uncovered with increasing regularity, and many of these are providing new insights into character distributions, as well as the tempo of avian evolution. Just 10 years ago, understanding of the early evolution of birds was based on a handful of fossils greatly separated temporally and phylogenetically (e.g., *Archaeopteryx* and a few derived ornithurines). Now, more than 50 individual taxa are known from throughout the Mesozoic (Chiappe and Dyke 2002), and from this new information it is now clear that feathers originated as a series of modifications early in the theropod radiation and that flight is a later innovation (reviewed in Chiappe and Dyke 2002, Xu et al. 2003). Numerous new discoveries of pre-neornithine fossils will undoubtedly provide alternative interpretations to character-state change throughout the line leading to modern birds (for summaries of pre-neornithine relationships, see Chiappe and Dyke 2002).

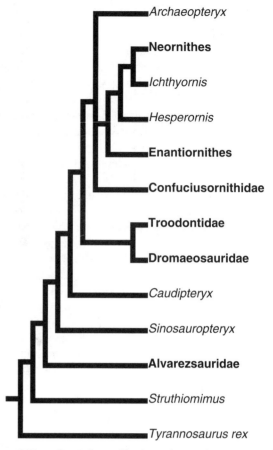

Figure 27.1. Relationships of birds to theropod dinosaurs (after Chiappe and Dyke 2002).

DNA Hybridization and Beyond

The DNA hybridization work of Sibley and Ahlquist (1990) has had a major impact on avian systematics. Their tree—the so-called "Tapestry" shown in figure 27.2—provided a framework for numerous evolutionary interpretations of avian biology. Avian systematists, however, have long noted shortcomings with the analytical methods and results of Sibley and Ahlquist (Cracraft 1987, Houde 1987, Lanyon 1992, Mindell 1992, Harshman 1994). Moreover, it is obvious that Sibley and Ahlquist, like many others before and after, designed their experiments with significant preconceived assumptions of group monophyly (again, many of which can be traced to Fürbringer 1888).

The spine of the DNA hybridization tree is characterized by a plethora of short internodes, which is consistent with the hypothesis of an early and rapid radiation (discussed more below). The critical issue, however, is that most of the deep internodes on Sibley and Ahlquist's (1990) tree were not based on a rigorous analysis of the data, and in fact the data are generally insufficient to conduct such analyses (Lanyon 1992, Harshman 1994). Relationships implied

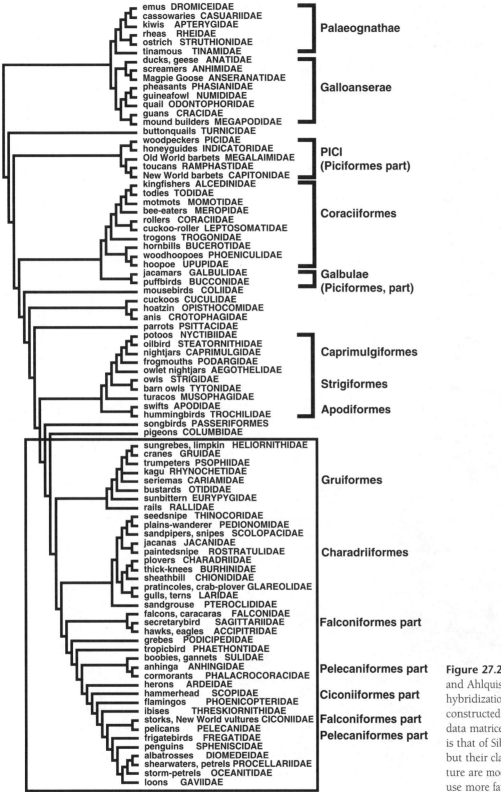

Figure 27.2. The "tapestry" of Sibley and Ahlquist (1990) based on DNA hybridization distances. The tree was constructed by hand from incomplete data matrices. The topology shown here is that of Sibley and Ahlquist (1990), but their classification and nomenclature are modified in some instances to use more familiar names.

by the tree therefore have ambiguous reliability. In addition, because of the manner in which experiments were designed, and possibly because of artifacts due to rate heterogeneity in hybridization distances, instances of incorrect rooting occur across the tree. Thus, although the DNA hybridization data have yielded insight about both novel and previously proposed relationships, they are difficult to interpret and compare with other results except as assertions of relationships.

The tree derived from DNA hybridization data postulated a specific series of relationships among taxa traditionally assigned ordinal rank, as well as among families. It is relevant here to summarize the overall structure of this tree as some of the major groupings it implies will be addressed in subsequent sections of this chapter. Suffice it to say at this point, the emerging morphological and molecular data confirm some of these relationships but not others, both among traditional "orders" but among families as well.

Among its more controversial claims, the DNA hybridization tapestry (fig. 27.2):

1. Recognizes a monophyletic Palaeognathae (ratites and tinamous) and Galloanserae (galliform + anseriform) but unites them, thus placing the neornithine root between them and all other birds: this rooting renders the Neognathae (all birds other than palaeognaths) paraphyletic, a conclusion refuted by substantial data (see below). Oddly, Sibley and Ahlquist (1990) contradicted this in their classification and grouped Galloanserae within their "Neoaves" (equivalent to Neognathae here).

2. Places Turnicidae (buttonquail), Pici (woodpeckers and their allies), and Coraciiformes (kingfishers, rollers, and allies) + Galbulae (traditionally united with the Pici) at the base of the Neoaves.

3. Identifies mousebirds, then cuckoos + Hoatzin, and finally parrots as sequential sister groups to the remaining neognaths.

4. Makes the large songbird (Passeriformes) assemblage the sister group to the remaining neognaths; this latter clade has the pigeons as the sister group of a large, mostly "waterbird," assemblage.

5. Depicts monophyly of Gruiformes (cranes, rails, and allies) and Charadriiformes (shorebirds, gulls, and allies) within the waterbirds: the falconiforms are also monophyletic, except that the New World vultures (Cathartidae) are placed in a family with the storks (Ciconiidae). Within the remainder of the waterbirds, the traditional orders Pelecaniformes (pelicans, gannets, cormorants) and Ciconiiformes (flamingos, storks, herons, ibises) are each rendered paraphyletic and interrelated with groups such as grebes, penguins, loons, and the Procellariiformes (albatrosses, shearwaters).

The Challenge of Resolving Avian Relationships

Initial optimism over the results of DNA hybridization has given way to a realization that understanding the higher level relationships of birds is a complex and difficult scientific problem. There is accumulating evidence that modern birds have had a relatively deep history (Hedges et al. 1996, Cooper and Penny 1997, Waddell et al. 1999, Cracraft 2001, Dyke 2001, Barker et al. 2002, Paton et al. 2002, *contra* Feduccia 1995, 2003) and that internodal distances among these deep lineages are short relative to the terminal branches (Sibley and Ahlquist 1990, Stanley and Cracraft 2002; the evidence is discussed below). To the extent these hypotheses are true, considerable additional data will be required to resolve relationships at the higher levels. This conclusion is supported by the results summarized here.

Although the base of Neoaves is largely unresolved at this time, recent studies are confirming some higher level relationships previously proposed, and others are resolving relationships within groups more satisfactorily than before (the songbird tree discussed below is a good example). At the same time, novel cladistic hypotheses are emerging from the growing body of sequence data (e.g., the proposed connection between grebes and flamingos; van Tuinen et al. 2001). So, even though our ignorance of avian relationships is still substantial, progress is being made, as this review will show.

In addition to summarizing the advances in avian relationships over the past decade (see also Sheldon and Bledsoe 1993, Mindell 1997), the following discussion of neornithine relationships is largely built upon newly completed studies from our various laboratories that emphasize increased taxon and character sampling for both molecular and morphological data. These studies include:

1. An analysis of the *c-myc* oncogene (about 1100 aligned base pairs) for nearly 200 taxa that heavily samples nonpasseriform birds (J. Harshman, M. J. Braun, and C. J. Huddleston, unpubl. obs.)

2. An analysis broadly sampling neornithines that uses 4800 base pairs of mitochondrial sequences in conjunction with 680 base pairs of the *PEPCK* nuclear gene (Sorenson et al. 2003)

3. An analysis of the *RAG-2* [recombination activating protein] nuclear gene for approximately 145 nonpasseriform taxa and a sample of passeriforms (J. Cracraft, P. Schikler, and J. Feinstein, unpubl. obs.)

4. A combined analysis of the RAG-2 data and a sample of 166 morphological characters for 105 family-level taxa (G. J. Dyke, P. Beresford, and J. Cracraft, unpubl. obs.)

5. A combined analysis of the *c-myc* and RAG-2 data for 69 taxa, mostly nonpasseriforms (J. Harshman, M. J. Braun, and J. Cracraft, unpubl. obs.)

6. A combined analysis of 74 "waterbird" taxa for 5300 base pairs of mitochondrial and RAG-2 gene sequences (S. Stanley, J. Feinstein, and J. Cracraft, unpubl. obs.)

7. An analysis of 146 passeriform taxa for 4108 base pairs of the RAG-1 and RAG-2 nuclear genes (F. K. Barker, J. F. Feinstein, P. Schikler, A. Cibois, and J. Cracraft, unpubl. obs.)

8. An analysis of 44 nine-primaried passeriforms ("Fringillidae) using 3.2 kilobases of mitochondrial sequence (Yuri and Mindell 2002).

Phylogenetic Relationships among Basal Neornithes

The Base of the Neornithine Tree

In contrast to the considerable uncertainties that exist regarding the higher level relationships among the major avian clades, the base of the neornithine tree now appears to be well corroborated by congruent results from both morphological and molecular data (fig. 27.3; summarized in Cracraft and Clarke 2001, García-Moreno et al. 2003; see below). Thus, modern birds can be divided into two basal clades, Palaeognathae (tinamous and the ratite birds) and Neognathae (all others); Neognathae, in turn, are composed of two sister clades, Galloanserae for the galliform (megapodes, guans, pheasants, and allies) and anseriform (ducks, geese, swans, and allies) birds, and Neoaves for all remaining taxa. This tripartite division of basal neornithines has been recovered using morphological (Livezey 1997a, Livezey and Zusi 2001, Cracraft and Clarke 2001, Mayr and Clarke 2003; see below) and various types of molecular data (Groth and Barrowclough 1999, van Tuinen et al. 2000, García-Moreno and Mindell 2000, García-Moreno et al. 2003, Braun and Kimball 2002, Edwards et al. 2002, Chubb 2004; see also results below). The DNA hybridization tree also recovered this basal structure, but the root, estimated by assuming a molecular clock without an outgroup, was placed incorrectly (fig. 27.2). In contrast, analyses using morphological or nuclear sequences have sought to place the root through outgroup analysis, and their results are consistent in placing it between palaeognaths and neognaths (Cracraft 1986, Groth and Barrowclough 1999, Cracraft and Clarke 2001; see also studies discussed below). Small taxon samples of mitochondrial data have also been particularly prone to placing the presumed fast-evolving passerine birds at the base of the neornithine tree (Härlid and Arnason 1999, Mindell et al. 1997, 1999), but larger taxon samples and analyses using better models of evolution (e.g., Paton et al. 2002) have agreed with the morphological and nuclear sequence analyses. Recent studies of nuclear short sequence motif signatures support the traditional hypothesis (Edwards et al. 2002), and

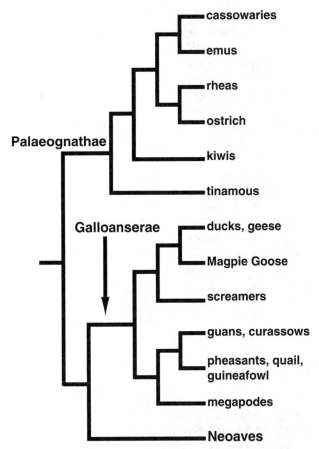

Figure 27.3. The basal relationships of modern birds (Neornithes). Relationships within Paleognathae are those based on morphology (Lee et al. 1997), which do not agree with results from molecular sequences. See text for further discussion.

it also worth noting that palaeognaths and neognaths are readily distinguished by large homomorphic sex chromosomes in the former and strongly heteromorphic chromosomes in the latter (Ansari et al. 1988, Ogawa et al. 1998).

Palaeognathae

Monophyly of palaeognaths is well corroborated, but relationships within the ratites remain difficult to resolve. The relationships shown in figure 27.3 reflect those indicated by morphology (Cracraft 1974, Lee et al. 1997, Livezey and Zusi 2001), and all the internodes have high branch support. Molecular data, on the other hand, have differed from this view and, in general, data from different loci and methods of analysis have yielded conflicting results. In most of these studies (Lee et al. 1997, Haddrath and Baker 2001, Cooper et al. 2001) the kiwis group with the emu + cassowaries, and the rhea and ostrich diverge independently at the base of the tree. When the extinct New Zealand moas are included in studies using most of the mitochondrial genome (Haddrath and Baker 2001, Cooper et al. 2001), they also tend to be placed toward the base of the tree. It can be noted that single gene trees often do not recover ratite monophyly with strong support, although these taxa generally group together.

Palaeognaths appear to exhibit molecular rate heterogeneity. Tinamous, in particular, and possibly rheas and ostriches appear to have higher rates of molecular evolution than do kiwis, emus, and cassowaries (Lee et al. 1997, van Tuinen et al. 2000, Haddrath and Baker 2001). Additionally, paleognath mitochondrial sequences, which have been the primary target of molecular studies, exhibit significant shifts in base composition, which have made phylogenetic interpretations difficult (Haddrath and Baker 2001). Thus, rate artifacts, nonstationarity, the existence of relatively few, deeply divergent species-poor lineages, and short internodal distances among those lineages all play a role in making the resolution of ratite relationships extremely difficult and controversial. Although palaeognath relationships may be solved with additional molecular and morphological data of the traditional kind, the discovery of major character changes in molecular sequences such as indels or gene duplications may also prove to be important.

Galloanserae

Despite occasional debates that galliforms and anseriforms are not sister taxa (Ericson 1996, 1997, Ericson et al. 2001), the predominant conclusion of numerous workers using morphological and/or molecular data is that they are (Livezey 1997a, Groth and Barrowclough 1999, Mindell et al. 1997, 1999, Zusi and Livezey 2000, Livezey and Zusi 2001, Cracraft and Clarke 2001, Mayr and Clarke 2003, Chubb 2004). Molecular studies questioning a monophyletic Galloanserae (e.g., Ericson et al. 2001) have all employed small taxon samples of mtDNA or nuclear DNA, but when samples are increased, or nuclear genes are used, Galloanserae are monophyletic and the sister group of Neoaves (Groth and Barrowclough 1999, García-Moreno and Mindell 2000, van Tuinen et al. 2000, García-Moreno et al. 2003, Chubb 2004; see also J. Harshman, M. J. Braun, and C. J. Huddleston, unpubl. obs.); three indel events in sequences from *c-myc* also support a monophyletic Galloanserae (fig. 27.4). The DNA hybridization tree of Sibley and Ahlquist (1990) recognized Galloanserae, but because the neornithine root was incorrectly placed, Galloanserae was resolved as the sister group of the palaeognaths. With respect to relationships within galliforms, a consistent pattern seems to have emerged (Cracraft 1972, 1981, 1988, Sibley and Ahlquist 1990, fig. 328, Harshman 1994, Dimcheff et al. 2000, 2002, Dyke et al. 2003; see also J. Harshman, M. J. Braun, and C. J. Huddleston, unpubl. obs.): (Megapodiidae (Cracidae (Numididae + Odontophoridae + Phasianidae))). The major questions remain centered around the relative relationships among the guinea fowl (numidids), New World quail (odontophorids), and pheasants (phasianids), as well as the phylogeny within the latter; recent studies suggest that the numidids are out-

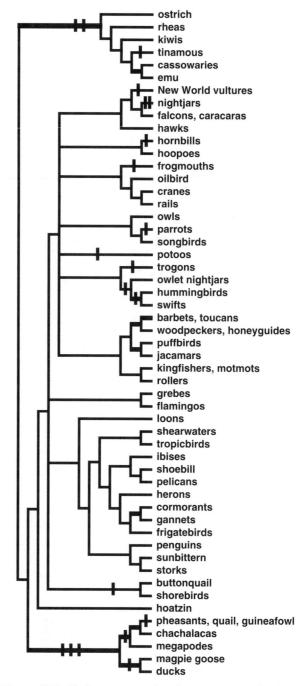

Figure 27.4. Phylogenetic tree based on approximately 1100 bases of the nuclear oncogene *c-myc*, including intron, exon coding, and 3' untranslated region sequence, for 170 taxa (J. Harshman, M. J. Braun, and C. H. Huddleston, unpubl. obs.). The tree shown is an unweighted parsimony majority rule bootstrap tree, plus other compatible branches. Thick branches have 70% or greater bootstrap support; thin branches may have very low support. Vertical tick marks represent phylogenetically informative indels. Most terminal branches represent several species, and all those are strongly supported, although for clarity the branches are not shown as thickened.

side quails and phasianids (Cracraft 1981, Dimcheff et al. 2000, 2002, Dyke et al. 2003).

Relationships among the basal clades of anseriforms are also not too controversial (Livezey 1986, 1997a,; Sibley and Ahlquist 1990: fig. 328 *contra* the "tapestry", Harshman 1994, Ericson 1997, Groth and Barrowclough 1999; for views of relationships within anatids, see Madsen et al. 1988, Livezey 1997b, Donne-Goussé et al. 2002). The screamers (Anhimidae) are the sister group to the magpie goose (Anseranatidae) + ducks, geese, and swans (Anatidae). We note, however, that the resolution of the basal nodes among screamers, magpie goose, and anatids has been difficult and that mitochondrial data sometimes unite the screamers and magpie goose (fig. 27.5), a grouping not suggested by nuclear, morphological, or combined data. The fact that both Livezey (1997a) and Ericson (1997) found the Late Cretaceous-Paleogene fossil *Presbyornis* to be the sister group of Anatidae (see also Kurochkin et al. 2002) is important because it sets the Late Cretaceous as the minimum time of divergence for the anatids and all deeper nodes.

Relationships within Neoaves

Relationships among the neoavian higher taxa have been discussed in a number of studies over the past several decades (e.g., Cracraft 1981, 1988, Sibley and Ahlquist 1990, Ericson 1997, Mindell et al. 1997, 1999, Feduccia 1999, van Tuinen et al. 2000, 2001, among others), and it is clear that relatively little consensus has emerged. The monophyly of many groups that have been accorded the taxonomic rank of "order" such as loons, grebes, penguins, parrots, cuckoos, and the large songbird group (Passeriformes) has not been seriously questioned but that of nearly all other higher taxa has. Thus, it is now broadly accepted that several traditional orders such as pelecaniforms, ciconiiforms, and caprimulgiforms are nonmonophyletic, whereas the status of others such as gruiforms, coraciiforms, piciforms, and falconiforms remains uncertain in the minds of many workers.

If one had to summarize the current state of knowledge, the most pessimistic view would see the neoavian tree as a "comb," with little or no resolution among most traditional families and orders. Short and poorly supported internodes among major clades of neoavians are characteristic of recent studies using nuclear (Groth and Barrowclough 1999, van Tuinen et al. 2000) or mitochondrial data sets (van Tuinen et al. 2000, 2001, Johnson 2001, Hedges and Sibley 1994, Johansson et al. 2001), and the data sets discussed here also illustrate this point. The trees discussed below will be interpreted within the framework of bootstrap resampling analyses that show sister lineages supported at the 70% level (heavy lines in the figures). Using this approach, relationships among the avian higher taxa can be interpreted as largely unresolved, producing the neoavian comb. Nevertheless, there are emerging similarities in phylogenetic pattern recovered across some of these different studies that suggest some commonality of phylogenetic signal. In these and other published cases, the

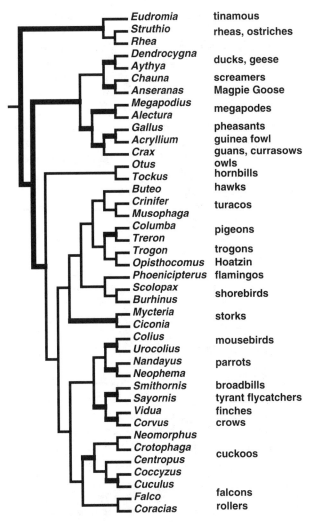

Figure 27.5. A phylogenetic hypothesis for 41 avian taxa based on about 4800 base pairs of mitochondrial sequence and 680 base pairs of *PEPCK* intron 9 nuclear gene using three paleognaths as the root (Sorenson et al. 2003). Nodes with bootstrap support values of 70% are shown in heavy black, based on maximum likelihood and maximum probability analyses of mitochondrial data and MP analyses of *PEPCK* intron 9.

primary reason for the neoavian comb is suspected to be insufficient character and/or taxon sampling. As noted above, current evidence suggests that many of these divergences are old and occurred relatively close in time. Thus, we are optimistic that most neoavian relationships will be resolved with additional data (see Discussion, below).

Phylogenetic Relationships among the "Waterbird Assemblage"

Over the years, many authors have suggested that some or all of the waterbird orders, in particular, seabirds (Procellariiformes), penguins (Sphenisciformes), loons (Gaviiformes), grebes (Podicipediformes), storks, herons, flamingos and allies (Ciconiiformes), pelicans, cormorants, and allies (Pelecaniformes), shorebirds and gulls (Charadriiformes), and cranes, rails, and allies (Gruiformes), are related to one another (see, e.g., Sibley and Ahlquist 1990, Hedges and Sibley 1994, Olson and Feduccia 1980a, 1980b, Cracraft 1988). Some authors have also linked various falconiform families to the waterbird assemblage (Jollie 1976–1977, Rea 1983), including a supposedly close relationship between New World vultures (Cathartidae) and storks (Ligon 1967, Sibley and Ahlquist 1990, Avise et al. 1994; but also see Jollie 1976–1977, Hackett et al. 1995, Helbig and Siebold 1996). As a consequence of these and newer molecular studies, it is now widely thought that several of the large traditional orders of waterbirds may not be monophyletic, and this is especially true of the pelecaniforms and ciconiiforms (Cottam 1957, Sibley and Ahlquist 1990, Hedges and Sibley 1994, Siegel-Causey 1997, van Tuinen et al. 2001).

The supposition that waterbirds are related to one another within neornithines as a whole is not well supported, although the available data are suggestive of a relationship among some of them (see above). Only the DNA hybridization tree of Sibley and Ahlquist (1990) covered all birds, and on their tree (fig. 27.2) the waterbirds and falconiforms are clustered together. Van Tuinen et al. (2001) recently re-evaluated waterbird relationships and compared new DNA hybridization data with results from about 4062 base pairs of mitochondrial and nuclear sequence data for 20 and 19 taxa, respectively. Their most general conclusion was there was relatively little branch support across the spine of the tree, indicating that relationships among waterbirds are still very much uncertain. They did, however, find support for several clades: (1) a grouping of (the shoebill *Balaeniceps* + pelicans) + hammerkop (*Scopus*), and these in turn to ibises and herons, (2) penguins + seabirds (Procellariiformes), and most surprisingly, (3) grebes + flamingos.

Previous studies have had insufficient taxon and character sampling, or both. Even though large taxon samples based on mitochondrial genes (fig. 27.6), or on the *c-myc* and *RAG-2* nuclear genes (figs. 27.4, 27.7A), are an improvement on previous work, by themselves or together (fig. 27.8), they are still inadequate to provide strong character support for most clades. Nevertheless, some congruence among these various studies is apparent. The *c-myc* data (fig. 27.4; J. Harshman, M. J. Braun, and C. J. Huddleston, unpubl. obs.), for example, recover (1) (cormorants + gannets) + frigatebirds, (2) (shoebills + pelicans) + ibises, (3) grebes + flamingos, and (4) buttonquails + shorebirds. At the same time, groups such as loons, tropicbirds, penguins, and storks do not show any clear pattern of relationships in the *c-myc* data or the nuclear/mitochondrial tree of van Tuinen et al. (2001). What is clear in the *c-myc* data is that New World vultures and storks are distantly removed from one another; New World vultures were not included in the van Tuinen et al. study. The *RAG-2* data (fig. 27.7; J. Cracraft, P. Schikler, and J. Feinstein, unpubl. obs.) also strongly support (1) a pelican/shoebill/hammerkop clade, (2) a cormorant/anhinga/gannet group-

ing, and (3) various clades within traditional charadriiforms and gruiforms. Both *c-myc* and *RAG-2* + morphology link frigatebirds to the sulids, phalacrocoracids, and anhingids.

In an attempt to address problems of sparse taxon sampling seen in previous studies, S. Stanley, J. Feinstein, and J. Cracraft (unpubl. obs.) examined 57 waterbird taxa for 5319 base pairs of mitochondrial and nuclear *RAG-2* sequences (fig. 27.6). When palaeognaths and Galloanserae are used as outgroups, the root of the waterbird tree was placed on one of the two gruiform lineages, thus suggesting, in agreement with Sibley and Ahlquist (1990) and van Tuinen et al. (2001), that gruiforms are outside the other waterbird taxa, although this is not strongly supported given available data. This larger analysis still provides little resolution for higher level relationships among waterbirds, but it does find support for a grebe + flamingo relationship, monophyly of charadriiforms, and the shoebill + pelicans + hammerkop clade, in agreement with van Tuinen et al. (2001) and the *c-myc* data (fig. 27.4).

The buttonquails (Turnicidae) have traditionally been considered members of the order Gruiformes. Recent molecular analyses, however, now place them decisively with the charadriiforms, and indeed they are the sister-group of the Lari (Paton et al. 2003). The *c-myc* data (fig. 27.4) are consistent with this topology and include a unique indel, uniting turnicids and chradriiforms.

The mitochondrial and *RAG-2* data also appear to contain phylogenetic signal for other clades even though they do not have high bootstrap values. Thus, when the data are explored using a variety of methods (e.g., transversion parsimony), the following groups are generally found (fig. 27.6): (1) an expanded "pelecaniform" clade that also includes taxa formerly placed in ciconiiforms (shoebill, hammerkop, ibises, and storks), (2) a grouping of grebes and flamingos with charadriiforms and some falconiforms, and (3) often a monophyletic Falconiformes (although the family Falconidae was not sampled), with no evidence of a relationship between storks and New World vultures. Tropicbirds (phaethontids) and herons (ardeids) represent divergent taxa that have no stable position on the tree. Some of these relationships are also seen in other data sets such as the *c-myc* data (fig. 27.4) and in the mitochondrial data of van Tuinen et al. (2001).

Phylogenetic Relationships among the Owls (Strigiformes), Swifts and Hummingbirds (Apodiformes), and Nightjars and Allies (Caprimulgiformes)

The DNA hybridization tree (fig. 27.2; Sibley and Ahlquist 1990) recognizes a monophyletic Caprimulgiformes that is the sister group of the owls; these two groups, in turn, are the sister group of the turacos (Musophagidae), and finally, all three are the sister clade of the swifts and hummingbirds (Apodiformes). There is now clear evidence that this hypothesis is not correct.

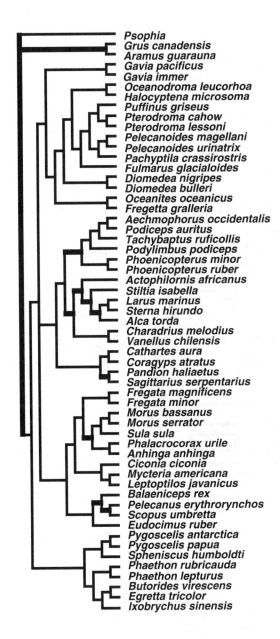

Psophia	**Psophiidae: trumpeters**
Grus canadensis	**Gruidae: cranes, limpkin**
Aramus guarauna	
Gavia pacificus	**Gaviidae: loons**
Gavia immer	
Oceanodroma leucorhoa	**Procellariidae: petrels,**
Halocyptena microsoma	**diving-petrels**
Puffinus griseus	
Pterodroma cahow	
Pterodroma lessoni	
Pelecanoides magellani	
Pelecanoides urinatrix	
Pachyptila crassirostris	
Fulmarus glacialoides	
Diomedea nigripes	**Diomediidae: albatrosses**
Diomedea bulleri	
Oceanites oceanicus	
Fregetta gralleria	
Aechmophorus occidentalis	**Podicipedidae: grebes**
Podiceps auritus	
Tachybaptus ruficollis	
Podylimbus podiceps	
Phoenicopterus minor	**Phoenicopteridae: flamingos**
Phoenicopterus ruber	
Actophilornis africanus	**Jacanidae: jacanas**
Stiltia isabella	**Glareolidae: pratincoles**
Larus marinus	**Laridae: terns, gulls**
Sterna hirundo	
Alca torda	**Alcidae: murres, auks**
Charadrius melodius	**Charadriidae: plovers**
Vanellus chilensis	
Cathartes aura	**Cathartidae: New World vultures**
Coragyps atratus	**Accipitridae: osprey**
Pandion haliaetus	**Sagittariidae: Secretarybird**
Sagittarius serpentarius	
Fregata magnificens	**Fregatidae: frigatebirds**
Fregata minor	
Morus bassanus	**Sulidae: boobies, gannets**
Morus serrator	
Sula sula	
Phalacrocorax urile	**Phalacrocoracidae: cormorants**
Anhinga anhinga	**Anhingidae: anhingas**
Ciconia ciconia	**Ciconiidae: storks**
Mycteria americana	
Leptoptilos javanicus	
Balaeniceps rex	**Balaenicipitidae: Shoebill**
Pelecanus erythrorynchos	**Pelecanidae: pelicans**
Scopus umbretta	**Scopidae: Hammerkop**
Eudocimus ruber	**Threskiornithidae: ibises**
Pygoscelis antarctica	**Spheniscidae: penguins**
Pygoscelis papua	
Spheniscus humboldti	
Phaethon rubricauda	**Phaethontidae: tropicbirds**
Phaethon lepturus	
Butorides virescens	**Ardeidae: herons**
Egretta tricolor	
Ixobrychus sinensis	

Figure 27.6. A phylogenetic hypothesis for "waterbird" higher taxa using 4164 base pairs of mitochondrial sequence (*cytochrome b, COI, COII, COIII*) and 1155 base pairs of the *RAG-2* nuclear gene (transversion weighted) using gruiform taxa as the root (S. Stanley, J. Feinstein, and J. Cracraft, unpubl. obs.). Thick branches represent interfamilial clades supported by bootstrap values greater than 70% (all families had high bootstrap values but are not shown for simplicity).

Both published and unpublished data have recently indicated that caprimulgiforms are not monophyletic. Instead of their traditional placement within caprimulgiforms, owlet-nightjars (Aegothelidae) are most closely related to the swifts and hummingbirds, a hypothesis first recognized in *c-myc* nuclear sequences (Braun and Huddleston 2001; fig. 27.4). This relationship is supported by morphological characters (Mayr 2002) as well as by combined morphological and *RAG-2* data (fig. 27.7B) and by combined *c-myc* and *RAG-2* data (fig. 27.8). Even with the aegothelids removed from the caprimulgiforms there is presently little support for the monophyly of the remaining families. The available molecular data for *c-myc*, *RAG-2*, or combined *c-myc/RAG-2* (figs. 27.4, 27.7A, 27.8) do not unite them, nor do combined *c-myc* and *RAG-1* fragments (Johansson et al. 2001) or morphology (Mayr 2002). The relationships of owls to various

caprimulgiform taxa are also not supported by available sequence data (figs. 27.4, 27.7, 27.8; Johansson et al. 2001, Mindell et al. 1997); however, one subsequent DNA hybridization study has supported this hypothesis, in addition to linking owls, caprimulgiforms, and apodiforms (Bleiweiss et al. 1994). Preliminary morphological data also suggest a relationship (Livezey and Zusi 2001).

Phylogenetic Relationships among "Higher Land Birds": Cuculiformes, Coraciiformes, Trogoniformes, Coliiformes, and Piciformes

Few avian relationships are as interesting as those associated with the "higher land bird" question, and it is a problem with important implications for the overall topology of the neornithine tree. Historically, groups such as the piciforms, coraci-

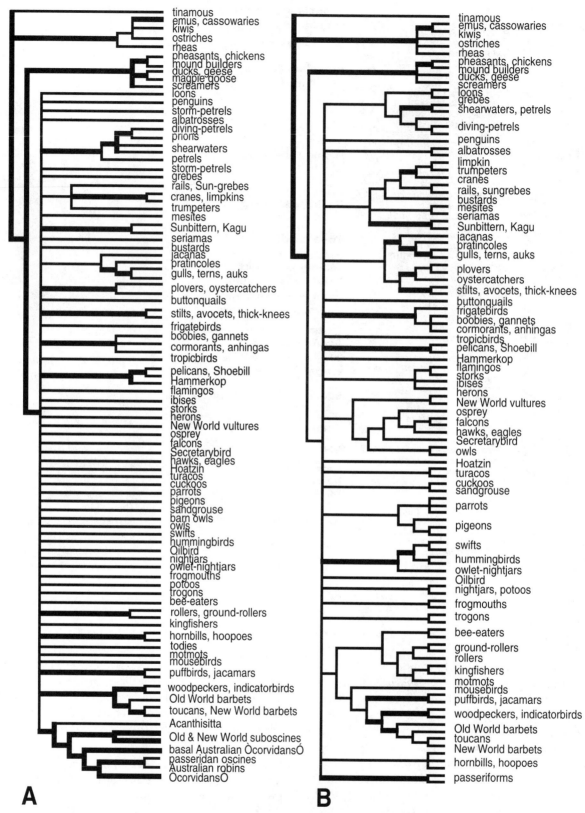

Figure 27.7. (A) A phylogenetic hypothesis for neoavian taxa using 1152 base pairs of the *RAG*-2 exon. (B) A phylogenetic tree based on 1152 base pairs of the *RAG*-2 exon and 166 morphological characters. Analyses are all unweighted parsimony. Thick branches have greater than 70% bootstrap support. Data from J. Cracraft, P. Schikler, J. Feinstein, P. Beresford, and G. J. Dyke (unpubl. obs.).

strong support from:

combined data and both data sets
combined data only
combined data and one separate data set

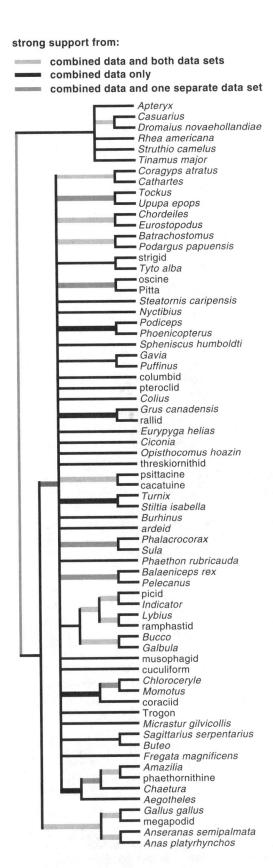

iforms, passeriforms, caprimulgiforms, and cuculiforms have been associated with one another in various classifications (e.g., Huxley 1867, Garrod 1874, Fürbringer 1888) and have been loosely called "higher land birds" (e.g., Olson 1985, Feduccia 1999, Johansson et al. 2001). Here we discuss the relationships within and among the coraciiform and piciform birds, their placement on the neornithine tree, and their relationships to the passeriforms.

Although the cuculiforms, coraciiforms, and piciforms have long been seen as "higher" neornithines and often closely related to passeriforms, this view was turned upside down by the DNA hybridization tree (Sibley and Ahlquist 1990), which postulated that all three groups were at the base of the neoavian tree (fig. 27.2). One of the two traditional groups of piciforms, Pici, was placed near the base of the neoavian tree adjacent to the turnicids, whereas the other, the jacamars and puffbirds (Galbulae), was placed as the sister group to a monophyletic "Coraciae," including traditional coraciiforms and trogons. The passeriforms were placed as the sister group to the entire waterbird assemblage but were not found to have any close relationship with either piciform or coraciiform taxa.

At present, none of these relationships can be confirmed or refuted. Available nuclear sequence data for *RAG-1* (Groth and Barrowclough 1999) as well as the *c-myc* and *RAG-2* data (figs. 27.4, 27.7) cannot resolve the base of Neoaves, indicating that the placement of these (or other) groups within neornithines remains an open question. Recent morphological and molecular studies, however, are identifying some well-supported clades within these groups. The two major clades of the piciforms, Pici and Galbulae, are each strongly monophyletic in all studies (see figs. 27.4, 27.7A,B, 27.8; Johansson et al. 2001), and evidence increasingly indicates that they are sister taxa. Some data, including *RAG-2* (fig. 27.7A) and fragments of *c-myc* and *RAG-1* (Johansson et al. 2001), cannot resolve this issue, but a monophyletic Piciformes is supported by morphology (Cracraft and Simpson 1981, Swierczewski and Raikow 1981, Raikow and Cracraft 1983, Mayr et al.

Figure 27.8. Phylogenetic hypothesis from combined *c-myc* and *RAG-2* data for 69 taxa, analyzed by unweighted parsimony. Branches with bootstrap support greater than 50% are shown. Thick branches have greater than 70% bootstrap support. To maximize the taxon overlap between data sets, equivalent species were combined, and this is reflected in the name given to the terminal node; for example, *Gallus gallus* was sequenced for both genes, but two different species were sequenced from *Aegotheles*, and species were sequenced from two different genera of megapodes. Data from J. Harshman, M. J. Braun, and J. Cracraft (unpubl. obs.).

2003), longer *c-myc* sequences (fig. 27.4), *RAG-2* + morphology (fig. 27.7B), combined *c-myc/RAG-2* data (fig. 27.8), and by other nuclear sequences (Johansson and Ericson 2003). Within Pici, it is now clear that the barbets are paraphyletic and that some or all of the New World taxa are more closely related to toucans (Burton 1984, Sibley and Ahlquist 1990, Lanyon and Hall 1994, Prum 1988, Barker and Lanyon 2000, Moyle 2004) than to other barbets; interrelationships within the barbet and toucan clade still need additional work.

DNA hybridization data were interpreted as supporting a monophyletic coraciiforms (Sibley and Ahlquist 1990). Although recent DNA sequences are insufficient to test coraciiform monophyly, they do show support for groups of families traditionally placed within coraciiforms. There is now congruent support, for example, for the monophyly of (1) hornbills + hoopoes/woodhoopoes (figs. 27.4, 27.7A,B, 27.8; Johansson et al. 2001), (2) motmots + todies (Johansson et al. 2001), and (3) kingfishers + motmots (figs. 27.4A, 27.7B, 27.8; Johansson et al. 2001), and support for (4) the kingfisher/motmot clade with the rollers (figs. 27.4, 27.7B, 27.8; Johansson et al. 2001).

Although they are clearly monophyletic (Hughes and Baker 1999), the relationships of the cuckoos are very uncertain, with no clear pattern across different studies. The distinctive Hoatzin (*Opisthocomus hoazin*) has been variously placed with galliforms (Cracraft 1981), cuculiforms (Sibley and Ahlquist 1990, Mindell et al. 1997), or turacos (Hughes and Baker 1999), yet there is no firm support in the *c-myc* (fig. 27.4), mitochondrial and *PEPCK* data (fig. 27.5), or in those from *RAG-2* and morphology (fig. 27.7B; see also Livezey and Zusi 2001) for any of these hypotheses. A relationship to galliforms at least can be rejected: hoatzins are clearly members of Neoaves, not Galloanserae (figs. 27.4, 27.5, 27.7A,B, 27.8; see also Sorenson et al. 2003).

Trogons and mousebirds are each so unique morphologically that they have been placed in their own order, but both have been allied to coraciiform and/or piciform birds by many authors (for reviews, see Sibley and Ahlquist 1990, Espinosa de los Monteros 2000). In recent years trogons have generally been associated with various coraciiforms on the strength of stapes morphology (Feduccia 1975), myology (Maurer and Raikow 1981), and osteology (Livezey and Zusi 2001). Mousebird relationships have been more difficult to ascertain, and no clear picture has emerged. In the mitochondrial-*PEPCK* data mousebirds group with parrots (fig. 27.5), whereas the *RAG-2* gene is uninformative. The study of Espinosa de los Monteros (2000) linked mousebirds with trogons and then that clade with parrots. The problem is that all these groups are old, divergent taxa with relatively little intrataxon diversity. Much more data will be needed to resolve their relationships.

Phylogenetic Relationships within the Perching Birds (Passeriformes)

The perching birds, order Passeriformes, comprise almost 60% of the extant species of birds. The monophyly of pas-

seriforms has long been accepted and is strongly supported by a variety of studies, including those using morphological or molecular data (Feduccia 1974, 1975, Raikow 1982, 1987; see also figs. 27.4, 27.5, 27.7, 27.8). Our current understanding of their basal relationships and biogeographic distributions strongly suggests that the group is old, with an origin probably more than 79 million years ago, well before the Cretaceous–Tertiary extinction 65 million years ago (e.g., Paton et al. 2002) and on a late-stage Gondwana (Cracraft 2001, Barker et al. 2002, Ericson et al. 2002). Recent molecular work using nuclear genes (Barker et al. 2002, Ericson et al. 2002) supports the hypothesis that the New Zealand wrens (Acanthisittidae) are the sister group to the remainder of the passerines, and that the latter clade can be divided into two sister lineages, the suboscines (Tyranni) and the oscines (Passeri). Resolving relationships within the suboscines and oscines has been complex, not only because of the huge diversity (about 1200 and 4600 species, respectively) but also because many of the traditional families are neither monophyletic nor related as depicted in Sibley and Ahlquist's (1990) tree. Nuclear gene sequences, however, are beginning to clarify phylogenetic patterns within this large group. The results presented here summarize some ongoing studies of the passerines, primarily using two nuclear genes (*RAG-1* and *RAG-2*; F. K. Barker, J. F. Feinstein, P. Schikler, A. Cibois, and J. Cracraft, unpubl. obs.) with dense taxon sampling, and represent the most comprehensive analysis of passeriform relationships to date (4126 aligned positions for 146 taxa).

The DNA hybridization data were interpreted by Sibley and Ahlquist (1990) as showing a division between suboscine and oscine passerines with the New Zealand wrens being the sister group to the remaining suboscines. Within the oscines, there were two sister clades, Corvida, which consisted of all Australian endemics and groups related to crows (the so-called "corvine assemblage"), and Passerida for all remaining taxa. The phylogenetic hypothesis shown in figure 27.9A, which is based on nuclear gene data (F. K. Barker, J. F. Feinstein, P. Schikler, A. Cibois, and J. Cracraft, unpubl. obs.), depicts a substantially different view of passeriform history. Thus, although the subdivision into suboscines and oscines is corroborated, the New Zealand wrens are the sister group of all other passerines. In addition, numerous taxa of the Australian "corvidans" are complexly paraphyletic relative to the passeridans and a core "Corvoidea."

The suboscine taxon sample is small, but these nuclear data are able to resolve a number of the major clades with strong support (fig. 27.9B). New World and Old World clades are sister groups (Irestedt et al. 2001, Barker et al. 2002). Within the Old World group, the data strongly support the pittas as being the sister group of the paraphyletic broadbills and the Malagasy asities (see also Prum 1993). The New World suboscines are divisible into two large clades. The first includes nearly 550 species of New World flycatchers, manakins, and cotingas; although this clade is strongly sup-

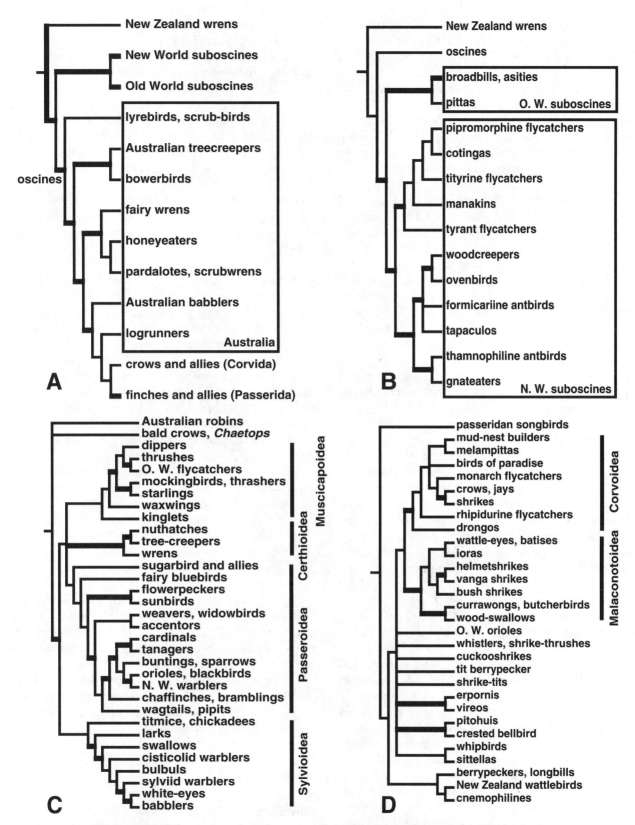

Figure 27.9. Phylogenetic analyses from an analysis of 146 passeriform taxa for 4126 base pairs of *RAG*-1 and *RAG*-2 exons using maximum parsimony. (A) Relationships among the basal lineages. (B) Relationships among the suboscine passeriforms. (C) Relationships among the passeridan songbirds. (D) Relationships among the basal oscines and corvidan songbirds. Data from F. K. Barker, J. F. Feinstein, P. Schikler, A. Cibois, and J. Cracraft (unpubl. obs.).

ported, relationships within the group are still uncertain (see also Johansson et al. 2002). The remaining 560 species of New World suboscines are split into the thamnophiline antbirds and their sister clade, the formicarinine antbirds and the ovenbirds and woodcreepers. The most thorough study of New World suboscine relationships to date is that of Irestedt et al. (2002), which examined more than 3000 base pairs of nuclear and mitochondrial sequences for 32 ingroup taxa of woodcreepers, ovenbirds, and antbirds; our results are congruent with those reported in their study.

As noted, the oscines, or songbirds, have been subdivided into two large assemblages, the Corvida and Passerida, based on inferences from DNA hybridization. This simple partition has been shown to be incorrect (Barker et al. 2002, Ericson et al. 2002), but we are now able to tell a much more interesting story because of a larger taxon sample. No fewer than five distinctive Australian "corvidan" clades are sequential sister groups to the core corvoid and passeridan clades (Barker et al. 2002; see also F. K. Barker, J. F. Feinstein, P. Schikler, A. Cibois, and J. Cracraft, unpubl. obs.): the lyrebirds (Menuridae), the bowerbirds and Australian treecreepers (Ptilonorhynchoidea), the diverse meliphagoid assemblage, the pomotostomine babblers, and the orthonychid logrunners (fig. 27.9A). This phylogenetic pattern firmly anchors the origin of the oscines in East Gondwana.

But the story of corvidan paraphyly is not yet exhausted. The passeridan clade has three basal clades (fig. 27.9C), one of which is the Australian robins (Eopsaltridae), included by DNA hybridization data within the corvidans. A second clade is the peculiar African genus *Picathartes*, the bald crows or rock-fowl, also placed toward the base of the passerines by hybridization data (see Sibley and Ahlquist 1990: 625–626), and its sister taxon, the rock-jumpers (*Chaetops*). Finally, there are the core passeridans (Ericson et al. 2000, Barker et al. 2002; see also F. K. Barker, J. F. Feinstein, P. Schikler, A. Cibois, and J. Cracraft, unpubl. obs.). It is not clear from the available data whether the *Picathartes* + *Chaetops* clade or the eopsaltrids is the sister group of the core passeridans, although present data suggest the robins are more closely related.

The basal relationships of the core passeridans are still unclear. There are four moderately well-defined clades within the group (fig. 27.9C; see also Ericson and Johansson 2003). The first, Sylvioidea, includes groups such as the titmice and chickadees, larks, bulbuls, Old World warblers, white-eyes, babblers, and swallows. The second, here termed Certhioidea, consists of the wrens, nuthatches, and treecreepers. The third is a very large group, Passeroidea, that includes various Old World taxa basally—the fairy bluebirds, sunbirds, flowerpeckers, sparrows, wagtails, and pipits—and the huge (almost 1000 species) so-called nine-primaried oscine assemblage (Fringillidae of Monroe and Sibley 1993), most of which are New World (Emberizinae: buntings, wood warblers, tanagers, cardinals, and the orioles and blackbirds; for recent discussions of relationships, see Groth 1998, Klicka et al.

2000, Lovette and Bermingham 2002, Yuri and Mindell 2002). The last group of core passeridans is Muscicapoidea, which encompasses the kinglets, waxwings, starlings, thrashers and mockingbirds, and the large thrush and Old World flycatcher clade of some 450 species.

With the elimination of the early "corvidan" clades discussed above (fig. 27.9A), the remainder of Sibley and Ahlquist's "Corvida" do appear to form a monophyletic assemblage, although it is not well supported at this time, and we restrict the name "Corvida" to this clade (fig. 27.9D). Although relationships among family-level taxa within this complex cannot be completely resolved with *RAG-1* and *RAG-2* sequences, these data do identify several well-defined clades, and they partition relationships more satisfactorily than previous work.

Two of the corvidan clades are well supported. The first we term here Corvoidea, which include the crows and jays (Corvidae) and their sister group, the true shrikes (Laniidae), the monarch and rhipidurine flycatchers, drongos, mud-nest builders (Struthidea, *Corcorax*), the two species of *Melampitta*, and the birds of paradise (Paradisaeidae). The second well-supported lineage of the corvidans we term Malaconotoidea. This "shrike-like" assemblage is comprised of the African bush-shrikes (Malaconotidae), the helmet shrikes (*Prionops*), *Batis*, the Asian ioras (Aegithinidae), and the vanga shrikes (Vangidae) of Madagascar. Also included in this clade are the wood-swallows (Artamidae) and their sister group the Australian magpies and currawongs (Cracticidae).

All other corvidans appear to be basal to the corvoids and malaconotoids but are, on present evidence, unresolved relative to these two clades. Most of these groups, including the pachycephalids, oriolids, campephagids, daphoenosittids, falcunculids, and other assorted genera are mostly Australasian in distribution, and presumably in origin. Also included in this melange are the vireos and their Asian sister group, *Erpornis zantholeuca*.

Outside of all these corvidan groups is a clade comprising some ancient corvidans that appear to be related: the New Zealand wattlebirds (Callaeatidae), the cnemophilines (formally placed in the birds of paradise), and the berrypeckers (Melanocharitidae). The basal position of these groups relative to the remaining Corvida provides persuasive evidence that the group as a whole had its origin in Australia (and perhaps adjacent Antarctica), further tying the origins of the oscine radiation to this landmass.

Discussion

Where We Are

To judge from the large numbers of papers reviewed above, research on the higher level relationships of birds has made significant progress over the last decade, yet it is obvious from the results of these studies that compelling evidence for relationships among most major clades of Neoaves is still lacking. Nevertheless, a function of this chapter is to serve as a benchmark of our current understanding of avian relationships, and one way expressing this progress is to propose a summary hypothesis that attempts to reflect the improvements in our knowledge of avian relationships, even though the underlying evidence may be imperfect. Different investigators, including the authors of this chapter, will disagree about what constitutes sufficient evidence for supporting the monophyly of a clade, and most would no doubt prefer to see a tree that is based on all avian higher taxa and a very large data set of molecular and morphological characters numbering in the tens of thousands. That ideal is 5–10 years away, however, yet it is still useful to examine how far have we come over the last decade.

Figure 27.10 depicts a summary phylogenetic hypothesis for the avian higher taxa. It represents an estimate of avian history at this point in time and is admittedly speculative in a number of places that we note below; it represents, moreover, a compromise among the authors. We therefore have no illusions that all of these relationships will stand the test of time and evidence, but a number will. The thick lines are meant to identify clades in which relatively strong evidence for their monophyly has been discovered in one or more individual studies. The thin lines depict clades that have been recovered in various studies, even though the evidence for these individual hypotheses may be weak. Congruence across studies suggests that with more data, many of these clades will gain increased support.

As already noted, the base of the neornithine tree is no longer particularly controversial, with palaeognaths and then Galloanserae being successive sister groups to Neoaves. Relationships within ratites are unsettled, however, because of conflict among the molecular data and with the morphological evidence.

Neoavian relationships, on the other hand, are decidedly uncertain, although new information becomes available with each new study. The base of the neoavian tree is a complete unknown, but within Neoaves evidence for relationships among a number of major groups is emerging. There is a suggestion that many of the traditional "waterbird" groups are related, although a monophyletic assemblage that includes all "waterbird" taxa itself is unlikely. Thus, some "waterbird" taxa are definitely related, others probably so, but other nonwaterbird taxa will almost certainly be found to be embedded within waterbirds. It now seems clear that some traditional groups such as Pelecaniformes and Ciconiiformes are not monophyletic, but many of their constituent taxa are related. Thus there is now evidence for a shoebill + pelican + hammerkop clade and for an anhinga + cormorant + gannet + (more marginally) frigatebird clade, and these two clades are probably related to each other, along with ibises, herons, and storks. Tropicbirds (phaethontids) are a real puzzle as this old, long-branch taxon is quite unstable on all trees.

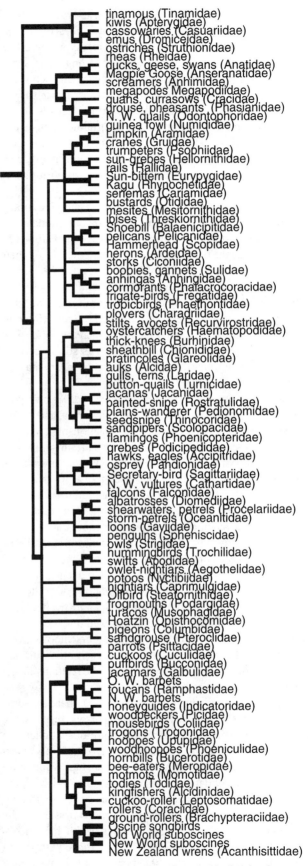

Figure 27.10. Summary hypothesis for avian higher level relationships (see discussion in text).

There is a core group of gruiform taxa with well supported relationships, including rails + sungrebes, on the one hand, and cranes + limpkins + trumpeters, on the other. Moreover, the kagu and sunbittern are strongly supported sister taxa. Aside from some morphological character data (e.g., Livezey 1998), there is little current evidence to support monophyly of traditional Gruiformes. This is an old group, with basal divergences almost certainly in the Cretaceous (Cracraft 2001) that cannot be resolved given the character data currently available; yet, there is no firm evidence that any of these groups is related to a nongruiform taxon, so we retain the traditional order.

Ongoing work in various labs is confirming the monophyly of the charadriiforms, including the buttonquails, often placed in gruiforms. Current sequence data (Paton et al. 2003) indicate the relationships shown in figure 27.10. There is also a suggestion in the molecular data presented earlier that charadriiforms are associated with flamingos + grebes, and possibly with some or all of the falconiforms. The latter group consists of three well defined clades (falcons, cathartids, and accipitrids + osprey + secretary bird), but whether these are related to each other is still uncertain. Morphology indicates that they are, but molecular data cannot yet confirm or deny this.

Relationships among the "higher land birds" remain controversial in many cases. Swifts, hummingbirds, and owlet-nightjars are monophyletic but their relationships to other taxa traditionally called caprimulgiforms are unsupported; as with gruiforms, we have no clear evidence that any of them are more closely related to other taxa, and so we retain the group. Whether owls cluster with these families is also uncertain. Again, all of these taxa are very old groups and resolution of their relationships will require more data.

The three "orders" Piciformes, Coraciiformes, and Passeriformes may or may not be related to one another, but in many studies subgroups of them are clustered together. More and more data sets are showing a monophyletic piciforms. Passeriforms are strongly monophyletic, and relationships among their basal clades are becoming well understood (see discussion above). Finally, the traditional coraciiforms group into two clades whose relationships to each other are neither supported nor refuted by our data. The relationships of mousebirds and trogons are also still obscure.

In contrast to many of the above groups, there are some highly distinctive taxa such as turacos, parrots, pigeons, the hoatzin, and cuckoos that have been notoriously difficult to associate with other groups using both molecular and morphological data. Deciphering their relationships will require larger amounts of data than are currently available.

Despite the appearance of substantial structure, the hypothesis of figure 27.10 could be interpreted pessimistically by examination of those clades subtended by thin branches—indicating insufficient support—versus those with thick-branched clades we judge to be either moderately or strongly supported. Seen in this way, the tree is mostly a polytomy and suggests we know very little about avian relationships. Viewed more optimistically, however, the tree is a working hypothesis that suggests progress is being made. Critically, this representation of our state of knowledge contradicts the false notion that the broad picture of avian phylogenetics has been drawn, and only the details remain to be filled (e.g., Mooers and Cotgreave 1994). Given the state of current activity in many laboratories around the world, we predict that in little more than five years a similar figure, whatever its configuration, will have a substantially larger proportion of well-supported clades.

The Future

These are exciting and productive times for avian systematists. We are witnessing the growth of molecular databases, containing sequences from homologous genes across most avian taxa. As recently as 10 years ago the availability of such comprehensive, comparative, discrete character data sets was little more than a dream. Within the next several years large data sets for both molecular and morphological data will be published that span all the major clades of nonpasseriform birds. At the same time, avian systematics is becoming increasingly collaborative with groups of researchers pooling resources and publishing together. These collaborations involve both molecular and morphological data and extend back across time through the incorporation of fossils.

All of these data will soon be publicly available on the Internet as a result of these collaborations, and these data should greatly accelerate avian systematic research. Discrete-character data sets lend themselves to continual growth and addition in a manner entirely absent from the early comprehensive work based on DNA hybridization distances (Sibley and Ahlquist 1990). These data sets will variously confirm, challenge, or overturn earlier hypotheses of avian phylogeny, and this may be expected to continue as both character and taxon sampling increase. We view the continued collection of comparative data as imperative not just for avian systematics, but for elaborating the insight into evolutionary history and processes at multiple hierarchical levels that only phylogeny can provide.

The Challenge

Just how difficult will it be to build a comprehensive avian Tree of Life (ATOL)? Several observations suggest it will be extremely so. First, there are about 20,000 nodes on the extant avian Tree of Life. Fossil taxa only add to that number. Then, there is the challenge presented by the history of birds itself. It is now evident that there have been many episodes of rapid radiation across the neoavian tree, perhaps involving thousands of nodes, and resolving these will require unprecedented access to specimen material (including anatomical preparations and fresh tissues) as well as large character sampling to establish relationships. Gone are the days when a single person or laboratory might hope to solve the problem of avian relationships. The problem is too difficult and complex for single laboratories in which time and money are limited. The scientific challenge presented by the avian Tree of Life will call for large taxon and character sampling, goals best achieved by a communitywide effort.

There are also conceptual roadblocks. One is the problem of uncertain knowledge. More taxa and characters may not guarantee a "satisfying" answer, by which we mean having resolution of nodes with sufficiently strong branch support that additional data will merely confirm what has already been found. The issue is that more taxa guarantee (some) uncertainty. More taxa are good, of course, but they also means more character data will likely be required to attain strong support for any particular node. Measuring phylogenetic understanding on very large trees such as the avian Tree of Life will also be a complex challenge. Measures of support are ambiguous in their own right, and whatever answer we get depends on the taxon and character sampling—that is, on the available data. Thus, what are the boundaries of a study? How will we know when to stop (because it has been determined we "know" relationships) and move on to an unresolved part of the tree? This is a nontrivial problem, but as we erect a scaffold that identifies strongly supported monophyletic groups, perhaps that will make it easier to circumscribe studies and resolve the tree more finely.

Another conceptual roadblock is the problem of investigator tenacity. It should be straightforward to build the scaffold of the avian Tree of Life. Systematists are doing that now. There will be—and already are—lots of trees that are moderately resolved but still have little satisfactory branch support (remember that the DNA hybridization tree was nearly "fully" resolved). So how much do we, the investigators, really want to know relationships? If the object is to publish more papers, then as more and more taxa are added, and if character sampling does not also increase, more and more nodes are likely to be supported rather poorly, especially across those parts of the tree representing rapid radiations (short internodes). Resolving these nodes with some measure of confidence will require substantial amounts of data (much more than is currently collected in typical studies). In the near future, this may not be an issue as technical innovations allow systematists to gather more data more rapidly. However, many investigators will not necessarily have easy access to these technologies, and it is already becoming apparent that being able to collect large volumes of data (genomes) does not necessarily mean that the data themselves are going to be phylogenetically useful for the problem at hand.

Although many phylogenetic problems in birds, at all taxonomic levels, will be quite difficult to resolve, we must be resolute. Resolving relationships is crucial for answering numerous questions in evolutionary biology, and to the extent that these questions are worth pursuing we should not settle for not knowing phylogeny. One result emerging from the studies discussed here illustrates this point. Evidence now indicates owlet-nightjars are the sister group of swifts and hummingbirds. Depending on the sister group of this clade, it implies either that adaptation to nocturnal lifestyle arose multiple times in aegothelids and other birds, or that nocturnal habits are primitive and swifts and hummingbirds are secondarily diurnal. Phylogeny thus provides important insight into understanding avian diversification.

Finally, our perspectives on avian evolution will not be—should not be—built on one kind of data. Tree topologies should reflect the most comprehensive description of character evolution over time, which means that all forms of character information—genetic, morphological, behavioral, and so forth—should be incorporated into analyses. They may not only contribute to phylogenetic resolution in their own right, but will give us a richer picture of the history of avian evolution.

Acknowledgments

F.K.B., G.J.D., S.S., P.B., and A.C. all received support from the AMNH F. M. Chapman Fund. Much of the research presented in this chapter is supported by the AMNH Monell Molecular Laboratory and Lewis B. and Dorothy Cullman Program for Molecular Systematics Studies. Work on the *c-myc* gene was aided by the able assistance of Chris Huddleston and a Smithsonian postdoctoral fellowship to J.H. M.S. acknowledges the help of Elen Oneal and support from the National Science Foundation. Work by J.G.-M., D.P.M., M.D.S., and T.Y. was supported by NSF grants DEB-9762427 and DBI-9974525.

Literature Cited

Ansari, H. A., N. Takagi, and M. Sasaki. 1988. Morphological differentiation of sex chromosomes in three species of ratite birds. Cytogenet. Cell Genet. 47:185–188.

Avise, J. C., W. S. Nelson, and C. G. Sibley. 1994. DNA-sequence support for a close phylogenetic relationship between some storks and New-World vultures. Proc. Natl. Acad. Sci. USA 91:5173–5177.

Barker, F. K., G. F. Barrowclough, and J. G. Groth. 2002. A phylogenetic hypothesis for passerine birds: taxonomic and biogeographic implications of an analysis of nuclear DNA sequence data. Proc. R. Soc. Lond. B 269:295–308.

Barker, F. K., and S. M. Lanyon. 2000. The impact of parsimony weighting schemes on inferred relationships among toucans and neotropical barbets (Aves: Piciformes). Mol. Phylogenet. Evol. 15:215–234.

Bleiweiss, R., J. A. W. Kirsch, and F.-J. Lapointe. 1994. DNA-DNA hybridization-based phylogeny for "higher" nonpasserines: reevaluating a key portion of the avian family tree. Mol. Phylogenet Evol. 3:248–255.

Braun, M. J., and C. J. Huddleston. 2001. Molecular phylogenetics of caprimulgiform nightbirds. P. 51 *in* Abstracts of the 119th Stated Meeting, American Ornithologists' Union, Seattle, WA, 16–19 August 2001.

Braun, E. L., and R. T. Kimball. 2002. Examining basal avian divergences with mitochondrial sequences: model complexity, taxon sampling, and sequence length. Syst. Biol. 51:614–625.

Burton, P. J. K. 1984. Anatomy and evolution of the feeding apparatus in the avian orders Coraciiformes and Piciformes. Bull. Br. Mus. Nat. Hist. Zool. 47(6):331–443.

Chiappe, L. M. 1995. The first 85 million years of avian evolution. Nature 378:349–355.

Chiappe, L. M. 2001. Phylogenetic relationships among basal birds. Pp. 125–139 in New perspectives on the origin and early evolution of birds (J. Gauthier and L. F. Gall, eds.). Peabody Museum of Natural History, Yale University, New Haven, CT.

Chiappe, L. M., and G. J. Dyke. 2002. The Mesozoic radiation of birds. Annu. Rev. Ecol. Syst. 33:91–124.

Chiappe, L. M., J. Shu'an, J. Qiang, and M. A. Norell. 1999. Anatomy and systematics of the Confuciusornithidae (Theropoda: Aves) from the late Mesozoic of northeastern China. Bull. Am. Mus. Nat. Hist. 242:1–89.

Chubb, A. L. 2004. New nuclear evidence for the oldest divergence aong neognath birds: the phylogenetic utility of ZENK (I). Mol. Phylogen. Evol. 30:140–151.

Cooper, A., C. Lalueza-Fox, S. Anderson, A. Rambaut, J. Austin, and R. Ward. 2001. Complete mitochondrial genome sequences of two extinct moas clarify ratite evolution. Nature 409:704–707.

Cooper, A., and D. Penny. 1997. Mass survival of birds across the Cretaceous-Tertiary boundary: molecular evidence. Science 275:1109–1113.

Cottam, P. A. 1957. The pelecaniform characters of the skeleton of the shoebill stork *Balaeniceps rex*. Bull. Br. Mus. Nat. Hist. 5:51–72.

Cracraft, J. 1972. The relationships of the higher taxa of birds: problems in phylogenetic reasoning. Condor 74:379–392.

Cracraft, J. 1974. Phylogeny and evolution of the ratite birds. Ibis 116:494–521.

Cracraft, J. 1977. John Ostrom's studies on *Archaeopteryx*, the origin of birds, and the evolution of avian flight. Wilson Bull. 89:488–492.

Cracraft, J. 1981. Toward a phylogenetic classification of the Recent birds of the world (Class Aves). Auk 98:681–714.

Cracraft, J. 1986. The origin and early diversification of birds. Paleobiology 12:383–399.

Cracraft, J. 1987. DNA hybridization and avian phylogenetics. Evol. Biol. 21:47–96.

Cracraft, J. 1988. The major clades of birds. Pp. 339–361 in The phylogeny and classification of the tetrapods, Vol. 1: Amphibians, reptiles, birds (M. J. Benton, ed.). Clarendon Press, Oxford.

Cracraft, J. 2001. Avian evolution, Gondwana biogeography and the Cretaceous-Tertiary mass extinction event. Proc. R. Soc. Lond. B 268:459–469.

Cracraft, J., and J. Clarke. 2001. The basal clades of modern birds. Pp. 143–156 in New perspectives on the origin and early evolution of birds (J. Gauthier and L. F. Gall, eds.). Peabody Museum of Natural History, Yale University, New Haven, CT.

Darwin, C. R. 1859. On the origin of species. John Murray, London.

Dimcheff, D. E., S. V. Drovetski, M. Krishnan, and D. P. Mindell. 2000. Cospeciation and horizontal transmission of avian sarcoma and leukosis virus *gag* genes in galliform birds. J. Virol. 74:3984–3995.

Dimcheff, D. E., S. V. Drovetski, and D. P. Mindell. 2002. Molecular evolution and systematics of tetraoninae and other Galliformes using mitochondrial 12S and ND2 genes. Mol. Phylogenet. Evol. 24:203–215.

Donne-Goussé, C., V. Laudet, and C. Hanni. 2002. A molecular phylogeny of Anseriformes based on mitochondrial DNA analysis. Mol. Phylogenet. Evol. 23:339–356.

Dyke, G. J. 2001. The evolution of birds in the early Tertiary: systematics and patterns of diversification. Geol. Jour. 36:305–315.

Dyke, G. J., B. E. Gulas, and T. M. Crowe. 2003. The suprageneric relationships of galliform birds (Aves, Galliformes): a cladistic analysis of morphological characters. Zool. J. Linn. Soc. 137:227–244.

Edwards, S. V., B. Fertil, A. Giron, and P. J. Deschavanne. 2002. A genomic schism in birds revealed by phylogenetic analysis of DNA strings. Syst. Biol. 51:599–613.

Ericson, P. G. P. 1996. The skeletal evidence for a sister-group relationship of anseriform and galliform birds—a critical evaluation. J. Avian Biol. 27:195–202.

Ericson, P. G. P. 1997. Systematic relationships of the Palaeogene family Presbyornithidae (Aves: Anseriformes). Zool. J. Linn. Soc. 121:429–483.

Ericson, P. G. P., L. Christidis, A. Cooper, M. Irestedt, J. Jackson, U. S. Johansson, and J. A. Norman. 2002. A Gondwanan origin of passerine birds supported by DNA sequences of the endemic New Zealand wrens. Proc. R. Soc. Lond. B 269:235–241.

Ericson, P. G. P., and U. S. Johansson. 2003. Phylogeny of Passerida (Aves: Passeriformes) based on nuclear and mitochondrial sequence data. Mol. Phylog. Evol. 29:126–138.

Ericson, P. G. P., U. S. Johansson, and T. J. Parsons. 2000. Major divisions in oscines revealed by insertions in the nuclear gene c-*myc*: a novel gene in avian phylogenetics. Auk 117:1069–1078.

Ericson, P. G. P., T. J. Parsons, U. S. Johansson. 2001. Morphological and molecular support for nonmonophyly of the Galloanserae. Pp. 157–168 in New perspectives on the origin and early evolution of birds (J. Gauthier and L. F. Gall, eds.). Peabody Museum of Natural History, Yale University, New Haven, CT.

Espinosa de los Monteros, A. 2000. Higher-level phylogeny of Trogoniformes. Mol. Phylogenet. Evol. 14:20–34.

Feduccia, A. 1974. Morphology of the bony stapes in New and Old World suboscines: new evidence for common ancestry. Auk 91:427–429.

Feduccia, A. 1975. Morphology of the bony stapes (columella) in the Passeriformes and related groups: evolutionary implications. Univ. Kans. Mus. Nat. Hist. Misc. Publ. 63:1–34.

Feduccia, A. 1995. Explosive evolution in Tertiary birds and mammals. Science 267:637–638.

Feduccia, A. 1999. The origin and evolution of birds. 2nd ed. Yale University Press, New Haven, CT.

Feduccia, A. 2002. Birds are dinosaurs: simple answer to a complex problem. Auk 119:1187–1201.

Feduccia, A. 2003. 'Big bang' for Tertiary birds? Trends Ecol. Evol. 18:172–176.

Fürbringer, M. 1888. Untersuchungen zur Morphologie und Systematik der Vögel. 2 vols. von Holkema, Amsterdam.

García-Moreno, J., and D. P. Mindell. 2000. Using homologous genes on opposite sex chromosomes (gametologs) in phylogenetic analysis: a case study with avian CHD. Mol. Biol. Evol. 17:1826–1832.

García-Moreno, J., M. D. Sorenson, and D. P. Mindell. 2003. Congruent avian phylogenies inferred from mitochondrial and nuclear DNA sequences. J. Mol. Evol. 57:27–37.

Garrod, A. H. 1874. On certain muscles of birds and their value in classification. Proc. Zool. Soc. Lond. 1874:339–348.

Gauthier, J. A.1986. Saurischian monophyly and the origin of birds. Pp. 1–55 in The origin of birds and the evolution of flight (K. Padian, ed.). Memoirs of the California Academy of Sciences 8. San Francisco, CA.

Groth, J. G. 1998. Molecular phylogenetics of finches and sparrows: consequences of character state removal in cytochrome *b* sequences. Mol. Phylogenet. Evol. 10:377–390.

Groth, J. G., and G. F. Barrowclough. 1999. Basal divergences in birds and the phylogenetic utility of the nuclear RAG-1 gene. Mol. Phylogenet. Evol. 12:115–123.

Hackett, S. J., C. S. Griffiths, J. M. Bates, and N. K. Klein. 1995. A commentary on the use of sequence data for phylogeny reconstruction. Mol. Phylogenet. Evol. 4:350–356.

Haddrath, O., and A. J. Baker. 2001. Complete mitochondrial DNA genome sequences of extinct birds: ratite phylo-genetics and the vicariance biogeography hypothesis. Proc. R. Soc. Lond. 268:939–945.

Haeckel, E. 1866. Generelle Morphologie der Organismen: allgemeine Grundzüge der organischen Formen-Wissenschaft, mechanisch begründet durch die von Charles Darwin reformirte Descendenz-Theorie. G. Reimer, Berlin.

Härlid, A., and U. Arnason. 1999. Analyses of mitochondrial DNA nest ratite birds within the Neognathae: supporting a neotenous origin of ratite morphological characters. Proc. R. Soc. Lond. 266:305–309.

Harshman, J. 1994. Reweaving the Tapestry: what can we learn from Sibley and Ahlquist (1990)? Auk 111:377–388.

Hedges, S. B., Parker, P. H., Sibley, C. G., and S. Kumar. 1996. Continental breakup and the ordinal diversification of birds and mammals. Nature 381:226–229.

Hedges, S. B., and C. G. Sibley. 1994. Molecules vs. morphology in avian evolution: the case of the "pelecaniform" birds. Proc. Natl. Acad. Sci. USA 91:9861–9865.

Helbig, A. J., and I. Seibold. 1996. Are storks and New World vultures paraphyletic? Mol. Phylogenet. Evol. 6:315–319.

Hennig, W. 1966. Phylogenetic systematics. University of Illinois Press, Urbana.

Holtz, T. R., Jr. 1994. The phylogenetic position of the Coelurosauria (Dinosauria: Theropoda). J. Paleontol. 68(5):1100–1117.

Holtz, T. R., Jr. 2001. Arctometatarsalia revisited: the problem of homoplasy in reconstructing theropod phylogeny. Pp. 99–122 in New perspectives on the origin and early evolution of birds (J. Gauthier and L. F. Gall, eds.). Peabody Museum of Natural History, Yale University, New Haven, CT.

Houde, P. 1987. Critical evaluation of DNA hybridization studies in avian systematics. Auk 31:17–32.

Hughes, J. M., and A. J. Baker. 1999. Phylogenetic relationships of the enigmatic hoatzin (*Opisthocomus hoazin*) resolved using mitochondrial and nuclear gene sequences. Mol. Biol. Evol. 16:1300–1307.

Huxley, T. H. 1867. On the classification of birds; and on the taxonomic value of the modifications of certain of the cranial bones observable in the class. Proc. Zool. Soc. Lond. 1867:415–472.

Huxley, T. H. 1868. On the animals which are most nearly intermediate between birds and reptiles. Geol. Mag. 5:357–365.

Irestedt, M., J. Fjeldsa, U. S. Johansson, and P. G. P. Ericson. 2002. Systematic relationships and biogeography of the tracheophone suboscines (Aves: Passeriformes). Mol. Phylogenet. Evol. 23:499–512.

Irestedt, M., U. S. Johansson, T. J. Parsons, and P. G. P. Ericson. 2001. Phylogeny of major lineages of suboscines (Passeriformes) analysed by nuclear DNA sequence data. J. Avian Biol. 32:15–25.

Johansson, U. S., and P. G. P. Ericson. 2003. Molecular support for a sister group relationship between Pici and Galbulae (Piciformes *sensu* Wetmore 1960). J. Avian Biol. 34:185–197.

Johansson, U. S., M. Irestedt, T. J. Parsons, and P. G. P. Ericson. 2002. Basal phylogeny of the Tyrannoidea based on comparisons of cytochrome *b* and exons of nuclear *c-myc* and RAG-1 genes. Auk 119:984–995.

Johansson, U. S., T. J. Parsons, M. Irestedt, and P. G. P. Ericson. 2001. Clades within the 'higher land birds', evaluated by nuclear DNA sequences. J. Zool. Syst. Evol. Res. 39:37–51.

Johnson, K. P. 2001. Taxon sampling and the phylogenetic position of Passeriformes: evidence from 916 avian cytochrome *b* sequences. Syst. Zool. 50:128–136.

Jollie, M. 1976–1977. A contribution to the morphology and phylogeny of the Falconiformes. Evol. Theory 1:285–298, 2:115–300, 3:1–141.

Klicka, J., K. P. Johnson, and S. M. Lanyon. 2000. New world nine-primaried oscine relationships: constructing a mitochondrial DNA framework. Auk 117:321–336.

Kluge, A. G., and J. S. Farris. 1969. Quantitative phyletics and the evolution of anurans. Syst. Zool. 18:1–32.

Kurochkin, E. N., G. J. Dyke, and A. A. Karhu. 2002. A new presbyornithid bird (Aves, Anseriformes) from the Late Cretaceous of southern Mongolia. Am. Mus. Nov. 3386:1–17.

Lanyon, S. M. 1992. Review of Sibley and Ahlquist 1990. Condor 94:304–307.

Lanyon, S. M., and J. G. Hall. 1994. Reexamination of barbet monophyly using mitochondrial-DNA sequence data. Auk 111:389–397.

Lee, K., J. Feinstein, and J. Cracraft. 1997. Phylogenetic relationships of the ratite birds: resolving conflicts between molecular and morphological data sets. Pp. 173–211 in Avian molecular evolution and systematics (D. P. Mindell, ed.). Academic Press, New York.

Ligon, J. D. 1967. Relationships of the cathartid vultures. Occ. Pap. Mus. Zool. Univ. Mich. 651:1–26.

Livezey, B. C. 1986 A phylogenetic analysis of recent anseriform genera using morphological characters. Auk 103:737–754.

Livezey, B. C. 1997a. A phylogenetic analysis of basal Anseriformes, the fossil *Presbyornis*, and the interordinal relationships of waterfowl. Zool. J. Linn. Soc. 121:361–428.

Livezey, B. L. 1997b. A phylogenetic classification of waterfowl (Aves: Anseriformes), including selected fossil species. Ann. Carnegie Mus. 66:457–496.

Livezey, B. C. 1998. A phylogenetic analysis of the Gruiformes (Aves) based on morphological characters, with an emphasis on rails (Rallidae). Philos. Trans. R. Soc. Lond. 353:2077–2151.

Livezey, B. C., and R. L. Zusi. 2001. Higher-order phylogenetics of modern Aves based on comparative anatomy. Neth. J. Zool. 51:179–205.

Lovette, I. J., and E. Bermingham. 2002. What is a wood warbler? Molecular characterization of a monophyletic Parulidae. Auk 119:695–714.

Macleay, W. S. 1819–1821. Horae entomologicae: or essays on the annulose animals. S. Bagster, London.

Madsen, C. S., K. P. McHugh, and S. R. D. Kloet. 1998. A partial classification of waterfowl (Anatidae) based on single-copy DNA. Auk 105:452–459.

Martin, L. D. 1983. The origin of birds and of avian flight. Curr. Ornithol. 1:106–129.

Maurer, D. R., and R. J. Raikow. 1981. Appendicular myology, phylogeny and classification of the avian order Coraciiformes (including Trogoniformes). Ann. Carnegie Mus. 50:417–434.

Mayr, E. 1942. Systematics and the origin of species: from the viewpoint of a zoologist. Columbia University Press, New York.

Mayr, E. 1969. Principles of systematic zoology. McGraw Hill, New York.

Mayr, E., and D. Amadon. 1951. A classification of recent birds. Am. Mus. Nov. 1946:453–473.

Mayr, G. 2002. Osteological evidence for paraphyly of the avian order Caprimulgiformes (nightjar and allies). J. Ornithol. 143:82–97.

Mayr, G., and J. Clarke. 2003. The deep divergence of modern birds: a phylogenetic analysis of morphological characters. Cladistics 19:527–553.

Mayr, G., A. Manegold, and U. S. Johansson. 2003. Monophyletic groups within "higher land birds"—comparison of molecular and morphological data. Z. Zool. Syst. Evol. Res. 41:233–248.

Meise, W. 1963. Verhalten der straussartigen Vögel und Monophylie der Ratitae. Pp. 115–125 *in* Proceedings of the 13th International Ornithological Congress (C. G. Sibley, ed.). American Ornithologists' Union, Baton Rouge, LA.

Mindell, D., ed. 1997. Avian molecular evolution and systematics. Academic Press, San Diego.

Mindell, D. P. 1992. DNA-DNA hybridization and avian phylogeny. Syst. Biol. 41:126–134.

Mindell, D. P., M. D. Sorenson, D. E. Dimcheff, M. Hasegawa, J. C. Ast, and T. Yuri. 1999. Interordinal relationships of birds and other reptiles based on whole mitochondrial genomes. Syst. Biol. 48:138–152.

Mindell, D. P., M. D. Sorenson, C. J. Huddleston, H. C. Miranda, Jr., A. Knight, S. J. Sawchuk, and T. Yuri. 1997. Phylogenetic relationships among and within select avian orders based on mitochondrial DNA. Pp. 213–247 *in* Avian

molecular evolution and systematics (D. P. Mindell, ed.). Academic Press, San Diego.

Monroe, B. L., and C. G. Sibley. 1993. A world checklist of birds. Yale University Press, New Heaven, CT.

Mooers, A. O., and P. Cotgreave. 1994. Sibley and Ahlquist's tapestry dusted off. Trends Ecol. Evol. 9:458–459.

Moyle, R. G. 2004. Phylogenetics of barbets (Aves: Piciformes) based on nuclear and mitochondrial sequence data. Mol. Phylogen. Evol. 30:187–200.

Norell, M. A., J. M. Clark, and P. J. Makovicky. 2001. Phylogenetic relationships among coelurosaurian theropods. Pp. 49–67 *in* New perspectives on the origin and early evolution of birds (J. Gauthier and L. F. Gall, eds.). Peabody Museum of Natural History, Yale University, New Haven, CT.

Ogawa, A., K. Murata, and S. Mizuno. 1998. The location of Z- and W-linked marker genes and sequence on the homomorphic sex chromosomes of the ostrich and the emu. Proc. Natl. Acad. Sci. USA 95:4415–4418.

O'Hara, R. J. 1988. Diagrammatic classifications of birds, 1819–1901: views of the natural system in 19th-century British ornithology. Pp. 2746–2759 *in* Acta XIX Congressus Internationalis Ornithologici (H. Ouellet, ed.). National Museum of Natural Sciences, Ottawa.

Olson, S. L. 1985. The fossil records of birds. Pp. 79–238 *in* Avian biology (D. S. Farner, J. King, and K. C. Parkes, eds.), vol. 8. Academic Press, New York.

Olson, S. L. 2002. Review of "New Perspectives on the Origin and Early Evolution of Birds. Proceedings of the International Symposium in Honor of John H. Ostrom." Auk 119:1202–1205.

Olson, S. L., and A. Feduccia. 1980a. *Presbyornis* and the origin of the Anseriformes (Aves: Charadriomorphae). Smithson. Contrib. Zool. 323:1–24.

Olson, S. L., and A. Feduccia. 1980b. Relationships and evolution of flamingos (Aves: Phoenicopteridae). Smithson. Contrib. Zool. 316:1–73.

Ostrom, J. H. 1976. Archaeopteryx and the origin of birds. Biol. J. Linn. Soc. 8:91–182.

Padian, K., and L. M. Chiappe. 1998. The origin and early evolution of birds. Biol. Rev. 73:1–42.

Paton, T. A., A. J. Baker, J. G. Groth, and G. F. Barrowclough. 2003. *RAG*-1 sequences resolve phylogenetic relationships within charadriiform birds. Mol. Phylogen. Evol. 29:268–278.

Paton, T. A., O. Haddrath, and A. J. Baker. 2002. Complete mitochondrial DNA genome sequences show that modern birds are not descended from transitional shorebirds. Proc. R. Soc. Lond. B 269:839–846.

Payne, R. B., and C. J. Risley. 1976. Systematics and evolutionary relationships among the herons (Ardeidae). Misc. Publ. Mus. Zool. Univ. Michigan 150:1–115.

Prum, R. O. 1988. Phylogenetic interrelationships of the barbets (Capitonidae) and toucans (Ramphastidae) based on morphology with comparisons to DNA-DNA hybridization. Zool. J. Linn. Soc. Lond. 92:313–343.

Prum, R. O. 1993 Phylogeny, biogeography, and evolution of the broadbills (Eurylaimidae) and asites (Philepittidae) based on morphology. Auk 110:304–324.

Prum, R. O. 2002. Why ornithologists should care about the theropod origin of birds. Auk 119:1–17.

Raikow, R. 1987 Hindlimb myology and evolution of the Old World suboscine passerine birds (Acanthisittidae, Pittidae, Philepittidae, Eurylaimidae). Am. Ornithol. Union Ornith. Monogr. 41:1–81.

Raikow, R. J. 1982. Monophyly of the Passeriformes: test of a phylogenetic hypothesis. Auk 99:431–445.

Raikow, R. J., and J. Cracraft. 1983. Monophyly of the Piciformes: a reply to Olson. Auk 100:134–138.

Rea, A. M. 1983. Cathartid affinities: a brief overview. Pp. 26–54 in Vulture biology and management (S. R. Wilbur and J. A. Jackson, eds.). University of California Press, Berkeley.

Sereno, P. C. 1999. The evolution of dinosaurs. Science 284:2137–2147.

Sheldon, F. H., and A. H. Bledsoe. 1993. Avian molecular systematics, 1970s to 1990s. Annu. Rev. Ecol. Syst. 24:243–278.

Sibley, C. G., and Ahlquist, J. E. 1990. Phylogeny and classification of birds: a study in molecular evolution. Yale University Press, New Haven, CT.

Sibley, C. G., J. E. Ahlquist, and B. L. Monroe. 1988. A classification of the living birds of the world based on DNA-DNA hybridization studies. Auk 105:409–423.

Sibley, C. G., and B. L. Monroe, Jr. 1990. Distribution and taxonomy of the birds of the world. Yale University Press, New Heaven, CT.

Siegel-Causey, D. 1997. Phylogeny of the Pelecaniformes: molecular systematics of a primitive group. Pp. 159–172 in Avian molecular evolution and systematics (D. P. Mindell, ed.). Academic Press, New York.

Simpson, G. G. 1961. Principles of animal taxonomy. Columbia University Press, New York.

Simpson, S. F., and J. Cracraft. 1981. The phylogenetic relationships of the piciformes (Class Aves). Auk 98:481–494.

Sorenson, M. D., E. Oneal, J. García-Moreno, and D. P. Mindell. 2003. More taxa, more characters: the hoatzin problem is still unresolved. Mol. Biol. Evol. 20:1484–1499.

Stanley, S. E., and J. Cracraft. 2002. Higher-level systematic analysis of birds: current problems and possible solutions. Pp. 31–43 in Molecular systematics and evolution: theory and practice (R. DeSalle, G. Giribet, and W. Wheeler, eds.). Birkhäuser Verlag, Basel.

Storer, R. W. 1960. Evolution in the diving birds. Pp. 694–707 in Proceedings of the XII International Ornithological Congress (G. Bergman, K. O. Donner, and L. von Haartman, eds). Tilgmannin Kirjapaino, Helsinki.

Stresemann, E. 1927–1934. Aves. Pp. 1–899 in Handbuch der Zoologie (W. Kükenthal and T. Krumbach, eds.), vol. 7, pt. 2. Walter de Gruyter, Berlin.

Stresemann, E. 1959. The status of avian systematics and its unsolved problems. Auk 76:269–280.

Strickland, H. E. 1841. On the true method of discovering the natural system in zoology and botany. Ann. Mag. Nat. Hist. 6:184–194.

Swierczewski, E. V., and R. J. Raikow. 1981. Hindlimb morphology, phylogeny, and classification of the Piciformes. Auk 98:466–480.

Tarsitano, S., and M. K. Hecht. 1980. A reconsideration of the reptilian relationships of Archaeopteryx. Zool. J. Linn. Soc. 69:149–182.

van Tuinen, M., D. B. Butvill, J. A. W. Kirsch, and S. B. Hedges. 2001. Convergence and divergence in the evolution of aquatic birds. Proc. R. Soc. Lond. B 268:1–6.

van Tuinen, M., C. G. Sibley, and S. B. Sibley. 2000 The early history of modern birds inferred from DNA sequences of nuclear and mitochondrial ribosomal genes. Mol. Biol. Evol. 17:451–457.

Waddell, P. J., Y. Cao, M. Hasegawa, and D. P. Mindell. 1999. Assessing the Cretaceous superordinal divergence times within birds and placental mammals using whole mitochondrial protein sequences and an extended statistical framework. Syst. Biol. 48:119–137.

Wallace, A. R. 1856. Attempts at a natural arrangement of birds. Ann. Mag. Nat. Hist. 18:193–216.

Wetmore, A. 1930. A systematic classification for the birds of the world. Proc. U.S. Natl. Mus. 76:1–8.

Wetmore, A. 1934. A systematic classification for the birds of the world/revised and amended. Smithson. Inst. Misc. Coll. 89(13):1–11.

Wetmore, A. 1940. A systematic classification for the birds of the world. Smithson. Inst. Misc. Coll. 99(7):1–11.

Wetmore, A. 1951. A revised classification for the birds of the world. Smithson. Inst. Misc. Coll. 117(4):1–22.

Wetmore, A. 1960. A classification for the birds of the world. Smithson. Inst. Misc. Coll. 139(11):1–37.

Wheeler, Q. D. 1995. The "old systematics": classification and phylogeny. Pp. 31–62 in Biology, phylogeny, and classification of Coleoptera: papers celebrating the 80th birthday of Roy A. Crowson (J. Pakaluk and S. A. Slipinski, eds.). Muzeum i Instytut Zoologii PAN, Warsaw.

Xu, X., M. A. Norell, X.-L. Wang, P. J. Makovicky, and X.-C. Wu. 2002. A basal troodontid from the Early Cretaceous of China. Nature 415:780–784.

Xu, X., Z. Zhou, X. Wang, X. Kuang, F. Zhang, and X. Du. 2003. Four-winged dinosaurs from China. Nature 421:335–340.

Yuri, T., and D. M. Mindell. 2002. Molecular phylogenetic analysis of Fringillidae, "New World nine-primaried oscines" (Aves: Passeriformes). Mol. Phylogenet. Evol. 23:229–243.

Zusi, R. L., and B. C. Livezey. 2000. Homology and phylogenetic implications of some enigmatic cranial features in galliform and anseriform birds. Ann. Carnegie Mus. 69:157–193.

Maureen A. O'Leary
Marc Allard
Michael J. Novacek
Jin Meng
John Gatesy

Building the Mammalian Sector of the Tree of Life

Combining Different Data and a Discussion of Divergence
Times for Placental Mammals

Mammals are species that comprise a clade formed by the common ancestor of marsupials, monotremes, and placentals and all of its living and extinct descendants. The oldest fossils that form part of Mammalia (specifically the "crown clade"; see below) are diminutive forms known as multituberculates (Rougier et al.1996, McKenna and Bell 1997, Luo et al. 2002: fig. 1) and members of a clade referred to as Australosphenida (Luo et al. 2002; see also Flynn et al.1999, Rauhut et al. 2002), both of which date to the Middle Jurassic period. Mammals inhabit all land masses of the world (Nowak 1999) and have invaded such a wide range of habitats that they currently can be found living in the air and the sea, on land and within it. To exploit these habitats different mammalian taxa have evolved into the largest animals ever to have inhabited the earth (the blue whale), some of the most intelligent forms of life based on the ratio of brain size to body size (e.g., humans and chimpanzees), and forms possessing such extraordinary behaviors as the ability to echolocate (e.g., certain bats) as a means of understanding their surroundings (Nowak 1999). Building the mammal part of the Tree of Life amounts to discovering the branching diagram (phylogenetic tree) that describes how fossil and living mammal species diversified from a common ancestor through time.

The living members of Mammalia possess a variety of anatomical characteristics, including mammary glands, a specialized skin gland that can produce milk to feed offspring. Most living mammals, and some extraordinarily well-preserved fossil mammals (e.g., Hu et al.1997, Meng and Wyss 1997, Ji et al. 2002) also have hair. Hair serves multiple functions in mammals, including insulation, camouflage, and display. The circulatory system of mammals is characterized by a four-chambered heart consisting of fully separated venous and arterial circulation, and the mammalian brain includes dramatically expanded areas of gray matter (Vaughan 1986), as well as highly developed centers for processing visual and olfactory stimulation. Bony features, such as the presence of three ear ossicles and a jaw joint, known as the dentary-squamosal joint (Olson 1944, 1959, Kermack and Mussett 1958, Simpson 1960, Crompton and Jenkins 1973, 1979, Kermack et al.1973, Hopson 1994, Crompton 1995, Cifelli 2001), also have served to diagnose mammals, particularly fossil mammals. Different subgroups of mammals also are famous for their diversity of dental specializations, such as the tribosphenic molar (see below).

Researchers investigating mammalian phylogenetics have pursued a range of questions from such focused tasks as understanding the relationships of several closely related species, to broad investigations of interordinal relationships, many of which have involved the interpretation of numerous key fossils. As is discussed throughout this volume, as part of the Tree of Life effort, simultaneous analyses of hundreds or even thousands of taxa are emerging, often referred to as supermatrix analyses. The last decade has been characterized by an enormous increase in the amount of molecular sequence data available for the study of mammal phylogenetics as well as the discovery of a number of very significant new fossils. As we discuss below, mammalian phylogenetics is now moving away from a pattern of investigating how and why there is incongruence between data partitions (e.g., "molecules vs. morphology") and toward large-scale integration of historically heterogeneous character data (e.g., osteology, histology, molecular sequences, behavior). This approach often facilitates the discovery of clades supported by characters that may come from many different aspects of the organism. The clade Mammalia is poised to become one of the first Linnaean classes to be examined using a global simultaneous analysis of molecular and phenotypic data because Mammalia is a clade of relatively low taxonomic diversity relative to examples like Insecta or Aves, and because it includes many species, including our own (*Homo sapiens*), which are particularly well characterized from both a molecular and a morphological standpoint.

In this chapter, we do not provide a historical or taxonomic review of work on mammal phylogenetics—this has been provided recently for fossil data (Cifelli 2001) and for molecular data (Waddell et al. 1999, and references therein). Instead, we describe current efforts to move toward simultaneous analysis of phylogenetic data for mammals as an exemplar clade that forms part of the Tree of Life. We discuss some of the methodological justification for simultaneous analysis and explore particular problems within mammalian phylogenetics, primarily focusing on questions of interordinal relationships, because these have historically been some of the most challenging and contentious problems. Finally, as an example of how phylogenies can be applied to other evolutionary questions, we discuss using phylogenies to determine the age of placental mammals.

How Many Mammals Are There?

Zoologists now have recognized more than 5000 extant species of mammals (Wilson and Reeder in press), but no contemporary tally of the number of extinct mammal species has been conducted. A count of genera, extinct and extant, can be obtained from the recent classification of McKenna and Bell (1997) and was reported by Shoshani and McKenna (1998) to be 1083 living genera and 4076 extinct genera. Not only are the majority of mammalian genera extinct, but for every one extant genus of mammals, there are almost four extinct genera (fig. 28.1). We estimate that a count of species that included both extinct and extant taxa might uncover, conservatively, 20,000 species. With so much extinction recorded by fossils, this diversity must be accounted for in building a phylogenetic tree for mammals. Put another way, a tree based on living mammals alone encompasses only a fraction of the known diversity of Mammalia.

Mammal Clades and Broad-Level Classification

Figure 28.2 illustrates the tripartite division of Mammalia into Monotremata (the echidnas and duck-billed platypuses; species that lay eggs rather than produce live young), Marsupialia (kangaroos, opossums, koalas, and relatives), and Placentalia (elephants, whales, primates, shrews, mice, dogs, bats, and relatives). We have organized this tree using crown clade and stem clade concepts (Jefferies 1979, Ax 1987, Rowe 1988, de Queiroz and Gauthier 1992, Wible et al.1995, Rougier et al.1996, McKenna and Bell 1997) as a means of defining Mammalia and the clades within it. The use of crown and stem clades as a final basis for classification remains controversial for a variety of reasons (e.g., McKenna and Bell 1997, Nixon and Carpenter 2000). Nonetheless, as we discuss below, certain issues in mammalian phylogenetics, such as the timing of the origin of Placentalia, have been muddled by the inconsistent use of clade names by different authors. Different authors often mean different species when they use the word "placental," particularly when referring to fossil taxa and their relationships to clades of living taxa. We use crown and stem clade concepts here because, lacking another unambiguous, widely agreed upon (or used) mammalian classification, these terms serve as an effective heuristic device for discussing recent problems in mammalian phylogenetics, such as the time of origin of various clades.

We can define the crown clade Placentalia as the common ancestor of *Elephas maximus* (an elephant), *Bos taurus* (domestic cow), and *Dasypus novemcinctus* (an armadillo), and all of its living and fossil descendants. Some of the descendants that form part of Placentalia include humans and the rest of the order Primates, as well as a number of other orders (fig. 28.1) containing such taxa as bats, anteaters, flying lemurs, whales, carnivores, elephants, hippos, and tree shrews, to name a few examples. Although crown clades are defined by, and include, many extant species, it is important to consider that crown clades also contain fossil species. For example, the entirely extinct clade Desmostylia is part of the crown clade Placentalia because desmostylians have been demonstrated to be closely related to such species as *Elephas maximus* and relatives (Domning et al.1986, Novacek and Wyss 1987, Novacek 1992a) and therefore constitute descendants of the common ancestor of *Elephas maximus*, *Bos taurus*, and *Dasypus novemcinctus*. Many crown clades include numerous fossils; in fact, fossils may be the majority of species in crown clades, as is the case for placental mammals (fig. 28.1).

The crown clade Marsupialia is defined as the common ancestor of *Didelphis virginiana* (an opossum) and *Macropus giganteus* (a kangaroo), and all of its living and fossil descendants. The crown clade Monotremata is defined as the common ancestor of *Ornithorynchus anatinus* (a platypus) and *Tachyglossus aculeatus* (an echidna), and all of its living and fossil descendants. These crown clades contribute to the

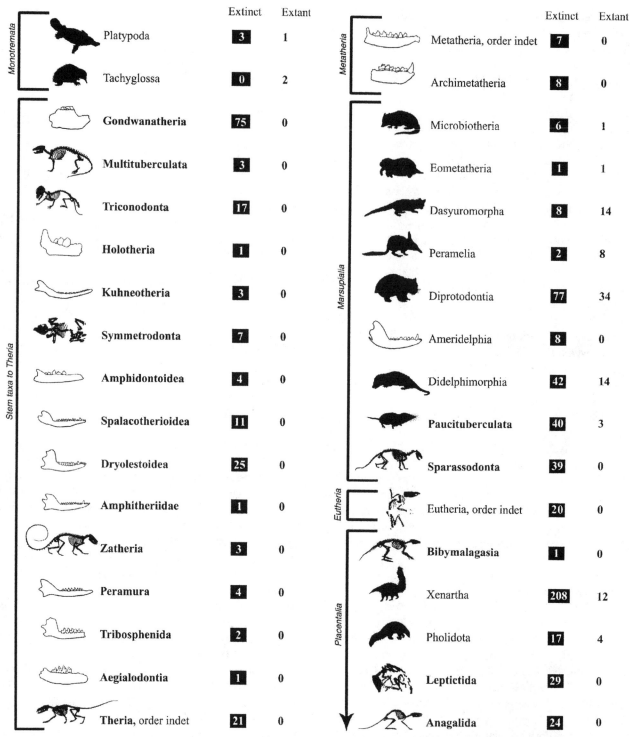

	Extinct	Extant		Extinct	Extant
Platypoda	3	1	Metatheria, order indet	7	0
Tachyglossa	0	2	Archimetatheria	8	0
Gondwanatheria	75	0	Microbiotheria	6	1
Multituberculata	3	0	Eometatheria	1	1
Triconodonta	17	0	Dasyuromorpha	8	14
Holotheria	1	0	Peramelia	2	8
Kuhneotheria	3	0	Diprotodontia	77	34
Symmetrodonta	7	0	Ameridelphia	8	0
Amphidontoidea	4	0	Didelphimorphia	42	14
Spalacotherioidea	11	0	**Paucituberculata**	40	3
Dryolestoidea	25	0	**Sparassodonta**	39	0
Amphitheriidae	1	0	Eutheria, order indet	20	0
Zatheria	3	0	**Bibymalagasia**	1	0
Peramura	4	0	Xenartha	208	12
Tribosphenida	2	0	Pholidota	17	4
Aegialodontia	1	0	**Leptictida**	29	0
Theria, order indet	21	0	**Anagalida**	24	0

Figure 28.1. A summary of generic-level extinction within Mammalia. Taxonomic categories listed within Mammalia are primarily, although not exclusively, the rank of Order (from McKenna and Bell 1997). We have summarized the numbers of extinct and extant genera contained within each category with a reconstruction of an example species on the left. A skeleton or a jaw represents categories with no living members; a silhouette of an example living species represents categories that have at least one living member. In general, jaws indicate relatively less documented taxa; however, many species in the categories represented here by jaws are also known from skulls and postcranial data. Note that the numbers of extinct genera far exceed the numbers of extant genera and that many categories with extant members contain a majority of fossil taxa. Higher groupings (Monotremata, Placentalia, Theria, Metatheria, Marsupialia, and Eutheria) follow figure 28.2 and are not necessarily those of McKenna and Bell (1997). For example, McKenna and Bell's (1997: 80–81) Placentalia of indeterminate order and Epitheria of indeterminate order are listed here as "Eutheria, order indet." McKenna and Bell (1997) classified some Cretaceous taxa within groups marked here as part of Placentalia

		Extinct	Extant
	Macroscelidea	8	4
	Mimotonida	3	0
	Lagomorpha	56	12
	Mixodontia	12	0
	Rodentia	743	430
	Cimolesta	128	4
	Creodonta	61	0
	Carnivora	354	108
	Lipotyphla	205	53
	Chiroptera	41	178
	Primates	218	54
	Scandentia	2	5
	Ungulata, order indet	1	0
	Tubulidentata	4	1
	Procreodi	29	0
	Condylarthra	59	0
	Arctostylopida	9	0

		Extinct	Extant
	Cete	238	40
	Artiodactyla	601	82
	Meridiungulata, order indet	10	0
	Litopterna	51	0
	Notoungulata	167	0
	Astrapotheria	16	0
	Xenungulata	2	0
	Pyrotheria	6	0
	Altungulata, order indet	1	0
	Perissodactyla	238	6
	Hyracoidea	18	3
	Embrithopoda	5	0
	Sirenia	29	2
	Desmostylia	6	0
	Proboscidea	42	2

(e.g., Ungulata order indet., Meridiungulata). *Contra* Springer et al. (2003), however, this is not tantamount to evidence that the crown clade Placentalia (fig. 28.2) contains Cretaceous taxa because the McKenna and Bell (1997) classification is not based on a phylogenetic analysis. It should not be assumed, therefore, that groupings described in McKenna and Bell (1997) are necessarily monophyletic because in many cases this remains to be explicitly tested. Taxa listed as Eutheria and Metatheria are stem clades to Placentalia and Marsupialia, respectively (fig. 28.2); the category "Stem taxa to Theria" has not been formally named. Figures redrawn from Sinclair (1906), Scott (1910), Riggs (1935), Simpson (1967), Clemens (1968), Kermack et al. (1968), Krebs (1971), Casiliano and Clemens (1979), Kielan-Jaworowska (1979), Jenkins and Krause (1983), Dashzeveg and Kielan-Jaworowska (1984), Rose (1987), Fox et al. (1992), Rougier (1993), Kielan-Jaworowska and Gambaryan (1994), Cifelli and DeMuizon (1997), Hu et al. (1997), Novacek et al. (1997), Cifelli (1999), Ji et al. (1999), Pascual et al. (1999), and DeMuizon and Cifelli (2000).

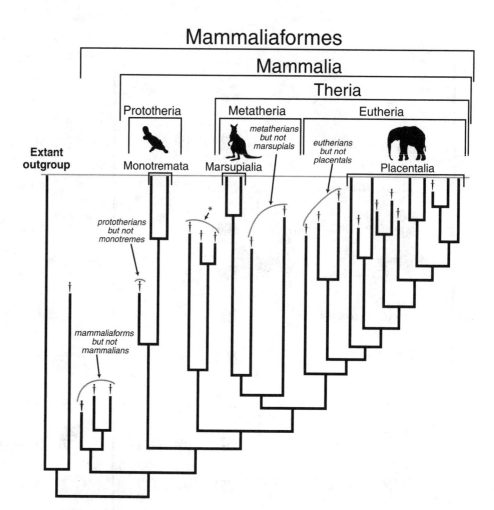

Figure 28.2. Simplified schematic of the tripartite division of mammals indicating crown clades (Placentalia, Marsupialia, Monotremata, Theria, and Mammalia) and stem taxa to these crown clades. The horizontal gray line indicates the Recent. Protatherians, eutherians, metatherians, and mammaliaforms are stem-based taxa (e.g., De Queiroz and Gauthier 1992). Different branch lengths are a reminder that the fossils are staggered in time throughout these clades. Lineages that are fully extinct are denoted by a dagger. The stem clade to the crown clade Theria that is indicated by an asterisk is currently unnamed. Note that these taxa would not be considered "therians."

definition of two larger crown clades: Theria, which is the common ancestor of *Elephas maximus* and *Didelphis virginiana* and all of its descendants; and Mammalia itself, which is the common ancestor of *Elephas maximus* and *Ornithorynchus anatinus* and all of its descendants. Theria is also a crown clade for which the majority of taxa are fossils.

The terms Prototheria, Metatheria, Eutheria, and Mammaliaformes represent stem-based taxa (De Queiroz and Gauthier 1992). They are defined as follows: Metatheria consists of all species more closely related to the marsupial *Didelphis virginiana* than to the placental *Elephas maximus*; Eutheria consists of all species more closely related to the placental *Elephas maximus* than to the marsupial *Didelphis virginiana*; Prototheria consists of all species more closely related to *Ornithorynchus anatinus* than to *Elephas maximus* (in other words, to any member of Theria); and Mammaliaformes consists of all taxa more closely related to *Elephas maximus* than to the clade Reptilia (*sensu* Gauthier et al. 1988).

It is important to recognize that this method of defining larger clades implies that all placentals are also eutherians, and alternatively that it is possible to be a eutherian without being a placental (fig. 28.2). Any stem species that falls outside the crown clade Placentalia would not be considered a

"placental" mammal; no matter how "placental-like" it is in terms of its characters. The same would apply to other stem species throughout the mammalian family tree. The recognition of both crown and stem taxa depends on the pattern of ancestry and descent, not on the characters that diagnose a particular taxon. It may sound counterintuitive to define a clade based on something other than anatomical characters; however, as discussed by Rowe (1988), defining a clade by common ancestry is consistent with evolutionary thinking and allows so-called defining traits to reverse without disqualifying a species' membership in a larger clade. For example, if we defined species as mammalian because they have extensive hair, then strictly speaking, we would be barred from considering whales (which virtually lack hair) as part of Mammalia. This, however, contradicts findings from phylogenetic analyses that indicate that whales are deeply nested within placental mammals. Likewise, there is evidence that extinct pterosaurs from the Mesozoic had hair (Padian and Rayner 1993, Unwin and Bakhurina 1994). The membership of pterosaurs within Mammalia has never emerged from any phylogenetic analysis because pterosaurs share a greater number of traits with the clade Archosauria (crocodilians, dinosaurs, birds, and relatives) than they do with Mammalia. As noted in McKenna and Bell (1997), this does not mean that crown and stem clades do not have diagnostic characters; they do. Because characters show homoplasy, however, the topology, not the character, is used to define the group.

Deciphering the interrelationships of placental orders (how placental species form larger clades) has remained a challenging problem for morphological systematists despite decades of study (e.g., Novacek 1992b, Shoshani and McKenna 1998). In the early 1990s several hypotheses of interordinal relationships, derived initially from anatomical data, were beginning to be intensively tested with new molecular data. The taxon within Placentalia that branched first was generally considered to be an edentate, and the following clades were thought to be monophyletic (although alternative arrangements certainly remained under consideration): Glires (rodents + rabbits); Paenungulata (hyraxes + elephants + sea cows); Archonta (primates, bats, tree shrews, and flying lemurs), and Ungulata (hoofed mammals) (discussed in Novacek 1992b). Nonetheless, the base of Placentalia remained rather bushlike with certain nodes appearing repeatedly but with other higher groupings remaining unstable and often supported by only a few synapomorphies.

The infusion of numerous molecular sequence characters into mammalian phylogenetics, particularly during the last decade, introduced a variety of new phylogenetic hypotheses. Controversies developed concerning the monophyly of several higher clades, including insectivorans (moles, shrews, and relatives), archontans, ungulates, glires, rodents (mice, voles, and close relatives), and artiodactylans (hoofed mammals with an even number of toes). Indeed, molecular data have even resulted in trees that did not support the fundamental tripartite division of mammals, instead associating monotremes and marsupials as sister taxa to the exclusion of placentals (Janke et al.1994, 1997, 2002; see also arguments in the morphological literature: Gregory 1947, Kühne 1973). New higher clades, notably Afrotheria (golden moles, elephant shrews, elephants, aardvarks, hyraxes, tenrecs, and manatees) and Laurasiatheria (whales, artiodactylans, carnivorans, perissodactylans, pangolins, bats, and several insectivorans) have been first proposed on the basis of entirely molecule-based analyses (e.g., Murphy et al. 2001, 2002, Madsen et al. 2001). These clades represent a fairly fundamental restructuring of the placental branching sequence that had not been previously supposed from an investigation of morphological data. New molecule-based hypotheses remind systematists that even long-held notions of relationship are hypotheses that can be overturned by new data. In many cases, however, the combined analyses of molecular and morphological data required to test these new hypotheses are only just emerging.

The Importance of Combining Data

Many ideas about the evolution of mammals have at their core an implicit or explicit hypothesis of genealogical history. Such commonly discussed topics as scenarios of adaptation or notions of the age or place of origin of a taxon are fundamentally dependent on a hypothesis that states how mammal species branched from each other through time. A weakly tested hypothesis of relationship is a shaky scaffolding for all other inferences placed on top of it, underscoring the importance of the Tree of Life to evolutionary study as a whole.

Phylogenetics, like many historical sciences, stands in contrast to experimental sciences because the hypotheses to be tested (e.g., the origin of primates) are not experiments that can be repeated (like the function of a particular enzyme in a human cell). Students of historical problems can, however, establish tests of a different kind: they can formulate a hypothesis of relationships about how species branched from each other through time, and test that hypothesis by looking for evidence to reject it. Phylogenetic hypotheses (e.g., trees) start as nothing more than a guess about the roadmap of evolution; these guesses are then tested against the biological and paleontological evidence available. They are rejected if they do not efficiently explain all of the data.

How do we know when we have the tree of mammals? Essentially, like any hypothesis, we can always continue to test it. When a hypothesis of relationship has been tested and retested by adding new data (e.g., molecular sequences, bones, soft tissues, behavior) to a global parsimony analysis, and the hypothesis remains unchanged, the hypothesis is the best explanation of all the data. We could then refer to such a hypothesis as robust or stable (Nixon and Carpenter 1996). It is important to appreciate that such well-tested trees may not necessarily have high support measures, such as bootstrap values or Bremer support. Once we establish a phylogenetic tree that is stable to the addition of new data (i.e., we add new data and the tree does not change), we can examine how the characters used to build the tree changed through time (using a method called optimization), as well as what the tree says about the age and place of origin of various clades.

A phylogenetic tree of mammals is tested by studying heritable, phylogenetically independent traits (molecular, morphological, or behavioral), called characters. Arguments that certain characters are "misleading" and should be eliminated before a tree is even tested are circular because if we do not know the tree (which is built from characters) then we do not know which characters are "good" and which are "bad" before we test them. Some character systems that have been described as "bad" include mitochondrial DNA (mtDNA; e.g., Naylor and Brown 1998, Luckett and Hong 1998), teeth (e.g., Naylor and Adams 2001, Cifelli 2000), and all morphological evidence (e.g., Hedges and Maxson 1996). Others that have been described as "good" include cranial characters (e.g., McDowell 1958), certain nuclear genes, and molecular markers known as SINES (e.g., Nikaido et al.1999, 2001). We would argue that empirical work is currently not extensive enough to substantiate such generalizations, something that is underscored by there being a number of exceptions to all the examples listed here as being "good" or "bad." But what if we have tested some characters; can we generalize on the basis of those results? If a character has been found to have a lot of homoplasy in worms, should we assume that it also has a lot of homoplasy in mammals? Because there are examples where such generalizations have failed, and because characters can be informative even if they show homoplasy, it seems premature to introduce such assumptions into phylogenetic analyses.

Also problematic for the rigorous testing of phylogenies is a tendency to assert a priori that a particular set of characters is correlated functionally or developmentally, and that these characters will promote a misleading phylogenetic signal if they are included in an analysis. Often this is enforced by constraining tree topology, and using that tree to evaluate the informativeness of new data. This is problematic for at least two reasons: (1) functional correlation and phylogenetic correlation are not necessarily the same thing, and one should therefore not be used as a basis for assuming the other (Farris 1969), and (2) a priori lumping of characters into functional complexes forces them to group together in phylogenetic analysis; breaking them into individual characters does not. The latter allows the data themselves to indicate whether the individual parts of the proposed "complex" even evolved at the same node.

Finally, we might ask which species are necessary to sample in order to find the tree of mammals. If our goal is to discover the Tree of Life, then ultimately we want to know where all species, fossil and living, fit on the tree. This does not mean that all analyses must include all species from the outset, but it does mean that we should think about the problem of mammalian phylogenetics on a large scale and cumulatively, with the results of each analysis open to testing by adding new species and characters. Historically, many of the earliest cladistic analyses of mammals contained relatively few taxa by contemporary standards and used higher taxa as operational taxonomic units (OTUs; e.g., Novacek 1982, 1992a, Rowe 1988). As computerized search algorithms have become increasingly powerful, more recent analyses have tended to use genera (e.g., Rougier et al.1998, Murphy et al. 2001a, Gatesy and O'Leary 2001, Meng et al. 2003). Malia et al. (2003) emphasize, for trees to be maximally accountable to the data, there ultimately should be a final shift toward sampling at the species level as analyses become increasingly exhaustive and algorithms become more powerful.

Supertrees and Supermatrices

Separating characters into groups such as different genes, genes and morphology, or types of morphology is known as data partitioning (Kluge 1989, Nixon and Carpenter 1996). This approach is sometimes explicitly preferred for phylogeny reconstruction, and trees from separate analyses may subsequently be combined into a summary tree. This approach has been formalized as "supertree" methods (Sanderson et al.1998), and supertrees have been constructed to investigate higher clades of mammals (e.g., Liu et al. 2001). Alternatively, data partitions can simply be combined in a single simultaneous analysis, or a "supermatrix" (e.g., Murphy et al. 2001, Gatesy et al. 2002, Malia et al. 2003), the analysis of which relies on character congruence or agreement among individual characters (Kluge 1989, Nixon and Carpenter 1996). Partitioning data matrices, determining the separate tree results, and then summarizing the shared topological patterns using consensus techniques was a sequence of operations originally called taxonomic congruence (Kluge 1989). The supertree approach is not the same as the taxonomic congruence approach in which traditional consensus techniques (e.g., Adams consensus, strict consensus, majority rules consensus) are used to summarize results. Rather than comparing shared clusters of taxa directly, commonly used supertree methods, such as matrix representation with parsimony (Baum 1992, Ragan 1992), recode separate phylogenetic trees into a new data matrix that represents these trees. Then the supertree matrix is analyzed using parsimony algorithms. Various implementations of matrix representation with parsimony differ in how they move from original tree topologies to a coded data matrix, how "characters" are weighted, and how conflicts are reconciled (Baum 1992, Ragan 1992, Baum and Ragan 1993, Purvis 1995, Ronquist 1996, Sanderson et al.1998). Salamin et al.'s (2002) recent supermatrix of grasses provides a review of these methods.

Kluge (1989) criticized the separate analysis of data partitions on numerous grounds (see also Nixon and Carpenter 1996, DeSalle and Brower 1997). Others support separate analyses of data partitions, and there have been several criticisms and rejoinders on this topic (Bull et al.1993, Miyamoto and Fitch 1995, Nixon and Carpenter 1996, Cunningham 1997). These arguments are primarily directed at separate data analysis, not supertrees specifically, but the general arguments are relevant to evaluating the usefulness of supertrees.

Perhaps the primary inadequacy of the supertree approach is its insensitivity to hidden support (Barrett et al.1991) among the characters of different matrices used in separate analyses (see Wilkinson et al. 2001, Pisani and Wilkinson 2002). The simplest way to avoid the problematic issues of weighting and redundancy in supertree analysis is to include each character state once in a supermatrix and let character congruence determine the best-supported tree topology (Farris 1983, Kluge 1989). There is no theoretical difference between the analysis of one mammal clade, or the analysis of all mammals, or the analysis of the Tree of Life. The only difference is one of scale. Therefore as we begin to solve computational problems that have limited the scale of phylogenetic analyses (e.g., the number of taxa that can be analyzed), the construction of supermatrices rather than supertrees becomes increasingly compelling.

Supermatrices of Extinct and Extant Taxa: Computational Issues and Missing Data

It is well known that as the number of taxa increases, so does the difficulty of the phylogenetic problem. For four taxa there are three unrooted networks possible; for 14 taxa there are more than 316 billion possible unrooted networks, and this number rapidly increases (Kitching et al.1998: 41). Under the existing paradigm of finding the most parsimonious solution for each problem, large matrices require an extremely large amount of computational power. Building a simultaneously analyzed tree for all living and extinct mammals, conservatively 20,000 species, represented by tens of thousands of characters, will pose substantial computational challenges.

Many large matrices have been examined using PAUP* (Swofford 2000); however, several investigators (Nixon 1999, Goloboff et al.1999, Goloboff and Farris 2001) have commented that PAUP* exhibits severe limitations regarding the rapidity of parsimony search strategies if a data matrix contains more than 400 taxa. Computational challenges such as these represent an active area of research (e.g., DeSalle et al. 2002, Janies and Wheeler 2002), and new search algorithms such as POY (Wheeler and Gladstein 2000), the parsimony ratchet (Nixon 1999) implemented through WinClada (Nixon 2002) and NONA (Goloboff 1994), and TNT (Goloboff, et al.1999, Goloboff and Farris 2001) have allowed investigators to execute some of the largest parsimony-based searches with relative efficiency. Both WinClada/NONA and TNT software, for example, regularly produced large trees from matrices containing more than 1700 OTUs (Allard et al. 2002). These new methods represent a considerable advance for the examination of large data sets.

Results of some of the largest phylogenetic analyses published to date were determined using these new rapid heuristic methods but still include fewer than 1000 taxa; examples include eukaryotes (440 taxa; Lipscomb et al.1998) and seed plants (500 taxa; Rice et al.1997, Nixon 1999, Janies and Wheeler 2001). An intraspecific analysis of humans that included 1771 individuals (Allard et al. 2002) was tested in a phylogenetic framework using WinClada/NONA. For mammals two examples of particularly large published matrices included 264 taxa (for molecular data using the program POY; Janies and Wheeler 2001) and 91 taxa (molecular and morphological data, using PAUP*; Gatesy et al. 2002).

A supermatrix of mammals, particularly one that combines molecular and nonmolecular (morphology, behavior) characters, will have substantial missing data. Missing data may occur because no investigator has scored a particular set of taxa and characters for a clade, because a feature has changed so much as to be absent, because a character is inapplicable, or because a feature has not been preserved for study (in a fossil, e.g., see discussion in Gatesy and O'Leary 2001, Kearney 2002). Operationally, some investigators create composite taxa (e.g., a combination of two or more species to make a genus level OTU) expressly to reduce missing data. A composite behaves in a tree search as a single taxon, the assumption being that it is monophyletic with respect to the other taxa in the analysis. Composite taxa (and the implicit monophyly assumptions they encode) can, however, have serious effects on the resulting topology. A recent review (Malia et al. 2003) of a mammalian supermatrix that included composite taxa (Madsen et al. 2001) showed that the construction of composite taxa did have dramatic and not necessarily beneficial effects on tree topology. Prendini (2001) and Malia et al. (2003) recommend breaking all composites above the species level, an approach consistent with recent arguments that missing data do not create false or misleading evidence (e.g., Kearney 2002).

Specific Problems in Mammal Phylogeny

Mammaliaformes

There are a number of interesting fossil taxa that fall outside the crown clade Mammalia but that are part of the clade Mammaliaformes (fig. 28.2). These fossils capture critical stages in the transition from an amniote sister taxon of mammals to the crown clade Mammalia and include such groups as Sinoconodontidae, Morganucodonta, Docodonta, and Haramiyoidea (McKenna and Bell 1997). These close mammal relatives are of generally small size and are known from fossils that date back to the Triassic and Jurassic. Support for the hypothesis that these taxa belong outside of crown clade Mammalia has been demonstrated in many different analyses, (Rowe 1988, Wible et al.1995, Rougier et al.1996, Hu et al.1997, Ji et al.1999, 2002, Luo et al. 2001a, 2001b, 2002, Wang et al. 2001, Rauhut et al. 2002).

As described above, Mammaliaformes contain not only these basal forms, but also the crown clades of monotremes, marsupials and placentals. Investigation of general mammaliaform relationships also concerns the diversification of fossils that are more highly nested. Some of the most researched mammaliaform problems include the position of: (1) multituberculates (mentioned above as critical for dating the entire mammal crown clade), (2) "triconodonts" and their relation to monotremes and therians, and (3) the relationship of monotremes to other Mesozoic mammals.

Parsimony analyses have commonly placed multituberculates within the crown clade Mammalia (Rowe 1988, Wible et al.1995, Rougier et al.1996, Hu et al. 1997, Ji et al. 1999, Luo et al. 2001b, 2002, Wang et al. 2001). Multituberculates may be either stem monotremes (Luo et al. 2001b, Wang et al. 2001), more closely related to therians (Luo et al. 2002, Rauhut et al. 2002), or part of an unresolved polytomy with therians and monotremes (Wible et al.1995). Phylogenetic investigations of "triconodonts" [Austrotriconodontidae, Amphilestidae, and Triconodontidae (McKenna and Bell 1997)], a group that may not be monophyletic, have shown them to be the sister group of the crown clade Mammalia (Hu et al.1997, Luo et al.

2001b, 2002, Ji et al. 2002) or a member of the crown clade Mammalia (Rowe 1988, Wible et al.1995, Rougier et al.1996, Luo et al. 2001a 2002) or have been unable to resolve their position with respect to crown Mammalia (Luo et al. 2001b, Wang et al. 2001). When "triconodonts" fall within Mammalia, they commonly form the sister group of the clade consisting of multituberculates and their relatives (Rowe 1988, Luo et al. 2002: fig. 1, Ji et al. 2002, Rauhut et al. 2002). It is clear that phylogenetic relationships of triconodonts and multituberculates are still unstable and that the resolution of this problem will affect the content and relationships of Mammalia as well as minimum estimates of the age of this clade.

Understanding the relationships of monotremes to other mammals requires discussion of the term tribospheny. Tribospheny refers to a shape of the molar teeth such that in occlusion there is both a crushing (*sphene*) and a shearing (*tribos*) component to the movement between the upper and lower teeth (Simpson 1936). This dental feature is present to varying degrees in the monophyletic group Theria (fig. 28.2) and their close relatives; the oldest fossils of these have generally been found in the Northern Hemisphere. Recent discoveries of Mesozoic mammals from southern continents (Rich et al.1997, Flynn et al.1999), however, have raised intriguing questions about the origin of tribospheny and the relationships of Mammalia. Several fossil taxa from southern continents show a complex molar pattern that is tribosphenic in shape. Rich et al. (1997, 1999, 2001a, 2001b) interpreted *Ausktribosphenos* and *Bishops*, fossil mammals from the Cretaceous period of Australia, as having tribosphenic molars and suggested that these fossils were closely related to hedgehoglike placentals. This affiliation for *Ausktribosphenos* and *Bishops*, however, has not yet been corroborated by comprehensive cladistic analyses (Kielan-Jaworowska et al.1998, Musser and Archer 1998, Archer et al.1999, Rich et al. 2001a, 2001b). Archer et al. (1999) interpreted the dentition of the Early Cretaceous taxon *Steropodon*, a fossil monotreme, as having a modified tribosphenic pattern. Monotremes are toothless in the living adult forms, prohibiting direct comparison of their adult dentition teeth with that of other mammals. The significance of theses observations is that depending on the phylogenetic hypothesis, monotremes may be descended from a taxon with tribospheny, a character that has long been thought to be more typical of therians. Recent phylogenetic analyses also suggest that the tribosphenic molar pattern, long thought to have evolved once, has evolved independently twice: once in an endemic southern (Gondwanan) clade that is survived by extant monotremes and again independently in a northern (Laurasian) clade composed of extant marsupials, placentals, and their extinct relatives (Luo et al. 2001a, 2002, Ji et al. 2002, Rauhut et al. 2002).

Glires

Within the diverse clades that form part of placental mammals, certain clades have been the focus of numerous inves-

tigations drawing on morphological, molecular, and fossil evidence. One such clade is Glires, which consists of two extant mammalian orders: Rodentia (rats, mice, and relatives) and Lagomorpha (hares and pikas). We define these here as crown clades (fig. 28.3). Together, lagomorphs and rodents constitute nearly half of extant mammalian species diversity (Nowak 1999). Stem taxa to Lagomorpha first appeared in the Paleocene of Asia (McKenna 1982; fig. 28.3). Crown clade Rodentia, by contrast, may contain some taxa collected in Paleocene rocks of North America (Wood 1962, Dawson et al.1984, Korth 1984, Dawson and Beard 1996), but recent large phylogenetic analyses (Meng et al. 2003) do not fully substantiate this (fig. 28.3).

Morphologists, including paleontologists, have extensively researched relationships within Glires and occasionally questioned its monophyly (for review, see Meng and Wyss 2001, Meng et al. 2003). During the last decade, molecular biologists have challenged both Glires monophyly and rodent monophyly (Graur et al.1991, 1996, Li et al.1992), arguing that the guinea pig is more closely related to primates than to other rodents and that rabbits are more closely related to primates than to rodents. As pointed out by several studies, however, early molecular results that indicated a nonmonophyletic Glires or Rodentia appear to have been artifacts of small data sets or other methodological problems (Allard et al.1991, Hasegawa et al.1992, Graur 1993, Novacek 1993, Catzeflis 1993, Sullivan and Swofford 1997, Halanych 1998). More recent molecular studies investigating this claim have included more taxa, more gene sequences, or both (Madsen et al. 2001, Murphy et al. 2001, 2002, Waddell et al. 2001). These analyses have corroborated morphological studies that support Glires monophyly.

Several recent morphological studies continue to support the monophyly of Glires (Li et al.1987, Novacek 1992a, 1992b, Luckett and Hartenberger 1993, Shoshani and McKenna 1998, Meng and Wyss 2001). Figure 28.3 is a phylogenetic tree superimposed on a stratigraphic distribution of Glires, which resulted from an analysis of 50 taxa and 227 morphological characters (Meng et al. 2003). It shows the sister group relationship of Rodentia and Lagomorpha as determined from morphological data (for support values, see Meng et al. 2003). Because of the current topological congruence between molecular and morphological data, both showing support for Glires, we anticipate that combined analyses that include fossils will continue to support this result.

Cetacea

One of the most debated problems in placental mammalian phylogenetics concerns the position of the order Cetacea (whales, dolphins, and porpoises). The question of cetacean affinities is worthy of special consideration here because it has been examined with diverse data sets, including combined (total evidence) analyses of multiple genes and morphology. Most of these studies have focused on identifying

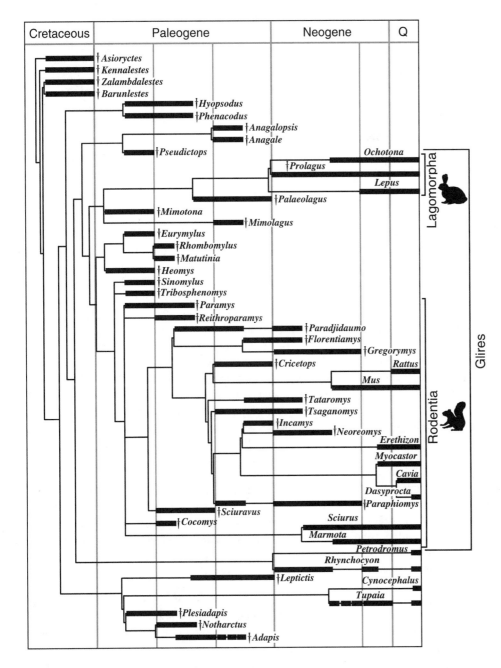

Figure 28.3. Tree of the relationships of Glires (a monophyletic clade of rodents and rabbits) and closely related taxa: strict consensus of 10 most parsimonious trees derived from osteological data (Meng et al. (2003) plotted against the stratigraphic record. Thick lines represent known durations of fossil lineages. Dashed lines indicate uncertain durations of lineages. Daggers indicate extinct taxa. The polytomy at the base of Rodentia makes designation of the membership of extinct taxa within crown clade Rodentia somewhat unclear; taxa such as *Paramys*, *Reithroparamys*, and *Cocomys* are not unambiguously part of crown Rodentia based on these data. Three crown clades are defined as follows: Rodentia, common ancestor of *Rattus* and *Marmota* and all of its descendants; Lagomorpha, common ancestor of *Ochotona* and *Lepus* and all of its descendants; Glires, common ancestor of *Ochotona* and *Marmota* and all of its descendants. See Meng et al. (2003) for support values.

the extant sister taxon of Cetacea and on determining the relationship of Cetacea to an extinct group of terrestrial mammals called Mesonychia, which are hoofed mammals that are part of the clade Paraxonia. Some of the alternative hypotheses under consideration include the following: (1) Cetacea are excluded from a monophyletic Artiodactyla (the order that includes the even-toed hoofed mammals, e.g., pigs, hippos, camels, and ruminants) and is the sister group of the extinct clade Mesonychia, (2) Cetacea are related to a subgroup of artiodactylans (i.e., hippos) and should be placed within Artiodactyla (thereby rendering the traditional concept of Artiodactyla paraphyletic), (3) Cetacea are the sister taxon of a monophyletic Artiodactyla to the exclusion of

Mesonychia, and (4) Cetacea are the sister taxon of Mesonychia and this clade is nested within Artiodactyla.

Phylogenetic analyses of Cetacea based on skeletal characters, anatomy of the digestive tract, transposons, amino acid sequences, DNA–DNA hybridization scores, and DNA sequences have been presented recently (reviewed in Gatesy and O'Leary 2001). Some studies have attempted to include as much published data as possible, or to qualify their conclusions based only on partial evidence. Other studies, many of which produced well-resolved phylogenetic trees, were obtained only after ignoring potentially contradictory published evidence (entire character data partitions or parts of them) and were not, therefore, the product of rigorous phylogenetic tests. Not surprisingly, the results of these partitioned analyses have rarely agreed.

One recent example is that of Thewissen et al. (2001). Although their phylogenetic analysis included previously undescribed morphological data, their study incorporated only a subset of previously published data. It excluded some osteological characters that happened to support a close relationship between mesonychians and Cetacea to the exclusion of artiodactylans, the relationship that Thewissen et al. (2001) claimed to reject. Similarly, Thewissen et al. (2001) also excluded all published molecular data, thereby barring molecules from influencing the phylogenetic placement of Cetacea. Although molecular sequence data have not been extracted from extinct mesonychians, certain analyses indicate that the thousands of informative molecular characters collected to date overwhelmingly contradict the results of Thewissen et al. (2001) and place Cetacea within a paraphyletic Artiodactyla, closest to hippopotamids (e.g., Gatesy et al.1999a, 1999b, Matthee et al. 2001). Exclusion of so much data in this way makes it impossible to assess the phylogenetic relevance of this study as presented.

Likewise, many other recent studies of whale phylogeny included analyses of very small subsets of data in isolation from the majority of published character evidence. For example, in a recent study of morphological characters of the digestive tract, Langer (2001) found that stomach morphology supported the monophyly of Artiodactyla to the exclusion of Cetacea. Langer (2001) then asserted that other morphological characters, which contradicted his hypothesis, were necessarily the result of convergent adaptation to an aquatic life and should be dismissed as phylogenetic evidence. Langer (2001) did not actually include these "convergent" characters in his matrix, but concluded that Artiodactyla was monophyletic regardless. Likewise, in a recent phylogenetic analysis of mtDNA and morphological data, Luckett and Hong (1998) argued that more than 90% of the approximately 1000 molecular characters they examined were too variable to be of any use and were eliminated from phylogenetic analysis. Putative aquatic adaptations, such as near-hairlessness in hippos and whales, also were not considered valid phylogenetic evidence. In this same tradition, Naylor and Adams (2001) hypothesized that dental traits, not aquatic specializations or molecular data, were just too homoplastic or correlated to include in phylogenetic analysis, leading them to present a preferred tree that excluded all dental evidence and to propose general arguments impugning the use of dental characters in mammalian phylogenetics (see O'Leary et al. 2003).

More inclusive studies of characters and taxa (Gatesy et al.1999a, 1999b, 2002, O'Leary 2001) have shown that results based on subsets of the total database are highly unparsimonious (e.g. O'Leary et al. 2003). We have compiled two large combined supermatrices of whales, artiodactylans, and close relatives. The first matrix includes 75 extant taxa and more than 37,000 characters from three morphological

data sets (Messenger and McGuire 1998, Geisler 2001, Langer 2001), a matrix of SINE transposon insertions (Nikaido et al. 1999, 2001), and 51 genes/gene products from the mitochondrial and nuclear genomes (fig. 28.4; Gatesy et al. 2002). This analysis includes most of the characters discussed in Luckett and Hong (1998), Naylor and Adams (2001), Langer (2001), and Thewissen et al. (2001) as well as tens of thousands of characters (mostly molecular) that were published prior to these papers. The supermatrix of extant taxa did not support topologies promoted by the authors of partitioned analyses (fig. 28.4, gray dots) and each of the partition-based hypotheses required at least 300 extra character steps beyond minimum tree length. Some groups supported by bootstrap scores of 100% and Bremer supports of more than 100 steps were not recovered in the more restricted analyses of Luckett and Hong (1998), Naylor and Adams (2001), Langer (2001), or Thewissen et al. (2001).

We also constructed a whale/artiodactylan supermatrix that included extinct taxa (fig. 28.5). The combined data set was composed of 50 extinct taxa, 18 extant taxa, and ~36,500 characters. Morphological data were primarily from Geisler (2001), and molecular characters (including alignment methodology) came from Gatesy et al. (2002). Per taxon, this supermatrix has much more missing character data, but should be a better test of the phylogenetic tree because it includes basal extinct species, such as primitive whales, early artiodactylans, and mesonychians. The strict consensus of minimum length topologies is not well resolved because of the instability of several taxa, including Mesonychia (fig. 28.5)

Use of a maximum agreement subtree (e.g., Cole and Hariharan 1996), which summarizes the maximum number of relationships that are supported by all minimum length topologies, helps clarify where character conflict is most pronounced. Instead of collapsing uncertain taxa to basal nodes as in an Adams (1972) consensus tree, the agreement tree excludes these taxa and just shows the relationships that are consistent with all of the equally short trees (fig. 28.6). This indicates that the differences among the most parsimonious trees are due to the instability of fossil taxa, not to alternative relationships for the living taxa (see also discussion of this in O'Leary 2001). Like the supermatrix for extant taxa only (fig. 28.4), the combined fossil/extant supermatrix (fig. 28.5) is consistent with a close relationship between hippopotamuses and whales, a result that was strictly contradicted in several analyses that used subsets of published data (Luckett and Hong 1998, Langer 2001, Thewissen et al. 2001). Furthermore, controversial relationships supported by the analysis of Naylor and Adams (2001), such as perissodactylan paraphyly and a grouping of Ovis and Camelus to the exclusion of Tragulus, were overwhelmingly rejected (figs. 28.4–28.6). In this analysis the fossil data have not altered the primary relationships of the extant taxa as determined from molecular data alone. We do not argue here in favor of a particular phylogenetic result, but instead suggest simply that more comparative work will be required to sort

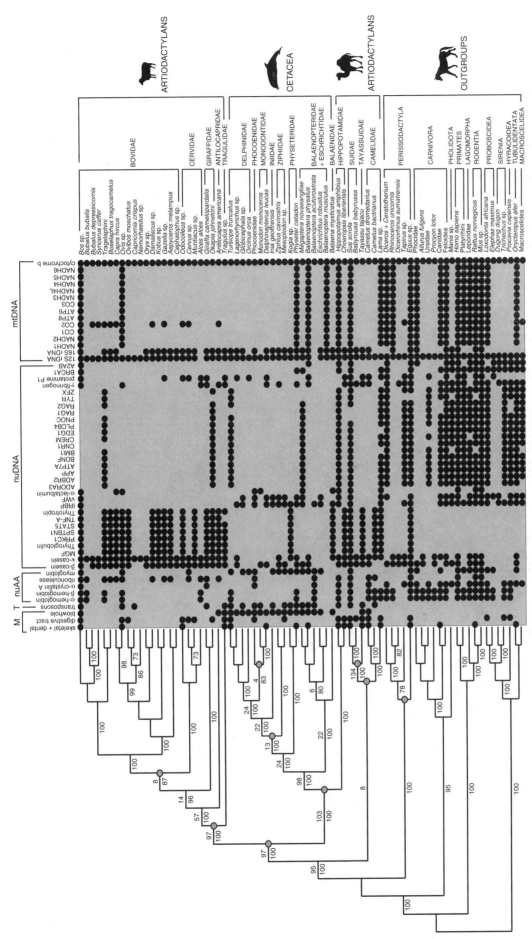

Figure 28.4. Single minimum length topology of 67,357 steps supported by parsimony analysis of the extant whale–artiodactylan supermatrix with all characters unordered. OTUs are shown to the right. Higher level taxa are in capitals and are delimited by brackets to the right of OTUs. Black circles indicate taxa sampled for these data sets; gray represents missing data in the supermatrix. One thousand random taxon addition replicates were used in each constrained heuristic search, but given the complexity of the supermatrix data set, these branch support scores may be lower than indicated. Bootstrap scores (Felsenstein 1985) that were greater than 69% are indicated below internodes. One thousand bootstrap replicates were done using heuristic searches of informative characters with simple taxon addition and tree bisection reconnection branch swapping (Swofford 2000). Gray circles at nodes mark clades that were inconsistent with the combined supertree analysis of Liu et al. (2001) and/or the restricted character analyses of Luckett and Hong (1998), Langer (2001), Naylor and Adams (2001), or Thewissen et al. (2001). The tree is rooted according to the hypotheses of Madsen et al. (2001) and Murphy et al. (2001).

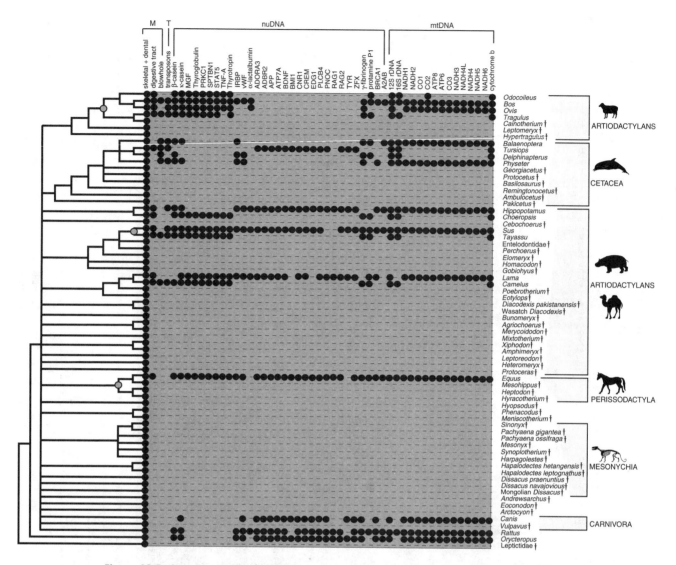

Figure 28.5. Strict consensus of 4522 minimum length topologies (32,613 steps) supported by parsimony analysis of the extinct + extant whale supermatrix with all characters unordered [130 random addition replicates in PAUP* beta version 10 (Swofford 2000)]. OTUs are shown to the right. Higher level taxa are in capitals and are delimited by brackets to the right of OTUs. Data sets are shown at the top of the figure (M, morphology; T, transposons; nuAA, nuclear amino acid sequences; nuDNA, nuclear DNA; mtDNA, mitochondrial DNA; for other abbreviations, see Gatesy et al. (2002), and taxonomic sampling for each data set is indicated by black circles as in figure 28.4. Gray circles at nodes mark clades that were inconsistent with the combined supertree analysis of Liu et al. (2001) and/or the analyses of Luckett and Hong (1998), Langer (2001), Naylor and Adams (2001), and Thewissen et al. (2001). The tree is rooted with Leptictidae (see Geisler 2001). Daggers indicate fossil taxa. Matrix available through TreeBASE.

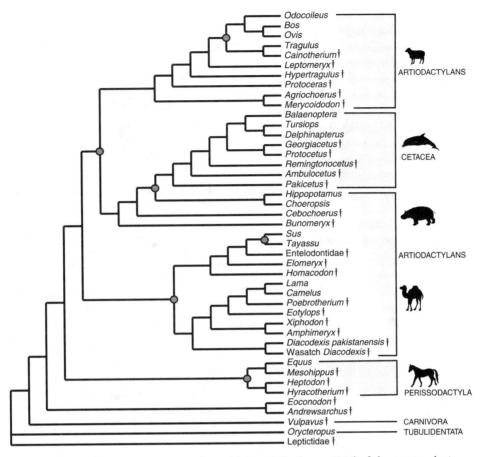

Figure 28.6. A maximum agreement subtree (Cole and Hariharan 1996) of shortest topologies found for the extinct + extant whale supermatrix (see strict consensus in fig. 28.5). This shows relationships that are stable among all most parsimonious trees. The phylogenetic positions of Mesonychia and some other taxa vary among minimum length trees and are therefore excluded from the agreement subtree. Gray circles at nodes mark relationships that are inconsistent with the combined supertree analysis of Liu et al. (2001) and/or the analyses of Luckett and Hong (1998), Langer (2001), Naylor and Adams (2001), and Thewissen et al. (2001). Daggers indicate fossil taxa.

out all the relationships among living whales and their extinct relatives.

Mammalian Supertrees and Supermatrices

A number of recently published large-scale molecule-based supermatrix analyses include increasingly greater numbers of taxa (>50) analyzed simultaneously. For example, Murphy et al. (2001) performed a simultaneous analysis of 64 mammal taxa using data from 18 different gene segments. Their results showed some variance in tree topology depending on the method of phylogenetic analysis (e.g., parsimony vs. maximum likelihood). These authors figured the maximum-likelihood tree, which showed the clades Glires, Xenarthra (sloths, anteaters, and armadillos), Afrotheria, and Laurasiatheria. The tree also supported results like artiodactylan

paraphyly. Importantly, however, Afrotheria was not supported in the parsimony analysis.

Liu et al. (2001) published a supertree analysis, which also resulted in a large mammalian tree, with 91 terminal taxa. This combined summary of previous morphological and molecular studies was largely congruent with traditional hypotheses of relationship based on morphology. In other words, the most basal clade was an edentate group, and the monophyly of Insectivora, Artiodactyla, Rodentia, and Glires was supported, but the monophyly of Afrotheria and Laurasiatheria was not. Liu et al. (2001) attempted to limit the overall redundancy of information in their supertree data set by using only the most recent and comprehensive published analysis for each gene, but there were still considerable duplications of evidence in the supertree data set. Reviews and assumptions of monophyly that were not based on primary data analysis also were included as evidence in the supertree data set. Gatesy et al. (2002) suggested that these duplications of evidence and other problems with supertree analysis led to phylogenetic results that were not supported by the underlying character data. Actual analysis of the characters (fig. 28.4) by those authors shows that just within Paraxonia (whales + artiodactylans), the supertree topology is more than 450 character steps less parsimonious than the minimum length tree supported by the data (i.e., the supermatrix analysis). The reanalysis by Gatesy et al. (2002) also supports monophyly of Rodentia and Glires, albeit with the minimum sample size.

Applying the Phylogeny of Mammals to the Determination of the Age of a Clade: Ghost Lineages and Molecular Clocks

Great interest has been focused on determining divergence times of different mammal clades and answering such questions as what is the oldest placental mammal, the age of the clade Placentalia (e.g., the basal split within Placentalia), and the age of the ancestral eutherian lineage leading to Placentalia. Calculating these dates is fundamentally related to phylogeny reconstruction. For familiar clades (like Placentalia) whose names have been in circulation under a variety of definitions, it is particularly important to employ explicit clade definitions (e.g., crown clades) when comparing divergence dates derived exclusively from fossil evidence with those derived from calibrated molecular evidence. Obviously (with the exception of occasional ancient DNA discoveries) the only divergence times that can be estimated using calibrated molecule-based divergence times are those between pairs of extant clades. Many other divergence dates can be assessed using the fossil record alone (e.g., the divergence of two extinct clades, the divergence of one extinct and one extant clade), but these cannot be compared directly with molecule-based divergence dates. It is key to compare explicit clade *branching points* regardless of the method of determining the dates.

One means of determining the age of a divergence time that relies on few assumptions is to compare the clade in question with the age of its sister taxon. This process was described by Marshall (1990) and formalized by Norell (1992) as ghost lineage analysis. Ghost lineage analysis entails simply the assumptions of phylogeny reconstruction. Ghost lineages are predicated on the idea that, if two monophyletic taxa are sisters, then they must have split at the same time. The oldest species among the taxa in either clade puts a minimum age on the split (Marshall 1990; fig. 28.7). For example, in figure 28.7, the split between clade B and clade A marks the most basal branching point within crown clade Placentalia (as defined above). To refer to a recent analysis (Springer et al. 2003), this would represent the split between the clade (Xenarthra + Boreoeutheria) and Afrotheria, for example. The oldest of these clades is B making it both the

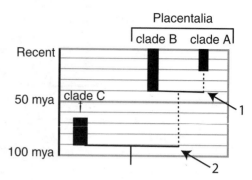

Figure 28.7. Schematic explaining the ghost lineage concept and how it can be applied to calculating the date of the basal split within the crown clade Placentalia and the age of the ancestral eutherian lineage leading to Placentalia using a hypothetical example. For the two members of Placentalia, clade A is younger than its sister, clade B. Although clade A's actual fossil record extends to only 20 Mya, clade A must have split from B at least 40 Myr old based on its phylogenetic relationships and the age of clade B. In other words, taxon A has a ghost lineage (dashed line 1) of 20 Myr. Because clade B is the oldest member of Placentalia its age puts a minimum divergence date on the basal split within Placentalia. This is the relevant date for comparison with molecule-based estimates of the origin of Placentalia. The sister taxon of Placentalia (clade C) is older than either taxon B or taxon A; therefore, there is a ghost lineage of 50 Myr (dashed line 2) extending the date of the ancestral eutherian lineage leading to Placentalia.

oldest placental and the clade that puts a minimum date of 40 Myr (million years) on the basal split within Placentalia in this hypothetical example. The segments of time for the lineage that are not recorded by fossils but which are dictated by the phylogeny are referred to as ghost lineages (fig. 28.7, dashed lines). Clade A has a ghost lineage of 20 Myr, during which time it had already split from B (but no fossils of clade A have been found in this interval). The split between Placentalia and its extinct eutherian sister taxon can also be calculated. If clade C is the eutherian sister taxon of Placentalia and is 90 Myr old, then ghost lineage logic dictates that the ancestral eutherian lineage leading to Placentalia must have split from clade C at the same time [90 million years ago (Mya)]. Ghost lineages can be calculated on any cladogram, even a cladogram of extant taxa alone. They are most effective, however, if the fossils that are part of the clade have been analyzed simultaneously with the extant taxa. If an older member of the clade is found, the phylogenetic analysis and the hypothesis of the age of the clade can be revised accordingly. Using a crown clade definition, the minimum age estimate for the origin of Placentalia is synonymous with the timing of the first split within Placentalia. The oldest species nested within crown clade Placentalia will determine the age of this divergence.

An alternative means of calculating the age of a clade is to use a molecular clock. This generally has been described as "an independent means of estimating times of origin for extinct clades" (Smith and Peterson 2002: 66) relative to the use of paleontological data. In its original formulation the molecular clock model of uniform rates of gene sequence change (Zuckerkandl and Pauling 1962, 1965) was adopted as a means of determining the absolute age of the divergence event between two lineages given a certain calibration (fig. 28.8). This differs from ghost lineages, which amount to a minimum estimate of divergence. The rate of divergence (e.g., the number of nucleotide changes that occurred in a given lineage since a splitting event) is not, however, a known quantity. The rate is derived from independent evidence used to calibrate the clock. Sometimes the calibration is a date of a selected fossil or fossils. Alternatively the date of a major geological event, such as the opening of the Atlantic or the separation of South America and Africa, has been equated with the time of separation of two taxa (reviewed in Smith and Peterson 2002). Once the calibration is established and the rate of divergence calculated, that rate is assumed to be accurate for all lineages compared (fig. 28.8). Figure 28.8 illustrates the basic equation for the calculation of divergence times using a molecular clock. This simple formula has been applied in numerous cases, but there have been criticisms of the very rationale for even applying the molecular clock (Novacek 1982, Goodman et al.1982, Ayala 1997, Ayala et al.1998). In its simplest incarnation, the molecular clock has been largely "discredited" (Smith and Peterson 2002).

Unlike ghost lineage analyses, molecular clock analyses entail not only the assumptions of phylogenetic analysis, but also at least two other important additional assumptions: (1) that the dates of origin of the fossils used to calibrate the rate of gene change ("clock") are accurate, and (2) that nucleotide changes (substitutions) occur at a uniform rate (fig. 28.8B; Zuckerkandl and Pauling 1965; see also Li 1997) or some "relaxation" of rate uniformity (methods reviewed in Smith and Peterson 2002: 75). Each of these assumptions introduces a separate set of problems.

Regarding calibration, Novacek (1982) noted that any error in the calibration of a divergence taken from a relevant fossil taxon (T in fig. 28.8) could grossly affect the estimate of divergence dates based on the clock model, a problem rediscovered by Lee (1999) and Alroy (1999). For example, molecular estimates of divergence are typically calibrated simply by a fossil's first appearance; however, this could be an underestimate of age if a well-tested phylogenetic hypothesis has not been taken into consideration. It would be appropriate to check the age of any taxon used as a calibration point against the age of its sister to see if the calibration point has a ghost lineage extending its age in time. This type of calibration has rarely been explicitly employed.

Second, and equally important, the rate (and thereby the date of a split between taxa) will also be miscalculated if a given gene has evolved at a faster rate in one of the two taxa

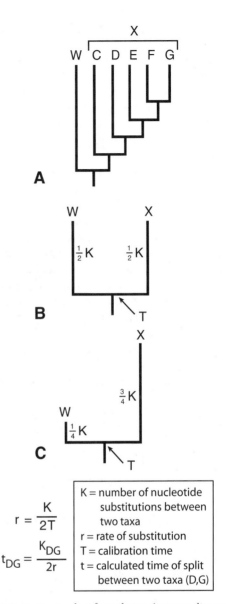

Figure 28.8. In its simplest formulation (average distance methods) a molecular clock is calibrated on the basis of the split between two taxa, for example, taxon W and taxon X (A), that is assumed to have occurred at a given date T. Using the number of nucleotide differences between X and W (e.g., 50 bp), the rate of nucleotide substitution, K, for other taxa either in the clade or outside of it can be calculated using the formula shown (C). Once this rate is established, the time elapsed between the split of taxon D and taxon G (t_{DG}), for example, can be calculated. In both the initial calculation of rate and in the calculation of the split between D and G, the assumption is that the rate of change is distributed equally down each lineage (B). It is entirely possible, however, that this assumption is violated such as is shown in (C), even for closely related taxa.

than in the other (fig. 28.8C). Enthusiasts of the molecular clock have responded to this criticism by abandoning the clock in a strict sense for a variety of different types of molecular estimates of divergence. These have been argued to be more robust because they either assess rate heterogeneity a priori or because they have a built-in ability to account for different rates of nucleotide evolution among taxa.

For example, investigators have applied relative-rate tests (Sarich and Wilson 1967; see also Tajima 1993) to compare the rates of substitution in a set of taxa, rejecting those genes that show significant rate heterogeneity. Many investigators (e.g., Kumar and Hedges 1998) have then gone on to apply the clock on genes that do not possess significant rate heterogeneity. Wu and Li (1985) used a relative rate test to compare rodent and human lineages using either artiodactylans or carnivorans as outside reference taxa. They concluded that the rate of synonymous substitution is about twice as high in the rodent lineage as in the human lineage. This did not prompt the authors to reject the molecular clock; however, instead they suggested that differences in rates could be tied to various biological parameters, arguing in this case that rodents with their short generation times would be expected to share a fairly uniform rate, but one much higher than that of humans and other primates. Similarly, Li and Graur (1991: 85) stated, "Although there is no global clock for the mammals, local clocks may exist for many groups of closely related species." There is, however, no particular evidence that gene rates are necessarily less heterogeneous between closely related taxa than between distantly related taxa, a matter that detracts from any justification for a distinction between "local" and "global" clocks. Furthermore, as discussed by Ayala et al. (1998) and Smith and Peterson (2002: 73), relative rate tests are not very powerful because they allow "considerable rate variation to go undetected" and "predicted times of origin [to be] wrong by as much as 50%" depending on the amount of unperceived rate heterogeneity. Thus, rate heterogeneity in molecular estimates of divergence times remains an important problem.

Complicating matters has been the observation that different genes also often provide different estimates of divergence. Hence, a number of workers have tried to correct the problem of rate heterogeneity by simply sampling more genes for the split in question and averaging the results. Here, it is argued that the large errors associated with estimating divergences based on the clock model can be minimized by incorporating data from many nucleotides in several genes (Fitch 1977, Li and Graur 1991). Such approaches draw on large sample sizes of sequence information and have been applied to estimates of divergence dates of many mammal and bird lineages (Hedges et al.1996, Kumar and Hedges 1998). These efforts to broaden nucleotide sampling in order to achieve supposedly more reliable estimates address only one dimension of uncertainty associated with these calculations—the possible heterogeneity in rates for *different genes*. They do not correct for the above noted problem

of variation in rates that exist *between* two lineages after a given splitting event (fig. 28.8C).

Modeling rate variation has become an alternative to the methods above as reviewed in Smith and Peterson (2002; see also Cutler 2000). Collectively these methods relax the strict assumption that rates of molecular evolution stay the same over time. The accuracy of the estimate of a divergence, however, depends on the reliability of the model of molecular substitution, which can be problematic given that the "actual patterns of amino acid or nucleotide substitution . . . are usually unknown" (Smith and Peterson 2002: 75). For example, the assumption that closely related species have similar rates of evolution described above, has found its way into certain model-based estimates of molecular divergence, where this is referred to as autocorrelation of rates (e.g., Sanderson 1997, Thorne et al.1998). However, as stated above, it is unclear that there is broad-based empirical evidence supporting this claim. Furthermore, model-based methods typically require a tree a priori because these methods require comparisons to be made topologically. Typically, these trees are derived not from combined data analyses but instead exclusively from molecular data. This tendency introduces potential shortcomings because it ignores the impact of nonmolecular data on the topology.

Divergence Times for Placentalia

Dating the radiation of Placentalia has become one of the most discussed topics in mammalian phylogenetics, in part because it has been promoted as a notorious "molecules versus morphology" debate in the scientific press. As noted above, the discussions have been complicated by pronounced variation in the definitions of such terms as "eutherian," "placental," and "therian" (e.g., compare Novacek 1999, Ji et al. 2002, Luo et al. 2002, Smith and Peterson 2002). Here we employ the stem and crown clade definitions outlined above (fig. 28.2) to explain the dating of clades using ghost lineages and as a basis for supporting our best assessment of the minimum ages of certain clades based on ghost lineages.

Calculating the minimum estimate for the age of the basal split within Placentalia using ghost lineages requires a tree that is a well-tested phylogenetic hypothesis of placental relationships that includes living and fossil species, in particular, fossil species from the Cretaceous. This permits discovery of the result that Cretaceous taxa belong within Placentalia. Preferably this tree would be derived from a combined (simultaneous) analysis of different data types (e.g., molecular and morphological) for both extant and extinct taxa. Global analyses of this kind for Placentalia, however, are only currently underway. For the purposes of illustrating the ghost lineage method, we discuss here how such a minimum age would be calculated using results from smaller phylogenetic analyses of Placentalia. Any minimum age calculations presented here would be open to testing by larger, more diverse total evidence analyses.

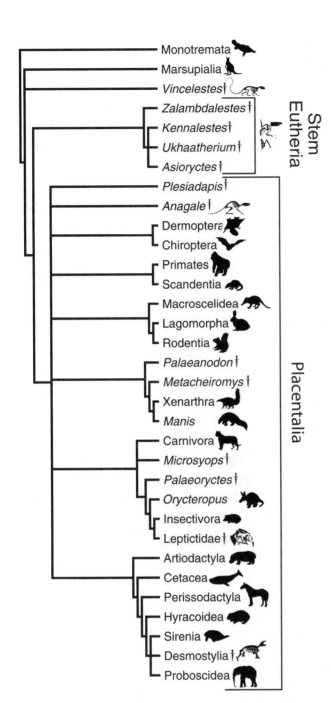

Figure 28.9. Strict consensus of eight minimum length topologies (180 steps) derived from a matrix originally analyzed by Novacek (1992a) with several taxa (e.g., *Ukhaatherium*) and characters added. Cretaceous taxa (*Zalambdalestes*, *Uhkaatherium*, *Kennalestes*, and *Asioryctes*) all fall outside of the crown clade Placentalia as its sister taxon. Fossil taxa included within Placentalia (e.g., *Palaeoryctes*) date to the early Paleocene. Using ghost lineages, this places an early Paleocene minimum divergence data on the basal split within Placentalia. Although the extinct taxa *Plesiadapis* and *Anagale* form part of a polytomy with other clades in Placentalia, in each minimum length topology these taxa fall within crown Placentalia and are therefore labeled accordingly here. The parsimony search (PAUP*, ver. 4.10) was heuristic with tree bisection and reconnection branch swapping, amb-option (internal branches collapsed if the minimal possible length of the branch is zero) in effect, multistate taxa treated as a polymorphism, all characters unordered (1000 random addition replicates). Tree rooted through Monotremata; images as in figure 28.1. Daggers indicate wholly extinct taxa; taxa without daggers also often contain a majority of extinct species. Consistency index = 0.6667; homoplasy index = 0.3333 (both of the former excluding uninformative); retention index = 0.7447; rescaled consistency index = 0.5151. Matrix available through TreeBASE.

Relevant analyses in the literature fall into two groups: (1) those that test the relationships of a number of extant placental taxa (more than two OTUs) and one or more extinct Cretaceous taxa, and (2) those that sample one representative extant placental crown clade member (OTU) and several extinct taxa from the Cretaceous. The second type of analysis obviously does not contain enough placental taxa (more than one) to permit the discovery of a Cretaceous taxon within Placentalia, but these types of analyses do contribute some information on the distribution of Cretaceous taxa within Theria (fig. 28.2).

Example analyses that fall into the first category are O'Leary and Geisler (1999) for Paraxonia (see also O'Leary

and Uhen 1999), Meng et al. (2003; see also fig. 28.3) for Glires, and an updated version of Novacek (1992a; fig. 28.9; see also Novacek 1999) that includes a number of newly discovered taxa [Shoshani and McKenna (1998) does not fit this category because it does not treat Cretaceous taxa as OTUs]. Inspection of the trees noted above for Glires and Paraxonia shows that the Cretaceous taxa included in each case fall outside the branching points between sampled members of Placentalia. An analysis across Placentalia (fig. 28.9) that includes the recently discovered and highly complete taxon *Ukhaatherium* (Novacek et al.1997) indicates that the Cretaceous taxa (*Zalambdalestes*, *Uhkaatherium*, *Kennalestes*,

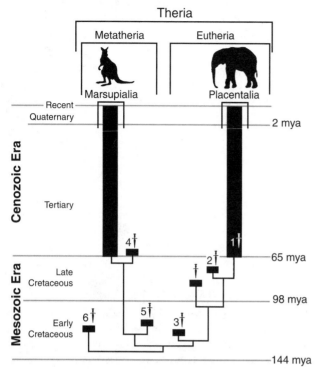

Figure 28.10. Schematic describing the age of the clade Placentalia and closely related clades (daggers indicate fossil taxa). (1) indicates a fossil taxon that falls within the crown clade Placentalia. (2) indicates a fossil that is the eutherian sister taxon of Placentalia. (3) indicates a fossil that is another Eutherian stem taxon to Placentalia (in this case the oldest member of Theria). Fossils that fall at position 2 or 3 are not directly relevant to calculating the minimum age of Placentalia. Current ghost lineage calculations indicate that all fossils within Placentalia (position 1) have a minimum age of approximately 64 Myr old (fig. 28.9). Taxa in position 2 have a minimum age of Late Cretaceous (77 Myr old; see fig. 28.9; see also Rougier et al. 1998, Rauhut et al. 2002; or 65 Myr old based on *Protungulatum*, Ji et al. 2002); taxa in position 3 have a minimum age of 125 Myr (based on *Eomaia*, Ji et al. 2002); neither of these is directly relevant to the age of Placentalia. Taxa in position 4 [*Pucadelphys* (Ji et al. 2002) or *Andinodelphys* (Rougier et al. 1998)] put a minimum age of Early Paleocene on the ancestral metatherian lineage leading to Marsupialia, and those in position 5 on the oldest member of Metatheria [Early Cretaceous based on *Kokopelia*, Ji et al. (2002) and Luo et al. (2002)]. *Eomaia*, if correctly dated at 125 Myr old, currently qualifies as the oldest known member of the crown clade Theria, but *contra* Ji et al. (2002), its discovery does not promote congruence between molecular and paleontological estimates of the basal split within Placentalia. The outgroup to Theria, position 6 (*Vincelestes*, fig. 28.9; *Slaughteria* or *Pappotherium*, Rougier et al. 1998; or *Kielantherium*, Rauhut et al. 2002, Ji et al. 2002) also dates to the Early Cretaceous or possibly Late Jurassic (*Peramus*; see Ji et al. 2002).

and *Asioryctes*) all fall outside the basal divergence within Placentalia. These taxa form a eutherian sister clade to Placentalia. This analysis overturns previous hypotheses that these Cretaceous taxa belonged within Placentalia (e.g., Novacek 1992b; see also Novacek et al.1997, Novacek 1999). Thus, based on this analysis, the basal split (here a polytomy; fig. 28.10) within Placentalia (fig. 28.10, position 1) is determined by the oldest taxon in the clade, *Palaeoryctes*, which dates to the Early Tertiary (specifically the Early Paleocene, ~64 Mya).

These ghost lineage calculations for minimum divergence times can be compared with recent molecular clock estimates. Sequence data representing many loci have been used under a clock assumption to determine molecular estimates of divergence for vertebrate groups, including placentals (e.g., Hedges et al.1996, Springer 1997, Kumar and Hedges 1998, Hedges and Poling 1999). Molecular estimates for the origin of placental clades have often been markedly older than those suggested by most calculations based on the fossil record. Hedges et al. (1996), for example, showed dates of more than 100 Myr for the origin of several lineages within crown Placentalia (e.g., primates, edentates, rodents, and artiodactylans). Because these clades are within Placentalia, these early dates, if corroborated, could pull back the dates of origin for many placental clades (depending on tree topology) well into the Cretaceous.

A second study by Kumar and Hedges (1998) greatly expanded coverage to 658 genes representing 207 vertebrate species and showed similarly ancient divergences, including a divergence time of 129 Myr, for certain members of Placentalia. These authors also estimated the split between Marsupialia and Placentalia to have occurred 173 Mya. Kumar and Hedges's (1998) analysis of mammalian divergence times has been revised, most notably by Eizirik et al. (2001), who analyzed 10,000 base pairs (bp) in 64 mammal taxa to arrive at somewhat more recent divergence times for most mammal orders, between 64 and 109 Myr, estimates that conformed more closely with those of Springer (1997). Most recently Springer et al. (2003) also contributed new dates based on 19 nuclear and three mitochondrial genes. These analyses employed model-based molecular estimates of divergence. In particular, the model of Springer et al. (2003) incorporated multiple fossil constraints on divergence times and allowed rates of molecular evolution to vary on different branches. They still obtained the result that not only were there several supraordinal divergences within the Cretaceous but also, and importantly, divergences *within* four placental orders (Lipotyphla, Rodentia, Primates, and Xenarthra) occurred prior to the Cretaceous–Tertiary boundary, as early as 74–77 Mya. Their estimate for the age of the basal split within Placentalia was 97–122 Mya. Clearly, these results disagree with the numbers presented above derived from ghost lineage calculations, which put the age of the basal split within Placentalia in the Early Tertiary at ~64 Mya. The fossil record (fig. 28.3) shows no evidence of a split within Rodentia on the order of 74 Mya or within Glires at greater than 80 Mya to match the ages of clade diversifications in Springer et al. (2003).

A number of published analyses or remarks suggest otherwise, that there is growing consensus between paleontological and molecular estimates for divergences within Placentalia that occurred well within the Cretaceous. For example, Springer et al. (2003: 1060) argued that "McKenna and Bell [1997] . . . recognized 22 genera from the Late Cretaceous and one genus from the Early Cretaceous as crown-group Placentalia," which would seem to lend paleontological support for results in Springer et al. (2003). The McKenna and Bell (1997) classification, which is a monumental literature review and synthesis of taxonomic work on Mammalia, is not, however, a classification based on a phylogenetic analysis. Strict phylogenetic readings of this classification may result in claims that are not necessarily based on analysis of character data.

Archibald (1996) and Archibald et al. (2001) also argued for the antiquity of some lineages within Placentalia based on new fossils known as zhelestids and zalambdalestids from Uzbekistan. These fossils are thought to be between 85 and 90 Myr old, and using a cladistic analysis of dental, jaw, and snout characters, these authors concluded that zhelestids were early members of a "superorder" of placental ungulates (hoofed mammals) and that zalambdalestids were associated with Glires. In other words, these authors hypothesized that their Cretaceous fossils fell in position 1 in the schematic in figure 28.10, within crown clade Placentalia. If supported, these proposals would obviously offer paleontological evidence for a much earlier origin of certain placental clades and, using ghost lineages, for the clade Placentalia as a whole.

Because of the taxon sampling, however, the Archibald et al. (2001) analysis did not amount to an explicit test of the affiliation of the new fossil taxa with Placentalia. Archibald et al. (2001) analyzed four extinct taxa (the Tertiary ungulates *Protungulatum* and *Oxyprimus* and the Tertiary Glires taxa *Tribosphenomys* and *Mimotoma*) that they argued were representative crown placentals. Although *Tribosphenomys* and *Mimotona* have subsequently been demonstrated to be members of Glires (Meng et al. 2003), and thus Placentalia, this is not the case for *Protungulatum*. Furthermore, ungulate phylogeny in general is very much in flux [e.g., compare Novacek (1992a, 1992b) with Springer et al. (2003) or Gatesy et al. (2002)]. Accordingly, robust tests of membership within any crown clade should include living members of that clade in the analysis.

Character sampling was also problematic in Archibald et al. (2001). Cranial and postcranial characters cited as evidence of the monophyly of Placentalia to the exclusion of forms like the zalambdalestids (Novacek et al.1997) were not considered. Instead of incorporating these characters into their data matrix, Archibald et al. (2001) excluded them, arguing that they did not occur universally within placentals. Such an operation implicitly suggests that some characters that may show homoplasy are more expendable than other characters, even though the authors themselves included many dental characters in their matrix that show homoplasy on their most parsimonious trees. A priori elimination of data that might produce a conflicting result is clearly not justified if the goal is to provide a robust test of alternative hypotheses of relationship (see similar problems in the above discussion on cetacean evolution).

Recent studies of some of the above taxa further suggest that Cretaceous forms such as zalambdalestids are stem groups outside crown placentals. When Meng et al. (2003) included Archibald et al.'s (2001) surrogate crown Glires and zalambdalestids in an extensive cladistic analysis of Glires, zalambdalestids did not emerge as a member of crown Glires or as its sister clade (fig. 28.3). Instead, zalambdalestids occupied a very basal position on the tree several nodes away from the other crown placental taxa in that analysis (e.g., tupaids, dermopterans, and macroscelideans). Wible et al. (2004), in the most comprehensive comparisons to date of zalambdalestid morphology, also found no clear evidence for a close affinity between zalambdalestids and Glires, or between zalambdalestids and another placental subclade.

A second paper claimed an emerging congruence between molecular and paleontological estimates of diversification for Placentalia; Ji et al. (2002: 816) identified a 125 Myr-old skeleton, *Eomaia*, from Northern China as a "eutherian (placental)" mammal and suggested that this discovery indicated a much more ancient date of origin for Placentalia than had been demonstrated previously from fossil evidence. The phylogenetic analysis in Ji et al. (2002), however, shows *Eomaia* several branches outside the basal split within Placentalia (fig. 28.10, position 3). Topologically, *Eomaia* is actually a very basal member of Eutheria on the stem to, and well outside of, Placentalia. Eomaia is of no direct relevance to molecular estimates of the basal split within Placentalia. Even if Ji et al 2002) were to use a stem-based definition of Placentalia, *Eomaia* would still not be relevant to the controversial molecule-based estimates of divergence within Placentalia. This is because the molecule-based estimates apply to the basal split within Placentalia and topologically *Eomaia* is far removed from that split.

Moreover, Ji et al. (2002) failed to point out a more relevant implication of the age of *Eomaia*—that, if correctly dated, it is one of the oldest known members of the crown clade Theria (figs. 28.2, 28.10). It provides paleontological evidence that the split between Marsupialia and Placentalia is at least 125 Myr old, a date that is still 50 Myr more recent than the estimate of 173 Myr for this divergence which emerged from the molecular clock analysis of Kumar and Hedges (1998). Contra Kumar and Hedges (1998: 917), it is not the case that "the molecular estimate for the marsupial-placental split, 173 Myr ago, corresponds well with the fossil based estimate (178–143 Myr ago)." Their characterization of the fossil-based estimate is too ancient and is not supported by ghost lineage analysis.

Clearly there remains a marked lack of agreement between minimum ages for the origin of Placentalia (and clades within it) as calculated using ghost lineages and dates of divergence derived from molecules. Our observations here corroborate those of Rodríguez-Trelles et al. (2002: 8112), who noted that "although data sets have become larger and methods of analysis considerably more sophisticated, the discrepancy between the fossil record and molecular dates has not disappeared." Indeed, several other problems with the molecular estimates can be noted. For example, the more conventional clock method employed by Kumar and Hedges (1998) does not actually require an a priori tree as some molecular estimates of divergence do. However, their results imply topologies some of which are discrepant with published trees based on character data (morphological, molecular or combined). The most conspicuous example is Glires (fig. 28.11), a clade that has been shown on the basis of morphological and molecular data to be monophyletic with respect to humans (e.g., Meng et al. 2003, Gatesy et al. 2002; see also figs. 28.3, 28.4). The topology implied by the Kumar and Hedges (1998) analysis is incongruent with the topologies of character-based analyses, because if rodents and rabbits are more closely related to each other than either is to humans, then the clock estimates should show humans splitting from rodents and rabbits at the same time. Similarly, the implied topology of Kumar and Hedges (1998) for ruminants, suids, and cetaceans (using the mean as an indicator of the sequence of divergence) is not consistent with published parsimony analyses based either on molecules, morphology, or both (see Gatesy and O'Leary 2001). Finally, the sequence of divergence of Paraxonia (sometimes referred to as Cetartiodactyla), Carnivora, and Perissodactyla is not corroborated by molecule-based phylogenetic analyses (Gatesy et al.1999a, 1999b, Murphy et al. 2001) or supermatrices (Gatesy et al. 2002). Presented with these conflicting results, we place greater importance on the tree topology because it introduces fewer assumptions.

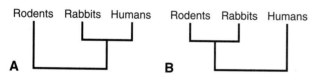

Figure 28.11. The topology implied by Kumar and Hedges' (1998) molecular clock divergences (A) contradicts that of published parsimony analyses of character data (B). Kumar and Hedges' (1998) molecular clock estimates state that a rodent (Sciurognathi) split from humans at 112 ± 3.5 Mya and that rabbits (Lagomorpha) split from humans 90.8 ± 2.0 Mya. Rodents and rabbits have been shown to be more closely related to each other than either taxon is to humans (Murphy et al. 2001). In order for the Kumar and Hedges's (1998) dates to be possible, rabbits would have to be more closely related to humans than they are to rodents (A), a topology that contradicts the tree generated from character data (B).

Certain recent authors (Smith and Peterson 2002, Springer et al. 2003) have argued that they find "convincing" (Smith and Peterson 2002: 82) the variety of molecular estimates, including "linearized tree methods that assume a single rate, quartet dating methods allowing two rates, and new Bayesian methods that allow rate variation across the topology" (Springer et al. 2003: 1061), because they all produce some intraordinal Cretaceous divergence dates for Placentalia. Less sanguine, however, are the observations of Rodríguez-Trelles et al. (2001, 2002). Rodríguez-Trelles et al. (2002: 8114) described a fundamental flaw inherent in clock-based estimates of divergence that "leads to dates that are systematically biased toward substantial overestimation of evolutionary times," especially with large samples of molecular sequence data. This is extremely problematic for molecular estimates of divergence, because large sample sizes were expected to improve these estimates. It remains unconvincing that the explanation for the incongruence between molecular and paleontological estimates is simply a poor fossil record.

Nonetheless, Smith and Peterson (2002: 65) insisted repeatedly that "a global rock bias" exists because "paleontological sampling in the Late Cretaceous is still too restricted geographically to draw any firm conclusions about the existence of a Pre-Tertiary record for modern orders [i.e., Placentalia]." But these authors did not address the arguments of Novacek (1999: 246), who noted that despite persistent geographic irregularities in the mammal fossil record, it remains "much enriched and much studied compared to other vertebrate groups" with many taxa documented from both the Cretaceous and the Tertiary periods. He argued that "[a]pparent patterns of mammalian distribution are not so easily ascribed to biases due to an impoverished record, as they might be for birds, amphibians, or other groups." This argument is consistent with the results of Foote et al. (1999), who showed that it was extremely unlikely statistically that members of Placentalia existed in the Cretaceous but simply have not been found as fossils. Smith and Peterson (2002: 71) doubted the Foote et al. (1999) results, based largely on North American and some Asian localities, could be generalized globally, because several "molecular phylogenies suggest a Gondwanan origin for many mammalian orders." The idea of a Gondwanan origin for Placentalia, however, remains untested by morphology or combined analyses and may not be substantiated once ancient placental fossils have been analyzed simultaneously with molecular sequences in phylogenetic analyses.

Thus, we fail to see why a convergence among molecular methods is a compelling validation of their results. All of these studies share the same premise that assessment of a large number of nucleotides somehow increases reliability of the molecular dates, but none fully addresses the possibility that substitution rates could differ markedly between any two related lineages. As long as this possibility remains insufficiently investigated and understood, reliance on molecular estimates for the timing of diversification in mammals and other groups seems unwarranted.

Conclusions

Discovering the mammalian section of the Tree of Life will require an enormous push for collection of both morphological and molecular data. We have outlined here a recommendation that these data should be assembled into supermatrices because this will create the strongest connection between the resulting tree topology and the underlying character data. We have also noted that advances in search algorithms make it increasingly straightforward to analyze thousands of taxa simultaneously, making a single supermatrix for Mammalia (combining extinct and extant taxa), a goal that is becoming increasingly within reach.

We have described how tree structure is extremely important for reconstructing the time of origin of a clade; one example of the many ways the mammalian sector of the Tree of Life can be applied to other evolutionary questions. Still other applications include understanding how characters have transformed through time (optimization), information that can even be used to reconstruct missing data (e.g., skin, behavior) in fossils (e.g., O'Leary 2001). Investigations of biogeographic area of origin and time of origin are also highly dependant on a well-tested underlying tree.

The last decade in particular has witnessed an enormous increase in the amount of molecular data available for phylogenetic analysis, and this new work is greatly enhanced by a sophisticated bioinformatics infrastructure, namely, the publicly supported molecular sequence database known as GenBank at the National Center for Biotechnology Information, which makes molecular sequence data quickly and freely available to investigators worldwide. This availability of raw data for molecule-based phylogenetic analyses makes the construction of molecular supermatrices relatively straightforward. New raw data can be quickly compared and combined with previously collected raw data for new phylogenetic analyses. The fact that this database now supports multiple alignments will make the synthesis even easier.

Morphological data, by contrast, currently are not supported by an equivalent centralized database within which raw observations from published morphological analyses are organized and archived for future phylogenetic analysis. As a result, systematists working with morphological data often find themselves in the position of "recollecting" data someone has amassed before. This is an unacceptable and wasteful repetition of effort that is in part responsible for restraining large-scale supermatrix analyses that combine molecular and morphological matrices. We believe that the databasing of morphological observations (homology statements) must be improved and that this is one of the most crucial modifications that must occur for a Tree of Life effort to be successful, not just for Mammalia but for all species. Our knowledge of extinct species, which far outnumber extant species in the mammalian clade, also comes almost exclusively from morphology. The full integration of molecular and morphological data so critical to resolving problems in mammal phylogeny will be most easily accomplished after the development of an appropriate bioinformatics infrastructure for archiving morphological data such as has been proposed as MorphoBank (2003).

The recent explosion of published phylogenetic analyses for many mammal clades includes contributions from such historically disparate fields as histology, paleontology, and molecular biology, challenges mammalian systematists to absorb data collected outside their field of specialization. Integration of these data will provide the greatest explanatory power because it will cast phylogenetic analysis not as a search for a subset of characters and taxa that will unlock phylogenetic truth, but as an accretionary synthesis of detailed comparative work across all phenotypic and genotypic systems and in all taxa.

Acknowledgments

We thank J. Cracraft and M. Donoghue for inviting us to contribute this article to the Tree of Life symposium at the American Museum of Natural History. For helpful discussion or comments on the manuscript, we thank K. de Queiroz, M. Malia, Jr., and one anonymous reviewer. Figure 28.1 was prepared by E. Heck with the assistance of L. Merrill; other figures were prepared by L. Betti-Nash or the authors. This work was supported by grants to M.A.O. (NSF DEB 9903964), M.A. (NSF DEB-9629319; the research participation program of the Oak Ridge Institute for Science and Education and the Counterterrorism and Forensic Science Research Unit at the FBI Academy), M.J.N. (NSF-DEB 9996172, NSF-DEB-0129031, the Antorchas Foundation, and the Frick Laboratory Endowment), J.G. (NSF-DEB 9985847), and J.M. (NSF-EAR-0120727 and Chinese National Science Foundation grant 49928202).

Literature Cited

Adams, E. N., III. 1972. Consensus techniques and the comparison of taxonomic trees. Syst. Zool. 21:390–397.

Allard, M. W., K. Miller, M. Wilson, K. L. Monson, and B. Budowle. 2002. Characterization of the Caucasian haplogroups present in the SWGDAM forensic mtDNA data set for 1771 human control region sequences. J. Forens. Sci. 47:1215–1223.

Allard, M. W., M. M. Miyamoto, and R. L. Honeycutt. 1991. Tests for rodent polyphyly. Nature 353:610–611.

Alroy, J. 1999. The fossil record of North American mammals: evidence for a Paleocene evolutionary radiation. Syst. Biol. 48:107–118.

Archer, M., R. Arena, M. Bassarova, K. Black, J. Brammall, B. Cooke, P. Creaser, K. Crosby, A. Gillespie, H. Godthelp, et al.1999. The evolutionary history and diversity of Australian mammals. Austral. Mammal. 21:1–45.

Archibald, J. D. 1996. Fossil evidence for a Late Cretaceous origin of "hoofed" mammals. Science 272:1150–1153.

Archibald, J. D, A. O. Averianov, and E. G. Ekdale. 2001. Late Cretaceous relatives of rabbits, rodents, and other extant eutherian mammals. Nature 414:62–65.

Ayala, F. J. 1997. Vagaries of the molecular clock. Proc. Natl. Acad. Sci. USA 94:7776–7783.

Ayala, F. J., A. Rzhetsky, and F. J. Ayala. 1998. Origin of the metazoan phyla: molecular clocks confirm paleontological estimates. Proc. Natl. Acad. Sci. USA 95:606–611.

Ax, P. 1987. The phylogenetic system: the systematization of organisms on the basis of their phylogenies. Wiley, New York.

Barrett, M., M. Donoghue, and E. Sober. 1991. Against consensus. Syst. Zool. 40:486–493.

Baum, B. 1992. Combining trees as a way of combining data sets for phylogenetic inference, and the desirability of combining gene trees. Taxon 41:3–10.

Baum, B., and M. A. Ragan. 1993. Comment on Baum method for combining phylogenetic trees—reply. Taxon 42:637–640.

Bremer, K. 1994. Branch support and tree stability. Cladistics 10:295–304.

Bull, J. J., J. P. Huelsenbeck, C. W. Cunningham, D. L. Swofford, and P. J. Waddell. 1993. Partitioning and combining data in phylogenetic analysis. Syst. Biol. 42:384–397.

Casiliano, M. L., and W. A. Clemens. 1979. Symmetrodonta. Pp. 151–161 in Mesozoic mammals: the first two-thirds of mammalian history (J. Lillegraven, Z. Kielan-Jaworowska, and W. Clemens, eds.). University of California Press, Berkeley.

Catzeflis, F. M. 1993. Mammalian phylogeny: morphology and molecules. Trends Ecol. Evol. 8:340–341.

Cifelli, R. L. 1999. Tribosphenic mammal from the North American Early Cretaceous. Nature 401:363–366.

Cifelli, R. L. 2000. Cretaceous mammals of Asia and North America. Paleontol. Soc. Korea Spec. Publ. 4:49–84.

Cifelli, R. L. 2001. Early mammalian radiations. J. Paleontol. 75:1214–1226.

Cifelli, R. L., and C. De Muizon. 1997. Dentition and jaw of Kokopellia juddi, a primitive marsupial or near-marsupial from the medial Cretaceous of Utah. J. Mammal. Evol. 4:241–258.

Clemens, W. A., Jr. 1968. A mandible of Didelphodon vorax (Marsupialia, Mammalia). Contrib. Sci. 133:1–11.

Cole, R., and R. Hariharan. 1996. An O(n log n) algorithm for the maximum agreement subtree problem for binary trees. Pp. 323–332 in Proceedings of the 7th annual ACM-SIAM Symposium on Discrete Algorithms (SODA96).

Crompton, A. W. 1995. Masticatory function in nonmammalian cynodonts and early mammals. Pp. 55–75 in Functional morphology in vertebrate paleontology (J. J. Thomason, ed.). Cambridge University Press, Cambridge.

Crompton, A. W., and F. A. Jenkins, Jr. 1973. Mammals from reptiles: a review of mammalian origins. Annu. Rev. Earth Planet. Sci. 1:131–155.

Crompton, A. W., and F. A. Jenkins, Jr. 1979. Origin of mammals. Pp. 59–73 in Mesozoic mammals—the first two-thirds of mammalian history (J. A. Lillegraven, Z. Kielan-Jaworowska, and W. A. Clemens, eds.). University of California Press, Berkeley.

Cunningham, C. W. 1997. Is congruence between data partitions a reliable predictor of phylogenetic accuracy? Empirically testing an iterative procedure for choosing among phylogenetic methods. Syst. Biol. 46:464–478.

Cutler, D. J. 2000. Estimating divergence times in the presence of an overdispersed molecular clock. Mol. Biol. Evol. 17:1647–1660.

Dashzeveg, D., and Z. Kielan-Jaworowska. 1984. The lower jaw of an aegialodontid mammal from the Early Cretaceous of Mongolia. Zool. J. Linn. Soc. 82:217–227.

Dawson, M. R., and K. C. Beard. 1996. New late Paleocene rodents (Mammalia) from Big Multi Quarry, Washakie Basin, Wyoming. Palaeovertebrata 25:301–321.

Dawson, M. R., C. Li, and T. Qi. 1984. Eocene ctenodactyloid rodents (Mammalia) of eastern central Asia. Spec. Publ. Carnegie Mus. Nat. Hist. 9:138–150.

De Muizon, C., and R. L. Cifelli. 2000. The "condylarths" (archaic Ungulata, Mammalia) from Early Palaeocene of Tiupampa (Bolivia): implications on the origin of the South American ungulates. Geodiversitas 22:47–150.

De Queiroz, K., and J. Gauthier. 1992. Phylogenetic taxonomy. Annu. Rev. Ecol. Syst. 23:449–480.

DeSalle, R., and A. V. Z. Brower. 1997. Process partitions, congruence, and the independence of characters: inferring relationships among closely related Hawaiian Drosophila from multiple gene regions. Syst. Biol. 46:751–764.

DeSalle, R., G. Giribet, and W. Wheeler (eds.). 2002. Techniques in molecular systematics and evolution. Birkhäuser Verlag, Boston.

Domning, D. P., C. E. Ray, and M. C. McKenna. 1986. Two new Oligocene Desmostylians and a discussion of tethytherian systematics. Smithson. Contrib. Paleobiol. 59:1–56.

Eizirik, E., W. J. Murphy, and S. J. O'Brien. 2001. Molecular dating and biogeography of the early placental mammal radiation. J. Hered. 92:212–219.

Farris, J. S. 1969. A successive approximations approach to character weighting. Syst. Zool. 18:374–385.

Farris, J. S. 1983. The logical basis of phylogenetic analysis. Pp.1–36 in Advances in cladistics (N. I. Platnick and V. A. Funk, eds.), vol. 2. Columbia University Press, New York.

Felsenstein, J. 1985. Confidence limits on phylogenies: an approach using the bootstrap. Evolution 39:783–791.

Fitch, W. M. 1977. The phyletic interpretation of macro-molecular sequence information: sample cases. Pp. 211–248 in Major patterns of vertebrate evolution (M. K. Hecht, P. C. Goody, and B. B. Hecht, eds.). Plenum Press, New York

Flynn J. J., J. M. Parish, B. Rakotosamimanana, W. F. Simpson, and A. R. Wyss. 1999. A middle Jurassic mammals from Madagascar. Nature 401:57–60.

Foote, M., J. P. Hunter, C. M. Janis, and J. J. Sepkoski. 1999. Evolutionary and preservational constraints on origins of biologic groups: divergence times of eutherian mammals. Science 283:1310–1314.

Fox, R. C., G. P. Youzwyshyn, and D. W. Krause. 1992. Post-Jurassic mammal-like reptile from the Palaeocene. Nature 358:233–235.

Gatesy, J., C. Matthee, R. DeSalle, and C. Hayashi. 2002. Resolution of a supertree/supermatrix paradox. Syst. Biol. 51:652–664.

Gatesy, J., M. Milinkovitch, V. Waddell, and M. Stanhope. 1999a. Stability of cladistic relationships between Cetacea and higher level artiodactyl taxa. Syst. Biol. 48:6–20.

Gatesy, J., P. O'Grady, and R. H. Baker. 1999b. Corroboration among data sets in simultaneous analysis: hidden support for phylogenetic relationships among higher level artiodactyl taxa. Cladistics 15:271–314.

Gatesy, J., and M. A. O'Leary. 2001. Deciphering whale origins with molecules and fossils. Trends Ecol. Evol. 16:562–570.

Gauthier, J., A. G. Kluge, and T. Rowe. 1988. Amniote phylogeny and the importance of fossils. Cladistics 4:105–209.

Geisler, J. H. 2001. New morphological evidence for the phylogeny of Artiodactyla, Cetacea, and Mesonychidae. Am. Mus. Nov. 3344:1–53.

GenBank. 2003. Molecular data archive. Available: www.ncbi. nlm.nih.gov. Last accessed 25 December 2003.

Goloboff, P. 1994. NONA: a tree search program, vers. 2 (program and documentation). Available: *www.cladistics. com*. Last accessed 8 December 2003.

Goloboff, P., and J. S. Farris. 2001. Methods for quick consensus estimation. Cladistics 17:s26–s34.

Goloboff, P., J. S. Farris, and K. Nixon. 1999. T.N.T.: tree analysis using new technology. Available: http://www. cladistics.com. Last accessed 8 December 2003.

Goodman, M., A. E. Romero-Herrara, H. Dene, J. Czelusniak, and R. E. Tashian. 1982. Amino acid sequence evidence on the phylogeny of primates and other eutherians. Pp. 115–191 in Macromolecular sequences systematic and evolutionary biology (M. Goodman, ed.). Plenum Press, New York.

Graur, D. 1993. Molecular phylogeny and the higher classi-fication of eutherian mammals. Trends Ecol. Evol. 8:141–147.

Graur, D., L. Duret, and M. Gouy. 1996. Phylogenetic position of the order Lagomorpha (rabbits, hares and allies). Nature 379:333–335.

Graur, D., W. A. Hide, and W. H. Li. 1991. Is the guinea-pig a rodent. Nature 351:649–652.

Gregory, W. K. 1947. The monotremes and the palimpsest theory. Bull. Am. Mus. Nat. Hist. 88:1–52.

Halanych, K. M. 1998. Lagomorphs misplaced by more characters and fewer taxa. Syst. Biol. 47:138–146.

Hasegawa, M., Y. Cao, J. Adachi, and T.-A. Yano. 1992. Rodent polyphyly. Nature 355:595.

Hedges, S. B., and L. R. Maxson. 1996. Re: molecules and morphology in amniote phylogeny. Mol. Phylogenet. Evol. 6:312–314.

Hedges, S. B., P. H. Parker, C. G. Sibley, and S. Kumar. 1996. Continental breakup and the ordinal diversification of birds and mammals. Nature 381:226–229.

Hedges, S. B., and L. L. Poling. 1999. A molecular phylogeny of reptiles. Science 283:998–1001.

Hopson, J. A. 1994. Synapsid evolution and the radiation of non-eutherian mammals. Pp. 190–219 in Major features of vertebrate evolution (R. S. Spencer, ed.). Paleontological Society, Knoxville, TN.

Hu, Y. M., Y. Q. Wang, Z. X. Luo, and C. K. Li. 1997. A new symmetrodont mammal from China and its implications for mammalian evolution. Nature 390:137–142.

Janies, D., and W. Wheeler. 2001. Efficiency of parallel direct optimization. Cladistics 17:S71–S82.

Janies, D., and W. Wheeler. 2002. Theory and practice of parallel direct optimization. Pp. 115–123 in Molecular systematics and evolution (R. DeSalle, W. Wheeler, and G. Giribet, eds.). Birkhauser Verlag, Boston.

Janke, A., G. Feldmaier-Fuchs, W. K. Thomas, A. von-Haeseler, and S. Pääbo. 1994. The marsupial mitochondrial genome

and the evolution of placental mammals. Genetics 137:243–256.

Janke, A., O. Magnell, G. Wieczorek, M. Westerman, and U. Arnason. 2002. Phylogenetic analysis of 18S rRNA and the mitochondrial genomes of the wombat, *Vombatus ursinus*, and the spiny anteater, *Tachyglossus aculeatus*: increased support for the Marsupionta hypothesis. J. Mol. Evol. 54:71–80.

Janke, A., X. Xu, and U. Arnason. 1997. The complete mitochondrial genome of the wallaroo (*Macropus robustus*) and the phylogenetic relationship among Monotremata, Marsupialia, and Eutheria. Proc. Natl. Acad. Sci. USA 94:1276–1281.

Jefferies, R. P. S. 1979. The origin of chordates—a methodological essay. Pp. 443–477 in The origin of major invertebrate groups (M. R. House, ed.). Systematics Association spec. vol. 12. Academic Press, New York.

Jenkins, F., Jr., and D. W. Krause. 1983. Adaptations for climbing in North American multituberculates (Mammalia). Science 220:712–715.

Ji, Q., Z. Luo, and S.-A. Ji. 1999. A Chinese triconodont mammal and mosaic evolution of the mammalian skeleton. Nature 398:326–330.

Ji, Q., Z.-X. Luo, C.-X. Yuan, J. R. Wible, and J.-P. Zhang, and J. A. Georgi. 2002. The earliest known eutherian mammal. Nature 416:816–822.

Kearney, M. 2002. Fragmentary taxa, missing data, and ambiguity: mistaken assumptions and conclusions. Syst. Biol. 51:369–381.

Kermack, D. M., K. A. Kermack, and F. Mussett. 1968. The Welsh pantothere *Kuehneotherium praecursoris*. J. Linn. Soc. Lond. Zool. 47:407–422.

Kermack, K. A., and F. Mussett. 1958. The jaw articulation of the Docodonta and the classification of Mesozoic mammals. Proc. R. Soc. Lond. B. 148:535–554.

Kermack, K. A., F. Mussett, and H. W. Rigney. 1973. The lower jaw of *Morganucodon*. Zool. J. Linn. Soc. 53:87–175.

Kielan-Jaworowska, Z., R. L. Cifelli, and Z. Luo. 1998. Alleged Cretaceous plcental from down under. Lethaia 31:267–268.

Kitching I. J., P. L. Forey, C. J. Humphries, and D. M. Williams. 1998. Cladistics (2nd ed.): the theory and practice of parsimony analysis. Systematics Association publ. no. 11. Oxford Science Publications, New York.

Kluge, A. G. 1989. A concern for evidence and a phylogenetic hypothesis of relationships among Epicrates (Boidae, Serpentes). Syst. Zool. 38:7–25.

Korth, W. W. 1984. Earliest Tertiary evolution and radiation of rodents in North America. Bull. Carnegie Mus. Nat. Hist. 24:1–71.

Krebs, B. 1971. Evolution of the mandible and lower dentition in dryolestids (Pantotheria, Mammalia). Pp. 89–102 in Early mammals (D. M. Kermack and K. A. Kermack, eds.). Zoological Journal of the Linnaean Society, vol. 50, suppl. 1.

Kühne, W. G. 1973. The systematic position of monotremes reconsidered. Z. Morphol. Tiere 75:59–64.

Kumar, S., and S. B. Hedges. 1998. A molecular timescale for vertebrate evolution. Nature 392:917–920.

Langer, P. 2001. Evidence from the digestive tract on phylogenetic relationships in ungulates and whales. J. Zool. Syst. Evol. Res. 39:77–90.

Lee, M. S. Y. 1999. Molecular clock calibrations and metazoan divergence dates. J. Mol. Evol. 49:385–391.

Li, C. K., R. W. Wilson, M. R. Dawson, and L. Krishtalka. 1987. The origin of rodents and lagomorphs. Pp. 91–108 in Current mammalogy, vol. 1 (H. H. Genoways, ed.). Plenum Press, New York.

Li, W.-H. 1997. Molecular evolution. Sinauer, Sunderland, MA.

Li, W.-H., and D. Graur. 1991. Fundamentals of molecular evolution. Sinauer, Sunderland, MA.

Li, W.-H., W. A. Hide, and D. Graur. 1992. Origin of rodents and guinea-pigs. Nature 359:277–278.

Lipscomb, D. L., J. S. Farris, M. Källersjö, and A. Tehler. 1998. Support, ribosomal sequences and the phylogeny of the Eukaryotes. Cladistics 14:303–338.

Liu, F. R., M. M. Miyamoto, N. P. Freire, P. Q. Ong, M. R. Tennent, T. S. Young, and K. F. Gugel. 2001. Molecular and morphological supertrees for eutherian (placental) mammals. Science 291:1786–1789.

Luckett, W. P., and J.-L. Hartenberger. 1993. Monophyly or polyphyly of the order Rodentia: possible conflict between morphological and molecular interpretations. J. Mammal. Evol. 1:127–147.

Luckett, W. P., and N. Hong. 1998. Phylogenetic relationships between the orders Artiodactyla and Cetacea: a combined assessment of morphological and molecular evidence. J. Mammal. Evol. 5:127–182.

Luo, Z.-X., R. L. Cifelli, and Z. Kielan-Jaworowska. 2001a. Dual origin of tribosphenic mammals. Nature 409:53–57.

Luo, Z.-X., A. W. Crompton, and A.-L. Sun. 2001b. A new mammaliaform from the Early Jurassic and evolution of mammalian characteristics. Science 292:1535–1540.

Luo, Z.-X., Z. Kielan-Jaworowska, and R. L. Cifelli. 2002. In quest for a phylogeny of Mesozoic mammals. Acta Palaeontol. Pol. 47:1–78.

Madsen, O., M. Scally, C. J. Douady, D. J. Kao, R. W. DeBry, R. Adkins, H. M. Amrine, M. J. Stanhope, W. W. de Jong, and M. S. Springer. 2001. Parallel adaptive radiations in two major clades of placental mammals. Nature 409:610–614.

Malia, M. J., Jr., D. L. Lipscomb, and M. W. Allard. 2003. The misleading effects of composite taxa in supermatrices. Mol. Phylogenet. Evol. 27:522–527.

Marshall, C. R. 1990. The fossil record and estimating divergence times between lineages: maximum divergence times and the importance of reliable phylogenies. J. Mol. Evol. 30:400–408.

Matthee, C. A., J. D. Burzlaff, J. F. Taylor, and S. K. Davis. 2001. Mining the mammalian genome for artiodactyl systematics. Syst. Biol. 50:367–390.

McDowell, S. B. 1958. The Greater Antillean insectivores. Bull. Am. Mus. Nat. Hist. 115:117–214.

McKenna, M. C. 1982. Lagomorpha interrelationships. Geobios. Mem. Spec. 6:213–223.

McKenna, M. C., and S. K. Bell. 1997. Classification of mammals above the species level. Columbia University Press, New York.

Meng, J., Y.-M. Hu, and C. Li. 2003. The osteology of Rhombomylus (Mammalia, Glires) and its implication for Glires systematics and evolution. Bull. Am. Mus. Nat. Hist. 275:1–247.

Meng, J., and A. Wyss. 1997. Multituberculate and other mammal hair recovered from Palaeogene excreta. Nature 385:712–714.

Meng, J., and A. R. Wyss. 2001. The Morphology of Tribosphenomys (Rodentiaformes, Mammalia): phylogenetic implications for basal Glires. J. Mammal. Evol. 8:1–71.

Messenger, S. L., and J. A. McGuire. 1998. Morphology, molecules, and the phylogenetics of cetaceans. Syst. Biol. 47:90–124.

Miyamoto, M. M., and W. M. Fitch. 1995. Testing species phylogenies and phylogenetic methods with congruence. Syst. Biol. 44:64–76.

MorphoBank. 2003. Available: www.morphobank.net.

Murphy, W. J., E. Eizirik, W. E. Johnson, Y.-P. Zhang, O. A. Ryder, and S. J. O'Brien. 2001. Molecular phylogenetics and the origins of placental mammals. Nature 409:614–618.

Murphy, W. J., E. Eizirik, S. J. O'Brien, O. Madsen, M. Scally, C. J. Douady, E. Teeling, O. A. Ryder, M. J. Stanhope, W. W. de Jong, and M. S. Springer. 2002. Resolution of the early placental mammal radiation using Bayesian phylogenetics. Science 294:2348–2351.

Musser, A. M., and M. Archer. 1998. New information about the skull and dentary of the Miocene platypus Obdurodon dicksoni, and a discussion of ornithorhynchid relationships. Philos. Trans. R. Soc. Lond. Biol. Sci. B 353:1063–1079.

Naylor, G. J. P., and D. Adams. 2001. Are the fossil data really at odds with the molecular data? Morphological evidence for Cetartiodactyla phylogeny reexamined. Syst. Biol. 50:444–453.

Naylor, G. J. P., and W. M. Brown. 1998. Structural biology and phylogenetic estimation. Nature 388:527–528.

Nikaido, M., F. Matsuno, H. Hamilton, R. L. Brownell, Jr., Y. Cao, W. Ding, Z. Zuoyan, A. M. Shedlock, R. E. Fordyce, M. Hasegawa, and N. Okada. 2001. Retroposon analysis of major cetacean lineages: the monophyly of toothed whales and the paraphyly of river dolphins. Proc. Natl. Acad. Sci. USA 98:7384–7389.

Nikaido, M., A. P. Rooney, and N. Okada. 1999. Phylogenetic relationships among cetartiodactyls based on insertions of short and long interspersed elements: hippopotamuses are the closest extant relatives of whales. Proc. Natl. Acad. Sci. USA 96:10261–10266.

Nixon, K. C. 1999. The parsimony ratchet, a new method for rapid parsimony analysis. Cladistics 15:407–414.

Nixon, K. C. 2002. WinClada, ver. 1.00.08. K. C. Nixon, Ithaca, NY. Available: http://www.cladistics.com. Last accessed 25 December 2003.

Nixon, K. C., and J. M. Carpenter. 1996. On simultaneous analysis. Cladistics 12:221–241.

Nixon, K. C., and J. M. Carpenter. 2000. On the other "phylogenetic systematics." Cladistics 16:298–318.

Norell, M. A. 1992. Taxic origin and temporal diversity: the effect of phylogeny. Pp. 89–118 in Extinction and phylogeny (M. J. Novacek and Q. D. Wheeler, eds.). Columbia University Press, New York.

Novacek, M. J. 1982. Information for molecular studies from anatomical and fossil evidence on higher eutherian phylogeny. Pp. 3–41 in Macromolecular sequences in systematic and evolutionary biology (M. Goodman, ed.). Plenum Press, New York.

Novacek, M. J. 1992a. Fossils, topologies, missing data, and the higher level phylogeny of eutherian mammals. Syst. Biol. 41:58–73.

Novacek, M. J. 1992b. Mammalian phylogeny: shaking the tree. Nature 356:121–125.

Novacek, M. J. 1993. Reflections on higher mammalian phylogenetics. J. Mammal. Evol. 1:3–30.

Novacek, M. J. 1999. 100 million years of land vertebrate evolution: the Cretaceous-Early Tertiary transition. Ann. Mo. Bot. Gard. 86:230–258.

Novacek, M. J., G. W. Rougier, J. R. Wible, M. C. McKenna, D. Dashzeveg, and I. Horovitz. 1997. Epipubic bones in eutherian mammals from the Late Cretaceous of Mongolia. Nature 389:483–486.

Novacek, M. J., and A. R. Wyss. 1987. Selected features of the desmostylian skeleton and their phylogenetic implications. Am. Mus. Nov. 2870:1–8.

Nowak, R. M. 1999. Walker's mammals of the world. 6th ed. Johns Hopkins University Press, Baltimore, MD.

O'Leary, M. A. 2001. The phylogenetic position of cetaceans: further combined data analyses, comparisons with the stratitgraphic record and a discussion of character optimization. Am. Zool. 41:487–506.

O'Leary, M. A., J. Gatesy, and M. J. Novacek. 2003. Are the dental data really at odds with the molecular data? Morphological evidence for whale phylogeny (re)reexamined. Syst. Biol. 52:853–864.

O'Leary, M. A., and J. H. Geisler. 1999. The position of Cetacea within Mammalia: phylogenetic analysis of morphological data from extinct and extant taxa. Syst. Biol. 48:455–490.

O'Leary, M. A., and M. D. Uhen. 1999. The time of origin of whales and the role of behavioral changes in the terrestrial-aquatic transition. Paleobiology 25:534–556.

Olson, E. C. 1944. Origin of mammals based upon cranial morphology of the therapsid suborders. Geol. Soc. Am. Spec. Pap. 55:1–136.

Olson, E. C. 1959. The evolution of mammalian characters. Evolution 13:344–353.

Padian, K., and J. M. V. Rayner. 1993. The wings of pterosaurs. Am. J. Sci. 293A:91–166.

Pascual, R., F. J. Goin, D. W. Krause, E. Ortiz-Jaureguizar, and A. A. Carlini. 1999. The first gnathic remains of *Sudamerica*: Implications for gondwanathere relationships. J. Vert. Paleontol. 19:373–382.

Pisani, D., and M. Wilkinson. 2002. Matrix representation with parsimony, taxonomic congruence, and total evidence. Syst. Biol. 51:151–155.

Prendini, L. 2001. Species or supraspecific taxa as terminals in cladistic analysis? Groundplans versus exemplars revisited. Syst. Biol. 50:290–300.

Purvis, A. 1995. A modification to Baum and Ragan's method for combining phylogenetic trees. Syst. Biol. 44:251–255.

Ragan, M. A. 1992. A phylogenetic inference based on matrix representation of trees. Mol. Phylogenet. Evol. 1:53–58.

Rauhut, O. W. M., T. Martin, E. Ortiz-Jaureguizar, and P. Puerta. 2002. A Jurassic mammal from South America. Nature 416:165–168.

Rice, K. A., M. J. Donoghue, and R. G. Olmstead. 1997. Analyzing large data sets: *rbc*L 500 revisited. Syst. Biol. 46:554–563.

Rich, T. H., T. F. Flannery, P. Trusler, A. L. Kool, N. van Klaveren, and P. Vickers-Rich. 2001a. Early Cretaceous mammals from Flat Rocks, Victoria, Australia. Rec. Queen Victoria Mus. 106:1–30.

Rich, T. H., P. Vickers-Rich, A. Constantine, T. F. Flannery, L. Kool, and N. van Klaveren. 1997. A tribosphenic mammal from the Mesozoic of Australia. Science 278:1438–1442.

Rich, T. H., P. Vickers-Rich, P. Trusler, T. F. Flannery, R. Ciffelli, A. Constantine, L. Kool, and N. van Klaveren. 2001b. Corroboration of the Garden of Eden hypothesis. Pp. 324–332 in Faunal and floral migrations and evolution in SE Asia-Australia (I. Metcalfe, J. M. B. Smith, M. Morwood, I. Davidson, and K. Hewison, eds.). A. A. Balkema, Lisse, the Netherlands.

Riggs, E. S. 1935. A skeleton of *Astrapotherium* [Argentina]. Field Mus. Nat. Hist. Geol. Ser. 6. 13:167–176.

Rodríguez-Trelles, F., R. Tarrío, and F. J. Ayala. 2001. Erratic overdispersion of three molecular clocks: GPDH, SOD, and XDH. Proc. Natl. Acad. Sci. USA. 98:11405–11410.

Rodríguez-Trelles, F., R. Tarrío, and F. J. Ayala. 2002. A methodological bias toward overestimation of molecular evolutionary time scales. Proc. Natl. Acad. Sci. USA 99(12):8112–8115.

Ronquist, F. 1996. Matrix representation of trees, redundancy and weighting. Syst. Biol. 45:247–253.

Rose, K. D. 1987. Climbing adaptations in the early Eocene mammal *Chriacus* and the origin of Artiodactyla. Science 236:314–316.

Rougier, G. W. 1993. *Vincelestes neuquenianus* Bonaparte (Mammalia, Theria) un primitivo mamifero del cretacio inferior de la cuenca Neuquina. 1–3. Ph.D. thesis, Universidad de Buenos Aires.

Rougier, G. W., J. R. Wible, and J. A. Hopson. 1996. Basicranial anatomy of *Priacodon fruitaensis* (Triconodontidae, Mammalia) from the Late Jurassic of Colorado, and a reappraisal of mammaliaform interrelationships. Am. Mus. Nov. 3183:1–38.

Rougier, G. W., J. R. Wible, and M. J. Novacek. 1998. Implications of *Deltatheridium* specimens for early marsupial history. Nature 396:459–463.

Rowe, T. 1988. Definition, diagnosis, and origin of Mammalia. J. Vert. Paleontol. 8:241–264.

Salamin, N., T. R. Hodkinson, and V. Savolainen. 2002. Building supertrees: an empirical assessment using the grass family (Poaceae). Syst. Biol. 51:136–150.

Sanderson, M. J. 1997. A nonparametric approach to estimating divergence times in the absence of rate constancy. Mol. Biol. Evol. 14:1218–1231.

Sanderson, M. J., A. Purvis, and C. Henze. 1998. Phylogenetic supertrees: assembling the trees of life. Trends Ecol. Evol. 13:105–109.

Sarich, V. M., and A. C. Wilson. 1967. Immunological time scale for hominid evolution. Science 158:1200–1203.

Scott, W. B. 1910. Litopterna of the Santa Cruz Beds. Rep. Princeton Univ. Exped. Patagonia, 1896–1899. 7:1–156.

Shoshani, J., and M. C. McKenna. 1998. Higher taxonomic relationships among extant mammals based on morphology, with selected comparisons of results from molecular data. Mol. Phylogenet. Evol. 9:572–584.

Simpson, G. G. 1936. Studies of the earliest mammalian dentitions. Dental Cosmos 6:1–24.

Simpson, G. G. 1960. Diagnosis of the Classes Reptilia and Mammalia. Evolution 14:388–392.

Simpson, G. G. 1967. The beginning of the age of mammals in South America, Part 2. Bull. Am. Mus. Nat. Hist. 137:1–260.

Sinclair, W. J. 1906. Marsupials of the Santa Cruz Beds. Rep. Princeton Univ. Exped. Patagonia. 1896–1899. 4:333–359.

Smith, A. B., and K. J. Peterson. 2002. Dating the time of origin of major clades: molecular clocks and the fossil record. Annu. Rev. Earth Planet. Sci. 30:65–88.

Springer, M. S. 1997. Molecular clocks and the timing of the placental and marsupial radiations in relation to the Cretaceous-Tertiary boundary. J. Mammal. Evol. 4:285–302.

Springer, M. S., W. J. Murphy, E. Eizirik, and S. J. O'Brien. 2003. Placental mammal diversification and the Cretaceous-Tertiary boundary. Proc. Natl. Acad. Sci. USA 100:1056–1061.

Sullivan, J., and D. L. Swofford. 1997. Are guinea pigs rodents? The importance of adequate models in molecular phylogenetics. J. Mammal. Evol. 4:77–86.

Swofford, D. L. 2000. PAUP*: phylogenetic analysis using parsimony (and other methods), vers. 4 beta. Sinauer, Sunderland, MA.

Tajima, F. 1993. Simple methods for testing the molecular clock hypothesis. Genetics 135:599–607.

Thewissen, J. G. M., E. M. Williams, and S. T. Hussain. 2001. Skeletons of terrestrial cetaceans and the relationships of whales to artiodactyls. Science 413:277–281.

Thorne, J. L., H. Kishino, and I. S. Painter. 1998. Estimating the rate of evolution of the rate of molecular evolution. Mol. Biol. Evol. 15:1647–1657.

Unwin, D. M., and N. N. Bakhurina. 1994. *Sordes pillosus* and the nature of the pterosaur flight apparatus. Nature 371:62–64.

Vaughan, T. A. 1986. Mammalogy. Saunders, Forth Worth, TX.

Waddell, P. J., H. Kishino, and R. Ota. 2001. A phylogenetic foundation for comparative mammalian genomics. Genome Informatics 12:141–154.

Waddell, P. J., N. Okada, and M. Hasegawa. 1999. Towards resolving the interordinal relationships of placental mammals. Syst. Biol. 48:1–6.

Wang, Y., Y. Hu, J. Meng, and C. Li. 2001. An ossified Meckel's cartilage in two Cretaceous mammals and origin of the mammalian middle ear. Science 294:357–361.

Wheeler, W. C., and D. Gladstein. 2000. POY: the optimization of alignment characters (program and documentation). American Museum of Natural History, New York. Available: www.cladistics.org. Last accessed 25 December 2003.

Wible, J. R., M. J. Novacek, and G. W. Rougier. 2004. New data on the skull and dentition in the Mongolian Late Cretaceous mammal *Zalambdalestes*. Bull. Am. Mus. Nat. Hist. 281:1–144.

Wible, J. R., G. W. Rougier, M. J. Novacek, M. C. Mckenna, and D. D. Dashzeveg. 1995. A mammalian petrosal from the Early Cretaceous of Mongolia: implications for the evolution of the ear region and mammaliamorph relationships. Am. Mus. Nov. 3149:1–19.

Wilkinson, M., J. Thorley, D. Littlewood, and R. Bray. 2001. Towards a phylogenetic supertree of Platyhelminthes? Pp. 292–301 in Interrelationships of the Platyhelminthes (D. Littlewood and R. Bray, eds.). Chapman-Hall, London.

Wilson, D. E., and D. M. Reeder. In press. Mammal species of the world. A taxonomic and geographic reference. 3rd ed. Smithsonian Institution Press, Washington, DC.

Wood, A. E. 1962. The early Tertiary rodents of the family Paramyidae. Trans. Am. Philos. Soc. Phil. 52:1–261.

Wu, C.-I., and W.-H. Li. 1985. Evidence for higher rates of nucleotide substitution in rodents than in man. Proc. Natl. Acad. Sci. USA 82:1741–1745.

Zuckerkandl, E., and L. Pauling. 1962. Molecular disease, evolution and genic heterogeneity. Pp. 189–225 in Horizons in biochemistry (M. Kasha and B. Pullman, eds.). Academic Press, New York.

Zuckerkandl, E., and L. Pauling. 1965. Evolutionary divergence and convergence in proteins. Pp. 97–166 in Evolving genes and proteins (V. Bryson and H. J. Vogel, eds.). Academic Press, New York.

Bernard Wood
Paul Constantino

Human Origins

Life at the Top of the Tree

This chapter describes the relationships and recent evolutionary history of *Homo sapiens*, or modern humans. By relationships, we mean the details of how modern humans are related to the other great apes, the living animals closest to modern humans. By recent, we mean the part of our evolutionary history that postdates our most recent common ancestor with one of the other living great apes.

Modern humans are singular in some important ways, yet in others we closely resemble the other great apes. Three of our singularities are noteworthy. First, our habitat is more extensive and varied than that of any other contemporary vertebrate, let alone any other large-bodied primate. Second, the size of the modern human population exceeds that of any other large undomesticated mammal, and we outnumber all the other great apes by many, many orders of magnitude. With respect to behavior, we are not unique in possessing culture (Whiten et al. 1999), but we are unique in terms of the complexity of that culture. As for our commonalties with higher primates and with other mammals, one of the triumphs of molecular biology has been the ways it is helping us document the details of our relatedness to the rest of the living world. The extent to which we share DNA with chimpanzees (~95–99% depending on how it is measured) is well known, but it is less known yet no less significant that it is estimated that we share 40% of our DNA with a banana. The magnitude of this molecular conservatism serves to emphasize that whatever we discover to be the genetic basis of the unique aspects of modern human

behavior (be they differences in the genes themselves, or in the intensity of their expression; e.g., Enard et al. 2002), the genetic differences between modern humans and the other great apes are quantitatively trivial compared with the overwhelming majority of our genome that we share with other life on Earth.

Terminology

In this chapter, we have tried to avoid using technical terms, but some are necessary. For reasons given below, we treat modern humans as one of the "great apes," the others being the two African higher primates, the chimpanzee (*Pan*) and the gorilla (*Gorilla*), and the orangutan (*Pongo*) from Asia. Linnaean taxonomic categories immediately above the level of the genus, that is, the family and the tribe, have vernacular equivalents that end in "id" and "in," respectively. Thus, members of the Hominidae, the family to which modern humans belong, are called "hominids" and members of the Hominini, the tribe that includes modern humans, are called "hominins."

Paleoanthropologists have differed, and still do differ, in the way they use the family and tribe categories with respect to the classification of the higher primates. In the past, *Homo sapiens* has been considered to be distinct enough to be placed in its own family, Hominidae, with all the other great apes grouped together in another family,

Pongidae. Thus, we and our close fossil relatives were referred to as hominids and the other great apes and their close fossil relatives were referred to as pongids (table 29.1). As we show below, this scheme is inconsistent with morphological and genetic evidence suggesting that one of the living pongids, the chimpanzee, is more closely related to modern humans (the only living "old-style" hominid) than it is to any other pongid (table 29.2).

In response to these developments, some researchers have advocated combining modern humans and chimps in the same genus (e.g., Page and Goodman 2001, Wildman et al. 2003). According to the rules of zoological nomenclature, the name for such a genus must be *Homo*. In this contribution we adopt a less radical solution: we lump all the great apes into the family Hominidae; within that grouping, we recognizes three living subfamilies, the Ponginae (or "pongines") for the orangutans, the Gorillinae (or "gorillines") for the gorillas, and the Homininae (or "hominines") for both modern humans and chimpanzees. Within the latter subfamily we recognize two tribes, the Panini (or "panins") for the chimpanzees and the Hominini (or "hominins") for modern humans. The latter is further broken down into two subtribes, one for all the extinct-only hominin genera (Australopithecina) and the other (Hominina) for the genus *Homo*, which includes the only living hominin taxon, *Homo sapiens*. Thus, in order of decreasing inclusivity, modern humans are hominids (family), hominines (subfamily), and then hominins (tribe and subtribe). Therefore, in the terminology used hereafter modern humans and all the fossil taxa judged to be more closely related to modern humans than to chimpanzees are referred to as hominins, with the chimpanzee equivalent being panin. We use the informal term "australopith" for members of the subtribe Australopithecina.

Table 29.1

A Traditional "Premolecular" Taxonomy of the Living Higher Primates (Boldface Indicates Extinct Taxa).

Superfamily Hominoidea (hominoids)

Family Hylobatidae
    Genus *Hylobates*
Family Pongidae (pongids)
    Genus *Pongo*
    Genus *Gorilla*
    Genus *Pan*
Family Hominidae (hominids)
    Subfamily Australopithecinae ("australopithecines")
    Genus *Ardipithecus*
    Genus *Australopithecus*
    Genus *Kenyanthropus*
    Genus *Orrorin*
    Genus *Paranthropus*
    Genus *Sahelanthropus*
    Subfamily Homininae (hominines)
    Genus *Homo*

Table 29.2

A Taxonomy of the Living Higher Primates that Recognizes the Close Genetic Links Between *Pan* and *Homo* (Boldface Indicates Fossil-Only Hominin Taxa).

Superfamily Hominoidea (hominoids)

Family Hylobatidae
    Genus *Hylobates*
Family Hominidae (hominids)
    Subfamily Ponginae
        Genus *Pongo* (pongines)
    Subfamily Gorillinae
        Genus *Gorilla* (gorillines)
    Subfamily Homininae (hominines)
        Tribe Panini
            Genus *Pan* (panins)
        Tribe Hominini (hominins)
            Subtribe Australopithecina (australopiths)
            Genus *Ardipithecus*
            Genus *Australopithecus*
            Genus *Kenyanthropus*
            Genus *Orrorin*
            Genus *Paranthropus*
            Genus *Sahelanthropus*
            Subtribe Hominina (hominians)
            Genus *Homo*

A Different Scale

Compared with other chapters in this book, we will deal with evolutionary history at a unique level of taxonomic detail, that of the species and genus. The species category is the lowest taxonomic level commonly used, and genera are composed of one, or more, species. For a group to qualify for the rank of genus, the taxa within it are generally taken to be both adaptively homogeneous and members of the same clade. To comply with the latter requirement, the genus must contain all the descendants of a common ancestor and its members must be confined to that clade. Species that are "adaptively similar" but belong to different clades do not qualify for the rank of genus.

At this level of taxonomic detail, differences in taxonomic philosophy (see below) significantly affect the way researchers of human evolution interpret the fossil evidence. These differences most importantly affect decisions about the numbers of species that are recognized in the human fossil record. Thus, this contribution considers nuances of taxonomy that would simply not be noticed in other chapters devoted to larger and more diverse sections of the Tree of Life.

Close Relatives

For much of the last century, the data available for reconstructing the phylogeny of the higher primates were effectively restricted to gross observations of the phenotype.

Numerically, these data were either dominated by, or confined to, observations made from the "hard tissues," that is, from the skeleton and dentition. In the older literature, the phenotypic and behavioral differences among the higher primates were interpreted as indicating a substantial gap, if not a gulf, between modern humans and the nonhuman higher primates. For close to 150 years (Huxley 1863), some researchers have suggested that modern humans are more closely related to the African apes as (*Homo* (*Pan*, *Gorilla*)) than they are to the orangutan. However, these researchers generally insisted on putting a respectful distance between modern humans and the last common ancestor we shared with the African apes. It is only relatively recently that data sets dominated by gross morphological observations of hard tissues have been interpreted as favoring a particularly close link between modern humans and chimpanzees [i.e., ((*Homo*, *Pan*) *Gorilla*); Groves 1986, Groves and Paterson 1991, Shoshani et al. 1996]. Soft-tissue data also support a (*Homo*, *Pan*) clade, but these data are presently dominated by observations about the gross anatomy of the limbs, especially information about muscles (Gibbs et al. 2000, 2002).

Developments in biochemistry and immunology during the first half of the 20th century allowed the focus of the search for better evidence about the nature of the relationships between humans and the apes to be shifted from traditional gross morphology to the morphology of molecules. The earliest attempts to use molecular morphology to determine the relationships among the higher primates used proteins such as albumin and hemoglobin. Proteins are made up of a string of amino acids. In many instances, one amino acid may be substituted for another without affecting the primary function of a protein, but the substitution can be detected by appropriate methods. Zuckerkandl (1963) used enzymes to break up the hemoglobin protein into its component peptides and then separated the components using a method called starch gel electrophoresis. The patterns made by the hemoglobins of modern humans, gorilla, and chimpanzee were indistinguishable (Zuckerkandl 1963). Morris Goodman (1963) used sensitive immunological techniques to investigate the affinities of the albumin protein of higher primates and showed that, with respect to this molecule, modern human and chimpanzee albumins were again indiscernible (Goodman 1963). In the 1970s Vince Sarich and Alan Wilson continued the exploitation of minor variations in protein structure, and they, too, concluded that modern humans and African apes were very closely related (Sarich and Wilson 1966, 1967).

The discovery of the genetic code by James Watson and Francis Crick demonstrated that the sequence of bases in the DNA molecule specifies the genes that determine the nature of the proteins manufactured within a cell. This meant that the affinities between organisms could be pursued at the level of the genome, thus potentially eliminating the need to rely on morphological "proxies" (be they traditional hard- and/or soft-tissue anatomy or the morphology of proteins) for information about relatedness. The DNA within the cell is located either within the nucleus as nuclear DNA, or within the mitochondria as mitochondrial DNA (mtDNA). Comparisons among the DNA of organisms can be made using two methods. In DNA hybridization, the entire DNA is compared but at a relatively crude level. In DNA sequencing, the base sequences of comparable sections of DNA are determined and then compared. In brief, DNA hybridization tells you "a little about a lot" of DNA, whereas, before the sequencing of whole genomes, the sequencing method told you "a lot about a little" of DNA. The results of both hybridization (e.g., Caccone and Powell 1989) and sequencing (e.g., Bailey et al. 1992, Horai et al. 1995; see reviews by Gagneux and Varki 2001, Wildman et al. 2002) studies of both nuclear DNA and mtDNA suggest that modern humans and chimpanzees are more closely related to each other than either is to the gorilla. When researchers calibrate these differences using paleontological evidence such as the split between the apes and the Old World monkeys or the split between the orangutans and the African great apes, then the neutral mutation theory suggests that the hypothetical ancestor of modern humans and the chimpanzee lived between about 5 and 8 Mya (million years ago, e.g., Shi et al. 2003). Other researchers using a different calibration point favor a substantially earlier date (10–14 Mya) for the *Pan/Homo* split (Arnason and Janke 2002).

Ancestral Differences

Although there are an impressive number of contrasts between the gross morphology of living chimpanzees and modern humans, differences between the earliest hominins and the ancestors of the chimpanzee are likely to have been more subtle. Some of the features that distinguish modern humans and chimpanzees, such as those linked to upright posture and bipedalism, can be traced far back into human prehistory. Other features and distinctive behaviors of modern humans, such as our relatively diminutive jaws and chewing teeth and complex language, were acquired more recently and thus cannot be used to identify early hominins, even if we had a reliable hard tissue marker that allowed researchers to identify a behavior such as language in the fossil record. At least two early hominin genera, *Australopithecus* and *Paranthropus*, had absolutely and relatively larger chewing teeth compared with later *Homo*. This "megadontia" of the premolars and molars may have been an important derived feature of early hominins, but it has apparently been reversed in later hominins. We do not know whether megadontia evolved just once, or in more than one clade, nor can we be sure it is confined to hominins. For example, a very preliminary analysis of extinct ape taxa (P. Andrews and B. Wood, pers. comm.) suggests that some of these taxa also have relatively enlarged chewing teeth. How, then, are we to tell an early hominin from the ancestors of the chimpanzees, or from the lineage that provided the common ancestor of chimpanzees and modern humans?

The conventional presumption is that both the common ancestor and panin taxa would have had a locomotor system adapted for life in the trees with the trunk held either horizontal or upright and with the forelimbs adapted for knuckle-walking on large branches or on the ground. This would have been combined with projecting faces accommodating elongated jaws bearing relatively small chewing teeth and large, sexually dimorphic canine teeth that are honed against the lower premolars. Early hominins, on the other hand, would have been distinguished by at least some skeletal and other adaptations for an upright posture and bipedal walking and running, linked with a masticatory apparatus that combined relatively larger chewing teeth and more modest-sized canines that do not project as far above the rest of the teeth.

A Third Way?

These proposed distinctions between hominins, panins, and their hypothetical common ancestor are working hypotheses that need to be reviewed and if necessary revised as the relevant fossil evidence is uncovered. Evidence of only one of the presumed distinguishing features of the hominins and panins may not be sufficient to identify a fossil as being in either the hominin or panin lineage. This is because there is evidence that the higher primates, like many other groups of mammals, are prone to homoplasy, which is the independent acquisition of morphological characters. This means that we cannot exclude the possibility that some of what many have come to regard as the "key adaptations" of the hominins (e.g., bipedalism) as well as those of the other great ape lineages may have arisen in more than one clade and more than once in the same clade (see below). If so, it would be very difficult on the basis of the inevitably fragmentary fossil record to distinguish the earliest members of the hominin and panin lineages between 5 and 10 Mya.

Lastly, if only for the historical reasons given below, we need to acknowledge the likelihood that a 5–10 Myr-old fossil ape taxon may be neither a hominin nor a panin. For example, for many years fossil great ape taxa known from African sites were interpreted as being ancestral to either the gorilla or the chimpanzee. Cladistic analysis has since shown that most of these taxa display derived morphology that probably precludes them from being a member of the extant African ape clades (Stewart and Disotell 1998). Thus, instead of assuming that a 5–10 Myr-old fossil taxon must be either an ancestral hominin, an ancestral panin, or their common ancestor, we need to entertain the possibility that it may belong to a hitherto unknown hominin or panin subclade or to an extinct sister group of the *Pan/Homo* clade. Colleagues must also realize that morphology that is primitive compared with later, undisputed, hominins can only make a taxon a *candidate* for the common ancestry of the hominin

clade; it cannot be used to prove it is *the* common ancestor. It is also very likely that 5–10 Myr-old fossil ape taxa are part of an adaptive radiation for which we have no satisfactory extant model. We should be prepared to find fossil apes in this and even later time ranges that display novel combinations of familiar features, as well as evidence of novel morphological features.

How Many Species of Fossil Hominin Should We Recognize in the Human Fossil Record?

It is easy to forget that statements about how many species are sampled in the hominin fossil record are hypotheses. There is lively debate about the definition of living species, so it is not surprising there is a spectrum of opinion about how the species category should be interpreted in the paleontological context. All species are individuals in the sense that they have a history. They have a beginning (the result of a speciation event), a middle that lasts as long as the species persists, and an end, which is either extinction or participation in another speciation event. Living species are caught in geological terms at an instant in their history, much as a single still photograph of a running race is only a partial record of that race. In the hominin fossil record that, albeit imperfectly, samples hundreds of thousands of years of time, the same species may be sampled several times. So to return to our metaphor, the hominin fossil record may be providing us with more than one photograph of the same running race.

Paleoanthropologists must devise strategies to ensure that the number of species they recognize in the fossil record is neither a gross underestimate nor an extravagant overestimate of the actual number. They must also take into account that they are working with fossil evidence that is largely confined to the remains of the hard tissues that make up the bones and teeth. We know from living animals that many "good" species are osteologically and dentally very difficult to distinguish (e.g., *Cercopithecus* species). Thus, there are good logical reasons to suspect that a hard tissue-bound fossil record will always underestimate the number of species.

When this attitude to estimating the likely number of species in the fossil record is combined with a "punctuated equilibrium" and cladogenetic interpretation of evolution then a researcher is liable to interpret the fossil record as containing more rather than fewer species (table 29.3A, fig. 29.1). Conversely, researchers who favor a more gradualistic, or anagenetic, interpretation of evolution that emphasizes morphological continuity rather than morphological discontinuity, and who see species as individuals that are longer lived and more prone to substantial changes in morphology through time, will tend to resolve the fossil record into fewer species (table 29.3B). For the reasons given above the taxonomic hypothesis favored in this contribution is one that recognizes more rather than fewer species.

Table 29.3
Alternate Hominin Taxonomies.

A. A more speciose (or more taxic) hominin taxonomy.
Primitive Hominins
    Genus *Ardipithecus*
        Ardipithecus ramidus
    Genus *Orrorin*
        Orrorin tugenensis
    Genus *Sahelanthropus*
        Sahelanthropus tchadensis
Australopiths
    Genus *Australopithecus*
        Australopithecus africanus
        Australopithecus afarensis
        Australopithecus bahrelghazali
        Australopithecus anamensis
        Australopithecus garhi
    Genus *Paranthropus*
        Paranthropus robustus
        Paranthropus boisei
        Paranthropus aethiopicus
    Genus *Kenyanthropus*
        Kenyanthropus platyops
Homo
    Genus *Homo*
        Homo sapiens
        Homo neanderthalensis
        Homo erectus
        Homo heidelbergensis
        Homo habilis
        Homo rudolfensis
        Homo antecessor
B. A less speciose hominin toxonomy.
Primitive hominins
    Genus *Ardipithecus*
        Ardipithecus ramidus
Australopiths
    Genus *Australopithecus*
        Australopithecus africanus
        Australopithecus afarensis
        Australopithecus garhi
    Genus *Paranthropus*
        Paranthropus robustus
        Paranthropus boisei
Homo
    Genus *Homo*
        Homo sapiens
        Homo erectus
        Homo habilis

Inventory of Fossil Hominin Taxa

In this section, we summarize the main taxa researchers have recognized in the hominin fossil record. Some researchers think a list this long recognizes too many species (see above). In this inventory the taxa are presented in three groups: taxa that are (or may be) primitive hominins, australopiths, and taxa that are conventionally included in the genus *Homo*. Within each of the three groups, the taxa are considered in the order of their formal introduction into the scientific lit-

erature. As recommended by the International Code of Zoological Nomenclature (ICZN; Ride et al. 1985), when a taxon has been moved from its initial genus, the original reference is given in parentheses, followed by the revising reference. Further details about most of the taxa and a more extensive bibliography can be found in Wood and Richmond (2000). Recent relevant reviews are also contained in Hartwig (2002).

Primitive Hominins

This group includes one taxon, *Ardipithecus ramidus*, that is probably a member of the hominin clade and two taxa, *Orrorin tugenensis* and *Sahelanthropus tchadensis*, which may be hominins. There are too few fossils as yet to be sure that the three taxa should be in different genera (or perhaps even different species), but until we have more evidence, the original genus designations have been retained.

Ardipithecus ramidus (*White et al. 1994*)
White et al. (1995)

    Type specimen.   ARA-VP-6/1—associated upper and lower dentition, Aramis, Middle Awash, Ethiopia 1993.
    Approximate time range.   ~4.5–5.7 Myr.
    History and context.   The initial evidence for this taxon was in the form of approximately 4.5-Myr-old fossils recovered from late 1992 onward at a site called Aramis in the Middle Awash region of Ethiopia. A second suite of fossils, including a mandible, teeth, and postcranial bones, was recovered in 1997 from five different localities in the Middle Awash that range in age from 5.2 to >5.7 Myr (Haile-Selassie 2001). One of the new localities is in the Aramis region, the other four are several kilometers to the west in exposures lying against the western margin of the East African Rift. With hindsight the remains from Aramis may not be the first evidence of this species to be found for the 5 Myr mandibular fragment (KNM-LT 329) from Lothagam, Kenya, may also belong to *A. ramidus*.
    Characteristics and inferred behavior.   The remains attributed to *A. ramidus* have some features in common with living species of *Pan*, others that are shared with the African apes in general, and, crucially, several dental and cranial features that are shared only with later hominins such as *Australopithecus afarensis*. Thus, the discoverers have suggested that the material belongs to a hominin species. They initially allocated the new species to *Australopithecus* (White et al. 1994), but subsequently the same researchers assigned it to a new genus, *Ardipithecus* (White et al. 1995), which they suggest is significantly more primitive than *Australopithecus*.
    The case White and his colleagues set forward to justify their initial taxonomic judgment centered on the cranial evidence, whereas Haile-Selassie (2001) focused on two features of the dentition and one of the postcranial skeleton. The former researchers claim that compared with *A. afarensis*, *A. ramidus* has relatively larger canines, first deciduous man-

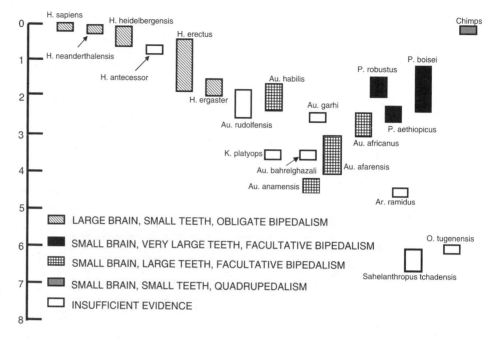

Figure 29.1. A proposed speciose taxonomy of hominins along with a depiction of their morphological–functional grade through time. See text for details.

dibular molars with less complex crowns, upper and lower premolar crowns that are more asymmetric (and thus more apelike), thinner enamel, and a flatter articular eminence. The researchers suggest that *A. ramidus* should be excluded from the apes because its upper central incisors are relatively small, its canine honing mechanism is poorly developed, the mandibular permanent molar crowns are too broad, the first deciduous mandibular molars have more complex crowns than those of *Pan*, and the foramen magnum is more anteriorly situated than it is in the apes. Haile-Selassie (2001) suggests that the relatively incisiform lower canines together with the dorsal orientation of the proximal joint surface of the proximal fourth pedal phalanx are further evidence of *A. ramidus* having more affinities with later hominins than with *Pan*.

Judging from the size of the shoulder joint *A. ramidus* weighed about 40 kg. Its chewing teeth were relatively small, and the position of the foramen magnum suggests that the posture and gait of *A. ramidus* were respectively more upright and bipedal than is the case in the living apes. The thin enamel covering on the teeth suggests that the diet of *A. ramidus* may have been closer to that of the chimpanzee than is the case for later hominins. The paleohabitat of both subsets of the *A. ramidus* hypodigm has been interpreted as predominantly woodland or grassy woodland (Woldegabriel et al. 2001). As yet we have no information about the size of the brain and only scant direct evidence from the limbs about the posture and locomotion (see above) of *A. ramidus*. The remains of a skeleton likely to belong to *A. ramidus* have been found at Aramis, and details are eagerly awaited.

Controversy. Although the evidence is far from conclusive, it is reasonable to regard *A. ramidus* as a primitive hominin until additional data suggest otherwise.

Orrorin tugenensis (*Senut et al. 2001*)

Type specimen. BAR 1000'00—fragmentary mandible, Kapsomin, Lukeino Formation, Tugen Hills, Baringo, Kenya 2000.

Approximate time range. ~6.0 Myr date is constrained by a 6.2 Myr underlying trachyte and a 5.6 Myr overlying sill.

History and context. The relevant remains come from four localities in the Lukeino Formation, Tugen Hills, Kenya. One of the 13 specimens recovered, a lower molar tooth crown, was discovered in 1974; the remaining 12 specimens were recovered in 2000.

Characteristics and inferred behavior. The Lukeino molar tooth has long been regarded as displaying a mixture of *Pan* and hominin morphology, but the researchers who recovered the more recent evidence claim that the BAR 1002'00 femur shows that *O. tugenensis* was "already adapted to habitual or perhaps even obligate bipedalism" (Senut et al. 2001). However, the grounds for interpreting its morphology as that of an obligate biped (presumably the shape and size of the head of the femur and the presence of a crestlike linea aspera on the posterior aspect of the shaft) are far from conclusive. A more detailed analysis of the external and internal morphology of three femora attributed to *O. tugenensis* (Pickford et al. 2002) is interpreted by the authors as confirming the locomotor mode as obligate bipedalism, but the computer-assisted tomographic scans of the femoral neck instead point to a more *Pan*-like regime of weight transmission. Otherwise, its discoverers admit that much of the critical dental morphology is "apelike" (Senut et al. 2001).

Controversy. In order to use the small size of the molar crowns of *Orrorin* as evidence of the latter's close link with *Homo*, parsimony dictates that all megadont early hominin fossil evidence must be placed in a large australopith subclade that is more distantly related to modern humans than is *O. tugenensis*. However, instead of belonging in the hominin clade, *O. tugenensis* may prove to belong to another part of the adaptive radiation that included the common ancestor of panins and hominins.

Sahelanthropus tchadensis *Brunet et al. 2002*

Type specimen.   TM266-01-060–1—an adult cranium, Anthracotheriid Unit, Toros-Menalla, Chad 2001.

Approximate time range.   ~6–7 Myr.

History and context.   The hypodigm was discovered during a survey of likely fossiliferous localities beyond the Koro Toro region in Chad. All the original specimens are from a single locality (Brunet et al. 2002). The dating is based on the match between the fauna in the Anthracotheriid Unit and the faunas known from Lukeino and from the Nawata Formation at Lothagam (Vignaud et al. 2002).

Characteristics and inferred behavior.   The cranium of *S. tchadensis* is chimp sized and displays a novel combination of primitive and derived features. Much about the cranial base and neurocranium is chimplike with the notable exception that the foramen magnum lies more anteriorly than is generally the case in chimps. Yet the presence of a supraorbital torus, relatively flat lateral facial profile, small, apically worn canines, low, rounded, molar cusps, relatively thick enamel, and relatively thick mandibular corpus are all features that would exclude *S. tchadensis* from any close relationship with the *Pan* clade and would place it in, or close to, the hominin clade. However, given the perils of inferring the characteristic morphology of a taxon from the evidence of a single individual, or even several individuals, these differences should be seen as indicative and not the final word about the taxonomy of this undoubtedly important late Miocene evidence (Wood 2002).

Australopiths

This group includes the fossil evidence assigned to all of the remaining hominin taxa that are not conventionally included in the genus *Homo*. As it is used in this and many other taxonomies, *Australopithecus* is almost certainly paraphyletic, but until we have more confidence that we can identify species from fragmentary hard tissue evidence and recover a reliable phylogeny from an incomplete fossil record there is little point in revising the generic terminology. In order to avoid more confusion than already exists, we have (with two exceptions) retained the original genus names. The exceptions are that *Zinjanthropus* and *Paraustralopithecus* are subsumed within the genus *Paranthropus*.

Australopithecus africanus *Dart 1925*

Type specimen.   Taung 1—a juvenile skull with partial endocast, Taung, now in South Africa 1924.

Approximate time range.   ~2.4–3 Myr.

History and context.   An early hominin child's skull found among the contents of a small cave exposed during mining at the Buxton Limeworks at Taungs (the name later changed to Taung) in southern Africa was referred by Raymond Dart to a new genus and species, *Australopithecus africanus*, which means literally, the "southern ape" of Africa. No other hominins have been recovered from the Buxton deposits.

Remains of hominins we now classify as *A. africanus* have been found at three other cave sites in southern Africa: Makapansgat, well to the northeast of Johannesburg, and at Sterkfontein and Gladysvale in the Blauuwbank Valley, close to Johannesburg. At these sites, as at Taung, early hominin fossils are mixed in with other animal bones in hardened rock and bone-laden cave fillings, or breccias. The cave sites in southern Africa can, at present, only be dated by relatively imprecise absolute physicochemical methods. More often, they have been dated by comparing the remains of the mammals found in the caves with mammalian fossils found at sites in East Africa that were dated using more precise and reliable absolute methods. In this and in other ways, the age of the *A. africanus*-bearing Sterkfontein Member 4 breccia has been estimated to be between 2.5 and 3 Myr. A hominin skeleton, StW 573, from Member 2 deep in the Sterkfontein cave may be somewhat older, ~4 Myr (Partridge et al. 2003), but it is too early to tell whether it belongs to *A. africanus* (Clarke 1998, 1999, 2002a). It has recently been suggested (Berger et al. 2002) that the Sterkfontein dates may be too old with 2.5 Myr being the upper and not the lower age limit of Member 4, but this reinterpretation has been contested (Clarke 2002b, Partridge 2002). The bones of the medium and large mammals found in the breccias of all the southern African hominin cave sites, as well as the hominins themselves, either were accumulated by predators or are there because the animals fell into and were then trapped in the caves. The other animal fossils and the plant remains found with *A. africanus* suggest that the immediate habitat was woodland with grassland beyond.

The first hominin to be recovered at Sterkfontein, TM 1511, was given the name *Australopithecus transvaalensis* Broom 1936 but was later transferred to a new genus, *Plesianthropus transvaalensis* (Broom 1936) Broom 1938. Raymond Dart allocated the Makapansgat fossil hominins to a new species, *Australopithecus prometheus* Dart 1948. However, after 1955 it became conventional to refer all the australopiths from southern Africa to a single genus, *Australopithecus*, and soon researchers and commentators subsumed both *A. transvaalensis* and *A. prometheus* into the species of *Australopithecus* with taxonomic priority, namely *A. africanus* Dart 1925.

Characteristics and inferred behavior.   The picture emerging from morphological and functional analyses suggests that, although *A. africanus* was capable of walking bipedally, it was probably not an obligate biped. It had relatively large chewing teeth, and apart from the reduced canines, the skull is relatively apelike. Its mean endocranial volume, a reasonable proxy for brain size, is ~450 cm^3. The Sterkfontein evidence suggests that males and females of *A. africanus* differed substantially in body size, but probably not to the degree they did in *A. afarensis* (see below).

Controversy. Some researchers have suggested that the *A. africanus* fossils recovered from Sterkfontein may sample more than one hominin species, but the case is not currently convincing enough (e.g., Lockwood and Tobias 1999) to abandon the existing single-species hypothesis as an explanation for the variation in that sample.

Paranthropus robustus Broom 1938

Type specimen. TM 1511—an adult, presumably male, cranium and associated skeleton, "Phase II Breccia," now member 3, Kromdraai B, South Africa 1938.

Approximate time range. ~2.0–1.5 Myr.

History and context. Evidence of *Paranthropus robustus* comes from Kromdraai, Swartkrans, Drimolen, and Cooper's caves in the Blauuwbank Valley, near Johannesburg, South Africa. Kromdraai and Swartkrans have been a focus of research since 1938 and 1948, respectively, with Members 1 and 2 at Swartkrans being the source of the main component of the *P. robustus* hypodigm. Research at Drimolen was only initiated in 1992 (Keyser et al. 2000),' yet already more than 80 hominin specimens have been recovered (Keyser 2000), and it promises to be a rich source of evidence about *P. robustus*.

Characteristics and inferred behavior. The brain, face, and chewing teeth of *Paranthropus robustus* are larger than those of *A. africanus*, yet the incisor teeth are smaller. Cranial and dental differences between the hominins recovered from Sterkfontein and Swartkrans have led to the suggestion that *P. robustus* was more herbivorous than *A. africanus*. Little is known about the postcranial skeleton of *P. robustus* except that the organization of the pelvis and the hip joint is much like that of *A. africanus*. It has been suggested that the thumb of *P. robustus* would have been capable of the type of grip necessary for stone tool manufacture, but this claim is not accepted by all researchers.

Controversy. Some workers point to differences between the hominins recovered from Swartkrans and Kromdraai and prefer to allocate the former material to a separate species, *Paranthropus crassidens* Broom 1949. However, most researchers treat the Swartkrans and Kromdraai evidence as a single species, and the Drimolen specimens apparently blur the distinction between the Kromdraai and Swartkrans hypodigms. For a time some researchers insisted that the *Australopithecus* and *Paranthropus* remains from southern Africa belonged to the same species, but the single species hypothesis has long since been abandoned.

Paranthropus boisei (Leakey 1959) Robinson 1960

Type specimen. OH 5—adolescent cranium, FLK, Bed I, Olduvai Gorge, Tanzania 1959.

Approximate time range. ~2.3–1.3 Myr.

History and context. The first evidence in East Africa of a hominin resembling *Paranthropus robustus* was two teeth found in 1955 at Olduvai Gorge. However, it was OH 5, a magnificent undistorted subadult cranium with a well-preserved dentition recovered by Louis and Mary Leakey in 1959, that convinced these researchers that these remains belonged to a new and distinctive hominin taxon *Zinjanthropus boisei* Leakey 1959. A fragmented cranium (OH 30) and several isolated teeth (OH 26, 32, 38, 46, and 60) have since been assigned to the same species. An ulna (OH 36) may also belong to it. Further evidence of *P. boisei* has since been recovered from the Peninj River on the shores of Lake Natron in Tanzania, the Omo Shungura Formation and Konso in Ethiopia, Chesowanja in the Chemoigut basin, at West Turkana in Kenya, and from Melema in Malawi. However, the site collection that has provided most of the evidence about *P. boisei* is that from Koobi Fora, on the eastern shore of Lake Turkana. The new species was initially included in a new genus, *Zinjanthropus*, but the generic distinction between *Zinjanthropus* and *Australopithecus* has long since been abandoned. It is now usual to refer to the taxon as either *Australopithecus boisei* or *Paranthropus boisei* (see below).

Characteristics and inferred behavior. Cranially *P. boisei* is presently the only hominin to combine a massive, wide, flat, face, massive premolars and molars, small anterior teeth, and a modest-sized neurocranium (~450 cm^3). The face of *P. boisei* is larger and wider than that of *P. robustus*, yet their brain volumes are similar. Cranial features of *P. boisei* include the complex overlap at the parietotemporal suture and the combination of an anteriorly situated foramen magnum and a modest-sized brain. The mandible of *P. boisei* has a larger and wider body or corpus than any other hominin (see *P. aethiopicus* below). The proportions of the dentition are very derived in that very large-crowned premolar and molar teeth are combined with small anterior (i.e., incisor and canine) teeth. The tooth crowns apparently grow at a faster rate than has been recorded for any other early hominin. There is, unfortunately, no postcranial evidence that can with certainty be attributed to *P. boisei*. The fossil record of *P. boisei sensu stricto* extends across about 1 Myr of time, during which there is little evidence of any substantial change in the size or shape of the components of the cranium, mandible, and dentition (Wood et al. 1994).

Paranthropus aethiopicus (Arambourg and Coppens 1968) Chamberlain and Wood 1985

Type specimen. Omo 18.18 (or 18.1967.18)—an edentulous adult mandible, locality 18, section 7, member C, Shungura Formation, Omo region, Ethiopia 1967.

Approximate time range. ~2.5–2.3 Myr.

History and context. Some researchers have suggested that the oldest of the East African evidence for *Paranthropus* should be taxonomically distinct and that the taxon name *Paraustralopithecus aethiopicus*, linked with a ~2.5–Myr-old mandible, would be available for such a taxon. Thus, when a distinctive 2.5–Myr-old *Paranthropus* cranium, KNM-WT 17000, was recovered from West Turkana, it was natural to consider whether this new specimen should also be assigned to the same taxon.

Characteristics and inferred behavior.   The mandible and the mandibular dentition of *Paranthropus boisei sensu lato* apparently become more derived about 2.3 Mya, and that shift forms part of the evidence for the interpretation that the "early" and "late" stages of *Paranthropus* in East Africa should be recognized taxonomically, with the former being referred to as *Paranthropus aethiopicus*. Among the differences between the two East African *Paranthropus* species are the more prognathic face, the less flexed cranial base and the larger incisors of *P. aethiopicus* compared with *P. boisei*.

Controversy.   When this taxon was introduced in 1968, it was the only megadont hominin in this time range. With the discovery of *A. garhi* (see below), it is apparent that robust mandibles with similar length premolar and molar tooth rows are associated with what are claimed to be two distinct forms of cranial morphology.

Australopithecus afarensis *Johanson et al. 1978*

Type specimen.   LH 4—adult mandible, Laetolil Beds, Laetoli, Tanzania 1974.

Approximate time range.   ~3–4 Myr.

History and context.   This taxon was established in 1978 for hominin fossils recovered from Laetoli in Tanzania and from Hadar in Ethiopia. Subsequently, evidence has come from other sites in Ethiopia, including two Middle Awash localities, Maka and Belohdelie, the sites of Fejej and White Sands in the Omo Region, and possibly from the Kenyan sites of Koobi Fora, Allia Bay, West Turkana, and Tabarin. *A. afarensis* is the earliest hominin to have a comprehensive fossil record that includes a skull, fragmented crania, many lower jaws, and sufficient limb bones to be able to attempt an estimation of stature and body mass. The collection includes a specimen, AL-288, that preserves just less than half of the skeleton of an adult female.

Characteristics and inferred behavior.   The range of body mass estimates is from 25 to >50 kg. The estimated brain volume of *A. afarensis* is between 400 and 500 cm^3. This is larger than the average brain size of a chimpanzee, but if the estimates of the body size of *A. afarensis* are approximately correct, then relative to estimated body mass, the brain of *A. afarensis* is not substantially larger than that of *Pan*. It has incisors that are much smaller than those of extant chimpanzees, but the premolars and molars of *A. afarensis* are relatively larger than those of the chimpanzee and the hind limbs of AL-288 are substantially shorter than those of a modern human of similar stature. Attempts to reconstruct the habitat of *A. afarensis* suggest that it was living in a more open woodland environment than that reconstructed for *A. ramidus*. The appearance of the pelvis and the relatively short lower limb suggest that, although *A. afarensis* was capable of bipedal walking, it was not adapted for long-range bipedalism. This indirect evidence for the locomotion of *A. afarensis* is complemented by the discovery at Laetoli of several trails

of fossil footprints. These provide very graphic direct evidence that a contemporary hominin, presumably *A. afarensis*, was capable of bipedal locomotion. The upper limb, especially the hand, retains morphology that most likely reflects a significant element of arboreal locomotion. The size of the footprints and the length of the stride are consistent with stature estimates based on the length of the limb bones of *A. afarensis*. These suggest that the standing height of adult individuals in this early hominin species was between 1.0 and 1.5 m. Recent analyses have shown that the dental and mandibular morphology of this taxon changed relatively little during its ~1 Myr time range.

Controversy.   When the classification of the material now referred to as *A. afarensis* was first discussed it was natural for researchers to consider its relationship to the remains of *Australopithecus africanus* Dart 1925. The results of morphological and cladistic analyses suggest that there are significant differences between the two hypodigms and that they are rarely sister taxa in cladistic analyses. The comparisons also emphasize that in nearly all the cranial characters examined *A. afarensis* displays a more primitive character state than does *A. africanus*. Despite the substantial range of estimated body mass and claims that the taxon subsumes a mix of upper limb morphology, most researchers continue to interpret this fossil evidence as representing one species.

Australopithecus bahrelghazali *Brunet et al. 1996*

Type specimen.   KT 12/H1—anterior portion of an adult mandible, Koro Toro, Chad 1995.

Approximate time range.   ~3.0–3.5 Myr.

History and context.   This taxon was established for Pliocene hominin remains recovered in Chad, north-central Africa.

Characteristics and inferred behavior.   The published evidence, a mandible and a maxillary premolar tooth, has been interpreted as being sufficiently distinct from *A. ramidus*, *A. afarensis* and *A. anamensis* to justify its allocation to a new species. Its discovers claim that its thicker enamel distinguishes the Chad remains from *A. ramidus*, that the more vertical orientation and reduced buttressing of the mandibular symphysis together with the more symmetrical crowns of the P$_3$ separate it from *A. anamensis*, and that its more complex mandibular premolar roots distinguish it from *A. afarensis*.

Controversy.   Not all researchers are convinced that these remains are sufficiently different from *A. afarensis* to justify their allocation to a new species.

Australopithecus anamensis *Leakey et al. 1995*

Type specimen.   KNM-KP 29281—an adult mandible with complete dentition, and a temporal fragment that probably belongs to the same individual, between the upper and lower pumiceous tuffs of the basal fluvial complex, Kanapoi, Kenya 1994.

Approximate time range.   ~4.0–4.5 Myr.

History and context.   The hypodigm of the new taxon, *Australopithecus anamensis*, centers on material recovered by Meave Leakey and her team from the site of Kanapoi, together with material recovered earlier from Allia Bay, northern Kenya (Leakey et al. 1995).

Characteristics and inferred behavior.   The main differences between *A. anamensis* and *A. afarensis* relate to details of the dentition. In some respects the teeth of *A. anamensis* are more primitive than those of *A. afarensis* (e.g., the asymmetry of the premolar crowns and the relatively simple crowns of the deciduous first mandibular molars), but in others (e.g., the low cross-sectional profiles and bulging sides of the molar crowns) they show similarities to more derived and temporally later *Paranthropus* taxa (see above). The upper limb remains are australopith-like, and a tibia attributed to *A. anamensis* has features associated with bipedality (Ward 2002). A useful detailed review of the fossil evidence has appeared recently (Ward et al. 2001).

Controversy.   Some researchers interpret *A. anamensis* not as a separate taxon, but as the more primitive, earlier segment of an effectively continuous hominin lineage including both *A. anamensis* and *A. afarensis*.

Australopithecus garhi *Asfaw et al. 1999*

Type specimen.   BOU-VP-12/130—a cranium from the Hata member, Bouri, Middle Awash 1997.

Approximate time range.   ~2.5 Myr.

History and context.   The evidence for this taxon comes from Bouri, in the Middle Awash of Ethiopia.

Characteristics and inferred behavior.   *Australopithecus garhi* combines a primitive cranium with large-crowned postcanine teeth. However, unlike *Paranthropus* (see above), the incisors and canines are large and the enamel lacks the extreme thickness seen in the latter taxon. A partial skeleton combining a long femur with a long forearm was found nearby but is not associated with the type cranium of *A. garhi* (Asfaw et al. 1999). Cut-marked animal bones found in nearby horizons of the same age suggest that either *A. garhi* or another contemporary hominin were defleshing animal bones, presumably with stone tools.

Controversy.   The discoverers of *A. garhi* interpret it as a probable ancestor of *Homo*, but it could equally well be the sister taxon of a *Homo*, *Paranthropus*, *A. africanus* clade. If future discoveries demonstrate that the mandibles of *P. aethiopicus* and *A. garhi* cannot be distinguished from each other, then the name *P. aethiopicus* would have priority for the hypodigm.

Kenyanthropus platyops *Leakey et al. 2001*

Type specimen.   KNM-WT 40000—cranium, Lomekwi, West Turkana, Kenya 1999.

Approximate time range.   ~3.3–3.5 Myr.

History and context.   Two specimens from West Turkana, KNM-WT 40000, a 3.5-Myr-old cranium and KNM-WT 38350 a 3.3-Myr-old maxilla, are respectively the holotype and the paratype of *Kenyanthropus platyops* (Leakey et al. 2001). The initial report lists 34 other potential members of the same hypodigm, but at this stage the researchers are reserving their judgment about the taxonomy of these remains, some of which have only recently been referred to *A. afarensis* (Brown et al. 2001).

Characteristics and inferred behavior.   The main reasons Leakey et al. (2001) did not assign KNM-WT 40000 and 38350 to *A. afarensis* are this material's reduced subnasal prognathism, anteriorly situated zygomatic root, flat and vertically orientated malar region, relatively small but thick-enameled molars, and the unusually small M^1 compared with the size of the P^4 and M^3. Some of the morphology of the new genus including the shape of the face is *Paranthropus*-like, yet it lacks the postcanine megadontia that characterizes *Paranthropus*. The authors note the face of the new material resembles that of *Homo rudolfensis*, but they rightly point out that the postcanine teeth of the latter are substantially larger than those of KNM-WT 40000. *K. platyops* displays a hitherto unique combination of facial and dental morphology.

Controversy.   White (2003) has argued (not persuasively, in our opinion) that KNM-WT 40000 is a cranium of *A. afarensis* and that its distinctive morphology is the result of pre- and postfossilization damage involving the infiltration of external matrix into cracks produced by weathering.

Homo

This group contains hominin taxa that are conventionally included within the *Homo* clade. One of us, along with others, have suggested that two of these taxa (*H. habilis* and *H. rudolfensis*) may not belong in the *Homo* clade (Wood and Collard 1999), but until we can generate sound phylogenetic hypotheses about the australopiths, it is not clear what their new generic attribution should be. Thus, for the purposes of this review, they are retained within *Homo*.

Homo sapiens *Linnaeus 1758*

Type specimen.   Linnaeus did not designate a type specimen.

Approximate time range.   ~150 Kyr (thousand years) to the present day.

History and context.   An early indication that modern humans were ancient enough to have a fossil record came when a series of skeletal remains were discovered by workmen at the Cro-Magnon rock shelter at Les Eyzies de Tayac, France, in 1868. A male skeleton, Cro-Magnon 1, was initially made the type specimen of a novel species, *Homo spelaeus* Lapouge 1899, but it was soon apparent that it was not appropriate to discriminate between this material and modern humans. Soon, more modern humanlike fossils were recovered from sites elsewhere in Europe, but the first African fossil evidence of populations that are difficult to distinguish from anatomically modern humans, from Singa in the

Sudan, did not come until 1924. Comparable evidence has since come from north, east, and southern Africa [e.g., Ethiopia (Dire-Dawa, 1933; Omo II, 1967; Herto, 1997); Morocco (Dar es-Soltan, 1937–1938), and Natal—now KwaZulu Natal (Border Cave, 1941–1942 and 1974)]. In the Near East, comparable fossil evidence has been recovered from sites such as Mugharet Es-Skhul (1931–1932) and Djebel Qafzeh (1933, 1965–1975). In Asia and Australasia, anatomically modern human fossils have been recovered from sites such as Wadjak, Indonesia (1889–1890), the Upper Cave at Zhoukoudian, China (1930 and thereafter), Niah Cave, Borneo (1958), Tabon, Philippines (1962), and the Willandra Lakes, Australia (1968 and thereafter). All this material has been judged to be within, or close to, the range of variation of living regional samples of modern human populations, and thus it is not appropriate to distinguish it taxonomically from *Homo sapiens*.

Characteristics and inferred behavior.   Paradoxically, it is easier to assemble information about the characteristic morphology of extinct hominin taxa than about the only living hominin species. For each morphological region what are the boundaries of living *H. sapiens* variation? How far beyond these boundaries, if at all, should we be prepared to go and still refer the fossil evidence to *H. sapiens*? These are simple questions to which one would have thought there would be ready answers. However, the morphological expression of modern humanness has proved to be complex and difficult to express. For example, spoken language is assumed to be a *sine qua non* of *H. sapiens*, but it is difficult if not impossible to determine language competence (as opposed to the potential for language) from the fossil record. It is claimed that the distinctive form of living and fossil *H. sapiens* crania can be reduced to two main influences, a retracted face and an expanded globular braincase (Lieberman et al. 2002), and the recently announced crania from Herto (White et al. 2003) are consistent with this prediction.

Controversy.   The origin of *H. sapiens* has been the subject of considerable debate. Most analyses have pointed to Africa ~100–200 Kyr ago as the source of modern human genetic variation (Relethford 2002; but see also Templeton 2002). The earliest evidence of anatomically modern human morphology in the fossil record comes from sites in Africa (e.g., Omo II and Herto) and the Near East (e.g., Qafzeh) listed above. It is also in Africa that there is evidence for a likely morphological precursor of anatomically modern human morphology. This takes the form of crania that are generally more robust and archaic-looking than those of anatomically modern humans yet which are not archaic enough to justify their allocation to *H. heidelbergensis*, or derived enough to be *H. neanderthalensis* (see below). Specimens in this category include Jebel Irhoud (Morocco, 1961 and 1963) from North Africa; Omo 2 (Kibish Formation) (Ethiopia, 1967); Laetoli 18 (Tanzania, 1976); Eliye Springs (KNM-ES 11693) (Kenya, 1985) and Ileret (KNM-ER 999 and 3884; Kenya, 1971 and 1976, respectively) from East Africa; and

Florisbad (Free State, 1932) and Cave of Hearths (Northern Province, 1947) in southern Africa. There is undoubtedly a gradation in morphology that makes it difficult to set the boundary between anatomically modern humans and *H. heidelbergensis*. However, it is clear that unless at least one boundary is set along this cline, morphological variation within *H. sapiens sensu lato* is so great that it strains credulity.

Homo neanderthalensis *King 1864*

Type specimen.   Neanderthal 1—adult calotte and partial skeleton, Feldhofer Cave, Elberfield, Germany 1856.

Approximate time range.   ~200–30 Kyr.

History and context.   The first evidence of Neanderthals to come to light was a child's skull found in 1829 from a site in Belgium called Engis. An adult cranium recovered in 1848 from Forbes' Quarry in Gibraltar also displays the distinctive Neanderthal morphology. However, the type specimen of *Homo neanderthalensis* King 1864 consists of an adult skeleton recovered in 1856 from the Feldhofer Cave in the Neander Valley, in Germany. Excavations were restarted at the Feldhofer Cave in 1997 and much of what was missing from the original skeleton plus the remains of other individuals have recently been recovered (Schmitz et al. 2002). After the initial recovery of hominins from the Feldhofer Cave it was some time before discoveries were made at other sites in Europe [e.g., Moravia (Sipka, 1880); Belgium (Spy, 1886); Croatia (Krapina, 1899–1906); Germany (Ehringsdorf, 1908–1925), and France (Le Moustier, 1908 and 1914; La Chapelle-aux-Saints, 1908; La Ferrassie, 1909, 1910, and 1912)]. The first evidence of Neanderthals beyond western Europe was recovered in 1924–26 at Kiik Koba in the Crimea. The first of many discoveries in the Near East was at Tabun (1929), and in 1938 the first fossils were recovered from Central Asia at Teshik-Tash. New Neanderthal localities continue to be discovered in Europe (e.g., St. Cesaire, 1979; Zaffaraya, 1983 and 1992; Moula-Guercy, 1991) and western Asia (Mezmaiskaya, 1993 and 1994). Thus, Neanderthal remains have been found throughout Europe, with the exception of Scandinavia, as well as in the Near East, the Levant, and western Asia. Many elements of the characteristic morphology of the Neanderthals can be seen in remains recovered from sites such as Steinheim and Reilingen (Germany) and Swanscombe (England) that date from ~200–300 Kyr. It is also said to be evident in precursor form in the remains that have been found in the Sima de los Huesos, a cave in the Sierra de Atapuerca, Spain (see *H. heidelbergensis*, below).

Characteristics and inferred behavior.   Features of the Neanderthal cranium include thick, double-arched brow ridges, a face that projects anteriorly in the midline, a large nasal skeleton, laterally projecting and rounded parietal bones and a rounded, posteriorly projecting occipital bone (i.e., an occipital "bun"). Estimates of brain size [means: female, 1286 cc. (*n* = 4); male, 1575 cc. (*n* = 7)] suggest that Neanderthal brains were as large, if not larger, than the brains of living

Homo sapiens, but they were perhaps slightly smaller relative to body mass. The Neanderthals were stout with a broad rib cage, a long clavicle, a wide pelvis, and limb bones that are generally robust with well-developed muscle insertions. The distal extremities tend to be short compared with most modern *H. sapiens*, but Neanderthals were evidently obligate bipeds. The generally well-marked muscle attachments and the relative thickness of long bone shafts have been interpreted as indicators of a strenuous lifestyle. The size and wear on the incisors suggest that the Neanderthals regularly used their anterior teeth as "tools" either for food preparation or to grip hide or similar material.

It is clear that the Neanderthals possessed the cognitive and manipulative abilities to create a sophisticated, versatile tool kit and possibly objects of symbolic value. Whether or not Neanderthals were capable of complex speech typical of modern humans remains unknown, largely because the neural adaptations that make speech possible do not preserve in the fossil record. Some reconstructions suggest that the Neanderthal vocal tract would have been capable of fewer differentiable vowel sounds than that of modern humans, but this hypothesis is difficult to test. Researchers have recently presented compelling evidence for deliberate defleshing (i.e., cannibalism) on the crania of ~100–Kyr-old Neanderthals from Moula-Guercy. Paleoenvironmental and anatomical data indicate that Neanderthals typically occupied cold, marginal habitats.

Controversy. In the past decade or so there has been an increasing acceptance that the Neanderthals are morphologically distinctive, so much so that many consider it unlikely that such a specialized form could have given rise to the morphology seen in modern humans. There is, however, another school of researchers who point to, and stress, the morphological continuity between the fossil evidence for *H. sapiens* and the remains others would attribute to *H. neanderthalensis*. Some have argued that morphologically intermediate specimens are evidence of admixture between Neanderthals and modern humans, but this interpretation has been challenged.

Recent developments. Recently researchers have been able to recover short fragments of mtDNA from the humerus of the Neanderthal type specimen (Krings et al. 1997, 1999). They were able to show that the fossil sequence falls well outside the range of variation of a diverse sample of modern humans, and they suggest that Neanderthals would have been unlikely to have made any contribution to the modern human gene pool. They conclude that this amount of difference points to 550–690 Kyr of separation. Subsequently, mtDNA has been recovered at two other Neanderthal sites, from rib fragments of a child's skeleton at Mezmaiskaya (Ovchinnikov et al. 2000) and from Vindija (Krings et al. 2000). The differences between the mtDNA fragments studied are similar to the differences between any three randomly selected African modern humans. The fragments of mtDNA that have been studied are short, but if the findings of the

three studies summarized in Krings et al. (1999) were to be repeated for other parts of the genome, then the case for placing Neanderthals in a separate species from modern humans on the basis of their skeletal peculiarities would be greatly strengthened (Knight 2003). There is disagreement about the influence that intentional burial may have had on the preservation of Neanderthal remains.

Homo erectus *(Dubois 1892) Mayr 1944*

Type specimen. Trinil 2—adult calotte, Trinil, Ngawi, Java (now Indonesia) 1891.

Approximate time range. ~1.8 Myr to 200 Kyr.

History and context. In 1890 Eugene Dubois discovered a mandible fragment in Java at a site called Kedung Brubus. Less than a year later, in 1891, at excavations on the banks of the Solo River at Trinil, workers unearthed a skullcap that became the type specimen of a new species. Dubois initially referred the skull cap to *Anthropopithecus erectus* Dubois 1892, but in 1894 he transferred the new species to *Pithecanthropus* (Dubois 1894), and since then others have transferred it to *Homo* (see below).

The focus for the next phase of the search for hominin remains in Java was upstream of Trinil where the Solo River cuts through the Plio-Pleistocene sediments of the Sangiran Dome. In 1936 a German paleontologist, Ralph von Koenigswald, recovered a cranium that resembled the distinctive shape of the Trinil skullcap, but the brain size, ~750 cm³, was even smaller than that of the Trinil calotte. In China in the early 1920s Gunnar Andersson and Otto Zdansky excavated for two seasons (1921 and 1923) at Locality 1 at Zhoukoudian (formerly Choukoutien) Cave, near Beijing. They recovered quartz artifacts, but apparently no fossil hominins. However, Zdansky subsequently realized that two "ape" teeth belonged to a hominin, and the next year they were assigned to a new hominin genus and species, *Sinanthropus pekinensis* Black 1927. The first cranium from Zhoukoudian was found in 1929, and excavations continued until their interruption by World War II. The fossils recovered from Locality 1 were consistent in their morphology and were similar in many ways to *Pithecanthropus erectus*, so much so that Ernst Mayr formerly proposed the taxa be merged and then subsumed into *Homo* as *Homo erectus* (Mayr 1944).

Since then, similar fossils have been found at other sites in China (e.g., Lantian, 1963–1964); southern Africa (Swartkrans, 1949 and thereafter); East Africa (Olduvai Gorge, 1960 and thereafter; West and East Turkana, 1970 and thereafter; Melka Kunture, 1973 and thereafter and also perhaps at Buia, Eritrea, 1995 and 1997); and North Africa (Tighenif, 1954–1955). Many also include the "Solo" remains from Ngandong, Indonesia, within *H. erectus*. Discoveries from East African sites have since provided crucial evidence about the postcranial morphology of *H. erectus* (e.g., OH 28).

Characteristics and inferred behavior. The crania of *H. erectus* have a low vault, a substantial more-or-less continuous torus above the orbits and a sharply angulated occipital region. The inner and outer tables of the cranial vault are thick. Cranial capacities vary from ~725 cm³ for OH 12, to ~1250 cm³ for the Solo V calotte from Ngandong. The greatest width of the face is in the upper part. The palate has similar proportions to those of modern humans, but the buttressing is more substantial. The body of the mandible is more gracile than that of the australopiths, but more robust than that of modern humans. The mandible lacks the well-marked chin that is a feature of modern humans. The tooth crowns are generally larger and the premolar roots more complicated than those of modern humans, and the third molars are usually smaller, or the same size, as the second molars. The dense cortical bone of the postcranial skeleton is generally thicker than is the case for modern humans. The limb bones are modern humanlike in their proportions, but they have more robust shafts, with the femoral and tibial shafts flattened from front to back (femur) and side to side (tibia) relative to those of modern humans.

All the dental and cranial evidence points to a modern humanlike diet for *H. erectus*, and the postcranial elements are consistent with a habitually upright posture and obligate, long-range bipedalism. There is no fossil evidence relevant to assessing the dexterity of *H. erectus*, but if *H. erectus* manufactured Acheulean artifacts then some dexterity would be implicit.

Controversy.   Over the years several authors have suggested that morphological continuity between *H. erectus* and later *H. sapiens* effectively invalidates the specific status of the former. This has resulted in the proposition that *H. erectus* be sunk into *H. sapiens* Linnaeus 1758. Recent advocates of this course of action include Wolpoff et al. (1994) and Tobias (1995).

Recent developments.   If the discoveries from Dmanisi, Georgia (Gabunia et al. 2000, Vekua et al. 2002) do prove to belong to early African *H. erectus* (see below), then their small brains and primitive cranial morphology would make *H. erectus sensu lato* a substantially different taxon.

Homo heidelbergensis *Schoetensack 1908*

Type specimen.   Mauer 1—adult mandible, Mauer, Heidelberg, Germany 1907.

Approximate time range.   ~600–100 Kyr.

History and context.   The Mauer mandible was considered distinctive because it has no chin and because the corpus is larger than those of the mandibles of modern humans living in Europe today. Cranial evidence from Zuttiyeh (Israel, 1925) has since been assigned to this group, as have fossils from Greece (Petralona, 1959); France (Arago, 1964–1969; Montmaurin, 1949); Hungary (Vértesszöllös, 1965); and Germany (Bilzingsleben, 1972–1977, 1983, and thereafter). Researchers responsible for the discovery and analysis of the large sample of ~400–600–Kyr-old (Bischoff and Shamp 2003) hominins from Sima de los Huesos, Sierra de Atapuerca, Spain, also assign that collection to *H. heidelbergensis*, but other researchers are more inclined to treat this evidence as an early form of *H. neanderthalensis* (see above).

The first relevant African evidence for *H. heidelbergensis*, or what some call "archaic" *H. sapiens*, came in 1921 with the recovery of a ~250–300 Kyr cranium from a cave in the Broken Hill Mine at Kabwe in what is now Zambia. Other morphologically comparable remains have been found from the same, or an earlier, time period in southern Africa (Hopefield/Elandsfontein, 1953 and thereafter), East Africa (Eyasi, 1935–1938; Ndutu, 1973), and North Africa (Rabat, 1933; Jebel Irhoud, 1961 and 1963; Sale, 1971; Thomas Quarry, 1969/72). The earliest evidence (~600 Kyr) of this African archaic group comes from Bodo (Ethiopia, 1976). Asian evidence for an archaic form of *Homo* comes from China (e.g., Dali, 1978; Jinniushan, 1984; Xujiayao, 1976/7, 1979; Yunxian, 1989/90) and possibly India (Hathnora, 1982). Most of these fossils are not reliably dated and their estimated ages range from 100 to 200 Kyr.

Characteristics and inferred behavior.   What sets this material apart from *H. sapiens* and *H. neanderthalensis* is the morphology of the cranium and the robusticity of the postcranial skeleton. Some brain cases are as large as those of modern humans, but they are always more robustly built with a thickened occipital region and a projecting face and with large separate ridges above the orbits, unlike the more continuous brow ridge of *H. erectus*. Compared with *H. erectus* (see above), the parietals are expanded, the occipital is more rounded, and the frontal bone is broader. The crania of *H. heidelbergensis* lack the autapomorphies of *H. neanderthalensis*, such as the anteriorly projecting midface and the distinctive swelling of the occipital region. The mean cranial capacity for this taxon, ~1200 cc, is substantially larger than the ~970 cc mean for *H. erectus*. However, the upper end of the range of *H. erectus* brain size overlaps the lower end of the range of *H. heidelbergensis*. *H. heidelbergensis* is the earliest hominin to have a brain as large as anatomically modern *H. sapiens*, and its postcranial skeleton suggests that its robust long bones and large lower limb joints were well suited to long-distance bipedal walking.

Controversy.   There are currently different views about the scope and phylogenetic relationships of *H. heidelbergensis*. Researchers who interpret the Steinheim, Swanscombe, and Sima de los Huesos remains as the beginnings of a distinctive Neanderthal taxon see insufficient "morphological space" for *H. heidelbergensis* and do not recognize it as a valid taxon (e.g., Stringer 1996). Instead, they advocate sinking *H. heidelbergensis* into *H. neanderthalensis*. Others have used an elaborate system of grades of "archaic *H. sapiens*" to accommodate the same fossil evidence or have taken to ignoring species-level classifications in favor of recognizing a larger number of paleo-, or p-demes (e.g., Howell 1999), which are defined as "local populations" of species. The researchers who do accept *H. heidelbergensis* as a valid taxon have different interpretations of it. Some researchers who recognize *H. heidelbergensis* interpret the taxon to include all non-Neanderthal "archaic" *Homo* fossils, whereas others interpret it as being confined to the European Middle Pleistocene. If there is to be a single species to cover the archaic material from Europe, Africa, and Asia, then the species name *H. heidelbergensis* Schoetensack 1908 has priority. However, if there was evidence that the non-European subset of the hypodigm sampled an equally good species, then the species name with priority is *H. rhodesiensis* Woodward 1921.

Homo habilis *Leakey et al. 1964*

Type specimen. OH 7—partial skull cap and hand bones, FLKNN, Bed I, Olduvai Gorge, Tanzania 1960.

Approximate time range. ~2.4–1.6 Myr.

History and context. In 1960 Louis and Mary Leakey recovered substantial parts of both parietal bones, six hand bones (OH 7), and "a large part of a left foot" (OH 8) from Bed I of Olduvai Gorge and in the next year or so further evidence of a "nonrobust" hominin came from both Beds I and II of Olduvai Gorge. In 1964, Leakey et al. set out the case for recognizing a new species for the nonrobust hominin from Olduvai and for accommodating it within the genus *Homo*. In due course additional specimens from Olduvai were added to the hypodigm of *H. habilis*, the most significant being the cranium OH 24 and the associated skeleton OH 62. Evidence of fossils resembling *H. habilis* from Koobi Fora includes a well-preserved skull (KNM-ER 1805), a well-preserved cranium (KNM-ER 1813), several mandibles, and some isolated teeth. Initially these specimens were not allocated to a species but were given the informal name "early *Homo*." Some of the hominin fossils recovered from members G and H of the Shungura Formation have also been assigned to *H. habilis*, as has a fragmentary cranium and some isolated teeth from member 5 at Sterkfontein, the cranium SK 847 from member 1 at Swartkrans and a maxilla from Hadar. Suggestions that *H. habilis* remains have been recovered from sites beyond Africa are as yet unsubstantiated (but see above the evidence recovered from Dmanisi).

Characteristics and inferred behavior. The endocranial volume of *H. habilis* as originally described (*H. habilis sensu stricto*) ranges from just less than 500 cm^3 to about 600 cm^3. All the crania are wider at the base than across the vault, but the face is broadest in its upper part. The only postcranial evidence that can with confidence be assigned to *H. habilis sensu stricto* are the postcranial bones associated with the type specimen, OH 7, and the associated skeleton, OH 62. If OH 62 is representative of *H. habilis sensu stricto*, the skeletal evidence suggests that its limb proportions and locomotion were australopith-like. The curved proximal phalanges and well-developed muscle markings on the phalanges of OH 7 also indicate the hand was used for more powerful grasping (such as would be needed for arboreal activities) than is the case in any other species of *Homo*. The inference that *H. habilis sensu stricto* was capable of spoken language was based on links between endocranial morphology and language comprehension and production that are no longer valid.

Controversy. The case for splitting *H. habilis sensu lato* (i.e., the Olduvai evidence plus crania such as KNM-ER 1470

and 1590) into two taxa, *H. habilis sensu stricto* (see above) and *Homo rudolfensis* (see below), has attracted broad support, but it is by no means universally accepted. As will be apparent from inferences about its locomotion and capacity for language set out above, in several ways *H. habilis sensu stricto* is adaptively more like the australopiths than later *Homo* taxa. This evidence combined with at best weak cladistic evidence (see below) for its inclusion in the *Homo* clade prompted Wood and Collard (1999) to suggest that both it and *H. rudolfensis* should be removed from the genus *Homo*. But what genus do those taxa properly belong to? The same authors recommended that until the phylogenetic relationships among the australopiths become clearer, they should be referred to *Australopithecus*, but that would make that taxon almost certainly paraphyletic. For the purposes of this review, we retain the conventional taxonomy of both taxa, at least until there is more consensus on this topic.

Homo ergaster *Groves and Mazák 1975*

Type specimen. KNM-ER 992, Area 3, Okote member, Koobi Fora Formation, Koobi Fora 1971.

Approximate time range. ~1.9–1.5 Myr.

History and context. This taxon was introduced in 1975 as part of a review of the taxonomy of the "early *Homo*" fossils from Koobi Fora. The type specimen is KNM-ER 992 an adult mandible that had been compared with, and by some workers referred with, *Homo erectus*. The paratypes include the skull KNM-ER 1805, but the only detailed analysis of KNM-ER 1805 has concluded that it should be referred to *H. habilis sensu stricto*. Any decision about whether *Homo ergaster* is a good taxon is dependent on researchers demonstrating that the type specimen KNM-ER 992 can be distinguished from *H. erectus* (see above). Similarities between the Koobi Fora component of the *H. ergaster* hypodigm and the juvenile skeleton, KNM-WT 15000 from West Turkana suggest that the latter should also be included in *H. ergaster*. More recently, it has been claimed that there is evidence for *H. ergaster* beyond Africa. Well-preserved crania and mandibles from Dmanisi, Republic of Georgia, in the Caucasus have been assigned to early African *H. erectus* (or *H. ergaster*) or to a new taxon, *Homo georgicus* (Gabunia et al. 2000, Vekua et al. 2002).

Characteristics and inferred behavior. The features claimed to distinguish *H. ergaster* from *H. erectus* fall into two categories. The first consists of the ways in which *H. ergaster* is more primitive than *H. erectus*. The best evidence in this category comes from details of the mandibular dentition and in particular the mandibular premolars. The second category consists of the ways in which *H. ergaster* is less specialized, or derived, in its cranial vault and cranial base morphology than is *H. erectus*. For example, it is claimed that *H. ergaster* lacks some of the more derived features of *H. erectus* cranial morphology such as thickened inner and outer tables and prominent sagittal and angular tori, but other researchers dispute the distinctiveness of this material (see below). *H.*

ergaster is the first large-bodied hominin taxon with a body shape that was closer to that of modern humans than to the australopiths (Wood and Collard 1999). It is also the first hominin to combine modern human-sized chewing teeth with a postcranial skeleton (e.g., long legs, large femoral head) committed to long-range bipedalism and to lack morphological features associated with arboreal locomotor and postural behaviors. The small chewing teeth of *H. ergaster* imply either that it was eating different food than the australopiths, or that it was preparing the same food extra-orally, probably by using tools and/or by cooking it.

Controversy.   Many researchers do not regard the *H. ergaster* hypodigm worthy of a separate species. They either dispute there are any consistent, or significant, morphological differences between the "early African" part of *H. erectus* (i.e., *H. ergaster*) and the main *H. erectus* hypodigm, or they acknowledge there are differences but suggest that they do not merit recognition at the level of the species.

Homo rudolfensis (*Alexeev 1986*) sensu *Wood 1992*

Type specimen.   Lectotype: KNM-ER 1470, Area 131, Upper Burgi member, Koobi Fora Formation, Koobi Fora, Kenya 1972.

Approximate time range.   ~1.8–1.6 Myr.

History and context.   In 1986 Alexeev suggested that differences between the cranium KNM-ER 1470 from Koobi Fora and *Homo habilis sensu stricto* from Olduvai Gorge justified referring the former to a different new species he named *Pithecanthropus rudolfensis*. Thus, if *Homo habilis sensu lato* does subsume more variability than is consistent with it being a single species and if KNM-ER 1470 is judged to belong to a *Homo* species other than *Homo habilis sensu stricto*, then *Homo rudolfensis* (Alexeev 1986) Wood 1992 is available as the name of a second early *Homo* taxon.

Characteristics and inferred behavior.   The main ways that *H. rudolfensis* differs from *H. habilis sensu stricto* are that they have different mixtures of primitive and derived, or specialized, features. For example, although the absolute size of the brain case is greater in *H. rudolfensis*, its face is widest in its mid-part, whereas the face of *H. habilis* is widest superiorly. Despite the absolute size of its brain (~750–800 cm³), when it is related to estimates of body mass the brain of *H. rudolfensis* is not substantially larger than those of the australopiths. The more primitive face of *H. rudolfensis* is combined with a robust mandible and mandibular postcanine teeth with larger, broader, crowns and more complex premolar root systems than those of *H. habilis*. At present no postcranial remains can be reliably linked with *H. rudolfensis*. The mandible and postcanine teeth are larger than one would predict for a generalized hominoid of the same estimated body mass, suggesting that its dietary niche made mechanical demands similar to those of the megadont australopiths.

Controversy.   The detailed case for dividing *Homo habilis sensu lato* into two species is set out in Wood (1991, 1992). A recent review of the cladistic and functional evidence for *H. rudolfensis* (Wood and Collard 1999) has concluded that there are few grounds for its retention in *Homo* and recommended that it (along with *H. habilis sensu stricto*) be transferred to *Australopithecus* as *Australopithecus rudolfensis* (Alexeev 1986 Wood and Collard 1999).

Homo antecessor *Bermudez de Castro et al. 1997*

Type specimen.   ATD6–5—mandible and associated teeth, Level 6, Gran Dolina, Spain 1994.

Approximate time range.   ~500–700 Kyr.

History and context.   The Gran Dolina (TD) site is a cave in the Sierra de Atapuerca that was exposed when a railway cutting was excavated a century ago. The fossils attributed to *H. antecessor* were recovered when a test excavation reached Level 6.

Characteristics and inferred behavior.   The authors of the initial report claim the combination of a modern humanlike facial morphology with the relatively primitive crowns and roots of the teeth is not seen in *H. heidelbergensis*, nor do the Gran Dolina remains have the derived *H. neanderthalensis* traits seen in *H. heidelbergensis*. It is the apparent lack of these derived features combined with differences from *H. ergaster* that led the authors to propose the new hominin species. They suggest that *H. antecessor* is probably the last common ancestor of Neanderthals and *H. sapiens*.

Controversy.   Many researchers question the grounds for excluding this material from *H. heidelbergensis*.

Phylogeny

There is a wide spectrum of opinion about phylogenetic relationships within the hominin clade. Most researchers are convinced that the existing methods are capable of recovering reliable phylogenetic relationships among fossil hominin taxa. However, a minority of researchers are less confident that reliable phylogenies can be extracted using traditional data obtained from the existing fossil record. One faction within this minority argues that until the selection of characters is better integrated with information about the molecular basis of development, character independence will never be assured (Lovejoy et al. 2000). Another faction within the minority suggests that even if character independence could be assured, much of the hard-tissue evidence provided by the fossil record may be so prone to various forms of homoplasy that the phylogenetic signal it retains is too weak and the homoplastic noise so strong that the former cannot be detected with any reliability (Corruccini 1994, Collard and Wood 2000). The introduction of new three-dimensional methods for capturing information about shape and size may improve the likelihood that phenetic information can be used to reconstruct phylogeny (Lockwood et al. 2002, Guy et al. 2003).

The phylogenetic tree in figure 29.2 is a consensus of recent attempts to recover the phylogeny of hominins. Some taxon hypodigms are so small that any phylogenetic hypothesis is speculative. Other hominin taxa are sufficiently well known (e.g., *P. boisei, A. afarensis, H. neanderthalensis*) that paucity of the fossil record per se is unlikely to be the reason for any ambiguity about their phylogenetic relationships. Two clades, later *Homo* and *Paranthropus*, are supported by nearly all phylogenetic reconstructions (e.g., Wood 1991, Skelton and McHenry 1992, Strait et al. 1997). Taxa that for many years have been regarded as human ancestors (e.g., *H. neanderthalensis* and late *H. erectus*) are almost certainly too derived to be directly ancestral to modern humans.

Conclusions

The living and fossil taxa within the (*Homo, Pan*) clade can be resolved into the four crude grades identified in figure 29.1. Many fossil taxa are excluded from this grade classification because they lack one or more of the necessary lines of evidence to infer brain size, relative tooth size, or locomotor pattern. Two of the grades coincide with major multitaxon clades and are coincident with *Homo* and *Paranthropus*, two of the five genera recognized within the (*Homo, Pan*) clade. Although the results of cladistic analyses of the hominin fossil record differ in detail (e.g., Strait et al. 1997, Wood and Collard 1999), nearly all agree about the robusticity of the *Homo sensu stricto* and *Paranthropus* clades.

A linear, sequential model is no longer tenable for the post-2.5-Myr period of human evolutionary history, but influential researchers continue to interpret the period between 5.0 and 3.0 Myr as a series of time-successive hominin species (Asfaw et al. 1999). Thus, they view *A. ramidus* as the direct ancestor of *A. anamensis* and the latter as the direct ancestor of *A. afarensis*. This simplistic interpretation was

always likely to be challenged by fresh fossil evidence, and this came in the form of a proposal to establish not just a new species but a new genus for fossil hominins discovered at West Turkana in 1998 and 1999. In that paper, Meave Leakey et al. (2001) make the case that *Kenyanthropus platyops* is a distinct taxon that shares some facial similarities with *Paranthropus* taxa without sharing the latter's distinctively large premolars and molars and thick enamel. The newly discovered and described *Sahelanthropus tchadensis* (Brunet et al. 2002) combines facial features hitherto considered apparently distinctive of advanced australopiths and *Homo* with a chimp-sized brain and a good many other cranial features seen only in *Pan*. All this suggests that the origins of the (*Homo, Pan*) clade and subsequent evolution within the hominin clade are a good deal more complex than many had anticipated (Wood 2002). It is truly remarkable that thus far no hominid fossil evidence in the 4–7 Myr time range has been interpreted as being more closely related to *Pan* than to *Homo*. Is this because none has yet been discovered? Or is it because we are aware of it but have misinterpreted it as belonging to the hominin and not the panin clade?

Acknowledgments

B.W. is grateful to the organizers of the Tree of Life meeting for their invitation to place modern humans in their proper context within the living world. The Henry R. Luce Foundation, the National Science Foundation, and The Leverhulme Trust have funded research by B.W. that is incorporated in this review. P.C. is supported by an NSF-IGERT graduate fellowship. Special thanks to Mark Collard for contributing to many of the ideas incorporated in this review, and to Sally Gibbs for carrying out the soft-tissue study.

Figure 29.2. A speculative phylogeny of the hominins over time. Solid lines indicate the authors' preferences. See text for details.

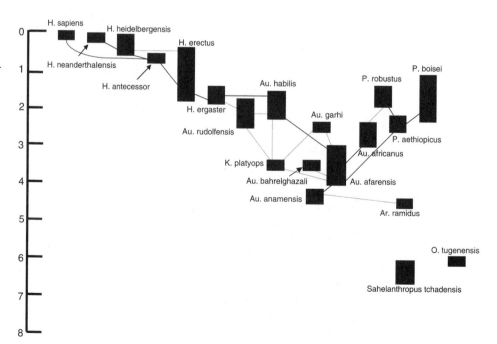

Literature Cited

Alexeev, V. 1986. The origin of the human race. Progress Publishers, Moscow.

Arambourg, C., and Y. Coppens. 1968. Decouverte d'un australopithecien nouveau dans les Gisements de l'Omo (Ethiopie). South African Journal of Science 64(2):58–59.

Arnason, U., and A. Janke. 2002. Mitogenomic analyses of eutherian relationships. Cytogenet. Genome Res. 96:20–32.

Asfaw, B., T. White, O. Lovejoy, B. Latimer, S. Simpson, and G. Suwa. 1999. *Australopithecus garhi*: a new species of early hominid from Ethiopia. Science 284:629–635.

Bailey, W. J., K. Hayasak, C. G. Skinner, S. Kehoe, L. U. Sieu, J. L. Slighthom, and M. Goodman. 1992. Reexamination of the African hominoid trichotomy with additional sequences from the primate beta-globin gene. Mol. Phylogenet. Evol. 1:97–135.

Berger, L. R., R. Lacruz, and D. J. de Ruiter. 2002. Revised age estimates of *Australopithecus*-bearing deposits at Sterkfontein, South Africa. 119:192–197.

Bermudez de Castro, J. M., J. L. Arsuage, E. Carbonell, A. Rosas, I. Martinez, and M. Mosquera. 1997. A hominid from the Lower Pleistocene of Atapuerca, Spain: possible ancestor to Neaderthals and modern humans. Science 276:1392–1395.

Bischoff, J. L., and D. D. Shamp. 2003. The Sima de los Huesos hominids date to beyond U/Th equilibrium (>350 kyr) and perhaps to 400–500 kyr: new radiometric dates. Journal of Archaeological Science 30:275–280.

Black, D. 1927. Further hominid remains of Lower Quaternary age from Chou Kow Tien deposit. Nature 120:954.

Broom, R. 1936. A new fossil anthropoid skull from South Africa. Nature 138:486–488.

Broom, R. 1938. The Pleistocene anthropoid apes of South Africa. Nature 142:377–379.

Broom, R. 1949. Anotehr new type of fossil ape-man (*Paranthropus crassidens*). Nature 163:57.

Brown, B., F. H. Brown, and A. Walker. 2001. New hominids from the Lake Turkana Basin, Kenya. J. Hum. Evol. 41:29–44.

Brunet, M., F. Guy, D. Pilbeam, H. T. Mackaye, A. Likius, D. Ahounta, A. Beauvilain, C. Blondel, H. Bocherens, J.-R. Boisserie, et al. 2002. A new hominid from the Upper Miocene of Chad, Central Africa. Nature 418:145–151.

Caccone, A., and J. R. Powell. 1989. DNA divergence among hominoids. Evolution 43:925–942.

Clarke, R. J. 1998. First ever discovery of a well-preserved skull and associated skeleton of *Australopithecus*. S. Afr. J. Sci. 94:460–463.

Clarke, R. J. 1999. Discovery of complete arm and hand of the 3.3 Mya *Australopithecus* skeleton from Sterkfontein. S. Afr. J. Sci. 95:477–480.

Clarke, R. J. 2002a. Newly revealed information on the Sterkfontein member 2 *Australopithecus* skeleton. S. Afr. J. Sci. 98:523–526.

Clarke, R. J. 2002b. On the unrealistic "revised age estimates" for Sterkfontein. S. Afr. J. Sci. 98:415–418.

Collard, M., and B. Wood. 2000. How reliable are human phylogenetic hypotheses? Proc. Natl. Acad. Sci. USA 97:5003–5006.

Corruccini, R. S. 1994. How certain are hominoid phylogenies? The role of confidence intervals in cladistics. Pp. 167–183 *in* Integrative paths to the past: paleoanthropological advances in honor of F. Clark Howell (R. S. Corruccini and R. L. Ciochon, eds.). Prentice Hall, Englewood Cliffs, NJ.

Dart, R. A. 1925. *Australopithecus africanus*: the man-ape of South Africa. Nature 115:195–199.

Dart, R. A. 1948. An *Australopithecus* from the Central Transvaal. South African Science 1:200–201.

Dubois, E. 1892. Palaeontologische onderzoekingen op Java. Versl. Mijnw. (Jakarta, Java) 3:10–14.

Dubois, E. 1894. *Pithecanthropus erectus*, eine menschenahnliche Uebergangsform aus Java. Batavia (Jakarta) Landesdruckerei.

Enard, W., P. Khaitovich, J. Klose, S. Zollner, F. Heissig, P. Giavalisco, K. Nieselt-Struwe, E. Muchmore, A. Varki, R. Ravid, G. M. Doxiadis, R. E. Bontrop, and S. Paabo. 2002. Intra- and interspecific variation in primate gene expression patterns. Science 296:340–343.

Gabunia, L., A. Vekua, D. Lordkipanidze, C. C. Swisher, III, R. Ferring, A. Justus, M. Nioradze, M. Tvalchrelidze, S. C. Anton, G. Bosinski, O. Joris, M.-A. de Lumley, G. Majsuradze, and A. Mouskhelishvili. 2000. Earliest Pleistocene hominid cranial remains from Dmanisi, Republic of Georgia: taxonomy, geological setting and age. Science 288:1019–1025.

Gagneux, P., and A. Varki. 2001. Genetic differences between humans and great apes. Mol. Phylogenet. Evol. 18:2–13.

Gibbs, S., M. Collard, and B. A. Wood. 2000. Soft-tissue characters in higher primate phylogenetics. Proc. Natl. Acad. Sci. USA 97:11130–11132.

Gibbs, S., M. Collard, and B. A. Wood. 2002. Soft-tissue anatomy of the extant hominoids: a review and phylogenetic analysis. J. Anat. 200:3–49.

Goodman, M. 1963. Man's place in the phylogeny of the primates as reflected in serum proteins. Pp. 204–234 *in* Classification and human evolution (S. Washburn, ed.). Aldine, Chicago.

Groves, C. P. 1986. Systematics of the great apes. Pp. 187–217 *in* Comparative primate biology (D. R. Swindler and J. Erwin, eds.). Liss, New York.

Groves, C. P., and V. Mazak. 1975. An approach to the taxonomy of the Hominidae: gracile Villafranchian hominids of Africa. Casopis pro mineralogii a geologii (Czech) 20(3):225–247.

Groves, C. P., and J. D. Paterson. 1991. Testing hominoid phylogeny with the PHYLIP programs. J. Hum. Evol. 20:167–183.

Guy, F. M. Brunet, M. Schmittbuhl, and L. Viriot. 2003. New approaches in hominoid taxonomy: morphometrics. Am. J. Phys. Anthropol. 121:198–218.

Haile-Selassie, Y. 2001. Late Miocene hominids from the Middle Awash, Ethiopia. Nature 412:178–181.

Hartwig, W. C. 2002. The primate fossil record. Cambridge University Press, Cambridge.

Horai, S., K. Hayasaka, R. Kondo, K. Tsugane, and N. Takahata. 1995. Recent African origin of modern humans revealed by complete sequences of hominoid mitochondrial DNAs. Proc. Natl. Acad. Sci. USA 92:532–536.

Howell, F. C. 1999. Paleo-demes, species clades, and extinctions in the Pleistocene hominin record. J. Anthropol. Res. 55:127–151.

Huxley, T. H. 1863. Evidence as to man's place in nature. Williams and Norgate, London.

Keyser, A. W. 2000. The Drimolen skull: the most complete australopithecine cranium and mandible to date. S. Afr. J. Sci. 96:189–193.

Keyser, A. W., C. G. Menter, J. Moggi-Cecchi, T. R. Pickering, and L. R. Berger. 2000. Drimolen: a new hominid-bearing site in Gauteng, South Africa. S. Afr. J. Sci. 96:193–197.

Knight, A. 2003. The phylogenetic relationship of Neandertal and modern human mitochondrial DNAs based on informative nucleotide sites. J. Human Evol. 44:627–632.

Krings, M., C. Capelli, F. Tschentscher, H. Geisert, S. Meyer, A. von Haeseler, K. Grossschmidt, G. Possnert, M. Paunovic, and S. Paabo. 2000. A view of Neandertal genetic diversity. Nat. Genet. 26:144–146.

Krings, M., H. Geisert, R. W. Schmitz, H. Krainitzki, and S. Paabo. 1999. DNA sequence of the mitochondrial hyper-variable region II from the Neandertal type specimen. Proc. Natl. Acad. Sci. USA 96:5581–5585.

Krings, M., A. Stone, R. W. Schmitz, H. Krainitzki, M. Stoneking, and S. Paabo. 1997. Neandertal DNA sequences and the origin of modern humans. Cell 90:19–30.

Leakey, L. S. B. 1959. A new fossil skull from Olduvai. Nature 184:491–493.

Leakey, L. S. B., and M. D. Leakey. 1964. Recent discoveries of fossil hominids in Tanganyika, at Olduvai and near Lake Natron. Nature 202:5–7.

Leakey, M. G., C. S. Feibel, I. McDougall, and A. Walker. 1995. New four-million-year-old hominid species from Kanapoi and Allia Bay, Kenya. Nature 376:565–571.

Leakey, M. G., F. Spoor, F. H. Brown, P. N. Gathogo, C. Kiarie, L. N. Leakey, and I. McDougall. 2001. New hominin genus from eastern Africa shows diverse middle Pliocene lineages. Nature 410:433–440.

Lieberman, D. E., B. M. McBratney, and G. Krovitz. 2002. The evolution and development of cranial form in *Homo sapiens*. Proc. Natl. Acad. Sci. USA 99:1134–1139.

Lockwood, C. A., J. M. Lynch, and W. H. Kimbel. 2002. Quantifying temporal bone morphology of great apes and humans: an approach using geometric morphometrics. J. Anat. 201:447–464.

Lockwood, C. A., and P. V. Tobias. 1999. A large male hominid cranium from Sterkfontein, South Africa, and the status of *Australopithecus africanus*. J. Hum. Evol. 36:637–685.

Lovejoy, C. O., M. J. Cohn, and T. D. White. 2000. The evolution of mammalian morphology: a developmental perspective. Pp. 41–55 in Development, growth and evolution (P. O'Higgins and M. Cohn, eds.). Academic Press, San Diego.

Mayr, E. 1944. On the concepts and terminology of vertical subspecies and species. Natl. Res. Coun. Committee Common Probl. Genet. Paleontol. Syst. Bull. 2:11–16.

Ovchinnikov, I. V., A. Gotherstrom, G. P. Romanova, V. M. Khritonov, K. Liden, and W. Goodwin. 2000. Molecular analysis of Neanderthal DNA from the northern Caucasus. Nature 404:490–493.

Page, S. L., and M. Goodman. 2001. Catarrhine phylogeny: noncoding evidence for a diphyletic origin of the mangabeys and for a human-chimpanzee clade. Mol. Phylogenet. Evol. 18:14–25.

Partridge, T. C. 2002. On the unrealistic "revised age estimates" for Sterkfontein. S. Afr. J. Sci. 98:418–419.

Partridge, T. C., D. E. Granger, M. W. Caffee, and R. J. Clarke. 2003. Lower Pliocene hominid remains from Sterkfontein. Science 300:607–612.

Pickford, M., B. Senut, D. Gommery, and J. Treil. 2002. Bipedalism in *Orrorin tugenensis* revealed by its femora. C. R. Palevol. 1:1–13.

Relethford, J. H. 2002. Genetics and the search for modern human origins. Wiley-Liss, New York.

Ride, W. D. L., C. W. Sabrosky, G. Bernardi, and R. V. Melville (eds.). 1985. International Code of Zoological Nomenclature. 1. British Museum of Natural History, London.

Robinson, J. T. 1960. The affinities of the new Olduvai australopithecine. Nature 186:456–458.

Sarich, V. M., and A. C. Wilson. 1966. Quantitative immunochemistry and the evolution of primate albumins. Science 154:1563–1566.

Sarich, V. M., and A. C. Wilson. 1967. Rates of albumin evolution in primates. Proc. Natl. Acad. Sci. USA 58:142–148.

Schmitz, R. W., D. Serre, G. Bonani, S. Feine, F. Hillgruber, H. Krainitzki, S. Paabo, and F. H. Smith. 2002. The Neadertal type site revisited: interdisciplinary investigations of skeletal remains from the Neader Valley, Germany. Proc. Natl. Acad. Sci. 99(20):13342–13347.

Schoetensack, O. 1908. Der Unterkiefer des *Homo heidelbergensis* aus den Sanden von Mauer bei Heidelberg. W. Engelmann, Leipzig, Germany.

Senut, B., M. Pickford, D. Gommery, P. Mein, K. Cheboi, and Y. Coppens. 2001. First hominid from the Miocene (Lukeino Formation, Kenya). C. R. Acad. Sci. Paris 332:137–144.

Shi, J., H. Xi, Y. Wang, C. Zhang, Z. Jiang, K. Zhang, Y. Shen, L. Jin, K. Zhang, W. Yuan, et al. 2003. Divergence of the genes on human chromosome 21 between human and other hominoids and variation of substitution rates among transcription units. Proc. Natl. Acad. Sci. USA 100:8331–8336.

Shoshani, J., C. P. Groves, E. L. Simons, and G. F. Gunnell. 1996. Primate phylogeny: morphological vs. molecular results. Mol. Phylogenet. Evol. 5:101–153.

Skelton, R. R., and H. M. McHenry. 1992. Evolutionary relationships among early hominids. J. Hum. Evol. 23:309–349.

Stewart, C. B., and T. R. Disotell. 1998. Primate evolution: in and out of Africa. Curr. Biol. 8:582–588.

Strait, D. S., F. E. Grine, and M. A. Moniz. 1997. A reappraisal of early hominid phylogeny. J. Hum. Evol. 32:17–82.

Stringer, C. B. 1996. Current issues in modern human origins. Pp. 115–134 in Contemporary issues in human evolution (W. E. Meikle, F. C. Howell, and N. G. Jablonski, eds.). California Academy of Science, San Francisco.

Templeton, A. 2002. Out of Africa again and again. Nature 416:45–51.

Tobias, P. V. 1995. The place of *Homo erectus* in nature with a critique of the cladistic approach. Pp. 31–41 in Human evolution in its ecological context (J. R. F. Bower and S. Sartono, eds.). *Pithecanthropus* Centennial Foundation, Leiden.

Vekua, A., D. Lordkipanidze, G. P. Rightmire, J. Agusti, R. Ferring, G. Maisuradze, A. Mouskhelishvili, M. Nioradze, M. Ponce de Leon, M. Tappen, M. Tvalchrelidze, and C. Zollikofer. 2002. A new skull of early *Homo* from Dmanisi, Georgia. Science 297:85–89.

Vignaud, P., P. Duringer, H. T. Mackaye, A. Likius, C. Blondel, J.-R. Boisserie, L. de Bonis, V. Eisenmann, M.-E. Etienne, D. Geraads, et al. 2002. Geology and palaeontology of the Upper Miocene Toros-Menalla hominid locality, Chad. Nature 418:152–155.

Ward, C. V. 2002. Interpreting the posture and locomotion of *Australopithecus afarensis*: where do we stand? Yrbk. Phys. Anthropol. 45:185–215.

Ward, C. V., M. G. Leakey, and A. Walker. 2001. Morphology of *Australopithecus anamensis* from Kanapoi and Allia Bay, Kenya. J. Hum. Evol. 41:255–368.

White, T. D. 2003. Early hominids-diversity or distortion? Science 299:1994–1997.

White, T. D., B. Asfaw, D. DeGusta, H. Gilbert, G. D. Richards, G. Suwa, and F. C. Howell. 2003. Pleistocene *Homo sapiens* from Middle Awash, Ethiopia. Nature 423:742–747.

White, T. D., G. Suwa, and B. Asfaw. 1994. *Australopithecus ramidus*, a new species of early hominid from Aramis, Ethiopia. Nature 371:306–312.

White, T. D., G. Suwa, and B. Asfaw. 1995. *Australopithecus ramidus*, a new species of early hominid from Aramis, Ethiopia. Nature 375:88.

Whiten, A., J. Goodall, W. C. McGrew, T. Nishida, V. Reynolds, Y. Sugiyama, C. E. G. Tutin, R. W. Wrangham, and C. Boesch. 1999. Cultures in chimpanzees. Nature 399:682–685.

Wildman, D., L. I. Grossman, and M. Goodman. 2002. Functional DNA in humans and chimpanzees shows they are more similar to each other than either is to other apes. Pp. 1–10 in Probing human origins (M. Goodman and A. S. Moffat, eds.). American Academy of Arts and Sciences, Cambridge, MA.

Wildman, D. E., M. Uddin, G. Liu, L. I. Grossman, and M. Goodman. 2003. Implications of natural selection in shaping 99.4% nonsynonymous DNA identity between humans and chimpanzees: enlarging genus *Homo*. Proc. Natl. Acad. Sci. USA 100:7181–7188.

Woldegabriel, G., Y. Haile-Selassie, P. R. Renne, W. K. Hart, S. H. Ambrose, B. Asfaw, G. Heiken, and T. White. 2001. Geology and palaeontology of the Late Miocene Middle Awash valley, Afar rift, Ethiopia. Nature 412:175–178.

Wolpoff, M. H., A. G. Thorne, J. Selinek and Z. Yinyun. 1994. The case for sinking *Homo erectus*: 100 years of *Pithecanthropus* is enough! Cour. Forschungsinst. Senckenb. 171:341–361.

Wood, B. A. 1991. Koobi Fora research project, Vol. 4: Hominid cranial remains. Clarendon Press, Oxford.

Wood, B. A. 1992. Origin and evolution of the genus *Homo*. Nature 355:783–790.

Wood, B. A. 2002. Hominid revelations from Chad. Nature 418:133–135.

Wood, B. A., and M. C. Collard. 1999. The human genus. Science 284:65–71.

Wood, B. A., and B. G. Richmond. 2000. Human evolution: taxonomy and paleobiology. J. Anat. 197:19–60.

Wood, B. A., C. Wood, and L. Konigsberg. 1994. *Paranthropus boisei*: an example of evolutionary stasis? Am. J. Phys. Anthropol. 95:117–136.

Woodward, A. S. 1921. A new cave man from Rhodesia, South Africa. Nature 108:371–372.

Zuckerkandl, E. 1963. Perspectives in molecular anthropology. Pp. 243–272 in Classification and human evolution (S. Washburn, ed.). Aldine, Chicago.

Perspectives on the Tree of Life

Edward O. Wilson

The Meaning of Biodiversity and the Tree of Life

It seems very likely, in accordance with the belief of many anthropologists, that the first words to emerge during the evolution of human speech were used to specify people, plants, animals, and other objects, a roster that proliferated rapidly thereafter. That step, which presumably occurred sometime during the transition from *Homo erectus* to *Homo sapiens* a half million years ago, can rightfully be considered the earliest roots of science. Accuracy and repeatability were vital for the sake of survival, then as now. Getting things by their right names, as the Chinese say, is the first step to wisdom.

And so it came to pass that the emergence of modern Western science included an effort to name the immense array of plant and animal species on Earth, and also to group them in a system that reflects their degree of similarity. That was an eighteenth-century achievement, culminating in the binomial nomenclatural system of the Swedish naturalist Carolus Linnaeus. Scientific taxonomy was followed by the notion of a genealogy of species, a nineteenth-century advance foreshadowed by the acceptance of evolution. In the twentieth century came the explanation of the process of species multiplication, one of the central achievements of the Modern Synthesis of evolutionary theory.

And now what? The answer, clearly, is a complete account of Earth's biodiversity, pole to pole, bacteria to whales, at every level of organization from genome to ecosystem, yielding as complete as possible a cause-and-effect explanation of the biosphere, and a correct and verifiable family tree for all the millions of species—in short, a unified biology. That vision, I presume, is widely shared, and why we are here.

Let me put this shared conception another way: we are here to reassert the rightful place of systematics in the mainstream of biology. In recent decades, as the molecular revolution swept over biology like a tidal wave, systematics sank in esteem. It was, in the view of the molecular triumphalists, old-fashioned biology. To many of them, its subject matter seemed spent, its practitioners dull and pedestrian. Professional taxonomists did not actually decline in population during this Dark Age, but their number, which is about 6000 worldwide today, fell sharply in relation to the total of scientists, of which perhaps half a million or more work in the United States alone. The total support given systematics research nationally from all sources, including museums, universities, and government agencies, is still a miserly $150 to $200 million annually.

But the problem with systematics, including primary descriptive taxonomy devoted to new species and monographs of those previously classified, was never obsolescence. The problem with systematics was the failure to recognize its true importance.

Consider, for example, the primary exploration of the biosphere. We do not know even to the nearest order of magnitude the number of living species on Earth. Estimates of the total number vacillate wildly according to method. They range from 3.6 million at the low end to more than 100 million at the high end. The estimated number of species of all kinds of organisms—plants, animals, and microorganisms—formally described with scientific names falls somewhere between 1.5 and 8 million, but a complete and careful census remains to be made. In short, we lack even an exact accounting of what we already know.

The following figures will give you an idea of how far we have to go in purely descriptive alpha taxonomy. About 69,000 species of fungi have been identified and named, but as many as 1.6 million are thought to exist. Of the nematode worms, making up four of every five animals on Earth—creatures so abundant that if all other matter on the surface of the planet were to disappear it is said you could still see the ghostly outline of most of it in nematodes—some 15,000 species are known but millions more may await discovery.

The truth is that we have only begun to explore life on Earth. The gap in knowledge is maximum in the case of the bacteria and the outwardly similar archaeans, the black hole of systematics, whose species could number in the tens of thousands or, with equal ease, in the tens of millions. Our ignorance of these microorganisms is epitomized by bacteria of the genus *Prochlorococcus*, arguably the most abundant organisms on the planet, and responsible for a large part of the organic production of the ocean, yet unknown to science until 1988. *Prochlorococcus* cells float passively in open water at 70,000–200,000 per milliliter, multiplying with energy captured from sunlight. Their extremely small size is what makes them so elusive. They belong to a special group called picoplankton, simple-celled organisms much smaller than conventional bacteria and barely visible at the highest optical magnification.

Even figures for the relatively well-studied vertebrates are spongy. Estimates for the living fish species of the world, including those both described and undescribed, range from 15,000 to 40,000. The global number of described and named amphibian species, including frogs, toads, salamanders, and the less familiar caecilians, has grown in the past 15 years by one-third, from 4000 to 5300 at this moment. In the same period of time the number of known mammals has also jumped from about 4000 to 5000. And similarly, the flowering plants, for centuries among the favorite targets of field biologists, contain significant pockets of unexplored diversity. About 272,000 species have been described worldwide, but the true number is certain to be more than 300,000, because each year about 2000 new species are added to the world list published in the standard *Index Kewensis* (available at http://www.ipni.org/).

You will recognize the following image in popular fiction: a scientist discovers a new species of animal or plant somewhere in the upper Amazon. At base camp the team celebrates and sends the good news back to the home institution. Mention of the event is made somewhere in the *New York Times*. The truth, I assure you, is radically different. Scientists expert in the classification of each of the most diverse groups, such as bacteria, fungi, and insects, are continuously burdened with new species almost to the breaking point. Working mostly alone and on minuscule budgets, they try desperately to keep their collections in order while eking out enough time to publish accounts of a small fraction of the novel life forms sent to them for identification.

Many systematists share this experience, of which my own example and those of fellow myrmecologists have been typical. About 11,000 species of ants have been named, but that number, we believe, is likely to double when tropical regions are more fully explored. While recently conducting a study of *Pheidole*, one of the world's two largest ant genera, I uncovered 340 new species, more than doubling the number in the genus and increasing the entire known fauna of ants in the Western Hemisphere by 20%. When my monograph was published in the spring of 2003, additional new species were still pouring in, mostly from collectors working in the tropics.

Why should we work so hard to complete the Linnaean enterprise? The answer is simple and compelling. To describe and to classify all of the surviving species of the world deserves to be one of the great scientific goals of the new century. In applied science, it is needed for effective conservation of natural resources, for bioprospecting (i.e., the search for new classes of pharmaceuticals and other natural products in wild species), and for impact studies of environmental change. In basic science, a complete biodiversity map is a key element in the advancement of ecology, including especially the understanding of ecosystem assembly and functioning. In reconstructing the Tree of Life, the new Linnaean enterprise is fundamental to genetics and evolutionary biology. Not least, it also offers an unsurpassable adventure: the exploration of a little-known planet.

Biodiversity exploration is the cutting edge of a still greater effort. Natural history remains far behind descriptive taxonomy. Of the named species—never mind those still undiscovered—fewer than 1% have been studied beyond the essentials of habitat preference and diagnostic anatomy. In addressing complex natural systems, ecologists and conservation biologists appear not to fully appreciate how thin is the ice on which they skate.

When large arrays of species are studied in depth for their intrinsic interest, the result is a surge in basic and applied research in other domains of biology. New phenomena are discovered and research agendas suggested that had never been conceived by researchers focused on favored single species such as *Escherichia coli* and *Homo sapiens*.

The complete census of Earth's biodiversity is no longer a distant dream. It is buoyed by the information revolution. New electronic technology, increasing exponentially in capacity and user-friendliness, is trimming the cost and time required for taxonomic description and data analysis. It promises to speed traditional systematics by a hundred times or more.

Within 10–20 years the combined methodology might work as follows: imagine an arachnologist making the first study of the spiders of an isolated rainforest in Ecuador.

He sits in a camp sorting newly collected specimens with the aid of a portable, internally illuminated microscope. After quickly sorting the material to family or genus, he enters the electronic keys that list character states for, say, 20 characters and pulls out the most probable names for each specimen in turn. Now the arachnologist consults monographs of the families or genera available on the World Wide Web, studying the illustrations, pondering the distribution maps and natural history recorded to date. If monographs are not yet available, he calls up digitized photographs from the central global biodiversity files of the most likely type specimens taken wherever they are—London, Vienna, São Paulo, anywhere photographic or electron micrographs have been made—and compares them with the fresh specimens by panning, rotating, magnifying, and pulling back again for complete views. Perhaps he feeds an automatic feature-matching program. Does this specimen belong to a new species? He records its existence (noting the exact location from his global positioning system receiver), habitat, web form, and other relevant information into the central files, and he states where the voucher specimens will be placed—perhaps later to become type specimens. Informatics has thus allowed the type specimens of Ecuadorian spiders to be electronically repatriated to Ecuador, and new data on its spider fauna to be made immediately and globally available.

The arachnologist has accomplished in a few hours what previously consumed weeks or months of library and museum research. He understands that biodiversity studies advance along three orthogonal axes. First are monographs, which treat all of the species across their entire ranges. Second are local biodiversity studies, which describe in detail the species occurring in a single locality, habitat, or region. When expanded to include more and more groups, local biodiversity studies may eventually cover all local plants, animals, and microorganisms, creating an all-taxa biotic inventory, a truly solid base for community ecology in its full complexity.

The next step in global biodiversity mapping can be expected to follow close behind, thanks to the swift advances occurring in genomics. Already on the order of 10,000 species from the major domains of organisms have been sequenced for their small subunit ribosomal genes. As the process accelerates, so will growth of these and other base pair data, and in a reasonably short time the sequences will become a standard tool for identification and phylogenetic reconstruction across all groups of organisms.

Next on the horizon and coming up fast are complete genomes and, in particular, those of functional genes. A method has recently been conceived, using parallel sequencing of single DNA or RNA strands through nanopores, that

if successful could read off the three billion base pairs of a human cell in hours or the thousand or so of a virus in seconds. Holes little more than a nanometer in width are punched through cell membrane with staphylococcus bacteria, forming channels just wide enough to thread single strands of nucleotides but not double strands. Electrical impulses force the strands through, and differences in conductance of the base pairs identify them after passage. The method is in an intermediate stage of development and may not in the end become operable, but at the very least it illustrates the potential of technologies, for example, those that include advances in the shotgunning method, poised to advance genomics and put it at the service of systematics and the rest of biology.

Ultrafast genomic mapping is not necessary for the identification of a butterfly or flowering plant. The larger and anatomically more complex eukaryotic organisms can be identified very swiftly by visual inspection of their diagnostic phenotypes, if not in the heads of experts then by the use of software that automatically scans specimens and their images with a capacity for near-instantaneous matching and identification. But rapid sequencing is crucial for viruses, bacteria, fungi, and many of the smaller soft-bodied animals.

When microorganisms can be quickly identified by their genomes, the impact on biology will be enormous. For the first time a comprehensive picture of their diversity and geography will emerge. Ambiguities concerning the root of the Tree of Life will diminish as the earliest stages in the evolution of life are more precisely defined. The origin and role of natural transgenes in the early evolution of higher organisms will be clarified. In ecology the effect will be truly revolutionary, because microorganisms are a large part of the foundation of ecosystems, yet to date are largely unstudied. It will be possible to enter undisturbed ecosystems at micro and nano levels, observe thousands of kinds of microorganisms in action in the same way we now observe animals and plants macroscopically, and from these miniature and still unexplored rainforests of the ultrasmall, collect colonies and individuals for rapid identification. I believe it safe to predict that within 10–20 years, microbial systematics and microbial ecology will become major industries of science.

In exploring large and microscopic organisms alike, the grail of a global all-taxon biological inventory (ATBI) also seems attainable within a matter of decades, say, in 20 years, if it is made a scientific priority. The time has come to treat the global ATBI as a near-horizon goal rather than, as traditional in the past, an eventual destination. Above all, it is rendered urgent by the accelerating worldwide destruction of natural ecosystems and extinction of species. Conservation biologists are in near-unanimous agreement that human activity has inaugurated a mass extinction spasm not equaled since the end of the Mesozoic era 65 million years ago. At the present rate of environmental degradation, as many as a quarter of the still-existing plant and animal species could be gone or committed to early extinction within 30 years, and half by the end of the 21st century. Biology is the only science whose subject matter is vanishing. Alerted to the technological advances that promise to empower the global ATBI, and realizing the importance of such a thorough survey for humanity, a dozen or so groups around the world have initiated ATBIs on a continental or global scale, and to varying degrees of resolution—with or without microorganisms, for example, or based on existing databases and museum specimens or not. One of the most ambitious is the Global Biodiversity Information Facility (GBIF for short), conceived within the Organisation for Economic Co-operation and Development (OECD) in 1999, headquartered this year in Copenhagen, and funded by pledges from 14 OECD member countries. In 2001, another, private organization, the All Species Foundation, was begun in California with the same goal as the GBIF. That fall the All Species Foundation hosted a summit meeting at Harvard of organizations engaged in continental and global all-taxon censusing. They included GBIF; the Association for Biodiversity Information, which has been newly created from the Natural Heritage Network of the Nature Conservancy; the Biodiversity Foundation for Africa; and others.

In time such organizations will try to work out a plan for concerted action, a timeline, a budget, a suite of methodologies, and a fund-raising program that raises all ships. I expect that a heavy emphasis will be put on the financial support and upgrading of basic systematics research, including straightforward alpha taxonomy, which, I trust you will agree, undergirds everything we accomplish and hope to accomplish in systematics generally.

The effort to complete a global biodiversity map is likely to follow the following stages:

- First and foremost is the high-resolution imaging of primary types of all species for which this is practicable or, in absence of types, other authenticated material.
- At the same time, or soon thereafter, with the supervision of expert systematists, the images, collection data, and bibliography references and synonymy will be placed on the Internet.
- Then this vastly more accessible database will be used to prepare monographs, field guides, and instructional manuals at a greatly speeded-up pace.
- In the longer term, field exploration will pick up to fill the gaps, yielding Internet diagnoses of new species and expansion of databases for already known species.
- Simultaneously, there will be ongoing phylogenetic reconstructions of species, updated as novelties and new data are added. The Tree of Life, including the interpretation of the evolutionary history of all living taxa and the antecedent taxa recoverable by cladistic inference and the fossil record, will emerge with constantly improving clarity.
- Finally, a true encyclopedia of life will be pieced together, transiting all levels of biological organization, genome to ecosystem, and enlarged continuously during the generations to come.

In visualizing the universal tree, the living species can be thought of as the growing tips of the twigs and leaves, and their antecedents the branches. The living species are monitored in organismic and evolutionary time, the intervals of which witness changes that can be observed within a human generation. The histories of the branches, in contrast, are reconstructed in evolutionary time, across intervals that in most cases extend deep into geological history.

Systematists who work on living species, the twigs and leaves of the Tree of Life, produce information increasingly vital to the rest of biology, from molecular and cell biology and the medical sciences to ecology and conservation biology. Those who work on phylogeny, the branching patterns across evolutionary time, provide the basis of a sound higher classification and our integrated picture of the history of life. Exploratory systematics and phylogenetic reconstruction are synergistic, reinforcing one another, illuminating biodiversity as it is in this instant of geological time and tracing its origins through deep geological time.

From the alpha taxonomy of species and geographical races to their phylogeny, modern systematics becomes at last a seamless web of rigorous science and cutting-edge technology. Applied to each level of biological organization in turn, it is the key to a unified biology.

In other chapters of this volume are dispatches from the front delivered by some of our leading authorities on the systematics and evolution of virtually the complete spread of biodiversity. They will make clear that in drawing the Tree of Life, from the still tangled and problematic trunk of bacteria and archaeans to the mind-boggling productions of the flowering plants and animals, a new biology is emerging. They will establish, I am confident, that systematics is what ties biology together. Implicit also will be the necessity of this knowledge for the preservation of Earth's fauna and flora, including that awkwardly bipedal, bulge-headed, tool-making, incessantly chattering Old World primate species, *Homo sapiens*. The universal ATBI and the unified Tree of Life are the conceptions that will surely fire the ambition and release the energies of those committed to evolutionary biology.

David B. Wake

A Tree Grows in Manhattan

When the first full genome for a microbe was published, I was teaching an evolution course, and as I read the article I was first surprised and then thrilled to learn that the discovery had such profound evolutionary significance. Along with many others, I realized that we were entering a new world, one in which evolutionary biologists such as I had new responsibility. We now could, and therefore must, build a Tree of Life. It has long been a dream of comparative biology to explain how life has evolved and what evolutionary relationships mean. It has been a personal dream to make evolutionary biology predictive. Because evolution seems to run in grooves, following avenues of least resistance, knowing something about one taxon gives one a very good sense of what a closely related taxon will be like. Why should this be so? Evidently there are rules to be discovered, generalities to be established. Genetics, especially as it relates to development, provides some inspiration. But imagine what we might learn if we knew the true Tree of Life! Such a tree would include vastly more than what I now have the courage to identify as "only" full genomic information, but even that would be a great start.

It has been nearly 20 years since my colleague Allan Wilson first told me about how it was possible to amplify and soon to sequence DNA. He thought it would be only a short time before systematists would be routinely sequencing DNA and using the data to frame and test evolutionary hypotheses. I thought he was optimistic, but he was right. About the time that these conversations were taking place, Marvalee

Wake and I bought our first personal computer (we actually thought it would be possible to share one!). Systematists everywhere were having such experiences, and before long we were armed with methods, techniques, machines, and most important, with an intellectual framework (coming out of the phylogenetics revolution starting with Hennig on the one hand and numerical methods on the other, in the 1960s). Rapid progress ensued, leading to the first inkling that we might try assembling a Tree of Life, envisioned in the Nobel Symposium in Sweden in 1988. But most of us toiled with our own taxa, which systematists have historically divided up so as to avoid direct confrontational competition. The organization of the systematics community into provincial societies (within the herpetological community alone there are three mainly North American societies and dozens more elsewhere in the world, most with their own journals) did not help bring groups together, but gradually, with the National Science Foundation playing a critically important role at several points along the way, we began to interact effectively, and the successful conference we have experienced is the most recent manifestation.

Not surprisingly, early attempts to develop a tree of all life began within the community of microbial biologists, not only because they had less (in the sense of organismal complexity) to work with and had to turn to molecules, but also because they already were familiar with many molecular biological techniques and were ready to move when the era of PCR (polymerase chain reaction) arrived. Perhaps more surprising is how rapidly the systematics community embraced molecular methods and approaches, not as a replacement for more traditional morphological approaches (which continued to develop methodologically, with a focus on building large character-based databases and analyzing them in diverse ways), but as an exceedingly important addition to our "tool kits."

The New York meeting was an unqualified success from my viewpoint. The oral presentations were uniformly outstanding—well prepared, well delivered, and designed for effective communication with a diverse audience. Remarkably, there was no dissent from the fundamental premise —that we want, need, and *can* produce a Tree of Life. Furthermore, in a field that has experienced intellectual warfare, what controversies arose in terms of data analysis and the like were downplayed in the interests of the general good. Perhaps we were all on good behavior because of the high degree of idealism expressed so beautifully by Ed Wilson in his inspiring opening address, and the symbolism of a remarkable address by Rita Colwell, the Director of the National Science Foundation and a person who thoroughly understands and appreciates the goal we have set for ourselves. For whatever reason, there was a wonderful sense of a common purpose, as well as of duty and responsibility. And in the background of it all was the intellectual imperative that the tools are at hand to accomplish our goal.

It is amazing to me how much comparative DNA sequence is accumulating and at what a high rate! Lacking such data, we would not even be talking about a Tree of Life initiative, but for taxon after taxon we witnessed the impact of molecular data. In some instances the goal of many systematists, a "total evidence" approach incorporating morphological and molecular data, integrated with fossil evidence, is

emerging (e.g., mammals). However, large molecular databases do not assure phylogenetic resolution, as we have learned in the case of birds. For some relatively large taxa (e.g., my own group, the amphibians, with about 5500 species), it may be possible to obtain sequence information for nearly all species, so as to put the "leaves" on the tree. But for microbes (astonishingly complex in the extent of paraphyly), despite an enormous accumulation of sequence data, the number of unsampled taxa is staggering and one wonders what the impact of as yet unsampled lineages will be.

I was struck by the estimates of one after another of the specialists that the numbers of taxa in their areas were vastly greater than previously thought. We remain in a phase of discovery, as we were reminded by the very recent description of a new order of insects. The number of species of amphibians is growing more than 3% per year, and vertebrates are supposed to be well known. Certainly at the level of basal taxa we have a great deal to learn, even for our best-known groups. So, the task is large, and if we are to accomplish it we will have to modify our publication strategy and streamline the process by which we describe taxa.

There will be more Tree of Life conferences and they will become increasingly inclusive, of researchers as well as taxa. We will work together not only because we stand to benefit from the interaction, but above all because we *must*. Information about what we have in the world will improve our chances of preserving biodiversity. Just knowing the Tree of Life will not assure its preservation, but for those of us for whom taxa count and trees count, having the requisite information will, we expect, enable us to more effectively act. We live in challenging and exciting times, but they are perilous as well, and it will take more than knowledge and wisdom to preserve the main structure of the Tree of Life on this planet.

David M. Hillis

The Tree of Life and the Grand Synthesis of Biology

In the 1980s, there was rapid growth of the field of phylogenetics. The developments were so extensive that at the 1988 Nobel symposium titled "The Hierarchy of Life" (Fernholm et al. 1989), one participant wondered aloud if young biologists could be attracted into the field given that "all the big questions have been answered." I doubted that pronouncement; from my view, the field of phylogenetics was still in its nascent stages. I thought most of the big and interesting questions, as well as the major challenges, awaited us in the future. Morris Goodman agreed, and he described his vision for "a new age of exploration that promises to bring to fruition Darwin's dream of reconstructing the true genealogical history of life" (Goodman 1989:43). In many ways, that symposium did represent a turning point for phylogenetics, and the symposium that represents the subject of this book shows just how far we have come since the 1980s. The advances in progress on the Tree of Life have been greater in the 1990s than in all previous years combined, and the prognosis for the future has never been brighter.

A few comparisons between the 1988 Nobel symposium and the present symposium, "Assembling the Tree of Life," demonstrate just how much progress we have made. The description of PCR (polymerase chain reaction) had only been published the year before the Nobel symposium (Mullis and Faloona 1987), and DNA sequencing data were just beginning to have a major impact on the field of phylogenetic analysis. Statistical analysis of phylogenetic trees was in its infancy in 1988, although several of the papers published in the proceedings of that symposium discussed emerging methods for assessing the strength of support for inferred trees. Even though data sets in 1988 were rather small by today's standards, computational resources (both software and hardware) were already limiting. Maximum likelihood analyses were virtually unmentioned at the 1988 symposium, and the computational constraints of such analyses made their application to large problems impractical. Therefore, systematists were severely limited by lack of data, weakly developed statistical methodology, and computational constraints. However, the stage was set for all of these bottlenecks to be removed or reduced.

In figure 32.1, I show an analysis of papers in the *Science Citation Index* for the past two decades (1982–2001). In 1982, there were 186 papers in the *Science Citation Index* that had the word "phylogeny" (or its derivative "phylogenetic") in the title, abstract, or key words. This means that it was possible to read about one paper every other day, and still read virtually all the literature on phylogenetics published worldwide. As I said above, the growth of the field through the 1980s was impressive: by the end of the decade, there had been more than a doubling of papers on phylogeny (393 papers in 1990), and in that year it would have been necessary to read more than a paper a day to read all the papers in the field. However, the real growth of the field of phylogenetics (at least in terms of number of papers published, and therefore in the number of phylogenetic trees presented) occurred throughout the 1990s.

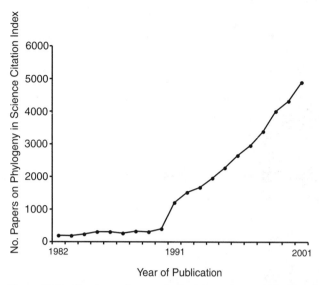

Figure 32.1. Numbers of papers in the *Science Citation Index* that include the words "phylogeny" or "phylogenetic" in the title, abstract, or key words, published from 1982 through 2001.

In 2001, almost 5000 papers were published on phylogeny. The total number of papers in the *Science Citation Index*, across all fields of science, was 999,618 in 2001. That means that a staggering 1 paper out of every 200 published in all fields of science was on phylogeny! Today, when I pick up a journal in almost any biological field, I expect to see some kind of phylogenetic analysis in at least one of the articles. If one wanted to attempt to read all the papers on phylogeny, that would require reading about 100 papers a week.

Recent progress on the Tree of Life has not resulted just because phylogenies are so much easier to infer now than they were a decade or so ago. The importance of understanding the relationships among the subjects of their studies finally became widely accepted (by biologists, of all fields) in the 1990s, as well. As phylogenies for many groups (as well as genes) became widely available, the power of comparative analyses became apparent in all areas of biology. Until phylogenies were widely available, biologists were likely to view objects of study in biology much as a chemist would view atoms in a chemical equation. Every hydrogen atom (of the same isotope) can be treated like all others. However, virtually nothing in biology is like hydrogen atoms. Every gene, every individual, every species, and every clade is more closely related (and more similar) to some genes, individuals, species, and clades than it is to others. This makes biology difficult, but not impossible. However, it does mean that every biologist must think at some level about phylogeny to put his or her work in the context of the rest of biology.

As I watched the presentations in this symposium, I was awed in two ways. First, the progress on reconstructing the Tree of Life has been nothing short of phenomenal. Our annual progress on understanding new relationships within the Tree of Life is now much greater than all the accumulated knowledge on relationships that we had in the late 1980s. The applications of the Tree of Life to problems as diverse as forensics, origins of new diseases, ecology, behavior, development, molecular evolution, and assessment of global biodiversity is astonishing, and it is hard to keep up with all the new developments. Second, and despite all the recent progress, I was struck with the view that we are on the brink of yet another turning point: as the Tree of Life becomes more complete, its applications are also expanding exponentially. A complete Tree of Life would allow analyses that we would never contemplate today. Even the goal of discovering all the species on Earth is much more likely to be achieved if we have a complete Tree of Life for all the known species. A complete Tree of Life would allow us to catalog and organize all the species we know about, greatly increasing the potential to automate the discovery and description of the remaining unknown species. Fields such as ecology could move from treating communities as unknown "black boxes" to understanding their complexity and differences, perhaps allowing ecology to emerge as a truly predictive science. With phylogeny as a framework, molecular biology could move from a largely descriptive science to a field of explanation and prediction. The Tree of Life would also allow us to organize, connect, and synthesize all the information on all the species of Earth. A grand, web-based "encyclopedia of life" would result, and the field of biology would be immediately transformed. After that point, any information that anyone collected on any species would contribute to the understanding of all of life. In short, the Tree of Life represents the first (and most critical) step in the Grand Synthesis of biology.

Will someone writing an overview of the 2022 Tree of Life Symposium see the trend shown in figure 32.1 continue? My guess is that the trend will continue for at least a few years, but perhaps not decades, if the phylogenetic revolution is to be truly successful. The term "phylogeny" is now emphasized in papers that use phylogenetic methods in part because the approach is still considered innovative in many fields. However, in the future, if the Tree of Life initiative is truly successful, people will not think to distinguish their papers in this way. If all of biology is connected through a Tree of Life, then studying biology in a phylogenetic context should become almost transparent. People will include phylogenetic analyses as a matter of ordinary operating procedure. So, the best measure of the success of the phylogenetic revolution will come when analyzing biological data in a phylogenetic context merits as much of an emphasis in a paper as using a computer to analyze data does today, namely, something that virtually everyone does as a matter of necessity. And as with computers, new students in biology won't even be able to imagine how we ever got along without phylogenetic analysis.

Literature Cited

Fernholm, B., K. Bremer, and H. Jörnvall (eds.). 1989. The hierarchy of life: molecules and morphology in phylogenetic analysis. Excerpta Medica (Elsevier Science), Amsterdam.

Goodman, M. 1989. Emerging alliance of phylogenetic systematics and molecular biology: a new age of exploration. Pp. 43–61 *in* The hierarchy of life: molecules and morphology in phylogenetic analysis (B. Fernholm, K. Bremer, and H. Jörnvall, eds.). Excerpta Medica (Elsevier Science), Amsterdam.

Mullis, K. B., and F. A. Faloona. 1987. Specific synthesis of DNA *in vitro* via a polymerase catalyzed chain reaction. Methods Enzymol. 155:335–350.

Michael J. Donoghue

Immeasurable Progress on the Tree of Life

In listening to the Assembling the Tree of Life (ATOL) symposium in New York, and in reading the manuscripts for this volume, I was overwhelmed by the enormous progress that we have made, over such a short time, on what Darwin so aptly called "the great Tree of Life." The word "immeasurable"—in the dictionary sense of "indefinitely extensive"—seems to apply perfectly to this situation. But what about the other, more literal, meaning of the word immeasurable? Is phylogenetic progress also "incapable of being measured"? This is the question I want to address. My sense is that there are many facets of "progress" that matter to us and that we would like to be able to measure. For some of these we can devise proper metrics, and we might even be able to provide concrete numbers. For others, as I'll argue, we aren't even entirely sure what we'd like to measure, and we're still a long way from being able to quantify how we are doing.

Let me back up, and ask, What are the ways we might think about expressing progress—to measure where we stand now in relation to where we were a decade ago and where we hope to end up? One possibility would be to tally the number of known species on Earth that have been included in bone fide phylogenetic analyses [in December 2003 there were almost 35,000 species represented in TreeBASE (available at http://www.treebase.org), but the real number might be more like 80,000], or maybe even the number that could potentially be included today if we harnessed all of the data in relevant databases [e.g., DNA sequences in GenBank (available at http://www.ncbi.nlm.nih.gov)]. Another possibility

would be to chart trends in the number of phylogenetic papers published over the years (e.g., Sanderson et al. 1993; Hillis, ch. 32 in this vol.).

These are certainly interesting measures, and the numbers, insofar as we know them, certainly do bolster the gut-level feeling that we're making lots of progress. They don't, however, capture much about the nature and the quality of what's being learned. Maybe we should also be gauging our coverage of the Tree of Life in terms of the number of major lineages represented by some reasonable number of exemplars, or perhaps we should somehow represent the size and the variety of the data sets that are being analyzed. Or, perhaps a metric is needed to reflect changing levels of confidence in the clades being identified. Another worthy measure, for very obvious purposes, would gauge how many phylogenetic studies have provided solutions to practical problems. Success stories along these lines abound—identifying the source of an emerging infectious disease, pointing the way toward crop improvement, orienting the search for new pharmaceuticals, and so on (see Yates et al., ch. 1 in this vol.; examples of the practical importance of phylogenetic research are also highlighted in a brochure sponsored by the National Science Foundation (Cracraft et al. 2002). But how do we attach a number to such achievements? Patents pending, perhaps, although this would record only a small fraction of the successes.

Ultimately, I think we would all like a measure that captures how phylogenetic studies have affected our understanding of life—how the living world is structured, how it works, and how it has come about. At first glance this truly does seems immeasurable, in the "not-capable-of-measurement" sense of the word. But on second thought, maybe there is a reasonably good proxy for this, which takes us back to Willi Hennig (e.g., Hennig 1966). What if we could faithfully tally up cases in which traditionally recognized taxonomic groups had been convincingly demonstrated to be paraphyletic? Paraphyletic groups are ones that contain an (inferred) ancestor and some, but not all, of its descendants. In practice, of course, paraphyly is "discovered" when a phylogenetic analysis identifies one or more new clades that unite some of the lineages previously assigned to the traditional group with one or more lineages placed outside of that group. In other words, the "negative" discovery of paraphyly is *precisely* the "positive" discovery of new "cross-cutting" clades.

Before we think about whether we could actually count up discoveries of paraphyly, let's contemplate why this might be a satisfying measure of phylogenetic progress. First of all, it's worth noting that this measure relates how changes in our knowledge of phylogenetic relationships have affected the application of taxonomic names, and as such, it can potentially be assessed everywhere in the Tree of Life, from the very base out to the tips, without needing to refer to particular groups or their characters. In this sense, it is a measure without units. Second, it registers a change in the language that we use to describe the structure of diversity, which can deeply (although often quite subtly) influence the way we perceive diversity, orient our research, and teach. Third, the discovery of paraphyly has immediate impacts on our understanding of character evolution. Some characters previously thought to have evolved convergently are seen instead to be homologous—to have evolved only once, in the inferred ancestor of a newly discovered cross-cutting clade. Even more generally, the recognition of paraphyly allows us to infer a sequence of evolutionary events, which helps fill in what appeared to be major gaps between traditional taxa. Often this is just the information we need to choose among competing evolutionary hypotheses about how and why major transitions occurred. In many of the same ways, of course, such discoveries also help us make sense of biogeography. Fourth, such discoveries generally change the way we perceive shifts in diversification, especially by accentuating differences in the number of species between sister groups.

Putting the third and fourth points together, my guess is that discoveries of paraphyly will eventually have even more profound impacts on how we view the connection between character change and diversification. In particular, I think we'll be forced to develop a more nuanced (and more productive) view of "key innovations." It will become increasingly natural to think from the outset about a series of changes culminating in a combination of traits that ultimately affected diversification. Rather than simply moving the causal explanation down a node or two in the phylogeny, this distributes the causation across a series of nodes and character

changes. Also, increasingly we'll focus on how apparently subtle changes early in such a chain rendered new morphological designs accessible, which in turn enabled the evolution of the traits that we most often associate with the success of clades, with ecological transitions, and so forth.

To illustrate these points, let's look at green plants. Figure 33.1 provides an overview of our present knowledge of phylogenetic relationships among the major lineages—highly simplified, of course, and consciously pruned (rendered pectinate) to serve my purposes (see O'Hara 1992 for a general discussion of such simplifications). Several widely known traditional groups are supported as monophyletic in all recent analyses, including the entire green plant clade (viridophytes), land plants (embryophytes), vascular plants (tracheophytes), seed plants (spermatophytes), flowering plants (angiosperms), and monocotyledons (monocots). A number of other traditionally recognized groups have repeatedly been determined to be paraphyletic, confirming suspicions that they represent grades of organization, diagnosed only by ancestral features of the more inclusive clades to which they belong. Specifically, "green algae," "bryophytes," "pteridophytes," "gymnosperms," and "dicotyledons" all appear to be paraphyletic. In each case, one or more new clades were discovered that linked some lineages traditionally assigned to the group to related taxa. So, for example, the streptophyte and charophyte clades (as circumscribed here; for an alternative, see Delwiche et al., ch. 9 in this vol.) include lineages that used to be assigned to the green algae (the Charophyta in the traditional sense) along with the land plant clade. Likewise, the euphyllophyte clade unites all extant lineages of seedless vascular plants, except the lycophytes, with the seed plants, and so on. In the case of the "bryophytes" and the "gymnosperms," names were proposed for new cross-cutting clades ("stomatophytes" and "anthophytes," respectively), but recent analyses have cast doubt on their existence (see Nickrent et al. 2000, Donoghue and Doyle 2000). Nevertheless, in both cases it remains quite clear that these traditional groups are paraphyletic (see Delwiche et al., ch. 9, and Pryer et al., ch. 10 in this vol.).

The impact of these discoveries on our understanding has been enormous. The most obvious and immediate effect was on our ability to dissect the evolutionary sequence of events surrounding the greatest transformations in plant history. For example, take the transition from living in water to living on land (see Graham 1993). Before we recognized the paraphyly of green algae and of bryophytes, this shift appeared to entail a large number of steps, which we had no real basis for putting in order. This implied either many extinctions and, consequently, gaps in our knowledge, or else some sort of wholesale transformation from one life form to another. Under these circumstances, alternative theories emerged and remained viable. What kind of environment did the immediate ancestors of the land plants live in, and what did they look like? After all, "green algae" live in saltwater or in freshwater; may be unicells, colonies, filaments, or more complex

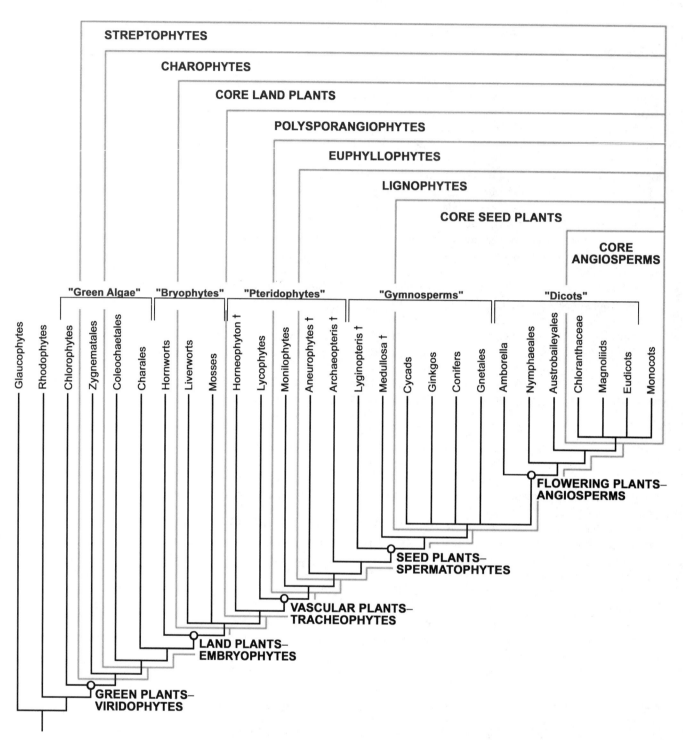

Figure 33.1. An overview of green plant phylogeny, illustrating progress through the recognition and abandonment of paraphyletic groups (e.g., "green algae" and "bryophytes") with the discovery of new major clades (e.g., streptophytes and euphyllophytes). For references to the primary literature, underlying evidence, levels of support, outstanding controversies, and additional evolutionary implications, see Kenrick and Crane (1997), Doyle (1998), Donoghue (2002), Judd et al. (2002, ch. 7), and chapters 9–11 in this volume. Note that Delwiche et al. (ch. 9 in this vol.; also Karol et al. 2001) use the name "Charophyta" for the clade here referred to as the streptophytes. The usage adopted here may better reflect original intentions (e.g., Bremer and Wanntorp 1981) and subsequent usage (e.g., Kenrick and Crane 1997); in any case, such nomenclatural problems highlight the desirability of providing explicit phylogenetic definitions for clade names.

forms; may or may not have cell walls separating the nuclei; and so on. And what about the evolution of the land plant life cycle—alternating between multicellular haploid (gametophyte) and diploid (sporophyte) phases? In short, the transition to land largely remained a mystery.

With the discovery of a series of intervening clades (fig. 33.1; Karol et al. 2001; see Delwiche et al., ch. 9 in this vol.), we're now able to infer a sequence of events from the first green plants through the transition to land. We can be quite certain that their immediate ancestors lived in freshwater, probably quite close to the shore; had rather complex parenchymatous construction; and bore eggs (and zygotes) on the parent plant in specialized containers. Likewise, we can finally put to rest the debate about the life cycle: the land plant life cycle originated through the intercalation of a multicellular diploid phase (through delayed meiosis) into an ancestral life cycle in which the diploid zygote underwent meiosis directly to yield haploid spores.

This example is meant only to illustrate the sorts of insights that can follow the discovery of paraphyly, and so to justify such a measure of progress. What can we say, then, about the number of these discoveries in recent years, or about our expectations in the future? In *The Hierarchy of Life* (Fernholm et al. 1989), the last major attempt to take stock of phylogenetic progress, Gareth Nelson remarked: "Paraphyly, it would seem, is the most common discovery of modern systematic research" (Nelson 1989: 326). This may well be true, but is there a way to put a number on it? Sadly, aside from asking experts on each major clade to come up with a list (or an account along the lines of fig. 33.1), we aren't really able to do this. We haven't been keeping track in any systematic way and, as I will argue, we haven't developed the necessary informatics tools.

Let us suppose that we wanted to be able to tally up those changes in knowledge of phylogeny that significantly changed our view of the world, and that for this purpose we wanted to focus on discoveries that changed the way that taxonomic names are used. Specifically, we would be looking for cases in which the name of a paraphyletic group had been abandoned altogether, or the circumscription had been adjusted so that the name again referred to a hypothesized clade. These are what might be called "meaningful" taxonomic changes, to distinguish them from other sorts of name changes. We would want to avoid, for example, changes only in the Linnaean taxonomic rank that a group is assigned (e.g., a shift from Family to Order). As things now stand, such rank assignments are fundamentally arbitrary, yet our nomenclatural codes are intimately tied to them, and in some cases a cascade of name changes can be required without any underlying advance in our knowledge of phylogeny. Also, it's important to note that quite a few clades are discovered and named that don't contradict the monophyly of any previously named taxon—instead, they resolve bits of the Tree of Life that were more or less unresolved and to which taxonomic names had not been applied. The point is that the problem

is not as "simple" as just tracking changes in the names being used in the taxonomic literature.

What we really are talking about is tracking changes in the relationship between taxonomic names and hypothesized clades. If we knew how taxonomic names mapped onto a tree at some initial time, we could see at a later time how many names applied to the same clades versus how many no longer applied to clades but to paraphyletic groups. To do this in practice, one would need, first of all, a database that recorded changes in our knowledge of phylogeny. TreeBASE is designed for this purpose, but unfortunately, it still isn't used consistently enough by the authors of phylogenetic papers. One presumes that this will improve (probably driven by more journals requiring the submission of phylogenetic data and results), in which case we will automatically develop the record we need to make solid tree comparisons over time.

But tracking trees is only one part of the problem. The other is to understand how names have been used at different times. Although for some groups of organisms there are databases that keep track of all the names that have ever been published (e.g., the International Plant Names Index, available at http://www.ipni.org), or even of the accepted names and synonyms (e.g., Species 2000, available at http://www.sp2000.org), it's hard to say exactly how these names correspond to hypothesized clades at any one time, much less at different times. The problem is that taxonomic names have not traditionally been defined in such a way that we can be sure whether they were even meant to refer to clades (sometimes, mostly in the past, names were knowingly applied to paraphyletic groups) or, if so, which lineages were intended to be included (even assuming complete agreement on phylogenetic relationships). Of course, we could get better about designating how names are meant to coincide with clades by, for example, consistently labeling clades in TreeBASE. This would be a step in the right direction, but it would be even better to adopt a nomenclatural system in which the connection between a taxonomic name and a hypothesized clade needed to be precisely defined at the outset. Here I am referring to "node-based" and "stem-based" definitions and other conventions discussed in relation to the PhyloCode (available at http://www.phylocode.org). Interestingly, taxonomic names under such a system tend to be maintained in the face of changes in phylogenetic knowledge, although with a different composition of lineages. Specifically, the name of a taxon discovered to be paraphyletic might well be retained for a more inclusive clade, unless it happened to become synonymous with a preexisting name. Overall, it is hard to say how the turnover of names would compare between the PhyloCode and our traditional nomenclature codes, where names are neither defined with respect to a tree nor fixed in terms of content.

The conclusion I draw from the above is that the actual abandonment of the names of paraphyletic groups is probably not going to be a very sensitive measure (under either traditional nomenclature or under the PhyloCode). Names can be retained and reconfigured in various ways, and in any case it would be hard to judge when a particular name had finally been dropped by the relevant taxonomic community. In the end, what we really want, regardless of "abandonment," is a database designed such that we can identify those phylogenetic discoveries that change how names map onto trees—whether a name refers to the same clade at different times or whether it can be made to refer to a clade only by changing the content to include lineages previously viewed as being outside the group. This would be a pretty sophisticated database, but I see no reason why it couldn't be developed.

My point is that it's time we attended to the business of naming clades and to the informatics issues surrounding the Tree of Life project. As Hennig stressed, "Investigation of the phylogenetic relationship between all existing species and the expression of the results of this research, in a form which cannot be misunderstood, is the task of phylogenetic systematics" (Hennig 1965: 97). Progress on the first of these goals—understanding phylogenetic relationships—has certainly been impressive. By comparison, progress on the second goal—expressing the results in a form that cannot be misunderstood—has been rather pathetic. Much of what we have learned about relationships has not been translated into the taxonomic language used to describe the diversity of Life. And much of what we have learned has not been properly incorporated into databases, so the effort is effectively wasted. I hope we have made real progress along these lines before we take stock again of the Tree of Life.

In summary, at this moment it strikes me that phylogenetic progress is immeasurable in both senses of the word—phylogenetic knowledge is expanding at a mind-boggling rate *and* we don't yet have the tools to measure this in the ways we would like. When we are eventually able to make measurements of the sort I have described, we will have achieved something truly monumental. We will certainly have charted much more of the Tree of Life, but we will also have changed the language we use to communicate about biological diversity and, therefore, how we think about the world. Perhaps most important, we will have rendered this knowledge widely accessible and prepared it for the queries that will propel the Tree of Life project to the next level. "Indefinitely extensive" will have become the only applicable meaning of "immeasurable."

Acknowledgments

I am grateful to Joel Cracraft for his leading role in organizing the symposium and editing the proceedings, and to the other speakers in the session on plants—Chuck Delwiche, Kathleen Pryer, and Pam Soltis. I have benefited from discussion of these issues with Susan Donoghue and Kevin de Queiroz. For their help with my presentation at the symposium and with figure 33.1, I am indebted to Brian Moore and Mary Walsh. Yale University, through Provost Alison Richard, generously supported the symposium and the participation of Yale students.

Literature Cited

Bremer, K., and H.-E. Wanntorp. 1981. A cladistic classification of green plants. Nord. J. Bot. 1:1–3.

Cracraft, J., M. Donoghue, J. Dragoo, D. Hillis, and T. Yates (eds.). 2002. Assembling the tree of life: harnessing life's history to benefit science and society. National Science Foundation. Available: http://ucjeps.berkeley.edu/tol.pdf. Last accessed 25 December 2003.

Donoghue, M. J. 2002. Plants. Pp. 911–918 in Encyclopedia of evolution (M. Pagel, ed.), vol. 2. Oxford University Press, Oxford.

Donoghue, M. J., and J. A. Doyle. 2000. Demise of the anthophyte hypothesis? Curr. Biol. 10:R106–R109.

Doyle, J. A. 1998. Phylogeny of the vascular plants. Annu. Rev. Ecol. Syst. 29:567–599.

Fernholm, B., K. Bremer, and H. Jörnvall (eds.). 1989. The hierarchy of life. Nobel Symposium 70. Elsevier, Amsterdam.

Graham, L. E. 1993. Origin of the land plants. Wiley, New York.

Hennig, W. 1965. Phylogenetic systematics. Annu. Rev. Entomol. 10:97–116.

Hennig, W. 1966. Phylogenetic systematics. University of Illinois Press, Champaign-Urbana.

Judd, W. S., C. S. Campbell, E. A. Kellogg, P. F. Stevens, and M. J. Donoghue. 2002. Plant systematics: a phylogenetic approach. 2nd ed. Sinauer, Sunderland, MA.

Karol, K. G., R. M. McCourt, M. T. Cimino, and C. F. Delwiche. 2001. The closest living relatives of land plants. Science 294:2351–2353.

Kenrick, P, and P. R. Crane. 1997. The origin and early diversification of land plants: a cladistic study. Smithsonian Institution Press, Washington, DC.

Nelson, G. 1989. Phylogeny of the major fish groups. Pp. 325–336 in The hierarchy of life (B. Fernholm, K. Bremer, and H. Jörnvall, eds.). Nobel Symposium 70. Elsevier, Amsterdam.

Nickrent, D., C. L. Parkinson, J. D. Palmer, and R. J. Duff. 2000. Multigene phylogeny of land plants with special reference to bryophytes and the earliest land plants. Mol. Biol. Evol. 17:1885–1895.

O'Hara, R. J. 1992. Telling the tree: narrative representation and the study of evolutionary history. Biol. Philos. 7:135–160.

Sanderson, M. J., B. G. Baldwin, G. Bharathan, C. S. Campbell, D. Ferguson, J. M. Porter, C. Von Dohlen, M. F. Wojciechowski, and M. J. Donoghue. 1993. The growth of phylogenetic information and the need for a phylogenetic database. Syst. Biol. 42:562–568.

Joel Cracraft

Michael J. Donoghue

Assembling the Tree of Life

Where We Stand at the Beginning of the 21st Century

Few endeavors in biology, or in all the sciences, can match our quest to understand the course of life's history on Earth, which stretches across billions of years and captures the descent of untold millions of species. The notion that scientific inquiry might achieve that goal is little more than a century and a half old, and yet surprisingly, most of the species that have appeared on the twigs of the Tree of Life (TOL) have been put there only in the last decade. The systematists who have contributed to the chapters in this volume have collectively contributed a significant step toward a grand vision of systematic biology: achieving a comprehensive picture of the TOL is finally within our grasp. Darwin, Haeckel, Huxley, and the other giants who convinced the world of life's long history of change, and built the first scaffold of that history, might very well say "finally . . . it's about time"!

That it has taken so long to get to this point is testimony to the fundamental conceptual and technical challenges that have faced systematic biologists over the years. For many decades systematists had no clear theoretical or methodological idea how to recover life's history in an objective way. That challenge, as many of the greatest in the sciences, was met by deceptively simple logic. Willi Hennig, and the phylogenetic principles he developed (1950, 1966), quickly formed the foundation for quantitative, objective methodologies for comparing the characters of organisms. The technical challenges, in turn, were met when it became easier to collect new kinds of data, primarily molecular, and as computational

software and hardware improved to make these comparisons faster and more efficient.

The last major summary of our knowledge of the TOL— compiled from the 1988 Nobel symposium titled "The Hierarchy of Life" (Fernholm et al. 1989)—establishes a point of comparison with which to understand the intense work of the past decade. The phylogenetic trees presented in that volume rarely included more than 15–20 taxa, and data sets hardly exceeded 100 or 200 characters, most far fewer than that. Perusal of the journals of that time paints a similar story.

The scientific work summarized here, in contrast, manifests a huge growth in phylogenetics research. Virtually all the chapters include taxon and character samples that were unheard of a mere 10 years ago. Yet, because the focus of the chapters in this volume is the relationships among the higher taxa, even these summaries cannot convey the vast increase in our knowledge that has taken place at all hierarchical levels. For that, the reader will have to go to specialized volumes—Benton (1988), Stiassny et al. (1996), Fortey and Thomas (1998), Littlewood and Bray (2001), and Judd et al. (2002) are but five examples that have been published in recent years—as well as to the numerous journals publishing phylogenetic results in every issue.

Having knowledge of the phylogenetic relationships of life is crucial if we are to advance societal well-being, including, importantly, building a sustainable world. In this volume, the chapters by Yates et al. (ch. 1), Colwell (ch. 2), and

Futuyma (ch. 3) describe numerous examples of the contributions that phylogenetic understanding has already made to science and society. Phylogenetic relationships establish the framework for all comparative analyses of biological data, and this hierarchical structure is also a predictive tool that leads us from those characteristics we now know about organisms to those we might expect to find in those less known or newly discovered. Such logic, whether expressed explicitly or not, underlies the expectation that certain organisms might harbor pharmacologically important compounds, might be pathogenetic or toxic, might express agriculturally important gene products, and so on. Indeed, the use of phylogenetic knowledge, including analytical methods that have been developed to solve phylogenetic problems, has grown so rapidly in recent years that even a single volume devoted to the subject could not be comprehensive.

The practical outcomes and applications of TOL research are certainly a clear reason why society should continue to support a better understanding of phylogenetic relationships (see ch. 1–3; see also Cracraft 2002). Yet, what drives many scientists engaged in this effort is the sheer wonder associated with knowing a chunk of life's history. To step back and attempt to grasp the entire history of life on Earth is itself an almost unimaginable task. Here we are, one species out of hundreds of millions that have existed since the diversification of life began several billion years ago, and we are attempting to see how that history has unfolded. It is difficult enough to see how we will build the TOL for the living species, let alone for all those that vanished over the course of time, but it is an exciting prospect. All people on the planet understand something about their "genealogical roots," and that serves as a crucial metaphor for seeing how human existence and origins fit into the bigger picture of life's diversity. This is a nontrivial exercise, for truly understanding that history is bound to influence the ethical picture people develop about the importance of life forms other than our own and how these have been inextricably linked to our own well-being over time. Obviously, it is not easy for us to step back from an anthropocentric view of the world, but a TOL can facilitate such a perspective.

Darwin's vision had a profound effect on people's understanding of themselves. Yet the understanding that went along with this change in thought is not universally appreciated even today, despite 150 years of evolutionary thought and science. The TOL will be a key element in advancing an expanded vision of life's history.

The Tree of Life: An Ongoing Synthesis

The chapters in this volume summarize our current understanding of the phylogenetic history of the major groups of organisms. It is time to stand back and see the big picture. Figure 34.1 presents a summary TOL that attempts to provide an estimate of the interrelationships among the extant clades of life. Its scope and depth, which is skewed toward the "higher" eucaryotes, is primarily a function of the coverage of the chapters in this book, which, in turn, generally reflect known, described taxonomic diversity. Clearly, many more groups could have been added to this tree, and numerous friends working on megadiverse taxa have suggested how their favorite groups could be expanded. Yet, the best way for this tree to serve an educational purpose is to limit detail and to include groups that are familiar to a wide audience.

Conceptually, the tree is constructed as a composite—constructed by piecing together the trees presented for the different groups. It is not derived from an analysis of a "supermatrix." It attempts to represent relationships that are moderately to well supported, yet there are unresolved nodes. Some will see the tree as too conservative and would recommend resolving certain nodes; others would prefer that more nodes be depicted as ambiguous. Because the tree is not built from a data matrix, it is not a rigorous phylogenetic hypothesis in a traditional sense. Rather, it is a summary of where we are now and a step in the continuous process of building a TOL. Importantly, it also stands as a framework for discussing some of the key problems and controversies raised in the individual chapters of this volume.

The Basal Clades of Life

It has been standard for a number of years now to recognize three major basal branches ("domains") of the TOL, the Bacteria, the Archaea, and the Eucarya (see Baldauf et al., ch. 4, and Pace, ch. 5, in this vol.), all of which are generally treated as monophyletic. A major impediment for understanding the nature of that monophyly and the relationships among these groups is, of course, the problem of where to place the root of the TOL. The present conventional wisdom is that the root lies along the branch between the Bacteria and the other two on the basis of evidence presented by duplicated genes (ch. 4). Some workers, on the other hand, have raised the issue of lateral gene transfer as possibly confounding the placement of the root (Doolittle, ch. 6), or that analytical artifacts such as long-branch attraction can lead to misleading relationships, which also could affect the placement of the root (Philippe, ch. 7). Philippe also argues that we have seen the evolutionary world as proceeding from the simple to the complex and thus have potentially overlooked the possibility that "prokaryotic"-like organisms could have been derived from eucaryotes by simplification. A major concern for all these scenarios, however, is that given the monophyly of these three groups, the placement of the root may be unsolvable because it remains a three-taxon problem.

The trailblazing work of Carl Woese, Norman Pace, and others to use the small subunit ribosomal RNA (rRNA) gene to reconstruct life's earliest branches can truly be said to have revolutionized our view of the TOL, and at the same time those data have shaped how the question of basal relationships has been approached. It is now clear that rRNA se-

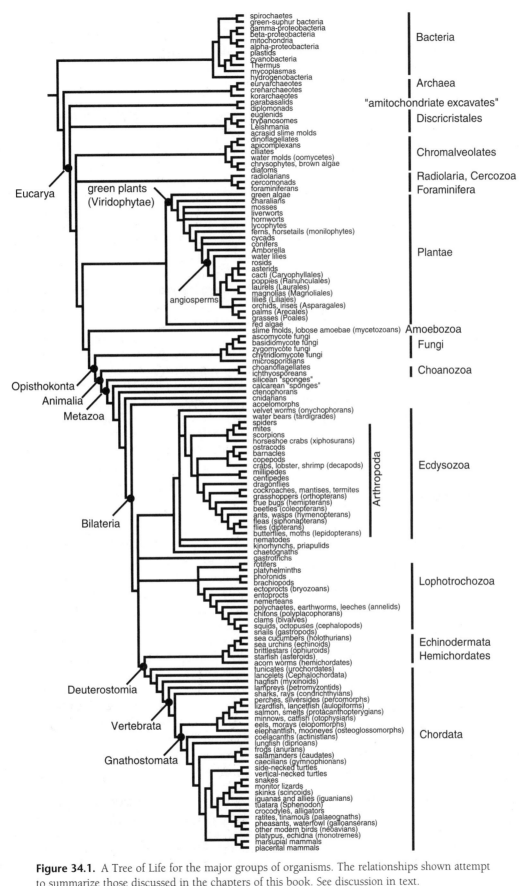

Figure 34.1. A Tree of Life for the major groups of organisms. The relationships shown attempt to summarize those discussed in the chapters of this book. See discussion in text.

quences alone cannot resolve the branching order among bacterial lineages to a convincing degree (Pace, ch. 5).

Bacterial relationships have been strongly influenced by decades of attempts to classify using phenetic data sets of a small number of key "characters" (e.g., gram-positive vs. gram-negative staining). This approach is bound to create some nonmonophyletic taxa. Bacterial systematists have also classified these taxa at high taxonomic rank (subkingdoms, divisions, phyla) on the basis of distinctiveness, and that tradition has continued as genetically distinct forms have been discovered from environmental samples. As Pace (ch. 5) describes, there are two main groups of Archaea, the crenarchaeotes and the euryarchaeotes. The third group shown on figure 34.1, the korarchaeotes, is only represented by environmental rRNA gene sequences and is of uncertain status (see also Baldauf et al., ch. 4).

Are viruses life, or not, and what has been their history? These are the subjects of chapter 8 by David Mindell and his colleagues. Although the topic of viral phylogeny was not the subject of a talk in the New York symposium Assembling the Tree of Life, its inclusion in this volume was deemed important for understanding the full panoply of biotic history. Mindell et al. show that viruses have arisen multiple times, and they summarize what we understand of their evolutionary relationships. Importantly, they also discuss how phylogenetics and its methodology can be applied to issues of human health.

Basal Eucarya

The base of the eucaryotic tree is very uncertain, with candidate groups being the parabasalid + diplomonad clade or discricristates, among others (Baldauf et al., ch. 4). Some would argue (Philippe, ch. 7) that the basal position of such taxa as parabasalids or diplomonads is probably a long-branch artifact. Their basal position seems reasonable at first glance, because it has been thought they branched off before the acquisition of the bacterial precursors of mitochondria. It is now known, however, that these "amitochondriate excavates" have some mitochondrial genes in their nuclear genomes. "Basal" eucaryotes remain one of the most unexplored regions of the TOL, and inasmuch as some groups are apparently very diverse, numerous candidates for the basal eucaryotic divergences are likely to emerge as new data are acquired.

There are three large monophyletic clades of eucaryotes, the green plants (upwards of 500,000 species), fungi (around 60,000 described), and animals (more than one million described). It is widely accepted that millions of species of fungi and animals remain to be discovered and described, whereas plant diversity has been more completely characterized. One of the more interesting phylogenetic findings of recent years is that the fungi and animals are sister taxa relative to other organisms (Opisthokonta; see Baldauf et al., ch. 4). It is important to note, however, that there are numerous single-

celled taxa whose relationships to these three clades are still unresolved; therefore, the tripartite division discussed here is certainly simplistic.

Plants

The overall backbone of plant phylogeny is moderately well supported (Donoghue, ch. 33, and Delwiche et al., ch. 9, in this vol.). The term "algae" has been applied to a diverse array of unrelated taxa possessing plastids, some of which lie at the base of the land plants, and indeed from the perspective of Delwiche et al., the land plants simply comprise a terrestrial lineage of green algae. Although the relationships among these algal groups need much further study, current molecular evidence identifies the Charales as the sister group of the land plants (embryophytes).

Within the embryophytes, the interrelationships among the three major groups of nonvascular plants—the liverworts, hornworts, and mosses—and the vascular plants (tracheophytes) are still a matter of controversy (Delwiche et al., ch. 9). The base of the tracheophyte tree is less controversial, with lycophytes being the sister group of the rest and then monilophytes (horsetails and various "fern" groups) being the sister group of the seed plants (Pryer et al., ch. 10). Relationships within the monilophytes, and especially at the base of the clade that includes the modern seed plants, are not entirely resolved. Within the latter group, which contains some 300,000 species, the angiosperms comprise the most diverse clade. The phylogenetic unity of the clade that includes the extant "gymnosperms" is still questionable, and the sister group of all the angiosperms has not yet been identified with confidence.

The angiosperms (flowering plants) are by far the dominant group of land plants, and their interrelationships have been the subject of a large number of morphological and molecular systematic studies over the last decade. Soltis and colleagues (ch. 11) have been important contributors to this effort. They note that relationships at the base of the angiosperms are moderately well understood. One of the more remarkable findings to emerge in recent years is that *Amborella trichopoda* of New Caledonia is the only living representative of the sister group of all other angiosperms, and the next branch contains the water lilies. The three largest clades within the core angiosperms—monocots, magnoliids, and the eudicots—are well defined, but their relationships to one another and to several other smaller clades remain unresolved (ch. 11).

Fungi

In recent years fungi have emerged as the sister group to the animals (see Baldauf et al., ch. 4, and Eernisse and Peterson, ch. 13, in this vol.). It is also becoming increasingly apparent that they will eventually be seen as one of the most diverse groups on Earth. The large-scale phylogenetic structure of the fungi has become clearer with the addition of sequence data, and it is now accepted that the two great groups of terrestrial fungi, the ascomycotes and basidiomycotes, are monophyletic and sister taxa (Taylor et al., ch. 12). As Taylor and colleagues note, relationships within these two diverse groups are still in need of considerable study. The base of the fungal tree is also poorly understood and is occupied by lineages usually assigned to two more obscure groups, the zygomycotans and chytridiomycotans, both of which may be nonmonophyletic.

Basal Animals

Animals are taken here to include the choanoflagellates and their sister group, the metazoans (see Eernisse and Peterson, ch. 13 in this vol.). Eernisse and Peterson review the evidence showing that animal and metazoan monophyly has become increasingly well established in recent years, but that relationships at the base of the Metazoa have been in a state of flux, particularly when it comes to those organisms typically called "sponges." Traditional classifications using morphological data recognized a monophyletic Porifera, but molecular data have led to the conclusion that siliceous sponges branched off first, followed by the calcareous sponges, the latter of which are the sister group to the eumetazoans (ch. 13). Relationships among the major clades of metazoans—ctenophorans, cniderians, placozoans, and eumetazoans—also remain uncertain because of conflicts among data sets (see ch. 13 for details)

Bilaterians

The monophyletic bilaterians are composed of three main groups, the ecdysozoans, lophotrochozoans, and deuterostomes, and more and more evidence is pointing to the conclusion that acoelomorph flatworms are their sister group (see Eernisse and Peterson, ch. 13, and Littlewood et al., ch. 14, in this vol.). Intense examination of the monophyly of these groups and the interrelationships of their included taxa has essentially revolutionized our view of bilaterian evolution over the last decade by eliminating the simplistic aceolomate to pseudocoelomate to coelomate description of phylogenetic history. Although the monophyly of ecdysozoans, lophotrochozoans, and deuterostomes—particularly the latter—is increasingly accepted (at least for the "core" taxa of the first two), their interrelationships are controversial, as is the placement of a number of small, morphologically disparate metazoan groups often classified at the phylum level (Littlewood et al., ch. 14, discuss no less than 15 "phyla"). Therefore, a major question is whether there exists an ecdysozoan + lophotrochozoan clade—thus implying the classical protostome–deuterostome dichotomy.

Lophotrochozoans

As reviewed by Eernisse and Peterson (ch. 13 in this vol.), the interrelationships among lophotrochozoan taxa are ex-ceedingly complex and contentious due to conflicts in data, especially morphological versus molecular. Several groups are regularly recognized: (1) the lophophorates, encompassing brachiopods and phoronids; (2) the trochozoans, including the annelids and mollusks, and their allies (see fig. 34.1); and (3) the platyzoans (rotifers, platyhelminths, and others). The latter two groups have traditionally been clustered in the Spiralia on the basis of possessing spiral cleavage and a trochophore larva, although it is entirely possible that lophophorates are within the trochozoans.

The two great groups of lophotrochozoans are sister taxa, the annelids (Siddall et al., ch. 15) and the mollusks (Lindberg et al., ch. 16). Within the former, leeches and earthworms are related, but the sister group of leeches within the earthworms is still uncertain. Morphological and molecular data conflict on annelid relationships, along with those of sipunculans, relative to the diverse marine polychaete worms (ch. 15). Clearly much more work will be required to resolve the history of these groups.

The interrelationships of the major clades of mollusks are moderately well accepted (Lindberg et al., ch. 16; see also fig. 34.1), with cephalopods and gastropods being sister taxa and related to bivalves and chitons at the base of the tree. All these groups have a deep evolutionary history, with considerable fossil diversity, and an integrated picture of their phylogeny will significantly advance paleontology. Not unexpectedly, the interrelationships of the recent molluscan biota are comparatively poorly understood given their extensive diversity.

Ecdysozoans

Different lines of evidence point to the ecdysozoans being a natural group (summarized in Eernisse and Peterson, ch. 13 in this vol.), yet many questions remain about their interrelationships, reflected in the unresolved tree in figure 34.1. Four ecdysozoan clades are now generally accepted (ch. 13 and 14): (1) the panarthropods; (2) nematodes and nematomorphs; (3) the kinorhynchs, priapulids, and loriciferans; and (4) chaetognaths. The latter two groups have low diversity, but the nematodes are thought to be the most numerically abundant metazoans on Earth, and they undoubtedly have a tremendous undescribed diversity greatly exceeding the 25,000 or so species already named. Littlewood and colleagues (ch. 14) briefly note recent progress on the phylogenetics of this group.

The arthropods—insects (Hexapoda); centipedes and millipedes (Myriapoda); crabs, crayfish, and their allies (Crustaceans); and the spiders and allies (Chelicerata)—include a number of megadiverse clades, especially the mites, spiders, and insects, and together they represent roughly 60% of the known species diversity on Earth. Wheeler and colleagues (ch. 17 in this vol.) describe the complex problem of deciphering relationships among the major groups of arthropods, the conflicting topologies implied by different data sets, and the fact that inclusion of fossil taxa in total evidence analyses often has dramatic effects on phylogenetic inferences. Although most of the evidence clusters crustaceans, myriapods, and hexapods together (as the Mandibulata because they possess mandibles) to the exclusion of the chelicerates, resolving relationships among the mandibulates has not been straightforward (ch. 17).

The higher level relationships of the chelicerates are moderately well supported, with mites and spiders being sister taxa and related to scorpions and their allies, and those three, in turn, are the sister group of the horseshoe crabs (fig. 34.1; Coddington et al., ch. 18). Over the past decade, relationships among the spiders have received considerable attention, and they are the best understood of the chelicerates, whereas relationships among the diverse clades of mites remain very poorly resolved (ch. 18).

As reviewed by Schram and Koenemann (ch. 19 in this vol.), the monophyly of the crustaceans has been contentious, with morphological data tending to support monophyly and some molecular data sets denying it. Even in this volume, differences of interpretation exist: Schram and Koenemann (ch. 19) question monophyly, whereas the analyses of Wheeler and colleagues (ch. 17) generally find a monophyletic Crustacea. Many of these differences, and those in the literature, come down to *apparent* conflicts between molecules and morphology, to alternative interpretations of morphological characters, especially those of fossils, and to which clade is to be called Crustacea. There is relatively little argument (see fig. 34.1; see also ch. 19), however, that the core crustacean clades are monophyletic and related to one another, especially the maxillopods (copepods, barnacles, ostracods) and the malacostracans (crabs, shrimps, and allies).

Arguably, the greatest challenge to the TOL—as we currently understand organic diversity—is the relationships within the hexapods, or insects and their allies. The vast diversity of forms creates multiple challenges for understanding insect history. Willmann (ch. 20 in this vol.) presents a summary of the complexities of hexapod phylogeny and how viewpoints have shifted over time, and Whiting (ch. 21) discusses phylogenetic relationships within the most diverse clade of hexapods, the holometabolic insects. Arguments over insect relationships exemplify the debates in other groups—molecules versus morphology, fossil versus extant taxa. The overall structure of the insect tree, however, is remarkably consistent from one study to the next (ch. 20): aside from a number of basal groups, most insects can be clustered into the Pterygota (those with wings), at the base of which are the mayflies and dragonflies (whether sister taxa or not is in dispute) and their great sister group, the Neoptera.

No less than 80% of the insects are found in the neopteran group, the Holometabola—those insects with complete metamorphosis. This generally accepted monophyletic group contains the most familiar of the insects—beetles, butterflies, wasps, flies, and so forth—yet the interrelationships of these well-defined groups have engen-

dered considerable debate (Whiting, ch. 21 in this vol.). The tree shown in figure 34.1 includes only five holometabolic clades whose relationships appear with regularity on trees using both morphology and/or molecules (ch. 21), but many other smaller clades are omitted. Clearly, the complexity of the vast taxonomic and morphological diversity of this group will feed controversy for many years, and it seems that considerable data will be required to resolve these long-standing phylogenetic questions.

Deuterostomia

The third great group of the bilaterians is the Deuterostomia (or Deuterostomata), which includes the echinoderms and hemichordates (ambulacrarians), on the one hand, and their sister group, the chordates. Until recently, the boundaries of the deuterostomes were ambiguous, but morphological and molecular work has clearly eliminated lophophorates, ectoprocts, and chaetognaths from the clade and established the remainder as a monophyletic group (see Eernisse and Peterson, ch. 13 in this vol.).

Ambulacraria

Smith and colleagues (ch. 22 in this vol.) review recent advances in hemichordate and echinoderm phylogenetics. The former group is small in terms of diversity, and relationships within the group have still not been deciphered satisfactorily. Echinoderms are also not especially diverse, having only about 6000 extant species, but they possess an extensive fossil record and are among the best known marine organisms. As Smith and colleagues detail, relationships among the major monophyletic groups are moderately well supported on both molecular and morphological grounds (fig. 34.1).

Chordata and Vertebrata

The overall pattern of chordate phylogeny is moderately well corroborated by both morphological and molecular data (fig. 34.1; see Rowe, ch. 23 in this vol.). The tunicates and lancelets are the successive sister taxa to the craniates (hagfish + vertebrates). For many years the hagfish and lampreys (fig. 34.1) were grouped together as the agnathans, but the preponderance of evidence does not favor this, especially the morphological and developmental data. Rowe notes in his review, however, that some molecular data find a monophyletic Agnatha; therefore, the problem needs further attention using combined data sets, and fossils as well as extant taxa.

Moving up the vertebrate tree, the next node subtends the sharks and allies (Chondrichthyes) and all other vertebrates (Osteichthyes), which are together termed the Gnathostomata. The Osteichthyes, in turn, are subdivided into the sarcopterygians (coelacanths, lungfish, and tetrapods) and the actinopterygian fishes. Stiassny and colleagues (ch. 24) lead us through the world of things called "fishes," in their case, chondrichthyans, actinopterygians, and the "fishlike" sarcopterygians. Chondrichthyans are easily divided into elasmobranchs (sharks, rays) and chimaeras, but relationships within the former clade are still uncertain. Morphological data recognize two basal sister taxa (galeomorphs and squalomorphs) and support a moderately resolved phylogeny within the latter; the conflict comes with some emerging molecular data that is said to question the monophyly of the rays and sharks (ch. 24). Within sarcopterygians, resolution of the coelacanth–lungfish–tetrapod trichotomy has been contentious. Stiassny et al. remain agnostic on this issue, whereas Rowe (ch. 23) resolves this in favor of lungfish + tetrapods while noting that the debate continues.

The actinopterygian fishes are the most diverse group of vertebrates and have a huge diversity of forms, so relationships have generally been difficult to resolve. Most of the actinopterygian nodes on figure 34.1 are based on morphological data, as Stiassny and colleagues (ch. 24) note, but new molecular data are being generated at a rapid rate. Although the interrelationships of these major groups might be generally accepted, phylogenetic understanding within most of them has a long way to go, especially given their high diversity.

It has long been accepted that amphibians are at the base of the tetrapod tree and are the sister group to all other vertebrates, which are grouped together as the Amniota (Rowe, ch. 23 in this vol.). Living amphibia are clearly monophyletic, and the relationships among the three clades have long been accepted (Cannatella and Hillis, ch. 25). Thus, caecilians are the sister group of the salamanders and frogs. Relationships within the three living taxa, especially within salamanders and frogs, are greatly unsettled.

The amniotes, so named because they share an amniote egg, are divided into two major clades, the Reptilia—including turtles, lepidosaurs (snakes, lizards, tuataras), and archosaurs (crocodiles and birds)—and the Mammalia (Rowe, ch. 23, and Lee et al., ch. 26, in this vol.). Higher level relationships within the reptiles have been particularly contentious. Crocodiles and birds go together on all trees, but the turtles, tuatarans, and snakes and lizards sort out in different ways depending on the data set. There are significant conflicts across and within data sets that leave these relationships unresolved. In contrast to this somewhat dismal situation, Lee and colleagues (ch. 26) show that higher level relationships within turtles and within the lizards and snakes, for example, are becoming better understood (fig. 34.1), although at lower taxonomic levels many gaps in our knowledge still exist.

Higher level relationships within living birds remain perhaps the least understood of all the major groups of tetrapods (Cracraft et al., ch. 27). The basal split between the tinamous and ratites (paleognaths) and all other birds (neognaths), and then within the neognaths between the galliforms–anseriforms and all others (Neoaves), are well supported by various data. Phylogenetic pattern among the traditional neoavian "orders," on the other hand, are largely

unresolved. The reason for this is pretty simple—lack of adequate character and taxon sampling—but that is rapidly changing.

Our understanding of mammalian interrelationships has made great strides in recent years because of the addition of very large morphological and molecular data sets. Yet, at the same time, as discussed by O'Leary and colleagues (ch. 28 in this vol.), there exists a great deal of conflict among data sets, and over their interpretation. All agree that monotremes are the sister group of the marsupials and placentals, but within the latter group there is considerable debate about how the traditional orders are related. The increasingly large molecular data sets appear to be converging on an answer, but morphological (and paleontological) data often conflict.

Finally, the symposium included a discussion of our current picture of hominid phylogenetics (Wood and Constantino, ch. 29 in this vol.)—for after all, it is a subject that generates great scientific and public interest and controversy. In contrast to other contributors, Wood and Constantino focus attention on the basal taxa of *Homo*—what systematists generally call species—because it is difficult to understand human evolution without delimiting those units. These authors come down on the "many species" side of the debate, as opposed to "just a few," and they argue that deciphering relationships among these taxa is challenging because so much of the fossil material is fragmentary and difficult to compare. They also demonstrate that debates over human origins—in the sense of which species is related to which—are likely to continue for quite some time.

Perspectives on the Tree of Life

A volume like this was not possible a decade or so ago, as a comparison with *The Hierarchy of Life* (Fernholm et al. 1989) makes clear. New analytical methods and new and more abundant data have transformed the field. But there has also been a sea change in biology's attitude toward systematics and TOL research. Our interest in life on Earth has accelerated, not only because it is rapidly disappearing, or is in our self-interest to find new ways to make money from it, or because increased understanding will contribute to the well-being of humanity, or it is intrinsically interesting. It is all these reasons. In his perspective, E. O. Wilson (ch. 30) makes the case for "a complete account of Earth's biodiversity, pole to pole, bacteria to whales, at every level of organization from genome to ecosystem, yielding as complete as possible a cause-and-effect explanation of the biosphere, and a correct and verifiable family tree for all the millions of species—in short a unified biology." Amen to that. Indeed, discovering and describing biodiversity and understanding the TOL go hand in hand, and both are increasingly seen as a foundation for all of biology. Importantly, TOL research has moved into mainstream experimental and molecular biology.

Growth in TOL research over the past decade, as David Wake (ch. 31) and David Hillis (ch. 32) observe, is readily apparent. Hillis also makes the important point that as TOL research expands, so do its applications to science and society. We are certainly on a roll, but how we might measure progress is not so straightforward, as Michael Donoghue notes (ch. 33). His tentative conclusion, discussed more below, is that it is the recognition and abandonment of paraphyletic groups that is perhaps the best measure of progress. Although it seems that some new paraphyletic groups will inevitably be created as more taxa are investigated, a successful war against paraphyly is the surest measure of success.

The Tree of Life: Progress Against Paraphyly

A survey of previous literature leads one to the conclusion that assembling the TOL must be an exceedingly complex problem because very few have attempted to resolve the whole tree (one of the few attempts has been in the popular literature; Tudge 2000). The present volume signals that we have entered a new era of research in phylogenetics. If we look back more than a decade ago, the overall state of knowledge discussed at the 1988 Nobel symposium might appear disappointing, and in today's terms, it was. If we compare, for example, the "summary tree" from that symposium (fig. 34.2) with the one discussed here (fig. 34.1), the con-

trast is striking. As noted above, it reflects a change not only in data and data analysis but also in attitude toward what we now know we can accomplish. The latter is not to be dismissed: a decade ago, not everyone was convinced a universal tree was at hand, or possible (even for relatively small chunks of the tree). Today, the attitude of systematists has changed. We will have a universal tree, and the operative questions are when, how well supported it will be, and how we are going to create a new field of phyloinformatics to tap the tree's benefits.

Concepts of monophyly, paraphyly, and polyphyly are not really associated with phylogeny per se but how relationships map to classification. When we say that the goal of TOL research is to discover and eliminate paraphyly, we mean eliminate named groups that are not natural groups. The practical manifestation of the chapters in this book is to rid systematic biology of nonmonophyletic groups, but this activity will be resisted by some. TOL research is caught to some extent in the language of the past, in which groups are ranked on the basis of distinctness. In the past, it was morphological distinctness, but today "genetic distinctness"—however that might be measured objectively—is increasingly an important criterion. The notion that distinctness should enter into hierarchical classifications through ranking has created paraphyletic groups in its wake and hindered phylogenetic progress. The plethora of high taxonomic ranks, such as domains, kingdoms, phyla, and the like, does nothing to clarify the phylogenetic history of life.

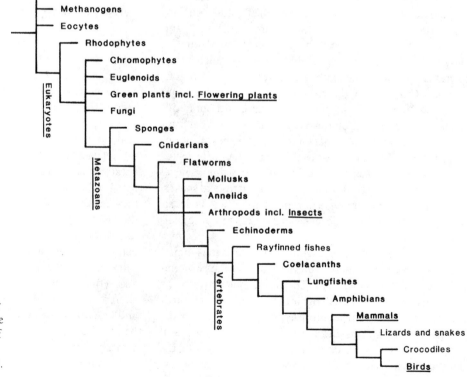

Figure 34.2. A "summary" tree (hierarchy) of life of selected major groups of organisms in which those taxa underlined were the subject of discussion at the 1988 Nobel symposium (Fernholm et al. 1989).

Although we can be "immeasurably" optimistic that progress on the TOL will continue unabated, those involved in research know the task is a difficult one. A theme of the 1988 Nobel symposium was molecules versus morphology. In the early years of molecular systematics, there was an abundance of exuberance that molecules were going to sweep away morphology in reconstructing the TOL. That has not exactly happened, if one is to judge by the myriad molecular data sets that conflict with one another. Indeed, as this volume attests, more and more workers are seeking to combine molecular and morphological data, and there is a growing realization that if we are truly to have a TOL, extinct life—at least 90% of all of it—must be included. The view here is that most of the conflicts we see among different data sets are more a matter of the selection of data, method of analysis, and lack of sufficient data than they are anything substantially "wrong" with a particular kind of data. Evidence is evidence, and we should, as scientists, bring all that is relevant to bear on a problem that we can. This view echoes Colin Patterson's (1989) closing remarks for the 1988 Nobel symposium: molecules allow us to gather large amounts of data quickly, but morphological data give us access to other dimensions of life—ontological, paleontological, temporal, and of form and function. Systematics needs all this. Biology needs all this.

Acknowledgments

All the participants and coauthors of the chapters in this book, especially Sandie Baldauf, Mark Siddall, Ward Wheeler, Pam Soltis, Kathleen Pryer, Timothy Rowe, Douglas Eernisse, Tim Littlewood, Max Telford, and John Taylor, provided expert advice and help in constructing figure 34.1.

Literature Cited

Benton, M. (ed.). 1988. The phylogeny and classification of the tetrapods, vols. 1 and 2. Clarendon Press, Oxford.

Cracraft, J. 2002. The seven great questions of systematic biology: an essential foundation for conservation and the sustainable use of biodiversity. Ann. Missouri Bot. Garden. 89:127–144.

Fernholm, B., K. Bremer, and H. Jörnvall (eds.). 1989. The hierarchy of life: molecules and morphology in phylogenetic analysis. Elsevier, Amsterdam.

Fortey, R. A., and R. H. Thomas. 1998. Arthropod relationships. Chapman and Hall, London.

Hennig, W. 1950. Grundzüge einer Theorie des phylogenetischen Systematik. Deutscher Zentraverlag, Berlin.

Hennig, W. 1966. Phylogenetic systematics. University of Illinois Press, Urbana.

Judd, W. S., C. S. Campbell, E. A. Kellogg, P. F. Stevens, and M. J. Donoghue. 2002. Plant systematics: a phylogenetic approach. 2nd ed. Sinauer, Sunderland, MA.

Kenrick, P., and P. R. Crane. 1997. The origin and early diversification of land plants: a cladistic study. Smithsonian Institution Press, Washington, DC.

Littlewood, D. T. J., and R. A. Bray (eds.). 2001. Interrelationships of the platyhelminthes. Taylor and Francis, London.

Patterson, C. 1989. Phylogenetic relations of major groups: conclusions and prospects. Pp. 471–488 *in* The hierarchy of life: molecules and morphology in phylogenetic analysis (B. Fernholm, K. Bremer, and H. Jörnvall, eds.). Elsevier, Amsterdam.

Stiassny, M. L. J., L. R. Parenti, and G. D. Johnson (eds.). 1996. Interrelationships of fishes. Academic Press, San Diego.

Tudge, C. 2000. The variety of life. Oxford University Press, Oxford.

Index